HANDBOOK OF ADVANCED METHODS AND PROCESSES IN

# OXIDATION CATALYSIS

From Laboratory to Industry

HANDBOOK OF ADVANCED METHODS AND PROCESSES IN

# OXIDATION CATALYSIS

From Laboratory to Industry

Editors

**Daniel Duprez**
University of Poitiers, France

**Fabrizio Cavani**
University of Bologna, Italy

ICP Imperial College Press

*Published by*

Imperial College Press
57 Shelton Street
Covent Garden
London WC2H 9HE

*Distributed by*

World Scientific Publishing Co. Pte. Ltd.
5 Toh Tuck Link, Singapore 596224
*USA office:* 27 Warren Street, Suite 401-402, Hackensack, NJ 07601
*UK office:* 57 Shelton Street, Covent Garden, London WC2H 9HE

**Library of Congress Cataloging-in-Publication Data**
Duprez, Daniel, 1945–
Handbook of advanced methods and processes in oxidation catalysis : from laboratory to industry / Daniel Duprez, University of Poitiers, France, Fabrizio Cavani, Universita di Bologna, Italy.
pages cm
Includes bibliographical references and index.
ISBN 978-1-84816-750-6 (hardcover : alk. paper)
1. Oxidation. 2. Catalysis. 3. Chemistry, Organic. I. Cavani, Fabrizio. II. Title.
QD281.O9D87 2014
660'.28443--dc23
2014017262

**British Library Cataloguing-in-Publication Data**
A catalogue record for this book is available from the British Library.



Typeset by Stallion Press
Email: enquiries@stallionpress.com

Printed in Singapore by B & Jo Enterprise Pte Ltd

# Preface

## Advanced Methods and Processes in Oxidation Catalysis From Laboratory to Industry

edited by **Daniel Duprez** *(University of Poitiers, France)* &
**Fabrizio Cavani** *(Università di Bologna, Italy)*

Since the discovery by Humphry Davy in 1817 of the flameless combustion of coal gas over Pt wires, tremendous progress has been made in the understanding of complex phenomena occurring in oxidation catalysis. In parallel, advanced technologies were developed to make these processes more efficient and safer. In the nineteenth century, researchers observed that hydrocarbon oxidation could lead to organic intermediates on noble metals. The huge demand from the chemical industry for new compounds prompted them to take advantage of the selective oxidation to synthesize oxygenated chemicals. Synthesis of new compounds required specific oxide catalysts much more selective than noble metals. Considerable progress was made during the twentieth century while the development of cleaner, greener and safer catalytic processes remains a permanent objective of the chemical industry today.

This book offers a comprehensive overview of the most recent developments in both total oxidation and combustion and also in selective oxidation. For each topic, fundamental aspects are paralleled with industrial applications. The book covers oxidation catalysis, one of the major areas of industrial chemistry, outlining recent achievements, current challenges and future opportunities. One distinguishing feature of the book is the selection of arguments which are emblematic of current trends in the chemical industry, such as miniaturization, use of alternative, greener oxidants, and innovative systems for pollutant abatement. Topics outlined are described in terms of both catalyst and reaction chemistry, and also reactor and process technology.

The book is presented in two volumes. The first ten chapters are devoted to total oxidation while the next eighteen chapters deal with selective oxidation.

Different aspects of total oxidation processes are reviewed in the first part of the book: hydrocarbon oxidation (Chapter 1) and soot oxidation (Chapter 2) for mobile applications while oxidation of volatile organic compounds (VOC) is treated in the next five chapters. Chapter 3 provides a general overview of VOC oxidation while chlorinated VOCs are specifically discussed in Chapter 4 and persistent VOC in Chapter 5. Plasma catalysis processes for VOC abatement are reviewed in Chapter 6. Finally, Chapter 7 gives the point of view of industry for the development and applications of catalysis for air depollution technologies. Total oxidation is also used for energy production by combustion processes exemplified in Chapter 8. The last two chapters are devoted to oxidation processes in liquid media by electrochemical techniques (Chapter 9) or more generally as "advanced oxidation processes" for water depollution (Chapter 10).

The part devoted to selective oxidation includes chapters aimed at providing an overview of oxidation technologies from an industrial perspective, with contributions from chemical companies such as eni SpA, Radici Chimica, Polynt, Sabic, DSM, and Clariant (Chapters 11–16). Then, Chapters 17–19 gives an updated view of experimental tools and techniques aimed at the understanding of catalyst features and interactions between catalysts and reactants/products. Chapters 20–23 are focussed on specific classes of homogenous and heterogeneous catalysts, such as vanadyl pyrophosphate, polyoxometalates, supported metals and metal complexes. Finally, Chapters 24–28 deal with classes of reactions, reactor configurations and process technologies used in selective oxidation, again offering a perspective on recent developments and new trends, such as oxidation of alkanes, oxidations under supercritical conditions, use of non-conventional oxidants, membrane and structured reactors.

Daniel Duprez and Fabrizio Cavani

# Contents

# Chapter 1

# Oxidation of CO and Hydrocarbons in Exhaust Gas Treatments

Jacques BARBIER Jr and Daniel DUPREZ*

The present chapter aims to describe the kinetics and mechanisms of CO and HC oxidation in exhaust gas treatments. Attention will be paid to reactions carried out on noble metal catalysts (Pt, Pd, Rh) usually employed in three-way catalysts (spark ignition engines) around stoichiometry. The effect of ceria, usually employed as an oxygen storage material, will also be reviewed.

## 1.1. Introduction

Since 1972 in the United States and 1989 in Europe, regulations have been imposed on the automobile industry to limit air pollution emitted by vehicles. Since these dates, legislation has been regularly reinforced with more and more severe regulations concerning four categories of pollutants: carbon monoxide, hydrocarbons (and other organics), nitrogen oxides (NO and $NO_2$) and soot particulates.[1–3] To achieve abatement of these pollutants, automotive catalytic converters were implemented on new cars to eliminate CO, HC and NOx, while particulate filters are intended to be mounted in the exhaust gas pipe of diesel engines. Oxidation of CO and hydrocarbons is an important process occurring over three-way catalysts. These catalysts are currently employed in the catalytic converters of gasoline engines (close-looped engines) while similar formulations are used in diesel oxidation converters. Three-way (TW) catalysts contain Pt, Pd and Rh deposited on a mixed oxide made typically of doped alumina (La, Ca, . . .) and an oxygen storage capacity component ($Ce_xZr_{1-x}O_2$ binary oxides or CeZrXOy ternary oxides, X being another rare earth element).[4–7] The term "oxygen storage capacity" was introduced by Yao and Yu Yao to qualify the ability of the catalyst to work in cycling conditions: the solid stores oxygen during the lean phases and releases it during the rich ones.[8] With this method, the noble metals continue to be fed with O species when the $O_2$ concentration significantly decreases in the gas phase.

*Laboratoire de Catalyse en Chimie Organique, UMR 6503 CNRS-Université de Poitiers, 40 avenue du Recteur Pineau, 86022 Poitiers cedex. France.

Oxidation of CO and hydrocarbons in conditions of exhaust gas conversion (concentrations around 1% for CO and less for HCs) has been widely studied since the implementation of catalytic converters. Yu Yao was one of the first authors to publish a systematic study of these reactions over Pt, Pd and Rh catalysts in $O_2$ excess.[9,10] Moreover, the effect of ceria was also investigated, making Yao and Yu Yao's reports a source of important information. Their results will be analyzed and summarized in the first part of this chapter. In the second part, more recent studies will be reviewed with special attention paid to investigation under cycling conditions.

## 1.2. The Pioneer Works (1970–1990)

In TW catalysis, an optimal conversion of all the pollutants (reducers like CO and HC, and oxidants like NO and $NO_2$) is achieved for an S ratio (defined by Schlatter)[11] close to unity. The S ratio is given in Eq. 1.1, in which chemical formulae represent the volume percentages of the gases.

$$S = \frac{2O_2 + NO + 2NO_2}{CO + H_2 + 3nC_nH_{2n} + (3n+1)C_nH_{2n+1}} \tag{1.1}$$

The numerator represents the number of O atoms available in the oxidants ($O_2$ and NOx) while the denominator represents the number of O atoms required for a total oxidation of the reducers: CO, HC (alkenes and alkanes) and $H_2$. The Schlatter equation may easily be extended to other HC (aromatics for instance) or oxygenated compounds. However, other gases such as $H_2O$ and $CO_2$ are not supposed to react with pollutants, which is not always observed (see Section 1.4). Yu Yao investigated CO and HC oxidation in $O_2$ excess ($S = 2$). Oxidation reactions were carried out over Pd, Pt and Rh catalysts of different metal loading and dispersions and at different temperatures. In the publications of Yu Yao,[9,10] the reactions were carried out over bulk metals (wires), alumina-supported catalysts and finally over metals supported on ceria-alumina. Specific rates (per gram of catalyst) were reported as well as activation energies, metal dispersions of supported catalysts or metal area of bulk catalysts. From this information it was possible to calculate turnover frequencies (TOF) extrapolated at a given temperature (the same for every metal catalyst).[12] Metal catalysts are compared on the basis of their TOF.

### 1.2.1. *Oxidation of carbon monoxide*

#### 1.2.1.1. *Effect of metal particle size*

Oxidation of carbon monoxide (Eq. 1.2) is a reaction which can be catalyzed by all the noble metals usually employed in TW converters but also by many oxides or

**Table 1.1.** Intrinsic activity of Pd, Pt and Rh for CO oxidation at 250°C ($s^{-1}$). Metal dispersion (%) is given in parentheses.

| Metal | Pd | Pt | Rh |
|---|---|---|---|
| Unsupported | 4.6 | 0.31 | 10.1 |
| Supported on $Al_2O_3$ | 2.9 (41%) | 0.24 (7%) | 1.8 (57%) |
| | 0.9 (67%) | 0.10 (87%) | 0.4 (69%) |

mixed oxides.[13]

$$CO + {}^1\!/\!_2O_2 \rightarrow CO_2 \tag{1.2}$$

Table 1.1 compares the turnover frequencies extrapolated at 250°C for $S = 2(0.5\%CO + 0.5\%O_2)$.

The three metals show a moderate sensitivity to the metal particle size, large particles having the highest turnover frequency. Platinum exhibits the lowest structure sensitivity while rhodium appears to have the highest sensitivity to metal dispersion. Oh *et al.* confirmed that rhodium was more active than platinum in $O_2$ excess.[14,15] However, Rh is more sensitive than Pt to the presence of NO,[14] or a hydrocarbon.[15] For instance, in a $0.5\%CO + 0.5\%O_2 + 500$ ppm NO or a $0.1\%CO + 1\%O_2 + 0.2\%CH_4$, Pt appears to be more active than Rh. Although the reaction was carried out in conditions far from those encountered in catalytic converters (silica support), the study by Cant *et al.* gives useful information about the reaction of the stoichiometry ($1.3\%CO + 0.65\%O_2$).[16] The results are given in terms of TOF (molecule $CO_2$ per metal atom per hour). At 127°C, the following ranking is observed: Ru; 250 > Pt; 30 > Rh; 23 > Pd; 4 > Ir; 0.4 while at 177°C, the same metal/silica catalysts exhibit the following activity: Ru; 5900 > Rh; 900 > Pt; 150 > Pd; 110 > Ir; 12. The changes between 127 and 177°C are due to the lowest activation energy of Pt (58 kJ $mol^1$) instead of 100 kJ $mol^1$ for the other metals. The very good behavior of Ru in CO oxidation is also observed in many other reactions involved in TW catalysts. Unfortunately, the volatility of Ru tetroxide made impossible the use of this metal in automotive converters.[17]

#### 1.2.1.2. *Effect of ceria*

In $O_2$ excess, ceria (20% in alumina) changes the activity of Pt and Pd very little but significantly increases that of Rh.[10] By contrast, the influence of ceria is much more marked around the stoichiometry ($S = 1$) at least for Pt and Rh.[18,19] It is clear that the beneficial effect of ceria can be observed mainly at low $O_2$ concentration and most probably in cycling conditions.

**Table 1.2.** Kinetic orders and activation energies for CO oxidation at 250°C over Pd, Pt and Rh catalysts. From Ref. 10.

| Metal | Pd | | | Pt | | | Rh | | |
|---|---|---|---|---|---|---|---|---|---|
| Support | None | $Al_2O_3^a$ | $CeO_2$-$Al_2O_3$ | None | $Al_2O_3^a$ | $CeO_2$-$Al_2O_3$ | None | $Al_2O_3^a$ | $CeO_2$-$Al_2O_3$ |
| m ($O_2$) | +1.0 | +0.9 | 0 | +1.0 | +1.0 | +0.5 | +1.0 | +1.0 | 0 |
| n (CO) | −1.0 | −0.9 | +1.0 | −1.0 | −0.9 | +0.3 | −1.0 | −0.8 | +0.2 |
| $E_a$ (kJ mol$^{-1}$) | 125 | 108–133 | 50 | 125 | 104–125 | 84 | 117 | 92–113 | 104 |

[a]Metal dispersion on alumina: 16–65% for Pd, 4–87% for Pt and 7–69% for Rh.

### 1.2.1.3. *Kinetics and mechanisms*

Kinetic data reported by Yu Yao[10] are summarized in Table 1.2. Rates were expressed according to the power law equation:

$$r = ke^{\left(-\frac{E}{RT}\right)} P_{O2}^{m} P_{CO}^{n} \tag{1.3}$$

Kinetic orders (m and n) and activation energies E were determined by varying the concentrations and the temperature around the conditions given in Table 1.1.

On unsupported metals and on alumina-supported catalysts, the kinetic orders with respect to $O_2$ are close to +1 while the reaction is auto-inhibited by CO (orders close to −1). Ceria has a dramatic effect on the reaction, with all the kinetic orders becoming nil or positive. Activation energies (close to 120 kJ mol$^{-1}$ on unsupported metals) decrease with the particle size and the presence of ceria.

The mechanism generally proposed for the reaction on unsupported metals and alumina-supported catalysts is a classical Langmuir–Hinshelwood mechanism with CO and $O_2$ competing for the same metal sites M.

*Adsorption of CO and $O_2$*

$$CO + M \Leftrightarrow CO - M: \quad \text{equilibrium constant } K_{co} \tag{1.4}$$

$$\tfrac{1}{2}O_2 + M \Leftrightarrow O - M: \quad \text{equilibrium constant } K_o \tag{1.5}$$

*Surface reaction (determining step)*

$$CO - M + O - M \rightarrow CO_2 + 2M: \quad \text{rate constant k} \tag{1.6}$$

which leads to the following rate equation:

$$r = k\frac{K_{CO}\,K_O\,P_{CO}\,P_O^{1/2}}{[1 + K_{CO}\,P_{CO} + K_O P_O^{1/2}]^2} \tag{1.7}$$

Although heat of $O_2$ adsorption is higher than that of CO on noble metals, CO coverage is always higher than that of oxygen (see Section 1.3). Under the conditions of CO oxidation, CO appears to be more strongly adsorbed than $O_2$ so that: $K_{CO}$ $P_{CO} \gg 1 + K_O P_O^{1/2}$. The rate equation Eq. 1.7 may then be simplified as:

$$r = k\frac{K_O P_O^{1/2}}{K_{CO} P_{CO}} \tag{1.8}$$

which explains the order −1 with respect to CO. Orders +1 in oxygen can be obtained by modifying the mechanism, supposing either that $O_2$ is not dissociatively adsorbed

$$O_2 + M \Leftrightarrow O_2 - M \tag{1.9}$$

or that the reaction proceeds via a Rideal mechanism between adsorbed CO and gaseous $O_2$:

$$CO - M + O_2 \rightarrow CO_2 + O - M \tag{1.10}$$

Nevertheless, in the latter case (Rideal mechanism), order in CO should be nil or slightly positive so that only the LH mechanism with non-dissociated $O_2$ can account for the experimental orders. Ceria would create new sites for $O_2$ adsorption. As CO and $O_2$ do not compete for the same sites, positive orders in CO and $O_2$ are observed.

### 1.2.2. *General trends in hydrocarbon oxidation*

The hydrocarbon reactivity depends on numerous factors: chain length, unsaturation, presence of cycles more or less distorted... For instance, Bart *et al.* showed that light alkanes and acetylene were particularly refractory to oxidation over a commercial Pt-Rh/$CeO_2$-$Al_2O_3$ catalyst.[20] Table 1.3 gives the light-off temperatures ($T_{50}$ required to reach a 50% conversion) of some hydrocarbons over this catalyst. In the alkane series, HC reactivity increases significantly with the chain length, methane being by far the most refractory hydrocarbon with a $T_{50}$ above 500°C. Alkenes and aromatics are relatively easy to oxidize, their $T_{50}$ being comprised between

**Table 1.3.** Light-off temperatures $T_{50}$ (50% conversion) for different hydrocarbons and alcohols over a Pt-Rh/$CeO_2$-$Al_2O_3$ commercial catalyst. The synthetic gas mixture contains 0.15% HC (in C1 equivalent) + 0.61%CO + 0.2%$H_2$ + 480 ppm NO + 10 $CO_2$ + 10% $H_2O$. It is at the stoichiometry (S = 1). Volumic space velocity was 50,000 $h^{-1}$. From Ref. 20.

| *n*-Alkanes | Alkenes, alkyne | Aromatics | Alcohols |
|---|---|---|---|
| Methane : 515°C | Ethylene : 205°C | Benzene : 205°C | Methanol : 195°C |
| Ethane : 435°C | Propene : 185°C | Toluene : 220°C | Ethanol : 200°C |
| Propane : 290°C | Acetylene : 285°C | *o*-Xylene : 225°C | Propanol : 205°C |
| Hexane : 195°C | | | Butanol : 210°C |

185 and 225°C. Bart *et al.* also investigated a series of alcohols whose oxidation seems very easy.

Some of the factors influencing the hydrocarbon reactivity have been recently reinvestigated on a series of 48 hydrocarbons.[21] The main results will be reported in Section 1.3.

### 1.2.3. *Oxidation of light alkanes*

#### 1.2.3.1. *Effect of metal particle size*

Yu Yao investigated the oxidation of C1–C4 alkanes over Pd, Pt and Rh catalysts[9] (Table 1.4). As for CO oxidation, TOF were calculated on the basis of specific activities and metal dispersions reported by Yu Yao. For each alkane, TOF were extrapolated to the same temperature (using the activation energy also reported by Yu Yao), which allowed a direct comparison between the three metals.

It was confirmed that oxidation rates strongly depend on the length of the molecule. Palladium was the most active metal for methane oxidation, the order of activity being: Pd $\gg$ Rh > Pt. It is still very active in ethane oxidation with an inversion between Pt and Rh (Pd > Pt > Rh). For C3–C4 hydrocarbons, platinum is definitely the most active catalyst (Pt $\gg$ Pd $\gg$ Rh). Whatever the alkane molecule, all the metals show high structure sensitivity in oxidation: the greater the particle size, the higher the TOF. As the specific activity $R_m$ (per gram of metal) is proportional to the product D × TOF, there exists a value of the metal dispersion D for which $R_m$ is maximal. Depending on the hydrocarbon, this optimal dispersion is between 15 and 40% for Pt, while it is somewhat higher for Pd (about 50%).

Similar size effects were observed by Hicks *et al.* for methane oxidation (6.5%$CH_4$ + 15%$O_2$) over Pt and Pd catalysts.[22,23] At 350°C, TOF of 0.005 $s^{-1}$

**Table 1.4.** Oxidation of C1–C4 alkanes over Pd, Pt and Rh catalysts (unsupported or supported on alumina). From Ref. 9.

| Metal | Disp.% ↓ | HC → T°C→ | $CH_4$ 400 | $C_2H_6$ 350 | $C_3H_8$ 250 | $C_4H_{10}$ 225 |
|---|---|---|---|---|---|---|
| Pd | Unsupported | | 5.4 | 3.6 | 0.25 | 0.19 |
| " | 16 | | 0.31 | 0.093 | 0.0072 | 0.0042 |
| " | 65 | | 0.012 | 0.030 | 0.0045 | 0.0014 |
| Pt | Unsupported | | 0.017 | 0.93 | 10.0 | 10.4 |
| " | 6 | | 0.0095 | 0.31 | 1.5 | 5.2 |
| " | 87 | | — | — | 0.16 | 1.75 |
| Rh | Unsupported | | 0.050 | 0.16 | 0.010 | 0.0076 |
| " | 7 | | 0.017 | 0.011 | 0.0006 | 0.0004 |
| " | 57 | | 0.0085 | 0.0095 | 0.0004 | 0.0004 |

and 0.008 $s^{-1}$ were found for well-dispersed and sintered Pt, respectively. They amounted to 0.02 $s^{-1}$ and 1.3 $s^{-1}$ for well-dispersed and sintered Pd catalysts. The preferential orientation along more active surfaces during sintering has been employed to interpret these results.[24,25]

### 1.2.3.2. *Effect of ceria*

Except for Rh, ceria has rather a negative effect in alkane oxidation (Table 1.5). This is essentially due to the fact that $O_2$ chemisorption is not a limiting factor in alkane oxidation (see Section 1.2.3.3). Rhodium was also the metal most sensitive to the presence of ceria for CO oxidation (see Section 1.2.1.2). There is certainly a specificity to the interaction of this metal with ceria.

### 1.2.3.3. *Kinetics and mechanisms*

Kinetic data relative to propane oxidation are reported in Table 1.6. Contrary to what was observed in CO oxidation, the kinetic orders strongly depend on the nature of the metal. They are nil or positive for Pd and Rh while, on Pt, a negative order with respect to $O_2$ and an order of +2 with respect to $C_3H_8$ are recorded. Relative close orders (around −1 in $O_2$ and +1 in HC) were reported by Yu Yao for methane oxidation over Pt, which tends to prove that the mechanism of oxidation is similar for both alkanes.

**Table 1.5.** Effect of ceria in alkane oxidation. Activity ratio (per g. of metal) between catalysts supported on 20%$CeO_2$-$Al_2O_3$ and catalysts supported on pure alumina (S = 2). From Ref. 9.

| Reaction | Pd (0.15%) | Pt (0.22%) | Rh (0.15%) |
|---|---|---|---|
| $CH_4 + O_2$ | 0.3 (400°C) | 0.05 (500°C) | 1 (500°C) |
| $C_3H_8 + O_2$ | 0.2 (350°C) | 0.5 (250°C) | 3 (400°C) |

**Table 1.6.** Oxidation of propane over Pd, Pt and Rh catalysts. Kinetic orders with respect to $O_2$ (m) and to $C_3H_8$ (n) and activation energies. From Ref. 9.

| Metal | Pd | | | Pt | | | Rh | | |
|---|---|---|---|---|---|---|---|---|---|
| Support | None | $Al_2O_3$ | $CeO_2$-$Al_2O_3$ | None | $Al_2O_3$ | $CeO_2$-$Al_2O_3$ | None | $Al_2O_3$ | $CeO_2$-$Al_2O_3$ |
| m ($O_2$) | 0 | +0.1 | +0.1 | −1 | −1 | −1 | +0.1 | 0 | +0.1 |
| n ($C_3H_8$) | +0.4 | +0.6 | +0.6 | +1.2 | +2 | +2 | +0.5 | +0.5 | +0.4 |
| $E_a$ (kJ $mol^{-1}$) | 96 | 66–96 | 63 | 92 | 84–105 | 96 | 92 | 100 | 84 |

The results of Table 1.6 show that oxygen is more strongly bound to the metals than propane. The difference is much more marked on Pt than on the other two metals, which leads to a negative order with respect to $O_2$. There is no mechanism unanimously accepted for alkane oxidation on Pt. For this reaction an "oxygenolysis" mechanism comparable to that of hydrogenolysis has been proposed with the following elementary steps:[12]

*Dehydrogenating adsorption of propane:*

$$C_3H_8 + Pt \rightarrow C_3H_{8-x} - Pt + x/2\,H_2 \tag{1.11}$$

*Dissociative adsorption of $O_2$:*

$$\tfrac{1}{2}O_2 + Pt \Leftrightarrow O - Pt \tag{1.12}$$

*C–C bond rupture in the adsorbed hydrocarbon species, the decomposition occurring either spontaneously:*

$$C_3H_{8-x} - Pt + Pt \rightarrow CH_y - Pt + CH_{8-x-y} - Pt \tag{1.13}$$

*or by reaction with $O_2$:*

$$C_3H_{8-x} - Pt + 2O - Pt \rightarrow CO - Pt + H_2O - Pt + C_2H_{6-x} - Pt \tag{1.14}$$

These reactions are rapidly followed by oxidation of HC fragments and hydrogen.

On Pt, only Eq. 1.14 is able to account for the kinetic observations. If mechanism Eq. 1.11–Eq. 1.12–Eq. 1.14 occurs, the kinetic derivation leads to the following rate equation:

$$r = k\frac{K_C K_O^2 P_C P_O}{[1 + K_C P_C + K_O P_O^{1/2}]^3} \tag{1.15}$$

in which $P_C$ and $P_O$ are the partial pressures in propane and $O_2$, $K_C$ and $K_O$ the equilibrium constants of steps Eq. 1.11, Eq. 1.12 and k, the rate constant of step Eq. 1.14, which is supposed to be the rate-determining step. On Pt (the most active metal for propane oxidation), oxygen is strongly adsorbed so that: $K_O P_O^{1/2} \gg 1 + K_C P_C$. The rate equation (Eq. 1.15) can then be simplified:

$$r = k\frac{K_C\,P_C}{K_O\,P_O^{1/2}} \tag{1.16}$$

As for CO oxidation (Eq. 1.8), order $-1$ in $O_2$ experimentally observed may suggest that the $O_2$ molecule could react before it is dissociated.

**Table 1.7.** Propene oxidation at 150°C: turnover frequencies ($s^{-1}$) of unsupported metals and metals supported on alumina. Metal dispersion is given in parentheses. Gas composition: 0.1% $C_3H_6$ + 1% $O_2$ + $N_2$. From Ref. 9.

| Metal | Pd | Pt | Rh |
|---|---|---|---|
| Unsupported | 0.56 | 0.35 | 0.95 |
| Supported on $Al_2O_3$ | 0.10 (40%) | 0.016 (13%) | 0.003 (7%) |
| | | 0.20 (87%) | 0.030 (69%) |

## 1.2.4. *Alkene oxidation*

### 1.2.4.1. *Effect of metal particle size and effect of ceria*

Specific activities of Pd, Pt and Rh catalysts in propene oxidation are reported in Table 1.7. Contrary to what was observed in alkane oxidation, propene oxidation is not very sensitive to the nature of the metal. Quite similar TOF were measured over unsupported metals, while Pt and Pd seemed to be slightly more active than Rh when supported on alumina. Propene oxidation is not very sensitive to metal particle size. However, intrinsic activity would be rather higher on small particles. As TOF are higher or much higher on unsupported metals, it seems that alumina could play a negative role in propene oxidation. The intermediary formation of partially oxidized compounds (acrolein, alcohols, ...) is not excluded. Alumina might store and stabilize these intermediates, slowing down the total oxidation.

Propene oxidation is much faster than propane oxidation over Pd and Rh. The reverse tendency would occur over Pt. However, propene is more strongly adsorbed on Pt than propane, which explains why, in the oxidation of $C_3H_6/C_3H_8$ mixtures, propene oxidizes first; propane oxidation starts when virtually all the propene is oxidized.[26]

Ceria has a moderate effect in propene oxidation. It is rather positive on Pt and Rh. The presence of 20% ceria in alumina can increase the activity by a factor of two or three on these metals. For Pd, the effect of ceria seems limited and rather negative.

### 1.2.4.2. *Kinetics and mechanisms*

Kinetic orders are very different to those observed for alkane oxidation (Table 1.8). They are rather close to those measured in CO oxidation at least for Pd and Pt, rhodium showing a different behavior.

Propene appears to be more strongly adsorbed than $O_2$ over Pt and Pd: kinetic orders are definitely positive in $O_2$ and negative in $C_3H_6$. This inhibiting effect of propene is not observed on Rh, on which $O_2$ appears to be more strongly adsorbed than propene.

**Table 1.8.** Propene oxidation over Pd, Pt and Rh catalysts. Kinetic orders with respect to $O_2$ (m) and to $C_3H_6$ (n) and activation energies. From Ref. 9.

| Metal | Pd | | | Pt | | | Rh | | |
|---|---|---|---|---|---|---|---|---|---|
| Support | None | $Al_2O_3$ | $CeO_2$-$Al_2O_3$ | None | $Al_2O_3$ | $CeO_2$-$Al_2O_3$ | None | $Al_2O_3$ | $CeO_2$-$Al_2O_3$ |
| m ($O_2$) | +1.5 | +1.5 | +0.7 | +1.8 | +2.0 | +1.5 | −1.3 | −0.8 | 0 |
| n ($C_3H_6$) | −0.6 | −0.5 | −0.3 | −0.8 | −1.0 | −0.6 | +1.3 | +0.9 | +0.5 |
| $E_a$ (kJ mol$^{-1}$) | 125 | 63–117 | 63 | 92 | 67–125 | 80 | 96 | 67–92 | 92 |

The mechanism of propene oxidation is undoubtedly different from that of alkanes. Propene adsorption does not require a C-H bond rupture, alkene molecules being adsorbed on most metals via the $\pi$ electrons of the C=C double bond. This adsorption would be strong on Pd and Pt and much weaker on Rh. It is interesting to note that Rh is the metal most sensitive to the presence of the support, its intrinsic activity being 30 to 300 times less when it is supported on alumina (Table 1.7). This suggests that the support could play a role in propene adsorption, tending to inhibit the reaction on the metal. Oxidation most likely goes further according to steps similar to those written for propane (Eqs 1.13 and 1.14).

As a rule, activation energies are close to those measured on alkanes. Again, ceria tends to decrease Ea for Pd and Pt while it is virtually unchanged for Rh.

### 1.2.5. *Overview of the behavior of Pd, Pt and Rh catalysts in CO and HC oxidation*

Activity of Pd, Pt and Rh catalysts for CO and HC oxidation and corresponding rate equations depend first on the relative adsorption equilibrium of the reducer and oxygen on the metals. From the kinetic data reported in Sections 1.2.2 to 1.2.4, the scheme represented in Fig. 1.1 can be drawn.

This scheme allows us to account for the general behavior of Pd, Pt and Rh catalysts in oxidation. Reducers whose adsorption constant is higher than that of $O_2$ (bars on the right of $O_2$) are strongly adsorbed and behave as inhibitors of the reaction (negative orders while that of $O_2$ is positive). Conversely, reducers whose adsorption constants are lower than that of $O_2$ (bars on the left of $O_2$) are weakly adsorbed: $O_2$ acts as an inhibitor of the reaction (negative orders while those of the reducers are positive). Ceria significantly changes this picture as it offers new sites for $O_2$ adsorption. Chlorine has a detrimental effect on propane and propene oxidation as it blocks hydrocarbon adsorption.[26] Fortunately, water produced during oxidation leads to progressive catalyst dechlorination, which helps in restoring activity.

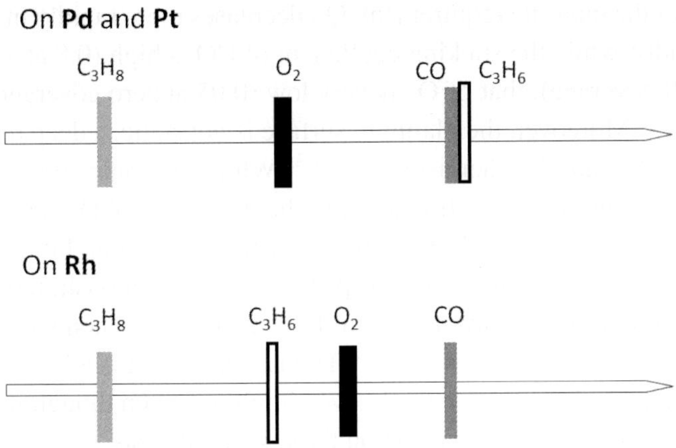


**Figure 1.1.** Relative adsorption constant of CO, propane, propene and $O_2$ on Pd, Pt and Rh catalysts (unsupported or supported on alumina).

## 1.3. Recent Investigations (After 1990)

### 1.3.1. *Oxidation of carbon monoxide*

The most recent advances in CO oxidation were detailed in the review paper of Royer and Duprez.[13] As far as noble metals are concerned, the kinetic data of Yao and Yu Yao,[8] reported in Section 1.2.1., remain valid. Nevertheless, the intensive works of Ertl and coworkers on this reaction allowed a more detailed and more exact picture of what really occurs at the metal surface. There is an apparent contradiction between the data of adsorption heats of $O_2$ and CO and the assumption that $O_2$ coverage is very low in the reaction. Heat of chemisorption of CO on noble metals was reported or reviewed in many papers by Engel and Ertl,[27] Nieuwenhuys,[28] Toyoshima and Somorjai,[29] Bradford and Vannice[30] and Ge *et al.*[31] For Pt, most of the available data show that the heat of chemisorption of CO, $Q_{CO}$, are comprised between 109 and 138 kJ mol$^{-1}$. Only Ge *et al.* reported higher values for $Q_{CO}$. Heats of chemisorption of $O_2$ ($Q_O$) were first measured by Brennan *et al.* for most metals used in catalysis.[32] For Pt they found a value of 265 kJ mol$^{-1}$, in agreement with the heat reported by Nieuwenhuys (230 kJ mol$^{-1}$).[28] The significant differences between $Q_O$ and $Q_{CO}$ on Pt prompted Yeo *et al.* to revisit the question on Pt (111).[33] Again, the same conclusions could be drawn from this new investigation with values of $Q_O$ and $Q_{CO}$ at zero coverage of 339 and 185 kJ mol$^{-1}$, respectively. The following question should then be addressed: why is oxygen coverage so low in CO oxidation in spite of its high heat of chemisorption? Several phenomena may explain this apparent discrepancy. First, $O_2$ should be dissociated upon chemisorption which requires two adjacent sites. Both $Q_{CO}$ and $Q_O$ decrease when CO and O coverage increases but, because of the dual site requirement, $Q_O$ decreases more rapidly in the presence of CO. Secondly, while the sticking coefficient of CO is high (0.8 at zero coverage and 0.2 at full coverage), that of $O_2$ is very low (0.05 at zero coverage and 0.02 at full coverage).[33] Moreover, the platinum surface is not static, a deep reconstruction occurring upon CO and $O_2$ chemisorption.[34,35] While CO chemisorption is not very perturbed by this surface reconstruction, the chemisorption of $O_2$ can be drastically altered. These phenomena lead to oscillating reactions with highly contrasted CO and O coverage changing with time and space, as clearly demonstrated by the group of Ertl.[36,37] For that reason, hysteresis in the reaction rate when CO or $O_2$ partial pressure is varied, it is often observed.[38] The Langmuir–Hinshelwood mechanism described in Section 1.2.1.3 is certainly oversimplified, even though it accounts for most of the kinetic observations made on three-way catalysts.

The role of ceria has also been studied in many papers published after 1990.[39–42] It is now accepted that two types of sites on ceria should be distinguished: those that are located at the metal-support interface and those that are located on the support, not in interaction with the metal. Direct evidence of these specific sites of ceria were obtained by Johansson *et al.* on model catalysts prepared by electron beam lithography.[43] Most authors concluded that sites located at the metal-support interface would be very active in CO oxidation. Serre *et al.* described these sites as a bridged oxygen ion bonded both to Pt atoms and Ce ions (Pt-O-Ce). Oxygen would be very labile with a high propensity of O vacancy formation during the rich phases. The doped catalyst loses a great part of its exceptional activity under prolonged oxidative medium ($S > 1$). This implies that the promotion by ceria is more marked in transient conditions around the stoichiometry than under $O_2$ excess.

The state of the metal during CO oxidation has also been debated in the literature. Though it is expected that Pt remains in the metallic form, the rhodium could be largely oxidized in reaction.[41] CO adsorption would occur on ionic Rh sites surrounded by O vacancies.

All these investigations have described the role of ceria, but the original assumption of Yu Yao[8] and Oh and Eickel[18] that new sites for $O_2$ chemisorption are created on $CeO_2$ remains valid for a good description of kinetic observations.

### 1.3.2. *Oxidation of hydrocarbons and alcohols*

#### 1.3.2.1. *Light hydrocarbons (C1–C6)*

One of the major results of Yu Yao was to demonstrate that oxidation of light alkanes was extremely sensitive to the particle size of metals, with turnover frequencies being much higher on big particles (see Table 1.4).[9] This behavior was confirmed by Gololobov *et al.* for oxidation of C1–C6 alkanes over Pt.[44] Their results are summarized in Table 1.9.

**Table 1.9.** Turnover frequency of Pt in oxidation of light alkanes (C1–C6).[a] Effect of particle size of metal. TOF values are given in $10^{-2}\,s^{-1}$. HC concentration: 5,000 ppm (in C1 equivalent) diluted in air; GHSV (gas hourly space velocity): 60,000 $h^{-1}$. From Ref. 44.

| | *n*-Alkane | $CH_4$ | $C_2H_6$ | $C_3H_8$ | $C_4H_{10}$ | $C_6H_{14}$ |
|---|---|---|---|---|---|---|
| $d_{Pt}$(nm) | T (°C) | 420 | 319 | 240 | 195 | 172 |
| $1.3 \pm 0.3$ | | 0.25 | 0.26 | 0.28 | 0.24 | 0.30 |
| $2.2 \pm 0.5$ | | 0.46 | 0.32 | 0.36 | 0.70 | 1.68 |
| $2.9 \pm 0.6$ | | 0.90 | 1.03 | 1.28 | 3.82 | 6.99 |
| $6.5 \pm 1.7$ | | 0.67 | 0.82 | 1.26 | 3.50 | 5.84 |
| $8.3 \pm 2.2$ | | 0.63 | 0.77 | 1.42 | 5.17 | 6.45 |
| $11.5 \pm 3.7$ | | 0.59 | 0.52 | 1.69 | 4.98 | 7.96 |

[a]Reaction temperatures were adjusted to obtain the same *n*-alkane conversion on the catalyst sample having a mean particle size of 1.3 nm.

It is confirmed that longer alkanes are easier to oxidize: the temperature at which the same oxidation rate can be observed decreases with the number of carbon atoms in the molecule (420°C for methane vs 172°C for hexane). The particle size sensitivity significantly increases with the chain length: the ratio between maximal and minimal TOF values is 3.6 for methane while it amounts to 26.5 for *n*-hexane. Within the domain of $d_{Pt}$ investigated here (1.3–11.5 nm), TOF values continuously increase with the particle size for *n*-hexane while a maximum is observed at 2.9 nm for methane and ethane.

Although alkanes are not strongly adsorbed over noble metals, it seems that ethane and propane may shift light-off temperatures for $CH_4$ oxidation to higher values. This behavior was clearly observed over Pd catalysts and ascribed to a change of the reduction state of Pd in the presence of ethane and propane.[45] Inhibition of methane oxidation by C2–C3 alkanes is virtually not observed over Pt, whereas strongly adsorbed hydrocarbons such as alkenes and acetylene may strongly affect alkane oxidation.[46]

Many authors have shown that the support could play a role, not only in changing particle size but also in modifying adsorption properties of the metals. Ceria could stabilize ionic species of platinum leading to a strong metal-support interaction. Bera *et al.* have compared the behavior of $Pt/CeO_2$ and $Pt/Al_2O_3$ in TW catalysis.[47] The enhanced activity observed in several reactions ($CO + O_2$, $CO + NO$ and $HC + O_2$, Table 1.10) has been attributed to the formation of new sites ($\text{-}O^{2-}Ce^{4+}\text{-}O^{2-}Pt^{n+}\text{-}O^{2-}$ with $n = 2$ or 4). Ceria-supported catalysts are more active than alumina ones for all the reactions. NO as an oxidant is more sensitive in nature to support than $O_2$. Moreover, ceria is a better promoter for oxidation of CO and propane than that of methane. Whatever the oxidant (NO or $O_2$), methane oxidation remains difficult with a modest promotion by ceria.

**Table 1.10.** Support effect on Pt and Pd catalysts for different reactions involved in three-way catalysis. The catalytic activity is given by the temperature at which a conversion close to 100% is observed. Partial pressures of reactants are in the range of 1–20 Torr with molar ratios close to the stoichiometry. From Ref. 47.

| Reaction | 1%$Pt/CeO_2$ | 1%$Pt/Al_2O_3$ | 1%$Pd/CeO_2$ | 1%$Pd/Al_2O_3$ |
|---|---|---|---|---|
| $NO + CO$ | 270 | 400 | 175 | 350 |
| $NO + NH_3$ | 225 | 280 | 275 | 330 |
| $NO + CH_4$ | 350 | 425 | 450 | 475 |
| $NO + C_3H_8$ | 325 | 475 | 330 | 525 |
| $CO + O_2$ | 180 | 230 | 175 | 230 |
| $CH_4 + O_2$ | 400 | 425 | 330 | 350 |
| $C_3H_8 + O_2$ | 110 | 175 | 230 | 300 |

The deep oxidation of $C_2$-$C_4$ alkanes has been studied by Garetto *et al.* over Pt catalysts supported on various oxides.[48] Pt/zeolite catalysts show better performances in oxidation than $Pt/Al_2O_3$ and Pt/MgO. Support acidity is not a major contributing factor for this enhancement of activity, the promotion effect being observed both on acidic (HY, ZSM5, Beta) and rather basic zeolites (KL). The following turnover frequencies ($h^{-1}$) were obtained for propane oxidation at 250°C ($C_3H_8/O_2/N_2$ molar ratio = 0.8/9.9/89.3):

Pt/Beta; 39,000 > Pt/ZSM5; 10,000 > Pt/HY; 9,000 > Pt/KL; 1,400 > $Pt/Al_2O_3$; 85 > Pt/MgO; 30.

It seems that the zeolites allow a better adsorption of alkanes by using a confinement effect. A similar effect was observed on hierarchical porous silica membranes with mesopores of 4 nm.[49] These mesopores allowed a better stabilization of Pt particles and preferential adsorption of the reactants.

Less conventional supports were also used for hydrocarbon oxidation. For example, Postole *et al.* investigated the performance of palladium deposited on boron nitride.[50] This PdO/BN catalyst showed relatively good performances in propene and methane oxidation even in the presence of moisture.

Acidic or basic promoters can affect the activity of Pt and Pd supported on alumina or zirconia.[51,52] Basic promoters (e.g. Na) improve CO and propene oxidation while they inhibit NO oxidation. By contrast, acidic promoters (e.g. sulfates) show the greatest effect in propane oxidation. Carbon monoxide and propene are too strongly adsorbed (with respect to $O_2$): the effect of basic promoters would be to weaken their adsorption on the metals and thereby to reinforce that of oxygen. The reverse is true for acidic promoters: they strengthen the adsorption of propane, too weakly adsorbed on Pt and Pd (see Section 1.2.5). The same result can be observed if sulfates are replaced by $SO_2$ directly injected with the propane/air mixture.[53] The effect of acidic promoters on $Rh/Al_2O_3$ is somewhat different. Lee *et al.* reported that sulfates increased both the reactivity of propene and propane.[54] This result can be explained by the relative strengths of $C_3H_8$, $C_3H_6$ and $O_2$ adsorption on Rh on which both the alkene and the alkane are weakly adsorbed (Fig. 1.1).

The state of palladium can change upon hydrocarbon oxidation. Under sub-stoichiometry of $O_2$, PdO is reduced before CO and alkene oxidation starts while it is reduced during the light-off oxidation of propane.[55,56] In methane combustion, even in $O_2$ excess, it was proven that the oxidation starts when some PdO sites are reduced by $CH_4$.[57,58] The active sites of palladium for methane oxidation would be created by association of oxidized $Pd^{2+}$ and reduced Pd. Finally, combining Pd with other metals like Ni can improve the catalytic activity in CO and propene oxidation.[59] Again, the support ($CeZrOx/Al_2O_3$) plays a decisive role in favoring interaction between the two metals.

### 1.3.2.2. *Heavy hydrocarbons (>C6)*

If oxidation of light hydrocarbons were still the subject of many studies in the 1990s, the tendency after 2000 has been to investigate oxidation of heavier hydrocarbons representing the fraction of HC emissions between gaseous hydrocarbons and soots.[21,60,61] Catalytic oxidation of 48 hydrocarbons from C6 to C20 was investigated by Diehl *et al.* over a well-dispersed $Pt/Al_2O_3$ catalyst.[21] From this study it was possible to draw some conclusions and rules about hydrocarbon reactivity as a function of their molecular structure.

#### 1.3.2.2a. Normal alkanes and alkane isomers

The tendency observed with light alkanes is confirmed up to *n*-C20: *n*-alkane oxidability increases with the chain length with a correlative decrease of the light-off temperature (temperature for a 50% conversion, $T_{50}$). Nevertheless, the decrease of $T_{50}$ is weak above 10 carbon atoms (Table 1.11). A correlation was found between

**Table 1.11.** Light-off temperature of normal alkanes over a 1%$Pt-Al_2O_3$ catalyst. Reaction conditions: HC concentration: 1,500 ppm C diluted in air; air flow rate: 20 $cm^3\ min^{-1}$; 35 mg catalyst. From Ref. 21.

| Hydrocarbon | $T_{50}$ (°C) | Hydrocarbon | $T_{50}$ (°C) |
|---|---|---|---|
| Propane | 292 | *n*-Decane | 171 |
| *n*-Hexane | 243 | *n*-Hexadecane | 165 |
| *n*-Octane | 202 | Eicosane (*n*-C20) | 164 |
| *n*-Nonane | 183 | | |

the light-off temperature and the ionization potential of hydrocarbon. Normal alkanes are adsorbed on Pt by C-H bond rupture and C-Pt bond formation. $O_2$ can form very reactive oxygen species (O, superoxides, peroxides) by an electron transfer from the metal to O atoms. A likely hypothesis is that these electrons may come from the adsorbed hydrocarbon: the easier the hydrocarbon ionization, the higher the oxygen reactivity.

Oxidation of alkane isomers strongly depends on the number of primary, secondary, tertiary or quaternary carbons in the molecule. Secondary carbons and, still more, tertiary carbons are easy to oxidize while primary and quaternary carbons are much more difficult to oxidize. An example is given in Fig. 1.2 for selected hydrocarbons containing eight carbon atoms. They can be ranked by decreasing oxidation rate: *n*-octane $\approx$ methyl-2-heptane > trimethyl-2,3,4-pentane $\gg$ dimethyl-2,2-hexane > trimethyl-2,2,4-pentane $\gg$ tetramethyl-2,2,3,3-butane. This latter hydrocarbon, which contains only primary and quaternary carbons, is extremely difficult to oxidize.

#### 1.3.2.2b. Normal alkenes and alkene isomers

The behavior of alkenes in oxidation significantly differs from that of alkanes. Normal alkenes are generally easy to oxidize over Pt, Pd and Rh. Contrary to normal alkanes, their reactivity depends little on the chain length for $7 < n < 10$ while the light-off temperature of shorter alkenes (C2–C6) slightly increases with the chain

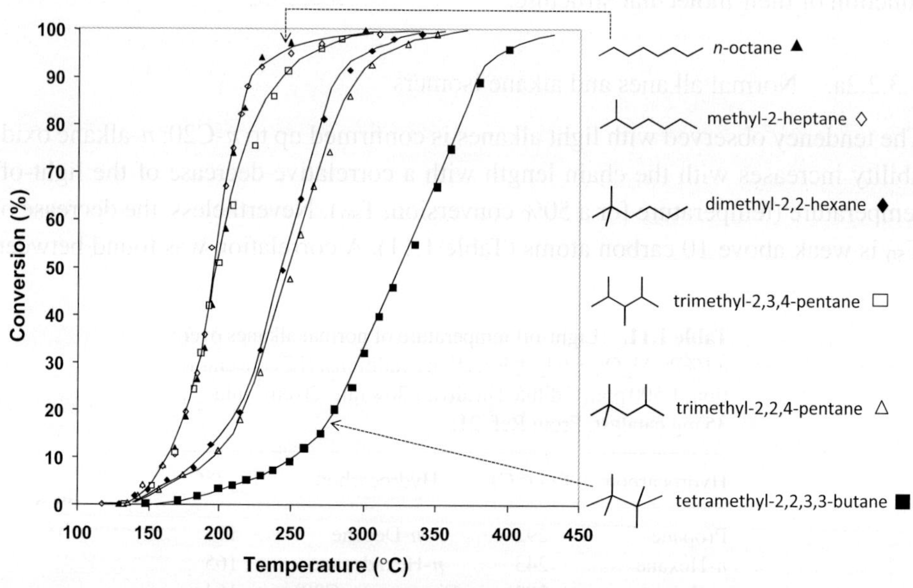


**Figure 1.2.** Oxidation of *n*-octane and C8 alkane isomers over $Pt/Al_2O_3$. Reactions conditions: see Table 1.11. Adapted from Ref. 21.

**Table 1.12.** Light-off temperatures of normal alkenes and alkene isomers with seven and eight carbon atoms. Reaction conditions: see Table 1.11. From Ref. 21.

| Formula | Name | Light-off temperature $T_{50}$(°C) |
|---|---|---|
|  | Hept-1-ene | 197 |
| 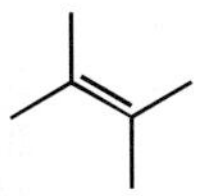 | Dimethyl-2,2-but-2-ene | 162 |
| 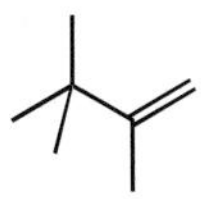 | Trimethyl-2,3,3-but-1-ene | 160 |
|  | Oct-1-ene | 198 |
| 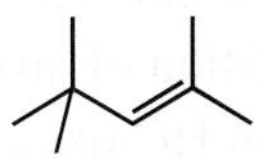 | Trimethyl-2,4,4-pent-2-ene | 166 |
| 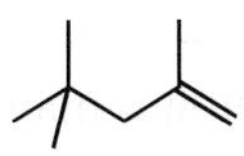 | Trimethyl-2,4,4-pent-1-ene | 176 |

length.[62] Light olefins (ethylene, propene, butenes) oxidize at much lower temperatures than the corresponding alkanes, while *n*-heptene and *n*-octene oxidize at around 200°C, the same temperature as *n*-heptane and *n*-octane. For longer hydrocarbons, the reverse situation is even observed: *n*-decene oxidizes at a higher temperature (200°C) than *n*-decane (167°C). For these hydrocarbons, the superiority of Pt for oxidation reactions is much less marked.[63,64]

Contrary to branched alkanes, alkene isomers are rather more reactive than the corresponding *n*-alkenes (Table 1.12). As a matter of fact, normal alkenes are too strongly adsorbed on the metals. Substituting H atoms by alkyl groups tends to decrease the adsorption heat of alkene isomers, which contributes to equilibrate the adsorption coefficients of hydrocarbons and oxygen.[65]

Octadienes were also studied by Diehl *et al.* who showed that the presence of a second double bond has little effect on the reactivity of the hydrocarbon. For instance, $T_{50}$ = 201°C for oct-1-ene, 208°C for octa-1,5-diene and 211°C for octa-1,7-diene.[21]

#### 1.3.2.2c. Monocyclic aromatics

Oxidation of monocyclic aromatics, especially benzene and toluene, has been the subject of many investigations in the last 20 years. As a rule, conversion efficiencies on a diesel oxidation catalyst are in the order alkenes > alkanes > aromatics.[66]

**Table 1.13.** Effect of CO on hydrocarbon oxidation ($T_{50}$ in °C). Reaction conditions: mixture of hydrocarbons (25 ppm each) + 0.6%$O_2$ + 1%CO (when present); Gas hourly space velocity: 30,000 h-1. Catalysts: metal/$Al_2O_3$/monolith crushed and sieved to 100 – 300 $\mu$m. Metal content (in the whole monolith): 0.306%Pt; 0.305%Pd; 0.246%Rh. From Ref. 63.

| | Platinum | | Palladium | | Rhodium | |
|---|---|---|---|---|---|---|
| HC | No CO | CO present | No CO | CO present | No CO | CO present |
| Hexene | 171 | 312 | 184 | 260 | 133 | 201 |
| Toluene | 186 | 326 | 200 | 279 | 254 | 202 |
| Benzene | 186 | 327 | 208 | 284 | 282 | 249 |
| Isooctane | 192 | 328 | 263 | 286 | 294 | 287 |

Platinum seems to be the best catalyst for benzene and toluene oxidation provided that there is no CO or alkenes in the gas mixture.[63,64,67,68] Rhodium, very active in CO and alkene oxidation, may help platinum to work in oxidation of aromatics free of inhibition.[64] An example of the inhibiting effect of CO on HC oxidation is presented in Table 1.13.

Other molecules present in the gas mixtures like alcohols or ketones may have a moderate effect on aromatic oxidation but the reverse (inhibition of alcohol oxidation by aromatics) is most often observed.[69] Different supports of Pt were used for toluene oxidation: $Al_2O_3$,[21,63,70,71] $Al_2O_3$/Al,[72] ZnO/$Al_2O_3$,[73] $TiO_2$,[69] mesoporous fibrous silica[74] or monoliths.[75] Zeolites, generally promoted by platinum, were shown to give excellent catalysts for aromatic oxidation.[76] Basic zeolites showed excellent performances in oxidation of *m*-xylene even in the absence of platinum.[77] Palladium catalysts, either supported on alumina or ceria-alumina, were also investigated for oxidation of benzene and several alkylbenzenes.[78,79]

The nature and the size of the alkyl substituent on the aromatic cycle may have a great effect on the conversion efficiency. Diehl *et al.* showed that the light-off temperatures of monoalkylbenzenes on Pt/$Al_2O_3$ increased with the number of carbon atoms in the alkyl group.[21] This tendency was observed up to four carbon atoms. For very long alkyl groups (as in *n*-decylbenzene), $T_{50}$ decreases and becomes close to that of benzene (Table 1.14). Diehl *et al.* also compared the oxidation of several polyalkylbenzenes. It was shown that the monomethylbenzene (toluene) oxidized at the highest temperature: increasing the number of methyl groups increases the hydrocarbon reactivity. A similar tendency is observed for branched alkyl groups (mono and ditertiobutyl).

1.3.2.2d. Polycyclic hydrocarbons

Polycyclic aromatic hydrocarbons (PAH) are generally difficult to oxidize. Naphthalene was generally chosen as a model of these PAHs. Pt/$Al_2O_3$,[80] Pd/Ce-$Al_2O_3$,[81]

**Table 1.14.** Effect of substituents (nature and number) on the reactivity of alkylbenzenes over $Pt/Al_2O_3$. Reaction conditions: see Table 1.11.

| Hydrocarbon | $T_{50}$ (°C) | Hydrocarbon | $T_{50}$ (°C) |
|---|---|---|---|
| Benzene | 191 | *m*-Xylene | 209 |
| Toluene | 216 | Mesitylene | 205 |
| Ethylbenzene | 227 | Hexamethylbenzene | 196 |
| *n*-Butylbenzene | 228 | Tertiobutylbenzene | 181 |
| *n*-Decylbenzene | 195 | Ditertiobutylbenzene | 176 |

polycrystalline ceria[82] or ceria-zirconia[83] were used as catalysts for oxidation of naphthalene or a mixture of naphthalene with other compounds.

Diehl *et al.* showed that naphthalene oxidized over $Pt/Al_2O_3$ at 240°C far over the light-off temperature of *n*-decane (171°C). However, when the bicyclic molecule is partially (tetraline) or totally (decaline) hydrogenated, its oxidation becomes much easier ($T_{50} = 230$°C for tetraline and 205°C for decaline).[21] Oxidation of polycyclic hydrocarbons is often not total at moderate temperatures. Two kinds of behavior can be observed:[21]

— Oxygenated intermediates are formed during the light-off tests and total conversion to $CO_2$ and $H_2O$ is achieved only at very high conversions. The case of fluorene is typical of this behavior. The selectivity to $CO_2$ exceeds 50% only for conversions greater than 80%. There is a significant formation of fluorenone all along the fluorene conversion.

— Dehydrogenated intermediates can be formed up to very high conversions. For instance, acenaphthene oxidation gives rise to acenaphthylene as an intermediate of reaction. The selectivity to acenaphthylene is still 20% at a 40% conversion of acenaphthene over Pt.

It thus seems very important to analyze both $CO_2$ and all these intermediates during the light-off tests of polycyclic hydrocarbon oxidation.

## 1.4. Conclusions

Oxidation reactions over Pt, Pd and Rh are generally structure sensitive with turnover frequencies higher on bigger particles. The structure sensitivity is very high for light alkanes and moderate for other hydrocarbons and CO.

Oxidation of CO is fast over noble metals. The presence of CO may inhibit the oxidation of the most reactive hydrocarbons.

The reactivity of normal alkanes increases with the chain length, methane being by far the most difficult hydrocarbon to oxidize. As a rule, alkane isomers are less reactive than normal alkanes with the same number of carbon atoms. Isomers with quaternary carbons are extremely difficult to oxidize. Alkenes are more reactive than alkanes and, contrary to these hydrocarbons, alkene isomers are easier to oxidize than normal alkenes.

Monocyclic aromatics are generally oxidized around 200°C over Pt. Nature, number and size of the alkyl groups can affect the reactivity of the benzene ring. Polycyclic hydrocarbons are more difficult to oxidize. The reaction is often complex and gives rise to oxygenated or dehydrogenated intermediates up to high conversions. As a matter of fact, the selectivity to $CO_2$ should be checked very carefully.

## References


1. Silver, R., Sawyer, J. and Summers, J. (1992). *Catalytic Control of Air Pollution. Mobile and Stationary Sources, ACS Symposium Series* 495.
2. Armor, J. (1994). *Environmental Catalysis, ACS Symposium Series* 552.
3. Heck, R., Farrauto, R. and Gulati, S. (2002). *Catalytic Air Pollution Control. Commercial Technologies*. 2nd Edition. J. Wiley & Sons, New York.
4. Collins, N. and Twigg, M. (2007). Three-way catalyst emissions control technologies for spark-ignition engines — recent trends and future developments, *Topics Catal.*, **42–43**, pp. 323–332.
5. Cuif, J., Blanchard, G., Touret, O., *et al.* (1997). *SAE Technical Paper* No. 97-0463.
6. Rohart, E., Larcher, O., Deutsch, S., *et al.* (2004). From Zr-rich to Ce-rich: thermal stability of OSC materials on the whole range of composition, *Topics Catal.*, **30–31**, pp. 417–423.
7. Matsutomo, S. (2004). Recent advances in automobile exhaust catalysts, *Catal. Today*, **90**, pp. 183–190.
8. Yao, H. and Yu Yao, Y. (1984). Ceria in automotive exhaust catalysts. 1. Oxygen storage, *J. Catal.*, **86**, pp. 254–265.
9. Yu Yao, Y. (1980). Oxidation of alkanes over noble metal catalysts, *Ind. Eng. Chem., Prod. Res. Dev.*, **19**, pp. 293–298.
10. Yu Yao, Y. (1984). The oxidation of CO and hydrocarbons over noble metal catalysts, *J. Catal.*, **87**, pp. 152–162.
11. Schlatter, J. (1978). *SAE Technical Paper* No. 780199.
12. Duprez, D. (1995). Review: catalytic converters for reducing pollution caused by traffic, *J. Chim. Phys.*, **92**, pp. 1952–1983.
13. Royer, S. and Duprez, D. (2011). Catalytic oxidation of carbon monoxide over transition metal oxides, *ChemCatChem.*, **3**, pp. 24–65.
14. Oh, S. and Carpenter, J. (1986). Role of NO in inhibiting CO oxidation over alumina-supported rhodium, *J. Catal.*, **101**, pp. 114–122.
15. Oh, S., Mitchell, P. and Siewert, R. (1991). Methane oxidation over alumina-supported metal catalysts with and without cerium additives, *J. Catal.*, **132**, p. 287.
16. Cant, N., Hicks, P. and Lennon, B. (1978). Steady state oxidation of carbon monoxide over supported noble metal catalysts with particular reference to platinum, *J. Catal.*, **54**, pp. 372–383.
17. Taylor, K. (1984). *Catalysis, Science and Technology*, J. Anderson and M. Boudart (eds), Vol. 5. Springer Verlag, Berlin, p. 119.

18. Oh, S. and Eickel, C. (1988). Effect of cerium addition on CO oxidation kinetics over alumina-supported rhodium catalysts, *J. Catal.*, **112**, pp. 543–555.
19. Barbier Jr, J. and Duprez, D. (1993). Reactivity of steam in exhaust-gas catalysis: 1. Steam and oxygen/steam conversions of CO and propane over Pt-Rh catalysts, *Appl. Catal. B: Environmental*, **3**, pp. 61–83.
20. Bart, J., Pentenero, A. and Prigent, M. (1992). *Catalytic Control of Air Pollution, ACS Symposium Series*, R. Silver, J. Sawyer, J. Summers (eds), **Vol. 495**. pp. 42–60 (Chapter 4).
21. Diehl, F., Barbier Jr, J., Duprez, D., *et al.* (2010). Catalytic oxidation of heavy hydrocarbons over Pt/Al2O3. Influence of the structure of the molecule on its reactivity, *Appl. Catal. B: Environmental*, **95**, pp. 217–227.
22. Hicks, R., Li, H., Young, M. *et al.* (1990). Structure sensitivity of methane oxidation over platinum and palladium, *J. Catal.*, **122**, pp. 280–294.
23. Hicks, R., Li, H., Young, M. *et al.* (1990). Effect of catalyst structure on methane oxidation over palladium on alumina, *J. Catal.*, **122**, pp. 295–306.
24. Garbowski, E., Feumi-Jantou, C., Mouaddib, N. *et al.* (1994). Catalytic combustion of methane over palladium supported on alumina catalysts: evidence for reconstruction of particles, *Appl. Catal.*, **109**, pp. 277–291.
25. Garbowski, E. and Primet, M. (1995). Catalytic combustion of methane over palladium supported on alumina catalysts: evidence for reconstruction of particles. Corrigendum, *Appl. Catal.*, **125**, pp. 185–187.
26. Marécot, P., Fakche, A., Kellali, B., *et al.* (1994). Propane and propene oxidation over platinum and palladium on alumina. Effect of chloride and water, *Appl. Catal. B: Environmental*, **3**, pp. 283–294.
27. Engel, T. and Ertl, G. (1979). Elementary steps on catalytic oxidation of carbon monoxide over platinum metals, *Adv. Catal.*, **28**, pp. 1–78.
28. Nieuwenhuys, B., (1983). Adsorption and reaction of CO, NO, $H_2$ and $O_2$ over Group VIII-metal surfaces, *Surf. Sci.*, **126**, pp. 307–336.
29. Toyoshima, I. and Somorjai, G. (1979). Heats of chemisorption of $O_2$, $H_2$, CO, $CO_2$ and $N_2$ on polycrystalline and single-crystal transition metal surfaces, *Catal. Rev. -Sci. Eng.*, **19**, pp. 105–159.
30. Bradford, M. and Vannice, M. (1996). Estimation of CO heats of chemisorption on metal surfaces from vibrational spectra, *Ind. Eng. Chem. Res.*, **35**, pp. 3171–3178.
31. Ge, Q., Kose, R. and King, D. (2000). Adsorption energetics and bonding from femtomole calorimetry and from first principles theory, *Adv. Catal.*, **45**, pp. 207–259.
32. Brennan, D., Hayward, D. and Trapnell, B. (1960). The calorimetric determination of the heats of adsorption of oxygen on evaporated metal films, *Proc. Roy. Soc. A*, **256**, pp. 81–105.
33. Yeo, Y., Vattuone, L. and King, D. (1997). Calorimetric heats for CO and $O_2$ adsorption and for the catalytic CO oxidation reaction on Pt{111}, *J. Chem. Phys.*, **106**, pp. 392–401.
34. Martin, R., Gardner, P. and Bradshaw, A. (1995). The adsorbate-induced removal of the Pt(100) surface reconstruction. 2. CO, *Surf. Sci.*, **342**, pp. 69–84.
35. Sharma, R., Brown, W. and King, D. (1998). The adsorption of CO on Pt{110} over the temperature range from 90 to 300K studied by RAIRS, *Surf. Sci.*, **414**, pp. 68–76.
36. Imbihl, R. and Ertl, G. (1995). Oscillatory kinetics in heterogeneous catalysis, *Chem. Rev.*, **95**, pp. 697–733.
37. Bar, M., Nettesheim, S., Rotermund, H. *et al.* (1995). Transition between front and spiral waves in a bistable surface reaction, *Phys. Rev. Lett.*, **74**, pp. 1246–1249.
38. Dicke, J. Rotermund, H. and Lauterbach, J. (2000). *Surf. Sci.*, **454**, pp. 352–357.
39. a) Serre, C., Garin, F., Belot, G., *et al.* (1993). Reactivity of Pt-$Al_2O_3$ and Pt-$CeO_2$-$Al_2O_3$ for the oxidation of carbon monoxide by oxygen. Catalyst characterization by TPR using CO as reducing agent, *J. Catal.*, **141**, pp. 1–8; b) Serre, C., Garin, F., Belot, G., *et al.* (1993). Reactivity

of $Pt-Al_2O_3$ and $Pt-CeO_2-Al_2O_3$ for the oxidation of carbon monoxide by oxygen. Influence of the pretreatment step on the oxidation mechanism, *J. Catal.*, **141**, pp. 9–20.
40. Nibbelke, R., Nievergeld, A., Hoebink, J., *et al.* (1998). Development of a transient kinetic model for the CO oxidation by $O_2$ over a $Pt/Rh/CeO_2/\gamma-Al_2O_3$ three-way catalyst, *Appl. Catal. B: Environmental*, **19**, pp. 245–259.
41. Manuel, I., Chaubet, J., Thomas, C., *et al.* (2004). Simulation of the transient CO oxidation over $Rh^0/SiO_2$ and $Rh^{x+}/Ce_{0.68}Zr_{0.32}O_2$, *J. Catal.*, **224**, pp. 269–277.
42. Shekhtman, S., Goguet, A., Burch, R., *et al.* (2008). CO multipulse TAP studies of 2% $Pt/CeO_2$ catalyst: Influence of catalyst pretreatment and temperature on the number of active sites observed, *J. Catal.*, **253**, pp. 303–311.
43. Johansson, S., Österlund, L. and Kasemo, B. (2001). CO oxidation bistability diagrams for Pt/CeOx and $Pt/SiO_2$ model catalysts prepared by electron-beam lithography, *J. Catal.*, **201**, pp. 275–285.
44. Gololobov, A., Bekk, I., Bragina, G., *et al.* (2009). Platinum Nanoparticle Size Effect on Specific Catalytic Activity in n-Alkane Deep Oxidation: Dependence on the Chain Length of the Paraffin, *Kin. Katal.*, **50**, p[. 830–836.
45. Demoulin, O., Le Clef, B., Navez, M., *et al.* (2008). Combustion of methane, ethane and propane and of mixtures of methane with ethane or propane on $Pd/\gamma-Al_2O_3$ catalysts, *Appl. Catal. A: General*, **344**, pp. 1–9.
46. Dubien, C., Schweich, D., Mabilon, G., *et al.* (1998). Three-way catalytic converter modelling: fast- and slow-oxidizing hydrocarbons, inhibiting species, and steam-reforming reaction, *Chem. Eng. Sci.*, **53**, pp. 471–481.
47. Bera, P., Patil, K., Vayaram, V., *et al.* (2000). Ionic dispersion of Pt and Pd on CeO2 by combustion method: Effect of metal-ceria interaction on catalytic activities for NO reduction and CO and hydrocarbon oxidation, *J. Catal.*, **196**, pp. 293–301.
48. Garetto, T., Rincón, E. and Apesteguia, C. (2007). The origin of the enhanced activity of Pt/zeolites for combustion of $C_2$-$C_4$ alkanes, *Appl. Catal. B: Environmental*, **73**, pp. 65–72.
49. Yacou, C., Ayral, A., Giroir-Fendler, A., *et al.* (2009). Hierarchical porous silica membranes with dispersed Pt nanoparticles, *Micr. Meso. Mater.*, **126**, pp. 222–227.
50. Postole, G., Bonnetot, B., Gervasini, A., *et al.* (2007). Characterisation of BN-supported palladium oxide catalyst used for hydrocarbon oxidation, *Appl. Catal. A: General*, **316**, pp. 250–258.
51. Wu, H., Liu, L. and Yang, S. (2001). Effects of additives on supported noble metal catalysts for oxidation of hydrocarbons and carbon monoxide, *Appl. Catal. A: General*, **211**, pp. 159–165.
52. Yentekakis, I., Tellou, V., Botzolaki, G., *et al.* (2005). A comparative study of the $C_3H_6+NO+O_2$, $C_3H_6+O_2$ and $NO+O_2$ reactions in excess oxygen over Na-modified Pt/gamma-$Al_2O_3$ catalysts, *Appl. Catal. B: Environmental*, **56**, pp. 229–239.
53. Corro, G., Montiel, R. and Vásquez, C. (2002). Promoting and inhibiting effect of $SO_2$ on propane oxidation over $Pt/Al_2O_3$, *Catal. Comm.*, **3**, pp. 533–539.
54. Lee, A., Seabourne, C. and Wilson, K. (2006). Sulphate-promotion and structure-sensitivity in hydrocarbon combustion over $Rh/Al_2O_3$ catalysts, *Catal. Comm.*, **7**, pp. 566–570.
55. Maillet, T., Solleau, C., Barbier Jr, J., *et al.* (1997). Oxidation of carbon monoxide, propene, propane and methane over a $Pd/Al_2O_3$ catalyst. Effect of the chemical state of Pd, *Appl. Catal. B: Environmental*, **14**, pp. 85–95.
56. Ferhat-Hamida, Z., Barbier Jr, J., Labruquere, S., *et al.* (2001). The chemical state of palladium in alkene and acetylene oxidation — A study by XRD, electron microscopy and TD-DTG analysis, *Appl. Catal. B: Environmental*, **29**, pp. 195–205.
57. Castellazzi, P., Groppi, G., Forzatti, P., *et al.* (2010). Role of Pd loading and dispersion on redox behaviour and $CH_4$ combustion activity of $Al_2O_3$ supported catalysts, *Catal. Today*, **155**, pp. 18–26.

58. Baylet, A., Marécot, P., Duprez, D., *et al.* (2011). In situ Raman and in situ XRD analysis of PdO reduction and Pd degrees of oxidation in Pd supported on gamma-Al2O3 catalyst under different atmospheres, *Phys. Chem. Chem. Phys.*, **13**, pp. 4607–4613.
59. Hungría, A., Calvino, J., Anderson, J., *et al.* (2006). Model bimetallic Pd-Ni automotive exhaust catalysts: Influence of thermal aging and hydrocarbon self-poisoning, *Appl. Catal. B: Environmental*, **62**, pp. 359–368.
60. Mathieu, O., Lavy, J. and Jeudy, E. (2009). Investigation of Hydrocarbons Conversion Over a Pt-Based Automotive Diesel Oxidation Catalyst: Application to Exhaust Port Fuel Injection, *Top. Catal.*, **52**, pp. 1893–1897.
61. Raux, S., Frobert, A. and Jeudy, E. (2009). Low Temperature Activity of Euro4 Diesel Oxidation Catalysts: Comprehensive Material Analyses and Experimental Evaluation of a Representative Panel, *Top. Catal.*, **52**, pp. 1903–1908.
62. Amon-Meziere, I., Castagna, F., Prigent, M., *et al.* (1995). *SAE Technical Paper* No. 950932.
63. Patterson, M., Angove, D. and Cant, N. (2000). The effect of carbon monoxide on the oxidation of four $C_6$ to $C_8$ hydrocarbons over platinum, palladium and rhodium, *Appl. Catal. B: Environmental*, **26**, pp. 47–57.
64. Patterson, M., Angove, D. and Cant, N. (2001). The effect of metal order on the oxidation of a hydrocarbon mixture over alumina-supported combined platinum/rhodium catalysts, *Appl. Catal. B: Environmental*, **35**, pp. 53–58.
65. Tsai, Y., Xu, C. and Koel, B. (1997). Chemisorption of ethylene, propylene and isobutylene on ordered Sn/Pt(111) surface alloys, *Surf. Sci.*, **385**, pp. 37–59.
66. Hasan, A., Leung, P., Tsolakis, A., *et al.* (2011). Effect of composite after treatment catalyst on alkane, alkene and monocyclic aromatic emissions from an HCCI/SI gasoline engine, *Fuel*, **90**, pp. 1457–1464.
67. Dryakhlov, A. and Kiperman, S. (1981). Kinetic of the thorough oxidation of small quantities of benzene in air on a platinum catalyst. 1. Principal kinetic laws, *Kinet. Katal.*, **22**, pp. 159–163.
68. Liotta, L. (2010). Catalytic oxidation of volatile organic compounds on supported noble metals *Appl. Catal. B: Environmental*, **100**, pp. 403–412.
69. Santos, V., Carabineiro, S., Tavares, P., *et al.* (2010). Oxidation of CO, ethanol and toluene over $TiO_2$ supported noble metal catalysts, *Appl. Catal. B: Environmental*, **99**, pp. 198–205.
70. Ordóñez, S., Bello, L., Sastre, H., *et al.* (2002). Kinetics of the deep oxidation of benzene, toluene, n-hexane and their binary mixtures over a platinum on gamma-alumina catalyst, *Appl. Catal. B: Environmental*, **38**, pp. 139–149.
71. dos Santos, A., Lima, K., Figueiredo, R., *et al.* (2007). Toluene deep oxidation over noble metals, copper and vanadium oxides, *Catal. Lett.*, **114**, pp. 59–63.
72. Burgos, M., Paulis, M., Antxustegi, M., *et al.* (2002). Deep oxidation of VOC mixtures with platinum supported on $Al_2O_3$/Al monoliths *Appl. Catal. B: Environmental*, **38**, pp. 251–258.
73. Kim, K. and Ahn, H. (2009). Complete oxidation of toluene over bimetallic Pt-Au catalysts supported on ZnO/$Al_2O_3$, *Appl. Catal. B: Environmental*, **91**, pp. 308–318.
74. Uchisawa, J., Kosuge, K., Nanba, T., *et al.* (2009). Effect of meso- and macropore structures of Pt-supported fibrous silica on the catalytic oxidation of toluene, *Catal. Lett.*, **133**, pp. 314–320.
75. Barresi, A., Cittadini, M. and Zucca, A. (2003). Investigation of deep catalytic oxidation of toluene over a Pt-based monolithic catalyst by dynamic experiments, *Appl. Catal. B: Environmental*, **43**, pp. 27–42.
76. Beauchet, R., Magnoux, P. and Mijoin, J. (2007). Catalytic oxidation of volatile organic compounds (VOCs) mixture (isopropanol/o-xylene) on zeolite catalysts, *Catal. Today*, **124**, pp. 118–123.
77. Beauchet, R., Mijoin, J., Batonneau-Gener, I., *et al.* (2010). Catalytic oxidation of VOCs on NaX zeolite: Mixture effect with isopropanol and o-xylene, *Appl. Catal. B: Environmental*, **100**, pp. 91–96.

78. Padilla, J., Del Angel, G. and Navarette, J. (2008). Improved Pd/$\gamma$-$Al_2O_3$–Ce catalysts for benzene combustion *Catal. Today*, **133–135**, pp. 541–547.
79. Kim, S., and Shim, W. (2009). Properties and performance of Pd based catalysts for catalytic oxidation of volatile organic compounds, *Appl. Catal. B: Environmental*, **92**, pp. 429–436.
80. Shie, J., Chang, C., Chen, J., *et al.* (2005). Catalytic oxidation of naphthalene using a Pt/$Al_2O_3$ catalyst, *Appl. Catal. B: Environmental*, **58**, pp. 289–297.
81. Klingstedt, F., Neyestanaki, A., Lindfors, L., *et al.* (2000). Hydrothermally stable catalysts for the removal of emissions from small-scale biofuel combustion systems, *React. Kinet. Catal. Lett.*, **70**, pp. 3–9.
82. Ndifor, E., Garcia, T., Solsona, B., *et al.* (2007). Influence of preparation conditions of nano-crystalline ceria catalysts on the total oxidation of naphthalene, a model polycyclic aromatic hydrocarbon, *Appl. Catal. B: Environmental*, **76**, pp. 248–256.
83. Bampenrat, A., Meeyoo, V., Kitiyanan, B., *et al.* (2008). Catalytic oxidation of naphthalene over CeO2-ZrO2 mixed oxide catalysts, *Catal. Comm.*, **9**, pp. 2349–2352.

# Chapter 2

# Soot Oxidation in Particulate Filter Regeneration

Junko UCHISAWA, Akira OBUCHI and Tetsuya NANBA*

This chapter deals with the mechanism of soot oxidation for the regeneration of particulate filters. Nature of catalysts and effect of process parameters (temperature, $NO_2$ partial pressure, intimacy of contact between soot and catalyst, ...) are reviewed. The role of oxygen mobility and of nature of oxygen species are also discussed.

## 2.1. Introduction

Generally, diesel engines are operated under lean-burn conditions, with excess air to fuel, so that the exhaust gases contain lower concentrations of CO and hydrocarbons than gasoline engines, which are operated near the stoichiometric air-to-fuel ratio. However, the liquid fuel is directly injected and burned in the cylinder; therefore, the air-to-fuel ratio may become very rich locally and part of the fuel is thermally decomposed to form solid carbonaceous particles, commonly referred to as soot. The soluble organic fraction (SOF) derived from fuel components with higher boiling points, lubricant oil, and sulfate (sulfuric acid mist) produced from sulfur in the fuel are later attached to this soot, which forms the main constituent of diesel particulate matter (PM).

Most of the PM mass from diesel engines ranges from 0.1 to 0.5 μm in diameter,[1] which is much less than the suspended particulate matter (SPM) standard prescribed in the Air Quality Standards of Japan (10 μm or less), and the PM2.5 (2.5 μm or less) standard prescribed by the United States Environmental Agency. To make matters worse, the majority of PM is made up of ultrafine particles with diameters of 0.005–0.1 μm. Ultrafine particles of 0.05 μm or smaller are called nanoparticles, and these cause concern with regard to invasion of respiratory organs such as bronchial tubes, alveolar cells, and further blood vessels, and are thus detrimental to human health.[2]

*National Institute of Advanced Industrial Science and Technology (AIST), Energy-saving System Team, Research Center for New Fuels and Vehicle Technology, 16-1 Onogawa, Tsukuba, 305-8569, Japan.

From the 1980s to early 2000s, control of PM was mainly accomplished using technologies that improved engine combustion, most successfully with the common-rail system, i.e., an electronically controlled high-pressure fuel injection system. However, as more severe emission regulations have been enforced worldwide,[3] the adoption of after-treatment systems that directly remove exhausted PM has become inevitable.

There are two types of PM after-treatment devices: diesel oxidation catalysts (DOC) and diesel particulate filters (DPF). DOCs involve a flow-through honeycomb washcoating of an oxidation catalyst, most typically Pt-Pd/$Al_2O_3$ (Fig. 2.1a), and are effective for the removal of the SOF in PM, in addition to CO and hydrocarbons in the gas phase, oxidizing them completely to $CO_2$ and $H_2O$. The SOF can be treated as it is still in a liquid state at high temperature when it passes through the DOC, so that it can easily contact the catalyst surface. Depending on the SOF content, PM concentration can be decreased by 20–30% using the DOC method. This level of performance was sufficient in the past when PM regulations were not so stringent. Pd, which has a lower oxidation activity, has been combined with Pt as an active component of the DOC because too much oxidation activity of Pt alone produces sulfate by the oxidation of $SO_2$ contained in the exhaust gas, which may increase PM emissions. However, sulfate production has been of little concern recently because many advanced countries are now restricting sulfur content in diesel fuel to very low levels, such as below 10 ppm.

The second after-treatment device, the DPF, basically filters the exhaust gas to separate and collect all PM, including soot. The wall-flow monolith-type DPF is the most commonly used structure and has excellent PM collection efficiencies, exceeding 95%. The structure is comprised of a flow-through type honeycomb with gas-permeable porous walls which alternately seal the end of each honeycomb channel, as schematically shown in Fig. 2.1b. Most commercially produced wall-flow DPFs are made of cordierite ($2MgO{\cdot}2Al_2O_3{\cdot}5SiO_2$) or silicon carbide (SiC), which have high heat resistance (1,000°C or more) and thermal shock resistance. The

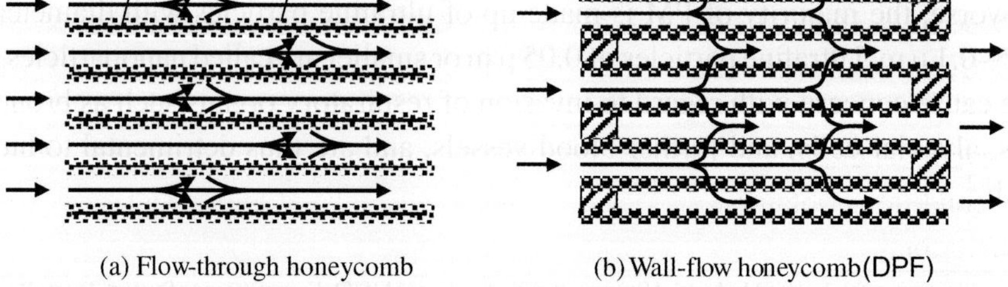


**Figure 2.1.** Schematic illustration of honeycomb substrates for diesel after treatment.

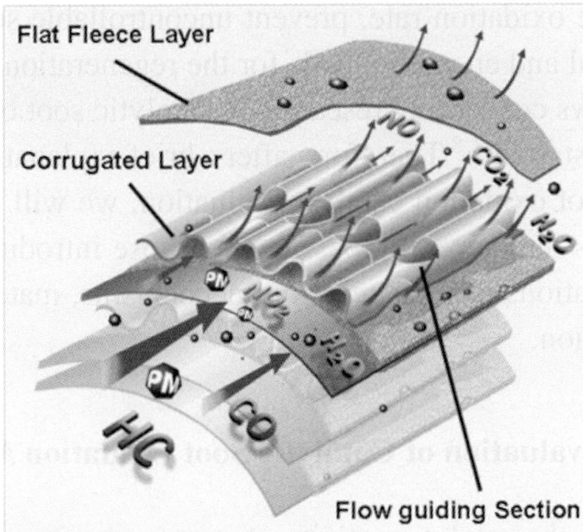


**Figure 2.2.** Example of a flow-through type DPF made by stacking corrugated metal plates (PM-METALIT® produced by Emitec).[6]

filter walls, which are approximately 0.3 mm thick, have interconnected pores of *ca.* 10 μm in diameter and PM is trapped in these walls. Most PM is collected near or over the surfaces of the inlet side of the DPF walls, which is known as shallow filtration. Other types of DPF include those made of woven alumina or SiC cloth and reinforced with wire net,[4] and wire net with fine cylindrically rolled meshes.[5] With these types of DPF, PM is collected in all of the filter material, which is known as deep filtration. More recently, flow-through type DPFs that stack corrugated metal plates have been developed by Emitec (Fig. 2.2),[6] in which the pressure drop remains low while collecting PM with moderate collection efficiencies. The shape of the corrugated plates is designed to create a turbulent flow of the exhaust gas to promote collision and trapping of PM on the wall surfaces of the narrow passages.

The PM collected in DPFs should be continuously or occasionally oxidized and removed from the filter to maintain filtration ability, which is referred to as regeneration. With DPF application, it is most important to realize smooth regeneration with high reliability while the vehicle is in operation. It is necessary to heat the DPF to 600°C or more, which is not attained during typical vehicle operation, to complete PM oxidation (burnout) within a few minutes or less. Artificial heating is achieved by the catalytic combustion of fuel supplied to a DOC placed upstream of the DPF, by retarded combustion in the engine, via post injection into the engine cylinders, or by direct injection into the exhaust pipe. Such heating is referred to as active regeneration, in contrast to that in a system that uses exhaust gas temperature conditions obtained in typical engine operations, which is known as passive regeneration.

In this chapter, the latest research and development in the oxidation of soot in PM emitted from diesel engines is reviewed. Catalysts are necessary when using

DPFs to increase the oxidation rate, prevent uncontrollable soot combustion, and decrease instrumental and energetic loads for the regeneration process. There have been excellent reviews concerning research on catalytic soot oxidation since 1980, when R&D of DPFs started.[7,8] Therefore, after a brief explanation of the commonly used methods for soot oxidation catalyst evaluation, we will classify the recently investigated catalysts into several types, mostly those introduced since 2000, and give detailed explanations of their reaction mechanisms, materials, and situations for practical application.

## 2.2. Method for Evaluation of Catalytic Soot Oxidation Activity

In laboratory studies, the catalytic activity of soot (carbon) oxidation is most commonly evaluated by temperature-programmed reaction (TPR). A typical experimental configuration for TPR is shown in Fig. 2.3. A sieved granular catalyst and commercially available carbon black (CB) with no metal impurities, used as a model diesel soot, is mixed together with a catalyst/CB ratio of 10:100 by weight and the mixture is installed in a tubular quartz reactor. Neeft *et al.* reported that the contact between the catalyst and PM in an actual DPF is as loose as that formed by rough mixing the materials with a spatula.[9] However, many laboratory-scale investigations have been carried out by mixing the catalyst and CB more thoroughly with a mortar or by instrumental ball milling, probably to highlight the differences in the activities

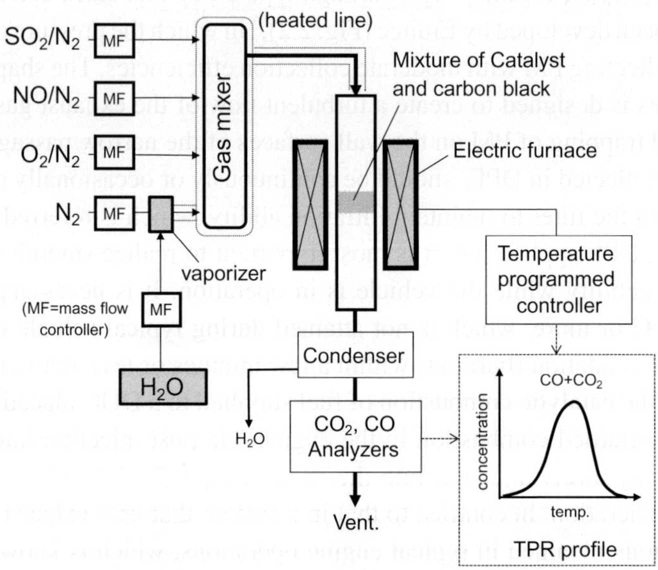


**Figure 2.3.** Experimental configuration of the temperature-programmed reaction (TPR) used to evaluate soot oxidation catalysts.

of catalyst samples. The contact formed with these methods is referred to as a tight contact. The reactant gas, a model diesel exhaust gas containing NO, $SO_2$, $H_2O$, and $O_2$ with $N_2$ balance, is passed through a mixture of the loose or tight contact and the reactor temperature is then ramped linearly at a rate of 1–10 K/min. The humidified reactant gas should be introduced after the sample temperature is above the dew point to prevent water condensation on the sample, which would instantly change the loose contact to a tight contact. The concentrations of product, especially $CO_2$ and CO, are analyzed continuously or periodically with a short interval, using, for instance, a non-dispersive infrared (IR) gas analyzer and a Fourier transform infrared (FTIR) spectrometer equipped with a multi-reflectance gas cell. The catalytic activity is evaluated from the $CO_2$ or $CO_2 + CO$ concentration profile. The peak ($T_p$) and partial oxidation ($T_x$) temperatures, at which a certain percentage (x%) of CB is oxidized during the TPR, are often used as an index of the catalytic activity (Fig. 2.4a). The method developed by Ozawa *et al.*[10] for the analysis of thermogravimetric data may be able to treat the TPR results of carbon oxidation more quantitatively but has not yet been applied to a great extent.

Under practical scale and reaction conditions using DPF test pieces loaded with catalysts and engine exhaust gases, the catalytic performance is often evaluated by the balance point temperature (BPT; Fig. 2.4b), which is determined from the changes in pressure drop ($\Delta$p) through the DPF with time at different temperatures. At low temperatures, $\Delta$p increases with time as PM accumulates on the DPF. As the temperature is raised, the PM oxidation rate increases and finally exceeds the PM accumulation rate at a certain point, which results in a decrease in pressure drop across the DPF with time. The BPT is defined as the temperature at which the rate of change of $\Delta$p is equal to zero.

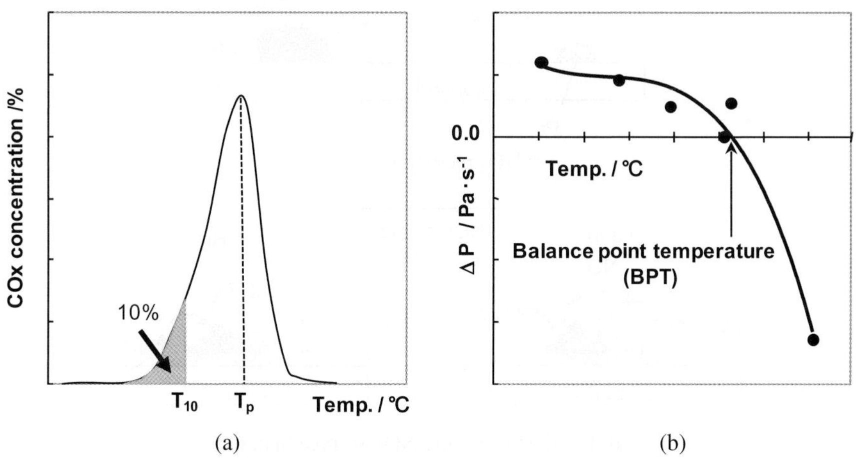


**Figure 2.4.** Indices of soot oxidation activity.

## 2.3. Classification of PM Oxidation Catalyst

Numerous soot oxidation catalysts have been reported since the 1980s, because soot oxidation is fundamentally a simple complete oxidation reaction (carbonaceous compounds $\rightarrow CO_2 + H_2O$), so that sophisticated catalysts with high selectivity are not required. However, there is a critical problem in establishing contact and interaction, directly or indirectly, between the reactant (soot) and the catalyst, both of which are solid materials. Therefore, soot oxidation catalysts reported to date can be classified according to the assumed working mechanism that solves this problem. In this review the authors classify the catalysts into the four types shown in Fig. 2.5, based on the mediator for the oxidation reaction that connects the active sites of catalyst and soot surfaces: mobile catalysts, mobile oxygen catalysts, $NO_2$ mediating

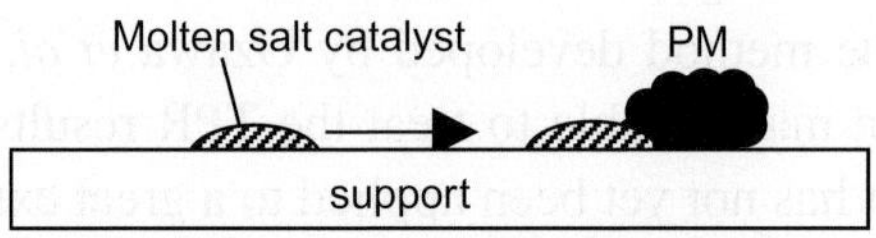


(a) Mobile catalyst

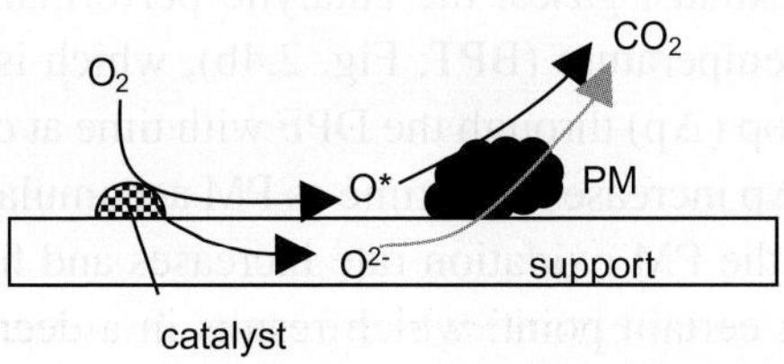


(b) Mobile oxygen catalyst

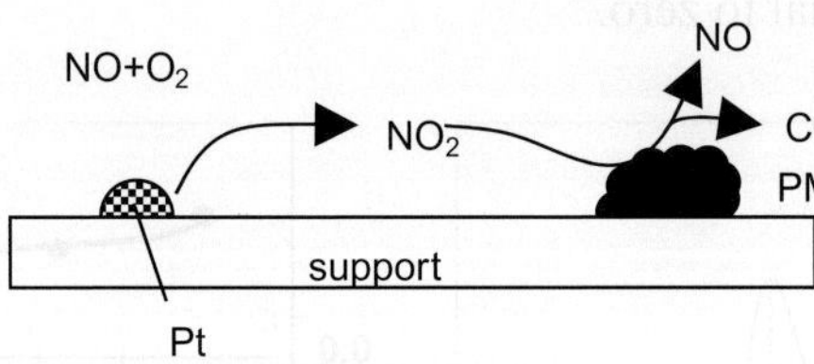


(c) $NO_2$ mediating catalyst

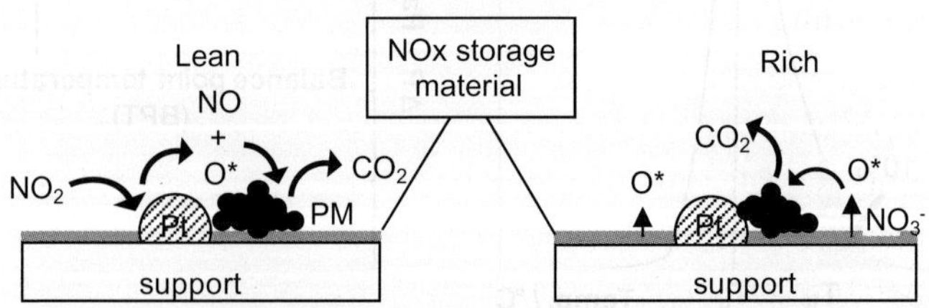


(d) Catalyst involving NOx storage material

**Figure 2.5.** Soot oxidation catalysts classified according to the working mechanism.

catalysts, and catalysts involving NOx storage materials. Similar classification was employed in Fino's recent review article.[8]

In most cases, the catalyst should be supported directly on a DPF where soot is collected. The mobile catalysts are characterized by having a relatively low melting point or a high vapor pressure, which provides high mobility and the ability to establish new contact points with soot under oxidation. The mobile oxygen catalysts, which are mainly composed of a simple or mixed-metal oxide, have surface or lattice oxygen species that are mobile due to spillover on the surface or by ionic conduction in the bulk, which promote the oxidation reaction by the supply of oxygen from the active sites of the catalyst to the soot surface. The $NO_2$ mediating catalysts oxidize NO that is typically present in the diesel exhaust gas to $NO_2$, a stronger oxidizer than $O_2$, on the catalyst surface. The $NO_2$ produced desorbs from the catalyst surface, diffuses into the gas phase, and then reaches and oxidizes the soot surface. $NO_2$ is a stable molecule, therefore this type of catalyst is not necessarily supported on a DPF but can be placed upstream of the DPF. Catalysts that involve NOx storage materials are often seen in catalysts that intend to simultaneously remove PM and NOx. The NOx storage material originally introduced for the removal of NOx under lean conditions can supposedly function as a mediator for soot oxidation.

It should be noted that in practical application, the working mechanism of a catalyst cannot be explicitly classified into one specific type, but rather is a combination of two or more of these types and in some cases other mechanisms may also be involved. Consequently, the following classification should be regarded as based only on the predominant working mechanism. In the next section, the reaction mechanisms of each catalyst type will be explained in more detail and promising examples of each catalyst type are given.

## 2.4. Mechanisms and Examples of Each Catalyst Type

### 2.4.1. *Mobile catalysts*

This catalyst may also be referred to as the molten salt catalyst, as referred to by Jelles *et al.*, according to the catalyst materials of this type reported to date.[11] It is well known that some of the transition metal oxides, alkaline, and alkaline-earth metal oxides promote carbon oxidation.[12] These oxides are solid and immobile at room temperature but become mobile on the surfaces of soot and support materials on a micrometer scale above certain temperatures, the melting point, or so-called Tamman temperature. In such a mobile state, the catalyst can maintain contact with the soot while the soot surfaces are continually excavated by oxidation.

However, high mobility at working temperatures is not the only requisite for a material to become a good catalyst of this type. It is also necessary to have high reversibility of the oxidation states, i.e., high redox properties. For example, $V_2O_5$ is assumed to supply oxygen to carbon surfaces by recycling between different redox states, e.g., $V_2O_5$ and $V_6O_{13}$.[13] Other transition metal compounds, especially those of Mo and Cu, also have both high mobility and redox properties, and exhibit high carbon oxidation activity.

Alkaline and alkaline-earth metal ions do not seem to have high redox properties because they have only one stable ionic valency, +1 and +2, respectively. However, thermodynamic considerations of the oxides of these metals predict that the most stable compounds at approximately 500°C in 0.2 atm of $O_2$ are $Li_2O$, $NaO_2$, $KO_2$, $RbO_2$, $CsO_2$, BeO, MgO, CaO, SrO, and $BaO_2$.[14] Speculating from these chemical formula, some metals may naturally generate peroxide ($O_2^{2-}$) or superoxide ($O_2^-$) ions. In addition, for Li, Mg, Ca, and Sr oxides, it was estimated that peroxides such as $Li_2O_2$, $MgO_2$, $CaO_2$, and $SrO_2$ are produced under certain conditions. The oxidation activity of alkaline and alkaline-earth metal oxides has been correlated to the electronegativity of the metals by Castoldi *et al.*, as shown in Fig. 2.6 (alumina-supported catalyst in tight contact, $T_{50}$: 50% oxidation temperature in TPR experiments).[15] Higher activities are evident for metals with lower electronegativities. It is speculated that as the electronegativity difference between the metal and oxygen increases, the valence electron of the metal is donated to a larger number of oxygen atoms than that in typical oxides ($O^{2-}$), which results in the formation of active oxygen species, such as $O_2^{2-}$, $O_2^-$, and $O^-$. In addition, compared with the corresponding carbonates and typical oxides, the melting point of these peroxides and superoxides are generally lower, from 400 to 600°C,[16] from which high mobility is expected.

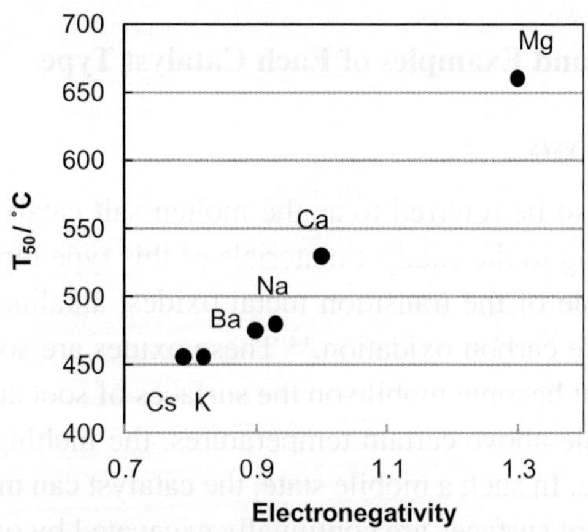


**Figure 2.6.** Correlation between the electronegativity of alkaline and alkalineearth metal catalysts and soot oxidation activity. (Reproduced from Ref. 15)

Mixed-metal oxides and eutectic mixtures based on $V_2O_5$ or $MoO_3$ have been well investigated from the viewpoint that these materials have relatively low melting points. $AgVO_3$ (m.p. = 470°C),[17] mixed oxides of V with K or Cs,[18,19] $Cs_2MoO_4$-$V_2O_5$ ($Cs_2O$:$MoO_3$:$V_2O_5$ = 0.33:0.33:0.34, m.p. = 352°C),[20] and K or Cs-Fe-V oxides supported on $Al_2O_3$[21] are among the reported active catalysts.

Mul *et al.* reported that lower melting points can be obtained in the form of chlorides or oxychlorides with accordingly higher soot oxidation activities.[22] However, these compounds have higher vapor pressures than the corresponding metal oxides, so that they are evaporated and lost during extended use, and are not considered practical.

Similarly, metal nitrates have been studied as candidates with low melting points and high oxidizing activities, most of which are alkaline metal nitrates, with $KNO_3$ (m.p. = 334°C) being the most commonly examined.[23–26] Gross *et al.* reported that the carbon oxidation mechanism of $KNO_3$ combined with $CeO_2$ is promoted by a synergistic effect of $KNO_3$ and superoxide ($O_2^-$) generated on $CeO_2$.[27] The nitrate ion is expected to be continuously supplied from $NO_2$ generated from NO in the exhaust gas. However, the problem is loss of $KNO_3$, as is the case for chlorides and oxychlorides. Wu *et al.* found that by using $CeO_2$-$ZrO_2$ as a support material instead of $CeO_2$ alone, $KNO_3$ becomes more stable on the surface.[28] Milt *et al.* also reported that by the addition of $Ba(NO_3)_2$ together with $KNO_3$, the change of $KNO_3$ to inert $K_2SO_4$ in the presence of $SO_2$ in the exhaust gas is retarded and the active period is prolonged because the formation of $BaSO_4$ suppresses that of $K_2SO_4$.[29,30]

### 2.4.2. *Mobile oxygen catalysts*

Lattice oxygen is often used for oxidation of hydrocarbons over various metal oxide catalysts (Mars–van Krevelen mechanism).[31] Among the metal oxides that exhibit such catalysis, $CeO_2$ is known to have especially high redox properties between $Ce^{3+}$ and $Ce^{4+}$ and release lattice oxygen into the gas phase. Utilization of this mobile oxygen has been attempted in many studies. In TPR experiments, the 50% oxidation temperature ($T_{50}$) of CB mixed with $CeO_2$ calcined at 1,000°C was *ca.* 570°C for tight contact.[32] Even after such high temperature calcinations, $CeO_2$ showed lower $T_{50}$ than the temperature of 620°C or more for non-catalytic CB oxidation. Bueno-Lopez *et al.* has provided evidence for the involvement of $CeO_2$ lattice oxygen in soot oxidation from temporal analysis of the products;[33] however, the details of the reaction mechanism are yet to be clarified. At present, the following three mechanisms have been proposed.

(1) Active oxygen is released from the lattice by redox reactions of $CeO_2$ and oxidizes the soot (Fig. 2.7, mechanism 1).[34,35] The redox capacity of $CeO_2$

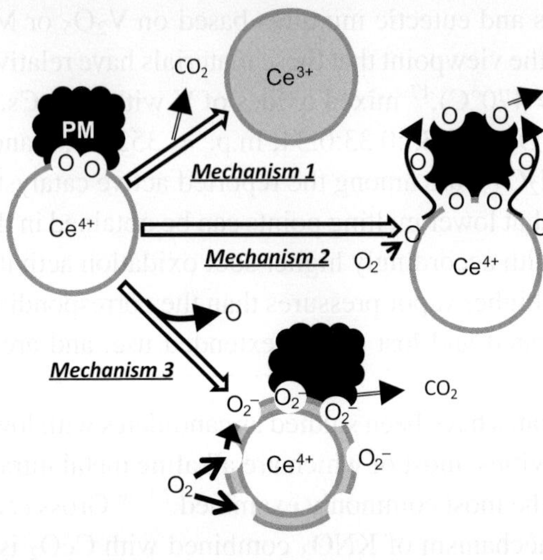


**Figure 2.7.** Three suggested mechanisms for soot oxidation on $CeO_2$, with different types of mobile oxygen.

is well known as the oxygen storage capacity (OSC). When this capacity is improved by the formation of a solid solution with Zr or rare earth metals (CeMOx), the solid solution exhibits higher activity for soot oxidation than $CeO_2$ alone.[36–38]

(2) Oxygen molecules are dissociatively adsorbed on $CeO_2$. The adsorbed active oxygen spills over the $CeO_2$ surface, moves onto the soot surface, and reacts at active sites on the soot surface (mechanism 2).[39] In this mechanism, the OSC does not affect the activity, because the migration of lattice oxygen is not involved.

(3) Active oxygen species, such as superoxide ($O_2^-$), are formed on surface oxygen vacancy sites generated by the reduction of $CeO_2$, and these oxidize the soot (mechanism 3).[40,41] The formation of superoxide species has been evidenced by electron paramagnetic resonance analysis.[42,43] In this reaction mechanism, OSC is also presumed to not influence the catalytic activity, although the reducibility of the surface oxygen has a significant effect on the formation of superoxide species.

In any of these mechanisms, the loading of metals with redox properties onto $CeO_2$ is effective for enhancement of the catalytic activity. Among the various metals attempted, Ag is especially attractive, for Ag/$CeO_2$ exhibited high soot oxidation activity.[44–49] Ag itself is known to have high catalytic activity for carbon oxidation, as observed by the behavior of small Ag particles on a graphite surface that penetrate into the graphite by oxidizing the contact point, as is the case for V, K, and Cu.[50]

However, the reaction mechanism proposed by Shimizu *et al.* for $Ag/CeO_2$ is that Ag does not directly oxidize soot, but promotes the reduction of $Ce^{4+}$ to $Ce^{3+}$, on which adsorbed oxygen is activated and oxidizes the soot.[51] Clarification of the detailed mechanism of soot oxidation on $CeO_2$, including the role of metal components, will make a significant contribution to the development of more active mobile oxygen type catalysts.

$CeO_2$ does not only generate active oxygen, but also promotes NO oxidation to $NO_2$, with which soot oxidation is promoted. This process is described later.

Perovskite-type oxides are another type of material that have mobile lattice oxygen and are active as soot oxidation catalysts. The perovskite composition is expressed as $ABO_3$, and substitution of either the A or B site metals with a different valency metal will increase the amount of oxygen vacancies and lattice oxygen mobility, and also improve the redox properties, thereby improving the soot oxidation activity.[52–54] Fino *et al.* proposed a mechanism in which soot oxidation proceeds through spillover of active oxygen adsorbed on perovskites.[55] The amount of active oxygen was dependent on the number of oxygen vacancies on the surface. However, the mechanism of soot oxidation over perovskite oxides is not yet fully understood.

### 2.4.3. *$NO_2$ mediating catalysts*

In typical diesel engine exhaust gas that contains nitrogen oxides (mainly NO) in addition to soot, the oxidation rate of soot can be substantially increased by using Pt catalysts such as $Pt/Al_2O_3$; the effect was first reported by Cooper *et al.* and Hawker *et al.* of Johnson Matthey.[56–58] The Pt catalyst and soot do not have to directly contact each other, or even be close to each other, because the effect is attributed to $NO_2$ produced from NO over the catalyst, which is a more active oxidizer than $O_2$, that can diffuse in the gas phase from the catalyst to the soot surfaces and promote soot oxidation at low temperatures. The reaction processes are expressed as follows:

$$2NO + O_2 \rightarrow 2NO_2 \tag{2.1}$$

$$2NO_2 + C(\text{soot}) \rightarrow 2NO + CO_2 + CO \tag{2.2}$$

Reaction (2.2) starts from as low as 250°C, so that the soot collected in the DPF is slowly and continuously oxidized under typical engine exhaust gas temperature conditions (180–300°C). As a result, Johnson Matthey refers to its developed DPF that uses this catalytic process as a continuously regenerating trap (CRT).

The authors have made detailed investigations of the effect of coexisting gas on reaction (2.2) and found that the presence of $H_2O$ (water vapor) and small concentrations of $SO_2$ substantially promote this reaction.[59] Figure 2.8 shows the results of TPR of a carbon black (model soot) in loose contact with $Pt/SiO_2$ under different

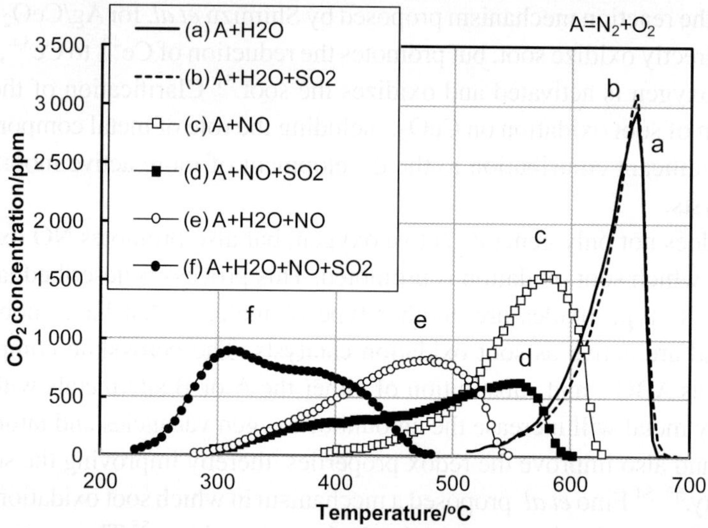


**Figure 2.8.** Effect of the reaction gas composition on the activity of $Pt/SiO_2$ for carbon oxidation. (Reproduced from Ref. 59)

atmospheres. The TPR profile in 10% $O_2$ + 7% $H_2O$ with $N_2$ balance had a peak temperature ($T_p$) around 660°C (Fig. 2.8a), with the total profile being almost the same as that under dry conditions (10% $O_2/N_2$, not shown). This suggests that $Pt/SiO_2$ itself has no remote oxidation catalysis effect to oxidize the soot. The TPR results show no significant change when $SO_2$ was added in addition to $O_2$ and $H_2O$ (Fig. 2.8b). $T_p$ shifted lower, to 580°C, for the first time when the mixture of $O_2$ (10%) and NO (1,000 ppm) was used as a reactant gas (Fig. 2.8c). This is explained as an example of the $NO_2$ mediating catalysis, although the effect is small. However, when $H_2O$ is added to this composition (reactant gas composed of $O_2$ + NO + $H_2O/N_2$), $T_p$ is substantially lowered to 480°C (Fig. 2.8e). Finally, when $O_2$, NO, $H_2O$, and $SO_2$ (100 ppm) are all present in the reactant gas, $T_p$ became 300°C and the light-off temperature was below 250°C (Fig. 2.8f). It is speculated that the high soot oxidation performance of Pt catalysts reported to date has always been realized under such gas mixing conditions.

It is unusual that carbon oxidation is promoted by $SO_2$, which in many cases hampers catalytic reactions. The authors have speculated on the $SO_2$ promotion effect as shown in Fig. 2.9.[60] The first attack of $NO_2$ to carbon (soot) under $H_2O$-deficient conditions may produce partially oxidizing groups, such as carbonyls and acid anhydrides, on the carbon surfaces, but the reaction may not proceed further because these groups are rather inert against complete oxidation because of their electron-accepting nature. However, when $H_2O$ and sulfuric acid ($H_2SO_4$ or $SO_3$) are present in the gas phase, the partially oxidized surface groups may be converted to

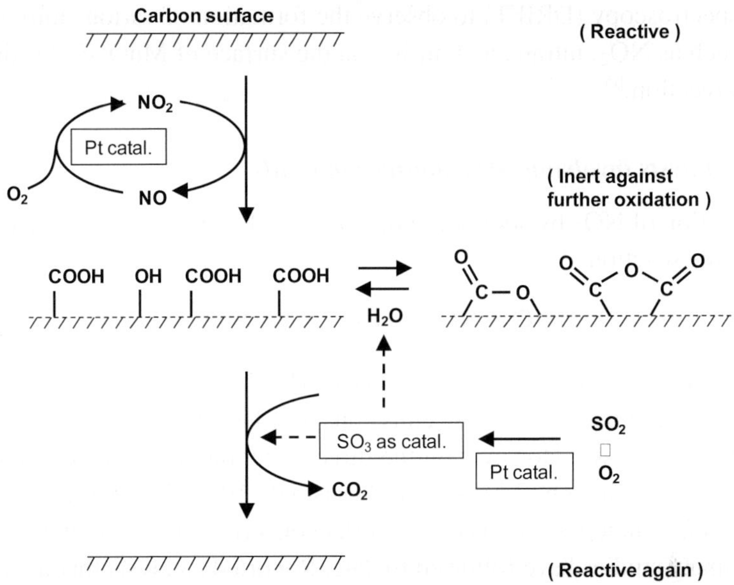


**Figure 2.9.** Suggested reaction scheme for carbon oxidation with Pt catalysts in the presence of NO, $SO_2$, and $H_2O$ in the reactant gas. (Reproduced from Ref. 60)

carboxyls by hydration and further decarbonated by sulfuric acid acting as a catalyst for these reactions. As a result, carbon in the top layer of the soot surface may be completely oxidized, revealing the next carbon layer so that oxidation reactions may continue. The effect of $H_2SO_4$ on the promotion of carbon oxidation in the presence of $NO_2$ was experimentally confirmed.[60] Taking this effect into account, the authors have attempted to optimize the Pt catalyst support for the promotion of soot oxidation at low temperatures.[61] However, the sulfur content in fuel is now reduced to extremely low levels to keep other catalytic processes active, which makes it difficult to apply this effect in practice.

Recently, $NO_2$ mediating type Pt catalysts have also been supported on DPFs with an active regenerating system. In such cases, the catalyst should have durability against sintering at temperatures as high as 700°C during repeated regeneration procedures. Pfeifer *et al.* have recently demonstrated that Pt and Pd-combined catalysts have higher durability against temperature than Pt catalysts.[62] In addition, a Pt-free catalyst composed of Pd-Au-Ag-Ni-Co alloy nanoparticles supported on mixed oxides of $TiO_2$ and $RuO_2$ has been reported to have high resistance to sulfur and an anchoring, i.e., an anti-sintering effect.[63] In recent years, non-Pt group catalysts have also been investigated, due to concerns regarding the shortage of precious metal natural resources. Tikhomirov *et al.* has reported low temperature oxidation of soot with $MnO_x$-$CeO_2$ as a substitute for Pt catalysts to oxidize NO, although the catalyst was weak against sulfur poisoning.[64] Wu *et al.* employed diffuse reflectance infrared spectroscopy (DRIFT) to observe the formation of various nitrogen oxide species, such as $NO_2$, nitrate, and nitrite, on the surface of $MnO_x$-$CeO_2$ during the oxidation reaction.[65]

### 2.4.4. *Catalysts involving NOx storage materials*

If the reduction of $NO_2$ by soot does not stop at NO, but proceeds further to $N_2$, as shown by the reaction

$$C + NO_2 \rightarrow CO_2 + 1/2N_2 \quad (2.3)$$

then both harmful substances, soot and NOx, can be simultaneously treated, although a stoichiometric imbalance of these emissions may be a problem for complete elimination of both species. However, in the direct reaction between carbon and $NO_2$, the selectivity for reaction (2.3) against (2.2) is only 10–15%.[66] Yoshida *et al.*[67] and Teraoka *et al.*[68] undertook pioneering works concerning this simultaneous reduction, and many studies have followed to date. Reported catalysts that are active for this reaction are perovskites, such as $La_{2-x}K_xCu_{1-y}V_yO_4$,[66] and spinels, such as $CuFe_2O_4$[69] and $CoCr_2O_4$,[70] Co and K supported on $La_2O_3$ or $CeO_2$,[71] K-doped $Fe_2O_3$,[72] and $MnO_x$-$CeO_2$.[64] Among these studies, Milt *et al.* have found high NOx storage and soot oxidation activities of K/$La_2O_3$ through its nitrate formation ability, and also the reducibility of nitrate to $N_2$ under reducing atmospheres.[73] Furthermore, they commented on the possibility of realizing the simultaneous reduction of soot and NOx by using this catalyst as a NOx storage material in the NOx storage reduction system (NSR) developed by Toyota.

Researchers of the Toyota group had earlier discovered this phenomenon.[74,75] The NSR is an exhaust gas treatment system used to reduce NOx from lean-burn gasoline engines that was developed in the early 1990s.[76,77] A Pt catalyst doped with base metals, such as Ba and K, is used under typically lean-burn (oxidizing) conditions with periodic, short time operation under rich-burn (reducing) conditions. Under lean conditions, NO in the exhaust gas is oxidized to $NO_2$ and absorbed as nitrates and nitrites. Under occasional rich conditions, these compounds are decomposed and emit $NO_2$ and NO, which are further reduced to $N_2$ by $H_2$ and CO that are abundant in the reducing atmosphere. Over the entire process, NOx is eliminated and the base material is restored to the oxide (such as BaO), hydroxide ($Ba(OH)_2$), and carbonate ($BaCO_3$), all of which have NOx storage ability.

The researchers of Toyota tested the performance of the NSR catalyst supported on a monolithic DPF under varying atmospheres between lean and rich conditions and found that soot oxidation is promoted under both conditions, in addition to the reduction of NOx, and a system they call DPNR (diesel particulate NOx reduction) was developed. From the detection of $O_2^-$ species by electron spin resonance (ESR), they suggested that during the course of nitrate and nitrite formation

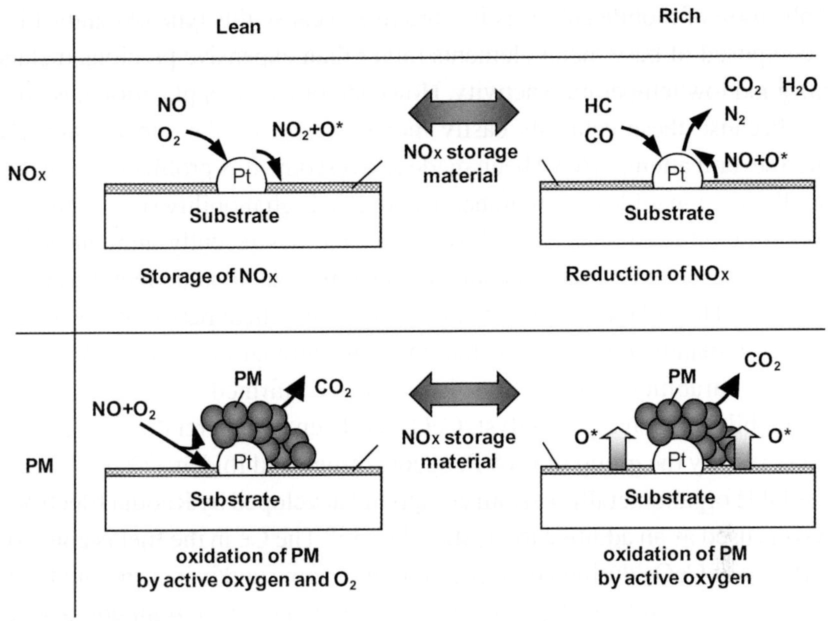


**Figure 2.10.** Suggested reaction mechanism for DPNR with the involvement of active oxygen produced by changes to the air-to-fuel ratio. (Reproduced from Ref. 74)

and decomposition, active oxygen is produced, which promotes soot oxidation as schematically shown in Fig. 2.10. Forzatti and co-workers investigated the reaction mechanism of DPNR in detail and reported that the nitrated NOx storage material, in addition to $NO_2$, directly oxidizes soot under lean conditions, while a small quantity of $CO_2$ emission (apparent soot oxidation) under rich conditions may result from $CO_2$ desorption from the carbonated NOx storage material prompted by $H_2O$, or soot oxidation by $NO_2$ produced from decomposition of the nitrated NOx storage material.[78] The same group also reported that the decomposition of nitrated NOx storage material is most prominent when both Pt and the storage material are present on the same supporting particles.[79] Furthermore, when K compound is used as a storage material, there seems to be another soot oxidation mechanism in which K is directly involved as a catalyst because the material promotes soot oxidation even in the absence of NO in the reactant gas.[80]

Krishna *et al.* similarly reported that an unknown species was produced from K added to $Pt/Al_2O_3$, which was mobile on the catalyst surface and promoted soot oxidation, although the species could not be identified.[81]

## 2.5. Practical Application and Improvement of Soot Oxidation Catalysts

Several examples of practical applications and improvement of soot oxidation catalysts, based on the mechanisms discussed, will be explained in this section.

Application of mobile catalysts is attractive because this type of catalyst is generally composed of base metal elements rather than expensive precious metals and exhibits good low temperature activity. However, practical application has not been realized because these catalysts easily move away from the support and DPF.[82] Although several attempts have been made to overcome this problem, it may be fundamentally incompatible to simultaneously obtain high mobility (i.e., high activity) and good durability in such a catalyst. Panasonic has recently announced a DPF system that uses a PM oxidation catalyst containing an alkaline metal as an active component.[83] They claim to have obtained almost identical performance at temperatures approximately 100°C lower than the conventional Pt catalyst. However, the stability and durability of the catalyst is yet to be confirmed.

As a mobile oxygen type catalyst, $CeO_2$ has been commercialized as a fuel born catalyst (FBC) by the group of PSA Peugeot Citroen and Ibiden.[84,85] In this system, an oil-soluble organometallic cerium compound developed by Rhodia Electronics & Catalysis is used as an additive to the diesel fuel.[86] The Ce in the fuel is converted to fine particles of $CeO_2$ during combustion in the engine and is incorporated with the soot particles.[87] Therefore, the soot and $CeO_2$ particles are in tight contact, so that high catalytic oxidation efficiency is expected. After trapping the soot containing $CeO_2$ on the DPF made of SiC until the back pressure increases to a certain level, additional fuel is supplied to the exhaust line by post injection with a common-rail system and then catalytically combusted in a DOC placed upstream of the DPF, and the temperature is raised sufficiently high (*ca.* 500°C) to accomplish regeneration, i.e., burnout of the soot. After regeneration, the $CeO_2$ particles are reported to mostly accumulate as agglomerates of relatively large particles in the flow channels of the DPF in the inlet side.

FBCs with other metal components, such as Pt,[88] Mn,[89] and Fe, have also been reported. The Rhodia group has developed and provided Fe and Ce-combined FBCs, and more recently a Fe-based FBC, which is no longer combined with Ce.[90] The use of Fe-FBC together with a catalyzed DPF is more effective than the Fe-FBC alone.

$CeO_2$ has also been used as a supported catalyst. Ibiden has developed a new structural SiC-DPF, which it calls the Octosquare asymmetric filter.[91] The cross section of the inlet channels is wider than that of the outlet channels and 300 nm $CeO_2$ particles are loaded in thin layer regions (*ca.* 10 μm thick) just below the surface of the filter walls. The catalyst location almost overlaps that of the soot collected in shallow filtration, so that good contact between the catalyst and soot is expected to improve the soot oxidation performance.

Mazda has developed and commercialized a catalyst with Pt supported on a mixed oxide of $CeO_2$ with Zr or Pr for light-duty diesel engine cars.[92] Isotopic kinetic experiments using $^{18}O_2$ revealed that formation of the mixed oxide increases the mobility of lattice oxygen and thereby soot oxidation.

As for the $NO_2$ mediating catalysts, Johnson Matthey and car manufacturers have applied CRT systems for used cars in Europe[93–95] and the US.[96] They are operated as passive regenerating systems and are not equipped with a device to artificially raise the DPF temperature. In order to prevent system failure caused by excess soot accumulation, the ratio of NOx to PM in the exhaust gas should be larger than a certain level.[97] A two-stage system composed of various types of DPF without loading catalysts, preceded by a flow-through honeycomb DOC with loaded Pt, has been developed based on the reaction mechanism. NO in the exhaust gas is converted to $NO_2$ in the first stage, which flows down to the second stage and oxidizes soot collected in the DPF. Engelhard developed another type of DPF called a catalyzed DPF (CDPF) or catalyzed soot filter (CSF), which is composed of only a DPF on which soot oxidation catalysts are directly supported.[98] More recently, a two-stage system composed of a DOC followed by a CDPF has been developed. In a comparison of DOC-DPF, CDPF, and DOC-CDPF, the latter performed best, for the BPTs were in the order of DOC-CDPF (250°C) $<$ DOC-DPF (265°C) $<$ CDPF (280°C).[99] Meanwhile, investigations to seek the advantage of a single-stage type filter are being conducted. Koltsakis *et al.* have reported that a single-stage type filter is cost effective and better under continuously regenerating conditions at relatively low temperatures because the DPF temperature can be higher.[100] In any of the above cases, NO is repeatedly oxidized to $NO_2$ and serves to oxidize soot by supporting the oxidation catalysts on the DPF. Generally, in a wall-flow type DPF, soot is collected in a very shallow layer near the inner surfaces of the filter wall, whereas the NO oxidation catalyst is supported on all parts of the wall. It might seem that $NO_2$ produced from a catalyst located downstream of the flow cannot reach the soot; however, a simulation study by Haralampous *et al.* demonstrated that a large portion of $NO_2$ diffuses back to the soot accumulation layer.[101] In addition, a CDPF contributes to the complete oxidation of gaseous pollutants such as CO and hydrocarbons.

Hino Motors has succeeded in the commercialization of the world's first new vehicle models equipped with a DOC-CDPF employing an active regeneration system in which the use is not restricted by the NOx/soot ratio in the exhaust gas or by the driving conditions.[102] The temperature of the CDPF is increased during the occasional application of the regeneration mode and is controlled by catalytic combustion of the fuel supplied by post injection with a common-rail. More recently, they have put a more advanced system into practice.[103] The system consists of a combination of an active regenerating DPF system and a urea-SCR, which meets the stringent 2010 regulations in Japan that restrict PM and NOx emissions to 0.01 and 0.7 g/kWh, respectively. The production of $NO_2$ in the DPF system not only serves to promote soot oxidation in a continuous regeneration mode but also promotes the subsequent SCR reaction, i.e., NOx reduction by $NH_3$ produced from urea.

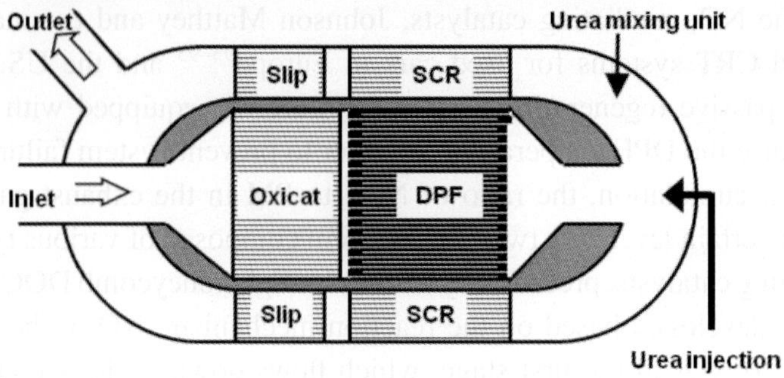


**Figure 2.11.** Compact combined system of a DPF and urea-SCR, developed by Johnson Matthey. (Reproduced from Ref. 104)

A group of Johnson Matthey and other companies has also developed a combined system of a DPF and urea-SCR in a compact form as shown in Fig. 2.11.[104] In the almost cylindrical-shaped system, the exhaust gas entering from the left side passes through a DOC and DPF situated near the central axis, turns back at the right end, passes through SCR and $NH_3$-slip prevention (Slip) catalysts, both surrounding the DOC and DPF system, and is finally emitted from an outlet located near the inlet port. The urea water is added to the system from the right end, mixed well with the exhaust gas, and hydrolyzed to $NH_3$ before the exhaust reaches the SCR honeycomb. This configuration realized a more compact system, and high performance for PM and NOx removal were confirmed both under steady-state and transient engine operations.

Toyota has commercialized a DPNR system for passenger cars[105] and light-duty trucks.[106] As shown in Fig. 2.12, the system consists of a newly introduced fuel injector in the exhaust pipe, a flow-through honeycomb supporting the NSR catalyst, a DPF supporting the NSR catalyst, and finally a flow-through honeycomb supporting an oxidation catalyst. The soot collected in the DPF is continuously eliminated under typical lean and occasional rich conditions. Active regeneration by temporal heating is applied using the combination of fuel injection in the exhaust pipe and catalytic combustion over the NSR catalyst, which also functions as an oxidation catalyst, in order to regenerate the NOx storage capacity of the sulfated NSR catalyst. Alkaline NOx storage materials are deactivated into sulfates, which can be decomposed at high temperatures with $SO_2$ desorption, and the alkalinity for NOx storage is restored. The occasional rich conditions necessary for NOx removal are achieved by fuel injection in the exhaust pipe while maintaining lean engine combustion. The Pt supported on a DPF will be sintered by repetitive active regeneration and catalytic activity will be lost. Pt catalysts using $CeO_2$-$Al_2O_3$ as a support have recently been developed to avoid this.[107] The Pt

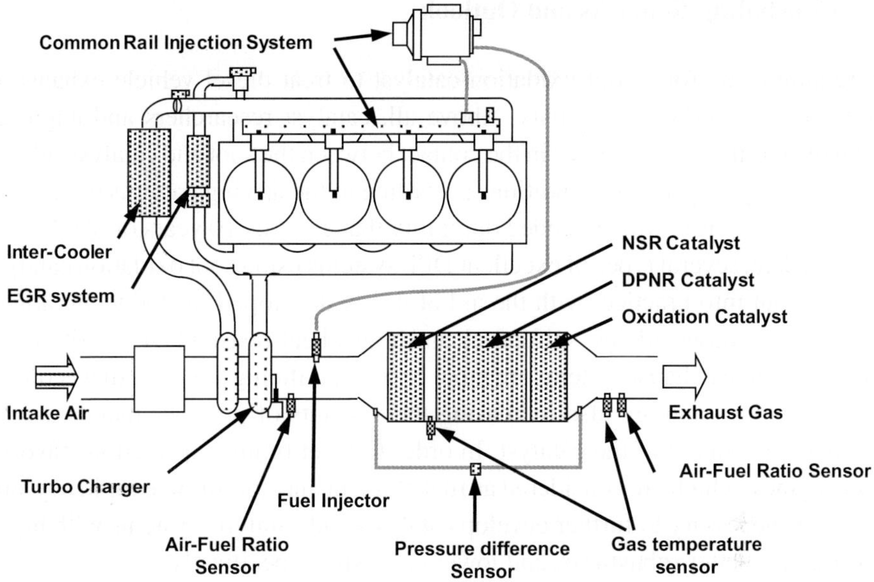


**Figure 2.12.** DPNR system developed by Toyota. (Reproduced from Ref. 105)

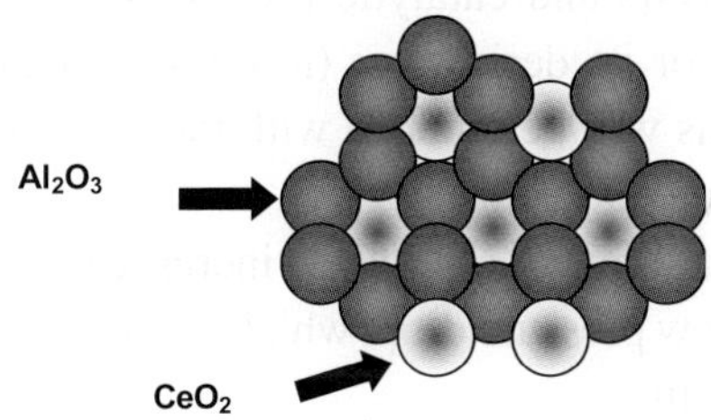


**Figure 2.13.** Schematic configuration of the $CeO_2$-$Al_2O_3$ support with Pt catalyst for the DPNR system with high durability against sintering. (Reproduced from Ref. 106)

particles are highly dispersed on $CeO_2$ particles and $Al_2O_3$ acts as a barrier to prevent sintering of the $CeO_2$ particles to each other, as schematically shown in Fig. 2.13.

Fuel consumption increases as active regeneration at high temperatures is more frequently applied; therefore, it is desirable to use diesel fuels with low sulfur content, at least below 10 ppm.[108] Even when the fuel sulfur content is decreased to such levels, it is effective to employ a sulfur trap containing alkaline K compounds upstream of the DPNR to maintain good performance for longer periods.[109] In addition, Kustov *et al.* have studied the feasibility of using Sr- or Ca-doped alumina,[110] or Sr-doped $ZrO_2$ as NOx storage materials.[111]

## 2.6. Concluding Remarks and Outlook

The requirements for a soot oxidation catalyst to treat diesel vehicle exhaust are many and very difficult to satisfy. Above all, catalyst researchers and engineers should overcome the difficulties in the contact between the soot and catalyst, address the widely varying reaction conditions according to the driving patterns of vehicles, and provide a system with extended durability that exceeds 10 years or 600,000 km of use. To date, several types of excellent DPF systems using soot oxidation catalysts have been put into practice, with the aid of newly developed related technologies, such as the common-rail, low sulfur fuel, and highly heat-resistant DPFs (SiC-DPF). However, engine exhaust systems are more or less sacrificing the good fuel economy of diesel engines and are also increasing the amounts of precious metals used as the active component of the catalyst. In order to meet future expectations favoring diesel engines, which are considered as fuel efficient and appropriate for heavy-duty trucks, it is necessary to further develop catalysts and catalytic systems with higher performance, better robustness, and lower cost. More specifically:

— development of catalysts that decrease or eliminate the use of Pt and other precious metals, resources which are limited;
— development of catalysts and catalytic processes that are not based upon the involvement of $NO_2$ or its derivatives (nitrates), because the NOx concentration in the exhaust gas will be reduced with the enforcement of more stringent emission regulations; and
— improvement and development of DPFs to increase the catalyst loading capacity while maintaining a low pressure drop, which will enable the integration of DPFs and DOCs into one unit.

Furthermore, as clearly demonstrated by the three-way catalytic system for gasoline vehicles and DPNR explained in this chapter, the automobile exhaust gas treatment catalysts do not function alone, but rather require a highly controlled engine system. It is therefore necessary to further develop technologies to functionally integrate engine and after-treatment devices in which the soot oxidation catalyst is a component.

## References

1. Kittelson, D. (1998). Engines and Nanoparticles: A Review, *J. Aerosol Sci.*, **29**, pp. 575–588.
2. Kumar, P., Robins, A., Vardoulakis, S., *et al.* (2010). A Review of the Characteristics of Nanoparticles in the Urban Atmosphere and the Prospects for Developing Regulatory Control, *Atmos. Env.*, **44**, pp. 5035–5052.
3. Johnson, T. (2008). Diesel Emission Control in Review, *SAE Technical Paper*, 2008-01-0069.

4. Goto, Y., Lee, J., Kawai, T., *et al.* (2004). Trapping Performance of Fine Particles from a Diesel Engine by Various Dpfs with Different Surface Structures, *SAE Technical Paper*, 2004-01-0598.
5. Khair, M. (2003). A Review of Diesel Particulate Filter Technologies, *SAE Technical Paper*, 2003-01-2303.
6. Brück, R., Müller-Haas, K., Holz, O., *et al.* (2009). Application of PM-METALIT® and SCRi® Systems, *ICPC*, pp. 1–14.
7. van Setten, B., Makkee, M. and Moulijn, J. (2001). Science and Technology of Catalytic Diesel Particulate Filters, *Catal. Rev.*, **43**, pp. 489–564.
8. Fino, D. (2007). Diesel Emission Control: Catalytic Filters for Particulate Removal, *Sci. Tech. Adv. Mater.*, **8**, pp. 93–100.
9. Neeft, J., Makkee, M. and Moulijn, J. (1996). Catalysts for the Oxidation of Soot from Diesel Exhaust Gases. I. An Exploratory Study, *Appl. Catal. B: Environmental*, **8**, pp. 57–78.
10. Ozawa, T. (1965). A New Method of Analyzing Thermogravimetric Data, *Bull. Chem. Soc. Japan*, **38**, pp. 1881–1886.
11. Jelles, S., van Setten, B., Makkee, M., *et al.* (1999). Molten Salts as Promising Catalysts for Oxidation of Diesel Soot: Importance of Experimental Conditions in Testing Procedures, *Appl. Catal. B: Environmental*, **21**, pp. 35–49.
12. McKee, D. (1981). The Catalyzed Gasification Reactions of Carbon, *Abstracts of Papers of the American Chemical Society*, **181**, pp. 1–118.
13. MacKee, D. (1983). Mechanism of the Alkali Metal Catalyzed Gasification of Carbon, *Fuel*, **62**, pp. 170–175.
14. From authors' calculations with "Outokumpu HSC Chemistry for Windows (ver. 4.0, 1999)", Outokumpu Research Of Information Service.
15. Castoldi, L., Matarrese, R., Lietti, L., *et al.* (2009). Intrinsic Reactivity of Alkaline and Alkaline-earth Metal Oxide Catalysts for Oxidation of Soot, *Appl. Catal. B: Environmental*, **90**, pp. 278–285.
16. Lide, D. (1990–1991). *Handbook of Chemistry and Physics*, 71st edition, CRC Press, Boca Raton, FL.
17. Domesle, R., Herbert, M. and Hanau V. (1984). Catalyst for Reducing the Ignition Temperature of Diesel Soot and Process for Making the Catalyst, *US Patent* No. 4455393.
18. Saracco, G., Badini, C., Russo, N., *et al.* (1999). Development of Catalysts Based on Pyrovanadates for Diesel Soot Combustion, *Appl. Catal. B: Environmental*, **21**, pp. 233–242.
19. Liu, J., Zhao, Z., Xu, C., *et al.* (2005). Diesel Soot Oxidation over Supported Vanadium Oxide and K-promoted Vanadium Oxide Catalysts, *Appl. Catal. B: Environmental*, **61**, pp. 36–46.
20. Setten, van B., Dijk, van R., Jelles, S., *et al.* (1999). The Potential of Supported Molten Salts in the Removal of Soot from Diesel Exhaust Gas, *Appl. Catal. B: Environmental*, **21**, pp. 51–61.
21. Neri, G., Rizzo, G., Galvagno, S., *et al.* (2003). K- and Cs-FeV/$Al_2O_3$ Soot Combustion Catalysts for Diesel Exhaust Treatment, *Appl. Catal. B: Environmental*, **42**, pp. 381–391.
22. Mul, G., Kapteijn, F. and Moulijn, J. (1997). Catalytic Oxidation of Model Soot by Metal Chlorides, *Appl. Catal. B: Environmental*, **12**, pp. 33–47.
23. Carrascull, A., Lick, I., Ponzi, E., *et al.* (2003). Catalytic Combustion of Soot with a $O_2$/NO Mixture. $KNO_3$/$ZrO_2$ Catalysts, *Catal. Commun.*, **4**, pp. 124–128.
24. Galdeano, N., Carrascull, A., Ponzi, M., *et al.* (2004). Catalytic Combustion of Particulate Matter Catalysts of Alkaline Nitrates Supported on Hydrous Zirconium, *Thermochimica Acta*, **421**, pp. 117–121.
25. Tikhomirov, K., Krocher, O. and Wokaun, A. (2006). Influence of Potassium Doping on the Activity and the Sulfur Poisoning Resistance of Soot Oxidation Catalysts, *Catal. Lett.*, **109**, pp. 49–53.
26. Zhang, Y., Qin, Y. and Zou, X. (2006). The $CO_2$ Absorption Characteristics and Catalytic Oxidation Activities of V/K and V/Ce/K-based Catalysts for Diesel Soot Oxidation, *Catal. Commun.*, **7**, pp. 523–527.

27. Gross, M., Ulla, M. and Querini, C. (2009). Catalytic Oxidation of Diesel Soot: New Characterization and Kinetic Evidence Related to the Reaction Mechanism on K/$CeO_2$ Catalyst, *Appl. Catal. A: General*, **360**, pp. 81–88.
28. Wu, X., Liu, D., Li, K., *et al.* (2007). Role of $CeO_2$-$ZrO_2$ in Diesel Soot Oxidation and Thermal Stability of Potassium Catalyst, *Catal. Commun.*, **8**, pp. 1274–1278.
29. Milt, V., Peralta, M, Ulla, M., *et al.* (2007). Soot Oxidation on a Catalytic NOx Trap: Beneficial Effect of the Ba–K Interaction on the Sulfated Ba,K/$CeO_2$ Catalyst, *Catal. Commun.*, **8**, pp. 765–769.
30. Peralta, M., Milt, V., Cornaglia, L., *et al.* (2006). Stability of Ba,K/$CeO_2$ Catalyst During Diesel Soot Combustion: Effect of Temperature, Water, and Sulfur Dioxide, *J. Catal.*, **242**, pp. 118–130.
31. Mars, P. and van Krevelen, D. (1954). Oxidations Carried out by Means of Vanadium Oxide Catalysts, *Special Supplement to Chemical Engineering Science*, **3**, pp. 41–59.
32. Atribak, I., Bueno-López, A. and García-García A. (2009). Role of Yttrium Loading in the Physico-chemical Properties and Soot Combustion Activity of Ceria and Ceria–zirconia Catalysts, *J. Mol. Catal. A: Chemical*, **300**, pp. 103–110.
33. Bueno-Lopez, A., Krishna, K., Makkee, M., *et al.* (2005). Active Oxygen from $CeO_2$ and its Role in Catalyzed Soot Oxidation, *Catal. Lett.*, **99**, pp. 203–205.
34. Aneggi, E., Boaro, M., Leitenburg, C., *et al.* (2006). Insights into the Redox Properties of Ceria-based Oxides and their Implications in Catalysis, *J. Alloys and Compounds*, **408–412**, pp. 1096–1102.
35. Issa, M., Petit, C., Brillard, A., *et al.* (2008). Oxidation of Carbon by $CeO_2$: Effect of the Contact Between Carbon and Catalyst Particles, *Fuel*, **87**, pp. 740–750.
36. Masui, T., Minami, K., Koyabu, K., *et al.* (2006). Synthesis and Characterization of New Promoters Based on $CeO_2$–$ZrO_2$–$Bi_2O_3$ for Automotive Exhaust Catalysts, *Catal. Today*, **117**, pp. 18–192.
37. Ishihara, T., Oishi, T. and Hamamoto, S. (2009). Praseodymium Oxide Doped with Bi for Diesel Soot Oxidation at Low Temperature, *Catal. Commun.*, **10**, pp. 1722–1724.
38. Reddy, B., Bharali, P., Thrimurthulu, G., *et al.* (2008). Catalytic Efficiency of Ceria–zirconia and Ceria–hafnia Nanocomposite Oxides for Soot Oxidation, *Catal. Lett.*, **123**, pp. 327–333.
39. Krishna, K., Bueno-López, A., Makkee, M., *et al.* (2007). Potential Rare-earth Modified Ceo$_2$ Catalysts for Soot Oxidation Part II: Characterization and Catalytic Activity with NO + $O_2$, *Appl. Catal. B: Environmental*, **75**, pp. 201–209.
40. Setiabudi, A., Chen, J., Mul, G., *et al.* (2004). $CeO_2$ Catalyzed Soot Oxidation: The Role of Active Oxygen to Accelerate the Oxidation Conversion, *Appl. Catal. B: Environmental*, **51**, pp. 9–19.
41. Zhang, D., Murata, Y., Kishikawa, K., *et al.* (2008). Synthesis of Large Surface Area MnOx-$CeO_2$ Using CTA and its Catalytic Activity for Soot Combustion, *J. Ceram. Soc. Japan*, **116**, pp. 230–233.
42. Machida, M., Murata, Y., Kishikawa, K., *et al.* (2008). On the Reasons for High Activity of $CeO_2$ Catalyst for Soot Oxidation, *Chem. Mater*, **20**, pp. 4489–4494.
43. Saab, E., Abi-Aad, E., Bokova, M., *et al.* (2007). EPR Characterization of Carbon Black in Loose and Tight Contact with $Al_2O_3$ and $CeO_2$ Catalysts, *Carbon*, **45**, pp. 561–567.
44. Li, K., Wang, H. and Wei, Y. (2009). Selective Oxidation of Carbon Using Iron-modified Cerium Oxide, *J. Phys. Chem. C*, **113**, pp. 15288–15297.
45. Harrison, P., Ball, I., Daniell, W., *et al.* (2003). Cobalt Catalysts for the Oxidation of Diesel Soot Particulate, *Chemical Engineering Journal*, **95**, pp. 47–55.
46. Weng, D., Li, J., Wu, X., *et al.* (2008). Promotional Effect of Potassium on Soot Oxidation Activity and $SO_2$-poisoning Resistance of Cu/$CeO_2$ Catalyst, *Catal. Commun.*, **9**, pp. 1898–1901.

47. Liu, J., Zhao, Z., Xu, C., *et al.* (2010). $CeO_2$-supported Vanadium Oxide Catalysts for Soot Oxidation: The Roles of Molecular Structure and Nanometer Effect, *J. Rare Earths*, **28**, pp. 198–204.
48. Aneggi, E., Llorca, J., Leitenburg, C., *et al.* (2009). Soot Combustion over Silver-supported Catalysts, *Appl. Catal. B: Environmental*, **91**, pp. 489–498.
49. Yuechang, W., Jian, L., Zhen, Z., *et al.* (2010). Preparation and Characterization of $Co_{0.2}/Ce_{1-X}zr_{x}o_2$ Catalysts and their Catalytic Activity for Soot Combustion, *Chinese J. Catal.*, **31**, pp. 283–288.
50. Murrell, L. L. and Carlin, R. T. (1996). Silver on Ceria: An Example of a Highly Active Surface Phase Oxide Carbon Oxidation Catalyst, *J. Catal.*, **159**, pp. 479–490.
51. Shimizu, K., Kawachi, H. and Satsuma, A. (2010). Study of Active Sites and Mechanism for Soot Oxidation by Silver-loaded Ceria Catalyst, *Appl. Catal. B: Environmental*, **96**, pp. 169–175.
52. Doggali, P., Kusaba, S., Teraoka, Y., *et al.* (2010). $La_{0.9}Ba_{0.1}CoO_3$ Perovskite Type Catalysts for the Control of CO and PM Emissions, *Catal. Commun.*, **11**, pp. 665–669.
53. Zhang, G., Zhao, Z., Liu, J., *et al.* (2009). Macroporous Perovskite-type Complex Oxide Catalysts of $La_{1-X}k_{x}co_{1-Y}fe_{y}o_3$ for Diesel Soot Combustion, *J. Rare Earths*, **27**, pp. 955–960.
54. Shimokawa, H., Kusaba, H., Einaga, H., *et al.* (2008). Effect of Surface Area of La–K–Mn–O Perovskite Catalysts on Diesel Particulate Oxidation, *Catal. Today*, **139**, pp. 8–14.
55. Fino, D., Russo, N., Saracco, G., *et al.* (2003). The Role of Superficial Oxygen in Some Perovskites for the Catalytic Combustion of Soot, *J. Catal.*, **217**, pp. 367–375.
56. Cooper, B., Jung, H. and Toss, J. (1990). Treatment of Diesel Exhaust Gases, *US Patent* No. 4902487.
57. Cooper, B. and Thoss, J. (1989). Role of NO in Diesel Particulate Emission Control, *SAE Technical Paper*, 890404.
58. Hawker, P., Myers, N., Huthwohl, G., *et al.* (1997). Experience with a New Particulate Trap Technology in Europe, *SAE Technical Paper*, 970182.
59. Uchisawa, J., Obuchi, A., Zhao, Z., *et al.* (1998). Carbon Oxidation with Platinum Supported Catalysts, *Appl. Catal. B: Environmental*, **18**, pp. L183–187.
60. Uchisawa, J., Obuchi, A., Ogata, A., *et al.* (1999). Effect of Feed Gas Composition on the Rate of Carbon Oxidation with $Pt/SiO_2$ and the Oxidation Mechanism, *Appl. Catal. B: Environmental*, **21**, pp. 9–17.
61. Uchisawa, J., Wang, S., Nanba, T., *et al.* (2003). Improvement of Pt Catalyst for Soot Oxidation Using Mixed Oxide as a Support, *Appl. Catal. B: Environmental*, **44**, pp. 207–215.
62. Pfeifer, M., Kögel, M., Spurk, P., *et al.* (2007). New Platinum/Palladium Based Catalyzed Filter Technologies for Future Passenger Car Applications, *SAE Technical Paper*, 2007-01-0234.
63. Fuc, P. (2008). Non Pt Catalyst Group in Active Part of New PM Filter, *SAE Technical Paper*, 2008-01-1551.
64. Tikhomirov, K., Kröcher, O., Elsener, M., *et al.* (2006). $MnO_x$-$CeO_2$ Mixed with Oxides for the Low-temperature Oxidation of Diesel Soot, *Appl. Catal. B: Environmental*, **64**, pp. 72–78.
65. Wu, X., Lin, F., Xu, H., *et al.* (2010). Effects of Adsorbed and Gaseous Nox Species on Catalytic Oxidation of Diesel Soot with $MnO_x$-$CeO_2$ Mixed Oxides, *Appl. Catal. B: Environmental*, **96**, pp. 101–109.
66. Fino, D., Fino, P., Saracco, G., *et al.* (2003). Studies on Kinetics and Reactions Mechanism of $La_{2-x}K_xCu_{1-y}V_yO_4$ Layered Perovskites for the Combined Removal of Diesel Particulate and NOx, *Appl. Catal. B: Environmental*, **43**, pp. 243–259.
67. Yoshida, K. Makino, S., Sumiya, S., *et al.* (1989). Simultaneous Reduction of Nox and Particulate Emissions from Diesel Engine Exhaust, *SAE Technical Paper*, 892046.
68. Teraoka, Y., Nakano, K., Kagawa, S., *et al.* (1995). Simultaneous Removal of Nitrogen Oxides and Diesel Soot Particulates Catalyzed by Perovskite-type Oxides, *Appl. Catal. B: Environmental*, **5**, pp. L181–185.

69. Shangguan, W., Teraoka, Y. and Kagawa, S. (1997). Kinetics of Soot-$O_2$, Soot-NO and Soot-$O_2$-NO Reactions over Spinel-type $CuFe_2O_4$ Catalyst, *Appl. Catal. B: Environmental*, **12**, pp. 237–247.
70. Fino, D., Russo, N., Saracco, G., *et al.* (2006). Catalytic Removal of Nox and Diesel Soot over Nanostructured Spinel-type Oxides, *J. Catal.*, **242**, pp. 38–47.
71. Pisarello, M., Milt, V., Peralta, M., *et al.* (2002). Simultaneous Removal of Soot and Nitrogen Oxides from Diesel Engine Exhausts, *Catal. Today*, **75**, pp. 465–470.
72. Kureti, S., Weisweiler, W. and Hizbullah, K. (2003). Simultaneous Conversion of Nitrogen Oxides and Soot into Nitrogen and Carbon Dioxide over Iron Containing Oxide Catalysts in Diesel Exhaust Gas, *Appl. Catal. B: Environmental*, **43**, pp. 281–291.
73. Milt, V., Pissarello, M., Miro, E., *et al.* (2003). Studies on Kinetics and Reactions Mechanism of $La_{2-x}K_xCu_{1-y}V_yO_4$ Layered Perovskites for the Combined Removal Of Diesel Particulate and NOx, *Appl. Catal. B: Environmental*, **43**, pp. 243–259.
74. Nakatani, K., Hirota, S., Takeshima, S., *et al.* (2002). Simultaneous PM and NOx Reduction System for Diesel Engines, *SAE Technical Paper*, 2002-01-0957.
75. Suzuki, J. and Matsumoto, S. (2004). Development of Catalysts for Diesel Particulate NOx Reduction, *Topics in Catal.*, **28**, 1–4, pp. 171–176.
76. Miyoshi, N., Matsumoto, S., Katoh, K., *et al.* (1995). Development of New Concept Three-way Catalyst for Automotive Lean-burn Engines, *SAE Technical Paper*, 950809.
77. Roy, S. and Baiker, A. (2009). NOx Storage-reduction Catalysis: From Mechanism and Materials Properties to Storage-reduction Performance, *Chem. Rev.*, **109**, pp. 4054–4091.
78. Castoldi, L., Matarrese, R., Lietti, L., *et al.* (2006). Simultaneous Removal of NOx and Soot on Pt–Ba/$Al_2O_3$ NSR Catalysts, *Appl. Catal. B: Environmental*, **64**, pp. 25–34.
79. Nova, I., Lietti, L., Castoldi, L., *et al.* (2006). New Insights in the NOx Reduction Mechanism with $H_2$ over Pt-Ba/$\gamma$-$Al_2O_3$ Lean NOx Trap Catalysts under Near-isothermal Conditions, *J. Catal.*, **239**, pp. 244–254.
80. Matarrese, R., Castoldi, L., Lietti, L., *et al.* (2009). Simultaneous Removal of NOx and Soot over Pt-Ba/$Al_2O_3$ and Pt-K/$Al_2O_3$ Catalysts, *Topics in Catal.*, **42–43**, pp. 293–297.
81. Krishna, K. and Makkee, M. (2006). Soot Oxidation over NOx Storage Catalysts: Activity and Deactivation, *Catal. Today*, **114**, pp. 48–56.
82. van Setten, B., Spitters, C., Bremmer, J., *et al.* (2003). Stability of Catalytic Foam Diesel-soot Filters Based on $Cs_2O$, $MoO_3$, and $Cs_2SO_4$ Molten-salt Catalysts, *Appl. Catal. B: Environmental*, **42**, pp. 337–347.
83. Inoue, M., Miyazaki, T., Tokubuchi N., *et al.* (2004). Exhaust Gas Purification Catalyst and Exhaust Gas Purification Material, *US Patent* No. 6696386.
84. Ohno, K., Shimato, K., Taoka, N., *et al.* (2000). Characterization of SiC-DPF for Passenger Car, *SAE Technical Paper*, 2000-01-0185.
85. Salvat, O., Marez, P. and Belot, G. (2000). Passenger Car Serial Application of a Particulate Filter System on a Common Rail Direct Injection Diesel Engine, *SAE Technical Paper*, 2000-01-0473.
86. Blanchard, G., Seguelong, T., Michelin, J., *et al.* (2003). Ceria-based Fuel-borne Catalysts for Series Diesel Particulate Filter Regeneration, *SAE Technical Paper*, 2003-01-0378.
87. de Sousa Filho, P., Gomes, L., de Oliveira, K., *et al.* (2009). Amphiphilic Cerium(III) $\beta$-Diketonate as a Catalyst for Reducing Diesel/Biodiesel Soot Emissions, *Appl. Catal. A: General*, **360**, pp. 210–217.
88. Shafer, M., Schauer, J., Copan, W., *et al.* (2006). Investigation of Platinum and Cerium from Use of a FBC, *SAE Technical Paper*, 2006-01-1517.
89. Guinther, G., Human, D., Miller, K., *et al.* (2002). The Role that Methylcyclopentadienyl Manganese Tricarbonyl (MMT®) Can Play in Improving Low-temperature Performance of Diesel Particulate Filters, *SAE Technical Paper*, 2002-01-2728.

90. Harié, V., Pitois, C., Rocher, L., *et al.* (2008). Latest Development and Registration of Fuel Borne Catalyst for DPF Regeneration, *SAE Technical Paper*, 2008-01-0331.
91. Ogyu, K., Oya, T., Ohno, K., *et al.* (2008). Improving of the Filtration and Regeneration Performance by the Sic-DPF with the Layer Coating of PM Oxidation Catalyst, *SAE Technical Paper*, 2008-01-0621.
92. Suzuki, K., Harada, K., Yamada, H., *et al.* (2007). Study on Low Temperature Oxidation of Diesel Particulate Matters by Oxygen Storage Component for the Catalyzed Diesel Particulate Filter, *SAE Technical Paper*, 2007-01-1919.
93. Allansson, R., Cooper, B., Thoss, J., *et al.* (2000). European Experience of High Mileage Durability of Continuously Regenerating Diesel Particulate Filter Technology, *SAE Technical Paper*, 2000-01-0480.
94. van Poppel, M., Stevens, M. and de Keukeleere, D. (2001). Performance of a Continuous Regenerating Trap on City Buses in Real Traffic Conditions, *SAE Technical Paper*, 2001-24-0060.
95. Bal, B., Hully, D., Lausseur, P., *et al.* (2004). Experience of Continuously Regenerating Particulate Traps on City Buses in Europe, *SAE Technical Paper*, 2004-01-00078.
96. Chatterjee, S., Conway, R., Lanni, T., *et al.* (2002). Performance and Durability Evaluation of Continuously Regenerating Particulate Filters on Diesel Powered Urban Buses at NY City Transit – Part II, *SAE Technical Paper*, 2002-01-0430.
97. Babu, K., Sudipto, B., Kang, B., *et al.* (2005). The Effect of NOx/Soot Ratio on the Regeneration Behavior of Catalysed Diesel Particulate Filters for Heavy Duty Applications, *SAE Technical Paper*, 2005-26-347.
98. Farrauto, R., Voss, K. and Heck, R. (1993). A Base Metal Oxide Catalyst for Reduction of Diesel Particulates, *SAE Technical Paper*, 932720.
99. Allansson, R., Blakeman, P., Cooper, B., *et al.* (2002). Optimising the Low Temperature Performance and Regeneration Efficiency of the Continuously Regenerating Diesel Particulate Filter (CR-DPF), *SAE Technical Paper*, 2002-01-0428.
100. Koltsakis, G., Haralampous, O., Dardiotis, C., *et al.* (2005). Performance of Catalyzed Particulate Filters without Upstream Oxidation Catalyst, *SAE Technical Paper*, 2005-01-0952.
101. Haralampous, O., Koltsakis, G., Samaras, Z., *et al.* (2004). Reaction and Diffusion Phenomena in Catalyzed Diesel Particulate Filters, *SAE Technical Paper*, 2004-01-0696.
102. Toorisaka, H., Minamikawa, J., Narita, H., *et al.* (2004). DPR Developed for Extremely Low PM Emissions in Production Commercial Vehicles, *SAE Technical Paper*, 2004-01-0824.
103. Hirabayashi, H., Furukawa, T., Koizumi, W., *et al.* (2011). Development of New Diesel Particulate Active Reduction System, *SAE Technical Paper*, 2011-01-1277.
104. Walker, P., Allansson, R., Blakeman, P., *et al.* (2003). The Development And Performance of the Compact SCR-trap System: A 4-Way Diesel Emission Control System, *SAE Technical Paper*, 2003-01-0778.
105. Ohki, H., Ishiyama, S. and Asano, A. (2003). Control Technology for a Passenger Car Diesel Engine Equipped with the DPNR System, *SAE Technical Paper*, 2003-01-1880.
106. Shoji, A., Kamoshita, S., Watanabe, T., *et al.* (2004). Development of a Simultaneous Reduction System of NOx and Particulate Matter for Light-duty Truck, *SAE Technical Paper*, 2004-01-0579.
107. Ohashi, N., Asanuma, T., Fukuma, T., *et al.* (2008). Development of Next-generation Nox Reduction System for Diesel Exhaust Emission, *SAE Technical Paper*, 2008-01-0065.
108. Asanuma, T., Hirota, S., Yanaka, M., *et al.* (2003). Effect of Sulfur-free and Aromatics-free Diesel Fuel on Vehicle Exhaust Emissions Using Simultaneous PM and NOx Reduction System, *SAE Technical Paper*, 2003-01-1865.
109. Nishioka, H., Yoshida, K., Asanuma, T., *et al.* (2010). Development of Clean Diesel NOx After-treatment System with Sulfur Trap Catalyst, *SAE Technical Paper*, 2010-01-0303.

110. Kustov, A. and Makkee, M. (2009). Application of NOx Storage/Release Materials Based on Alkali-earth Oxides Supported on $Al_2O_3$ for High-temperature Diesel Soot Oxidation, *Appl. Catal. B: Environment*, **88**, pp. 263–271.
111. Kustov, A., Ricciardi, F. and Makkee, M. (2009). NOx Storage and High Temperature Soot Oxidation on Pt–Sr/$ZrO_2$ Catalyst, *Topics in Catalysis*, **52**, pp. 2058–2062.

# Chapter 3

# The Catalytic Oxidation of Hydrocarbon Volatile Organic Compounds

Tomas GARCIA,* Benjamin SOLSONA† and Stuart H. TAYLOR‡

Hydrocarbons are present in many VOCs whose abatement is required for environmental reasons. After having presented the main processes used for air clean-up, the catalytic systems developed for hydrocarbon oxidation (alkanes, olefins, aromatics) are reviewed.

## 3.1. Introduction

The term "volatile organic compound" (VOC) refers to a chemically diverse and wide-ranging class of compounds which can be difficult to define, and in fact many definitions currently exist. VOCs are defined by the US Environmental Protection Agency as[1]

> . . . any compound of carbon, excluding carbon monoxide, carbon dioxide, carbonic acid, metal carbides or carbonates and ammonium carbonates which participates in atmospheric photochemical reactions . . .

This definition is non-specific, and does not focus on chemical nature or functionality, hence any organic compound with a vapour pressure exceeding 0.1 mmHg under standard conditions (25°C and 760 mmHg) could be regarded as a VOC. VOCs are emitted from a wide variety of natural and anthropogenic sources. Natural sources include volcanic activity, swamps, vegetation, animals and insects; control of emissions from these sources is not generally practical. Emissions from anthropogenic sources are also widespread, originating from manufacturing and processing industries, processes and products using organic solvents, combustion processes and vehicle exhaust to name just a few. However, for the VOCs emitted

*Department d'Enginyeria Química, Universitat de València, C/ Dr. Moliner 50, 46100 Burjassot, Valencia, Spain.
†Instituto de Carboquímica (CSIC), C/Miguel Luesma, 50018 Zaragoza, Spain.
‡Cardiff Catalysis Institute, School of Chemistry, Cardiff University, Main Building, Park Place, Cardiff, CF10 3AT, UK.

from these anthropogenic sources the possibility exists to control their release into the atmosphere.

The release of VOCs into the environment has widespread environmental implications. Pollution by VOCs has been linked to the increase in photochemical smog[2] and ozone depletion.[3] In addition, many VOCs are themselves toxic and/or carcinogenic. The US Clean Air Act of 1990[4] was one of the first measures to call for a 90% reduction in the emissions of 189 toxic chemicals, with 70% of these classed as VOCs, by 1998. Hence, in recent years, the development of effective technologies for the removal of VOCs from the atmosphere has increased in importance with the introduction of legislation to control their release. Various methods have been proposed, and one of the best is heterogeneous catalytic oxidation. This has the advantage over the more common original thermal oxidation process, since it requires less supplementary fuel and is therefore a less expensive process. However, the characteristics of the catalyst selected for this process are of vital importance for successful operation, and potential problems such as lifetime and deactivation must be solved if catalytic oxidation is to be employed universally. Catalysts currently in use include noble metals, notably platinum and palladium, and those based on metal oxides, however, irrespective of the type of catalyst, the most important characteristics are activity and selectivity for total oxidation.

The development of noble metal catalysts and transition metal oxides for catalytic oxidation of VOCs has been widely reported in the literature.[5–9] The review paper published in 1987 by Spivey presents a good overview of catalytic combustion of VOCs.[5] More recent reviews, focusing on the catalytic combustion of a wide range of VOCs by a wide variety of catalysts[6] and on chlorinated VOCs,[7] were published in 2004. In the last two years, two more reviews have been published. These reviews focused on the development of non-noble metal oxide catalysts for catalytic combustion of VOCs[8] and on catalytic combustion catalysts for the removal of polycyclic aromatic hydrocarbons.[9] This review is not intended to be an exhaustive account, but should provide an overview of the current state of research for catalysts used for alkane and aromatic total oxidation. The aim is also to identify the types of catalysts that are likely to be of use in the future, and the obstacles that must be overcome to produce viable catalysts. The development of a catalyst that may be used for the combustion of all classes of compounds under the general term VOC presents a major challenge for future research, as this has not yet been achieved.

## 3.2. Technology Options for VOC Abatement

A number of technologies are currently used for the abatement of VOCs into the atmosphere. In general, these can be divided into two types; those that remove VOCs from aerial effluent, but do not destroy them, and technologies which remove them by destroying them. The former options include adsorption, absorption and condensation. Adsorption often uses an adsorbent, such as activated carbon or a zeolite, whilst absorption more often makes use of a scrubber containing an appropriate liquid tuned to capture the VOCs. Both processes are relatively straightforward to operate and do not require significant additional energy input, but they do not remove VOCs to very low levels and once the capacity of the adsorption/absorption media has been reached it must be replaced and the contaminated media requires disposal. Condensation is an attractive option and can be used for higher VOC concentrations ($>1\%$) as it allows recovery of the VOCs which may have an economic value. However, it is often necessary to carry out cryogenic condensation, which is a costly process, and may not be offset by the value of the recovered compounds.

Thermal and catalytic oxidation are also well-established treatment technologies, which result in the destruction of VOCs, mainly to carbon dioxide and water, but this depends on the chemical composition of the VOCs. The release of carbon dioxide into the environment may not be entirely satisfactory, but it is more benign than VOCs and generally has a lower impact as a greenhouse gas. Thermal combustion or incineration requires temperatures in excess of 1,000°C. Whilst it is a simple and often effective method of control, the high temperatures required result in a relatively fuel intensive process, with limited control over the ultimate products. The latter is particularly problematic and can result in an incomplete oxidation of the waste stream and the formation of toxic by-products such as dioxins, dibenzofurans and oxides of nitrogen, if conditions are not carefully controlled. Alternatively, heterogeneous catalytic oxidation offers many potential advantages. The use of a catalyst in the oxidative destruction of VOCs significantly lowers the process operating temperature, which is typically in the range 300–600°C. This reduction in temperature is advantageous, as the supplementary fuel demand to sustain oxidation is reduced. Furthermore, there may be some legislative advantages as catalytic oxidation is no longer regarded as an incineration process, eliminating certain regulatory requirements. In addition, catalytic oxidation offers a much greater degree of control over the reaction products and can operate with dilute effluent streams, which cannot be treated as easily by thermal combustion. Hence, catalytic oxidation may be considered as the most appropriate method for end-of-pipe VOC pollution control.

## 3.3. Operational Parameters Affecting the Catalytic Combustion of VOCs

The catalytic oxidative destruction of VOCs offers some significant advantages over other current VOC abatement methods, as it enables complete destruction at relatively low temperatures with a high volumetric throughput. Disregarding the nature of the catalyst, a variety of operational parameters is known to influence the efficiency of the process. These include operating temperature, pre-heating of the system, space velocity and the nature and concentration of VOCs in the effluent stream. Examples of the effect of these operating parameters will be briefly considered.

### 3.3.1. *Temperature*

It is obvious that temperature has an influence on catalytic oxidation efficiency; however, in general the temperature required for complete oxidation of a VOC cannot be used independently as a controlling factor. This is because the reaction temperature varies according to the VOCs present, their concentrations, and the catalyst employed. The use of high temperatures will increase the efficiency of destruction of VOCs, but it can also accelerate catalyst deactivation, resulting in a reduction of activity. When applied industrially, relatively low temperatures are preferred so that operational energy costs can be minimised.

According to Prasad *et al.*[10] the evolution of the total oxidation rate with increasing temperature follows some general characteristics. At lower temperatures no activity is observed, but by increasing the temperature initially, a reaction confined to the surface of the catalyst takes place. In this regime the reaction rate increases exponentially with temperature. A further increase of the temperature results in the onset of limitations imposed by heat and mass transfer, in spite of the reaction still being confined to the catalyst surface. Finally, a further increase of the temperature can result in the initiation of homogeneous gas phase reactions, and these can become predominant, reducing the influence of the catalyst.

### 3.3.2. *Influence of pre-heating*

The efficiency of a catalytic oxidation system may be enhanced by pre-heating the effluent gas prior to catalytic combustion. If pre-heating to a sufficiently high temperature is achieved, the process effectively combines the methods of thermal and catalytic oxidation. The combination with the thermal oxidation process significantly enhances the effectiveness of the subsequent catalytic oxidation. For example, Tichenor and Palazzolo[11] have determined the relative contribution to destruction efficiency for a pre-heater operated at a range of temperatures. The destruction of a mixture of isopropanol, methyl ethyl ketone, ethyl acetate, benzene and *n*-hexane, oxidised at a space velocity of 50,000 $h^{-1}$ in the temperature range 300–450°C over a bimetallic Pt-Pd catalyst supported on a ceramic monolith was investigated. The pre-heating stage contributed significantly to the oxidation efficiency, particularly at higher temperatures. Hence, pre-heated VOC containing effluent, which could arise from various sources, could reduce the overall fuel requirement necessary to sustain the combustion. In addition, the products of thermal oxidation may influence the activity of the catalyst. This has been demonstrated in studies by Zięba *et al.*[12] indicating that thermal oxidation during pre-heating caused 5–20% oxidation of the VOC content. The effects of pre-heating were investigated for the oxidation of ethylene, methane and toluene, over industrial combustion catalysts (0.1% Pt/alumina and copper/cobalt/manganese oxides supported on alumina) with temperatures in the range 77–477°C. The results for the system incorporating pre-heating of the VOCs were compared with those for a conventional system with no pre-heating; when pre-heated, an increased oxidation efficiency of 5–30% was observed. It was proposed that thermal oxidation was acting as a source of radicals, consisting of hydrogen, oxygen, hydroxyl and organic species, and these enhanced the production of radicals in the subsequent catalytic oxidation step. The production of radicals in thermal oxidation has been reported previously,[13] and their importance in gas phase oxidation reactions is well established.[14] Supplementary fuel provided to sustain combustion was also found to act as a source of radicals. However, it should also be noted that water vapour may be formed in the pre-heating thermal oxidation step, which may inhibit catalytic oxidation. The addition of 5% water vapour to the effluent gas resulted in a 5–10% reduction in activity of the catalytic oxidation of toluene.[12] It was proposed that water formed in the pre-heater could also suppress oxidation activity by competing for adsorption sites on the catalyst surface.

### 3.3.3. *Effect of space velocity*

As with many catalytic reactions, the space velocity has a significant effect on the activity of the total oxidation of VOCs. As the space velocity increases, destruction efficiency generally decreases, as would be expected as normal behaviour in heterogeneously catalysed reactions. Not surprisingly, for commercial applications, a catalyst capable of achieving high levels of total oxidation activity at relatively high space velocities with no reduction in specificity towards total oxidation products is preferred. This would allow the catalyst to be used in simple end-of-pipe applications, and would also reduce the amount of catalyst required to achieve complete destruction, thus reducing expenditure on both the catalyst and capital engineering costs.

The effects of increasing space velocity have been demonstrated by Vassileva *et al.*[15] The combustion of benzene over a 0.5 wt% Pd/30% $V_2O_5/Al_2O_3$ catalyst for space velocities of 330, 2,000 and 5,000 $h^{-1}$ at a constant oxygen to benzene molar ratio of ca. 7.5 have been studied. Results demonstrated that higher space velocities showed the highest initial activities, but conversion levelled off at less than 100%. However, the activity at a space velocity of 330 $h^{-1}$ rapidly increased and became constant at a significantly higher conversion close to 100%. In the region in which catalytic activity is constant for each space velocity, activity markedly decreased as space velocity increased.

### 3.3.4. *Effect of VOC concentration*

Catalytic oxidation is ideally suited to the destruction of low concentrations of VOCs, which is a major advantage for this abatement technique since it allows removal of low VOC levels from waste streams; this is essential if industry is to comply with current air pollution legislation. The applicability of various abatement techniques to differing VOC concentrations is shown in Table 3.1.[1] A catalyst capable of efficiently destroying a wide range of concentrations is desired, so that compliance with legislation concerning their release can be achieved, regardless of the actual concentration in the waste stream. The specific effects of VOC concentrations may vary for different VOCs, as it will also depend on factors such as heating characteristics and chemical composition, as this will influence their enthalpy of combustion and will impact catalytic oxidation efficiency. As an example, the effects of increasing VOC concentration have been considered by Tichenor and Palazzolo[11] for a bimetallic platinum-palladium catalyst used for the oxidation of a mixture of hydrocarbons at total concentrations of 1,200 and 6,000 vppm. The higher VOC concentration resulted in higher oxidation activity, a result which was more evident at lower temperatures (305°C) than at higher temperatures (400°C).

### 3.3.5. *Type of VOC*

Individual VOCs are combusted at a specific temperature according to their chemical composition, type of catalyst and the reaction conditions used. The ease of destruction of VOCs by catalytic combustion can generally be correlated to the

**Table 3.1.** Suitability of various VOC abatement technologies for VOC concentration; data adapted from Ref. 1.

| Abatement method | Minimum concentration/ppm | Maximum concentration/ppm |
|---|---|---|
| Thermal oxidation (no heat recovery) | 20 | 1,000 |
| Thermal oxidation (with heat recovery) | 1,000 | 30,000 |
| Catalytic oxidation | 50 | 10,000 |
| Adsorption | 20 | 20,000 |
| Absorption | 1,000 | 20,000 |
| Condensation | 6,000 | 10,000 |
| Biofiltration | 500 | 2,000 |
| Membrane technology | 1 | 1,000 |
| UV oxidation | 1 | 30,000 |

chemical class of the compound, such that a general order for the ease of oxidation can be observed. For example, Tichenor and Palazzolo[11] determined such an order for a Pt/Pd bimetallic catalyst on a ceramic honeycomb monolith. The inlet temperature and space velocity of the system were varied in the ranges 260–425°C and 15,000–80,000 $h^{-1}$, to give 98–99% conversion, from which the following ranking was obtained:

$$\text{alcohols} > \text{aldehydes} > \text{aromatics} > \text{ketones} > \text{acetates} > \text{alkanes} > \text{chlorinated hydrocarbons}$$

All the compounds were totally oxidised, with the exception of chlorinated hydrocarbons, which were seen to partially deactivate the catalyst. Chlorinated hydrocarbons are frequently difficult to destroy, with both chlorinated reagents and products acting as catalyst poisons and thus causing catalyst deactivation, resulting in a decrease of activity. Similar behaviour is also often observed for fluorinated VOCs.

Comparable orders for the ease of oxidation may be obtained for other catalysts used in VOC abatement, and activities vary specifically according to the stability of the class of compound and the ability of the compound and/or its oxidised products to act as catalyst poisons. Specific compounds within these general classes may have higher or lower destruction efficiencies depending on their exact nature and on the composition of the catalyst used.

### 3.3.6. *Effect of VOC mixtures*

Industrial gas effluents rarely consist of a single VOC, hence it is essential to determine the effects caused by any interaction between components of a VOC mixture. The effects of using different mixture compositions are rarely predictable and many studies only concentrate on streams containing a single component. If there is little or no interaction between the components of a mixture, it would be expected that the mixture would be combusted with a similar efficiency to the pure components. This is rarely seen. For example, over a Pt/Pd catalyst at 305°C, 90% conversion of hexane was observed, whereas only 75% conversion was evident under the same conditions when hexane was present in a mixture with isopropanol, methyl ethyl ketone, ethyl acetate and benzene.[11] However, these effects are not universal, as the same study determined that the destruction of ethyl acetate is greater when present in a mixture of hydrocarbons than when present alone, and there was no observable difference in the destruction efficiency of benzene when oxidised alone or in mixtures.

A number of general statements may be made about the interactions between classes of compounds and the subsequent effects on combustion. Aliphatic hydrocarbons are usually combusted with greater efficiency alone, with significantly lower activity observed for these compounds in mixtures with aromatics. In contrast, esters are frequently destroyed with greater efficiency when present in mixtures, although this is probably due to the involvement of homogeneous gas phase reactions.[16]

A decrease in conversion of a VOC when present in a mixture, as compared to the activity of the pure compound, is generally attributable to the existence of competitive mechanisms for the oxidation of the individual components. This has been observed by Papenmeier and Rossin[17] for chloroform and dichloromethane oxidised by a 3% Pt/alumina catalyst, with each chloro-organic suppressing the reactivity of the other, compared to the pure compounds alone. The combustion of both compounds occur by similar mechanisms, involving adsorption of the chloro-organic onto an oxygen-covered platinum surface and subsequent decomposition, and both are also inhibited by the formation of HCl. As both chloro-organics have similar adsorption equilibrium constants, competitive adsorption effects are expected in the two-component mixture. Increasing the concentration of one chloro-organic relative to the other resulted in increased surface coverage, hence fewer sites are available for the adsorption and oxidation of the other component, thus decreasing its oxidation. Papenmeier and Rossin[17] state that competitive adsorption effects occurring in mixed feed streams may result in the reversal of the order in which individual species are combusted relative to the order observed for pure compounds. This is illustrated by a study of the combustion of lean mixtures (200–2000 ppm) of aromatic hydrocarbons over a Pt/alumina catalyst, in the temperature range 100–350°C.[18] Relative activity for combustion, both alone and in two, three and four component mixtures, were determined and strengths of adsorption of each compound on the catalyst surface were calculated. The reactivity for the pure compounds decreased in the order:

benzene > toluene > ethylbenzene > *o*-xylene > styrene

In mixtures, this order was reversed, as the relative strengths of adsorption of the aromatic compounds dictated the extent of surface coverage and hence reactivity. Strongly adsorbed compounds will block catalytic sites, and thus reduce activity. For example, styrene, the most strongly adsorbed of these compounds, will decrease the adsorption of the other components in a mixture onto the catalyst surface, and thus inhibits their oxidation. If these results are applied to all VOCs, it can be proposed that inhibition is due to competition for adsorption sites. Compounds that show low oxidation activity tend to be strongly adsorbed on the catalyst and thus will inhibit the oxidation of more reactive and hence weakly adsorbed compounds. Therefore, in mixtures of VOCs of differing reactivities, it can be expected that the more reactive species will not be oxidised to the same extent as the less active compounds.

In order to ensure that all components of a mixture of VOCs are completely oxidised it is often necessary to increase the reaction temperature from that which would be required to combust the components of the mixture separately. Industrially this increases the cost, so it would be beneficial to develop catalysts in which interaction between VOCs and competition for sites on the catalyst surface are minimised.

## 3.4. Review of VOC Oxidation Catalysts

Catalytic oxidation has been established as one of the most appropriate technologies for VOC abatement. An assessment of the suitability of catalytic oxidation for hydrocarbon control, along with competing processes, is given in Table 3.2. In the literature there are many studies focusing on the catalytic oxidation of VOCs, however, it is beyond the scope of this work to comprehensively review these studies. Rather we will concentrate on the catalytic total oxidation of simple short-chain alkanes and aromatic compounds as illustrative examples of VOC abatement.

### 3.4.1. *Catalytic oxidation of alkanes*

Among the different groups of VOCs, alkanes represent one of the most interesting from a practical point of view. The study of the catalytic deep oxidation of alkanes is of outstanding importance since they are emitted from a number of different industries and automotive vehicles. Generally speaking, and taking this statement with caution, alkanes are the least reactive among volatile organic compounds, and therefore the most demanding conditions are required for their total oxidation. This usually means high temperatures and low space velocities. Often, milder conditions are sufficient to remove other VOCs, such as alcohols, aldehydes and ketones, and the same reaction conditions applied to alkanes will result in little or no conversion.

Methane is the most stable alkane, and although it is not strictly a VOC, the abatement by catalytic oxidation is extremely important. In fact it is very common that VOCs are grouped into methane and other non-methane VOCs. Methane, in addition to the drawbacks of most VOCs, is also a powerful greenhouse gas with a potential of ca. 30 times higher than that of carbon dioxide. Conversely, it has not been linked, as have most VOCs, to the production of tropospheric ozone. Methane is released into the atmosphere through several sources, including industries, gas power plants and even natural gas powered vehicles.[19] In recent years natural gas has been increasingly used as a fuel due to its clean combustion, high energy to $CO_2$ released ratio and improved transportation methods (liquefied natural gas and gas pipelines). The production/consumption of natural gas has doubled over the last 30 years and proved reserves have followed a similar trend.[20] Hence, the increasing use of natural gas also provides a driving force for improving catalysts to control emissions.

**Table 3.2.** Hydrocarbons emission control technology ratings.

| Parameter | Technology for VOC control | | | | | | | | | |
|---|---|---|---|---|---|---|---|---|---|---|
| | Condensation | Absorption | Adsorption | Incineration | Regenerative oxidation | Catalytic oxidation | Flameless oxidation | BIF combustion | Biofiltration | Flares |
| *Type of VOC* | | | | | | | | | | |
| Hydrocarbon gases | F | B | D | A | A | A | A | A | B | A |
| Hydrocarbons condensed | A | A | A | A | A | A | A | A | B | A |
| *Flow rate/ concentration* | | | | | | | | | | |
| Low/Low | B | A | A | B | D | A | B | A | C | F |
| High/Low | C | A | A | C | A | A | C | A | C | B |
| Low/High | A | D | D | A | D | D | A | C | B | F |
| High/High | A | D | D | A | D | D | A | C | B | B |
| *Process type* | | | | | | | | | | |
| Continuous process | A | A | A | A | A | A | A | A | A | A |
| Batch or variable process | A | A | A | D | D | D | B | D | C | A |

Rating: A = Excellent; B = Good; C = Satisfactory; D = Poor; F = Unacceptable
Table adapted and shortened from Refs. 21, 22.

Due to the relatively inert character of methane, the thermal combustion in the absence of a catalyst usually requires high temperatures, typically over 1200°C, which means high energy consumption and the increased probability of the formation of undesired NOx. Therefore, catalysed combustion should provide the potential to operate at significantly lower temperatures. The total oxidation of other short-chain alkanes is also important. For example, propane is abundant and it is present in both oil and natural gas. A relatively high concentration of propane is present in liquefied petroleum gas (LPG), which is mainly comprised of propane and butanes. In natural gas, propane is present as a minor component but with typical concentrations of ca. 1 wt%. Propane is emitted into the atmosphere due to incomplete combustion by a range of industries, and also in the increasingly used LPG vehicles, as a substitute for gasoline and diesel.

The reactivity of an alkane can be related inversely to the energy of its constituent C–H bonds. More specifically, the reactivity will be linked to the energy of the weakest C–H bond. The C–H bond energy of tertiary carbons is lower than those of secondary carbons and these are lower than primary carbons (Table 3.3). Consequently, $n$-butane, which contains two secondary carbons, is more reactive than propane, which only contains one, and propane is more reactive than ethane, which has two primary carbon atoms.[23] Finally methane is the least reactive, since it has the strongest C–H bond of the alkanes.[24]

Although it is always difficult to generalise, olefins are easier to oxidise than alkenes, but still more difficult than oxygenates. The relationship between the energy of the weakest C–H bond and reactivity is also applicable to olefins. Thus ethylene only presents vinylic C–H bonds, meanwhile propylene and butylenes have both vinylic and allylic C–H bonds. Since the allylic C–H bonds (Table 3.3) are weaker than vinylic C–H bonds, it is reasonable to assume that the oxidation of propylene and butylenes is more facile than that of ethylene, and indeed it is.

The activation of a C–H bond is the first, and rate-determining, step in the combustion of most hydrocarbons, especially for alkanes. Once the first C–H bond is broken then the formation of carbon dioxide takes place reasonably easily. Burch *et al.*,[25] studying the catalytic combustion of various hydrocarbons on metal and metal oxide-based catalysts, proposed that the initial activation of the substrate may occur through either homolytic or heterolytic scission on the surface of the catalyst.

**Table 3.3.** Typical bond energies for C–C and C–H bonds.

| | C–C | Primary C–H | Secondary C–H | Tertiary C–H | Allylic C–H | Vinylic C–H |
|---|---|---|---|---|---|---|
| Energy/kJ mol$^{-1}$ | 376 | 420 | 401 | 390 | 361 | 445 |

Related to this, Hodnett and co-workers[26,27] surveyed a range of gas-phase selective oxidation reactions using alkanes and olefins as raw materials. A strong correlation between the strength of the weakest C–H bond energy and reaction products was reported. In the case of total oxidation reactions, the desired final product is in all cases the same, $CO_2$, which does not present any C–H bonds and it is therefore not possible to establish such a relationship with selectivity, but the catalytic activity will mainly be determined by the weakest C–H bond of the reactant.

In addition to the characteristics of the substrates and the final reaction products, it would be very interesting to try to predict the catalytic activity of a catalyst from its physical properties. It is well known that for pure metal oxides a correlation exists between catalytic activity and the heat of formation of the metal oxide. Moro-Oka and co-workers[28,29] demonstrated for several substrates (propylene, isobutylene, acetylene, ethylene and propane) that the lower the heat of formation of the catalyst oxide, divided by the number of oxygen atoms in the oxide, then the greater the activity for total oxidation. From the metal oxides tested, those of palladium and platinum were the most active, which corresponds to the elements with the most unstable oxides.[5]

Golodets[24] grouped the metal oxides together based on the stability of the oxides. Thus the metals which do not form stable bulk oxides remain as reduced metals during oxidation reactions at low temperatures. This group of unstable oxides is composed of the noble metal oxides, such as palladium, platinum, gold and rhodium, which in general are the most reactive for total oxidation reactions. The oxides with an intermediate stability are comprised of the oxides of iron, cobalt and nickel. Meanwhile the most stable oxides are those corresponding to alkali, alkali-earths, rare earths and actinides, and usually the most stable oxides present the lowest activity.

As noble metals seem to be the preferred catalysts for total oxidation we will emphasise these, especially palladium, platinum and gold. In fact, noble metal catalysts, such as platinum and palladium, dispersed on a high area metal oxide, are the commercial catalysts of choice due to their high intrinsic oxidation activity. Table 3.4 presents some advantages and disadvantages of the different catalysts used for the total oxidation of alkanes. Data from the table must be taken with caution, but it can be useful for a quick comparison of the different types of catalysts employed.

#### 3.4.1.1. *Noble metal-basedx catalysts*

Some specific reviews on the total oxidation of short-chain alkanes have been published previously[19,30,31] and these focus particularly on catalysts based on palladium and platinum. Accordingly, in this chapter only a few more general points concerning these catalysts will be commented on. It is well known that the catalysts of choice are those based on palladium and platinum, since they generally present the highest activity. Many key points regarding the catalytic performance of palladium

**Table 3.4.** Catalytic systems for combustion of alkanes and olefins.

| System | Reactivity | Thermal stability | Structure sensitive? | Price | Sulfur tolerant? |
|---|---|---|---|---|---|
| Pt-supported | Very high | Good | Very likely The larger the crystallite, the higher the activity | Very high | Medium |
| Pd-supported | Very high | Good | Very likely The larger the crystallite, the higher the activity | Very high | Low |
| Au-supported | Variable (highly depends on the support) | Poor | Yes The smaller the crystallite, the higher the activity | High | Low |
| Bulk cobalt oxide | High | Poor | Yes The smaller the crystallite, the higher the activity | Low | — |
| Perovskite | High | Very good | Yes The smaller the crystallite, the higher the activity | Medium | — |

and platinum catalysts have been addressed in a wealth of detail, although there still remains a degree of debate around some key questions. One key area is: what is the relationship between the catalytic activity and the particle size of the noble metals? In addition, is total alkane oxidation a structure-sensitive reaction?

Several authors consider that over a certain range of particle sizes, there is an increase in turnover frequency as the metal particle size increases. Some explanations have been proposed for this, for example, that small PdOx crystallites are in closer contact with the support and contain stronger Pd–O bonds which are required for the rate-determining C–H activation step, hence the activity is higher when large crystallites are present.[32] Other suggestions include that the oxygen adsorbed on the surface of the catalyst is more or less reactive depending on the crystallite size.[33,34] Other authors could not find this relationship between crystallite size and turnover frequency,[35] obtaining different reaction rates in catalysts with different crystallite sizes, but without a significant trend. Similarly, for platinum catalysts, some authors have proposed structure sensitivity of the reaction,[36] and others have concluded that it is structure insensitive.[37]

Another question is: which is the most appropriate metal for short alkane total oxidation, platinum or palladium? This is not necessarily an easy question to answer. Palladium and platinum each have advantages and disadvantages associated with their use. Palladium catalysts have been more widely studied in the literature since it is generally considered by many to be more active than platinum. However, this is not always the case, as, depending on the reaction conditions and the substrate selected, platinum catalysts can exhibit better performance. For example, a catalyst operating in fuel-rich conditions means that the catalyst is in a more reduced state when compared with one operating in an oxygen-rich atmosphere. Since platinum is more active if it is not fully oxidised, and inversely, more fully oxidised palladium is the most active form, a platinum-based catalyst will be preferable under hydrocarbon-rich conditions.[35] Although generally for methane oxidation, palladium is the most active catalyst, platinum-based catalysts can be better catalysts for the oxidation of higher hydrocarbons.[33,38] Additionally, there is some evidence that platinum seems to be more resistant to sulfur than palladium,[39] and it is also better than Pd in complex catalysts and as a component of multioxide catalysts.[19] The advantages of working with catalysts containing both platinum and palladium have also been reported. Narui and co-workers[40] demonstrated when using alumina as a support for methane combustion, that palladium-platinum catalysts, apart from being more active than simple palladium catalysts, were more stable with time-on-line. This effect was attributed to the improved dispersion of palladium and to the prevention of sintering of palladium oxide particles when platinum was present. Similarly, Yamamoto and co-workers[36] showed the same effect.

There is also debate surrounding the optimal oxidation state for palladium-based catalysts. The oxidation of short-chain alkanes on Pd-based catalysts takes place through a Mars–van Krevelen mechanism. Therefore palladium species are continuously oxidising and reducing during the catalytic runs and therefore the oxidation states of the active metals are essential for this reaction to take place appropriately. Either PdO or a core-shell $PdO/Pd^0$ system have been reported to be the active phase. Burch and Urbano[41] propose that the fully oxidised form of palladium (PdO) is the active site. These authors demonstrated that metallic palladium or chemisorbed oxygen on palladium present low activity, but a catalyst in which the surface of the palladium has been oxidised to the equivalent of 3–4 monolayers of oxygen is highly active. In contrast, Hicks *et al.*[34] and Oh *et al.*[42] proposed that PdO situated on Pd crystallites were remarkably more active for methane oxidation than PdO. Fully oxidised PdO did not lead to high activities. Oh *et al.*[42] also proposed that bulk PdO was completely inactive.

The modification of palladium-based catalysts by addition of various promoters and additives, usually metals or metal oxides, has been investigated. Studies have shown improved catalytic performance for the total oxidation of light alkanes, usually leading to higher conversions and lower deactivation. The reason for this promotion is still under discussion, since the metal oxide additives alone usually show relatively low activity for alkane oxidation over the range of reaction temperatures. Alloying phenomena, modification of the properties of the support, modification of the PdO particle size, variations in the Pd oxidation states or an enhanced reduction–reoxidation cycle are considered as the most likely factors for the enhancement of activity. For example, if $Pd/Al_2O_3$ catalysts are modified with titania, a decrease in the Pd–O bond strength and a facilitation of the reduction/decomposition of PdO to metallic palladium takes place.[43] In a comprehensive study many metal oxides were tested as promoters of $Pd/Al_2O_3$ for methane oxidation.[44] All the promoted catalysts showed better activity than the unmodified $Pd/Al_2O_3$, although the catalytic behaviour depended on the nature of the metal oxide added. Thus, adding oxides of nickel, copper and rhodium resulted in the most active catalysts, decreasing the temperature of the light-off curves by more than 100°C. The enhanced behaviour of NiO-promoted catalysts was explained in terms of the increased temperature required to transform palladium oxide to reduced palladium.[44]

Another of the most commonly studied additives is vanadium. The enhanced catalytic activity of vanadium-promoted palladium-alumina catalysts has been related to a range of different effects, depending on the study referred to. For example, according to some authors[45] the improvement is due to a modification in the palladium reducibility and particle size. On the other hand, Escandón *et al.*[46] showed that the activity increase could be produced by palladium-vanadium interactions or a modification of the support properties. Yazawa *et al.*[47] suggested that for propane combustion, although both the dispersion and the oxidation states of palladium affect the catalytic activity on a series of supports, the oxidation state of palladium affects the catalytic activity to a greater extent than the dispersion. Accordingly the partially-oxidized palladium showed the highest catalytic activity for propane combustion. Similarly Garcia and co-workers[48] observed a large promotion of the total oxidation of $C_1-C_4$ alkanes when vanadium was added to a $Pd/TiO_2$ catalyst. This enhancement was explained by the generation of a more facile reduction-oxidation cycle which increased the reaction rate. In the case of niobium-oxide modified $Pd/TiO_2$ catalysts[49] the addition of niobium significantly modified the nature of the palladium and niobium species. It was observed that after the addition of niobium, a marked increase in the oxygen mobility takes place. This could not only promote the presence of palladium species in a totally oxidized state but also resulted in the formation of new and very easily reducible species identified by subambient temperature programmed reduction (TPR). It was concluded that the niobium promotional effect in propane oxidation was due to the presence of these active oxygen species.

In the case of platinum catalysts the addition of some metal oxides has also resulted in enhanced catalytic activity, which has been attributed to the modification of platinum crystallite size, and especially to the modification of the oxidation state of platinum.[50] It has been reported that promoters with large electronegativities, such as molybdenum, vanadium, tungsten and niobium, enhance the catalytic activity compared with the unpromoted $Pt/Al_2O_3$, since more electronegative promoters present a higher resistance to oxidation, and platinum remains less oxidised than those with electropositive promoters, such as alkaline and alkaline-earth metals, which are even less reactive than the unpromoted $Pt/Al_2O_3$ catalyst.[51]

The emission of methane into the atmosphere takes place mainly as a component of natural gas, which also contains ethane and propane in different concentrations. Since methane is the least reactive it could be expected that the presence of ethane or propane does not affect the methane oxidation rate. However, Ruiz and co-workers[52] have demonstrated over a $Pd/\gamma\text{-}Al_2O_3$ catalyst that, in the presence of ethane and propane, both enhancement and suppression of methane oxidation can take place. Hence, the exact composition of the natural gas will be important in determining the final activity of the catalysts. This effect is not expected to be isolated to a palladium-based catalyst and highlights the wide variety of conditions over which a VOC catalyst must be expected to operate.

Palladium and platinum are well known as oxidation catalysts and have been widely employed in three-way car exhaust catalysts, for example. Conversely, for many years gold was largely viewed as inert for many reactions including oxidation. The pioneering work of Hutchings[53] and Haruta[54] demonstrated that gold can be catalytically active, and gold has become the most fashionable catalytic element with a large number of published studies focusing on it. As discussed above, in the case of palladium and platinum catalysts, it seems that generally the larger the crystallite size, the higher the catalytic activity for alkane total oxidation. Conversely, for gold catalysts the catalytic activity for methane oxidation apparently often decreases with an increase of the gold particle size. Nieuwenhuys *et al.*,[55] working with $Au/Al_2O_3$ catalysts, observed an inverse relationship between the catalytic performance and the average size of the gold particles. This trend was very clear when $Au/Al_2O_3$ catalysts were doped with alkali or alkali-earth metal oxides.[56] However, if the catalyst was promoted with transition metal oxides the intrinsic activity of the transition elements must also be considered.[57] Waters and co-workers[58] conducted a very detailed study of methane combustion over transition metal oxide supported gold catalysts prepared by co-precipitation, and concluded that the best catalytic performance was obtained with $Co_3O_4$ as the support. They justified these results with the oxidation state of gold on the surface of the catalysts, concluding that active catalysts comprised both reduced and oxidised gold, and that activity increased with increasing surface concentration of the oxidised form. Haruta and co-workers also found that $Co_3O_4$ was the best support for gold in the oxidation of alkanes.[59] Longer chain alkanes have also been investigated, and Gasior *et al.*,[60] for example, have studied propane oxidation over gold catalyst supported on different metal oxides, such as those of magnesium, silicon, tin, iron, titanium and cerium. No correlation between the activity of the catalysts for propane total oxidation and the gold particle size or the reducibility of the catalysts was evident. Therefore, it was concluded that the catalytic activity of gold supported catalysts for VOC combustion was strongly dependent on the nature of the support.

In a more wide-ranging study Solsona *et al.*[61] investigated the total oxidation of methane, ethane and propane over gold supported on a range of metal oxides. The addition of Au reduced the light-off temperature when compared with the activity of the metal oxide support alone, and generally catalysts prepared by co-precipitation were more active than catalysts prepared by impregnation. The catalytic behaviour of gold catalysts was mainly determined by the characteristics of the support, but the presence of gold systematically decreased the light-off temperature compared to the support alone. The incorporation of gold by an appropriate method improved the reducibility of the metal oxide support, and in the particular case of $Co_3O_4$ it also increased the number of oxygen vacancies,[62] accounting for the observed improvement.

Miao and Deng[63] reported high catalytic activity for $Au/Co_3O_4$ catalysts in methane combustion and showed that the activity could be improved by adding small amounts of platinum. Surprisingly, the addition of palladium to $Au/Co_3O_4$ did not result in any enhancement of the total oxidation activity.

### 3.4.1.2. *Metal oxide-based catalysts*

Metal oxides of non-noble metals are usually less active, but present the advantage of comparatively lower prices. Several studies have investigated transition metal oxide catalysts and demonstrated that $Co_3O_4$ is the most active for alkane total oxidation.[64,65] As long ago as 1927 the catalytic oxidation of methane in the air was studied over a large range of metals, metal oxides and composites. Catalysts investigated include uranium oxide, thorium oxide, cerium oxide, platinum, nickel, platinum and nickel, copper and cobalt, cobalt oxide and nickel oxide amongst others. From the wide range of catalysts $Co_3O_4$ was the most active.[66] Another study also focused on the catalytic activity of non-noble metal oxides for $C_1-C_3$ alkanes and the following trend was established:[61]

$$Co_3O_4 > Mn_2O_3 > CuO > Fe_2O_3 > CeO_2 > TiO_2$$

In another study the following trend of activity was also obtained:[67]

$$Co_3O_4 > CuO > NiO > Mn_2O_3 > Cr_2O_3$$

Bulk $Co_3O_4$ can show activity higher than that of supported palladium and platinum catalysts, especially if the cobalt oxide presents a high surface area.[68] Liu and co-workers[69] tested bulk nanocrystalline $Co_3O_4$ prepared by a soft reactive grinding (SRG) procedure and obtained excellent activities for propane catalytic combustion, showing complete conversion at 240°C. These authors proposed that a high concentration of superficial electrophilic oxygen ($O^-$) species was important for achieving high activity.

A major disadvantage of using a $Co_3O_4$ catalyst is that at higher temperatures serious deactivation can take place. This is due to the transformation of the active $Co_3O_4$ phase into the less active CoO phase,[70] or to the sintering and agglomeration of $Co_3O_4$ particles.[68,71] Therefore, the use of $Co_3O_4$ as a catalyst for the oxidation of particularly stable alkanes, such as methane, may be limited by stability issues. In order to overcome this limitation, promoters, such as niobium have been added to increase the stability and prevent this sintering,[72] although the activity is adversely affected. If working with more reactive substrates, such as propane, such high temperatures are not required and, therefore, good stability with time-on-line has been demonstrated.[68,69]

Supporting $Co_3O_4$ is a possible strategy to improve the catalytic performance. Cobalt oxide has been supported on several different types of alumina[68] and tested for propane total oxidation. It was observed that once supported there was a decrease in the overall catalytic activity and the activity per cobalt site when compared with the bulk oxide. Supporting cobalt oxide also led to the formation of Co-Al-O species with low reducibility and consequently a lower activity. A limited range of alumina supports were studied and scope still remains for developing alumina-supported cobalt-oxide catalysts.

Liotta and co-workers[70] showed that $CeO_2$, in spite of presenting low reactivity for methane combustion, is a very good support and promoter for $Co_3O_4$. Adjusting the compositional mixture of $CeO_2$ and $Co_3O_4$ can result in more active catalysts than pure $Co_3O_4$. More importantly, if it is calcined in an appropriate manner, it does not suffer deactivation over a range of higher reaction temperatures. The improved performance of the $CeO_2/Co_3O_4$ catalyst has been explained on the basis of a higher mobility of the lattice oxygen.[70] The increased stability has been attributed to the increasingly more difficult transformation of $Co_3O_4$ to CoO when $CeO_2$ is added.[70]

Manganese oxide is probably, after cobalt oxide, the most promising metal oxide for the total oxidation of short-chain alkanes. Busca and co-workers[73] showed that $\alpha$-$Mn_2O_3$ and $Mn_3O_4$ presented high activity, much higher than that of an $\alpha$-$Fe_2O_3$ catalyst, for propane and propene combustion. The spinel-type compound $Mn_3O_4$ was more active than $Mn_2O_3$ and mixtures of $Mn_2O_3$–$Fe_2O_3$ for both propene and propane oxidation. Only for some specific compositions and only for propane oxidation some $Mn_2O_3$–$Fe_2O_3$ catalysts showed comparable activity to that of $Mn_3O_4$. In another study of the total oxidation of methane, supported manganese oxide catalysts were less active than bulk manganese oxide,[74] and alumina was shown to be a better support than $TiO_2$ and $SiO_2$. On the other hand Zaki and co-workers[75] studied methane total oxidation over manganese oxide catalysts supported on $ZrO_2$, $TiO_2$, $TiO_2$-$Al_2O_3$ and $SiO_2$-$Al_2O_3$. They found that $SiO_2$-$Al_2O_3$ was the best support, and the supported catalyst was more active than bulk $\alpha$-$Mn_2O_3$, emphasising the importance of the strong acidity of the silica-alumina support in achieving good performance. Apparently the C–H activation, which is the rate-determining step for alkane oxidation, was favoured by a strongly acidic surface such as that of silica-alumina.

Surprisingly, uranium oxide has been shown to be highly active for the deep oxidation of several VOCs.[76] It showed moderate conversion for propane and butane oxidation and only a low activity for methane and ethane.[77] Inversely to the situation observed with cobalt oxide, when $U_3O_8$ was supported on silica the activity increased compared with the bulk oxide. Supporting the uranium oxide resulted in modification of the structure and chemistry of the oxide leading to an increase in the defective structure, which resulted in an increased oxidation activity.

A range of mixed-metal oxides have also been studied for total oxidation of short-chain alkanes, and one particular group of oxides is the perovskites. Perovskites have the formula $ABX_3$, which has a cubic structure in which each A cation is coordinated to twelve X anions and each B cation to six X anions. It has been common to study the partial substitution of the A or B cations, as this structure can be achieved with multiple A and B compositions. Catalysts with this structure have been shown to be very active for combustion reactions, although the catalytic activity varies largely depending on the preparation methods and especially on the composition. In spite of the high catalytic activity of perovskites, they are less active than typical platinum or palladium catalysts, although they present the advantage of being able to tolerate higher reaction temperatures.[78] The high capacity for total oxidation of perovskite-type oxides can be explained on the basis of the particular characteristics of the structure, which can present a number of different types of defects.[79] Thus, depending on the particular composition, perovskites can exhibit unique properties, such as high electronic and ionic conductivity, an excellent capacity for reversible oxygen sorption and the ability to stabilise mixed valences of several active metals.[80] It has been reported that the reactivity of perovskites is mainly determined by the characteristics of the B cation,[81] whilst the role of the A cation is more structural, especially when it is partially substituted by a cation of different valency. It is this effect which defines the formation of crystal lattice vacancies, which are able to stabilise unusual oxidation states of the B cation.[82] Baiker *et al.*[83] showed that there was no significant influence of the A-site cations on the catalytic activity during methane total oxidation. Typical perovskite compositions employed in the literature are based on lanthanum-cobalt and lanthanum-manganese oxides due to their high reactivity, as cobalt and manganese oxides are the most active amongst simple oxides.

Due to the high reactivity, versatility and thermal stability of perovskites many studies have been devoted to the combustion of different alkanes. As long ago as 1985, Seiyama and co-workers[81] showed the high propylene oxidation activity of several perovskites with different compositions, and demonstrated that the activity of the perovskite was mainly determined by the characteristics of the B cation in the $ABX_3$ structure. In other work Arai *et al.*[84] studied the catalytic oxidation of methane over various perovskite-type oxides, including partially cation-substituted lanthanum, iron, lanthanum-manganese and lanthanum-cobalt oxides, showing, in the best cases, comparable activity to that of platinum-alumina catalysts.

One of the disadvantages of perovskites is their relatively low surface area, which limits potential activity. Thus, it has been demonstrated that the higher the surface area and the lower the particle size, generally the higher the total oxidation activity.[19,85] In order to increase the surface area, a mesostructured $LaCoO_3$ perovskite prepared by a nanocasting method was tested for the combustion of methane.[86,87] The catalytic results demonstrated higher activity than the conventional bulk $LaCoO_3$ perovskite prepared by the conventional citrate method. Apart from the higher surface area, the enhanced activity was associated with the presence of high valent +4 cobalt ions.[86]

Supporting perovskites could also be another alternative to increase conversion, in addition to improving the mechanical properties. Supported $LaCoO_3$ with different loadings was prepared using $Ce_{0.8}Zr_{0.2}O_2$ as a support[88] and showed that intermediate compositions are far more reactive than pure $LaCoO_3$ perovskites or $Ce_{0.8}Zr_{0.2}O_2$ alone. A structured perovskite was found to be more reactive than the highly dispersed oxide elements on the support, thus demonstrating that rather than just the effect of supporting the perovskite, there is a synergetic effect between $LaCoO_3$ and $Ce_{0.8}Zr_{0.2}O_2$.

Another strategy to enhance the activity is to incorporate or impregnate a noble metal into the perovskite structure. This approach can then provide a catalyst with two kinds of active sites. Palladium-containing perovskite-type oxides with the formula $LaTi_{0.5}Mg_{0.5-x}Pd_xO_3$, (where $0 \leq x \leq 0.10$) were tested for methane total oxidation.[89] Although the perovskite structure was formed across the composition range, at least some of the palladium was not incorporated into the perovskite structure, and remained as a separate phase. These authors propose that, conversely to more conventional $Al_2O_3$-supported palladium catalysts, the incorporation of palladium into the perovskite matrix and the interaction of PdO-$Pd^0$ means that smaller palladium particles are more active at higher temperatures. $Mn_{1-x}Fe_xO_3$ (where $x = 0.4, 0.5, 0.6, 1$) perovskite catalysts promoted with palladium, either fresh or pre-treated with $SO_2$, have also been tested for methane total oxidation.[90] Catalysts with low palladium loadings (2–2.5%) were more active, although less tolerant to $SO_2$, than high palladium loadings (10%). Similarly Cimino *et al.*,[91] working with Pd-$LaMnO_3$ monolithic catalysts (using different palladium loadings) for the combustion of methane, found that the different reactivity of palladium active sites, and those associated with the perovskite, allowed the catalyst to operate over a wider range of operating conditions than typical palladium and $LaMnO_3$ catalysts alone. On the other hand Russo *et al.*[92] demonstrated that the incorporation of palladium into the La-Mn-O perovskite structure decreased the reaction temperature required to achieve a given methane conversion by more than 50°C compared with the palladium-free catalyst. The higher activity of the palladium catalyst was explained by the capability to deliver a higher amount of intrafacial oxygen than the palladium-free catalyst. The authors propose that this palladium-modified perovskite was as active as the typical commercial Pd/$\gamma$-$Al_2O_3$ catalyst, but with a palladium content four times lower. In recent work Eyssler *et al.*[93] compared the catalytic performance for methane oxidation for Pd/$LaFeO_3$, with palladium on the surface of $LaFeO_3$, and $LaFe_{0.95}Pd_{0.05}O$, with palladium incorporated into the perovskite structure. Palladium was in different coordination and oxidation states in both catalysts. Thus $Pd^{3+}$ in a distorted octahedral coordination was predominant in $LaFe_{0.95}Pd_{0.05}O_3$, while $Pd^{2+}$ was mainly in a square planar coordination in Pd/$LaFeO_3$. These different characteristics modified the catalytic performance. Palladium Pd/$LaFeO_3$ was most active, shifting the light-off curve by ca. 100°C to a lower temperature compared with the Pd-free $LaFeO_3$ catalyst. $LaFe_{0.95}Pd_{0.05}O_3$ only presented an activity comparable to that of the Pd-free catalyst.

### 3.4.2. *Catalytic oxidation of aromatics*

Aromatic hydrocarbons are among the most toxic VOCs, and therefore, their removal is of outstanding importance. In spite of containing double bonds these compounds are relatively stable, presenting reactivity lower than that of similar non-aromatic hydrocarbons containing double bonds. This stability, which hinders their removal via catalysis, has been associated with the delocalisation of electrons in the aromatic rings. The simplest and most common aromatic VOC is benzene, which has significant toxicity and is highly carcinogenic. To corroborate this, one classification of VOCs, the German TA-Luft system, identifies three groups of VOCs:

(i) extremely hazardous to health;
(ii) class A compounds, causing harm to the environment;
(iii) class B compounds, with low environmental impact.

In this classification benzene is included in the first group, which is that of the highly hazardous compounds. Consequently in most countries the use of benzene is restricted to those processes where it is essential as a reactant, or originates from natural products, such as petroleum derivatives and gasoline. Other aromatic compounds, such as toluene and xylene, also present a relatively high toxicity, but not as much as that of benzene and they do not exhibit strong carcinogenic properties.

Aromatic VOCs are released into the atmosphere from a wide range of sources. It is surprising that these emissions still originate partly from gasoline-fuelled vehicles. This is because benzene and other aromatics remain in gasoline due to their high octane number. Emissions from stationary sources are also abundant and come from a variety of industries, such as chemicals, petrochemicals, paints, coatings and steel manufacture. Another important characteristic of aromatics is their high capacities as solvents; in fact they are considered as the most powerful solvents of the hydrocarbon family. Although uses are restricted for environmental reasons, aromatics are still being utilised in many industries, especially the chemical and coating industries.

The removal of aromatics via catalytic oxidation has been reported to be a good choice. Supported noble metals and metal oxides have been proposed as efficient catalysts for the total oxidation of aromatics. Generally speaking, noble metal-based catalysts have been reported to be the most active catalysts, and they also demonstrate good stability. Metal oxide-based catalysts are generally not as active as noble metals, but they are more tolerant to the presence of sulfur or chlorides. Hence, there are a great number of studies involving the catalytic combustion of aromatic VOCs in the literature. The behaviour of catalysts based on noble and transition metals have all been widely reported. For example, Au, Ir, Pd, Pt and Rh are generally supported on silica,[94] alumina,[95,96] $TiO_2$,[97] activated carbon[98–100] and zeolites or mesoporous silicates.[101,102]

#### 3.4.2.1. *Noble metal-based catalysts*

According to many studies reported in the literature metallic catalysts require a high active surface area, which means that small metal particles with high dispersion are necessary. Due to the fact that small metal particles tend to sinter at relatively low temperatures, they should ideally be strongly anchored on a support, which itself is thermally stable and maintains a high specific surface area at high temperatures. Consequently, it has been reported recently that the sol–gel method applied to $Pd/SiO_2$ catalysts allows the stabilisation of palladium species by the silica network, leading to more active catalysts for the total oxidation of aromatics. Although, by this preparation method, the metallic palladium particles are located within the $SiO_2$ particles, they are readily accessible via the microporosity of the structure.[94] Determination of the catalyst performance showed that they were very active, and total benzene conversion was obtained at 200°C.

The structure sensitivity and *in situ* activation of benzene combustion on $Pt/\gamma$-$Al_2O_3$ catalysts of different platinum and chlorine loadings have been studied.[96] The light-off curves shifted to lower temperatures with increasing platinum particle size, suggesting that benzene total oxidation is a structure-sensitive reaction. Total benzene conversion was obtained at ca. 300°C. It has been proposed that benzene oxidation on platinum proceeds via a Langmuir–Hinshelwood mechanism, which involves the rapid and strong adsorption of benzene on metallic platinum and assumes that the rate constant of oxygen adsorption is very low compared with the rate constant of the surface reaction.[96] On the other hand, Ordoñez *et al.*[103]

reported that the kinetics of benzene and toluene catalytic combustion can be fitted to a Mars–van Krevelen mechanism. A modified Mars–van Krevelen mechanism was proposed for mixtures of the aromatic VOCs and this accounted for the effects of competitive adsorption.

Catalytic oxidation of aromatic VOCs has also been investigated over a 1 wt% Pd/$\gamma$-$Al_2O_3$ catalyst.[96] It was found that the increasing VOC conversion with time-on-line was dependent on the oxidation state of palladium and the growth of palladium particles on the catalyst, suggesting that it is a structure-sensitive reaction. The order of oxidation activity for a variety of aromatics over the supported palladium catalyst was:

$$o\text{-xylene} \sim \text{toluene} > \text{benzene}$$

Complete conversion was attained at 240°C for benzene, at 190°C for toluene and at 190°C for *o*-xylene over the pre-reduced catalyst. For all the three VOCs conversion increased as reaction time increased due to the formation of larger palladium particles.

It has been reported recently that the addition of platinum to a Pd/$\gamma$-$Al_2O_3$ catalyst resulted in an increase of catalytic activity.[104] Moreover, it was effective in preventing the deactivation of the catalysts for benzene combustion. On the contrary, the addition of platinum beyond a certain amount decreases activity, because the palladium active sites block the platinum active sites. It was reported that the activity of the catalysts was related to the oxidation state of the metal, Pd/Al ratio and particle size. Complete benzene oxidation over Pt-Pd bimetal catalyst supported on $\gamma$-$Al_2O_3$ was also affected by the Pt-Pd ratio and the formation of small particles with a uniform size distribution was suggested to increase the activity.[104] The full benzene conversion was obtained at 250°C.

It is generally accepted that a suitable VOC oxidation catalyst should satisfy at least two criteria; lower temperature activity and high thermal stability. Thus, a novel, very active and remarkably stable 0.01%Pt–0.02%Pd catalyst was prepared by using stainless steel as the support.[105] The stainless steel was pre-treated by an anodic oxidation process. Total oxidation of toluene was achieved at 210°C, despite the fact that the catalyst was calcined at 1,000°C. The average diameter of the platinum and palladium particles was 1–2 nm, showing that the anodized dielectric film has a considerable affinity for dispersing active phases of platinum and palladium, even after the high calcination temperature employed. The results indicate that the anodic oxidation process in the support preparation was quite effective in increasing activity and maintaining a stable catalyst and it is worthy of further study to assess more fully the potential as a support.

The oxidation of toluene has also been investigated over a range of noble metal catalysts (Pt, Pd, Ir, Rh and Au) supported on $TiO_2$.[97] The catalysts were prepared by liquid phase reduction deposition (LPRD) and by incipient wetness impregnation (IMP). LPRD is based on the adsorption of metal ions or complexes onto the surface followed by reduction. Highly dispersed particles on the support can be produced, overcoming limitations of conventional methods. In the total oxidation of toluene the following order of activity was observed:

$$Pt/TiO_2 > Pd/TiO_2 > Rh/TiO_2 \approx Ir/TiO_2 \ggg Au/TiO_2$$

The same order was observed for both preparation methods, although the LPRD method always led to more active catalysts. Toluene oxidation over the platinum and palladium catalysts, was found to be a structure-sensitive reaction.

Vanadium-promoted silver and palladium catalysts supported on $Al_2O_3$ or $TiO_2$ were found to be highly active in the catalytic combustion of benzene.[106–108] A synergistic effect was observed when vanadium was added to palladium-supported $TiO_2$ or $Al_2O_3$ catalysts. The high catalytic activity of vanadium-promoted catalysts for benzene oxidation was related to the presence of two catalytic sites. These were readily reducible vanadium sites, since the presence of palladium increases the reducibility of vanadium species,[106] and palladium sites are in a higher oxidation state and have a larger particle size.[107] Thus, the promoting effect can be linked to a better activation of oxygen on the metal particles, which enables the reverse oxidation of $V^{4+}$ and leads to an equilibrium in the redox process.[108]

A very rigorous study has been carried out by Adreeva and co-workers on benzene catalytic combustion on a vanadia-promoted gold-supported catalyst. Gold-vanadia catalysts on different supports, such as $\gamma$-$Al_2O_3$,[109] $CeO_2$,[110] mesoporous titania and zirconia[111] have been evaluated. The combination of gold and vanadia leads to a significant increase in the catalytic activity with a decrease in the temperature for 100% conversion of benzene. Generally, it was observed that the presence of gold influenced the process of $V^{5+}$ reduction to $V^{4+}$ improving the redox transfer during the catalytic operation. From these studies it can be seen that the lowest temperature for total benzene conversion is obtained in the case of a gold-vanadia catalyst supported on $CeO_2$, as this showed complete conversion at 200°C. It was concluded that the high activity was due to an optimum of "the surface catalysts structure design", in combination with the available free ceria surface area. It is worth highlighting that the temperature needed for full benzene conversion in vanadia-promoted gold catalysts is much lower than that needed to achieve complete conversion for other novel gold-supported catalysts, e.g. those supported on $CeO_2$,[110] $Al_2O_3$-$CeO_2$,[110] $Fe_2O_3$[112] and $TiO_2$ or $TiO_xN_y$,[113] where temperatures higher than 350°C are required.

In the last decade, different authors have studied the use of different washcoats, which act as a host for the active palladium component, to prepare monolithic catalysts for toluene total oxidation. $Y_2O_3$,[114] $Ce_xY_{1-x}O_2$,[115] $Ce_xZr_{1-x}O_2$,[116] $Ce_{0.8}Zr_{0.15}La_{0.05}O\delta$,[117] $Ce_{0.8}Zr_{0.2}O_2$[117] and $Ce_{0.4}Zr_{0.4}Y_{0.1}Mn_{0.1}O\delta$[118] have all been evaluated. For the different washcoats the $Pd/Ce_{0.8}Zr_{0.15}La_{0.05}O\delta$ had the best catalytic activity, reaching 95% toluene conversion at a temperature as low as 190°C. It was observed that doping $La^{3+}$ into the $CeO_2$-$ZrO_2$ solid solution could generate more oxygen vacancies, and it could also inhibit the sintering of the $CeO_2$-$ZrO_2$ solid solution. Furthermore, the $Ce_{0.8}Zr_{0.15}La_{0.05}O\delta$ washcoat had much better redox properties than the others.

It can be difficult to perform complete oxidation of aromatic VOCs at very low temperatures. Actually, in most cases, studies show that reaction temperatures of at least 200°C are necessary to achieve total oxidation of toluene when more conventional catalysts are employed. A remarkably lower oxidation temperature was reported by Masui *et al.*[119] The authors prepared $Pt/CeO_2$–$ZrO_2$–$Bi_2O_3/\gamma$-$Al_2O_3$ ($Pt/CZB/Al_2O_3$) catalysts by a wet impregnation method in the presence of polyvinyl pyrolidone (PVP). By the optimisation of the amount of platinum, complete oxidation of toluene was achieved at a temperature as low as 120°C on a 7 wt% Pt/16 wt% $Ce_{0.64}Zr_{0.15}Bi_{0.21}O_{1.895}/\gamma$-$Al_2O_3$ catalyst. The high oxidation activity observed for the catalyst was attributed to the concerted effect of platinum and $Ce_{0.64}Zr_{0.16}Bi_{0.20}O_{1.90}$, and this effect contributed to the high mobility of lattice oxygen in this catalyst. Despite the complex nature of the catalyst and the high precious metal content, the high activity at a relatively low temperature is significant.

Recently, the promoting effect of manganese oxide on the catalytic combustion of aromatic VOCs has also been observed. Very active manganese-promoted $Pt/Al_2O_3$ combustion catalysts have been reported by Aguero *et al.*[120] A transition phase $\theta$-$\delta$-$Al_2O_3$ was used as support. The transitional alumina was prepared following a very interesting method, as a relatively high specific surface area was achieved (103 $m^2\,g^{-1}$), despite the high calcination temperature (1,000°C). For these catalysts, complete toluene conversion was reached at a very low temperature of ca. 75°C. It was reported that at least two different platinum active sites were found to exist on the surface and that the modification of the acid-base properties of the support affects the oxidation state of the platinum particles and their dispersion. In addition, it was established that the catalytic activity increased with increasing platinum dispersion. This catalyst is one of the most active ever reported for the catalytic combustion of toluene.

It has been published that hydrophobic-activated carbons can be suitable supports for noble metal species active for total oxidation. The catalytic behaviour of platinum[98] and palladium[99] supported on carbon-based monoliths was studied in the low temperature catalytic combustion of benzene, toluene and *m*-xylene, and compared with the corresponding behaviour of Pt-supported on $\gamma$-$Al_2O_3$ coated monoliths. Carbon-based monoliths showed much better catalytic performance, which was ascribed to the fact that the carbon surface is more hydrophobic than the $\gamma$-$Al_2O_3$, and the poisoning effect of water molecules produced during the combustion was reduced. It was stated that the noble metal dispersion depends on both the porous texture and surface chemistry of the support. Fortunately, no gasification of the carbon-coated monoliths was observed during the catalytic combustion of aromatic VOCs in the temperature range studied and it was observed that the platinum catalysts were more active than palladium. In addition, whilst the palladium catalysts with smaller palladium particle sizes were more active, in the case of platinum catalysts, the opposite was observed, which might be due to a structure-sensitivity effect. It is also worth highlighting that catalysts supported on carbon-based monoliths were very active, reaching total conversions at temperatures ranging from 150 to 200°C depending on both the type of VOCs and the nature of the catalyst. Although the low temperatures used in the work avoided the destruction of the support, it is well known that the relatively low resistance to oxidation of carbon supports is the main drawback for their use in the catalytic oxidation of VOCs. Interestingly, it has been reported that the presence of phosphorus compounds on the carbon surface has an inhibiting effect on the oxidation reaction at moderate temperatures.[99] Specifically, impregnation with phosphoric acid produces C–O–P bonds that block the active carbon sites and limit the oxidation of the carbon.

The performance of other high porosity materials has also been evaluated as supports for noble metals in the catalytic combustion of aromatic VOCs. A series of platinum-supported MCM-41 and ZSM-5 catalysts prepared by impregnation were studied by Xia *et al.*[101] The most hydrophobic catalyst of platinum supported on MCM-41, which had a large surface area and pore size, was the most active for the total oxidation of toluene in air, and activity was maintained even in the presence of added water vapour. Complete toluene conversion was reached at a temperature lower than 200°C. However, platinum supported on ZSM-5, which is a microporous and hydrophilic material, showed much lower catalytic activity than Pt/MCM-41. The authors stated that the high oxidation activity of the catalyst depended mainly on the high hydrophobicity and partly on its large pore size and high platinum loading.

The effect of support pore size and shape on the catalytic activity of palladium catalysts supported on FAU and MOR zeolites and MCM-41 and KIT-1 mesoporous materials have been studied by Ryoo *et al.*[102] Generally, it can be concluded that noble metal catalysts supported on mesoporous materials showed higher activity, whilst the low activity of palladium catalysts supported on microporous zeolites was ascribed to mass transfer limitations. In order to avoid these limitations, supported platinum on sepiolite catalysts with high macroporosities have recently been reported.[121] These catalysts were fabricated as conformed ceramic extrudates by a single-step synthesis route, using activated carbon as a templating agent and also for *in situ* reduction of the metal salt. The choice of sepiolite as support for the metal active phase was motivated by the excellent rheological properties of pastes, which after subsequent heat treatment led to conformed bodies with high mechanical strength, thermal resistance and large external surface areas. It was reported that the use of an impregnated carbon procedure as a preparation pathway to supported noble metal catalysts inhibits metal particle agglomeration, especially at high metal loadings. This macroporous support has been shown to significantly improve the catalytic activity for toluene combustion, reaching full toluene conversion at a temperature lower than 225°C.

Supported ruthenium catalysts have been examined for the catalytic combustion of toluene. A range of different supports have been evaluated[122,123] such as $\gamma$-$Al_2O_3$, $CeO_2$, $SnO_2$ and $ZrO_2$. Ru/$CeO_2$ showed the highest activity for all tests, regardless of whether or not it was pre-treated in a hydrogen atmosphere. The catalytic activity of Ru/$SnO_2$ was significantly reduced by reduction treatment, whereas the activity of Ru/$ZrO_2$ and Ru/$\gamma$-$Al_2O_3$ were enhanced due to the formation of ruthenium in the metallic state. In the case of Ru/$SnO_2$, the formation of an intermetallic compound with a core-shell structure was confirmed, and resulted in the deterioration of catalytic activity. The catalytic activity was strongly related to the ability of the ruthenium species to be easily oxidised and reduced at low temperatures.[122,123] Such ruthenium species were present on $CeO_2$ in a highly dispersed state, resulting in the highest activity.

#### 3.4.2.2. *Metal oxide-based catalysts*

As presented above, noble metal supported catalysts have high activities and selectivities at low temperature, lower than 200°C in some cases, but they are intrinsically relatively expensive and can be unstable in the presence of chloride and/or sulfur compounds.[124] Therefore, metal oxides, commonly of the transition elements, have also been intensively studied and successfully used for the catalytic combustion of aromatic VOCs, as they offer advantages on the basis of cost and potential stability.[8] They are supported over porous oxides or used as bulk catalysts without a support. The catalytic combustion of benzene, toluene and *p*-xylene by various transition metal oxide species supported on $\gamma$-$Al_2O_3$ has been studied recently.[125] It was observed that CuO/$\gamma$-$Al_2O_3$ was the most active of the seven metal oxides supported on $\gamma$-$Al_2O_3$ which were investigated. On the other hand, CuO was impregnated onto four different supports, $CeO_2$, $\gamma$-$Al_2O_3$, $TiO_2$ and $V_2O_5$, in order to define the optimal combination. CuO/$CeO_2$ was the most active catalyst, followed by CuO/$\gamma$-$Al_2O_3$. Finally, it was reported that the activity of CuO/$CeO_2$ with respect to the type of VOC was in the following order:

$$\text{toluene} > p\text{-xylene} > \text{benzene}$$

Various bulk metal oxide catalysts, including $CeO_2$, CuO, $Fe_2O_3$, $V_2O_5$, $ZrO_2$, $\gamma$-$Al_2O_3$ and $TiO_2$ (rutile), have been evaluated for the catalytic incineration of aromatic hydrocarbons. It was concluded that $CeO_2$ was the most active catalyst, and complete oxidation of toluene was achieved at 240°C. A similar temperature for complete conversion was found in the case of *p*-xylene, whereas benzene was more refractory and complete conversion was not reached even at 300°C. From the data it was concluded that it was not the surface area, but the redox properties that led to the different activities of the metal oxide catalysts.[126] As was observed for other metal oxides, bulk $CeO_2$ was more active than alumina-supported $CeO_2$,[127] which was only able to reach full toluene conversion at a temperature slightly lower than 300°C under similar conditions.

The use of cobalt oxide in the form of $Co_3O_4$ can be considered as an alternative to noble metal-based catalysts for the catalytic combustion of aromatic VOCs, since this metal oxide offers the advantage of both high reactivity and a relatively low price when compared with precious metals. The negative aspect of using cobalt oxide is the low stability during the catalytic reactions at very high temperatures. However, by optimising the characteristics of the metal oxide throughout the preparation method in the case of bulk metal oxides,[128,129] or supporting the $Co_3O_4$ on different metal oxides such as $Al_2O_3$ or $CeO_2$,[128,130,131] it is possible to prevent the sintering of the catalyst and facilitate the re-oxidation of cobalt species[128] during the catalytic combustion of aromatic VOCs. The latter is an important concept as the reaction takes place at intermediate temperatures and re-oxidation could be relatively slow. In the last decade, efforts have focused on modifying the characteristics of the cobalt oxide properties to improve activity. In the case of a supported metal oxide catalyst, it has been generally assumed that the most active cobalt-supported species are $Co_3O_4$-supported crystallites, loosely or moderately interacting with the support surface, and catalysts have attained complete benzene conversion at a temperature slightly higher than 250°C.[128,129]

In the case of bulk metal oxides, different strategies have been developed for the preparation of $Co_3O_4$ catalysts,[132,133] which have led to metal oxides with beneficial characteristics. In this context it is worth mentioning the development of the nanocasting technique, which uses a hard template and allows the preparation of highly ordered bulk metal oxides with high surface areas. Interestingly, during the procedure high calcination temperatures are often used, which could improve the stability of the catalysts. Mesoporous $Co_3O_4$ replicas have been reported in the literature[129,134] and present good preparative reproducibility, which is likely to be related to the fast transportation and to the low energy of cubic $Co_3O_4$ crystallites. Mesoporous cobalt oxides have been successfully prepared by a nanocasting route using mesoporous KIT-6 silica as a hard template for the catalytic combustion of toluene. These materials present extremely high surface areas although the synthesis conditions need to be properly adjusted. Thus, if optimally synthesised, these cobalt oxides can reach surface areas of ca. 175 $m^2\,g^{-1}$, in spite of having been calcined at a temperature as high as 550°C.[129] These materials have shown very high catalytic activity for the total oxidation of toluene, reaching full conversion at 225°C. Nanocast cobalt catalysts exhibit a catalytic activity much higher than that shown by a $Co_3O_4$ catalyst prepared by conventional techniques. The results of nanocast catalysts are likely to be related to both the high surface area and high oxygen mobility. Equally as important as the high catalytic activity obtained is the high catalytic stability that these catalysts present at moderate temperatures, regardless of the VOC employed, which is presumably a result of the high calcination temperature used, and is an advantage of the preparation method.

The nanocasting method has also been used to synthesise ordered chromium metal oxides. Mesoporous chromia is known to perform well for the combustion of VOCs. For example, by adopting a neutral templating strategy using poly (alkylene oxide) as a template, Sinha and Suzuki[135] generated 3D mesoporous chromium oxide with a surface area of 78 $m^2\ g^{-1}$ after calcination at 500°C. They found excellent catalytic activity over the mesoporous material for the oxidation of toluene, and related the performance to the multivalency of chromium ions. Furthermore, using KIT-6 as a template and chromium nitrate as the metal source, Wang *et al.*[136] fabricated ordered mesoporous $Cr_2O_3$ with a surface area of 82 $m^2\ g^{-1}$ after calcination at 400°C. They observed 100% conversion of toluene at 350°C and 30,000 $h^{-1}$ space velocity. Finally, Xia *et al.*[137] reported that with a high surface area of 106 $m^2\ g^{-1}$ and multivalent $Cr^{3+}$, $Cr^{5+}$, and $Cr^{6+}$ ions, mesoporous ordered chromia, fabricated in an autoclave through a novel solvent free route using KIT-6 as the hard template, was able to catalyse total toluene oxidation at 300°C.

Manganese oxide, as supported and as bulk metal oxide, has also been studied for the catalytic combustion of aromatic VOCs.[138, 139] The performance of an $\theta$-$\delta$-$Al_2O_3$ supported manganese catalyst has recently been evaluated for the abatement of aromatic VOCs.[138] As may be expected, the reactivity of the catalysts for toluene combustion was roughly correlated with the reducibility of the surface manganese oxide species, which was linked to the existence of an intrafacial mechanism. Full toluene conversion was reached at temperatures higher than 300°C. The catalytic activity of alumina-supported manganese catalysts has been improved by the addition of $CeO_2$[140] or $CeZrO_2$[141] to the alumina support. In these materials it was concluded that cerium improved the catalytic role of manganese in toluene oxidation. Oxygen mobility was also promoted in a redox mechanism in which manganese species served as the active sites. It was also observed that the partial substitution of $Ce^{4+}$ with $Zr^{4+}$ into the lattice of $CeO_2$ can form solid solutions, which results in the improvement of oxygen storage capacity, oxidation properties, thermal stability, and the catalytic activity at lower temperature. In addition, it was reported that the $MnO_X$ dispersion was greatly promoted by the support surface. Using these catalysts complete benzene conversion was attained at 290°C.

Recently, the performance of a very active manganese catalyst has been reported.[142] Manganese-zirconium and manganese-titanium supported oxide catalysts were prepared by means of reactions in molten sodium and potassium nitrate fluxes at 500°C. Alkaline metal nitrate salts can be conveniently used as solvents due to their low melting points, e.g. only 120°C for the $LiNO_3$–$NaNO_3$–$KNO_3$ eutectic, and the ability of the nitrate anion to be a donor of oxide ions ($O^{2-}$) or $O^-$ species. It was reported that $ZrO_2$-supported catalysts prepared using a $KMnO_4$ precursor showed the highest performance for the combustion of both benzene and toluene with $T_{50}$ values of 236 and 167°C, respectively. This catalyst was remarkably more active than $LaCoO_3$ and $Pt/Al_2O_3$ reference catalysts tested using the same experimental conditions. Thus, it was concluded that the molten salt technique has advantages for the preparation of manganese catalysts. The crucial point is probably the favourable chemical form of manganese species together with high specific surface areas.

Bulk manganese oxide catalysts, $Mn_3O_4$, $Mn_2O_3$ and $MnO_2$, and promoted manganese oxide catalysts with alkaline and alkaline-earth metals have been evaluated.[139] The catalytic activities were in the order:

$$Mn_3O_4 > Mn_2O_3 > MnO_2$$

$Mn_3O_4$ completely oxidised toluene at ca. 280°C. The activity was correlated directly with the surface area and oxygen mobility, and it was observed that the addition of potassium, calcium and magnesium also enhanced activity. The performance of manganese metal oxides has also been improved by the formation of solid solutions with other metal oxides, such as iron.[143] Formation of a $Fe_2O_3$–$Mn_2O_3$ solid solution is associated with a change of the cubic structure of $Mn_2O_3$, in which a proportion of $Mn^{3+}$ is replaced by smaller $Fe^{3+}$ ions. In these materials, the existence of structural defects favours the adsorption of oxygen. These oxygen species are very reactive and they notably improve the catalytic activity for toluene combustion.

Different groups[106,144,145] have studied benzene catalytic combustion over $VO_X/TiO_2$ catalysts. A two-step mechanism has been suggested where the first step is an adsorption of the aromatic ring on the catalyst via nucleophilic attack, and the second step is electrophilic substitution of the adsorbed species. Partial oxidation products were formed in both the presence and the absence of gas-phase oxygen, indicating that the surface oxygen was involved in the oxidation process. It was observed that there was a direct correlation between increasing vanadium loading and catalytic activity. Recently, the impact of doping with molybdenum and tungsten oxides has been reported. A significant increase in activity was observed when molybdenum or tungsten was added.[145]

In the last decade, some authors have studied the application of perovskite catalysts to the oxidation of VOCs. Transition metal perovskites based on lanthanum ($LaMO_3$, where M = Mn or Co) have demonstrated that they are very good oxidation catalysts for the removal of aromatic compounds.[146–149] The redox properties of the M cation, the availability of weakly-bonded oxygen at the surface and the presence of lattice defects are suggested as being responsible for the catalytic activity. It is generally accepted that these factors can be determined by the conditions used for catalyst preparation, such as temperature and time of the hydrothermal treatment, nature of the metal precursor, and base strength of the solution.[147] Redox titrations have shown that cobalt is present in $LaCoO_3$ exclusively in the 3+ oxidation state, whereas $LaMnO_3$ contains a considerable amount of $Mn^{4+}$. Generally in the literature it can be found that $LaMnO_3$ was more active than $LaCoO_3$. Since cobalt is present only in the 3+ oxidation state, oxygen is strongly anchored on the surface and consequently difficult to remove. In addition, it has also been reported that the partial replacement of $La^{3+}$ by $Sr^{2+}$ has often resulted in a considerable increase in combustion activity,[148,149] both in cobalt and manganese perovskite catalysts. For these materials it has been reported that aromatic VOCs can be totally removed by oxidation at temperatures ranging from 200 to 300°C, depending on the experimental conditions, such as VOC concentration, space velocity and the nature of the aromatic VOC.

It is well known that the major limitation of the application of perovskites as combustion catalysts is their lower surface area and their increased tendency to sinter. One solution to increase the contact surface between the VOC and the perovskite is to disperse it on a large surface area and thermally stable support. Thus, supported $LaCoO_3$ perovskites on $CeZrO_2$ have been studied recently.[150] The use of a $CeZrO_2$ support for lanthanum cobalt perovskites promoted the catalytic activity with respect to the corresponding bulk perovskites, decreasing the temperature for complete toluene oxidation by more than 50°C. The increased activity was related to two factors: (i) the larger exposed surface and (ii) the composition of the support which provided the increased oxygen mobility of the catalyst.

Several studies have been carried out in order to investigate the properties of transition metal exchanged or impregnated zeolite catalysts in the partial and deep oxidation of aromatic hydrocarbons. The availability of zeolites with a variety of porous structures, different composition and degree of hydrophobicity, as well as the possibility to control the acidic properties and location of exchanged cations, have contributed to the increased use of zeolites.[151] The results have shown that the catalytic behaviour depends on the reducibility and acidity of the catalysts, and on the oxygen carrier capacities.[152,153] However, it has been reported that a very high temperature was needed for benzene catalytic combustion, ca. 550°C, in the case of NaX, CaA and ZSM-5 zeolites in their parent and protonated forms.[152] The formation of carbonaceous deposits (coke) inside the pores or on the outer surface of zeolites is the main cause of their low activity and remarkable deactivation during the transformation of organic reactants.[154] Transition metal exchange zeolites improve the activity by increasing both the zeolite acidity (by cation hydrolysis), and oxygen chemisorption.[153] Thus, HY, NaY and HMFI zeolites exchanged with copper and caesium have been studied.[153,154] The addition of caesium leads to a decrease of the light-off temperature by 50°C. In these solids, the position and geometry of the copper ions in the zeolite matrix are of great significance for the redox behaviour and activity for toluene oxidation. The increase of the copper content of CuNaHY zeolites, from 1 to 8 wt%, caused a decrease of about 100°C for the temperature of complete toluene oxidation to $CO_2$ and reduces the temperature required to oxidise coke. However, temperatures higher than 350°C are still needed to reach complete toluene conversion to $CO_2$. Moreover, some selected oxides ($Mn^{2+}$, $Co^{2+}$, $Fe^{3+}$ and $Cu^{2+}$) supported on clinoptilolite were tested for the catalytic incineration of toluene. Manganese oxide on clinoptilolite was found to be the most active and durable of all the catalysts tested,[151] demonstrating total toluene conversion at a temperature slightly higher than 300°C.

Finally, it has been indicated that the use of mesoporous ordered silica materials as supports for metal oxides exhibited higher catalytic activity compared to microporous zeolite catalysts for catalytic combustion of toluene in the presence of excess oxygen, and furthermore coke formation could be effectively minimised.[155] Li *et al.*[155] reported that the activity of a copper-manganese oxide supported catalyst followed the order:

Cu-Mn/MCM-41 > Cu-Mn/ZSM-5 > Cu-Mn/BETA > Cu-Mn/porous silica

In recent years different mesoporous silica supports have been used and these have been mainly SBA-15[156–159] and MCM-41.[156,160] It can be observed that SBA-15 is the most widely used support. This support has a high surface area and uniform pore-size distribution, allowing higher dispersions of active phases and better control over the particle size compared with conventional amorphous silica or MCM-41. Amongst the supported metal oxides, CuO and $Co_3O_4$ supported on SBA-15 are found to have the highest activity for benzene oxidation.[156,157] The conversion of benzene over these catalysts reaches 100% at about 250°C. The activity is in the order:

$$CuO \sim Co_3O_4 > MnO > FeO > NiO$$

SBA-15 alone shows very low activity for benzene oxidation. The copper and cobalt oxide catalysts have the best redox properties, which is one of the reasons why they exhibit the highest activity for the catalytic combustion of benzene.[158] For the CuO/MCM-41 catalyst, the temperature required for 80% conversion of benzene was at least 500°C, which is 200°C higher than that for the CuO/ SBA-15 catalyst.

## 3.5. Conclusions

Catalytic oxidation as a method for VOC abatement is now a well-established technology, and one which offers significant advantages over other abatement methods. A wide variety of catalysts have been studied for the total oxidation of common

VOCs, such as short-chain alkanes and aromatics, and undoubtedly advances in terms of activity and stability have been achieved. Scope still exists to develop more active and stable catalysts for the total oxidation of more demanding VOCs, such as halogenated compounds and those containing heteroatoms like sulfur. There is also significant potential to improve our understanding of the catalysis at a molecular level and this endeavour should help to develop better total oxidation catalysts, and Impact on other areas like selective oxidation.

## References

1. Mukhopadhyay, N. and Moretti, E. (1993). *Current and Potential Future Industrial Practices for Controlling Volatile Organic Compounds*, American Institute of Chemical Engineers, Center for Waste Control Management, New York.
2. Jennings, M., Palazzolo, M., Krohn, N., *et al.* (1985). *Catalytic Incineration for the Control of Volatile Organic Compound Emission*, Pollution Technology Review, No. 121. Noyes Publications, New Jersey.
3. Molina, M. and Rowland, F. (1974). Stratospheric Sink for Chlorofluoromethanes-Chlorine Atom Catalyzed Destruction of Ozone, *Nature*, **249**, pp. 810–812.
4. Environmental Protection Agency, US Clean Air Act, 1990, USA.
5. Spivey, J. (1987). Complete Catalytic Oxidation of Volatile Organics, *Ind. Eng. Chem. Res.*, **26**, pp. 2165–2180.
6. Heneghan, C., Hutchings, G. and Taylor, S. (2004). Destruction of Volatile Organic Compounds by Heterogeneous Catalytic Oxidation, *Catalysis*, **17** (Royal Society of Chemistry, London, eds J. Spivey, G. Roberts), pp. 105–151.
7. Everaert, K. and Baeyens, J. (2004). Catalytic Combustion of Volatile Organic Compounds, *J. Hazard. Mater.*, **B109**, pp. 113–139.
8. Li, W., Wang, J. and Gong, H. (2009). Catalytic Combustion of VOCs on Non-noble Metal Catalysts, *Catal. Today*, **148**, pp. 81–87.
9. Ntainjua, N. and Taylor, S. (2009). The Catalytic Total Oxidation of Polycyclic Aromatic Hydrocarbons, *Top. Catal.*, **52**, pp. 528–541.
10. Prasad, R., Kennedy, L. and Ruckenstein, E. (1984). Catalytic Combustion, *Catal. Rev. Sci. Eng.*, **26**, pp. 1–58.
11. Tichenor, B. and Palazzolo, M. (1987). Destruction of Volatile Organic Compounds via Catalytic Incineration, *Environ. Prog.*, **6**, pp. 172–76.
12. Zięba A., Banaszak, T. and Miller, R. (1995). Thermal-catalytic Oxidation of Waste Gases, *Appl. Catal. A: Gen.*, **124**, pp. 47–57.
13. Banaszak, T., Miller, R. and Zembrzuski, M. (1987). The Influence of Flame-generated Free Radicals on the Thermal Oxidation of Waste Gases. *JAPCA*, **37**, pp. 1434–1438.
14. Mackie, J. (1991). Partial Oxidation of Methane: The Role of the Gas Phase Reactions, *Rev. Catal. Sci. Eng.*, **33**, pp. 169–240.
15. Vassileva, M., Andreev, A., Dancheva, S. *et al.* (1989). Complete Catalytic Oxidation of Benzene Over Supported Vanadium Oxides Modified by Palladium, *Appl. Catal. A: Gen.*, **49**, pp. 125–141.
16. Musialik-Piotrowska, A. and Syczewska, K. (1989). Destruction of Volatile Organic Mixtures by Catalytic Combustion, *Environ. Prot. Eng.*, **15**, pp. 117–126.
17. Papenmeier, D. and Rossin, J. (1994). Catalytic Oxidation of Dichloromethane, Chloroform and Their Binary Mixtures over a Platinum Alumina Catalyst, *Ind. Eng. Chem. Res.*, **33**, pp. 3094–3103.

18. Barresi, A. and Baldi, G. (1994). Deep Catalytic Oxidation of Aromatic Hydrocarbon Mixtures: Reciprocal Inhibition Effects and Kinetics. *Ind. Eng. Chem. Res.*, **33**, pp. 2964–2974.
19. Choudhary, T., Banerjee, S. and Choudhary, V. (2002). Catalysts for Combustion of Methane and Lower Alkanes, *Appl. Catal. A: Gen.*, **234**, pp. 1–23.
20. BP Statistical Review of World Energy. (2009).
21. Wani, H., Branion, R. and Lau, A. (1997). Biofiltration: A Promising and Cost-effective Control Technology for Odors, VOCs and Air Toxics, *J. Environ. Sci. Health*, **A32**, 7, pp. 2027–2055.
22. VOC Control: Technology Selection Criteria and Commercial Systems Review. (1995). *Air Poll. Consultant*, **5,** 4.1–4.18.
23. Kung, H. (1994). Oxidative Dehydrogenation of Light (C2 to C4) Alkanes, *Adv. Catal.*, **40**, pp. 1–38.
24. Golodets, G. (1983). *Heterogeneous Catalytic Reactions Involving Molecular Oxygen*, Elsevier, New York.
25. Burch, R., Crittle, D. and Hayes, M. (1999). C-H Bond Activation in Hydrocarbon Oxidation on Heterogeneous Catalysts, *Catal. Today*, **47**, pp. 229–234.
26. Batiot, C. and Hodnett, B. (1996). The Role of Reactant and Product Bond Energies in Determining Limitations to Selective Catalytic Oxidations, *Appl. Catal. A: Gen.*, **137** pp. 179–191.
27. Cassidy, F. and Hodnett, B. (1998). Selective Oxidation Catalysts: An Evaluation of the Discriminating Capacity of Active Sites on Oxide Catalysts with Molecular Oxygen as Oxidant, *CATTECH*, **2**, pp. 173–180.
28. Moro-Oka, Y., Morikawa, Y. and Ozaki, A. (1967). Regularity in the Catalytic Properties of Metal Oxides in Hydrocarbon Oxidation, *J. Catal.*, **7**, pp. 23–32.
29. Moro-Oka, Y. and Ozaki, A. (1966). Regularity in the Catalytic Properties of Metal Oxides in Propylene Oxidation, *J. Catal.*, **5**, pp. 116–124.
30. Gelin, P. and Primet, M. (2002). Complete Oxidation of Methane at Low Temperature over Noble Metal Based Catalysts: a review, *Appl. Catal. B : Environ.*, **39**, pp. 1–37.
31. Ciuparu, D., Lyubovsky, M., Altman, E., *et al.* (2002) Catalytic Combustion of Methane Over Palladium-based Catalysts, *Catal. Rev.*, **44**, pp. 593–649.
32. Fujimoto, K., Ribeiro, F., Avalos-Borja M., *et al.* (1998). Structure and Reactivity of PdOx/ZrO2 Catalysts for Methane Oxidation at Low Temperatures. *J. Catal.*, **179**, pp. 431–442.
33. Hicks, R., Qi, H., Young, M., *et al.* (1990). Structure Sensitivity of Methane Oxidation over Platinum and Palladium, *J. Catal.*, **122**, pp. 280–294.
34. Hicks, R., Qi, H., Young, M., *et al.* (1990). Effect of Catalyst Structure on Methane Oxidation over Palladium on Alumina, *J. Catal.*, **122**, pp. 295–306.
35. Baldwin, T. and Burch, R. (1990). Catalytic Combustion of Methane over Supported Palladium Catalysts: I. Alumina Supported Catalysts, *Appl. Catal. B: Environ.*, **66**, pp. 337–358.
36. Yamamoto, H. and Uchida, H. (1998). Oxidation of Methane over Pt and Pd Supported on Alumina in Lean-burn Natural-gas Engine Exhaust, *Catal. Today*, **45**, pp. 147–151.
37. Burch, R. and Loader, P. (1994). Investigation of $Pt/Al_2O_3$ and $Pd/Al_2O_3$ Catalysts for the Combustion of Methane at Low Concentrations, *Appl. Catal. B: Environ.*, **5**, pp. 149–164.
38. Ribeiro, F., Chow, M. and Dalla Betta, R. (1994). Kinetics of the Complete Oxidation of Methane over Supported Palladium Catalysts, *J. Catal.*, **146**, pp. 537–544.
39. Corro, G., Fierro, J. and Vázquez, O. (2005). Strong Improvement on $CH_4$ Oxidation over $Pt/\gamma\text{-}Al_2O_3$ Catalysts, *Catal. Commun.*, **6**, pp. 287–292.
40. Narui, K., Yata, H., Furuta, K., *et al.* (1999). Effects of Addition of Pt to $PdO/Al_2O_3$ Catalyst on Catalytic Activity for Methane Combustion and TEM Observations of Supported Particles, *Appl. Catal. A: Gen.*, **179**, pp. 165–173.
41. Burch, R. and Urbano, F. (1995). Investigation of the Active State of Supported Palladium Catalysts in the Combustion of Methane, *Appl. Catal. A: Gen.*, **124**, pp. 121–138.

42. Oh, S., Mitchell, P. and Siewert, R. (1991). Methane Oxidation over Alumina-supported Noble Metal Catalysts with and without Cerium Additives, *J. Catal.*, **146**, pp. 287–301.
43. Wang, C., Lin, H. and Ho, C. (2002). Effects of the Addition of Titania on the Thermal Characterization of Alumina-supported Palladium, *J. Mol. Catal. A: Chem.*, **180**, pp. 285–291.
44. Ishihara, T., Shigematsu, H., Abe, Y., *et al.* (1993). Effects of Additives on the Activity of Palladium Catalysts for Methane Combustion, *Chem. Lett.*, **3**, pp. 407–410.
45. Neyertz, C. and Volpe, M. (1998). Preparation of Binary Palladium-vanadium Supported Catalysts from Metal Acetylacetonates, *Colloids Surf. A*, **136**, pp. 63–69.
46. Escandón, L., Ordóñez, S., Diez, F., *et al.* (2003). Methane Oxidation over Vanadium-modified Pd/$Al_2O_3$ Catalysts, *Catal. Today*, **78**, pp. 191–196.
47. Yazawa, Y., Yoshida, H., Takagi, N., *et al.* (1999). Acid Strength of Support Materials as a Factor Controlling Oxidation State of Palladium Catalyst for Propane Combustion, *J. Catal.*, **187**, pp. 15–23.
48. Garcia, T., Solsona, B., Murphy, D., *et al.* (2005). Deep Oxidation of Light Alkanes over Titania-supported Palladium/Vanadium Catalysts, *J. Catal.*, **229**, pp. 1–11.
49. Taylor, M., Ntainjua Ndifor, E., Garcia, T., *et al.* (2008). Deep Oxidation of Propane Using Palladium–Titania Catalysts Modified by Niobium, *Appl. Catal. A: Gen.*, **350**, pp. 63–70.
50. Yazawa, Y., Takagi, N., Yoshida, H., *et al.* (2002). The Support Effect on Propane Combustion over Platinum Catalyst: Control of the Oxidation-resistance of Platinum by the Acid Strength of Support Materials, *Appl. Catal. A: Gen.*, **233**, pp. 103–112.
51. Yazawa, Y., Yoshida, H., Komai, S., *et al.* (2002). The Additive Effect on Propane Combustion over Platinum Catalyst: Control of the Oxidation-resistance of Platinum by the Electronegativity of Additives, *Appl. Catal. A: Gen.*, **233**, pp. 113–124.
52. Demoulin, O., Le Chef, B., Navez, M., *et al.* (2008). Combustion of Methane, Ethane and Propane and of Mixtures of Methane with Ethane or Propane on Pd/$\gamma$-$Al_2O_3$ Catalysts, *Appl. Catal. A: Gen.*, **344**, pp. 1–9.
53. Hutchings, G. (1985). Vapor Phase Hydrochlorination of Acetylene: Correlation of Catalytic Activity of Supported Metal Chloride Catalysts, *J. Catal.*, **96,** pp. 292–295.
54. Haruta, M., Yamada, N., Kobayashi, T., *et al.* (1989). Gold Catalysts Prepared by Coprecipitation for Low-temperature Oxidation of Hydrogen and of Carbon Monoxide, *J. Catal.*, **115**, pp. 301–309.
55. Grisel, R., Kooyman, P. and Nieuwenhuys, B. (2000). Influence of the Preparation of Au/$Al_2O_3$ on $CH_4$ Oxidation Activity, *J. Catal.*, **191**, pp. 430–437.
56. Gluhoi, A., Bogdanchikova, N. and Nieuwenhuys, B. (2005). Alkali (earth)-doped Au/$Al_2O_3$ Catalysts for the Total Oxidation of Propene, *J. Catal.*, **232**, pp. 96–101.
57. Gluhoi, A. and Nieuwenhuys, B. (2007). Catalytic Oxidation of Saturated Hydrocarbons on Multicomponent Au/$Al_2O_3$ Catalysts Effect of Various Promoters, *Catal. Today*, **119**, pp. 305–310.
58. Waters, R., Weimer, J. and Smith, J. (1995). An Investigation of the Activity of Coprecipitated Gold Catalysts for Methane Oxidation, *Catal. Lett.*, **30**, pp. 181–188.
59. Tsubota, S., Ueda, A., Sakur, H., *et al.* (1994). Application of Supported Gold Catalysts in Environmental Problems, *Environ. Catal.*, **34**, pp. 420–458.
60. Gasior, M., Grzybowska, B., Samson, K., *et al.* (2004). Oxidation of CO and C-3 Hydrocarbons on Gold Dispersed on Oxide Supports, *Catal. Today*, **91**, pp. 131–135.
61. Solsona, B., Garcia. T., Jones, C., *et al.* (2006). Supported Gold Catalysts for the Total Oxidation of Alkanes and Carbon Monoxide, *Appl. Catal. A: Gen.*, **312**, pp. 67–76.
62. Solsona, B., Garcia, T., Hutchings, G., *et al.* (2009). TAP Reactor Study of the Deep Oxidation of Propane Using Cobalt Oxide and Gold-containing Cobalt Oxide Catalysts, *Appl. Catal. A: Gen.*, **365**, pp. 222–230.

63. Miao, S. and Deng, Y. (2001). Au–Pt/$Co_3O_4$ Catalyst for Methane Combustion, *Appl. Catal. B: Environ.*, **31**, pp. L1–L4.
64. Germain, J. (1967). *Catalytic Conversion of Hydrocarbons*, Academic Press, New York.
65. Boreskov G. (1982). In J. Anderson, and M. Boudart (eds), *Catalysis, Science and Technology*, Vol. 3, Springer Verlag, New York, p. 39.
66. Yant, W. (1927). The Activity of Various Metals and Metal Oxide Catalysts in Promoting the Oxidation of Methane by Air, *J. Am. Chem. Soc.*, **49**, pp. 1454–1460.
67. McCarthy, J., Chang, Y., Wong, V., *et al.* (1997), Kinetics of High Temperature Methane Combustion by Metal Oxide Catalysts, Symp. Catalytic Combustion, San Francisco, *Am. Chem. Soc., Div. Petrol. Chem.*, **42**, pp. 158–165.
68. Solsona, B., Davies, T., Garcia, T., *et al.* (2008). Total Oxidation of Propane Using Nanocrystalline Cobalt Oxide and Supported Cobalt Oxide Catalysts, *Appl. Catal. B: Environ.*, **84**, pp. 176–184.
69. Liu, Q., Wan, L., Chen, M., *et al.* (2009). Dry Citrate-precursor Synthesized Nanocrystalline Cobalt Oxide as Highly Active Catalyst for Total Oxidation of Propane, *J. Catal.*, **263**, pp. 104–113.
70. Liotta, L., Di Carlo, F., Pantaleo, G., *et al.* (2006). $Co_3O_4/CeO_2$ Composite Oxides for Methane Emissions Abatement: Relationship between $Co_3O_4$–$CeO_2$ Interaction and Catalytic Activity, *Appl. Catal. B: Environ.*, **66**, pp. 217–227.
71. Trimm, D. (1995). Materials Selection and Design of High Temperature Catalytic Combustion Units, *Catal. Today*, **26**, pp. 231–238.
72. Trigueiro, F., Ferreira, C., Volta, J., *et al.* (2006). Effect of Niobium Addition to Co/$\gamma$-$Al_2O_3$ Catalyst on Methane Combustion, *Catal. Today*, **118**, pp. 425–432.
73. Baldi, M., Sanchez-Escribano, V., Gallardo-Amores, J., *et al.* (1998). Characterization of Manganese and Iron Oxides as Combustion Catalysts for Propane and Propene, *Appl. Catal. B, Environ.*, **17**, pp. L175–L182.
74. Hu, J., Chu, W. and Shi, L. (2008). Effects of Carrier and Mn Loading on Supported Manganese Oxide Catalysts for Catalytic Combustion of Methane, *J. Nat. Gas Chem.*, **17**, pp. 159–164.
75. Zaki, M., Hasan, M., Pasupulety, L., *et al.* (1999) CO and $CH_4$ Total Oxidation over Manganese Oxide Supported on $ZrO_2$, $TiO_2$, $TiO_2$-$Al_2O_3$ and $SiO_2$-$Al_2O_3$ Catalysts, *New J. Chem.*, **23**, pp. 1197–1202.
76. Hutchings, G., Heneghan, C. and Taylor, S. (1996). Uranium-oxide-based Catalysts for the Destruction of Volatile Chloro-organic Compounds, *Nature*, **384**, pp. 341–343.
77. Taylor, S. and O'Leary, S. (2000). A Study of Uranium Oxide Based Catalysts for the Oxidative Destruction of Short Chain Alkanes, *Appl. Catal. B: Environ.*, **25**, pp. 137–149.
78. Machida, M., Eguchi, K. and Arai, H. (1990). Effect of Structural Modification on the Catalytic Property of Mn-substituted Hexaaluminates, *J. Catal.*, **123**, pp. 477–485.
79. Tejuca, L., Fierro, J. and Tascón, J. (1989). Structure and Reactivity of Perovskite-Type Oxides, in D. Eley, H. Pines, P. Weisz (eds.), *Advances in Catalysis*, Vol. 36, Academic Press, New York, p. 237.
80. Alifanti, M., Kirchnerova, J., Delmon B., *et al.* (2004) Methane and Propane Combustion over Lanthanum Transition-metal Perovskites: Role of Oxygen Mobility, *Appl. Catal. A: Gen.*, **262**, pp. 167–176.
81. Seiyama, T., Yamazoe, N. and Eguchi, K. (1985). Characterization and Activity of Some Mixed Metal Oxide Catalysts, *Ind. Eng. Chem. Prod. Res. Dev.*, **24**, pp. 19–27.
82. Yamazoe, N. and Teraoka, Y. (1990). Oxidation Catalysis of Perovskites, Relationships to Bulk Structure and Composition, *Cat. Today*, **8**, pp. 175–199.
83. Baiker, A., Marti, P., Keusch, P., *et al.* (1994). Influence of the A-site cation in $ACoO_3$ (A = La, Pr, Nd, and Gd) Perovskite-type Oxides on Catalytic Activity for Methane Combustion, *J. Catal.*, **146**, pp. 268–276.

84. Arai, H., Yamada, T., Eguchi, K., *et al.* (1986). Catalytic Combustion of Methane over Various Perovskite-type Oxides, *Appl. Catal. A: Gen.*, **26**, pp. 265–276.
85. Zhong, Z., Chen, K., Ji, Y., *et al.* (1997). Methane Combustion over B-site Partially Substituted Perovskite $LaFeO_3$, *Appl. Catal. A: Gen.*, **156**, pp. 29–41.
86. Wang, Y., Wang, Y., Liu, X., *et al.* (2009). Nanocasted Synthesis of the Mesostructured $LaCoO_3$ Perovskite and Its Catalytic Activity in Methane Combustion, *J. Nanosci. Nanotechno.*, **9**, pp. 933–936.
87. Leanza, R., Rossetti, I., Fabbrini, L., *et al.* (2000). Perovskite Catalysts for the Catalytic Flameless Combustion of Methane – Preparation by Flame-hydrolysis and Characterisation by TPD-TPR-MS and EPR. *Appl. Catal. B: Environ.*, **28**, pp. 55–64.
88. Alifanti, M., Blangenois, N., Florea, M., *et al.* (2005). Supported Co-based Perovskites as Catalysts for Total Oxidation of Methane, *Appl. Catal. A: Gen.*, **280**, pp. 255–265.
89. Petrović, S., Karanović, L., Stefanov, P., *et al.* (2005). Catalytic Combustion of Methane over Pd Containing Perovskite Type Oxides, *Appl. Catal. B: Environ.*, **58**, pp. 133–141.
90. Koponen, M., Venalainen, T., Suvanto, M., *et al.* (2006). Methane Conversion and $SO_2$ Resistance of $LaMn1\text{-}xFexO_3$ ($x = 0.4, 0.5, 0.6, 1$) Perovskite Catalysts Promoted with Palladium, *J. Mol. Catal. A: Chem.*, **258**, pp. 246–250.
91. Cimino, S., Casaletto, M., Lisi, L., *et al.* (2007). Pd-LaMnO3 as Dual Site Catalysts for Methane Combustion, *Appl. Catal. A: Gen.*, **327**, pp. 238–246.
92. Russo, N., Palmisano, P. and Fino, D. (2009). Pd Substitution Effects on Perovskite Catalyst Activity for Methane Emission Control, *Chem. Eng. J.*, **154**, pp. 137–141.
93. Eyssler, A., Mandaliev, P., Winkler, A., *et al.* (2010). The Effect of the State of Pd on Methane Combustion in Pd-Doped LaFeO3, *J. Phys. Chem. C*, **114**, pp. 4584–4594.
94. Lambert, S., Cellier, C., Gaigneaux, E., *et al.* (2007). $Ag/SiO_2$, $Cu/SiO_2$ and $Pd/SiO_2$ Cogelled Xerogel Catalysts for Benzene Combustion: Relationships Between Operating Synthesis Variables and Catalytic Activity, *Catal. Commun.*, **8**, pp. 1244–1248.
95. Kim, S. and Shim, W. (2009). Properties and Performance of Pd-based Catalysts for Catalytic Oxidation of Volatile Organic Compounds, *Appl. Catal. B: Environ.*, **92**, pp. 429–436.
96. Garetto, T. and Apesteguıa, C. (2001). Structure Sensitivity and In Situ Activation of Benzene Combustion on $Pt/Al_2O_3$ Catalysts. *Appl. Catal. B: Environ.*, **32**, pp. 83–94.
97. Santos, V., Carabineiro, S., Tavares, P., *et al.* (2010). Oxidation of CO, Ethanol and Toluene over $TiO_2$ Supported Noble Metal Catalysts, *Appl. Catal. B: Environ.*, **99**, pp. 198–205.
98. Perez-Cadenas, A., Morales-Torres, S., Maldonado-Hodar, F., *et al.* (2009). Carbon-based Monoliths for the Catalytic Elimination of Benzene, Toluene and *m*-Xylene, *Appl. Catal. A: Gen.*, **366**, pp. 282–287.
99. Bedia, J., Rosas, J., Rodrıguez-Mirasol, J., *et al.* (2010). Pd-supported on Mesoporous Activated Carbons with High Oxidation Resistance as Catalysts for Toluene Oxidation, *Appl. Catal. B: Environ.*, **94**, pp. 8–18.
100. Perez-Cadenas, A., Morales-Torres, S., Kapteijn, F., *et al.* (2008). Carbon-based Monolithic Supports for Palladium Catalysts: The Role of the Porosity in the Gas-phase Total Combustion of *m*-Xylene, *Appl. Catal. B: Environ.*, **77**, pp. 72–277.
101. Xia, Q., Hidajat, K. and Kawi, S. (2001). Adsorption and Catalytic Combustion of Aromatics on platinum-supported MCM-41 Materials, *Catal. Today*, **68**, pp. 255–262.
102. Ryoo, M., Chung, S., Kim, J., *et al.* (2003). The Effect Of Mass Transfer on the Catalytic Combustion of Benzene and Methane over Palladium Catalysts Supported on Porous Materials, *Catal. Today*, **83**, pp. 131–139.
103. Ordóñez, S., Bello, L., Sastre, H., *et al.* (2002). Kinetics of Deep Oxidation of Benzene, Toluene, N-Hexane and their Binary Mixtures over a Platinum Catalyst, *Appl. Catal. B: Environ.*, **38**, pp. 139–149.

104. Kim, H., Kim, T., Koh, H., *et al.* (2005). Complete Benzene Oxidation over Pt-Pd Bimetal Catalyst Supported On $\gamma$-Alumina: Influence of Pt-Pd Ratio on the Catalytic Activity, *Appl. Catal. A: Gen.*, **280**, pp. 125–131.
105. Chen, M., Ma, Y., Li, G., *et al.* (2008). Support Effect, Thermal Stability, and Structure Feature of Toluene Combustion Catalyst. *Catal. Commun.*, **9**, pp. 990–994.
106. Garcia, T., Solsona, B., Cazorla-Amorós, D., *et al.* (2006). Total Oxidation of Volatile Organic Compounds by Vanadium Promoted Palladium-titania Catalysts: Comparison of Aromatic and Polyaromatic Compounds, *Appl. Catal. B: Environ.*, **62**, pp. 66–76.
107. Ferreira, R., de Oliveira P. and Noronha, F. (2004). Characterization and Catalytic Activity of Pd/$V_2O_5$/$Al_2O_3$ Catalysts on Benzene Total Oxidation, *Appl. Catal. B: Environ.*, **50**, pp. 243–249.
108. Vassileva, M., Andreev, A. and Dancheva, S. (1991). Complete Catalytic Oxidation of Benzene over Supported Vanadium Oxides Modified by Silver, *Appl. Catal. A: Gen.*, **69**, pp. 221–234.
109. Andreeva, D., Nedyalkova, R., Ilieva, L., *et al.* (2004). Gold–vanadia Catalysts Supported on Ceria–alumina for Complete Benzene Oxidation, *Appl. Catal. B: Environ.*, **52**, pp. 157–165.
110. Andreeva, D., Nedyalkova, R., Ilieva, L., *et al.* (2003). Nanosize Gold-ceria Catalysts Promoted by Vanadia for Complete Benzene Oxidation, *Appl. Catal. A: Gen.*, **246**, pp. 29–38.
111. Idakieva, V., Ilieva, L., Andreeva, D., *et al.* (2003). Complete Benzene Oxidation over Gold-vanadia Catalysts Supported on Nanostructured Mesoporous Titania and Zirconia, *Appl. Catal. A: Gen.*, **243**, pp. 25–39.
112. Albonetti, S., Bonelli, R., Delaigle, R., *et al.* (2010). Catalytic Combustion of Toluene over Cluster-derived Gold/Iron Catalysts, *Appl. Catal. A: Gen.*, **372**, pp. 138–146.
113. Centeno, M., Paulis, M., Montes, M., *et al.* (2005). Catalytic Combustion of Volatile Organic Compounds on Gold/Titanium Oxynitride Catalysts, *Appl.Catal. B: Environ.*, **61**, pp. 177–183.
114. Lingyun, J., Mai, H., Jiqing, L., *et al.* (2007). Preparation and Catalytic Performance of Pd Monolithic Catalysts Supported by Y2O3 Washcoat, *Chin J. Catal.*, **28**, pp. 635–640.
115. Lingyun, J., Mai, H., Jiqing, L., *et al.* (2008). Palladium Catalysts Supported on Novel CexY1–xO Washcoats for Toluene Catalytic Combustion, *J. Rare Earths*, **26**, pp. 614–619.
116. Zhang, Q., Zhao, L. and Yue, L. (2008). Toluene Combustion over Pd-based Monolithic Catalysts with a Novel CexZr1-$O_2$ Washcoat, *React. Kinet. Catal. Lett.*, **93**, pp. 27–33.
117. Lei, Y., Leihong, Z., Qingbao, Z., *et al.* (2009). Catalytic Combustion of Toluene over Pd-based Monolithic Catalysts with a Novel Washcoat Ce0.8Zr0.15La0.05O, *J. Rare Earths*, **27**, pp. 733–738.
118. Liu, Z., Wang, J., Zhonga, J., *et al.* (2007). Catalytic Combustion of Toluene over Platinum Supported on Ce–Zr–O Solid Solution Modified, *J. Hazard. Mat.*, **149**, pp. 742–746.
119. Masui, T., Imadzu, H., Matsuyama, N., *et al.* (2010). Total Oxidation of Toluene on Pt/$CeO_2$–$ZrO_2$–$Bi_2O_3$/ -$Al_2O_3$ Catalysts Prepared in the Presence of Polyvinyl Pyrrolidone, *J. Hazard. Mat.*, **176**, pp. 1106–1109.
120. Aguero, F., Barbero, B., Fernando, M., *et al.* (2009). Mixed Platinum-Manganese Oxide Catalysts for Combustion of Volatile Organic Compounds, *Ind. Eng. Chem. Res.*, **48**, pp. 2795–2800.
121. Blanco, J., Petre, A., Yates, M., *et al.*, (2007). Tailor-made High Porosity VOC Oxidation Catalysts Prepared by a Single-step Procedure, *Appl, Catal, B: Environ.*, **73**, pp. 128–134.
122. Aouad, S. Abi-Aad, E. and Aboukaıs, A. (2009). Simultaneous Oxidation of Carbon Black and Volatile Organic Compounds over Ru/$CeO_2$ Catalysts, *Appl. Catal. B: Environ.*, **88**, pp. 249–256.
123. Mitsui, T., Tsutsui, K., Matsui, T., *et al.* (2008). Support Effect on Complete Oxidation of Volatile Organic Compounds over Ru Catalysts, *Appl. Catal. B: Environ.*, **81**, pp. 56–63.

124. Ordóñez, S., Hurtado, P., Sastre, H., *et al.* (2004). Methane Catalytic Combustion over Pd/$Al_2O_3$ in Presence of Sulphur Dioxide: Development of a Deactivation Model, *Appl. Catal. A: Gen.*, **259**, pp. 41–48.
125. Wang, C., Lin, S., Chen, C., *et al.* (2006). Performance of the Supported Copper Oxide Catalysts for the Catalytic Incineration of Aromatic Hydrocarbons, *Chemosphere*, **64**, pp. 503–509.
126. Wang, C. and Lin, S. (2004). Preparing an Active Cerium Oxide Catalyst for the Catalytic Incineration of Aromatic Hydrocarbons, *Appl. Catal. A: Gen.*, **268**, pp. 227–233.
127. Del Angel, G., Padilla, J., Cuauhtemoc, I., *et al.* (2008). Toluene Combustion on g-$Al_2O_3$–$CeO_2$ Catalysts Prepared from Boehmite and Cerium Nitrate, *J. Mol. Catal. A: Chem.*, **281**, pp. 173–178.
128. Solsona, B., Davies, T., García, T., *et al.* (2008). Total Oxidation of Propane Using Nanocrystalline Cobalt Oxide and Supported Cobalt Oxide Catalysts, *Appl. Catal. B: Environ.*, **84**, pp. 176–184.
129. Garcia, T., Agouram, S., Sánchez-Royo, J., *et al.* (2010). Deep Oxidation of Volatile Organic Compounds Using Ordered Cobalt Oxides Prepared by a Nanocasting Route, *Appl. Catal. A: Gen.*, **386**, pp. 16–27.
130. Liotta, L., Ousmane, M., Di Carlo, G., *et al.* (2009). Catalytic Removal of Toluene over $Co_3O_4$–$CeO_2$ Mixed Oxide Catalysts: Comparison with Pt/$Al_2O_3$, *Catal. Lett.*, **127**, pp. 270–276.
131. Ataloglou, T., Vakrosa, J., Bourikas, K., *et al.* (2005). Influence of the Preparation Method on the Structure–Activity of Cobalt Oxide Catalysts Supported on Alumina for Complete Benzene Oxidation, *Appl. Catal. B: Environ.*, **57**, pp. 299–312.
132. Radwan, N., El-Shall, M. and Hassan, H. (2007). Synthesis and Characterization of Nanoparticle $Co_3O_4$, Cuo and Nio Catalysts Prepared by Physical and Chemical Methods to Minimize Air Pollution, *Appl. Catal. A: Gen.*, **331**, pp. 8–18.
133. Cao, J., Zhao Y. and Yan, G. (2003). Sol–gel Preparation and Characterization of $Co_3O_4$ Nanocrystals, *J. Univ. Sci. Technol. Beijing*, **10**, pp. 54–57.
134. Lu, A., Zhao, D. and Wan, Y. (2010). *Nanocasting: A Versatile Strategy for Creating Nanostructured Porous Materials*, RSC Publishing, RSC Nanoscience and Nanotechnology, Cambridge.
135. Sinha, A. and Suzuki, K. (2007). Novel Mesoporous Chromium Oxide for VOCs Elimination, *Appl. Catal. B: Environ.*, **70**, pp. 417–422.
136. Wang, Y., Yuan, X., Liu, X., *et al.* (2008). Mesoporous Single-crystal $Cr_2O_3$: Synthesis, Characterization, and its Activity in Toluene Removal, *Solid State Sci.*, **10**, pp. 1117–1123.
137. Xia, Y., Dai, H., Jiang, H., *et al.* (2009). Mesoporous Chromia with Ordered Three-dimensional Structures for the Complete Oxidation of Toluene and Ethyl Acetate, *Environ. Sci. Technol.*, **43**, pp. 8355–8360.
138. Agüero, F., Scian, A., Barbero, B., *et al.* (2009). Influence of the Support Treatment on the Behavior of MnOx/$Al_2O_3$ Catalysts used in VOC Combustion, *Catal. Lett.*, **128**, pp. 268–280.
139. Kim, S. and Shim, W. (2010). Catalytic Combustion of VOCs over a Series of Manganese Oxide Catalysts, *Appl. Catal. B: Environ.*, **98**, pp. 180–185.
140. Kim, H., Choi, S. and Inyang, H. (2008). Catalytic Oxidation of Toluene in Contaminant Emission Control Systems using Mn-Ce/$\gamma$-$Al_2O_3$, *Environ. Technol.*, **29**, pp. 559–569.
141. Yan, S., Wang, J., Zhong, J., *et al.* (2008). Effect of Metal Doping into Ce0.5Zr0.5$O_2$ on Catalytic Activity of MnOx/Ce0.5–xZr0.5–xM0.2xOy/$Al_2O_3$ for Benzene Combustion, *J. Rare Earths*, **26**, pp. 841–845.
142. Raciulete, M. and Afanasiev, P. (2009). Manganese-containing VOC Oxidation Catalysts Prepared in Molten Salts, *Appl. Catal. A: Gen.*, **368**, pp. 79–86.
143. Duran, F., Barbero, B., Cadus, L., *et al.* (2009). Manganese and Iron Oxides as Combustion Catalysts of Volatile Organic Compounds, *Appl. Catal. B: Environ.* **92**, pp. 194–201.

144. Lichtenberger, J. and Amidiris, M. (2004). Catalytic Oxidation of Chlorinated Benzenes over $V_2O_5/TiO_2$ Catalysts. *J. Catal.*, **223**, pp. 296–308.
145. Debecker, D., Delaigle, R., Bouchmella, K., *et al.* (2010). Total Oxidation of Benzene And Chlorobenzene with $MoO_3$- and $WO_3$-promoted $V_2O_5/TiO_2$ Catalysts Prepared by a Nonhydrolytic Sol–gel Route, *Catal. Today*, **157**, pp. 125–130.
146. Spinicci, R., Faticanti, M., Marini, P., *et al.* (2003). Catalytic Activity of $LaMnO_3$ and $LaCoO_3$ Perovskites Towards VOCs Combustion, *J. Mol. Catal. A: Chem.*, 197, pp. 147–155.
147. Deng, J., Zhang, L., Dai, H., *et al.* (2009). A Study on the Relationship Between Low-Temperature Reducibility and Catalytic Performance of Single-Crystalline La0.6Sr0.4$MnO_3$+d Microcubes for Toluene Combustion, *Catal. Lett.*, **130**, pp. 622–629.
148. Deng, J., Zhang, L., Daia, H., *et al.* (2009). Hydrothermally Fabricated Single-crystalline Strontium-substituted Lanthanum Manganite Microcubes for the Catalytic Combustion Of Toluene, *J. Mol. Catal. A: Chem.*, **299**, pp. 60–67.
149. Huang, H., Liu, Y., Tang, W., *et al.* (2008). Catalytic Activity of Nanometer La1 xSrx$CoO_3$ (x = 0, 0.2) Perovskites Towards VOCs Combustion La1 xSrx$CoO_3$ (x = 0, 0.2) Perovskites Prepared by a Co-precipitation Method, *Catal. Commun.*, **9**, pp. 55–59.
150. Alifanti, M., Florea, M. and Parvulescu, V. (2007). Ceria-based Oxides as Supports for $LaCoO_3$ Perovskite; Catalysts for Total Oxidation of VOC, *Appl. Catal. B: Environ.*,**70**, pp. 400–405.
151. Pozan Soylu, G., Özçelik, Z. and Boz, I. (2010). Total Oxidation of Toluene over Metal Oxides Supported on a Natural Clinoptilolite-type Zeolite, *Chem. Eng. J.*, **162**, pp. 380–387.
152. Dıaz, E., Ordoñez, S., Vega, A., *et al.* (2005). Evaluation of Different Zeolites in their Parent and Protonated Forms for the Catalytic Combustion of Hexane and Benzene. *Microp. and Mesop. Mat.*, **83**, pp. 292–300.
153. Ribeiro, M., Silva, J., Brimaud, S., *et al.* (2007). Improvement of Toluene Catalytic Combustion by Addition of Cesium in Copper Exchanged Zeolites, *Appl. Catal. B: Environ.*, **70**, pp. 384–392.
154. Antunes, A., Ribeiro, M., Silva, J., *et al.* (2001). Catalytic Oxidation of Toluene over Cunahy Zeolites Coke Formation and Removal, *Appl. Catal. B: Environ.*, **33**, pp. 149–164.
155. Li, W., Zhuang, M., Xiao, T., *et al.* (2006). MCM-41 Supported Cu-Mn Catalysts for Catalytic Oxidation of Toluene at Low Temperatures, *J. Phys. Chem. B*, **110**, pp. 21568–21571.
156. Yang, J., Jung, W., Lee, G., *et al.* (2008). Catalytic Combustion of Benzene over Metal Oxides Supported on SBA-15, *J. Ind. Eng. Chem.*, **14**, pp. 779–784.
157. Mu, Z., Li, J., Duan, M., *et al.* (2008). Catalytic Combustion of Benzene on Co/$CeO_2$/SBA-15 and Co/SBA-15 Catalysts, *Catal. Commun.*, **9**, pp. 1874–1877.
158. Yang, J., Jung, W., Lee, G., *et al.* (2010). Effect of Pretreatment Conditions on the Catalytic Activity of Benzene Combustion over SBA-15-Supported Copper Oxides, *Top. Catal.*, **53**, pp. 543–549.
159. Deng, J., Zhang, L., Dai, H., *et al.* (2009). In situ Hydrothermally Synthesized Mesoporous $LaCoO_3$/SBA-15 Catalysts: High Activity for the Complete Oxidation of Toluene and Ethyl Acetate, *Appl. Catal. A: Gen.*, **352**, pp. 43–49.
160. Li, W., Zhuang, M. and Wang, J. (2008). Catalytic Combustion of Toluene on Cu-Mn/MCM-41 Catalysts: Influence of Calcination Temperature and Operating Conditions on the Catalytic Activity, *Catal. Today*, **137**, pp. 340–344.

# Chapter 4

# Catalytic Oxidation of Volatile Organic Compounds: Chlorinated Hydrocarbons

Juan R. GONZÁLEZ-VELASCO,* Asier ARANZABAL,* Beñat PEREDA-AYO,* M. Pilar GONZÁLEZ-MARCOS,* and José A. GONZÁLEZ-MARCOS*

This chapter deals with the catalytic oxidation of chlorinated volatile organic compounds (Cl-VOCs). The intended oxidation products are $CO_2$ and $H_2O$, which are accompanied by HCl when VOCs are chlorinated. Uncompleted combustion can form CO, with the presence of water displacing the $HCl/Cl_2$ equilibrium, and other intermediate products may also be formed. Many catalysts have been reported in the literature for this purpose, including noble metals, transition metal oxides, mixed oxides, zeolites and perovskite-based catalysts. Our intention is to give a global and integrated view of all results presented in the literature, concentrating on the understanding of catalytic behaviours and mechanisms in the abatement of chlorinated volatile organic compounds with single or complex feed streams, and in the presence of water, hydrogen-supplying molecules and other co-pollutants, such as nitrogen oxides. The catalysts must exhibit adequate thermal and chemical stability with a remarkable resistance to deactivation, mainly by chlorine, which is also reviewed here.

## 4.1. Introduction

Among the chemicals emitted into the atmosphere, volatile organic compounds (VOCs) are classified worldwide as hazardous air pollutants. Although no widely supported definition of a VOC exists, the available definitions are mostly related to their vapour pressure, photochemical reactivity and/or effects on air quality and health. According to EC Directive 1999/13/EC, VOCs are functionally defined as organic compounds having a vapour pressure of 0.01 kPa or more at 293.15 K, or having a corresponding volatility under particular conditions of use. Methane, ethane, CO, $CO_2$, organometallic compounds and organic acids are excluded from this definition. An important group of VOCs are organochlorines, which are widely used in industry, in applications such as cleaning agents and degreasers, chemical

*Department of Chemical Engineering, Faculty of Science and Technology, University of the Basque Country, P.O. Box 644, ES-48080 Bilbao, Spain.

extractants, additives for paints, inks and adhesives, raw materials in the synthesis of drugs, pesticides and polymers, solvents for chemical reactions and the stripping agents of paints. These compounds are emitted into the atmosphere and pose a significant health hazard due to three main reasons: high volatility and persistence in the environment, ability to travel great distances from their point of release and the ability to transform, in the atmosphere, into other compounds which are toxic or hazardous to humans and wildlife.

Because their concentrations are usually low and the symptoms slow to develop, VOCs are typically not acutely toxic but have chronic effects and are suspected of causing cancer. VOCs also play a significant role in the formation of photochemical smog, since they react with nitrogen oxides and sunlight to form ozone. For this reason, the US Environmental Protection Agency has determined that controlling VOCs is an effective method for minimizing ozone levels.[1]

Among the strategies to control VOC release, the most desirable are those that improve processes so that emissions are minimized, however, there are many situations in which it is impractical or impossible to avoid the production of some waste. The main stationary sources of chlorinated VOC emissions into the atmosphere can be classified into three groups according to the volume of their release:

- — heavy chemical plants manufacturing halogenated hydrocarbons for the synthesis of plastics,[2] insecticides, anaesthetics, etc.;
- — finishing processes, especially those based on the use of volatile solvents;
- — clean-up processes, such as composting plants or the air-strippers of contaminated wastewater, groundwater and land.[3]

The increasingly stringent environmental regulations limiting emissions of pollutants and the growth of social awareness of environmental protection have forced manufacturers to develop technology to remove VOCs efficiently from waste streams. For many years, thermal incineration has been considered one of the most effective control devices. The effluent is burned with air in a furnace or torch, using fuel gas as an energy source, since the low concentration of VOCs (1,000 ppm) does not allow for self-sustained combustion. In the case of Cl-VOCs, chlorine atoms are flame retardants. In addition, due to their chemical stability, these compounds require high temperatures (800–1,000°C) for complete destruction and the presence of chlorine tends to yield a large volume of highly toxic products of incomplete combustion, such as phosgene, dibenzofurans and/or dioxins.[4] In addition, nitrogen oxides are formed at high temperatures.

In recent decades, catalytic oxidation of Cl-VOCs has received increased attention due to its energy and efficiency advantages over a wide range of operating conditions. Catalytic oxidation refers to the burning of pollutants with the aid of a catalyst. The catalyst opens up a different reaction pathway with lower activation energy than gas-phase combustion and allows oxidation to proceed at lower temperatures. This results in lower energy requirements and a lower production of $NO_x$. Capital costs due to more compact incinerators and insulation needs are also lower.[5] The approximate ranges of operating variables are low to moderate temperatures (100–500°C), atmospheric pressure, high space velocity ($10^3$–$10^5$ $h^{-1}$), and low organic reactant concentration ($10^2$–$10^3$ ppm) in the air.

Among the wide range of Cl-VOCs used in industrial and commercial applications, a selected group of compounds has been used for catalytic oxidation studies: 1,2-dichloroethane (DCE), which is the main pollutant in flue gases of vinyl chloride monomer plants;[5] trichloroethylene (TCE) and dichloromethane (DCM), which are common cleaning solvents in dry-cleaning, metal degreasing and semiconductor manufacturing and are found in the off-gases of groundwater and soil remediation processes;[6–8] and chlorobenzene (ChB), which has been used as a representative of polychlorinated dibenzo dioxins and/or furans.[9–11] Other Cl-VOCs (e.g. tetrachloroehylene, carbon tetrachloride, chloroform) have occasionally been used. The reactivity of these compounds can vary significantly. As a general term, saturated chlorinated hydrocarbons are oxidized more readily than unsaturated chlorinated organic compounds.[12] Their reactivity is correlated with their adsorption capacities. The adsorption capacity of chlorinated ethylenes is lower for compounds containing a larger number of chlorine atoms in the molecule.[13] However, there is no agreement between the catalytic reactivity of each oxidized compound and the calculated values of atomic excitation energies.[14]

The purpose of this chapter is to review the literature dealing with the heterogeneous catalytic oxidation of Cl-VOCs, emphasizing the knowledge which leads to the process design for the abatement of such compounds. The technical problem of depollution by chemical reaction in the catalytic domain consists basically of three different aspects: activity, selectivity and durability. One of the main goals in designing a catalyst is to find a composition that lowers the temperature required for the conversion of pollutants. Furthermore, the reaction pathway over the catalyst should lead to the complete oxidation of products, i.e. $CO_2$, $H_2O$ and HCl. The presence of HCl is preferable to $Cl_2$, since it can be readily scrubbed downstream of the catalytic oxidizer preventing its exit through the stack. Catalyst stability and durability are as important as their activity and selectivity, as these properties have to remain constant with reaction time in order to fulfil international environmental regulations and be economically more attractive.

On the other hand, the appropriate design of a reactor for catalytic oxidation requires the knowledge of kinetic parameters and transport parameters, and an appropriate mathematical description of the optimal reactor behaviour. However, modelling the process by duplicating the conditions found in real waste streams containing VOC mixtures may be complex, since "mixture effects" may result in inhibition or enhancement of the oxidation of a given compound.

## 4.2. Catalysts for Chlorinated VOC Oxidation

Heterogeneous catalytic oxidation of hydrocarbons has been extensively studied for the last century, since a large fraction of high value chemical feed stocks, such as organic aldehydes, acids, ketones, alcohols, etc., are produced by the catalytic partial oxidation of light alkanes and alkenes. Historically, oxidation reactions have always been carried out over either noble metals or transition metal oxides.[15–18] Noble metals have been almost exclusively applied to complete oxidations, whereas transition metal oxides have been used for both selective and complete oxidations.

The first investigations on the catalytic oxidation of chlorinated VOCs started in the 1970s, when awareness of air pollution was increasing,[19–21] but it was in the 1990s that research interest in this topic increased noticeably, peaking in 2001–2005 (Fig. 4.1). In this period, noble metal and mixed oxide-based catalysts were most investigated, and interest in alternative materials for improving the performance of catalysts started, focusing specifically on metal oxides and protonic metal zeolites. In later years efforts were focused on catalyst formulation with enhanced redox properties and oxygen storage capacity, such as the combination of ceria, zirconia, titania and vanadia.

Table 4.1 highlights the activity of some recently reported catalysts. Typically, catalytic activity has been characterized by monitoring the rise in conversion as a function of temperature for a particular Cl-VOC in some given test condition. A characteristic curve referred to as the light-off or ignition curve is obtained. $T_{50}$ and $T_{90}$

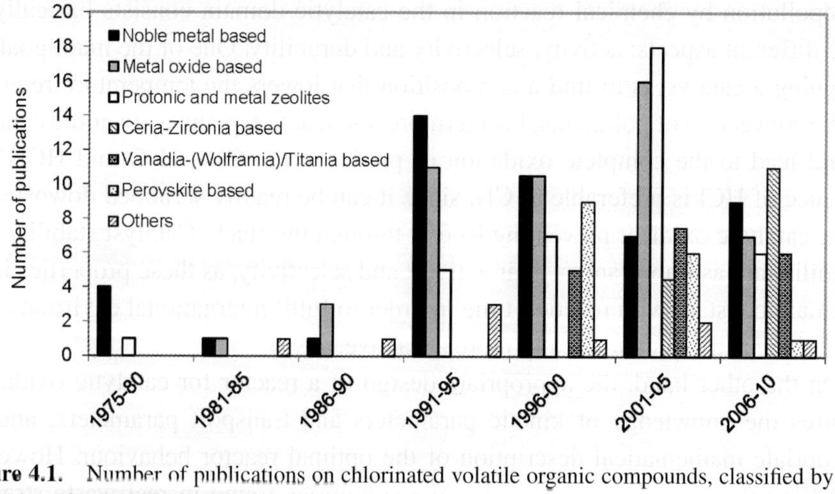


**Figure 4.1.** Number of publications on chlorinated volatile organic compounds, classified by type of catalyst, from 1975 to 2010.

**Table 4.1.** Catalytic conversion for Cl-VOC combustion reported in the recent literature[a].

| Catalyst | VOC concentration, %vol. | GSHV, $h^{-1}$ | $T_{50}$,°C | $T_{90}$,°C | Authors |
|---|---|---|---|---|---|
| 0.06% Pt/Fiber-glass | Cl-VOC mixture (0.55%) | 10, 000 | 485 | | Paukshtis *et al.*[22] |
| $MnO_x$ (1:4)/$TiO_2$-CNT[a] | ChB (100 ppm) | 36, 000 | | 150 | Tian *et al.*[23] |
| 5% $Mn_2O_3$/ $\gamma$-$Al_2O_3$ | DCE (100 ppm) | 80, 000 | 295 | 377 | Tseng *et al.*[24] |
| $CeO_2$ | DCE / TCE (1,000 ppm) | 30, 000 | 320/425 | 390/490 | de Rivas *et al.*[25] |
| $ZrO_2$ | DCE / TCE (1,000 ppm) | 30, 000 | 360/440 | 400/500 | de Rivas *et al.*[25] |
| $Mn_2O_3$ | DCE / TCE (1,000 ppm) | 30, 000 | 345/365 | 420/420 | de Rivas *et al.*[25] |
| $Ce_{0.5}Zr_{0.5}O_2$ | DCE / TCE (1,000 ppm) | 30, 000 | 295/390 | 395/465 | de Rivas *et al.*[25] |
| $Mn_{0.4}Zr_{0.6}O_2$ | DCE / TCE (1,000 ppm) | 30, 000 | 305/315 | 380/380 | de Rivas *et al.*[25] |
| Pt/H-ZSM-5 | DF (2,400 ppm) | 150, 000 | 280 | 310 | Komatsu *et al.*[10] |
| $MnCuO_x/TiO_2$ | ChB (500 ppm) | 5, 000 | 275 | 320 | Vu *et al.*[26] |
| $MnO_x$-$CeO_2$ | ChB (1,000 ppm) | 15, 000 | 175 | 220 | Xingyi *et al.*[27] |
| $CeO_2$ | TCE (1,000 ppm) | 15, 000 | 162 | 190 | Dai *et al.*[28] |
| Pd/$LaMnO_3$ | ChB (1,000 ppm) | 17, 800 | 243 (401[b]) | 342 | Giraudon *et al.*[29] |
| $V_2O_5$-$WO_3/TiO_2$ | ChB (100 ppm) | 37, 000 | 185 | 235 | Debecker *et al.*[30] |
| Si-MCM-41 | TCE (1,000 ppm) | 15, 000 | >550 | >550 | Wang *et al.*[31] |
| Pt/Si-MCM-41 | TCE (1,000 ppm) | 15, 000 | 419 | 485 | Wang *et al.*[31] |
| Pt/Ce-Si-MCM-41 | TCE (1,000 ppm) | 15, 000 | 380 | 445 | Wang *et al.*[31] |
| $SiO_2$ | TCE (1,000 ppm) | 15, 000 | 508 | | López-Fonseca *et al.*[32] |
| $ZrO_2$ | TCE (1,000 ppm) | 15, 000 | 480 | | López-Fonseca *et al.*[32] |
| H-BETA | TCE (1,000 ppm) | 15, 000 | 450 | | López-Fonseca *et al.*[32] |

*(Continued)*

**Table 4.1.** (*Continued*)

| Catalyst | VOC concentration, %vol. | GSHV, $h^{-1}$ | $T_{50}$,°C | $T_{90}$,°C | Authors |
|---|---|---|---|---|---|
| H-ZSM-5 | TCE (1,000 ppm) | 15,000 | 470 | 520 | López-Fonseca *et al.*[33] |
| H-MOR | TCE (1,000 ppm) | 15,000 | 450 | 500 | López-Fonseca *et al.*[33] |
| H-Y | TCE (1,000 ppm) | 15,000 | 480 | 530 | López-Fonseca *et al.*[33] |
| Pt/H-BETA | TCE (1,000 ppm) | 15,000 | 380 | 460 | López-Fonseca *et al.*[32] |
| Pd/H-BETA | TCE (1,000 ppm) | 15,000 | 310 | 380 | López-Fonseca *et al.*[34] |
| $Pd/Al_2O_3$ | TCE (1,000 ppm) | 15,000 | 410 | 475 | González-Velasco *et al.*[35] |

[a] Note: Carbon nanotube (CNT), 1,2-dichloroethane (DCE), trichloroethlyene (TCE), chlorobenzene (ChB), dibenzofuran (DF).
[b] Characteristic temperatures of the perovskite alone.

(temperatures at which 50% and 90% conversion are attained) are used as an indication of the relative reactivity of the catalysts.

### 4.2.1. *Noble metals*

In general terms, noble metals (Pt, Pd) show higher specific activity and are readily prepared in a highly dispersed form on supports of highly specific surfaces ($\gamma$-$Al_2O_3$, $SiO_2$, etc.). Small amounts (0.1–0.5 wt%) are enough for the production of good combustion catalysts.[36] Although activity is increased by increasing metal loading, no marginal advantage was observed above 1 wt% loadings.[19,35] Pt is the most used, but Pd-based catalysts have exhibited higher activity in the combustion of chlorinated hydrocarbons than Pt-based ones.[34,35] However, Pt promotes full oxidation of chlorohydrocarbons to $CO_2$, whereas Pd yields CO to a higher extent.[35,37] Due to the excess of oxygen in the reaction medium, noble metals, and especially Pd, are progressively oxidized and chlorinated during the catalytic test into, presumably, oxide and chloride metal species.[29,38–40] These species have been claimed to be potential active chlorination sites, especially in the oxidation of hydrogen-lean Cl-VOCs, such as TCE,[39,41–44] chloroform[37] and chlorobenzene.[29,38,45]

### 4.2.2. *Metal oxides*

The use of platinum and palladium is limited by cost and sensitivity of the noble metal catalysts to poisoning, especially by chlorine/chloride products. Thus, non-noble metal oxide catalysts have been considered as low-cost alternatives. The most common catalysts in this group are oxides of V, Cr, Mn, Fe, Co, Ni, Cu, Zn, Nb and Mo.[23,24,40,46–48] These are characterized by the high mobility of electrons and positive oxidation states. They may be prepared as uniform mass catalysts or supported on a highly specific surface area and cheaper oxide, such as $\gamma$-$Al_2O_3$. Since their specific activity is lower, metal loading of supported oxide catalysts is typically superior (5–20%), exhibiting more resistance to poisoning. Chromium oxide is reported to be the most active metal oxide[49,50] and has been extensively studied as supported on $Al_2O_3$,[51,52] porous carbon,[53,54] pillared clay,[55] zeolites,[13,56] $SnO_2$,[57] $TiO_2$ and $SiO_2$.[58] The degree of polymerization of chromate species increases with Cr loading, which leads to stronger redox ability on the surface.[59] Feijen-Jeurissen *et al.*[52] reported that 2 wt% chromia/alumina was more active than 2 wt% Pd/alumina in the oxidation of TCE. Over the chromia catalyst, a small amount of tetrachloroethylene (perchloroethylene, PCE) is formed under dry circumstances, but partial oxidation to CO was considerable, as reported earlier.[60,61] Nevertheless, the use of chromium can be restricted as it forms extremely toxic residues, such as chromium oxychloride, at low temperatures.[62]

In the last decade, manganese oxide has received increased attention as a cheap and environmentally friendly and active catalyst.[23,24] The optimal loading of Mn is related to its dispersion, oxidation state and oxygen chemisorption capacity.[23,63,64] Undesirable CO is also produced,[65] although selectivity to $CO_2$ is promoted with increasing manganese content.[64] Manganese oxide is also believed to be chlorinated during reaction,[48] leading to quick deactivation.

Alternatively, vanadium oxide catalysts, especially when supported on $TiO_2$, exhibit high activity — as high as chromium — and better stability in the oxidation of chloro-organics.[9,66–70] Although $TiO_2$-based $V_2O_5$-$WO_3$ catalysts were originally designed for the removal of nitrogen oxides ($NO_x$) by selective catalytic reduction (SCR),[71] they have also been used for the combined destruction of dioxins and $NO_x$.[72]

Catalyst manufacturing companies are currently marketing specific catalysts for the oxidation of chlorinated VOCs, mostly based on noble metals, such as Envicat-HHC (Süd-Chemie), Halocat® (Johnson Matthey), and HeraPur™ K-02130 and K-02134 (Heraeus). Alternatively, Haldor Topsøe's catalysts (CK305, CK306, CK395) are based on Cr, CrPd and Mn, respectively. The CRI Catalyst Company is offering a metal-supported titanium oxide catalyst for the decomposition of gaseous dioxin emissions. A systematic comparison between noble metal-based and chromia-based commercial catalysts concluded that the apparent catalytic activity of both types of catalysts for pure Cl-VOCs was in the same order of magnitude.[73,74] Heck *et al.*[75] provide a review of commercial catalytic technology for Cl-VOC abatement.

Mixed-metal oxides have also been used in order to improve the catalyst by taking advantage of the best properties of each metal.[26,76–78] Among them, perovskite-type oxides with the general formula $ABO_3$ (where "A" represents a lanthanide and/or alkaline-earth metal ion and "B" a transition metal ion, such as Mn, Co, Fe, Ni, etc.) have been found to be effective catalysts.[79] Important properties of perovskites in their catalytic applications include the stability of mixed oxides and unusual valence states of the transition metal ions in their structure, the presence of defect sites, and the high mobility of oxygen ions. Cr, Co and Mn perovskites are found to be the most active.[29,80,81] The kind of lanthanide in the perovskite did not significantly influence the catalytic activity and the formation of by-products.[82] Polychlorinated by-product formation was rather low as compared with noble metals.[29]

In the last decade, the redox properties of $CeO_2$ have been intensively analysed. The mechanism of VOC oxidation reactions over ceria is generally considered a redox-type mechanism, in which the key steps are the supply of oxygen by the readily reducible oxide and its re-oxidation by oxygen. $CeO_2$ catalysts prepared by thermal decomposition of cerium nitrate were found to be very active, but not stable[28] in the oxidation of TCE. Modifying $CeO_2$ with other metal oxides, for instance by the partial substitution of $Ce^{4+}$ with $Zr^{4+}$ in the lattice structure, can improve the oxygen storage capacity (OSC), redox properties and thermal resistance of the catalyst, and enhance catalytic activity at low temperatures.[25,83]

### 4.2.3. *Zeolites*

Not only the nature of the metal, but also the nature of the support (hydrophobicity, acidity, pore structure and redox properties) has a great influence in the oxidation of chloro-organics.[12,59,63] Many papers have reported that the order of activity of a series of mixed oxide catalysts was in close agreement with the strong acidity of the samples.[34,84] Particularly, remarkable effects have been attributed to Brønsted acid sites of protonic zeolites (H-Y, H-MOR, H-ZSM-5 and H-BETA), as being efficient sites for Cl-VOC chemisorption.[85,86] Moreover, surface acidity significantly inhibited the selectivity to molecular chlorine and polychlorinated subproducts[87] in favour of the desired chlorinated deep combustion product, HCl. Adversely, selectivity to the incomplete combustion product CO is significant.[88] Hence, the combination of acidic and oxidizing characteristics makes metal-doped zeolites good catalysts for Cl-VOC abatement.[40,89]

## 4.3. Kinetic Studies

Most assessments of activity, selectivity and deactivation are based on the comparison of ignition curves under the same reaction conditions. The temperature at which a certain level of conversion is reached ($T_{50}$ or $T_{90}$) has also been widely used for selecting the best catalyst and for optimization of the preparation methods.

Several efforts have been undertaken in order to obtain kinetic model equations for the catalytic decomposition of chlorinated volatile organic compounds. Adequate information on the reaction mechanisms determines the rules for building the kinetic model. Some consideration involving the mass and heat transfer phenomena during the reaction has also been undertaken. As the inlet concentration of Cl-VOC is very low and the enthalpy changes are moderate, no significant temperature gradient occurs due to the transport effect during reactions.[90] Under typical reaction conditions, the low heat generation due to exothermicity allows for the consideration of the isothermal reactor assumption in fixed beds, monoliths and metal fibre structures. Likewise, negligible concentration gradients have been calculated theoretically[91] and confirmed experimentally.[90,92]

### 4.3.1. *Potential reaction rate models*

Catalytic oxidation of a chlorinated VOC in the presence of oxygen has been mainly assumed to fit the simple potential model:

$$-r_{\rm V} = -\frac{dC_{\rm V}}{d(W/F_{\rm V_0})} = kC_{\rm O_2}^{m}C_{\rm V}^{n} \tag{4.1}$$

where $m$ and $n$ are the reaction order for oxygen and Cl-VOC, respectively. Under excess oxygen, as usual in VOC oxidation, the oxygen concentration can be considered practically unchanged[46] so that it can be introduced in the apparent reaction constant $k'$ shifting Equation (4.1) to

$$-r_{\rm V} = -\frac{dC_{\rm V}}{d(W/F_{\rm V_0})} = k'C_{\rm V}^{n} \tag{4.2}$$

Equation (4.2) is also compatible with the Mars–van Krevelen mechanism[93] when oxygen incorporation into the catalyst occurs faster than Cl-VOC decomposition.

Taking into account the definition of Cl-VOC conversion, $X = 1 - (C_{\rm V}/C_{\rm V_0})$, (Eq. 4.2) can also be expressed in terms of conversion as

$$C_{\rm V_0}\frac{dX}{d(W/F_{\rm V_0})} = k'C_{\rm V_0}^{n}(1 - X)^{n} \tag{4.3}$$

or rearranged for subsequent integration as

$$\frac{dX}{(1 - X)^{n}} = k'C_{\rm V_0}^{n-1}d(W/F_{\rm V_0}) \tag{4.4}$$

Considering $n = 1$ and after integration, Equation (4.4) can be linearized to

$$\ln(1 - X) = -k'(W/F_{\rm V_0}) \tag{4.5}$$

when representing $\ln(1 - X)$ vs $W/F_{V_0}$, which has been verified for a huge variety of Cl-VOCs and catalysts. The apparent kinetic constant $k'$ has been determined by linear regression at different temperatures and subsequently the apparent activation energy was obtained. The first order reaction has been proved in the catalytic destruction of chlorinated benzenes on $V_2O_5$–$WO_3$/$TiO_2$/metal fibres.[91,92] Calculated activation energies resulted in values ranging between 50 and 37 kJ mol$^{-1}$, decreasing when the number of Cl-substituents increased.

Similar results have been achieved during DCM oxidation on PtNaY-zeolites with an excess of water,[94] with $E_a = 82$ kJ mol$^{-1}$; in $CCl_4$ oxidation on Cr-Ce/$Al_2O_3$/monoliths,[95] with $E_a = 79.5$ kJ mol$^{-1}$; for TCE in humid air on K-CuO catalysts,[96] with $E_a = 66$–$75$ kJ mol$^{-1}$; for DCE on $CrO_3$ catalysts;[97] for TCE on Ru/$Al_2O_3$,[98] with $E_a = 54.5$ kJ mol$^{-1}$; and for mixed DCM/TCE feed stream over various $Cr_2O_3$-supported catalysts,[74] with $E_a = 44$ kJ mol$^{-1}$.

Equation 4.5 can be written for the conversion of 50% and 90% at specific operational conditions as Equations (4.6) and (4.7), the temperature then corresponding to $T_{50}$ and $T_{90}$, respectively.

$$\ln(1 - 0.50) = -k'_0 \exp\left(\frac{-E_a}{RT_{50}}\right)\left(\frac{W}{F_{V_0}}\right) \tag{4.6}$$

$$\ln(1 - 0.90) = -k'_0 \exp\left(\frac{-E_a}{RT_{90}}\right)\left(\frac{W}{F_{V_0}}\right) \tag{4.7}$$

Dividing Equation (4.6) by Equation (4.7) and isolating $E_a$

$$E_a = \frac{1.2\mathrm{R}}{(1/T_{50} - 1/T_{90})} \tag{4.8}$$

which allows us to determine the apparent activation energy from the values of $T_{50}$ and $T_{90}$ deduced from the corresponding light-off curves.

We have used Equation (4.8) to estimate the apparent activation energy for most experimental data in the literature, obtaining values between 18 and 110 kJ mol$^{-1}$. Obviously, the obtained values of the activation energy depend on the particular Cl-VOC to be oxidized and the type of catalyst used.

The influence of other species in the reaction rate has also been investigated. For example, Gervasini *et al.*[99] studied the destruction of $CCl_4$ in mixtures with different hydrocarbons over Cu-Cr catalysts concluding in first order kinetics for both $CCl_4$ and the hydrocarbon. Tseng *et al.*[24] fitted conversion data directly to the general potential model (Eq. 4.1) for the oxidation of DCE over $Mn_2O_3$/$\gamma$-$Al_2O_3$ catalysts and concluded orders of 0.49 and 0.42 for DCE and oxygen, respectively.

### 4.3.2. *Mechanistic reaction rates*

Although the potential models are useful for the development and comparison of catalyst activity during the design phase, more complete models should be used to expand the scope for their use in reactor simulation and design. Evidence of different steps involved in the oxidation of Cl-VOC has been collected:

— adsorption and adsorbed species;[13,100]
— effect of the concentration of different species on the observed reaction rate;[9]
— selectivity towards the desired products and detection of intermediates.[52,90,101]

The oxidation reaction occurs first by the reactant adsorption on the active sites and subsequent loss of light molecules (HCl, $H_2O$, $Cl_2$) until its total oxidation. Sometimes more stable intermediates can be detected when the extent of the reaction is only partial.[29,35,50,102] Corella *et al.*[103] proposed the use of mechanistic equations including the equilibrium constants of the adsorbed species. Other authors[104,105] have proposed Langmuir–Hinshelwood kinetic equations involving the adsorption of oxygen and Cl-VOC mainly for $Pt/Al_2O_3$ catalysts

$$-r_V = \frac{k p_V p_{O_2}}{1 + K_1 p_V + K_2 p_{O_2}} \tag{4.9}$$

Wang *et al.*[106] and Lou and Lee[37] assumed the reaction occurred between oxygen on the active site and Cl-VOC from the gas phase as in the Rideal–Eley mechanism. This is coincident with the equation proposed by the Mars–van Krevelen model

$$-r_V = \frac{k_O k_r p_V p_{O_2}}{k_O p_{O_2} + \upsilon k_r p_V} \tag{4.10}$$

Miranda *et al.*[50] studied the kinetics of TCE oxidation using four different models: Langmuir–Hinshelwood (reaction between the two adsorbed species was the rate-controlling step), Rideal–Eley (assuming that oxygen reacts from the gas phase), Mars–van Krevelen (Eq. 4.10), and a $Cl_2$ inhibition model (Eq. 4.11). The last of these resulted in the best fit with the experimental laboratory data

$$-r_V = \frac{k p_V}{1 + K_{Cl_2} p_{Cl_2} + K_V p_V} \tag{4.11}$$

Very few kinetic studies have considered intermediate products formed during Cl-VOC oxidation, such as vinyl chloride in the case of DCE or tetrachloroethylene in the case of TCE. Aranzabal *et al.*[90] proposed for DCE oxidation a mechanism with dehydrochlorination as a first step, leading to the formation of vinyl chloride (Eq. 4.12), followed by the direct oxidation of $C_2H_3Cl$ to CO and $CO_2$

(Eqs. 4.13 and 4.14).

$$C_2H_4Cl_2 \rightarrow C_2H_3Cl + HCl \qquad r_1 = A_1 \exp\left(\frac{-E_{a1}}{RT}\right) p_{DCE} \tag{4.12}$$

$$C_2H_3Cl + 3/2O_2 \rightarrow 2CO + HCl + H_2O \qquad r_2 = A_2 \exp\left(\frac{-E_{a2}}{RT}\right) p_{VC} \tag{4.13}$$

$$C_2H_3Cl + 5/2O_2 \rightarrow 2CO_2 + HCl + H_2O \qquad r_3 = A_3 \exp\left(\frac{-E_{a3}}{RT}\right) p_{VC} \tag{4.14}$$

$$CO + 1/2O_2 \rightarrow CO_2 \qquad r_4 = A_4 \exp\left(\frac{-E_{a4}}{RT}\right) p_{CO} \tag{4.15}$$

Formation of $Cl_2$ can also be expected by the Deacon reaction due to the formation of HCl in the presence of a high oxygen concentration in the reaction environment,

$$2HCl + 1/2O_2 \leftrightarrows Cl_2 + H_2O \quad r_D = A_D \exp\left(\frac{-E_D}{RT}\right) p_{HCl}^2 \tag{4.16}$$

Table 4.2 shows the values of the kinetic parameters for each reaction stage, within the 95% confidence intervals, obtained from the minimization of the sum of residues

$$RSS = \sum_{i}^{\text{comp.}} \sum_{j}^{\text{expt.}} [p_{expt}(i,j) - p_{calc}(i,j)]^2 \tag{4.17}$$

To our knowledge, kinetic models based on a more comprehensive testing to predict the reaction rate composition in streams near the industrial operation parameters with a greater number of components have not yet been established.

## 4.4. Influence of Water Vapour and Co-Pollutants in Feed Streams

In practical applications the flue gas stream carries chloro-organics with a relatively low concentration in a mixture with water vapour (moisture) whose concentration

**Table 4.2.** Calculated kinetic parameters for the complete oxidation of DCE.

| Reaction | ln (A, mol $s^{-1}$ $kg^{-1}$ $Pa^{-1}$) | $E_a$, kJ $mol^{-1}$ |
|---|---|---|
| $C_2H_4Cl_2 \rightarrow C_2H_3Cl + HCl$ | $-5.3 \pm 1.0$ | $61.3 \pm 6.5$ |
| $C_2H_3Cl + 3/2O_2 \rightarrow 2CO + HCl + H_2O$ | $-8.9 \pm 2.1$ | $52.2 \pm 17.9$ |
| $C_2H_3Cl + 5/2O_2 \rightarrow 2CO_2 + HCl + H_2O$ | $15.5 \pm 3.6$ | $163.0 \pm 32.0$ |
| $CO + 1/2O_2 \rightarrow CO_2$ | $9.8 \pm 38.8$ | $108.2 \pm 107.3$ |
| $2HCl + 1/2O_2 \leftrightarrows Cl_2 + H_2O$ | $-20.2 \pm 0.7$ | $37.7 \pm 13.2$ |

Mean errors in the prediction for each compound, Pa: 2.6 (DCE), 4.0 (CO), 6.6 ($CO_2$), 3.7 (VC), 4.5 (global).

**Table 4.3.** Composition of vinyl chloride combustion wastes.

| Dichloroethane distillation residue | | Vinyliden chloride distillation residue | |
|---|---|---|---|
| Substance | Mass % | Substance | Mass % |
| 1,2-Dichloroethane | 43.7 | Trichloroethylene | 32.5 |
| 1,1,2-Trichloroethane | 36.7 | Vinyliden chloride | 24.8 |
| 1,1,3-Trichloropropane | 4.5 | Dichloroethane | 24.7 |
| 1,1,2-Trichloropropane | 4.4 | Trans-dichloroethylene | 6.9 |
| 1,2-Dicloropropane | 8.7 | Trichloroethane | 6.4 |
| Mono-chlorobenzene | 2.0 | Trans-trichloroethilene | 4.7 |

can exceed 10,000 ppm, e.g. in the case of gas streams from air stripping of Cl-VOCs in contaminated water.[7,107] Also, non-chlorinated VOCs, such as aliphatic or aromatic compounds, may frequently be present. Therefore, it is of great importance to analyse the impact of these molecules on the behaviour of Cl-VOC oxidation catalysts.

Actual chloro-organic wastes from plants producing vinyl chloride are an industrial example of complex mixtures of Cl-VOCs.[22] These wastes (heavy residues from dichloroethane and vinylidene chloride distillation) are characterized by a complex composition, particularly complicated by the presence of a significant amount of heavily oxidized polychloroolefins with an insufficient amount of hydrogen for the stoichiometric formation of HCl. Typical compositions of these wastes are given in Table 4.3.

It should also be considered that industrial exhaust gases contain $NO_x$, CO and $H_2O$ together with Cl-VOCs. The usual concentration of $NO_x$ and CO in, for example, cogeneration and combustion exhaust gases are, respectively, in the range 100–1,000 and 1,000–50,000 ppm.[108–110] From a practical point of view, it is thus crucial to assess whether designed catalysts still succeed in oxidizing Cl-VOCs efficiently in the presence of co-pollutants or if the latter induce any kind of deactivation.

Thus, the scope of this section is to elucidate how the presence of hydrogen-supplying compounds such as water or non-halogenated VOCs, the mutual effects of several Cl-VOCs on one another, and the presence of co-pollutants such as CO and $NO_x$ can affect catalytic activity and product selectivity for the optimization and selection of effective catalysts.

### 4.4.1. *Effect of water vapour*

The effect of water vapour on VOC oxidation has always needed careful investigation. Firstly, water is present in all exhaust gases due to its production in all incineration processes. Secondly, water is a product of the VOC oxidation itself. Thirdly, water is commonly regarded as a poison for the catalytic combustion of VOCs.[38,56,85] However, in the case of the combustion of aromatic Cl-containing VOCs, water could play an additional interesting role, namely by removing the $Cl^-$ remaining on the surface after the oxidation of the aromatic ring.[47] Finally, water could also react with chlorine to produce HCl by the Deacon reaction ($2Cl_2+2H_2O4HCl+O_2$) and change the balance of selectivity for $HCl/Cl_2$.[39] It was reported that the Cl-VOC adsorption capacity and the Cl-COV destruction activity are generally decreased upon contact with a high concentration of water vapour in the gas stream,[111–113] although a low concentration of water vapour is sometimes helpful for the complete destruction of hydrogen-lean Cl-VOCs, such as TCE or $CCl_4$.

It is not easy to compare the oxidative behaviour of different catalysts reported in the literature, as reaction conditions in which the catalysts were tested differ. In our laboratory various types of catalysts have been tested for Cl-VOC oxidation under similar experimental conditions, namely atmospheric pressure, a flow rate through the reactor of 500 $cm^3$ $min^{-1}$ and a gas hourly space velocity (GHSV) of 15,000 $h^{-1}$. The concentration of each chlorohydrocarbon in the gas stream was 1,000 ppm. Table 4.4 shows a comparison of $T_{90}$ and selectivity to the desired products — HCl and CO — for TCE oxidation experiments under dry and humid conditions. In addition, results from other researchers are discussed in the following section.

The activity and product distribution of alumina-supported platinum and palladium[35,39] and chromia and palladium[52] catalysts for the oxidation of 1,2-dichloroethane and trichloroethylene have been investigated in the absence and presence of water. In the case of TCE, over a 2 wt% $Pd/Al_2O_3$, 50% and 95% conversion was reached at 360 and 420°C, respectively.[52] González-Velasco *et al.*[39] reported the effect of water concentration on the destruction of TCE over 0.42 wt% $Pd/Al_2O_3$ and 0.44 wt% $Pt/Al_2O_3$ (Table 4.4). With palladium, the reaction proceeded to almost the same extent in both the absence or the presence of water in all temperature ranges, even when increasing water vapour concentration from 1,000 to 15,000 ppm. However, TCE oxidation light-off curves over $Pt/Al_2O_3$ showed that catalyst activity was enhanced by water at low temperatures (<400°C); at 400°C conversion was nearly the same in dry and humid streams; but above 400°C the presence of water inhibited the activity for TCE oxidation, i.e. $T_{90}$ of 500 vs 530°C without water vapour. This inhibition effect of TCE oxidation activity, especially above 400°C, is quite general for all the catalysts reported in the literature, as can be seen for those shown in Table 4.4.

Concerning the distribution of products, HCl, $Cl_2$ and $C_2Cl_4$ (PCE) were found to be the chlorine-containing products in the TCE catalytic oxidation whatever the type of catalyst.[32,33,35,39,40,43,52,102,114] In general, the amount of PCE increased with temperature, reached a maximum around 450–500°C and was significantly reduced at 550°C to very low levels. For example, Fig. 4.2 shows the evolution of

**Table 4.4.** Temperatures for 90% TCE-conversion and selectivity of desired products (HCl and CO) for different catalysts investigated in the literature, alone and in the presence of water vapour.

| | TCE (1,000 ppm) | | | TCE + water (15,000 ppm) | | |
|---|---|---|---|---|---|---|
| | $T_{90}$,°C | Selectivity | | $T_{90}$,°C | Selectivity | |
| Catalyst | | HCl[a] | $CO_2^b$ | | HCl[a] | $CO_2^b$ |
| 0.44% $Pt/Al_2O_3^{(1)}$ | 490 | 31% | 100% | 500[c] | 63%[c] | 100%[c] |
| 0.42% $Pd/Al_2O_3^{(1)}$ | 490 | 39% | 47% | 460[c] | 78%[c] | 100%[c] |
| H-Y(2) | 530 | 54% | 66% | 540 | 95%[c] | 72% |
| H-ZSM-5(2) | 510 | 52% | 63% | 530 | 94%[c] | 70% |
| H-MOR(2) | 500 | 51% | 69% | 525 | 98%[c] | 74% |
| H-BETA(3) | 450 | 32% | 27% | 510 | 99% | 28% |
| PdO/H-BETA(3) | 385 | 15% | 35% | 355 | 98% | 51% |
| Pt/H-BETA(4) | 450 | 20% | 68% | 430 | 74% | 99% |
| $CeO_2^{(5)}$ | 490 | 31% | 89% | 490 | 80% | 92% |
| $Ce_{0.68}Zr_{0.32}O_2^{(5)}$ | 475 | 41% | 79% | 500 | 84% | 81% |
| $Ce_{0.5}Zr_{0.5}O_2^{(5)}$ | 465 | 43% | 75% | 495 | 86% | 79% |
| $Ce_{0.15}Zr_{0.85}O_2^{(5)}$ | 450 | 47% | 73% | 490 | 88% | 78% |
| $ZrO_2^{(5)}$ | 500 | 51% | 62% | 530 | 94% | 70% |

[a]At 550°C [b]At 350°C [c]7,500 ppm of water vapour
(1)González-Velasco *et al.*[39]
(2)López-Fonseca *et al.*[33]
(3)López-Fonseca *et al.*[32]
(4)López-Fonseca *et al.*[40]
(5)de Rivas *et al.*[102]

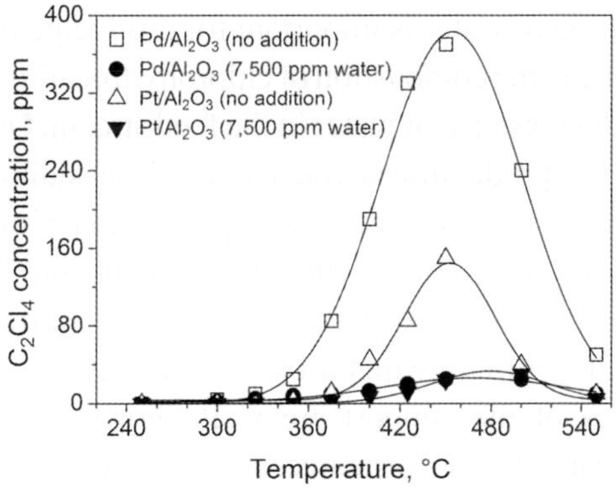


**Figure 4.2.** Evolution of the intermediate tetrachloroethylene with temperature in the oxidation of TCE over noble metal-based catalysts, in the absence and presence of water (adapted from González-Velasco *et al.*[39]).

PCE concentration with temperature for $Pt/Al_2O_3$ and $Pd/Al_2O_3$, in both the absence and the presence of water vapour.[39]

In the absence of water, PCE concentration over $Pd/Al_2O_3$ was more than double that for over $Pt/Al_2O_3$. In the presence of 1,000 ppm of water, PCE formation was reduced by a factor of four over $Pd/Al_2O_3$ and by a factor of two over $Pt/Al_2O_3$. Higher volumes of water (7,500 and 15,000 ppm) led to similar PCE concentrations (around 25 and 15 ppm) on $Pd/Al_2O_3$ and slightly higher concentrations over $Pt/Al_2O_3$ (34 and 26 ppm). The noble metals were responsible for the formation of PCE by chlorination of adsorbed TCE, through metal (oxy) chlorides. In this context, palladium exhibited a higher chlorination ability than platinum, which resulted in a higher selectivity to PCE in the former. A similar trend was found by Feijen-Jeurisen *et al.*;[39] the addition of water to the feed did not affect the conversion of TCE but resulted in an important decrease of the amount of PCE from 38 to 4%. Over the 2 wt% $CrO_x/Al_2O_3$ catalyst, a small amount of PCE was formed in a dry environment. The addition of water to the feed prevented the formation of PCE and the oxidation reaction over $CrO_x$ was inhibited.

Selectivity towards the desired oxidized products, i.e. HCl and $CO_2$, is shown in Table 4.4 for the TCE oxidation over different type of catalysts, at 550°C for HCl (as PCE has been almost totally oxidized at this temperature) and 350°C for $CO_2$. In general, the presence of water vapour in the feed stream greatly lowered the production of molecular chlorine in favour of HCl formation. For example, in the runs over $Pd/Al_2O_3$, when 1,000 ppm water was added, chlorine production was half of that produced under dry conditions and 10 times lower than in an excess of water (1.5%). Platinum showed higher selectivity to chlorine, hence, the effect of water was slightly lower.[39] The evolution of HCl and $CO_2$ selectivities with temperature is represented in Fig. 4.3 for the H-BETA and Pt/H-BETA catalysts as reported in the literature.[32] The addition of water is important for producing the desired HCl rather than $Cl_2$ as a product from the combustion of chlorohydrocarbons. Also, selectivity to $CO_2$ is enhanced by the presence of water in the feed stream. On the other hand, with metal-free H-BETA, CO production is much higher. The addition of Pt improves this situation, but at the expense of the concomitant production of molecular chlorine and highly chlorinated by-products (PCE in the TCE combustion). Water was also found to promote complete oxidation of CO, through the water-shift reaction, resulting in 100% selectivity to $CO_2$ for noble metal-based catalysts above 400°C.

The characteristics of the support are well known for playing a significant role in determining catalytic performance,[115–117] as already seen from the examples given above. Studies on the applicability of H-type zeolites (H-Y, H-ZSM5, H-MOR) as potential catalysts for chlorocarbon destruction demonstrated that acid sites (mainly Brønsted-type sites) acted as efficient chemisorption sites for chlorinated molecules.[33,85,86] The order of activity for TCE oxidation was found to

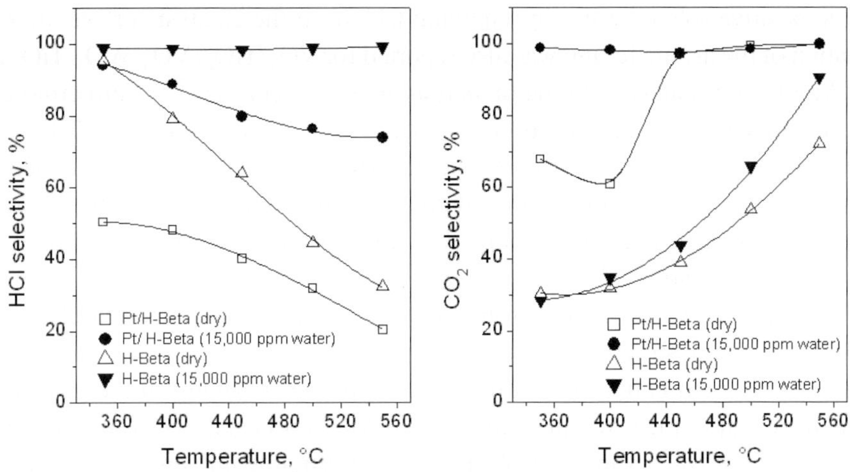


**Figure 4.3.** Evolution of the desired products with temperature in the oxidation of TCE over noble metal-based catalysts, in the absence and presence of water (data from López-Fonseca *et al.*[32]).

be: H-MOR > H-ZSM5 > H-Y. The addition of water to the feed stream did not alter the activity order observed under dry air. The excess of water led to increased conversion over H-MOR and H-ZSM5 at lower temperatures (partial conversion), meaning that both oxygen and water are involved in the key step(s) of TCE destruction. Zeolite H-Y was insensitive to added water. Also, the addition of water resulted in a substantial improvement for deep oxidation selectivity. It suppressed the generation of $Cl_2$ while increasing the selectivity to HCl and led to a much lower (intermediate) production of PCE. The selectivity to $CO_2$ was not complete but was somewhat better in the presence of water.

The combustion of DCE and TCE over ceria-zirconia mixed oxides with varying Ce/Zr ratios was investigated by de Rivas *et al.*[83] Under humid conditions (15,000 ppm of water vapour), catalytic activity was negatively impacted. The inhibiting effect was more perceptible for ceria-zirconia oxides and pure zirconia; by contrast, pure ceria was almost unaffected. This slight inhibition reflected the disfavoured adsorption of the chlorinated compounds at the catalytic sites in the presence of water molecules. Ce-rich oxides showed a lower affinity for water. These results reveal that this adsorption capacity was a key property of an adequate Ce/Zr catalyst for Cl-VOC abatement in industrial waste gas streams, with $Ce_{0.2}Zr_{0.8}O_2$ showing the highest activity. The addition of water to the feed stream induced beneficial effects on the deep oxidation catalytic selectivity, appreciably enhancing the yields of HCl and $CO_2$ since it promoted the reverse Deacon reaction ($4HCl+O_2 \leftrightharpoons 2Cl_2+2H_2O$). Also the formation of chlorinated by-products was significantly suppressed.

The positive role of water in removing $Cl^-$ from the catalyst surface after the oxidation of the aromatic ring was also reported for $VO_x/TiO_2$, $VO_x$-$WO_x/TiO_2$ and $VO_x$-$MoO_x/TiO_2$ catalysts in the oxidation of chlorobenzene.[47,93] Furthermore, in the catalytic oxidation of 4,4,6-trichlorophenol (TCP) over $V_2O_5$-$WO_3/TiO_2$ catalysts, the presence of water affected the adsorption capacity, steady-state activity and product distribution, due to the competitive adsorption of TCP and water, the surface reactions of hydroxyl groups with $Cl^-$ species and adsorbed chlorinated hydrocarbon derivatives.[68]

The effect of water vapour on the activity of MCM-48 supported chromium catalysts for the destruction of TCE was reported by Kawi and Te.[56] The results showed that a high concentration of water vapour (8,000 ppm) in the gas stream was detrimental to the catalytic activity of Cr/MCM-48 catalysts, whereas a lower concentration of water vapour (1,000 ppm) had negligible influence. These authors also showed that, although Cr/Si-MCM-48 showed quite similar stable activity at low concentration of water vapour in the gas stream, it had significantly higher TCE adsorption and hydrophobicity compared to Cr/MCM-48.

Abdullah *et al.*[76] prepared the following single and bimetallic catalysts: $Cr_{1.5}$/$SiCl_4$-Z, $Cu_{1.5}/SiCl_4$-Z and $Cr_{1.0}Cu_{0.5}/SiCl_4$-Z (where Z represents H-ZSM-5), and analysed the effect of water vapour in the combustion of DCM, TCM (trichloromethane) and TCE. In the presence of 9,000 ppm of water vapour in the feed, conversions of DCM and TCM dropped from 99% to 86%, compared to a drop from 94% to 88% for TCE. Despite inhibiting the conversion of Cl-VOC, water increased the carbon dioxide yield by being directly involved in the combustion reaction as a hydrolysis agent. The effect was more noticeable in TCE combustion, as it involved the formation of relatively unstable vinyl carbocation compared with DCM and TCM, which gave rise to alkyl carbocation. In addition, the role of water as a hydrogen-supplying agent decreased by-product formation by suppressing chlorine-transfer reactions.

### 4.4.2. *Effect of other VOCs as $H_2$-supplying agents*

Off-gases streams containing mixtures of organic molecules present an additional concern since not only a high conversion of each pollutant must be achieved, but the formation of unwanted reaction products must also be limited. As a result, the optimization and selection of effective catalysts does not appear to be an easy task, owing to the large variety of VOC molecules and the complex nature of VOC mixtures encountered in practice.

This section is specifically devoted to the analysis of the mutual mixture effects found in feed streams that simultaneously contain chlorinated and non-chlorinated compounds. Among the chlorinated hydrocarbons emitted in gaseous industrial waste streams, 1,2-dichloroethane and trichloroethylene have usually been chosen as model compounds, the latter with a H/Cl ratio below 1. On the other hand, *n*-hexane (HEX) and toluene (TOL) have been selected as model compounds of the category of non-chlorinated aliphatic hydrocarbons. HEX, TOL, DCE and TCE were used because alkanes and chlorinated ethanes and ethylenes have been ranked intermediate and low, respectively, on their ease of destruction. The usual operating conditions selected for most studies are as follows: contact time about 0.12 s, VOC concentration around 1,000 ppm in air.

#### 4.4.2.1. *Combustion of Cl-VOC binary mixtures*

The mutual effects of the simultaneous decomposition of two chlorinated compounds (DCE+DCM, DCE+TCE and DCM+DCE) have been investigated by López-Fonseca *et al.*[101,118] over a series of H-type zeolites (H-ZSM-5, H-MOR and chemically dealuminated H-Y zeolite). Competition among the Cl-VOCs for adsorption sites in binary mixtures led to a significant increase in ignition temperatures with a pattern of higher activity for DCE than DCM and TCE. The topology of the structure appeared to play a key role in controlling the catalytic behaviour of H-MOR since it could eventually be severely deactivated by coke formation during the combustion of 1,2-dichloroethane. The effects observed on selectivity when simultaneously oxidizing two chlorinated VOCs were remarkable. On one hand, the formation of chlorinated by-products or reaction intermediates was suppressed. On the other hand, the selectivity towards deep oxidation chlorinated product (HCl) was promoted. The favoured formation of HCl was particularly noticeable when DCE was a component of the mixture, since the water molecules formed as a result of the total decomposition promoting HCl formation by the Deacon reaction.[39,119]

#### 4.4.2.2. *Mixtures with non-chlorinated VOCs*

Among the large number of Cl-VOCs which are discharged into the environment, those containing more chlorine atoms than hydrogen atoms within their molecules (e.g. TCE or $CCl_4$), cannot be totally converted by air to the most desirable HCl, which is easily scrubbed in an alkaline medium, and therefore even more toxic products such as $Cl_2$ and $COCl_2$ are formed. To improve selectivity to HCl, either a hydrogen-rich fuel or water vapour may be added to the feed stream. Indeed, in real flue gases, Cl-VOCs are present in a matrix of non-chlorinated organic compounds, where the inhibition or enhancement of oxidation of a given compound may occur.[92,120]

The presence of non-chlorinated VOC was found to enhance the destruction of unsaturated chlorocarbons on alumina-based catalysts; chlorocarbon light-off temperatures were lower. The magnitude of the shift was dependent on the VOC species and on the relative concentration of VOCs with respect to unsaturated chlorocarbons. Simultaneously, VOC light-offs were shifted to moderately higher temperatures. Unsaturated and aromatic VOCs produce a strong effect towards the enhancement of Cl-VOC oxidation, while saturated VOCs shows little enhancement. For example, the effectiveness of VOCs for enhanced destruction of TCE was found to follow the order: ethane < toluene < ethylene.[115]

González-Velasco *et al.*[39] evaluated the presence of hexane (1,000 ppm) and toluene (1,000 ppm) in the oxidation of TCE (1,000 ppm) over supported low-loading Pt and Pd catalysts, as a way of providing hydrogen in order to convert all the chlorine to HCl and to minimize undesirable by-products ($C_2Cl_4$ and CO). The $T_{50}$ for TCE shifted to lower values in the presence of non-halogenated VOC admixtures. The magnitude of this shift depends on the VOC nature: hexane produced a stronger effect towards the enhancement of TCE conversion than toluene, especially over $Pd/Al_2O_3$ (Fig. 4.4). On the other hand, the product composition can be observed in Fig. 4.5, in particular those corresponding to the desired products, i.e. HCl and $CO_2$, and the undesired intermediate $C_2Cl_4$. Hexane and toluene showed effectiveness in promoting selectivity to HCl, and hence inhibiting $Cl_2$ and $C_2Cl_4$ formation. This improvement in HCl selectivity with the addition of hydrogen-supplying compounds by removing chlorine attached to the catalyst surface has also been observed for most catalysts reported in the literature, mainly in the oxidation of hydrogen-lean feed molecules. Thus, Gervasini *et al.*[99] studied the oxidative destruction of pure $CCl_4$ and admixtures with hexane and toluene, over Cu-Cr and $MnO_2$ catalysts both dispersed on alumina supplied by Engelhard, Italy. Enhancement of $CCl_4$ conversion was observed in the presence of both hydrocarbons. The amount of hydrogen supplied, i.e. the H/Cl ratio in the feed stream, was found to be the

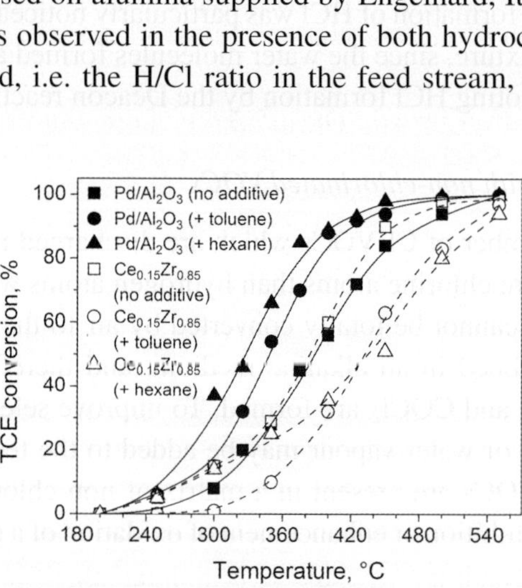


**Figure 4.4.** Light-off curves for the TCE oxidation over $Pd/Al_2O_3$ and $Ce_{0.15}Zr_{0.85}O_2$ catalysts in the absence and presence of toluene and *n* hexane as hydrogen supplying compounds (adapted from González-Velasco *et al.*[39] and de Rivas *et al.*[102]).

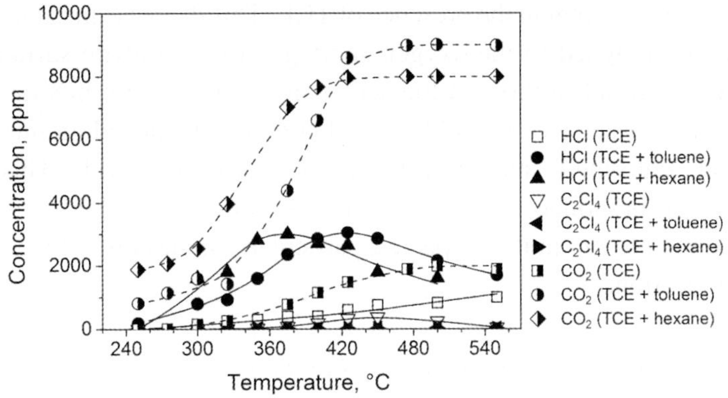


**Figure 4.5.** Evolution of the intermediate tetrachloroethylene with temperature in the oxidation of TCE over noble metal-based catalysts, in both the presence and absence of water (adapted from González-Velasco *et al.*[39]).

most important parameter in increasing $CCl_4$ conversion. With noble metal-based catalysts the main C-containing product of combustion is $CO_2$; in fact, at temperatures above 400°C only $CO_2$ was obtained in the experiments reported in Fig. 4.5, namely 9,000 ppm when toluene was added to the TCE feed stream, and 8,000 ppm in the case of *n*-hexane. Such deep oxidation to $CO_2$ has not been observed when using catalysts without metallic active sites.

The behaviour of a Pt-based catalyst on a metallic monolith support washcoated with alumina, with the addition of lanthanum and cerium, was studied by Musialik-Piotrowska and Mendyka.[121] The activity of the catalyst was tested in the oxidation of ChB and DCE alone and in two-component mixtures with toluene, *n*-hexane, acetone, ethanol, and ethyl acetate. The influence of non-chlorinated compounds on ChB oxidation differed from one compound to another. Over the whole range of reaction temperatures, ethanol enhanced the conversion of ChB by 10%. The addition of both hydrocarbons also slightly improved ChB destruction, while DCE conversion was inhibited in the presence of each non-chlorinated compound that was added. Both chlorinated hydrocarbons not only inhibited catalytic destruction of each of the non-chlorinated compounds added, but also increased the reaction selectivity and concentration of the intermediate yielded, the first of which was acetaldehyde.

Ceria-based mixed oxides have more recently been examined in the literature.[102, 114] The simultaneous oxidative decomposition of DCE and TCE, and hexane and toluene, were studied over $Ce_xZr_{1-x}O_2$ mixed oxides. As for activity, competitive adsorption played an important role in the mutual inhibition detected (Fig. 4.4). The competition and interference between reactants led to an increase in the temperature at which complete removal of each VOC was achieved. TCE was more affected by the presence of toluene, while the oxidation of toluene was simultaneously inhibited to a larger extent in the presence of TCE. This decrease in catalytic activity was essentially assigned to the oxygen shortage at the catalytic surface and to a largely disfavoured adsorption on the active sites due to steric hindrance. On the other hand, the presence of non-chlorinated compounds played an important role in minimizing chlorine formation in favour of hydrogen chloride. This beneficial effect was more pronounced for TCE combustion due to the higher reactivity of hydrogen atoms present in these species, particularly in hexane, which can reverse the Deacon reaction.

#### 4.4.2.3. *Effect of non-hydrocarbon co-pollutants*

There is some indication in the literature that SCR catalysts containing vanadia supported on titania are also able to oxidize chlorinated hydrocarbons and dioxine-like molecules to carbon dioxide, water and hydrochloric acid in the presence of nitrogen oxides and ammonia, and that they might therefore be suitable for the control of $NO_x$ and Cl-VOCs in incinerator flue gases.[9,72,122]

Jones and Ross[122] examined the possibility of using a series of vanadia supported on different oxides for the complete oxidation of chlorinated hydrocarbons (ethyl chloride or ChB) in the presence of NO and $NH_3$. They showed that vanadia supported on an alumina-modified zirconia is capable of achieving 100% conversion of the chlorinated molecules while also giving 100% conversion of the NO over a relatively wide range of temperatures (320–410°C). This material also showed resistance to poisoning by HCl, but it is slowly poisoned by $SO_2$. The addition of $WO_3$ to the formulation has no significant effect on the catalytic behaviour towards NO and ethyl chloride but it improves resistance to $SO_2$.

The influence of NO on the performance of $VO_x/TiO_2$, $VO_x$-$WO_x/TiO_2$ and $VO_x$-$MoO_x/TiO_2$ was investigated in the combustion of ChB.[9,72] NO proved to induce an increase in ChB conversion, however, this only occurred if $O_2$ was present and was maximized when the catalyst contained W or Mo. The suggested mechanism for this effect was for NO to first be oxidized to $NO_2$, mainly on $WO_x$ and $MoO_x$; then, $NO_2$ replaces or assists $O_2$ in the re-oxidation step of the $VO_x$ phase (as described by Mars and van Krevelen[123]), thus speeding up the oxidation cycle, which macroscopically corresponds to the increase in ChB conversion. The reaction scheme is represented in Fig. 4.6. It appears that NO is definitively not a poison; on the contrary, it acts as a dopant of chlorinated aromatics combustion on $VO_x$-based catalysts.

## 4.5. Chlorinated VOC Catalyst Deactivation and Regeneration

As has already been reported throughout this chapter, an adequate catalyst for the destruction of chlorinated VOCs must show high oxidation activity at low

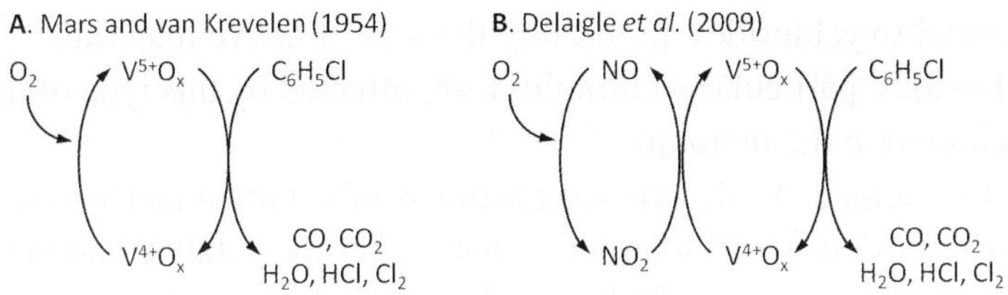


**Figure 4.6.** Scheme of the chlorobenzene oxidation in the absence (A) and in the presence (B) of NO over $VO_X/TiO_2$ catalysts (schemes from Bertinchamps *et al.*[72]).

temperatures along with high selectivity to carbon dioxide and HCl. Furthermore, the catalyst should maintain its activity even when the operating conditions are altered, i.e. in the presence of water vapour, after the addition of Cl-VOC binary mixtures or in the presence of a non-hydrocarbon co-pollutant. Apart from activity and selectivity, it is worth noting that the catalyst must show a remarkable resistance to deactivation.

Catalyst deactivation is a great impediment to industrial applications.[26] In fact, catalyst stability and durability are as important as activity and selectivity, so these properties have to keep constant with reaction time in order to fulfil international environmental regulations and be economically attractive.[124] Indeed, over the past three decades, the science of catalyst deactivation has been steadily developing, while the literature addressing this topic has expanded considerably. One of the pioneer works dealing with catalyst deactivation was presented by Spivey and Butt.[62]

The deactivation pathway may change depending on the nature of the catalyst (supported noble metals, transition metal oxides, zeolites, mixed oxides) and could be caused by several factors, both physical and chemical.[62] Apart from the catalyst itself, the specific characteristics of the process, such as temperature, space velocity or type of chlorinated feed, appear to be major factors influencing the type and extent of deactivation.

In general, for the catalytic oxidation of Cl-VOCs there are several reasons for catalyst deactivation.[26] These include:

— volatilization of the active phase;
— poisoning by modification of the catalyst structure, mainly caused by strong adsorption of chlorine and/or hydrogen chloride;
— formation of a coke deposit that can block the porous structure; and
— thermal degradation, which can sinter metal particles or destroy the catalyst structure.

### 4.5.1. *Volatilization of the active phase*

The interaction between chlorine formed during the catalytic combustion of Cl-VOCs and metals present on the catalyst surface could lead to the formation of volatile metal oxychlorides, promoting the loss of active material.[73,76,125] Transition metal oxides, particularly chromium, are affected by this type of deactivation as is well reported in the literature.[73,125–127]

In order to elucidate the deactivation pathway of commercial Cr-based catalysts, Padilla *et al.*[73] carried out a "long-term" test. After the catalyst had been used for dichloromethane oxidation for 50 hours, the authors noticed the formation of thin red deposits on the inner wall of the reactor exit along with some light-green or yellow condensates. Although the exact loss of the active phase from the catalyst was not measured, the presence of chromium in the condensates was confirmed. Consequently, the activity decrease was related to the loss of chromium, probably via the formation of $CrO_2Cl_2$ species due to the attack of chlorine.

The type of chlorinated VOC has been reported to be a key factor affecting the loss of chromium during the reaction. Rachapudi *et al.*[125] followed the activity decrease over time on a stream of Cl-VOCs, in humid air, over a Cr-zeolite catalyst at 500°C and 1%. A notable difference in the extent of the deactivation was observed depending of the Cl-VOC feed. When TCE was used as the Cl-VOC source, the conversion decreased from 77% to 35% after 50 hours, whereas when vinyl chloride (VC) was used, the activity decrease was much lower, from 98% to 93%. As the characterization of the catalyst after the aging process revealed, the different deactivation patterns were in accordance with the amount of Cr lost during the reaction. That is, a larger amount of chromium was lost during the TCE oxidation, nearly half of the Cr present in the fresh catalyst, whereas only 17% was lost during VC oxidation.

In agreement with this fact, Padilla *et al.*[73] concluded that for low Cl-contents and/or high H/Cl ratios (as in the case of VC, with only one chlorine atom/molecule), chromium-based catalysts can be recommended due to their high activity, but for relatively high Cl-contents or low H/Cl ratios (as in the case of TCE, with three chlorine atoms/molecules) chromium is lost from the catalyst resulting in its deactivation. Thus, the overall amount of Cl and the H/Cl ratio in the flue gas are key factors in determining the loss of chromium and consequently the life of the catalyst.

Further to this, Rachapudi *et al.*[125] proposed a reaction pathway to describe the volatilization of chromium cations. The interaction of chromium cations (i.e. the $CrO_3$-like compound) with a chlorine-containing species is suggested to occur by one or more of the following reactions:

$$CrO_3(s) + 2HCl(g) \rightarrow CrO_2Cl_2(g) + H_2O(g) \tag{4.18}$$

$$CrO_3(s) + Cl_2(g) \rightarrow CrO_2Cl_2(g) + 1/2O_2(g) \tag{4.19}$$

$$CrO_3(s) + C_2HCl_3(g) + 1/2O_2 \rightarrow CrO_2Cl_2(g) + HCl(g) + 2CO \tag{4.20}$$

These reactions are listed in order of decreasing probability based on $\Delta G_g$, which range from about −60 to −40 kJ at 500°C. Note that in the presence of water, the reversible Deacon reaction will further reduce the concentration of chlorine, and thus the likelihood of Reaction (4.19) between $CrO_3$ and $Cl_2$.

The primary product, $CrO_2Cl_2$, is volatile (m.p. −96.5°C, b.p. 117°C) at reaction conditions and thus becomes the likely vehicle for chromium cation mobility and loss. Furthermore, it is probable that much of the $CrO_2Cl_2$ lost from the catalyst would have decomposed in the cooler portions of the reactor according to

$$2CrO_2Cl_2(g) \rightarrow Cr_2O_3(s) + 2Cl_2(g) + 1/2O_2(g) \tag{4.21}$$

which is consistent with the dark green residues (believed to be $Cr_2O_3$) found by Padilla *et al.*[73] and Rachapudi *et al.*[125] after the reaction section, especially during TCE aging experiments, due to lower H/Cl ratios.

The catalyst deactivation due to active phase loss has also been reported for other catalysts apart from Cr-based ones. For example, the deactivation of the $V_2O_5/SiO_2$-$TiO_2$ catalyst during TCE oxidation[127] was attributed to the loss of vanadium (5.0 wt% before reaction vs 1.7 wt% after). As reported by Kießling *et al.*[126] a perovskite-type catalyst ($LaCoCO_3$) can also be deactivated due to active phase loss. They observed that the catalyst structure can be modified or even totally destroyed after chloromethane oxidation at 500°C. The loss of active phase was evidenced by the blue ($CoCl_2$) and dark ($Co_3O_4$) crystal deposits at the end of the reactor tube, leading to an irreversible deactivation.

In contrast, the volatilization of noble metals does not appear to be the most probable deactivation pathway. Corella *et al.*[103] performed a deactivation study over Pt- and Pd-based commercial catalysts for the oxidation of dichloromethane for a period of 120 h (1,000 ppm of DCM, 10,000 $h^{-1}$, 1 vol% steam, 400°C). The analysis by ICP-MS of the condensates obtained in these tests showed the absence of Pt and Pd, and consequently catalyst deactivation could not be assigned to the volatilization of the active phase. More recently Miranda *et al.*[128] also confirmed no volatilization of the active phase when using different noble metals such as Ru, Pd, Rh and Pt supported over alumina for TCE oxidation.

### 4.5.2. *Poisoning by modification of catalyst structure*

The interaction of catalysts with chlorine is the main problem met in the design of catalysts for the combustion of chlorinated compounds.[129] Almost all types of catalysts, to varying extents, are influenced by chlorine poisoning, which leads in most cases to catalyst structure modification and therefore to deactivation.

The presence of water vapour plays a predominant role in limiting the catalyst deactivation due to chlorine poisoning when supported noble metals are used as catalysts. Guillemot *et al.*[130] studied the effect of the addition of water in the deactivation process over 1.2% Pt/HY for the oxidation of PCE. They observed that the catalyst deactivation resulted much more quickly in the absence of steam. In fact, the $CO_2$ yield fell from near 100% to 5% after four hours of reaction at 500°C in the absence of water, whereas in the presence of $H_2O$ the $CO_2$ yield was only reduced to 95%. In that case, the deactivation was not attributed to the formation of coke as its content was almost identical for both experiments, with and without $H_2O$. In contrast, notable differences were observed in the chlorine content: 5.15% after reaction in the absence of water vs 0.34% in the presence of water. As the oxidation of PCE does not produce water due to the absence of hydrogen atoms in the molecule, if water steam is not added, the catalyst is rapidly deactivated by the chlorine deposited on the platinum oxide forming the $PtCl_4$ species. On the other hand, it can be assumed that water acts as a catalyst surface cleaner, limiting chlorine deposition onto the noble metal and thus decreasing catalyst deactivation. Consequently, the exceptional stability of a Pt-based commercial catalyst tested by Corella *et al.*[103] during dichloromethane oxidation can be assigned to the excess of steam (1 vol%) added during the reaction. Along the same lines, the water production due to TCE oxidation could be the reason for the slight deactivation observed over Ru-, Pd-, Pt- and Rh-based catalysts.[128]

As has been well reported in the literature, when noble metals are supported over zeolites, both metallic and cationic sites play a role in Cl-VOC transformation.[32,130] Consequently, it is reasonable to consider that the deactivation may affect not only the noble metal, as previously reported, but also the cationic sites. Aranzabal *et al.*[124] found that loss of acidity was more severe in the presence of additional HCl (around 70–75%), and it occurred mainly over strong acid sites of H-MOR, H-ZSM-5 and H-BETA. In the deactivation study carried out by Guillemot *et al.*,[130] three different Pt-exchanged zeolites (1.2% Pt/HY, 1% Pt/NaX and 0.65% Pt/NaY) were used during PCE oxidation. After four hours of reaction, the chlorine content measured by elemental analysis showed that NaX-based catalysts were more sensitive to chlorine poisoning with a total loading of 4.2%, in contrast to 0.4 and 0.3% observed for NaY and HY, respectively. This result was assigned to the location of sodium cations in the zeolite structure. As deactivation and chlorine content on the 1% Pt/NaX catalyst is higher, it seems that the $Na^+$ cations of this zeolite react more easily with Cl to form NaCl. Formation of NaCl clusters not only decreases catalyst activity as fewer cations are still active, but probably also blocks the access of PCE molecules to other active sites, such as metallic ones located inside supercages, leading to a high deactivation. On the contrary, the chlorine content observed over the NaY zeolite was much lower, which means that $Na^+$ cations located on this zeolite are more resistant to chlorination.

Although transition metal oxides are in general less catalytically active than noble metals for the destruction of chlorinated VOCs, they can largely resist the deactivation by chlorine poisoning.[28,131] This fact has positioned supported transition metal oxides as the potential substitutes for noble metal-based catalysts.

In order to determine the nature and extent of the deactivation of the transition metal oxides, Vu *et al.*[26] performed a complete characterization of Mn-Cu mixed oxide supported over $TiO_2$ after reaction with ChB for five days. When the reaction was carried out at 350°C, ChB was totally converted and no deactivation was observed, whereas at 300°C the conversion decreased from near 100% to 75% during the first hour, and then the catalyst performance stabilized at this value until the end of the test. Catalyst characterization after the reaction by scanning electron microscopy (SEM) revealed the presence of chlorine homogeneously distributed all over the $MnCuO_x/TiO_2$ catalyst surface, indicating that chlorinated species are retained during ChB oxidation at low temperatures. Previous studies[63] have reported that under working conditions the Mn-Cu mixed oxide may become an oxychlorinated compound ($MnCuO_{x-a}Cl_{2a}$), leading to catalyst deactivation. However, these chlorinated manganese compounds remain active for the ChB oxidation, leading to a stable performance after the initial activity drop, which can be related to the active phase chlorination.

Recently, cerium oxide has attracted much attention in environmental catalysis, either as an effective promoter or as a supporting material based on its high OSC and facile $Ce^{4+}/Ce^{3+}$ redox cycle. Although $CeO_2$-based catalysts were found to be very active in Cl-VOCs catalytic combustion, they deactivated quickly due to the strong adsorption of HCl or $Cl_2$.[28,131] Whether under dry or humid conditions, Dai *et al.*[131] reported that the conversion of TCE drops rapidly after about 10 h of operation at 350°C. In order to understand the deactivation of $CeO_2$ catalysts, several techniques were employed to characterize the catalysts after reaction. On the one hand, EDS (energy-dispersive X-ray spectroscopy) and TG (thermogravimetry) results indicated no coke deposition on the catalyst surface and the Raman spectrum indicated that the bands located at 177,208 $cm^{-1}$ and 119,327 $cm^{-1}$ corresponding to $CeCl_3$ and CeOCl, respectively, were not observed. Thus, the deactivation of the $CeO_2$ catalyst does not result from coke deposition or from the formation of non-active species such as $CeCl_3$ and CeOCl. On the other hand, the XPS spectrum confirmed the presence of a large abundance of chlorine species on the deactivated $CeO_2$ catalyst. Therefore, the authors concluded that the catalyst deactivation was probably due to $Cl_2$ strongly adsorbed on the catalyst. The result is that the active sites for TCE catalytic combustion are blocked and the surface oxygen species (which assist the catalytic oxidation) cannot be rapidly compensated by gas-phase oxygen, which leads to the catalyst deactivation.

Wang *et al.*[78,132] improved the catalytic performance and also the resistance to deactivation of $CeO_2$ catalysts including manganese oxide. They found that the incorporation of Mn increases the amount of active oxygen in the structure of the $MnO_x$-$CeO_2$ catalyst and consequently improves the activity in ChB catalytic combustion. Furthermore, by increasing the Mn content, the stability of the catalyst is promoted due to the larger availability of active oxygen which facilitates the removal of chlorine deposits.

The temperature at which the reaction is carried out is another key parameter in the deactivation by chlorine poisoning over transition metal oxides and $CeO_2$-based catalysts.[26,78,131–133] As the literature reveals, deactivation can be limited to some extent by increasing the reaction temperature. In that way, surface oxygen, which at low temperatures is not active, may become activated and favour the continuous removal of strongly adsorbed Cl species as the reaction proceeds.

### 4.5.3. *Formation of coke deposits*

Carbonaceous deposits on the catalyst often occur with the oxidation of VOCs and lead to the loss of catalyst activity.[134] Although the mechanisms of formation of carbonaceous deposits are very complex, involving catalytic and non-catalytic reactions, it has been suggested that unsaturated chlorinated compounds are the promoters for coke formation.[124,128]

During Cl-VOCs oxidation under noble metal-based catalysts, a significant volume of unsaturated chlorinated by-products — such as polychlorinated benzenes during chlorobenzene oxidation[38,129] or tetrachloethylene (PCE) during TCE oxidation[128] — are produced. Depending on the choice of the noble metal, Miranda *et al.*[128] observed notable differences in the product distribution. While a $Ru/Al_2O_3$ catalyst produced saturated polychloromethanes as the main chlorinated by-products, Pd and Pt favoured the production of unsaturated chlorinated by-products such as PCE. The result confirmed that unsaturated compounds are possible promoters for coke formation, as Pt- and Pd-based catalysts produced the greatest amount of coke. On the other hand, $Ru/Al_2O_3$ showed a very stable performance, as unsaturated products were not produced; thus leading to a very low amount of carbon deposition.

Acid catalysts such as zeolites are also affected by coke deposition. The adsorption of reactants over acid sites is very strong, and thus the contact time is long enough to promote the large number of successive reactions which lead to the formation of coke from reactants or from by-products such as olefins.[124] Furthermore, the coking rate is believed to be related to the number of acid sites, that is, the greater the number of acid sites the higher the coking rate,[135] and the increase of surface acidity caused by the HCl generated in the reaction also enhances the formation of these carbonaceous deposits.[128]

It is generally believed that coking mainly occurs inside the zeolite pores and affects the activity in two different ways: site coverage (active sites poisoned by coke adsorption) and pore blockage (active sites inaccessible to reactants). The pore blockage due to coke deposition is more evident in zeolites with a one-dimensional pore structure, such as H-MOR. Aranzabal *et al.*[124] showed that a small amount of coke was enough to avoid the diffusion of reactants over H-MOR, whereas the stability of H-BETA was enhanced due to its three-dimensional pore system, in spite of the higher amount of coke deposited. They also reported that the coke weight profile decreased along the catalytic bed with the distance from the inlet, suggesting that coking results from a reactant or an intermediate formed early in the oxidation of DCE, i.e. vinyl chloride, according to a series-parallel reaction mechanism:

$$C_2H_4Cl_2 \rightarrow C_2H_3Cl \begin{array}{l} \nearrow CO, CO_2, HCl \\ \searrow CH_X \quad (Coke) \end{array} \tag{4.22}$$

The addition of water during Cl-VOC oxidation over zeolites considerably reduces the amount of coke deposited due to the steam gasification reaction.[136] However, the preferential adsorption of water molecules over acid sites resulted in a higher deactivation of the catalysts.[124] Thus, the effect of competitive adsorption on acid sites between $H_2O$ and Cl-VOC is stronger than the effect of lower coke formation.

The deactivation of other types of catalysts such as $CeO_2$,[28,131] $MnO_x$-$CeO_2$ mixed oxides[78,132] or Mn-Cu mixed oxides due to coke deposition during the oxidation of Cl-VOCs is very limited. These catalysts are very selective to $CO_2$ and consequently the formation of other chlorinated or non-chlorinated carbon-containing by-products, which are considered as coke precursors, is limited and thus their deactivation is not related to coke deposition.

### 4.5.4. *Thermal degradation*

When the catalyst is exposed to high temperatures during long operation times, thermal degradation may cause changes in the catalyst structure, morphology and physico-chemical characteristics, which leads to catalyst deactivation. The long-term operation has usually been simulated, especially in automotive catalysis,[137] by exposing the catalysts to high temperature excursions.

In considering the oxidation of chlorinated VOCs, Dai *et al.*[131] and de Rivas *et al.*[83] studied the changes in the physico-chemical characteristics of thermally-aged cerium oxide and Ce/Zr mixed oxides, respectively. In general, increasing the calcination temperature of the catalysts resulted in lower surface areas, higher crystallinities (larger particle sizes), lower total acidity and lower oxygen vacancies. The authors clearly observed that the increase in calcination temperature led to a progressive decrease in catalytic activity.

When metals are deposited on the catalyst surface, active phase migration to less accessible sites[125] or the loss of metal dispersion[129] could be the result of catalyst aging and consequently deactivation.

### 4.5.5. *Catalyst regeneration*

Catalyst regeneration in Cl-VOC oxidation is a relatively recent topic, and little work has been reported in the literature to date. The deactivation of catalysts due to active phase volatilization or physico-chemical changes during reaction (surface area, crystallinity, acidity, etc.) cannot be recovered and consequently this type of deactivation is irreversible. However, as has already been reported, in some cases the activity of the catalyst can be partially recovered when the deactivation is related to chlorine poisoning.

The most usual protocol to regenerate a catalyst deactivated by chlorine poisoning is to flow dry or wet air through the catalyst at high temperatures and in the absence of chlorinated compounds. Flowing dry air at 350°C for 5 h, Vu *et al.*[26] reported that the initial activity of a Mn-Cu mixed-oxide catalyst could be almost completely recovered. SEM analysis performed after regeneration revealed that the intensity of the peak at 2.6 keV due to the presence of chlorine in the catalyst surface decreased notably, although complete removal of chlorine was not achieved. Dai *et al.*[131] concluded that the removal of chlorine from the catalyst is a slow step, but they inferred that, if the flow of humid air goes on for long enough, complete removal of chlorine could be achieved and therefore the activity of the regenerated $CeO_2$ may be recovered.

## 4.6. Outlook and Conclusions

Heterogeneous catalytic oxidation of Cl-VOCs has received much attention in recent decades because of its energy and efficiency advantages for many different chloride molecules, over a wide range of operating conditions. The approximate ranges of operating variables are low to moderate temperatures (100–500°C), atmospheric pressure, high space velocities ($10^3$–$10^5$ $h^{-1}$) and low reactant concentration ($10^2$–$10^3$ ppm) in the air. The reaction pathway over the catalyst should lead to a complete oxidation product, i.e. $CO_2$, $H_2O$ and HCl, which is preferable to $Cl_2$.

In the 1990s and the first part of the 2000s, noble metal and mixed oxide-based catalysts were mostly investigated, with interest first focused on alternative materials, especially metal oxides and protonic and metal zeolites. More recently, efforts have been focused on formulations with enhanced redox properties and oxygen storage capacity, such as combinations of ceria, zirconia, titania and vanadia. Noble metals (Pt, Pd) show higher specific activity when they are prepared in a highly dispersed form on supports with a high specific surface area ($\gamma$-$Al_2O_3$, $SiO_2$); in this case a small amount of metal (0.1–0.5 wt%) was sufficient for good combustion. These species have been shown as active chlorination sites, especially in the oxidation of hydrogen-lean Cl-VOCs, such as trichloroethylene, chloroform and chlorobenzene.

The use of noble metals is limited by cost and sensitivity to poisoning, especially by chlorine/chloride products. Thus, oxides of V, Cr, Mn, Fe, Co, Ni, Cu, Zn, Nb and Mo have been considered as low-cost alternatives. Their specific activity is lower compared to noble metals; so that metal loadings of supported oxide catalysts are typically superior (5–20%) and they exhibit more resistance to poisoning. Chromium oxide has been reported as the most active metal oxide and has been extensively studied as supported on $Al_2O_3$, porous carbon, pillared clay, zeolites, $SnO_2$, $TiO_2$ and $SiO_2$. One limitation is that over metal oxide catalysts the undesired partial oxidation to CO was considerable. Also, the use of chromium is restricted owing to the formation of extremely toxic residues, such as chromium oxychloride, at low temperatures.

Alternatively, $V_2O_5/TiO_2$ exhibited activity as high as that of chromium and better stability in the oxidation of chloro-organics. Although $TiO_2$-based $V_2O_5$-$WO_3$ catalysts were originally designed for $NO_x$ removal by SCR, they have been successfully used for the combined destruction of dioxins and $NO_x$. Cr-, Co- and Mn-perovskites have also been found to be effective catalysts for the oxidation of Cl-VOCs due to their high stability and the unusual valence states of the transition metal ion in their structure, the presence of defect sites, and the high mobility of oxygen ions. Polychlorinated by-product formation was rather low compared with noble metals. In the last decade, $CeO_2$ has been intensively studied. The mechanism of Cl-VOC oxidation over $CeO_2$ is generally considered a redox-type mechanism, in which the key steps are the supply of oxygen by the readily reducible oxide and its re-oxidation by oxygen.

Concerning the nature of the support, it has been extensively reported that the order of activity is in close agreement with the strong acidity of the material. In particular, remarkable effects have been attributed to Brønsted acid sites of protonic zeolites (H-Y, H-MOR, H-ZSM-5 and H-BETA) as efficient sites for Cl-VOC chemisorption. In addition, surface acidity significantly inhibited production of $Cl_2$ and polychlorinated subproducts. Adversely, production of CO by incomplete combustion was significant. However, it is suggested that the combination of acidic and oxidizing characteristics make metal-doped zeolites good catalysts for Cl-VOC abatement.

Most assessments of the activity, selectivity and deactivation have been based on the comparison of the light-off curves under the same reaction conditions. Several efforts have been made in order to obtain kinetic model equations for catalytic decomposition of Cl-VOCs. In general, potential first order kinetics have been deduced, with the apparent activation energy ranging between 18 and 110 kJ mol$^{-1}$, depending on the particular Cl-VOC and the type of catalyst. In addition, some mechanistic reaction rate equations have been developed which define the different steps in the Cl-VOC oxidation, including the equilibrium constants of the adsorbed species. The oxygen can be first adsorbed on the catalyst surface (Langmuir–Hinshelwood or Mars–van Krevelen) or can react directly from the gas phase (Riedel–Eley).

The effect of water vapour on VOC oxidation has always needed careful investigation, as water is present in all exhaust gases and is also a product of the VOC oxidation itself. The Cl-VOC adsorption capacity and the Cl-COV destruction activity generally decrease upon contact with a high concentration of water vapour in the gas stream, although a low concentration of water vapour is sometimes helpful for complete destruction of hydrogen-lean Cl-VOC, such as TCE or $CCl_4$. The presence of water vapour in the feed stream greatly decreased the production of molecular chlorine, in favour of HCl formation, and it led to a much lower production of intermediate by-products, such as PCE, in TCE combustion. Also, the addition of water resulted in a substantial improvement for deep oxidation selectivity; in general, selectivity to $CO_2$ was improved to practically 100% with noble metal-based catalysts.

Cl-VOCs containing more chlorine than hydrogen atoms within their molecules (e.g. TCE, $CCl_4$) require hydrogen-rich fuel or water vapour added to the feed stream in order to achieve higher HCl production. Thus, the presence of non-chlorinated VOCs (ethane, ethylene, *n*-hexane, toluene, etc.) in the environment enhanced the destruction of hydrogen-lean chloro-organics by removing chlorine attached to the catalyst surface.

Laboratory studies, like most of those reviewed here, are usually conducted with a single feed stream composition making it possible to determine the effect of each individual co-pollutant potentially present in real effluents. On the other hand, the performance of catalysts in real plant conditions occurs in the presence of complex mixtures with a wide range of varying parameters at the same time, e.g. temperature fluctuations, co-pollutant concentration or space velocity, which makes the analysis difficult. However, effort should be made in order to gain a more integrated understanding of what happens in the case of complex mixtures.

Catalyst deactivation is a great impediment to industrial applications. In general, for Cl-VOC oxidation there are several factors that lead to catalyst deactivation, such as volatilization of the active phase, poisoning by strong adsorption of $Cl_2$/HCl, formation of a coke deposit on the porous structure, and metal particle sintering caused by the temperature. Transition metal oxides, particularly chromium and vanadia, are deactivated by the volatilization of metal oxychlorides formed on the catalyst surface. But no volatilization of the active phase when using noble metals such as Ru, Pd, Rh and Pt has been confirmed for TCE oxidation.

The interaction of catalysts with chlorine is the main problem met in the design of catalysts for the combustion of Cl-VOCs. Transition metal oxides can resist the deactivation by chlorine poisoning to a greater degree than noble metals, although they are, in general, less catalytically active in the destruction of Cl-VOCs. On the contrary, ceria-based catalysts are active in the catalytic combustion, but they are deactivated quickly due to the strong adsorption of $HCl/Cl_2$. The beneficial role of the presence of water for decreasing catalyst poisoning has already been stated above. This is the only type of deactivation that can be considered reversible, i.e. catalysts can be regenerated. The most usual protocol to regenerate the catalyst is by flowing dry or wet air through the catalyst at high temperatures and in the absence of chlorinated compounds.

Unsaturated chlorinated intermediates/by-products are mainly responsible for coke deposition and deactivation, thus Pt- and Pd-based catalysts produce the highest amount of carbon deposition. Acid catalysts, such as zeolites, are also affected by coke deposition because of the strong adsorption of reactants over the acid sites, increasing the contact time to promote a large number of successive reactions that lead to coke formation from reactant or by-products.

Raw predictions can be made on the impacts of the concomitant presence of several co-pollutants in the gas stream, along with target pollutants of different natures and catalyst deactivation factors on industrial applications. However, the difficulty of obtaining a real, clear picture of the interconnected effects of these parameters should encourage researchers to perform experiments in conditions increasingly closer to the target application.

## Acknowledgements

The authors would like to acknowledge the financial support received from the Basque government (Consolidated Research Group, GIC 07/67-JT-450-07). One of the authors (BPA) wants also to acknowledge the PhD Research Grant from the Spanish Science and Innovation Ministry.

## Nomenclature

$A$ Frequency factor, same as kinetic constant
$C_i$ Concentration of component i, mol $cm^{-3}$
$E_a$ Activation energy, kJ $mol^{-1}$
$K_i$ Equilibrium constant for an elementary step
$k$ Kinetic constant

| | |
|---|---|
| $n, m$ | Reaction order, dimensionless |
| $p_i$ | Partial pressure of component $i$, atm |
| $R$ | Ideal gas constant, 8.314 J $mol^{-1}$ |
| $-r_V$ | Reaction rate of VOC oxidation, (mol V) (g cat)$^{-1}$ $s^{-1}$ |
| $T$ | Temperature, K |
| $T_{50}$ ($T_{90}$) | Temperature at which the reaction conversion is 50% (90%), K |
| $W/F_{V_0}$ | Space-time, (g cat) s (mol V)$^{-1}$ |
| $X$ | Conversion of Cl-VOC |
| *Greek symbol* | |
| $\upsilon$ | Oxygen stoichiometric coefficient in the oxidation reaction |
| *Indexes* | |
| 0 | Initial (inlet) value |
| expt | Experimental |
| calc | Calculated |
| *Chemical molecules* | |
| ChB | Chlorobenzene |
| DF | Dibenzofurane |
| Cl-VOC | Chlorinated Volatile Organic Compound |
| DCE | 1,2-Dichloroethane |
| DCM | Dichloromethane |
| HEX | *n*-Hexane |
| PCE | Perchloroethylene, same as tetrachlorethylene |
| TCE | Trichloroethylene |
| TCM | Trichloromethane |
| TCP | 4,4,6-Trichlorophenol |
| TOL | Toluene |
| VOC | Volatile Organic Compound |

## References


1. US EPA (United States Environmental Protection Agency). Requirements for Preparation, Adoption, and Submittal of Implementation Plans: Definitions. Code of Federal Regulations 40CFR51.100(s). 2003.
2. Weber H, Dimmling W, Möller K. Environmental Protection in the Production of Vinyl Chloride Monomer (VCM). *Dechema Monogr* 1976; 80: 57–76.
3. Kosusko M, Nunez C. Destruction of Volatile Organic Compounds Using Catalytic Oxidation. *J Air Waste Manag Assoc* 1990; 40: 254–259.

4. Yasuhara A, Morita M. Formation of Chlorinated Compounds in Pyrolysis of Trichloroethylene. *Chemosphere* 1990; 21: 479–486.
5. Müller H, Deller K, Despeyroux B, Peldszus E, Kammerhofer P, Kühn W, *et al.* Catalytic Purification of Waste Gases Containing Chlorinated Hydrocarbons with Precious Metal Catalysts. *Catal Today* 1993; 17: 383–390.
6. Agarwal S, Spivey J, Butt J. Catalyst Deactivation During Deep Oxidation of Chlorohydrocarbons. *Appl Catal A: Gen* 1992; 82, 259–275.
7. Gavaskar AR, Kim BC, Rosansky SH, Ong SK, Marchand EG. Crossflow Air Stripping and Catalytic Oxidation of Chlorinated Hydrocarbons From Groundwater. *Environ Progress* 1995; 14: 33–40.
8. Dries J, Bastiaens L, Kuypers S, Springael D., Agathos N, Diels L. Combined Removal of Chlorinated Ethenes and Heavy Metals by Zerovalent Iron in Batch and Continuous Flow Column Systems. *Environ Sci Technol* 2005; 39: 8460–8465.
9. Delaigle R, Debecker DP, Bertinchamps F, Gaigneaux EM. Revisiting the Behaviour of Vanadia-based Catalysts in the Abatement of (Chloro)-Aromatic Pollutants: Towards an Integrated Understanding. *Top Catal* 2009; 52: 501–516.
10. Komatsu T, Ooshima R. Catalytic Combustion of Dioxin Analogue Compounds on Pt Supported Zeolite. *J Jap Petrol Inst* 2009; 52: 332–340.
11. Hetrick C, Patcas F, Amiridis M. Effect of Water on the Oxidation of Dichlorobenzene over $V_2O_5/TiO_2$ Catalysts. *Appl Catal B: Environ* 2011; 101: 622–628.
12. Windawi H, Wyatt M. Catalytic Destruction of Halogenated Volatile Organic Compounds: Mechanisms of Platinum Catalyst Systems. *Platinum Met Rev* 1993; 37: 186–193.
13. Chintawar P, Greene H. Adsorption and Catalytic Destruction of Trichloroethylene in Hydrophobic Zeolites. *Appl Catal B: Environ* 1997; 14: 37–47.
14. Koyer-Golkowska A, Musialik-Piotrowska A, Rutkowski J. Oxidation of Chlorinated Hydrocarbons over Pt-Pd-based Catalyst: Part 1. Chlorinated Methanes. *Catal Today* 2004; 90: 133–138.
15. Golodets, GI. *Heterogeneous Catalytic Reactions Involving Molecular Oxygen*, Stud Surf Sci Catal Vol. 15. Amsterdam: Elsevier; 1983.
16. Spivey J. Complete Catalytic Oxidation of Volatile Organics. *Ind Eng Chem Res* 1987; 26: 2165–2180.
17. Albonetti S, Cavani F, Trifiro F. Key Aspects of Catalyst Design for the Selective Oxidation of Paraffins. *Catal Rev–Sci Eng* 1996; 38; 413–438.
18. Farrauto R, Bartholomew C. *Fundamentals of Industrial Catalytic Processes*. Chichester UK: John Wiley & Sons; 2003.
19. Bond G, Sadeghi N. Catalyzed Destruction of Chlorinated Hydrocarbons. *J Appl Chem Biotechnol* 1975; 25: 241–248.
20. Fevrier D, Mignon P, Vernet J. Reactivity of Some Halogenated Alkanes on 13X Molecular Sieve. *J Catal* 1977; 50: 390–399.
21. Pope D, Walker D, Moss R. Evaluation of Platinum-honeycomb Catalysts for Destructive Oxidation of Low Concentrations of Odorous Compounds in Air. *Atmosphere Environ* 1978; 12: 1921–1927.
22. Paukshtis EA, Simonova LG, Zagoruiko AN, Balzhinimaev BS. (2010). Oxidative Destruction of Chlorinated Hydrocarbons on Pt-containing Fiber-glass Catalysts. *Chemosphere* 2010; 79: 199–204.
23. Tian W, Fan X, Yang H, Zhang X. Preparation of $MnO_x/TiO_2$ Composites and their Properties for Catalytic Oxidation of Chlorobenzene. *J Hazard Mater* 2010; 177: 887–891.
24. Tseng TK, Wang L, Ho CT, Chu H. The Destruction of Dichloroethane over a $\gamma$-Alumina Supported Manganese Oxide Catalyst. *J Hazard Mater* 2010; 178: 1035–1040.
25. De Rivas B, López-Fonseca R, Gutiérrez-Ortiz MA, Gutiérrez-Ortiz JI. Catalytic Performance of Chlorinated Ce/Zr Mixed Oxides for Cl-VOC Oxidation. In: Zamorano M, Brebbia CA,

Kungolos, Popov V, Itoh H. (eds.) *Waste Management and The Environment IV WIT Trans Ecol Environ* 2008; 109: 857–866.

26. Vu, V., Belkouch, J., Ould-Dris, A., *et al.* (2009). Removal of Hazardous Chlorinated VOCs over Mn-Cu Mixed Oxide Based Catalyst, *J. Hazard. Mater.*, **169**, pp. 758–765.
27. Xingyi W, Qian K, Dao L. Catalytic combustion of chlorobenzene over $MnO_x$-$CeO_2$ mixed oxide catalysts. *Appl Catal B: Environ* 2009; 86: 366–375.
28. Dai Q, Wang X, Lu G. Low-temperature Catalytic Destruction of Chlorinated VOCs over Cerium Oxide. *Catal Commun* 2007; 8: 1645–1649.
29. Giraudon J, Elhachimi A, Leclercq G. Catalytic Oxidation of Chlorobenzene over Pd/Perovskites. *Appl Catal B: Environ* 2008; 84: 251–261.
30. Debecker DP, Delaigle R, Eloy P, Gaigneau EM. Abatement of model molecules for dioxin total oxidation on $V_2O_5$-$WO_3$/$TiO_2$ catalysts: The case of substituted oxygen-containing VOC. *J Mol Catal A: Chem* 2008; 289: 38–43.
31. Wang XY, Dai QG, Zheng Y. Low temperature catalytic combustion of thichloroethylene over La, Ce, and Pt catalysts supported on MCM-41. *Chin J Catal* 2006; 27: 468–470.
32. López-Fonseca R, Gutiérrez-Ortiz JI, Gutiérrez-Ortiz MA, González-Velasco JR. Catalytic oxidation of aliphatic chlorinated volatile organic compounds over Pt/H-BETA zeolite catalyst under dry and humid conditions. *Catal Today* 2005; 107–108: 200–207.
33. López-Fonseca R, Aranzabal A, Gutiérrez-Ortiz JI, Álvarez-Uriarte JI, González-Velasco JR. Comparative Study of the Oxidative Decomposition of Trichlorethylene over H-type Zeolites under Dry and Humid Conditions. *Appl Catal B: Environ* 2001; 30: 303–313.
34. López-Fonseca R, Gutiérrez-Ortiz JI, González-Velasco JR. Noble Metal Loaded Zeolites for the Catalytic Oxidation of Chlorinated Hydrocarbons. *React Kinet Catal Lett* 2005; 86: 127–133.
35. González-Velasco JR, Aranzabal A, Gutiérrez Ortiz JI, López-Fonseca R, Gutiérrez-Ortiz MA. Activity and Product Distribution of Alumina Supported Platinum and Palladium Catalysts in the Gas-phase Oxidative Decomposition of Chlorinated Hydrocarbons. *Appl Catal B: Environ* 1998; 19: 189–197.
36. Zwinkels MFM, Järaas SG, Menon PG, Griffik TA. Catalytic Materials for High-temperature Combustion. *Catal Rev–Sci Eng* 1993; 35: 319–358.
37. Lou J, Lee S. Destruction of Trichloromethane with Catalytic Oxidation. *Appl Catal B: Environ* 1997; 12: 111–123.
38. Van den Brink R, Louw R, Mulder P. Formation of Polychlorinated Benzenes During the Catalytic Combustion of Chlorobenzene Using a Pt/$\gamma$-$Al_2O_3$ Catalyst. *Appl Catal B: Environ* 1998a; 16: 219–226.
39. González-Velasco JR, Aranzabal A, López-Fonseca R, Ferret R, González-Marcos JA. Enhancement of the Catalytic Oxidation of Hydrogen-lean Chlorinated VOCs in the Presence of Hydrogen-supplying Compounds. *Appl Catal B: Environ* 2000; 24: 33–43.
40. López-Fonseca R, Gutiérrez-Ortiz JI, González-Velasco JR. Catalytic Combustion of Chlorinated Hydrocarbons Over H-BETA and PdO/H-BETA Zeolite Catalysts. *Appl Catal A: Gen* 2004; 271: 39–46.
41. Yu T, Shaw H, Farrauto R. Catalytic Oxidation of Trichloroethylene over PdO Catalyst on $\gamma$-$Al_2O_3$. *ACS Symp. Series* 1992; 495: 141–152.
42. Shaw H, Wang Y, Yu T, Cerkanowick AE. Catalytic Oxidation of Trichloroethylene and Methylene Chloride. *ACS Symp Series* 1992; 518: 358–379.
43. Bickle G, Suzuki T, Mitarai Y. Catalytic Destruction of Chlorofluorocarbons and Toxic Chlorinated Hydrocarbons. *Appl Catal B: Environ* 1994; 4: 141–153.
44. Park J, Lee CW, Chang J, Park S, Shin, C. Catalytic Oxidation of Trichloroethylene over Pd-loaded Sulfated Zirconia. *Bull Korean Chem Soc* 2004; 25: 1355–1360.
45. Becker L, Forster H. Oxidative Decomposition of Chlorobenzene Catalyzed by Palladium Containing Zeolite Y. *J Catal* 1997; 170: 200–203.

46. Lago RM, Green MLH, Tsang SC, Odlyha M. Catalytic Decomposition of Chlorinated Organics in Air by Copper Chloride Based Catalysts. *Appl Catal B: Environ* 1996; 8: 107–121.
47. Krishnamoorthy S, Rivas J, Amiridis M. Catalytic Oxidation of 1,2-Dichlorobenzene over Supported Transition Metal Oxides. *J Catal* 2000; 193: 264–272.
48. Döbber D, Kießling D, Schmitz W, Wendt G. $MnO_x/ZrO_2$ Catalysts for the Total Oxidation of Methane and Chloromethane. *Appl Catal B: Environ* 2004; 52: 135–143.
49. Yim SD, Chang K, Koh DJ, Nam I, Kim YG. Catalytic Removal of Perchloroethylene (PCE) over Supported Chromium Oxide Catalysts. *Catal Today*, 2000; 63: 215–222.
50. Miranda B, Díaz E, Ordóñez S, Vega A, Díez FV. Oxidation of Trichloroethene over Metal Oxide Catalysts: Kinetic Studies and Correlation with Adsorption Properties. *Chemosphere*, 2007; 66: 1706–1715.
51. Ramanathan K, Spivey J. Catalytic Oxidation of 1,1-Dichloroethane. *Comb Sci Technol* 1989; 63: 247–255.
52. Feijen-Jeurissen MMR, Jorna JJ, Nieuwenhuys BE, Sinquin G, Petit C, Hindermann J. Mechanism of Catalytic Destruction of 1,2-Dichloroethane and Trichloroethylene Over $\gamma$-$Al_2O_3$ and $\gamma$-$Al_2O_3$ Supported Chromium and Palladium Catalysts. *Catal Today* 1999; 54: 65–79.
53. Petrosius SC, Drago RS, Young V, Grunewald GC. Low-temperature Decomposition of Some Halogenated Hydrocarbons Using Metal Oxide/Porous Carbon Catalysts, *J Am Chem Soc* 1993; 115: 6131–6137.
54. Kang M, Lee C. Methylene Chloride Oxidation on Oxidative Carbon-supported Chromium Oxide Catalyst. *Appl Catal A: Gen* 2004; 266: 163–172.
55. Storaro L, Ganzerla R, Lenarda M, Zanoni R, Jiménez López A, Olivera-Pastor P, *et al.* Catalytic Behavior of Chromia and Chromium-doped Alumina Pillared Clay Materials for the Vapor Phase Deep Oxidation of Chlorinated Hydrocarbons. *J Mol Catal A: Chem* 1997; 115: 329–338.
56. Kawi S, Te M. MCM-48 Supported Chromium Catalyst for Trichloroethylene Oxidation. *Catal Today* 1998; 44: 101–109.
57. Solymosi F, Rasko J, Papp E, Oszkó A, Bánsági T. Catalytic Decomposition and Oxidation of $CH_3Cl$ on $Cr_2O_3$-doped $SnO_2$. *Appl Catal A: Gen* 1995; 131: 55–72.
58. Yim Sd, koh Dj, Nam IS, Kim YG, Effect of the Catalyst Supports on the Removal of Perchloroethylene (PCE) over Chromium Oxide Catalysts. *Catal Lett* 2000; 64: 201–207.
59. Yim S, Nam I. Characteristics of Chromium Oxides Supported on $TiO_2$ and $Al_2O_3$ for the Decomposition of Perchloroethylene. *J Catal* 2004; 221: 601–611.
60. Manning M. Fluid Bed Catalytic Oxidation: An Underdeveloped Hazardous Waste Disposal Technology. *Hazard Waste*, 1984; 1: 41–65.
61. Weldon J, Senkan S. Catalytic Oxidation of $CH_3Cl$ by $Cr_2O_3$. *Comb Sci Technol* 1986; 47: 229–237.
62. Spivey J, Butt J. Literature-review: Deactivation of Catalysts in the Oxidation of Volatile Organic Compounds. *Catal Today* 1992; 11: 465–500.
63. Liu Y, Luo M, Wei Z, Xin Q, Ying P, Li C. Catalytic Oxidation of Chlorobenzene on Supported Manganese Oxide Catalysts. *Appl Catal B: Environ* 2001; 29: 61–67.
64. Gutiérrez-Ortiz J, López-Fonseca R, Aurrekoetxea U, González-Velasco JR. Low-temperature Deep Oxidation of Dichloromethane and Trichloroethylene by H-ZSM-5-Supported Manganese Oxide Catalysts. *J Catal* 2003; 218: 148–154.
65. Hung S, Barresi A, Pfefferle L. Flow Tube Reactor Studies of Catalytically Stabilized Combustion of Methyl Chloride. *Proceedings of the 23$^{rd}$ International Symposium on Combustion*. Orleans, France: Combustion Institute; 1991. p. 909–915.
66. Krishnamoorthy S, Baker J, Amiridis M. Catalytic Oxidation of 1,2-Dichlorobenzene over $V_2O_5/TiO_2$-based Catalysts. *Catal Today* 1998; 40: 39–46.
67. Weber R, Sakurai T. Low Temperature Decomposition of PCB by $TiO_2$-based $V_2O_5/WO_3$ Catalyst: Evaluation of the Relevance of PCDF Formation and Insights into the First

Step of Oxidative Destruction of Chlorinated Aromatics. *Appl Catal B: Environ* 2001; 34: 113–127.
68. Lomnicki S, Lichtenberger J, Xu Z, Waters M, Kosman J, Amiridis MD. Catalytic Oxidation of 2,4,6-Trichlorophenol over Vanadia/Titania-based Catalysts. *Appl Catal B: Environ* 2003; 46: 105–119.
69. Lichtenberger J, Amiridis MD. Catalytic Oxidation of Chlorinated Benzenes over $V_2O_5/TiO_2$ Catalysts. *J Catal* 2004; 223: 296–308.
70. Bertinchamps F, Grégoire C, Gaigneaux EM. Systematic Investigation of Supported Transition Metal Oxide Based Formulations for the Catalytic Oxidative Elimination of (Chloro)-Aromatics. Part II: Influence of the Nature and Addition Protocol of Secondary Phases to $VO_x/TiO_2$. *Appl Catal B: Environ* 2006; 66: 10–22.
71. Armor J. Environmental Catalysis. *Appl Catal B: Environ* 1992; 1: 221–256.
72. Bertinchamps F, Treinen M, Blangenois N, Mariage E, Gaigneaux EM. Positive Effect of $NO_x$ on the Performances of $VO_x/TiO_2$-based Catalysts in the Total Oxidation Abatement of Chlorobenzene. *J Catal* 2005; 230: 493–498.
73. Padilla A, Corella J, Toledo J. Total Oxidation of Some Chlorinated Hydrocarbons with Commercial Chromia Based Catalysts. *Appl Catal B: Environ* 1999; 22: 107–121.
74. Toledo J, Corella J, Sanz A. Noble Metal-based Catalysts for Total Oxidation of Chlorinated Hydrocarbons. *Environ Progress* 2001; 20: 167–174.
75. Heck R, Farrauto R, Gulati S. *Catalytic Air Pollution Control. Commercial Technology*, 3rd edition. Hoboken, New Jersey: John Wiley & Sons; 2009.
76. Abdullah A, Bakar M, Bhatia S. Combustion of Chlorinated Volatile Organic Compounds (VOCs) Using Bimetallic Chromium-copper Supported on Modified H-ZSM-5 Catalyst. *J Hazard Mater* 2006; 129: 39–49.
77. Gutiérrez-Ortiz JI, De Rivas B, López-Fonseca R, Martín S., González-Velasco JR. Structure of Mn-Zr Mixed Oxides Catalysts and their Catalytic Performance in the Gas-phase Oxidation of Chlorocarbons. *Chemosphere* 2007; 68: 1004–1012.
78. Wang X, Kang Q, Li D. Catalytic Combustion of Chlorobenzene Over $MnO_x$-$CeO_2$ Mixed Oxide Catalysts. *Appl Catal B: Environ* 2009; 86: 166–175.
79. Schneider R, Kießling D, Wendt G. Cordierite Monolith Supported Perovskite-type Oxides: Catalysts for the Total Oxidation of Chlorinated Hydrocarbons. *Appl Catal B: Environ* 2000; 28: 187–195.
80. Poplawski K, Lichtenberger J, Keil FJ, Schnitzlein K, Amiridis MD. Catalytic Oxidation of 1,2-Dichlorobenzene over $ABO_3$-type Perovskites. *Catal Today* 2000; 62: 329–336.
81. Sinquin G, Petit C, Libs S, Hindermann JP, Kiennemann A. Catalytic Destruction of Chlorinated $C_1$ Volatile Organic Compounds (CVOCs) Reactivity, Oxidation and Hydrolysis Mechanisms. *Appl Catal B: Environ* 2000; 27: 105–115.
82. Stephan K, Hackenberger M, Kießling D, Wendt G. Total Oxidation of Methane and Chlorinated Hydrocarbons on Zirconia Supported $A_{1-x}Sr_xMnO_3$ Catalysts. *Chem Eng Technol* 2004; 27: 687–693.
83. De Rivas B, López-Fonseca R, Sampedro C, Gutiérrez-Ortiz JI. Catalytic Behaviour of Thermally Aged Ce/Zr Mixed Oxides for the Purification of Chlorinated VOC-containing Gas Streams. *Appl Catal B: Environ* 2009; 90: 545–555.
84. Imamura S. Catalytic Decomposition of Halogenated Organic Compounds and Deactivation of the Catalysts. *Catal Today* 1992; 11: 547–567.
85. López-Fonseca R, Aranzabal A, Steltenpohl P, Gutiérrez-Ortiz, González-Velasco JR. Performance of Zeolites and Product Selectivity in the Gas-phase Oxidation of 1,2-Dichloroethane. *Catal Today* 2000; 62: 367–377.
86. López-Fonseca R, De Rivas B, Gutiérrez-Ortiz JI, Aranzabal A, González-Velasco JR. Enhanced Activity of Zeolites by Chemical Dealumination for Chlorinated VOC Abatement. *Appl Catal B: Environ* 2003; 41: 31–42.

87. Scire S, Minico S. The Role of the Support in the Oxidative Destruction of Chlorobenzene on Pt/Zeolite Catalysts: An FT-IR Investigation. *Catal Lett* 2003; 91: 199–205.
88. González-Velasco JR, López-Fonseca R, Aranzabal A, Gutiérrez-Ortiz JI, Steltenpohl P. Evaluation of H-Type Zeolites in the Destructive Oxidation of Chlorinated Volatile Organic Compounds, *Appl. Catal. B: Environ.*, 2000; 24: 233–242.
89. Chatterjee S, Greene H. Effects of Catalyst Composition on Dual Site Zeolite Catalysts Used in Chlorinated Hydrocarbon Oxidation. *Appl Catal A: Gen* 1993; 98: 139–158.
90. Aranzabal A, González-Marcos JA, Ayastuy JL, González-Velasco JR. Kinetics of Pd/Alumina Catalysed 1,2-Dichloroethane Gas-Phase Oxidation, *Chem. Eng. Sci.*, 2006; 61: 3564–3576.
91. Everaert K, Mathieu M, Baeyens J, Vansant E. Combustion of Chlorinated Hydrocarbons in Catalyst-coated Sintered Metal Fleece Reactors. *J Chem Technol Biotechnol* 2003; 78: 167–172.
92. Everaert K, Baeyens J. Catalytic Combustion of Volatile Organic Compounds. *J Hazard Mater* 2004; B109: 113–139.
93. Bertinchamps F, Attianese A, Mestdagh MM, Gaigneaux EM. Catalysts for Chlorinated VOCs Abatement: Multiple Effects of Water on the Activity of $VO_x$ Based Catalysts for the Combustion of Chlorobenzene. *Catal Today* 2006; 112: 165–168.
94. Pinard L, Mijoin J, Magnoux P, Guisnet M. Oxidation of Chlorinated Hydrocarbons over Pt Zeolite Catalysts. 1. Mechanisms of Dichloromethane Transformations over PtNaY Catalysts. *J Catal* 2003; 215: 234–244.
95. Tanilmis T, Atalay S, Alpay HE, Atalay FS. Catalytic Combustion of Carbon Tetrachloride. *J Hazard Mater* 2002; 90: 157–167.
96. Berty J, Stenger Jr HG, Buzan GE, Hu K. Oxidation and Removal of Chlorinated Hydrocarbons. *Stud Surf Sci Catal* 1993; 75: 1571–1574.
97. Bonny R, Lenfant C, Thyrion F. Catalytic Oxidation and Decomposition of $CH_2Cl_2$ on Supported $CrO_3$ at Low Temperature. *Int J Environ Stud* 1997; 53: 75–85.
98. Miranda B, Díaz E, Ordóñez S, Díez FV. Catalytic Combustion of Trichoroethene over $Ru/Al_2O_3$: Reaction Mechanism and Kinetic Study. *Catal Commun* 2006; 7: 945–949.
99. Gervasini A, Pirola C, Ragaini V. Destruction of Carbon Tetrachloride in the Presence of Hydrogen-supplying Compounds with Ionization and Catalytic Oxidation. *Appl Catal B: Environ* 2002; 38: 17–28.
100. Avdeev VI, Kovalchuk VI, Zhidomirov GM, d'Itri JL. DFT Analysis of the Mechanisms of 1,2-Dichloroethane Dichlorination on Supported Cu-Pt Bimetallic Catalysts. *J Struct Chem* 2007; 48: S171–S183.
101. López-Fonseca R, Gutiérrez-Ortiz JI, Ayastuy JL, Gutiérrez-Ortiz MA, González-Marcos JR. Gas-phase Catalytic Combustion of Chlorinated VOC Binary Mixtures. *Appl Catal B: Environ* 2003; 45: 13–21.
102. De Rivas B, López-Fonseca R, Gutiérrez-Ortiz MA, Gutiérrez-Ortiz JI. Role of Water and Other H-rich Additives in the Catalytic Combustion of 1,2-Dichloroethane and Trichloroethylene. *Chemosphere* 2009; 75: 1356–1362.
103. Corella J, Toledo J, Padilla A. On the Selection of the Catalyst Among the Commercial Platinum-based Ones for Total Oxidation of Some Chlorinated Hydrocarbons. *Appl Catal B: Environ* 2000; 27: 243–256.
104. Klinghoffer A, Rossin J. Catalytic Oxidation of Chloroacetonitrile over 1% Platinum Alumina Catalyst. *Ind Eng Chem Res* 1992; 31: 481–486.
105. Rossin J, Farris M. Catalytic Oxidation of Chloroform over a 2% Platinum Alumina Catalyst. *Ind Eng Chem Res* 1993; 32: 1024–1029.
106. Wang Y, Shaw H, Farrauto R. Catalytic Oxidation of Trace Concentration of Trichloroethylene over 1.5% Platinum on $\gamma$-Alumina. *ACS Symp Series* 1992; 495: 125–140.
107. Horsley, J. *Catalyst for the Elimination of Volatile Organic Compounds. Halogenated Compounds*. Catalytica Studies Division. Environmental Report, 1992.

108. Heck R. Catalytic Abatement of Nitrogen Oxides: Stationary Applications. *Catal Today* 1999; 53: 519–523.
109. Sanger M, Werther J, Ogada T. $NO_x$ and $N_2O$ Emission Characteristics from Fluidised Bed Combustion of Semi-dried Municipal Sewage Sludge. *Fuel* 2001; 80: 167–177.
110. Rogaume T, Auzanneau M, Jabouille F, Goudeau JC, Torero JL. The Effects of Different Air-flows on the Formation of Pollutants during Wastes Incineration. *Fuel* 2002; 81: 2277–2288.
111. Imamura S, Tarumoto H, Ishida S. Decomposition of 1,2-Dichloroethane on Titanium Dioxide/Silica. *Ind Eng Chem Res* 1989; 28: 1449–1452.
112. Haber J, Machej T, Janik R, Krysciak J, Sadowska H. On the Mechanism of Catalytic Oxidation of $CH_2Cl_2$ on $\gamma$-$Al_2O_3$ and its Oscillatory Behavior. *Z Phys Chem* 1996; 197: 97–112.
113. Van den Brink RW, Mulder P, Louw R, Sinquin G, Petit C, Hindermann JP. Catalytic Oxidation of Dichloromethane on $\gamma$-$Al_2O_3$: A Combined Flow and Infrared Spectroscopic Study. *J Catal* 1998; 180: 153–160.
114. De Rivas B, Gutiérrez-Ortiz JI, López-Fonseca R, González-Velasco JR. Analysis of the Simultaneous Catalytic Combustion of Chlorinated Aliphatic Pollutants and Toluene over Ceria-zirconia Mixed Oxides. *Appl Catal A: Gen* 2006; 314: 54–63.
115. Windawi H, Zhang Z. Catalytic Destruction of Halogenated Air Toxins and the Effect of Admixture with VOCs. *Catal Today* 1996; 30: 99–105.
116. Scire S, Minico S, Crisafulli C. Catalytic Combustion of Volatile Organic Compounds on Gold/Cerium Oxide Catalysts. *Appl Catal B: Environ* 2003; 45: 117–125.
117. Pinard L, Magnoux P, Ayrault P, Guisnet M. Oxidation of Chlorinated Hydrocarbons over Zeolite Catalysts. 2. Comparative Study of Dichloromethane Transformation over NaX and NaY Zeolites. *J Catal* 2004; 221: 662–665.
118. López-Fonseca R, Gutiérrez-Ortiz JI, Gutiérrez-Ortiz MA, González-Velasco JR. Dealuminated Y Zeolites for Destruction of Chlorinated Volatile Organic Compounds. *J Catal* 2002; 209: 145–150.
119. Aranzabal A, González-Marcos JA, López-Fonseca R, Gutierrez-Ortiz MA, González-Velasco JR. Deep Catalytic Oxidation of Chlorinated VOC Mixtures from Groundwater Stripping Emissions. *Stud Surf Sci Catal* 2000; 130: 1229–1234.
120. Tichenor B, Palazzolo M. Destruction of Volatile Organic Compounds via Catalytic Incineration. *Environ Progress* 1987; 6: 172–176.
121. Musialik-Piotrowska A, Mendyka B. Catalytic Oxidation of Chlorinated Hydrocarbons in Two-components Mixtures with Selected VOCs. *Catal Today* 2004; 90: 139–144.
122. Jones J, Ross J. The Development of Supported Vanadia Catalysts for the Combined Catalytic Removal of the Oxides of Nitrogen and of Chlorinated Hydrocarbons from Flue Gases. *Catal Today* 1997; 35: 97–105.
123. Mars P, van Krevelen D. Oxidations Carried out by Means of Vanadium Oxide Catalysts. *Chem Eng Sci* 1954; 3: 41–59.
124. Aranzabal A, González-Marcos JA, Romero-Sáez M, González-Velasco JR, Guillemot M, Magnoux P. Stability of Protonic Zeolites in the Catalytic Oxidation of Chlorinated VOCs (1,2-Dichloroethane). *Appl Catal B: Environ* 2009; 88: 533–541.
125. Rachapudi R, Chintawar P, Greene H. Aging and Structure/Activity Characteristics of Cr-ZSM-5 Catalysts During Exposure to Chlorinated VOCs. *J Catal* 1999; 185: 58–72.
126. Kießling D, Schneider R, Kraak P, Haftendorn M, Wendt G. Perovskite-type Oxides: Catalysts for the Total Oxidation of Chlorinated Hydrocarbons. *Appl Catal B: Environ* 1998; 19: 143–151.
127. Kulazynski M, van Ommen JG, Trawczynski J, Walendziewski J. Catalytic Combustion of Trichloroethylene Over $TiO_2$-$SiO_2$ Supported Catalysts. *Appl Catal B: Environ* 2002; 36: 239–247.

128. Miranda B, Díaz E, Ordóñez S, Vega A, Díez FV. Performance of Alumina-supported Noble Metal Catalysts for the Combustion of Trichloroethene at Dry and Wet Conditions. *Appl Catal B: Environ* 2006; 64: 262–271.
129. Van den Brink, RW, Louw R, Mulder P. Increased Combustion Rate of Chlorobenzene on Pt/$\gamma$-$Al_2O_3$ in Binary Mixtures with Hydrocarbons and with Carbon Monoxide. *Appl Catal B: Environ* 2000; 25: 229–237.
130. Guillemot M, Mijoin J, Mignard S, Magnoux P. Mode of Zeolite Catalysts Deactivation During Chlorinated VOCs Oxidation. *Appl Catal A: Gen* 2007; 327: 211–217.
131. Dai Q, Wang X, Lu G. Low-temperature Catalytic Combustion of Trichloroethylene over Cerium Oxide and Catalyst Deactivation. *Appl Catal B: Environ* 2008; 81: 192–202.
132. Wang X, Kang Q, Li D. Low-temperature Catalytic Combustion of Chlorobenzene over $MnO_x$-$CeO_2$ Mixed Oxide Catalysts. *Catal Commun* 2008; 9: 2158–2162.
133. Wu M, Wang X, Dai Q, Wu Y, Li D. Low Temperature Catalytic Combustion of Chlorobenzene ver Mn-Ce-O/$\gamma$-$Al_2O_3$ Mixed Oxides Catalyst. *Catal Today* 2010; 158: 336–342.
134. Li W, Wang J, Gong H. Catalytic Combustion of VOCs on Non-noble Metal Catalysts. *Catal Today* 2009; 148: 81–87.
135. Guisnet M, Magnoux P. Fundamental Description of Deactivation and Regeneration of Acid Zeolites. *Stud Surf Sci Catal* 1994; 88: 53–68.
136. McMinn T, Moates F, Richardson J. Catalytic Steam Reforming of Chlorocarbons: Catalyst Deactivation. *Appl Catal B: Environ* 2001; 31: 93–105.
137. González-Velasco JR, Gutiérrez-Ortiz MA, Botas J, Bernal S, Gatica JM, Pérez-Omil JA. HREM and XRD Characterisation of Thermal Ageing of Pd/$CeO_2$/$Al_2O_3$ Automotive Catalysts. *Stud Surf Sci Catal* 1999; 126: 187–194.

# Chapter 5

## Zeolites as Alternative Catalysts for the Oxidation of Persistent Organic Pollutants

Stéphane MARIE-ROSE,* Mihaela TARALUNGA,† Xavier CHAUCHERIE,† François NICOL,† Emmanuel FIANI,‡ Thomas BELIN,* Patrick MAGNOUX* and Jérôme MIJOIN*

This chapter is focussed on the development at laboratory level of a new way of catalytic persistent organic pollutants (POPs) oxidation based on zeolites to reduce emissions and meet stringent environmental regulations.

### 5.1. Introduction

Polychlorinated dibenzo-p-dioxins (PCDD/Fs) and polycyclic aromatic hydrocarbons (PAHs) are considered to fulfil the criteria for persistent organic pollutants (POPs) under the United Nations Economic Commission for Europe (UN-ECE) Protocol (Aarhus Protocol), revised in 2009 as ECE/EB.AIR/2009/14; the UN Environment Programme's POPs Convention (Stockholm Convention); and the recent Industrial Emission Directive 2010/75/EU from the European Commission for integrated pollution prevention and control. These pollutants are mainly by-products of thermal processes and solid waste combustion is one of numerous POPs emission sources.[1,2] As legislation in France and Europe grows in stringency, the development of new pollutant-control techniques is necessary. In this context, this research project was focussed on POPs catalytic oxidation with regard to the promising activity of faujasite zeolites for the oxidation of PAHs/PCDD/Fs.

*Université de Poitiers, Institut de Chimie des Milieux et Matériaux de Poitiers (IC2MP), UMR CNRS 7285, 4 rue Michel Brunet, 86022 Poitiers Cedex, France.

†VEOLIA Environnement Recherche & Innovation (VERI), 10 rue Jacques Daguerre, 92500 Rueil-Malmaison, France.

‡ADEME, 20 avenue du Grésillé BP 90406, 49004 Angers Cedex 01, France.

### 5.1.1. *Persistent organic pollutants: Environmental legislation, sources and formation mechanisms from solid waste combustion*

Persistent organic pollutants are toxic chemicals, which adversely affect human health and the environment due to their capacity to accumulate, persist and pass from wildlife species into the food chain. POPs represent a large range of organic products. Dioxins (PCDDs) and furans (PCDFs), polychlorobenzenes (PCBzs), polychlorinated biphenyls (PCBs) and polycyclic aromatic hydrocarbons (PAHs) are quite well known in terms of formation mechanisms and toxicity. Solid waste combustion has been identified as one source of POPs formation.[1–4] In Europe, emissions of POPs from solid waste combustion are regulated by the European Directive 2000/76/CE and in France several 'regional' laws also apply. Dioxin and furan emissions are thus restricted to 0.1 ng I-TEQ/Nm$^3$ (at 11% vol. $O_2$). Considering the best available techniques (BATs) applied to the energy-from-waste (EfW) industry, dioxin and furan emission levels are mostly between 0.01 and 0.1 ng I-TEQ/Nm$^3$ (at 11% vol. $O_2$). The toxicity of PAHs is recognised and concentrations in ambient air are regulated by the European Directive 2004/107/EC, however no emission limit values have so far been set up. In France, following the first proposal of 'Grenelle de l'environnement' presented in 2008, the authorities encouraged the EfW industry to improve the PAHs emissions monitoring and also to consider reduction techniques.

#### 5.1.1.1. *Dioxins and furans: Formation mechanisms from solid waste combustion*

Dioxins and furans form in virtually every combustion process of solid materials if chlorine, oxygen and organic matter are present within an appropriate temperature range.[1,5–7] In waste combustion conditions, two possible ways for dioxins/furans to form are described in the literature (Fig. 5.1).[8–14] The first mechanism (homogenous route) occurs between 200–800°C downstream of the furnace and consists of a

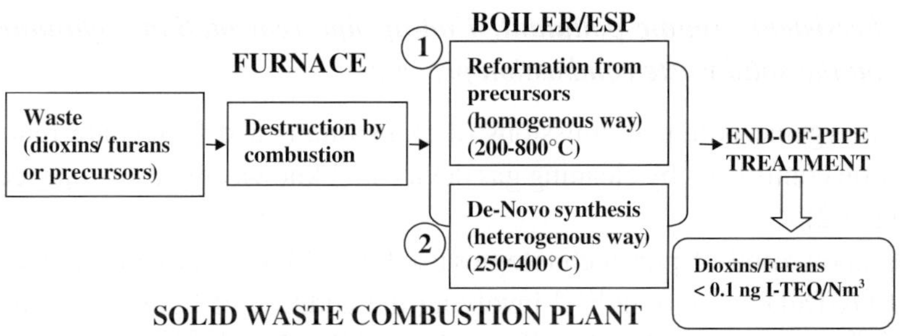


**Figure 5.1.** Dioxin/furan formation and destruction pathways in a waste combustion plant.

homogenous gas phase reaction from precursors (benzene, chlorobenzene, PAHs, chlorophenols, etc...).[2,9,15]

The second way (*de novo* synthesis) contributes significantly to overall PCDD/F concentrations found before end-of-pipe gas cleaning systems in waste combustion plants. It consists of a heterogeneous gas-solid phase reaction taking place on the ash particles surface. Inorganic chlorides (NaCl, HCl) form $Cl_2$ through the Deacon reaction in the presence of $O_2$ and with inorganic compounds acting as catalysts ($CuCl_2$, $FeCl_3$, ...); subsequently, $Cl_2$ reacts with aromatics in the flue gas and ash which contains the carbonaceous fractions required to form chlorinated organic compounds and organic fractions, which leads to the formation of dioxins/furans. Many studies on dioxin formation mechanisms have highlighted that the kinetics of dioxin/furan synthesis is modified by different factors such as temperature, oxygen, $SO_2$ and water concentrations, and catalyst nature.[2,3,7,9,11,12,16] Deacon's reaction kinetics seem then to control the dioxin/furan formation mechanism in the combustion processes. Avoiding the formation/presence of chlorine is therefore considered one of the primary methods of reducing dioxin/furan content.

#### 5.1.1.2. *Polycyclic organic aromatics: Formation mechanisms from solid waste combustion*

Polycyclic organic aromatics are generally produced during incomplete combustion of organic compounds and their formation is increased at high temperatures and low $O_2$ concentrations. Five PAHs products, derived from benz(a)pyren (BaP) and known as BaPs, are classified as hazardous (carcinogenic, mutagenic and/or toxic for reproduction) by the 2455/2001/CE1 decision: benzo(a)pyrene, benzo(b)-fluoranthene, benzo(ghi)perylene, benzo(k)fluoranthene and indeno(1,2,3-cd)pyrene. Different studies[17–20] have showed that prior to the end-of-pipe gas cleaning units, total PAHs emissions can vary from 0.3 to 2.9 mg/t of waste, while BaPs emission comprised between 0.1 and 0.8 mg/t of waste.

### 5.1.2. *Persistent organic pollutant: End-of-pipe remediation techniques on the solid waste combustion process*

Persistent organic pollutant emissions (dioxins/furans and PAHs) are reduced by combustion control and by cleaning gas devices[8,21] known as end-of-pipe technologies (Fig. 5.2).

Most known end-of-pipe technologies used for PAHs and dioxin/furan reduction are DeNOx/DeDiox systems[22–24] involving the addition of lignite coke, activated carbon[25,26] or an activated carbon/lime mixture (Fig. 5.3).[27–29]

The activated carbon adsorption efficiency on PCDD/Fs is generally between 40 and 70% within the temperature range of 140–220°C, and this process does not

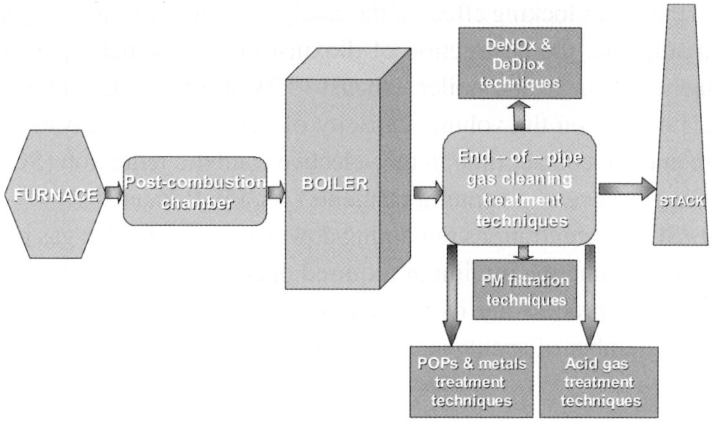


**Figure 5.2.** Waste combustion plant process scheme.

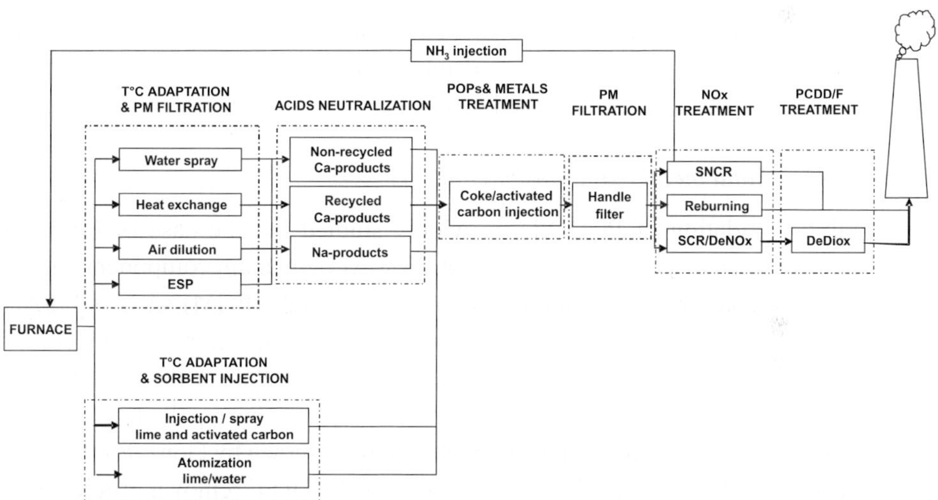


**Figure 5.3.** Reduction techniques for end-of-pipe pollutants in a waste combustion plant.

guarantee a complete removal of dioxins.[21,24] POPs removal efficiency depends largely on the injection system, the specific surface area, the sorbents — flue gas mixing, the flue gas temperature (recommended 140–170°C) — the type of filtering devices as well as the mass flow rate of the sorbents and the contact time between sorbents and the flue gas.

Beside adsorption techniques, an 85% control efficiency of dioxins/furans can be achieved by the employment of the DeDiox system.[2,30] The catalysts used are mostly composed of Ti, V and W oxides. Additionally, Pt and Au oxides supported on silica-boria-alumina were also found to be effective at 200°C.[31]

To avoid the pore blocking effect of the catalysts with coarse fly ash particles and ammonium sulphate, the destruction of dioxins/furans is usually performed after the cleaning stages or at the boiler exhaust (350–400°C).[28] The POPs reduction rate depends mainly on the volume capacity of the installed catalyst, the reaction temperature and the contact time in the selective catalytic reduction (SCR) reactor, as well as the presence of poisoning elements (As, Pb, P, K, etc. . .).

The $SO_2/SO_3$ concentrations remaining downstream of the flue gas pre-cleaning components determine the minimum required operational temperature. The major advantages of this process are its relative operational simplicity with no residue, except a small number of spent catalysts (after 8–10 years on average), which can be recycled.

### 5.1.3. *Overview of catalytic oxidation process for POPs removal: Laboratory solutions*

Nowadays, most of the catalysts investigated for the catalytic destruction of persistent organic pollutants are categorised into two broad classes: metal or metal oxide-based catalysts and zeolite-based catalysts.

#### 5.1.3.1. *Metal and metal oxide-based catalysts*

The catalytic oxidation of PCDD/Fs is widely studied but the case of PAHs in waste processes has not been well discussed. However, the main components in the catalysts investigated belong to the platinum group metals (PGM) such as $Pd/Al_2O_3$[32] and $Pd\text{-}Rh/CeO_2/Al_2O_3$[33] which have the interesting characteristics of stability and low volatility.[34] Platinum is the most studied noble metal for the oxidation of hydrocarbons[35–40] and is highly selective towards $CO_2$. However, palladium is more efficient with chlorinated compounds than platinum, as shown by Gonzalez-Velasco *et al.*[41] Those catalysts can achieve as high as 90% PCDD/F removal efficiency at 573 K[42] and a high PAH conversion (over 95%) is obtained with a $Pt/Al_2O_3$ catalyst at moderate temperature.

Diehl *et al.*[43] studied the deep oxidation of various hydrocarbons over $1\%Pt/Al_2O_3$ including PAHs. It appears that the reactivity of PAHs is related to several factors such as the loss of aromaticity (partial hydrogenation of the cycles) and the propensity of the bicyclic aromatics to form partially oxidised compounds as intermediates. Catalytic studies over $Rh/Al_2O_3$ catalysts in steam reforming can somehow explain the reactivity of the PAHs; most results are attributed to the metal particles or to their dispersion on the support.[44,45] Indeed, better degradations are obtained on metal particles, which have a more marked electron-donor effect. Using various supports with equal Pt loading (0.5 wt%) for the complete oxidation of naphthalene, Ntainjua *et al.*[46] have shown that $SiO_2$ permitted a higher efficiency (90%

at 473 K) than all other supports. These results were attributed to the metal dispersion, Pt oxidation state and the strong interaction between metal and support. Total oxidation of PAHs was reported over various metals supported on $\gamma$-$Al_2O_3$ catalysts by Zhang *et al.*[47] Ru and Co catalysts were less $CO_2$ selective and Mo or W less active, while Pt and Pd, to a lesser extent, were found to be the most active. However, the application of PGM is usually limited because of their relatively high cost, low stability toward the produced HCl and $Cl_2$[48,49] and low selectivity towards HCl (Pt/$Al_2O_3$). Moreover, formation of polychlorinated compounds is favoured on noble metal-based catalysts.[50–54]

Potentially, metal oxide catalysts could be more resistant to deactivation but the total oxidation activity is usually lower than metal-based catalysts. Compared to 0.5%Pt/$Al_2O_3$, a ceria catalyst is more efficient for naphthalene oxidation, as demonstrated by Garcia *et al.*[55] Various metal oxides ($CoO_x$, $MnO_x$, CuO, ZnO, $Fe_2O_3$, $TiO_2$) were synthesised and compared to $CeO_2$, which nevertheless achieves better performance and remains stable at 473 K for 50 h.[56] $VO_x$, $CrO_x$, $MnO_x$ and $FeO_x$ are also commonly used as catalysts,[48,57–59] with the highest activities toward PCDD/F decomposition obtained with $V_2O_5$ while $FeO_x$ catalysts show the lowest activities. Using commercial $V_2O_5$-$WO_3$/$TiO_2$ catalysts and a gas mixture containing PCBs, PCDD/F and PAHs, Weber *et al.*[60] have demonstrated that the non-chlorinated PAHs were almost entirely oxidised at 423 K ($>$ 90%). However, low efficiencies were observed in the case of polychlorinated aromatics since these compounds were mainly adsorbed on the surface without being oxidised.

The presence of water (as steam in the exhaust emissions) has no inhibitory effect on Pt/$Al_2O_3$ for temperatures below 673 K[61] and reduces the chlorine formation. However, a study on the catalytic decomposition of dioxins over $V_2O_5$-$WO_3$/$TiO_2$-based catalysts has shown both the important roles of water content,[62] characterised by an inhibition through competitive adsorption, as well as surface chlorine removal.

#### 5.1.3.2. *Zeolite-based catalysts*

Chlorobenzenes are well known as important precursors of PCDD/Fs and are suitable model compounds for the complete oxidation of chlorinated POPs. Noble metal-based catalysts such as Pt/$Al_2O_3$ show high efficiencies but promote the formation of polychlorinated benzenes.[40] Zeolites have unique properties for the deep oxidation of chlorinated compounds thanks to their well-defined framework and the presence of acid sites or transition metal cations.[63] As stated by Corma,[64] zeolites present interesting properties of reactant/product partitioning and of molecules preactivated by the molecular confinement effect. Moreover, their adsorptive properties can be modulated by modifying the nature of the extra framework cations and the Si/Al ratio.

In the case of the oxidative combustion of 1,2-dichloroethane and trichloroethylene, the protonic zeolites (H-FAU, H-MOR, H-ZSM5) exhibit high rates of activity.[65–67] The combination of a metal and zeolites leads not only to better catalytic activities but also to a reduction in the quantity of polychlorinated compounds produced.[54,68] As an example, Cr-Y and Pd-Y are respectively active for the catalytic oxidation of dichloromethane[69] and chlorobenzene. On the Pt/H-type zeolites systems, the deep oxidation of chlorobenzene appears to be strongly related to the type of zeolite: Pt/H-Y and Pt/H-β[70] are more active than Pt/H-ZSM5 or Pt/H-ferrierite.[54,68,71] An explanation of these results is based on the acidity of the support, which is involved in the activation of the reactant molecule. Moreover, the zeolite structure plays a key role in the product shape selectivity effect: the lower the size of the channels, the lower the amount of polychlorinated by-products.[68]

Similar to the observation of the supported metal-based catalysts, benzofuran is more easily oxidised than chlorinated compounds.[72] Having a greater activity than $Pt/Al_2O_3$, the protonated form of this catalyst (Pt/HY) appears to be more interesting for the oxidation of dioxin/furan mixtures. These observations are also confirmed with tetrachloroethylene by Guillemot *et al.*[73,74] Again, proof of the important role of the support is given by the quantity of by-products formed, which increases in the following order: $Pt/HFAU < Pt/Al_2O_3 < Pt/SiO_2$.[40]

The activity of the zeolite catalyst can be improved by adjusting the number of acid sites as demonstrated by Lopez-Fonseca *et al.*[75] with the 1,2-dichloroethane oxidation over a HFAU zeolite. Brønsted acidic sites are reported to favour 1,2-dichlorobenzene destruction, through a concerted multicentre mechanism. Over Brønsted protonic sites, two dichlorobenzene molecules react to give chlorobenzene, HCl, and a second dichlorobenzene molecule adsorbed on the zeolite oxygen framework. It is proposed that these strongly adsorbed species be oxidised by oxygen and water to produce $CO_X$ and HCl, and to restore protonic zeolite sites.[72,76] Thus, USHY zeolites, which possess a low Si/Al ratio and a high number of acid sites, are appropriate catalysts. These features were successfully used to destroy PAHs such as 1-methylnaphtalene with no aromatic by-products.[77,78]

The presence of water in the feed sometimes affects the catalytic behaviour in a detrimental way.[77,79,80] However, the amount of water does not change the order of the catalytic activities of zeolites.

## 5.2. Preliminary Study on POP Precursors

Chlorobenzenes and furans are known to be precursors of dioxins during solid waste combustion processes[81,82] and are often used as model molecules to imitate the structure of dioxins. During this work, the selective combustion of chlorobenzenes

**Table 5.1.** Characterization of the catalysts. Platinum dispersion and the number of accessible platinum atoms $n_{Pt}$.

| Catalysts | Pt dispersion (%) | nPt ($\times 10^{18}$ atoms.g$^{-1}$) |
|---|---|---|
| 1%$PtAl_2O_3$ | 78 | 23.8 |
| 1.2%$PtSiO_2$ | 50 | 18.5 |
| 1.1%PtHFAU(5) | 51.5 | 17.5 |
| 0.8%PtHFAU(5) | 48 | 12 |
| 0.6%PtHFAU(5) | 8 | 1.5 |
| 0.3%PtHFAU(5) | 10 | 1 |
| 0.1%PtHFAU(5) | 40 | 1.7 |
| HFAU(5) | 0 | 0 |

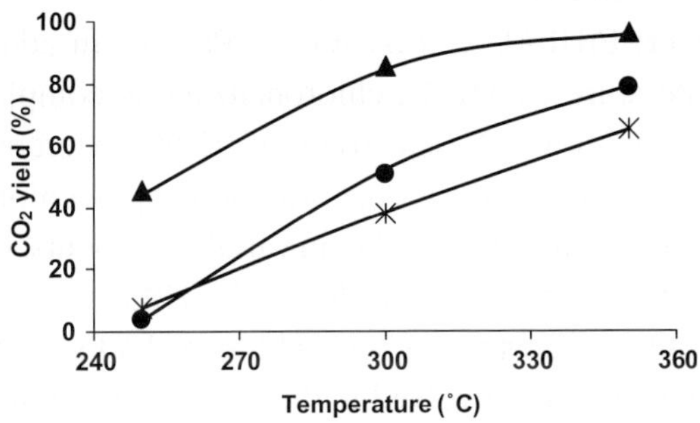


**Figure 5.4.** $CO_2$ yield of chlorobenzene conversion after 4 h of reaction versus temperature over 1%$PtAl_2O_3$ (•), 1.2%$PtSiO_2$ (∗) and 1.1%PtHFAU(5) (▲) catalysts.

and benzofuran on platinum loaded zeolites was studied in order to explain the roles of the support and that of the metal in the catalytic combustion process.

### 5.2.1. *Chlorobenzene catalytic combustion over zeolite catalysts*

Three types of platinum catalysts were compared in this study: 1%$PtAl_2O_3$, 1.2%$PtSiO_2$ and a series of PtHFAU zeolitic catalysts with a Si/Al ratio of 5 (Table 5.1).

Before the reaction, the catalysts were calcined *in situ* under dry air at 400°C for six hours. The reactions were carried out at atmospheric pressure under wet air (1.03% water corresponding to 53% relative humidity), with a gas hourly space velocity (GHSV) of 18,000 $h^{-1}$ and for temperatures varying between 250 and 350°C.

Figure 5.4 shows that whatever the reaction temperature, 1.1%PtHFAU(5) is the most active and the most selective catalyst for this reaction, despite its lower

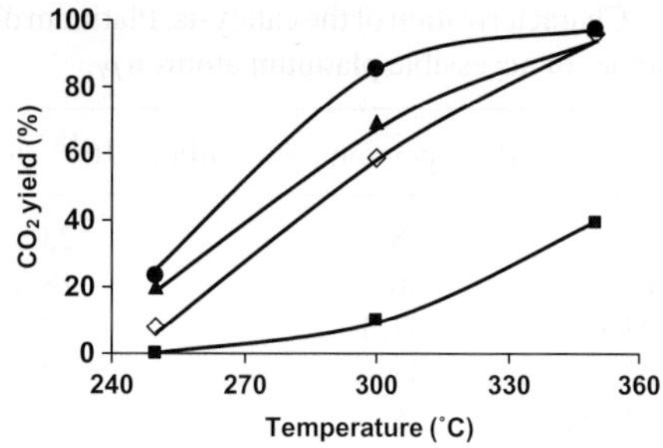


**Figure 5.5.** $CO_2$ yield of chlorobenzene conversion after 4 h of reaction versus temperature over 0% (■), 0.2% (◇), 0.3% (▲) and 1.1%PtHFAU(5) (•) catalysts.

number of accessible platinum atoms (nPt). The activity of tested catalysts decreased in the following order: PtHFAU > $PtAl_2O_3$ > $PtSiO_2$. In addition, it should be noted that at a temperature of 350°C, chlorobenzene is completely destroyed on the PtHFAU catalysts with a $CO_2$ selectivity of 97.5%. Moreover, $Cl_2$ was never detected when using a zeolite catalyst. The formation of polychlorinated compounds (PhClx) was observed over all catalysts, but in variable amounts: over 1.1%PtHFAU, PhClx represents only 3.5 ppm (0.5% yield) in the effluent against 14 ppm (2.1% yield) over 1.2%$PtSiO_2$ and 25 ppm (3.8% yield) over 1%$PtAl_2O_3$. It is indeed well known that platinum is directly involved in the formation of PhClx according to its localization and oxidation state.[50–54]

The influence of the platinum content and/or the number of accessible platinum atoms per gram of catalyst (nPt) was studied in the oxidation of chlorobenzene at different temperatures (250–350°C) on a series of PtHFAU(5) catalysts (Table 5.1). The $CO_2$ yield reported after four hours of reaction, increased simultaneously with increasing temperature and the amount of platinum deposited on the zeolite (Fig. 5.5).

Whatever the platinum content (from 0.2 to 1.1%), the PtHFAU(5) catalyst is able to completely oxidise chlorobenzene at 350°C with very good $CO_2$ selectivity (from 90 to 99%). Moreover, at this temperature, 40% of the chlorobenzene is converted into $CO_2$ on the sole zeolitic support (0 wt% Pt) with a $CO_2$ selectivity of about 80% and with only traces of CO and benzene (0.5 ppm) as by-products. At 300°C, the $CO_2$ yield increased with the amount of platinum, and for 0.6% Pt a quasi-plateau was observed (Fig. 5.6).

This behaviour is generally characteristic of a bifunctional mechanism; beyond a turning point value of the platinum concentration, the activity of the support becomes the limiting step. Therefore, chlorobenzene initially undergoes a partial transformation on the acid sites of the zeolite, and the reaction is then completed on the metal particles. On the platinum loaded catalysts, polychlorinated products (PhClx) were

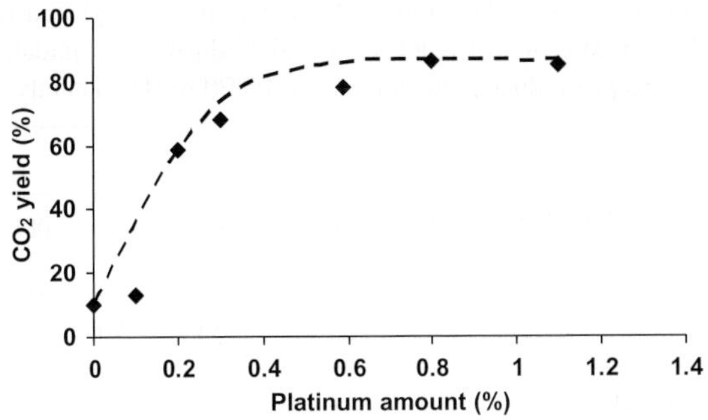


**Figure 5.6.** $CO_2$ yield of chlorobenzene conversion after 4 h of reaction at 300°C over PtHFAU(5) catalyst as a function of the amount of platinum.

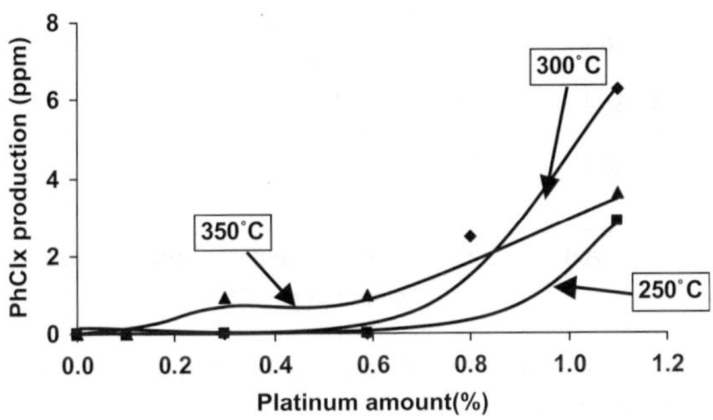


**Figure 5.7.** Polychlorobenzenes production after 4 h of reaction at 250°C (■), 300°C (◆) and 350°C (▲) as a function of the amount of platinum over PtHFAU(5) catalysts.

also observed mainly from 300°C and whatever the reaction temperature, PhClx production increases with the platinum amount (Fig. 5.7).

According to the literature, the platinum state does not seem to be a key factor for the catalytic oxidation of chlorobenzene.[51,53,63,68,71,83] Certain studies have used platinum in a reduced state[54] and others have used platinum in its oxidised state.[50,52] However, studies carried out on the oxidation of VOCs showed that reduced Pt ($Pt^0$) deposited on zeolites was the most active species for the oxidation of aromatic hydrocarbons and ketones.[36,84,85] In this context, the 1.1%PtHFAU(5) catalyst formerly reduced *in situ* under hydrogen for six hours at 450°C was tested in the oxidation of chlorobenzene at 300°C. The particular effect of this treatment was to slightly increase the total conversion of chlorobenzene but with a much higher number of polychlorinated compounds (from 6.3 to 33.8 ppm) and amount of coke deposited on the catalyst after reaction (from 0.4 to 1.17%) (Table 5.2).

Such results reveal that a prereduction of the PtHFAU(5) catalysts is not crucial since the slightly higher conversions obtained are detrimental to $CO_2$ selectivity.

**Table 5.2.** Chlorobenzene conversion, $CO_2$ selectivity ($S_{CO2}$) and polychlorinated compound production after 4 h of reaction at 300°C during the oxidation of 667 ppm chlorobenzene over a pre-reduced and calcined 1,1%PtHFAU(5) catalyst.

| | | | PhClx | | $C_6H_6$ | | |
|---|---|---|---|---|---|---|---|
| Catalyst | $C_6H_5Cl$ Conv. (%) | $S_{CO_2}$ (%) | ppm | % | ppm | % | %C |
| 1.1%PtHFAU | 87.3 | 97.8 | 6.3 | 1.1 | 0.6 | 0.1 | 0.4 |
| 1.1%PtHFAU[a] | 94.2 | 85.2 | 33.8 | 5.1 | — | — | 1.17 |

[a] catalyst reduced under hydrogen at 450°C for 6 h.

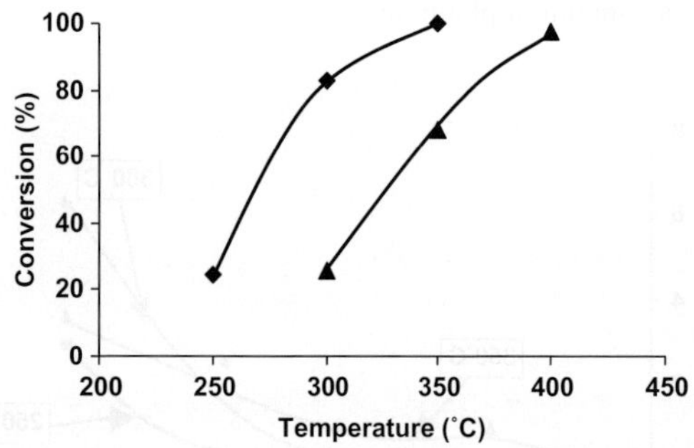


**Figure 5.8.** Conversion of chlorobenzene (♦) and 1,2-dichlorobenzene (▲) over 0.6%PtHFAU(5) after 4 h of reaction versus temperature.

### 5.2.2. *Oxidation of polychlorinated hydrocarbons*: 1, 2 *dichlorobenzene*

The results obtained on 0.6%PtHFAU(5) showed a 50°C increase of the light-off curve of 1,2-dichlorobenzene (1,2-$PhCl_2$) in comparison with the chlorobenzene curve (Fig. 5.8).

The reduction in the $CO_2$ selectivity during the oxidation of the 1,2-$PhCl_2$ compared to that of chlorobenzene was mainly due to the formation of polychlorinated compounds as well as CO traces and slightly higher coke formation at low temperatures (Table 5.3).

The effect of the 1,2-$PhCl_2$ concentration (from 133 to 1,333 ppm) was studied on the 0.8%PtHFAU(5) catalyst for temperatures varying from 250 to 400°C. The oxidation of the 1,2-$PhCl_2$ is complete at 400°C (for 133 and 333 ppm) and decreases slowly with increasing concentration (Fig. 5.9).

Kinetic studies, carried out in parallel, showed that it was possible to achieve total conversion of $PhCl_2$ at 300°C even without the addition of platinum for concentrations of about 10 ppm, which are much higher than those of dioxin precursors observed in industrial processes.

**Table 5.3.** Chlorobenzene conversion, $CO_2$ ($S_{CO_2\%}$) and PhClx ($S_{PhClx\%}$) selectivities, and amount of coke (%C) obtained after 4 h of reaction over 0.6Pt%HFAU(5) at various temperatures.

| | PhCl | | | | 1,2-$PhCl_2$ | | | |
|---|---|---|---|---|---|---|---|---|
| T°C | Conversion (%) | $S_{CO_2}$ (%) | $S_{PhClx}$ (%) | %C | Conversion (%) | $S_{CO_2}$ (%) | $S_{PhClx}$ (%) | %C |
| 250 | 24.5 | 80 | — | — | — | — | — | — |
| 300 | 82.7 | 94.5 | — | < 0.1 | 25.1 | 72 | 1.1 | 0.73 |
| 350 | 100 | 96.5 | 0.1 | 0.6 | 60.8 | 87.5 | 7.2 | 0.77 |
| 400 | — | — | — | — | 97.5 | 87.2 | 8.1 | 0.15 |

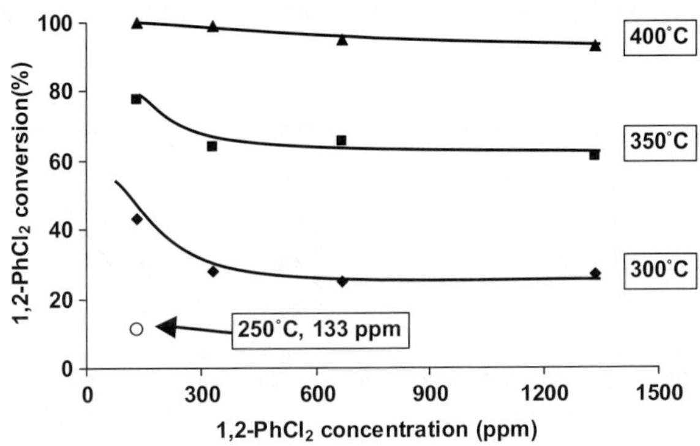


**Figure 5.9.** 1,2-dichlorobenzene conversion after 4 h of reaction as a function of concentration over 0.8%PtHFAU(5) catalyst at 250°C (○), 300°C (♦), 350°C (■) and 400°C (▲).

### 5.2.3. *Catalytic combustion of mixed polychlorinated and oxygenated cyclic hydrocarbons: 1,2-dichlorobenzene and benzofuran*

The 0.6%PtHFAU(5), 1%Pt$Al_2O_3$ and 1.2%PtHFAU(5) catalysts were tested at 300°C, in the presence of a benzofuran (200 ppm)/1,2-$PhCl_2$ (667 ppm) mixture.

It is worth noting that at 300°C benzofuran was completely converted on all the catalysts. The 1%Pt$Al_2O_3$ catalyst, which contains more platinum than the 0.6%PtH-FAU(5) catalyst, is, however, not the most active, and the 1.2%PtHFAU(5) catalyst, which has the same number of accessible platinum atoms, is twice as active and leads to a 1.6 times higher $CO_2$ yield (Table 5.4). This clearly shows that the presence of acid sites has a strong influence on the oxidation process of the mixture; PtHFAU(5) catalysts are then more adapted to the oxidation than polychlorinated aromatics and zeolite-based catalysts are suitable for the destruction of chlorinated aromatic compounds. In addition, the production of polychlorinated compounds (PhClx) is

**Table 5.4.** 1,2-$PhCl_2$ global and in $CO_2$ conversion (in mixture), polychlorobenzenes production (ppm) after 4 h at 300°C, the number of accessible Pt atoms (nPt) and coke level over catalysts.

| Catalysts | 0.6%PtHFAU(5) | 1.2%PtHFAU(5) | 1%$PtAl_2O_3$ |
|---|---|---|---|
| nPt (x $10^{18}$atoms.$g^{-1}$) | 1.45 | 21 | 23.85 |
| Conv. 1,2-$PhCl_2$ (%) | 42 | 70.1 | 35.8 |
| Conv. 1,2-$PhCl_2$ in $CO_2$ (%) | 37.8 | 52 | 32.2 |
| PhClx | 7.5 | 19.6 | 17.7 |
| % coke (%C) | 5.4 | 1.34 | 0.5 |

more important on the 1%$PtAl_2O_3$ and 1.2%PtHFAU(5) catalysts than on 0.6%PtHFAU(5) (Table 5.4). The mixture effect potentially induced by benzofuran was then evaluated.

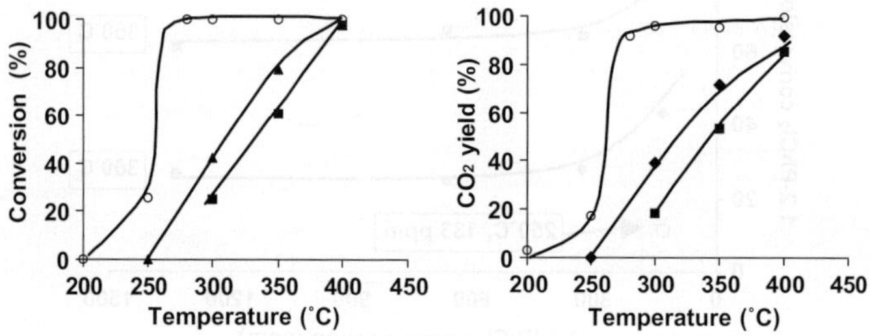


**Figure 5.10.** Global conversion of sole benzofuran (○), sole 1,2-$PhCl_2$ (■), 1,2-$PhCl_2$ in mixture (▲) (a) and corresponding $CO_2$ yields (b) after 4 h of reaction vs temperature over 0.6%PtHFAU(5).

On 0.6%PtHFAU(5) and from 300°C, the presence of benzofuran in the mixture significantly increased the total conversion of the 1,2-$PhCl_2$ as well as the $CO_2$ yield (Fig. 5.10a and 5.10b). The presence of benzofuran with 1,2-$PhCl_2$ is thus beneficial to the oxidative destruction of the compounds in the mixture. Moreover, a positive effect was also found in terms of catalyst stability. The 1.2%PtHFAU(5) catalyst was tested over four days in the catalytic oxidation of the benzofuran/1,2-dichlorobenzene mixture at 300°C in usual reaction conditions (Fig. 5.11). No decrease in the conversion of benzofuran was observed during the four-day reaction; it remained at 100% thoughout. This was not the case for the conversion of 1,2-dichlorobenzene and the conversion of the benzofuran/1,2-dichlorobenzene mixture into $CO_2$, which decreased during the first 24 hours before stabilizing thereafter.

Deactivation was much more pronounced for the production of polychlorobenzenes and led to an increase in $CO_2$ selectivity (Table 5.5). However, the number of Brønsted acid sites able to retain pyridine at 150°C fell by 20% after four hours of

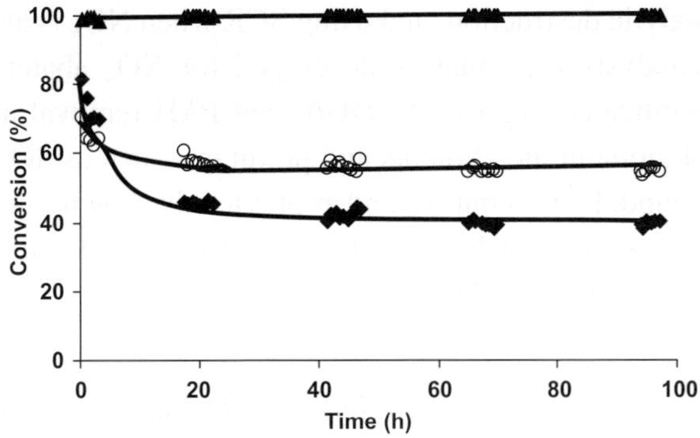


**Figure 5.11.** 1,2-Ph$Cl_2$ (♦) and benzofuran (▲) conversions as well as $CO_2$ yield (○) at 300°C during four days of testing over a 1.2%PtHFAU(5) catalyst.

**Table 5.5.** Physico-chemical characterization of a 1.2%PtHFAU(5) catalyst during four days of reaction at 300°C. $CO_2$ selectivity($S_{CO_2}$).

| Time (h) | 0.5 | 4 | 24 | 70 | 97 |
|---|---|---|---|---|---|
| D (%) | 56.7 | — | — | — | 6.2 |
| $n_B$ ($\mu$ mol.g$^{-1}$) | 776 | 615 | — | — | 447 |
| $n_L$ ($\mu$ mol.g$^{-1}$) | 243 | 253 | — | — | 317 |
| $S_{CO_2}$(%)$^b$ | 73.9 | 74.2 | 87.5 | 93.7 | 92.6 |

D: platinum dispersion.
$n_B$: number of Brønsted acid sites determined at 150°C by pyridine adsorption followed by Fourier transform infrared spectroscopy (FTIR).
$n_L$: number of Lewis acid sites determined at 150°C by pyridine adsorption followed by FTIR.

reaction, and after five days of reaction, a loss of 42% is observed. These results highlight the initial modification of the zeolite (partial collapse of the zeolitic structure) by dealumination which also leads to a higher number of Lewis acid sites associated with the extra framework aluminium species formation (Table 5.5). However, this process mostly takes place at the beginning of the reaction and after this transitory period the catalyst stability is not affected.

## 5.3. Advanced Study: Oxidation of PAHs in the Presence of a Complex Pollutants Matrix

Although no international emission control has to date been set up for PAHs, in 2008 the French parliament introduced better controls for the impact of PAH emissions in existing waste combustion facilities. Several efficient POP removal air pollution control technologies have been developed and include adsorption followed by dedusting systems[86] and catalytic destruction, and using SCR when $NO_x$ is also removed.[30,87] However, SCR catalysts were mainly developed for $NO_x$ abatement. Therefore, new catalysts specifically targeting PCDD/F and PAH removal are needed. POP molecules are not alone in the effluents and pollutants such as $SO_x$, $NH_3$, HCl and $NO_x$ can also be found. In this context, and as previously shown, zeolite-based materials are attractive catalysts for POP oxidation or for gas effluents similar to those associated with solid waste combustion.[40,42,72,76–78,88,89]

### 5.3.1. *Oxidation of 1-methylnaphthalene (1-MN) over faujasite zeolite-based catalysts*

Oxidation of 1-MN (900 ppm) was carried out at various temperatures (300 to 450°C), at high GHSV (18,000 $h^{-1}$) and in wet air (50% relative humidity) over a USHY zeolite, which is a readily available and thermally stable zeolite.

1-MN was completely transformed at 450°C with the carbon dioxide yields increasing continuously until they reached 100%. At 400 and 300°C, the initial conversion of 1-MN was total, however, after a few minutes of reaction a deactivation occurred (Fig. 5.12a). As the reactant initially disappeared, the yield into carbon dioxide was not 100% (Fig. 5.12b).

The difference between the disappearance of 1-MN and the carbon dioxide yield was due to the adsorption of the reactant and/or coke formation. The carbon contents measured after six hours of reaction over the samples were 11.0, 9.9 and 4.1 wt% for 300, 400 and 450°C, respectively. The carbon balance was then close to 100%, considering that the difference between 1-MN transformation and carbon dioxide yield is due to coke formation (or 1-MN adsorption). It must be noted that carbon monoxide was never detected. This observation is inconsistent with the results of Moljord *et al.*[90] who found a $CO/CO_2$ ratio of 0.47 at 400°C during coke combustion. The

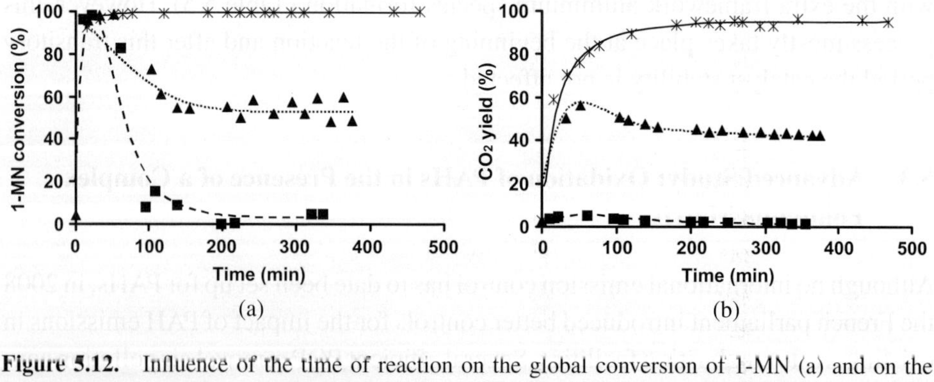


**Figure 5.12.** Influence of the time of reaction on the global conversion of 1-MN (a) and on the carbon dioxide yield (b) over a USHY zeolite at 300°C (■), 400°C (▲) and 450°C (♦).

absence of CO could be explained by the low concentration of aromatic compounds (900 ppm) in the feed.

The characterization of catalysts after reaction by infrared spectroscopy was carried out in order to determine the nature of the adsorbed species in the pores of the zeolite, which were mainly constituted of oxygenated aromatics compounds ($C_xH_yO_z$) as shown in the literature during xylene oxidation over Pd- and Pt-based zeolite catalysts (Scheme 5.1).[77,91]

$$1\text{-}MN \xrightarrow{v_1} [C_XH_YO_Z] \xrightarrow{v_2} CO_2 + H_2O$$

**Scheme 5.1.** 1-MN transformation over zeolite catalysts

1-MN is transformed into $C_xH_yO_z$ compounds before its complete oxidation into carbon dioxide. At a low temperature (300°C), the reaction rate $\upsilon_2$ is very low and only $C_xH_yO_z$ compounds are formed and retained initially in the pores of the zeolite. When the temperature increases, the reaction rate $\upsilon_2$ increases since the USHY zeolite possesses active acid sites able to transform $C_xH_yO_z$ into carbon dioxide. These sites, which are able to transform aromatic compounds into carbon dioxide, were specifically attributed to strong acid sites.

### 5.3.2. *Oxidation of 1-MN in the presence of mixed contaminants*

The catalytic combustion of 1-MN was first studied in the absence and presence of $NH_3$ over a USHY zeolite (Fig. 5.13). In the absence of $NH_3$, a conversion of 100% was achieved at 400°C (Fig. 5.13a). As previously mentioned, the USHY zeolite was active at 300 and 350°C. In the presence of 100 ppm $NH_3$, the activity of the

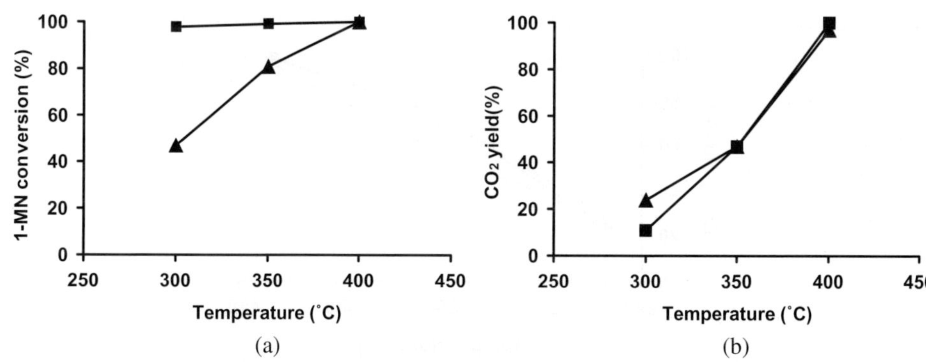


**Figure 5.13.** 1-MN conversion (a) and carbon dioxide yield (b) taken after 7 h of reaction as a function of temperature over USHY zeolite in the absence (▲) and presence (■) of $NH_3$.

catalyst increased and 1-MN was totally transformed at 300°C. The transformation of 1-MN on the USHY zeolite did not lead to any aromatics by-products and the only detected reaction product was carbon dioxide.

Whatever the operating conditions, in the presence or absence of $NH_3$, the CO yield increased with the reaction temperature and a 100% yield was obtained at 400°C (Fig. 5.13b). However, at 300°C, while 1-MN was fully transformed in the presence of $NH_3$, only 10% carbon dioxide was produced, mainly because of the formation of oxygenated compounds retained in the zeolite pores ('coke'),[77,78] which was favoured by $NH_3$ as revealed by elemental analysis showing a larger amount of carbon after the reaction. As for the coke oxidation reaction,[92] it was shown that 1-MN oxidation into carbon dioxide required strong Brønsted acid sites.[78] In our case, the basic character of $NH_3$ favours its adsorption at low temperatures on strong acid sites, which are able to transform 1-MN into carbon dioxide. Indeed, at 300°C the carbon dioxide yield was close to 20% in the absence of $NH_3$ as opposed to 10% in the presence of $NH_3$. When the reaction was carried out with $NH_3$, only the weaker sites were able to work, but these sites seemed to be active only in the conversion of 1-MN into intermediate oxygenated compounds that remained adsorbed on the solid surface.

The transformation of 1-MN was then studied in the presence of a $SO_2/NH_3$ mixture. The concentrations of $SO_2$ and $NH_3$ present in the feed were 180 and 54 ppm, respectively. In the mixture, 1-MN was fully transformed from 300°C in the presence of 100 ppm of $NH_3$ (Fig. 5.13a). Figure 5.14 represents a comparison of the carbon dioxide yield obtained after a seven hour reaction which was carried out with only 1-MN, a 1-MN/$SO_2$ mixture and a 1-MN/$NH_3$/$SO_2$ mixture as a function of the reaction temperature.

Figure 5.14 shows that when the reaction was carried out at 350°C with 180 ppm of $SO_2$ the $CO_2$ yield was enhanced. The promoting effect of $SO_2$ could be explained

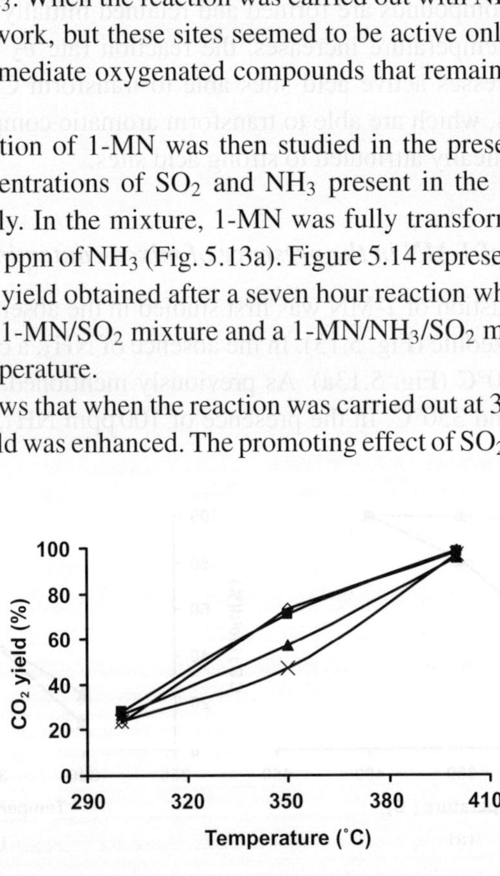


**Figure 5.14.** Carbon dioxide yield as a function of the temperature of the reaction. 1-MN only (×), 1-MN/$SO_2$ mixture (■), 1-MN/$SO_2$/$NH_3$ mixture (▲), 1-MN/$SO_2$/$NH_3$/HCl over USHY (◊).

by its oxidation into $SO_3$ (a reaction carried out in the presence of $O_2$), which could react with steam present in the feed to form sulfuric acid ($H_2SO_4$). The latter could contribute to the transformation of the oxygenated intermediate formed from 1-MN into more oxygenated intermediate molecules via a hydrolysis reaction.[93–95] When 54 ppm of $NH_3$ was added to the previous mixture, this promoting effect was partially cancelled. This decrease in $CO_2$ yield could be explained by a partial conversion of $SO_2$ into $NH_4HSO_4$ or $(NH_4)_2SO_4$ during the reaction. Indeed, the decrease of the promoting effect (about 14% loss in $CO_2$ yield) is correlated with the amount of $SO_2$ potentially consumed by $NH_3$ to form ammonium sulphate. As reported in the literature,[92,96–98] the formation of ammonium sulphate could lead to the deactivation of the catalyst. However, in our operating conditions no deactivation was observed during the seven hours, showing that the USHY zeolite seems not to be affected by the formation of ammonium sulphates.

When 217 ppm of NO and 750 ppm of HCl were added to the previous mixture, the total conversion of 1-MN was obtained at 300°C and a $CO_2$ yield of 100% was obtained at 400°C. The addition of 750 ppm of HCl in the 1-MN/$SO_2$/$NH_3$ mixture cancelled the inhibitory effect observed when the reaction was carried out in the presence of a $SO_2$/$NH_3$ mixture at 350°C; the results were identical to the ones obtained with sole 1-MN (Fig. 5.14).

This behaviour is undoubtedly the result of the neutralization (thus annihilating its effect) of $NH_3$ by HCl to produce ammonium chloride. Elemental analysis carried out on used samples showed a Cl concentration in accordance with this assumption. Once $NH_3$ was neutralised, the HCl remaining in the feed could also contribute to the hydrolysis of adsorbed species, however its effect should be limited since a 100% $CO_2$ yield was still obtained at 400°C. Moreover, it is worth noting that a significant DeNOx effect was observed as around 75% of the initial $NO_X$ was eliminated.

## 5.4. Conclusion

Future legislative reinforcement will prompt the industry to develop more effective POP reduction techniques as the existing ones suffer because the limits are not well adapted to the reduction of specific POPs such as PAHs.

The work presented in this chapter shows that the use of zeolites could be an interesting and complementary alternative to the current end-of-pipe gas cleaning techniques. Indeed, these catalysts present good properties towards the elimination of PCDD and PCDF precursors. If the addition of noble metals improves their efficiency, the kinetic studies showed that with concentrations in the range observed in the industry, non-doped zeolites are able to easily eliminate their precursors in the SCR temperature range. In regard to PAH reduction, acid zeolites are also very efficient when considering a representative solid waste combustion gas matrix, as they allow a significant $NO_x$ concentration reduction of up to 75%.

Further studies should however be carried out to better evaluate the effects of dust, higher water concentrations and catalyst shape.

## References


1. Cunliffe, A. and Williams, P. (2009). De-novo formation of dioxins and furans and the memory effect in waste incineration flue gases, *Waste Management*, **29**, pp. 739–748.
2. Hartenstein, H. (2003). Dioxin and Furan Reduction Technologies for Combustion and Industrial Thermal Process Facilities, *The Handbook of Environ. Chem.*, Vol. 3, Part O, pp. 389–423.
3. Wu, Y., Lin, L., Hsieh, L. *et al.* (2009). Atmospheric dry deposition of polychlorinated dibenzo-p-dioxins and dibenzofurans in the vicinity of municipal solid waste incinerators, *J. Hazard. Mat.*, **162**, pp. 521–529.
4. Grosso, M., Cernuschi, S., Palini, E., *et al.* (2004). PCDD/Fs release during normal and transient operation of a full scale MSWI plant, *Organohalogen Compounds*, **66**, pp. 1243–1249.
5. Hunsinger, H., Jay, K. and Vehlow, J. (2002). Formation and destruction of PCDD/F inside a grate furnace, *Chemosphere*, **46**, 9–10, pp. 1263–1272.
6. Ba, T., Zheng, M., Zhang, B., *et al.* (2009). Estimation and characterization of PCDD/Fs and dioxin-like PCBs from secondary copper and aluminum metallurgies in China, *Chemosphere*, **75**, 9, pp. 1173–1178.
7. Huang, H. and Buekens, A. (1995). On the mechanisms of dioxin formation in combustion processes, *Chemosphere*, **31**, 9, pp. 4099–4117.
8. Zimmermann, R., Blumenstock, M., Heger, H., *et al.* (2001). Emission of Nonchlorinated and Chlorinated Aromatics in the Flue Gas of Incineration Plants during and after Transient Disturbances of Combustion Conditions: Delayed Emission Effects, *Environ. Sci. & Tech.*, **35**, pp. 1019–1030.
9. Wilhelm, J., Stieglitz, L., Dinjus, E., *et al.* (2001). Mechanistic studies on the role of PAHs and related compounds in PCDD/F formation on model fly ashes, *Chemosphere*, **42**, pp. 797–802.
10. Stieglitz, L., Vogg, H., Zwick, G. *et al.* (1991). On formation conditions of organohalogen compounds from particulate carbon of fly ash, *Chemosphere*, **23**, pp. 1255–1264.
11. Everaert, K. and Baeyens, J. (2001). Correlation of PCDD/F Emissions with Operating Parameters of Municipal Solid Waste Incinerators, *J. Air & Waste Manag. Assoc.*, **51**, pp. 718–724.
12. McKay, G. (2002). Dioxin characterisation, formation and minimisation during municipal solid waste (MSW) incineration: review, *Chem. Eng. Journal*, **86**, pp. 343–368.
13. Stanmore, B. (2004). The formation of dioxins in combustion systems, *Combustion & Flamme*, **136**, pp. 398–427.
14. Khachatryan L. and Dellinger, B. (2003). Formation of chlorinated hydrocarbons from the reaction of chlorine atoms and activated carbon, *Chemosphere*, **52**, pp. 709–716.
15. Weber, R., Behnisch, P., Brouwer, A., *et al.* (2006). Contemporary relevance of dioxin and dioxin like compound contaminations in residues from recycling of HCH waste, *Organohalogen Compd.*, **68**, pp. 905–910.
16. Shao, K., Yan, J., Li, X., *et al.* (2010). Effects of $SO_2$ and $SO_3$ on the formation of polychlorinated dibenzo-p-dioxins and dibenzofurans by de novo synthesis, *Zhejiang Univ.-Sci. A* (*Appl. Phys. & Eng.*), **11**, 5, pp. 363–369.
17. Chen, S., Hsieh, L. and Chiu. S. (2003). Characteristics of the PAH emissions from the incineration of livestock wastes with/without APCD, *Environ. International*, **28**, 7, pp. 659–668.

18. Xiaofang, Y. (2008). Polycyclic aromatic hydrocarbon (PAH) emission from co-firing municipal solid waste (MSW) and coal in a fluidized bed incinerator, *Waste Manage.*, **28**, pp. 1543–1551.
19. Chung, T., Liao, C., Chang-Chien, G. (2010). Distribution of polycyclic aromatic hydrocarbons and polychlorinated dibenzo-p-dioxins/dibenzofurans in ash from different units in a municipal solid waste incinerator, *Waste Manage. Research*, **28**, 9, pp. 789–799.
20. Singh, S. and Vinit, P. (2007). The effect of temperature on PAHs emission from incineration of acrylic waste, *Environ. Monit. Assess.*, **127**, pp. 73–77.
21. Buekens, A. and Huang, H. (1998). Comparative evaluation of techniques for controlling the formation and emission of chlorinated dioxins/furans in municipal waste incineration, *J. Hazard. Mater.*, **62**, pp. 1–33.
22. Goemans, M., Clarysse, P., Joannes, J., *et al.* (2003). Catalytic NOx reduction with simultaneous dioxin and furan oxidation, *Chemosphere* **50**, pp. 489–497.
23. Weber, R. (2004). Relevance of PCDD/PCDF Formation for the Evaluation of POPs Destruction Technologies. – Necessity and Current Status, *Organohalogen Compd.*, **66**, pp. 1270–1280.
24. Parizek, T., Bébar, L. and Stehlik, P. (2008). Persistent pollutants emission abatement in waste-to-energy systems, *Clean Techn. Environ. Policy*, **10**, pp. 147–153.
25. Inoue, K. and Kawamoto, K. (2005). Fundamental Adsorption Characteristics of Carbonaceous Adsorbents for 1,2,3,4-Tetrachlorobenzene in a Model Gas of an Incineration Plant, *Environ. Sci. & Tech.*, **39**, pp. 5844–5850.
26. Inoue, K. and Kawamoto, K. (2008). Adsorption characteristics of carbonaceous adsorbents for organic pollutants in a model incineration exhaust gas, *Chemosphere*, **70**, pp. 349–357.
27. Le Cloirec, P. and Laplanche, A. (2005). Réduction des dioxines, furannes et polychlorobiphenyls, *Les techniques de l'ingénieur*, **J3935**, pp. 1–10.
28. Kulkarni, P., Crespo, J. and Afonso, C. (2008). Dioxins sources and current remediation technologies — A review, *Environ. International*, **34**, pp. 139–153.
29. Wielgosinski, G. (2010). The Possibilities of Reduction of Polychlorinated Dibenzo-P-Dioxins and Polychlorinated Dibenzofurans Emission, *Intern. J. Chem. Eng.*, **2010**, pp. 1–11.
30. Goemans, M., Clarysse, P., Joannès, J. *et al.* (2004). Catalytic NOx reduction with simultaneous dioxin and furan oxidation, *Chemosphere*, **54**, pp. 1357–1365.
31. Everaert, K. and Baeyens, J. (2004). Catalytic combustion of volatile organic compounds, *J. Hazard. Mater.*, **109**, pp. 113–139.
32. Wey, M., Chen, J., Huang, H., *et al.* (2003). Oxidation of organic pollutants in incineration flue gas by a fluidized palladium catalyst, *Combust. Sci. Technol.*, **175**, pp. 1211–1236.
33. Chen, J., Wey, M., Yeh, C., *et al.* (2004). Simultaneous treatment of organic compounds, CO, and NOx in the incineration flue gas by three-way catalyst, *Appl. Catal. B: Environ.*, **48**, pp. 25–35.
34. Hermia, J. and Vigneron, S. (1993). Catalytic incineration for odour abatement and VOC destruction, *Catal. Today*, **17**, pp. 349–358.
35. Whang, X., Shen, S., Hidajat, K., *et al.* (2004). Naphthalene Oxidation over 1%Pt and 5%Co/$\gamma$-$Al_2O_3$ Catalysts: Reaction Intermediates and Possible Pathways, *Catal. Lett.*, **96**, pp. 87–96.
36. Tsou, J., Pinard, L. and Magnoux, P. (2003). Catalytic oxidation of volatile organic compounds (VOCs): Oxidation of o-xylene over Pt/HBEA catalysts, *Appl. Catal. B: Environ.*, **46**, pp. 371–379.
37. Ordonez, S., Bello, L., Sastre, H., *et al.* (2002). Kinetics of the deep oxidation of benzene, toluene, n-hexane and their binary mixtures over a platinum on $\gamma$-alumina catalyst, *Appl. Catal. B: Environ.*, **38**, pp. 139–149.
38. Papaefthimiou, P., Ioannides, T. and Verykios, X. (1997). Combustion of non-halogenated volatile organic compounds over group VIII metal catalysts, *Appl. Catal. B: Environ.*, **13**, pp. 175–184.
39. Shie, J., Chang, C., Chen, J., *et al.* (2005). Catalytic oxidation of naphthalene using a Pt/$Al_2O_3$ catalyst, *Appl. Catal. B: Environ.*, **58**, pp. 289–297.

40. Taralunga, M., Mijoin, J. and Magnoux, P. (2005). Catalytic destruction of chlorinated POPs — Catalytic oxidation of chlorobenzene over PtHFAU catalysts, *Appl. Catal. B: Environ.*, **60**, pp. 163–171.
41. Gonzalez-Velasco, J., Aranzabal, A., Gutierrez-Ortiz, J., *et al.* (1998). Activity and product distribution of alumina supported platinum and palladium catalysts in the gas-phase oxidative decomposition of chlorinated hydrocarbons, *Appl. Catal. B: Environ.*, **19**, pp. 189–197.
42. Taralunga, M., Mijoin, J. and Magnoux, P. (2004). Catalytic oxidation of chlorobenzene, a model compound for dioxin, over Pt/zeolite catalysts, *Organohalogen Compounds*, **66**, pp. 1160–1166.
43. Diehl, F., Barbier Jr, J., Duprez, D., *et al.* (2010). Catalytic oxidation of heavy hydrocarbons over Pt/$Al_2O_3$. Influence of the structure of the molecule on its reactivity, *Appl. Catal. B: Environ.*, **95**, s pp. 217–227.
44. Delahay, D. and Duprez, D. (1989). Effects of dispersion and partial reduction on the catalytic properties of Rh$Al_2O_3$ catalysts in the steam reforming of mono- and bicyclic aromatics, *J. Catal.*, **115**, pp. 542–550.
45. Duprez, D. (1992). Selective steam reforming of aromatic compounds on metal catalysts, *Appl. Catal. A*, **82**, pp. 111–157.
46. Ntainjua, N., Carley, A. and Taylor, S. (2008). The role of support on the performance of platinum-based catalysts for the total oxidation of polycyclic aromatic hydrocarbons, *Catal. Today*, **137**, pp. 362–366.
47. Zhang, X., Shen, S., Yu, L., *et al.* (2003). Oxidative decomposition of naphthalene by supported metal catalysts, *Appl. Catal. A: Gen.*, **250**, pp. 341–352.
48. Taylor, S., Heneghan, C., Hutchings, G., *et al.* (2000). The activity and mechanism of uranium oxide catalysts for the oxidative destruction of volatile organic compounds, *Catal. Today*, **59**, pp. 249–259.
49. Krishnamoorthy, S., Baker, J. and Amiridis, M. (1998). Catalytic oxidation of 1,2-dichlorobenzene over $V_2O_5/TiO_2$-based catalysts, *Catal. Today*, **40**, pp. 39–46.
50. van den Brink, R., Louw, R. and Mulder, P. (1998). Formation of polychlorinated benzenes during the catalytic combustion of chlorobenzene using a Pt/$\gamma$-$Al_2O_3$ catalyst, *Appl. Catal. B: Environ.*, **16**, pp. 219–226.
51. van den Brink, R., Mulder, P. and Louw, R. (1999). Catalytic combustion of chlorobenzene on Pt/$\gamma$-$Al_2O_3$ in the presence of aliphatic hydrocarbons, *Catal. Today*, **54**, pp. 101–106.
52. van den Brink, R., Krzan, M., Feijen-Jeurissen, M.,*et al.* (2000). The role of the support and dispersion in the catalytic combustion of chlorobenzene on noble metal based catalysts, *Appl. Catal. B: Environ.*, **24**, pp. 255–264.
53. van den Brink, R., Louw, R., Mulder, P. (2000). Increased combustion rate of chlorobenzene on Pt/$\gamma$-$Al_2O_3$ in binary mixtures with hydrocarbons and with carbon monoxide, *Appl. Catal. B: Environ.*, **25**, pp. 229–237.
54. Scirè, S., Minico, S., Cristafulli, C. (2003). Pt catalysts supported on H-type zeolites for the catalytic combustion of chlorobenzene, *Appl. Catal. B: Environ.*, **45**, pp. 117–125
55. Garcia, T., Solsona, B. and Taylor, S. (2005). Nano-crystalline Ceria Catalysts for the Abatement of Polycyclic Aromatic Hydrocarbons, *Catal. Lett.*, **105**, pp. 183–189.
56. Garcia, T., Solsona, B. and Taylor, S. (2006). Naphthalene total oxidation over metal oxide catalysts, *Appl. Catal. B: Environ.*, **66**, pp. 92–99.
57. Liu, Y., Wei, Z., Feng, Z., *et al.* (2001). Destruction of Chlorobenzene and o-Dichlorobenzene on a Highly Active Catalyst: MnOx/$TiO_2$–$Al_2O_3$, *Catal.*, **202**, pp. 200–204.
58. Cho, C. and Ihm, S. (2002). Development of New Vanadium-Based Oxide Catalysts for Decomposition of Chlorinated Aromatic Pollutants, *Environ. Sci. Technol.*, **36**, pp. 1600–1606.
59. Yim, S., Koh, D. and Nam, I. (2002). A pilot plant study for catalytic decomposition of PCDDs/PCDFs over supported chromium oxide catalysts, *Catal. Today*, **75**, pp. 269–276.

60. Weber, R., Sakurai, T. and Hagenmaier, H. (1999). Low temperature decomposition of PCDD/PCDF, chlorobenzenes and PAHs by $TiO_2$-based $V_2O_5$–$WO_3$ catalysts, *Appl. Catal. B: Environ.*, **20**, pp. 249–256.
61. Rossin, J. and Farris, M. (1993). Catalytic oxidation of chloroform over a 2% platinum alumina catalyst, *Ind. Eng. Chem. Res.*, **32**, pp. 1024–1029.
62. Yang, C., Chang, S., Hong, B., *et al.* (2008). Innovative PCDD/F-containing gas stream generating system applied in catalytic decomposition of gaseous dioxins over $V_2O_5$–$WO_3$/$TiO_2$-based, *Chemosphere*, **73**, pp. 890–895.
63. Becker, L. and Förster, H. (1997). Oxidative Decomposition of Chlorobenzene Catalyzed by Palladium-Containing Zeolite Y, *J. Catal.*, **170**, pp. 200–203.
64. Corma, A. (2003). State of the art and future challenges of zeolites as catalysts, *J. Catal.*, **216**, pp. 298–312.
65. Gonzalez-Velasco, J., Lopez-Fonseca, R., Aranzabal, A., *et al.* (2000). Evaluation of H-type zeolites in the destructive oxidation of chlorinated volatile organic compounds, *Appl. Catal. B: Environ.*, **24**, pp. 233–242.
66. Lopez-Fonseca, R., Gutierrez-Ortiz, J., Gutierrez-Ortiz, M., *et al.* (2002). Dealuminated Y Zeolites for Destruction of Chlorinated Volatile Organic Compounds, *J. Catal.*, **209**, pp. 145–150.
67. Imamura, S. (1992). Catalytic decomposition of halogenated organic compounds and deactivation of the catalysts, *Catal. Today*, **11**, pp. 547–567.
68. Sciré, S. and Minico, S. (2003). The Role of the Support in the Oxidative Destruction of Chlorobenzene on Pt/Zeolite Catalysts: An FT-IR Investigation, *Catal. Letters*, **91**, 3–4, pp. 199–205.
69. Chatterjee, S. and Greene, H. (1991). Oxidative catalysis of chlorinated hydrocarbons by metal-loaded acid catalysts, *J. Catal.*, **130**, pp. 76–85.
70. Lopez-Fonseca, R., Gutierrez-Ortiz, J., Gutierrez-Ortiz, M., *et al.* (2005). Catalytic oxidation of aliphatic chlorinated volatile organic compounds over Pt/H-BETA zeolite catalyst under dry and humid conditions, *Catal. Today*, **107**, pp. 200–207.
71. Scire, S., Minico, S., Crissafulli, C., *et al.* (2002). Catalytic combustion of chlorobenzene over Pt/zeolite catalysts, Studies in Surface Science and Catalysis, *Stud. Surf. Sci. Catal.*, **142**, pp. 1023–1030.
72. Taralunga, M., Innocent, B., Mijoin, J., *et al.* (2007). Catalytic combustion of benzofuran and of a benzofuran/1,2-dichlorobenzene binary mixture over zeolite catalysts, *Appl. Catal. B: Environ.*, **75**, pp. 139–146.
73. Guillemot, M., Mijoin, J., Mignard, S., *et al.* (2007). Mode of zeolite catalysts deactivation during chlorinated VOCs oxidation, *Appl. Catal. A: Gen.*, **327**, pp. 211–217.
74. Guillemot, M., Mijoin, J., Mignard, S., *et al.* (2008). Adsorption of tetrachloroethylene (PCE) in gas phase on zeolites of faujasite type: Influence of water vapour and of Si/Al ratio, *Micro. Meso. Mater.*, **111**, pp. 334–342.
75. Lopez-Fonseca, R., de Rivas, B., Gutierrez-Ortiz, J., *et al.* (2003). Enhanced activity of zeolites by chemical dealumination for chlorinated VOC abatement, *Appl. Catal. B: Environ.*, **41**, pp. 31–42.
76. Taralunga, M., Mijoin, J. and Magnoux, P. (2006). Catalytic destruction of 1,2-dichlorobenzene over zeolites, *Catalysis Communications*, **7**, pp. 115–121.
77. Marie-Rose, S., Belin, T., Mijoin, J., *et al.* (2009). Catalytic combustion of polycyclic aromatic hydrocarbons (PAHs) over zeolite type catalysts: Effect of water and PAHs concentration, *Appl. Catal. B: Environ.*, **90**, pp. 489–496.
78. Marie Rose, S., Belin, T., Mijoin, J., *et al.* (2009). Destruction of PAH and dioxin precursors using selective oxidation over zeolite catalysts. Influence of the presence of ammonia in the flue gas, *Appl. Catal. B: Environ.*, **93**, pp. 106–111.

79. Ribeiro, F., Chow, M. and Dallabetta, R. (1994). Kinetics of the Complete Oxidation of Methane over Supported Palladium Catalysts, *J. Catal.*, **146**, pp. 537–544.
80. Cullis, C. and Willatt, B. (1984). The inhibition of hydrocarbon oxidation over supported precious metal catalysts, *J. Catal.*, **86**, pp. 187–200.
81. Altarawneh, M., Dlugogorski, B., Kennedy, E., *et al.* (2009). Mechanisms for formation, chlorination, dechlorination and destruction of polychlorinated dibenzo-p-dioxins and dibenzofurans (PCDD/Fs), *Progress in Energy and Combustion Science*, **35**, pp. 245–274.
82. Everaert, K. and Baeyens, J. (2002). The formation and emission of dioxins in large scale thermal processes, *Chemosphere*, **46**, pp. 439–448.
83. de Jong, V., Cieplik, M., Reints, W.,*et al.* (2002). A Mechanistic Study on the Catalytic Combustion of Benzene and Chlorobenzene, *J. Catal.*, **211**, pp. 355–365.
84. Dégé, P., Pinard, L., Magnoux, P., *et al.* (2001). Catalytic oxidation of volatile organic compounds (VOCs). Oxidation of o-xylene over Pd and Pt/HFAU catalysts, *C. R. Acad. Sci. Ser. Iic*: *Chem.*, **4**, pp. 41–47.
85. Dégé, P., Pinard, L., Magnoux, P., *et al.* (2000). Catalytic oxidation of volatile organic compounds: II. Influence of the physicochemical characteristics of Pd/HFAU catalysts on the oxidation of o-xylene, *Appl. Catal. B*: *Environ.*, **27**, pp. 17–26.
86. Chang, S., Yeh, J., Chein, H., *et al.* (2008). PCDD/F Adsorption and Destruction in the Flue Gas Streams of MWI and MSP via Cu and Fe Catalysts Supported on Carbon, *Environ. Sci. & Tech.*, **42**, pp. 5727–5733.
87. Finocchio, E., Busca, G. and Notaro, M. (2006). A review of catalytic processes for the destruction of PCDD and PCDF from waste gases, *Appl. Catal. B*: *Environ.*, **62**, pp. 12–20.
88. Marie Rose, S. Mijoin, J., Magnoux, P., *et al.* (2008). Influence of $NH_3$ during the catalytic oxidation of a 1-methylnaphthalene / 1,2-dichlorobenzene mixture over Pt/zeolite catalysts, *Organohalogen Compounds*, **70**, pp. 2216–2219.
89. Scire, S., Minico, S., Crissafulli, C., *et al.* (2002). Catalytic combustion of chlorobenzene over Pt/zeolite catalysts, Studies in Surface Science and Catalysis, *Stud. Surf. Sci. Catal.*, **142**, pp. 1023–1030.
90. Moljord, K., Magnoux, P. and Guisnet M. (1995). Coking, aging and regeneration of zeolites XV. Influence of the composition of HY zeolites on the mode of formation of coke from propene at 450°C, *Appl. Catal. A*: *Gen.*, **122**, pp. 21–32.
91. Guisnet, M., Dégé, P. and Magnoux, P. (1999). Catalytic oxidation of volatile organic compounds 1. Oxidation of xylene over a 0.2 wt% Pd/HFAU(17) catalyst, *Appl. Catal. B*: *Environ.*, **20**, pp. 1–13.
92. Ham, S. and Nam, I. (2002). Selective Catalytic Reduction of Nitrogen Oxides by Ammonia, *Catalysis*, **16**, pp. 236–271.
93. Hinz, A., Skoglundh, M., Fridell, E., *et al.* (2001). An Investigation of the Reaction Mechanism for the Promotion of Propane Oxidation over $Pt/Al_2O_3$ by $SO_2$, *J. Catal.*, **201,** pp. 247–257.
94. Oi-Uchisawa, J., Obuchi, A., Ogata, A., *et al.* (1999). Effect of feed gas composition on the rate of carbon oxidation with Pt/SiO2 and the oxidation mechanism, *Appl. Catal. B*: *Environ.*, **21**, pp. 9–17.
95. Yao, H., Stepien, H. and Gandhi, H. (1981). The effects of $SO_2$ on the oxidation of hydrocarbons and carbon monoxide over $Pt\gamma$-$Al_2O_3$ catalysts, *J. Catal.*, **67**, pp. 231–236.
96. Ham, S., Choi, H., Nam, I., *et al.* (1995). Effect of Copper Contents on Sulfur Poisoning of Copper Ion-Exchanged Mordenite for NO Reduction by $NH_3$, *Ind. Eng. Chem. Res.*, **34**, pp. 1616–1623.
97. Ham, S., Choi, H., Nam, I., *et al.* (1996). Effect of oxygen on selective catalytic reduction of NO by $NH_3$ over copper ion exchanged mordenite-type zeolite catalyst, *Catal. Lett.*, **42**, pp. 35–40.
98. Janssen, F. (1999). Environmental Catalysis — Stationary Sources, In G. Ertl, H. Knozinger, J. Weitkamp (eds), *Environmental Catalysis*, Wiley-VCH Verlag GmbH, Weinheim, pp. 119–179.

# Chapter 6

# Plasma Catalysis for Volatile Organic Compounds Abatement

J. Christopher WHITEHEAD*

This chapter deals with the combination of plasma with catalysis. It offers an effective solution to the environmental clean-up of pollutant VOCs in waste gas streams. Abatement of the halomethanes and hydrocarbons at low temperatures is detailed. The role of ozone in plasma catalysis remediation is also reviewed.

## 6.1. Introduction

Non-thermal, atmospheric pressure plasma has been extensively investigated for the removal of volatile organic compounds (VOCs) from waste gas streams and other environments for almost 20 years.[1] A wide range of organic as well as halo-organic and organo-sulfur species[2,3] have been studied using a variety of plasma reactors such as dielectric barrier, corona, surface, gliding arc and packed-bed discharges.[4,5] In these discharges, there are very high energy electrons with effective temperatures in excess of 10,000 K or kinetic energies of $>10$ eV which create excited state species and radicals in the gas without significantly raising its temperature ($T_{gas} \sim 300$–500 K). This high degree of non-equilibrium characterises a non-thermal plasma and can provide an oxidative environment containing reactive species such as O atoms and OH radicals at a gas temperature that remains close to ambient, minimising corrosion effects that might occur in the much higher operating temperatures of pyrolysis or thermal catalytic oxidation. In addition, non-thermal plasma systems have the advantage of very rapid response, producing oxidative reagents almost instantly without the need for high temperature heating, and allowing sophisticated control systems with intelligent feedback that increase energy efficiency. Working at atmospheric pressure eliminates the need for expensive pumping systems also making the technique cost effective. In general terms, plasma technology has been found to be most effective for the removal of low concentrations ($<1{,}000$ ppm) of VOCs from air or other gas streams and is a viable alternative to the commonly

*School of Chemistry, The University of Manchester, Oxford Road, Manchester M13 9PL, UK.

used abatement techniques such as adsorption, thermal incineration and catalytic oxidation.

Ideally, the products of plasma treatment of waste streams containing VOCs should be $CO_2$ and $H_2O$, representing complete hydrocarbon oxidation with inorganic oxides or acids such as $SO_2$ for sulfur-containing VOCs, and hydrogen halides or oxides for halo-organic species. However, it has been found that plasma treatment can often lead to the formation of various unwanted by-products which are mainly organic intermediates and CO due to incomplete oxidation and $NO_x$ formation when the gas stream is air. Hence, in order to improve and increase the efficiency of VOC decomposition, an innovative technology has been proposed where a plasma reactor is coupled with a catalyst. Such a combination helps overcome some of the disadvantages of both catalytic and plasma treatments, often with synergistic benefits. Plasma activation of a catalyst can often occur at much lower temperatures than thermal catalyst activation and can, in many cases, reduce problems due to coking, for example, poisoning by sulfur or sintering at high temperatures. The general aim of the hybridisation of plasma and catalyst is to increase the degree of removal of the VOC and to improve the selectivity for product formation, thereby reducing unwanted by-products and increasing energy efficiency. There have been several review articles on plasma catalysis for the removal of VOCs which form the background to this chapter.[6–9]

## 6.2. Plasma Catalyst Interactions

When combining plasma with a catalyst, there are two configurations that are generally used. These are illustrated in Fig. 6.1. Either the catalyst can be located directly in the discharge region (A), which is called a one-stage arrangement, or it can be placed downstream from the plasma (B) and called a two-stage arrangement.

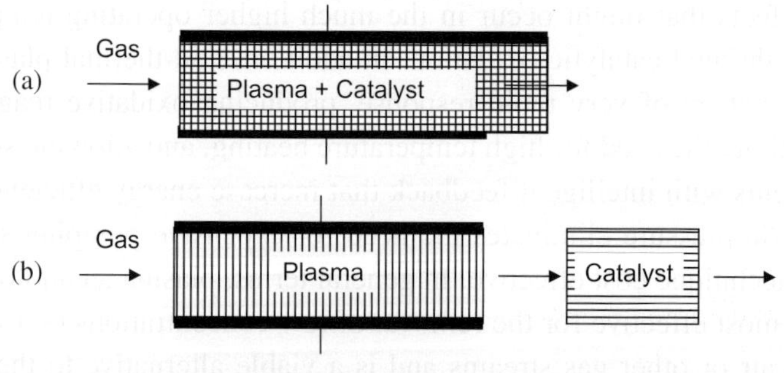


**Figure 6.1.** A schematic diagram of the different plasma catalyst configurations. Configuration (A) places the catalyst directly into the discharge region — a one-stage arrangement. Configuration (B) has the catalyst downstream from the discharge — a two-stage arrangement.

Compared with a thermal catalysis experiment where energy comes from supplied heat, in non-thermal plasma-activated catalysis the energy comes from the electrical discharge. This produces reactive and energetic species in the gas-phase which may then come into contact with the surface of the catalyst. The primary species produced in a plasma are generally short-lived and consist of electrons, photons, ions, excited-state atoms or molecules and radicals. It is these species that will interact with the catalyst in a one-stage plasma catalyst arrangement. In contrast, it will only be relatively long-lived species exiting from the plasma that are in contact with the downstream catalyst in a two-stage plasma catalyst configuration. These will be the end-products, by-products and long-lived reactive intermediates of the plasma processing (of which ozone is an important and common example) and, possibly, vibrationally excited species. Vibrational energy can be a significant mode for efficiently activating surface adsorption and desorption processes.

Another important aspect that must be considered in using the one-stage configuration is that the action of plasma-generated species may modify the condition or state of the catalyst surface.[10] Plasma is quite routinely used to prepare catalyst surfaces prior to their use, giving them increased effectiveness by changing their activity or surface properties. For example, a plasma has been used to prepare platinum nanoclusters on carbon nanotubes as a catalyst[11] and a thermal plasma spraying process can provide a durable coating for nickel catalysts for biomass gasification.[12] In addition, the properties of the catalyst may affect the physical characteristics of the discharge, especially its electrical properties where the packing material (i.e. the catalyst) may modify the electrical conductivity, changing the electrical breakdown voltage and the maximum attainable plasma power. This can change the range of reactive species and their concentrations and modify the nature of the discharges, which might change from being predominantly filamentary discharges in the gas-phase to surface discharges along the surface of the catalyst.[13] Very importantly, the presence of a catalyst in the discharge may add adsorbent material which can prolong the lifetime of species increasing their reaction times. Chen's recent review[8] presents a useful distinction between the chemical and physical interactions that might arise in plasma catalysis.

## 6.3. Plasma Catalysis for the Abatement of Halomethanes

### 6.3.1. *CFC-12*

The destruction of chlorofluorocarbons (CFCs) which are ozone-depleting substances banned by the Montreal Protocol represents a particular technical challenge. Very high temperatures (>1,250°C) provided by high-temperature incineration and thermal plasma torches are needed to destroy the CFCs safely without the formation of extremely hazardous dioxins. The strength of the carbon-halogen bonds in CFCs makes them extremely energy-intensive species to destroy. A study of the destruction of CFC-12, $C_2F_2Cl_2$, at a concentration of ~400 ppm in an atmospheric pressure air stream was performed using only a plasma discharge.[14] The plasma reactor was a dielectric packed-bed using barium titanate as the packing material. The percentage destruction of CFC-12 in air ranged between ~8-40% depending on the CFC-12 concentration and the residence (and hence reaction) time in the plasma reactor. An enhancement of a factor of greater than two in the destruction could be achieved by using a leaner oxygen mixture (~0.02% $O_2$ in $N_2$). The products of destruction are CO, $CO_2$, $COF_2$, $F_2$ and $Cl_2$ with significant by-product formation of $NO_x$ (NO and $NO_2$) which is a common occurrence associated with plasma processing in air. In an air gas stream, NO is initially formed by the recombination of oxygen and nitrogen atoms produced in the plasma. $NO_2$ is then formed from the NO through a variety of steps including reaction with ozone and oxygen atoms. The initial step in the destruction of the CFC-12 is thought to involve electron-induced dissociation of $CF_2Cl_2$, probably involving initial electron attachment to the highly electronegative molecule.

By incorporating a catalyst in the form of a fine powder into the barium titanate packing material, it is possible to investigate the plasma-catalytic destruction of CFC-12 in a one-stage configuration. A range of catalytic materials including γ-alumina, $TiO_2$ and HZSM-5 and NaX zeolites were investigated.[15] Figure 6.2 shows the effect of these materials on the overall destruction of 500 ppm of CFC-12 in an air stream. It can be seen that there is only a modest increase in destruction in plasma catalysis except for the $TiO_2$ catalyst which increases the percentage CFC-12 destruction from ~12% to ~27%.

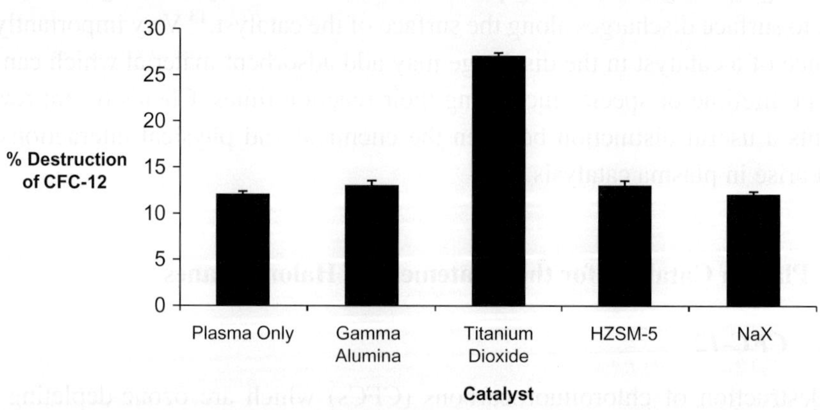


**Figure 6.2.** Chart showing the destruction of ~500 ppm CFC-12 in an atmospheric pressure air stream using plasma catalysis with a range of catalysts in a one-stage plasma catalysis configuration. The plasma is a barium titanate packed-bed. Adapted from.[15]

The effectiveness of titanium dioxide in plasma catalysis arises from a direct interaction of the electrons in the plasma discharge with the catalyst that is analogous to the well-known properties of $TiO_2$ as a photocatalyst. $TiO_2$ has Schottky defects giving n-type semiconductor properties and a bandgap of 3.2 eV which can be activated by light at a wavelength of <385 nm. This excitation creates electron hole-pair states giving radical cations, $OH^-$ and $O_2^-$, from chemisorbed moisture in air-based systems, with oxidative activity attributed mostly to valence-band holes, $O_2$, $HO_2$ and $H_2O_2$.[16] In the plasma catalysis system, it is thought[15] that the electrons in the discharge, which typically have a mean energy of ~3.5 eV, create electron hole-pair states in an analogous manner to UV excitation

$$TiO_2 + e^-(> 3.2\,\text{eV}) \rightarrow h^+ + e^-$$

It should be noted that the photon energy and flux emitted from these discharges are not sufficient to explain the enhancement in destruction by photocatalysis. While $TiO_2$ is found to be effective in more than doubling the CFC-12 destruction efficiency, the NaX zeolite catalyst which has no effect in enhancing the CFC-12 destruction (Fig. 6.2) reduces the $NO_x$ formation by more than a factor of two compared with the use of a plasma discharge alone. Ideally in plasma catalysis, we wish to achieve both enhancement in destruction and minimisation of unwanted by-products.

### 6.3.2. *Dichloromethane*

Dichloromethane, $CH_2Cl_2$, (DCM) is a common solvent used in a wide range of industrial processes and is a known hazardous air pollutant. Its destruction by plasma methods has been studied widely by ourselves and others.[2] We have extended these studies to the integration of plasma and catalysts with the aim of enhancing destruction whilst minimising by-product formation.[17] Again we used a barium titanate packed-bed plasma reactor. Figure 6.3 shows the results of using a range of zeolite catalysts in a two-stage configuration where the catalyst is downstream from the plasma. There is only a modest increase in DCM destruction brought about by the catalyst (<37%). However, we see that the NaX catalyst is able to reduce the $NO_x$ emissions by more than a factor of two as we observed for CFC-12. The sodium zeolites are more basic forms and this makes them good de-$NO_x$ catalysts in plasma reactors at low temperatures (<350°C).[18]

To gain further insight into how the plasma and catalyst interact in this system, we performed some characterisation of the catalysts after a period of treatment in the plasma environment using BET (Brunauer, Emmett and Teller) surface area analysis, SEM (scanning electron microscopy) investigation of the surface morphology, elemental analysis and ATR-IR (attenuated total reflection Infrared) spectroscopy

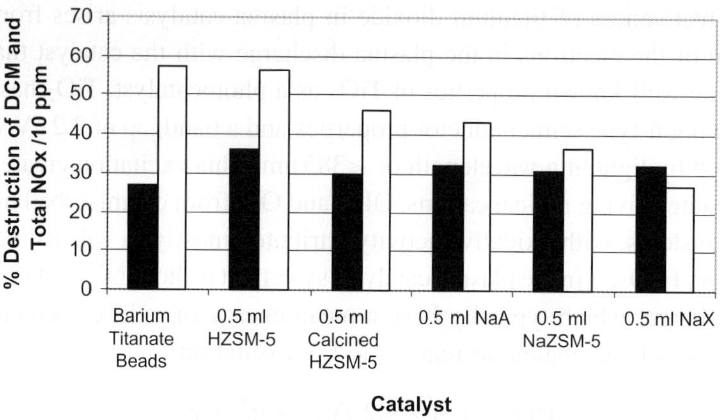


**Figure 6.3.** The percentage destruction of DCM (solid bars) and the $NO_X$ formation (open bars) for various zeolite materials used in a two-stage plasma catalyst configuration for the remediation of 500 ppm DCM in an atmospheric pressure air stream. Adapted from.[17]

to determine the changes in active sites on the catalyst surface. The surface areas of the catalysts were unchanged or only slightly reduced (<10%) after processing and there was no change in the surface morphology as evidenced by SEM nor was there any deposition of carbon on the catalysts. This indicates that their stability during plasma-activated catalysis is superior to that of thermal catalysis where, for example, in the catalytic oxidation of DCM on $\gamma$-$Al_2O_3$ at 600 K it is found that activity is lost by carbon deposition.[19] In addition, in thermal catalysis, it is found that the surface becomes chlorinated whereas in plasma catalysis there is only a small uptake of chlorine (<2% by weight).

Figure 6.4 shows the results of ATR-IR spectroscopy for $\gamma$-alumina used in plasma catalysis of DCM in air in both a one- and two-stage configuration. The strong absorbances <1,000 $cm^{-1}$ are associated with Al-O vibrations of solid alumina and are significantly reduced by plasma processing, suggesting a high degree of modification of the alumina. This spectral region also includes any possible Al-Cl vibrations but their existence cannot be confirmed as they are masked by the alumina absorbances. Thus it is not possible to determine whether the chlorine in the catalyst is in the form of $AlCl_3$ or not. The peaks at around 3000–2900 $cm^{-1}$ which increase upon processing, more for one-stage than for two-stage, correspond to adsorbed DCM and its decomposition products.[19] The peaks at 3,800–3,600 $cm^{-1}$ represent surface hydroxyl and are largely unaffected by the plasma processing. The bands at 1,400, 1,320 and 1,250 $cm^{-1}$ come from C-H stretching and bending vibrations of DCM by-products rather than DCM itself. The peak at ~1,070 $cm^{-1}$ indicates C-O vibrations typical of a methoxy group.

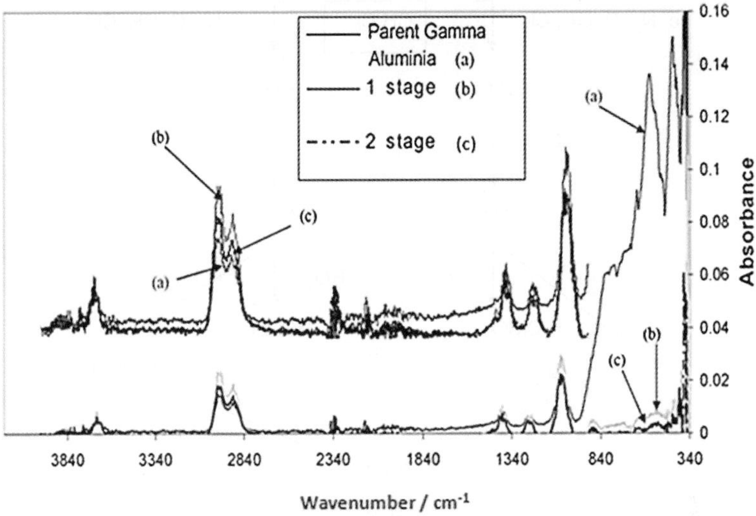


**Figure 6.4.** ATR-FTIR spectrum of the γ-alumina catalyst before use (a); after use in one-stage and two-stage plasma catalysis arrangements, (b) and (c), respectively. The region from 900–4,000 $cm^{-1}$ is also shown in enlarged form. Adapted from.[17]

The products of the plasma-catalysis processing are predominantly CO and phosgene, HCOCl, as well as small amounts of $CO_2$ indicating incomplete oxidation during the plasma-catalytic processing. Plasma decomposition of the air also gives oxides of nitrogen. A simplified mechanism (not including $NO_x$ formation) which derives from the gas-phase plasma reduction but also gives insight into the role of the catalyst is shown in Fig. 6.5. The plasma-produced species are O, $N_2$* and electrons. OH and Cl are produced in subsequent reaction steps but provide a rapid propagation route for the decomposition of DCM and the subsequent abatement chemistry. Key in this mechanism is the role played by chloroperoxy radicals and formaldehyde and phosgene. The alumina surface provides a source of hydroxyl radicals to which the DCM is bound by the displacement of a chlorine as HCl to form a surface-bound chloromethoxy species which can be transformed into phosgene or formaldehyde which can desorb or be converted into CO.[19] The intermediate HCl may then chlorinate the alumina surface or desorb.

Further insight into the mechanism of plasma-assisted catalysis can be gained by studying the temperature dependence of the oxidation process and comparing the decomposition of the DCM as a function of temperature for both plasma catalysis and thermal catalysis[20] The catalyst is incorporated into the barium titanate packed-bed plasma reactor in a one-stage configuration. The reactor can be heated to 400°C with an electric heater. Figure 6.6 shows the temperature variation for the decomposition of DCM with a catalyst.

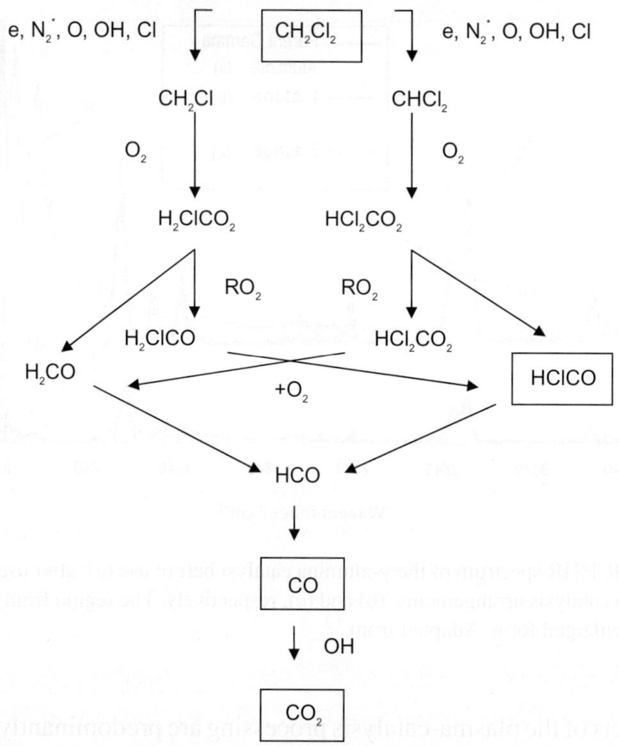


**Figure 6.5.** Simplified mechanism of dichloromethane transformation in a plasma reactor.

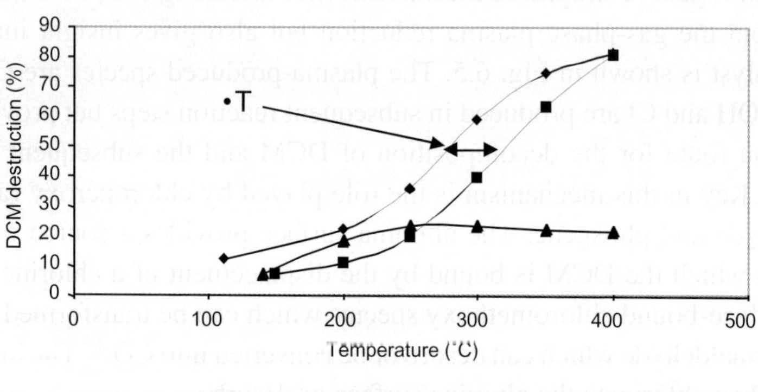


**Figure 6.6.** The percentage destruction of 500 ppm of dichloromethane in atmospheric pressure air studied as a function of heater temperature for plasma alone (▲), plasma catalysis (◆) and thermal catalysis (■). The catalyst is titanium dioxide, $TiO_2$, and the plasma is a barium titanate packed-bed.

Three experiments were undertaken. In the first, only a plasma was used to destroy the DCM. After an initial rise in the percentage destruction, the destruction remained approximately constant at ~20% for temperatures >250°C. For conventional thermal catalysis, there is little destruction for temperatures <200°C when the destruction rises steadily to reach ~80% at 400°C, the maximum possible in the experiment. Plasma catalysis shows a similar form of temperature dependence to that of thermal catalysis but it starts from a higher baseline and is displaced to lower temperatures. This displacement corresponds to about 50°C at 50% destruction of DCM. As the electrical power required to sustain the plasma discharge is very low (1 W) compared with that for the electrical heating (~50 W), this corresponds to a significant reduction in the energy requirement (~34%) achieved by the combination of plasma and catalyst.

## 6.4. Plasma Catalysis for the Abatement of Hydrocarbons

### 6.4.1. *Propane and propene*

Because of the strength of the C-H bond in saturated hydrocarbons, these molecules are difficult to destroy by plasma methods alone. Unsaturated hydrocarbons, by contrast, are much easier to remediate as is demonstrated in Fig. 6.7. The destruction of propene requires significantly less plasma energy density (<100 J per litre of processed gas) to achieve ~70% destruction than is needed to achieve the same destruction of propane (>1,000 J litre$^{-1}$).[21] The chemistry of the removal of both propane and propene in the plasma is initiated by oxygen atoms generated in the discharge by electron impact dissociation of molecular oxygen, but there are many more possible pathways available in the case of propene than there are for propane. These pathways generally involve the formation of OH radicals by H atom abstraction, which are then able to rapidly advance the further destruction of the hydrocarbon. These initial steps are significantly more rapid for atomic oxygen reaction with

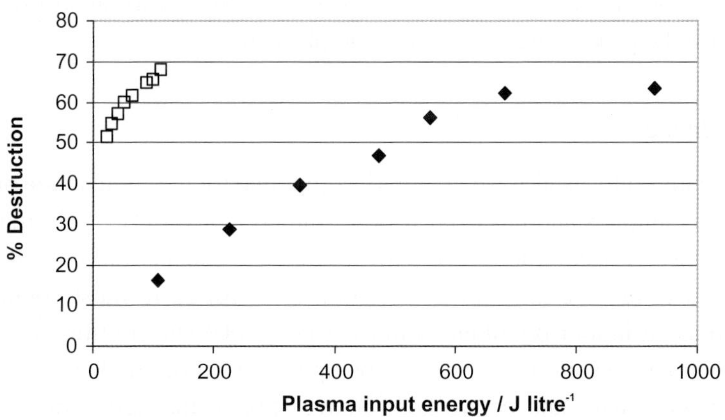


**Figure 6.7.** The percentage destruction of 100 ppm propane (•) and propene (•) in an atmospheric air stream using a surface discharge plasma reactor as a function of plasma energy density. Adapted from.[21]

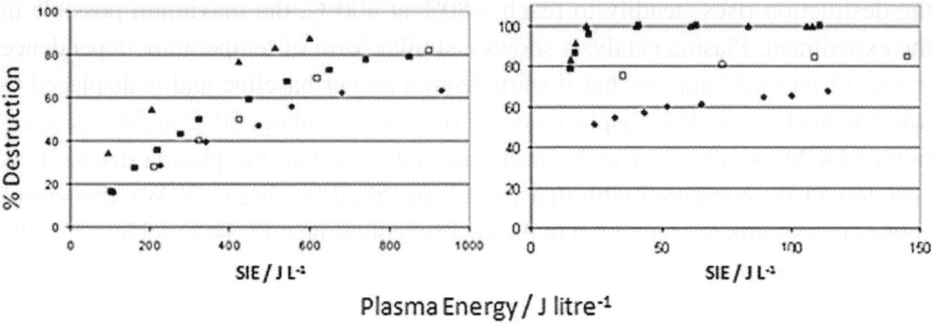


**Figure 6.8.** Effect of catalyst on 100 ppm propane (left hand panel) and propene (right hand panel) destruction in air as a function of plasma energy density. •plasma only (no catalyst), •$TiO_2$, •γ-$Al_2O_3$ and •NaA.

propene than with propane (by about a factor of 100) and this is reflected in the different energy input requirement seen in Fig. 6.7. The subsequent chemistry is also complex and by-product formation of $CH_2O$ and HCOOH is observed for the plasma treatment of propene in addition to CO and $CO_2$, but these by-products are not observed for the destruction of propane.

The use of plasma catalysis can increase the energy efficiency of the abatement process for both propane and propene, but the effect of the addition of a catalyst is modest in the case of propane processing but quite dramatic for propene, as shown in Fig. 6.8, where 100% destruction can be achieved at a plasma energy density of 20 J litre$^{-1}$.[22] The equivalent destruction for propene in the absence of a catalyst at this energy would be only 50%. In both cases, the most effective material was found to be γ-alumina which may result from reactions involving surface OH species as was seen for DCM and also the effectiveness of γ-alumina as an absorbent which will increase the processing time within the plasma.

### 6.4.2. *Toluene*

Toluene is commonly used as a benchmark compound for the comparison of different abatement techniques and has been widely studied in a variety of plasma catalysis configurations. It is a common solvent in the chemical and pharmaceutical industries and its release into the atmosphere is strictly controlled in most countries. It is also found as an indoor pollutant from tobacco smoke and furniture and cleaning products, with a concentration that may be up to 2–10 times higher than that found in outdoor air.[23] Kim and co-workers[24] have studied the destruction of a range of aromatics including toluene at atmospheric pressure using a surface discharge plasma reactor packed with $Ag/TiO_2$ catalyst. They achieved complete destruction of the toluene with a plasma input energy density of 130 J litre$^{-1}$ giving CO and $CO_2$

as the main decomposition products and formic acid, HCOOH, as an intermediate at low energy densities ($<40\,\text{J litre}^{-1}$). There were no other ring-cleavage or ring-retaining products detected indicating that the toluene completely decomposed to small fragments. The selectivity for $CO_2$ formation over CO is $\sim$75% but it was not possible to totally eliminate CO by increasing the plasma energy density. The energy density is an important processing parameter as the amount of $NO_x$ by-product produced will depend on its magnitude. Achieving optimum processing conditions by minimising $NO_x$ while maximising VOC destruction is a key goal that might be achieved by judicious selection of the catalyst.

Metal oxide catalysts are also commonly used in plasma catalysis for toluene abatement. Guo and researchers[25,26] found that a manganese oxide/alumina catalyst in a dielectric barrier discharge (DBD) plasma reactor was the most effective for toluene removal, with an increase by a factor of two in removal compared to the use of plasma alone. They also found that it gave the best conversion of toluene to $CO_2$. Van Durme[23] found that a $CuOMnO_2/TiO_2$ catalyst in a two-stage configuration, downstream from a corona plasma reactor, increased toluene destruction by up to 40 times and it was also effective in reducing $NO_x$ levels. Guo and co-workers[25] proposed a mechanism in which the oxygen on the catalyst surface (MO) would attack the toluene releasing the $CH_3$ group.

$$C_6H_5CH_3 + MO \rightarrow C_6H_5 - MO + CH_3$$

Oxygen atom attack will initiate the surface decomposition of the phenyl radical

$$C_6H_5 - MO + O \rightarrow CO + CO_2 + H_2O + MO$$

The methyl radical can also be catalytically converted to CO and $CO_2$

$$CH_3 - MO \rightarrow CH_2 - MO + H$$

$$CH_2 - MO \rightarrow CO + CO_2 + H_2O + MO$$

Surface analysis shows that the oxidation state of the manganese oxide changes from $Mn_2O_3$ to $Mn_3O_4$ in the discharge and that this form has a better oxidising capacity for toluene.

When we study the temperature dependence for the destruction of toluene in atmospheric pressure air by plasma catalysis using a one-stage, packed-bed plasma system,[27] we see a different type of behaviour to that previously described for DCM. The results are shown in Fig. 6.9. Processing in the absence of a catalyst gives low destruction with essentially no temperature dependence as we saw for DCM, and the use of $TiO_2$ as a catalyst shows a similar dependence as it did with DCM. However, the use of $Ag/\gamma\text{-}Al_2O_3$ and $Ag/TiO_2$ catalysts show very different behavior with temperature. At 257°C, there is 68% toluene destruction using the $Ag/TiO_2$ catalyst, in comparison to only 19% for $TiO_2$ alone. By 400°C, 100% conversion is

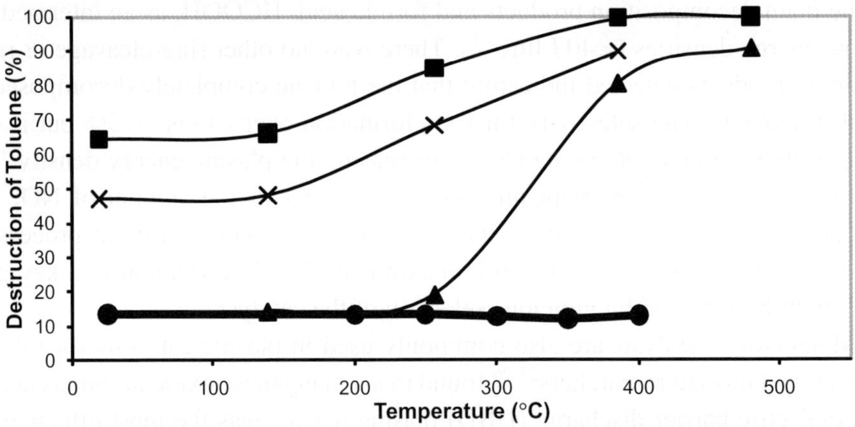


**Figure 6.9.** Comparison of catalyst performance for the plasma-catalytic destruction of 500 ppm of toluene in atmospheric pressure air as a function of temperature, using (■) Ag/$\gamma$-$Al_2O_3$, (▲) $TiO_2$ and (×) Ag/$TiO_2$ catalysts and supports, in a one-stage configuration. The thick black line (•) shows results for plasma alone, with no catalyst. The plasma is a barium titanate packed-bed. Adapted from.[27]

achieved with the plasma-Ag/$TiO_2$ catalyst combination. A similar but less efficient behavior is noted for plasma-catalytic destruction of toluene with the Ag/$\gamma$-$Al_2O_3$ catalyst.

From these data, we can distinguish two different types of behavior exhibited in the temperature dependence for plasma catalysis in a one-stage configuration. The first is highly efficient and only weakly dependent on temperature (e.g. Ag/$TiO_2$) in contrast to the behavior observed in conventional thermal catalysis (e.g. Fig. 6.6), which shows a steadily increasing rapid rate of removal of the pollutant above a well-defined temperature threshold. However, there are also cases where plasma catalysis shows a temperature profile similar to that of thermal catalysis, such as the plasma-catalytic destruction of DCM and toluene using $TiO_2$ (Figs 6.6 and 6.9). It is generally observed, however, that plasma-activated catalysis is more efficient than thermal catalysis at all temperatures until the two curves converge at very high temperatures. These two different temperature profiles for plasma-activated catalysis may indicate that there are two contributing mechanisms for plasma-activated catalysis; the first being virtually temperature independent whereas the second is temperature dependent, with a threshold at a certain temperature in a similar manner to thermal catalysis, but generally shifted to lower temperatures. The first, which is only slightly temperature dependent, reflects the behavior of the plasma alone where activation is caused predominantly by gas-phase chemistry. The production of active species, such as atomic oxygen or ozone, can then adsorb onto the catalyst surface possibly by some plasma-induced activation of the catalyst by electrons or photons created directly by the discharge. Electron-induced processes and production of electrons have only a slight temperature dependence.[28] For the strongly temperature dependent curve, thermal activation of the catalyst becomes more important, overtaking the contribution of the plasma-activated processes. The relative contribution of these two mechanisms determines the observed temperature profile for the destruction.[29]

From Fig. 6.9, we can also identify a synergistic effect of combining plasma with the Ag/$TiO_2$ catalyst by focussing on the data at a low temperature, 25°C, where the plasma alone gives a dissociation of $\sim$12% compared with $<$8% from thermal catalysis. If the two effects were additive this would give a combined destruction of $<$20%. The actual extent of destruction achieved by plasma catalysis at 25°C is 65%; a synergistic gain of more than a factor of three. However, at higher temperatures there is a much less marked difference between the destruction achieved thermally and by plasma catalysis and this synergy disappears. No synergistic effects were observed in our study of the destruction of DCM in air using plasma catalysis at any temperature. Clearly, the common perception of plasma catalysis being a universally synergistic phenomenon is false.

Another way in which information about the mechanism of plasma-assisted catalysis can be derived is to use the temperature dependent data to extract kinetic information via an Arrhenius-type analysis. Destruction of toluene in an atmospheric pressure air stream was studied using thermal catalysis and plasma-assisted catalysis for Ag/$\gamma$-$Al_2O_3$ and $Mn_2O_3$/$\gamma$-$Al_2O_3$in both a one- and two-stage system.[30] The results show that the increase in toluene destruction with temperature that is observed for both catalysts is achieved in very different ways. It was found that there is no difference between the measured activation energy for thermal catalysis and the two-stage (or downstream) plasma catalysis arrangement. This would suggest that there is no activation of the catalysts in the downstream position caused by the migration of long-lived species produced in the plasma. On the other hand, it was found that plasma could activate a catalyst placed inside the discharge (one-stage configuration). Plasma treatment decreases the activation energy ($E_a$) for the silver-alumina catalyst in a one-stage arrangement but does not increase the number of active centres on the surface (as indicated by the Arrhenius $A$ parameter). In the case of $MnO_2$/alumina, the converse is found: plasma activation does not change the activation energy but it does form additional active centres (an increase in $A$). It is suggested that these active centres could be explained by a change in the oxidation state of manganese ions under plasma action as noted by Guo *et al.*[25] Alternatively, the interaction of active molecules (ozone, oxygen atoms, and radicals) with the catalyst surface can also promote electron transfer from the Mn sites.

## 6.5. The Role of Ozone in Plasma Catalysis for VOC Abatement

Ozone is commonly produced in discharges in air. Although its decomposition to oxygen is thermodynamically favoured, it remains stable at ambient temperatures. Ozone in the outlet of a plasma system is extremely undesirable and catalysts are often used to remove it. However, it is well established that ozone on a catalytic surface can induce secondary oxidation processes.[31] Due to its long lifetime, ozone may play an important role in the plasma-catalytic oxidation of VOCs. In a two-stage configuration, ozone may still be present in significant concentrations after the discharge, and oxidation of the VOC may take place with ozone either in the gas-phase or on the surface of a downstream catalyst. In general, the rates of gas-phase reactions of ozone with VOCs are very small, but ozone can be adsorbed and decompose on catalytic surfaces leading to the formation of strongly oxidising species such as atomic oxygen.[32,33] These species can then oxidise any adsorbed VOC on the catalyst surface and enhance the conversion of CO to $CO_2$.

Using a plasma source of ozone (~750 ppm), it is possible to inject ozone into an air stream containing toluene (70 ppm) downstream of a $MnO_2$ catalyst and achieve 100% destruction of the toluene and complete removal of the ozone by the action of the catalyst.[34] The products of the processing were $CO_2$ and CO. Repeating the experiment without the catalyst gave no decomposition of toluene indicating that there is no gas-phase destruction of toluene by ozone. The nature of the catalyst is important in this decomposition, as Ye *et al.* report that mixing ozone with toluene on a ceramic cordierite did not give rise to any measurable toluene decomposition despite removal of the ozone on the ceramic.[35] Similar experiments with a $MnO_2$ catalyst were effective, however.[36] It has been reported that experiments using a $CuO/MnO_2/TiO_2$ catalyst were effective at removing toluene in an air stream when ozone was mixed with the stream before the catalyst, but that in a humid air stream the extent of toluene removal was reduced because of the blocking of active sites on the catalyst by water.[23] It is likely that the injection of plasma-generated ozone into a VOC waste stream before an ozone-decomposing catalyst, such as $MnO_2$, may be a general method for the oxidation of VOCs and warrants further investigation.

## 6.6. Cycled Systems for Plasma Catalytic Remediation

Many of the plasma catalysis schemes described in this chapter are not optimised in a way that is suitable for processing large volumes of waste gas in an efficient manner on a larger scale. Plasma reactors are generally limited to rather low gas flows compared to normal waste gas emissions and are presently best suited to niche applications of small flows with low concentrations of pollutants. A more energy-efficient configuration can be constructed where the pollutant is initially adsorbed

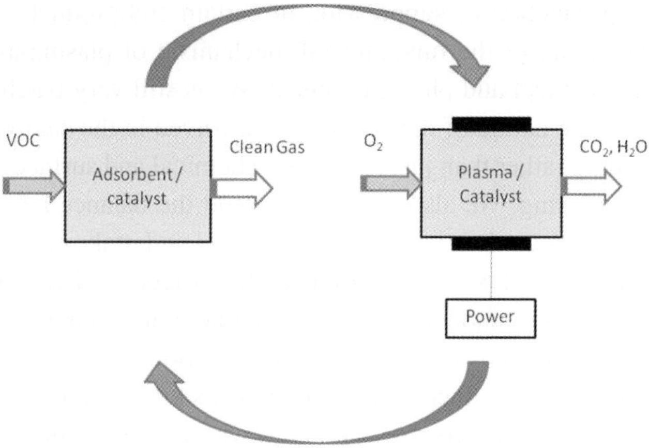


**Figure 6.10.** Schematic diagram of a cycled plasma catalyst/adsorbent system. Adapted from.[37]

onto a suitable material and concentrated for later treatment by plasma, which can then be discharged in contact with the adsorbent material using an oxidising gas. This provides a plasma-catalytic treatment of the adsorbed gas, which can destroy the pollutant VOC with a lower oxidising gas flow for a short period compared to the large gas flows and longer time over which adsorption takes place. A schematic view of the arrangement is shown in Fig. 6.10. The cycle consists of adsorption followed by plasma regeneration of the adsorbed VOC. This can take place while another adsorbent system is in operation.

Kim *et al.* used $TiO_2$, γ-alumina and zeolites to adsorb benzene and toluene with a surface-discharge plasma reactor with oxygen to regenerate the adsorbent.[37] Kuroki *et al.* used a similar arrangement with a honeycomb zeolite as an adsorbent in a dielectric barrier plasma to treat toluene (17 ppm) in air streams. The adsorbent was regenerated in the plasma stage using either $O_2$ or $N_2$ to desorb and treat the VOC.[38] Typically, this type of process can achieve greater energy efficiency and cope with larger gas loads than the single pass plasma catalysis arrangement and can be expected to feature in future experiments and demonstrator projects on a larger scale.

## 6.7. Conclusions

The combination of plasma with catalysis offers an effective solution to the environmental clean-up of pollutant VOCs in waste gas streams. The hybridisation of the two techniques can bring about improved efficiency by increasing destruction of the pollutant and reducing the operating temperature and energy requirement for the processing. Additionally, the use of certain catalysts can give increased selectivity in terms of the production or suppression of certain end-products. However, our detailed understanding of the fundamental mechanism of plasma-assisted catalysis from both a chemical and physical perspective is still very patchy. We need to know more about the nature of the active sites involved in the catalysis using real-time, *in situ* probing rather than just relying on chemical and surface analysis of the catalyst after processing. We also must understand the balance between homogeneous and heterogeneous chemistry taking place and undertake effective theoretical modelling of plasma catalysis to develop realistic surface- and gas-phase chemical mechanisms. We need to understand more about how the catalyst affects the electrical and gaseous dynamics of the discharge. Additionally, real gaseous waste is rarely composed of well-defined single components and we must understand how plasma catalysis works for mixtures of VOCs whose composition may vary with time. Issues of complex interacting chemistry that have been observed in plasma-only systems with $NO_x$ in VOC remediation[39] and in reaction promotion in mixtures of hydrocarbons,[40] will have their analogues in plasma catalysis where there will be competition for active sites on the catalyst surface that will affect reactivity and selectivity. If we can devise experiments that provide answers to these issues, then we may come closer to the goal of being able to design or select catalysts specifically for plasma-assisted catalysis to engineer effective solutions for VOC abatement.

## Acknowledgments

It is a pleasure to acknowledge financial support from the UK Engineering and Physical Sciences Research Council and the contributions from members of the Manchester Plasma Chemistry Group whose work is presented here: Tarryn Blackbeard, Vladimir Demidyuk, Helen Gallon, Alice Harling, Stuart Fischer, Sarah Hill, Claire Ricketts, Xin Tu, Anna Wallis and Kui Zang.

## References

1. Penetrante, B.M. and Schultheis, S.E. (1993). *Non-Thermal Plasma Techniques for Pollution Control*. Parts A & B. Berlin: Springer-Verlag.
2. Fitzsimmons, C., Ismail, F., Whitehead, J.C. and Wilman, J.J. (2000). The Chemistry of Dichloromethane Destruction in Atmospheric-Pressure Gas Streams by a Dielectric Packed-Bed Plasma Reactor. *J. Phys. Chem. A*, **104**, pp. 6032–6038.
3. Ahmad, I.K., Wallis, A.E. and Whitehead, J.C. (2003). The plasma destruction of odorous molecules: Organosulphur compounds. *High Temp. Mater. Proc.*, **7**, pp. 487–499.
4. Yamamoto, T. (1997). VOC Decomposition by nonthermal plasma processing — a new approach. *J. Electrostat.*, **42**, pp. 227–238.
5. Chen, H.L., Lee, H.M., Chen, S.H. and Chang, M.B. (2008). Review of Packed-Bed Plasma Reactor for Ozone Generation and Air Pollution Control. *Ind. Eng. Chem. Res.*, **47**, pp. 2122–2130.

6. Kim, H.-H., Ogata, A. and Futamura, S. (2006). Applications of Plasma-Catalyst Hybrid Processes for the Control of $NO_X$ and Volatile Organic Compounds, in *Trends in Catalysis Research*, Bevy, L.B. (ed.) Nova Science Publishers Inc.: New York. pp. 1–50.
7. Van Durme, J., Dewulf, J., Leys, C. and Van Langenhove, H. (2008). Combining non-thermal plasma with heterogeneous catalysis in waste gas treatment: A review. *Appl. Catal. B Environ*, **78**, pp. 324–333.
8. Chen, H.L., Lee, H.M., Chen, S.H., Chang, M.B., Yu, S.J. and Li, S.N. (2009). Removal of Volatile Organic Compounds by Single-Stage and Two-Stage Plasma Catalysis Systems: A Review of the Performance Enhancement Mechanisms, Current Status, and Suitable Applications. *Environ. Sci. Technol.*, **43**, pp. 2216–2227.
9. Whitehead, J.C. (2010). Plasma catalysis: A solution for environmental problems. *Pure Appl. Chem.*, **82**, pp. 1329–1336.
10. Gallon, H.J., Tu, X., Twigg, M.V. and Whitehead, J.C. (2011). Plasma-Assisted Methane Reduction of a NiO Catalyst — Low Temperature Activation of Methane and Formation of Carbon Nanofibres. *Appl. Catal. B Environ.*, **106**, pp. 616–620.
11. Kim, S., Jung, Y. and Park, S.J. (2008). Preparation and electrochemical behaviors of platinum nanocluster catalysts deposited on plasma-treated carbon nanotube supports. *Colloid. Surface A*, **313**, pp. 189–192.
12. Ronkkonen, H., Klemkaite, K., Khinsky, A., Baltusnikas, A., Simell, P., Reinikainen, M., Krause, O. and Niemela, M. (2011). Thermal plasma-sprayed nickel catalysts in the clean-up of biomass gasification gas. *Fuel*, **90**, pp. 1076–1089.
13. Tu, X., Gallon, H.J., Twigg, M.V., Gorry, P.A. and Whitehead, J.C. (2011). Dry reforming of methane over a Ni/Al(2)O(3) catalyst in a coaxial dielectric barrier discharge reactor. *J. Phys. D: Appl. Phys.*, **44**, pp. 274007/1 – 274007/10.
14. Ricketts, C.L., Wallis, A.E., Whitehead, J.C. and Zhang, K. (2004). A mechanism for the destruction of CFC-12 in a nonthermal, atmospheric pressure plasma. *J. Phys. Chem. A*, **108**, pp. 8341–8345.
15. Wallis, A.E., Whitehead, J.C. and Zhang, K. (2007). Plasma -Assisted Catalysis for the Destruction of CFC-12 in an atmospheric pressure gas streams using $TiO_2$. *Catal. Lett.*, **113**, pp. 29–31.
16. Chen, C., Lei, P., Ji, H., Ma, W. and Zhao, J. (2004). Photocatalysis by titanium dioxide and polyoxometalate/TiO2 cocatalysts. Intermediates and mechanistic study. *Environ. Sci. Technol.*, **38**, pp. 329–337.
17. Wallis, A.E., Whitehead, J.C. and Zhang, K. (2007). The Removal of DCM from Atmospheric Pressure Air Streams using Plasma-assisted Catalysis. *Appl. Catal. B Environ.*, **72**, pp. 282–288.
18. Sun, Q., Zhu, A.M., Yang, X.F., Niu, J.H. and Xu, Y. (2003). Formation of $NO_X$ from N-2 and O-2 in catalyst-pellet filled dielectric barrier discharges at atmospheric pressure. *Chem. Commun.* pp. 1418–1419.
19. van den Brink, R.W., Mulder, P., Louw, R., Sinquin, G., Petit, C. and Hindermann, J.P. (1998). Catalytic oxidation of dichloromethane on gamma-Al2O3: A combined flow and infrared spectroscopic study. *J. Catal.*, **180**, pp. 153–160.
20. Harling, A.M., Wallis, A.E. and Whitehead, J.C. (2007). The effect of temperature on the removal of DCM using non-thermal, atmospheric-pressure plasma-assisted catalysis. *Plasma Proc. Polym.*, **4**, pp. 463–470.
21. Hill, S.L., Kim, H.H., Futamura, S. and Whitehead, J.C. (2008). The destruction of atmospheric pressure propane and propene using a surface discharge plasma reactor. *J. Phys. Chem. A*, **112**, pp. 3953–3958.
22. Hill, S.L. (2007). Non-Thermal Plasma Processing of Propane and Propene, Ph.D. thesis, The University of Manchester.
23. Van Durme, J., Dewulf, J., Sysmans, W., Leys, C. and Van Langenhove, H. (2007) Efficient toluene abatement in indoor air by a plasma catalytic hybrid system. *Appl. Catal. B Environ.*, **74**, pp. 161–169.

24. Kim, H.H., Ogata, A. and Futamura, S. (2005). Atmospheric plasma-driven catalysis for the low temperature decomposition of dilute aromatic compounds. *J. Phys. D: Appl. Phys.*, **38**, pp. 1292–1300.
25. Guo, Y.F., Ye, D.Q., Chen, K.F. and He, J.C. (2007). Toluene removal by a DBD-type plasma combined with metal oxides catalysts supported by nickel foam. *Catal. Today*, **126**, pp. 328–337.
26. Guo, Y.F., Liao, X.B., He, J.H., Ou, W.J. and Ye, D.Q. (2010). Effect of manganese oxide catalyst on the dielectric barrier discharge decomposition of toluene. *Catal. Today*, **153**, pp. 176–183.
27. Harling, A.M., Demidyuk, V., Fischer, S.J. and Whitehead, J.C. (2008). Plasma-catalysis destruction of aromatics for environmental clean-up: Effect of temperature and configuration. *Appl. Catal., B*, **82**, pp. 180–189.
28. Harling, A.M., Kim, H.H., Futamura, S. and Whitehead, J.C. (2007). Temperature dependence of plasma-catalysis using a nonthermal, atmospheric pressure packed bed; the destruction of benzene and toluene. *J. Phys. Chem. C*, **111**, pp. 5090–5095.
29. Blackbeard, T., Demidyuk, V., Hill, S.L. and Whitehead, J.C. (2009). The effect of temperature on the plasma-catalytic destruction of propane and propene: a comparison with thermal catalysis. *Plasma Chem. Plasma P.*, **29**, pp. 411–419.
30. Demidyuk, V. and Whitehead, J.C. (2007). Influence of temperature on gas-phase toluene decomposition in plasma-catalytic system. *Plasma Chem. Plasma P.*, **27**, pp. 85–94.
31. Einaga, H. and Futamura, S. (2004). Catalytic oxidation of benzene with ozone over alumina-supported manganese oxides. *J. Catal.*, **227**, pp. 304–312.
32. Roland, U., Holzer, F. and Kopinke, F.D. (2002). Improved oxidation of air pollutants in a non-thermal plasma. *Catal. Today*, **73**, pp. 315–323.
33. Subrahmanyam, C., Renken, A. and Kiwi-Minsker, L. (2007). Novel catalytic dielectric barrier discharge reactor for gas-phase abatement of isopropanol. *Plasma Chem. Plasma P.*, **27**, pp. 13–22.
34. Harling, A.M., Glover, D.J., Whitehead, J.C. and Zhang, K. (2009). The role of ozone in the plasma-catalytic destruction of environmental pollutants. *Appl. Catal. B Environ.*, **90**, pp. 157–161.
35. Ye, D.Q., Huang, H.B., Chen, W.L. and Zeng, R.H. (2008). Catalytic decomposition of toluene using various dielectric barrier discharge reactors. *Plasma Sci. Technol.*, **10**, pp. 89–93.
36. Huang, H.B., Ye, D.Q. and Guan, X.J. (2008). The simultaneous catalytic removal of VOCs and $O_3$ in a post-plasma. *Catal. Today*, **139**, pp. 43–48.
37. Kim, H.H., Ogata, A. and Futamura, S. (2008). Oxygen partial pressure-dependent behavior of various catalysts for the total oxidation of VOCs using cycled system of adsorption and oxygen plasma. *Appl. Catal. B Environ.*, **79**, pp. 356–367.
38. Kuroki, T., Fujioka, T., Okubo, M. and Yamamoto, T. (2007). Toluene concentration using honeycomb nonthermal plasma desorption. *Thin Solid Films*, **515**, pp. 4272–4277.
39. Harling, A.M., Whitehead, J.C. and Zhang, K. (2005). $NO_X$ formation in the plasma treatment of halomethanes. *J. Phys. Chem. A*, **109**, pp. 11255–11260.
40. Demidyuk, V., Hill, S.L. and Whitehead, J.C. (2008). Enhancement of the Destruction of Propane in a Low-Temperature Plasma by the Addition of Unsaturated Hydrocarbons: Experiment and Modeling. *J. Phys. Chem. A*, **112**, pp. 7862–7867.

# Chapter 7

# Catalytic Abatement of Volatile Organic Compounds: Some Industrial Applications

Pascaline TRAN,* James M. CHEN* and Robert J. FARRAUTO*,†

The objective of this chapter is to illustrate the point of view of industry with some applications in volatile organic compound (VOC) abatement: (i) vent gases from purified terephthalic acid (PTA) plants, (ii) nitrogen-containing VOC (acrylonitrile), (iii) regenerative oxidation catalyst technology.

## 7.1. Introduction

Volatile organic compounds (VOCs) is a general term designating organic molecules, primarily hydrocarbons, but also oxygenates, aromatics and organic heteroatom materials (i.e. halide-containing). They primarily originate from uncontrolled chemical and combustion processes such as power plants, food processors, pharmaceutical operations, etc. The structure and composition of the VOC determines the degree to which they photo-chemically react with oxides of nitrogen ($NO_x$) forming hazardous ozone, also referred to as smog, in a series of complicated atmospheric reactions. Consequently they are regulated by local and federal laws. They are also referred to as reactive organic compounds (ROCs). Their approximate reactivity as a function of molecular type is indicated below.

oxygenates > aromatics > olefins > saturated molecules (large > small)

Methane is excluded from current US regulations due to its low photo-chemical reactivity. However, this must be reconsidered for future regulations given its contribution to greenhouse effects.

In general, many VOC generating processes operate in lean (excess air) environments and thus abatement involves oxidation. When large amounts are present it is sometimes economical to recover them for recycling since many are used as solvents. For chemical plants such as painting or coating operations, alcohols, paraffinic and

*BASF Corporation, 25 Middlesex-Essex Turnpike, Iselin, New Jersey.
†Earth and Environmental Engineering Department, Columbia University in the City of New York.

aromatic solvents are commonly used to disperse pigments. After spraying operations they are evaporated but cannot be emitted into the atmosphere due to local regulations. Thermal incineration can be used, however it requires high temperatures (must be in the flammable regime) and often excessive fuel costs to maintain a temperature sufficiently high to thermally combust the VOCs.[1] Further, more undesirable by-products can often form such as dioxins. Catalysts reduce activation energies by changing the rate-limiting step to a less energy intensive one thereby permitting reactions to occur at lower temperatures.[2] Furthermore, catalytic processes can operate outside (below) the flammability range allowing small concentrations of pollutants to be oxidized into harmless products. Consequently catalytic oxidation is a method of choice for many VOC abatement applications.

The catalytic components, either precious or base metals, are selected based on their specific activity for abating the molecules of interest. Catalytic components are highly dispersed (high catalytic surfaces areas) onto carriers (also referred to as the support) to maximize the number of sites upon which the reactants can interact (chemisorb) to facilitate the catalytic process. The greater the number of available sites the greater the catalytic kinetics and the rate of reaction. Also, the carrier which disperses the metals must be resistant to certain contaminants in the exhaust, such as $SO_x$, to avoid deactivation. The catalytic components are impregnated from solutions onto a high surface area particulate material (sphere, extrudate, tablet) or a washcoat (catalytic component deposited on a high surface area powdered carrier) deposited onto the walls of a ceramic or metal monolith, as shown in Fig. 7.1. The monolith offers many advantages relative to particulates such as low pressure drop, flexibility in orientation and absence of attrition. For these and other reasons monoliths are exclusively used in catalytic converters for the automobile.[3]

In some VOC applications, particulate catalysts are more suitable than monoliths. The basic hydrocarbon VOC abatement reaction is

$$C_xH_y + (x + y/2)O_2 \rightarrow xCO_2 + yH_2O \tag{7.1}$$

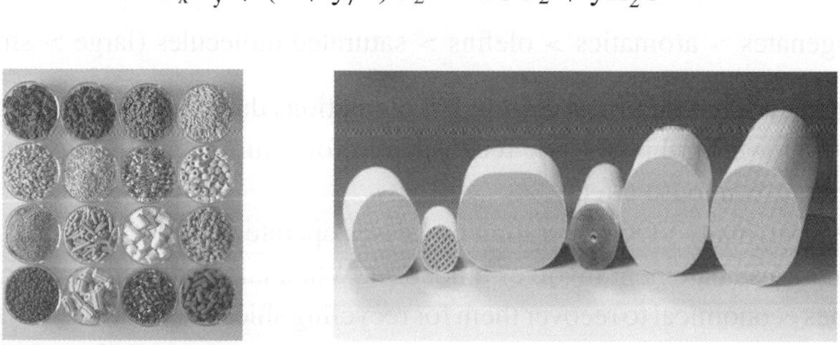

(a) (b)

**Figure 7.1.** Particulate and monolith structures. (a) Extrudates, tablets and spheres. (b) Structured reactors (monoliths or honeycombs).

CO, although not a VOC, often accompanies the VOC and thus it too is oxidized

$$CO + {}^1\!/_2\, O_2 \rightarrow CO_2 \tag{7.2}$$

For hetero-atom VOCs such as halide compounds (X)

$$CH_3X + O_2 \rightarrow 3CO_2 + 4H_2O + HX + X_2 \tag{7.3}$$

For perfluoro compounds

$$C_2F_6 + 3H_2O \rightarrow CO + CO_2 + 6HF \tag{7.4}$$

For oxygenates

$$C_xH_yO + (x + 1/4y - 1/2)O_2 \rightarrow xCO_2 + y/2H_2O \tag{7.5}$$

The abatement system is usually designed for maximum heat recovery, as shown in the diagram in Fig. 7.2. The igniter is needed only to pre-heat the pollutant-laden inlet gas to initiate the oxidation reaction. If the inlet gas temperature becomes sufficiently higher than that necessary for "light-off" no external heat source is needed to sustain catalytic oxidation. The catalyst shown is a precious metal(s) deposited on a high surface area carrier such as $\gamma$-$Al_2O_3$ washcoated onto the walls of the monolith structure.

Monoliths are almost always washcoated with precious metal (Pt and or Pd) due to their high activity density relative to less active base metals such as oxides of Cu, Mn, Co, Cr, etc. The inlet air containing pollutants is pre-heated through the exit heat exchanger. If the exotherm for the combustion reaction is sufficiently large a second heat exchanger, in series, may be used to export it for other uses such as the generation of process heat.

This chapter will focus on three examples of industrial systems used to abate emissions from various VOC sources. Precise details on the composition and nature

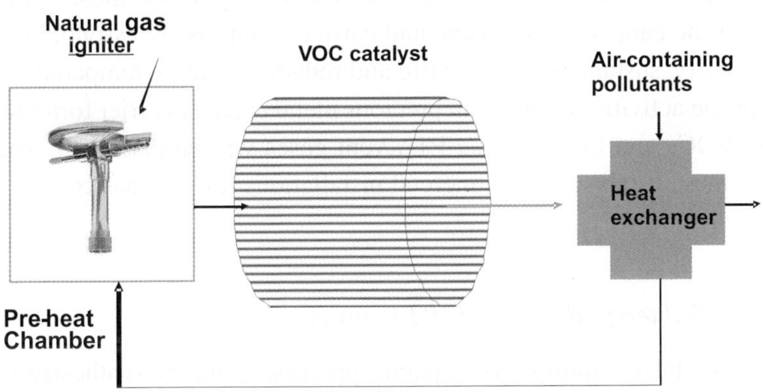


**Figure 7.2.** Catalytic monolith abatement system with heat recovery.

of the catalyst will not be given due to company confidentiality requirements. System designs, catalyst deactivation examples and regeneration methods[1] are available. A recent general review of precious metal catalyst candidates used for VOC abatement is also available.[4]

## 7.2. Case #1: Catalytic Oxidation of Purified Terephthalic Acid

### 7.2.1. *Introduction*

Terephthalic acid (TA) produced in a purified form (PTA) is used almost exclusively in the manufacture of polyethylene terephthalate (PET polyester) fibers. A smaller percentage of PTA is used for the manufacture of polyester films, polybutylene terephthalate resins and barrier resins for carbonated beverage bottles. Between 1994 and 1998, global PTA capacity increased by 62%, from 10 million to 17 million metric tons. Approximately 70% of worldwide PTA capacity is located in the Asia-Pacific region.[5]

The vent gases from purified terephthalic acid (PTA) plants contain CO, methyl bromide (MeBr) and various VOCs from the partial oxidation reaction of *p*-Xylene to terephthalic acid. Catalytic oxidation has been accepted as the most effective technology to control these emissions. Various catalytic components and carriers (or supports) were investigated in order to develop catalysts capable of controlling emissions at lower temperatures to reduce energy consumption.

Precious metal catalysts such as Pt and Pd along with proprietary promoters were deposited on high surface area carriers washcoated onto the walls of a ceramic monolith. These generic formulations have been shown to be the most active, stable and economical and are therefore widely used commercially for these applications. The key issue is the durability or life of the catalyst due to potential catalyst poisoning effects as a result of the presence of bromine-containing compounds. Optimal combinations of the catalytic component and carrier materials are essential to enhance catalyst activity, sustain the catalyst life and reduce operating temperatures. In this case study, the activities of different precious metal catalyst/carrier formulations for removing VOCs and CO from the PTA vent gases will be discussed. The performances of these catalysts in commercial installations for vent gas control will also be discussed.

### 7.2.2. *Purified terephthalic acid (PTA) process*

Production of PTA requires two separate processes, one to synthesize crude TA and one to purify crude TA. The predominant process for crude TA manufacture

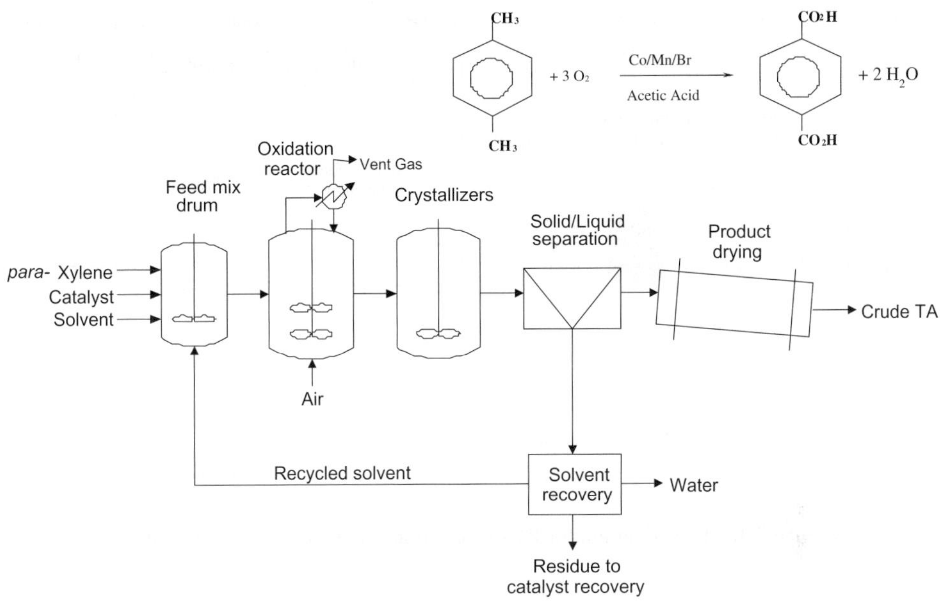


**Figure 7.3.** Flow diagram for crude TA production using the Amoco process.

is the liquid-phase air oxidation of *p*-xylene developed by Amoco, as shown in Fig. 7.3.[6] Fresh and recycled acetic acid, *p*-xylene and a catalyst (normally manganese or cobalt acetate and sodium bromide) are continuously fed to the reactor operating at 175–230°C and 220–435 psi. Air is added in greater than stoichiometric amounts to facilitate partial oxidation of xylene to TA. After oxidation, the conversion of *p*-xylene exceeds 95%, and the yield of TA is at least 90%.[7] The crude TA, which includes impurities such as toluic acid and carboxybenzaldehyde from the partial oxidation of *p*-Xylene, is then sent to the purification stage (see Fig. 7.4) to make PTA that is suitable for polyester production. This is accomplished via the catalytic hydrogenation of carboxybenzaldehyde to the more water-soluble *p*-toluic acid which remains in solution upon the subsequent recrystallization of TA.

Air emissions from PTA production mostly come from the reactor vent gas of the TA production stage. There are no VOC emissions associated with the purification of TA. After passing through aqueous adsorbers, the reactor vent gas still contains the VOCs of *p*-xylene, acetic acid, methyl acetate, toluene, benzene, methyl bromide and CO. With increasingly stringent environmental regulations combined with the expansion of PTA capacity, more and more PTA plants are required to further reduce air emissions from the adsorber vent gas.

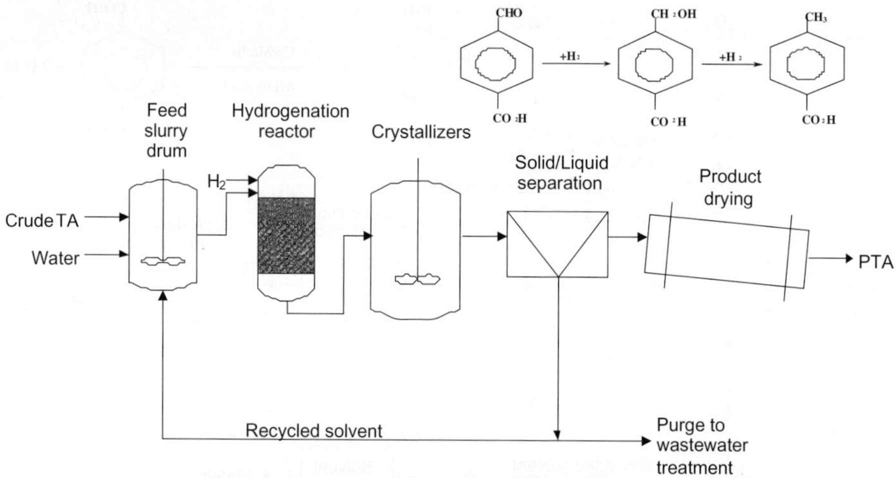


**Figure 7.4.** Flow diagram for PTA production using the Amoco purification.

### 7.2.3. *Emission control technologies of vent gases from PTA purification processes*

The use of carbon adsorption, which can achieve 97% VOC but has no effect on CO emission, may be inadequate for most plants. Thermal oxidation can reduce VOC and CO levels in the vent gases by 99% and more. However, considerable supplemental fuel may be required to achieve the necessary operating temperature (about >800°C).[6] Catalytic oxidation, which can also simultaneously control VOC and CO, but at lower temperatures (<500°C), offers a unique benefit for controlling these vent gases with significant energy savings. A catalytic oxidizer (CATOX) can also easily be retrofitted as an integral part of the production process. It can be placed before pressure let-down (operating under pressure) or after pressure let-down (operating under ambient pressure) of the reactor vent.

Figure 7.5 shows a simple flow scheme of a catalytic oxidizer. The reactor vent gas is first pre-heated by passing through a recuperative heat exchanger to absorb heat from the outlet gas stream. The feed gas is then heated to a design inlet temperature by a gas burner before entering the catalyst. Across the catalyst, VOC and CO emissions are oxidized to $CO_2$ and $H_2O$ while MeBr is converted to $CO_2$, water and $HBr/Br_2$. The clean gas passes through the heat exchanger for heat recovery before being vented.

In the early 1990s, Engelhard (now BASF) commercialized a $Pt/Al_2O_3$ catalyst, VOCat 300H, for this application. The catalyst successfully achieved >98% removal efficiencies for all organics at a PTA installation operating at 375°C inlet with 420°C outlet (Fig. 7.6). The catalyst has been in operation in excess of five years. With

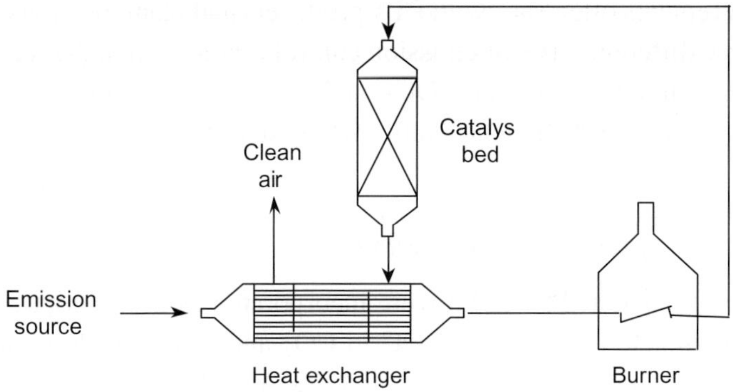


**Figure 7.5.** Flow scheme of a typical catalytic oxidizer.

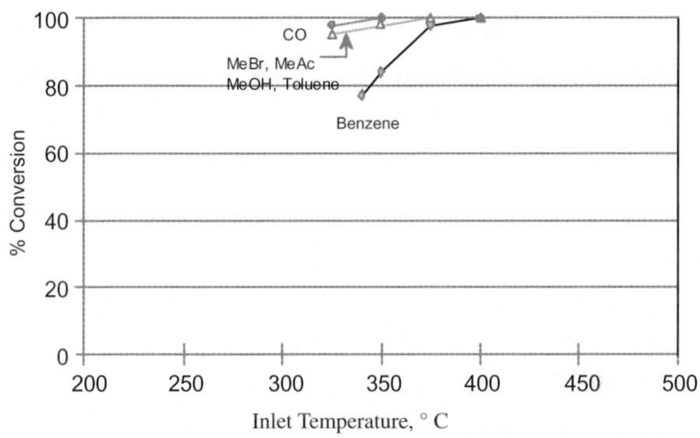


**Figure 7.6.** VOCat300H achieves 98%+ CO and VOC conversions at 375°C+.

successful experiences in over 25 commercial installations, catalytic oxidation has been widely accepted as the best control technology for PTA vent gas treatment.

The energy requirement to operate CATOX depends on the operating temperature, which depends on the catalytic activity and VOC conversion efficiencies required. Reducing the fuel requirement is a major incentive for developing catalysts that can operate at lower temperatures. The catalyst development work and performance of improved commercial catalysts for different vent gas control requirements are presented.

The vent gas flow rate for a PTA unit ranges from 20,000 to 80,000 standard cubic feet per minute (SCFM). Typical PTA vent gas composition is 3,000 to 7,000 ppm CO, 5 to 50 ppm MeBr and 500 to 1,000 ppm of other VOCs, 500 ppm $CH_4$, 3 to 5% $O_2$, 2% $H_2O$ and balanced $N_2$. The VOCs are xylene, toluene, acetic acid, methyl and ethyl acetates, methanol, and benzene (5 to 50 ppm). Depending on the CO and VOC concentrations, the heat release across the oxidation catalyst is 20–80°C. Emission control requirements often vary with PTA producers and plant locations. In general, there are three different types of emission control requirements: the removal of CO only; the removal of CO and total VOCs; and the removal of CO, VOCs and MeBr. The required destruction efficiency varies from 90 to 98%.

### 7.2.3.1. *Catalyst preparation and testing*

Precious metals such as Pt and Pd were incorporated as key catalytic ingredients to promote complete oxidation of VOC to $CO_2$ and water, while minimizing the formation of partially oxidized products and CO. A number of catalyst formulations were prepared by having Pt, Pd and other additives supported on various oxides such as Al, Ce, Mn, Ti, Zr, Si, La, W, etc., and their mixtures. These supports were then washcoated onto ceramic cordierite monolith (honeycomb) substrates. Commercially, ceramic monoliths having a cell density of 200–400 cells per square inch (CPSI) are often used. Since monolith configurations have lower pressure drop and higher geometric area than extrudates, a catalytic oxidizer using honeycomb catalysts can be constructed with a smaller reactor cross section.

Catalytic performance was determined by measuring the conversions of feed VOCs, methyl bromide and CO across a catalyst sample. These tests were conducted in a fixed-bed reactor. Two test units were designed to measure the performances for high pressure and ambient pressure applications, respectively. For the low pressure unit, a catalyst sample (1″D × 3″L) was placed inside a one-inch I.D. quartz reactor residing in a vertical furnace (Fig. 7.7).

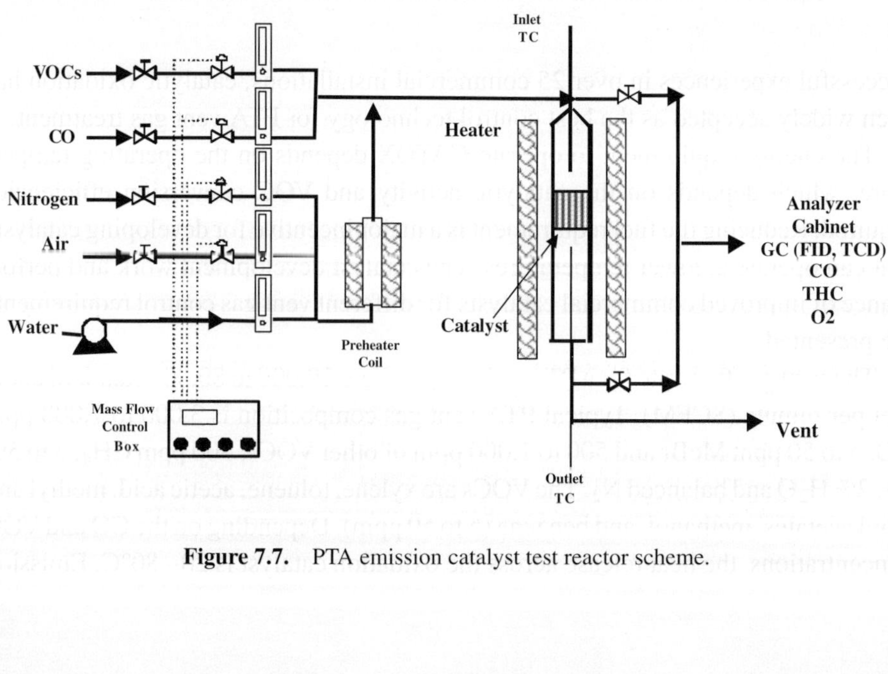


**Figure 7.7.** PTA emission catalyst test reactor scheme.

Different feed gases, such as CO, methyl bromide, $O_2$, toluene, xylene, acetic acid, benzene and balanced $N_2$, were supplied by feed cylinders. Water was introduced by a water pump. Flow rates of the feed streams were regulated by flow controllers. The flows of $O_2$, $N_2$ and water were pre-heated and combined with VOC streams before entering the reactor. Thermocouples were located ½″ above and below the catalyst sample to record the reaction temperatures. Gas samples taken immediately before and after the catalyst were sent to a GC equipped with a thermal conductivity detector (TCD) and a flame ionization detector (FID) to measure the concentrations of CO and various hydrocarbon compounds. The test unit was also equipped with continuous $O_2$, CO and total hydrocarbon analyzers. A schematic of the test unit is shown in Fig. 7.7. For the high pressure unit, the tests were conducted up to 150 psig total pressure. The baseline feed composition was 7,000 ppm CO, 50 ppm toluene, 50 ppm benzene, 50 ppm methyl bromide, 3% $O_2$, 2% $H_2O$ and balanced $N_2$. The catalyst temperature varied from 150–450°C.

### 7.2.3.2. *Catalyst parameters*

The catalytic behavior under PTA vent gas environments is quite different from many other conventional VOC control operations. Figures 7.8 and 7.9 show the conversion responses of a $Pt/Al_2O_3$ catalyst for CO, toluene and methyl bromide. When these gases were present individually, these compounds were oxidized at relatively low temperatures (less than 320°C), as shown in Fig. 7.8. However, when they were present together, as in PTA vent gas environments, the oxidation rates for all compounds were substantially suppressed and the catalyst was required to operate at significantly higher temperatures (e.g. > 370°C) in order to achieve high conversions (see Fig. 7.9).

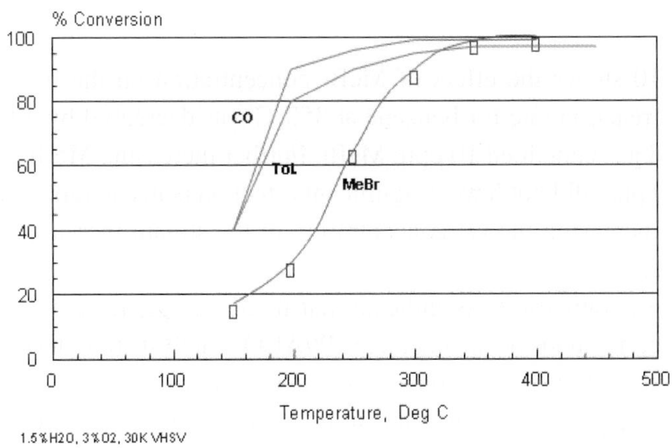


**Figure 7.8.** Performances of $Pt/Al_2O_3$ when CO, VOC and MeBr are present individually.

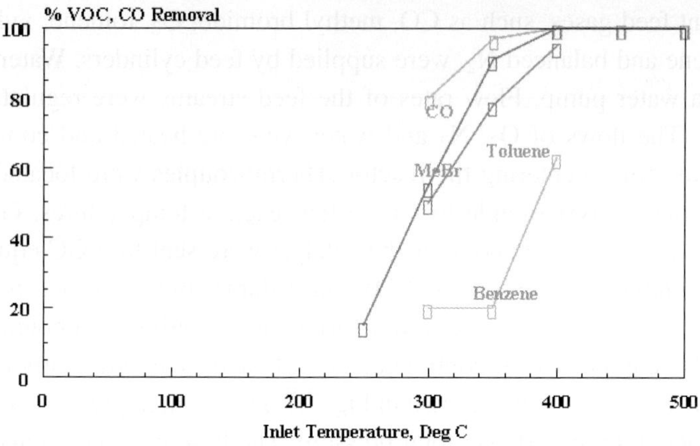


**Figure 7.9.** Performances of $Pt/Al_2O_3$ when CO, VOC and MeBr are present together.

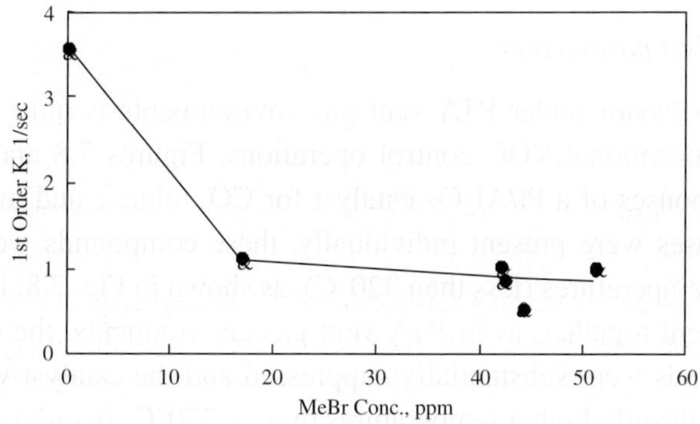


**Figure 7.10.** Effect of methyl bromide concentration on benzene removal rate constant K for $Pt/Al_2O_3$.

Figure 7.10 shows the effect of MeBr concentration on the benzene removal kinetics. The reaction rate for benzene at 350°C was decreased by a factor of four when the feed gas contained 10 ppm MeBr. Further increasing MeBr concentration from 10 to 50 ppm did not have a significant effect on benzene removal. It was also verified that increasing the $O_2$ concentration in the stream increased the catalytic oxidation rate.

The above parametric tests indicate that there was a strong mutual inhibition among these compounds occurring on the $Pt/Al_2O_3$ catalyst. To reduce the oxidation temperature, the mutual inhibition effect may need to be reduced.

Different catalyst support materials were investigated to determine if such inhibition could be affected by support properties. Table 7.1 compares the support surface area and surface acidity of two experimental catalysts. Both catalysts have high

**Table 7.1.** Comparison of catalyst properties.

| Catalyst Ingredient | Pt | Pt |
|---|---|---|
| Support Type | A (High Acidity) | B (Low Acidity) |
| BET Surface Area, $m^2/g$ | 150 | 208 |
| Acidity (TPD), mmoles $NH_3/g$ | 0.07 | 0.02 |

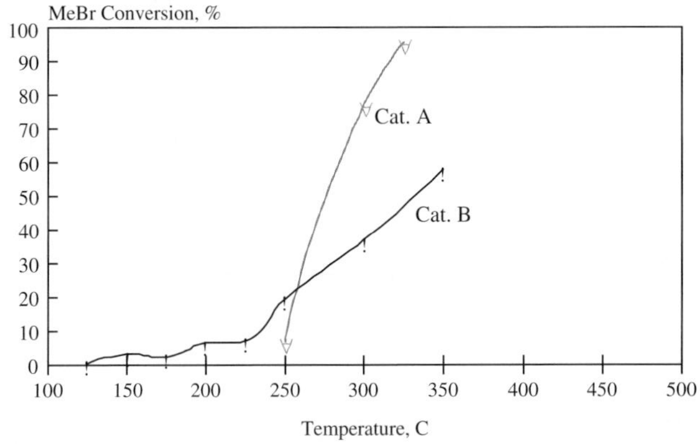


**Figure 7.11.** Comparison of MeBr oxidation activity of Pt catalysts with different support properties.

surface for high Pt dispersion. Catalyst A is an Al-based support while catalyst B is a Si-based support. Catalyst B has lower surface acidity than catalyst A, as shown from ammonia titration measurements.[8] Figure 7.11 shows that MeBr conversion for catalyst B is substantially lower than for catalyst A.

Figure 7.12 shows the effect of MeBr on CO conversion performance for these two supports which had the same Pt loading. These tests were conducted at 30,000 1/hr SV. It can be seen that for catalyst A, the CO conversion rates were significantly suppressed by the presence of 50 ppm MeBr.

On the other hand, the CO oxidation activity for catalyst B was much higher, and was not greatly affected by the presence of MeBr. Also, unlike the results obtained for the Pt/$Al_2O_3$ catalyst, the CO activity for catalyst B was not affected by varying $O_2$ concentrations (from 1.5 to 10 vol%). From these parametric tests, it becomes quite apparent that the properties of a catalyst support can have a significant impact on the performance of a Pt catalyst.

In addition to the support materials, the effect of precious metal content on the catalyst performance was also studied. Pt loading has a strong impact on the catalyst activity when oxidizing methyl bromide and benzene. For other emissions, such as CO, methyl acetate, xylene, etc., lowering Pt content has little effect on the conversion.

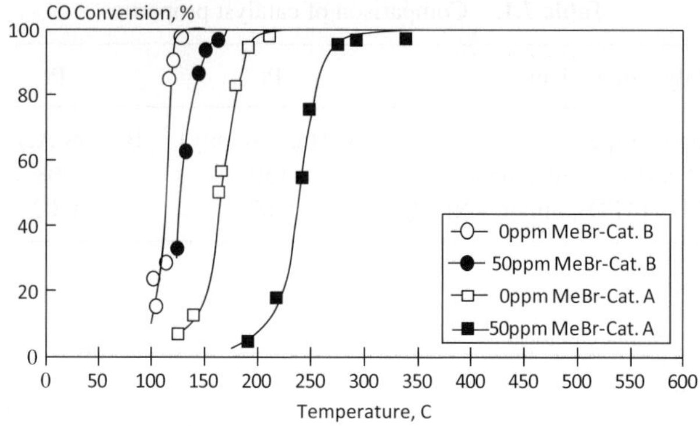


**Figure 7.12.** Impact of MeBr on CO activity for Pt catalysts with different support properties (30000 VHSV, 7000 ppm CO, 1.5% water).

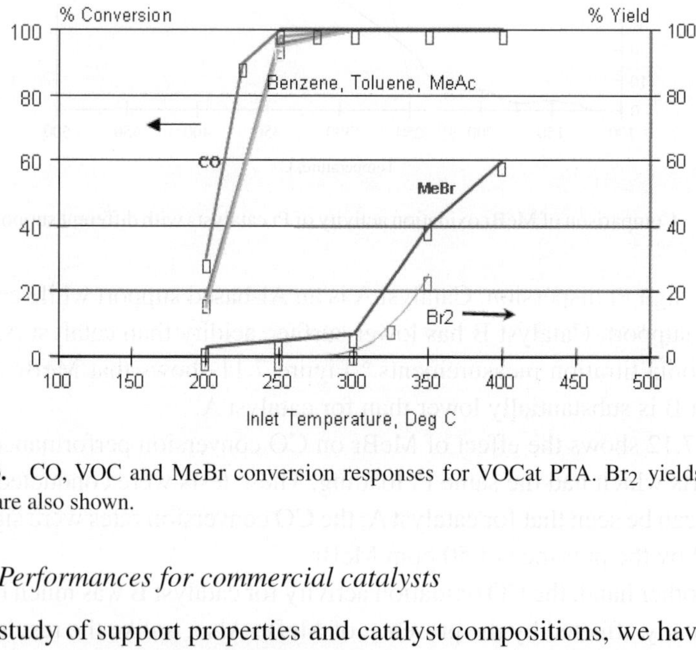


**Figure 7.13.** CO, VOC and MeBr conversion responses for VOCat PTA. $Br_2$ yields from MeBr conversion are also shown.

### 7.2.3.3. *Performances for commercial catalysts*

From the study of support properties and catalyst compositions, we have commercialized two new catalyst formulations for PTA vent gas control. The performances of these two catalysts are shown in Figs. 7.13 and 7.14, respectively. The VOCat PTA catalyst has been developed to selectively abate CO and most VOCs at temperatures as low as 225°C without significantly converting MeBr. VOCat 450H has been developed to remove all these compounds within a temperature range of 250−300°C. As compared with the $Pt/Al_2O_3$ catalyst, the improved formulation of the VOCat 300H can reduce the operating temperature by about 75°C. With this catalyst improvement, a net annual fuel saving of \$100,000 to \$150,000 can be realized for a PTA plant operating at a 50,000 SCFM flow rate.

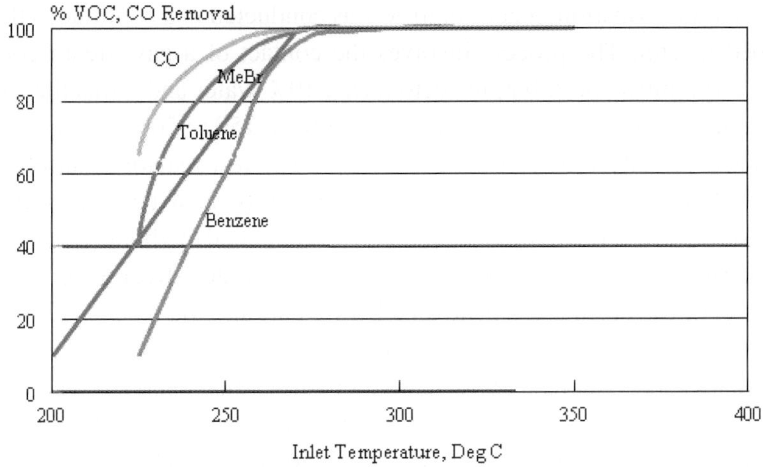


**Figure 7.14.** CO and VOC conversion responses for VOCat450H.

## 7.3. Case #2: Oxidation of Nitrogen-Containing VOCs: Precious Metal Catalysts vs Base Metal Catalysts

### 7.3.1. *Introduction*

The emissions from the manufacture of many plastics, for example, ABS plastics, contain highly hazardous VOCs such as acrylonitrile. Removal of these nitrogen-containing VOCs by thermal oxidation at elevated temperatures often results in the formation of the undesirable nitrogen oxides ($NO_x$). Catalytic oxidation lowers the oxidation temperature and provides a possible means of controlling VOCs without generating $NO_x$. However, the formation of $NO_x$ is greatly influenced by the catalyst type and the temperature regime in which the oxidation reaction will occur. The purpose of this investigation is to study the activity and selectivity of precious metal and base metal catalysts with respect to acrylonitrile oxidation.[9,10,11]

### 7.3.2. *Catalysts and test conditions*

The study includes the use of both honeycomb and pelleted catalysts. The precious metal catalysts (PM catalysts) are platinum (Pt) and platinum/palladium (Pt/Pd), and the base metal catalysts are manganese, manganese copper and chromium. The platinum catalysts consist of one with a high surface area (150 $m^2/g$) and one with a low surface area (25 $m^2/g$). The platinum high surface area catalyst will be referred as the Pt (HS) catalyst in the text and the one with low surface area will be the Pt (LS) catalyst.

The catalytic oxidation of acrylonitrile was conducted in a continuous flow with a fixed-bed reactor. The process involves the contact of a gas stream containing 250 ppm acrylonitrile or 250 ppm acetonitrile, 10% water and air as the oxidizing agent with the catalyst at temperatures ranging from 150 to 500°C. The volumetric hourly space velocity (VHSV) used is 20,000. The test reactor is similar to that used for PTA-VOC conversion (Fig. 7.7). The unit consists of a one-inch I.D. quartz reactor residing in a vertical furnace. The VOC is contained in a bomb that is placed in a controlled temperature bath. The VOC and water saturated streams were combined with the rest of the air in a pre-heat zone before entering the reactor. Flow rates of the feed streams were maintained using MKS mass flow controllers. The gas exiting the reactor was analyzed using a Rosemount hydrocarbon analyzer. The formation of $NO_x$ was monitored with a $NO_x$ analyzer. The performance of the catalyst at different temperatures is measured as % conversion of the VOC.

### 7.3.3. *Activity of precious metal catalysts*

The surface area of the support greatly affected the performances (VOC conversion and $NO_x$ selectivity) of the catalyst. The catalyst in which the precious metal was dispersed onto a high surface area carrier (150 $m^2/g$) showed complete conversion at around 225°C, while the catalyst with a low surface area (25 $m^2/g$) showed complete conversion at 300°C (Fig. 7.15).

Figure 7.16 shows the effect of surface area on the $NO_x$ selectivity for the platinum catalyst (% $NO_x$ formed = 100*ppm $NO_x$/ppm acrylonitrile converted). It can be seen that the low surface material makes a slightly higher amount of $NO_x$ than the high surface area catalyst.

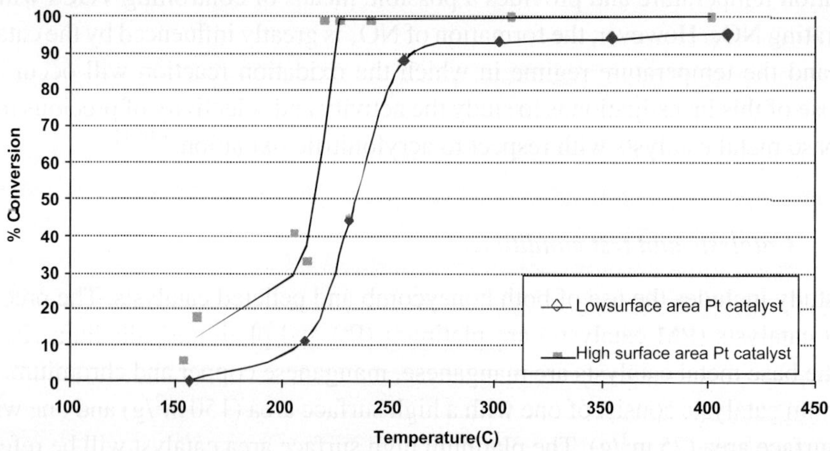


**Figure 7.15.** Effect of surface area of support on Pt catalyst activity for the destruction of acrylonitrile.

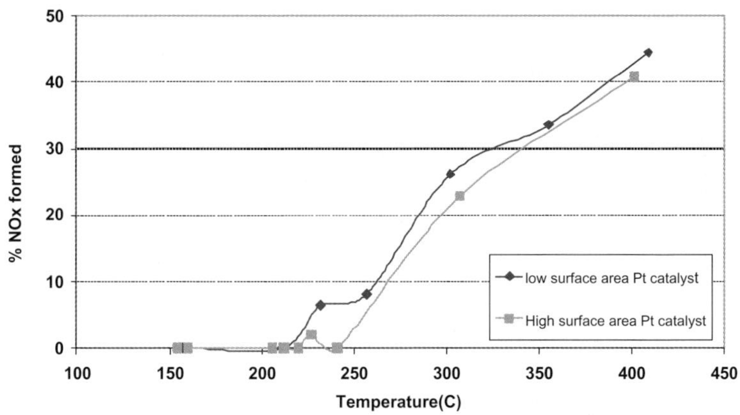


**Figure 7.16.** Effect of surface area on $NO_X$ formation selectivity.

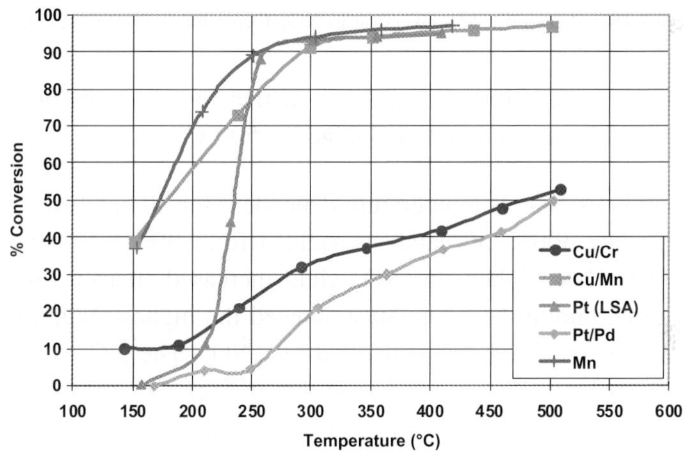


**Figure 7.17.** Acrylonitrile conversion over pellet catalysts (PM and base catalysts).

## 7.3.4. *Precious metal vs base metal (pellet type)*

The performance of a base metal catalyst in pellet form was evaluated with acrylonitrile and results indicated that the manganese-based catalyst was more active than the chromium-based catalyst. Both manganese and manganese doped with copper showed complete conversion of acrylonitrile at around 260°C, while the chromium catalyst showed only about 50% conversion at 500°C, as shown in Fig. 7.17. When the performance of the precious metal catalysts was compared to that of the base metal catalysts it was found that:

(1) Manganese-based catalysts showed the same performance as a Pt (LS) catalyst.
(2) Manganese-catalysts are more active than a Pt/Pd catalyst.

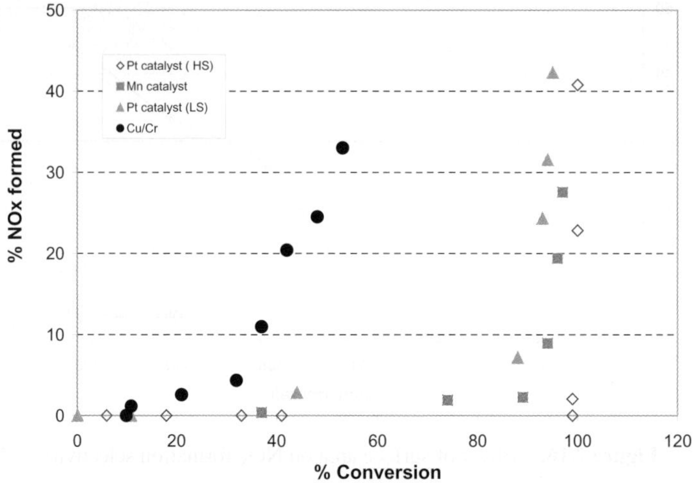


**Figure 7.18.** Acrylonitrile conversion vs $NO_X$ formation for different catalysts.

(3) Chromium-based catalysts are not very active for nitrogen-containing VOC. Only about 60% conversion was obtained at 500°C.

The formation of $NO_x$ is temperature dependent for all catalysts evaluated. At temperatures below 300°C all catalysts yielded from 0% to less than 5% $NO_x$. However, at 300°C or higher, the $NO_x$ yield increased sharply for all catalysts evaluated. The chromium-based catalyst showed the highest $NO_x$ formation. All of the acrylonitrile oxidized was converted into $NO_x$. The $NO_x$ formation vs the percentage of acrylonitrile converted is shown in Fig. 7.18. At 100% conversion only 2.5% $NO_x$ was formed for the Pt (HS) catalyst, while 20% and 25% $NO_x$ were formed for the manganese catalyst and for the Pt (LS) catalyst, respectively. This is due to the fact the Pt (HS) catalyst is much more active than both the Pt (LS) and the manganese catalyst, and therefore can reach complete conversion at a lower temperature, hence the lower $NO_x$ conversion. As the temperature increases it can be seen that the $NO_x$ yield increases rapidly.

## 7.4. Case #3: Regenerative Catalytic Oxidation Catalysts

### 7.4.1. *Introduction*

Regenerative catalytic oxidation (RCO) is a VOC abatement technology that is gaining acceptance in plants where energy costs are high and hours of operation are long. By combining the features of regenerative thermal efficiency and catalytic oxidation, RCO technology provides an efficient and cost-effective solution to air pollution problems in a variety of industries where hours of operation reach 4,000 annually and the challenge is to reduce energy and operating costs, yet comply with increasingly stringent VOC control regulations.[12] These industries may include printed circuit board fabrication (laminate manufacturing), printing (lithography and flexography), coating (cans, coils and fabrics), forest products (production of woods ranging from plywood to medium- and high-density fiberboard), automotive spray-paint booths and component manufacturing (spray-painting of parts, resin components, adhesive components, miscellaneous metalworking applications).

VOC abatement involves collecting the air streams discharged from factory exhaust stacks and processing them through an abatement system. Of the wide range of approaches to controlling the low concentrations of VOCs in industrial air stream emissions, the most frequently and consistently used method is oxidation. In the oxidation reaction, hydrocarbons, or VOCs, are combined with oxygen at elevated temperatures to yield carbon dioxide ($CO_2$) and water vapor, which can then be safely discharged into the atmosphere. Complete oxidation requires time, temperature and turbulence. The primary operating cost for abatement equipment is the energy needed to raise the gas temperature to facilitate the oxidation reaction. Thus, oxidation technologies that are developed to minimize the energy requirement will have low overall operating costs.

The energy cost can be reduced by two different approaches. One approach is to increase the thermal efficiency of the system. By doing so, more of the energy that is added to facilitate the oxidation is retained in the equipment instead of being discharged with the treated air into the atmosphere. Among various heat exchange designs, regenerative heat recovery is considered the most thermally efficient, as it can achieve up to 95% thermal efficiency. Because of its energy saving feature, regenerative thermal oxidation (RTO) has become the most common oxidation technology employed in the United States in recent years. The basic principle of thermal oxidation is to raise the VOC-laden process exhaust to its oxidation temperature, 800°C, thereby converting the VOCs to $CO_2$ and water through incineration.[13]

RTO systems use multiple heat sinks of thermally stable ceramic material to alternately collect and release heat from the process exhaust stream. In a basic system, two chambers of heat sink material are installed adjacent to an oxidation chamber. As the VOC-laden air stream enters the RTO system, it passes through the heat sink material. The heat sink, which is hot, gives its thermal energy to the incoming process air, pre-heating it prior to its entry into the oxidation chamber. As the oxidized air stream leaves the combustion chamber, it passes through another chamber of heat sink material, which is cool. The heat sink absorbs the thermal energy from the air stream, which is subsequently discharged into the atmosphere only moderately warmer than when it arrived. During the thermal transfer cycle in the regenerative heat sink chambers, the inlet heat sink becomes cool and the outlet heat sink becomes hot. At the end of this cycle, air control valves change positions so that once again the incoming air stream passes through the hot heat sink and leaves through the cool heat sink. This cyclical absorption and rejection of thermal energy results in highly efficient thermal energy recovery, as high as 95%. Depending on their configuration, RTOs can have VOC destruction efficiencies as high as 99%. To be effective, a typical RTO system needs to raise the process exhaust to its oxidation temperature, approximately 800°C, and maintain it at that temperature with sufficient oxygen for at least half a second. The VOCs are oxidized to $CO_2$ and water, which can be discharged harmlessly into the atmosphere.

The other approach to reduce the energy cost is to lower the oxidation temperature. This is normally achieved by employing a catalyst to accelerate the oxidation reaction. Instead of 800°C for thermal oxidation, commercial catalytic systems typically operate at temperatures of 300 to 550°C to achieve the required VOC destruction efficiency. To further reduce energy requirements most catalytic abatement systems include a recuperative heat exchanger that enables recovery of up to 65% of the thermal energy from the exhaust gases.

RCO combines both energy saving approaches to achieve the best overall energy efficiency for an oxidation abatement system. RCO systems can provide the high heat recovery efficiency of an RTO system and with special catalyst-coated media installed on top of the heat sink material; RCO systems can also provide VOC destruction at lower temperatures than those associated with catalytic oxidation to further reduce energy requirements.

Figure 7.19 shows a schematic diagram of an RCO system. The basic operation of an RCO system is essentially the same as an RTO unit, with the only difference being lower oxidation temperatures. Thus, essentially all RTO units can be converted to RCO simply by placing a catalyst layer on top of the heat sink material. Also, since oxidation reactions occur at the catalyst rather than in the oxidation chamber, the volume of the oxidation chamber in an RCO unit can be significantly smaller than that in an RTO unit.[14]

Table 7.2 compares the energy cost of an RTO, an RCO and a catalytic oxidation unit. RCO has the lowest operating cost among these technologies. Relative to an RTO unit, the fuel cost for RCO is about 45% lower. Additionally, there are electrical savings of 20% due to a lower pressure drop across the bed. For a 60,000 SCFM unit, a yearly net saving of $75,000 can be realized by converting RTO to RCO.

### 7.4.2. *RCO catalyst properties*

The RCO operating temperature relies on the stability and activity of the RCO catalyst to oxidize VOCs in the process streams. With the Clean Air Act Amendments of 1990, many industries are required to reduce VOC emissions. Since the exhaust

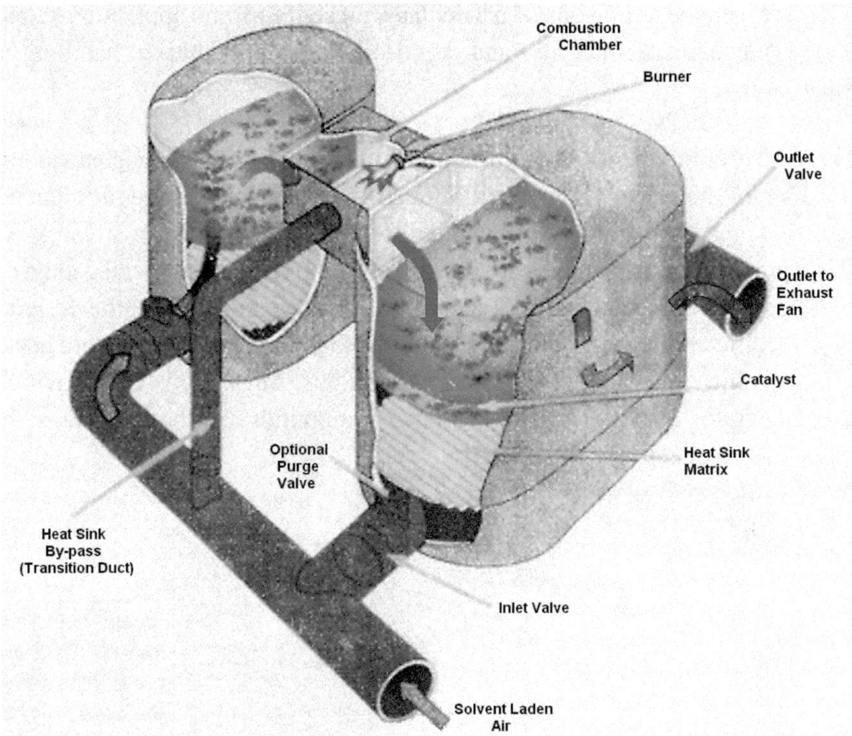


**Figure 7.19.** RCO system schematic.

**Table 7.2.** Fuel operating cost comparison for different oxidizer systems (flow rate: 20,000 SCFM, inlet temperature: 21°C, fuel cost: US \$3.50/MMBTU for natural gas @ 1,000 BTU/$ft^3$).

| Oxidizer System | No VOC | 50 lb/hr VOC |
|---|---|---|
| Thermal Oxidizer | \$186.89/hr | \$182.36/hr |
| Recuperative Thermal Oxidizer (65%)* | \$41.36/hr | \$38.07/hr |
| Regenerative Thermal Oxidizer (95%)* | \$6.51/hr | \$3.82/hr |
| Catalytic Oxidizer | \$77.13/hr | \$73.47/hr |
| Recuperative Catalytic Oxidizer (65%)* | \$23.33/hr | \$20.22/hr |
| Regenerative Catalytic Oxidizer (95%)* | \$3.38/hr | \$0.71/hr |

*Thermal efficiency.

gas stream in each industry contains its own unique emissions, different catalysts may be needed to provide the most cost-effective solutions for different industries. A suitable catalyst formulation must meet the following criteria:

(1) For a typical exhaust stream, 95 to 99% VOC reduction efficiency at 450°C;
(2) Complete oxidation to $CO_2$ and water;

(3) Good resistance to poisons or particulates present in some applications; and
(4) Good mechanical integrity and abrasion resistance during handling and operation.

The active components of the RCO catalyst are chosen dependent upon the composition of the exhaust. Engelhard (now BASF) developed three different commercial RCO catalysts tailored to handle different types of exhaust streams. RCO 5000 is a precious metal catalyst, RCO 6000 is a mixed precious metal catalyst and RCO 7000 is a base metal catalyst. Each catalyst is designed to meet the design criteria for specific applications and can be made into forms for both structure packing (e.g. honeycomb) and random packing (e.g. saddles, rings, etc.). The activities of these catalysts for converting different VOC compounds are shown in Figs. 7.20, 7.21 and 7.22, respectively.

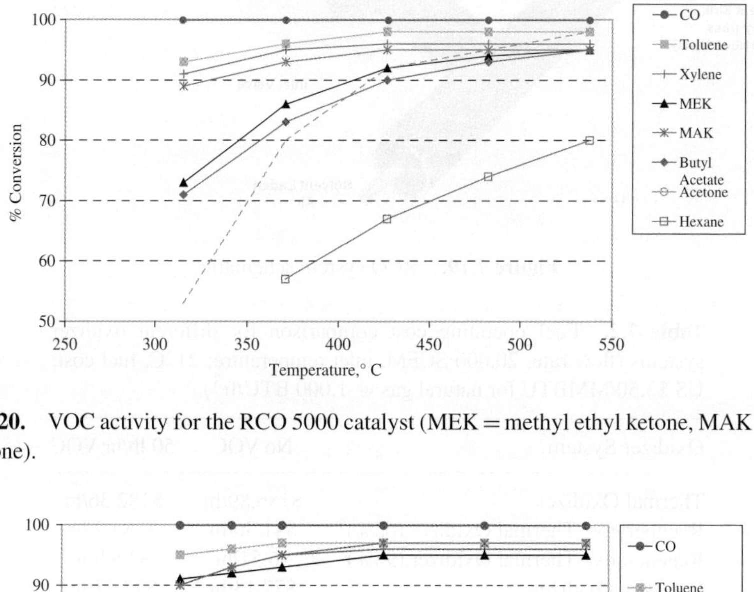


**Figure 7.20.** VOC activity for the RCO 5000 catalyst (MEK = methyl ethyl ketone, MAK = methyl amyl ketone).

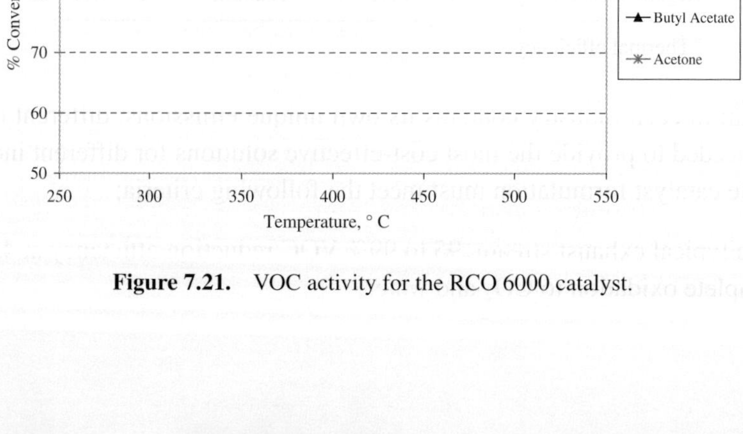


**Figure 7.21.** VOC activity for the RCO 6000 catalyst.

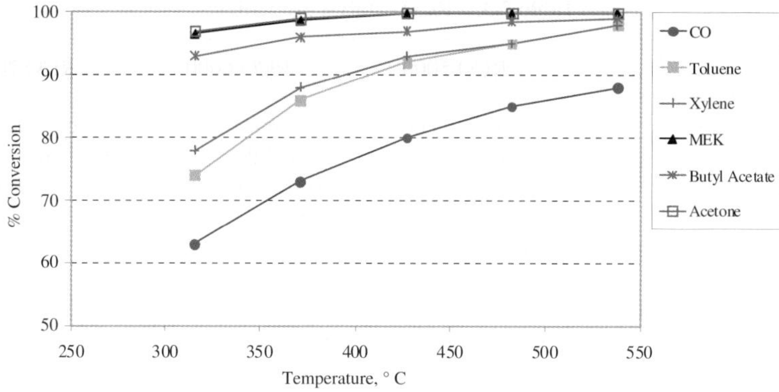


**Figure 7.22.** VOC activity for the RCO 7000 catalyst.

RCO 5000 is designed for air emissions containing mainly aromatics (toluene), alkenes (pinene, propylene) and carbon monoxide (CO). This catalyst has excellent thermal stability, is very resistant to poisons such as sulfur and can be regenerated by washing. RCO 5000 has operated successfully for more than two years in forest product applications such as oriented strand board (OSB) and plywood manufacturing processes. The air emissions from these processes contain primarily pinene, toluene, CO and methanol. In dust-laden environments, particulate matters in the exhaust stream can accumulate on the surface of the catalyst causing masking of the active component. When this occurs, the catalyst can be regenerated by washing, which can be carried out *in situ* or the catalyst can be removed and sent back to BASF.

For applications where the exhaust stream contains aromatics, alkenes, CO and oxygenated compounds, RCO 6000 is preferred (Fig. 7.21). This catalyst can operate at temperatures as low as 350°C, has excellent thermal stability and can also be regenerated with washing. RCO 6000 has been successfully used in automotive applications (paint booths and oven processes).

RCO 7000 is preferred for exhaust streams containing mainly oxygenated compounds such as acetone or butyl acetate. This catalyst can destroy these oxygenated VOCs at 350°C. However, it is not very active for aromatics and would need temperatures higher than 850°C to completely destroy them. Additionally, since it is a base metal catalyst, RCO 7000 is not effective for CO and has been found to form CO when used with streams containing methyl ethyl ketone.

For pharmaceutical applications where the exhaust stream contains mainly chlorinated compounds, a combination precious and base metal catalyst was developed. This catalyst has not yet been commercialized.

Table 7.3 summarizes the characteristics of each catalyst.

**Table 7.3.** RCO catalyst properties.

| Catalyst | RCO 5000 | RCO 6000 | RCO 7000 |
|---|---|---|---|
| Catalytic Ingredients | Precious metal | Precious metal | Base metal |
| Shape | 1″ saddle | 1″ saddle | 1/2″ ring |
| Good Conversion | Aromatics<br>Alkenes<br>Alkynes | Aromatics<br>Alkenes<br>Alkynes<br>Oxygenated VOCs | Oxygenated VOCs |
| Fairly Good Conversion | Oxygenated VOCs | | Aromatics |
| Poor Conversion | C-6 paraffins | C-6 paraffins | C-6 paraffins |
| CO Removal | Excellent | Excellent | Poor |
| Poisons | Organosilicates | Organosilicates<br>Sulfur | Organosilicates<br>Sulfur |
| Alkaline/Acid Washing | Yes | Yes | No |
| Recommended Applications | Forest products | Automotive | Automotive |
| Recommended Operating Temperature (°C) | 350 | 350 | 420 |

- Aromatic VOCs: toluene, xylene, benzene, ethylbenzene, etc.
- Oxygenated VOCs: ketones, acetates, etc.
- Paraffins: methane, ethane, propane, butane, pentane, hexane

### 7.4.3. *Commercial applications*

One application for the RCO catalyst which will provide a substantial cost reduction is the RTO to RCO conversion. One alternative is to put a layer of proprietary precious metal catalyst on top of the existing bed in the regenerative chamber of the RTO. This will reduce the operating temperature required to oxidize the VOCs to about 450°C. Another alternative is to remove some ceramic saddles before installing the RCO catalyst on top of the catalyst bed. These approaches can be made to meet the customer's needs for maximizing fuel or electrical savings.

There are several important issues that need to be considered before conversion to RCO is attempted. Each VOC stream needs to be examined to ensure there are no significant quantities of catalyst poisons such as silicon, phosphorus, arsenic and other heavy metals that might deactivate the catalyst. In addition, the catalyst's performance could be affected by masking and/or fouling from particulates present in the gas stream. However, if this occurs, the catalyst can be regenerated relatively easily, first by washing with a chemical solution, then by reheating to between 450 and 500°C for one to two hours. It is important to discuss individual air streams with prospective process emission suppliers before making any decisions on VOC abatement.

Table 7.4 gives the list of commercial experiences of RCO catalysts. The experiences include applications for printing (lithography and flexography), coating (cans,

**Table 7.4.** Examples of BASF (formerly Engelhard) RCO installations.

| Location | Allocations | Units | Flow Rate SCFM ($Nm^3$/hr) |
|---|---|---|---|
| Alabama | Forest Products | 1 | 43,000 (73,100) |
| Mississippi | Forest Products | 1 | 43,000 (73,100) |
| Oklahoma | Forest Products | 1 | 86,000 (146,200) |
| Arkansas | Forest Products | 1 | 86,000 (146,200) |
| Michigan | Forest Products | 2 | 86,000 each (146,200) |
| West Virginia | Forest Products | 2 | 86,000 (146,200) |
| Tennessee | Printing | 1 | 7,000 (11,900) |
| Michigan | Auto Paint | 3 | 604,000 total (1,027,000) |

coils and fabrics), forest products (production of woods ranging from plywood to medium- and high-density fiberboard) and automotive OEM (spray-paint booths). Examples of RCO performances in some of the applications are given below.

A plywood manufacturing plant was required to remove at least 95% of the VOCs emitted from the drying operation. A two-chamber RCO unit was installed to handle the exhaust flow of 43,000 SCFM. The emissions were primarily CO, pinene, methanol, toluene and aldehydes. With the RCO 5000 catalyst, the oxidation chamber was controlled at 450°C to achieve the required overall destruction efficiency. The system has been in operation for over two years. Catalyst samples are periodically removed for activity measurements in the laboratory. Over the period, the catalyst has maintained its fresh activity with no performance degradation observed. During the two years of operation, the system has gone through a number of bake-out operations to remove the condensable organic compounds.

For OSB manufacturing, the emissions contain VOCs and particulates of carbonates, fatty acid salts and sulfates. These particulates, present in the process streams, have been shown to cause binding of heat transfer media. A wet electrostatic precipitator is installed before an RTO or RCO oxidizer to remove these particulates to prevent plugging. Under normal operations, the RCO 5000 catalyst maintained its performance. However, it was noticed that the particulate matters in the exhaust stream could accumulate on the surface of the catalyst, causing masking of the active component. When this occurred, the catalyst was regenerated by washing.

A three-chamber RTO unit is used to control VOC emissions from a clear coat process of an automotive coating application. The unit was converted to RCO by removing the ceramic saddles and loading the RCO 6000 catalyst on top of the bed. The stream contains methyl ethyl ketone, butyl acetate, xylene, trimethylbenzene and ethylbenzene. After conversion to operate as an RCO, the system was measured to achieve 99.3% VOC destruction efficiency at an oxidation temperature of 375°C. The oxidizer continues to demonstrate greater than 99% destruction efficiency after 10 months of service.

## 7.5. Conclusions

The vent gases from crude PTA production contain emissions of CO, methyl bromide and other VOCs that can be effectively abated by catalytic oxidation. In the presence of bromine-containing compounds, the activities of precious metal catalysts were strongly influenced by the support materials. The successful development of new catalytic materials with optimal combinations of the catalyst composition and support material has enabled the oxidizer operation to proceed at a substantially lower temperature and with a lower fuel cost.

Platinum and manganese catalysts are more active than palladium- and chromium-based catalysts when oxidizing nitrile compounds. A Pt (HS) catalyst is more active than a manganese catalyst and Pt (LS) catalyst. The Pt high surface area catalyst reaches complete conversion at 225°C while the manganese catalyst does not reach complete conversion until 300°C. Because $NO_x$ formation is temperature dependent, the Pt high surface area catalyst can present an advantage. With this catalyst, $NO_x$ formation can be minimized because it can be operated at low temperature.

RCO combines the benefits of high heat recovery and low catalytic oxidation temperature to effectively reduce VOC emission control costs. RCO units have been successfully commercialized to control VOC emissions. With three distinct RCO catalyst formulations that have been developed to handle different process streams, BASF can provide end users with a highly energy efficient and cost-effective solution to many of their VOC emission problems.

It should be understood that each catalytic system must be carefully designed to meet the percent abatement required to adhere to local regulations. It is necessary for the design to satisfy the allowable pressure drop, life expectancy and space available in the vent or stack of the process.

## References

1. Heck, R. and Farrauto, R. (2009). *Catalytic Air Pollution Control: Commercial Technology*, 3rd edition, Wiley and Sons, Hoboken, New Jersey, Chapter 11.
2. Heck, R. and Farrauto, R. (2009). *Catalytic Air Pollution Control: Commercial Technology*, 3rd edition, Wiley and Sons, Hoboken, New Jersey, Chapter 1.
3. Heck, R. and Farrauto, R. (2009). *Catalytic Air Pollution Control: Commercial Technology*, 3rd edition, Wiley and Sons, Hoboken, New Jersey, Chapter 6.
4. Liotta, L. (2010). Catalytic Oxidation of Volatile Organic Compounds on Supported Noble Metals, *Appl. Catal. B: Environ.*, **100**, pp. 403–412.
5. Ainsworth, S. (1994). Purified Terephthalic Acid Capacity Takes Off, *C&E News*, **72**, pp. 19–20.
6. Davis, W. (2000). Air Pollution Engineering Manual, in *Air & Waste Management Association*, 2nd Edition, John Wiley & Sons, Inc., pp. 434–440.

7. Kirk, R. and Othmer, D. (1978). *Encyclopedia of Chemical Technology*, 3rd Edition, Vol. 17, John Wiley & Sons, New York.
8. Chen, M. and Nguyen, P. (1997). Catalytic Method and Device for Controlling VOC, CO and Halogenated Organic Emissions, *US Patent 5,643,545*.
9. Yao Li, S., Lian Li, S. and Lu Li., B. (1997). Reactions of Organic Exhausts and the Thermal Stability of Catalysts, *React. Kinet. Catal. Lett.*, **62**, pp. 89–95.
10. Vanin, G., Noskov, A., Popova, G. *et al.* (1993). The Industrial Plant for Unsteady State Purification of Flue-gases from Acrylonitrile and Cyanic Acid, *Catal. Today*, **17**, pp. 251–260.
11. Vlasenko, V. *et al.* (1992). Development and Model tests of a Method of Catalytic Purification of Gases from Acrylonitrile Impurities, *Khim. Promst+*, **24**, pp. 273–275.
12. Chen, J. (1996). Lower Operating Temperatures Oxidize VOCs, *Pollution Engineering.*, 12, pp. 42–44.
13. Cloud, R. (1996). Control VOC Emissions, *Chem. Process.*, 35, pp. 72–75.
14. Gribbon, S. (1996). *Regenerative Catalytic Oxidation*, Special Conference for VOC and Air Toxic Control, Air and Waste Management Association: Clearwater, FL, pp. 88–97.

# Chapter 8

# Hydrocarbon Processing: Catalytic Combustion and Partial Oxidation to Syngas

Unni OLSBYE*

Depending on the reaction conditions (nature of the co-reactants: air, steam of carbon dioxide; temperature,...), hydrocarbon processing may produce syngas by partial oxidation or heat by combustion. Though noble metals are widely used for total oxidation, non-noble metals, especially nickel catalysts, offer a good compromise between activity and cost for syngas production. Catalyst deactivation is an essential feature of these high temperature processes. Reactor concepts and designs are also reviewed.

## 8.1. Introduction

The title reactions of this chapter share a special feature compared to other catalytic oxidation processes. Given the proper set of reaction conditions, the target products of each process are the most thermodynamically stable combination of molecules. Unless other criteria exist (and we will come back to this later), the main role of the catalyst is therefore to reduce the activation energy of the process, to make it proceed under milder conditions.

Table 8.1 shows the main reactions involved in each of the processes (Eqs. 8.1–8.9). Industrial catalytic combustion (Eq. 8.1) is a means of energy production from hydrocarbons, whereas the target product of catalytic partial oxidation (Eq. 8.2), synthesis gas (syngas; $CO + H_2$), is the first intermediate in the production of valuable chemicals from natural gas. Such processes are expected to become increasingly important in the coming decades, since oil reserves are depleting and the increasing global population still depends on liquid, fossil fuel based energy carriers for the transportation sector and will need to replace oil as a raw material even for the production of everyday goods such as building materials, PC covers, clothes etc.[1] Typical downstream processing of syngas units uses the Fischer–Tropsch process

*Centre of Material Science and Nanotechnology, Department of Chemistry, University of Oslo, P. O. Box 1126 Blindern, Oslo 0318, Norway.

**Table 8.1.** Main reactions involved in catalytic combustion and catalytic partial oxidation of methane (thermodynamic data from Ref. 4).

| Reaction | Eq. | Standard reaction enthalpy |
|---|---|---|
| $CH_4 + 2O_2 \rightarrow CO_2 + 2H_2O$ | (8.1) | $\Delta H_{298K} = -802.3$ kJ/mol |
| $CH_4 + 0.5O_2 \rightarrow CO + 2H_2$ | (8.2) | $\Delta H_{298K} = -35.7$ kJ/mol |
| $CH_4 + H_2O = CO + 3H_2$ | (8.3) | $\Delta H_{298K} = 206.2$ kJ/mol |
| $CH_4 + CO_2 = 2CO + 2H_2$ | (8.4) | $\Delta H_{298K} = 247.3$ kJ/mol |
| $CO + H_2O = CO_2 + H_2$ | (8.5) | $\Delta H_{298K} = -41.1$ kJ/mol |
| $CH_4 = C + 2H_2$ | (8.6) | $\Delta H_{298K} = 74.9$ kJ/mol |
| $2CO = C + CO_2$ | (8.7) | $\Delta H_{298K} = -\ 172.4$ kJ/mol |
| $nCO + 2nH_2 = C_nH_{2n} + nH_2O$ | (8.8) | |
| $CO + 2H_2 = CH_3OH$ | (8.9) | |

(Eq. 8.8) and the methanol process (Eq. 8.9); both require a $H_2$: CO ratio close to two, in agreement with the stoichiometric ratio of the catalytic partial oxidation (CPO) reaction of methane, the main component of natural gas (Eq. 8.2). If higher hydrocarbons are used, the $H_2$: CO ratio may be adjusted by the water-gas shift reaction (Eq. 8.5) by adding either steam or $CO_2$ to the feed.

Both title processes may be run without a catalyst.[2,3] Adding a catalyst to the process has the advantage of decreasing the activation energy of the reaction, so the processes may be run under milder conditions. However, since catalysts are susceptible to poisoning (by sulfur, for example), additional upstream gas cleaning may be required in the catalytic processes.

Figure 8.1 shows the thermodynamic equilibrium composition of the effluent gas under stoichiometric conditions for the partial oxidation of methane to synthesis gas ($CH_4 : O_2 = 2 : 1$) versus temperature at 1 atm (a) and 4 atm (b) total pressure. At 1 atm pressure, full methane conversion and 100% CO selectivity is

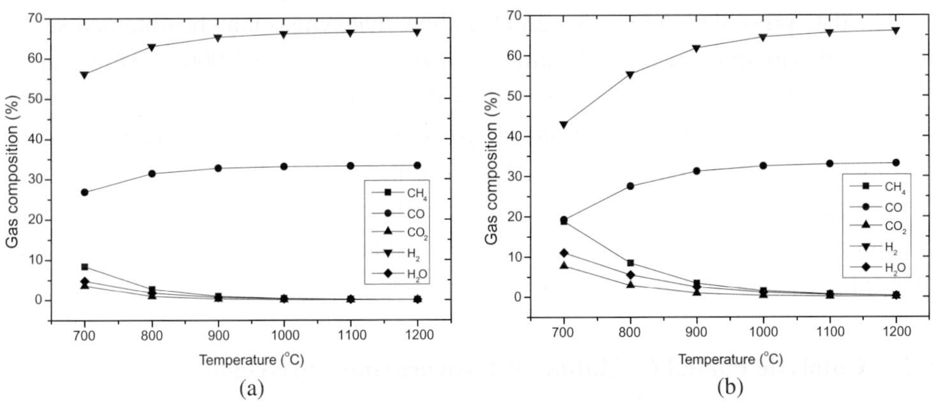


**Figure 8.1.** Equilibrium gas composition for methane oxidation at $CH_4 : O_2 = 2$, 800°C and (a) 1 atm total pressure (b) 4 atm total pressure.

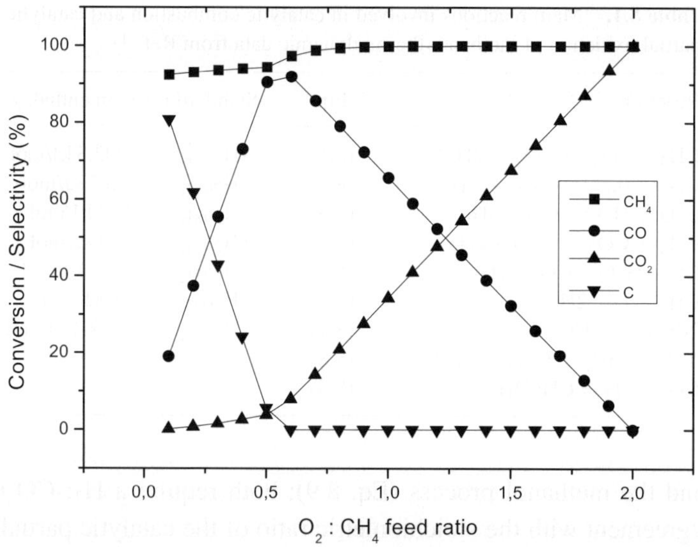


**Figure 8.2.** Equilibrium $CH_4$ conversion and product selectivity versus $O_2$ : $CH_4$ feed ratio at 800°C and 1 atm total pressure.

obtained at temperatures higher than 1,000°C. When increasing the pressure to 4 atm, the temperature for full methane conversion and 100% CO selectivity increases to 1.100°C. When using the stoichiometric composition for methane combustion ($CH_4 : O_2 = 1 : 2$) in the calculations (not shown), full conversion of methane to $CO_2$ and $H_2O$ was obtained over the whole temperature range (700–1,200°C) at both pressures.

Figure 8.2 shows the equilibrium $CH_4$ conversion and product selectivities versus $O_2$ : $CH_4$ feed ratio at 800°C and 1 atm. Decreasing the $O_2$ : $CH_4$ ratio improves selectivity to synthesis gas over the complete combustion products, but introduces another by-product, carbon deposits, here represented by graphite.

The data presented in Figs. 8.1 and 8.2 show that syngas may be obtained selectively by thermodynamic equilibrium at temperatures around 1,000°C. However, in order to produce syngas selectively at less than 1,000°C, a catalyst with a kinetic selectivity for the partial oxidation of hydrocarbons to synthesis gas needs to be developed. The status on this approach is described in Sec. 8.2.2.

The data presented in Figs. 8.1 and 8.2 further show that under fuel-rich (non-stoichiometric) conditions, catalytic combustion may lead to partial oxidation, rather than full oxidation products, at equilibrium.

## 8.2. Catalytic Partial Oxidation of Hydrocarbons to Syngas

The first report on catalytic partial oxidation of hydrocarbons to synthesis gas was published by Liander in 1929,[5] followed by Padovani and Franchetti in 1933[6] and

Prettre *et al.* in 1946.[7] Ni-based catalysts were used in each study, and equilibrium conversion to syngas was reported for temperatures above 850°C. Despite the promising results, the CPO reaction did not receive further attention until the late 1980s.

### 8.2.1. *The catalyst*

#### 8.2.1.1. *Non-noble metal catalysts*

Nickel-based catalysts have been the subject of many studies of the CPO reaction during the past two decades, confirming the activity of this metal for the title reaction (see for example Refs. 8–24). It is now generally accepted that the reduced metal has superior activity for methane activation compared with the corresponding metal oxide (see Sec. 8.2.2 below). As a CPO catalyst, Ni has three main advantages and two main shortcomings. The main advantages are its high CPO activity, high abundance and low cost, while its main shortcomings are the low free energy of oxidation and its significant activity for coke formation compared to noble metals.[8] Lunsford *et al.* studied CPO over an alumina-supported Ni catalyst, and found that the catalyst was divided into three zones: first $NiAl_2O_4$ and $NiO/Al_2O_3$, active for methane combustion, and then $Ni/Al_2O_3$, active for methane reforming.[9] Many attempts have been made to improve the performance of Ni-based catalysts by adding promoters or changing the catalyst support. Use of rare earth supports or the addition of slightly basic oxides to the support may restrict coke formation.[10–17] MgO-based supports (including hydrotalcite-derived Mg(Al)O) have received particular attention, and it is believed that the advantageous effect of MgO on catalyst activity and stability is related to a combination of high Ni dispersion (due to solid solution Ni-Mg-O formation) and suppression of coke formation by the slightly basic MgO.[18–24] It should, however, be noted that the formation of solid solutions, or mixed oxide phases, such as Mg(Ni)O, $NiAl_2O_4$ and $NiLaAl_{11}O_{19}$, not only promotes Ni dispersion, but also renders Ni more difficult to reduce.[25,26] This means that the reaction must be run at a higher (C+H)/O feed ratio in order to maintain Ni in the reduced state in the reforming section.

Co and Fe, which are well-known catalysts for the Fischer–Tropsch reaction (in principle the reverse reaction of Eqs. 8.3 and 8.4) have also been tested as CPO catalysts. However, Co, and especially Fe, are more difficult to reduce than Ni, and in general show lower activity for the CPO reaction than Ni.[27] CoO and $Fe_3O_4$ have further been reported to have significant activity for the complete oxidation of methane.[27–29] Enger *et al.* recently tested $Co/Al_2O_3$ catalysts for the CPO reaction and reported that the addition of promoters such as Fe, Cr, Re, Mn, W, Mo, V and Ta oxides dramatically reduced the conversion capacity of Co, while Ni promotion enhanced it.[30]

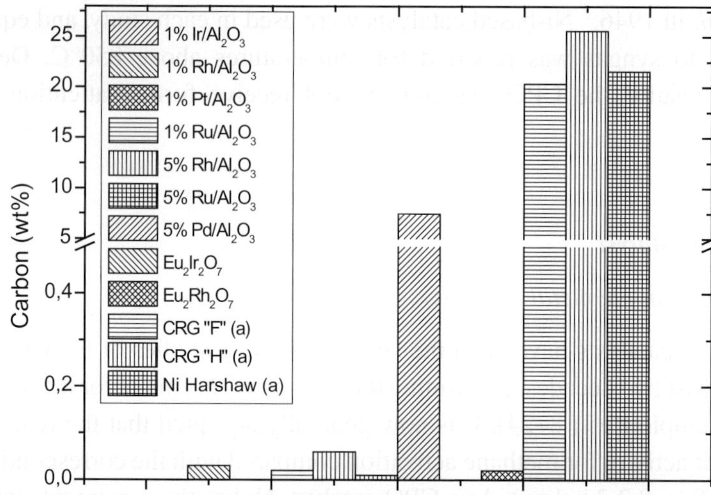


**Figure 8.3.** Comparison of various catalysts for carbon deposition during the catalytic partial oxidation of methane to synthesis gas. m(cat) = 50 mg, T = 777°C, $CH_4 : O_2 = 2$, flow rate = 8 ml/min, time = 24 h. (a) The British Gas "CRG "F" and "CRG "H" nickel catalysts and the nickel Harshaw catalyst were left on stream for 150 min, 80 min and 60 min respectively. Data from Ref. 8

### 8.2.1.2. *Noble metal catalysts*

Vernon *et al.* tested a series of noble metal catalysts, either Ru, Rh, Pt, Pd or Ir supported on $Al_2O_3$, or rare earth ruthenium pyrochlore materials.[31] All materials yielded the equilibrium CPO conversion of methane to syngas at 777°C and 1 atm. After the reaction, it was found that Ru had been reduced out of the pyrochlore structures. Claridge *et al.* studied coke formation over alumina-supported or pyrochlore-derived catalysts and reported that the coke forming rates decreased in the order: Ni > Pd > Rh > Ir, as illustrated in Fig. 8.3.[32]

More recently, various support materials for Ir were studied by Nakagawa *et al.*, who reported that among the support materials – $TiO_2$, $ZrO_2$, $Y_2O_3$, $La_2O_3$, MgO, $Al_2O_3$ and $SiO_2$ – $TiO_2$ yielded the highest resistance towards carbon deposition.[33]

### 8.2.1.3. *Non-metallic catalysts*

Ceria is one of the best known oxygen storage materials and is, for example, used in three-way catalysts.[34,35] Several recent works describe ceria as an oxidizer for the conversion of $CH_4$ to syngas.[36–40] Perovskite-type oxides ($ABO_3$) represent another class of reducible oxides with the potential to be partial oxidation catalysts.[41] In general, however, non-metal catalysts are found to have too low an activity for the CPO reaction to be commercially interesting. Therefore, studies of these classes of materials for CPO have focused on high temperature applications (above 800°C).[42–46] Most recent works combine such materials with metals for use in the CPO process.

For instance, ceria combined with platinum, and/or chemically modified with, for example, zirconium, samarium or bismuth to adjust oxygen diffusivity and capacity;[40,47,48] Pt or Rh supported on $CeO_2$;[38] Pt or Ru supported on $Ce_{1-x}Zr_xO_2$,[39] and Rh or Pt supported on $La_{0.75}Sr_{0.25}Fe_{0.6}Co_{0.15}Al_{0.25}O_{3-\delta}$perovskite.[47]

### 8.2.2. *The reaction mechanism*

Two reaction mechanisms have been proposed for the CPO reaction. One is the indirect mechanism, in which part of the methane undergoes combustion and the remaining methane subsequently reforms with $CO_2$ and $H_2O$ to yield the equilibrium product. The other is the direct mechanism, in which CO and $H_2$ are formed selectively from methane and oxygen without further oxidation. Kinetic studies of the reaction are complicated by the strong exo- and endo-thermicity of the individual reaction steps (Table 8.1), possibly leading to uncontrolled temperature profiles in the reactor. Therefore, most mechanistic studies of the CPO reaction have been performed by pulsing reactants over a shallow catalyst bed in a vacuum in temporal analysis of products (TAP) reactors.

Au and Wang[49] and Hu and Ruckenstein[50,51] studied CPO over $Ni/SiO_2$ by an isotopic pulse technique and observed that over oxidized $NiO/SiO_2$, methane reacted directly with NiO, without previous dissociation, yielding $CO_2$ and $H_2O$ as the only products. Over reduced $Ni/SiO_2$, methane dissociation was fast, and isotopic exchange in the effluent methane showed that this was not the rate-limiting step of the reaction. Reduced $Ni/SiO_2$ produced mainly CO and $H_2$. Hu and Ruckenstein further studied $Ni/La_2O_3$ by a pulse technique, and observed that the initial activity of the catalyst for the conversion of $CH_4/O_2$ depended on whether it was pre-reduced ($Ni/La_2O_3$, high activity) or not ($NiO/La_2O_3$, low activity). After ten pulses, the catalyst reached a state of partial reduction, giving an intermediate activity.[52,53] Transient responses suggested that the partially reduced catalyst produced CO as the primary product, followed by $CO_2$. In contrast, Verykios *et al.* studied the kinetics of CPO over $Ni/La_2O_3$ and reported that the reaction was dominated by complete combustion followed by reforming to CO and $H_2$. High CO selectivity at low contact time was observed only at very low oxygen partial pressures ($CH_4 : O_2 = 25$).[54] Several studies reported indirect indications of a combustion-reforming mechanism over Ni-based catalysts, i.e. a steep temperature increase followed by a temperature decrease in the flow direction of fixed-bed reactors.[55–57]

Buyevskaia *et al.* studied the reaction mechanism over a $Rh/Al_2O_3$ catalyst in a TAP reactor and found that the reaction pathway depended to a large degree on surface reduction. A high degree of surface reduction was required for hydrogen formation to take place.[58] Later, Buyevskaia studied Rh black by the same method, and reported that the active species for methane dissociation was neither metallic

Rh nor rhodium oxide, but probably $Rh^+$.[59] Mallens *et al.* studied the response of pulsing methane, oxygen or a mixture of the two over Rh sponge and reported that CO and $H_2$ were the primary products, from transient response curves where these products appeared before $CO_2$ and $H_2O$.[60] Similar studies and results are reported for several noble metal catalysts on various supports (see, for example, Ref. 57 and the references therein). However, even for noble metal catalysts, steep temperature gradients, indicating a combustion-reforming reaction scheme, are observed for reactants under atmospheric pressure.[61]

The studies exemplified above do not give a clear answer regarding whether a direct pathway from methane to syngas exists. However, mounting evidence suggests that the indirect mechanism dominates under process conditions of practical significance.

### 8.2.3. *Catalyst deactivation*

Given an indirect reaction mechanism, the oxidation and reforming zones of the CPO reactor represent different challenges with respect to catalyst deactivation, which can be the result of oxidation, sintering and (possibly) melting of the active phase in the hot oxidation zone, and coke formation in the reforming zone.

Metal particle sintering during reaction is a common feature of CPO catalysts (see, for example, Ref. 57 and the references therein). All metals shown to be active for the CPO reaction have melting points well above the reaction temperature.[62] However, since the melting points of particles drop dramatically with decreasing particle size, in the nanometer range, especially for diameters less than 10 nm (see, for example, Refs. 63–65), active metals could be expected to be more or less melted under typical CPO reaction conditions, thus facilitating metal sintering. Indications exist that supports which form solid solutions or mixed oxide phases with Ni tend to stabilize smaller Ni particles.[25,26]

Coke formation is a kinetic as well as a thermodynamic issue. Starting with thermodynamics, coke formation (represented by graphite) is favored at low H : C and O : C ratios, as illustrated by Eqs. (8.6) and (8.7) in Table 1, and by Fig. 8.2. A more extensive illustration of the thermodynamic coke-forming regime is given by Bartholomew[66] and Rostrup-Nielsen *et al.*[67] Several authors have reported the coke-free operation of noble metal catalysts under conditions which would thermodynamically result in graphite formation.[68] This result is probably related to kinetic selectivity. Consider the following simplified set of reactions between gas phase species and a surface site S:

$$CH_4 + S \rightarrow C - S + 2H_2 \quad (8.10)$$

$$C - S + \text{“O”} \rightarrow CO + S \quad (8.11)$$

Kinetically, a low steady-state concentration of carbon-containing adsorbates (which are potential coke precursors) is favored when Eq. 8.10 ($CH_4$ decomposition) is slow compared to Eq. 8.11 (C oxidation). Attempts to avoid deactivation by coke formation therefore focus on either increasing the rate of the gasification reaction, or slowing down the rate of the hydrocarbon decomposition reaction. As such, Besenbacher *et al.* added Au to a Ni catalyst and demonstrated that the catalyst maintained its activity with time on stream under butane steam reforming conditions, while the corresponding Ni catalyst deactivated rapidly due to coke formation.[69] Scanning tunneling microscopy (STM) investigations and density functional theory (DFT) modeling suggested that Au altered the electronic properties of Ni, making it less active for $CH_4$ activation.[69,70] An additional interpretation, which has also been used to explain the effect of adding $H_2S$ to a methane dry reforming feed (i.e. hindering coke formation in the so-called sulphur-passivated reforming (SPARG) process),[71] is that a certain number of neighboring sites is required to form aromatic graphite precursors from carbon-containing adsorbates, and that too many of those sites are covered by $H_2S$(ads) (or Au) for this reaction to take place.

Attempts to reduce coke formation by facilitating carbon oxidation are mainly focusing on Ni-based catalysts and are related to the use of basic carrier materials (see Sec. 8.2.1.1. above). Various types of coke may be formed, the main types being encapsulating coke, which covers the catalyst surface, and filamentous coke, which makes the metal particle propagate from the catalyst surface, forming long carbon whiskers with the metal particle at the tip. It has been suggested that filamentous coke forms by diffusion of carbon through the metal particle (see, for example, Ref. 72 and the references therein). However, recent transmission electron microscopy (TEM) studies coupled with DFT calculations indicated that the growth of carbon filaments on Ni particles proceeds by the surface diffusion of carbon and nickel atoms.[73] It is not yet fully understood why noble metals are more resistant towards coke formation than Ni (cf. Fig. 8.3). Rostrup-Nielsen recently suggested that it might be due to low diffusion rates in the subsurface layers, or that steps are not created as easily on these metals as on Ni.[68]

### 8.2.4. *Reactor concepts*

The high operating temperature and moderate exothermicity of the overall CPO reaction (Eq. 8.2) is ideal for autothermal reactor operation. For catalysts supporting the indirect reaction scheme, however, heat management becomes an issue. Simulations of an adiabatic fixed-bed reactor containing $Ni/Al_2O_3$ showed that a hot spot temperature as high as 1,500°C could be obtained with a $CH_4 : O_2$ ratio of 1.67, when assuming that methane combustion and reforming took place in series.[74] In the following sections, three reactor concepts which address this challenge will be presented.

Another issue of CPO is the cost and explosion hazard associated with cryogenic air separation. Two reactor concepts addressing this challenge will also be presented. Finally, a short description of the autothermal reformer, which is today the preferred reactor choice for the industrial scale conversion of hydrocarbons with oxygen to syngas, will be given.

#### 8.2.4.1. *Fluid-bed reactor*

Fluid-bed reactors (FBR) offer a potential solution to heat management, due to efficient mixing of solids in the catalyst bed. Olsbye *et al.* reported smooth temperature profiles in a lab-scale fluid-bed reactor when performing CPO over a $Ni/Al_2O_3$ catalyst (9°C in the fluid-bed compared to 150°C in the fixed-bed reactor under the same CPO conditions).[75] Mleczko and Wurzel reported another advantage of FBR compared to fixed-bed reactors; fluid-bed operation allowed the $Ni/Al_2O_3$ catalyst to remain in a reduced state, active for the reforming reactions.[76] Bharadwaj and Schmidt tested Pt, Rh and Ni supported on $Al_2O_3$, and found similar activity for Rh and Ni catalysts, and inferior activity for Pt, during fluid-bed methane CPO.[77] Mleczko *et al.*[76] Bharadwaj and Schmidt[77] and Santos *et al.*[78] all reported limited or no coke formation during FBRCPO operation, and ascribed this observation to carbon gasification due to the continuous recirculation of the catalyst between oxygen-rich and oxygen-deficient regions.

A potential downside of fluid-bed reactors is mechanical attrition due to collisions between catalyst particles, leading to the formation of catalyst fines, which may be carried out of the reactor's isothermal zone and cause back formation of methane in colder zones (with less advantageous equilibrium conversion, see Fig. 8.1).[75,78] Exxon patented a fluid bed based CPO process with $Ni/Al_2O_3$ catalyst and reported that an intermediate catalyst activity, with 0.5–2.5 wt% Ni, would limit the problem of methane back formation.[79] Their findings were later confirmed by Mleczko *et al.*[76]

#### 8.2.4.2. *Heat-integrated wall reactor*

The heat-integrated wall reactor (HIWR) was first presented by Ioannides and Verykios.[80] It consists of two tubular reactors, one centered inside the other. The walls of the inner tube are covered with catalyst and gas travels from the end of the inner combustion chamber to the outer reforming chamber, in a counter-current manner. Heat produced in the inner tube will be transported through the wall and provide heat to the reforming reaction. Experimental studies using a $Rh/Al_2O_3$ catalyst showed an increased conversion (e.g. 87% vs 56%) and smoother temperature curve (e.g. $dT = 190°C$ vs 290°C) for a HIWR reactor compared to a conventional fixed-bed reactor when operated under the same temperature and flow conditions. The potential of the HIWR concept was later confirmed by modeling studies.[81]

The main challenges in future developments of this reactor concept are to restrict the rate of the oxidation reaction in order to avoid the steep temperature gradient in the first part of the catalytic zone (e.g. by gradual oxygen feed as suggested by Veser *et al.*),[81] and to improve the heat transfer properties of the reactor walls.

### 8.2.4.3. *Short contact time monolith reactor*

The short contact time monolith reactor was first introduced by Hickman and Schmidt as a direct partial oxidation concept, using either a Pt- or a Rh-covered ceramic monolith.[82] Complete methane conversion and $> 90\%$ selectivity to CO and $H_2$ was obtained over the Rh-coated monolith for, for example, an inlet temperature of 460°C with air as the oxidant, $CH_4/O_2 = 1.6$ and 10 ms contact time. The Pt-coated monolith gave inferior methane conversion and $H_2$ selectivity, and the selectivity difference was explained by a sequence of elementary steps showing that the activation energy for OH formation was far inferior on Pt than on Rh.[82] Later, Heitnes *et al.* studied Ni- and Pt-coated ceramic monoliths under similar reaction conditions and found steep temperature gradients in the reactor, as well as gas compositions corresponding to equilibrium at the monolith outlet temperature.[83] In a follow-up study of Ni- or Pd-coated monoliths and Pt gauze, the temperature profiles again suggested combustion-reforming for the monolith and gauze reactors.[84]

Even if the reaction is not selective, monolith reactors may be interesting CPO reactors. Coating the reactor walls instead of using packed-bed reactors offers a reduced pressure drop over the catalytic zone, and use of metal-based monoliths with high heat transfer coefficients may reduce the temperature gradients in the reactor. Such an effect was reported by Aartun *et al.* when using a $Rh/Al_2O_3$/Fecralloy monolith reactor for the CPO of propane. They observed a temperature gradient of approximately 100°C at full oxygen conversion with a $C:O:N$ ratio of $0.8:1:4$.[85] Recent work on short contact time reactors further comprises packed $Rh/Al_2O_3$ reactors, where kinetic studies again supported an indirect route to syngas formation.[86] Furthermore, Tanaka *et al.* reported that a Rh/Ce/MgO catalyst gave lower temperature gradients than the corresponding Rh/MgO catalyst in short contact time CPO, and ascribed this effect to an enhanced reducibility of Rh in the presence of Ce, enabling reforming activity in the first part of the reactor.[87]

### 8.2.4.4. *Cyclic reactor with oxygen carrier material*

One possible approach to avoid cryogenic air separation is the cyclic oxidation process, in which a reducible metal oxide is used as an oxygen source (Eq. 8.12) and subsequently reoxidized in an air separation unit (Eq. 8.13).

$$yCH_4 + MO_x = yCO + 2yH_2 + MO_{x-y} \quad (8.12)$$

$$MO_{x-y} + 1/2yO_2 = MO_x \quad (8.13)$$

Such a split oxidation process was previously evaluated in 1949 by Lewis *et al.* for the conversion of $CH_4$ to syngas using CuO as an oxygen source in a fluid-bed reactor at 900°C.[88] They suggested a two-step reaction; a first step where copper oxide is rapidly reduced and part of the $CH_4$ is totally oxidized, and a second slower step where $CH_4$ was reformed with the formed carbon dioxide and steam to syngas. More recently, Otsuka *et al.* investigated ceria, $CeO_2$, mainly impregnated with Pt, as an oxidant in the $CH_4$ partial oxidation to syngas.[36,89] Other researchers have studied Pt and Rh supported on ceria,[38] or Pt or Ru supported on $Ce_{1-x}Zr_xO_2$.[39] Another group of materials which has been studied as cyclic oxidation materials is perovskite-based oxides.[90,91] An increase in syngas selectivity with a degree of oxide reduction has generally been observed, and it has tentatively been suggested that there are two types of oxygen on the oxide surface, one selective and one unselective for syngas formation.[36,39,89,90] Very recently, however, thermogravimetric and catalytic studies of Pt/Ce(Zr)$O_2$ as well as Rh or Pt supported on (La,Sr)(Fe,Co,Al)$O_{3-\delta}$ perovskites at 600°C showed that syngas selectivity was in either case directly correlated with the thermodynamic redox properties of the reducible oxide. CPO selectivity is thus thermodynamically, not kinetically, controlled even in the cyclic CPO process.[47,48] A main challenge for future studies of this reactor concept is to develop materials which are stable during multiple redox cycles and have a large oxygen release capacity in the oxygen partial pressure range suited for selective syngas formation (see Fig. 8.2).

#### 8.2.4.5. *Membrane reactor*

Continuous air separation by an oxygen-conducting membrane which constitutes the wall of a CPO reactor is another approach which has received much interest, also from industry.[92] Two types of membrane materials have been studied: zirkonia-based membranes, which are efficient oxygen ion conductors but require electrodes to transfer electrons to the reduction interface, and perovskites (of general formula $ABO_3$, with dopants in the A and/or B site), which are mixed ionic/electronic conductors (MIEC).[92]

Recently, focus has been set mainly on MIEC materials, and several materials with high ionic and electronic conductance have been reported, such as $La_{1-x}A_xCo_{1-y}B_yO_{3-\delta}$ (A = Sr, Ba or Ca and B = Fe, Cu or Ni),[93–96] $SrFe_{1-x}Cu_xO_{3-\delta}$[97,98] and $La_{0.2}Sr_{0.8}Fe_{0.8}Cr_{0.2}O_{3-\delta}$ perovskites.[99] In a study of $La_{1-x}A_xBa_{0.8}Co_{0.2}O_{3-\delta}$ perovskites, it was observed that the packing of a Ni/$Al_2O_3$

catalyst on the reaction-side surface of the membrane led to a fivefold increase in oxygen permeation and a fourfold increase in methane conversion.[100,101] The main challenges in ongoing studies of MIEC is to reduce their brittleness while maintaining their stability towards large oxygen gradients.

#### 8.2.4.6. *Autothermal reforming*

The autothermal reformer reactor (ATR) is not a pure CPO concept, since the oxidation reaction proceeds in a flame. The reason for its inclusion here is its industrial use, while the concepts described above are still at the development stage (see, for example, Ref. 102). The ATR has a compact design consisting of a burner, a combustion chamber and a catalyst bed. Part of the methane is combusted with oxygen in the burner. The hot effluent is then reformed with additional methane over a conventional reforming catalyst, giving equilibrium syngas yield at the reactor outlet. Steam is added to the feed to prevent soot formation.[102] Reducing the amount of steam would give syngas production closer to the desired $H_2 : CO = 2 : 1$ ratio (see Fig. 8.1 above), and this challenge is subject to further development.[103]

## 8.3. Catalytic Combustion

Energy production by combustion is a well-established process, including natural gas and methane feeds.[104] From thermodynamic considerations, a stoichiometric combustion process is advantageous at all temperatures. However, at temperatures above 1600°C, the direct combination of $N_2$ and $O_2$ in air becomes thermodynamically favorable, and this temperature therefore represents an upper boundary for the process.[3]

### 8.3.1. *The catalyst*

The main role for the combustion catalyst is to lower the initiation temperature of the oxidation process. Once the reaction is initiated, the reactor temperature increases, and ignition (i.e. the point at which the temperature of the fuel mixture is sufficiently high to maintain the temperature without external heating) takes place.[105]

Ignition is particularly challenging for methane, which is quite an inert molecule and requires high initiation temperatures. Once the reaction starts, there is considerable heat production, and the temperature may approach, or even surpass, 1,600°C. Natural gas mixtures give easier temperature control, because the $C_2{-}C_4$ components will oxidize at lower temperatures than methane, and provide heat for the methane oxidation reaction.[3] Veser and Schmidt studied the catalytic ignition of $C_1{-}C_4$ alkanes over a platinum foil, and reported that the ignition temperature decreased as: $CH_4 > C_2H_6 > C_3H_8 \sim C_4H_{10}$.[106] They further observed a decrease in ignition temperature with an increasing fuel-to-air ratio, in particular for methane, and ascribed this observation to competitive adsorption of alkanes versus oxygen on the Pt surface.

As expected, reports on catalyst activity for combustion follow the order of reactivity of partial oxidation (see Sec. 8.2 and, for example, Ref. 105 and the references therein). Schmidt and co-workers reported that the ignition temperature for the combustion of ethane decreased in the order: Pt < Pd < Rh < Ir < Ni.[107,108] Beck *et al.* recently studied methane combustion in a gradientless flow-circulation reactor over $Pt/Al_2O_3$ with Pt particles ranging from 1.3–10 nm in size, at 380–500°C, in a large excess of oxygen.[109] An important structure sensitivity was observed, with a bell-shaped turnover frequency curve peaking at a Pt particle diameter of 2 nm. X-ray absorption near edge structure (XANES) spectra of the catalysts indicated that for the optimum Pt particle diameter, the particles consisted partly of $PtO_x$ and partly of Pt, while samples with smaller and larger diameters consisted predominantly of either $PtO_x$ or $Pt^0$, respectively. The bell-shaped curve was interpreted as a combination of an increased surface tension at decreasing particle diameter, and an optimum $Pt^0/PtO_x$ combination for the combustion reaction.

### 8.3.2. *Reactor concepts*

Chemical looping combustion (CLC) was first introduced as a means of increasing the efficiency of catalytic combustion, and in recent years, also as a means of $CO_2$ capture. CLC is typically carried out in a dual fluid-bed reactor system, in which a metal, typically dispersed on an inert support material, is oxidized by air in the first reactor (Eq. 8.14), and then transferred to the second reactor where it makes contact with the hydrocarbon to produce $CO_2$ and $H_2O$ (Eq. 8.15). An additional feature of CLC is that air is separated from the oxidation zone, thereby avoiding the possibility of $NO_x$ formation.

$$M + 0.5O_2 \rightarrow M - O \qquad (8.14)$$

$$CH_4 + 4M - O \rightarrow CO_2 + 2H_2O + 4M \qquad (8.15)$$

The selection of a suitable oxygen carrier material is a key issue for the further development of CLC. The main requirements are sufficient oxygen storage capacity, high reactivity under oxidizing and reducing conditions, high selectivity for the complete combustion of the hydrocarbon, low tendency for carbon formation and high attrition resistance. Metals which have been reported as promising oxygen carriers comprise Ni, Cu, Mn, Co and Fe (see, for example, Ref. 110 and the references therein). However, the process has still to demonstrate its industrial potential.

## References


1. International Energy Agency: World Energy Outlook 2010.
2. Wilhelm, D., Simbeck, D., Karp, A., *et al.* (2001). Syngas production for gas-to-liquids applications: technologies, issues and outlook, *Fuel Proc. Techn.*, **71**, (1–3), pp. 139–148.
3. Lee, J. and Trimm D. (1995). Catalytic combustion of methane, *Fuel Proc. Techn.*, **42**, (2–3), pp. 339–359.
4. HSC Chemistry. (2000). Chemical Reaction and Equilibrium Software with Extensive Thermochemical Database, Version 4.1, Outokumpu Research Oy.
5. Liander H. (1929). The utilisation of natural gases for the ammonia process, *Trans. Faraday Soc.*, **25**, pp. 462–471.
6. Padovani, C. and Francetti Giorn, P. (1933). Incomplete Oxidation of Methane With Oxygen and Air, *Giorn. chim. ind. applicata*, **15**, pp. 429–432.
7. Prettre, M. Eichner, C. and Perrin, M. (1946). The catalytic oxidation of methane to carbon monoxide and hydrogen, *Trans. Faraday. Soc.*, **42**, pp. 335–340.
8. Claridge, J., Green, M., Tsang, S., *et al.* (1993). A study of carbon deposition on catalysts during the partial oxidation of methane to synthesis gas, *Catal. Letters*, **22**, pp. 299–305.
9. Dissanayake, D., Rosynek, M., Kharas, K., *et al.* (1991). Partial oxidation of methane to carbonmonoxide and hydrogen over a Ni/$Al_2O_3$ catalyst, *J. Catal.*, **132**, pp. 117–127.
10. Zhu, T. and Flytzani-Stephanopoulos, M. (2001). Catalytic partial oxidation of methane to synthesis gas over Ni-$CeO_2$, *Appl. Catal. A*, **208**, pp. 403–417.
11. Chu, W., Yan, Q., Liu, X., *et al.* (1998). Reactivity of Pt/$Al_2O_3$ and Pt/$CeO_2$/$Al_2O_3$ catalysts for partial oxidation of methane to syngas, *Stud. Surf. Sci. Catal.*, **119**, pp. 855–860.
12. Chu, W., Yan, Q., Liu, S., *et al.* (2000). Improvements of ceria promoted nickel catalysts for natural gas oxidation to syngas, *Stud. Surf. Sci. Catal.*, **130**D, pp. 3573–3578.
13. Tsipouriari, V. and Verykios, X. (1998). Kinetic study of the catalytic partial oxidation of methane to synthesis gas over Ni/La2O3 catalyst, *Stud. Surf. Sci. Catal.*, **119**, pp. 795–800.
14. Lu, Y., Liu, Y. and Shen, S. (1998). Design of stable Ni catalysts for partial oxidation of methane to synthesis gas, *J. Catal.*, **177**, pp. 386–388.
15. Choudhary, V., Rajput, A. and Mamman, A. (1998). NiO alkaline earth oxide catalysts for oxidative methane-to-syngas conversion: Influence of alkaline earth oxide on the surface properties and temperature-programmed reduction/reaction by $H_2$ and methane, *J. Catal.*, **178**, pp. 576–585.
16. Slagtern, A. and Olsbye, U. (1994). Partial oxidation of methane to synthesis gas using La-M-O catalysts, *Appl. Catal. A*, **110**, 99–108.
17. Chu, Y., Li, S., Lin, J., *et al.* (1996). Partial oxidation of methane to carbon monoxide and hydrogen over NiO/$La_2O_3$/gamma-$Al_2O_3$ catalyst, *Appl. Catal. A*, **134**, pp. 67–80.
18. Choudhary, V., Mamman, A. and Sansare, S. (1992). Selective oxidation of methane to CO and $H_2$ over Ni/MgO at low temperatures, *Angew. Chem., Int. Ed. Engl.*, **31**, pp. 1189–1190.
19. Choudhary, V., Ramarjeet, R. and Rane, V (1992). Low temperature catalytic selective partial oxidation of methane to CO and $H_2$ over Ni/$Yb_2O_3$, *J. Phys. Chem.*, **96**, pp. 8686–8688.
20. Choudhary, V., Rajput, A. and Prabhakar, B. (1993). Nonequilibrium oxidative conversion of methane to CO and $H_2$ with high selectivity and productivity over Ni/$Al_2O_3$ at low temperatures, *J. Catal.*, **139**, pp. 326–328.
21. Ruckenstein, E. and Hu, Y. (1999). Methane partial oxidation over NiO MgO solid solution catalysts, *Appl. Catal. A: Gen.*, **183**, pp. 85–92.
22. Hu, Y. and Ruckenstein, E. (1998). Catalyst temperature oscillations during partial oxidation of methane, *Ind. Eng. Chem. Res.*, **37**, pp. 2333–2335.
23. Santos, A., Menendez, M., Monzon, A., *et al.* (1996). Oxidation of methane to synthesis gas in a fluidized bed reactor using MgO-based catalysts, *J. Catal.*, **158**, pp. 83–91.

24. Basini, L., D'Amore, M., Fornasari, G., *et al.* (1998). Ni/Mg/Al anionic clay derived catalysts for the catalytic partial oxidation of methane - Residence time dependence of the reactivity features, *J. Catal.*, **173**, pp. 247–256.
25. Slagtern, Å., Olsbye, U., Blom, R., *et al.* (1997). Characterization of Ni on La modified $Al_2O_3$ catalysts during $CO_2$ reforming of methane, *Appl. Catal. A*, **165**, (1–2), pp. 379–390.
26. Råberg, L., Jensen, M., Olsbye, U., *et al.* (2007). Propane dry reforming to synthesis gas over Ni based catalysts. Influence of support and operating parameters on catalyst activity and stability, *J. Catal.*, **249**,(2), pp. 250–260.
27. Slagtern, Å., Swaan, H., Olsbye, U., *et al.* (1998). Catalytic partial oxidation of methane over Ni-, Co- and Fe-based catalysts, *Catal. Today*, **46**, (2–3), pp. 107–115.
28. Choudhary, V., Rajput, A., Prabhakar, B., *et al.* (1998). Partial oxidation of methane to CO and $H_2$ over nickel and/or cobalt containing $ZrO_2$, $ThO_2$, $UO_2$, $TiO_2$ and $SiO_2$ catalysts, *Fuel*, **77**, pp. 1803–1807.
29. Slagtern, A. and Olsbye, U. (1994). Partial oxidation of methane to synthesis gas using La-M-O catalysts, *Appl. Catal. A*: *Gen.*, **110**, pp. 99–108.
30. Enger, B., Lødeng, R. and Holmen A. (2009). Modified cobalt catalysts in the partial oxidation of methane at moderate temperatures, *J. Catal.*, **262**, pp. 188–198.
31. Vernon, P., Green, M., Cheetham, A., *et al.* (1990). Partial oxidation of methane to synthesis gas, *Catal. Lett.*, **6**, pp. 181–186.
32. York, A., Xiao, T. and Green, M. (2003). Brief overview of the partial oxidation of methane to synthesis gas, *Topics in Catal.*, **22**, (3–4), pp. 345–358.
33. Nakagawa, K., Anzai, K., Matsui, N., *et al.* (1998). Effect of support on the conversion of methane to synthesis gas over supported iridium catalysts, *Catal. Lett.*, **51**, pp. 163–167.
34. Funabiki, M., Yamada, T. and K. Kayano, K. (1991). Auto exhaust catalysts, *Catal. Today*, **10**, pp. 33–43.
35. Kacimi, S., Barbier, J., Taha, R., *et al.* (1993). Oxygen storage capacity of promoted $Rh/CeO_2$ catalysts — Exceptional behaviour of $RhCu/CeO_2$, *Catal. Lett.*, **22**, pp. 343–350.
36. Otsuka, K., Wang, Y., Sunada, E., *et al.* (1998). Direct partial oxidation of methane to synthesis gas by cerium oxide, *J. Catal.*, **175**, pp. 152–160.
37. Otsuka, K., Wang, Y. and Nakamura, M. (1999). Direct conversion of methane to synthesis gas through gas-solid reaction using $CeO_2$-$ZrO_2$ solid solution at moderate temperature, *Applied Catal. A*, **183**, pp. 317–324.
38. Fathi, M., Bjorgum, E., Viig, T., *et al.* (2000). Partial oxidation of methane to synthesis gas: Elimination of gas phase oxygen, *Catal. Today*, **63**, pp. 489–497.
39. Pantu, P., Kim, K. and Gavalas, G. (2000). Methane partial oxidation on $Pt/CeO_2$-$ZrO_2$ in the absence of gaseous oxygen, *Appl. Catal. A*, **193**, pp. 203–214.
40. Sadykov, V., Kuznetsova, T., Alikina, G., *et al.* (2004). Fuel-rich methane combustion: Role of the Pt dispersion and oxygen mobility in a fluorite-like complex oxide support, *Catal. Today*, **117**, pp. 475–483.
41. Choudhary, T., Banerjee, S. and Choudhary, V. (2002). Catalysts for combustion of methane and lower alkanes, *Appl. Catal. A*, **234**, pp. 1–23.
42. Zeng, Y., Tamhankar, S., Ramprasad, N., *et al.* (2003). A novel cyclic process for synthesis gas production, *Chem. Eng. Sci.*, **58**, pp. 577–582.
43. Dai, X., Wu, Q., Li, R., *et al.* (2006). Hydrogen production from a combination of the water-gas shift and redox cycle process of methane partial oxidation via lattice oxygen over LaFeO3 perovskite catalyst, *J. Phys. Chem. B*, **110**, (51), pp. 25856–25862.
44. Dai, X., Yu, C., Li, R., *et al.*, (2008). Synthesis gas production using oxygen storage materials as oxygen carrier over circulating fluidized bed, *Journal of Rare Earths*, **26**, pp. 76–80.
45. Dai, X., Yu, C. and Wu, Q. (2008). Comparison of $LaFeO_3$, $La_{0.8}Sr_{0.2}FeO_3$, and $La_{0.8}Sr_{0.2}Fe_{0.9}Co_{0.1}O_3$ perovskite oxides as oxygen carrier for partial oxidation of methane, *J. Nat. Gas Chem.*, **17**, pp. 415–418.

46. Kuhn, J. and Ozkan, U. (2008). Effect of Co content upon the bulk structure of Sr- and Co-doped $LaFeO_3$, *Catal. Lett.*, **121**, pp. 179–188.
47. Mudu, F., Arstad, B. and Bakken, E. (2010). Perovskite-type oxide catalysts for low temperature, anaerobic catalytic partial oxidation of methane to syngas, *J. Catal.*, **275**, pp. 25–33.
48. Mudu, F., Arstad, B., Fjellvåg, H., *et al.* (2011). Thermodynamic Control of Product Formation During the Reaction Between $CH_4$ and Pt Promoted Ceria-zirconia Solid Solutions, *Catal. Lett.*, **141**, pp. 8–14.
49. Au, C. and Wang, H. (1996). Pulse study of methane partial oxidation to syngas over $SiO_2$-supported nickel catalysts, *Catal. Lett.*, **41**, pp. 159–163.
50. Hu, Y. and Ruckenstein, E. (1998). Isotopic GC-MS study of the mechanism of methane partial oxidation to synthesis gas, *J. Phys. Chem. A*, **102**, pp. 10568–10571.
51. Hu, Y. and Ruckenstein, E. (1998). Broadened pulse-step change isotopic sharp pulse analysis of the mechanism of methane partial oxidation to synthesis gas, *J. Phys. Chem. B*, **102**, pp. 230–233.
52. Hu, Y. and Ruckenstein, E. (1995). Pulse MS study of the partial oxidation of methane over $Ni/La_2O_3$ catalyst, *Catal. Lett.*, **34**, pp. 41–50.
53. Hu, Y. and Ruckenstein, E. (1996). Transient kinetic studies of partial oxidation of $CH_4$, *J. Catal.*, **158**, pp. 260–266.
54. Tsipouriari, V., Zhang, Z. and Verykios, X. (1998). Catalytic partial oxidation of methane to synthesis gas over Ni-based catalysts - I. Catalyst performance characteristics, *J. Catal.*, **179**, pp. 283–291.
55. Prettre, M., Vermeiren, W., Blomsma, E., *et al.* (1992). Catalytic and thermodynamic approach of the oxyreforming reaction of methane, *Catal. Today*, **13**, pp. 427–436.
56. York, A., Xiao, T. and Green, M. (2003). Brief overview of the partial oxidation of methane to synthesis gas, *Topics in Catalysis*, **22**, (3–4), pp. 345–358.
57. Enger, B., Lødeng, R. and Holmen, A. (2008). A review of catalytic partial oxidation of methane to synthesis gas with emphasis on reaction mechanisms over transition metal catalysts, *Appl Catal. A*, **346**, pp. 1–27.
58. Uchijima, T., Nakamura, J., Sato, K., *et al.* (1994). Production of synthesis gas by partial oxidation of methane and reforming of methane with carbon dioxide, *Stud. Surf. Sci. Catal.*, **81**, pp. 325–327.
59. Buyevskaya, O., Wolf, D. and Baerns, M. (1994). Rhodium-catalyzed partial oxidation of methane to CO and $H_2$ — transient studies on its mechanism, *Catal. Lett.*, **29**, pp. 249–260.
60. Mallens, E., Hoebink, J. and Marin, G. (1997). The reaction mechanism of the partial oxidation of methane to synthesis gas: A transient kinetic study over rhodium and a comparison with platinum, *J. Catal.*, **167**, pp. 43–56.
61. Li, B., Maruyama, K., Nurunnabi, M., *et al.*, (2004). Temperature profiles of alumina-supported noble metal catalysts in autothermal reforming of methane, *Appl. Catal. A*, **275**, pp. 157–172.
62. Haynes, W. (ed) (2010–2011), *Handbook of Chemistry and Physics*, $91^{st}$ Edition, CRC Press, Boca Raton, FL.
63. Allen, G., Bayles, R., Gile, W., *et al.* (1986). Small particle melting of pure metals, *Thin Solid Films*, **144**, (2), pp. 297–308.
64. Buffat, P. and Borrel, J. (1976). Size effect on melting temperature of gold particles, *Phys. Rev. A*, **13**, pp. 2287–2298.
65. Sun, J. and Simon, S. (2007). The melting behavior of aluminum nanoparticles, *Thermochimica Acta*, **463**, pp. 32–40.
66. Bartholomew, C. (1982). Carbon deposition in steam reforming and methanation, *Catal. Rev. Sci. Eng.*, **24**(1), pp. 67–112.
67. Rostrup-Nielsen, J., Sehested, J. and Nørskov, J. (2002). Hydrogen and synthesis gas by steam- and $CO_2$ reforming, *Adv. Catal.*, **47**, pp. 65–139.

68. Rostrup-Nielsen J. (2008). Steam Reforming, in G. Ertl, H. Knözinger, F. Schüth, *et al.*(eds), *Handbook of Heterogeneous Catalysis,* $2^{nd}$ Edition, Vol. 6, Wiley-VCH, Weinheim, Chapter 13.11, pp. 2882–2903.
69. Besenbacher, F., Chorkendorff, I., Clausen, B., *et al.* (1998). Design of a surface alloy catalyst for steam reforming, *Science*, **279**, pp. 1913–1915.
70. Jacobsen, J., Pleth Nielsen, L., Besenbacher, F., *et al.* (1995). Atomic-scale determination of misfit dislocation loops at metal-metal interfaces, *Phys. Rev. Lett.*, **75**, pp. 489–492.
71. Rostrup-Nielsen, J. (1984). Sulfur-passivated nickel catalysts for carbon-free steam reforming of methane, *J. Catal.*, **85**, pp. 31–43.
72. Snoeck, J., Froment, G. and Fowles, M. (1997). Filamentous carbon formation and gasification: Thermodynamics, driving force, nucleation, and steady-state growth, *J. Catal.,* **169**, pp. 240–249.
73. Helveg, S., López-Cartes, C., Sehested, J., *et al.* (2004). Atomic-scale imaging of carbon nanofibre growth, *Nature*, **427**, pp. 426–429.
74. de Groote, A. and Froment, G. (1996). Simulation of the catalytic partial oxidation of methane to synthesis gas, *Appl. Catal. A*, **138**, pp. 245–264.
75. Olsbye, U., Tangstad, E. and Dahl Stud, I. (1994). Partial oxidation of methane to synthesis gas in a fluidized-bed reactor, *Surf. Sci. Catal.*, **81**, pp. 303–308.
76. Marshall, K. J. and Mleczko, L. (1999). CFO modelling of an internally circulating fluidized-bed reactor, *Chem. Eng. J.*, **54**, (13–14), pp. 2085–2093.
77. Bharadwaj, S. and Schmidt, L. (1994). Synthesis gas formation by catalytic oxidation of methane in fluidized bed reactors, *J. Catal.*, **146**, pp. 11–21.
78. Santos, A., Mendez, M. and Santamaria, J. (1994). Partial oxidation of methane to carbon monoxide and hydrogen in a fluidized bed reactor, *Catal. Today*, **21**, pp. 481–488.
79. Eberly, pp., Goetsch, D., Say, G., *et al.* (1989). Method and catalyst for synthesis gas production. Exxon Research and Engineering Company, EP 0335 668 A3. 10 ppp.
80. Ioannides, T. and Verykios, X. (1998). Development of a novel heat-integrated wall reactor for the partial oxidation of methane to synthesis gas, *Catal. Today*, **46**, pp. 71–81.
81. Veser, G., Frauhammer, J. and Friedle, U. (2000). Syngas formation by direct oxidation of methane — Reaction mechanisms and new reactor concepts, *Catal. Today,* **61**, pp. 55–64.
82. Hickman, D. and Schmidt, L. (1993). Production of syngas by direct catalytic oxidation of methane, *Science*, **259**, pp. 343–346.
83. Heitnes, K., Lindberg, S., Rokstad, O., *et al.* (1994). Catalytic partial oxidaiton of methane to synthesis gas using monolithic reactors, *Catal. Today*, **21**, pp. 471–480.
84. Heitnes, K., Lindberg, S., Rokstad, O., *et al.* (1995). Catalytic partial oxidation of methane to synthesis gas, *Catal. Today*, **24**, pp. 211–216.
85. Aartun, I., Venvik, H., Holmen, A., *et al.* (2005). Temperature profiles and residence time effects during catalytic partial oxidation and oxidative steam reforming of propane in metallic microchannel reactors. *Catal. Today*, **110**, pp. 98–107.
86. Donazzi, A., Beretta, A., Groppi, G., *et al.* (2008). Catalytic partial oxidation of methane over a 4% Rh/alpha $Al_2O_3$ catalyst Part I: Kinetic study in annular reactor, *J. Catal.*, **255**, pp. 241–258.
87. Tanaka, H., Kaino, R., Okumura, K., *et al.* (2009). Catalytic performance and characterization of Rh-CeO2/MgO catalysts for the catalytic partial oxidation of methane at short contact time, *J. Catal.*, **268**, pp. 1–8.
88. Lewis, W., Gilliland, E. and Reed, W. (1949). Reaction of methane with copper oxide in a fluidized bed, *Ind. Eng. Chem.*, **41**, pp. 1227–1237.
89. Otsuka, K., Ushiyama, T. and Yamanaka, I. (1993). Partial oxidation of methane using the redox of cerium oxide, *Chem. Lett.*, **22**, pp. 1517–1520.
90. Dai, X., Li, R., Yu, C., *et al.*, (2006). nsteady-state direct partial oxidation of methane to synthesis gas in a fixed-bed reactor using $AFeO_3$ (A = La, Nd, Eu) perovskite-type oxides as oxygen storage, *J. Phys. Chem. B*, **110**, pp. 22525–22531.

91. Kharton, V., Patrakeev, M., Waerenborgh, J., *et al.* (2005). Methane oxidation over perovskite-related ferrites: Effects of oxygen nonstoichiometry, *Solid State Sci.*, **7**, pp. 1344–1352.
92. Hu, Y. and Ruckenstein, E. (2004). Catalytic conversion of methane to synthesis gas by partial oxidation and $CO_2$ reforming, *Adv. Catal.*, **48**, pp. 297–345.
93. Teraoka, Y., Zhang, H., Furukawa, S., *et al.* (1985). Oxygen permeation through perovskite-type oxides, *Chem. Lett.*, **14**, pp. 1743–1746.
94. Teraoka, Y., Nobunaga, T. and Yamazoe, N. (1988). Effect of cation substitution on the oxygen semipermeability of perovskite-type oxides, *Chem. Lett.*, **17**, pp. 503–506.
95. Teraoka, Y., Nobunaga, T., Okamoto, K., *et al.* (1991). Influence of constituent metal-cations in substituted $LaCoO_3$ on mixed conductivity and oxygen permeability, *Solid State Ion.*, **48**, pp. 207–212.
96. Buwmeister, H. and Burggraf, A. (1996). Dense ceramic membranes for oxygen separation In A. Burggraf and L. Cot (eds), *Fundamentals of Inorganic Membrane Science and Technology*, Elsevier, Amsterdam, pp. 435–460.
97. Balachandran, U., Dusek, J., Mieville, R., *et al.* (1995). Dense ceramic membranes for partial oxidation of methane to syngas, *Appl. Catal. A*, **133**, pp. 19–29.
98. Balachandran, U., Dusek, J., Maiya, P., *et al.* (1997). Ceramic membrane reactor for converting methane to syngas, *Catal. Today*, **36**, pp. 265–272.
99. Mazanec, T., Cable, T., Frye, J., *et al.* (1994). Solid multi-component membranes, electrochemical reactor components, electrochemical reactors and use of membranes, reactor components, and reactor for oxidation reactions, *US Patent 5,306,411*.
100. Wang, H. and Ruckenstein, E. (1999). Catalytic partial oxidation of methane to synthesis gas over gamma-$Al_2O_3$-supported rhodium catalysts, *Catal. Lett.*, **59**, pp. 121–127.
101. Mallens, E., Hoebink, J. and Marin, G. (1995). An investigation on the reaction mechanism for the partial oxidation of methane to synthesis gas over platinum, *Catal. Lett.*, **33**, pp. 291–304.
102. Aasberg-Petersen, K., Christensen, T., Stub Nielsen, C., *et al.* (2003). Recent developments in autothermal reforming and pre-reforming for synthesis gas production in GTL applications, *Fuel Processing Technology*, **83**, pp. 253–261.
103. Dybkjær, I. and Christensen, T. (2001). Syngas for large scale conversion of natural gas to liquid fuels, *Stud. Surf. Sci. Catal.*, **136**, pp. 435–440.
104. Sarofim, A. and Flanagan, R. (1976). $NO_x$ control for stationary combustion sources, *Prog. Energy Combust. Sci.*, **2**, pp. 1–25.
105. Hohn, K., Huang, C. and Cao, C. (2009). Catalytic ignition of light hydrocarbons, *J. Nat. Gas Chem.*, **18**, pp. 115–123.
106. Veser, G. and Schmidt, L. (1996). Ignition and extinction in the catalytic oxidation of hydrocarbons over platinum, *AIChE Journal*, **42**, pp. 1077–1087.
107. Veser., G., Ziauddin, M. and Schmidt, L. (1999). Ignition in alkane oxidation on noble-metal catalysts, *Catal. Today*, **47**, pp. 219–228.
108. Ziauddin, M., Veser, G. and Schmidt, L. (1997). Ignition-extinction of ethane-air mixtures over noble metals, *Catal. Lett.*, **46**, pp. 159–167.
109. Beck, I., Bukhtiyarov, V., Pakharukov, I., *et al.* (2009). Platinum nanoparticles on $Al_2O_3$: Correlation between the particle size and activity in total methane oxidation, *J. Catal.*, **268**, pp. 60–67.
110. Dueso, C., Abad, A., García-Labiano, F., *et al.* (2010). Reactivity of a $NiO/Al_2O_3$ oxygen carrier prepared by impregnation for chemical-looping combustion, *Fuel*, **89**, pp. 3399–3409.

# Chapter 9

# Oxygen Activation for Fuel Cell and Electrochemical Process Applications

Christophe COUTANCEAU and Stève BARANTON*

This chapter aims at describing the main issues for the activation of molecular and atomic oxygen in electrochemical processes. The electroreduction of molecular oxygen has important applications in the energy conversion domain (i.e. fuel cells and metal-air batteries), in electrosynthesis reactors and in the chlor-alkali process. The electro-oxidation of organic compounds, for the electrosynthesis process, fuel cell or sensor applications, necessitates an efficient activation of the oxygen atom of water, which will be the second aspect of this chapter.

## 9.1. Introduction

The electrocatalytic activation of oxygen is of great importance in several electrochemical systems and processes such as fuel cells,[1] metal-air batteries,[2,3] the chlor-alkali process,[4] electrosynthesis reactors,[5] etc.

In a proton exchange electrolyte fuel cell (PEMFC) working with hydrogen as the fuel, as do metal-air batteries and the chlor-alkali process, the activation of oxygen is a necessary step for the dioxygen reduction reaction at the cathode of the electrochemical reactor. This reaction limits the efficiency of these systems,[6] as its kinetics are rather low in comparison with, for instance, the reaction kinetics of hydrogen oxidation in a fuel cell.[7,8] This leads to a high overpotential for the oxygen reduction reaction (ORR) to occur, close to 0.23 V under standard conditions.

In fuel cells working with a liquid fuel, usually an alcohol such as methanol (a direct methanol fuel cell – DMFC), ethanol (a direct ethanol fuel cell – DEFC), glycerol (a direct glycerol fuel cell – DGFC), etc., in addition to the necessity to activate the ORR at the cathode, the alcohol oxidation reaction at the anode also involves a high overpotential.[9] This high overpotential is mainly due to the formation, after dissociative adsorption of the alcohol at the catalyst surface,[10] of poisoning species which block the catalytic surface; the main one adsorbed is carbon monoxide.[5,11,12]

*Laboratoire de Catalyse en Chimie Organique, UMR 6503 CNRS-Université de Poitiers, 40 avenue du Recteur Pineau, 86022 Poitiers cedex. France.

In order to remove such poisoning species from the catalytic surface, extra oxygen atoms have to be provided to complete the oxidation reaction to $CO_2$; these extra oxygen atoms, from water, have to be activated at the catalytic surface. This phenomenon is called the bifunctional theory of electrocatalysis or, more simply, the bifunctional mechanism.[13,14]

The same problem is encountered during electrosynthesis by the oxidation of a chemical compound.[5] In an electrochemical reactor, the oxidation reaction has to be counterbalanced with a reduction reaction in order to close the electrical circuit. Under these conditions, it is better for industrial applications to use oxygen from air, which is free, as the oxidative agent. Such a system then becomes very close to a fuel cell system, apart from the oxidation reaction that has to be controlled here, whereas the complete oxidation of alcohol into $CO_2$ is sought in direct alcohol fuel cells (DAFCs). For this reason, fuel cell systems will be considered in this chapter to illustrate the important problem of oxygen activation for electrochemical processes.

## 9.2. Thermodynamics

As an example, the working principle of a DEFC is illustrated in Fig. 9.1. The electrochemical cell consists of two electronic conducive electrodes, an anode and a cathode separated by an ionic conductive solid electrolyte (a proton exchange membrane generally of Nafion type).[15–17] At the anode the electro-oxidation of alcohol takes place as follows:

$$\begin{gathered} CH_3CH_2OH + 3H_2O \rightarrow 2CO_2 + 12H^+ + 12e^- \\ E^0_{CO_2/C_2H_5OH} = 0.085\ \text{V vs SHE} \end{gathered} \tag{9.1}$$

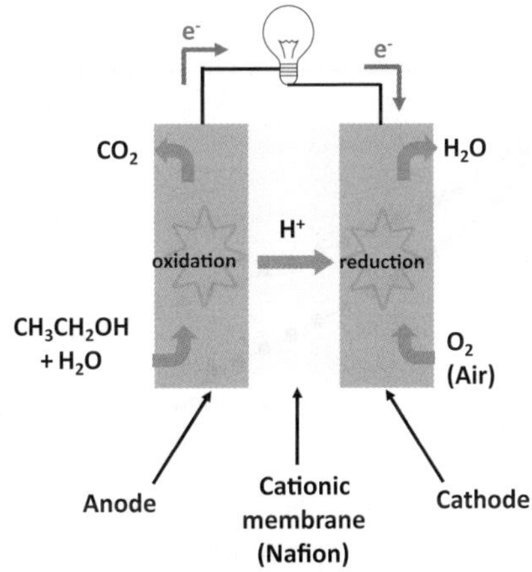


**Figure 9.1.** Working principle of a direct ethanol fuel cell.

whereas at the cathode, the oxygen reduction reaction occurs according to:

$$O_2 + 4H^+ + 4e^- \rightarrow 2H_2O$$
$$E^0_{O_2/H_2O} = 1.229 \text{ V vs SHE} \quad (9.2)$$

where $E^0_{CO_2/EtOH}$ and $E^0_{O_2/H_2O}$ are the standard electrode potentials and SHE is the standard hydrogen electrode.

Equation 9.1 clearly indicates that the extra atoms of oxygen have to be provided, from water, and therefore activated, to perform the complete oxidation of ethanol into carbon dioxide. The overall combustion reaction is then:

$$CH_3CH_2OH + 3O_2 \rightarrow 2CO_2 + 3H_2O \quad (9.3)$$

The $\Delta G^0_R$ and $\Delta H^0_R$ of ethanol combustion into $CO_2$ and $H_2O$ under standard conditions are equal to $-1325$ kJ mol$^{-1}$ and 1366 kJ mol$^{-1}$, respectively,[18] so that the following equality is respected:

$$U^0_{eq.} = -\frac{\Delta G^0_R}{nF} = E^0_{O_2/H_2O} - E^0_{CO_2/EtOH} = 1.144 \text{ V} \quad (9.4)$$

with $F = 96485$ C mol$^{-1}$, the Faraday constant, and $n = 12$, the number of electrons exchanged per molecule for complete oxidation to $CO_2$.

The energy efficiency under reversible standard conditions of the cell is then:

$$\varepsilon_{rev.} = -\frac{\Delta G^0_R}{\Delta H^0_R} = \frac{1325}{1366} = 0.97 \quad (9.5)$$

Figure 9.2 shows typical polarization curves recorded in a DAFC with different anodic catalysts. Even with the best recognized catalysts for ethanol

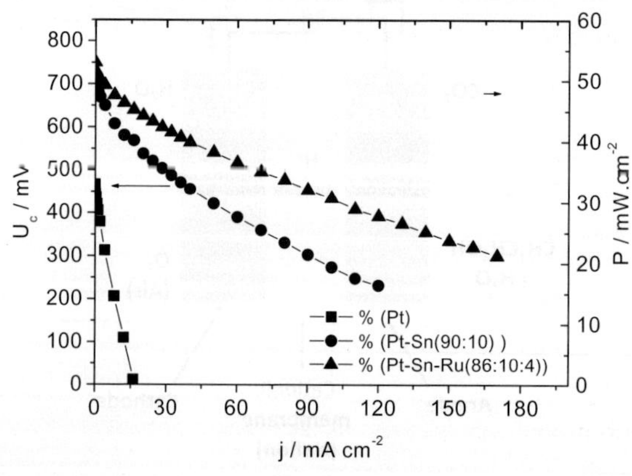


**Figure 9.2.** Polarization curves recorded at 80°C in a DEFC fitted with different anode catalysts.

electro-oxidation,[19,20] it appears that the open circuit voltage (OCV) is lower than the standard equilibrium voltage which, under the present experimental temperature and with the approximation that ΔH (enthalpy of reaction) and ΔS (entropy of reaction) are constant over the considered temperature range, is close to:

$$U_{eq.}(T) = \frac{\Delta G_R(T)}{nF} = \frac{\Delta H_R^0 - T\Delta S^0}{nF} = \frac{1366.10^3 - 353 \times 161}{12 \times 96485} = 1.131\ V \qquad (9.6)$$

Moreover, when a current density j is provided by the cell its voltage U(j) decreases greatly. In the first approximation, these effects result mainly from three limiting factors: the charge transfer overpotentials $\eta_a$ and $\eta_c$ at the anode and at the cathode, respectively, due to reaction rates of the electrochemical processes, the ohmic drop $R_e j$ in the electrolyte and interfaces, and mass transfer limitations for reactants and products.[21] The cell voltage can then be expressed as follows:

$$U(j) = E_c(j) - E_a(j) = E_c^0 + \eta_c - \left(E_a^0 + \eta_a\right) - R_e j = U_{eq.}^0 - \eta_a + \eta_c + R_e j \qquad (9.7)$$

The overpotentials $\eta_a$ and $\eta_c$ correspond to $(E_a - E_a^0 > 0)$ and $(E_c - E_c^0 < 0)$, where $E_a$ and $E_c$ are the anode and cathode potentials at a given current density, respectively, $E_a^0$ and $E_c^0$ are the reaction equilibrium potentials at the anode and at the cathode, respectively. The overpotentials $\eta_a$ and $\eta_c$ take into account both the slow kinetics of the electrochemical reactions (activation polarization) and the limiting rate of mass transfer (concentration polarization). Here, the fuel crossover effect, i.e. the depolarization of the cathode by the alcohol coming from the anode through the membrane which shifts its potential towards more negative potentials, thus increasing the absolute value of the cathode overpotential,[22] is neglected as its impact on the cell voltage is lower than that of other limitations.[23] Considering the voltage limitations occurring in an operating cell, the voltage efficiency of the cell can be determined as follows:

$$\varepsilon_U = \frac{U(j)}{U_{eq.}^0} \qquad (9.8)$$

If the ethanol oxidation reaction is complete, leading to $CO_2$, twelve moles of electrons are exchanged per mole of ethanol consumed. But, this reaction can stop at stages of the mechanism, leading to the formation of acetic acid or aceltaldehyde, for example, with a transfer of only four moles or two moles of electrons involved per mole of ethanol consumed, respectively. In that case, a faradic efficiency will appear, expressed as follows:

$$\varepsilon_F = \frac{n_{exp.}}{n_{th.}} \tag{9.9}$$

where $n_{exp.}$ is the experimental number of exchanged electrons per ethanol molecule and $n_{th.}$ is the number of electrons exchanged for the complete oxidation of ethanol into $CO_2$ ($n_{th.} = 12$).

From these three efficiencies, the overall energy efficiency of the cell can be expressed as follows:

$$\varepsilon_{cell} = \varepsilon_{rev.} \times \varepsilon_U \times \varepsilon_F = \frac{\Delta G^0}{\Delta H^0} \times \frac{U(j)}{U^0_{eq.}} \times \frac{n_{exp.}}{n_{th.}}$$

$$= \frac{n_{th.}FU^0_{eq.}}{\Delta H^0_R} \times \frac{U(j)}{U^0_{eq.}} \times \frac{n_{exp.}}{n_{th.}} = \frac{n_{exp.}FU(j)}{\Delta H^0_R} \tag{9.10}$$

From this equation, it appears that the better way to significantly increase the overall energy efficiency is to increase $\varepsilon_E$ (the potential efficiency) and $\varepsilon_F$ (the faradic efficiency), since $\varepsilon_{rev.}$ (the reversible efficiency) is given by the thermodynamics (it can be slightly increased by changing the pressure and temperature operating conditions).

The decrease of the cell global overvoltage $|\eta|$ is directly related to the increase of the rate of the electrochemical reactions occurring at both electrodes. Both the electrode potential and the catalytic electrode material will synergistically increase the reaction rate V. Indeed, the current intensity j is proportional to the rate of reaction V. For a first order electrochemical reaction, the rate is proportional to the reactant concentration $C_i$ and the current density can be expressed as follows:

$$j = \frac{nFV_i}{A} = nFk(T, E)C_i = nFC_i e^{-\frac{\Delta \breve{G}^+}{RT}} \tag{9.11}$$

where n is the number of exchanged electrons, F is the Faraday constant (96485 C $mol^{-1}$), A is the geometric surface area of the electrode, k(T, E) the kinetics constant of the reaction, R the perfect gas constant, T the temperature and $\Delta\breve{G}^+$ the electrochemical activation energy. In electrocatalysis, two essential activation terms have to be considered; the first comes from the electrode potential activation energy and the second comes from the chemical activation energy. Considering an electrode, the cathode where the reduction reaction of oxygen has to be activated, as an example, the electrochemical activation energy $\breve{G}^+$ is:

$$\Delta\breve{G}^+ = \Delta G_0^+ - \alpha_c nFE^+ \tag{9.12}$$

where $\Delta G_0^+$ is the chemical energy of activation, $\alpha_c nFE^+$ is the electrical part of the activation energy, $\alpha_c$ the transfer coefficient ($0 < \alpha_c < 1$) corresponding to the fraction of the total electric energy used for the reaction activation, and $E^+$ is the cathode potential. The cathodic current density $j_c$ can then be expressed as:

$$j_c = nFk_0C_{O_2}e^{-\frac{\Delta G_0^+}{RT}}e^{\frac{\alpha_c nFE^+}{RT}} = j_{0,c}e^{\frac{\alpha_c nFE^+}{RT}} \quad (9.13)$$

Thus the increase of the current density at a given potential involves the increase of the exchange current density $j_{0,c}$, which is the result of the increased electro-catalytic activity of the electrode, i.e. a decrease of the chemical activation energy $\Delta\breve{G}^+$. Such a goal can be reached by modifying the nature and structure of the electrode material.

## 9.3. Molecular Oxygen Electroreduction

Oxygen can undergo two different reduction reactions, one following a four-electron process to form water as presented in Eq. 9.2, but also a second one following a two-electron process as shown in the following equation:

$$O_2 + 2H^+ + 2e^- \rightarrow H_2O_2 \quad E^0{}_{O_2/H_2O} = 0.695\text{ V vs SHE} \quad (9.14)$$

Although thermodynamic data suggest a high instability of hydrogen peroxide in an acid medium, the kinetics of its decomposition is very low in aqueous solution. Tarasevich *et al.* proposed that the more likely reaction mechanism for its decomposition involved a redox disproportionation reaction:[24]

$$H_2O_2 + 2H^+ + 2e^- \rightarrow 2H_2O \quad (9.15)$$

$$\underline{H_2O_2 \rightarrow O_2 + 2H^+ + 2e^-} \quad (9.16)$$

$$2H_2O_2 \rightarrow O_2 + 2H_2O \quad (9.17)$$

Under these conditions, the transfer of several electrons leading to unstable intermediates involves an increase of the free energy for the breaking of the O-O bond, which could explain the low kinetics of $H_2O_2$ decomposition.

The complete electroreduction of molecular oxygen in water involving the exchange of four electrons (a four-electron process) as presented in Eq. 9.2 is highly irreversible and the theoretical thermodynamic potential cannot be achieved at room temperature in aqueous medium, even using platinum electrodes, without working under very special experimental conditions. For example, the reversible potential of the $O_2/H_2O$ redox system was determined by Bockris and Huq[25] in an $O_2$-saturated 0.01 N $H_2SO_4$ electrolyte after a very long and thorough purification of the electrolytic solution, using a platinum electrode which had undergone a very thorough oxidation of its surface by treatment in a concentrated solution of $HNO_3$ and $H_2SO_4$ followed by exposure to pure oxygen at 500°C for two hours. Hoare[26,27] has also reported obtaining the reversible potential of the $O_2/H_2O$ redox system on the face of a platinized membrane. But, in general, the potential achieved for the $O_2/H_2O$ redox system under an oxygen pressure of 1 bar and at room temperature does not exceed 1.1 V vs SHE.

The development by Frumkin of the rotating ring-disc and the related hydrodynamic theory[28,29] led to important progresses for the determination of the oxygen reaction reduction mechanism. In particular, several reaction pathways have been evidenced.[30] In such a technique the oxygen reduction reaction is performed on a disc electrode, the potential of which is linearly varied between two limits, whereas a ring electrode is maintained at a potential high enough to allow the oxidization of the hydrogen peroxide, formed at the disc electrode, into oxygen (Fig. 9.3).

From data obtained by the rotating ring-disc electrode technique, several models for oxygen reduction in aqueous solution were proposed;[31–34] the general scheme (Scheme 9.1) describing the parallel and serial reactions involving oxygen and hydrogen peroxide is one of them.

As shown in Scheme 9.1, $O_{2(bulk)}$, $O_{2(surf.)}$ and $O_{2(ads.)}$ correspond to molecular oxygen in the solution, adjacent to the electrode surface and adsorbed on a catalytic site, respectively.

According to the reaction pathway involved in the oxygen reduction reaction, the serial or parallel route electrode materials can be classified into two groups. The first group corresponds to materials favouring the formation of the hydrogen peroxide intermediate in the course of the oxygen reduction reaction; gold,[34] graphite[35] and mercury[24] belong to this group. The second involves the parallel mechanisms, meaning the direct four-electron reduction process of oxygen into water at the same

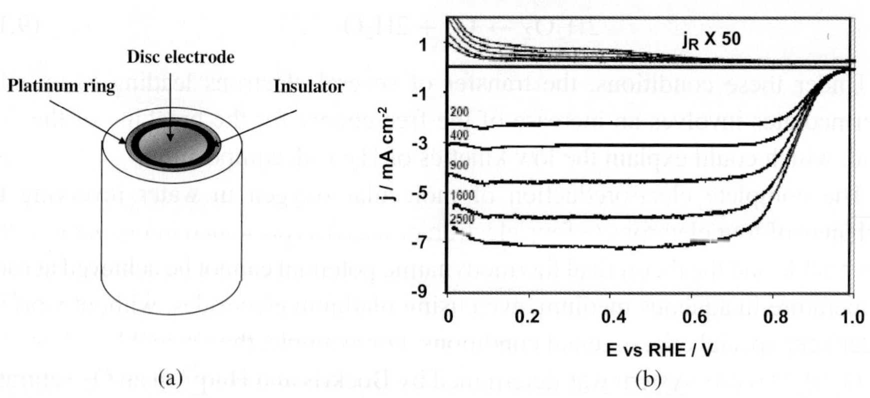


**Figure 9.3.** (a) Scheme of a rotating ring-disc electrode. (b) Polarization curves of ORR recorded at a Pt (40 wt%)/C disc electrode and at a Pt ring electrode with different electrode rotation rates from 200 to 2,500 revolutions per minute recorded in $O_2$-saturated 0.1 $HClO_4$ electrolyte. ($T = 20°C$, $v = 5\,mV\,s^{-1}$, $E_{ring} = 1.2\,V$ vs reversible hydrogen electrode (RHE)).

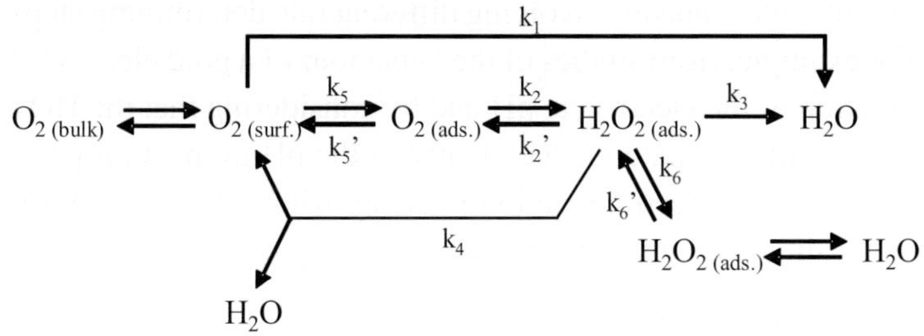


**Scheme 9.1.** Parallel and serial reactions of the oxygen reduction.

time as the two-electron process via the formation of hydrogen peroxide, and is represented by platinum[36] and silver.[24]

### 9.3.1. *Oxygen reduction on first group materials*

When considering materials of the first group, such as gold and graphite, most authors propose that the reduction reaction of oxygen involves the formation of a s-$O_2^-$ species (where s is an adsorption site), even if the proposed mechanism could differ. For example, Taylor and Humffray[37] proposed the following mechanism for the oxygen reduction into hydrogen peroxide, where the rate determining step (rds) is the step described in Eq. 9.18 for pH $< 10$:

$$s + O_2 \rightarrow s - O_{2(ads.)} \qquad \text{(rds)} \tag{9.18}$$

$$s - O_{2(ads.)} + e^- \rightarrow s - O_2^- \tag{9.19}$$

$$s - O_2^- + H_2O + e^- \rightarrow s + HO_2^- + OH^- \tag{9.20}$$

In the mechanism proposed by Morcos and Yeager,[38] the two first steps are identical to those presented above but the last one becomes:

$$2s - O_2^- + H_2O + e^- \rightarrow 2s + HO_2^- + OH^- + O_2 \tag{9.21}$$

whereas Zurilla *et al.*[34] found no evidence for the adsorption step to be rate determining before the first electron transfer, and described the steps seen in Eqs 9.18 and 9.19 as a single step:

$$s + O_{2(ads.)} + e^- \rightarrow s - O_2^- \tag{9.22}$$

These mechanisms could explain the fact that, on such surfaces, the kinetics of the oxygen electroreduction reaction into hydrogen peroxide is higher in an alkaline medium than in an acid medium. Indeed, if the s-$O_2^-$ species is likely to be stable in an alkaline medium, the free energy for the formation of this species is certainly higher in an acid medium. The formation of such an entity must involve the simultaneous transfer of an electron and a proton, which affects the kinetics of the process.

However, other mechanisms involving different rate determining steps were also proposed. For example, from studies of the behaviour of a gold electrode toward the ORR as a function of the electrolyte pH and by considering that the $HO_2^-$ ion operates as a reaction intermediate independently of the pH in most proposed kinetics, Bonnemay *et al.*[39] developed a reaction pathway where the rate determining step consisted of the simultaneous exchange of two electrons:

$$O_2 + 2e^- \rightarrow O_2^{2-} \qquad \text{(rds)} \qquad (9.23)$$

$$O_2{}^{2-} + H^+ \rightarrow HO_2^- \qquad (9.24)$$

$$HO_2^- + H^+ + e^- \rightarrow H_2O_2^- \qquad \text{(slow step)} \qquad (9.25)$$

$$H_2O_2^- + 2H^+ + e^- \rightarrow 2H_2O \qquad (9.26)$$

The works carried out by Genshaw *et al.* showed that the ORR on gold followed different mechanisms according to the potential range, which were characterized by different Tafel slopes.[40] In the region of low current density (lower reduction overpotentials) the complete reduction of $O_2$ in water via the four-electron process dominated, whereas in the high current density region (higher reduction overpotentials) $O_2$ is mainly reduced into hydrogen peroxide via the two-electron process. The existence of different mechanisms of $O_2$ reduction depending on the potential range was confirmed by Strbac *et al.* on an Au(100) surface in an alkaline medium.[41] They found three potential regions each involving a different reaction pathway: in the low current density region, where $AuOH^{(1-\lambda)-}$ species are present at the gold surface (coming from a partial electron transfer: $Au + OH^- \rightarrow AuOH^{(1-\lambda)-} + \lambda e^-$), the four-electron producing water process is favoured; at more negative potentials, the two-electron process producing hydrogen peroxide becomes prominent; and finally at very high overpotentials, the four-electron process occurs again. However, the first group materials are generally less active towards ORR than those of the second group; in particular, platinum is known to be the most active monometallic catalyst for this reaction. This fact is illustrated in Fig. 9.4, where the activity of bulk gold and bulk platinum electrodes are compared in an acid medium.

### 9.3.2. *Oxygen reduction on second group materials*

Platinum, which belongs to the second group of materials,[36,42] is considered to have the best catalytic activity for the ORR in an acid as well as in an alkaline medium. This metal possesses surface properties which facilitate the adsorption of molecular oxygen, and thus it is able to catalyse the direct reduction of oxygen into water. Moreover, it has a greater stability in an $O_2$-saturated acid medium, a very oxidant medium, than less noble materials and its surface properties are less altered, or with a lower kinetics. However, despite these properties, the ORR on the platinum surface

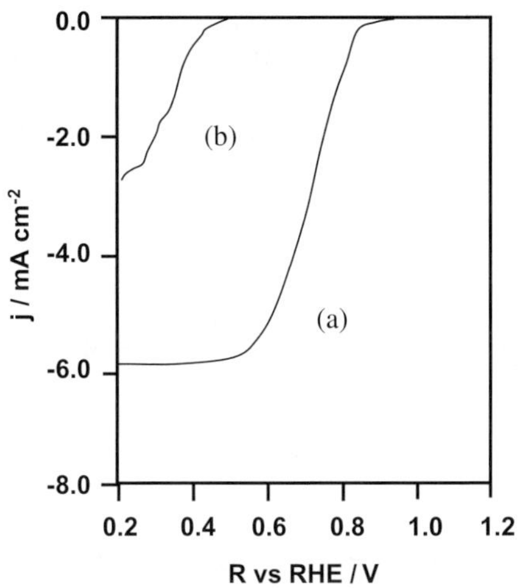


**Figure 9.4.** Polarization curves of ORR on (a) bare platinum and (b) bare gold electrodes recorded in $O_2$-saturated 0.5 M $H_2SO_4$ electrolyte (T = 20° C, v = 2 mV $s^{-1}$, $\Omega$ = 2500 rpm).

remains an irreversible reaction occurring with low kinetics. As shown in Fig. 9.4, the reaction starts at a potential of around 1.0 V vs RHE, which corresponds to an overpotential of at least 0.2 V. Moreover, the great irreversibility of the reaction makes it difficult to determine the kinetics constants; the exchange current densities are generally in the order of $10^{-4}$–$10^{-7}$ mA $cm^{-2}$,[43] which avoids the possibility of measuring current densities in the neighbourhood of the equilibrium potential.[44] The kinetics then has to be determined in a potential range where the current densities are high enough to be measured; such measurements are then realized under non-equilibrium conditions and are analysed using simplified models. The first step in the oxygen reduction mechanism on platinum can be described by three different possible models leading to three different reaction pathways as shown in Fig. 9.5.[45]

The adsorption of oxygen on platinum involving the Bridge model and the Griffiths model can lead to the direct reduction in water via the four-electron process, whereas the adsorption mode involving the Pauling model leads to the formation of a hydrogen peroxide intermediate which can either remain adsorbed and further reduced (four-electron process) or can desorb producing hydrogen peroxide (two-electron process).

The Griffith mechanism and the Bridge mechanism involve lateral interactions between the dioxygen and the metal. The interaction between the π orbital of dioxygen and the $d_z$ orbital of platinum is a bonding one, whereas the non-bonding interaction is realized via the partially filled $d_{xz}$ or $d_{yz}$ orbitals of platinum and the orbital π* of dioxygen;[48,49] Figure 9.6 illustrates these interactions. The bonding lateral interactions are very strong and tend to lengthen and weaken the O-O bond until it

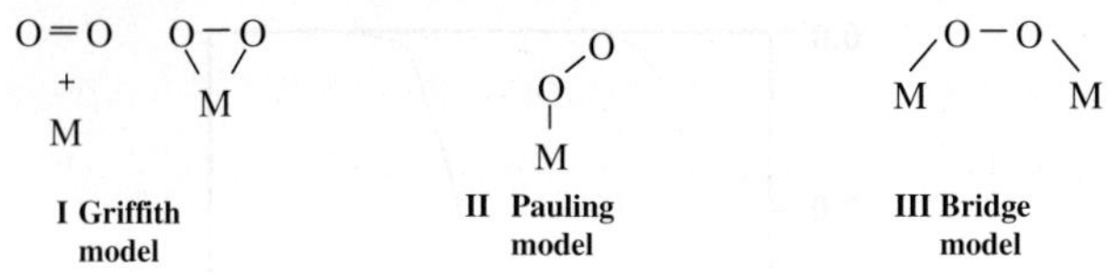


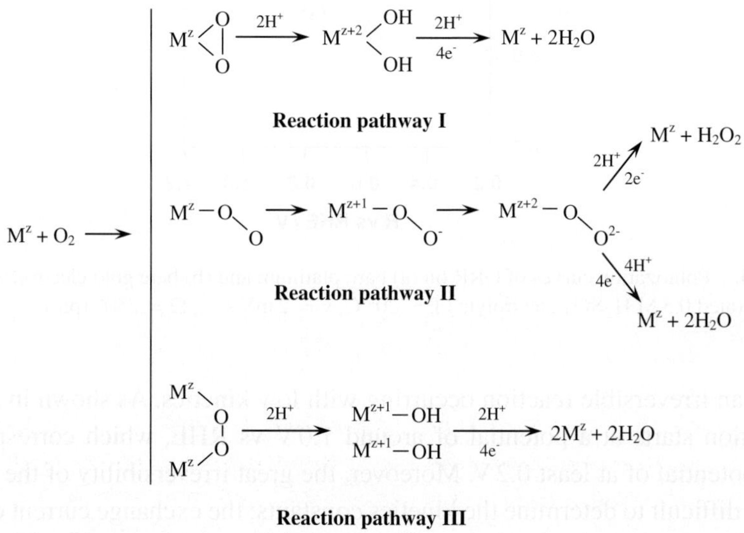


**Figure 9.5.** Possible reaction pathways as a function of the oxygen adsorption mode on platinum[45] (Pauling adsorption model[46] and Griffith adsorption model[47]).

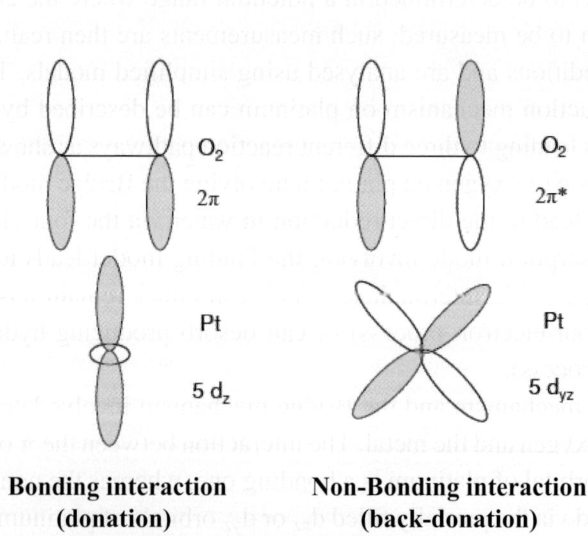


**Figure 9.6.** Interactions platinum-oxygen in the case of Griffith and Bridge adsorption models.

breaks. The formation of adsorbed hydroxyl species on platinum is then described by both mechanisms. The reduction of the Pt-OH species in the presence of protons will lead to the formation of water and the regeneration of the catalytic site. These kinds of adsorption modes then lead preferentially to the four-electron process with water production. Concerning the Pauling adsorption mode, the formation of water is penalized because it is expected to involve radicals $O^{-\bullet}$ or $OH^{\bullet}$ as intermediates.[45]

It appears that the mechanism of oxygen reduction involves different characteristics of the catalytic material, which may change its activity and selectivity:

(i) the d orbitals of platinum involved in the reaction can be modified by electronic effects[49] induced by the material structure (surface atom coordination, defect density, etc.);
(ii) the Bridge adsorption mode requires the presence of two adjacent platinum atoms, so that a geometric effect[50] can be involved (crystal cell parameters, exposed surface domains, etc.);
(iii) the dioxygen adsorption will be influenced by the local environment of the surface (presence of surface oxides, presence of surface charges, etc.).

The change of the nature of the catalytic material can indeed allow the modification of these characteristics, but also the control of the structure and of the morphology of platinum – in terms of crystallite size, exposed surface domains, internal strains, etc. For these reasons, and also in order to decrease the amount of the noble metal, platinum is used for electrocatalysis as nanoparticles dispersed on a high surface area electronic conductive carbon powder (carbon nanograins,[51] carbon nanotubes,[52] carbon nanofibers,[53] etc.), as shown in Fig. 9.7.

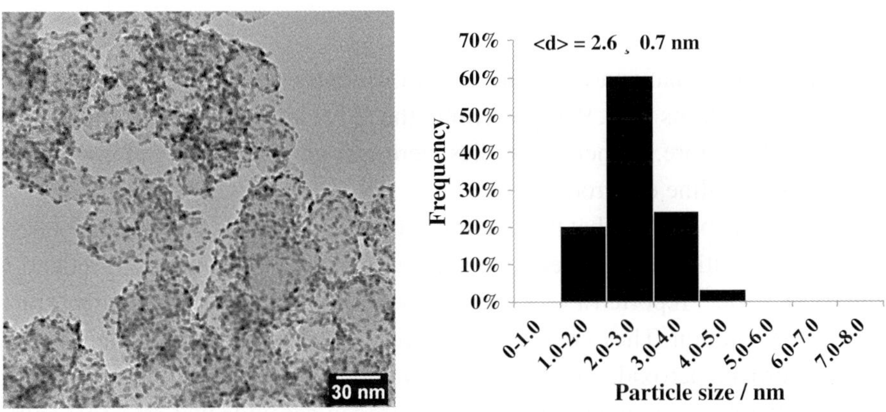


(a) (b)

**Figure 9.7.** (a) Transmission electron microscopy (TEM) image of Pt/C catalysts synthesized by a microwave-assisted polyol process and (b) related histograms of the size distribution.[51]

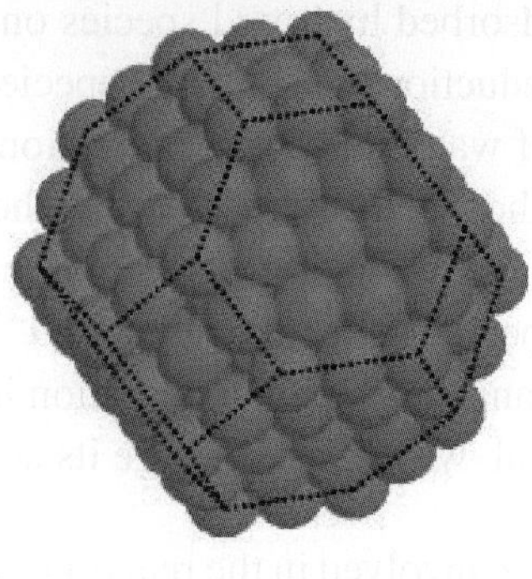

**Figure 9.8.** Representation of a perfect truncated octahedron.

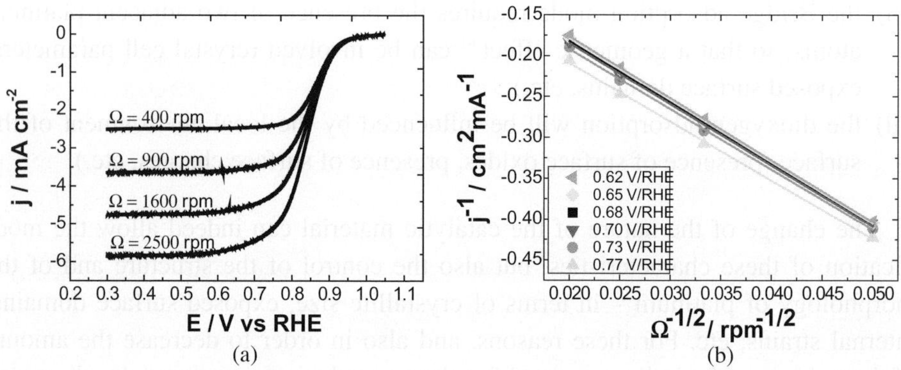


**Figure 9.9.** (a) j(E) polarization curves at different electrode rotation rates (Ω) recorded on Pt(40wt%)/C catalyst prepared via a microwave-assisted polyol synthesis method in a $O_2$-saturated 0.5M $H_2SO_4$ electrolyte. (T = 20°C, scan rate = 1 mVs$^{-1}$); (b) Koutecky–Levich plots determined from (a) at different potentials.

These platinum nanoparticles are generally represented as cuboctahedrons or truncated octahedrons (Fig. 9.8) presenting the (111) and (100) surface domains of platinum.[54] Therefore, numerous studies were carried out on ORR at low index Pt (hkl) monocrystalline electrodes.[55,56]

It should also be noted that the platinum surface is very sensitive to the presence of species in solution and to electrode pre-treatments (anodization, pre-reduction). Damjanovic *et al.*[36] reported a very strong dependence of the reaction pathway on the purity of the solution. They concluded that the oxygen reduction reaction occurred without hydrogen peroxide intermediate formation on a pre-reduced platinum electrode, and therefore that the production of hydrogen peroxide was effective only on sites affected by the presence of adsorbed impurities.

The oxygen reduction reaction on platinum is often studied using the rotating disc electrode as shown in Fig. 9.9a in the case of a Pt/C catalyst prepared by a microwave-assisted polyol method.[51] To access kinetics data, a mathematic treatment of the data is generally performed using the Koutecky–Levich equation:[57,58]

$$\frac{1}{j} = \frac{1}{j_k} + \frac{1}{j_l^{diff.}} \tag{9.27}$$

where j is the absolute value of the oxygen reduction current density at an electrode potential E, $j_k$ is the kinetic current density and $j_l^{diff.}$ is the diffusion limiting current density. The number of exchanged electrons can be drawn from the diffusion limiting current density given by the Levich equation:[59,60]

$$j_l^{diff.} = 0.2nF\nu^{-1/6}D_{O_2}^{2/3}C_{O_2}\Omega^{1/2} = nB\sqrt{\Omega} \tag{9.28}$$

where nF is the Faraday constant, $\nu$ is the kinematic viscosity of the solution ($1.07 \times 10^{-2}\,cm^2\,s^{-1}$ in 0.5 M $H_2SO_4$), $D_{O_2}$ the diffusion coefficient of molecular oxygen in 0.5 M $H_2SO_4$ ($2.1 \times 10^{-5}\,cm^2\,s^{-1}$), $C_{O_2}$ the oxygen concentration in a saturated 0.5 M $H_2SO_4$ electrolyte ($1.03 \times 10^{-6}\,mol\,cm^{-3}$)[61] and $\Omega$ the rotation rate expressed in revolutions per minute (rpm). Then, the slope of the 1/j vs $1/\Omega^{1/2}$ straight lines (Fig. 9.9b) allows the determination of the number of exchanged electrons as a function of the potential during ORR.

Considering the following mechanism proposed by Tarasevich *et al.*,[24] a more detailed Koutecky–Levich equation could be used:

$O_{2(sol.)} \rightarrow O_{2(surf.)}$ (diffusion coefficient in bulk electrolyte: $j_l^{diff.}$)

$O_{2(surf.)} \rightarrow O_{2(cata.)}$ (diffusion of $O_2$ in the catalytic film: $j_l^{film}$)

$O_{2(cata.)} \rightarrow O_{2(ads.)}$ (adsorption process: $j_l^{ads.}$)

$O_{2(ads.)} + 2e^- \rightarrow [O_{2(ads.)}]^{2-}$ (electron transfer rds: $j_0$, Tafel slope b)

$[O_{2(ads.)}]^{2-} + xH^+ + ye^- \rightarrow$ products ($x = 2, y = 0$ for $H_2O_2$; $x = 4, y = 2$ for $H_2O$)

Assuming that the electron transfer step is the rate determining step, Eq. 9.27 can be expressed as follows:[62]

$$\begin{aligned}\frac{1}{j} &= \frac{1}{j_l^{diff.}} + \frac{1}{j_l^{film}} + \frac{1}{j_l^{ads.}} + \frac{1}{j_0\left(\frac{\theta}{\theta_e}\right)\exp\left(\frac{\alpha nF}{RT}|\eta|\right)} \\ &= \frac{1}{j_l^{diff.}} + \frac{1}{j_l^{film}} + \frac{1}{j_l^{ads.}} + \frac{1}{j_0\left(\frac{\theta}{\theta_e}\right)\exp(|\eta|/b)}\end{aligned} \tag{9.29}$$

where $\eta = E - E_{eq.}$ is the overpotential, and $\theta$ and $\theta_e$ are the degree of coverage of the platinum surface by oxygen containing species at potential E and at equilibrium potential $E_{eq.}$, respectively. Because it is assumed that the electron transfer is the rate determining step, the adsorption process of oxygen is also assumed to be more rapid and then $\theta$ can be considered equal to $\theta_e$ at all potentials.

The film diffusion limiting current density and the adsorption limiting current density are both independent of disk electrode rotation rates and applied potential (E), thus it is impossible to dissociate them and Eq. 9.29 can be written:

$$\frac{1}{j} = \frac{1}{j_l^{diff.}} + \frac{1}{j_l^{film}} + \frac{1}{j_l^{ads.}} + \frac{1}{j_0 \exp(|\eta|/b)} = \frac{1}{j_l^{diff.}} + \frac{1}{j_l} + \frac{1}{j_0 \exp(|\eta|/b)}$$

$$\text{with } \frac{1}{j_l} = \frac{1}{j_l^{film}} + \frac{1}{j_l^{ads.}} \tag{9.30}$$

then, by mixing Eqs. 9.26 and 9.29:

$$\frac{1}{j} = \frac{1}{j_l^{diff.}} + \frac{1}{j_l} + \frac{1}{j_0 \exp(|\eta|/b)} = \frac{1}{j_l^{diff.}} + \frac{1}{j_k} \tag{9.31}$$

where

$$\frac{1}{j_k} = \frac{1}{j_l} + \frac{1}{j_0 \exp(|\eta|/b)} \tag{9.32}$$

From Eqs 9.30 and 9.31, it is possible to draw $1/j_k$ as a function of the potential E to obtain $1/j_l$ because at high overpotentials $1/j_k$ tends toward $1/j_l$, as observed on Fig. 9.10a. Then, Eq. 9.31 can be transformed as follows:

$$\eta = b \ln\left(\frac{j_k}{j_l - j_k} + \ln\frac{j_l}{j_0}\right) \tag{9.33}$$

In the present case, from the Koutecky–Levich plots in Fig. 9.9b, four electrons were exchanged over the whole considered potential range; an exchange current

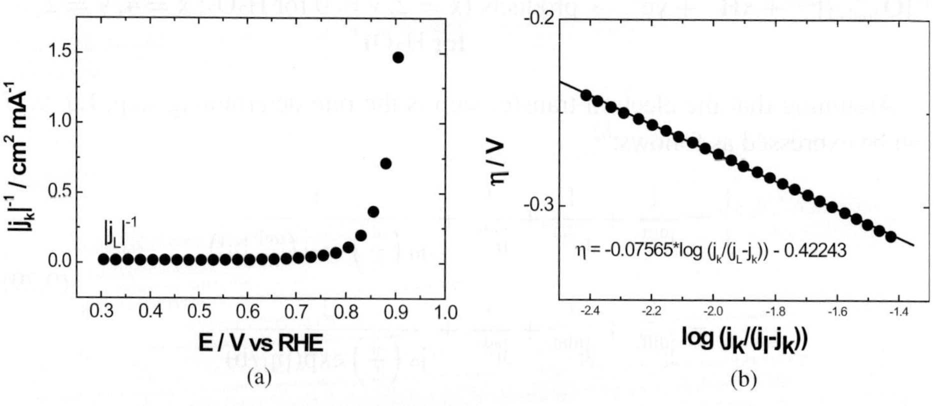


**Figure 9.10.** (a) Plot of $1/j_k$ as a function of the electrode potential drawn from data obtained in Figure 9.9; (b) related Tafel plot.

density value of $1.3 \times 10^{-7}$ A cm$^2$ and a Tafel slope of 75.7 mV decade$^{-1}$ were obtained (Fig. 9.10b). Such a value of j is consistent with that obtained by different authors with Pt/C catalysts prepared by a classical polyol method ($2.1 \times 10^{-7}$ A cm$^2$ at 70°C)[63] or by a microwave-assisted polyol process ($1.25 \times 10^{-7}$ A cm$^{-2}$ at 50°C).[64]

### 9.3.3. *Alternative catalysts to pure platinum for ORR*

Numerous studies have shown that platinum-based binary alloy electrocatalysts such as PtFe, PtCo, PtNi and PtCr exhibit a higher catalytic activity for ORR in an acid electrolyte than pure platinum.[49,65–67] Figure 9.11 displays the polarization curves of the oxygen reduction reaction recorded in an $O_2$-saturated $H_2SO_4$ electrolyte with different $Pt_3M_1$/C catalysts prepared via a colloidal method (Bönnemann method).[68,69] All bimetallic catalysts allow the achievement of higher activity towards ORR as higher current densities are recorded at high potentials in comparison with pure platinum. On such bimetallic catalysts, the number of exchanged electrons remains close to four, as indicated by the value of the limiting current density at high overpotentials, which are equal to that obtained with the pure platinum catalyst.

The observed electrocatalytic enhancement was generally interpreted either by an electronic factor, i.e. the change of the d-band vacancy in Pt upon alloying, and/or geometric effects (Pt coordination number and Pt–Pt distance).

For the geometric effect, the transition metals used for the modification are miscible in platinum to form alloy structures or a solid solution, at least for bulk platinum alloys.[70,71] X-ray diffraction (XRD) measurements given in the literature indicate that such a property is conserved for nanoparticles with Fe, Ni and Co.[72] In the cases presented above, alloys are platinum rich and the bimetallic materials display the same face centered cubic (fcc) structure as pure platinum, as indicated

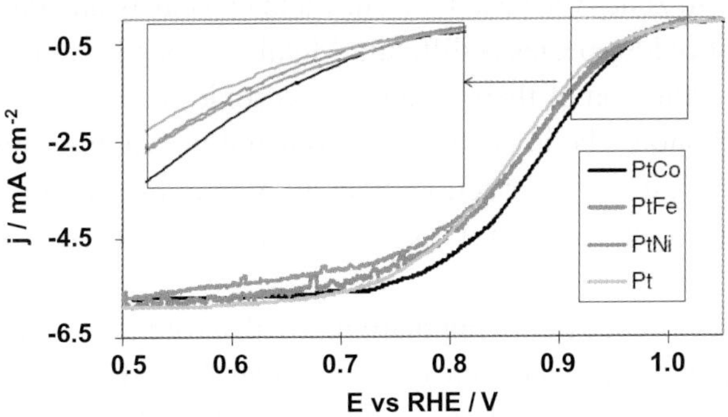


**Figure 9.11.** Polarization curves of the ORR recorded using different bimetallic $Pt_3M_1$ (40 wt%)/C catalysts in $O_2$-saturated $H_2SO_4$ at T = 20°C, v = 3 mV s$^{-1}$ and $\Omega$ = 2,500 rpm.

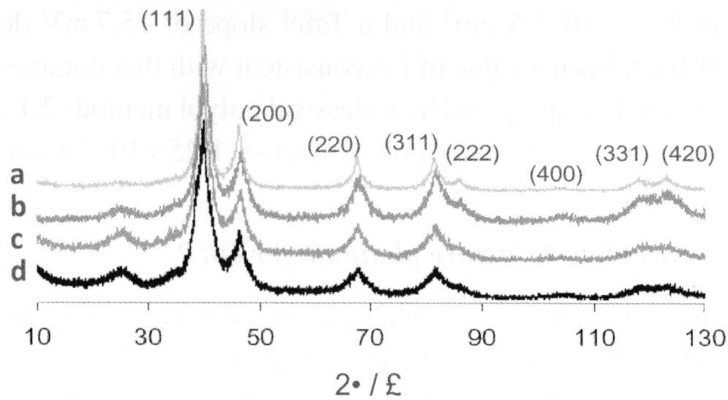


**Figure 9.12.** XRD patterns of (a) Pt (40 wt%)/C, (b) $Pt_3Ni$(40wt%)/C, (c) $Pt_3Fe$ (40 wt%)/C and (d) $Pt_3Co$ (40 wt%)/C prepared by the Bönnemann method.

by the XRD patterns in Fig. 9.12. However, a small contraction of the cell parameter could be detected (i.e. a shift of the diffraction peaks towards higher $2\theta$ values) for the bimetallic compounds. The distortion of the platinum cell leads to the modification of the inter-atomic distance and further to the adsorption step of oxygen; the activity toward the oxygen reduction reaction is then changed.

For the electronic effect, the addition of a foreign transition metal to platinum generates a modification of the vacancy orbital energy levels of platinum.[49,73] The density of vacancy d orbitals influences the ability of the catalytic material to adsorb oxygen and to desorb water. Moreover, Wang and Balbuena calculated the change in free energy of oxygen adsorption in order to better understand the role of the second metal.[74] They showed a strong link between the oxygen adsorption energy and the electronic structure of the metal. They compared several metals (Fig. 9.13) by plotting the difference in the Gibbs free energy for the reduction of an oxide and hydroxide with that obtained with platinum ($\Delta\Delta G_4$) as a function of the difference in the Gibbs free energy of oxygen adsorption with that obtained using platinum ($\Delta\Delta G_1$). Metals in zone A have a d-vacancy and can easily adsorb oxygen, but the reduction of their oxide species is difficult. Metals in zone B are less able to adsorb oxygen, but the reduction of their oxides or hydroxides species is easier. The best catalyst for ORR should be in zone C; but no metal lies in this zone. However, Co and Ni are the closest and their action in an alloy with platinum is to facilitate the adsorption of molecular oxygen without significantly penalizing the metal oxide reduction.

Lastly, another effect by alloying platinum with a foreign transition metal is discussed in the literature. The galvanic effect[73] comes from the fact that the transition metals are more easily oxidisable than platinum. They allow for the protection of the platinum surface, preventing it from oxidation, and only metallic platinum is active for the oxygen reduction reaction.

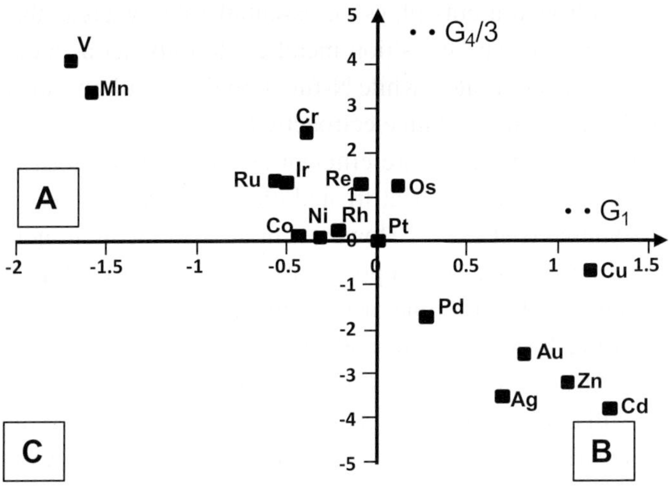


**Figure 9.13.** Difference in Gibbs free energy for the formation and the reduction of oxide species of different metals compared with those of platinum.[74]

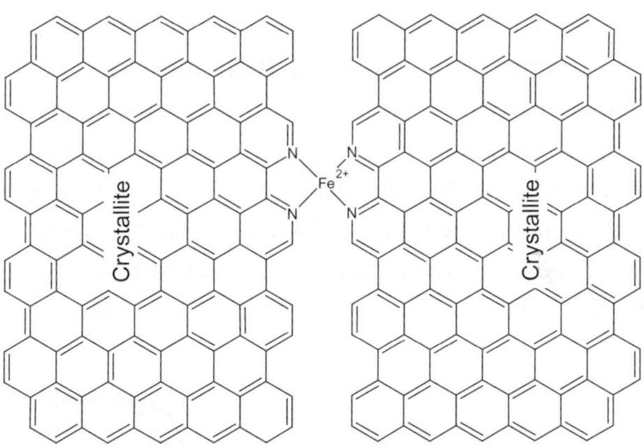


**Figure 9.14.** Schematic representation of the catalytic site after pyrolysis.[80]

Several non-platinum-based catalysts have also been studied, including transition metal chalcogenides[75] and macrocycles, heat treated[76,77] or not,[78,79] in order to improve the oxygen reduction reaction. Very recently, Lefèvre *et al.*[80] proposed a catalyst of $CN_xFe$ type with a comparable activity toward ORR as that of platinum, at least in a certain potential range. This extremely promising result opens the route for the development of a new class of active materials for electrochemical reactions. However, although the authors proposed the structure presented in Fig. 9.14 for the active site, the real nature of the active site, which is the key point for the understanding of its activity, is not yet clearly identified. It could be that the pyridinic configuration of nitrogen atoms plays an essential role, whereas that of the iron cations remains obscure. The transition metal could only act as a catalyst for the formation of the $CN_x$ active sites, while N-functionalized carbon substrates without metal present also contain a certain electroactivity.[81]

Transition metal macrocycles are efficient as a cathode catalyst in an alkaline fuel cell;[82] the most active non-heat treated electrocatalyst for oxygen reduction is iron phthalocyanine (FePc).[83,84] However, in an acid medium, non-heat treated metallophthalocyanines have shown poor stability, leading to a drastic decrease of activity towards the ORR with time or voltammetric cycles.[78] Metallophthalocyanines were then often heat treated over 500°C under an inert atmosphere to improve their stability.[48,85] But the nature and structure of the active centre after pyrolysis of macrocycles is still controversial.[86–89] The fact that the active centre of the non-heat treated FePc is clearly identified is a good point to study the activity of this material in an acid medium. It is known to reduce molecular oxygen mainly into water via a four-electron process in an acid medium, in contrast to CoPc, which reduces molecular oxygen mainly into hydrogen peroxide ($H_2O_2$) via a two-electron process.[78]

The electroactivity of FePc is dramatically related to the crystalline structure of the macrocycle particles. Iron phthalocyanine exists under two stable phases: α- and β-FePc. According to the works of Ballirano *et al.*[90] and Kirner *et al.*[91] on metal phthalocyanines, these structures differ by the relative position of the macrocycles in the cell (in parallel stack for the α phase, in perpendicular stack for the β phase) and hence, by the Me-Me distance in the crystallographic cell, which is higher in the β phase than in the α phase. As a consequence, differences exist in the electrocatalytic activity and selectivity of each catalyst towards oxygen reduction, as shown in Fig. 9.15. The α-FePc leads to an onset of the reduction wave at a

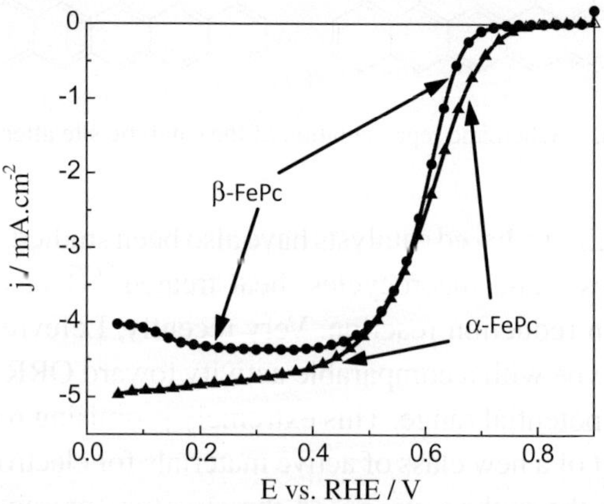


**Figure 9.15.** Polarization curves of the ORR recorded at different FePcC catalysts in $O_2$-saturated $H_2SO_4$ electrolyte at T = 20°C, v = 5 mV s$^{-1}$ and Ω = 2,500 rpm.

potential 100 mV higher than that of β-FePc and to higher absolute values of current densities in the diffusion plateau. The higher overpotential for the ORR at β-FePc compared with that obtained at α-FePc was explained by the structure of β-FePc in a perpendicular stack in the cell which makes oxygen absorption difficult, due to either interplanar spacing or accessibility.[92] In the case of cofacial dicobalt porphyrins, Collman *et al.*[93] showed that the higher onset potential of the oxygen reduction wave, higher current densities and higher selectivity towards water production, were achieved for a Co–Co distance close to 0.4 nm, which is a value very close to that of the Fe–Fe distance in α-FePc. Normally, the reduction of oxygen in a cobalt macrocycle occurs via a two-electron process to form $H_2O_2$ as the main product. To explain the selectivity into water in this configuration, the oxygen adsorption was proposed to occur via the formation of the μ-peroxo Co–$O_2$–Co species. Yeager[94] proposed the formation of a FePc dimer species at high potentials, involving two different reduction mechanisms according to the adsorption mode of $O_2$, M–$O_2$–M or M–O–M–$O_2$. Baranton *et al.*,[92] on the basis of Tafel slopes determination and using the Parsons equation,[95,96] proposed the following mechanism as the most probable in the low overpotential region (E > 0.7 V vs RHE):

$$2Fe^{II}Pc + O_2 \rightarrow PcFe^{III} - O_2Fe^{III}Pc \qquad \text{rds} \qquad (9.34)$$

$$Fe^{III} - O_2 - Fe^{III} + H^+ + e^- \rightarrow Fe^{II}Pc - OH + Fe^{III}Pc - O^\bullet \qquad (9.35)$$

$$Fe^{II}Pc - OH + Fe^{III}Pc - O^\bullet + H^+ + e^- \rightarrow 2Fe^{II}Pc - OH \qquad (9.36)$$

$$(2x)Fe^{II}Pc - OH + H^+ + e^- \rightarrow Fe^{II}Pc + H_2O \qquad (9.37)$$

In the high overpotential region (E < 0.7 V vs RHE), an *in situ* reflectance spectroscopy technique allowed the determination of several oxygenated species formed at the FePc electrodes: an absorption band located at 880 $cm^{-1}$ assigned to the vibration of the O-O bond of the hydrogen peroxide molecule,[97,98] one at 1,130 $cm^{-1}$, which corresponds to a vibration of the O-O bond in the $O_2^-$ adsorbed species,[99] and a band of lower intensity located at 1050 $cm^{-1}$ assigned to the vibration of the O-O bond in the adsorbed $O_2H$ species.[98,99] It was then proposed that FePc was able to reduce oxygen to $H_2O$ and $H_2O_2$ at 0.5 V vs RHE. The oxygen reduction in that potential range has then to occur via an adsorbed hydrogen peroxide species and can lead to either water or hydrogen peroxide as the main product, as it was suggested by Zagal *et al.*[84]

## 9.4. Atomic Oxygen Activation: Alcohol Electro-Oxidation

The electro-oxidation in a fuel cell of low molecular mass alcohols, such as methanol and ethanol, appears particularly convenient for two main reasons: they are liquids (allowing easy storage compared with hydrogen) and their theoretical mass energy density is rather high (6.1 and 8.0 kWh $kg^{-1}$ for methanol and ethanol, respectively), close to that of gasoline (10.5 kWh $kg^{-1}$).[100] Moreover, it has recently been proposed that fuel cells can be used as electrochemical reactors for oxidizing heavier alcohols or polyols (ethylene glycol, glycerol, etc.) for the cogeneration of energy and high added value chemicals.[5] In the first case, the complete oxidation of alcohol into $CO_2$ is sought, whereas in the second case, the selective oxidation of the alcohol groups into carbonyl or carboxyl functions without the C-C bond cleavage has to be performed. Considering as an example the electro-oxidation of ethanol in an acid medium, three different products can be obtained according to the number of exchanged electrons in the reaction:

$$CH_3CH_2OH + 3H_2O \rightarrow 2CO_2 + 12H^+ + 12e^- \quad (9.38)$$

$$CH_3CH_2OH + H_2O \rightarrow CH_3CO_2H + 4H^+ + 4e^- \quad (9.39)$$

$$CH_3CH_2OH \rightarrow CH_3CHO + 2H^+ + 2e^- \quad (9.40)$$

For a DEFC application, the complete oxidation of ethanol in $CO_2$ leads to the recovery of the maximal mass energy density, whereas its oxidation in acetic acid leads to a third, and the formation of acetaldehyde to a sixth, of this maximal energy density (moreover, acetaldehyde is a toxic compound, whereas acetic acid has a commercial application). However, the breaking of the C-C bond is difficult to perform at low temperatures, and the main reaction products are acetaldehyde and acetic acid or acetate,[101,102] which leads to a low faradic efficiency (17–33% of the theoretical energy). Moreover, as a consequence of the acidic environment of the ionomeric conducting membrane (Nafion) and of the low working temperatures of DAFCs (60–120°C), the use of platinum is impossible to avoid,[10] owing to its catalytic properties to activate C-H bond cleavage during the first adsorption steps, although this leads to rather poor electro-oxidation kinetics. However, strongly bonded species are formed on platinum after the dissociative adsorption step of alcohols, which limits the number of accessible active sites for further reaction. For example, methanol is adsorbed, with the formation of adsorbed carbon monoxide as a poisoning species.[103] With ethanol and polyols, even if the C-C bond breaking is difficult, adsorbed CO is also observed by *in situ* infrared reflectance spectroscopy.[5,12,104] In each case, the formation of such poisoning species leads to poor activity, and the challenge is to enhance the activity of Pt. In order to improve the reaction kinetics, a deep understanding of the mechanisms of the electrocatalytic reactions is a key issue. As several different steps are generally necessary to carry out the complete electrocatalytic reaction, an optimized catalyst should be multifunctional. The composition of the catalysts (nature and proportion of the metals involved)[9,105] and the structure (size of particles,[106] atomic arrangement,[11] superficial structure,[107–109] etc.) are crucial.

On a pure Pt/C catalyst, Rousseau *et al.*[101] showed that the electro-oxidation of ethanol at the anode of a DEFC working at 80°C mainly led to the formation of acetaldehyde, acetic acid and carbon dioxide, with chemical yields of 47.5%, 32.5% and 20.0%, respectively. By comparing the mass yield and the faradic yields, they concluded that no other products were formed in a significant amount. This result confirms that Pt is able to break the C-C bond to some extent. *In situ* infrared measurements on ethanol adsorption and electro-oxidation at platinum electrodes have clearly shown that the adsorbed CO species are formed from 0.3 V vs RHE at the platinum surface;[10,12] moreover Iwasita and Pastor[110] found some traces of $CH_4$ at potentials lower than 0.4 V vs RHE. Previous studies showed that the initial steps of ethanol adsorption and oxidation on Pt can follow two different modes:[110–112]

$$Pt + CH_3 - CH_2OH \rightarrow Pt - OCH_2 - CH_3 + e^- + H^+ \qquad (9.41)$$

$$Pt + CH_3 - CH_2OH \rightarrow Pt - CH_2 - CH_2OH + e + H^+ \qquad (9.42)$$

It was shown by Hitmi *et al.*[113] that acetaldehyde was formed at potentials lower than 0.6 V vs RHE. From long-term electrolysis experiments on a Pt catalyst coupled with high performance liquid chromatography (HPLC) analyses, Vigier *et al.* detected acetaldehyde at potentials as low as 0.35 V vs RHE whereas no acetic acid was detected in this potential range.[19] It could then be proposed that at $E < 0.6\,V$ vs RHE the ethanol electro-oxidation occurs following the mechanism:

$$Pt - OCH_2 - CH_3 \rightarrow Pt + CH_3 - CHO + H^+ + e^- \qquad (9.43)$$

$$Pt - CH_2 - CH_2OH \rightarrow Pt + CH_3 - CHO + e^- + H^+ \qquad (9.44)$$

At $E < 0.4\,V$ vs RHE, where acetaldehyde is formed, it can adsorb on platinum and form a Pt-$CH_3$-CO species according to:

$$Pt + CH_3 - CHO \rightarrow Pt - CH_3 - CHO \qquad (9.45)$$

$$Pt - CH_3 - CHO \rightarrow Pt - CO - CH_3 + e^- + H^+ \qquad (9.46)$$

As it was shown by SNIFTIRS measurements (subtractively normalized interfacial Fourier transform infrared spectroscopy measurements) that adsorbed CO species exist from 0.3 V vs RHE on a Pt surface:[12]

$$Pt - CO - CH_3 + Pt \rightarrow Pt - CO + Pt - CH_3 \qquad (9.47)$$

and at $E < 0.4$, where Pt is able to adsorb hydrogen to form Pt-H species:

$$Pt - CH_3 + Pt - H \rightarrow Pt + CH_4 \qquad (9.48)$$

At potentials higher than 0.6 V vs RHE, the dissociative adsorption of water occurring on a platinum surface according to Eq. 9.49 provides OH adsorbed species, allowing the catalyst to further oxidize the adsorption residues of ethanol. Then, oxidation of adsorbed CO species (Eq. 9.50) occurs, which is in agreement with *in situ* Fourier transform infrared (FTIR) measurements where $CO_2$ starts to be detected from ca. 0.65 V vs RHE[10,12] and CO stripping experiments at a platinum surface, where CO is removed from ca. 0.6 V vs RHE.[12] Acetaldehyde can also be oxidized following Eq. 9.51:

$$Pt + H_2O \rightarrow Pt - OH + H^+ + e^- \quad (E > 0.6\,V \text{ vs RHE}) \tag{9.49}$$

$$Pt - CO + Pt - OH \rightarrow 2Pt + CO_2 + H^+ + e^- \tag{9.50}$$

$$Pt - CH_3 - CHO + Pt - OH \rightarrow Pt + CH_3 - CO_2H + H^+ + e^- \tag{9.51}$$

In this mechanism, supported by *in situ* infrared measurements, product distribution analyses by HPLC and gas chromatography (GC), and CO stripping experiments, it appears that one of the limiting steps is the activation of water by the catalytic surface providing extra oxygen atoms to complete the oxidation reaction. This fact is more drastic in the case of methanol where the poisoning of platinum by adsorbed CO species completely blocks the catalytic surface preventing any reaction to occur and any oxidation current density below 0.6 V vs RHE, whereas in the case of ethanol some current densities are recorded from lower potentials due to the oxidation of ethanol to form acetaldehyde which does not necessitate the occurrence of any extra oxygen atoms. In the case of ethanol electro-oxidation, the simplified mechanism in Fig. 9.16 can be proposed.

It appears clearly that the limiting step for achieving higher faradic and potential efficiency is the activation of water molecules to provide the extra oxygen atoms in the oxidation process allowing the formation of $CO_2$ and acetic acid.

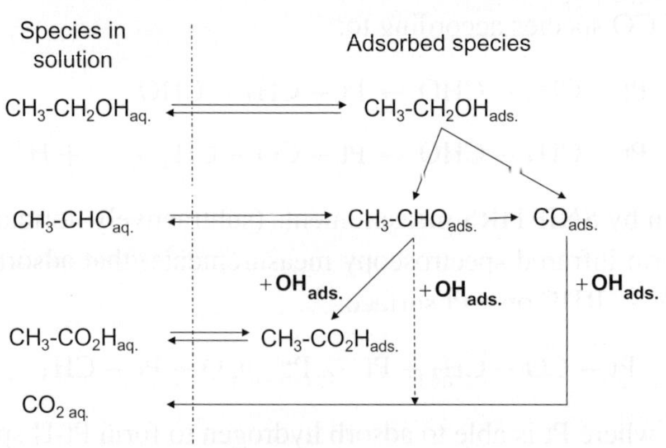


**Figure 9.16.** Schematic mechanism of ethanol electrooxidation at a Pt/C electrode.

In the case of methanol, the same problem arises for the removal of adsorbed CO coming from the dehydrogenation reaction of the alcohol according to:

$$CH_3OH \rightarrow CH_2OH_{ads.} + H^+ + e^- \quad (9.52)$$

$$CH_2OH_{ads.} \rightarrow CHOH_{ads.} + H^+ + e^- \quad (9.53)$$

$$CHOH_{ads.} \rightarrow CHO_{ads.} + H^+ + e^- \quad (9.54)$$

then either

$$CHO_{ads.} + OH_{ads.} \rightarrow CO_2 + 2H^+ + 2e^- \quad (9.55)$$

or

$$CHO_{ads.} \rightarrow CO_{ads.} + H^+ + e^- \quad (9.56)$$

$$CO_{ads.} + OH_{ads.} \rightarrow CO_2 + H^+ + e^- \quad (9.57)$$

From such a mechanism, the oxidation current density can be expressed as in Eq. 9.60, considering that two organic species from methanol are adsorbed on the platinum surface ($CO_{ads}$ and $CHO_{ads}$), as was suggested in the previous mechanism:

$$j = nFk_1\theta_{CO}\theta_{OH}\exp\left(\frac{\alpha_1 n_1 F}{RT}E\right) + nFk_2\theta_{CHO}\theta_{OH}\exp\left(\frac{\alpha_2 n_2 F}{RT}E\right) \quad (9.58)$$

$$j = nF\theta_{OH}\left[k_1\theta_{CO}\exp\left(\frac{\alpha_1 n_1 F}{RT}E\right) + k_2\theta_{CHO}\exp\left(\frac{\alpha_2 n_2 F}{RT}E\right)\right] \quad (9.59)$$

From the proposed mechanism and Eq. 9.59, it can be deduced that higher current densities will be obtained in the case of $\theta_{OH} = \theta_{CHO} = 0.5$ and $\theta_{CO} = 0$. Moreover, higher potential efficiency will be achieved if the OH species could be adsorbed at lower potentials than for platinum. However, *in situ* infrared spectroscopy clearly demonstrated that the dissociative adsorption of both methanol and ethanol leads to the formation of strongly adsorbed CO species at low potentials.[103,114,115] Indeed, in both cases, the existence of an absorption band located close to 2,050 $cm^{-1}$ is clearly visible on the SNIFTIR spectra (Fig. 9.17). Therefore, bifunctional catalysts are able to activate two different reaction steps (alcohol and water adsorption, and surface reaction between adsorbed species via a Langmuir–Hinshelwood mechanism), and so exhibiting active sites with different properties is necessary. As an example, investigations into the possibility of enhancing activity toward methanol electro-oxidation with PtRu-based[11,107] and ethanol electro-oxidation with PtSn-based[10,19,101] electrodes are of great interest with regard to improving the electrical efficiency of DMFCs and DEFCs. The second metals are recognized as being able to activate water and adsorb OH species at lower potentials than platinum.

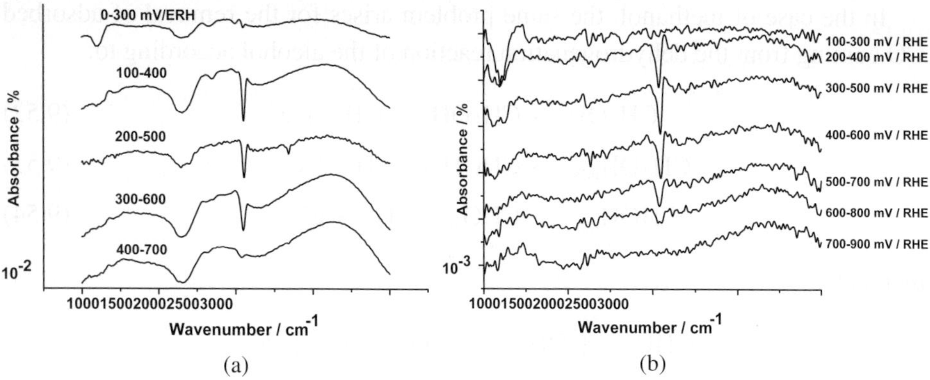


**Figure 9.17.** SNIFTIR spectra recorded in 0.1 M $HClO_4$ in the presence of (a) 0.1 M methanol and (b) 0.1 M ethanol on a Pt/C electrode at various potentials.

Several approaches have been considered to optimize the electrochemical behaviour of bimetallic catalysts. For example, some authors have studied the effect of PtRu catalyst structure on methanol and CO electro-oxidation,[11,107–109] whereas other authors[116–119] have focused on the optimal Pt/Ru atomic ratio.

Dubau *et al.*[11] showed that, for a given Pt/Ru atomic ratio, alloying Pt and Ru did not necessarily lead to the most active catalyst for the methanol oxidation reaction, but that dispersing Pt and Ru metals in strong interactions on the same carbon grains (decoration of platinum particles by smaller ruthenium particles) led to higher current densities for the electro-oxidation of methanol for potentials lower than 0.5 V vs RHE. An electrocatalytic enhancement of methanol oxidation at Pt particles decorated by Ru compared with the alloy compounds of the same composition was also observed by other authors.[107–109]

The differences in activity as a function of Ru content are due to the balance between the initial step of adsorption–dehydrogenation of methanol at Pt sites and the following step of adsorbed CO species oxidation.[120] An atomic ratio of 50/50 is proposed by some authors as optimal for methanol electro-oxidation,[118] whereas other authors favour a Pt/Ru atomic ratio closer to 80/20.[105,116,120] The discrepancies in the literature most likely come from a lack of knowledge of the surface composition. Studies on catalysts, allowing accurate determination of the platinum coverage, prepared by spontaneous deposition of Ru at Pt nanoparticles forming nanosized Ru islands of monoatomic height, indicated that the best activity with regard to methanol oxidation was found for a Ru coverage close to 40–50% at 0.3 V vs RHE and 0.5 V vs RHE.[121,122]

According to the bifunctional theory of electrocatalysis for the complete oxidation of methanol, the large amount of Ru may explain the higher activity of the $Pt_{0.5}Ru_{0.5}$ catalyst at low potentials. At potentials above 0.5 V vs RHE, the catalytic surface is blocked by adsorbed oxygen species,[13,14] which makes the adsorption of organic species more difficult. While Ru adsorbs oxygen species at lower potentials, increasing the content of Ru causes a decrease of the limiting current. Moreover, pure Pt displays higher activity at temperatures lower than 60°C compared with PtRu alloys for potentials higher than 0.5 V vs RHE.[119] Both effects involve the decrease in oxidation limiting current with the increase in the Ru atomic ratio. Electrochemical nuclear magnetic resonance (NMR) studies[122] have shown that CO diffusion at a pure Pt surface was slow compared with that of nearby PtRu islands. Then, the slow diffusion of adsorbed CO from Pt sites away from Ru to Pt sites close to Ru, where the oxygenated species present on Ru oxidize the CO species, can be proposed as the rate determining step.[119]

If PtRu bimetallic materials are considered as the most adapted catalysts for methanol electro-oxidation,[116,118,123,124] PtSn bimetallic catalysts showed higher electrocatalytic activity toward ethanol electro-oxidation.[125–128] Modification of platinum by tin greatly enhances the electro-oxidation rate of ethanol and may change the reaction mechanism. As discussed earlier in this chapter, the formation of adsorbed CO species is a consequence of the dissociative adsorption of methanol and ethanol (note that $CO_{ads}$ was also detected with ethylene glycol and glycerol).[5,104] Beyond being a poisoning species, CO is then also a reaction intermediate of alcohol oxidation. For this reason, numerous studies concerning the electro-oxidation of carbon monoxide on bimetallic catalysts have been carried out.[12,14,129–131] As a main result, it appeared that PtSn catalysts displayed better activity towards CO oxidation than PtRu catalysts, since lower onset potentials of the CO oxidation wave were recorded with the former material.[129] This paradoxical electrochemical behaviour of PtSn-based catalysts makes their study very controversial. Nevertheless, the catalytic enhancement of the electro-oxidation of adsorbed CO on PtSn catalysts is generally accepted. To explain the difference in activity towards methanol, ethanol and CO electro-oxidation, Wang *et al.*[132] proposed that, at the $Pt_3Sn$ alloy surface, a unique state of adsorbed CO was formed at high CO coverage on this surface. This adsorption state is unique in the sense that it can be obtained only from dissolved CO, not from methanol or ethanol due to platinum atom dilution by tin, and only on PtSn alloys, not on PtRu alloys. Therefore, the bifunctional mechanism could not be invoked alone to explain the catalytic enhancement.

The existence of an electronic effect (ligand effect) has been proposed. For example, Tong *et al.*[122] reported, on the basis of a combination of solid state NMR and electrochemical methods, a correlation between the $2\pi^*$ Fermi level local density of state and the steady state current of CO oxidation. The presence of Ru could weaken the Pt-CO bond, leading to an enhanced CO oxidation rate. The electronic effect of Ru was also reported to explain the higher rate of CO adsorption from methanol on a PtRu black catalyst by the $^{13}CO$ labelling method.[107] However, electrochemical infrared reflection-absorption spectroscopy (EC-IRRAS) data obtained by Park *et al.*[133] showed that the degree of electronic modification of Pt by neighbouring Ru was not significant enough to involve a change in catalytic activity.

In the case of PtSn catalysts, no evidence of a ligand effect was observed from an *in situ* FTIR study on $Pt_3Sn(110)$ bulk alloy[134] and PtSn nanoparticles supported on carbon.[135] It was proposed that the bifunctional mechanism was mainly involved in the oxidation process. The fact that the transition from positive to negative Stark shift of infrared $\nu$(CO) frequency during CO oxidation was much more pronounced on a PtSn/C catalyst than on the Pt/C catalyst was interpreted in terms of the different ways in which $OH_{ads}$ (necessary to oxidize CO) nucleates on each catalyst.

Finally, trimetallic compounds have been developed to enhance the electrocatalytic activity of Pt-based catalysts, for either methanol or ethanol electro-oxidation. In 1965, Shropshire showed that the adsorption of molybdate ($Na_2MoO_4$) on a Pt black electrode before adding the fuel (HCHO or $CH_3OH$) resulted in a decrease of 0.3 V of the oxidation onset potential with respect to pure Pt.[136] More recently, it was shown that adding molybdenum to PtRu catalysts increased the activity toward methanol electro-oxidation at low potentials.[137,138] An increase of ca. 0.1 V of the OCV of a direct methanol single cell was observed, which can be explained by a decrease in surface poisoning or by an effect on the composition of chemisorbed species.[139,140] Moreover, the higher capacity of this third metal to form oxy-hydroxyl species adsorbed at low potentials was often proposed as an explanation for the enhancement of methanol oxidation. In the case of ethanol, a trimetallic $Pt_{0.86}Sn_{0.1}Ru_{0.04}$/C catalyst[101] led to a DEFC performance at 80°C, twofold higher than that obtained with a $Pt_{0.9}Sn_{0.1}$/C catalyst under the same experimental conditions. The OCV remained similar for both catalysts (moreover, the presence of Ru was shown to involve no noticeable change in the product distribution of the anode outlet).[101] Here, the role of Ru may be to limit strong adsorption of surface poisons by diluting adsorption sites,[141] or to provide extra oxygen atoms from adsorbed OH species as soon as tin starts to form higher oxide species that are not catalytically active.

## 9.5. Conclusion

Oxygen activation is involved in many electrochemical processes such as fuel cell cathodes, metal-air batteries, chlor-alkali processes, electrosynthesis reactors, etc. It also has an important role in the field of metallic corrosion. The knowledge of the adsorption and charge transfer mechanism of oxygenating species is then of great importance either for improving the efficiency of the electrochemical devices (electric generators, synthesis reactors, etc.) or to protect metal from oxidation. In this chapter we have discussed the role of oxygen (molecular and atomic) activation in electrochemical processes taking as an example the reaction mechanisms of oxygen reduction and alcohol oxidation involved in low temperature fuel cells. These reactions are very complicated, involving multiple electron transfers and several elementary steps. Even the electro-oxidation of CO, which involves only a two-electron transfer mechanism, is relatively difficult to perform, leading to high overvoltage, since the adsorption of CO is very strong and blocks the active sites of Pt-based electrocatalysts. It is necessary to determine a detailed reaction mechanism, where all the adsorbed species and intermediate products are clearly identified, and where the rate determining step is known, in order to point out the reaction steps to be preferentially activated. In the case of the oxygen reduction reaction, the mechanism is not yet completely elucidated although this reaction has been studied for a long time. This reaction involves the transfer of several electrons which makes it very difficult to study, and moreover several parallel and successive steps may occur, leading to two different reaction intermediates and products. For alcohol electro-oxidation, even if the reaction intermediates and products could be detected by *in situ* spectroscopic and analytical methods (gas chromatography, high precision liquid chromatography, differential electrochemical mass spectroscopy, FTIR spectroscopy, etc.), the mechanisms are not totally clear; for example, the paradoxical behaviour of tin toward the oxidation of CO, methanol and ethanol is difficult to explain.

This chapter also shows that enhanced oxygen activation is obtained by modifying platinum with a foreign transition metal. The role of the second metal is not the same for the oxygen reduction reaction and for the water adsorption and activation; in the first case, geometric and electronic effects where the second metal leads to a change in the Pt cell parameter and in the electronic density of state in the d orbitals of platinum favouring the adsorption of $O_2$ are mainly involved, whereas in the second case, the bifunctional mechanism in which the second metal is able to provide extra oxygen atoms by adsorbing water at lower potentials than platinum, seems to be the main effect.

## References


1. Kadjo, A., Brault P., Caillard, A., *et al.* (2007). Improvement of Proton Exchange Membrane Fuel Cell Electrical Performance by Optimization of Operating Parameters and Electrodes Preparation, *J. Power Sources*, **172**, pp. 613–622.
2. Wang, X., Sebastian, P., Smit, M., *et al.* (2007). Studies on the Oxygen Reduction Catalyst for Zinc–air Battery Electrode, *J. Power Sources*, **124**, pp. 278–284.
3. Rudd, E. and Gibbons, D. (1994). High Energy Density Aluminum/Oxygen Cell, *J. Power Sources*, **47**, pp. 329–340.

4. Chatenet, M., Aurousseau, M. and Durand, R. (2000). Electrochemical Measurement of the Oxygen Diffusivity and Solubility In Concentrated Alkaline Media on Rotating Ring-disk and Disk Electrodes: Application to Industrial Chlorine-soda Electrolyte, *Electrochim. Acta*, **45**, pp. 2823–2827.
5. Simões, M., Baranton, S. and Coutanceau, C. (2010). Electro-oxidation of Glycerol at Pd Based Nano-catalysts for an Application in Alkaline Fuel Cells for Chemicals and Energy Cogeneration, *Appl. Catal. B: Environ.*, **93**, pp. 354–362.
6. Ralph, T. and Hogarth, P. (2002), Catalysis for Low Temperature Fuel Cells, Part I: The Cathode Challenges, *Platinum Metals Review*, **46**, pp. 3–14.
7. Gasteiger, H., Gu, W., Makharia, R., *et al.* (2003). Beginning-of-life MEA Performance: Efficiency Loss Contributions, in W. Vielstich, A. Lamm, H. Gasteiger (eds), *Handbook of Fuel Cells: Fundamentals, Technology and Applications*, Vol. 3, Wiley, Chichester, UK, pp. 593–610.
8. Gasteiger, H., Kocha, S., Sompalli, B., *et al.* (2005). Activity Benchmarks and Requirements for Pt, Pt-Alloy, and Non-Pt Oxygen Reduction Catalysts for PEMFCs, *Appl. Catal. B: Environ.*, **56**. pp. 9–35.
9. Coutanceau, C., Brimaud, S., Lamy, C., *et al.* (2008). Review of Different Methods for Developing Nanoelectrocatalysts for the Oxidation of Organic Compounds, *Electrochim. Acta*, **53**, pp. 6865–6880.
10. Léger, J., Rousseau, S., Coutanceau, C., *et al.* (2005). How Bimetallic Electrocatalysts Does Work for Reactions Involved in Fuel Cells?: Example of Ethanol Oxidation and Comparison to Methanol, *Electrochim. Acta*, **50**, pp. 5118–5125.
11. Dubau, L., Hahn, F., Coutanceau, C., *et al.* (2003). On the Structure Effects of Bimetallic PtRu Electrocatalysts Towards Methanol Oxidation, *J. Electroanal. Chem.*, **554–555**, pp. 407–415.
12. Vigier, F., Coutanceau C., Hahn, F., *et al.* (2004), On the Mechanism of Ethanol Electro-oxidation on Pt and PtSn Catalysts: Electrochemical and In Situ IR Reflectance Spectroscopy Studies, *J. Electroanal. Chem.*, **563**, pp. 81–89.
13. Watanabe, M. and Motoo, S. (1975). Electrocatalysis by Ad-atoms: Part II. Enhancement of the Oxidation of Methanol on Platinum by Ruthenium Ad-atoms, *J. Electroanal. Chem.*, **60**, pp. 267–273.
14. Watanabe, M. and Motoo, S. (1975). Electrocatalysis by Ad-atoms: Part III. Enhancement of the Oxidation of Carbon Monoxide on Platinum by Ruthenium Ad-atoms, *J. Electroanal. Chem.*, **60**, pp. 275–283.
15. Aricò, A., Cretì, P., Antonucci, P., *et al.* (1998). Optimization of Operating Parameters of a Direct Methanol Fuel Cell and Physico-chemical Investigation of Catalyst-electrolyte Interface, *Electrochim. Acta*, **24**, pp. 3719–3729.
16. Hikita, S., Yamane, K. and Nakajima. Y. (2001). Measurement of Methanol Crossover in Direct Methanol Fuel Cell, *J. S. A. E. Review*, **22**, pp. 151–156.
17. James, D. and Pickup, P. (2010). Effects of Crossover on Product Yields Measured for Direct Ethanol Fuel Cells, *Electrochim. Acta*, **55**, pp. 3824–3829.
18. Lamy, C., Lima, A., LeRhun, V., *et al.* (2002). Recent Advances in the Development of Direct Alcohol Fuel Cells (DAFC), *J. Power Sources*, **105**, pp. 283–296.
19. Vigier, F., Coutanceau, C., Perrard, A., *et al.* (2004). Development of Anode Catalysts for a Direct Ethanol Fuel Cell, *J. Appl. Electrochem.*, **34**, pp. 439–446.
20. Zhou, W., Li, W., Song, S., *et al.* (2004). Bi- and Tri-metallic Pt-based Anode Catalysts for Direct Ethanol Fuel Cells, *J. Power Sources*, **131**, pp. 217–223.
21. Ralph, T. and Hogarth, P. (2002). Catalysis for Low Temperature Fuel Cells, Part III: Challenges for the Direct Methanol Fuel Cell, *Platinum Metals Review*, **46**, pp. 146–164.
22. Scott, K., Taama, K., Argyropoulos, W., *et al.* (1999). The Impact of Mass Transport and Methanol Crossover on the Direct Methanol Fuel Cell, *J. Power Sources*, **83**, pp. 204–216.

23. Thomas, S., Ren, X., Gottesfeld, S., *et al.* (2002). Direct Methanol Fuel Cells: Progress in Cell Performance and Cathode Research, *Electrochim. Acta*, **47**, pp. 3741–3748.
24. Tarasevich, M., Sadwoski, A. and Yeager, E. (1983). Oxygen Electrochemistry, in J. Conway, J. Bockris, E. Yeager, *et al.* (eds.), *Comprehensive Treatise of Electrochemistry*, Vol. 7, Plenum Press, New York, pp. 301–398.
25. Bockris, J. and Huq, A. (1956). The Mechanism of the Electrolytic Evolution of Oxygen on Platinum, *Proc. R. Soc. London Ser. A*, **237**, pp. 277–296.
26. Hoare, J. (1963). The Normal Oxygen Potential on Bright Platinum, *J. Electrochem. Soc.*, **110**, pp. 1019–1021.
27. Hoare, J. (1979). Some Aspects of the Reduction of Oxygen at a Platinum-oxygen Alloy Diaphragm, *J. Electrochem. Soc.*, **126**, pp. 1502–1504.
28. Frumkin, A. and Nekrasov, L. (1959). On the Ring-disk Electrode, *Dokl. Akad. Nauk. SSSR*, **126**, pp. 115–118.
29. Stradins, J. (1997). Alexander N. Frumkin and the Electrochemistry of the 20th Century, *Electrochim. Acta*, **42**, pp. 731–736.
30. Kinoshita, K. (1992). Oxygen Electrochemistry, in K. Kinoshita (ed.), *Electrochemical Oxygen Technology*, John Wiley and Sons, Inc., New York, pp. 19–112.
31. Damjanovic, A., Genshaw, M. and Bockris J. (1966). Distinction between Intermediates Produced in Main and Side Electrodic Reactions, *J. Chem. Phys.*, **45**, pp. 4057–4059.
32. Wroblowa, H., Pan, Y. and Razumney, G. (1976). Electroreduction of Oxygen: A New Mechanistic Criterion, *J. Electroanal. Chem.*, **69**, pp. 195–201.
33. Appleby, A. and Savy, M. (1978). Kinetics of Oxygen Reduction Reactions Involving Catalytic Decomposition of Hydrogen Peroxide: Application to Porous and Rotating Ring-disk Electrodes, *J. Electroanal. Chem.*, **92**, pp. 15–30.
34. Zurilla, R., Sen, R. and Yeager E. (1978). The Kinetics of the Oxygen Reduction Reaction on Gold in Alkaline Solution, *J. Electrochem. Soc.*, **125**, pp. 1103–1109.
35. Yeager, E., Krouse, P. and Rao, K. (1964). The Kinetics of the Oxygen-peroxide Couple on Carbon, *Electrochim. Acta*, **9**, pp. 1057–1070.
36. Damjanovic, A., Genshaw, M. and Bockris, J. (1967). The Mechanism of Oxygen Reduction at Platinum in Alkaline Solutions with Special Reference to $H_2O_2$, *J. Electrochem. Soc.*, **114**, pp. 1107–1112.
37. Taylor, R. and Humffray, A. (1975). Electrochemical Studies on Glassy Carbon Electrodes: II. Oxygen Reduction in Solutions of High pH (pH $>$ 10), *J. Electroanal. Chem.*, **64**, pp. 63–84.
38. Morcos, I. and Yeager, E. (1970). Kinetic Studies of the Oxygen: Peroxide Couple on Pyrolytic Graphite, *Electrochim. Acta*, **15**, pp. 953–975.
39. Bonnemay, M., Bernard, C., Magner, G., *et al.* (1971). Étude en Fonction du pH de la Réduction de L'oxygène sur L'or, *Electrochim. Acta*, **16**, pp. 537–548.
40. Genshaw, M., Damjanovic, A. and Bockris. J. (1967). Hydrogen Peroxide Formation in Oxygen Reduction at Gold Electrodes: I. Acid Solution, *J. Electroanal. Chem.*, **15**, pp. 163–172.
41. Strbac, S., Anastasijevic, N. and Adžić, R. (1992). Oxygen Reduction on Au (100) and Vicinal Au (910) and Au (11, 1, 1) Faces in Alkaline Solution: A Rotating Disc-ring Study, *J. Electroanal. Chem.*, **323**, pp. 179–195.
42. Huang, J., Sen, R. and Yeager, E. (1979). Oxygen Reduction on Platinum in 85% Orthophosphoric Acid, *J. Electrochem. Soc.*, **126**, 786–792.
43. Coutanceau, C., Croissant, M., Napporn, T., *et al.* (2000). Electrocatalytic Reduction of Dioxygen at Platinum Particles Dispersed in a Polyaniline Film, *Electrochim. Acta*, **46**, pp. 579–588.
44. Sepa, D., Vojnovic, M., Vracar, Lj. M., *et al.* (1987). Different Views Regarding the Kinetics and Mechanisms of Oxygen Reduction at Pt and Pd Electrodes, *Electrochim. Acta*, **32**, pp. 129–134.
45. Yeager, E. (1981). Recent Advances in the Science of Electrocatalysis, *J. Electrochem. Soc.*, **128**, pp. 160C–171C.

46. Pauling, L. (1964). Nature of the Iron–oxygen Bond in Oxyhæmoglobin, *Nature*, **203**, pp. 182–183.
47. Griffith, J. S. (1956). On the Magnetic Properties of Some Haemoglobin Complexes, *Proc. R. Soc. Lond. A*, **235**, pp. 23–36.
48. Yeager, E. (1986). Dioxygen Electrocatalysis: Mechanisms in Relation to Catalyst Structure, *J. Mol. Catal.*, **38**, pp. 5–25.
49. Toda, T., Igarashi, H., Uchida, H., *et al.*, (1999). Enhancement of the Electroreduction of Oxygen on Pt Alloys with Fe, Ni, and Co, *J. Electrochem. Soc.*, **146**, pp. 3750–3756.
50. Shim, J., Yoo, D. and Lee. J. (2000). Characteristics for Electrocatalytic Properties and Hydrogen–oxygen Adsorption of Platinum Ternary Alloy Catalysts in Polymer Electrolyte Fuel Cell, *Electrochim. Acta*, **45**, pp. 1943–1951.
51. Lebègue, E., Baranton, S. and Coutanceau, C. (2011). Polyol Synthesis of Nanosized Pt/C Electrocatalysts Assisted by Pulse Microwave Activation, *J. Power Sources*, **196**, pp. 920–927.
52. Sun, X., Li, R., Villers, D., *et al.* (2003). Composite Electrodes Made of Pt Nanoparticles Deposited on Carbon Nanotubes Grown on Fuel Cell Backings, *Chemical Physics Letters*, **379**, pp. 99–104.
53. Celebi, S., Nijhuis, T., van der Schaaf, J., *et al.* (2011). Carbon Nanofiber Growth on Carbon Paper for Proton Exchange Membrane Fuel, *Carbon*, **49**, pp. 501–507.
54. Markovic, N. and Ross, P. (2000). Electrocatalysts by Design: From the Tailored Surface to a Commercial Catalyst, *Electrochim. Acta*, **45**, pp. 4101–4115.
55. Marković, N. and Ross, P. (2002), Surface Science Studies of Model Fuel Cell Electrocatalysts, *Surf. Sci. Rep.*, **45**, pp. 117–229.
56. Marković, N., Gasteiger, H. and Ross, P. (1997). Kinetics of Oxygen Reduction on Pt(hkl) Electrodes: Implications for the Crystallite Size Effect with Supported Pt Electrocatalysts, *J. Electrochem. Soc.*, **144**, pp. 1591–1597.
57. Paulus, U., Schmidt, T., Gasteiger, H., *et al.* (2001). Oxygen Reduction on a High-surface Area Pt/Vulcan Carbon Catalyst: A Thin-film Rotating Ring-disk Electrode Study, *Electroanal. Chem.*, **495**, pp. 134–145.
58. Paulus, U., Wokaum, A., Scherer, G., *et al.* (2002). Oxygen Reduction on High Surface Area Pt-based Alloy Catalysts in Comparison to Well Defined Smooth Bulk Alloy Electrodes, *Electrochim. Acta*, **47**, pp. 3787–3798.
59. Bughun, I. and Anson, F. (1997). Adsorption on Graphite and Catalytic Reduction of 0 2 by the Macrocyclic Complex of Cobalt(II) with 2, 3, 9, 10-tetraphenyl- 1, 4, 8, 11 tetraazacyclotetradeca- 1, 3, 8, 10-tetraene, *J. Electroanal. Chem.*, **430**, pp. 155–161.
60. Convert, P., Coutanceau, C., Crouigneau, P., *et al.* (2001). Electrode Modified by Electrodeposition of CoTAA Complexes as Selective Oxygen Cathode in a Direct Methanol Fuel Cell, *J. Appl. Electrochem.*, **31**, pp. 945–952.
61. Jakobs, R., Janssen, L. and Barendrecht, E. (1985). Oxygen Reduction at Polypyrrole Electrodes: I. Theory and Evaluation of the rrde Experiments, *Electrochim. Acta*, **30**, pp. 1085–1091.
62. Napporn, W., Léger, J.-M. and Lamy, C. (1996). Electrocatalytic Oxidation of Carbon Monoxide at Lower Potentials on Platinum-based Alloys Incorporated in Polyaniline, *J. Electroanal. Chem.*, **408**, pp. 141–147.
63. Oh, H., Oh, J., Hong, Y., *et al.*, (2007). Investigation of Carbon-supported Pt Nanocatalyst Preparation by the Polyol Process for Fuel Cell Applications, *Electrochim. Acta*, **52**, pp. 7278–7285.
64. Liu, Z., Gan, L., Hong, L., *et al.* (2005). Carbon-supported Pt Nanoparticles as Catalysts for Proton Exchange Membrane Fuel Cells, *J. Power Sources*, **139**, pp. 73–78.
65. Paffet, M., Beery, J. and Gottesfeld, S. (1988). Oxygen Reduction at $Pt_{0.65}Cr_{0.35}$, $Pt_{0.2}Cr_{0.8}$ and Roughened Platinum, *J. Electrochem. Soc.*, **135**, pp. 1431–1436.

66. Beard, B. and Ross, P. (1990). The Structure and Activity of Pt-Co Alloys as Oxygen Reduction Electrocatalysts, *J. Electrochem. Soc.*, **137**, pp. 3368–3374.
67. Toda, T., Igarashi, H. and Watanabe, M. (1999). Enhancement of the Electrocatalytic $O_2$ Reduction on Pt–Fe Alloys, *J. Electroanal. Chem.*, **460**, pp. 258–262.
68. Bönnemann, H., Brijoux, W., Brinkmann, R., *et al.* (1991). Formation of Colloidal Transition Metals in Organic Phases and Their Application in Catalysis, *Angew. Chem. Int. Engl.*, **30**, pp. 1312–1314.
69. Vogel, W., Britz, P., Bönnemann, H., *et al.* (1997). Structure and Chemical Composition of Surfactant-stabilized PtRu Alloy Colloids, *J. Phys. Chem. B*, **101**, pp. 11029–11036.
70. Massalski, T., Baker, H., Bennett, L., *et al.* (1986), in T. Massalski (ed.), *Binary Alloy Phase Diagrams*, American Society for Metals, Metals Park, OH.
71. Leroux, C., Cadeville, M., Pierron-Bohmes, V., *et al.* (1988). Comparative Investigation of Structural and Transport Properties of L10 NiPt and CoPt Phases: The Role of Magnetism, *J. Phys. F: Metal Physics*, **18**, pp. 2033–2052.
72. Antolini, E., Salgado, J., da Silva, R., *et al.* (2007). Preparation of Carbon Supported Binary Pt-M Alloy Catalysts (M = First Row Transition Metal) by Low-medium Temperature Methods, *Materials Chemistry and Physics*, **101–102**, pp. 395–403.
73. Mukerjee, S., Srinivasan, S., and Soriaga. M. (1995). Role of Structural and Electronic Properties of Pt and Pt Alloys on Electrocatalysis of Oxygen Reduction: An In Situ XANES and EXAFS Investigation, *J. Electrochem. Soc.*, **142**, pp. 1409–1422.
74. Wang, Y. and Balbuena, P. (2005). Design of Oxygen Reduction Bimetallic Catalysts: Ab-initio-derived Thermodynamic Guidelines, *J. Phys. Chem. B*, **109**, pp. 18902–18906.
75. Alonso-Vante, N., Cattarin, S. and Musiani, M. (2000). Electrocatalysis of $O_2$ Reduction at Polyaniline + Molybdenum-doped Ruthenium Selenide Composite Electrodes, *J. Electroanal. Chem.*, **481**, pp. 200–207.
76. Lalande, G., Côté, R., Tamizhmani, G., *et al.* (1995). Physical, Chemical and Electrochemical Characterization of Heat-treated Tetracarboxylic Cobalt Phthalocyanine Adsorbed on Carbon Black as Electrocatalyst for Oxygen Reduction in Polymer Electrolyte Fuel Cells, *Electrochim. Acta*, **40**, pp. 2635–2646.
77. Lalande, G., Faubert, G., Côté, R., *et al.* (1996). Catalytic Activity and Stability of Heat-treated Iron Phthalocyanines for the Electroreduction of Oxygen in Polymer Electrolyte Fuel Cells, *J. Power Sources*, **61**, pp. 227–237.
78. Coutanceau, C., El Hourch, A., Crouigneau, P., *et al.* (1995). Conducting Polymer Electrodes Modified by Metal Tetrasulfonated Phthalocyanines: Preparation and Electrocatalytic Behaviour Towards Dioxygen Reduction in Acid Medium, *Electrochim. Acta*, **40**, pp. 2739–2748.
79. Coutanceau, C., Crouigneau, P., Léger, J., *et al.* (1994). Mechanism of Oxygen Electroreduction at Polypyrrole Electrodes Modified by Cobalt Phthalocyanine, *J. Electroanal. Chem.*, **379**, pp. 389–397.
80. Lefèvre, M., Proietti, E., Jaouen, F., *et al.* (2009). Iron-based Catalysts with Improved Oxygen Reduction Activity in Polymer Electrolyte Fuel Cells, *Science*, **324**, pp. 71–74.
81. Shao, Y., Sui, J., Yin, G., *et al.* (2008). Nitrogen-doped Carbon Nanostructures and their Composites as Catalytic Materials for Proton Exchange Membrane Fuel Cell, *Appl. Cat. B Environ.*, **79**, pp. 89–99.
82. Jasinski, R. (1965). Cobalt Phthalocyanine as a Fuel Cell Cathode, *J. Electrochem. Soc.*, **112**, pp. 526–528.
83. Savy, M., Andro, P., Bernard, C., *et al.* (1973). Etude de la Reduction de L'oxygene sur les Phtalocyanines Monomeres et Polymeres. I. Principes Fondamentaux, Choix de L'ion Central, *Electrochim. Acta*, **18**, pp. 191–197.
84. Zagal, J., Bindra, P. and Yeager, E. (1980). A Mechanistic Study of $O_2$ Reduction on Water Soluble Phthalocyanines Adsorbed on Graphite Electrodes, *J. Electrochem. Soc.*, **127**, pp. 1506–1517.

85. Biloul, A., Coowar, F., Contamin, O., *et al.* (1990). Oxygen Reduction in Acid Media on Supported Iron Naphthalocyanine: Effect of Isomer Configuration and Pyrolysis, *J. Electroanal. Chem.*, **289**, pp. 189–201.
86. Gruening, G., Wiesener, K., Gamburtsev, S., *et al.* (1983). Investigations of Catalysts from the Pyrolyzates of Cobalt-containing and Metal-free Dibenzotetraazaannulenes on Active Carbon for Oxygen Electrodes in an Acid Medium, *J. Electroanal. Chem.*, **159**, pp. 155–162.
87. Scherson, D., Gupta, S., Fierro, C., *et al.* (1983). Cobalt Tetramethoxyphenyl Porphyrin-emission Mossbauer Spectroscopy and $O_2$ Reduction Electrochemical Studies, *Electrochim. Acta*, **28**, pp. 1205–1209.
88. Biloul, A., Gouérec, P., Savy, M., *et al.* (1996). Oxygen Electrocatalysis Under Fuel Cell Conditions: Behaviour of Cobalt Porphyrins and Tetraazaannulene Analogues, *Appl. Electrochem.*, **26**, pp. 1139–1146.
89. Van Veen, J., Colijn, H. and Van Baar, J. (1988). On the Effect of a Heat Treatment on the Structure of Carbon-supported Metalloporphyrins and Phthalocyanines, *Electrochim. Acta*, **33**, pp. 801–804.
90. Ballirano, P., Caminiti, R., Ercolani, C., *et al.* (1998). X-ray Powder Diffraction Structure Reinvestigation of the α and β Forms of Cobalt Phthalocyanine and Kinetics of the α → β Phase Transition, *J. Am. Chem. Soc.*, **120**, pp. 12798–12807.
91. Kirner, J., Dow, W. and Scheidt, W. (1976). Molecular Stereochemistry of Two Intermediate-spin Complexes. Iron(Ii) Phthalocyanine and Manganese(Ii) Phthalocyanine, *Inorg. Chem.*, **15**, pp. 1685–1690.
92. Baranton, S., Coutanceau, C., Garnier, E., *et al.* (2006). How Does α-FePc Catalysts Dispersed onto High Specific Surface Carbon Support Work Towards Oxygen Reduction Reaction (orr)?, *J. Electroanal. Chem.*, **590**, pp. 100–110.
93. Collman, J., Denisevich, P., Konai, Y., *et al.* (1980). Electrode Catalysis of the Four-electron Reduction of Oxygen to Water by Dicobalt Face-to-face Porphyrins, *J. Am. Chem. Soc.*, **102**, pp. 6027–6036.
94. Yeager, E. (1984). Electrocatalysts for $O_2$ Reduction, *Electrochim. Acta*, **29**, pp. 1527–1537.
95. Damjanovic, A. and Bockris, J. (1966). The Rate Constants for Oxygen Dissolution on Bare and Oxide-covered Platinum, *Electrochim. Acta*, **11**, pp. 376–377.
96. Gnanamuthu, D. and Petrcelli, J. (1967). A Generalized Expression for the Tafel Slope and the Kinetics of Oxygen Reduction on Noble Metals and Alloys, *J. Electrochem. Soc.*, **114**, pp. 1036–1041.
97. Holtze, R. (1988). Electrochemical and Surface Raman Spectroscopic Studies of an Iron Porphyrine Adsorbed on an Electrode, *Electrochim. Acta*, **33**, pp. 1619–1627.
98. Milligan, D. and Jacox, M. (1963). Infrared Spectroscopic Evidence for the Species $HO_2$, *Chem. Phys.*, **38**, pp. 2627–2631.
99. Anastasijevic, N., Vesovic, V. and Adžić, R. (1987). Determination of the Kinetic Parameters of the Oxygen Reduction Reaction Using the Rotating Ring-disk Electrode: Part I. Theory, *J. Electroanal. Chem.*, **229**, pp. 305–316.
100. Lamy, C. and Léger, J.-M., (1994). Les Piles à Combustible: Application au Véhicule Électrique, *J. Phys. IV*, **4**, pp. 253–278.
101. Rousseau, S., Coutanceau, C., Lamy, C., *et al.* (2006). Direct Ethanol Fuel Cell (DEFC): Electrical Performances and Reaction Products Distribution Under Operating Conditions with Different Platinum-based Anodes, *J. Power Sources*, **158**, pp. 18–24.
102. Bambagioni, V., Bianchini, C., Marchionni, A., *et al.* (1990). Pd and Pt–Ru Anode Electrocatalysts Supported on Multi-walled Carbon Nanotubes and their Use in Passive and Active Direct Alcohol Fuel Cells with an Anion-exchange Membrane (Alcohol = Methanol, Ethanol, Glycerol), *J. Power Sources*, **190**, pp. 241–251.

103. Beden, B., Juanto, S., Léger, J.-M., *et al.* (1987). Infrared Spectroscopy of the Methanol Adsorbates at a Platinum Electrode: Part III. Structural Effects and Behaviour of the Polycrystalline Surface, *J. Electroanal. Chem.*, **238**, pp. 323–331.
104. Demarconnay. L., Brimaud. S., Coutanceau, C., *et al.* (2007). Ethylene Glycol Electrooxidation in Alkaline Medium at Pluri-metallic Pt Based Catalysts, *J. Electroanal. Chem.*, **601**, pp. 169–180.
105. Dubau. L., Coutanceau. C., Garnier, E., *et al.* (2003). Electrooxidation of Methanol at Platinum–Ruthenium Catalysts Prepared from Colloidal Precursors: Atomic Composition and Temperature Effects, *J. Appl. Electrochem.*, **33**, pp. 419–429.
106. Mayrhofer, K., Arenz, M., Blizanac, B., *et al.* (2005). CO Surface Electrochemistry on Pt-nanoparticles: A Selective Review, *Electrochim. Acta*, **50**, pp. 5144–5154.
107. Waszczuk, P., Wieckowski, A., Zelenay, P., *et al.* (2001). Adsorption of CO Poison on Fuel Cell Nanoparticle Electrodes from Methanol Solutions: A Radioactive Labeling Study, *J. Electroanal. Chem.*, **511**, pp. 55–64.
108. Brankovic, S., Wang, J., Zhu, Y., *et al.* (2002). Electrosorption and Catalytic Properties of Bare and Pt Modified Single Crystal and Nanostructured Ru Surfaces, *J. Electroanal. Chem.*, **524/525**, pp. 231–241.
109. Brankovic S., Marinkovic, N., Wang, J., *et al.* (2002). Carbon Monoxide Oxidation on Bare and Pt-modified Ru(10$\bar{1}$0) and Ru(0001) Single Crystal Electrodes, *J. Electroanal. Chem.*, **532**, pp. 57–66.
110. Iwasita, T. and Pastor, E. (1994). A DEMS and FTIR Spectroscopic Investigation of Adsorbed Ethanol on Polycrystalline Platinum, *Electrochim. Acta*, **39**, pp. 531–537.
111. Iwasita, T. and Pastor, E. (1994). D/H Exchange of Ethanol at Platinum Electrodes, *Electrochim. Acta*, **39**, pp. 547–551.
112. Rightmire, R., Rowland, R., Boos, D., *et al.* (1964). Ethyl Alcohol Oxidation at Platinum Electrodes, *J. Electrochem. Soc.*, **111**, pp. 242–247.
113. Hitmi, H., Belgsir, E., Léger, J.-M., *et al.* (1994). A Kinetic Analysis of the Electro-oxidation of Ethanol at a Platinum Electrode in Acid Medium, *Electrochim. Acta*, **39**, pp. 407–415.
114. Perez, J., Beden, B., Hahn, F., *et al.* (1989). *In situ* Infrared Reflectance Spectroscopic Study of the Early Stages of Ethanol Adsorption at a Platinum Electrode in Acid Medium, *J. Electroanal. Chem.*, **262**, pp. 251–261.
115. Iwasita, T. and Nart, F. (1991). Identification of Methanol Adsorbates on Platinum: An *in situ* FT-IR Investigation, *J. Electroanal. Chem.*, **317**, pp. 291-298.
116. Kabbabi, A., Faure, R., Durand, R., *et al.* (1998). *In situ* FTIRS Study of the Electrocatalytic Oxidation of Carbon Monoxide and Methanol at Platinum–Ruthenium Bulk Alloy Electrodes, *J. Electroanal. Chem.*, **444**, pp. 41–53.
117. Iwasita, T., Hoster, H., John-Anaker, A., *et al.* (2000). Methanol Oxidation on PtRu Electrodes. Influence of Surface Structure and Pt-Ru Atom Distribution, *Langmuir*, **16**, pp. 522–529.
118. Dinh, H., Ren, X., Garzon, F., *et al.* (2000). Electrocatalysis in Direct Methanol Fuel Cells: In situ Probing of PtRu Anode Catalyst Surfaces, *J. Electroanal. Chem.*, **491**, pp. 222–233.
119. Gasteiger, H., Markovic, N., Ross, P., *et al.* (1994). Temperature-dependent Methanol Electro-oxidation on Well-characterized Pt-Ru alloys, *J. Electrochem. Soc.*, **141**, pp. 1795–1803.
120. Gasteiger, H., Markovic, N., Ross, P., *et al.* (1993). Methanol Electrooxidation on Well-characterized Platinum–ruthenium Bulk Alloys, *J. Phys. Chem.*, **97**, pp. 12020–12029.
121. Waszczuk, P., Solla-Gullón, J., Kim, H., *et al.* (2001). Methanol Electrooxidation on Platinum/Ruthenium Nanoparticle Catalysts, *J. Catal.*, **203**, pp. 1–6.
122. Tong, Y., Kim, H., Babu, P., *et al.* (2002). An NMR Investigation of CO Tolerance in a Pt/Ru Fuel Cell Catalyst, *J. Am. Chem. Soc.*, **124**, pp. 468–473.
123. Schmidt, T., Gasteiger, H. and Behm, R. (1999). Methanol Electrooxidation on a Colloidal PtRu-alloy Fuel-cell Catalyst, *Electrochem. Commun.*, **1**, pp. 1–4.

124. Hamnett A. (1999). Mechanism of Methanol Oxidation, in A. Wieckowski (ed.). *Interfacial Electrochemistry: Theory, Experiments and Applications*, Marcel Dekker, New York, pp. 843–883.
125. Lamy, C., Rousseau, S., Belgsir, E., *et al.* (2004). Recent Progress in the Direct Ethanol Fuel Cell: Development of New Platinum–tin Electrocatalysts, *Electrochim. Acta.*, **49**, pp. 3901–3908.
126. Delime, F., Léger, J.-M. and Lamy, C. (1999). Enhancement of the Electrooxidation of Ethanol on a Pt–PEM Electrode Modified by Tin. Part I: Half Cell Study, *J. Appl. Electrochem.*, **29**, pp. 1249–1254.
127. Song, S., Zhou, W., Zhou, Z., *et al.* (2005). Direct Ethanol PEM Fuel Cells: The Case of Platinum Based Anodes, *Int. J. Hydrogen Energy*, **30**, pp. 995–1001.
128. Tsiakaras, P. (2007). PtM/C (M = Sn, Ru, Pd, W) Based Anode Direct Ethanol–PEMFCs: Structural Characteristics and Cell Performance, *J. Power Sources*, **171**, pp. 107–112.
129. Nappom, W., Léger, J.-M and Lamy, C. (1996). Electrocatalytic Oxidation of Carbon Monoxide at Lower Potentials on Platinum-based Alloys Incorporated in Polyaniline, *J. Electroanal. Chem.*, **408**, pp. 141–147.
130. Morimoto, Y. and Yeager, E. (1998). CO Oxidation on Smooth and High Area Pt, Pt-Ru and Pt-Sn Electrodes, *J. Electroanal. Chem.*, **441**, pp. 77–81.
131. Massong, H., Wang, H., Samjeské, G., *et al.* (2007). The Co-catalytic Effect of Sn, Ru And Mo Decorating Steps of Pt(111) Vicinal Electrode Surfaces on the Oxidation of CO, *Electrochim. Acta*, **46**, pp. 701–707.
132. Wang, K., Gasteiger, H., Markovic, N., *et al.* (1996). On the Reaction Pathway for Methanol and Carbon Monoxide Electrooxidation on Pt-Sn Alloy Versus Pt-Ru Alloy Surfaces, *Electrochim. Acta*, **41**, pp. 2587–2593.
133. Park, S., Wieckowski, A. and Weaver, M. (2003). Electrochemical Infrared Characterization of CO Domains on Ruthenium-decorated Platinum Nanoparticles, *J. Am. Chem. Soc.*, **125**, pp. 2282–2290.
134. Stamenković, V., Arenz, M., Blizanac, B., *et al.* (2005). In situ CO Oxidation on Well Characterized $Pt_3Sn$(hkl) Surfaces: A Selective Review, *Surf. Science*, **576**, pp. 145–157.
135. Arenz, M., Stamenković, V., Blizanac, B., *et al.* (2005). Carbon-supported Pt–Sn Electrocatalysts for the Anodic Oxidation of $H_2$, CO, and $H_2$/CO Mixtures: Part II: The Structure–activity Relationship, *J. Catal.*, **232**, pp. 402–410.
136. Shropshire, J. (1965). The Catalysis of the Electrochemical Oxidation of Formaldehyde and Methanol by Molybdates, *J. Electrochem. Soc.*, **112**, pp. 465–469.
137. Lima, A., Coutanceau, C., Léger, J.-M., *et al.* (2001). Investigation of Ternary Catalysts for Methanol Electrooxydation, *J. Appl. Electrochem.*, **31**, pp. 379–386.
138. Jusys, Z., Schmidt, T., Dubau, L., *et al.* (2002). Activity of PtRuMeOx (Me = W, Mo or V) Catalysts Towards Methanol Oxidation and their Characterization, *J. Power Sources*, **105**, pp. 297–304.
139. Podlovchenko, B., Petrii, O., Frumkin, A., *et al.* (1966). The Behaviour of a Platinized-platinum Electrode in Solutions of Alcohols Containing More than One Carbon Atom, Aldehydes and Formic Acid, *J. Electroanal. Chem.*, **11**, pp. 12–25.
140. Smirnova, N., Petrii, O. and Grzejdziak, A. (1988). Effect of Ad-atoms on the Electro-oxidation of Ethylene Glycol and Oxalic Acid on Platinized Platinum, *J. Electroanal. Chem.*, **251**, pp. 73–87.
141. Adžić, R. (1984). Electrocatalytic Properties of Surfaces Modified by Foreign Metal Adatoms, in H. Gerischer and C. Tobias (eds), *Advances in Electrochemistry and Electrochemical Engineering*, Vol. 13, Wiley-Interscience, New York, pp. 159–260.

# Chapter 10

# Advanced Oxidation Processes in Water Treatment

Gabriele CENTI and Siglinda PERATHONER*

The use of advanced oxidation processes (AOPs) to remove pollutants in water treatment applications has been widely studied and applied industrially, but it is still an area of active development. New problems derive from toxic, refractory and xenobiotic micropollutants and the increasing requirements in terms of energy efficiency and quality of water in several industrial wastewater streams. This chapter introduces the AOP technologies and discusses the possibilities offered by using catalysts in these methods, with selected examples regarding the industrial applicability of these AOPs and open questions about their further development.

Water is an essential compound for life on Earth and its quality is crucial for the future of humanity. Human alterations to the water cycle combined with direct and indirect pollution have a profound effect on water availability and quality. Consequently, not only is the availability of water resources decreasing, but the quality is also worsening.[1] Only a small fraction of the available surface water meets the quality necessary for human and industrial use of water, and there is an increasing need for improved technologies to treat wastewater, remediate polluted water resources and eliminate micropollutants from water.

A variety of organic compounds from wastewater sources, such as pharmaceuticals, additives of personal care products, household chemicals, and industrial chemicals are widely recognized as a potential threat to aquatic ecosystems and to human health.[2] Even if present in low concentration ranges, they accumulate in the environment as they are largely non-biodegradable. Major concerns regarding the risk micropollutants pose to aquatic ecosystems and to human health involve endocrine disrupting effects, chemosensitizing effects, possible interactions of contaminant mixtures and chronic effects from long-term exposure.[3,4]

*Dipartimento di Chimica Industriale ed Ingegneria dei Materiali and CASPE/INSTM V.le F. Stagno D'Alcontres 31, 98166 Messina, Italy.

The term micropollutants is used to cover a range of substances found in water, of natural or synthetic origin, that includes those discharged directly as well as those that form inadvertently (e.g. chlorine disinfection by-products), or during treatment through degradation processes, and that may persist despite treatment. Endocrine disrupting chemicals (EDCs) are exogenous agents interfering with the synthesis, secretion, transport, binding, action and elimination of natural hormones in the body, which are responsible for the maintenance of homeostasis, reproduction, development and behavior.

These micropollutants are hardly removed using conventional water treatment technologies, but can often be converted using tertiary water treatment technologies, such as those based on advanced oxidation processes (AOPs), which will be discussed in this chapter. In addition, the use of these tertiary treatment technologies allows the impact of wastewater from industrial sources upon the environment to be reduced. Also, their integration with conventional water treatment technologies may often result in more economic and more efficient processes. An example is the pre-treatment of industrial wastewater by AOPs to remove toxic chemicals such as phenols and pesticides before biological treatment (active sludge).[5,6]

Industry and municipalities use about 10% of the globally accessible run-off and generate a stream of wastewater containing numerous chemical compounds. About 300 million tons of synthetic chemicals are discharged annually in wastewaters[2] and this excessive amount highlights the dimension of the problem. In addition, water withdrawal with re-emission (as wastewater) at a different location may often cause over exploitation of aquifers and a lowering of water tables, in some cases below the level for intrusion of salty water into groundwater. Increasing the recycling of wastewater is thus becoming a necessity to decrease the negative environmental impact of production, but may also become an opportunity to improve water management and process economics, as exemplified by the case of rinse water from the electronics industry.[7,8] The integration of AOP methods in the water technology train to recycle industrial wastewater is often necessary.[9]

Tertiary treatment systems currently used in wastewater treatment plants include membrane technologies (microfiltration, ultrafiltration, reverse osmosis, etc.), activated carbon adsorption and other methods.[9] Often these treatment methods are not fully effective in producing water with acceptable levels of the most persistent pollutants (e.g. phenols, pesticides, solvents, household chemicals and drugs, etc.), whereas different AOPs are recommended when wastewater components have a high chemical stability and/or low biodegradability. With respect to membranes, which are also effective in removing these persistent pollutants, AOPs may be more cost-effective and not sensitive to fouling.

In addition to the treatment of wastewaters and raw waters for human consumption and industrial uses, remediation of contaminated sites is another relevant area for the use of AOPs.[10] The pump and treat method is still the most common for a fast remediation of contaminated water resources. When non-volatile chemicals are present, the use of AOP is often the necessary option to eliminate the pollutants before water reintroduction. In addition, in some cases AOPs have been adapted for *in situ* remediation (for example, Fenton processes; see below).

AOPs are thus a water treatment technology of increasing relevance for the cleaning of waste and contaminated water with persistent residue. These methodologies are able to effectively degrade and remove specific pollutants, which would otherwise have been extremely difficult to eliminate with conventional processes, since many of these compounds are not biodegradable. Although known for many years, relevant progress is still possible with AOPs, in terms of efficient use of oxidants and integration with other technologies. The use of solid catalysts is one of the relevant options that will be emphasized here to further promote these methodologies.[11,12]

## 10.1. Advanced Oxidation Processes

Advanced oxidation processes indicate an ensemble of various technologies, based on the use of different oxidations, which are able to give complete mineralization (transformation to $CO_2$) of most organic chemicals. The common aspect is that all these methods are based on the formation and reaction of $^{\bullet}OH$ (hydroxyl) radicals, which are much more reactive than all other oxidizing species used in oxidative pollution abatement in drinking water and wastewater.[13,14] Depending on the type of oxidant and mechanism of reaction, other oxidizing species may also be involved in the conversion of organic pollutants.[15]

The first distinction between the different AOP methods is based on the different classes of processes: photochemical degradation processes ($UV/O_3$, $UV/H_2O_2$), photocatalysis ($TiO_2/UV$, photo-Fenton) and chemical oxidation processes ($O_3$, $O_3/H_2O_2$, $H_2O_2/Fe^{2+}$). A further distinction is based on the presence of homogeneous or heterogeneous reactions, the latter based on the use of solid catalysts (Fig. 10.1). Homogeneous processes can be further subdivided into processes that use an external input of energy (ultraviolet radiation, ultrasound, electrical or thermal energy) and processes that do not use energy.

### 10.1.1. *AOPs using UV radiation and/or chemical oxidation*

UV radiation is used to produce hydroxyl radicals (for example, by homolytic splitting of hydrogen peroxide) and in some cases for the degradation of compounds sensitive to UV radiation. When colored chemicals are present (dyes, for example) solar light could be used, with the dye also acting as photosensitizer. The costs of UV radiation treatment processes are largely dependent on the absorption properties

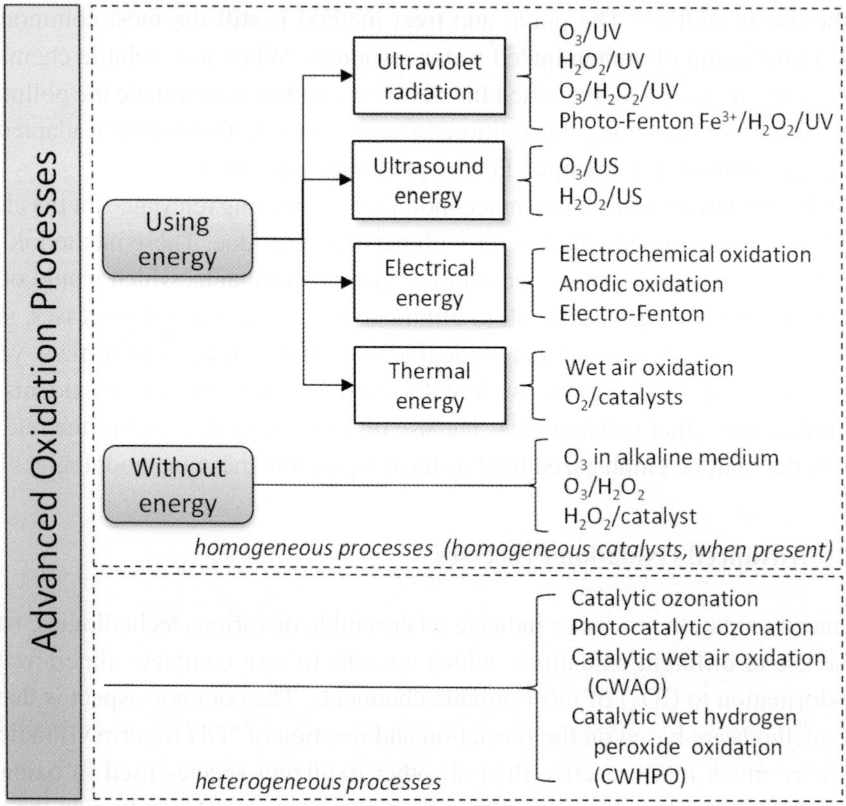


**Figure 10.1.** Classification of advanced oxidation processes (AOPs). Elaborated from Ref. 13.

of the compounds to be eliminated. Compounds that absorb light at lower wavelengths such as pentachlorphenol and N-nitrosodimethylamine (NDMA) could be cost-effectively photodegradated.[16]

The first step in combining ozone ($O_3$) with ultraviolet radiation is the ozone photolysis, which gives rise to the formation of hydroxyl radicals according to the following reaction:

$$H_2O + O_3 \quad \square h\nu \rightarrow 2^{\bullet}OH + O_2 \tag{10.1}$$

Two hydroxyl radicals may recombine to give hydrogen peroxide ($H_2O_2$) which can also be homolitically cleaved by UV radiation to give back two hydroxyl radicals. Generally, the aqueous medium saturated with ozone is irradiated with ultraviolet light at a wavelength of 254 nm, because for this wavelength the extinction coefficient for gas-phase ozone (3,300 $M^{-1}$ $cm^{-1}$) is two orders of magnitude greater than that of $H_2O_2$ (18.6 $M^{-1}$ $cm^{-1}$).[17] The method is limited by the low solubility of ozone, but is effective in the treatment of low concentrations of hard to remove chemicals. An example is the degradation of dinitrotoluene (DNT) and trinitrotoluene (TNT), present in the wastewater of explosive as well as various pharmaceutical and dye productions. Chen *et al.*[18] reported a high mineralization (94% removal of the total organic carbon – TOC) with a UV radiation of 96 W and an ozone dosage of 3.8 g/h. The use of UV radiation increases about 76% of the TOC removal. Another example is the elimination from wastewater of the industrial solvent N-methyl-2-pyrolidone (NMP). Using an $O_3$/UV combination, it is possible in a few hours to completely mineralize this compound using a relatively low radiation intensity ($1.5 \times 10^{-6}$ Einstein/L$\cdots$).[19] In general, ozone-based AOPs, which also include various combinations with UV irradiation and hydrogen peroxide, have the highest performances for degrading and detoxifying persistent pesticides in water bodies,[20] a problem of increasing relevance in many wells in rural areas. An example of industrial use is the treatment of effluents from the paperboard industry,[21] where many toxic phenolic-type products derived from lignin are present. The method is effective in their removal and allows wastewater to be sent for conventional biological treatment. The $O_3$/UV combination is also effective in bacterial elimination. An example is the inactivation of bacteria in the water from fish culture tanks.[22] However, Bustos *et al.*[23] reported recently in studying the ozone, UV and $O_3$/UV processes for the reuse of effluents in sewage treatment plants, that none of the processes achieved the disinfection levels required for human usage. A comprehensive review of the applications of ozone for industrial wastewater treatment was reported by Rice,[24] while Ried *et al.*[25] more recently reviewed the integration of ozone systems in biological treatment steps.

The $H_2O_2$/UV radiation process is based on the direct homolytic photolysis of hydrogen peroxide to give two hydroxyl radicals:

$$H_2O_2 \quad \square h\nu \rightarrow 2^{\bullet}OH \tag{10.2}$$

Although the use of $H_2O_2$ is preferable to ozone, due to the much higher solubility and the possibility of addressing the issue of wastewater with higher concentrations of pollutants, there are two main limitations. In aqueous solution, the two $^{\bullet}OH$ radicals generated in the photolytic process are not separated quickly due to the solvent cage effect, and thus the recombination to $H_2O_2$ is relatively fast, i.e. the quantum efficiency of the process is limited (around 50%). In addition, due to the high reactivity, long diffusional paths are not possible. The second problem is that $H_2O_2$ only absorbs below 280 nm and its absorption coefficient at 254 nm (the main emission of the low-pressure Hg-arc lamps) is only about 19 $M^{-1}$ $cm^{-1}$. For an efficient absorption of the UV light high $H_2O_2$ concentrations are necessary, but this gives rise to increased $^{\bullet}OH$ scavenging by $H_2O_2$.

The photolysis rate increases in alkaline conditions because the absorption coefficient of the peroxide anion ($HOO^-$) is higher than that of $H_2O_2$ (240 $M^{-1}$ $cm^{-1}$ as compared to about 19 $M^{-1}$ $cm^{-1}$). However, at basic pH, decomposition of $H_2O_2$ to water and oxygen occurs. Several metal ions present as impurities in the wastewater also catalyze the $H_2O_2$ decomposition. Many parameters, as in the $O_3$/UV case, thus determine the rate and effectiveness of $H_2O_2$/UV operations, such as the intensity of UV radiation, the ratio of oxidant/UV doses, the concentration of pollutants and other elements in the wastewater, the design of the photoreactor, etc.

An interesting recent example is the use of this method for the elimination of pharmaceutical compounds in municipal wastewater. For example, carbamazepine, clofibric acid, diazepam and diclofenac present in concentrations between 0.006 and 1.9 $\mu g\ L^{-1}$ in the wastewater of the Aachen-Soers region (Germany) could be effectively removed by $H_2O_2$/UV or $O_3$/UV treatment, while conventional biological treatments or membrane bioreactors are not effective.[26]

Another example is the elimination of naproxen, a non-steroidal anti-inflammatory drug commonly used for the reduction of pain. Felis *et al.*[27] reported a conversion > 93% after a 3 min treatment with $H_2O_2$/UV, at pH 6, a naproxen concentration of 1.06 $\mu$g/L, and 1 mg/L of hydrogen peroxide.

Various works have also compared $H_2O_2$/UV with possible alternatives, from $O_3$/UV to Fenton (Fe(II)/$H_2O_2$) and photo-Fenton process (Fe(II)/$H_2O_2$/UV). The effectiveness depends on the specific type of wastewater needing treatment, but a general trend is present. Jamil *et al.*[28] recently reported a comparative study among different photochemical oxidation processes to enhance the biodegradability of paper mill wastewater. The main results are summarized in Table 10.1 in terms of mineralization after 1 h (residual carbon oxygen demand — COD), increase of biodegradability (biological/carbon oxygen demand ratio — $BOD_5$/COD ratio), dechlorination (formation of $Cl^-$ ions deriving from the conversion of organochlorine compounds; hydroxyl radicals may react with them forming HOCl which

**Table 10.1.** Effectiveness of different AOPs in the treatment of paper mill wastewater. Elaborated from the results in Ref. 28.

| AOPs | Initial | UV | $H_2O_2$/UV | Fenton $H_2O_2$/Fe(II) | Photo-Fenton $H_2O_2$/Fe(II)/UV |
|---|---|---|---|---|---|
| *Mineralization* residual COD, mg $O_2$/L | 10,300 | — | 9,270 | 5,870 | 1,960 |
| *Biodegradability* $BOD_5$/COD ratio | 0.21 | 0.25 | 0.45 | 0.55 | 0.70 |
| *Residual free $Cl^-$* mmol/L | 75 | 10 | 22 | 25 | 36 |
| *Suspended solids (TSS)* mg/L | 5,950 | — | 600 | 250 | 200 |

COD = carbon oxygen demand. $BOD_5$ = Biological oxygen demand (after 5 days); TSS = Total suspended solids. *Conditions of testing*: UV/$H_2O_2$, $H_2O_2$ (5 g/L); Fenton reaction, $H_2O_2$ (2 g/L), Fe(II) (0.75 g/L); photo-Fenton, $H_2O_2$ (1.5 g/L), Fe(II) (0.5 g/L), pH = 3 for the all processes.

inhibits the further oxidation reaction), and volume of total suspended solids after reaction. All AOP methods are effective in treating this difficult industrial wastewater stream, but the photo-Fenton process shows a significant enhancement in the biodegradability of the wastewater, removal of COD and total suspended solids (TSS). The results in Table 10.1 are under the optimal conditions for each AOP, i.e. for different concentrations of $H_2O_2$ (see Table 10.1 legend).

The photo-Fenton process, in spite of the higher complexity and the use of both iron ions and a UV lamp, allows a better utilization of $H_2O_2$, which is a significant component of the overall cost.

An aspect often underestimated is the analysis of the toxicity of the effluents after AOPs. An interesting example was reported by de Luis *et al.*[29] who investigated the toxicity (Microtox test) of phenol solutions treated with $H_2O_2$/UV and $H_2O_2$/Fe AOPs. In the $H_2O_2$/Fe(II) system, with a $H_2O_2$ dosage between 5 and 8 mol per mole of phenol, acceptable levels of toxicity (values less than 30) and complete primary degradation of phenol were achieved.

At greater oxidant dosages, toxicity was largely due to intermediate-type carboxylic acids, but below 5 mol of $H_2O_2$ per mol of phenol, aromatic-type compounds were mainly responsible for toxicity, where the increase of catalyst concentration had a negative effect.

The $H_2O_2$/UV system behaves differently. After about 15 min of reaction, the primary degradation was achieved in most cases, and the minimum value of toxicity was reached. For longer times, there is an increase in the toxicity associated with the appearance of intermediates such as carboxylic acids. Aromatic-type compounds were primarily responsible for toxicity in the initial phase before reaching the minimum toxicity. Moreover, increasing the dosage of oxidants caused a decline in the value of the minimum toxicity.

There is thus no relation between the conversion of phenol and the degree of mineralization and toxicity of the effluents. In addition, there is a great dependence on the type of treatment and operative conditions. Without these tests, it is not possible to arrive at realistic conclusions especially when the AOP method is a pre-treatment for consecutive biological treatments.

When $H_2O_2$ is used in an $O_3$/UV process, it accelerates the decomposition of ozone and increases the generation of $^{\bullet}OH$ radicals, and also the cost of the process. Table 10.2 compares the costs of the different UV-based processes, but it should be remembered that the costs may depend on a number of parameters. As a rule of thumb, however, $O_3/H_2O_2$/UV should be applied only when the other methods do not provide the necessary level of removal.

The reaction of $O_3$ with $H_2O_2$ is called the peroxone process. The reaction of $H_2O_2$ with $O_3$ is slow, but that of its anion, $HO_2^-$, is fast. It is thus necessary to work at pH values below the p$K_a$ of $H_2O_2$. According to Sein *et al.*[31] the rate

**Table 10.2.** Operating costs (excluding labor costs) of AOPs. Adapted from Ref. 30.

| AOPs | Treatment cost ($\$/m^3$) |
|---|---|
| $O_3$ | 5.3 |
| $O_3$/UV[a] | 8.7 |
| $H_2O_2$/UV[a] | 4.6 |
| $O_3/H_2O_2$/UV[a] | 11.2 |

[a]Optimum lamp life: 2,000 h. Operating costs calculated on the basis of 90% COD removal.

determining step is the addition of $O_3$ to $HO_2^-$ to generate the $HOOOOO^-$ adduct which may lose $O_2$ or break down into $HO_2^\bullet$ and $O_3^{\bullet-}$ radicals. The latter is a short-lived, rather inert intermediate that decays into $O_2$ and $OH^-$. The $HO_2^\bullet$ radical, when not recombined with other radicals or reacting with organic species, may undergo a further series of consecutive reactions giving rise to hydroxyl radicals and other radical or radical anion species. The presence of UV radiation further complicates this already complex reaction mechanism. The $^\bullet$OH yield is about 50% with respect to $O_3$ consumed.

The $O_3/H_2O_2$/UV is suitable for the treatment of dye in textile industry effluents,[30,32,33] because the $H_2O_2$/UV process shows a slower rate of decolorization than ozonation, which is more effective in TOC removal. The combination of the two methods allows both a high rate of decolorization and TOC removal.

The combination with Fenton-type catalysis may further enhance the performance in difficult cases of industrial wastewater. An interesting example is the treatment of wastewater from the production of terephthalic acid (TPA), a large-scale chemical used in the production of plastics, fibers, dyes, pesticides, perfumes, medicines, etc. Water effluents from the TPA manufacturing process contain high concentrations of TPA along with other toxic chemicals such as isophthalic acid (IPA) and benzoic acid (BA). The COD is very high, typically around 35 g/L, and the water has a dark reddish-brown color. Treatment methods based on biological degradation and other methods have a limited efficiency. Chandrasekara Pillai *et al.*[34] investigated the use of different combinations of ozone, hydrogen peroxide, UV radiation and $Fe^{2+}$ ions in the treatment of these effluents. The combined use of $O_3/H_2O_2/Fe^{2+}$/UV yielded the highest COD degradation rate, with better COD removal, higher than 90% at 4 h, and complete TPA, IPA and BA conversion in about 2 h. Using tert-butyl alcohol as an OH scavenger, Chandrasekara Pillai *et al.*[34] also differentiated in the overall rate of oxidation (COD elimination, $k_{tot}$) between two contributions: (i) the direct molecular oxidation ($k_{\rm mol}$) and (ii) the indirect free radical oxidation ($k_{\rm rad}$). The results are summarized in Table 10.3. Ozone added to

**Table 10.3.** Treatment of wastewater from terephthalic acid production: rate constants of individual contributions, molecular reaction ($k_{mol}$) or radical reaction ($k_{rad}$), and overall reaction ($k_{tot}$) for the optimized ozonation systems. Elaborated from the results in Ref. 34.

| Rate ($10^{-3}$ min$^{-1}$) | $O_3$ | $O_3$/UV | $O_3/H_2O_2$/UV | $O_3$/Fe(II) | $O_3/H_2O_2$/Fe(II)/UV |
|---|---|---|---|---|---|
| $k_{tot}$ | 4.98 | 5.26 | 8.05 | 7.28 | 9.59 |
| $k_{mol}$ | 2.39 | 3.21 | 2.52 | 2.72 | 3.32 |
| $k_{rad}$ | 2.59 | 2.05 | 5.53 | 4.56 | 6.28 |

water can react directly with organic and inorganic compounds or indirectly after decay by the formation of hydroxyl radicals (Eqs. 10.3a and 10.3b). The $OH^-$ ion initiates the decay of ozone in water; high pH values therefore enhance this reaction. The formed hydroxyl radicals react very quickly with other water constituents.

$$O_3 + OH^- \rightarrow HO_2^- + O_2 \tag{10.3a}$$

$$O_3 + HO_2^- \rightarrow {}^{\bullet}OH + O_2^{\bullet -} + O_2 \tag{10.3b}$$

The direct reaction of ozone with organic and inorganic compounds is very selective but slower than the indirect reaction. Ozone reacts with the functional groups of organic compounds, the double-bond structures and aromatic structures. The indirect reaction is quicker and non-selective. The results in Table 10.3 confirm that the rate of direct molecular oxidation is nearly constant in the various systems, while both the presence of $H_2O_2$ and Fe(II) ions increase the rate of radical reaction due to the more efficient creation of hydroxyl radicals. Generally, the direct ozonation is important if the radical reaction is inhibited or the water contains many substances, which terminate the radical chain reaction (scavengers). The main scavengers are $HCO_3^-$, $CO_3^{2-}$, $PO_3^{4-}$ and humic acids.[35] In general, compounds without aromatic and double-bond structures, like methyl tert-buthyl ether (MTBE), have a low conversion rate by ozone alone.

MTBE, being a widely used gasoline additive to increase octane number, and having low biodegradability, is another common surface and groundwater contaminant. Ozone treatment leads to about 30–40% MTBE removal, while the combination of $O_3$ and $H_2O_2$ leads to a reduction in concentration of MTBE between 37 and 70% with an ozone dosage of 4 mg/L and 1.36 mg/L $H_2O_2$.[36] In general, MTBE then atrazine are the most stable water contaminants against AOPs. Table 10.4 gives an overview of the removal efficiency of main trace contaminants in surface and groundwater.[37]

Energy consumption and formation of by-products have to be taken into account as possible limitations for practical implementation. Energy demand for these oxidation processes ranges between 0.2–0.7 kWh/m$^3$, depending on the AOP and feed conditions. $O_3/H_2O_2$ is the most energy efficient AOP, but still requires more energy

**Table 10.4.** Removal efficiency of main trace contaminants in surface and groundwater. $O_3$ dosing = 1 mg $O_3$/mg dissolved organic carbon (DOC). Adapted from Ref. 37.

| Transformation | $O_3$ | $O_3$/UV | $O_3/H_2O_2$ |
|---|---|---|---|
| >90% | Carbamazepine, Diclofenac, Primidone, Diuron, Sulfamethoxazole, Bentazone | Carbamazepine, Diclofenac | Carbamazepine, Diclofenac |
| >50% | Linuron, 1,5-naphthalenedisulfonic acids (NDSA), 1,7-NDSA, 2,1-NDSA | Atrazine, Iopamidol | Atrazine, Iopamidol |
| <50% | Atrazine, MTBE, Iopamidol | MTBE | MTBE |
| Uncertain | Primidone (AOP), Amidotrizoic acid, Desethylatrazine | | |

than ozonation, while the use of UV radiation lowers energy efficiency. $H_2O_2$/UV requires 5–10 times more energy (up to 0.66 kWh/m$^3$) compared to ozone alone (about 0.04–0.09 kWh/m$^3$ for 5–6 mg/L $O_3$). The use of solar energy may reduce these costs.

At the Solar Platform at Almería (www.psa.es) a photo-Fenton process ($H_2O_2$/Fe(II) + solar light) has been applied to the pre-treatment of water used to wash fertilizer containers, in particular for the total degradation of methylphenylglycine (MPG). In about 3 h, a complete conversion of MPG (initial concentration of 500 mg/L) occurs, with a TOC removal of 38% using a concentration of $Fe^{2+}$ ($FeSO_4 \cdot 7H_2O$) of 20 mg/L and $H_2O_2$ concentration of 400 mg/L.[38]

In addition to a higher energy demand, the main drawback of oxidation technologies is the possible formation of toxic by-products. The by-product of most concern is the bromated ion ($BrO_3^-$), a potential human carcinogen, which is formed by applying oxidation technologies to bromide ions ($Br^-$) containing waters. A second by-product of concern is the carcinogenic NDMA, which forms during the ozonation of N,N-Dimethylsulfamid (DMS) present in the groundwater of some rural areas (DMS is a metabolite of the pesticide Tolyfluanid). The integration of ozonation (still the most used AOP method, due to the lower energy requirements) with other AOPs or the combination of ozone with $H_2O_2$, Fe(II) and/or UV may minimize these by-products, but, in general, it increases the process costs. The presence of radical scavengers in the water may influence the performances.

An interesting example on the question of by-products is provided by the surface and groundwater contamination by pesticides. Triazines are among the most abundant pesticides and since 1997, atrazine has been found in more than 50% of samples of both surface water and groundwater, while its degradation product, desethylatrazine, is the most frequently found (about half of all analyses).[39] Other frequently found herbicides include diuron and isoproturon (urea substitute family), with diuron found in approximately 35% of surface water samples and 5% of groundwater samples since 1997, with corresponding values of 20% and 5% for isoproturon. The reaction of $O_3$ and atrazine is slow (7 $M^{-1}s^{-1}$), but if $H_2O_2$ is added then atrazine is rapidly oxidized ($3 \times 10^9$ $M^{-1}s^{-1}$), with atrazine removal over 95%.[39] Ozonation produces deisopropylatrazine (DIA) and desethylatrazine (DEA) as its main by-products. The concentration of by-products is dependent on the $O_3$ concentration, the contact time and the use of $H_2O_2$.[39] For atrazine removal ranging from 40 to 80%, DEA production varies from between 20 to 50% of the atrazine removed and DIA from 8 to 22%. Based on these results, the French authorities have not approved the use of AOPs to remove pesticides.[39] However, by using $H_2O_2$/UV, it is possible to efficiently remove these by-products.[40]

The alternative is to use Fe(II) or Cu(II) ions to promote the hydroxyl radical formation (Fenton mechanism):

$$Fe^{2+} + H_2O_2 \rightarrow Fe^{3+} + OH^- + {}^\bullet OH \tag{10.4}$$

As mentioned before, the method may integrate with UV radiation (photo-Fenton). The Fenton is probably the oldest AOP, but has the disadvantages of generating iron hydroxide sludge and requiring a very low pH value. Table 10.5

**Table 10.5.** Examples of the use of the Fenton reagent ($H_2O_2/Fe^{2+}$) to reduce COD in industrial wastewater (TS = total solids, MMA = methyl methacrylate, PMMA = poly methyl methacrylate). Adapted from Ref. 39.

| Industry | Influent flow rate $m^3/d$ | Influent COD mg/L | Treatment condition | Treated water COD mg/L | Quantity sludges discharges |
|---|---|---|---|---|---|
| Washing water abrasive paper | 10–20 | 10,000 (500 mg/L phenolic) | Batch T = 50°C $FeSO_4$ | 900 (<0.5 mg/L phenolics) | 80 t/y 45% TS |
| De-inked pulp and newspaper | 50,000 | 200 | Continuous no heating $FeCl_2$ | <100 | On site incineration |
| Wastewater fine chemicals production | 20 | 9,000 | Batch T = 80°C 2 bar, $FeSO_4$ | 1,600 | 30 t/y 50% TS |
| MMA and PMMA process water | 3,000 | 800–1,000 | Continuous no heating $FeSO_4$ | <500 | — |
| Textile industry wastewater | 100 | >2,500 | Continuous 100°C, 3 bar $FeSO_4$ | ~500 | — |
| Explosive manufacture process | 200 | up to 100,000 10,000 mg/L in $Cl^-$ | Continuous 100°C, 3 bar mixture of $M^{n+}$ | <100 | 70 t/y 45% TS |

reports some examples of the use of the Fenton ($H_2O_2/Fe(II)$) process to treat industrial wastewater. The Fenton reagent requires a large amount of acidic and alkaline chemicals, the ideal pH being about 2.5. It is usually applied to wastewater with a COD in the range 1–100 g/L having poor biodegradability. In the textile industry case, de-sizing baths with a non-biodegradable sizing agent and exhausted dye baths are treated by oxidation in a special reactor at 100–130°C and a pressure of approximately 3 bar. Compared with the dark Fenton reaction, the photo-Fenton (i.e. combining the Fenton process with light irradiation) process has various advantages, such as the increase of the degradation rate, minimization of sludge generation and the use of solar energy, among others. However, it requires additional costs for the photoreactor and energy, because even using solar energy requires recirculating large volumes of wastewater. The industrial application of photo-Fenton is generally limited.

Another problem is related to the salt counterion. Inorganic anions ($Cl^-$, $HPO_4^{2-}$,$SO_4^{2-}$, etc. already present in wastewater or added as reagents) have a significant effect on the reaction rate in the case of the Fenton process. They may: (i) complex Fe(II) or Fe(III), affecting iron species reactivity and distribution; (ii) induce precipitation leading to a decrease of the active dissolved Fe(III); (iii) act as a scavenger of hydroxyl radicals; and (iv) oxidize iron ions influencing the redox mechanism present in the Fenton process. Chloride ions, for example, show a strong inhibitory effect for oxidation reactions.[41]

Wet air oxidation (WAO), e.g. the oxidation of organic compounds by oxygen at high temperatures and under pressure in autoclave reactors, is another widely used technology, especially in the treatment of industrial wastewater or sludges. Use of catalysts may promote the reaction performance,[12,42–47] as will be discussed later. The use of WAO as a pre-treatment before the biological process (active sludge) has been previously discussed.[48] This appears as one of the preferable solutions in several examples of industrial wastewater and sludge treatments. Table 10.6 gives an overview of relevant WAO processes for the treatment of industrial wastewater, with some examples of the combination of WAO with Fenton.

WAO processes typically operate under an oxygen pressure of 5–200 bar (when using air, about four times higher pressure is necessary) and at elevated temperatures (125–320°C). Residence times range from 15 to 120 min, but longer times are often necessary for high organic loadings or high levels of COD removal (typically ranging between 75 and 90%). In contrast to supercritical water oxidation (SCWO), a complete mineralization of the waste stream is often impossible by WAO, since some low molecular weight oxygenated compounds (especially acetic and propionic acids, methanol, ethanol, and acetaldehyde) are resistant to oxidation. For instance, removal of acetic acid is usually minimal at temperatures lower than 300°C. Organic nitrogen compounds are easily transformed into ammonia, which is also very stable in WAO conditions. In addition, there are gaseous emissions that should also be

**Table 10.6.** WAO processes for the treatment of industrial wastewater. Adapted from Ref. 47.

| Process | Application | Reactor | Temperature range, °C | Pressure range, MPa | Retention time, min | Catalyst/ oxidant |
|---|---|---|---|---|---|---|
| Zimpro | Preliminary detoxification, COD reduction | Co-current bubble column | 150–325 | 2–12 | 20–240 | — |
| Wetox | Pre-treatment biological | Horizontal, compartmentalized | 200–250 | 4 | 30–60 | — |
| Vertech | General industrial wastewater | Subsurface vertical reactor | 180–280 | 8–11 | 60 | — |
| Kenox | Pre-treatment biological | Two concentric shells, ultrasonic probe | 200–240 | 4–5 | 40 | — |
| Oxyjet | Pharmaceutical, chemical, and wood waste | Jet mixer and tubular reactor | 140–300 | | <5 | — |
| LoProx Bayer | Chemical/ Pharma waste | Multistage bubble column | 120–200 | 0.3–2 | | $Fe^{2+}$+ co-cat-/$O_2$ |
| Ciba-Geigy | Chemical/ Pharma waste | Multi-tubular | ~300 | ~15 | | $Cu^{2+}$/air |
| Athos | Residual sludges | | 235–250 | 4.4–5.5 | | $Cu^{2+}/O_2$ |
| WPO | Aquifer decontamination | | 90–130 | 0.1–0.5 | | Fe-Cu-Mn/$H_2O_2$ |
| Orcan | Refractory waste pre-treatment | | 120 | 0.3 | | $Fe^{2+}$/air + $H_2O_2$ |

treated. Therefore, WAO is a pre-treatment of liquid wastes that requires additional treatments of the process liquid and gas streams. Over 100 plants are in operation today, mostly to treat waste streams from petrochemical, chemical, and pharmaceutical industries as well as residual sludge from wastewater treatment. The use of catalysts allows the use of milder reaction conditions, and promotes conversion of the reaction intermediates (for example, acetic acid and ammonia) which are very difficult to convert in the absence of catalysts.

### 10.1.2. *AOPs using other energy sources*

Instead of UV radiation, other sources of energy may also be used to promote the generation of hydroxyl radicals. Ultrasound (US) generates local bubbles where very high temperatures and pressure conditions can be reached that give rise of the formation of $^{\bullet}OH$ radicals, even directly from water. Although in principle cheaper, scale-up of this technology is difficult; however, it may be combined with other oxidation processes. He *et al.*[49] reported an interesting example of the combination of ozonation and ultrasound for the conversion of p-aminophenol (PAP), an intermediate in the production of pharmaceutical products and dyes. For an initial PAP concentration of 10 mmol/L, a degradation efficiency of 99% at 30 min was obtained in addition to a TOC removal of 77%. The ozone dose used in the process was 5.3 g/h at a temperature of 25°C with a pH value of 11 and with an ultrasonic energy density of 0.3 W/mL. The presence of scavengers of the hydroxyl radicals decreased the removal efficiency.

The $O_3$/US method is effective for biological sludge anaerobic degradation.[50] The ultrasound induces the disruption of sludge flocs and microbial cell walls, which can then be effectively attached by ozone. It could also be used to improve the removal efficiency of ozone in treating wastewater containing high concentrations of pollutants such as phenolic compounds.[51] It may be noted, however, that the synergistic effect of combining ozonation with ultrasonic irradiation is observed only when the free radical attack is the controlling mechanism and the rate of generation of free radicals, due to ultrasonic action alone, is somewhat lower (at lower frequencies of operation and power dissipation levels).

Combining ozone and/or hydrogen peroxide with ultrasound leads to better utilization of both the oxidants and hence higher degradation rates due to the dissociation of ozone and hydrogen peroxide under the action of ultrasound. The mass transfer resistance, which is a major limiting factor for the application of ozone or hydrogen peroxide alone, is also eliminated due to the enhanced turbulence generated by ultrasound. The operating frequency is a crucial factor for the synergism. In general, it should not be increased beyond 500 kHz.[51] There are also optimum concentrations of ozone or hydrogen peroxide beyond which the beneficial effects of combining the two techniques are not observed. The rates of combination techniques are severely hampered due to the presence of radical scavengers and hence pre-treatment in terms of pH adjustment and/or adsorption with activated carbon should be undertaken to adjust the concentration of these scavengers below a certain value.

The use of additives may result in enhancement of the free radicals by two mechanisms.[51] Solid particles may increase the number of cavitation events due to creating instabilities in the system and facilitating the easy generation of cavities.

Some additives such as iron or $CCl_4$ intensify the degradation process by way of generating additional oxidizing mechanisms in the system. Use of dissolved salts such as NaCl also results in the enhancement of the degradation process by altering the distribution of the pollutants in the system (more concentration of pollutants at the cavity implosion sites). Many other factors determine the effectiveness of the operations, such as the presence of catalysts, the geometry of the reactor, power dissipation into the system, and operating frequency for the target pollutant. Therefore, a specific optimization and scale-up of the operations is rather difficult but is necessary.

### 10.1.3. *AOPs using heterogeneous catalysts*

The use of solid catalysts in AOPs offers several potential advantages:[12] (i) operation in milder reaction conditions, (ii) better engineering of the process, (iii) the possibility of selective conversion of specific target chemicals, and (iv) the combined removal by adsorption with catalytic regeneration in a separate step. A different case is instead that of the photocatalytic technologies, where semiconductors such as titania are required to generate, by photo-induced processes, the active species (hydroxyl radicals) responsible for the degradation of the pollutants. The use of catalysts is also often indispensable in converting recalcitrant contaminants, such as EDCs. AOP methods, in combination with solid catalysts, are the most effective technologies for the treatment of wastewaters from the production of pharmaceuticals and personal care products (PPCPs), and especially EDCs.[52]

The use of solid catalysts in AOP methods requires different characteristics of the solid with respect to those typically used in environmental catalysis applications. Limited attempts have been made to tailor solid catalysts for water treatment applications. There are problems of: (i) pore structure design (intraparticle diffusivity is usually an issue determining the catalytic performances); (ii) leaching (often driven from the reaction products and thus depending on the progress of the reaction); and (iii) stability (related, but not only depending on the possible leaching) and selectivity (typically, quite complex mixtures should be treated in practical applications). A more effective catalyst design in water treatment technologies requires taking advantage of the new possibilities offered from using microstructured and nanostructured materials (fibers, nanotubes, multilayer composite materials) in multiphase reactors.[53]

The efficient integration between catalyst and reactor design in wastewater treatments is a critical element in increasing the use of solid catalysts in these technologies. A further general issue in the design of solid catalysts for AOP methods is the fact that the effective oxidizing species are very reactive and short-lived, such as radical species ($H^\bullet$, $HO^\bullet$, $HO_2^\bullet$, etc.). Knowledge regarding the catalyst design necessary to optimize performances in the presence of these very reactive species is still quite limited. For example, many studies have reported the effects of catalyst preparation, catalyst stability, deactivation, and catalyst reuse and regeneration in WAO.[47] However, specific studies on the optimal pore geometry and wetting character of the solid catalysts are limited, while these two parameters are critical in homogeneous/heterogeneous radical pathways.

Optimal reactor design is critical for the effectiveness and economic viability of AOPs. The WAO process poses significant challenges to chemical reactor engineering and design, due to the: (i) multiphase nature of WAO reactions; (ii) temperatures and pressures of the reaction; and (iii) radical reaction mechanism. In multiphase reactors, complex relationships are present between parameters such as chemical kinetics, thermodynamics, interphase/intraphase intraparticle mass transport, flow patterns, and hydrodynamics influencing reactant mass transfer. Complex models of WAO are necessary to take into account the influence of catalyst wetting, the interface mass-transfer coefficients, the intraparticle effective diffusion coefficient, and the axial dispersion coefficient.[54,55]

In WAO with solid catalysts, three-phase reactors are used: trickle bed, bubble slurry column, and bubble fixed-bed (monolith) or three-phase fluidized-bed reactors. When the catalyst is present in the liquid phase (homogeneous) or absent, two-phase reactors such as bubble columns, jet-agitated reactors, and mechanically stirred reactor vessels are used. The limitations and advantages of these reactors for the application to WAO are listed in Table 10.7.

#### 10.1.3.1. *Catalytic ozonation*

The effect of catalysts on the ozonation rate is due to the acceleration of ozone decomposition with the production of active free radicals or the acceleration of molecular ozone reactions. The first effect promotes the reaction with respect to ozonation alone, but there is generally a strong dependence on the pH value of the solution. The presence of radical scavengers in the treated water can result in a significant reduction of the efficiency of contaminant removal due to the rapid reaction of these compounds with hydroxyl radicals.[56] This situation is common in the case of wastewaters containing suspended material. A common catalytic element is iron that operates according to Eq. 10.5.[57]

$$Fe^{2+} + O_3 \rightarrow FeO^{2+} + O_2 \qquad (10.5a)$$

$$FeO^{2+} + H_2O \rightarrow Fe^{3+} + {}^{\bullet}OH + OH^- \qquad (10.5b)$$

Co-based catalysts, such as $CoO_x$-$CeO_2$ mixed oxides, $CoO_x$ supported over alumina, and CoMgAl and CoNiAl hydrotalcites are also active in the reaction. Between these, the CoNiAl material shows the highest activity for the conversion

**Table 10.7.** Advantages and limitations of multiphase reactors used in catalytic wet air oxidation. Adapted from Ref. 47.

| Reactor type | Advantages | Limitations |
|---|---|---|
| Trickle bed | — High conversion as both gas and liquid flow regimes approach plug flow<br>— Low liquid hold-up<br>— High catalyst loading<br>— Low pressure drop | — Poor liquid phase distribution<br>— Often only partial wetting of the catalyst<br>— High intraparticle resistance<br>— Poor radial mixing<br>— Low mass-transfer coefficient<br>— Temperature control can be difficult |
| Slurry phase and three-phase fluidized-bed reactor | — high external mass transfer (gas–liquid, liquid–solid)<br>— Low intraparticle resistance<br>— Ease of catalyst addition/regeneration<br>— Ease of thermal management | — Catalyst separation<br>— high axial mixing<br>— Low catalyst load<br>— High liquid-to-solid ratio |
| Bubble fixed-bed reactor | — High gas–liquid mass transfer (better gas–liquid interaction)<br>— High liquid hold-up<br>— Well-wetted catalyst<br>— Channeling eliminated<br>— Good temperature control | — High axial backmixing<br>— Lower conversion compared to three-phase reactor<br>— High pressure drop<br>— Flooding problems |

of phenol in the presence of ozone.[58] After 4 h of reaction the TOC removal was 90%, but the leaching of $Ni^{2+}$ in solution was considerable (17 mg/L). Other catalyst types, however, show higher stability. For example, Fe-Mn and Mn-Ce oxides show a stable behavior without leaching in the depuration of phenolic wastewater with ozone.[59] The use of the catalyst allows a significant increase in the $BOD_5$/COD ratio and biodegradability after treatment with respect to ozone alone. However, the impact of the use of the catalyst in terms of improved ecotoxicity of the treated wastewater was less clear, with a complex behavior with time on stream.

Nanosized cerium oxide also shows excellent properties.[60,61] Oxalic and oxamic acid, aniline, and reactive dyes (C.I. Reactive Blue 5) could be effectively converted. The use of the catalyst significantly enhances the reaction rate and a synergetic effect between cerium oxide and carbon (used as the support) was noted. In addition, carbon is also active in the catalytic ozonation, and multiwall carbon nanotubes were shown to have high efficiency in the catalytic ozonation of oxalic and oxamic acids.[62]

Various other catalysts have been investigated in catalytic ozonation processes. Among the most widely used catalysts in heterogeneous catalytic ozonation are[63]: (i) metal oxides ($MnO_2$, $TiO_2$, $Al_2O_3$, FeOOH, and $CeO_2$); (ii) metals (Cu, Ru,

Pt, Co) on supports ($SiO_2$, $Al_2O_3$, $TiO_2$, $CeO_2$, and activated carbon); (iii) zeolites modified with metals; and (iv) activated carbon.

Adsorption of ozone and/or an organic molecule on its surface is the first requirement, but the exact nature of the sites which catalyze the generation of hydroxyl radicals after ozone adsorption is still debated.[63] The suggestions for ozone decomposition sites are: (i) surface redox sites; (ii) Lewis centers of metal oxides ($Al_2O_3$, $TiO_2$, $ZrO_2$, etc.); (iii) non-dissociated hydroxyl groups of metal oxides; and (iv) basic centers of the catalyst.

Many parameters also influence the reaction rate, very important among them is:

(i) the pH of the solution which influences the charge of surface (active) centers on the catalyst, the charge of ionic or ionizable organic molecules and also the charge of possible catalyst poisons;
(ii) the rate of both ozone and organic molecule adsorption, and the presence of competitive effects;
(iii) the influence of inorganic ions such as $F^-$, $PO_4^{3-}$, $SO_4^{3-}$, $HCO_3^-$, $CO_3^{2-}$; and
(iv) natural organic matter which acts as radical scavengers and/or site blockers.

They can thus largely influence the efficiency of the catalytic ozonation processes. In regard to the last aspect, it should be noted that most of the studies are still performed using model compounds and solutions, and thus the application to real wastewater, where the cited inorganic ions and natural organic matter are often present, may result in completely different behavior.

Considering the few studies using industrial wastewater, the use of supported Mn/Co for the catalytic ozonation of industrial wastewater containing chloro and nitro aromatics may be recommended.[64] At neutral pH almost complete decolorization and removal of chloro and nitro pollutants was achieved in 80 min. Catalytic ozonation significantly increased the biodegradability of industrial wastewater together with a decrease in the ecotoxicity.

Thiruvenkatachari *et al.*[65] instead used Fe and $TiO_2$ to treat the effluents contaminated with TPA, a known endocrine disruptor. In the study, catalytic ozonation with Fe (55 mg/L of $Fe_2(SO_4)_3$) as well as $TiO_2$ (90 mg/L) achieved the complete degradation of 50 mg/L of TPA in 10 min. They also investigated various alternative AOP methods and evaluated the operating costs based on the electrical energy requirement in kW and reagent costs. The results are summarized in Table 10.8.

Although the UV/$H_2O_2$/Fe/$O_3$ system was found to have the best organic degradation performance (with minimum degradation time), catalytic ozonation ($O_3$/Fe) seemed to show a satisfactory organic degradation performance and to be an economically more viable choice for the degradation of terephthalic acid.

**Table 10.8.** Overall pseudo first order kinetic rate constants ($k_{overall}$), and operating cost evaluation for TPA destruction (TPA 50 mg/L, pH 6) using different AOPs. Adapted from Ref. 65.

| AOP | $k_{overall}$, $min^{-1}$ | Operating cost, \$ $m^{-3}$ |
|---|---|---|
| $UV/H_2O_2/Fe/O_3$ | 0.607 | 2.80 |
| $H_2O_2/Fe/O_3/TiO_2/UV$ | 0.442 | 3.01 |
| $O_3/Fe$ | 0.406 | 1.02 |
| $O_3/TiO_2$ | 0.332 | 1.04 |
| $H_2O_2/Fe/UV$ | 0.055 | 2.12 |
| $H_2O_2/UV$ | 0.033 | 2.90 |
| $TiO_2/UV$ | 0.004 | 1.70 |

### 10.1.3.2. *Heterogeneous photocatalysis*

Many excellent reviews have discussed in detail the use of heterogeneous photocatalysis (mainly $TiO_2$) for the purification of wastewater[66–76] and for this reason we will recall here only a few aspects. Many papers have demonstrated that heterogeneous photocatalysis using semiconductor catalysts ($TiO_2$, ZnO, $Fe_2O_3$, CdS, GaP, and ZnS) can efficiently degrade a wide range of persistent organics into readily biodegradable compounds, and eventually mineralize them to innocuous carbon dioxide and water. However, from a practical perspective, the reaction rate is one to two orders of magnitude lower than that in other AOPs, as shown also in Table 10.8. The rate falls a further order of magnitude using solar light instead of UV radiation (in the case of $TiO_2$), compensating for the advantage of a lower energy cost for the UV lamp. Therefore, heterogeneous photocatalysis has been applied to niche applications and generally only for low volume wastewater.[77]

Among the semiconductor catalysts, titanium dioxide ($TiO_2$) has received the greatest interest from research and development departments investigating photocatalysis technology. $TiO_2$ is the most active photocatalyst under the photon energy of 300 nm $< \lambda <$ 390 nm and remains stable after repeated catalytic cycles, whereas Cds or GaP are degraded to produce toxic products.[74] Furthermore, $TiO_2$ has several functional properties such as chemical and thermal stability, resistance to chemical attack, strong mechanical properties, etc., making it unique for application in photocatalytic water treatment.

The application of $TiO_2$ photocatalysts to water treatment still faces many technical challenges. The post-separation of the $TiO_2$ semiconductor catalyst after water treatment remains the major obstacle towards the practicality of their use as an industrial process. The fine particle size of the $TiO_2$, together with their large surface area-to-volume ratio and surface energy, creates a strong tendency for catalyst agglomeration during operation. Such particle agglomeration is highly detrimental in view of particle size preservation, surface area reduction, and its reusable lifespan.

Other technical challenges include the development of catalysts with a broader photoactivity range (i.e. activity in the visible region) and its integration with a feasible photocatalytic reactor system. In addition, the understanding of the theory behind the common reactor operational parameters and their interactions is also inadequate and presents a difficult task for process optimization.

In general, the mechanism of action is known. Photons of energy ($h\nu$) greater than or equal to the band gap energy of $TiO_2$ (i.e. 3.2 eV for $TiO_2$anatase) and 3.0 eV for $TiO_2$rutile) are adsorbed from the solid; one electron ($e^-$) is photoexcited to the empty conduction band in femtoseconds. The light wavelength for such photon energy usually corresponds to $\lambda < 400$ nm. The photonic excitation leaves behind a positive hole ($h^+$) in the valence band, and thus creates the electron-hole pair ($e^-$–$h^+$). This primary event is followed by a series of chain oxidative/reductive reactions that occur at the photon-activated surface:[74]

(i) charge-carrier trapping of $e^-$ and $h^+$ by conduction and valence bands, respectively;
(ii) trapping of $e^-$ and $h^+$ at surface sites, or alternatively recombination with energy dissipation;
(iii) reaction of photoexcited electrons with surface electron scavengers. The presence of electron scavengers is vital for prolonging the recombination and successful functioning of photocatalysis. The presence of oxygen prevents the recombination of electron-hole pairs with the formation of superoxide radicals ($O_2^{\bullet -}$). This radical can be further protonated to form the hydroperoxyl radical ($HO_2^{\bullet}$) and subsequently $H_2O_2$. The $HO_2^{\bullet}$ radical also has scavenging properties and thus a positive effect on the recombination time of trapped $h^+$. Photocatalysis thus requires the presence of dissolved oxygen, but also of water molecules, because without solvation the highly reactive hydroxyl radicals would react with each other and not with the organics;
(iv) oxidation of hydroxyls ($OH^-$) by reaction with holes ($h^+$) to form the hydroxyl radical ($^{\bullet}OH$) which is the most active species for the oxidation of organic species;
(v) photodegradation of the organic species by $^{\bullet}OH$ and direct reaction of photoholes ($h^+$) with organics. The trapped holes are powerful oxidants (+1.0 to +3.5 V against NHE), while trapped electrons are good reducing species (+0.5 to –1.5 V against NHE), depending on the type of catalysts and oxidation conditions. Some simple organic compounds (e.g. oxalate and formic acid) can be mineralized by direct electrochemical oxidation where the trapped $e^-$ is scavenged by metal ions in the system;
(vi) protonation of superoxides to form $HOO^{\bullet}$ which may react with $e^-$ to form $HO_2^-$ which, in the presence of protons, forms $H_2O_2$.

Due to the presence of a radical chemically induced from surface trapped electrons and holes, all substances which may react with these species have a large effect on the reaction rates. $Cu^{2+}$, $Fe^{2+}$, $Al^{3+}$, $Cl^-$, and $PO_4^{3-}$ at certain levels may decrease photo-mineralization reaction rates, but Cu(II) and Fe(II) may also catalyze the Fenton and photo-Fenton reactions. For low amounts, there is thus a promotion, but larger amounts may be negative. The addition of oxyanion oxidants such as $ClO_2^-$, $ClO_3^-$, $IO_4^-$, $S_2O_8^-$, and $BrO_3^-$ increase photoreactivity by scavenging conduction-band electrons and reducing the charge-carrier recombination. The presence of salts also modifies the double layer and thus influences the adsorption of organics and oxygen, with regard to the mobility of radical species. Predicting the behavior of real wastewater from model solutions (on which most of the studies have been focused) may result in incorrect expectations.

Many mechanistic studies on organic compounds (e.g. phenol, chlorophenol, oxalic acid) have extensively investigated the photodegradation over a $TiO_2$ surface. Aromatic compounds can be hydroxylated by the reactive $^{\bullet}OH$ radical that leads to successive oxidation/addition and eventually ring opening. The resulting intermediates, mostly aldehydes and carboxylic acids, will be further carboxylated to produce carbon dioxide and water.

To date, the most widely applied photocatalyst in the research of water treatment is the Degussa P-25 $TiO_2$ catalyst. This catalyst is used as a standard reference for comparisons of photoactivity under different treatment conditions. The fine particles of the Degussa P-25 $TiO_2$ have always been applied in a slurry form. Immobilization of catalysts on a flat support (as necessary for practical operations) reduces the number of catalyst active sites and enlarges the mass transfer limitations. With the slurry $TiO_2$ system, an additional process step would need to be incorporated for post-separation of the catalysts. This separation process is crucial to avoid the loss of catalyst particles and related contamination of treated water.

Many studies have been dedicated to catalyst immobilization strategies that are suitable for use in slurry reactors or membrane reactors or both.[74] Photocatalyst modification and doping ion order, with a view to enlarging the photonic activation of catalysts to the visible region, is another area of very active recent development.

In conclusion, semiconductor photocatalytic technology using either UV light or solar radiation has been extensively investigated. It could be applicable to a large range of water contaminants, ranging from hazardous contaminants of pesticides, herbicides, and detergents to pathogens, viruses, coliforms, and spores. Photocatalytic treatment of wastewater is a viable alternative for treating biodegradable chemicals that are present only in very low amounts and very low volume streams. For higher pollutant concentrations and/or larger wastewater volumes, other AOPs are preferable. They may also integrate with photocatalytic technology. An alternative photocatalytic technology may be an effective pre-treatment step.

Promoting the feasibility of photocatalytic water treatment technology requires addressing several key technical constraints ranging from catalyst development to reactor design and process optimization.[74] These include:

(i) improvement of the photo-efficiency of the catalyst in order to fully utilize solar light;
(ii) immobilization of the catalyst for cost-effective solid–liquid separation;
(iii) extension of the photocatalytic operation to a wider pH range to minimize additives;
(iv) effective design of photocatalytic reactor systems or use of parabolic solar collectors for higher utilization of solar energy to reduce the electricity costs; and
(v) integration with other AOPs.

An example of the latter point is the integration of $TiO_2$ with $UV/H_2O_2$ and $Fe(II)/H_2O_2$ systems.[78] This approach allows the efficient treatment of real effluents from the textile industry, as the wastewater is composed of various dyes (yellow, red, and blue). Over 90% COD removal could be achieved. The added peroxide is totally consumed, and the solution shows low toxicity.

Similar results were obtained by Riga *et al.*[79] and Soutsas[80] who compared the $TiO_2$/UV method with various alternative AOPs in the conversion of various dyes used in the textile industry.

#### 10.1.3.3. *Catalytic wet air oxidation*

The use of catalysts in wet air oxidation (indicated with the acronyms CWAO or WACO) has already been implemented in various industrial processes (see Table 10.6), but in the form of homogeneous transition metals salts which favors the redox activation of oxygen. Two examples of industrial processes using solid catalysts are:

(i) *Osaka Gas Process* operates at about 250–320°C (70 bar) using a Fe–Co–Ni–Ru–Pd–Pt–Cu–Au–W catalyst supported on $TiO_2$ and $ZrO_2$ (either as spherical particles or as honeycomb structures). It uses a bubble column reactor. This process is similar to the non catalytic Zimpro process (Table 10.6). The process is used for the treatment of coal gasifier effluents, wastewater from coke ovens, concentrated cyanide wastewater from the nitridation of steel, and sewage sludge and residential wastes.[81]
(ii) *Nippon Shokubai Kagaku* operates in the temperature range of 160–270°C and at pressures in the 0.9–8.0 MPa range, with typical residence times in the region of about 1 h, using a trickle-bed reactor and a solid catalyst based on Pt- or Ru-supported on $TiO_2/CeO_2$.[82]

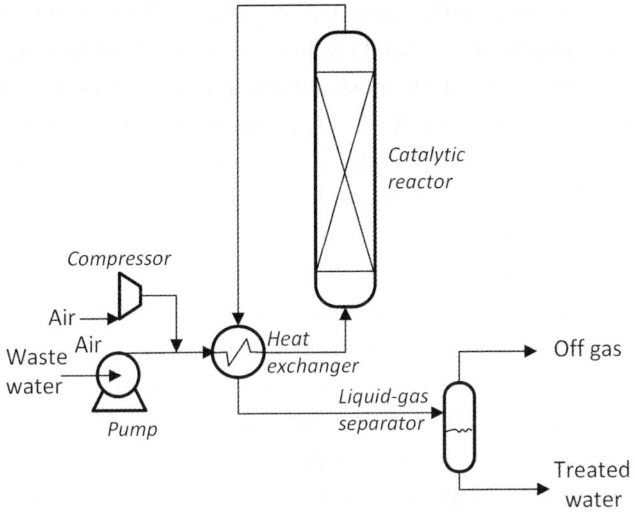


**Figure 10.2.** Simplified flow chart of the catalytic wet air oxidation process.

The simplified flow of the operations of these two CWAO processes is recorded in Fig. 10.2. CWAO is cost-effective for highly concentrated effluents (COD in the 10–100 g/L) and/or for wastewater containing components that are not readily biodegradable or are toxic to biological treatment systems. In fact, a main cost element is that the process requires high temperatures and pressures. The process becomes autogenic at COD levels of about 10,000 mg/L, at which the system will require external energy only at start-up. CWAO process plants also offer the advantage that they can be highly automated, have relatively small plant footprints, and are able to deal with variable effluent flow rates and compositions. Another issue is the corrosion-resistant materials required for the reactor (for example, Ti-lined autoclave), because acetic acid is a main by-product of the reaction.

The use of solid catalysts is potentially attractive for the possibility of lowering the reaction temperatures and pressures, and allowing easier separation with respect to homogeneous catalysts, thus reducing both fixed and running plant costs. Several reviews are dedicated to CWAO.[42–47,83–91] The main problem related to the use of solid catalysts is stability, which is lowered by the leaching of the active component (even though a number of stable catalysts have been developed), and the poisoning of active sites or fouling of the catalyst surface following deposition of reaction intermediates. Various heterogeneous catalysts including noble metals, metal oxides, and mixed oxides have been prepared and tested for the CWAO of model compounds and real wastewaters.

In the CWAO of phenol, noble metals such as Ru, Rh, Pd, and Pt generally show higher catalytic activity and higher resistance to metal leaching than base metal oxide catalysts. They are usually supported on $\gamma$-$Al_2O_3$, $TiO_2$, $CeO_2$, $ZrO_2$, and carbon materials with less than 5% metal loading. Among the noble metals used for the CWAO of phenol, Ru is one of the most active catalysts, especially when supported over ceria.[92] In the presence of a $Ru/CeO_2$ catalyst, almost complete conversion of phenol was achieved after 3 h run under 160°C and 20 bar of oxygen. The introduction of $ZrO_2$ into $CeO_2$ increased the mechanical strength, specific surface area, and adsorption capacity of pelletized Ru catalysts. With the 2 wt% $Ru/ZrO_2$-$CeO_2$ catalyst, phenol and TOC removal stabilized at approximately 100% and 96%, respectively, in a continuous packed-bubble column reactor for 100 h at 140°C and 4 MPa of air.[93] For the conversion of *o*-chlorophenol $Ru/Ce_xZr_{1-x}O_2$ catalysts showed better performances.[94] The conversion of *o*-chlorophenol increases with the initial pH value due to the acceleration of dechlorination, while pH values which are too high limit the total mineralization of *o*-chlorophenol by preventing the adsorption of reaction intermediates. Mixed oxides are instead typically not stable in long-term runs, especially due to leaching. For example, 4 wt% $Cu/CeO_2$ catalysts show good performances for the CWAO of phenol, *o*-chlorophenol, and *p*-nitrophenol.[95] At 160°C and 1 MPa, complete conversion of phenol and *o*-chlorophenol is achieved after 100 min and 130 min, respectively, while only about 60% of *p*-nitrophenol is degraded after 200 min. However, significant amounts of $Cu^{2+}$ ions were detected in the treated solution.

In the CWAO of carboxylic acids, noble metals give the best performances and stability. Ruthenium supported over ceria-zirconia shows the best performance.[96] These catalysts are also preferable for CWAO of N-containing compounds such as aniline.[92] Over $Ru/CeO_2$, ammonium ions formed in the reaction are selectively oxidized into molecular nitrogen in the temperature range of 180 to 200°C, but above 200°C nitrite and nitrate ions form.

Little work has been focused on the feasibility of solid catalysts for the treatment of real wastewater. Some of the studies on the CWAO of industrial wastewaters are summarized in Table 10.9.

**Table 10.9.** Summary of studies on CWAO of industrial wastewaters. Adapted from Ref. 46.

| Catalyst | Industrial wastewater | Reaction conditions | Ref. |
|---|---|---|---|
| $Ru/TiO_2$ and $Ru/ZrO_2$ | Kraft bleaching plant effluents Olive oil mill wastewater | 190°C, 5.5 MPa of air<br>140–190°C, 5.0–7.0 MPa of air | 97, 98 |
| Eggshell $Ru/TiO_2$ | Coke plant wastewater | 250°C, 4.8 MPa of air | 99 |
| $Ce$-$Cu/Al_2O_3/TiO_2$ | Coke plant wastewater | 140–220°C, 1.2 MPa of $O_2$ | 100 |
| Cu/CNF | Textile wastewater | 120–160°C, 0.63–0.87 MPa of $O_2$ | 101 |
| Ru on AC-ceramic sphere | Resin effluent | 160–240°C, 1.5 MPa of $O_2$ | 102 |

Very large amounts of water are used during processing in the pulp and paper industry. In particular, the bleaching process produces refractory organic compounds including lignin and polysaccharide fragments, organic acids, aliphatic alcohols, etc., which are hardly degraded by biological treatment. $Ru/TiO_2$ and $Ru/ZrO_2$ catalysts show high activity in treating these effluents.[97] More than 99% of TOC removal was achieved within 8 h of reaction at 190°C and 5.5 MPa of air. The reaction is characterized by a fast initial step with rapid fragmentation of large molecules to short organic acids, followed by a slow reaction step as these acids, especially acetic acid, tend to be resistant to further oxidation. Acetic acid can be eliminated by consecutive biological treatment.

Olive oil mill wastewater (OMW), deriving from olive oil production, has a high load of organic and antibacterial phenolic compounds and thus poses a severe environmental threat. The TOC and phenolic compounds content of OMW is effectively eliminated by CWAO (190°C and 70 bar of air) using Pt and Ru supported on titania and zirconia.[98] Toxicity towards *Vibrio fischeri* and phytotoxicity decreased simultaneously. In particular, the CWAO over the Ru catalysts considerably reduced the total phenolic contents of OMW and thus produced an effluent suitable for further treatment by anaerobic biological treatment. The Ru catalysts are stable over a long period of operation in a trickle-bed reactor.

Coke-plant wastewater is generated from coal coking, coal gas purification, and the by-product recovery processes of coke factories. Eggshell $Ru/TiO_2$ catalyst with 0.25 wt% Ru loading are active in treating this wastewater.[99] At 250°C and 4.8 MPa, COD removal and ammonia/ammonium compounds ($NH_3$-N) removal were 96% and 93%, respectively. The eggshell configuration allows an enhanced activity. Han *et al.*[100] investigated the CWAO of high-strength organic coking wastewater over the Ce-Cu and Ce-Mn catalysts. The Cu-Ce (2:1) catalyst supported on $\gamma$-$Al_2O_3/TiO_2$ showed high levels of activity and stability. With this catalyst, 95.2% of COD was removed after 60 min of reaction at 180°C and 1.2 MPa of $O_2$, while the concentration of leached copper ions was only 5.81 mg/L.

The effluents from the dye and textile industry are another serious problem for aquatic ecosystems. Carbon nanofibers (CNFs) can be effectively used as a catalyst support for the CWAO of textile washing wastewater.[101] In the presence of a 3 wt% Cu/CNF catalyst, color reduction and TOC removal were close to 97% and 74.1%, respectively, after 180 min of reaction under 140°C and 8.7 bar of oxygen pressure. The 43% toxicity was simultaneously reduced, and was evaluated by the acute toxicity results based on bioluminescence in *Vibrio fisheri*.

Effluents from resin production have a high COD (19,500 mg/L) and contain phenol, formaldehyde, and methanol. They could be treated by CWAO using a 3 wt% Ru on ceramic spheres coated with carbon.[102] With this catalyst, the conversion of COD was about 92% at 200°C and 1.5 MPa of $O_2$. The catalyst was stable for about 30 days.

Although limited, there are various examples showing that real effluents can be effectively treated by the CWAO approach.

### 10.1.3.4. *Wet hydrogen peroxide catalytic oxidation*

Wet hydrogen peroxide oxidation (WHPCO; alternative names are catalytic wet peroxide oxidation (CWPO), and wet peroxidation or wet oxidation by hydrogen peroxide) uses $H_2O_2$ as an oxidant instead of oxygen. It operates at significantly milder reaction conditions (temperature lower than 100°C, atmospheric pressure) and does not require costly autoclave reactors, e.g. significantly lower operative and fixed capital costs. However, $H_2O_2$ (HP) is a more costly reactant than oxygen (air). It is based on Fenton chemistry, where transition metals such as iron and copper are able to react with HP to generate hydroxylradicals ($^{\bullet}OH$), which are the effective oxidizing agents. The original Fenton studies, and many investigations and applications, are still based on the use of homogeneous catalysts ($Fe^{2+}$ and $Cu^{+}$ salts), as discussed in the previous sections, but the use of solid catalysts containing iron or copper is increasing (heterogeneous Fenton).[102–110] UV radiation is also able to catalyze the formation of these hydroxyl radicals from hydrogen peroxide, as discussed earlier, and thus the combination of solid catalysts with UV radiation is used (photo-Fenton).[110,111]

The Fenton oxidation, particularly using solid catalysts, is especially suitable for: (a) domestic wastewater treatment, (b) wastewater reclamation and reuse (destruction and removal of harmful substances from the water streams to substantially promote the reuse of safe, treated water in industry or in agriculture), (c) sludge production reduction, and (d) water disinfection. The use of solid catalysts allows some of the drawbacks of homogeneous Fenton catalysts to be overcome, including:[104]

(i) operations occuring in a limited pH range around 3.5;
(ii) difficulty in the recovery or elimination of iron or copper after the reaction; and
(iii) the complexation of iron ions by reaction products such as oxalic acid which leads to a progressive lowering of the reaction rate and decreased efficiency in $H_2O_2$ use as a function of the conversion.

In general, WHPCO is preferable to WACO in the following situations, where capital costs predominate over running costs:

(i) pre-treatment of streams (before sending to biological units) which requires the selective conversion of hardly-biodegradable or toxic chemicals, but with low levels of TOC removal (for example, in the pre-treatment of streams from the agro-food industry); and
(ii) treatment of wastewater having a medium-low organic content (1–30 gC/L), and low volumes of effluents (1–100 $m^3$/day).

Few reviews have discussed the use of solid Fenton-type catalysts for the WHPCO reaction.[104,112–119] The Fenton reaction consists essentially of the oxidation of $Fe^{2+}$ to $Fe^{3+}$ by $H_2O_2$ (Eq. 10.4). The Fe(III) may then be reduced again to Fe(II) through the following reaction:

$$Fe^{3+} + H_2O_2 \rightarrow Fe^{2+} + H^+ + HOO^\bullet \qquad (10.6)$$

Side reactions are the following:

$$^\bullet OH + H_2O_2 \rightarrow HOO^\bullet + H_2O \qquad (10.7a)$$

$$^\bullet OH + Fe^{3+} \rightarrow Fe^{2+} + OH^- \qquad (10.7b)$$

Besides iron salts, other transition metals that exhibit at least two oxidation states such as copper, ruthenium, cerium, and manganese can also promote the generation of hydroxyl radicals from $H_2O_2$. However, due to economic and environmental concerns, iron- or copper-based catalysts are usually preferred.

In simplified terms, the difference between WHPCO and WACO mechanisms can be explained as follows. In WHPCO, the rate of reaction depends on the redox reaction of $H_2O_2$ with iron (or other redox metals) to form the active radical species. In WACO, the rate depends on the electrophilic character of the catalyst, e.g. its rate of generation of surface radical species. Although this property also depends on the presence of redox sites, the Fenton mechanism is much more effective to close the cycle.

The main classes of catalysts used for heterogeneous WHPCO reaction are: clays and anionic clays (hydrotalcites), metal-ion exchanged zeolites and mesoporous silica containing transition metals, and doped metal oxides. Although some other transition metals have been also used (Mn, V), most catalysts contain iron and/or copper as the active elements. Leaching of the active metal is also a significant problem in this case. While different types of catalysts have been reported, only a few of them have been effectively proven to have a stable activity in long-term continuous experiments or at least in several repeated batch tests. Between the stable catalysts, Fe- and Cu-PILC (pillared clays) materials[112–114,120] have the best combination of activity and stability. However, the limited quantity of active elements (around 2% wt. of iron or copper) necessary to achieve stable performances, limits the overall activity.

Clays are naturally abundant minerals. The most common clays used for Fenton catalysis are montmorillonites, bentonites, and saponites. Synthetic clays such as beidellite have also been used, although they have a higher cost and not significantly different properties. These clays have a layered structure constituted by aluminosilicate sheets having an excess negative charge that interacts strongly, by electrostatic forces, with charge balancing cations. The interlamelar space is generally not accessible to organic substrates due to the strong electrostatic interaction between sheets and charge balancing cations. For their use, it is thus necessary to make the interlamelar space of layered clays accessible, for example, by putting large cations ("pillars") between the sheets. The pillars are generally polycationic species of a large size, so that being located in the interlamelar spaces increases the distance between the clay sheets, and make the intergallery space accessible to molecules.

Montmorillonites, for example, are natural layered clays whose structure is formed by $AlO_6$ octahedra sandwiched by $SiO_4$ tetrahedra. The presence of Al in the framework introduces negative charges that require the presence of charge balancing cations occupying interlamelar spaces. Polyaluminium cations can be suitably used to create stable pillars (pillared clays). Accessibility of non-cationic substrates to the intergallery space of pillared clays considerably increases the catalytic activity of pillared clays with respect to the original montmorillonite. Copper or iron ions may then be introduced by ion exchange or partial substitution of Al ions in the pillars. One general problem is that many aspects in the preparation of the pillars such as pH, concentration of metal ions, temperature, aging, etc., influence the composition and structure of the pillars as well as the location of the active metal. The difficulty in having pillars constituted by a single species or with a homogeneous distribution of polycations[121] is a major limit for understanding and making reliable comparisons of the literature data. In particular, the side reaction of the decomposition of $H_2O_2$, which is very sensitive to metal impurities, can be highly dependent on a specific sample, but has scarcely been investigated. However, under strictly controlled preparation conditions there is good reproducibility in the preparation.

Barrault *et al.*[122] (see review in Ref. 114 for more details and specific references) were among the first to report the good performances of mixed (Al–Cu)-pillared clays prepared from a crude bentonite sample in phenol oxidation with hydrogen peroxide. They evidenced that the introduction of copper in the pillaring position resulted in more effective catalysts, although the amount of copper introduced was only about 0.5 wt%. They also evidenced that copper introduced in this position resulted in stability against leaching and the catalyst is reusable in successive batch reactor tests. Later, they were also among the first to report the excellent behavior of mixed (Al-Fe)-pillared clays (FAZA) in the WHPCO conversion of phenol.[123] In mild reaction conditions (70°C, atmospheric pressure), about 80% of the initial amount of phenol was transformed into $CO_2$ in about 2 h. They demonstrated that the reaction is heterogeneous and that the TOC abatement obtained with the FAZA catalyst is much higher than that observed with homogeneous iron species in the same reaction conditions. They also proposed[124] that three different iron species were present:

(i) isolated iron species in highly distorted octahedral symmetry, located in a layer of the clay, probably the substitution of Al atoms;

(ii) iron species belonging to oxide clusters; and
(iii) isolated iron species in octahedral symmetry, present only in the catalyst pillared by Al-Fe mixed complexes, probably belonging to the pillars as extra framework species or by substituting the Al atoms of the pillars.

The latter species are those active in the total phenol oxidation by hydrogen peroxide. These species are also rather stable against leaching. The authors reported long-time (350 h) catalytic experiments in a continuous flow reactor that showed the high stability of the Al-Fe-PILC. However, stability in continuous flow reactor experiments has not been reported in other studies of WHPCO reactions. This is the critical problem for the industrial development of the process.

Many other authors have investigated these clay-type catalysts, as well as other catalysts (hydrotalcites, mixed oxides, micro- and meso-porous materials) in the WHPCO reaction,[112–115,120] but similar results and no peculiar performances have been reported. In several cases, leaching of the active metal was observed, while critical parameters such as the efficiency of use of $H_2O_2$ have not often been analyzed. $H_2O_2$ is a costly reagent determining to a large extent the process cost, hence efficient use is key. However, this issue has been frequently ignored and the disappearance of the model pollutant and/or reduction of total organic carbon have been the only parameters reported. $H_2O_2$ optimization should be considered as one of the most important parameters when ranking the efficiency of solid catalysts.

The sensitivity to pH is another aspect not always considered. Homogeneous Fenton processes need to operate in a narrow pH range (2.5–4.0) and this is a main process issue, as it requires the addition of large amounts of salts to the wastewater for pH control. The use of solid Fenton-type catalysts allows, in principle, operation at natural pH, without the need to adjust the pH, with better process economics. In addition, solid Fenton-type catalysts avoid the need for precipitation of iron and remove the iron sludges after treatment. This is a relevant process issue, especially when the operation is a pre-treatment before further stages such as biological treatments.

Fe-ZSM-5 has been reported to operate over a larger pH range with respect to homogeneous iron ions,[103] but usually the effect of pH has not been systematically investigated. A main problem in Fe-ZSM-5 is the leaching of iron. Although it may be reused several times in batch reactor tests, in long-term experiments deactivation occurs.

Most studies of the WHPCO reaction have been performed using model compounds. Between the main classes of chemicals investigated are phenol and phenol-derivatives. For example, tyrosol as a model of polyphenolic compounds in wastewater from olive oil milling,[125] *p*-coumaric acid,[126] cinnamic acid,[127] and

4-nitrophenol.[128] In addition, some dyes (for example, monoazo dye acid chrome dark-blue[129] and orange II[130]), MTBE,[131] and a few others were studied.

Few results have been reported for the WHPCO treatment of real wastewater streams. Kim *et al.*[132,133] investigated real dye house effluents in both batch reactors and continuous flow pilot plant-scale reactors by using $Cu/Al_2O_3$ and Al–Cu-PILC catalysts. The removal of TOC and color was strongly related to the consumption of $H_2O_2$. Copper components in the catalysts, especially in the Al–Cu-PILCs, showed successful activity toward complete removal of TOC and color. In addition, Al–Cu-PILC catalysts were extremely stable against copper leaching.

Caudo *et al.* have investigated copper-based and iron-based pillared clays (Cu-PILC, Fe-PILC) and compared them with the WHPCO of both model phenolic compounds (*p*-coumaric and *p*-hydroxybenzoic acids) and real OMW.[134] These two catalysts showed comparable performances in all these reactions, although they showed some differences in the rates of the various steps of reaction. In particular, Cu-PILC showed a lower formation of oxalic acid (main reaction intermediate) with respect to Fe-PILC. Neither catalyst showed any leaching of the transition metal that was different from other copper-based catalysts prepared by wetness impregnation on oxides (alumina, zirconia) or ion exchange of clays (bentonite) or zeolite ZSM-5. No relationship was observed between copper reducibility in the catalyst and the performance in WHPCO, or between the rate of copper leaching and catalytic behavior. Cu-PILC showed comparable activity to dissolved $Cu^{2+}$ ions, although the turnover number was lower assuming that all copper ions in Cu-PILC were active. Cu-PILC shows a high resistance to leaching and a good catalytic performance, which was attributed to the presence of copper, essentially in the pillars of the clay. A highly efficient use of $H_2O_2$ in the first hour of reaction with the participation of dissolved $O_2$ in solution was shown. For longer reaction times, however, the efficiency of $H_2O_2$ use considerably decreased.

Cu-PILC was used in the treatment of various real wastewaters from agro-food production[135] including wastewaters derived from citrus juice production, the extracted concentrated polyphenolics fraction from OMW, and OMW derived from three different sources. In the latter cases, tests were performed both in a lab-scale reactor and in a larger volume (about 10 L) reactor. The results showed that Cu-PILC might be used to treat real wastewater from agro-food production, and not only simple model chemicals. In all cases, using a semi-batch slurry-type reactor with a continuous feed of $H_2O_2$, the behavior in both TOC and polyphenol abatement may be described using pseudo first order reaction rates. Using real wastewater the rate constants are one to two orders of magnitude lower than using model molecules and a decrease in the ratio between the rate constant of phenols conversion and the rate constant of TOC abatement is observed. However, this ratio maintains over one in all cases. A typical value is around two, but the composition of wastewater and reaction conditions influence this ratio. Scaling-up to a larger volume semi-continuous slurry-type reactor causes a further lowering by one order of magnitude in the rate constants of TOC and polyphenols depletion. This is due to fouling of the catalyst related to the preferential coupling of the organic radicals and deposition over the catalyst with respect to their further degradation by hydroxyl radicals generated from $H_2O_2$ activation on the copper ions of the catalyst.

Al-Fe-PILC and $H_2O_2$ were used by Molina *et al.*[136] to study the treatment of 4-chlorophenol and an industrial wastewater deriving from cosmetics manufacture. Experiments were carried out at 90°C and atmospheric pressure and the influence of Fe load, catalyst concentration, and $H_2O_2$/COD ratio (between 0.5 and 2 times the stoichiometric ratio) were analyzed. Higher values of these parameters favor COD reduction. The Fe leaching in all cases was relatively low (1.2 mg/L), but not enough to justify a high stability as indicated by the authors. The same authors[137] have also investigated $Fe_2O_3$/SBA-15 nanocomposite catalysts for the WHPCO of a wastewater coming from a pharmaceutical plant (continuous operations). At 80°C using an initial oxidant concentration corresponding to twice the theoretical stoichiometric amount for complete carbon depletion and an initial pH of about 3, TOC removal of around 50% after 200 min was obtained. Using the catalyst in the form of extruded pellets for continuous operations in a fixed-bed reactor, about 60% TOC mineralization with good stability (for 55 h) was observed. The $BOD_5$/COD ratio increased from 0.20 to 0.30 after treatment.

Finally, Britto *et al.*[138] used a Cu/resin catalyst (which was then carbonized) for phenol abatement in industrial wastewater from a naphtha cracking unit of a petrochemical plant. The wastewater contains about 60 mg $L^{-1}$ phenol, was oil and grease free, and various contaminants were present: phosphate, sulfate, sulfide, $NH_3$, and traces of BTX. A nearly total phenol conversion at around 40°C (pH close to neutral) was observed, while the COD abatement was 35%, although the contact time and which part of phenol was adsorbed was not indicated. Copper leaching (10 mg $L^{-1}$) was also present.

## 10.2. Conclusions

AOP methods have been used over a long time, but there is still the need for further development, as highlighted in this chapter. Interest has been fostered in relation to the newly emerging problems associated with these methods, including:

(i) hardly biodegradable chemicals, such as pharmaceuticals, endocrine disrupting compounds, or some chemicals present in industrial wastewater;
(ii) the need for high-quality water for recycling in many industries; and

(iii) the possibility of reducing the cost of water treatment by combining AOP methods (such as pre-treatment) with conventional ones (such as biological treatment or anaerobic digestion) in order to remove toxic chemicals (such as polyphenols).

A general tendency is to combine many AOPs to improve performance (using different oxidants, UV radiation, etc.), but this is not always reflected in better cost-effectiveness, as discussed in relation to some of the examples.

In addition, there are often constraints which are not always considered. The Fenton reaction, for example, cannot be applied in the treatment of drinking water, as it only works below pH 3. Waters that contain strongly UV absorbing material may also not be treatable with UV/$H_2O_2$ (inner filter effect). Often the use of solid catalysts may offer viable solutions to improve performance, eliminating some of the complexity and costs in combining multiple AOPs, or bypassing some constraints present in using some AOP methods.

We have presented here a general view of the different AOPs and discussed the possibility of their improvement using solid catalysts, selecting some representative examples regarding the industrial applicability of these advanced oxidation processes and highlighting unanswered questions to be addressed in their further development. However, not all the possible technologies and their combinations have been discussed, due to space constraints. In particular, the use of electrochemical methods (from anodic oxidation to electro-Fenton, electrocatalytic, and photoelectrocatalytic processes) has been not discussed. Some consideration of these can be found in reviews.[13,15,106]

At present, the application of solid catalysts in AOP methods is mainly limited to laboratory/model waste or small-scale use. Although still limited, we have reported a number of successful examples of their use with industrial wastewater, even if the commercial application is essentially limited to a few examples in the area of catalytic wet air oxidation. It is thus necessary to develop new and/or improved catalysts, and to work on the engineering side, from reactor design to process optimization and scale-up. On the catalyst side, it has been remarked that a rational design, in relation to the complex chemistry present in AOP methods, which is a combination of redox and radical chemistry, is still missing. However, the design is translated from the use of solid catalysts in gas-solid reactions, which does not correspond to the need for AOP methods. Due to the formation of short-living and highly reactive oxidizing species, such as the hydroxyl radical, the catalytic chemistry is quite different from that present in "conventional" liquid–solid catalytic reactions. This aspect, and the relevance for the design of improved catalysts, has received limited attention in the literature.

We have also discussed, in relation to several examples, that often the focus of research has been on the conversion of the pollutant and COD/TOC removal, but this is often not the critical aspect, in particular when referring to the three sectors of application discussed above. When treating hardly biodegradable or toxic chemicals, attention should be given to the increase in the $BOD_5$/COD ratio, and the ecotoxicity of the treated solution. Similarly, this should be observed when used as a pre-treatment to biological operations. Integration with other technologies, such as the use of membranes, is the critical aspect for water reuse. The leaching of metals from the catalyst is a critical parameter in terms of both water quality and stability of the catalyst.

Continuous operations, instead of batch reactor tests as in the majority of the literature data, have to be used. Finally, cost is the conditioning factor for industrial development. In this respect, a parameter very few have considered is the efficiency of the use of oxidants ($O_3$, $H_2O_2$) or energy. More cost-effective processes are needed to ensure the wide-scale development of these water treatment technologies.

## References

1. Escobar, I. and Schäfer, A. (2009). *Sustainable Water for the Future*, Elsevier, The Netherlands.
2. Schwarzenbach, R., Escher, B., Fenner, K., *et al.* (2006). The Challenge of Micropollutants in Aquatic Systems, *Science*, **313**, pp. 1072–1077.
3. Musolff A., Leschik S., Reinstorf F., *et al.* (2010). Micropollutant Loads in the Urban Water Cycle, *Environ. Sci. Technol.*, **44**, pp. 4877–4883.
4. Food and Agriculture Organization of the United Nations (FAO), Statistical Database, Rome, 2006. http://faostat.fao.org/. Accessed on 29 August 2012.
5. Mantzavinos, D. and Psillakis, E. (2004). Enhancement of Biodegradability of Industrial Wastewaters by Chemical Oxidation Pre-treatment, *J. Chem. Techn. and Biotechn.*, **79**, pp. 431–454.
6. Walid, K. and Al-Qodah, Z. (2006). Combined Advanced Oxidation and Biological Treatment Processes for the Removal of Pesticides from Aqueous Solutions, *J. Hazardous Materials*, **B137**, pp. 489–497.
7. Centi, G. and Perathoner, S. (1999). Recycle Rinse Water: Problems and Opportunities, *Catal. Today*, **53**, pp. 11–21.
8. Centi, G., Gotti, M., Perathoner, S., *et al.* (2000). Rinse Water Purification Using Solid Regenerable Catalytic Adsorbents, *Catal. Today*, **55**, pp. 51–60.
9. (a) Centi, G. and Perathoner, S. (2001). Sustainable Water Use and Technologies, *Chim. Ind. (Milan)*, **1–2**, pp. 1–8.
   (b) Moreno Escobar, B., Gomez Nieto, M. and Hontoria García, E. (2005). Simple Tertiary Treatment Systems, *Water Science and Techn. Water Supply*, **5**, pp. 35–41.
10. Annable, M., Teodorescu, M., Hlavinek, P., *et al.* (2008). *Methods and Techniques for Cleaning-up Contaminated Sites*. NATO Science for Peace and Security Series C: Environmental Security, Springer, Heidelberg, Germany.
11. Centi, G. and Perathoner, S. (2003). Remediation of Water Contamination Using Catalytic Technologies, *Appl. Catal. B: Environ.*, **41**, pp. 15–29.

12. Centi, G. and Perathone, S. (2005). Use of Solid Catalysts in Promoting Water Treatment and Remediation Technologies, *Catalysis*, **18**, pp. 46–71.
13. Poyatos, J., Muñio, M., Almecija, M., *et al.* (2010). Advanced Oxidation Processes for Wastewater Treatment: State of the Art, *Water Air Soil Pollut.*, **205**, pp. 187–204.
14. Robertson P. (2011). *Advanced Oxidation Technologies for Waste and Potable Water Treatment*. John Wiley & Sons Inc., New York.
15. von Sonntag, C. (2008). Advanced Oxidation Processes: Mechanistic Aspects, *Water Science and Techn.*, **58**, pp. 1015–1021.
16. Vogelpohl, A. (2007). Applications of AOPs in Wastewater Treatment, *Water Science and Techn.*, **55**, pp. 207–211.
17. Guittoneau S., Duguet J., Bonnel C., *et al.* (1990). Oxidation of Parachloronitrobenzene in Dilute Aqueous Solution by $O_3$+UV and $H_2O_2$+UV: A Comparative Study, *Ozone Science and Eng.*, **12**, pp. 73–94.
18. Chen, W., Juan, C. and Wei, K. (2007). Decomposition of Dinitrotoluene Isomers and 2, 4, 6-Trinitrotoluene in Spent Acid from Toluene Nitration Process by Ozonation and Photo-ozonation, *J. Hazardous Materials*, **147**, pp. 97–104.
19. Muruganandham, M., Chen, S. and Wu, J. (2007). Mineralization of N-methyl-2-purolidone by Advanced Oxidation Process, *Separation and Purification Techn.*, **55**, pp. 360–367.
20. Keisuke, I. and El-Din, M. (2005). Aqueous Pesticide Degradation by Ozonation and Ozone-based Advanced Oxidation Processes: A Review (Part II), *Ozone: Science & Eng.*, **27**, pp. 173–202.
21. Amat, A., Arques, A., Miranda, M., *et al.* (2005). Use of Ozone and/or UV in the Treatment of Effluents from Board Paper Industry, *Chemosphere*, **60**, pp. 1111–1117.
22. Sharrer, M. and Summerfelt, S. (2007). Ozonation Followed by Ultraviolet Irradiation Provides Effective Bacteria Inactivation in a Freshwater Recirculating System, *Aquacultural Eng.*, **37**, pp. 180–191.
23. Bustos, Y., Vaca, M., Lopez, R., *et al.* (2010). Disinfection of a Wastewater Flow Treated by Advanced Primary Treatment using $O_3$, UV and $O_3$/UV Combinations, *J. Env. Science and Health, Part A: Toxic/Hazardous Substances & Env. Eng.*, **45**, pp. 1715–1719.
24. Rice, R. (1996). Applications of Ozone for Industrial Wastewater Treatment: A Review, *Ozone: Science and Eng.*, **18**, pp. 477–515.
25. Ried, A., Mielcke, J., Wieland, A., *et al.* (2007). An Overview of the Integration of Ozone Systems in Biological Treatment Steps, *Water Science and Techn.*, **55**, pp. 253–258.
26. Jose, H., Gebhardt, W., Moreira, R., *et al.* (2010). Advanced Oxidation Processes for the Elimination of Drugs Resisting Biological Membrane Treatment, *Ozone: Science & Eng.*, **32**, pp. 305–312.
27. Felis, E., Marciocha, D., Surmacz-Gorska, J., *et al.* (2007). Photochemical Degradation of Naproxen in the Aquatic Environment, *Water Science and Techn.*, **55**, pp. 281–286.
28. Jamil, T., Ghaly, M., El-Seesy, I., *et al.*, (2011). A Comparative Study Among Different Photochemical Oxidation Processes to Enhance the Biodegradability of Paper Mill Wastewater, *J. Hazardous Materials*, **185**, pp. 353–358.
29. De Luis, A., Lombraa, J., Menndez, A., *et al.* (2011). Analysis of the Toxicity of Phenol Solutions Treated with $H_2O_2$/UV and $H_2O_2$/Fe Oxidative Systems, *Ind. Eng. Chem. Res.*, **50**, pp. 1928–1937.
30. Yonar, T., Yonar, G., Kestioglu, K., *et al.* (2005). Decolorisation of Textile Effluent Using Homogeneous Photochemical Oxidation Processes, *Coloration Techn.*, **121**, pp. 258–264.
31. Sein, M., Golloch, A., Schmidt, T., *et al.* (2007). No Marked Kinetic Isotope Effect in the Peroxone ($H_2O_2$/$D_2O_2$ + $O_3$) Reaction: Mechanistic Consequences, *ChemPhysChem*, **8**, pp. 2065–2067.

32. Shu, H. and Chang, M. (2005). Pre-ozonation Coupled with UV/$H_2O_2$ Process for the Decoloration and Mineralization of Cotton Dyeing Effluent and Synthesized C. I. Direct Black 22 Wastewater, *J. Hazardous Materials*, **121**, pp. 127–133.
33. Shu, H. (2006). Degradation of Dyehouse Effluent Containing C. I. Direct Blue 199 by Processes of Ozonation, UV/$H_2O_2$ and Sequence of Ozonation with UV/$H_2O_2$, *J. Hazardous Materials*, **133**, pp. 92–98.
34. Chandrasekara, P., Kwon, T. and Moon, I. (2009). Degradation of Wastewater from Terephthalic Acid Manufacturing Process by Ozonation Catalyzed with $Fe^{2+}$, $H_2O_2$ and UV Light: Direct Versus Indirect Ozonation Reactions, *Appl. Catal. B: Environ.*, **91**, pp. 319–328.
35. Gottschalk, C., Libra, J. and Saupe, A. (2009). Ozonation of Water and Wastewater. *A Practical Guide to Understanding Ozone and its Application*, 2nd Edition. Wiley-VCH, Germany.
36. Acero, J., Haderlein, S., Schmidt, T., *et al.* (2001). MTBE Oxidation by Conventional Ozonation and the Combination Ozone/Hydrogen Peroxide: Efficiency of the Processes and Bromate Formation, *Environ, Sci. Technol.*, **35**, pp. 4252–4259.
37. Miehe, U., Hinz, C., Hoa E., *et al.* (2010). DOC and Trace Organic Removal via Ozonation and Underground Passage – Expected Benefit and Limitations, Report of the Project OXIRED. Kompetenzzentrum Wasser Berlin, Germany.
38. Oller, I., Malato, S., Sánchez Pérez, J., *et al.* (2007). Advanced Oxidation Process-biological System for Wastewater Containing a Recalcitrant Pollutant, *Water Science and Techn.*, **55**, pp. 229–235.
39. Suty, H., De Traversay, C. and Cost, M. (2004). Applications of Advanced Oxidation Processes: Present and Future, *Water Science and Techn.*, **49**, pp. 227–233.
40. Beltràn, F., Gonzàlez, M., Rivas, F., *et al.* (1996). Aqueous UV Radiation and UV/$H_2O_2$ Oxidation of Atrazine First Degradation Products: Deethylatrazine and Deisopropylatrazine, *Env. Toxicology and Chem.*, **15**, pp. 868–872.
41. Orozco, S., Bandala, E., Arancibiam C., *et al.* (2008). Effect of Iron Salt on the Color Removal of Water Containing the Azo-dye Reactive Blue Using Photo-assisted Fe(II)/$H_2O_2$ and Fe(III)/$H_2O_2$ Systems, *J. Photochem. and Photobiology A: Chem.*, **198**, pp. 144–149.
42. Levec, J. and Pintar, A. (1995). Catalytic Oxidation of Aqueous Solutions of Organics. An Effective Method for Removal of Toxic Pollutants from Waste Waters, *Catal. Today*, **24**, pp. 51–58.
43. Oliviero, L., Barbier, J. and Duprez, D. (2003). Wet Air Oxidation of Nitrogen-containing Organic Compounds and Ammonia in Aqueous Media, *Appl. Catal. B: Environ.*, **40**, pp. 163–184.
44. Luck, F. (1999). Wet Air Oxidation: Past, Present and Future. *Catal. Today*, **53**, pp. 81–91.
45. Matatov-Meytal Y., Sheintuch M. (1998). Catalytic Abatement of Water Pollutants, *Ind. Eng. Chem. Res.*, **37**, pp. 309–326.
46. Kim, K. and Ihm, S. (2011). Heterogeneous Catalytic Wet Air Oxidation of Refractory Organic Pollutants in Industrial Wastewaters: A Review, *J. Hazardous Materials*, **186**, pp. 16–34.
47. Bhargava, S., Tardio, J., Prasad, J., *et al.* (2006). Wet Oxidation and Catalytic Wet Oxidation, *Ind. Eng. Chem. Res.*, **45**, pp. 1221–1258.
48. Mantzavinos, D., Sahibzada, M., Livingston, A., *et al.* (1999). Wastewater Treatment: Wet Air Oxidation as a Precursor to Biological Treatment, *Catal.Today*, **53**, pp. 93–106.
49. He, Z., Song, S., Ying, H., *et al.* (2007). p-Aminophenol Degradation by Ozonation Combined With Sonolysis: Operating Conditions Influence and Mechanism, *Ultrasonics Sonochemistry*, **14**, pp. 568–574.
50. Xu, G., Chen, S., Shi, J., *et al.* (2010). Combination Treatment of Ultrasound and Ozone for Improving Solubilization and Anaerobic Biodegradability of Waste Activated Sludge, *J. Hazardous Materials*, **180**, pp. 340–346.

51. Gogate, P. (2008). Treatment of Wastewater Streams Containing Phenolic Compounds Using Hybrid Techniques Based on Cavitation: A Review of the Current Status and the Way Forward, *Ultrasonics Sonochemistry*, **15**, pp. 1–15.
52. Esplugas, S., Bila, D., Krause, L., *et al.* (2007). Ozonation and Advanced Oxidation Technologies to Remove Endocrine Disrupting Chemicals (EDCs) and Pharmaceuticals And Personal Care Products (PPCPs) in Water Effluents, *J. Hazard. Mater.*, **149**, pp. 631–642.
53. Centi, G. and Perathoner, S. (2003). Novel Catalyst Design for Multiphase Reactions, *Catal Today,* **79–80**, pp. 3–13.
54. Eftaxias, A., Larachi, F. and Stüber, F. (2003). Modelling of Trickle Bed Reactor for the Catalytic Wet Air Oxidation of Phenol, *Can. J. Chem. Eng.*, **81**, pp. 784–794.
55. Iliuta, I. and Larachi, F. (2001).Wet Air Oxidation Solid Catalysis Analysis of Fixed And Sparged Three-Phase Reactors, *Chem. Eng. Process.*, **40**, pp. 175–185.
56. Legube, B. and Karpel Vel Leitner, N. (1999). Catalytic Ozonation: A Promising Advanced Oxidation Technology for Water Treatment, *Catal. Today*, **53**, pp. 61–72.
57. Momani, F. (2007). Degradation of Cyanobacteria Anatoxin-A by Advanced Oxidation Processes, *Separation and Purification Techn.*, **57**, pp. 85–93.
58. Gruttadauria, M., Liotta, L., Carlo, G., *et al.* (2007). Oxidative Degradation Properties of Co-based Catalysts in the Presence Of Ozone, *Appl. Catal. B: Environ.*, **75**, pp. 281–289.
59. Martins, R. and Quinta-Ferreira, R. (2011). Phenolic Wastewaters Depuration and Biodegradability Enhancement by Ozone over Active Catalysts, *Desalination*, **270**, pp. 90–97.
60. Orge, C., Órfão, J., Pereira, M., *et al.* (2011). Ozonation of Model Organic Compounds Catalysed by Nanostructured Cerium Oxides, *Appl. Catal. B: Environ.*, **103**, pp. 190–199.
61. Orge, C., Órfão, J. and Pereira M. (2011).Catalytic Ozonation of Organic Pollutants in the Presence of Cerium Oxide-carbon Composites, *Appl. Catal. B: Environ.*, **102**, pp. 539–546.
62. Gonalves, A., Figueiredo, J., Órfão, J., *et al.* (2010). Influence of the Surface Chemistry of Multi-walled Carbon Nanotubes on their Activity as Ozonation Catalysts, *Carbon*, **48**, pp. 4369–4381.
63. Nawrocki, J. and Kasprzyk-Hordern, B. (2010). The Efficiency and Mechanisms of Catalytic Ozonation, *Appl. Catal. B: Environ.*, **99**, pp. 27–42.
64. Li, B., Xu, X., Zhu, L., *et al.* (2010). Catalytic Ozonation of Industrial Wastewater Containing Chloro and Nitro Aromatics using Modified Diatomaceous Porous Filling, *Desalination*, **254**, pp. 90–98.
65. Thiruvenkatachari, R., Kwon, T., Jun, J., *et al.* (2007). Application of Several Advanced Oxidation Processes for the Destruction of Terephthalic Acid (TPA), *J. Hazardous Materials*, **142**, pp. 308–314.
66. Chong, M., Jin, B., Chow, C., *et al.* (2010). Recent Developments in Photocatalytic Water Treatment Technology: A Review, *Water Research*, **44**, pp. 2997–3027.
67. Zhang, W., Zou, L. and Wang, L. (2010). Photocatalytic $TiO_2$/Adsorbent Nanocomposites Prepared via Wet Chemical Impregnation for Wastewater Treatment: A Review, *Appl. Catal. A: General*, **371**, pp. 1–9.
68. Friedmann, D., Mendive, C. and Bahnemann, D. (2010). $TiO_2$ for Water Treatment: Parameters Affecting the Kinetics and Mechanisms of Photocatalysis, *Appl. Catal., B: Environ.*, **99**, pp. 398–406.
69. Rizzo, L. (2010). Water and Wastewater Treatment by Heterogeneous Photocatalysis: A Review, in G. Castello (ed), *Handbook of Photocatalysts: Preparation, Structure and Applications,* Nova Science Publishers, New York, Ch. 7, pp. 271–296.
70. Kuwahara, Y., Kamegawa, T., Mori, K., *et al.* (2010). Design of New Functional Titanium Oxide-based Photocatalysts for Degradation of Organics Diluted in Water and Air, *Current Organic Chem.*, **14**, pp. 616–629.
71. Hernandez-Alonso, M., Fresno, F., Suarez, S., *et al.*, (2009). Development of Alternative Photocatalysts to $TiO_2$: Challenges and Opportunities, *Energy & Env. Science*, **2**, pp. 1231–1257.

72. Zaviska, F., Drogui, P., Mercier, G., *et al.* (2009). Advanced Oxidation Processes for Waters and Wastewaters Treatment: Application to Degradation of Refractory Pollutants, *Revue des Sciences de l'Eau*, **22**, pp. 535–564.
73. Akpan, U. and Hameed, B. (2009). Parameters Affecting the Photocatalytic Degradation of Dyes Using $TiO_2$-based Photocatalysts: A Review, *J. Hazardous Materials*, **170**, pp. 520–529.
74. Malato, S., Fernández-Ibáñez, P., Maldonado, M., *et al.* (2009). Decontamination and Disinfection of Water by Solar Photocatalysis: Recent Overview and Trends, *Catal. Today*, **147**, pp. 1–59.
75. Han, F., Kambala, V., Srinivasan, M., *et al.* (2009). Tailored Titanium Dioxide Photocatalysts for the Degradation of Organic Dyes in Wastewater Treatment: A Review, *Appl. Catal. A: General*, **359**, pp. 25–40.
76. Gaya, U. and Abdullah, A. (2008). Heterogeneous Photocatalytic Degradation of Organic Contaminants over Titanium Dioxide: A Review of Fundamentals, Progress and Problems, *J. Photochem. Photobiol. C: Photochem. Reviews*, **9**, pp. 1–12.
77. Wang, L. and Lu, G. (2008). Titanium Oxide Based Photocatalysts: From Research to Applications, *Recent Patents on Materials Science*, **1**, pp. 165–175.
78. García, J., Oliveira, J., Silva, A., *et al.* (2007). Comparative Study of the Degradation of Real Textile Effluents by Photocatalysis Reactions Involving $UV/TiO_2/H_2O_2$ and $UV/Fe^{2+}/H_2O_2$ Systems, *J. Hazardous Materials*, **147**, pp. 105–110.
79. Riga, A., Soutsas, K., Ntampegliotis, K., *et al.* (2007). Effect of System Parameters and of Inorganic Salts on the Decoloration and Degradation of Procion H-exl Dyes. Comparison of $H_2O_2/UV$, Fenton,UV/Fenton, $TiO_2/UV$ and $TiO_2/UV/H_2O_2$ Processes, *Desalination*, **211**, pp. 72–86.
80. Soutsas, K., Karayannis, V., Poulios, I., *et al.* (2010). Decolorization and Degradation of Reactive Azo Dyes via Heterogeneous Photocatalytic Processes, *Desalination*, **250**, pp. 345–350.
81. Okada, N., Nakanishi, Y. and Harada, Y. (1978). Treatment of Wastewaters Containing Ammoniacal Nitrogen, Japanese Patent JP 53 020 663 A.
82. Ishii, T., Mitsui, K., Sano, K., *et al.* (1991). Method for Wet Oxidation of Wastewater, Eur. Pat. Appl., EP 431932 A1 19910612.
83. Debellefontaine, H. and Foussard, J. (2000). Wet Air Oxidation for the Treatment of Industrial Wastes. Chemical Aspects, Reactor Design and Industrial Applications in Europe, *Waste Managem.*, **20**, pp. 15–25.
84. Roy, S., Vashishtha, M. and Saroha, A. (2010) Catalytic Wet Air Oxidation of Oxalic Acid Using Platinum Catalysts in Bubble Column Reactor: A Review, *J. Eng. Science and Technology Review*, **3**, pp. 95–107.
85. Pruesse, U., Thielecke, N. and Vorlop, K. (2008). Catalysis in Water Remediation, in G. Ertl (ed), *Handbook of Heterogeneous Catalysis*, 2nd Edition, Vol. 5, Wiley-VCH: Weinheim, Germany, pp. 2477–2500.
86. Bhargava, S., Tardio, J., Jani, H., *et al.* (2007). Catalytic Wet Air Oxidation of Industrial Aqueous Streams, *Catalysis Surveys from Asia*, **11**, pp. 70–86.
87. Levec, J. and Pintar, A. (2007). Catalytic Wet-air Oxidation Processes: A Review, *Catal. Today*, **124**, pp. 172–184.
88. Zou, L., Li, Y. and Hung, Y. (2007). Wet Air Oxidation for Waste Treatment, in L. Wang, Y. Hung, N. Shammas (eds.) *Handbook of Environmental Engineering*, Vol. 5 (Advanced Physicochemical Treatment Technologies), Springer, New York, pp. 575–610.
89. Cybulski, A. (2007). Catalytic Wet Air Oxidation: Are Monolithic Catalysts and Reactors Feasible?, *Ind. & Eng. Chem. Research*, **46**, pp. 4007–4033.
90. Kolaczkowski, S., Plucinski, P., Beltran, F., *et al.* (1999). Wet Air Oxidation: A Review of Process Technologies and Aspects in Reactor Design, *Chem. Eng. J.*, **73**, pp. 143–160.

91. Luck, F. (1996). A Review of Industrial Catalytic Wet Air Oxidation Processes. *Catal. Today*, **27**, pp. 195–202.
92. Barbier, J., Oliviero, L., Renard, B., *et al.* (2005). Role of Ceria-supported Noble Metal Catalysts (Ru, Pd, Pt) in Wet Air Oxidation of Nitrogen and Oxygen Containing Compounds, *Topics Catal.*, **33**, pp. 77–86.
93. Wang, J., Zhu, W., Yang, S., *et al.* (2008). Catalytic Wet Air Oxidation of Phenol with Pelletized Ruthenium Catalysts, *Appl. Catal. B: Environ.* **78**, pp. 30–37.
94. (a) Li, N., Descorme, C. and Besson, M. (2007). Catalytic Wet Air Oxidation of Aqueous Solution of 2-Chlorophenol over Ru/zirconia Catalysts, *Appl. Catal. B: Environ.* **71**, pp. 262–270. (b) Li, N., Descorme, C. and Besson, M. (2007) Catalytic Wet Air Oxidation of 2-Chlorophenol over Ru Loaded $Ce_xZr_{1-x}O_2$ Solid Solutions, *Appl. Catal. B: Environ.* **76**, pp. 92–100.
95. Posada, D., Betancourt, P., Liendo, F., *et al.* (2006). Catalytic Wet Air Oxidation of Aqueous Solutions of Substituted Phenols, *Catal. Lett.*, **106**, pp. 81–88.
96. Wang, J., Zhu, W., He, X., *et al.* (2008). Catalytic Wet Air Oxidation of Acetic Acid over Different Ruthenium Catalysts, *Catal. Commun.*, **9**, pp. 2163–2167.
97. Pintar, A., Besson, M. and Gallezot, P. (2001). Catalytic Wet Air Oxidation of Kraft Bleaching Plant Effluents in the Presence of Titania and Zirconia Supported Ruthenium, *Appl. Catal. B: Environ.*, **30**, pp. 123–139.
98. Minh, D., Gallezot, P., Azabou, S., *et al.* (2008). Catalytic Wet Air Oxidation of Olive Oil Mill Effluents: 4. Treatment and Detoxification of Real Effluents, *Appl. Catal. B: Environ.*, **84**, pp. 749–757.
99. Yang, M., Sun, Y., Xu, A., *et al.* (2007). Catalytic Wet Air Oxidation of Coke-plant Wastewater on Ruthenium-based Eggshell Catalysts in a Bubbling Bed Reactor, *Bull. Environ. Contam.Toxicol.*, **79**, pp. 66–70.
100. Han, L., Zhu, J., Kang, J., *et al.* (2009). Catalytic Wet Air Oxidation of High-strength Organic Coking Wastewater, *Asia-Pac. J. Chem. Eng.*, **4**, pp. 624–627.
101. Rodríguez, A., Ovejero, G., Romero, M., *et al.* (2008). Catalytic Wet Air Oxidation of Textile Industrial Wastewater Using Metal Supported on Carbon Nanofibers, *J. Supercrit. Fluids* **46**, pp. 163–172.
102. Liu, W., Hu, Y. and Tu, S. (2010). Active Carbon-ceramic Sphere as Support of Ruthenium Catalysts for Catalytic Wet Air Oxidation (CWAO) of Resin Effluent, *J. Hazard. Mater.*, **179**, pp. 545–551.
103. Centi, G., Perathoner, S. and Romeo, G. (2001). Fe/MFI as a New Heterogeneous Fenton-type Catalyst in the Treatment of Wastewater from Agroindustrial Processes, *Stud. Surf. Sci. Catal.*, **135**, pp. 5156–5163.
104. Centi, G., Perathoner, S., Torre, T., *et al.* (2000). Catalytic Wet Oxidation with $H_2O_2$ Of Carboxylic Acids on Homogeneous and Heterogeneous Fenton-type Catalysts, *Catal Today*, **55**, pp. 61–69.
105. Perathoner, S. and Centi, G. (2005). Wet Hydrogen Peroxide Catalytic Oxidation (WHPCO) of Organic Waste in Agro-food and Industrial Streams, *Topics Catal.*, **33**, pp. 207–224.
106. Chakinala, A., Gogate, P., Burgess, A., *et al.* (2009). Industrial Wastewater Treatment Using Hydrodynamic Cavitation and Heterogeneous Advanced Fenton Processing, *Chem. Eng. J.*, **152**, pp. 498–502.
107. Li, D. and Qu, J. (2009). The Progress of Catalytic Technologies in Water Purification: A Review, *J. Environ. Sci.*, **21**, pp. 713–719.
108. Guélou, E., Barrault, J., Fournier, J., *et al.* (2003). Active Iron Species in the Catalytic Wet Peroxide Oxidation of Phenol over Pillared Clays Containing Iron. *Appl Catal B: Environ.*, **44**, pp. 1–8.
109. Carriazo, J., Guelou, E., Barrault, J., *et al.* (2003). Catalytic Wet Peroxide Oxidation of Phenol over Al–Cu Or Al–Fe Modified Clays, *Appl Clay Sci.*, **22**, pp. 303–308.

110. Sirtori, C., Zapata, A., Oller, I., *et al.* (2009). Solar Photo-Fenton as Finishing Step for Biological Treatment of a Pharmaceutical Wastewater. *Environ. Sci. Technol.*, **43**, pp. 1185–1191.
111. Pérez, M., Torrades, F., Domènech, X., *et al.* (2002). Fenton and Photo-Fenton Oxidation of Textile Effluents, *Water Res.*, **36**, pp. 2703–2710.
112. Bauer, R. and Fallmann, H. (1997). The Photo-Fenton Oxidation: A Cheap and Efficient Wastewater Treatment Method, *Res. Chem. Intermediates*, **23**, pp. 341–354.
113. Perathoner, S. and Centi, G. (2010). Catalytic Wastewater Treatment Using Pillared Clays, in A. Gil, S. Korili, R. Trujillano, *et al.* (eds), *Pillared Clays and Related Catalysts*, Springer: Heidelberg, Germany, pp. 167–200.
114. Herney-Ramírez, J. and Madeira, L. (2010). Use of Pillared Clay-based Catalysts for Wastewater Treatment Through Fenton-like Processes, in A. Gil, S. Korili, R. Trujillano, *et al.* (eds), *Pillared Clays and Related Catalysts*, Springer, Heidelberg, Germany, pp. 129–166.
115. Guélou, E., Tatibouët, J. and Barrault, J. (2010). Fe–Al-Pillared Clays: Catalysts for Wet Peroxide Oxidation of Phenol, in A. Gil, S. Korili, R. Trujillano, *et al.* (eds), *Pillared Clays and Related Catalysts*, Springer, Heidelberg, Germany, pp. 201–224.
116. Navalon, S., Alvaro, M. and Garcia, H. (2010). Heterogeneous Fenton Catalysts Based on Clays, Silicas and Zeolites, *Appl. Catal. B: Environ.*, **99**, pp. 1–26.
117. Pignatello, J., Oliveros, E. and MacKay, A. (2006). Advanced Oxidation Processes for Organic Contaminant Destruction based on the Fenton Reaction and Related Chemistry, *Critical Reviews in Env. Science and Techn.*, **36**, pp. 1–84.
118. Hartmann, M., Kullmann, S. and Keller, H. (2010). Wastewater Treatment with Heterogeneous Fenton-type Catalysts based on Porous Materials, *J. Materials Chem.*, **20**, pp. 9002–9017.
119. Herney-Ramirez, J., Vicente, M. and Madeira, L. (2010). Heterogeneous Photo-Fenton Oxidation with Pillared Clay-based Catalysts for Wastewater Treatment: A Review, *Appl. Catal. B: Environ.*, **98**, pp. 10–26.
120. Garrido-Ramirez, E., Theng, B. and Mora, M. (2010). Clays and Oxide Minerals as Catalysts and Nanocatalysts in Fenton-like Reactions: A Review, *Applied Clay Science*, **47**, pp. 182–192.
121. Centi, G. and Perathoner, S. (2008). Catalysis by Layered Materials: A Review, *Micropor. Mesopor. Mater.*, **107**, pp. 3–15.
122. Carriazo, J., Molina, R., Moreno, S. (2008). A Study on Al and Al-Ce-Fe Pillaring Species and their Catalytic Potential as they are Supported on a Bentonite, *Appl. Catal. A: General*, **334**, pp. 168–172.
123. Barrault, J., Bouchoule, C., Echachoui, K., *et al.* (1998). Catalytic Wet Peroxide Oxidation (CWPO) of Phenol over Mixed (AlCu)-pillared Clays, *Appl Catal B: Environ*, **15**, pp. 269–274.
124. Barrault, J., Abdellaoui, M., Bouchoule, C., *et al.* (2000). Catalytic Wet Peroxide Oxidation Over Mixed (Al–Fe) Pillared Clays, *Appl Catal B: Environ*, **27**, pp. L225–L230.
125. Guélou, E., Barrault, J., Fournier, J., *et al.* (2003). Active Iron Species in the Catalytic Wet Peroxide Oxidation of Phenol over Pillared Clays Containing Iron, *Appl Catal B: Environ*, **44**, pp. 1–8.
126. Achma, R., Ghorbel, A., Dafinov, A., *et al.* (2008). Copper-supported Pillared Clay Catalysts for the Wet Hydrogen Peroxide Catalytic Oxidation of Model Pollutant Tyrosol, *Appl. Catal. A: General*, **349**, pp. 20–28.
127. Caudo, S., Centi, G., Genovese, C., *et al.* (2006). Homogeneous Versus Heterogeneous Catalytic Reactions to Eliminate Organics from Waste Water Using $H_2O_2$, *Topics Catal.*, **40**, pp. 207–219.
128. Tabet, D., Saidi, M., Houari, M., *et al.* (2006). Fe-pillared Clay as a Fenton-type Heterogeneous Catalyst for Cinnamic Acid Degradation, *J. Env. Management*, **80**, pp. 342–346.
129. Chirchi, L. and Ghorbel, A. (2002). Use of Various Fe-modified Montmorillonite Samples for 4-Nitrophenol Degradation by $H_2O_2$, *Applied Clay Science*, **21**, pp. 271–276.

130. Timofeeva, M., Khankhasaeva, S., Chesalov, Y., *et al.* (2009). Synthesis of Fe,Al-pillared Clays Starting from the Al,Fe-polymeric Precursor: Effect of Synthesis Parameters on Textural and Catalytic Properties, *Appl. Catal. B-Environ.*, **88**, pp. 127–134.
131. Herney-Ramirez, J., Lampinen, M., Vicente, M., *et al.* (2008). Experimental Design to Optimize the Oxidation of Orange II Dye Solution Using a Clay-based Fenton-like Catalyst, *Ind. Eng. Chem. Res.* **47**, pp. 284–294.
132. Gonzalez-Olmos, R., Roland, U., Toufar, H., *et al.* (2009). Fe-zeolites as Catalysts for Chemical Oxidation of MTBE in Water with $H_2O_2$, *Appl. Catal. B-Environ.*, **89**, pp. 356–364.
133. Kim, S., Kim, D., Oh, S., *et al.* (2002). Catalytic Wet Oxidation of Dyehouse Effluents with $Cu/Al_2O_3$ and Cu–Al pillared Clay, *Stud. Surf. Sci. Catal.*, **145**, pp. 355–358.
134. Kim, S. and Lee, D. (2004). Preparation of Al–Cu Pillared Clay Catalysts for the Catalytic Wet Oxidation of Reactive Dyes, *Catal Today*, **97**, pp. 153–158.
135. Caudo, S., Centi, G., Genovese, C., *et al.* (2007). Copper- and Iron-pillared Clay Catalysts for the WHPCO of Model and Real Wastewater Streams from Olive Oil Milling Production, *Appl Catal B: Environ.*, **70**, pp. 437–446.
136. Caudo, S., Genovese, C., Perathoner, S., *et al.* (2008). Copper-pillared Clays (Cu-PILC) for Agro-food Wastewater Purification with $H_2O_2$, *Micropor. Mesopor. Mater.*, **107**, pp. 46–57.
137. Molina, C., Zazo, J., Casas, J., *et al.* (2010). CWPO of 4-CP and Industrial Wastewater with Al-Fe Pillared Clays, *Water Science and Techn.*, **61**, pp. 2161–2168.
138. Melero, J., Martinez, F., Botas, J., *et al.* (2009). Heterogeneous Catalytic Wet Peroxide Oxidation Systems for the Treatment of an Industrial Pharmaceutical Wastewater, *Water Research*, **43**, pp. 4010–4018.
139. Britto, J., Botelho de Oliveira, S., Denilson, R., *et al.* (2008). Catalytic Wet Peroxide Oxidation of Phenol from Industrial Wastewater on Activated Carbon, *Catal. Today*, **133–135**, pp. 582–587.

# Chapter 11

## Selective Oxidation at SABIC: Innovative Catalysts and Technologies

Edouard MAMEDOV* and Khalid KARIM†

The paper overviews research carried out at SABIC Company in the last 15–20 years in the field of selective oxidation. Using different approaches, a number of effective catalysts were developed by proposing new or improving existing catalytic systems. On some of them reaction network and kinetics were studied that in combination with reaction engineering allowed elaborate process technology. The most advanced development is ethane direct oxidation to acetic acid which was commercialized at one of the SABIC plants.

In the early 1990s SABIC management launched a research program on the conversion of low-weight paraffins and olefins to value-added monomers and chemicals. Since that time SABIC researchers have been actively involved in the development and study of catalysts for different selective oxidation reactions. Some of the oxidation reactions and catalyst compositions developed by them are listed in Table 11.1 along with the yields and selectivities of target products.

The most advanced development is the direct oxidation of ethane to acetic acid, which has been commercialized at one of SABIC's plants. The catalyst for this process is based on mixed V and Mo oxides which were shown in the 1970s to catalyze mostly oxidative dehydrogenation of ethane to ethylene and produce much less acetic acid.[9] The addition of niobium results in the formation of the $(VNbMo)_5O_{14}$ solid solution, in which V sites activating paraffin and Mo sites catalyzing oxidation of intermediate olefin are isolated from each other by low active niobium. The presence of this structure in combination with the $MoO_3$ phase in the MoVNb oxide catalyst provides a significantly higher combined selectivity to ethylene and acetic acid.[10] This catalyst, however, also produces more ethylene than acetic acid. According to

*SABIC Technology & Innovation Center, 1600 Industrial Boulevard, Houston, TX 77478, USA.
†SABIC Technology & Innovation Complex,, P.O. Box 42503, Riyadh 11551, Saudi Arabia.

**Table 11.1.** Selective oxidation catalysts developed at SABIC company.

| Oxidation reaction | Catalyst | Y(%) | S(%) | Ref. |
|---|---|---|---|---|
| $C_2H_6 \rightarrow CH_3COOH$ | $MoVLaNbPdO_x$ | 30 | 60 | 1 |
| $C_2H_4 + CH_3COOH \rightarrow CH_3COOHC_2H_3$ | $Pd\text{-}Au/SiO_2$ | 600* | 90 | 2 |
| $C_3H_8 \rightarrow CH_2CHCH_3$ | $VSbAlNbO_x$ | 26 | 55 | 3 |
| $C_3H_8 + NH_3 \rightarrow CH_2CHCN$ | $VSbAlWBO_x$ | 13 | 61 | 4 |
| $C_3H_8 + NH_3 \rightarrow CH_2CHCN$ | $VWBiAlO_x$ | 7 | 65 | 5 |
| $i\text{-}C_4H_8 \rightarrow CH_2C(CH_3)CHO$ | $MoBiWFeCoNiSbCsMgZnO_x$ | 87 | 89 | 6,7 |
| $CH_2C(CH_3)CHO \rightarrow CH_2C(CH_3)COOH$ | $MoPVCuBiSbCsBO_x$ | 76 | 86 | 8 |

*Space-time yield as grams of vinyl acetate produced per hour per liter of catalyst.

**Table 11.2.** Effect of adding phosphorus on the activity and selectivity of the $Mo_{2.5}V_{1.0}Nb_{0.32}O_x$ catalyst at temperature 260°C, ethane-to-air ratio 15/85 and contact time of 3.2 seconds.

| | Conversion | | Product selectivity (%) | | | |
|---|---|---|---|---|---|---|
| P/V (atomic) | (%) | (%/m$^2$) | ($C_2H_4$ + AA) | $C_2H_4$ | AA | $CO_x$ |
| 0 | 65.0 | 0.85 | 57.5 | 27.0 | 30.5 | 42.5 |
| 0.01 | 63.2 | 1.11 | 59.1 | 25.3 | 33.8 | 40.9 |
| 0.03 | 62.2 | 1.21 | 60.7 | 24.1 | 36.6 | 39.3 |
| 0.04 | 53.3 | 0.99 | 60.4 | 10.5 | 49.9 | 39.6 |
| 0.06 | 33.6 | 1.29 | 64.9 | 14.7 | 50.2 | 35.4 |

the reaction network proposed in a paper by Rahman and colleagues,[11] acetic acid forms mainly by oxidation of the intermediate ethylene:

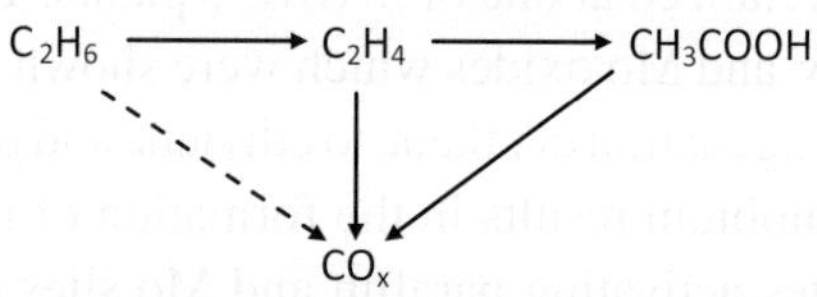


Ethylene conversion to acetic acid can be increased in at least two ways. One is making the catalyst surface more acidic, which will strengthen ethylene adsorption on the surface to give more time for its oxidation to acetic acid. A more acidic surface will also facilitate desorption of acetic acid thus reducing its surface over-oxidation to carbon oxides. This approach has been realized by adding small amounts of P, B and Te to the MoVNb oxide catalyst.[12] Table 11.2 presents the results of adding phosphorus.

**Table 11.3.** Effect of palladium on the activity and selectivity of the $Mo_{2.5}V_{1.0}Nb_{0.32}O_x$ catalyst at temperature 260°C, ethane-to-air ratio 15/85 and contact time of 3.2 seconds.

| | Conversion | | Product selectivity (%) | | | |
|---|---|---|---|---|---|---|
| P/V (atomic) | (%) | (%/m$^2$) | ($C_2H_4$ + AA) | $C_2H_4$ | AA | $CO_x$ |
| 0 | 65.0 | 0.85 | 57.5 | 27.0 | 30.5 | 42.5 |
| 0.00013 | 49.1 | 0.70 | 55.3 | 7.5 | 47.8 | 44.7 |
| 0.00048 | 49.7 | — | 56.9 | 3.9 | 53.1 | 43.6 |
| 0.00068 | 49.2 | 0.58 | 57.6 | 1.9 | 55.7 | 42.4 |

Increasing the amount of phosphorus introduced into the MoVNb oxide catalyst decreased ethane conversion because of lower specific surface areas of P-promoted catalysts. Ethane conversion normalized to the catalyst surface area, as is seen from Table 11.2, did not change much. Selectivity to acetic acid (AA) increased and selectivity to ethylene decreased significantly with the amount of phosphorus in the catalyst, while their combined selectivity stayed roughly unchanged. These results imply that the phosphorus added to the MoVNb oxide catalyst indeed increased conversion of the intermediate ethylene to acetic acid.

Another way to intensify ethylene oxidation to acetic acid is based on building extra active sites for this reaction. This has been done by doping the catalyst with palladium which is known to be catalytically active in the gas-phase oxidation of ethylene to acetic acid. The effect of palladium can be seen in Table 11.3 which lists the data from reference.[13]

Compared to the base catalyst, Pd-doped catalysts displayed somewhat lower overall activity in terms of both absolute and normalized conversion of ethane and essentially higher selectivity to acetic acid. As the Pd amount in the catalyst increased, selectivity to acetic acid increased at the expense of selectivity to ethylene. Combined selectivity to carbon oxides practically did not change. The only difference was that ethane oxidation on the base catalyst produced roughly equal amounts of CO and $CO_2$ while the reaction on Pd-promoted catalysts produced only $CO_2$. This result looks reasonable if one takes into account that Pd is a good catalyst for CO oxidation to $CO_2$.

The comparison of P- and Pd-promoted catalysts shows that Pd-promoted catalysts are more selective towards the formation of acetic acid. Other advantages are that they do not produce CO as a reaction product and display better redox behavior. A study of catalyst redox property revealed that Pd facilitated both the reduction of the MoVNb oxide catalyst with ethane and its re-oxidation with oxygen.[14] This is very important because the reaction occurs via the redox mechanism of a Mars–van Krevelen type. Such a conclusion comes from the data presented in Fig. 11.1. With increasing the amount of oxygen removed from the surface of the MoVNbPd

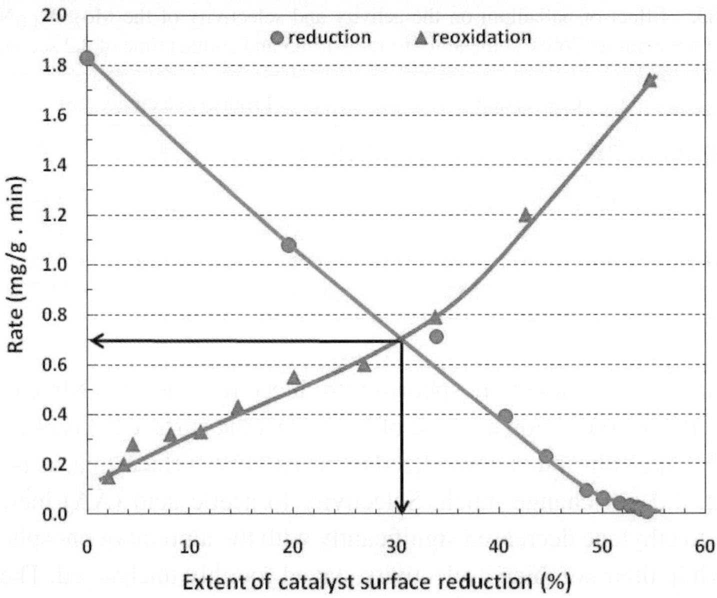


**Figure 11.1.** Rates of catalyst reduction with ethane and re-oxidation with oxygen at 260°C as a function of the extent of catalyst surface reduction.

oxide catalyst, the rate of catalyst reduction with ethane decreased and the rate of its re-oxidation with oxygen increased, so they became equal to each other at the extent of a catalyst surface reduction of about 30%. This number was close to the reduction extent of 25% determined for the steady-state catalyst by means of *in situ* gravimetric technique. The rate of ethane catalytic oxidation at the steady state was also close to the rates of catalyst reduction/re-oxidation at the intersection point.[15]

Selectivity to acetic acid on the MoVNbPd oxide catalyst was further improved by co-feeding water vapor.[11] As is seen in Fig. 11.2, increasing water vapor concentration in the feed from 0 to 20% enhanced selectivity to acetic acid from 60 to 80% at the expense of selectivities to ethylene and carbon oxides. Overall conversion of ethane practically did not change.

For the reasons listed above, a Pd-doped MoVNb oxide system has been picked as the basis for the development of industrial catalysts. This catalyst composition has been successfully scaled up in collaboration with a catalyst manufacturer and was commercialized in 2005 at the Ibn Rushd complex in Yanbu, Saudi Arabia. A block flow diagram for the ethane oxidation process is shown in Fig. 11.3.

The facility has a capacity of 30,000 metric tons per year and is currently operating. The catalyst, which is a calcined mixture of oxides of Mo, V, Nb, La and Pd, provides both high selectivity and a high yield of acetic acid. When an ethane–air system is employed, the selectivity to acetic acid is 60% at per pass conversion of

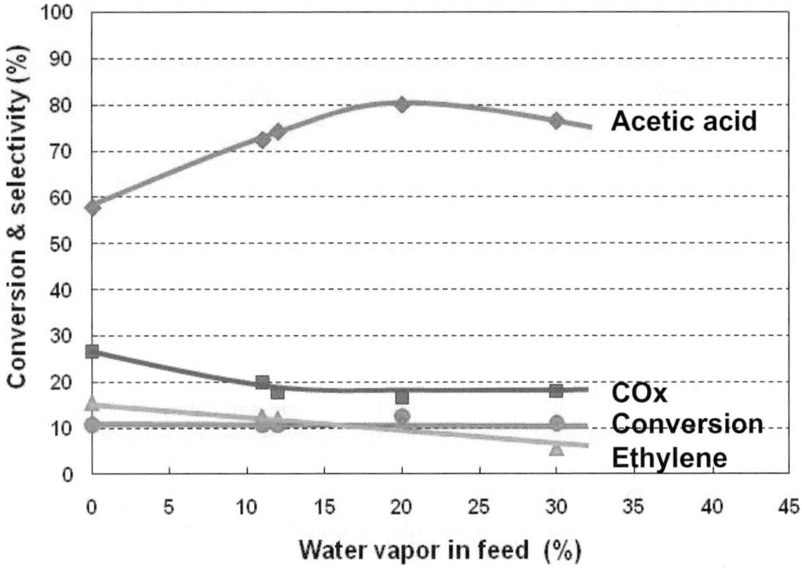


**Figure 11.2.** Effect of co-feeding water on ethane conversion and product selectivity on MoVNbPd oxide catalyst at temperature 260°C and pressure 15 bars.

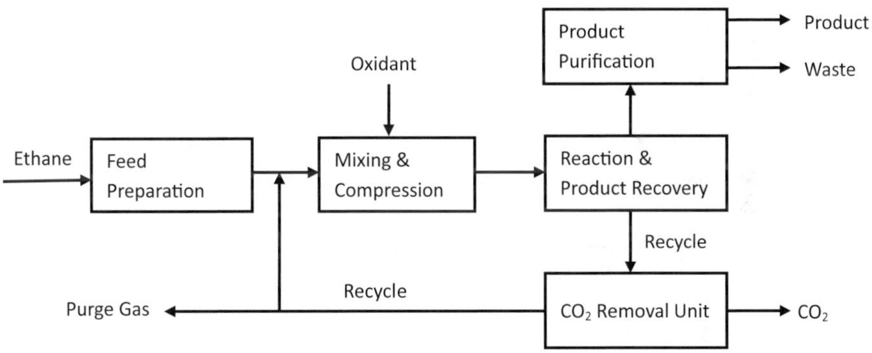


**Figure 11.3.** Block flow diagram for the SABIC process of direct ethane oxidation to acetic acid.

ethane of about 50% and at 100% conversion of oxygen. When using pure oxygen as an oxidant, the selectivity to acetic acid reaches as high as 67% at per pass conversions of about 14% and 100% of ethane and oxygen, respectively. In this case, the recycling of unconverted ethane may be beneficial.

Another SABIC achievement was a significant increase of activity attained for the shelled Pd-Au/$SiO_2$ catalyst in the oxidative conversion of ethylene and acetic acid to vinyl acetate.[2,16,17] This has been done by slightly increasing Au loading and by modifying the catalyst preparation method. In particular, prior reduction of the impregnated support which has been treated with a flow of nitrogen at temperatures

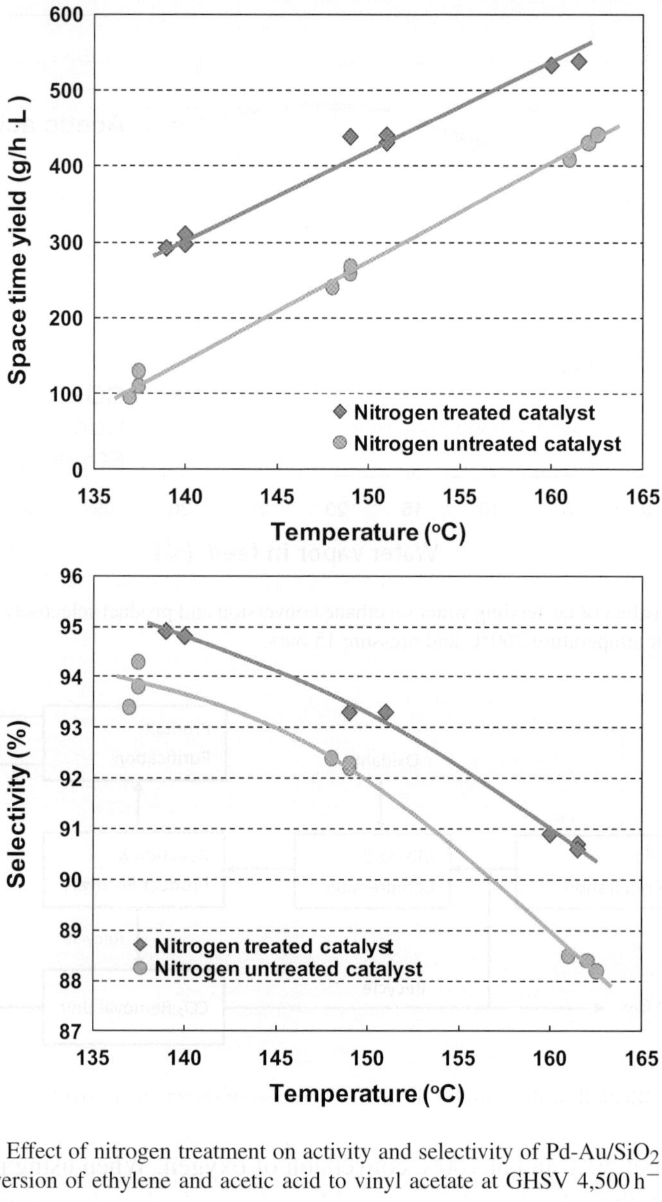


**Figure 11.4.** Effect of nitrogen treatment on activity and selectivity of Pd-Au/$SiO_2$ catalyst in the oxidative conversion of ethylene and acetic acid to vinyl acetate at GHSV 4,500 $h^{-1}$ and pressure 50 psig.

rising from room to 150°C. This procedure considerably increased both activity and selectivity of the Pd-Au/$SiO_2$ catalyst (see Fig. 11.4).

Other preparation steps, such as loading the support with Pd and Au, washing, drying, reducing and promoting with potassium acetate, have also been optimized to contribute to the improvement of catalyst performance. The catalysts prepared by this modified method have a uniform distribution of highly dispersed Pd and Au which are mainly concentrated in a thin shell of 0.05–0.20 mm. In terms of catalytic behavior, they are more active than the catalysts prepared by the prior art method. At 90% selectivity, the SABIC catalyst produces at least 600 grams of vinyl acetate per hour per liter catalyst, which is 20–30% higher than the catalysts prepared by conventional methods. The catalyst preparation has been scaled up and the scaled batch has been tested side by side with commercial catalysts made by different companies to confirm its superior performance.

The approach, based on "dilution" of catalytically active components with inert or low active material, has been employed to improve the performance of VSb-based oxide catalysts in reactions of the selective oxidation of propane. In particular, the incorporation of low active Al into the body of NiVSb, FeVSb, BiVSb and other mixed-metal oxides significantly improved their selectivity in the oxidative dehydrogenation of propane to propylene.[3] As is seen in Fig. 11.5, the addition of Al to the $Fe_{0.3}V_{1.0}Sb_{0.6}O_x$ catalyst increased selectivity to propylene by 2–3 times. The selectivity of an Al-diluted catalyst was further improved by adding small amounts of alkali and alkaline earth metals. Figure 11.5 illustrates the effect of potassium which, along with increasing catalyst intrinsic selectivity, also decreased propylene over-oxidation. This improvement in selectivity has been related to the lower acidity

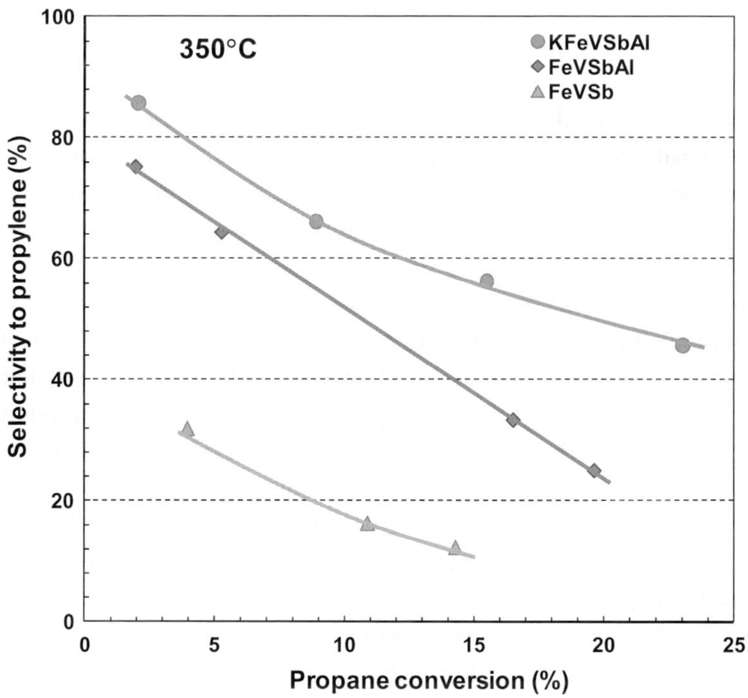


**Figure 11.5.** Oxidative dehydrogenation of propane on FeVSb-based oxide catalysts at 350°C.

of the K-promoted catalyst that facilitated propylene desorption to reduce its surface over-oxidation.

Solid-state dilution also positively affected the performance of the $V_1Sb_9O_x$ catalyst in the ammoxidation of propane to acrylonitrile. The insertion of low active Al and Mg into the body of this catalyst essentially improved selectivity to acrylonitrile.[4] The base catalyst comprised $VSbO_4$ as a major phase which was absent in the Al- and Mg-promoted catalysts. Instead they contained respectively $AlSbO_4$ and $MgSb_2O_6$ phases along with isolated tetrahedral $V^{5+}$ species.[18] Assuming that V-O sites dehydrogenate propane to propylene which then undergoes ammoxidation to acrylonitrile on the Sb-NH sites, one may suggest that Sb-NH sites neighboring low active Al or Mg were more selective in the ammoxidation of propylene than the Sb-NH sites neighboring active V.

Propane ammoxidation on Al- and Mg-diluted VSb oxide catalysts occurs according to the following network:

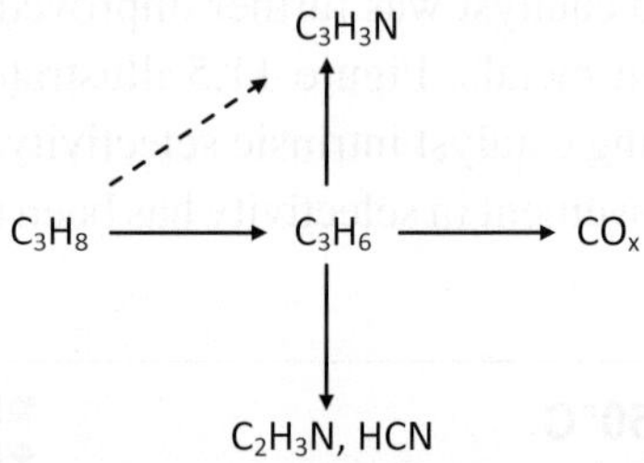


Acrylonitrile is mainly produced by ammoxidation of the intermediate propylene. The contribution of the direct conversion of propane to acrylonitrile has been estimated to be less than 5%. In general, the propylene ammoxidation pathway can be intensified in different ways. One of them is increasing the population of ammonia adsorbed species on the catalyst surface. Suggesting more ammonia was needed to adsorb onto the acidic surface, tungsten was put on the catalyst surface by an impregnation technique.[4] This procedure increased the number of acid sites without altering their strength. Due to this change in acidity, the W-promoted catalyst adsorbed more ammonia that reduced degradation and intensified ammoxidation of the intermediate propylene. As a result, selectivity to acrylonitrile increased considerably at the expense of selectivity to carbon oxides.[18]

Further improvement of selectivity to acrylonitrile has been achieved by co-feeding $CO_2$. According to the data reported in the Mamedov patent,[19] increasing the $CO_2$ percentage in the feed from 0 to 40 enhanced selectivity to acrylonitrile on the WVSbAl oxide catalyst from 60 to 72%. The positive effect of $CO_2$ was reversible and catalyst specific. For instance, in the presence of $CO_2$ selectivity of the WVSbMg oxide catalyst improved by 8%, while the selectivity of the VSbSnTi oxide catalyst developed by BP Chemical increased by only 2%.

A novel catalyst system containing neither Mo nor Sb has been proposed and developed for the ammoxidation of propane to acrylonitrile. It is based on mixed VWBi oxides.[5] The highest acrylonitrile selectivity of 55% was obtained on the catalyst that had a nominal composition $V_1W_{0.8}Bi_{1.6}O_x$ and comprised $BiVO_4$ and $Bi_2WO_6$ as major phases. On this catalyst acrylonitrile formation also occurred mainly by ammoxidation of the intermediate propylene. Herein it was suggested that propane dehydrogenation to propylene occured on the $BiVO_4$ phase which contained paraffin-activating V-O sites. The ammoxidation of propylene to acrylonitrile took place on the $Bi_2WO_6$ phase that held both the $\alpha$-H abstraction Bi-O sites and the nitrogen insertion W-NH sites. Of many elements tested as additives to the $VWBiO_x$, only aluminum and niobium considerably improved its catalytic behavior. As is seen in Fig. 11.6, Al- and Nb-promoted catalysts showed higher activity and higher selectivity than the unpromoted catalyst.

The enhanced performance of the Nb-containing catalyst has been attributed to the presence of the $Bi_3W_1Nb_9O_{30}$ phase in which active bismuth and tungsten sites are isolated from each other by low active niobium species. Poor crystallization has not allowed the relation between behavior and phase composition for the Al-promoted catalyst to be traced. It should be emphasized that the addition of molybdenum and antimony, which are present in many selective oxidation

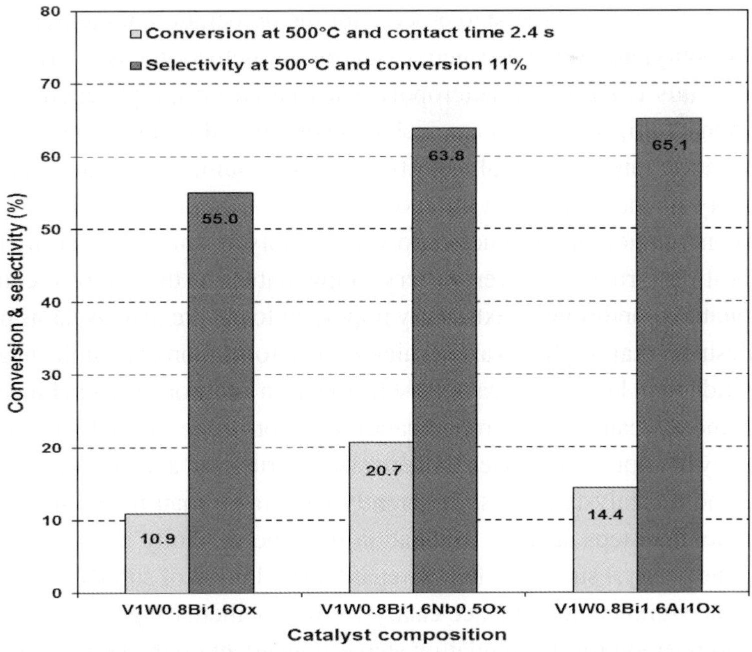


**Figure 11.6.** Effect of Al and Nb on the behavior of VWBi oxide catalysts in propane ammoxidation.

catalysts, negatively affected catalytic performance of the $VWBiO_x$. Both elements dramatically decreased the selectivity to acrylonitrile.

As part of SABIC's effort to develop a three-step process for producing methyl methacrylate from isobutylene, catalysts for the oxidation of isobutylene to methacrolein have been developed and studied. The best catalysts for this reaction are mixed-metal oxides based primarily on bismuth molybdate but containing a complex mixture of other elements, including Fe, Co, Ni and Zn to increase catalyst activity, and alkali metals such as Cs to moderate the oxidation activity of the most active and least selective sites. Properly controlling the relative amounts of these elements, especially the ratio of iron and bismuth and the ratio of bismuth and cesium, as well as the ratio of all of the metals relative to molybdenum, is the key to achieving a balance of activity and selectivity. In addition, obtaining the best catalyst structure requires careful dissolution of the component elements and proper control of catalyst precipitation, including pH, temperature and mixing, as well as careful optimization of drying, supporting and calcining the catalyst. Optimizing catalyst performance for the specific process needs is required to achieve good yields for the process as a whole. The highest methacrolein yield of 87% at selectivity of 89% has been obtained on the catalyst of the following elemental composition: $Mo_{12}Bi_{1.0}W_{0.3}Fe_{2.4}Co_{2.0}Ni_{4.0}Sb_{0.7}Cs_{0.6}Mg_{0.5}Zn_{0.5}O_x$.[6,7] This catalyst also displayed stable behavior in terms of both activity and selectivity during the lifetime test which lasted approximately 900 hours.

The second step in a three-step process for the preparation of methyl methacrylate from isobutylene is methacrolein oxidation to methacrylic acid. The effective catalysts for this reaction are heteropoly acids based on molybdenum, phosphorous and vanadium, with elements such as bismuth and copper added to provide higher oxidation rates, and alkali metals, such as cesium, used to neutralize much of the acidity of the underlying compound. As with the mixed-metal oxide catalysts, the composition of these heteropoly acid catalysts must be carefully tuned to give the best performance. Given the crystalline nature of the catalyst, controlling the precipitation conditions is extremely important to ensure that, as far as possible, only the desired catalyst phases are obtained, since formation of significant amounts of metal oxides can lead to low catalyst selectivity. In addition, proper choice of synthesis parameters can give improved catalyst performance through enhancements to purely physical properties such as pore size distribution and surface area. These heteropoly acid catalysts are also inherently less stable than mixed-metal oxides, so post-treatment steps such as calcination must be carefully controlled to avoid damaging the catalyst structure. In this regard, the addition of supports and stability modifiers can significantly enhance catalyst value by increasing the lifetime. Since the feed composition for this oxidation step is dependent on the performance of the catalyst for isobutylene oxidation, it is important to optimize catalyst performance to maximize overall process yield. Taking all these factors into account, a proprietary catalyst of nominal composition $Mo_{12}P_{1.5}V_{0.5}Cu_{0.1}Bi_{1.0}Sb_{0.8}Cs_{1.0}B_{0.5}O_x$ has been developed.[8] Under optimized reaction conditions it gave a 77% yield of methacrylic acid at a selectivity of 86%.

## References


1. Karim, K., Mamedov, E., Al-Hazmi, M., *et al.* (2001). Catalysts Methods for Producing Acetic Acid from Ethane Oxidation Using Mo, V, Pd and Nb Based Catalysts, Processes of Making Same and Methods of Using Same, *US Patent 6,310,241*.
2. Khanmamedova, A. (2004). Highly Selective Shell Impregnated Catalyst of Improved Space Time Yield for Production of Vinyl Acetate, *US Patent 6,825,149*.
3. Mamedov, E. and Shaikh, S. (2001). Catalyst System for Oxidative Dehydrogenation of Paraffins, *US Patent 6,235,678*.
4. Mamedov, E., Bethke, K., Shaikh, S., *et al.* (2004). Catalyst Compositions for the Ammoxidation of Alkanes and Olefins, Methods of Making and of Using Same, *US Patent 6,710,011*.
5. Mamedov, E., Shaikh, S. and Araujo, A. (2008). Catalyst Composition without Antimony or Molybdenum for Ammoxidation of Alkanes, a Process of Making and a Process of Using Thereof, *US Patent 7,449,426*.
6. Stevenson, S. and Liang, W. (2005). Preparation of Mixed Metal Oxide Catalysts for Catalytic Oxidation of Olefins to Unsaturated Aldehydes, *US Patent 6,946,422*.
7. Liang, W., Stevenson, S., Kauffman, J., *et al.* (2008). Mixed Metal Oxide Catalysts for the Production of Unsaturated Aldehydes from Olefins, *US Patent 7,361,791*.
8. Liang, W., Stevenson, S., McGuffey, A., *et al.* (2010). Catalyst for the Oxidation of a Mixed Aldehyde Feedstock to Methacrylic Acid and Methods for Making and Using Same, *US Patent 7,649,111*.
9. Thorsteinson, E., Wilson, T., Young, F., *et al.* (1978). The Oxidative Dehydrogenation of Ethane over Catalysts Containing Mixed Oxides of Molybdenum and Vanadium, *J. Catal.*, **52**, pp. 116–132.
10. Roussel, M., Bordes-Richard, E., Karim, K., *et al.* (2005). Oxidation of Ethane to Ethylene and Acetic Acid by MoVNbO Catalysts, *Catal. Today*, **99**, pp. 77–87.
11. Rahman, F., Loughlin, K., Al-Saleh, M., *et al.* (2010). Kinetics and Mechanism of Partial Oxidation of Ethane to Ethylene and Acetic Acid over MoV Type Catalysts, *Appl. Catal. A*, **375**, pp. 17–25.
12. Karim, K., Al-Hazmi, M. and Mamedov, E. (2000). Catalysts for the Oxidation of Ethane to Acetic Acid, Processes of Making Same and Processes of Using Same, *US Patent 6,013,597*.
13. Karim, K., Mamedov, E., Al-Hazmi, M., *et al.* (2000). Catalysts for Producing Acetic Acid from Ethane Oxidation, Processes of Making Same and Methods of Using Same, *US Patent 6,030,920*.
14. Roussel, M., Barama, S., Karim, K., *et al.* (2009). MoV-based Catalysts in Ethane Oxidation to Acetic Acid: Influence of Additives on Redox Chemistry, *Catal. Today*, **141**, pp. 288–293.
15. Fierro, J., Karim, K. and Mamedov, E. (1997). Unpublished data.
16. Khanmamedova, A. (2002). Highly Selective Shell Impregnated Catalyst of Improved Space Time Yield for Production of Vinyl Acetate, *US Patent 6,420,308*.
17. Khanmamedova, A., Li, B., Bates, R., *et al.* (2004). Highly Selective Shell Impregnated Catalyst of Improved Space Time Yield for Production of Vinyl Acetate, *US Patent 6,794,332*.
18. Shaikh, S., Bethke, K. and Mamedov, E. (2006). Propane Ammoxidation on Bulk, Diluted and Supported VSb Oxides, *Top. Catal.*, **38**, pp. 241–249.
19. Mamedov, E., Bethke, K., Shaikh, S. *et al.* (2007). Process for the Ammoxidation of Alkanes and Olefins, *US Patent 7,186,670*.

## Chapter 12

# Development of Selective Oxidation Catalysts at Clariant

Gerhard MESTL*

The chapter presents a brief overview of the current research on $V_2O_5/TiO_2$ catalysts for o-xylene oxidation to phthalic anhydride at Clariant. Phthalic anhydride is produced in tubular, salt-cooled reactors with a capacity of about 5 Mio to per annum. There is a rather broad variety of different process conditions realized in industry in terms of feed composition, air flow rate, as well as reactor dimensions which the phthalic anhydride catalyst portfolio has to match. Catalyst active mass compositions have been optimized at Clariant for these differently realized industry processes utilizing artificial neural networks trained on high-throughput data. Fundamental pilot reactor research unravelling new details of the reaction network of the o-xylene oxidation led to an improved kinetic reactor model which allowed further optimizing of the state of the art multi-layer catalyst system for maximum phthalic anhydride yields.

## 12.1. Introduction

As one of the world's leading specialty chemical companies, Clariant contributes to value creation with innovative and sustainable solutions for customers from many industries. The business units are divided into four Business Areas: Care Chemicals, Catalysis & Energy, Natural Resources and Plastics & Coatings. The product portfolio of Clariant provides competitive and innovative solutions to customers and research and development is focused on addressing the key trends of our time. These include energy efficiency, renewable raw materials, emission-free mobility and conserving finite resources.

Clariant's "Catalysts" business unit has its roots in the Süd-Chemie company which was founded by Justus von Liebig in Upper Bavaria close to Munich. This unit has a history stretching back over more than 150 years. Süd-Chemie's catalyst activities were established in 1962. Fifty years on, it is the world's leading manufacturer

*Head of Department Oxidation Catalysis, Clariant R&D, Waldheimer Str. 13, 83052 Bruckmühl, Germany.

of process catalysts for the chemical, petrochemical and refinery industry. The product portfolio is focused on value creation for customers and comprises, i.e. catalysts for the production of chemical products and for replacing oil with natural gas (gas-to-liquid, GTL), coal (coal-to-liquid, CTL) and biomass (biomass-to-liquid, BTL). The catalyst division is divided into three business segments: Syngas, Petrochemical and Specialty Catalysts. Selective partial oxidation catalysts belong to the Specialty Catalysts segment, with the main catalysis research centers located in Heufeld, Germany; Louisville, KY; and Toyama, Japan.

A little more than a decade ago, Clariant decided to venture into the field of partial oxidation reactions. Catalysts for selective partial oxidation reactions thus belong to the most recent developments in the catalyst portfolio of Clariant. The portfolio of oxidation catalysts currently comprises SulfoMax® for sulfuric acid production, FAMAX® for formaldehyde production, PHTHALIMAX® for phthalic anhydride production, SynDane® for maleic anhydride production, and most recently VAM$^2$ax® for vinyl acetate production.

The Department of Oxidation Catalysis belongs to the R&D Center located at Heufeld, Germany. Its mission is to direct customer needs, sales strategies and technical production requirements into highly efficient catalyst development. To this end the Department of Oxidation Catalysis is structured into task force units focusing on inorganic synthesis, catalyst formation, high throughput and bench-scale testing, pilot testing and kinetics, and process simulations. Thus, all aspects of catalyst research are covered by the department, such as design of active mass, shaping and forming, and the influence of reaction conditions, i.e. customer's plant conditions as well as long-term catalyst behavior.

## 12.2. Research in Oxidation Catalysis

An example of the general research strategy in the department is the development of PHTHALIMAX®, currently the most selective catalyst for phthalic anhydride (PA) production, which will be described in this chapter. In 2003, Clariant decided to start the development of its own PA catalyst. One year later, in 2004, the most innovative production technology was developed and installed. Since the first customer at the end of 2004, Clariant could win more than 80 catalyst installations globally.

Nowadays the accepted research strategy in heterogeneous catalysis follows a pyramidal hierarchy of four development scales.[1] High throughput experimentation, with hundreds of samples in the mg range tested in rather short time periods, is often the basis of the identification of new leads or new promoters. Clariant has dedicated a specialized laboratory to this important task of lead identification and promoter

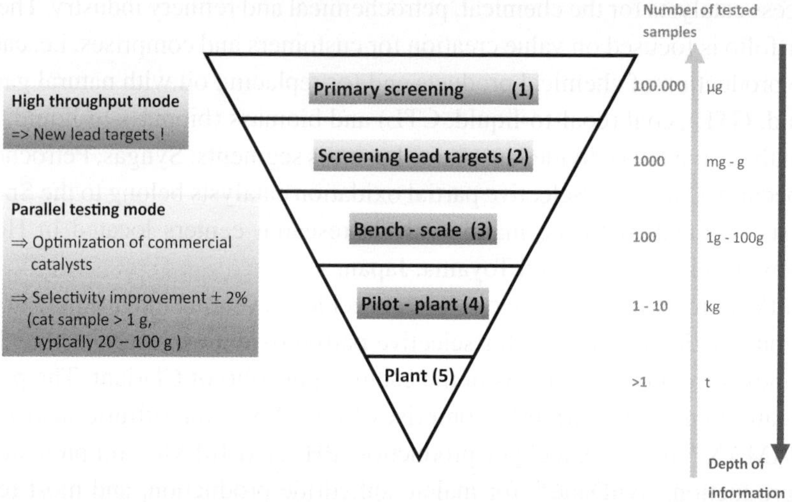


**Figure 12.1.** Schematic drawing of the research and development strategy realized in Clariant on five levels of continuous catalyst improvement.

optimization. This dedicated laboratory is fully equipped with state-of-the-art high throughput equipment, such as parallel synthesis robots for impregnation, parallel autoclaves, evaporation/drying stations, high throughput calcination ovens, multi-channel gas-phase or liquid-phase reactors, etc.; 48-channel, 12- or 6-channel reactor units, for catalyst testing at the $\mu$g-primary and mg-secondary scale, are available for catalytic gas-phase reactions of all kinds, for instance, selective partial oxidations. Promoter screening for oxidation catalysts is very efficiently conducted at this dedicated laboratory (Fig. 12.1).

Parallel bench-scale testing in the range of a few tenths of a gram of already formed catalyst shapes follows at the next level of catalyst development. In this way the R&D of catalyst formulation is not decoupled from the optimization of catalyst shapes together with developing the optimized pore structure, which is often crucial especially for selective partial oxidation.

In order to deduce fundamental information on intrinsic catalyst performance it is important to reduce the influence of the chosen reactor set-up on catalyst performance to a minimum. The first reactor requirement is ideal isothermal operation conditions.[2] The second requirement is continuously operated ideal plug flow without axial backmixing,[3] this being identical to a series of infinitesimally small, continuously stirred tank reactors each fulfilling the stationary concentration requirement. The realization of such an optimum reactor concept is not trivial, and in 1969 Temkin and Kul'kova[4] developed a concept in which actual-size catalyst bodies could be tested under ideal conditions. Catalyst spheres and inert cylinders are alternately placed in a tube with a diameter slightly bigger than the catalyst spheres. Inert cylinders and catalyst spheres are fixed by three wires. Excellent heat transport

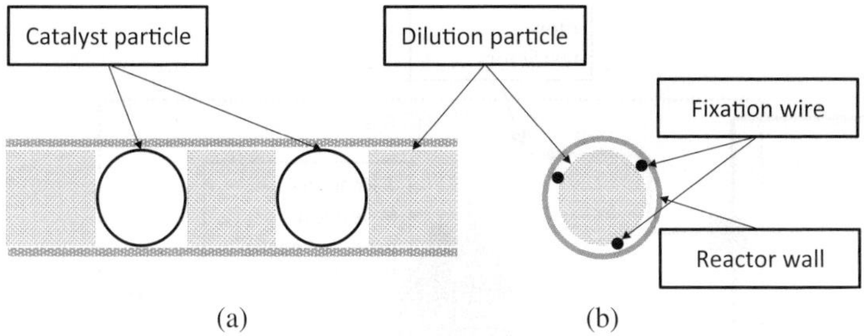


**Figure 12.2.** Schematic drawing of the original reactor concept of Temkin and Kul'kova[4]. a) side view of reactor, b) top view along reactor axis.

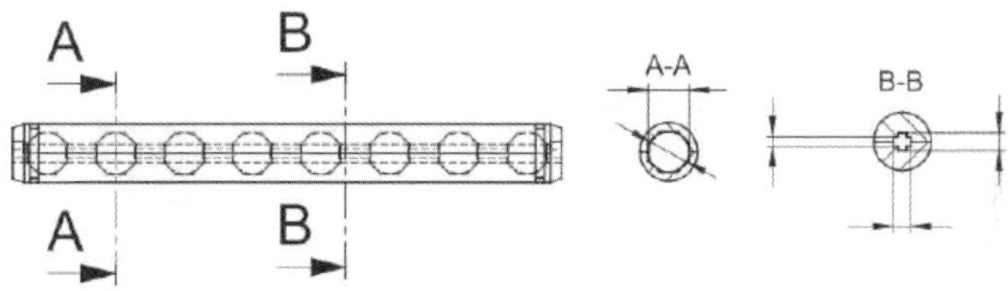


**Figure 12.3.** Schematic drawing of Temkin modules according to Clariant design.[5]

is guaranteed by the tiny slit between the inert spheres and reactor walls. Temkin and Kul'kova considered the space around each catalyst sphere as a continuously stirred tank reactor and could show that the overall reactor behavior only deviates by about 1% from an ideal serial connection of 30 continuously stirred tank reactors (CSTRs) and a targeted conversion of 50 mol % (Fig. 12.2).

This concept was more or less unappreciated for 40 years. Clariant has recently improved the fundamental Temkin concept and has extended it to a high throughput operation mode. The current design comprises reactor modules with eight spherical cavities which are connected via cross-shaped channels (Fig. 12.3).

The cavities can carry catalyst spheres, cylinders or rings with diameters of about 2–8 mm. The very narrow slit between the catalyst body and reactor wall leads to linear gas velocities of 0.2–3.2 m/s in the cavities, which is fully comparable with linear gas velocities in industrial reactors. The high gas flow in the connecting channels leads to optimum mixing in the following cavity. Four modules are charged per reactor of this seven-tube parallel unit.[5]

This improved reactor concept is characterized by the advantage that a comparably small number of technical catalyst bodies can be tested under the realistic contact time conditions of the respective industrial process. However, the considerably improved heat transfer properties of this reactor concept allow fully isothermal test conditions which cannot be achieved by conventional laboratory bench-scale units (high dilution is needed, which leads to bypassing).[5]

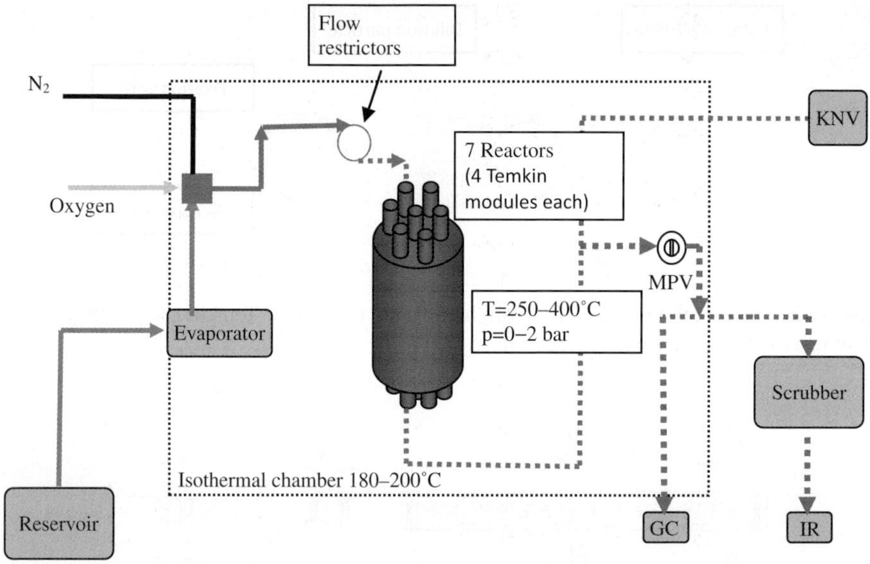


**Figure 12.4.** Schematic drawing of the high throughput Temkin unit.

Another advantage of this reactor lies in its minor axial backmixing, due to the micromachined channels, combined with optimum mixing in each cavity. The Bodenstein number (Bo) of this reactor was determined to be >120 under reaction conditions, thus characterizing a very narrow residence time distribution and hence ideal plug flow between the cavities (Fig. 12.4).

The next level of catalyst screening is then performed at the bench-scale level to determine the best catalyst shapes and active mass compositions for extended time periods. The bench-scale units used at Clariant are parallelized at 4 to 10 reactor tubes. At this level of testing, the reactor diameters typically used are fully comparable to industrial scale, only the reactor length is reduced to about 1.5 m. At this stage of testing, industrially-used salt bath cooling/heating is applied so as to remove the heat of reaction, similar to that found in an industrial operation (Fig. 12.5).

As in the Temkin unit at the previous development stage, the bench-scale units provide information on the intrinsic catalyst performance, i.e. activity constants, and conversion/selectivity relationships are determined. About 100 g of catalysts are typically charged into the reactors with dilution ratios between 1:5 and 1:15 to ensure a pseudo-isothermal operation. The units are equipped with fully automated moving thermocouples which allow the precise registration of the temperature profile and hence the average catalyst bed temperature along the reactor axis. Together with the fully automated on-line reaction gas analysis, a 24/7 operation is guaranteed. Typical test durations are in the range of 2 to 4 weeks depending on test

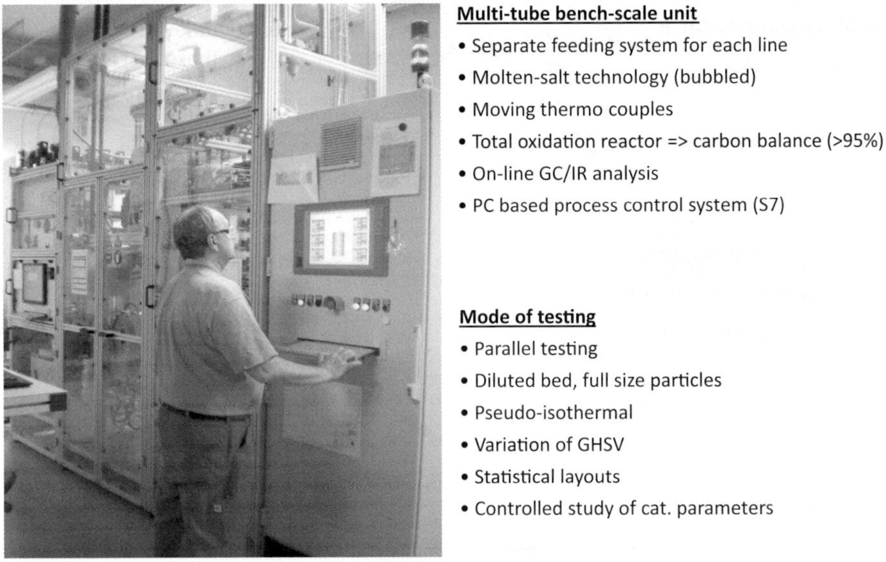


**Figure 12.5.** Multi-tube bench-scale unit.

protocol details, i.e. whether a fast deactivation stress test is combined with a standard routine.

Statistical layouts, e.g. response surface designs, are standard at this level of testing to fine tune all parameters for optimum catalyst performance, i.e. active mass compositions, promoter levels, textural properties, etc. As an example, Fig. 12.6 shows the response surface, i.e. catalyst activity constant as a function of two decisive parameters which were determined in a six-parameter at a time optimization program.

Data obtained in such a way for over 400 catalyst formulations has initially been used to train artificial neural networks, which in turn allow the prediction of catalyst performance of yet unprepared catalyst compositions. This approach is nowadays a standard tool when catalysts have to be designed for the special process conditions of a given customer. This broad database combined with artificial neural networks allows Clariant to customize its PHTHALIMAX® catalyst for phthalic anhydride production for each customer process.

The catalyst formulations with the optimum activity, identified as being most selective for given industrial conditions, are typically tested at the next level of R&D in pilot-scale reactors closely mimicking industrial plant conditions. Pilot tests are mandatory for further optimization of long-term behavior, and in addition reveal minor details of catalyst performance which escape bench-scale testing, i.e. the formation of minor by-products. Pilot reactor tests are performed under industrial operation conditions, i.e. gas hourly space velocities (GHSV) used in

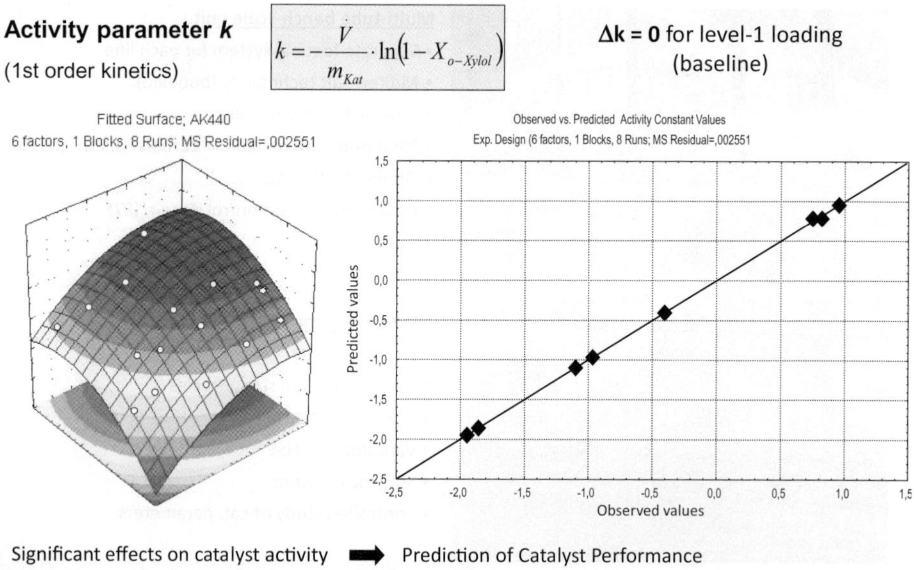


**Figure 12.6.** Response surface analysis of a six-parameter-at-a-time optimization program, left side: calculated response surface, white dots show experimental points, right side parity plot of observed vs. predicted catalyst activity levels.

customer plants, *o*-xylene loadings, air flow rates and plant pressure levels (reactor head pressure as well as pressure drop). Such tests provide important information for customers, such as the development of hot spots as a function of the operation conditions and the shut down as well as restart behavior of the catalyst. Moreover "soft" facts about the catalysts are determined which are also important for customers, such as dusting during loading and discharging, which would be hazardous for the loading team and hence must be avoided whenever possible. The completely dust-free PHTHALIMAX®, for example, can easily be loaded using the Clariant proprietary loading technology.

Clariant has installed a series of pilot units at the Department of Oxidation Catalysis, each equipped with reactor tubes of the different, industrially-realized diameters. The pilot reactors have been designed and built by DWE, Germany, one of the most experienced reactor construction companies in operation. The units are fully controlled by a Siemens S7 DCS system and equipped with mobile or multipoint thermocouples and on-line gas analysis allowing a 24/7 operation, comparable to an industrial plant. Each reactor unit is connected to twin-switch condensers for product separation, mimicking an industrial environment. The oxidation of *o*-xylene is industrially processed in fixed-bed tubular reactors with up to 30,000 tubes. Reaction temperatures range from 300 to 450°C with *o*-xylene feed concentrations between 0.5 vol% and 1.8 vol% in air. The reaction is conducted at nearly

**Figure 12.7.** Photograph showing Clariant's pilot reactor units for phthalic anhydride production.

atmospheric pressure and the cooling temperature, typically adjusted by a molten saltbath, ranges from 340°C to 390°C.[6,7] PA yields, from on-line reaction gas as well as from off-line crude PA analysis, and the by-product made, which is important for product quality, are precisely determined under customer conditions. With this information, catalysts for selective partial oxidation reactions can be further fine-tuned and optimized for the industrial demands (Fig. 12.7).

Because of the high economic relevance of the *o*-xylene conversion to phthalic anhydride, PA is a significant commodity in the chemical industry with an annual production of 4.5 million tons in 2005;[8] many attempts have been made in the past to determine the details of this reaction.[9–21] Prerequisite for the economic success story of PHTHALIMAX® was the fundamental understanding of the reaction network and kinetics, the optimization of active mass composition for each of the current four catalyst layers, as well as the tailoring of the management of the four layers which have to cooperate in an optimum way (Fig. 12.8).

To this end, one of the pilot units was designed as a sample port reactor equipped with 14 sample ports along the reactor axis and all ports being connected to the automated on-line reaction gas analysis. By using this sample port pilot plant, a detailed analysis can be made of conversion levels, intermediate and product concentrations as a function of operational conditions. Figure 12.9 shows an example of the *o*-xylene conversion with contact time as well as the concentrations determined for the main products.

The (by-)product formation could be studied in numerous experiments under varying operational conditions. Thus, temperature, reactor pressure, GHSV, air rate

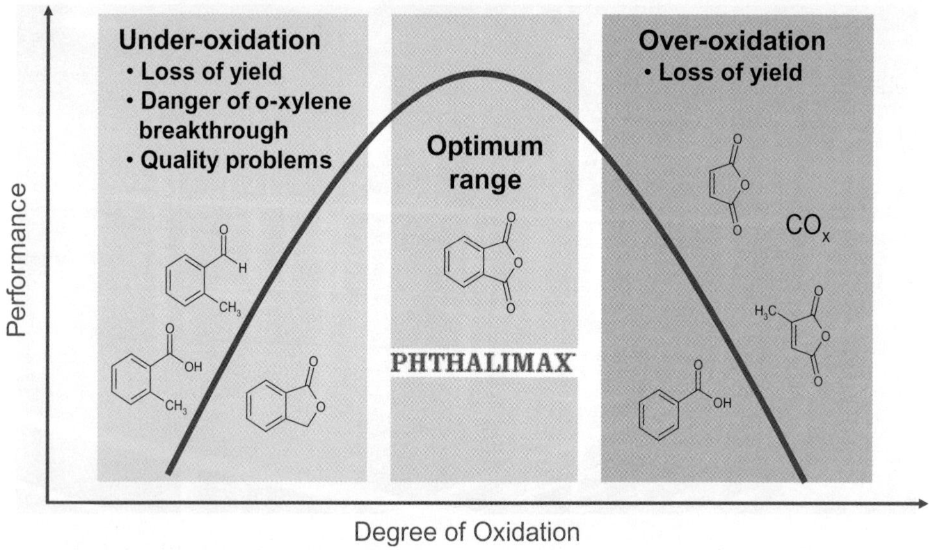


**Figure 12.8.** Schematic drawing of the window for optimum performance of a phthalic anhydride catalyst, left side: loss of yield and product quality issues due to under-oxidation by-products, right side: loss of yield and product quality issues due to over-oxidation by-products.

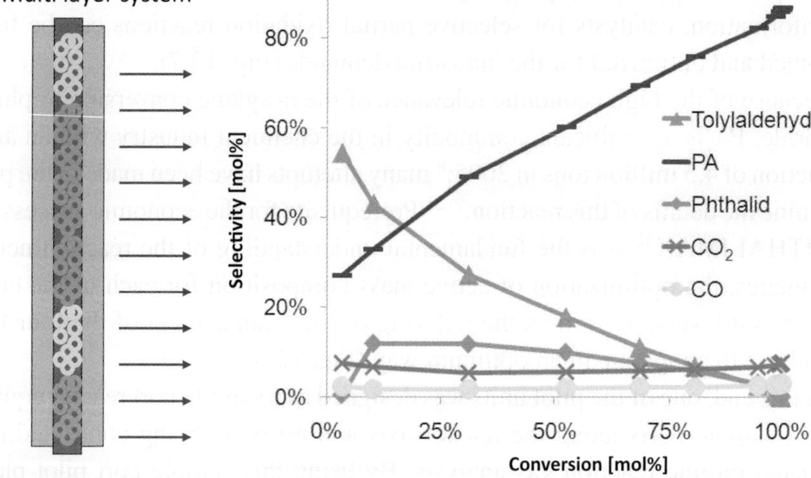


**Figure 12.9.** Left side: schematic drawing of the multi layer catalyst system as used in industrial processes, right side conversion-selectivity diagram as determined along the axis of the sample port pilot reactor.

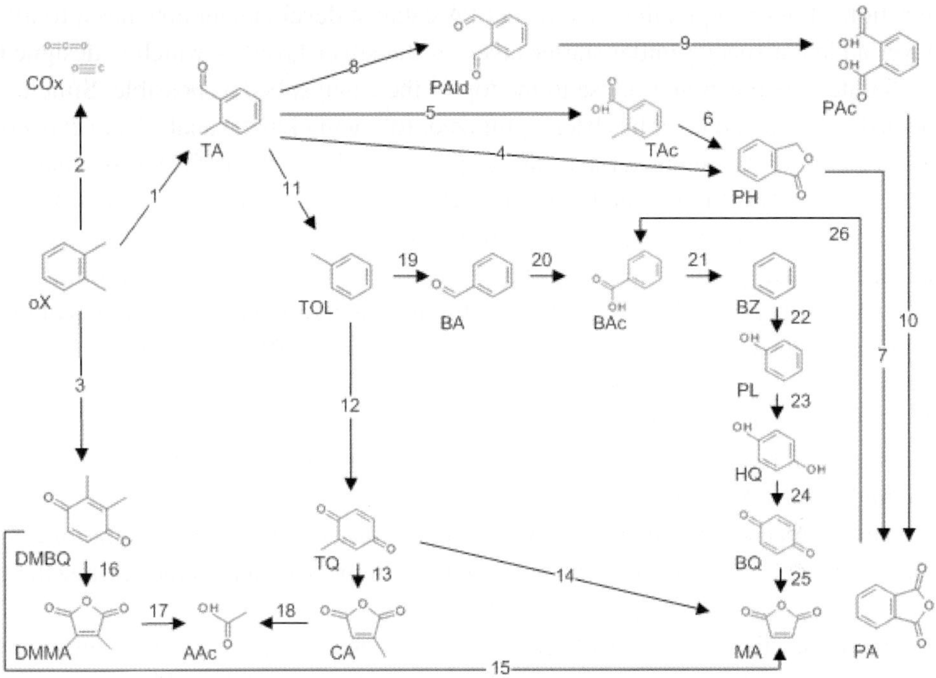


**Figure 12.10.** Scheme showing the improved reaction network as determined for the oxidation of *o*-xylene to phthalic anhydride.

and *o*-xylene loadings have been varied using statistical layouts for the determination of all intermediates and their reaction rates. Intermediates previously unknown in the literature were identified by gas chromatography combined with mass spectrometry (GC-MS) coupling and a new and improved reaction scheme for the catalytic oxidation of *o*-xylene could be proposed (Fig. 12.10).[22]

According to this new reaction network, the starting material is converted to the wanted product PA by three parallel routes: via phthalic aldehyde and phthalic acid (reactions 8, 9, 10), via tolylic acid and phthalide (reactions 5, 6, 7) and via phthalide (reactions 4 and 7). Phthalic anhydride when formed is very stable but it is converted in part via bencoic acid (reaction 26) over a rather complex reaction scheme to maleic anhydride (MA). Maleic anhydride is formed directly from *o*-xylene via tolylic aldehyde (reaction 1) and toluene (reaction 11) by two routes and via dimethylbenzochinone (DMBQ) (reaction 15). Toluene and DMBQ are converted over a series of reaction steps to acetic acid. The main by-products, CO and $CO_2$, are predominantly formed directly from *o*-xylene according to this mechanistic study.

Detailed kinetic studies using the sample pilot port allowed for catalyst and layer management optimization. The industrial phthalic anhydride catalyst system comprises at least three layers which have to be tuned to their task in the sequence of reactions. For example, the most recent PA catalyst development now has a fourth layer added, a special initial starter layer — the silver layer — which is designed to "ignite" the reaction as close to the top of the catalyst bed as possible. Simultaneously, the second layer has been optimized, following kinetic analysis, to convert as much of the starting material as fast as possible. This system of catalysts guarantees long catalyst lifetime and maximum PA yield. The third layer is necessary to selectively convert the remaining unconverted *o*-xylene to PA, while the fourth layer is necessary for product polishing, i.e. the conversion of unwanted by-products to $CO_x$. According to the kinetic studies it is highly important to convert as much of the *o*-xylene in the reactor as early as possible for maximum PA yield. Figure 12.11(a) shows the *o*-xylene conversion in a kinetic simulation and Fig. 12.11(b) shows the MA produced, it being the second main by-product after $CO_x$. As one can clearly see, the earlier *o*-xylene is converted in the reactor the less MA is produced, hence the higher the PA yield.

The detailed kinetics that has now been determined allows the simulations of customer plants. Operational parameter studies can help customers in the optimization of their operation with respect to market demands.

Moreover, the study of pilot reactor performance data is very important in the understanding of the long-term stability of an industrial catalyst. Plant operation often requires a catalyst lifetime of four years without losing too much PA yield during this period of time. PHTHALIMAX®, for example, loses about 0.5–1% PA yield per year. To study aging, catalysts are subjected to special stress tests on the bench-scale, as well as pilot units under high temperature and load conditions for

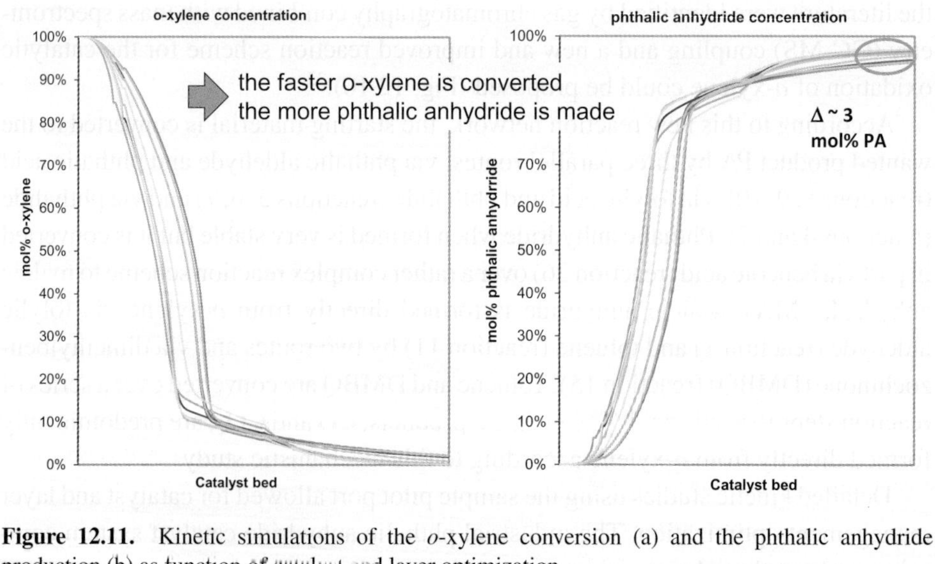


**Figure 12.11.** Kinetic simulations of the *o*-xylene conversion (a) and the phthalic anhydride production (b) as function of catalyst and layer optimization

longer time periods. Catalysts discharged after having exhibited a certain loss in performance during such stress tests are scrutinized using the modern characterization tool box of catalytic research.

Vanadia-titania catalysts have been the subject of research for a long time due to their outstanding industrial importance in (amm) oxidation reactions and for selective catalytic reduction of $NO_x$ (SCR-De$NO_x$), and there is an abundance of literature on this system.[23–38]

Among all physico-chemical characterization tools, Raman spectroscopy has frequently been used to characterize the vanadia-titania system.[39–50] Figure 12.12, for instance, shows a series of Raman spectra of the terminal metal-oxygen stretching vibration regime recorded for the pristine, freshly activated and equilibrated, as well as a high temperature-stressed catalyst from the hotspot zone in the reactor. The spectra have been normalized to the main band of $V_2O_5$ in this regime for better comparison, with the exception of the spectrum of the deactivated catalyst sample. While the pristine catalyst is characterized by Raman bands of $TiO_2$ and $V_2O_5$, bands which do not change much during the different calcination steps necessary for full catalyst activity, the Raman spectrum recorded for the catalyst under operation shows distinct differences! For one, the band assigned to the terminal V=O stretching vibration at

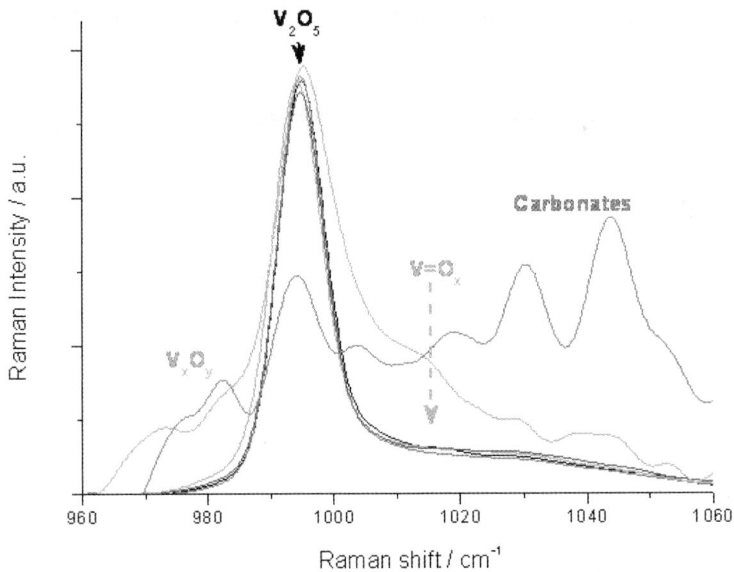


**Figure 12.12.** Recorded Raman spectra of the terminal metal-oxygen vibration regime. Spectra have been normalized to the intense band of $V_2O_5$ at 990 $cm^{-1}$ for better visual comparison with the exception of the spectrum of the deactivated catalyst. Color code: green and blue = pristine catalyst; black = after calcination at 300°C; red = after calcination at 410°C for 3 days; light blue = after standard operation for 3 months; pink = catalyst deactivated after high temperature stress test for 1 month.

995 $cm^{-1}$ has considerably broadened, indicating a loss in the crystallinity of $V_2O_5$, and this, in association with parallel X-ray diffraction studies (data not shown), revealed a complete loss of $V_2O_5$ reflections. Secondly, additional bands and shoulders are identified between 960 and 990 $cm^{-1}$, and between 1,000 and 1,020 $cm^{-1}$, hinting at a polymeric vanadium-oxygen surface species exhibiting terminal vanadyl groups, which might be related to the catalytically active surface phase.[28] Interestingly, Raman spectroscopy still shows the band assigned to crystalline $V_2O_5$, confirming once more the higher sensitivity of Raman for nanocrystalline phases as compared to X-ray diffraction (XRD). Hence, it can be concluded in full accordance with the literature[51,52] that the vanadia species has more or less completely spread over the titania support after activation in the catalytic reaction during the start-up.

The Raman spectrum recorded for the deactivated sample suffered a dramatic loss in the spectral intensity of nano-$V_2O_5$-related bands. Bands attributed to polymeric surface compounds seem to be less affected by the aging process but did also exhibit a loss of intensity. At the same time, new bands appear at still higher wavenumbers which can be assigned to surface carbonates due to carbonaceous deposits.[53,54] Hence, it can be concluded from the Raman spectra that the catalyst surface has at least been partly covered by surface coke.

X-ray photoelectron spectroscopy (XPS) has also been used frequently in the past to characterize vanada-titania catalysts.[55–58] Figure 12.13 displays the surface

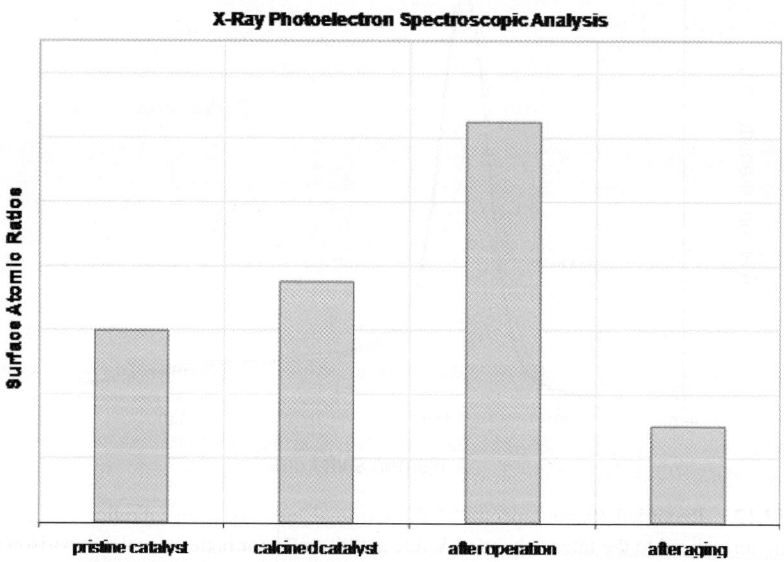


**Figure 12.13.** X-ray photoelectron spectroscopic analysis of the surface of V/Ti catalysts: atomic ratios of fresh, calcined, used and aged catalysts indicating a change in the surface concentration of the active component.

atomic ratios of V and Ti as determined by XPS for the differently treated catalyst surfaces. It can be seen from this quantitative data analysis that the vanadium dispersion increases from the fresh catalysts, over the calcined to the one under operation, indicating increased coverage of the $TiO_2$ surface by vanadia species, presumably the polymeric vanadium oxide species identified by Raman spectroscopy, which is formed during the solid-solid wetting process.[51,52] However, aging under the thermal stress test leads to a considerable reduction of the V/Ti ratio indicating that an aged catalyst exposes less catalytically active vanadium surface but more of the unreactive $TiO_2$. This loss of active vanadium surface can be attributed by a comparison with a quantitative bulk chemical analysis to either the diffusion of vanadium into $TiO_2$, as often described in the literature, or to agglomeration and crystallization into bigger $V_2O_{5-x}$ particles. A loss of active species by evaporation can be ruled out as the total vanadium concentration remained unchanged. A loss of vanadium species into the bulk of $TiO_2$ seems to be corroborated by XRD (data not shown) revealing partial rutilization[59,60] with lattice parameters different from pure rutile, which can be simulated assuming the well-described incorporation of $V^{3+/4+}$ ions into the $TiO_2$ lattice.[61]

High resolution transmission electron microscopy (TEM) was further used to confirm the results described above. Figure 12.14 shows the TEM images of the catalyst after stable operation and after the aging stress test. While the catalyst after stable operation only shows the lattice fringes of anatase covered by a layer of amorphous vanadium oxide surface species, the stressed catalyst shows a more complex structure. The core of the support oxide particle still shows the lattice fringes of anatase, but lattice distances of rutile can be seen in a surface-near layer around

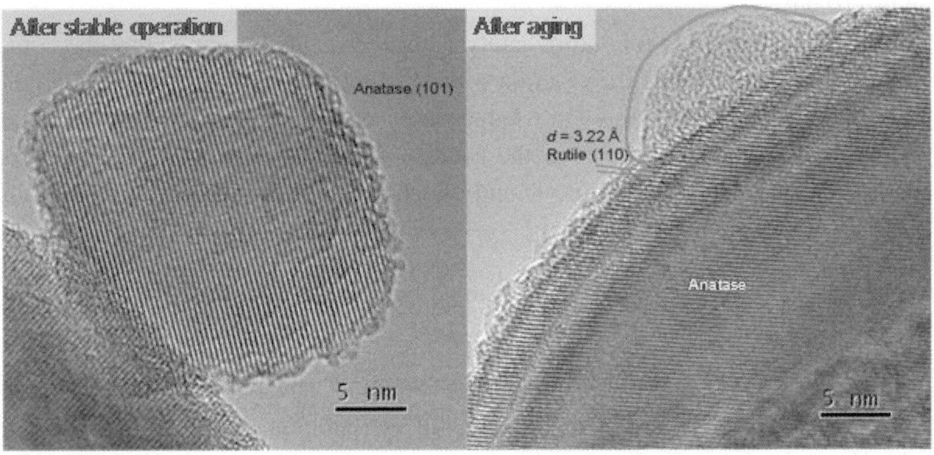


**Figure 12.14.** High resolution TEM images of a catalyst after operation (left image) and a catalyst after thermal aging (right image).

the anatase core. The catalyst particle is still covered by the amorphous vanadium oxide surface species, but a bigger particle of amorphous material has formed on the catalyst surface. Spatially resolved energy dispersive X-ray analysis (EDX) (data not shown) reveals that this particle is formed from carbonaceous deposits thus confirming the Raman results above.

This information helped the developers at the Clariant R&D laboratories to improve catalyst resistance against thermal stress.

Recently the PHTHALIMAX® family has been extended by a special catalyst for adiabatic post-reactor application, PHTHALIMAX®-F. This catalyst is designed to operate at rather low temperatures and with a high selectivity to convert under oxidation intermediates to PA and over-oxidation intermediates to $CO_x$ without combusting PA. One of the pilot units is equipped with an adiabatic post-reactor which is used for the development of this class of catalysts. The use of an adiabatic post-reactor with the correct operation of the Clariant post-catalyst can increase the overall plant PA yield by 1 wt%.

A new addition to Clariant's PA catalyst portfolio is the NAPHTHALIMAX® catalyst series, which has been especially developed for mixed *o*-xylene/naphthalene as well as 100% naphthalene feed streams. For the development of this catalyst type, a bench-scale unit as well as a pilot reactor had to be built which allowed the evaporation of naphthalene and its variable mixture with *o*-xylene streams at all mixing ratios. These two reactor units are not equipped with twin-switch condensers for safety reasons, but with a direct combustion unit for the product stream in order to reduce the potential hazard of exposure to the highly toxic naphtoquinone, a by-product of the mixed-feed process. This new family member shows very promising performance results at the pilot scale, with reaction gas yields up to 110 wt%.

Having set the basis for future catalyst improvements by unraveling the complex reaction networks of the reaction of *o*-xylene to phthalic anhydride, Clariant will continue to broaden this knowledge based on a profound physico-chemical understanding of the processes at the catalyst surface during its life cycle. This enables Clariant to further improve and widen the applicability of this family of PHTHALIMAX® catalysts.

## Acknowledgements

Prof. Dr-Ing. Th. Turek, Technical University Clausthal-Zellerfeld for the many fruitful discussions in our successful cooperation, Prof. Dr R. Schlögl, Fritz-Haber-Institute, Berlin, for the thorough catalyst characterizations, and the project managers, Dr Ch. Gückel, Dr H.-J. Wölk, R. Marx, Ch. Bäumler, and F. Schulz and their teams.

## References


1. Hagemeyer, A., Strasser P. and Volpe Jr, T. (2004). *High Throughput Screening in Heterogeneous Catalysis*, Wiley-VCH, Weinheim.
2. Wang, Y., Zheng, W., Chen, F., *et al.* (2008). Mechanistic Study of Propylene Oxidation over a Bi-molybdate Catalyst by in situ DRIFTS and Probe Molecules, *Appl. Catal. A: Gen.*, **351**, pp. 75–81.
3. Hagen, J. (2004). *Chemiereaktoren – Auslegung und Simulation*, Wiley-VCH, Weinheim.
4. Temkin, M. and Kul'kova, N. (1969). Laboratory Reactor with Ideal Displacement, *Kinet. Katal.*, **10**, pp. 461–463.
5. Hagemeyer, A., Mestl, G., Claus, P., *et al.* (2009). Temkin Screening Reaktor, *German Patent DE 20 2009 003* 014.8.r
6. Gütlhuber, F. (2002). Reactor Arrangement for Carrying Out Catalytic Gas Phase Reactions, Especially to Obtain Phthalic Anhydride, *WO2003022418, issued in Germany.*
7. Takada, M., Uhara, H. and Sato, T. (1978). Verfahren zur katalytischen Dampfphasenoxidation mit entsprechendem Reaktor, *German Patent DE2830765.*
8. Arpe, H. (2003). *Industrial Organic Chemistry*, 6th Edition, Wiley-VCH, Weinheim.
9. Bernardini, F. and Ramacci, M. (1966). Oxidation Mechanism of *o*-Xylene to Phthalic Anhydride, *Chim. Ind.*, **48**, pp. 9–17.
10. Herten, J. and Froment, G. (1968). Kinetics and Product Distribution in the Oxidation of *o*-Xylene on a Vanadium Pentoxide Catalyst, *Ind. Eng. Chem., Process Des. Develop.*, **7**, pp. 516–526.
11. Vanhove, D. and Blanchard, M. (1971). Catalytic Oxidation of *o*-Xylene on Catalysts of the Vanadium(V) Oxide-titanium (IV) Oxide Series, *Bull. Soc. Chim. Fr.*, **9**, pp. 3291–3295.
12. Saleh, R. and Wachs, I. (1987). Reaction Network and Kinetics of *o*-Xylene Oxidation to Phthalic Anhydride over $V_2O_5/TiO_2$ (Anatase) Catalysts, *Appl. Catal. A: Gen.*, **31**, pp. 87–98.
13. Ballarini, N., Brentari, A., Cavani, F., *et al.* (2009). A Revision of the Mechanism of *o*-Xylene Oxidation to Phthalic Anhydride with V/Ti/O Catalysts, and the Role of the Promoter Cs, *Catal. Today*, **142**, pp. 181–184.
14. Dias, C., Portela, M. and Bond, G. (1996). Oxidation of *o*-Xylene to Phthalic Anhydride over $V_2O_5/TiO_2$ Catalysts. Part 4. Mathematical Modelling Study and Analysis of the Reaction Network, *J. Catal.*, **164**, pp. 276–287.
15. Dias, C., Portela, M. and Bond, G. (1997). Synthesis of Phthalic Anhydride: Catalysts, Kinetics, and Reaction Modeling, *Catal. Rev. Sci. Eng.*, **39**, pp. 169–207.
16. Nikolov, V., Klissurski, D. and Anastasov, A. (1991). Phthalic Anhydride from *o*-Xylene Catalysis: Science and Engineering, *Catal. Rev. Sci. Eng.*, **33**, pp. 319–374.
17. Boag, I., Bacon, D. and Downie, J. (1975). Analysis of the Reaction Network for the Vanadia-catalyzed Oxidation of *o*-Xylene, *J. Catal.*, **38**, pp. 375–384.
18. Bernardini, F., Ramacci, M. and Paolacci, A. (1965). Analysis of Products of the Vapor Phase Oxidation of *o*-Xylene, *Chim. Ind.*, **47**, pp. 485–489.
19. Bernardini, F. and Ramacci, M. (1966). Calculation of Conversion Yield of *o*-Xylene to Phthalic Anhydride in an Industrial Reaction, *Chim. Ind.*, **48**, pp. 37–38.
20. Blanchard, M. and Vanhove, D. (1971). Vanadium Oxide-catalyzed Oxidation of Methyl Carbon-14 Labeled *o*-Xylene, *Bull. Soc. Chim. Fr.*, **11**, pp. 4134–4137.
21. Orozco, G., Gomez, J., Sanchez, O., *et al.* (2010). Effect of Kinetic Models on Hot Spot Temperature Prediction for Phthalic Anhydride Production in a Multitubular Packed Bed Reactor, *Canadian J. of Chem. Eng.*, **88**, pp. 224–231.
22. Marx, R., Wölk, H., Mestl, G., *et al.* (2011). Reaction Scheme of *o*-Xylene Oxidation on Vanadia Catalyst, *Appl. Catal. A: Gen.*, **398**, pp. 37–43.

23. Wainwright, M. and Hoffman, T. (1977). The Oxidation of *o*-Xylene on Vanadium Pentoxide Catalysts. I. Transient Kinetic Measurements, *Canadian J. Chem. Eng.*, **55**, pp. 552–556.
24. Bond, G. and Brückman, K. (1981). *Selectivity in Heterogeneous Catalysis*, Faraday Discussion No. 72, Royal Society of Chemistry, London.
25. Inomata, M., Mori, K., Miyamoto, A., *et al.* (1983). Structures of Supported Vanadium Oxide Catalysts. 1. Vanadium(V) Oxide/Titanium Dioxide (Anatase), Vanadium(V) Oxide/Titanium Dioxide (Rutile), and Vanadium(V) Oxide/Titanium Dioxide (Mixture of Anatase with Rutile), *J. Phys. Chem.*, **87**, pp. 754–761.
26. Gasior, M., Haber, J. and Machej, T. (1987). Evolution of $V_2O_5$-$TiO_2$ Catalysts in the Course of the Catalytic Reaction, *Appl., Catal.*, **33**, pp. 1–14.
27. Deo, G., Wachs, I. and Haber, J. (1994). Physical and Chemical Characterization of Surface Vanadium Oxide Supported on Titania: Influence of the Titania Phase (Anatase, Rutile, Brookite), *J. Crit. Rev. Surf. Chem.*, **4**, pp. 27–42.
28. Wachs, I. and Weckhuysen, B. (1997). Structure and Reactivity of Surface Vanadium Oxide Species on Oxide Supports, *Appl. Catal. A: Gen.*, **157**, pp. 67–90.
29. Bond, G., Sarkany, A. and Parfitt, G. (1979). The Vanadium Pentoxide-titanium Dioxide System. Structural Investigation and Activity for the Oxidation of Butadiene, *J. Catal.*, **57**, pp. 476–493.
30. Dias, C., Portela, M. and Bond, G. (1995). Oxidation of *o*-Xylene to Phthalic Anhydride over $V_2O_5/TiO_2$ I. Influence of Catalyst Composition, Preparation Method and Operating Conditions on Conversion and Product Selectivities, *J. Catal.*, **157**, pp. 344–352.
31. Dias, C., Portela, M. and Bond, G. (1995). Oxidation of *o*-Xylene to Phthalic Anhydride over $V_2O_5/TiO_2$ Catalysts II. Transient Catalytic Behaviour, *J. Catal.*, **57**, pp. 353–358.
32. Bond, G. (1997). Preparation and Properties of Vanadia/Titania Monolayer Catalysts, *Appl. Catal. A: Gen.*, **157**, pp. 91–103.
33. Wachs, I., Chan, S. and Saleh, R. (1985). The Interaction of $V_2O_5$ with $TiO_2$(Anatase) : II. Comparison of Fresh and Used Catalysts for *o*-Xylene Oxidation to Phthalic Anhydride, *J. Catal.*, **91**, pp. 366–369.
34. Kijenski, J., Baiker, A., Glinski, M., *et al.* (1986). Monolayers and Double Layers of Vanadium Pentoxide on Different Carriers: Preparation, Characterization and Catalytic Activities, *J. Catal.*, **101**, pp. 1–11.
35. Bond, G., Tahir, S. and Flamerz, S. (1991). Vanadium Oxide Monolayer Catalysts: Preparation, Characterization and Catalytic Activity, *Appl. Catal.*, **71**, pp. 1–31.
36. van Hengstum, A., Pranger, J., van Hengstum-Nijhuis, S., *et al.* (1986). Infrared Study of the Selective Oxidation of Toluene and o-Xylene on Vanadium Oxide/$TiO_2$, *J. Catal.*, **101**, pp. 323–330.
37. Deo, G. and Wachs, I. (1994). Effect of Additives on the Structure and Reactivity of the Surface Vanadium Oxide Phase in $V_2O_5/TiO_2$ Catalysts, *J. Catal.*, **146**, pp. 335–345.
38. Chary, K., Kishan, G., Bhaskar, T., *et al.* (1998). Structure and Reactivity of Vanadium Oxide Catalysts Supported on Anatase $TiO_2$, *J. Phys. Chem. B*, **102**, pp. 6792–6798.
39. Saleh, R., Wachs, I., Chan, S., *et al.* (1986). The Interaction of $V_2O_5$ with $Ti0_2$(anatase): Catalyst Evolution with Calcination Temperature and *o*-Xylene Oxidation, *J. Catal.*, **98**, pp. 102–114.
40. Chan, S., Wachs, I., Murrell, L., *et al.* (1984). In Situ Laser Raman Spectroscopy of Supported Metal Oxides, *J. Phys. Chem.*, **88**, pp. 5831–5835.
41. Went, G., Leu, L. and Bell, A. (1992). Quantitative Structural Analysis of Dispersed Vanadia Species in $TiO_2$(anatase)-supported Vanadium Pentoxide, *J. Catal.*, **134**, pp. 479–491.
42. Went, G., Oyama, S. and Bell, A. (1990). Laser Raman Spectroscopy of Supported Vanadium Oxide Catalysts, *J. Phys. Chem.*, **94**, pp. 4240–4246.
43. Centi, G. (1996). Nature of Active Layer in Vanadium Oxide Supported on Titanium Oxide and Control of its Reactivity in the Selective Oxidation and Ammoxidation of Alkylaromatics, *Appl. Catal. A: Gen.*, **147**, pp. 267–298.

44. Brückner, A. (2003). Monitoring Transition Metal Ions (TMI) in Oxide Catalysts during (Re)action: The Power of Operando EPR, *Phys. Chem. Chem. Phys.*, **5**, pp. 4461–4472.
45. Ramis, G., Busca, G., Bregani, F., *et al.* (1990). Fourier Transform-infrared Study of the Adsorption and Coadsorption of Nitric Oxide, Nitrogen Dioxide and Ammonia on Vanadia-Titania and Mechanism of Selective Catalytic Reduction., *Appl. Catal. B: Environ.*, **64**, pp. 259–278.
46. Grzybowska-Swierkosz, B. (1007). Vanadia-titania Catalysts for Oxidation of *o*-Xylene and Other Hydrocarbons, *Appl. Catal. A: Gen.*, **157**, pp. 263–310.
47. Topsoe, N. (1994). Mechanism of the Selective Catalytic Reduction of Nitric Oxide by Ammonia Elucidated by In Situ Online Fourier Transform Infrared Spectroscopy, *Science*, **265**, pp. 1217–2119.
48. Amiridis, M., Wachs, I., Deo, G., *et al.* (1996). Reactivity of $V_2O_5$ Catalysts for the Selective Catalytic Reduction of NO by $NH_3$: Influence of Vanadia Loading, $H_2O$, and $SO_2$, *J. Catal.*, **161**, pp. 247–253.
49. Si-Ahmed, H., Calatayud, M., Minot, C., *et al.* (2007). Combining Theoretical Description with Experimental in situ Studies on the Effect of Potassium on the Structure and Reactivity of Titania-supported Vanadium Oxide Catalyst, *Catal. Today*, **126**, pp. 96–102.
50. Bulushev, D., Reshetnikov, S., Kiwi-Minsker, L., *et al.* (2001). Deactivation Kinetics of V/Ti-Oxide in Toluene Partial Oxidation, *Appl. Catal. A: Gen.*, **220**, pp. 31–39.
51. Haber, J., Machej, T., Serwicka, E., *et al.* (1995). Mechanism of Surface Spreading in Vanadia-titania System, *Catal. Lett.*, **32**, pp. 101–114.
52. Knözinger, H. and Taglauer, E. (1993). Toward Supported Oxide Catalysts via Solid-solid Wetting, *Catal.*, **10**, pp. 1–40.
53. Bulushev, D., Kiwi-Minsker, L. and Renken, A. (2000). Vanadia/Titania Catalysts for Gas Phase Partial Toluene Oxidation. Spectroscopic Characterization and Transient Kinetics Study, *Catal. Today*, **57**, pp. 231–239.
54. Dias, C., Portela, M. and Bond, G. (1996). Oxidation of *o*-Xylene to Phthalic Anhydride over $V_2O_5/TiO_2$ Catalysts III. Study of Organic Residue Formed on the Catalyst Surface, *J. Catal.*, **162**, pp. 284–294.
55. Bond, G., Zurita, J. and Flamerz, S. (1986). Structure and Reactivity of Titania-supported Oxides. Part 2. Characterisation of Various Vanadium Oxide on Titania Catalysts by X-ray Photoelectron Spectroscopy, *Appl. Catal. B: Environ.*, **27**, pp. 353–362.
56. Nag, N. and Massoth, F. (1990). ESCA and Gravimetric Reduction Studies on Vanadium/Alumina and Vanadium/Silica Catalysts, *J. Catal.*, **124**, pp. 127–132.
57. Chiarello, G., Robba, D., De Michele, G., *et al.* (1993). An X-ray Photoelectron Spectroscopy Study of the Vanadia-titania Catalysts, *Appl. Surf. Sci.*, **64**, pp. 91–96.
58. Bond, G. and Flamerz, S. (1989). Structure and Reactivity of Titania-supported Oxides. IV. Characterization of Dried Vanadia/Titania Catalyst Precursors, *Appl. Catal. B: Environ.*, **46**, pp. 89–102.
59. Centi, G., Giamello, E., Pinelli, D., *et al.* (1991). Surface Structure and Reactivity of Vanadium-titanium Oxide Catalysts Prepared by Solid-state Reaction. 1. Formation of a Vanadium(IV) Interacting Layer, *J. Catal.*, **130**, pp. 220–237.
60. Centi, G., Pinelli, D., Trifiro, F., *et al.* (1991). Surface Structure and Reactivity of Vanadium-titanium Oxide Catalysts Prepared by Solid-state Reaction. 2. Nature of the Active Phase Formed During *o*-Xylene Oxidation, *J. Catal.*, **130**, pp. 238–256.
61. Grzybowska-Swierkosz, R. and Haber, J. (1984). *Vanadia Catalysts for Processes of Oxidation of Aromatic Hydrocarbons*, Polish Scientific Publishers, Warsaw.

# Chapter 13

# The Industrial Oxidation of KA Oil to Adipic Acid

Stefano ALINI* and Pierpaolo BABINI*

This chapter describes the main features of the industrial process for the synthesis of adipic acid by means of oxidation of the KA Oil mixture, with special focus on the reaction mechanism, reactor technology, safety aspects and materials. Aspects here examined include the various technical solutions which have been implemented during the last 60 years, with the aim of both improving the process performance and decreasing its environmental impact.

## 13.1. Introduction

Adipic acid is best known in the history of chemistry for its role in the invention of nylon 6,6 by W. H. Carothers. However, today, adipic acid is also an important link in the chemistry of intermediates because of its characteristics, price and market availability.

Although most of this product[1] (about 60%) is used in the manufacture of polyamide 6,6, which is formed by a reaction with hexamethylenediamine, the growth in the use of adipic acid during the last 20 years is also attributable to the expansion of polyurethane and polyester resins. After nylon, these product sectors, together with PVC plasticizers and synthetic lubricants, consume most of the "extra-nylon" adipic acid produced today.

The expansion of these sectors and the development of new markets have strongly contributed to the rise in the worldwide capacity to produce adipic acid, which in 2010 reached about 3.2 million tons. However, as we can see in Fig. 13.1, throughout the years, the geographical distribution of world capacity has undergone some changes. In fact, even though Europe and the United States still maintain their dominant position, in the last three years, the production capacity in China has recorded a huge increase. The newest adipic acid production plants in the world have been built in China by local manufacturers.

*Radici Chimica SpA, Via G. Fauser 50, 28100 Novara, Italy.

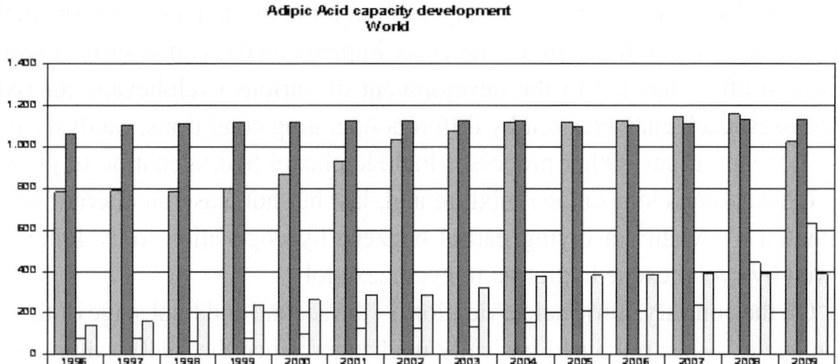


| | 1996 | 1997 | 1998 | 1999 | 2000 | 2001 | 2002 | 2003 | 2004 | 2005 | 2006 | 2007 | 2008 | 2009 |
|---|---|---|---|---|---|---|---|---|---|---|---|---|---|---|
| Europe | 781 | 791 | 786 | 801 | 871 | 931 | 1.036 | 1.074 | 1.119 | 1.121 | 1.126 | 1.146 | 1.159 | 1.024 |
| Americas | 1.065 | 1.105 | 1.112 | 1.119 | 1.120 | 1.125 | 1.125 | 1.125 | 1.125 | 1.100 | 1.105 | 1.110 | 1.130 | 1.130 |
| China | 73 | 73 | 64 | 79 | 101 | 126 | 126 | 134 | 154 | 214 | 214 | 235 | 442 | 630 |
| Asia s.à China | 140 | 160 | 204 | 233 | 266 | 286 | 286 | 321 | 371 | 379 | 384 | 389 | 389 | 389 |

**Figure 13.1.** Adipic acid production capacity in the world.

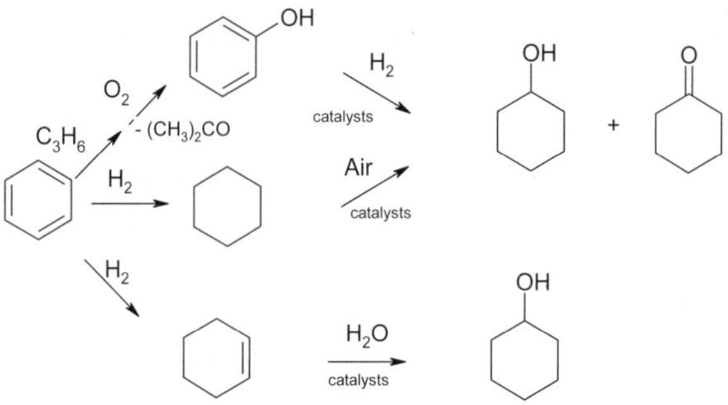


**Figure 13.2.** Summary of the main industrial processes for the production of KA oil or cyclohexanol starting from benzene.

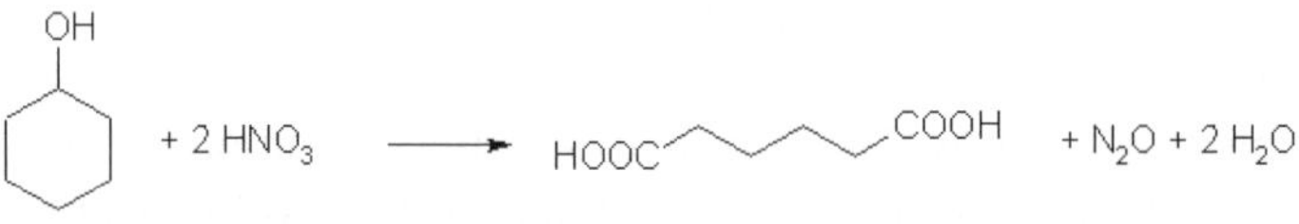


**Figure 13.3.** Oxidation of cyclohexanol into adipic acid.

The production cycle of adipic acid from hydrocarbons — benzene or cyclohexane — is a two-stage process: (i) the production of the raw material for the synthesis of adipic acid, consisting of cyclohexanol alone or, more often, a mixture of cyclohexanol and cyclohexanone (KA oil), and (ii) the production of adipic acid by nitric acid oxidation in the presence of copper and vanadium catalysts (see Figs. 13.2 and 13.3).

Most of the process research work has focused on the first stage, which ends with the production of the ketone-alcohol mixture. Much effort has gone into improving the overall yield of this reaction stage, through the introduction of radical changes in the chemistry of the various processes implemented on an industrial scale. This research effort has led to the development of various cyclohexane air oxidation processes, each characterized by different operating conditions, catalysts and catalyst concentrations. Other processes include phenol hydrogenation to produce the cyclohexanol/cyclohexanone mixture and, last but not least, an alternative process patented by Asahi[2] involving partial benzene hydrogenation to cyclohexene, followed by cyclohexene hydration to cyclohexanol.

On the contrary, nitric acid oxidation — the second and final stage of the production cycle — has remained basically unchanged since the 1940s, although claims have been made in scientific and patent literature that different operating conditions — reaction temperature, type of catalyst and catalyst concentration — could yield economic advantages by improving the selectivity or decreasing the consumption of nitric acid. Thus, we can assume that the various manufacturers have concentrated their efforts on improving the technology for the second step of the adipic acid synthesis process by identifying technical, and at times exclusive, solutions aimed at achieving objectives that were the priorities at particular times throughout the years: the safety of the production process, the quality of the final product, direct operating costs — particularly of energy — and investments, the need for increased production capacity, the reuse of by-products, production process automation and lastly, but only in terms of chronological order, the environmental footprint of the chemical process and the production plant.

Therefore, although there are no important new developments to report from the point of view of the chemical reaction, we have decided to dedicate this chapter to a thorough study of the more relevant aspects of the production process of this important intermediate.

The technological evolution occurring over the last 60 years of industrial history has also had an impact on chemical processes in general, and on the adipic acid production process in particular. This analysis offers us an opportunity to retrace over time the many difficult issues encountered in the development of a complex chemical process, such as the one at hand, and allows us to appreciate the importance of the history of the contributions made and the motivations underlying the changes introduced. Learning about this history will help us to avoid repeating the same mistakes and to move forward on a trend of constant improvement.

## 13.2. Nitric Acid Oxidation of a Cyclohexanol/Cyclohexanone Mixture to Produce Adipic Acid

Adipic acid (AA) is produced almost exclusively by nitric acid oxidation of a cyclohexanol/cyclohexanone[3] mixture in the presence of copper and vanadium catalysts in a homogeneous phase with the reaction mixture.

Despite the relatively simple stoichiometry, which can be represented theoretically by Eqs. 13.1 and 13.2 below, the reaction mechanism is much more complex. In fact, Eqs. 13.1 and 13.2 imply, but do not reveal, two different reaction pathways for the formation of adipic acid, one from cyclohexanol and the other from cyclohexanone, each consuming a different quantity of nitric acid, which is reduced to nitrous oxide.

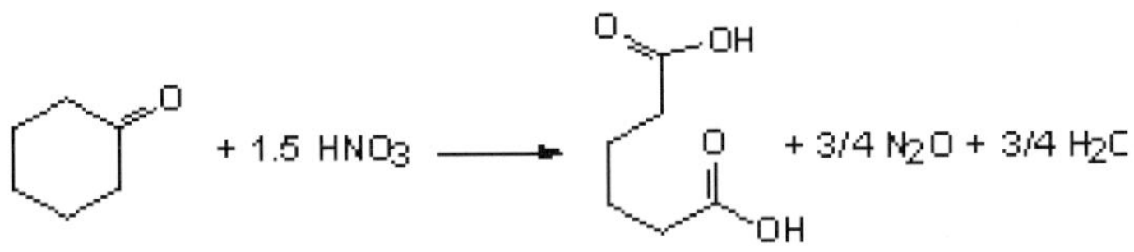


(13.1)

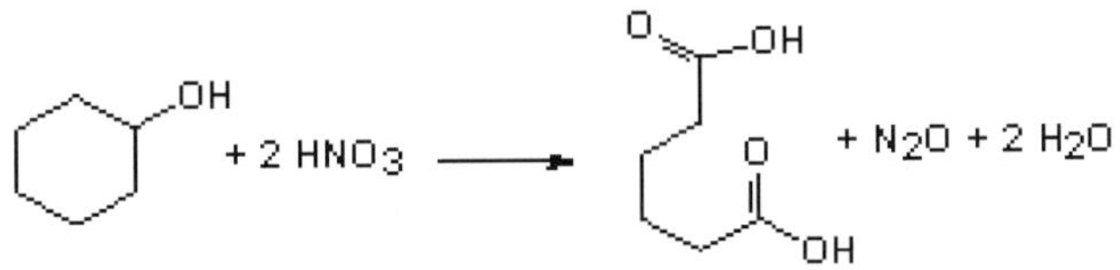


(13.2)

The nitric acid oxidation initiation mechanism is different for cyclohexanol and cyclohexanone: the reaction proceeds faster for cyclohexanol, which easily converts to cyclohexanone even at low temperatures, and proceeds at a slower rate for cyclohexanone. Used in pure form, cyclohexanone would require much more extreme conditions (higher temperatures and higher nitric acid concentrations) than those routinely used for the synthesis process in an industrial setting. Even when cyclohexanone is used in a mixture, it becomes necessary to create, chemically or physically, a nitrogen oxide-saturated environment, for example, by absorption of nitrous gasses in the solution fed into the reactor. The formation of nitrous acid via nitric acid reduction in the cyclohexanol to cyclohexanone oxidation and the consequent formation of higher nitrogen oxides ($NO + NO_2$), due to the instability of nitrous acid in an acid environment, ensure that cyclohexanone also becomes part of the reaction mechanism leading to the formation of adipic acid, both because it is part of the starting mixture and because it is produced in the first step of the cyclohexanol oxidation reaction.

The selectivity of the KA oil mixture oxidation reaction is strongly dependent on the complex activity of the nitric acid, which has acidifying, nitrating, nitrosating and oxidizing effects. These properties are, in turn, dependent on the process conditions (temperature, pressure and nitric acid concentration), as well as on the presence of catalysts and impurities in the reaction mass.[4] Good selectivity can be achieved by appropriate simultaneous modulation of all these properties and by choosing and maintaining the correct nitric acid concentration, and controlling reaction temperature and pressure. The great variety in the chemical properties of nitric acid makes this oxidizing agent unique. In fact, despite the amount of research conducted to find alternative synthesis routes that would avoid the use of nitric acid, as of today no competitive process has been found.[5] In order to appreciate the

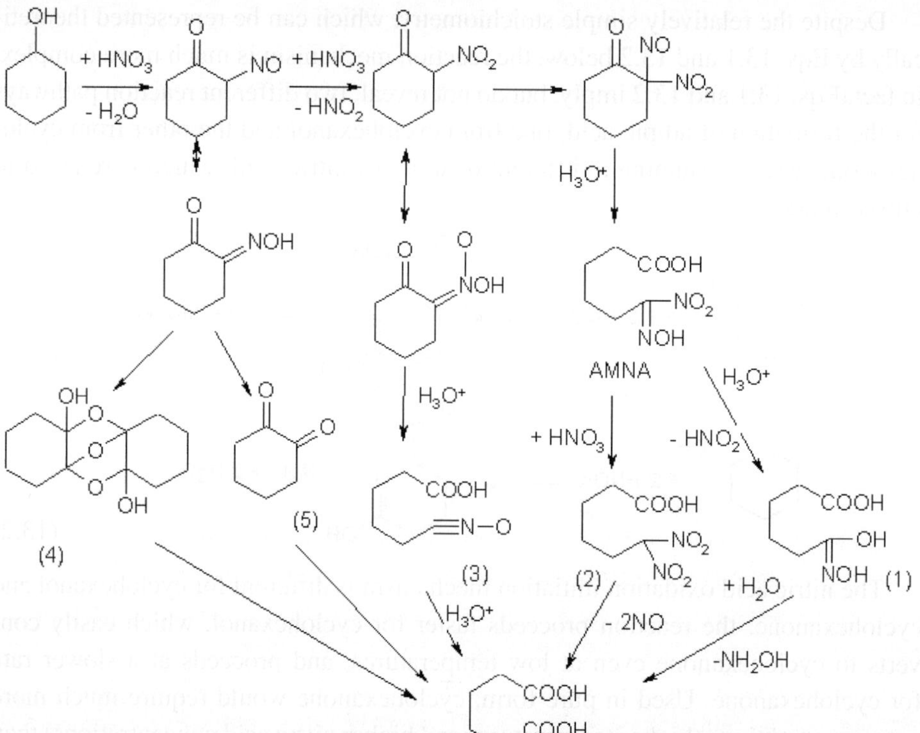


**Figure 13.4.** Main reactions and proposed routes involved in the cyclohexanol oxidation to adipic acid. (Elaborated from refs[4,5,7]).

truly unique properties of this reagent, we need to delve deeper into the reaction mechanism.

The reaction mechanism was first reported and discussed by van Asselt and van Krevelen[6] in 1963 and was further described in more depth in numerous papers[7] that continued to appear up to the beginning of the 1990s. This body of research generated a complete and detailed analysis of all the possible reactions that can occur depending on the related operating conditions.[4] The reaction pathway diagram in Fig. 13.4 shows at least five reaction paths leading to adipic acid.

The key intermediate in all the proposed mechanisms is 2-nitrosocyclohexanone, and thus the most significant $HNO_3$ property utilized is its nitrosating power. In fact, nitric acid acts as an oxidizing agent only in secondary reactions that occur only in the presence of previously nitrosated compounds.[8] Oxidation via the formation of adipomononitrolic acid (AMNA, paths 1 and 2) requires an oxidative step to 2-nitrocyclohexanone and a second nitrosating step to 2-nitroso 2-nitro cyclohexanone. This seems to be the favourite path in the presence of concentrated nitric acid solutions rich in higher nitrogen oxides (NOx), which increase the oxidizing power of the solution. In all industrial processes, the implemented operating conditions favour this synthesis path with increased selectivity to adipic acid, although it is also associated with greater nitric acid consumption and the formation of nitrous oxide. If the raw material is predominantly cyclohexanol, such as in the Asahi cyclohexene process, or if KA oil is produced from cyclohexanol via the boric acid process, this will be the predominant path to adipic acid and, all other conditions being equal, will produce adipic acid with high selectivity and with an almost stoichiometric formation of nitrous oxide (1 mol per mol of adipic acid). As the concentration of cyclohexanone in the KA oil mixture increases, a significant role is also played by other paths (2 and 3) that require lower nitric acid consumption — and thus involve less production of nitrous oxide — but lead to the increased formation of by-products (glutaric, succinic and oxalic acids), with a corresponding decrease in selectivity towards adipic acid.

Thus, the reaction path is strongly dependent on the operating conditions (concentration of Cu and V catalysts, their ratio, concentration of nitric acid, temperature of the oxidizing mother liquor, reactant ratio) and also on the composition of the KA oil mixture, in which the cyclohexanone concentration can vary from a few percentage points up to 50–60%, depending on the upstream process.

The reaction can be carried out in one or more reaction stages, each at a different temperature in a range from 60 to 100°C and with a residence time varying from 10 to 30 minutes per stage depending on the industrial process. In some cases, an additional stage for the completion of the reaction is added. For the reaction to occur, a large excess of nitric acid (at 50–60% concentration) is needed (the $HNO_3$: KA oil molar ratio must be no less than 7:1), as well as the presence of copper and vanadium catalysts.[9] Pressure can vary from 1 to 4 atm. Reaction selectivity is usually higher than 93–95%, depending on the KA oil mixture composition. The primary reaction by-products are glutaric acid (selectivity of 3–4%) and succinic acid (2–3%).

In accordance with the above reaction mechanism and under the simplifying assumption that cyclohexanol is the only raw material used, for each mol of alcohol, two moles of nitric acid are reduced to nitrous oxide, with a specific consumption of 0.863 kg $HNO_3$/kg AA and 100% selectivity. In practice, however, reaction selectivity is lower because of the formation of glutaric and succinic acids and $CO_2$, which raises the specific consumption above the stoichiometric value (for 95% selectivity, the consumption is 0.908 kg $HNO_3$/kg AA). This is the actual consumption, since the quantity of nitric acid used in the reaction is in reality greater. However, the excess nitric acid leads to the formation of higher nitrogen oxides (NO and $NO_2$), which can be recovered by absorption in water and converted back to nitric acid, which can then be recycled into the process.

The other routes to the formation of adipic acid shown in the reaction mechanism diagram are favoured by a more dilute solution of nitric acid and by lower reaction temperatures. These routes, however, require reaction conditions that increase process safety risk, since they lead to high concentrations of reaction intermediates. Furthermore, they could lead to an increase in by-product production. Thus, despite several attempts made in the past to change the operating conditions in order to favour other routes with the objective of decreasing the consumption of nitric acid (use of cyclohexanone-rich raw material, use of more dilute nitric acid at low reaction temperatures and slightly higher reaction times), no alternative conditions have been identified that would be sufficiently safe for a commercial production process.

## 13.3. Development of Reactors for Adipic Acid Synthesis

Additional factors that characterize the process are the high exothermicity of the reaction — a fact that one can appreciate by looking at the reaction enthalpy values (cyclohexanone to AA = $\Delta H^\circ - 172$ kcal/mole, cyclohexanol to AA = $\Delta H^\circ - 215$ kcal/mole) — the gases generated (the main gaseous compounds generated in the reaction, besides $N_2O$ as mentioned above, are $CO_2$, NO, $NO_2$ and $N_2$) and the operating conditions causing the presence of particularly corrosive fluids. These aspects have a decisive influence on the choice of technologies adopted and on the equipment design and engineering. By way of example, although we are dealing with a homogeneous catalysis reaction which requires no stirring once the reactants are well mixed, it is essential to keep the mass in turbulent regime in order to ensure that the heat exchange coefficients are adequate to compensate for the heat released in the reaction and to keep the temperature under control.

For the reasons just stated, it is clear that the development of continuous reactors for nitric acid oxidation to adipic acid has been a crucial aspect in the safety and economics of this process. The first reactors were relatively small-sized vessels equipped with an agitator, where the KA oil mixture was fed in near the agitator to ensure rapid dispersion of the mixture. The heat of the reaction was removed by means of external shell and tube heat exchangers. One can easily understand the difficulties in managing the process, mainly related to the frequent clogging of the coolers and to monitoring and controlling the temperature inside the reactors. When the crystallization of the adipic acid begins, the heat exchange inside the tubes decreases, making it necessary to increase the cooling water flow, which in turn worsens the thickening of the adipic acid layer and causes clogging. To this, we must add that there is very inefficient use of electric power in the case of agitators and recycling pumps. One of the first technological improvements was to move the heat exchange system inside the reactor itself, which made the heat exchange process more efficient. This new arrangement reduces the formation of encrustations on the heat exchanger surface and improves reaction temperature control. A deeper understanding of the reaction mechanism has allowed for optimizing the process operating conditions and designing more efficient and effective reactors that can take advantage of the strongly exothermic nature of the reaction to achieve benefits in terms of energy and safety.

KA oil conversion to 6-hydroxyimino-6-nitro hexanoic acid (adipomononitrolic acid, AMNA) proceeds very fast and is an exothermic process, while the conversion of AMNA to adipic acid is a relatively slow reaction and gives a small contribution to the total heat generated in the reaction. Thus, the latter process is in fact the rate determining step. In view of the high reaction rate, it is essential that the dispersion of the organic phase (KA oil) in the water reaction mass be quick, uniform and effective. The great difference in density and viscosity between KA oil and nitric acid makes the dispersion process quite difficult, and thus this aspect has to be researched with particular care in order to prevent safety problems. Furthermore, to make the situation worse, there is also a problem related to the evolution of the gas generated from the reaction mass, mostly nitrogen oxides (NOx and $N_2O$) and small quantities of carbon dioxide released in the decarboxylation of dicarboxylic acids. The solutions implemented in the various production plants are based on different engineering and technological approaches developed over the course of the years.

The need for increased production capacity has led to the design of plant engineering solutions that optimize investment costs and operating costs, particularly energy consumption costs. In order to avoid building plants with a large number of equipment units (reactors, heat exchangers, pumps, instruments) new reactors with high production capacity units have been developed.

In the late 1960s, Imperial Chemical Industries (ICI) studied a plug flow reactor;[10] an example consisting of a series of seven stages is shown in Fig. 13.5. All the nitric acid needed for the process is fed into the first stage, while KA oil is fed into each stage through internal distributors positioned right beneath the cooling pipe coil. The reactor does not require an agitator or circulating pumps to keep the reaction mass homogeneous. As a matter of fact, the arrangement was designed to take advantage of the reaction phenomenology itself. As said above, the reaction is characterized by a high rate of gas generation. The pressure inside a bubble submerged in a static fluid can build up so high as to cause violent vibrations that are capable of stirring the reaction mass. Observations have shown that by generating bubbles in the proximity of the nitric acid boundary layer, where the reaction occurs, it is possible to achieve very high heat transfer coefficients that allow for the removal of a large quantity of heat without causing the acid to crystallize. The reaction mass flows from stage to stage by overflow and the KA oil is fed into the lower part of the reactor.

As designed by ICI, the reaction can be conducted in multiple reactors in series with temperatures increasing from 70 to 100°C. In this arrangement, the nitric acid

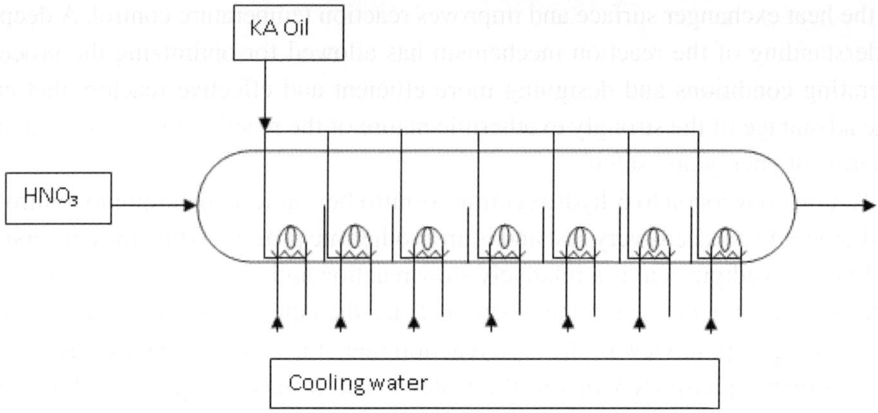


**Figure 13.5.** Ideal plug flow reactor for adipic acid synthesis.

and catalysts solution is fed only into the first reactor in the series, while the KA oil is fed into each reactor. The quantity of KA oil fed into each reactor is the amount needed to raise the temperature to the required value. Since the nitric acid is fed only into the first reactor, the nitric acid strength decreases progressively from reactor to reactor, allowing for improved total process yield.

BASF[11] opted for a single reactor with no mechanical stirring and no cooling system. The two reactants, nitric solution and cyclohexanol, are fed into the reactor in a ratio of approximately 200:1. In this arrangement, the exothermicity of the reaction is controlled by slightly increasing the temperature of the reaction mass. Stirring action is provided by the gas generated during the reaction and by the oxidizing fluid feeding rate, which is kept very high (60–70 $m^3$/h per $m^3$ of reactor volume). In this arrangement, reaction times are extremely low (1–3 minutes).

Regardless of the reactor design, an extremely important factor is the mixing of KA oil in the nitric solution. However, good mixing is very difficult to achieve because of the differences in the density and viscosity of the two reactants. Droplet feeding is the best method for achieving good dispersion in the reaction mass, an essential prerequisite to attain optimal reaction control and avoid leakage of partially reacted product into the downstream process stages, which were not designed to deal with the gas that would be generated. In this respect, the connecting pipes are the most critical elements, especially long horizontal stretches of pipe.

## 13.4. Safety Aspects

As described above, the main route to adipic acid involves the formation of AMNA. Laboratory studies have determined the heat of the reaction at each individual step.

The only intermediate that is isolable and quantitatively measurable using specific analytical methods is nitrolic acid, which under typical plant operating conditions has a concentration in the range of 1–2 g/l. The heat released in the decomposition of the nitrolic acid makes up about 10% of the total exothermicity of the entire reaction. The consecutive nature of the various steps involved leads to the consideration that an increase in the concentration of AMNA could lead to a significant increase in the potential energy of the whole system, thus creating a potentially dangerous condition. An in-depth knowledge of the reaction mechanisms and of the rules that govern such mechanisms is the key to setting safety blocks for the reaction parameters that would keep the system from getting out of control.

Under normal plant operating conditions, heat release is determined by the cyclohexanol/cyclohexanone mix flow, which is instrument-controlled in order to keep the correct mix-to-nitric acid ratio. By cutting off the KA oil feed, the reaction stops and heat release ceases. In most industrial reactors, the water cooling system inside the reactor is regulated by the temperature of the reaction mix. Redundant alarms and emergency stops are installed to automatically cut off the flow of the cyclohexanol/cyclohexanone mix in case of a malfunction. Some of the conditions and chains of events that could lead to potentially dangerous situations include:

(i) A drop in the reaction temperature (<50°C) favouring the formation of nitro- and nitroso-derivative compounds under the action of the nitric acid (at low temperatures the nitration reaction is favoured over the oxidation reaction). The accumulation of a significant excess quantity or concentration of these compounds, which could result from the failure to cut off the feed or from malfunctions in the drainage system, may create a danger of explosive decomposition due to the instability of these compounds.

(ii) The accumulation of unreacted cyclohexanol/cyclohexanone mix (e.g. resulting from the failure to adequately stir or mix the reaction mass). If the accumulated quantity exceeds a few tens of kilograms and the initiation of the reaction is delayed, heat and gas are produced and accumulate in an uncontrolled fashion, with explosion-like effects.

(iii) A cooling failure, due to a malfunction or breakdown in the reactor water system, leading to a temperature increase with a consequent rise in the reaction rate. In such a case, however, the heat released is limited, thanks to the fact that the flow of the cyclohexanol/cyclohexanone mixture stays constant.

In order to prevent these conditions from occurring, redundant alarm and emergency systems are installed. If the temperature falls below 65°C or rises above certain safety limits, or in the case of a lack of air for instruments, the emergency safety system will cut off the reactant feed to the reactor, thus removing the cause of the malfunction itself and allowing the system to return to safe operating conditions. For safe system operation, it is important to ensure suitable residence times and temperatures so that the oxidation reaction can be completed before the product recovery phases are started. This is particularly important in the crystallization stage, during which the mass needs to be cooled down almost to room temperature, a process associated with the increased risk of the build-up of dangerous compounds.

## 13.5. Materials

The severe reaction conditions — due to the presence of nitric acid at concentrations above 50%, together with a gas phase with significant concentrations of higher nitrous oxides (NO and $NO_2$) and water vapour, as well as high average temperatures (70–100°C) — create a highly corrosive environment that requires the use of high quality stainless steel and the adoption of manufacturing procedures, particularly for welding, in compliance with standards that include an adequate corrosion allowance.

These precautionary measures become absolutely necessary wherever the corrosion resistance of the material may have been reduced by heat treatments at high temperatures, such as welding. Corrosion in a nitric environment is due to the precipitation of insoluble chromium carbides, which deplete the chromium in the area around the weld, thus weakening the weld itself and making it more susceptible to attacks by nitric acid. In particular the intergranular corrosion[12] is a form of relatively rapid and localized corrosion associated with a defective microstructure known as carbide precipitation. When austenitic steels have been exposed for a period of time in the range of approximately 425–850°C, or when the steel has been heated to higher temperatures and allowed to cool through that temperature range at a relatively slow rate (such as occurs after welding or air cooling after annealing), the chromium and carbon in the steel combine to form chromium carbide particles along the grain boundaries throughout the steel. Formation of these carbide particles in the grain boundaries depletes the surrounding metal of chromium and reduces its corrosion resistance, allowing the steel to corrode preferentially along the grain boundaries. Steel in this condition is said to be "sensitized."

It should be noted that carbide precipitation depends upon carbon content, temperature and the length of time at that temperature. The most critical temperature range is around 700°C, at which 0.06% carbon steels will precipitate carbides in about 2 minutes, whereas 0.02% carbon steels are effectively immune from this problem.

It is possible to reclaim steel which suffers from carbide precipitation by heating it above 1,000°C, followed by water quenching to retain the carbon and chromium in solution and so prevent the formation of carbides. Most structures which are welded or heated cannot be given this heat treatment and therefore special grades of steel have been designed to avoid this problem. These are the stabilized grades 321 (stabilized with titanium) and 347 (stabilized with niobium). Titanium and niobium each have much higher affinities for carbon than chromium and therefore titanium carbides, niobium carbides and tantalum carbides form instead of chromium carbides, leaving the chromium in solution and ensuring full corrosion resistance.

Another method used to overcome intergranular corrosion is to use the extra low carbon grades such as grades 316L and 304L; these have extremely low carbon levels (generally less than 0.03%) and are therefore considerably more resistant to the precipitation of carbide.

Many environments do not cause intergranular corrosion in sensitized austenitic stainless steels, for example, glacial acetic acid at room temperature, alkaline salt solution such as sodium carbonate, potable water and most inland bodies of fresh water. For such environments, it would not be necessary to be concerned about sensitization. There is also generally no problem in light gauge steel since it usually cools very quickly following welding or other exposure to high temperatures.

The adoption of precautionary measures will help slow down the corrosion process but will not be able to stop it completely, especially when temperatures rise above 90°C and the concentration of nitric acid exceeds 50%.

In some cases, when it is of critical importance to the process operating conditions, one should weigh the economic benefits of using titanium. Titanium is more expensive but lighter than stainless steel and, because of its higher resistance to the corrosive action of nitric acid solutions in the presence of copper and vanadium catalysts, it is more reliable over the long run. This latter aspect is quite surprising, since corrosion standards show that titanium is, in fact, incompatible with nitric acid at high concentrations and high temperatures. But, nitric acid passivation by the copper and vanadium catalysts present in the reaction environment makes titanium stable.

## 13.6. Conclusion

We have dedicated this chapter mostly to the discussion of the chemical and technological aspects of adipic acid production. From an industrial application perspective, however, there are other factors that are strategic and fundamental in the development and success of the technology over time. Among such factors, the availability and cost of raw materials certainly play a primary role. The decision made in the 1950s to rely on cyclohexane as the feedstock for the process — in spite of the low yield of cyclohexane to KA oil transformation and the complexity of the process — together with the choice to use nitric acid as the oxidizing agent is still a winning solution, provided that technological development can solve the environmental problems and reduce operating costs. During the last 60 years, solutions have been developed and implemented for environmental problems related to the release of emissions into the air, discharges into water, recovery and reuse of by-products, solid waste reduction, safety and automation. Furthermore, all the ancillary operations for the recovery of the finished product, catalysts and nitric oxides have been optimized. The solutions to these issues were found over a relatively long period of time during which society was slowly evolving and new standards were being set for industrial production. In contrast, the new sustainability parameters set today cannot be met by progressive evolution. Any new industrial process would have to meet these ever more stringent environmental requirements from the very start. Research and development work on all these aspects in a lab environment or in a pilot plant would be a long-term project requiring a very large investment and, in any case, would not provide definitive information on the effects of scaled-up commercial production. The competitive advantage of the current technological process in terms of knowledge and know-how makes it practically irreplaceable in the short run, save for inventions that are capable of introducing radical changes.

Patents on this process have expired and the know-how acquired by each individual company is an important and intangible fixed asset that is difficult to protect. Adipic acid is still manufactured today primarily for nylon production. Nylon, which used to be a high value-added product, is now a commodity used in numerous high-volume applications. There are only a few niche markets today that allow for maintaining prices that provide acceptable margins. Consequently, any expenditure for research and development is becoming more difficult to justify, and industrial players are investing less and less in developing alternative processes. Research on new synthesis routes is confined to university research centres and is focused on the use of renewable resources.

## References

1. PCI Nylon Yellowbook. (2010). World PA6 & 66 Supply/Demand Report.
2. Mitsutani, A. and Kumano, S. (1997). Evaluation of New One-step Adipic Acid MFG Process, Special Evaluation Report N° 29, Nippon Cemtech Consulting Inc.; K. Yamashita, M. Iwasaki (1995). Production of Cyclohexene, *Japanese Patent JP7196538* (Applicant Asahi Chemical Industries).
3. Davis, D. (1997). Adipic Acid, in D. Danly and C. Campbell (eds) *Ullmann's Encyclopedia of Industrial Chemistry*, Vol A1, J. Wiley & Sons, New York, pp. 269–278; also in J. Kroschwitz and A. Seidel (eds) *Kirk–Othmer Encyclopedia of Chemical Technology*, 3rd Edition. Vol. 1 (1978). J. Wiley & Sons, New York, p. 510.
4. Castellan, A., Bart, J. and Cavallaro, S. (1991). Nitric Acid Reaction of Cyclohexanol to Adipic Acid, *Catal. Today*, **9**, pp. 255–283.

5. Cavani, F. and Alini, S. (2009). Synthesis of Adipic Acid: On the Way of More Sustainable Production, in F. Cavani (ed.) *Sustainable Industrial Processes*, Wiley-VCH Verlag GmbH & Co. KGaA, Weinheim, pp. 367–425.
6. Van Asselt, W. and van Krevelen, D. (1963). Preparation of Adipic Acid by Oxidation of Cyclohexanol and Cyclohexanone with Nitric Acid, *Rec. Trav. Chim. Pays Bas.*, **82**, pp. 51–67, 429–449; Van Asselt, W. and van Krevelen, D. (1963). Adipic Acid Formation by Oxidation of Cyclohexanol and Cyclohexanone with Nitric Acid, *Chemical Engineering Science*, **18**, pp. 471–483.
7. Godt Jr, H. and Quinn, J. (1956). A Study of the Nitric Acid Oxidation of Cyclohexanol to Adipic Acid, *J. Am. Chem. Soc.* **78**, p. 1461–1464; Ya. I., Lubyanitskii, R. Minati *et al.*, (1962). Kinetics conversion of 6-nitro-6-hydroxyiminohexanoic acid to adipic acid, *Zh. Fiz. Khim.*, **36**, p. 567–574; C.A. 57, 3279c.; Simmons, T., Love, R. and Kreuz, K. (1966). The Structure of $\alpha$-Nitro Ketones, *J. Org. Chem.*, **31**, p. 2400–2401; Parshall, G. (1978). Industrial Applications of Homogeneous Catalysis: A Review, *Mol. Catal.*, **4**, p. 243–270.
8. Topchiev, A. (1959). *Nitration of Hydrocarbons and Other Organic Compounds*, Pergamon Press, New York, p. 87.
9. Castellan, A., Bart, J. and Cavallaro, S. (1991). Industrial Production and Use of Adipic Acid, *Catal. Today*, **9**, pp. 237–254.
10. Hearfield, F. (1980). Adipic Acid Reactor Development with Benefits in Energy and Safety, *Chemical Engineer*, **316**, pp. 625–633.
11. Riegelbauer, G. and Wegerich, A. (1967). Continuous Production of Alkane Dicarboxylic Acids, Great Britain *GB 1,092,603* (Applicant BASF).
12. Tsai, N. S. and Eagar, T. W. (1984). The Size of the Sensitization Zone in 304 Stainless Steel Welds, *J. Materials for Energy Systems*, **6**, pp. 33–37.

# Chapter 14

# Selective Oxidation Reactions in Polynt: An Overview of Processes and Catalysts for Maleic Anhydride

Mario NOVELLI,* Maurizio LEONARDI* and Carlotta CORTELLI*

Maleic anhydride (MA) is a very versatile molecule thanks to both the double bond and the anhydride group; its worldwide annual production capacity is close to two million tonnes and it is expected to remain almost constant for the constant for the next decade all over the world.

The technologies utilized and available for MA production are different when starting from n-butane or benzene, and include fixed- or fluid-bed reactors and aqueous or solvent recovery systems.

The latest updates in the production from benzene include: the Polynt's high load technology, which consists of the retrofitting of the existing multi-tubular reactors, introducing an improved heat exchange capacity, a new benzene evaporator and an air mixing unit and the use of a specific catalyst based on Vanadium and Molybdenum Oxides.

The production from butane had been successfully applied beginning in the 1970s. Nowadays the available technologies use both fixed- and fluidised-bed reactors. The ALMA process, which is using a fluid bed reactor and solvent-based MA recovery, was developed in the 1980s and improvements have since been made in both the process and the catalyst.

There are several fixed bed technologies and catalysts. Polynt is active in this field, developing catalysts with different shapes and densities inside the tubes, in order to cover the various reaction technologies available on the market.

## 14.1. Introduction

The three "pillars" of Polynt's business are maleic anhydride (MAN), phthalic anhydride (PA) and trimellitic anhydride (TMA) which are produced through processes internally developed during the last 50 years, starting from the laboratory scale, up to the industrial production.

All these molecules are synthesized by selective oxidation reactions, which can be gas-phase heterogeneously catalyzed (maleic and phthalic anhydride) or

*Polynt S.p.A. Via Enrico Fermi 51, 24020 Scanzorosciate (BG), Italy.

homogeneously catalyzed (trimellitic anhydride). Since the beginning the research and development activities in Polynt have been focused on both the process (from the chemical and the engineering point of view) and on the catalyst formulation.

Phthalic anhydride is produced by partial oxidation of *o*-xylene or naphthalene. Recently the path from naphthalene has been dramatically reduced due to a lack of availability of the raw material. Polynt has been producing PA since its foundation in 1955 based on proprietary technology and catalysts.

Trimellitic anhydride is quite a "young" molecule for Polynt. Its industrial production started in the Scanzorosciate site in Italy in 1994 after a six-year project devoted to develop the technology and catalyst formulation. From 2007 a similar technology and catalyst have been used in Polynt's Changzhou plant in China, where, as in Italy, TMA is further transformed into trimellitate esters, utilized as special plasticizers.

Maleic anhydride is a more consolidated and quite "old" molecule, which is produced by Polynt both from benzene since 1959 and from *n*-butane (fluid bed) since 1994. Moreover, Polynt is the only one company in the world which covers all the catalysts used for the industrial production of MAN. Polynt licenses and sells not only the two catalysts utilized for captive use, but also the catalyst for *n*-butane fixed-bed technology.

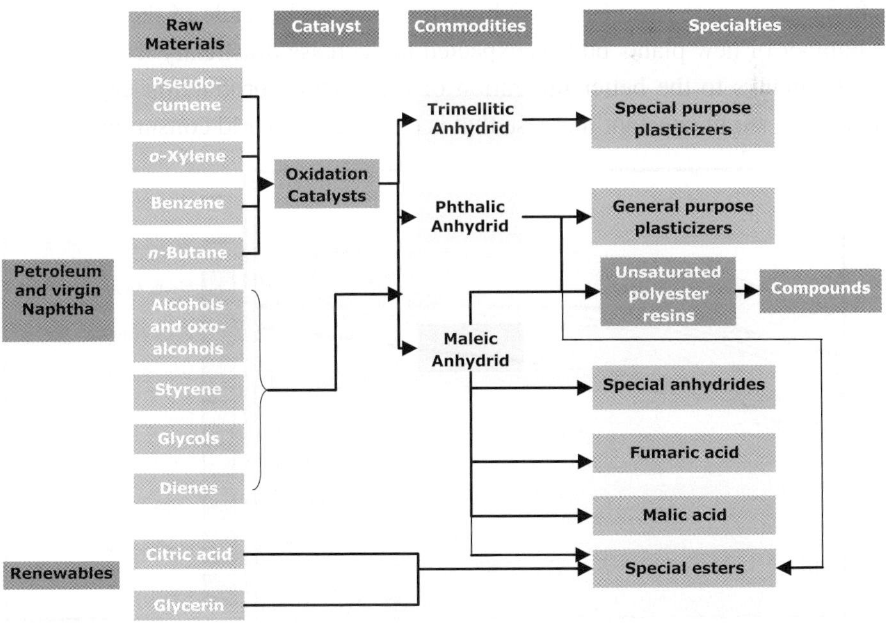


**Figure 14.1.** Polynt's integrated business, highlighting some applications for maleic, phthalic and trimellitic anhydrides.

In this chapter a short review of the history and the most recent developments of maleic anhydride production are given.

Before getting into the detail of this chapter, the scheme shown in Fig. 14.1 describes the integrated Polynt business, to give the reader an overview of some of the main applications for maleic, phthalic and trimellitic anhydrides.

It should be noted that all the technologies and catalysts in use in Polynt have been developed internally by its R&D and engineering teams. The historical traditions of a technology-oriented company, and the outstanding R&D know-how and facilities available in its laboratories, allow for continuous improvement of products and plants.

## 14.2. Maleic Anhydride Market Trends and Production

Maleic anhydride is industrially produced by selective oxidation of *n*-butane or selective oxidation of benzene. In 2011 the expected worldwide capacity was approximately 2.0 million tons of MAN, 80% of which was produced from *n*-butane (Fig. 14.2); the remaining 20% was produced from benzene.

It is important to underline that almost all the production from benzene is located in Asia (particularly in China), where the raw material is a derivative of coke, while in the rest of the world the use of *n*-butane is preferred for safety and economic considerations.

It is easy to note from Fig. 14.2 how, in the second decade of the 21st century, the number of new plants built is expected to decrease drastically if not disappear entirely, thanks to the better utilization of the existing production capacity. This matches the slight constant increase that is expected of world consumption.

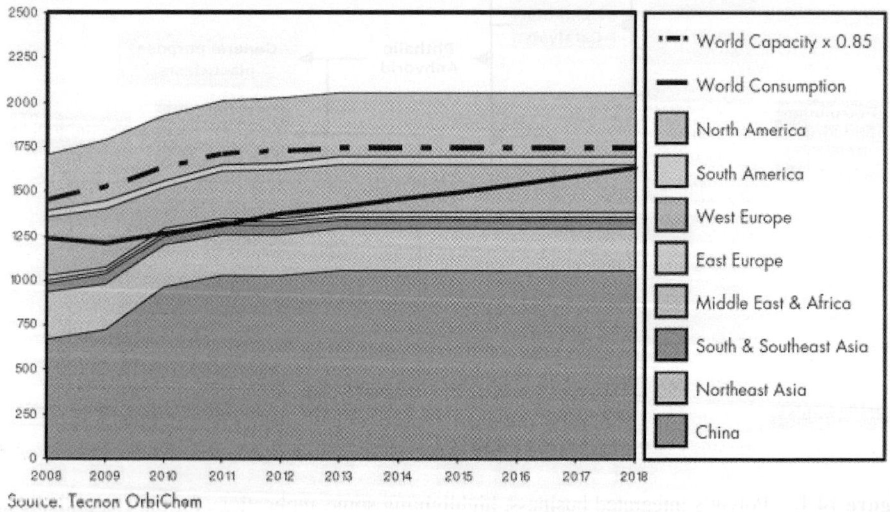


**Figure 14.2.** Worldwide production capacity of MAN.[1]

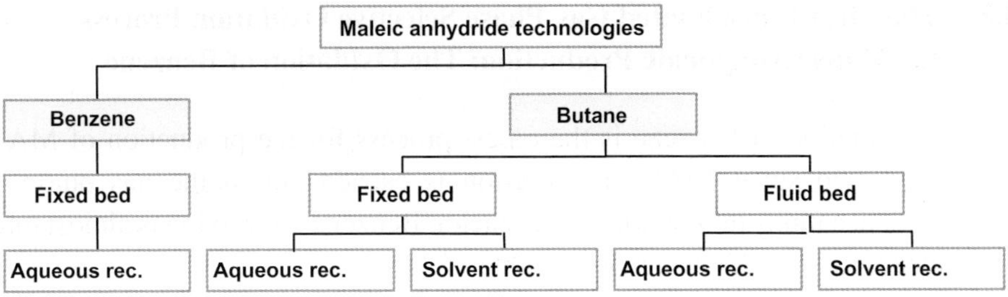


**Figure 14.3.** Technologies available for the production of maleic anhydride – raw materials, reactors and recovery sections.

The reactions involved in the two processes are:

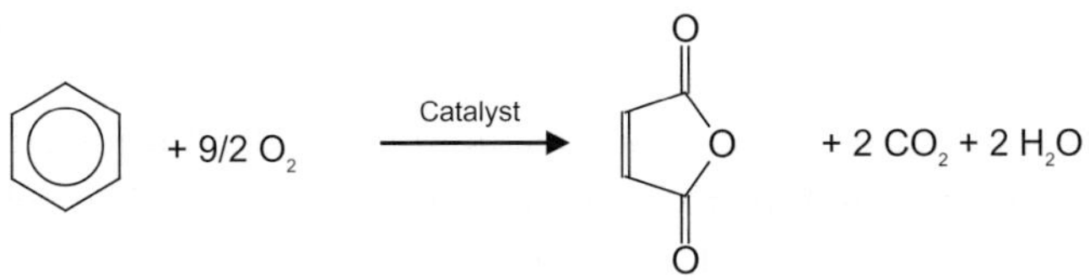


(14.1)

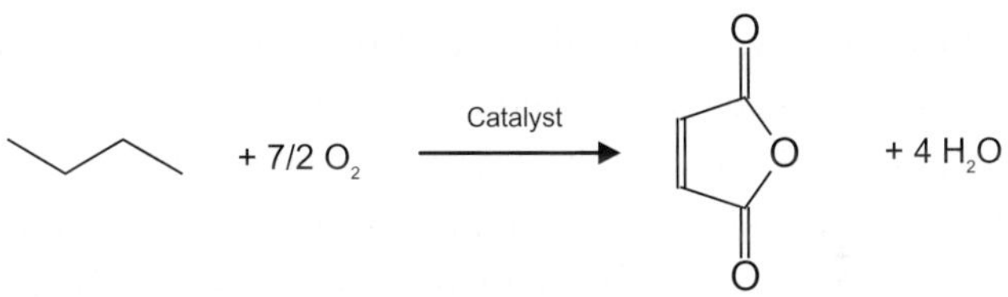


(14.2)

The technologies available to produce MAN are summarized in Fig. 14.3, where it is possible to note that the two raw materials can be transformed with two types of reactors, fixed or fluid bed, and two different kinds of recovery systems can be used, based on an aqueous or organic stream.

Despite the molar yield and hydrocarbon conversion being higher for the benzene process, the weight yield is higher in the case of the *n*-butane feed due to a better atomic efficiency for the C4 reaction, as reported in Table 14.1. Also shown are some typical industrial data for both C6 and C4 processes (fixed bed).

**Table 14.1.** Typical industrial data for *n*-butane and benzene processes, using fixed-bed technology.

| | | *n*-Butane | Benzene |
|---|---|---|---|
| Conversion | % | 80–82 | >97 |
| Selectivity | % | 72–74 | 72–74 |
| Molar yield | % | 58–60 | 70–72 |
| Weight yield | %wt | 95–100 | 88–92 |
| Catalyst life | years | >5 | 3–4 |

## 14.3. The Most Consolidated Gas-Phase Selective Oxidation Process for Maleic Anhydride Production: The Oxidation of Benzene

Selective oxidation of benzene is the oldest process for the production of MAN, industrially used since 1933. The reaction is carried out in the gas phase, at 400–450°C, in a multi-tubular plug flow reactor. Benzene conversion is almost total, while selectivity to MAN can reach a value of 74%.

Today almost 0.4 million tons of MAN are produced from benzene, the majority in China and the rest in Brazil, Mexico, India, Indonesia, Japan, Poland and Italy. Before the world crisis of 2009, several new plants were built every year in China for the Asian market, while in the rest of the world plants were shut down or converted to *n*-butane. Today, even in China, the new availability of *n*-butane from the northwest province of Xinjiang, has pushed a transformation of MAN production towards *n*-butane.

The modern catalyst is based on a vanadium-molybdenum mixed oxide, with a variety of promoters which increase activity and selectivity. The active phase is supported on an inert carrier, which allows the efficient removal of the reaction heat.

The history of Polynt's process and catalyst is described in Fig. 14.4, confirming the consolidated technology. Despite the age of this process, in recent years some improvements were introduced for both process technology and catalyst, to improve the MAN productivity and decrease the environmental impact.

The key point of this technology (as for the majority of oxidation reactions) is the heat removal. The reaction is largely exothermic ($\Delta H^\circ = -450$ kcal/mol) and the first processes (so called low load) permit the feeding of a quite diluted hydrocarbon

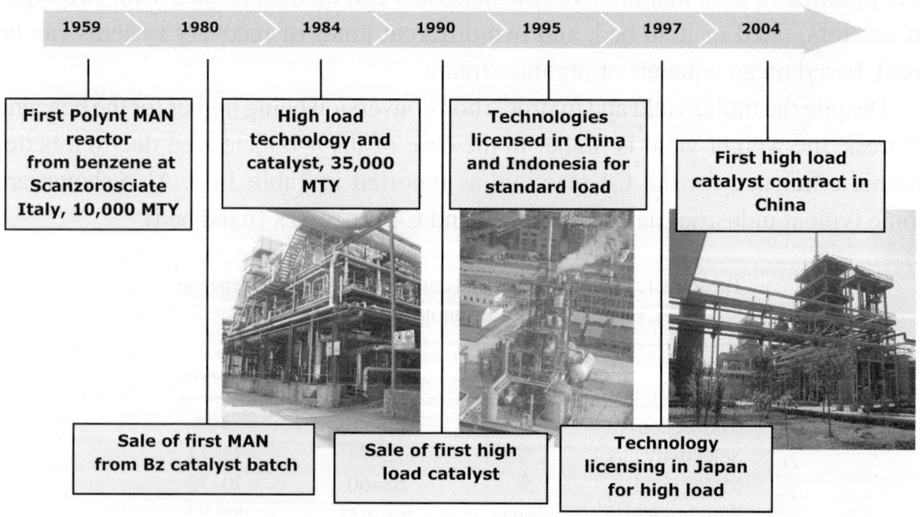


**Figure 14.4.** History of Polynt's catalysts and technologies referring to maleic anhydride.

stream (<1.2%). Today, thanks to a more efficient design of the heat exchanger and feeding system, in addition to a new catalyst shape and formulation, it is possible to reach a feed of 2% benzene in air.

Since this concentration is inside the flammability limits (1.2% of benzene in air is the lower limit, while 8% is the upper limit), the benzene evaporator and reactor should be protected using appropriate safety devices.

Figure 14.5 shows the simplified flow for the Polynt process using benzene as feedstock, the so-called high load technology, which allows reactor operation up to 210 g/h/tube (2% mol concentration) of benzene. This technology, which includes an innovative benzene feeding system and a new catalyst generation, has been widely applied in Polynt plants (two reactors of 13,000 tubes were retrofitted in the 1980s) and in other large reactors for MAN production from benzene. The reaction gases are cooled down by means of gas coolers and crude maleic anhydride is partially recovered by condensation and partially as a maleic acid (MAC) solution in a water scrubber. Off-gases leaving the scrubber are sent to a catalytic incinerator.

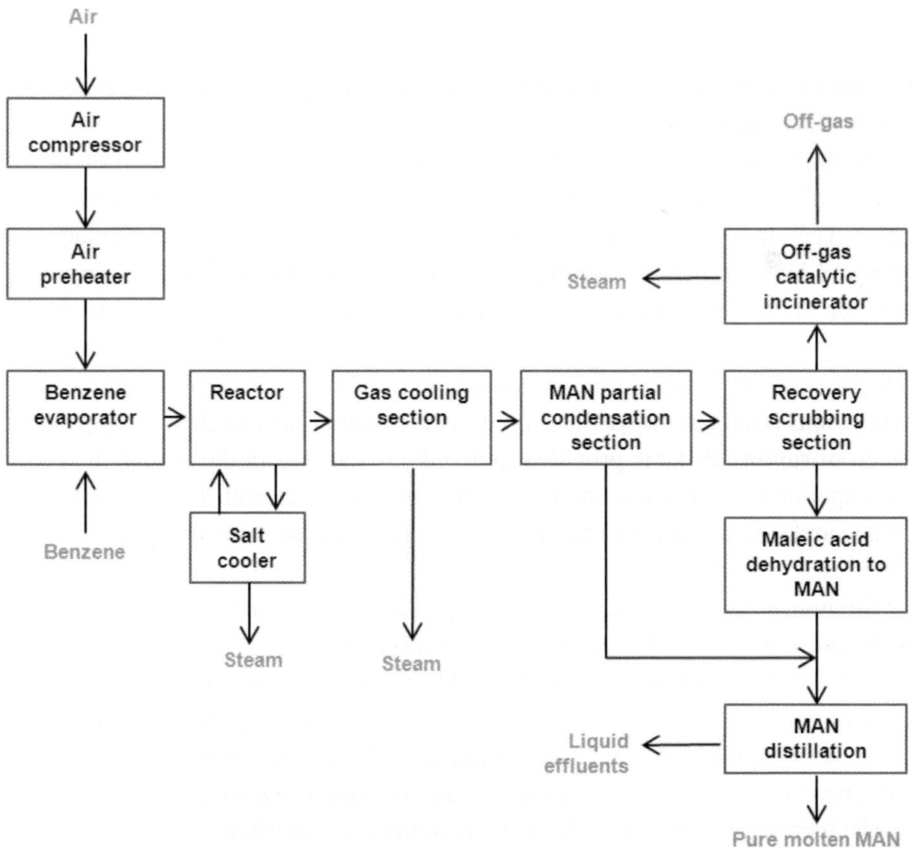


**Figure 14.5.** Simplified flow for the maleic anhydride technology from benzene.

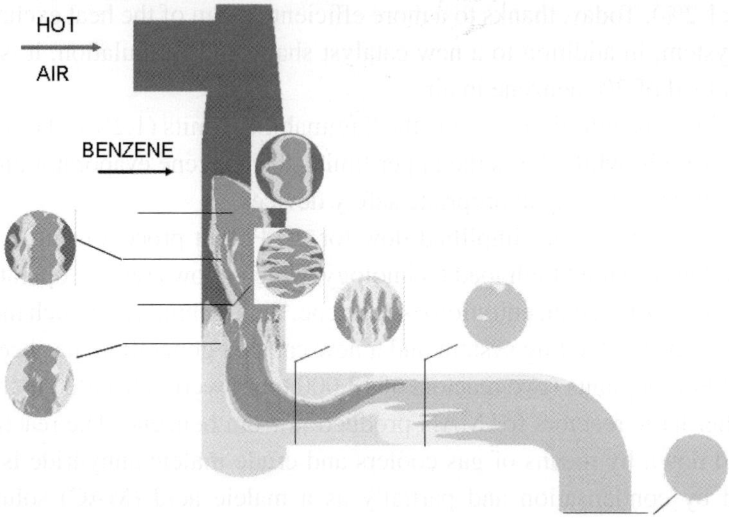


**Figure 14.6.** Benzene evaporator and air mixture device designed to guarantee a homogeneous hydrocarbon concentration in the gas entering the fixed-bed reactor. Colours are a function of the concentration.

Pure maleic anhydride is obtained in a batchwise-operated azeotropic dehydration/distillation column.

The high load technology previously mentioned was developed by Polynt with the intention of increasing the productivity of the existing plants in Scanzorosciate (Italy). The already existing multi-tubular reactors were retrofitted, introducing an improved heat exchange capacity, a new benzene evaporator and an air mixing unit able to assure a homogeneous blend of the aromatic with air (Fig. 14.6). The new evaporator is key equipment because it is able to avoid the risk of extra high benzene spots or droplets that, in contact with the hot surface of the first catalyst layer, could overheat the gas and ignite the mixture. Nevertheless an appropriate design of rupture disks is provided to avoid any damage in the oxidation section. The experience acquired in half a century of operation with benzene allowed a reliable and safe operation of the units with successful licensing activities in several countries.

The high load technology was developed not only thanks to process improvements, but also to a new catalyst shape and formulation.

It is well known that catalysts for benzene oxidation are formed by an active phase on an inert support. The active phase consists of a vanadium-molybdenum mixed oxide with a high content of vanadium; obviously other elements are present to improve activity or selectivity (such as tin, phosphorous, potassium and silicon).

The inert carrier is usually made by tabletting or extrusion of steatite, alumina, silicon carbide or other compounds, and its role is to allow a good heat exchange.

**Table 14.2.** Different supports for maleic anhydride from benzene catalyst.

| Support | Dimension (mm) | Geometric surface area in the reaction tube ($m^2$) | Benzene mass flow rate (g/h*tube) |
|---|---|---|---|
| Rings | 5×2×5 (external diameter, internal diameter, length) | 1.13 | 146 |
| Half rings | 7×4×5 (external diameter, internal diameter, length) | 1.17 | 151 |
| Cylinders | 5×5 (diameter, length) | 0.93 | 120 |
| Balls | 5 (diameter) | 0.78 | 101 |

The typical catalyst shapes for the low load catalyst are cylinders or balls. The hollow cylinder form (rings) is preferred to obtain a lower pressure drop with respect to balls or other shapes; this lower pressure drop leads to a lower energy consumption for the air compression.

The high load catalyst is, on the contrary, supported on a half cylinder support, which has a higher geometric specific surface area with respect to the conventional carriers; this means that it is possible to introduce a larger amount of active mass into the reactor, without increasing the thickness of the catalytic layer, as reported in Table 14.2.[2]

Today, the higher cost of benzene versus *n*-butane, discourages new plants building. Only in China, where coking benzene has a reasonable cost in comparison with n-butane, are new plants still under projecting and construction today. In the rest of the world only an exceptional situation, such as the unavailability of paraffin, a totally depreciated plant or market peculiarity, would still give a reason for the maintenance of old benzene-based plants. Moreover, even in China where the cheaper coking benzene is available, the *n*-butane-based technology is definitely gaining ground, drastically reducing the number of new reactors based on the aromatic feed.

The loss of interest in the production of MAN ex benzene in the Western world has prevented any significant development of the technology after the 1980s. Only China is expressing a real interest in cost optimization for the existing technology. Domestic study and investment were allocated in an optimized semi-continuous water distillation unit, while producers are wishing for a solvent recovery section coupled with a total continuous distillation unit.

## 14.4. Selective Oxidation of *n*-Butane for Maleic Anhydride Production

The processes from *n*-butane have been used industrially since the 1970s and the key factor for the success of this technology has been the development of the right catalyst (a vanadium-phosphorous mixed oxide), that is still the only option available in the industry.

Depending on the employed technologies (fixed or fluid bed) the *n*-butane selectivity varies from 65 to 53% molar, and MA conversions for both technologies does not exceed 90%.

The use of *n*-butane instead of benzene leads to many advantages:

(i) lower cost of the raw material: *n*-butane is present in the natural gas and it is also produced by the steam cracking of oil;
(ii) better atomic efficiency: starting from benzene two $CO_2$ molecules are co-produced;
(iii) lower environmental impact: benzene is a recognized carcinogenic substance;
(iv) better quality of MA due to more efficient separation and purification steps utilizing organic solvents as the recovery media.

The oxidation of *n*-butane to MAN was the first industrialized reaction converting paraffin to a large tonnage product and has continuously inspired researchers to improve and further understand the mechanism of the reaction, trying to translate this knowledge to other saturated organic molecules. The catalyst has been extensively studied and current research is very active on this topic. In the last 30 years many patents have been published by different companies and universities concerning modifications and improvements that increase the yield of the process.

It is well known that the only industrially used catalyst for the selective oxidation of *n*-butane to maleic anhydride is a vanadium/phosphorous mixed oxide, with a particular crystalline structure, the vanadyl pyrophosphate $(VO)_2P_2O_7$. (usually this catalyst is called VPO). Different preparation methods (inorganic or organic, in the last case using different solvents) are described in the literature. However, the final active phase is obtained by similar steps, which are: (i) synthesis of the vanadyl pyrophosphate precursor: $VOHPO_4 \cdot 0{,}5H_2O$; (ii) thermal decomposition of hemihydrate vanadyl acid orthophosphate, with total or partial loss of crystallization water; (iii) catalyst forming to reach better mechanical resistance as a function of the reactor type (fixed or fluid bed); and (iv) catalyst activation to obtain good and constant catalytic performances. In the available literature a variety of information and reviews can be found on the catalyst.[3–8]

Very important characteristics of the final catalyst, which can have a big impact on performance, are: (i) P/V ratio, (ii) crystallinity, (iii) oxidation state and (iv) presence of dopants; usually every catalyst formulation is designed and optimized for the final reactor and technology.

The processes can be very different in terms of technology, depending on the type of reactor (fixed or fluid bed, reactor height and diameter for the fixed bed) and type of MA recovery. Different companies have developed their own processes, as summarized in Table 14.3.

**Table 14.3.** Processes and licensors for MAN production from *n*-butane.

| Reactor typology | MAN recovery | Licensing company |
|---|---|---|
| Fixed bed | Solvent | Huntsman |
| — tube height: 3 or 6 m | Aqueous | Scientific Design — Conser |
| — tube diameter: 21 mm | | Polynt |
| — butane concentration: 1.6 to 2.0% mol. | | |
| Fluid bed | Solvent | CBI — Polynt |
| — space velocity: from 0.3 to 0.5 m/s | Aqueous | Ineos |
| — pressure: from 1.0 to 2.5 barg | | Mitsubishi |
| — butane concentration: 3.0 to 4.5% mol. | | |

## 14.5. Gas-Phase Selective Oxidation of *n*-Butane to Maleic Anhydride: The ALMA Process

In the 1980s Polynt (formerly Alusuisse), after developing a Vanadylpyrophosphate (VPO)fixed-bed catalyst and a MAN recovery solvent together with CBI (formerly ABB Lummus, an engineering company expert in fluid beds), jointly developed and successfully licensed a fluid-bed catalyst and technology named ALMA, an acronym for Alusuisse Lummus Maleic Anhydride (Fig. 14.7). Subsequently Du Pont, Ineos (at that time BP) and Mitsubishi developed their own fluid-bed technologies, but were not successfully licensed. The ambitious target to develop a MAN fluid-bed reactor was driven by the economic target to reduce investment and fixed costs for large single-train production capacities. The solvent recovery section

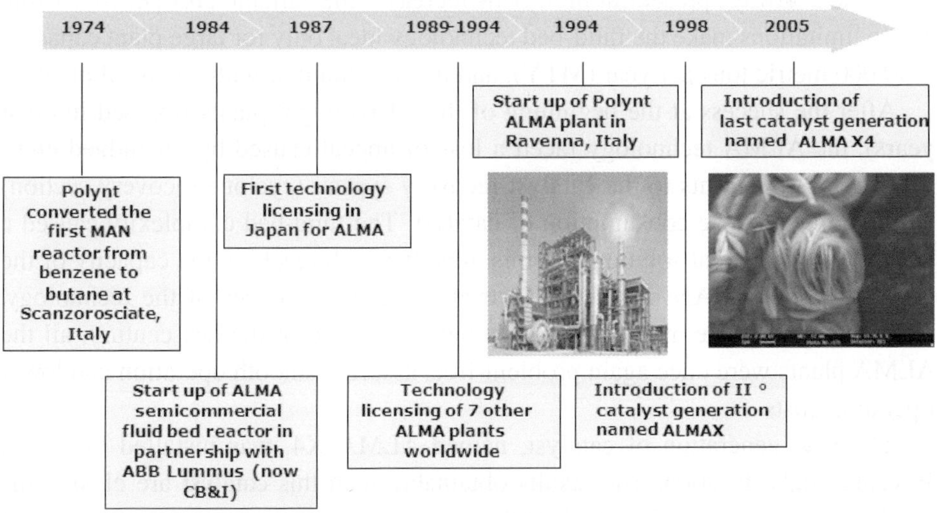


**Figure 14.7.** The history of the ALMA process.

and continuous distillation process, developed and patented by Polynt, was a clear breakthrough for MAN technology. In the 1990s almost all of the new MAN plants ex *n*-butane erected were ALMA plants.

The advantages of a fluid-bed reactor over a fixed-bed reactor are:

1. the possibility of feeding a higher hydrocarbon concentration within explosion limits;
2. excellent heat transfer, which is a key point in case of strong exothermic reaction, as this reaction is;
3. increased steam production suitable for conversion to power without fuel addition;
4. reduced power consumption for the air compressor;
5. improvements in catalyst handling and downtime;
6. large single-train capacity.

The advantages of a solvent recovery section coupled with a continuous distillation over an aqueous and batch purification section are:

1. elimination of the by-product formation of acids;
2. reduced steam consumption;
3. reduced liquid effluent;
4. superior MAN quality.

Beside these advantages, we should mention that the fluid-bed reactor is a complex technology that requires skilled engineers. Furthermore, the lower yield, a consequence of the extremely higher butane concentration and the backmixing phenomena always present in fluid-beds increases the variable cost per MAN ton. These limitations make the fluid-bed technology ideal only for large plant capacity, >50,000 metric tons per year (MTY), and in areas familiar with chemical plants.

After the success at the beginning of the 1990s (eight plants licensed in eight years), the ALMA technology faced a loss of appeal caused by misjudged engineering improvements to the catalyst recovery system (cyclone recovery section) causing an excessive consumption of catalyst. The fluid-bed complexity caused a delay in the technical solution and this, together with reaching the capacity of the Western world's MAN production, prevented further licences of the technology. After the appropriate modification, just before the end of the last century, all the ALMA plants were once again problem-free, assuring smooth operation and lower operating costs.

The new generation of catalyst, named ALMA X4, was installed in Polynt Ravenna, Italy, in 2005. The results obtainable with this catalyst are closing the gap between a fixed and fluid bed. At the same time the market request for new large MAN plants (up to 100,000 MTY) will assure new lifeblood for ALMA technology.

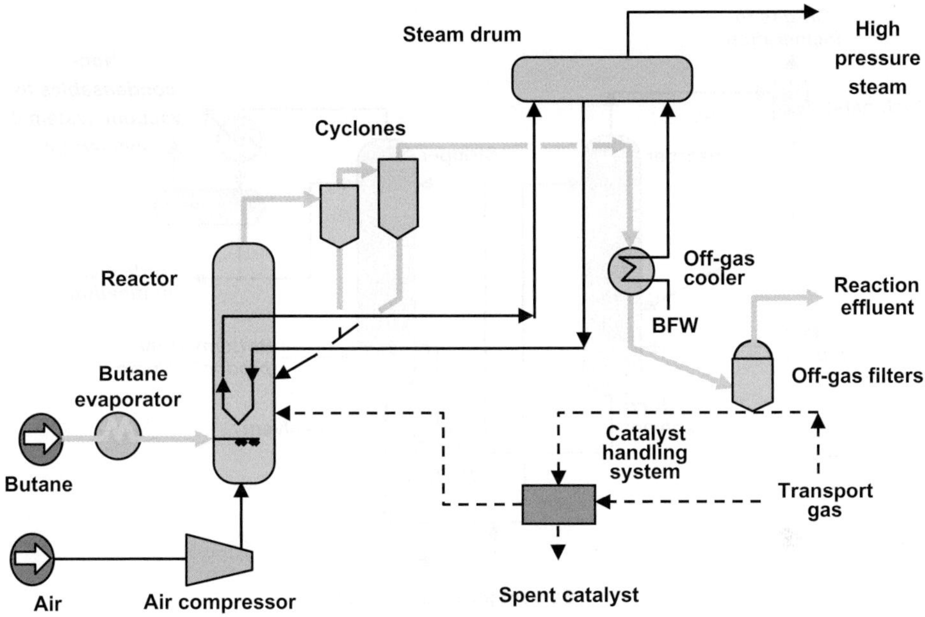


**Figure 14.8.** ALMA maleic anhydride reaction area.

Figure 14.8 represents the oxidation section for ALMA technology where air and $n$-butane, in concentration within the explosion limits, are mixed inside the fluid bed. Paraffin passing through the bed along the reactor is converted to MAN through the well-known Mars–van Krevelen mechanism. The reaction heat is removed with horizontal cooling coils maintaining an extremely homogeneous bed temperature profile, generating high pressure steam in the steam drum. The large amount of steam generated is further superheated in the post combustor and utilized in several ways, such as driving the air compressor or producing electrical power by a steam turbine and power generator, other than for plant needs. The high amount of steam per ton of MAN produced makes the ALMA reactor an ideal partner for downstream energivorous plants, e.g. 1–4 butanediol (BDO). The product, reactants and fluid-bed catalyst leaving the reactor are separated in a two-stage cyclone system where the catalyst is recovered and sent back to the fluid bed, while the gases, after cooling and a final catalyst fine filtration, are sent to the solvent recovery section.

In the reactor an appropriate gas space velocity (0.4–0.5 m/s) together with a proprietary-designed catalyst, having a particle size distribution of 50 microns as a mean diameter and density $>1{,}000\,kg/m^3$, allows a residence time inside the bed of tenths of seconds so that $n$-butane, at concentrations of up to 4.5 mol% can be converted at 80–86%, with an operating temperature of 410–430°C. The fluid bed catalyst acts as a flame arrestor maintaining the safety inside of the oxygen rich

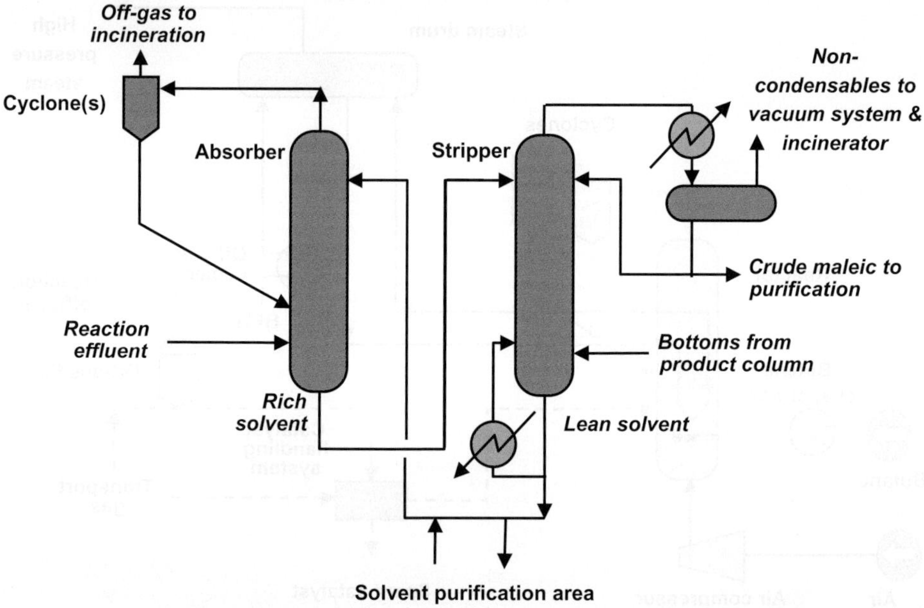


**Figure 14.9.** ALMA maleic anhydride recovery area.

atmosphere inside of the reactor. Through the reaction it is possible to have a gas mixture in the outlet that is below the minimum oxygen for combustion (MOC) value. The gas leaving the reactor is therefore safe for handling and not explosive due to the lack of oxygen.

The solvent recovery section (see Fig. 14.9) is constituted of a continuous loop of a patented non-toxic solvent that absorbs the MAN from the reactor gas in the absorber column. The MAN is later stripped from the rich solvent in the stripper column under vacuum and goes to the purification section, while the solvent is mostly recirculated back to the absorber, and a minor stream is sent to purification.

Subsequently the crude MAN is distilled in a continuous two-column train where the light ends (acrylic and acetic acids) are removed from the first column, while from the second column an outstanding MAN quality is obtained (Fig. 14.10).

The ALMA fluid-bed catalyst was a modification of the fixed-bed catalyst developed and sold by Polynt for the first time in the 1970s, when a MAN from benzene reactor was converted to *n*-butane.

The catalysts developed in the Italian laboratories were based on vanadyl pyrophosphate ($(VO)_2P_2O_7$) obtained by the calcination of vanadyl hydrogen

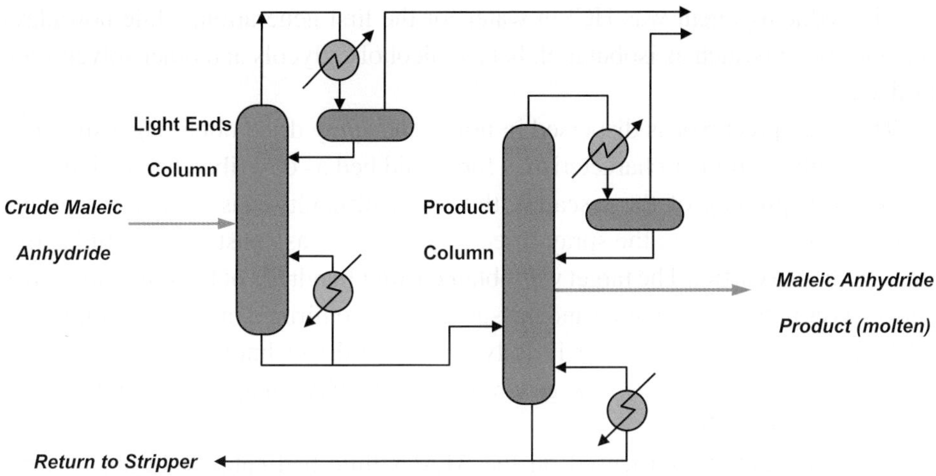


**Figure 14.10.** ALMA maleic anhydride purification area.

**Figure 14.11.** Scanning electron microscope image of a typical VPO catalyst defined as a “rosette”.

phosphate emihydrated ($VOHPO_4 \bullet 0.5H_2O$) (Fig. 14.11); this oxide is still the main component of modern catalysts.

The reaction can be described as follows:

$V_2O_5 + H_3PO_4 +$ reducing agent $\rightarrow VOHPO_4 \bullet 0.5H_2O +$ oxidized compounds
$VOHPO_4 \bullet 0.5H_2O \rightarrow (VO)_2P_2O_7$ with a thermal treatment.

The reducing agent was HCl in water for the first generation, while nowadays organic solvents such as isobutanol, benzyl alcohols, glycols and other solvents are utilized.

The dried precursor is dispersed in liquid and spray dried obtaining a spherical powder with the proper characteristics for a fluid bed as described by Geldart.[9]

At the beginning of the research, the major difficulty was to give the proper mechanical resistance to the spray-dried powder that was constituted of full active mass without a carrier. The target was obtained after hundreds of tests in a laboratory spray-dryer in the United State and the success was confirmed in a semi-commercial ALMA plant of 1.0 m diameter in Italy, where all the technological solutions for ALMA technology and the new generations of ALMA catalyst were and still are solved experimentally.

The R&D activities focused on the ALMA fluid-bed catalyst began in 1982 and are still on-going in Polynt's laboratories. Several patents have recently been published, covering the modifications introduced to improve yield and stability for the different ALMA catalyst generations.[10–14]

In 2001 ALMA technology was awarded by the Kirkpatrick Honour Award for being one of the five best technologies of the year, and in 1994 it was awarded by the European Better Environment Awards for Industry (EBEAFI) for "clean technology".

## 14.6. Some Recent Developments in the Fixed-Bed Process for Gas-Phase Selective Oxidation of *n*-Butane to Maleic Anhydride

To complete our discussion of the maleic anhydride technologies available in the world and in Polynt, we still need to describe the *n*-butane fixed-bed technology and catalyst. The origins date back to the 1970s when Polynt converted the first MAN from benzene reactor to *n*-butane (Fig. 14.12).

The oxidation area is identical to the benzene-based technology, excluding the paraffin evaporator and air mixer, having *n*-butane a lower boiling point than benzene. The recovery system used can be an aqueous system or as in the new plants, a solvent system, as described in Fig. 14.13.

After the pre-mixing of butane in air, the gases enter the multi-tubular reactor cooled with a mixture of nitrite and nitrate of sodium and potassium, where each tube has a 21 mm diameter and is 3.5 or 6 metres high. The catalyst, a VPO precursor, is tabletted into different shapes and activated to transform the $VOHPO_4 \cdot 0.5H_2O$ in $(VO)_2P_2O_7$. Nowadays the most modern generation of VPO catalysts are activated outside the reactors and are equilibrated so that the metal oxide loaded in the tubes is ready to work from the first instance of the reactants feed. It then ramps up to the

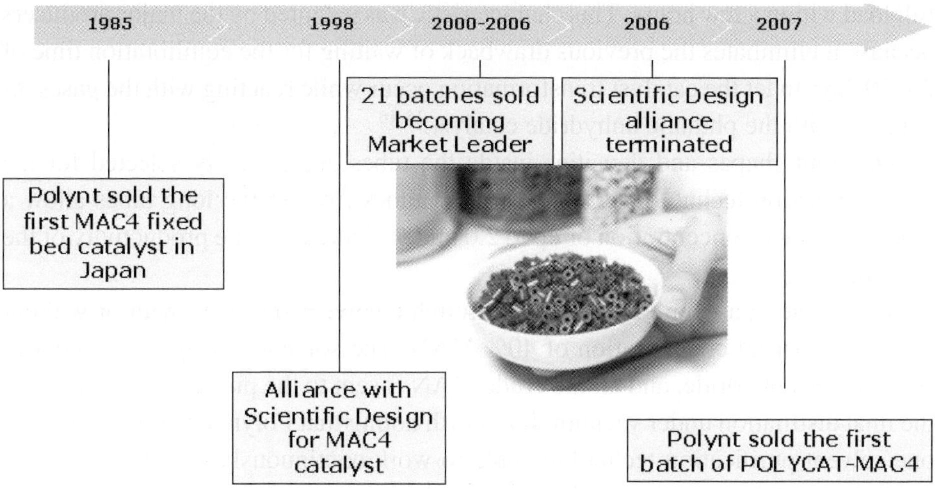


**Figure 14.12.** Maleic anhydride ex butane technology in a Polynt fixed bed.

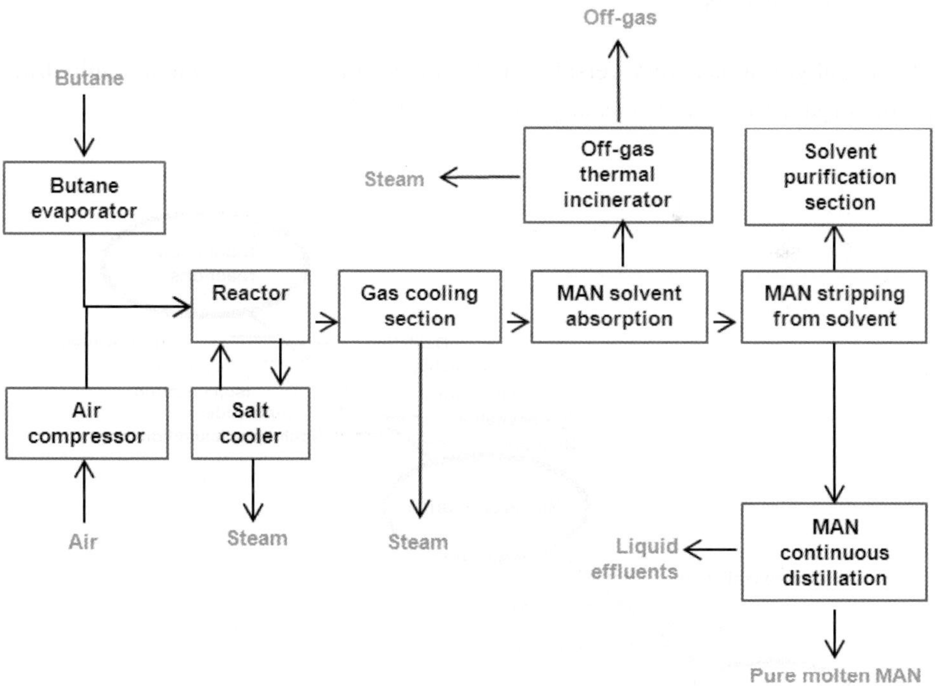


**Figure 14.13.** Simplified block diagram of maleic anhydride ex butane fixed-bed technology, solvent recovery.

full load within a few hours. This characteristic was patented by the major producers because it eliminates the previous drawback of waiting for the equilibration time of 20–30 days to let the catalyst transformation occur while reacting with the gases, as happens with the phthalic anhydride catalyst.[15,16]

Different shapes and densities inside the tubes are correctly selected for the various reaction technologies (long or short tubes), where the long tubes allow a higher *n*-butane concentration of up to 2.0 mol.%, increasing the productivity of the fixed bed.

The reacted gases are then passed through a solvent scrubber, with or without a previous partial condensation of 40% MAN. The solvent is stripped in order to separate the anhydride, and all the crude MAN is sent to the purification section for the final distillation under vacuum. It is worth noting that Polynt developed the only one solvent purification technology able to work continuously, avoiding cleaning downtime and the generation of wastes.

## 14.7. Conclusions

Maleic anhydride is a very versatile molecule with its double bond and anhydride group. Figure 14.14 shows its large reaction capability.

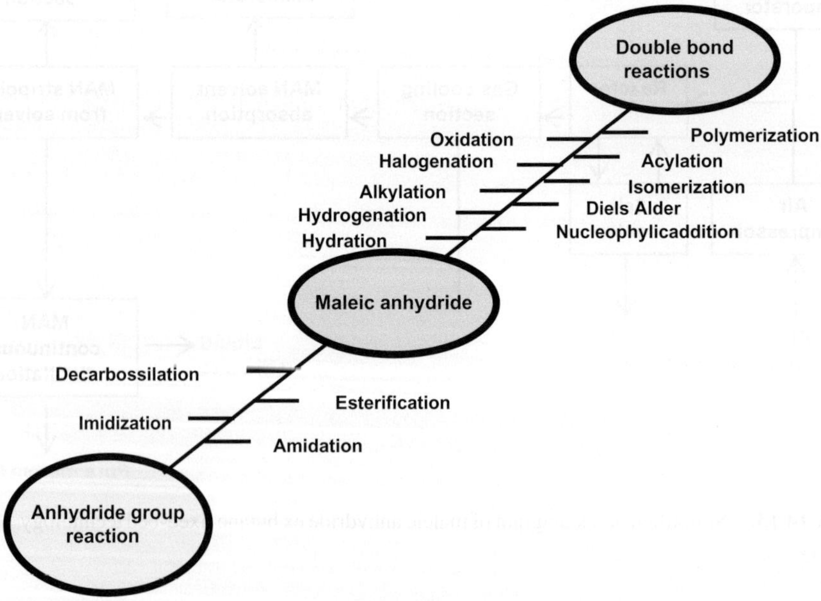


**Figure 14.14.** Reactivity of maleic anhydride.

MAN is a typical commodity meaning that transportation and selling cost have a very high impact. It brought to the installation of production capacity around the world. It is also important to take into consideration that the MAN consumption growth is strictly linked to the GDP increase. If the MAN capacity increased in the United States and Europe in the 1970s and 1980s, the 1990s were the years of increased capacity in Japan and Asia more broadly, while the new century has definitely been the period of growth in China specifically.

The technologies utilized and available are different when starting from *n*-butane or benzene, and include fixed- or fluid-bed reactors and aqueous or solvent recovery systems. Polynt is the only company in the market able to supply all the catalysts for these technologies.

The choice of the right technology for a new plant is linked to several factors such as: raw material availability and price, plant capacity, value of the energy produced (as steam or electric power), downstream integration with other products, etc.

Nowadays the worldwide annual production capacity is close to two million tonnes and is expected to remain almost constant for the next decade all over the world.

## References

1. Technon OrbiChem Nov. 2010.
2. Di Ciò A, Vitali A. (1981). Catalyst for the preparation of maleic anhydride. European Patent EP0037020B1 (assigned to Alusisse).
3. Torardi, C. and Calabrese, J. (1984). Ambient- and Low-temperature Crystal Structure of Vanadyl Hydrogen Phosphate, *Inorg. Chem.*, **23**, pp. 1308–1310.
4. Bordes, E. (1993). Nature of the Active and Selective Sites in Vanadyl Pyrophosphate, Catalyst of Oxidation of *n*-Butane, Butene and Pentane to Maleic Anhydride, *Catal. Today*, **16**, pp. 27–38.
5. Cavani F. and Ferruccio, F T. (1994). Catalyzing Butane Oxidation to Make Maleic Anhydride. *Chem. Tech.*, **24**, pp. 18–25.
6. Hutchings, G., Desmartin-Chomel, A., Olier, R., *et al.* (1994). Role of the Product in the Transformation of a Catalyst to its Active State, *Nature*, **368**, pp. 41–45.
7. Ballarini, N., Cavani, F., Cortelli, C., *et al.* (2006). VPO Catalyst for *n*-Butane Oxidation to Maleic Anhydride: A Goal Achieved, or a Still Open Challenge? *Top. Catal.*, **38**, pp. 147–156.
8. Guliants, V., Holmes, S., Benzinger, J., *et al.* (2001). In Situ Studies of Atomic, Nano- and Macroscale Order During $VOHPO_4{\cdot}0.5H_2O$ Transformation to $(VO)_2P_2O_7$, *J. Molec. Catal.*, **172**, pp. 265–276.
9. Geldart, D. (1986). *Gas Fluidization Technology*, Wiley-Interscience Publication, New York.
10. *Patent IT1237356B* (assigned to Lummus and Alusuisse).
11. Fumagalli, C., Stefani, G. (1991). Catalyst for oxidation ractions and process for its preparation, EP0182078B1, (assigned to Alusuisse).
12. Mazzoni, G., Cavani, F., and Stefani, G. (1997) Process for the tranformation of a vanadium/Phosphorous Mixed Oxide catalyst precursor into the active catalyst for the production of maleic anhydride, EP0804963B1 (assigned to Lonza).
13. Albonetti, S., Cavani, F., Ligi, S., and Mazzoni G. Vanadium/Phosphorous mixed oxide precursor and catalyst and their preparation, (2002) EP1183101B1 (assigned to Lonza).

14. Ghelfi, F., Mazzoni, G., Fumagalli, C., Cavani, F., and Pierelli, F. (2009). Niobium-doped vanadium/phosphorus mixed oxide catalyst US7638457B1 (assigned to Polynt).
15. Bortinger, A., Mazzoni, G., and Monti, T. (2005). *Phosphorous/Vanadium catalyst preparation* US6858561B2 (assigned to Lonza and Scientific Design).
16. Forkner, M. (2012). Maleic Anhydride catalyst and method for its preparation US8143461 (B2).

# Chapter 15

## Selective Oxidations at Eni

Roberto BUZZONI,* Marco RICCI,* Stefano ROSSINI† and Carlo PEREGO*

The marriage of green oxidant agents, like $H_2O_2$, and well-developed catalysts, such as TS-1, heteropolyanions, etc., is able to drive strong improvements in industrial selective oxidations. Classical approaches, key concepts, and new routes for hydrogen peroxide production are described with the aim of facing off the huge increase in $H_2O_2$ industrial needs, in an integrated way.

More challenging approaches to selective oxidation, attempting direct selective oxidation in the gas phase by oxygen or air, have also been explored using concepts like reactivity of lattice oxygen combined with periodic operations, also offering opportunities for producing hydrogen in an innovative manner.

### 15.1. Introduction

This chapter attempts to connect research on different chemical reactions and catalysts with a common thread. We will highlight a few short examples from this fascinating research field.

Having worked with TS-1, a titanium-containing molecular sieve, we will describe some selective oxidations by $H_2O_2$. Necessarily, when considering the importance of $H_2O_2$ in selective and sustainable oxidations, attention will be devoted to hydrogen peroxide production and to studies of its direct synthesis. Alternative approaches to selective oxidation, attempting direct selective oxidation in the gas phase by oxygen/air, involve passing through the redox and periodic operations concept that then drives us through to one-step hydrogen production. The role of hydrogen in the clean fuels strategy is fundamental; hydrogen demand in the refinery industry is growing dramatically and is projected to keep growing into the next decade as a result of clean fuels regulations and changing refinery product demand. New technology advancements targeted at improving the utilization and

*Eni S.p.A. – Research Center for Non-Conventional Energy - Istituto eni Donegani, via Fauser, 4 – 28100 Novara, Italy.

†Eni S.p.A. – Exploration & Production division, Via Unione Europea, 3 - 20097 San Donato Milanese (Mi), Italy.

lowering the cost of hydrogen will play a central role in future refineries. Moreover, another route towards clean fuels employs oxidative desulfurization, for example by the UOP/Eni oxidative desulfurization process by using heteropolyacids (HPAs) in which hydroperoxide is directly produced in the reaction medium. HPAs can also be used for selective oxidations with $H_2O_2$ which is an example of very attractive and intriguing chemistry.

## 15.2. TS-1 and Related Materials: A Materialized Dream

One of the most important industrial achievements in the field of liquid-phase selective oxidation has been the discovery of titanium silicalite-1 (TS-1) at Eni in the late 1970s, by Taramasso and co-workers.[1] Thanks to the unique reactivity properties of TS-1, several innovative processes that use hydrogen peroxide are now industrial processes, and others are under advanced development.

Now, about 30 years after TS-1 discovery, industrial catalysts are well developed and present in the technology portfolio of Polimeri Europa (since 2012, Versalis),[2] a petrochemical company wholly owned and controlled by Eni S.p.A. It is important to underline the key role of the most innovative proprietary catalysts, which have been developed as "fundamental elements", of some of the most advanced technologies.

A complete description of the peculiarities of TS-1 material and catalysts based upon it is not the objective of this chapter. An abundance of high-quality papers concerning TS-1 are available, as are papers on related materials, syntheses, characterizations and applications.[3–9] However, a few words are needed. Incorporation of titanium, in a tetrahedral coordination, into high-silica molecular sieve frameworks is the basis of the exceptional catalytic properties of the material. By using a mental model, if an all-silica MFI framework (mordenite framework inverted, eg. silicalite-1, S-1) is the host of atomically dispersed titanium, we are in the presence of a TS-1 material.

TS-1 is obtained by hydrothermal synthesis, from the progressive condensation of $SiO_4$ and $TiO_4$ tetrahedra around the template ion $(i\text{-}C_3H_7)_4N^+$ (from tetrapropylammonium hydroxide).

The crucial problem in the synthesis procedure is to find the proper conditions to obtain an isolated, tetrahedrally coordinated titanium species, since titanium greatly prefers an octahedral coordination.

The preparation first reported[1] is based on the controlled hydrolysis of aqueous solutions containing tetraethylorthosilicate, tetraethylorthotitanate and tetrapropylammonium hydroxide: this procedure, known as "mixed alkoxide", allows a fine control of synthesis parameters giving crystallization of high quality. Many other methods are reported in the literature.[10] One of the last most important improvements, from the point of view of application, was a new process which, by operating at a particularly low dilution, with suitable molar ratios between water and titanium in the reaction mixture, allows TS-1 materials to be prepared, in pure phase, with a high crystallization yield, using reduced reaction volumes and, at the same time, obtaining a high productivity.[11]

## 15.3. Selective Oxidation with Hydrogen Peroxide by TS-1 and Related Materials

In principle, from both the economic and the environmental point of view, the best way to perform oxidation reactions should be to use molecular oxygen. Unfortunately, this is not so easy. Thermodynamically, oxygen is indeed a powerful oxidant, but kinetically it is quite sluggish. The reason for that is the diradical (triplet) nature of molecular oxygen: since most organic substrates are closed shell (i.e. singlet) molecules, their reaction with oxygen is spin-forbidden and consequently quite slow. This situation is actually a lucky one, since it allows us not to be burned quickly in our oxygen-rich atmosphere. On the other hand, it forces chemists to develop alternative routes for many interesting oxidations.

A very successful approach is the use of oxygen donors, closed shell molecules XO able to react with an organic substrate S transferring an oxygen atom to it and affording the desired oxidized product SO and the co-product X arising from the reduction of XO (Reaction 15.1):

$$S + XO \rightarrow SO + X \tag{15.1}$$

Several single oxygen donors have been used including hydrogen peroxide ($H_2O_2$), hydroperoxides, organic peracids, nitrous oxide ($N_2O$), hypochlorites, etc. Among them, hydrogen peroxide is by far the most versatile and the easiest to use. Furthermore, its price is relatively low, its active oxygen content (i.e. the amount of oxygen which can be delivered to the substrate S) is very high (47%) and the co-produced water has no environmental or safety drawbacks.

For these reasons, hydrogen peroxide has been, and still is, an attractive reagent for selective oxidations on both the laboratory and industrial scale.

The availability of an industrial process for producing TS-1 allowed the development of several selective oxidations, all using hydrogen peroxide as a versatile and environmentally friendly oxidizing agent which gives water as the only co-product. The epoxidation of propene (Hyprox process), the phenol hydroxylation and the ammoximation of cyclohexanone are now $H_2O_2$-based industrial processes; benzene hydroxylation to phenol is currently under advanced evaluation.

### 15.3.1. *Propene epoxidation*

Propene oxide is particularly important, since it is a basic intermediate with a production capacity of around 7 • $10^6$ t/y. Its classical industrial production route is based on the propene epoxidation with chlorine (the chlorohydrin route) or with a hydroperoxide, usually tert-butyl hydroperoxide, (1-phenyl)ethyl hydroperoxide (ethylbenzene hydroperoxide) and cumyl hydroperoxide. Although these methods are satisfactory with regard to the synthesis of the epoxide, they do have a major drawback: the formation of a co-product, respectively a chloride salt (most usually calcium chloride), tert-butanol, 1-phenylethanol and cumyl alcohol. At the end of the reaction, the co-product must be recovered and recycled or, even worse, disposed of. In all cases, additional costs arise. Particularly in the chlorohydrin route, which is still in use to produce more than $2 \bullet 10^6$ t/y of the epoxide, 2.1 t of calcium chloride, polluted by chloro-organics, are produced per t of propylene oxide. For this reason, the epoxidation of propene with hydrogen peroxide, made possible by the TS-1 catalyst,[12,13] is particularly attractive: the only co-product is water, which does not need to be recycled, and no chlorinated by-products are formed.

Paparatto *et al.*[14] reported that in a slurry reactor, at temperatures below 60°C and under a little pressure, the reaction of propene with a suspension of TS-1 in aqueous methanol, into which aqueous $H_2O_2$ is fed, affords propene epoxide with very high selectivity, and both hydrogen peroxide decomposition and methanol oxidation are negligible. The only by-products are propene glycol and its two isomeric monomethyl ethers, all formed by the nucleophilic attack of water or methanol on the epoxide ring. Controlling the acidity with traces of buffering agents allows the achievement of selectivities up to 98% with respect to propene oxide.[14,15] At the end of the reaction, the unreacted propene, the epoxide and the methanol are recovered by distillation, whereas the residual aqueous solution can be treated in an ordinary biological plant.

No leakage of titanium has been observed under the reaction conditions. On the other hand, some deposition of heavy organic by-products on the TS-1 can occur, but the catalytic activity can be restored by combustion with air or by washing with a solvent at temperatures higher than 100°C. In 2001, the process was verified in a prototype plant with a capacity of *ca.* 6 t/day built in Ferrara, Italy.[16]

In 2008, Dow and BASF successfully started up the first commercial-scale production plant based on the novel BASF/Dow-developed HPPO (hydrogen peroxide for propylene oxide) technology at BASF's Antwerp, Belgium, facility. A second plant based on this technology was scheduled to begin production in Map Ta Phut, Thailand, in 2011. The two companies reported that, compared with traditional technologies, the epoxidation with hydrogen peroxide has a very low environmental impact, a simpler process layout and reduced investment costs.[17]

Since summer 2008, the South Korean company SKC has been producing propene oxide in Ulsan using the innovative HPPO process developed by Evonik/Uhde. The HPPO facility has a capacity of 100,000 metric t/y.[18]

### 15.3.2. *Phenol hydroxylation*

Hydroquinone and catechol are interesting intermediates used in the syntheses of several antioxidants. Furthermore, each of them also has specific uses: hydroquinone in the photographic industry, and catechol in the production of synthetic vanillin and other artificial aromas. In 2002 the production capacities were put at *ca.* 50,000 and 32,000 t/y respectively.

Historically, both hydroquinone and catechol were produced by cumbersome sequences of stoichiometric reactions aimed at the transformation of functional groups already present on the aromatic ring. In particular, the production of hydroquinone has often been used as an example of old stoichiometric methods to be replaced by catalytic ones.[19] A major improvement occurred in the 1970–1980s with the development of the hydroxylation of phenol with hydrogen peroxide catalyzed either by iron salts (Brichima process) or by strong mineral acids ($H_2SO_4$, $H_3PO_4$; Rhone–Poulenc process). In both cases, a mixture of hydroquinone and catechol was obtained. The ratio between them could be adjusted, to some extent, modifying the reaction conditions: catechol/hydroquinone ratio was 2.0 : 2.3 in the Brichima process and 1.2 : 1.5 in the Rhone–Poulenc process.

In 1986, a new process for the phenol hydroxylation by hydrogen peroxide marked the first industrial application of the TS-1 catalyst. At that time, the chemical company of the Eni group was EniChem, which built a plant in Ravenna, Italy for producing 10,000 t/y of the catechol/hydroquinone mixture.

The reaction is run at reflux in aqueous acetone or methanol, in a slurry reactor. The catechol/hydroquinone ratio is in the range 1.1 : 1.2 and is strongly affected by the choice of solvent. The TS-1 catalyst provides good yields, thus making possible the operation at a phenol conversion of around 30%, significantly higher than that of the Brichima or the Rhone–Poulenc processes (9 and 5%, respectively), still maintaining selectivities higher than 80–85 and 90% based on $H_2O_2$ and phenol, respectively. The high conversion implies good economics, due to low costs for phenol separation and recycling.[20,21]

### 15.3.3. *Benzene hydroxylation*

Phenol is one of the most important intermediates of the chemical industry. The global capacity for its production was around 10 Mt/y in 2008, with an actual production around 8.7 Mt. About 40% of the produced phenol is used for the synthesis of bisphenol A, a monomer for polycarbonates. Another 30% is consumed in the production of phenolic resins. The most important route for the industrial production of phenol is, by far, the cumene process, which accounts for 98% of the installed capacity. The cumene process is based upon the researches of Heinrich Hock on the formation and decomposition of cumyl hydroperoxide[22] and is sometimes referred to as the Hock process. It comprises the Friedel–Crafts alkylation of benzene with propene to afford cumene (iso-propylbenzene), followed by cumene oxidation with oxygen to cumyl hydroperoxide and by the cleavage of cumyl hydroperoxide in acidic medium to get phenol and acetone (reaction 15.2):

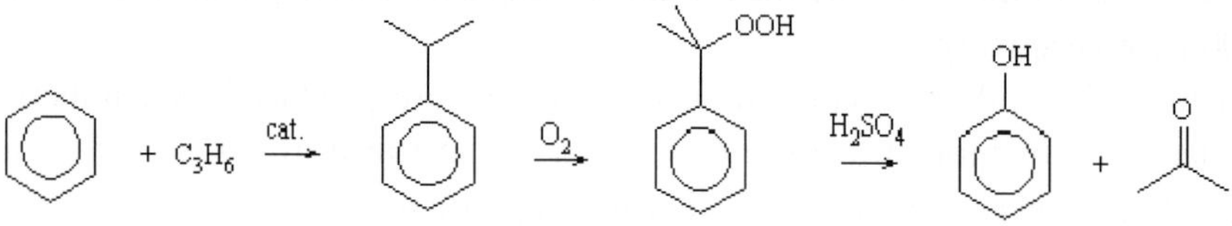


(15.2)

The cumene process is fully satisfactory in many aspects; in particular, selectivities to acetone and phenol, based on both benzene and propene, are very high and the environmental impact of the process is quite minimal.

However, even with the most competitive alkylation technologies using zeolite-based catalysts, the overall process is affected by huge recycling.[23] Furthermore, the cumene process suffers due to the co-production of about 0.6 t of acetone per t of phenol. Since the growth of the phenol market is significantly higher than that of acetone, much effort is currently being devoted to decouple phenol and acetone production and, particularly, to develop effective processes for the direct oxidation of benzene to phenol.

The latter oxidation, however, is a difficult reaction affected by poor selectivity due to the lack of kinetic control. Indeed, phenol is more reactive toward oxidation than benzene itself. So, a number of consecutive reactions occur, with substantial formation of over-oxidized products like catechol, hydroquinone, benzoquinones and tars.

A first success in the direct of benzene oxidation was provided by the Solutia process, discovered by Panov and co-workers at the Boreskov Institute of Catalysis in Novosibirsk, Russia and developed in close cooperation with Monsanto. In this process, the oxidant is nitrous oxide ($N_2O$) and an iron-containing zeolite is the catalyst.[24–26] Nevertheless, despite its brilliant results, it is unlikely that the Solutia process can become a major source of phenol since nitrous oxide availability is quite limited and its deliberate production would be expensive.

A promising alternative to nitrous oxide is provided by hydrogen peroxide and, in 1997, a project was started by Polimeri Europa to check if TS-1 might be a suitable catalyst for the oxidation of benzene to phenol.

At the beginning, however, the activity of TS-1 in the oxidation of benzene appeared to be very poor. TS-1 does not perform very well in two-phase systems and only solvents able to homogenize the hydrophobic substrate and the aqueous hydrogen peroxide could be used. Even so, in methanol, acetone, acetonitrile or tert-butanol, the selectivity to phenol rapidly dropped to values lower than 50%

at benzene conversion as low as 5%. Unexpectedly, however, the use of sulfolane allowed the conversion of benzene to reach close to 8%, maintaining the selectivity to phenol at higher than 80%. Detected by-products were catechol (7%), hydroquinone (4%), 1,4-benzoquinone (1%) and tars (5%). Even better results were obtained by a post-synthesis treatment of TS-1 with both hydrogen peroxide and ammonium hydrogen fluoride ($NH_4HF_2$). Upon such treatment ($H_2O_2$/F/Ti = 10/2.5/1; 60°C; 4 hours), a substantial amount of titanium (up to 75% of the initial value) was removed. Nevertheless, the crystalline structure of the zeolite remained unchanged and the catalytic activity did not decrease. On the contrary, it actually increased since the turnover frequency of residual titanium atoms rises from 31 to 80 $h^{-1}$. Even more important, at 8.6% benzene conversion, the selectivities, for both benzene and hydrogen peroxide, also increased from 83 to 94% with the formation of catechol (4%) and hydroquinone (2%) as the only by-products, without any evidence of further oxidation reactions.[27,28] The treated catalyst has been named TS-1B and has a quite peculiar UV-VIS spectrum: the absorption at 48,000–50,000 $cm^{-1}$ (typical of pure TS-1) was strongly reduced and a new one appeared, at 40,000 $cm^{-1}$, due to a new titanium species formed upon the treatment. No formation of amorphous extra framework titanium oxide, absorbing at 30,000–35,000 $cm^{-1}$, was observed.

Why is sulfolane, and only sulfolane, so effective in improving selectivities? In 1963, on the basis of the IR solution spectra, Russel Drago suggested that sulfolane forms a labile complex with phenol via a hydrogen bond.[29] More recently, Drago's suggestion was confirmed by *ab initio* calculations. Thus, it is reasonable that the improved selectivity observed while running the benzene oxidation in sulfolane is due to the formation of this complex, too large to enter the titanium silicalite pores, thus allowing phenol to remain relatively protected towards further oxidation.

This effect was confirmed by the calculation, using the software Sorption (Cerius 2), of the loading of both the free phenol molecule and the phenol-sulfolane complex, expressed as the number of loaded molecules in a crystal elementary unit of TS-1. The loadings turned out to be 13.6 and 0.8, respectively. Alternatively, the protective effect exerted by sulfolane can be evaluated by measuring the reaction rate, expressed as turnover frequency (TOF: moles of reacted substrate/moles of Ti × hour) for the oxidation of benzene and phenol, carried out separately in acetone and sulfolane as co-solvents. In the case of acetone, the phenol oxidation (TOF = 190) was 10 times faster with respect to that measured for benzene (TOF = 19); conversely, operating in sulfolane the rate measured for phenol (TOF = 51) was just 1.6 times higher than that measured for benzene (TOF = 31), according to the higher value of the observed selectivity.

To further increase the overall yield of the process, a second step can be added in which dihydroxylated by-products, hydroquinone and catechol are treated with

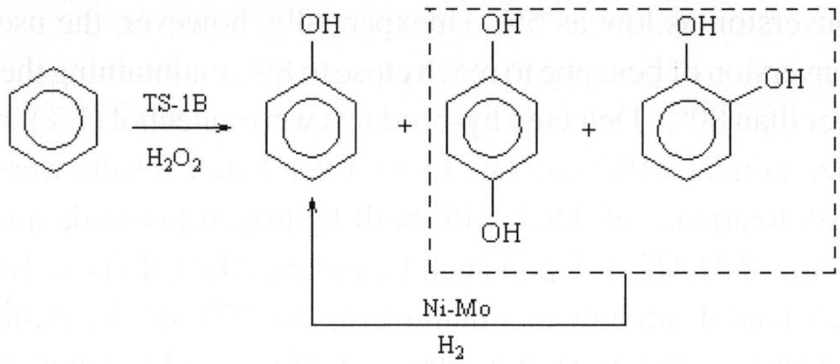


**Figure 15.1.** Polimeri Europa process for the direct oxidation of benzene to phenol.

hydrogen and partially deoxygenated to phenol, which is recycled back into the process (Fig. 15.1).[30]

The hydrodeoxygenation reaction (HDO) is carried out in the gas phase in a fixed-bed reactor (400°C, 25 bar of hydrogen), using commercial nickel and molybdenum oxides supported on alumina as catalysts. The HDO allows a quantitative conversion of both catechol and hydroquinone with a 96% selectivity to phenol.[31] The main by-products are heavy condensed polycyclic aromatic hydrocarbons.

The overall process performances, including the HDO section and the recycles, are:

Benzene conversion: 100%
$H_2O_2$ conversion: 100%
Selectivity on benzene: 97.7%
(Moles of produced phenol/moles of reacted benzene × 100)
Selectivity on $H_2O_2$: 71.0%
(Moles of produced phenol/moles of reacted $H_2O_2$ × 100)

Preliminary economic evaluations suggest that direct oxidation of benzene by hydrogen peroxide is not yet competitive compared with the traditional cumene process, particularly if the acetone recycling is taken into account, but also that it could become profitable should the acetone price fall close to its fuel value.[23]

### 15.3.4. *Cyclohexanone ammoximation*

Still in the phenol industrial cycle, after its reduction to cyclohexanone, a real breakthrough was achieved in the synthesis of ε-caprolactam, an intermediate produced at *ca.* 4 Mt/y which, in turn, is mainly used in nylon-6 production.

In the traditional approach to ε-caprolactam (Fig. 15.2), cyclohexanone reacts with hydroxylamine sulfate to obtain cyclohexanone oxime. The latter is then subjected to an oleum-catalyzed Beckmann rearrangement, affording the desired ε-caprolactam. This complex approach is penalized by its complexity, the necessity to avoid emissions of nitrous oxides ($NO_x$) and sulfur oxides ($SO_x$), and mainly by

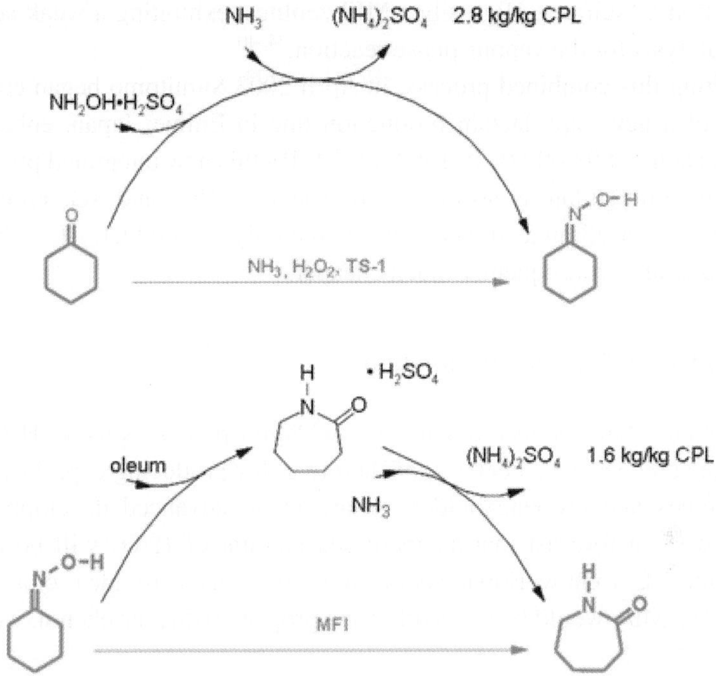


**Figure 15.2.** Comparison of the classical (black arrows) and green (green arrows) approaches to ε-caprolactam synthesis. In the ammoximation step 2.8 kg/kg ammonium sulfate is co-produced, and is comprised of the salt produced during hydroxylamine sulfate preparation (Raschig process) and cyclohexanone oxime synthesis.

the huge co-production of ammonium sulfate, which is formed well in excess of ε-caprolactam: about 2.8 kg/kg in the cyclohexanone oxime synthesis, and a further 1.6 kg/kg in the Beckmann rearrangement. So, the overall ammonium sulfate co-production, whose commercial value has become lower and lower in recent years, amounts to approximately 4.4 kg per kg of ε-caprolactam.[32]

An integrated process, which combines catalytic EniChem TS-1, catalyzed direct ammoximation of cyclohexanone and Sumitomo Chemical vapour-phase Beckmann rearrangement, both exploiting MFI based zeolite-like materials, is now industrially used for greener caprolactam production from cyclohexanone without co-producing any ammonium sulfate (Fig. 15.2).[33]

The first step of the integrated process, the ammoximation of cyclohexanone to cyclohexanone oxime, is based on a reaction, catalyzed by TS-1, between cyclohexanone, ammonia and hydrogen peroxide. It completely eliminates, on the one hand, the problems linked with the production and use of hydroxylamine and, on the other, the co-production of sulfates.

The second step, the Beckmann rearrangement in the vapour phase, is based on the transposition of cyclohexanone oxime to the corresponding lactam, without the co-production of sulfates. High silica MFI zeolites, exhibiting a weak acidity, are suitable catalysts for the vapour-phase reaction.[34–40]

Exploiting this combined process, in April 2003 Sumitomo began commercial operation of a new caprolactam production line in Ehime, Japan, enhancing the existing capacity of 93,000 t/y to 160,000.[41,42] By this new integrated process, low-value ammonium sulfate is no longer co-produced, $NO_x$ and $SO_x$ emissions are greatly reduced or eliminated, and water is the only co-product. Also, the plant is much simpler and, consequently much cheaper.

## 15.4. Hydrogen Peroxide Production

As new greener bulk industrial selective oxidation processes using $H_2O_2$ and Ti zeolite-like catalysts have been established (e.g. for producing caprolactam, dihydroxybenzenes and propene oxide), or are in an advanced development stage (phenol), we can forecast that an increased amount of $H_2O_2$ will be needed in the near future. Current worldwide demand for hydrogen peroxide is close to 2 Mt/y and about 0.2 Mt/y would be needed for one propene oxide, or phenol, world-scale plant.

A huge increase in $H_2O_2$ availability is a goal that can be reached in different ways:

(i) improvement of the classical process;
(ii) new production method(s), e.g. from carbon monoxide, oxygen and water, or a one-pot direct synthesis process from hydrogen and oxygen; or
(iii) $H_2O_2$ *in situ* generation from hydrogen and oxygen.

Dow/BASF chose an improved classical production approach further developed by Solvay with its high yield technology for the HPPO plants in Antwerp, Belgium and in Map Ta Phut, Thailand.[43]

Polimeri Europa focused its research on both the synthesis from carbon monoxide, oxygen and water and the direct synthesis from hydrogen and oxygen, developing the new DISY process ($H_2O_2$ production technology through direct synthesis).[44]

The *in situ* generation of $H_2O_2$ from hydrogen and oxygen is, in turn, a field of growing interest, pursued by many groups, for propene oxide and phenol production in particular.

### 15.4.1. *Current anthraquinone process*

At the industrial scale, hydrogen peroxide is produced almost exclusively by the alternate oxidation and reduction of alkylanthraquinone derivatives. This anthraquinone process, or AO (from autoxidation) process, was originally developed in Germany by IG Farbenindustrie between 1935 and 1945 but was scaled up to commercial scale only after the Second World War.

The process exploits the capability of a number of anthrahydroquinone derivatives to undergo an autoxidation reaction, eventually forming hydrogen peroxide. Practically, an organic working solution containing an alkylanthraquinone is subjected to cyclic hydrogenation and oxidation (Reactions 15.3 and 15.4):

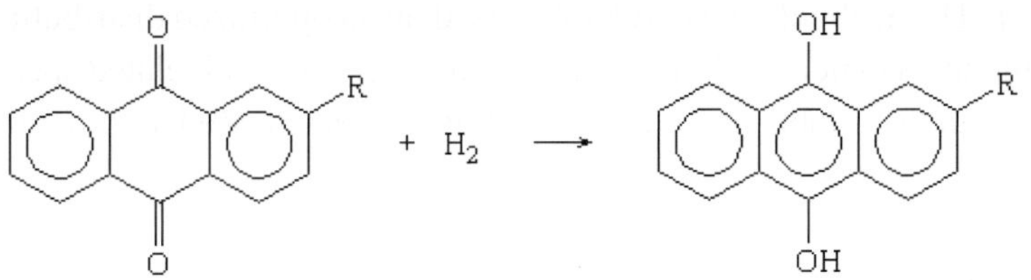


(15.3)

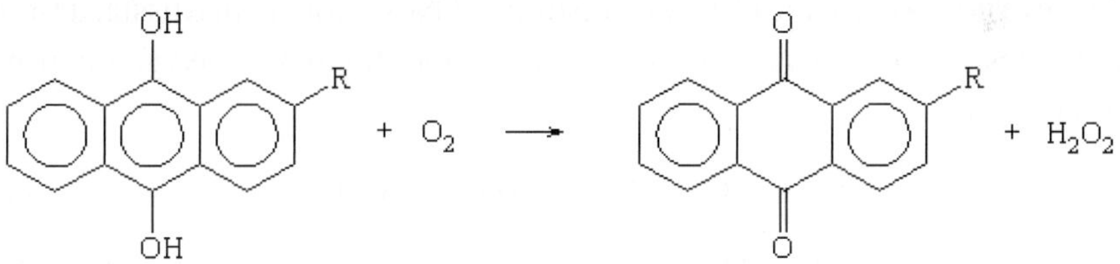


(15.4)

Hydrogen peroxide is formed, recovered by extraction with water, purified from all the organic impurities and finally concentrated by partial water removal under vacuum. The aqueous, commercial solution thus obtained is 60–70% by weight.

The process suffers from several drawbacks. For instance, the solvent must be able to dissolve both the apolar anthraquinone and the more polar hydroquinone and, at the same time, to provide immiscibility with water in order to allow the recovery of hydrogen peroxide. Complex mixtures of solvents are used, usually formed by an aromatic compound to dissolve the anthraquinone (toluene, methylnaphthalene) and a polar solvent to dissolve the hydroquinone (organic esters of phosphoric acid, diisobutylcarbinol, etc.). Even so, some derivatives of the quinone formed under the reaction conditions, particularly tetra- and octa-hydro derivatives are poorly soluble in the working solution.

Overall, the process is rather complex and is usually operated at relatively small plants (40–50 kt/y); few plants have a capacity that approaches 100 kt/y and many others are under 20 kt/y. So, the process is basically unable to provide the large amounts of cheap hydrogen peroxide that would be required should its use be developed for the synthesis of large volume intermediates, such as propene oxide or phenol.

For these reasons, considerable efforts have been made in recent years to identify and develop alternative routes to hydrogen peroxide and to improve existing ones. A new high-productivity/high-yield process has been developed by Solvay; this process doubles the productivity of the hydrogen peroxide plant. New hydrogen peroxide at a "mega-scale" single-train production unit in Antwerp, Belgium, the largest in the world (at 230 kt/y in a single line), is incorporating the new Solvay technology. Another plant of the same type and even higher capacity was planned for Thailand in 2011. This technology is probably based on an optimized distribution of alkyl anthraquinones (isomers of 2-amyl anthraquinone, AAQ, and related species), to be compared with the established working solution based on ethyl anthraquinone (EAQ).[45]

### 15.4.2. *Synthesis from carbon monoxide, oxygen and water*

In 1979 Yuri Yermakov, at the Catalysis Institute of Novosibirsk, in Russia, described a new synthesis of hydrogen peroxide from carbon monoxide, oxygen and water (Reaction 15.5):

$$O_2 + CO + H_2O \rightarrow H_2O_2 + CO_2 \qquad (15.5)$$

The reaction was catalyzed by palladium acetate and triphenylphosphine. However, despite favourable thermodynamics ($\Delta G^\circ_{298} = -134$ kJ/mol), the turnover number achieved (i.e. the moles of hydrogen peroxide formed per mole of palladium) was just five, mainly due to the oxidation of the triphenylphosphine ligand and the consequent precipitation of palladium metal.[46]

Some 20 years later it had become clear that nitrogen ligands, more stable towards oxidation than the phosphorus-based ones, could replace phosphines in a number of processes catalyzed by late transition metals. As a result, several nitrogen ligands, especially phenanthrolines and bipyridines, were tested for their capability to catalyze efficiently using the Yermakov reaction. The reactions were carried out in biphasic systems with the catalyst mostly dissolved into the organic phase, while the hydrogen peroxide was readily extracted in the aqueous one. In this way, the contact between the catalyst and the produced hydrogen peroxide was minimized, thus reducing the oxidation of the ligand and possible palladium-catalyzed decomposition of hydrogen peroxide. At the same time, separation between catalyst and product can be simply achieved by decantation, thus allowing recovery and reuse of the former. The reaction turned out to be significantly accelerated by small amounts of alcohols in the solvent medium: primary and secondary alcohols, however, were slowly oxidized, whereas tertiary alcohols were unaffected by the reaction conditions and proved to be the alcohols of choice. As far as nitrogen ligands are concerned, the best results were obtained with phenanthrolines with moderately bulky groups close to both nitrogen atoms: 2,9-dimethylphenanthrolines behave as excellent catalyst precursors. On the contrary, more hindered ligands, such as 2,9-di-*n*-butylphenanthroline and 2,9-diphenylphenanthroline, were unable to avoid precipitation of the palladium metal. In order to be sure that most of the palladium was actually confined to the organic phase, the organophilic 2,9-dimethyl-4,7-diphenyl-1,10-phenanthroline (bathocuproine) was selected as the ligand of choice. Small amounts of an organic acid were also needed.

Eventually, the best conditions turned out to be as follows: a toluene/2-methyl-2-butanol/water 35/25/40 biphasic mixture as solvent; palladium(II) acetate/bathocuproine/*p*-toluenesulfonic acid molar ratios 1/4/40; palladium concentration in the organic phase $9 \times 10^{-4}$ M; oxygen and carbon monoxide partial pressures 6,500 and 600 kPa, respectively; 70°C; reaction time 1 h. It should be noted that an excess of oxygen was required in the feeding gas in order to avoid any precipitation of the palladium metal, an important drawback under more reducing conditions. At the same time, the selected $O_2$/CO ratio provides a non-flammable mixture. Under the described conditions, it was possible (and easy) to work in a continuous operation mode. Simple phase-separation of the liquid allowed the recovery of an 8% aqueous solution of hydrogen peroxide, corresponding to a turnover number higher than 1,200.[47–49]

Regarding the reaction mechanism, in the original paper it was proposed that the reaction proceeded via reduction of a possible (bathocuproine)$Pd^{II}(AcO)_2$ complex by carbon monoxide through a hydride intermediate, affording a (bathocuproine)Pd° species. In order to close the catalytic cycle, it was already known that palladium(0) complexes (actually, with phosphines rather than phenanthrolines as ligands) could react with oxygen forming $Pd^{II}$ peroxo species and that the latter undergo hydrolysis forming hydrogen peroxide and restoring, at the same time, the starting $Pd^{II}$ species.

A few years later, this suggestion received substantial confirmation by the work of Shannon Stahl at the University of Wisconsin. The simple addition of two equivalents of bathocuproine to tris(dibenzylideneacetone)dipalladium(0) allowed the isolatation of the (bathocuproine)Pd°(dibenzylideneacetone) complex (I) which, apart from the dibenzylideneacetone moiety, is perfectly analogous to the Pd° species proposed as an intermediate in the catalytic cycle. Complex (I), fully characterized by $^1H$ and $^{13}C$ nuclear magnetic resonance (NMR), IR, UV-VIS spectroscopies and single-crystal X-ray crystallography, was then reacted with oxygen affording the expected peroxo complex (bathocuproine)$Pd^{II}(\eta^2 - O_2)$ (II), also characterized by NMR and IR spectroscopies and X-ray crystallography. Furthermore, it was found that the rate of the formation of complex (II) was fully comparable with the overall rate observed in our $H_2O_2$ synthesis. Finally, it was also confirmed that complex (II), treated with acetic acid, formed hydrogen peroxide and a complex (bathocuproine)$Pd^{II}AcO_2$ which is the starting species of the proposed mechanism.[50]

### 15.4.3. *Direct synthesis from hydrogen and oxygen*

The direct synthesis of hydrogen peroxide from hydrogen and oxygen (Reaction 15.6) is in principle the simplest method by which to form hydrogen peroxide and is forecast to offer lower production costs and also require quite low levels of investment.

$$H_2 + O_2 \rightarrow H_2O_2 \quad \Delta H = -135.8\ \text{kJ/mol} \qquad (15.6)$$

Although the catalytic liquid-phase oxidation of hydrogen for the production of hydrogen peroxide has been known since 1914, and a lot of patents have been issued, this process has not yet been put into practice.

There are three major drawbacks to the direct synthesis of hydrogen peroxide. First, hydrogen/oxygen mixtures are explosive over a wide range of concentrations, so the ratio of hydrogen to oxygen needs to be carefully controlled. Second, it is not easy to combine good results (productivity, selectivity, catalyst lifetime) and process conditions useful for $H_2O_2$ production and further industrial $H_2O_2$ purification and/or use. Third, despite its apparent simplicity, the reaction scheme is complex because of the occurrence of simultaneous or consecutive reactions, all of which are thermodynamically favoured and highly exothermic, that are working to lower selectivity:

(i) the formation of water

$$H_2 + 0,5O_2 \rightarrow H_2O \quad \Delta H = -241.6\ \text{kJ/mol} \qquad (15.7)$$

(ii) the decomposition of hydrogen peroxide

$$H_2O_2 \rightarrow 0,5O_2 + H_2O \quad \Delta H = -105.8\ \text{kJ/mol} \qquad (15.8)$$

(iii) the reduction of $H_2O_2$

$$H_2O_2 + H_2 \rightarrow 2H_2O \quad \Delta H = -211.5\ \text{kJ/mol} \qquad (15.9)$$

Each reaction can be favoured by experimental conditions (catalyst, promoters, acidity, solvent, etc.); a detailed description of the factors affecting conversion, selectivity and decomposition is available in published reviews.[51–53]

Based on the above premises and looking at the innovative industrial oxidation routes where hydrogen peroxide is used in combination with its proprietary TS-1 catalyst, Polimeri Europa focused on developing the new DISY process ($H_2O_2$ production technology through direct synthesis).

Development up to the pilot scale (5 l/h) of the direct synthesis of hydrogen peroxide from hydrogen and oxygen in methanol, carried out in very safe conditions, has been reported in patent applications[44,54,55] and presented at international congresses.[56,57]

By using continuous slurry-type reactors with recycled gas and gaseous feed make-up, $O_2$, $H_2$, $N_2$, liquid feed (solvent + promoters) and liquid effluent drawn through a sintered candle-type filter, concentrations of hydrogen peroxide higher than 7wt % and a molar selectivity based on hydrogen higher than 75% were obtained.

A productivity in the range of 120–200 kg/($m^3$*h) was also attained and a catalyst life of more than 2,000 hours, with typical reaction conditions as follows:

(i) $H_2$ (inlet) < 3.5%v; $O_2$ (inlet) < 12%v; $N_2$ to balance;
(ii) reaction temperature 40–50°C; reaction pressure 50–100 bar; and
(iii) solvent (feed): $CH_3OH$ + water; catalyst Pd-Pt/C; promoters: $H_2SO_4$, HBr.

Some of the key points reached and demonstrated, following the Polimeri Europa DISY approach, are the following:

(i) The gaseous mixtures employed lies well outside the explosion limits and process safety was witnessed by over 15,000 hours of trouble-free pilot operation (Fig. 15.3).
(ii) Optimization of solvent, promoters and reaction conditions and a deliberately developed heterogeneous catalyst, based on palladium and platinum as active

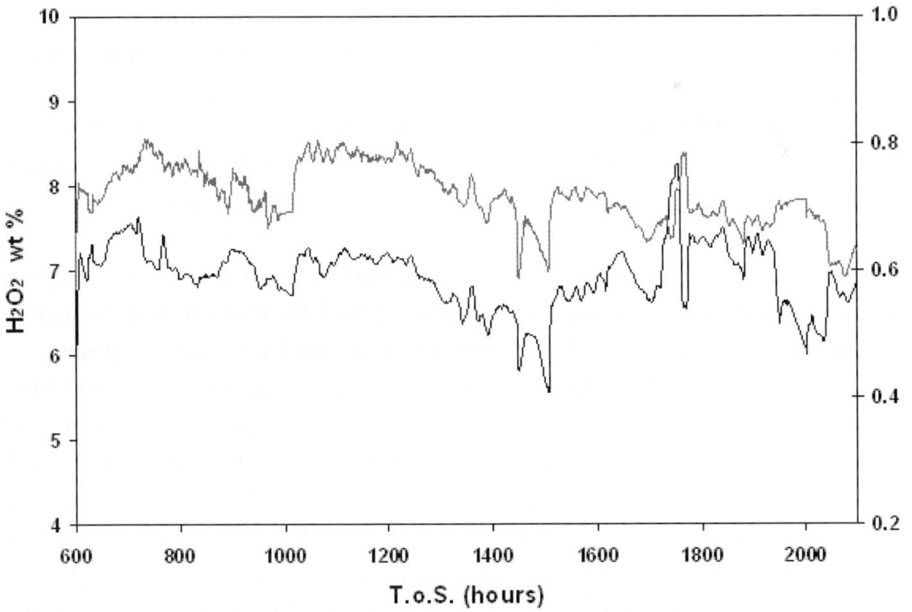


**Figure 15.3.** Direct synthesis of $H_2O_2$ from hydrogen and oxygen (Polimeri Europa DISY process). A typical pilot run: hydrogen peroxide concentration and hydrogen-based selectivity vs time on stream (T.o.S.). All values, including those measured during change in reaction parameters and plant stop-and-go procedures, are shown (smoothed by a moving average line on 10 periods).

components, allowed the achievement of low corrosion, long catalyst life and product stability.

(iii) Both integrated selective oxidation processes[54] as well as production of commercial-grade high-concentration aqueous hydrogen peroxide[55] are easily affordable.

## 15.5. Other Oxidations

As previously shown, the marriage of green oxidant agents, like $H_2O_2$, and well-developed catalysts, such as TS-1, is able to drive strong improvement in industrial chemistry.

Moreover, we have to cope with one of the biggest dreams/nightmares of industrial chemists: the high productivity AND selectivity AND safe oxidation of organic substrates by oxygen/air in the presence of suitable oxidation catalysts. Usually, this approach does not give fully acceptable results either with respect to the intrinsic safety of the process, or to its performances: quite often deep oxidation takes place or difficult safety constraints must be considered.

In the last 20 years, a number of approaches have been investigated.[58] The most promising are:

- A proper design of catalytically active sites, based on the principle of "site isolation",[59,60] coupled with appropriate catalyst morphology and shape. In this approach a limited array of oxidizing sites are structurally and chemically separated from the neighbouring active sites (e.g. mixed oxides and polyoxometalates).
- The cyclic approach in redox catalysis, also with the help of adapted special configurations of the catalytic reactors. From a theoretical point of view we can forecast: i) higher selectivity due to the reduced contribution of non-selective oxidation in the gas-phase or at the catalyst surface by hyper-reactive oxygen species; ii) improved safety since the approach avoids the formation of flammable mixtures in the gas phase; and iii) the handling of smaller volumes of gases.
- The choice of the best oxidant. If possible, the preparation of the "right oxidant" in line or *in situ* is considered economically promising, especially if a high level of process integration is considered, and must be carefully evaluated. Typical examples are: i) the use of alkylhydroperoxides in the oxidative desulfurization of gas-oil; ii) the continuous production of hydro-alcoholic solutions of hydrogen peroxide and their direct use in oxidation processes[61] for the in line approach; and iii) the hydro-oxidation of benzene[62] or propene with oxygen and hydrogen[63] for the *in situ* approach.

Often these approaches do not stand alone but are carefully connected; optimized catalysts are used in *ad hoc* technology configurations, by using proper oxidants.

A few examples follow that focus on the redox or depletive approach, the heteropolyacids/peroxide systems, and on oxidative desulfurization.

### 15.5.1. *Periodic operations: the redox approach*

A completely different approach to selective oxidation, with respect to the $H_2O_2$/TS-1 systems already described, was based on depletive, or redox, systems. This approach is obtained in atmospheres lacking an oxidant in the gas phase: the oxidation of the organic substrate takes place through lattice oxygen atoms of a, usually, multi-metal oxide pseudo-catalyst (or cataloreactant), via a Mars–van Krevelen type mechanism, and is followed by the re-oxidation of the reduced oxide in a separate step, spatially or temporary, thereby formally closing a catalytic cycle (Reactions 15.10 and 15.11).

$$\text{Substrate} + \text{Cat}_{\text{Ox}} \rightarrow \text{Product} + \text{Cat}_{\text{Red}} \quad (15.10)$$

$$\text{Cat}_{\text{Red}} + \text{Oxidant} \rightarrow \text{Cat}_{\text{Ox}} \quad (15.11)$$

As a general rule, this scheme is compatible with the concept of periodic operation in catalytic reactors.[64–67]

This approach can be conveniently used in order to enhance the efficiency in the selective oxidation of hydrocarbons by the spatially- or time-resolved separation of the reduction and re-oxidation step of the catalyst, by the application of the riser-regenerator concept (commercially realized by DuPont for the oxidation of butane to maleic anhydride)[68] or by a periodic or cyclic reactor operation (Fig. 15.4). This oxidation approach should:

(i) allow the maintenance of a very low, or even zero concentration of oxygen during the substrate oxidation, thus avoiding the formation of flammable mixtures in the gas phase and also, hopefully, allowing increased selectivity thanks to the suppression of undesired secondary reactions;
(ii) further improve the process safety requiring the handling of smaller volumes of gases;
(iii) allow the catalyst oxidation and reduction steps to be separately tuned, which can be carried out at different temperatures and conditions, an ideal situation to optimize both the reactions and the overall process performances; and
(iv) segregate discharge streams, with easier product separation.

Principal drawbacks of the method are due, in general, to the need for careful control of the inputs subjected to the periodic operations: for example, for circulating catalyst systems, a large mass of catalyst must be circulated per mass of product. The content of selective oxygen in the catalyst, catalyst shape and dimension, and attrition resistance are fundamental parameters to be fine- tuned. Furthermore, subjecting the

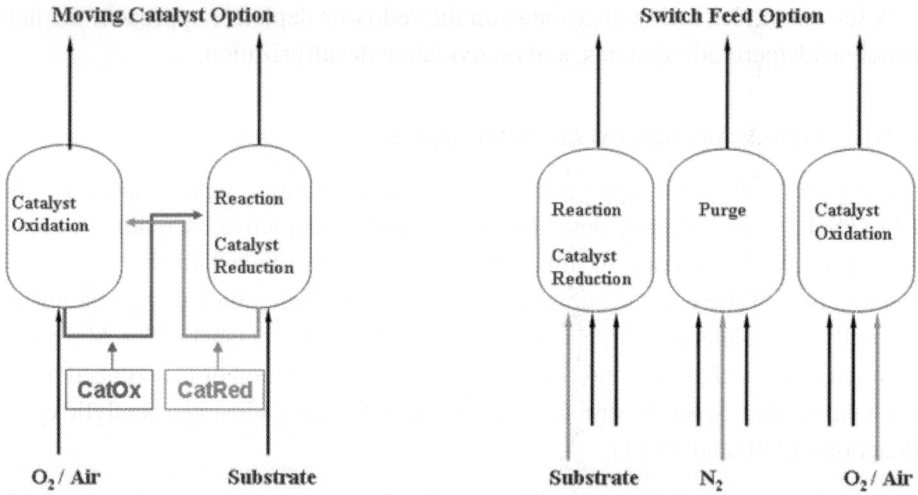


**Figure 15.4.** Schemes for the spatially-resolved separation of the reduction and re-oxidation steps, by moving catalyst or switching feed. In the first option (left), the substrate oxidation and the catalyst re-oxidation are spatially separated, in the second one (right), they are temporarily decoupled.

catalyst to transport, often at high speed, leads to an important attrition loss. Usually, re-oxidation of the catalyst isn't considered a key problem if it is reversible and fast enough; however, this point must be carefully evaluated at the lab stages of the research, with the available equivalent of useful oxygen per gram of catalyst, and researchers should not be surprised by the design of monster reactors or by the high speed required in the catalyst recirculation step.

Processes that have been studied with different redox approaches include:

- One-step, gas-phase, benzene to phenol direct oxidation over $Bi_{(1-x/3)}V_{(1-x)w}Nb_{(1-x)(1-w)}Mo_xO_4$ and $Cu_{(1-z)}Zn_zW_{(1-y)}Mo_yO_4$. Crystalline materials with a scheelite crystalline habit have been reported to be promising redox catalysts for this reaction.[69]
- Oxydehydrogenation of propane to propene, by vanadium oxide-based catalysts.[70,71]
- Dehydrogenation of alkylaromatics or paraffins to alkenylaromatics (in particular, styrene),[72,73] or olefins, on supported bismuth-vanadium oxides.
- Methane chemical looping combustion (CLC) over Ni-based mixed oxide materials.[74]
- Production of hydrogen over catalytic systems based on iron oxides.[75–78]

The approach has been also proposed as an energy conversion process with good potential to provide $CO_2$ capture via a chemical looping combustion cycle[74] as well as a process for the one-step hydrogen production by water splitting with intrinsic $CO_2$ separation.[75]

### 15.5.2. *The redox approach: the OSD (one step decarbonization) process*

Hydrogen is essential in the hydro-treatment processes used in refineries, or in complex bio-refineries, for the production of high-quality fuels with low environmental impact (e.g. desulfurization and intense dearomatization) and for the conversion of heavy crude oil and by-products into middle distillates. The importance of $H_2$ is also well known in petrochemistry (the synthesis of methanol, dimethyl ether, ammonia or hydrocarbons via Fischer–Tropsch). It also has potential as an energy vector for its clean fuel characteristics.

One-step hydrogen relates to a process for the production of hydrogen of high purity, which basically involves subjecting a solid to oxidation and treating the oxidized form thus produced with a hydrocarbon: the overall reaction leads to the formation of hydrogen or a species which can be easily transformed into hydrogen, as well as $CO_2$ which is obtained in a stream at a high concentration, which can be eliminated in exhaust reservoirs.

New processes for $H_2$ production, able to guarantee centralized production and $CO_2$ ready to be confined, are fundamental to drive towards total barrel conversion and the growing demand for high-quality fuels.

The OSD process is a highly innovative technology able to produce hydrogen from natural gas with simultaneous $CO_2$ capture. It belongs to the group of "chemical looping technologies" employing circulating fluidized beds and metal oxide particles for oxygen transfer, offering significantly reduced efforts and costs for gas separation. The OSD process is based on the use of a circulating "redox" solid material that can be oxidized via water splitting, thereby producing $H_2$, and reduced by a carbon-containing stream, typically hydrocarbons, producing $CO_2$. Readily sequestrable $CO_2$ is obtained within the redox loop after condensing water from the effluent stream purely composed of $CO_2$ and $H_2O$.

Water is the ideal raw material for hydrogen production, avoiding all the issues related to the exploitation of fossil fuels, so the OSD technology is aimed at producing hydrogen from water while reducing the capture of co-produced $CO_2$ to economically acceptable cost levels.

The general basic stoichiometry of the cycle is reported as follows:

$$4MeO_x + 4H_2O \rightarrow 4H_2 + 4MeO_{x+1} \tag{15.12a}$$

$$4MeO_{(x+1)} + CH_4 \rightarrow CO_2 + 2H_2O + 4MeO_x \tag{15.12b}$$

The overall reaction (15.12) is endothermic ($\Delta H = 39.5$ kcal/mol) and heat supply is needed to close the energy balance.

$$CH_4 + 2H_2O \rightarrow 4H_2 + CO_2 \tag{5.12}$$

Heat can be supplied by burning a small amount of hydrogen (15.12c)

$$x(H_2 + 0.5O_2 \rightarrow H_2O) \tag{15.12c}$$

so the overall reaction (15.12) becomes a weighted Eq. (15.13) where the energy balance depends upon the x value.

$$CH_4 + (2 - x)H_2O + 0.5xO_2 \rightarrow (4 - x)H_2 + CO_2 \tag{15.13}$$

An iron oxides system was identified as the best solid to support the cycle. The iron species involved are: $Fe_2O_3$ (hematite), where the iron has the maximum oxidation state ($Fe^{3+}$); $Fe_3O_4$ (magnetite), a structure where part of the iron is formally in a $Fe^{2+}$ state; FeO (wuestite), where all the iron atoms are formally in the $Fe^{2+}$ state; and $Fe^0$ (metallic iron).

In terms of overall material balance, the three consecutive reaction steps, in a quite generalized and simplified manner, can be written as follows:

$$8FeO + 8/3H_2O \rightarrow 8/3H_2 + 8/3Fe_3O_4 \quad Q = -47.4 \text{ Kcal} \tag{15.14a}$$

$$4Fe_2O_3 + 1CH_4 \rightarrow 8FeO + 1CO_2 + 2H_2O \quad Q = +96.39 \text{ Kcal} \tag{15.14b}$$

$$8/3Fe_3O_4 + 2/3O_2 \rightarrow 4Fe_2O_3 \quad Q = -90.2 \text{ Kcal} \tag{15.14c}$$

With the following overall reaction:

$$1CH_4 + 2/3H_2O + 2/3O_2 \rightarrow 8/3H_2 + 1CO_2 \quad Q = -40 \text{ Kcal} \tag{15.4}$$

The decoupling of the overall reaction (15.14) into three separate reactions allows us to point out that:

(i) the process meets the heat balance and the overall reaction heat is even negative, because the super-oxidation of the whole magnetite to hematite phase has a remarkable negative reaction heat; and
(ii) the fuel uptakes oxygen from the higher oxidation state such as hematite (15.14a) mainly forming a stream of $CO_2/H_2O$ with the negligible presence of reforming products.

The best solution to convert the elegant and fascinating chemistry of equations into a process is the choice of the appropriate reactor: a circulating fluidized-bed reactor (CFBR) configuration was considered the optimal choice. So the solid needs to have the appropriate mechanical resistance to support the stress of the circulation.

The amount and the rate of the exchanged oxygen per unit of mass of the oxygen carrier are both of paramount importance, because they dictate the space-time

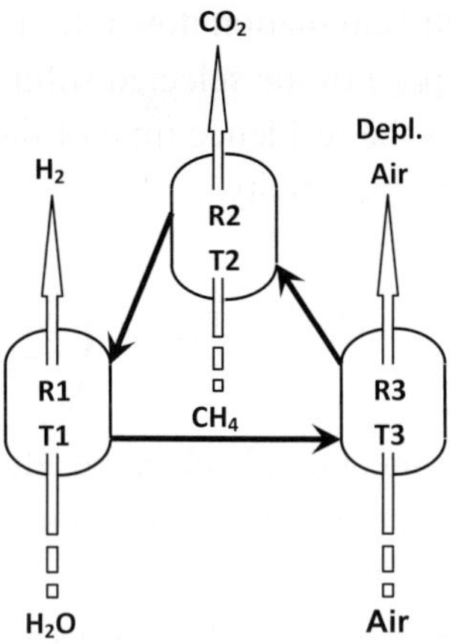


**Figure 15.5.** Concept scheme of the CFBR three vessels loop of the OSD process: R1 is a water-splitting reactor in which a partially reduced cataloreactant is used; R2 is a flameless combustor using the cataloreactant in the oxidized form; R3 can be a thermal unit in which part of the produced hydrogen can be used as fuel. T1, T2, and T3 represent different operative conditions (e.g. temperatures) of the reactors.

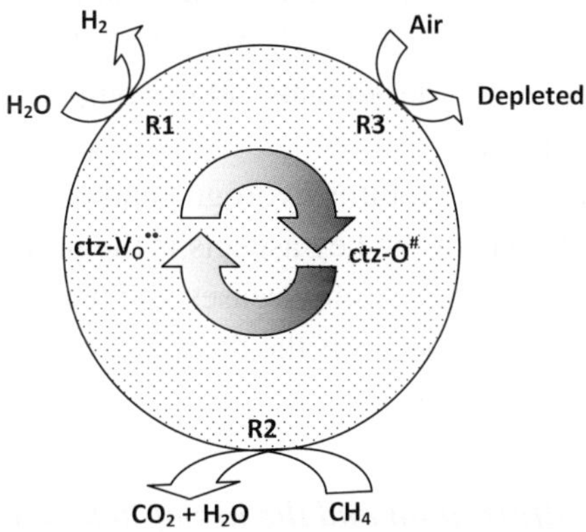


**Figure 15.6.** Concept scheme of the chemical evolution of the gas and solid reactants in the OSD process. The catalyst in oxygen rich and poor forms is represented as ctz-$O^{\#}$ and ctz-$V_O^{\bullet\bullet}$, respectively.

yield of the hydrogen production and because they govern the heat released during oxidation.

Figures 15.5 and 15.6 visually corroborate what is taking place in each vessel. It should be remembered that the solid is also the heat carrier and this adds additional constraints.

The sum of the adiabatic $\Delta T_i$ of reactions is dictated by the reaction kinetics, and the sum of reaction heats, in the unit time, ($\sum$j=R1-R3 Qj) should be balanced (15.15) with the solid circulation flow rate ($F_{OC}$) of the solid (kg/h) multiplied by the specific heat ($Cp_{OC}$) of the selected solid oxygen carrier (kcal/kg °C). The solid circulation controls the residence time of the solid in the reactors whose size depends on the required productivity.

$$\sum_{j} = R1 - R3 \quad Q_j = Cp_{OC} \times F_{OC} \times \left( \sum_{j} = R1 - R3\Delta T_j \right) \qquad (15.15)$$

If the heat release exceeds an optimal value, the adiabatic temperature ($\Delta T_i$) can be excessive, and an optimal temperature profile through the cycle should be kept to reach the highest efficiency. The efficiency of the system is a key parameter when evaluating a technology. The typical cold gas efficiency ($\eta_{cg}$) is given in (15.16) where $QH_2$ and $QCH_4$ are the molar fluxes of the hydrogen produced and fuel transformed.

$$h_{cg} = QH_2 \times DH_2 / QCH_4 \times CH_4 \qquad (15.16)$$

The preliminary data obtained returned an efficiency for the OSD process ($\eta_{OSD}$) in the order of 72% versus a literature value of the steam methane reforming (SMR) ranging from 73% (no steam export valorization) to 80% (maximum steam export valorization). Furthermore, in order to make a fair comparison, the SMR should be penalized for the $CO_2$ capture.

OSD is a unique process that returns hydrogen and pure carbon dioxide without any further management. One of the key aspects is the proprietary "multifunction" iron-based material: it not only reversibly ceases and picks up oxygen but it is also the heat carrier. The second key factor is the three vessels loop inherently returning the separated products.

### 15.5.3. *Oxidative desulfurization and the use of heteropolyacids*

In order to meet the new regulations about automotive fuels, deep desulfurization of oil fractions, towards near-zero sulfur liquid fuels, is mandatory.[79] It is generally achieved by hydrodesulfurization (HDS) which, can be very expensive (due to high capital costs, increased hydrogen consumption and shorter catalyst cycle) when applied to some recalcitrant higher-molecular-weight sulfur compounds (e.g. dibenzothiophenes).

An interesting new way to reduce the sulfur content in diesel below 10 ppm (ULSD, ultra-low sulfur diesel) is oxidative desulfurization (ODS) technology, which can be applied to pre-hydro-treated diesel.

The key feature of this technology is its complementarity with hydrodesulfurization: compounds which are less reactive with hydrogen are more reactive with

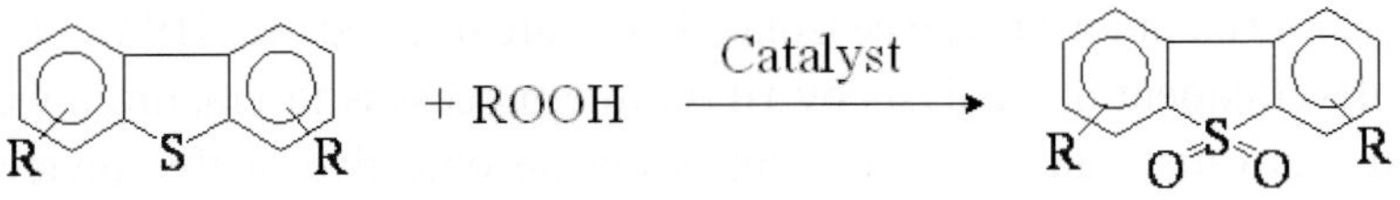


**Figure 15.7.** Alkyl-dibenzothiophene oxidation to the corresponding sulfone.

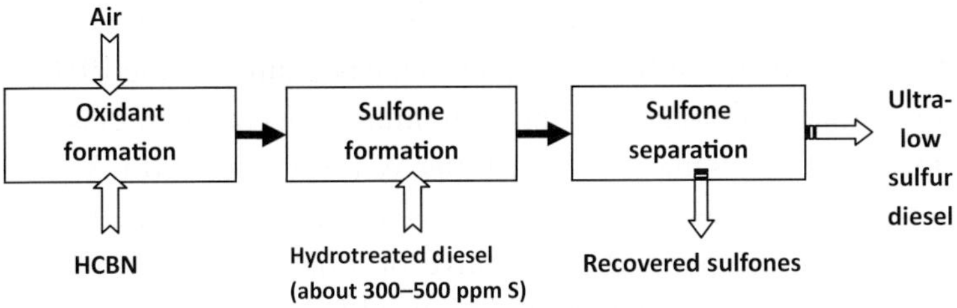


**Figure 15.8.** UOP/Eni oxidative desulfurization process scheme. HCBN (hydrocarbon) stands for any suitable hydrocarbon refinery stream. Adapted from Molinari *et al.*[81]

oxygen (e.g. dibenzothiophenes). These compounds can be oxidized under mild conditions by peroxides, and can be transformed to sulfoxides and sulfones (Fig. 15.7), highly polar compounds with physical properties quite different from those of gas-oil hydrocarbons, thus allowing their easy separation.

Another great advantage of the ODS process is its low reaction temperature and pressure, and that expensive hydrogen is not used in the process.

The oxidants used in this reaction are hydrogen peroxide, peracids or organic peroxides,[80] but the high cost of $H_2O_2$ or organic hydroperoxides makes the economics unfavourable in comparison with the traditional hydro-desulfurization process. In order to overcome this limitation, Eni/UOP jointly developed a new oxidative desulfurization process in which the hydroperoxide is directly produced in the reaction medium and in this way ULSD production can become economically convenient. A concept scheme of the process is given in Fig. 15.8.

Various oxidation catalysts proved to be effective in this new process, with heteropolyacids demonstrated to be amongst the most efficient.[58,82]

Molybdenum and tungsten in their highest oxidation state (VI) show an almost unique capability to form, along with the simple molybdate and tungstate anions, several polymolybdate and polytungstate species. Particularly upon acidification of molybdate or tungstate solutions also containing heteroatoms as other oxy anions (e.g. phosphate) or metal ions, the resulting polymolybdates and polytungstates can include such heteroatoms forming an extremely rich variety of heteropolymolybdates and heteropolytungstates. These heteropolyanions differ in stoichiometry and structure. In particular, the ratio of molybdenum or tungsten atoms to the heteroatoms is usually between 6 and 12. The free acids of many of these heteropolyanions are stable in acidic medium and can be isolated: they are referred to as HPAs. Despite the historic development of catalysis by HPAs (the process is fascinating), we cannot go deeply into the details as this item is not an objective of this review. However special issues[83,84] and very agreeable reviews[85–88] of this charming field are available.

The use of heteropolyanions as catalysts, however, was not new for the Eni researchers involved in oxidation chemistry. Some very peculiar heteropolyanions had been discovered at the Istituto Donegani at the beginning of the 1980s. Basically, they were quaternary ammonium (or phosphonium) salts of a new class of peroxidic tungsto-phosphates or arsenates in which the W (or Mo) to P (or As) ratio was as low as four: the best known, a typical example of this class, is the $PW_4O_{24}^{3-}$ anion, which was also the first peroxidic derivative of a heteropolyacid whose structure was solved.[89] Both the free acid and the reduced forms of these heteropolyanions are unstable, and it was not possible to isolate them.

The $PW_4O_{24}^{3-}$ anion turned out to be a valuable catalyst for a number of oxidations of organic substrates with hydrogen peroxide, including the epoxidation of olefins[90,91] and their cleavage, e.g. to get adipic acid from cyclohexene[92,93] or from the intermediate 1,2-cyclohexanediol.[94]

## References

1. Taramasso, M., Perego, G. and Notari, B. (1982). Preparation of Porous Crystalline Synthetic Material Comprised of Silicon and Titanium Oxides, *US Patent US4410501*. Snamprogetti Assignee.
2. Polimeri Europa Licensing Technology Brochure. (2009). Titanium Silicalite (TS-1) Zeolite Based Proprietary Catalyst. Available at: http://www.eni.com/en_IT/attachments/azienda/attivita-strategie/petrolchimica/licensing/TS1-A4-lug09.pdf (Accessed 11 February 2011).
3. Perego, C., Carati, A., Ingallina, P., *et al.* (2001). Production of Titanium Containing Molecular Sieves and their Application in Catalysis, *Appl. Catal. A: Gen.*, **221**, pp. 63–72.
4. Notari, B. (1996). Microporous Crystalline Titanium Silicates, *Adv. Catal.*, **41**, pp. 253–334.
5. Millini, R., Massara, E., Perego, G., *et al.* (1992). Framework Composition of Titanium Silicalite-1, *J. Catal.*, **137**, pp. 497–503.
6. Carati, A., Flego, C., Berti, D., *et al.* (1999). Influence of Synthesis Media on the TS-1 Characteristics, *Stud. Surf. Sci. Catal.*, **125**, pp. 45–52.
7. Bonino, F., Damin, A., Ricchiardi, G., *et al.* (2004). Ti-Peroxo Species in the TS-1/$H_2O_2$/$H_2O$ System, *J. Phys. Chem. B*, **108**, 11, pp. 3573–3583.
8. Clerici, M. (2001). The Role of the Solvent in TS-1 Chemistry: Active or Passive? An Early Study Revisited, *Top. Catal.*, **15**, pp. 257–263.
9. Parker Jr, W. and Millini, R. (2006). Ti Coordination in Titanium Silicalite-1, *J. Am. Chem. Soc.*, **128**, pp. 1450–1451.
10. Perego, G., Millini, R. and Belluss, G. (1998). Synthesis and Characterization of Molecular Sieves Containing Transition Metals in the Framework, in H. Karge and J. Weitkamp (eds), *Molecular Sieves Science and Technology*, Vol. 1, Springer, Berlin, pp. 187–228.

11. Carati, A., Berti, D., Millini, R., *et al.* (2010). Process for the Preparation of TS-1 Zeolites, *US Patent US2010/331576.* Polimeri Europa S.p.A. Assignee.
12. Clerici, M., Bellussi, G. and Romano, U. (1991). Synthesis of Propylene Oxide from Propylene and Hydrogen Peroxide Catalyzed by Titanium Silicalite, *J. Catal.*, **129**, pp. 159–167.
13. Clerici, M. (2006). TS-1 and Propylene Oxide, 20 Years Later, *Oil Gas-European Magazine*, **32**, pp. 77–82.
14. Paparatto, G., Forlin, A., De Alberti, G., *et al.* (2004). Integrated Process for the Preparation of Olefin Oxides, *US Patent US6888013.* Polimeri Europa S.p.A. Assignee.
15. Paparatto, G., Forlin, A. and Tegon, P. (2001). Process for the Preparation of Epoxides, *European Patent EP1072600.* EniChem S.p.A. Assignee.
16. Romano U. (2001). Ossido di Propilene. Nuova Tecnologia Produttiva. *Chim. Ind.* (*Milan*), **83**, pp. 30–31.
17. Dow Propyleneoxide News. (2009). Dow and BASF Win IChemE Award for Jointly Developed HPPO Technology. Available at http://www.dow.com/propyleneoxide/news/20091105a.htm (Accessed 13 May 2011).
18. Evonik Download Center. (2010). Evonik in South Korea. Available at: http://corporate.evonik.de/sites/dc/Downloadcenter/Evonik/Corporate/en/Locations/evonik-in-south-korea.pdf (Accessed 13 May 2011).
19. Sheldon, R. (1990). Catalytic Oxidations in the Manufacture of Fine Chemical, in G. Centi and F. Trifirò, (eds), *New Developments in Selective Oxidations*, Elsevier, Amsterdam, pp. 1–30.
20. Romano, U., Esposito, A., Maspero, F., *et al.* (1990). Selective Oxidation with Ti-silicalite, *Chim. Ind.* (*Milan*), **72**, pp. 610–616.
21. Bellussi, G. and Perego, C. (2008). Phenol Hydroxylation and Related Oxidations, in G. Ertl, H. Knozinger, F. Schüth, *et al.* (eds), *Handbook of Heterogeneous Catalysis,* $2^{nd}$ Edition, Vol. 8, Wiley-VCH, Weinheim, p. 3538.
22. Hock, H. and Lang, S. (1944). Autoxydation von Kohlenwasserstoffen, IX. Mitteil.: Über Peroxyde von Benzol-Derivaten, *Ber. Dtsch. Chem. Ges.* B, **77**, pp. 257–264.
23. Ricci, M., Bianchi, D. and Bortolo, R. (2009). Towards the Direct Oxidation of Benzene to Phenol, in F. Cavani, G. Centi, S. Perathoner, *et al.* (eds), *Sustainable Industrial Processes*, Wiley-VCH, Amsterdam, pp. 507–528.
24. Panov, G. (2000). Advances in Oxidation Catalysis: Oxidation of Benzene to Phenol by Nitrous Oxide, *Cattech*, **4**, pp. 18–32.
25. Uriarte, A., Rodkin, M., Gross, M., *et al.* (1997). Direct Hydroxylation of Benzene to Phenol by Nitrous Oxide, *Stud. Surf. Sci. Catal.*, **110** (Proceedings of the 3rd World Congress on Oxidation Catalysis), pp. 857–864.
26. Dubkov, K., Ovanesyan, N., Shteinman, A., *et al.* (2002). Evolution of Iron States and Formation of $\alpha$-Sites upon Activation of FeZSM-5 Zeolites, *J. Catal.*, **207**, pp. 341–352.
27. Balducci, L., Bianchi, D., Bortolo, R., *et al.* (2003). Direct Oxidation of Benzene to Phenol with Hydrogen Peroxide over a Modified Titanium Silicalite, *Angew. Chem. Int. Ed.*, **42**, pp. 4937–4940.
28. Bianchi, D., Balducci, L., Bortolo, R., *et al.* (2007). Oxidation of Benzene to Phenol with Hydrogen Peroxide Catalyzed by a Modified Titanium Silicalite (TS-1B), *Adv. Synth. Catal.*, **349**, pp. 979–986.
29. Drago, R., Wayland, B. and Carlson, R. (1963). Donor Properties of Sulfoxides, Alkyl Sulfites, and Sulfones, *J. Am. Chem. Soc.*, **85**, pp. 3125–3128.
30. Bianchi, D., Bortolo, R., Buzzoni, R., *et al.* (2004). Integrated Process for the Preparation of Phenol from Benzene with Recycling of By-products, *European Patent EP1424320.* Polimeri Europa S.p.A. Assignee.

31. Dalloro, L., Cesana, A., Buzzoni, R., *et al.* (2004). Process for the Preparation of Phenol by Means of the Hydrodeoxygenation of Benzene-diols, *European Patent EP1411038*. Polimeri Europa S.p.A. Assignee.
32. Clerici, M., Ricci, M. and Rivetti, F. (2006). Oxidation Processes with Hydrogen Peroxide and Hydroperoxides, in M. Beccari and U. Romano (eds) *Encyclopaedia of Hydrocarbons Vol. 2: Refining and Petrochemicals*, Eni & Istituto dell'Enciclopedia Italiana G. Treccani, Rome, pp. 661–686. Available at http://www.treccani.it/export/sites/default/Portale/sito/altre_aree/Tecnologia_e_Scienze_applicate/enciclopedia/inglese/inglese_vol_2/615-686_ING3.pdf (Accessed 13 May 2011).
33. Ichihashi, H. and Kitamura, M. (2002). Some Aspects of the Vapor Phase Beckmann Rearrangement for the Production of ε-Caprolactam over High Silica MFI Zeolites, *Catal. Today*, **73**, pp. 23–28.
34. Sato, H., Hirose, K., Ishii, N., *et al.* (1987). Production of ε-Caprolactam, *European Patent EP0234088*. Sumitomo Chemical CO Assignee.
35. Kitamura, M., Ichihashi, H. and Tojima, H. (1992). Process for Producing ε-Caprolactam and Activating Solid Catalysts Thereof, *European Patent EP0494535*. Sumitomo Chemical CO Assignee.
36. Heitmann, G., Dahlhoff, G. and Hölderich, W. (1999). Catalytically Active Sites for the Beckmann Rearrangement of Cyclohexanone Oxime to ε-Caprolactam, *J. Catal.*, **186**, pp. 12–19.
37. Bordiga, S., Ugliengo, P., Damin, A., *et al.* (2001). Hydroxyls Nests in Defective Silicalites and Strained Structures Derived upon Dehydroxylation: Vibrational Properties and Theoretical Modelling, *Top. Catal.*, **15**, pp. 43–52.
38. Ichihashi, H. and Kitamura, M. (2002). Some Aspects of the Vapor Phase Beckmann Rearrangement for the Production of ε-Caprolactam over High Silica MFI Zeolites, *Catal. Today*, **73**, pp. 23–28.
39. Flego, C. and Dalloro, L. (2003). Beckmann Rearrangement of Cyclohexanone Oxime over Silicalite-1: An FT-IR Spectroscopic Study, *Micropor. Mesopor. Mat.*, **60**, pp. 263–271.
40. Cesana, A., Palmery, S., Buzzoni, R., *et al.* (2010). Silicalite-1 Deactivation in Vapour Phase Beckmann Rearrangement of Cyclohexanone Oxime to Caprolactam, *Catal. Today*, **154**, pp. 264–270.
41. Sumitomo Chemical (2004). *Sumitomo Annual Report 2004*. Available at http://www.sumitomo-chem.co.jp/english/ir/library/annual_report/docs/ar2004_e.pdf (Accessed 16 May 2011).
42. Sumitomo Chemical (2003). *Sumitomo Annual Report 2003*. Available at http://www.sumitomo-chem.co.jp/english/ir/library/annual_report/docs/ar2003.pdf (Accessed 16 May 2011).
43. SOLVAY live (2007). High-yield Hydrogen Peroxide Production in Asia. *Solvay live*, **253**, p. 5. Available at http://www.solvaysites.com/sites/corporate/EN/NewsPress/Solvaygroupmagazine/Archives/Documents/SL253_Magazine_EN.pdf (Accessed 20 May 2011).
44. Paparatto G., D'Aloisio R., De Alberti G., *et al.* (2001) Catalyst and Process for the Direct Synthesis of Hydrogen Peroxide, *European Patent EP1160196*. Eni SpA-Polimeri Europa S.p.A. Assignee.
45. Chemsystems (2009). *Hydrogen Peroxide Report Abstract*, PERP 07/08-3, available at http://www.chemsystems.com/reports/search/docs/abstracts/0708_3_abs.pdf (Accessed 13 May 2011).
46. Zudin, V., Likholobov, V. and Yermakov, Y. (1979). Catalytic Synthesis of Hydrogen Peroxide from Oxygen and Water in the Presence of Carbon Monoxide and Phosphine Complexes of Palladium, *Kinet. Katal.* (*USSR*) (*Engl. Transl.*), **20**, pp. 1324–1325.
47. Bianchi, D., Bortolo, R., D'Aloisio, R., *et al.* (1999). Biphasic Synthesis of Hydrogen Peroxide from Carbon Monoxide, Water, and Oxygen Catalyzed by Palladium Complexes with Bidentate Nitrogen Ligands, *Angew. Chem. Int. Ed.*, **38**, pp. 706–708.
48. Bianchi, D., Bortolo, R., D'Aloisio, R., *et al.* (1999). Synthesis of Hydrogen Peroxide from Carbon Monoxide, Water, and Oxygen Catalysed by Palladium Complexes: A Study of the Catalyst Stabilisation, *Stud. Surf. Sci. Catal.*, **126** (Catalyst Deactivation), pp. 481–484.

49. Bianchi, D., Bortolo, R., D'Aloisio, R., *et al.* (1999). A Novel Palladium Catalyst for the Synthesis of Hydrogen Peroxide from Carbon Monoxide, Water And Oxygen, *J. Mol. Catal.*, **150**, pp. 87–94.
50. Stahl, S., Thorman, J., Nelson, R., *et al.* (2001). Oxygenation of Nitrogen-coordinated Palladium(0): Synthetic, Structural, and Mechanistic Studies and Implications for Aerobic Oxidation Catalysis, *J. Am. Chem. Soc.*, **123**, pp. 7188–7189.
51. Campos-Martin, J., Blanco-Brieva, G. and Fierro, J. (2006). Hydrogen Peroxide Synthesis: An Outlook Beyond the Anthraquinone Process, *Angewandte Chemie International Edition*, **45**, pp. 6962–6984.
52. Samanta, C. (2008). Direct Synthesis of Hydrogen Peroxide from Hydrogen and Oxygen: An Overview of Recent Developments in the Process, *Appl. Catal. A: Gen.*, **350**, pp. 133–149.
53. Centi, G. and Perathoner, S. (2009). One-step $H_2O_2$ and Phenol Syntheses: Examples of Challenges for New Sustainable Selective Oxidation Processes, *Catal. Today*, **143**, pp. 145–150.
54. Paparatto, G., D'Aloisio, R., De Albert, G., *et al.* (2000). New Catalyst, Process for the Production of Hydrogen Peroxide and its Use in Oxidation Processes, *European Patent EP0978316*. Polimeri Europa S.p.A. Assignee.
55. Paparatto G., Rivetti F., Andrigo P., *et al.* (2002). Process for the Production of Hydrogen Peroxide, *International Application WO0214217*. EniChem S.p.A.-Eni S.p.A.
56. Paparatto G., Buzzoni R. and Rivetti F. (2009). DISY: Toward a New Process for the Direct Synthesis of Hydrogen Peroxide, *6th World Congress on Oxidation Catalysis – ORAL COMMUNICATIONS Book of Abstracts*, O15-2A, pp. 96–97.
57. Buzzoni, R. and Perego, C. (2010). Strong Improvements and New Needs: A Two-way Connection in Industrial Catalysis, *The Sixth Tokyo Conference on Advanced Catalytic Science and Technology & The Fifth Asia Pacific Congress on Catalysis. Book of Abstracts*, IL-A02, pp. 22–23.
58. Cavani, F., Ballarini, N. and Luciani, S. (2009). Catalysis for Society: Towards Improved Process Efficiency in Catalytic Selective Oxidations, *Top. Catal.*, **52**, pp. 935–947.
59. Grasselli, R. (2002). Fundamental Principles of Selective Heterogeneous Oxidation Catalysis, *Top. Catal.*, **21**, pp. 79–88.
60. Grasselli, R. (2005). Selectivity Issues in (Amm) oxidation Catalysis, *Catal. Today*, **21**, pp. 23–31.
61. Paparatto, G., Rivetti, F., Andrigo, P., *et al.* (2001). Process for the Continuous Production of Hydrogen Peroxide in Organic Solvents, Using a Hydrogen Concentration Smaller than 4.5%Vol. and an Oxygen Concentration Smaller than 21.5%Vol., *European Patent EP1160195*. Eni S.p.A-EniChem S.p.A. Assignee.
62. Yamanaka, I. (2002). Reductive Activation of $O_2$ and Monooxygenation of Hydrocarbons by Eu Catalyst, *Catalysis Surveys from Japan*, **6**, pp. 63–72.
63. Dhingra, S., Schroden, R., Watson, K., *et al.* (2010). Hydro-oxidation Process Using a Catalyst Prepared from a Gold Cluster Complex, *US Patent US20100076208*. Dow Chemical Company Assignee.
64. Emig, G. and Liauw, M. (2002). New Reaction Engineering Concepts for Selective Oxidation Reactions, *Top. Catal.*, **21**, pp. 11–24.
65. Silveston, P., Hudgins, R. and Renken, A. (1995). Periodic Operation of Catalytic Reactors — Introduction and Overview, *Catal. Today*, 1995, **25**, pp. 91–112.
66. Boreskov, G. and Matros, Y. (1983). Flow Reversal of Reaction Mixture in a Fixed Catalyst Bed: A Way to Increase the Efficiency of Chemical Processes, *Appl. Catal.*, **5**, pp. 337–343.
67. Matros, Y. and Bunimovich, G. (1996). Reverse-flow Operation in Fixed-bed Catalytic Reactors, *Catal. Rev.*, **38**, pp. 1–68.
68. Contractor, R., Garnett, D., Horowitz, H., *et al.* (1994). A New Commercial-scale Process For *n*-Butane Oxidation to Maleic-anhydride Using a Circulating Fluidized-bed Reactor, *Stud. Surf. Sci. Catal.*, **82**, pp. 233–242.

69. Dalloro, L., Cesana, A., Buzzoni, R., *et al.* (2008). Process for the Preparation of Phenol, *US Patent US7868210.* Polimeri Europa S.p.A Assignee.
70. Ballarini, N., Cavani, F., Cericola, A., *et al.* (2004). Supported Vanadium Oxide-based Catalysts for the Oxidehydrogenation of Propane under Cyclic Conditions, *Catal. Today*, **91–92**, pp. 99–104.
71. Sanfilippo, D. and Miracca, I. (2006). Dehydrogenation of Paraffins: Synergies between Catalyst Design and Reactor Engineering, *Catal. Today*, **111**, pp. 133–139.
72. Ingallina, P., Carluccio, L., Bellussi, G., *et al.* (2000). Catalytic System and Process for the Oxidative Dehydrogenation of Alkylaromatics or Paraffins to the Corresponding Alkenylaromatics or the Corresponding Olefins, *European Patent EP1057530.* EniChem S.p.A.-EniTecnologie S.p.A. Assignee.
73. Ingallina, P., Carluccio, L., Bellussi, G., *et al.* (2001). Process for the Preparation of Catalytic Systems for the Oxidative Dehydrogenation of Alkylaromatics or Paraffins, *European Patent EP1160011.* EniChem S.p.A.-EniTecnologie S.p.A. Assignee.
74. Villa, R., Cristiani, C., Groppi, G., *et al.* (2003). Ni Based Mixed Oxide Materials for $CH_4$ Oxidation under Redox Cycle Conditions, *J. Mol. Catal. A: Chem.*, **204–205**, pp. 637–646.
75. Sanfilippo, D., Paggini, A., Piccoli, V., *et al.* (2001). Process for the Production of Hydrogen, *US Patent US687541.* Snamprogetti S.p.A. Assignee.
76. Cornaro, U. and Sanfilippo, D. (2004). Catalytic System and Process for the Production of Hydrogen, *US Patent US2004152790.* Eni S.p.A - Snamprogetti S.p.A. Assignee.
77. Sanfilippo, D., Miracca, I., Cornaro, U., *et al.* (2004). One-step-hydrogen: A New Direct Route by Water Splitting to Hydrogen with Intrinsic $CO_2$ Sequestration, *Stud. Surf. Sci. Catal.*, **147**, pp. 91–96.
78. Mizia, F., Rossini, S., Cozzolino, M., *et al.* (2009). One Step Decarbonization, in L. Eide (ed.), *Carbon Dioxide Capture for Storage in Deep Geologic Formations*, Vol. 3, CPL Press and BP, Thatcham, UK, pp. 201–220.
79. Barbara, P., Navarro, R., Campos-Martin, J., *et al.* (2011). Towards Near Zero-sulfur Liquid Fuels: A Perspective Review, *Catal. Sci. & Technol.*, **1**, pp. 23–42.
80. Campos-Martin J., Capel-Sanchez M., Perez-Presas P., *et al.* (2010). Oxidative Processes of Desulfurization of Liquid Fuels, *J. Chem. Technol. Biot.*, **85**, pp. 879–890.
81. Molinari, D., Baldiraghi, F. and Gosling, C. (2005). Oxidative Desulfurization in ULSD Strategies. *Oxidation and Functionalization: Classical and Alternative Routes and Sources*, DGMK-Conference, Milan, abstract available at http://www.dgmk.de/petrochemistry/abstracts_content13/Molinari.pdf (Accessed 26 May 2011).
82. de Angelis, A., Pollesel, P., Molinari, D., *et al.* (2007). Heteropolyacids as Effective Catalysts to Obtain Zero Sulfur Diesel, *Pure Appl. Chem.*, **79**, pp. 1887–1894.
83. AA.VV. (2003). *Applied Catalysis A: General*, **256**, Heteropoly Acids Special Issue, I. Kiricsi (ed.), pp. 1–327.
84. AA.VV. (1998). *Chemical Review*, **98**, Polyoxometalates, Hill, C.L. (ed.), pp. 1–390.
85. Misono, M. (2001). Unique Acid Catalysis of Heteropoly Compounds (Heteropolyoxometalates) in the Solid State, *Chem. Commun.*, **13**, pp. 1141–1152.
86. Timofeeva, M. (2003). Acid Catalysis by Heteropoly Acid, *Appl. Catal. A: Gen.*, **256**, pp. 19–35.
87. Jentoft, F., Klokishner, S., Kröhnert, J., *et al.* (2003). The Structure of Molybdenum-heteropoly Acids under Conditions of Gasphase Selective Oxidation Catalysis: A Multi-Method In Situ Study, *Appl. Catal. A: Gen.*, **256**, pp. 291–317.
88. Song, I. and Barteau, M. (2004). Redox Properties of Keggin-type Heteropolyacid (HPA) Catalysts: Effect of Counter-cation, Heteroatom, and Polyatom Substitution, *J. Mol. Catal. A: Chem.*, **212**, pp. 229–236.
89. Venturello, C., D'Aloisio, R., Bart, J., *et al.* (1985). A New Peroxotungsten Heteropoly Anion with Special Oxidizing Properties: Synthesis and Structure of Tetrahexylammoium Tetra(diperoxotungsto)phosphate (3-), *J. Mol. Catal.*, **32**, pp. 107–110.

90. Venturello, C., Alneri, E. and Ricci, M. (1983). A New, Effective Catalytic System for Epoxidation of Olefins by Hydrogen Peroxide under Phase-Transfer Conditions, *J. Org. Chem.*, **48**, pp. 3831–3833.
91. Venturello, C. and D'Aloisio, R. (1988). Quaternary Ammonium Tetrakis(diperoxotungsto) phosphates(3-) as a New Class of Catalysts for Efficient Alkene Epoxidation with Hydrogenperoxide, *J. Org. Chem.*, **53**, pp. 1553–1557.
92. Venturello, C. and Ricci, M. (1984). Carboxylic Acids from Olefins or Vicinal Dihydroxy Compounds, *European Patent EP 122804*. Montedison S.p.A. Assignee.
93. Antonelli, E., D'Aloisio, R., Gambaro, M., *et al.* (1998). Efficient Oxidative Cleavage of Olefins to Carboxylic Acids with Hydrogen Peroxide Catalyzed by Methyltrioctylammonium Tetrakis(oxodiperoxotungsto)phosphate(3-) under Two-phase Conditions. Synthetic Aspects and Investigation of the Reaction Course, *J. Org. Chem.*, **63**, pp. 7190–7206.
94. Venturello, C. and Ricci, M. (1986). Oxidative Cleavage of 1,2-diols to Carboxylic Acids by Hydrogen Peroxide, *J. Org. Chem.*, **51**, pp. 1599–1602.

# Chapter 16

# Selective Oxidation in DSM: Innovative Catalysts and Technologies

Paul L. ALSTERS,* Jean-Marie AUBRY,† Werner BONRATH,‡ Corinne DAGUENET,§ Michael HANS,‡ Walther JARY,** Ulla LETINOIS,‡ Véronique NARDELLO-RATAJ,† Thomas NETSCHER,‡ Rudy PARTON,§ Jan SCHÜTZ,‡ Jaap VAN SOOLINGEN,§ Johan TINGE§ and Bettina WÜSTENBERG‡

Selective oxidations are key transformations for the industrial production of a wide range of chemicals. Within DSM, oxidations reactions are used in the synthesis of vitamins and carotenoids, aroma compounds and polymer intermediates. Here we describe the use of both chemocatalysts and biocatalysts used in the industrial oxidation of aliphatic and aromatic systems, alcohols and poly hydroxyl-compounds.

## 16.1. Polyhydroxy Compounds: Ascorbic Acid

### 16.1.1. *Industrial production of vitamin C*

L-ascorbic acid or vitamin C (Fig. 16.1) is a water-soluble carbohydrate acid with antioxidant properties and is an essential nutrient for humans. It was first isolated from animal and plant sources in 1928.[1] Five years later, the molecular structure of crystallised ascorbic acid was elucidated.[2]

The strong reducing behaviour of vitamin C is based on its enediol structure (Fig. 16.1). The oxidation of ascorbic acid is a two-step process in which semi-dehydroascorbic acid, a strong acid (dimmer form), is involved.[3] This intermediate acts as a radical scavenger. The autoxidation of ascorbic acid is catalysed by transition metal ions[4] and dehydroascorbic acid is formed in the presence of oxygen.[5]

*DSM Innovative Synthesis B.V., A Unit of DSM Pharma Chemicals B.V., P.O. Box 18, 6160 MD Geleen, The Netherlands.

†Université Lille 1, Sciences et Technologie, LCOM, Equipe "Oxydation et Physico-chimie de la Formulation", CNRS UMR 8009, Bât. C6, 59655 Villeneuve d'Ascq Cedex, France.

‡Research and Development, DSM Nutritional Products, P.O. Box 2676, CH-4002 Basel, Switzerland.

§DSM Research, Industrial Chemicals, P.O. Box 18, 6160 MD Geleen, The Netherlands.

**DSM Fine Chemicals Austria, P.O. Box 296, 4021 Linz, Austria.

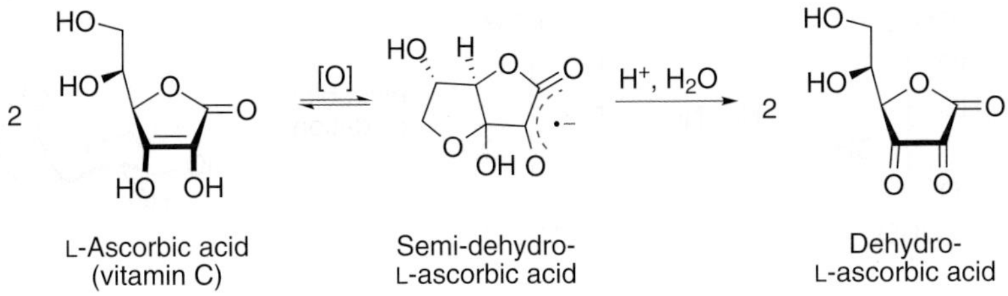


**Figure 16.1.** Antioxidant properties of L-ascorbic acid.

The reversible oxidation of ascorbic acid is the basis of its physiological activity. The reduction of dehydroascorbic acid, stable at pH 2.5–5.5 at 4°C for days, can be easily achieved by cysteine.[6]

The antioxidant and health-related properties of ascorbic acid offer a wide range of industrial applications, e.g. as a food and feed additive, and in the pharmaceutical, polymer, photographic and cosmetic industries.[7,8] L-ascorbic acid is by far the vitamin with the largest production volume; approximately 110,000 tons are manufactured globally each year.

Four Chinese producers are today supplying the majority of the global demand for this vitamin: (1) Weisheng Pharmaceutical Co (CSPC, Shijiazhuang), (2) North China Pharmaceutical Group Corp. (NCPC, Shijiazhuang), (3) Northeast Pharmaceutical Group Company Ltd. (NPGC, Shijiazhuang), and (4) Jiangshan Pharmaceutical Company (JSPC, Suzhou). As the sole remaining Western manufacturer, DSM Nutritional Products (formerly Roche Vitamins) has positioned itself in the premium segment of the vitamin C market (producing Quali-C™).

For more than 70 years the majority of commercially synthesised L-ascorbic acid has been produced by a variety of processes, which are generally variations of the Reichstein process. This process is a mixed chemical/fermentation synthesis method, which was developed by Reichstein and Grüssner in 1933.[9] In the classical procedure D-glucose is converted in five steps (four chemical steps plus one microbial step) into L-ascorbic acid (Fig. 16.2). In the first step D-glucose is hydrogenated to D-sorbitol. This catalytic hydrogenation is accomplished at high pressures and elevated temperatures over a nickel-alloy catalyst. In the second step D-sorbitol is regiospecifically oxidised to L-sorbose under aerobic conditions using *Gluconobacter* strains as the biocatalyst (formerly described as *Acetobacter*). This fermentative reaction is catalysed by sorbitol dehydrogenase (SLDH). In an acid catalysed reaction L-sorbose is then treated with acetone to yield 2,3:2,4-di-*O*-acetone-$\alpha$-L-sorbofuranose. Oxidation of this protected carbohydrate in high yield and selectivity results in diacetone-2-keto-L-gulonic acid, which can be deprotected to 2-keto-L-gulonic acid, followed by rearrangement under basic conditions or directly by acid treatment into L-ascorbic acid.[10]

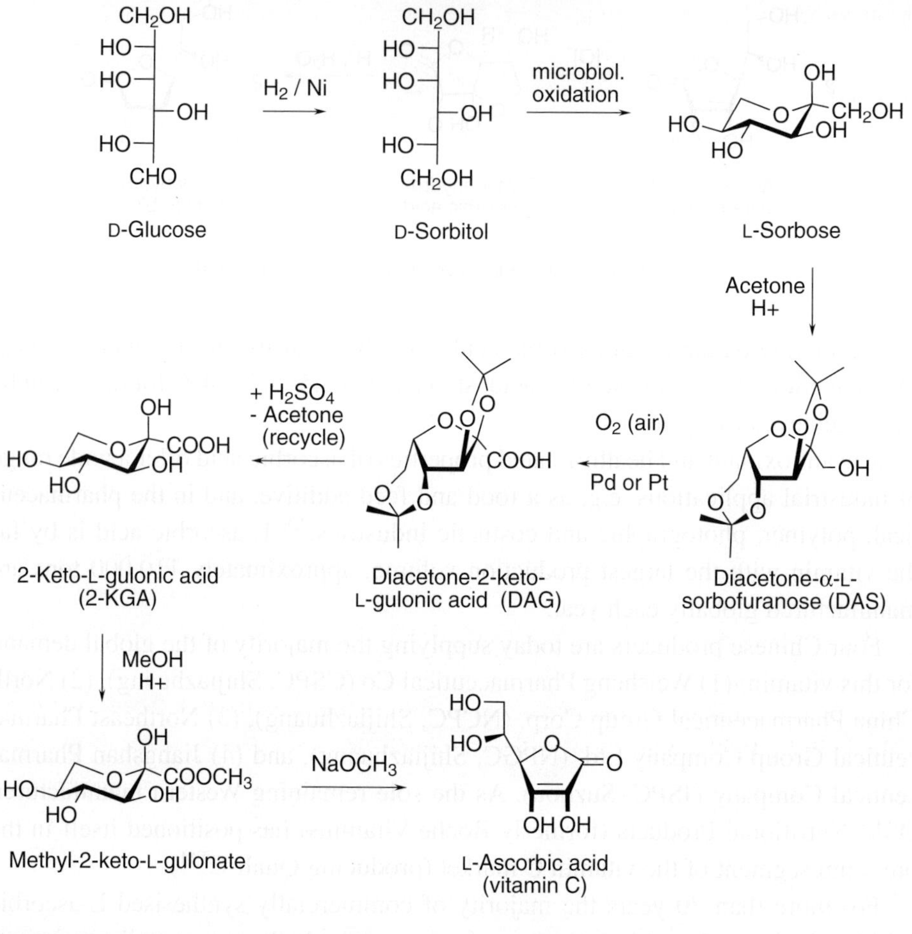


**Figure 16.2.** Schematic of the Reichstein process for the technical synthesis of L-ascorbic acid.

Historically, the oxidation reaction of diacetone-$\alpha$-L-sorbofuranose was carried out by the application of hypochlorite and a Ni-salt, electrochemistry,[8,11] or air oxidation in the presence of supported Pd or Pt catalysts (Fig. 16.2).[12,13]

A recent topic of oxidation reactions in the field of L-ascorbic acid is the direct oxidation of L-sorbose to 2-keto-L-gulonic acid.[14] In the past, several research groups have investigated this route with limited success.[15–17] Nowadays, oxidation of polyhydroxy compounds with an Au catalyst shows promising results. These new types of catalysts may solve the problem of selectivity during sorbose oxidation.[18,19]

While the first industrial trials using the Reichstein process allowed a yield of only 15–20% L-ascorbic acid, it is today possible to achieve a yield based on D-glucose of approximately 60%. However, the process is still characterised by high energy consumption, the need for high temperatures and pressures as well as the use of organics and water. As a result an increasing interest in the development of sustainable microbial alternatives to the chemical process has arisen in recent decades. With the application of modern technologies and novel scientific tools (e.g. genetic engineering and bioprocess design), it has become possible to develop processes which replace some of the chemical steps involved in the Reichstein synthesis of vitamin C.

### 16.1.2. *Microbial production of 2-KGA from D-sorbitol and/or L-sorbose*

The various production processes are usually classified according to: (i) the number of stages and (ii) the use of pure or mixed cultures. In this context a stage is described as a process unit applied for the conversion of a certain substrate into a certain product. This can sometimes be misleading because a process separation into different stages is either carried out to split the individual reaction steps performed by different micro-organisms or to improve the conversion of the same reaction by a sequential (cascade) process operation. Also, the term "mixed culture" can describe two different scenarios: the combined cultivation of a production strain with a growth facilitating strain or two individual production strains.

A two-stage 2-KGA-process starting from D-sorbitol was developed by Chinese scientists.[20,21] Similar to the Reichstein process, the 2-KGA-process starts with the conversion of D-glucose via D-sorbitol to L-sorbose and ends with the oxidation/rearrangement of 2-KGA to L-ascorbic acid. However, the two chemical reaction steps for the oxidation of L-sorbose to 2-KGA are replaced by a second mixed culture fermentation step with *Ketogulonicigenium* and a "helper strain" (e.g. *Bacillus megaterium*). Two dehydrogenases from *Ketogulonicigenium* are involved in the oxidation steps: (1) L-sorbose is oxidised to L-sorbosone (for chemical structures, see Figs. 16.2 and 16.3) by sorbose dehydrogenase (SDH) and

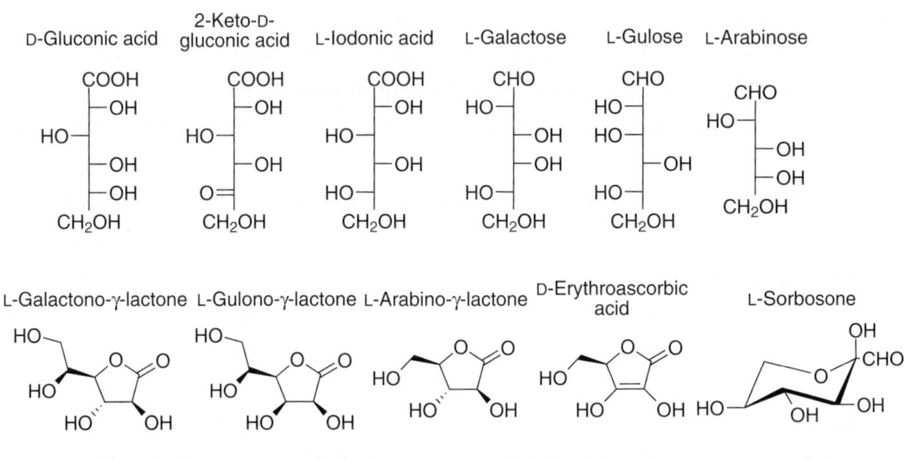


**Figure 16.3.** Chemical structures of various substrates and key intermediates.

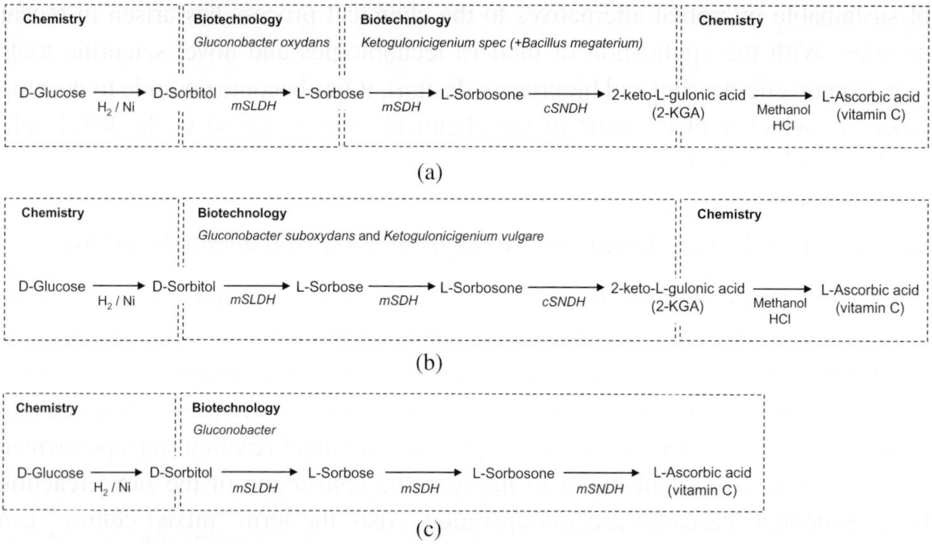


**Figure 16.4.** Industrial production of vitamin C (a) by a two-stage 2-KGA process, (b) by a single-stage 2-KGA process and (c) by a single-stage ascorbic acid process from D-sorbitol.

(2) L-sorbosone is oxidised to 2-KGA by sorbosone dehydrogenase (SNDH). The concomitant bacterium is required to enhance the growth of *Ketogulonicigenium*. The stimulating growth factor has not been identified so far, but recent observations suggest a mechanism based on secreted proteins.[22] Applying the two-stage process, 2-KGA was produced with a final concentration of up to 70 g/L and a yield of 90% based on L-sorbose (Fig. 16.4a). Today, this procedure is applied by all Chinese vitamin C producers.[23]

During the last four decades many attempts have been made to discover improved microbial methods for the production of 2-KGA from various substrates, including D-glucose, D-sorbitol, L-sorbose, L-sorbosone and L-iodonic acid (for chemical structures, see Figs. 16.2 and 16.3). The list of employed micro-organisms consists of *Gluconobacter*, *Ketogulonicigenium*, *Erwinia*, *Corynebacterium* and *Pseudomonas* strains.

At DSM Nutritional Products Ltd the main research activities were focused on processes employing *Gluconobacter* and *Ketogulonicigenium* strains. In 1990, an improved 2-KGA production was described with mutants of *Gluconobacter melanogenus* starting from D-sorbitol or L-sorbose.[24] With a yield on substrates of ~60%, a final 2-KGA concentration of 60 g/L was achieved within 90 h of cultivation.

Four years later a single-stage 2-KGA process starting from D-sorbitol was patented by DSM.[25] In this procedure the three oxidation reactions from D-sorbitol to 2-KGA are performed in a single process step by a mixed culture of *Gluconobacter suboxydans* IFO 3255 and *Ketogulonicigenium vulgare* DSM 4025 (Fig. 16.4). Advantages of this process concept are that: (i) the co-cultivation with a helper strain required in the two-stage process can be omitted and (ii) the formation of by-products derived from D-sorbitol (namely D-glucose, D-gluconic acid and 2-keto-D-gluconic acid) is substantially reduced.

More recently a single-stage continuous fermentation process for the conversion of L-sorbose to 2-KGA by *Ketogulonicigenium vulgare* DSM 4025 was developed. Here the requirement for a second helper micro-organism was superseded by supplementing the medium with baker's yeast.[26] In this system, 2-KGA was continuously produced from L-sorbose with a molar yield of 91.3%, a productivity of up to 4.80 g/L/h and a process duration of 110 h. However, a drawback of this method is the large amount of expensive baker's yeast needed (7.5% in the initial and feeding medium). Finally, a continuous two-stage fermentation process was developed for the conversion of L-sorbose to 2-KGA by mixed cultures of the production strain *Ketogulonicigenium vulgare* DSM 4025 and either *Bacillus megaterium* or *Xanthomonas maltophila* as the helper micro-organism.[27] By establishing a sophisticated process design for a continuous mixed culture, the costs for the production medium compared with the previously described single-stage pure culture approach could be reduced by >50%. However, this process still requires the conversion of D-sorbitol to L-sorbose in a separate step. As a conclusion it can be stated that both of the latter described approaches have shown the potential to evolve into a commercial application.

### 16.1.3. *Direct microbial production of vitamin C*

The direct synthesis by using micro-organisms remains the most challenging approach of L-ascorbic acid production. DSM Nutritional Products is indeed heading towards a technological platform by which this vitamin is produced in one fermentation step. The main advantages of this development are the minimisation of raw material consumption (glucose, other chemicals and solvents) and a more efficient energy use, which would result in a lower environmental impact.

Numerous attempts have been made to exploit new routes for the direct production of L-ascorbic acid from inexpensive feedstocks in a commercially attractive way. The list of putative production organisms contains microalgae (*Chlorella*, *Prototheca*), yeasts (*Candida*, *Saccharomyces*, *Zygosaccharomyces*), and (recombinant) prokaryotes (*Gluconobacter*, *Ketogulonicigenium*, *Xanthomonas*).

#### 16.1.3.1. *Microalgae*

Similar to higher plants, unicellular microalgae (microphytes) are able to produce L-ascorbic acid from D-glucose via L-galactose. In this pathway the direct precursor of L-ascorbic acid is L-galactono-$\gamma$-lactone (L-GalL). It has been shown that mutants of *Chlorella pyrenoidosa* are able to produce 2 g/L of L-ascorbic acid from 80 g/L of D-glucose in a one-step fermentation process.[28] However, most of the product remained associated with the biomass which is detrimental for the subsequent purification. A minor accumulation of L-ascorbic acid in the fermentation medium was achieved by using *Prototheca moriformis*.[29] However, the low productivities, together with the low growth rates of microalgae compared with bacterial or yeast cultures, make it rather unlikely that such microalgae cell systems will become commercially attractive.

#### 16.1.3.2. *Yeast*

Fungi of the genera *Zygomycetes*, *Ascomycetes* and *Basidiomycetes* synthesise D-erythroascorbic acid, a C5 analogue of L-ascorbic acid, from D-arabinose via the intermediate D-arabinono-$\gamma$-lactone. For *Candida albicans* and *Saccharomyces cerevisiae* it has been shown that the two involved enzymes, D-arabinose dehydrogenase (D-Ara-DH) and D-arabinono-$\gamma$-lactone oxidase (D-AL-Ox), are also capable of converting L-galactose to L-galactono-$\gamma$-lactone (for chemical structures, see Figs. 16.2 and 16.3) and finally L-ascorbic acid.[30–32] In recombinant yeast strains overexpressing these two enzymes led to an improved production of L-ascorbic acid from L-galactose, which was further enhanced by the additional expression of the L-galactose dehydrogenase from *Arabidopsis thaliana*.[33] Currently, research activities are focusing on strategies to provide the rather expensive carbon source L-galactose in an economical manner.

#### 16.1.3.3. *Prokaryotes*

There are only a few reports on direct production of L-ascorbic acid from carbohydrates by bacteria. In 1995, a new vitamin C producing enzyme (L-gulono-$\gamma$-lactone dehydrogenase) from *Ketogulonicigenium vulgare* DSM 4025 was characterised by Sugisawa *et al.*[34] However, there is currently no inexpensive access to the substrate L-gulono-$\gamma$-lactone. Later it was shown that the same organism is able to produce L-ascorbic acid from D-sorbitol, L-sorbose, L-gulose and L-sorbosone (for chemical structures, see Figs. 16.2 and 16.3).[35] Subsequently, a novel enzyme, PQQ-dependent L-sorbosone dehydrogenase (SNDH1), which directly converts L-sorbosone to L-ascorbic acid, was isolated and characterised.[36,37] As described for the previous direct fermentation approaches, the current research activities at DSM are focusing on the development of a single-stage fermentation process (Fig. 16.4c) that is economically competitive with the above mentioned Chinese two-stage 2-KGA process.

## 16.2. Aromatic Oxidations

### 16.2.1. *Oxidation of 2,3,5- and 2,3,6-trimethylphenol to 2,3,5-trimethylbenzoquinone*

In large-scale industrial syntheses of $\alpha$-tocopherol, the most important compound of the vitamin E group, 2,3,6-trimethylphenol (2,3,6-TMP, **3**) or 2,3,5-trimethylphenol (2,3,5-TMP, **4**) are important starting materials. They are oxidised to trimethylquinone (TMQ, **5**) using oxygen or peroxides (Fig. 16.5). Subsequent catalytic hydrogenation gives 2,3,5-trimethylhydroquinone (TMHQ, **1**), which is condensed with isophytol (**2**) to $\alpha$-tocopherol.[38]

In the oxidation of 2,3,6-TMP (**3**) or 2,3,5-TMP (**4**) inorganic acids or salts are used as oxidants as well as molecular oxygen, air or $H_2O_2$ combined with homogeneous or heterogeneous catalysts. In processes where inorganic acids and salts are used, **4** can be efficiently oxidised, whereas in catalytic protocols, 2,3,6-TMP **3** is more easily converted to **5** than **4**. Unlike the oxidation of 2,6-dimethylphenol, where the oxidative coupling to the corresponding diphenoquinone is the prevailing reaction, the oxidation of **3** leads to **5** and 2,2′,3,3′,6,6′-hexamethylbiphenyl-4,4′diol (**6**) as the hexamethyl diphenoquinone is not stable (Fig. 16.6).[39]

Chinese manufactures produce TMQ on an industrial scale by the oxidation of TMP using stoichiometric amounts of inorganic salts as oxidants, and thus generating stoichiometric amounts of inorganic waste. For example 2,3,6-TMP (**3**) is

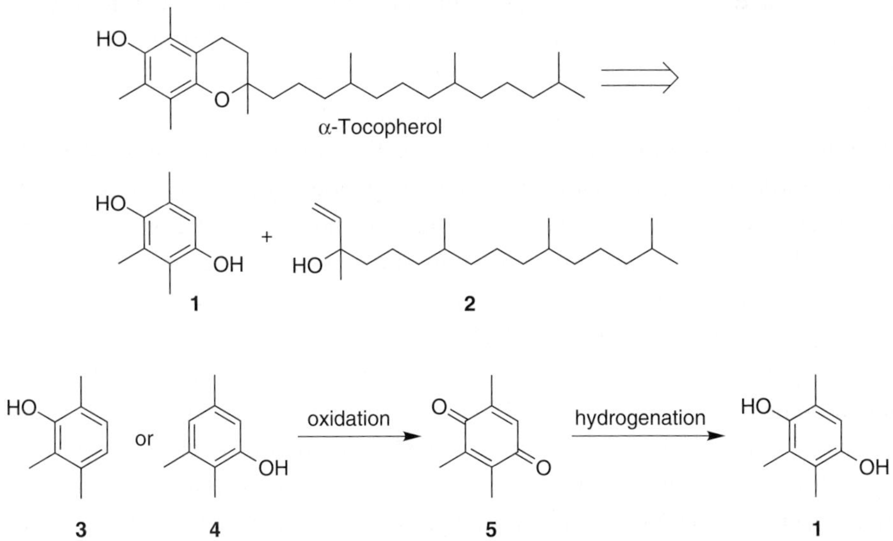


**Figure 16.5.** Synthesis concept for $\alpha$-tocopherol.

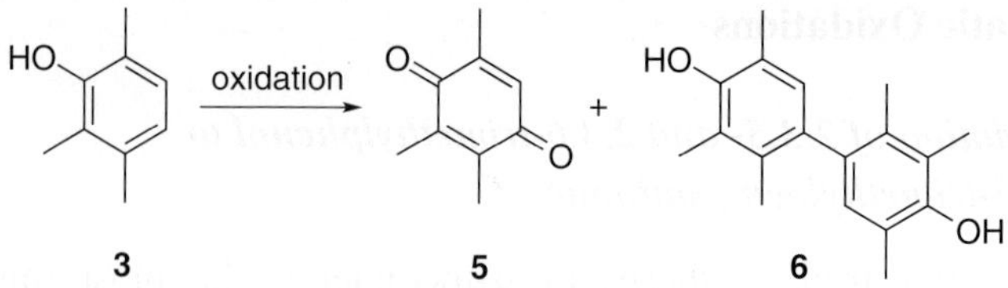


**Figure 16.6.** Main products in the oxidation of 2,3,6-TMP.

sulfonylated to 4-sulfonyl-2,3,6-trimethylphenol, followed by oxidation with $MnO_2$ to TMQ (**5**).[40]

Processes using $H_2O_2$, oxygen or air as oxidants in the presence of catalysts have the advantage that less inorganic waste is generated. Such methods are of great interest for the efficient production of economically important compounds in the life science industry.[41]

Homogeneous catalyst systems include transition metals, e.g. the cobalt Schiff base complex salcomine.[42–44] However, the quite expensive Schiff base cannot be recovered and recycled, which is a drawback of this method. Copper salts, e.g. $CuCl_2$ and $CuBr_2$, are widely used for the oxidation of TMP with molecular oxygen at larger scales.[45] In addition, applications of the halides of Cr, Mn, Fe, Ni and Zn as catalysts have been described.[46] Copper halides have been combined with earth metal halides, e.g. magnesium chloride.[47]

In general, high loadings of copper salt are required for an acceptable conversion due to deactivation of the catalyst. The addition of stabilising agents such as complexing agents, hydroxylamines, oximes or amines are beneficial to lower the amount of copper salt while maintaining a high conversion of TMP and selectivity to TMQ.[48] In addition, the use of ionic liquids as stabilising agents serves this goal.[49]

The major drawback in using oxygen as an oxidant is the risk of explosion of the highly flammable organic solvents at reaction temperatures above their flash point. Hazardous reaction conditions can be suppressed by using a biphasic reaction medium consisting of water and a longer chain alcohol and/or an aromatic solvent[50] or in a biphasic reaction mixture consisting of a neo-carboxylic acid with a carbon chain of 8–11 carbon atoms and water.[51] Carboxylic acids are also used as activators for the hydroxylation of TMP to TMHQ.[52]

Heteropolyacids and the corresponding polyoxometalates have been frequently applied in the oxidation of 2,3,6-TMP (**3**) with molecular oxygen or $H_2O_2$. Kholdeeva *et al.* studied the kinetics of 2,3,6-TMP oxidation over heteropolyacids $H_{3+n}PMo_{12-n}V_nO_{40}$ with molecular oxygen. They showed that the catalytically active species is the $VO_2^+$ ion.[53] Fujibayashi *et al.* showed that the activity of molybdovanadophosphate was greatly enhanced when supported on charcoal,[54] while 2,3,6-TMP is oxidised with molecular oxygen over catalytic amounts of heteropolyoxometalates containing molybdenum and vanadium in a solvent mixture of acetic acid and water. The heteropolyacid can be easily recycled using a biphasic system containing a water-acetic acid phase and an organic solvent non-miscible with water.[55]

Ti(IV) monosubstituted Keggin-type polyoxometalates (Ti-POMs) are stable towards the hydrolysis of Ti-O-W bonds and oxidative degradation. The catalytic oxidation of **3** using $H_2O_2$ as the oxidant over Ti-POMs yields 2,3,5-TMHQ (**1**) and the biphenol (**6**) as main products (Fig. 16.6). The product distribution depends significantly on the TMP/Ti-POM molar ratio.[56]

Another variant of homogeneous catalysis towards TMQ makes use of methyltrioxorhenium (MTO), which efficiently activates $H_2O_2$.[57] Dimethylcarbonate has been shown to be an efficient medium for this transformation, as MTO dissolves very well in this solvent.[58] Photo-oxidation of **3** has been described using porphyrins or metallophthalocyanins.[59]

The major advantage of heterogeneous catalysis consists of the simple separation of catalyst and product as well as in the recyclability of the catalyst. Spinel compounds have been reported for both the heterogeneous oxidation of 2,3,6-TMP to TMQ and for the one-step hydroxylation of 2,3,6-TMP to TMHQ. Hong *et al.* studied a one-step procedure consisting of the direct hydroxylation of the aromatic ring using spinel-type magnesium ferrite.[60] In two hours, 2,3,6-TMP was converted to 67% TMHQ and 32% TMQ over $MgFe_2O_4$ using $H_2O_2$ as the oxidant. The spinel $CuCo_2O_4$ shows an efficient catalytic activity using an excess of 30% hydrogen peroxide.[61] The combination of $CuCo_2O_4$ with hypocrellins, which are naturally occurring photosensitisers, mediated the oxidation under irradiation ($<400$ nm) and air.[62]

Copper hydroxyphosphate also allows the direct hydroxylation of 2,3,6-TMP, leading to selectivities for TMHQ (**1**) of $>80\%$. The main by-product is TMQ (**5**).[63] Heterogeneous catalysts which have been used for the oxidation of **3** include zeolites, mesoporous materials and molecular sieves containing transition metals.[64] Molecular sieves containing transition metals such as vanadium or copper in the framework can simply be mixed with 2,3,6-TMP and $H_2O_2$ in acetonitrile for the oxidation.[65]

Titanium containing mesoporous mesophase material, which is prepared under weakly alkaline conditions, shows high activity in the $H_2O_2$ based oxidations of bulky organic substrates. Hydrolytic instability, however, led in many cases to structure collapse during the oxidation processes and to a decrease in catalytic activity upon recycling. Improvement has been found upon preparation of the catalyst under moderate acidic conditions.[66] In contrast to titanium containing mesoporous mesophase silicate or amorphous $TiO_2$-$SiO_2$ mixed oxides, titanium leaching does not occur. Titanium dispersion and its accessibility were found to be crucial factors

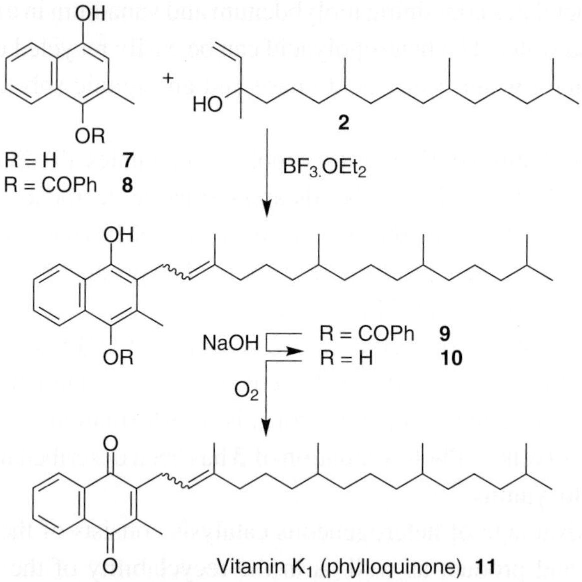


**Figure 16.7.** Synthesis of vitamin $K_1$.

determining the catalytic properties. $TiO_2$-$SiO_2$ aerogels containing 1.7–6.5 wt% of titanium allowed selectivities for oxidation to **5** up to 98%.[67]

### 16.2.2. *Vitamin K*

K-vitamins are of considerable economical interest for human and animal nutrition. Vitamin $K_1$ (phylloquinone, **11**, Fig. 16.7) and the water-soluble forms of menadione (**13**, Fig. 16.8) are used in pharmaceutical applications and as additives for poultry, pig and other animal feed.[68–70] The industrial process for the synthesis of the most important representative of this group, vitamin $K_1$ (**11**), is based on the work performed by research groups at Roche and Merck who used monoacylated starting materials.[71,72] The monobenzoate **8** derived from menadiol (menaquinol, **7**) can be alkylated with isophytol (**2**) in good yield to the crystalline dihydro-vitamin $K_1$ derivative **9** using $BF_3$ etherate as the catalyst.[73] After recrystallisation for enrichment of the *E*-isomer, **9** is saponified to hydroquinone **10** and subsequently oxidised with oxygen to vitamin $K_1$ (**11**). Through this industrial process, synthetic vitamin $K_1$ was introduced by Roche in 1953 under the brand name Konakion®.

Menaquinol (**7**), the key building block for the production of K vitamins, is obtained by the catalytic hydrogenation of menadione (vitamin $K_3$, **13**, Fig. 16.8). Commonly used processes for the preparation of **13** are based on the treatment of 2-methylnaphthaline (**12**) with (over-)stoichiometric amounts of strong

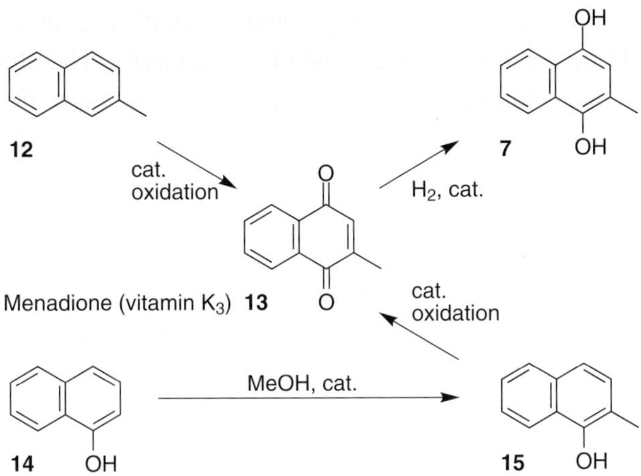


**Figure 16.8.** Preparation and use of menadione (**13**).

oxidants,[69,70] for example $CrO_3$ in sulfuric or acetic acid,[74,75] or hydrogen peroxide in acetic acid.[76] Alternative procedures use iron(III) chloride/$H_2O_2$, or nitric acid.[69] Many attempts have been made to achieve such industrially important oxidation reactions in satisfactory yield and regio- as well as chemo-selectivity by the use of more environmentally benign catalytic procedures, avoiding the problems of toxic waste.

Narayanan *et al.* claimed to obtain menadione (**13**) by oxidation of **12** with an excess of 30% hydrogen peroxide in an acetic acid solution at 100°C, in selectivities over 90% without using a catalyst, thus avoiding mineral acids and heavy metals like chromium. Kholdeeva *et al.* reported the surprising observation that **15** was not only oxidised by hydrogen peroxide, *tert*-butyl hydroperoxide and oxygen when using various solid supported catalysts, but also by non-catalysed treatment with molecular oxygen, with even superior selectivity.[77]

The efficient use of methyltrioxorhenium as a catalyst for this transformation has been described and considerably improved by the groups of Adam and Herrmann.[78] Under optimised conditions, less than 1 mol% of the catalyst is needed, and regioselectivities (1,4- vs 5,8-quinone) of above 85% can be obtained. The use of concentrated (85%) $H_2O_2$ in mixtures of acetic acid and acetic anhydride is, however, necessary, and the concentration of the acid is critical for the outcome of the reaction.

Ruthenium catalysts have also been applied as effective catalysts. The selective oxidation of **12** could be achieved by using terpyridine-derived ruthenium complexes, with catalyst loadings below 1 mol% and the addition of a phase transfer catalyst (PTC; an ammonium, phosphonium, or sulfonium salt) in biphasic aqueous systems, but also in methanol without PTC, and without the need for adding a mineral acid.[79] Practically 30% $H_2O_2$ could be employed, and yields of up to 60% of **13** could be achieved with around 60% selectivity. The ability of iron(III) salts to activate cheap oxidants such as hydrogen peroxide was noticed by Kowalski *et al.* and a practical procedure was developed in the group of Beller recently.[80] With an *in situ* catalyst prepared from inexpensive $FeCl_3$ hexahydrate, pyridine-2,6-dicarboxylic acid, and an amine, **13** could be obtained in a yield of 55% with 30% $H_2O_2$ as the oxidant in an alcoholic solvent.

Investigations of Anunziata *et al.* indicated the possibility of developing a competitive heterogeneous catalyst system; oxidation of **12** to **13** with 30% $H_2O_2$ in acetonitrile as the solvent delivered, after optimisation by using experimental design tools, menadione in 60% yield at a selectivity of around 90% with Ti-MCM-41.[81] An interesting alternative for the synthesis of **13** appeared in patent applications on the electrochemical oxidation of 2-methylnaphthalene (**12**).[82] The Cr(VI)/pyridine catalysed oxidation in an aqueous sulfuric acid/acetic acid mixture as mentioned by Machowska *et al.* was considerably improved by Harrison *et al.* from Hydro-Quebec by using the Ce(III)-Ce(IV) redox system in the presence of Cr(VI) as a catalyst, in mixtures of aqueous methanesulfonic acid and an organic solvent. High yields and selectivities at almost quantitative conversion were claimed for batch experiments in electrochemical cells.

For the preparation of **13**, 2-methyl-1-naphthol (**15**) can also be employed as a starting material (Fig. 16.8).[83] Compound **15** can be obtained from 1-naphthol (**14**) by catalytic alkylation with methanol, and further oxidised to menadione (**13**). Polyoxometallates (Keggin-type heteropoly compounds) containing phosphorus, molybdenum, tungsten, vanadium and oxygen were used as oxidants, as well as aqueous hydrogen peroxide in the presence of a niobium-based heterogeneous catalyst system. With 35% $H_2O_2$, oxidation to **13** occurs even in the absence of a catalyst with high selectivity. Recently, the use of *tert*-butyl hydroperoxide as an oxidant with supported iron tetrasulfophthalocyanine as the catalyst was investigated through kinetic studies by means of labelling experiments and spectroscopic studies.

## 16.3. Oxidations in Monoterpene Chemistry

Linalool and citral are important building blocks for the synthesis of isoprenoid natural products, e.g. vitamins A, E and K, carotenoids and a broad variety of fragrances.[38,70,84] In view of changing starting materials in future decades due to limited amounts of fossil resources, pinenes might be interesting sources for the fine chemical industry to produce linalool via pinane-2-hydroperoxide. The synthesis of linalool and citral from natural sources is discussed herein.

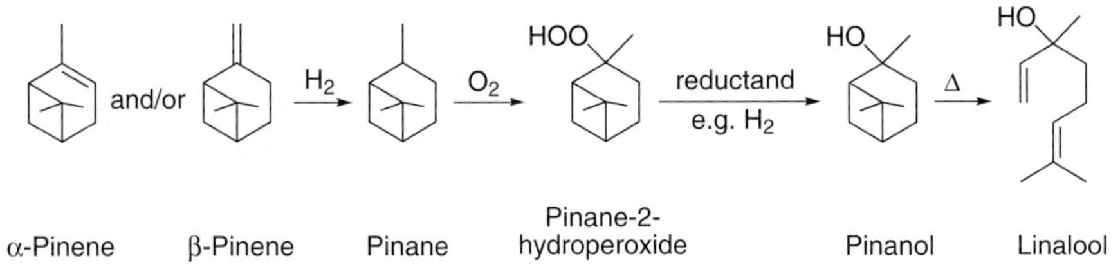


**Figure 16.9.** Preparation of linalool from pinenes.

### 16.3.1. *Oxidation of pinane to pinane hydroperoxide*

Linalool can be obtained from pinene extracts by hydrogenation, subsequent oxidation to the respective hydroperoxides, reduction to the respective alcohols and pyrolysis to linalool (Fig. 16.9).[85] Pinene extracts mostly contain mixtures of $\alpha$- and $\beta$-pinene. After hydrogenation a mixture of cis- and trans-pinane is obtained.

Pinanes were oxidised by applying a gas flow of air or oxygen in the presence of a catalyst,[86,87] or without a catalyst.[85,88] In both cases cis-pinane is more reactive than trans-pinane because of the steric effect of the gem-dimethyl group, which hinders the free radical attack on the tertiary C-H bond in the 2-position.

The catalyst applied was $Co(OAc)_2/Mn(OAc)_2/NH_4Br$ (12.8/1.5/5.0 mol%).[86] The authors oxidised a mixture of cis- and trans-pinane and needed only one catalytic step from pinane to pinanol. The formation of cis-pinanol is favoured over trans-pinanol because the attack of molecular oxygen at the 2-pinanyl radical occurs trans to the gem-dimethyl group due to steric reasons. The best conditions gave 17% conversion and selectivity of 54% to cis-pinanol and 17% to trans-pinanol. Without catalyst and applying the same reaction conditions 17% conversion and selectivity of 52% to cis-pinanol and 10% to trans-pinanol were obtained. The difference in cis/trans selectivity is probably due to the interaction of the transition metal with the 2-pinanyl radical.

When $Co(OAc)_2$ or $Mn(OAc)_2$ were used alone as catalysts and less oxygen was applied (instead of pure oxygen, air was applied and a solvent with a lower oxygen dissolution capacity was used) the reaction stopped at pinane-2-hydroperoxide.[87] The authors used enriched cis-pinane (>96%) and reported 37% as the highest yield to pinane-2-hydroperoxide.

The non-catalysed oxidation of pinane is an auto-oxidation.[85] Oxidation of cis-pinane at 100°C resulted in 15 wt% pinane-2-hydroperoxide which was converted to 67% cis-pinanol, 17% trans-pinanol and other by-products. Oxidation of trans-pinane at 100°C and subsequent reduction resulted in 17% cis-pinanol, 5% trans-pinanol and various by-products.[89]

Free radicals are generated either by the interaction between pinane and $O_2$ or by monomolecular decomposition of pinane-2-hydroperoxide. The recombination of

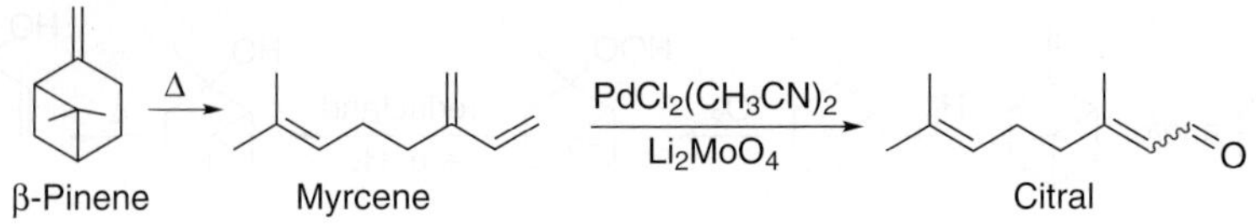


**Figure 16.10.** Preparation of citral from β-pinene.

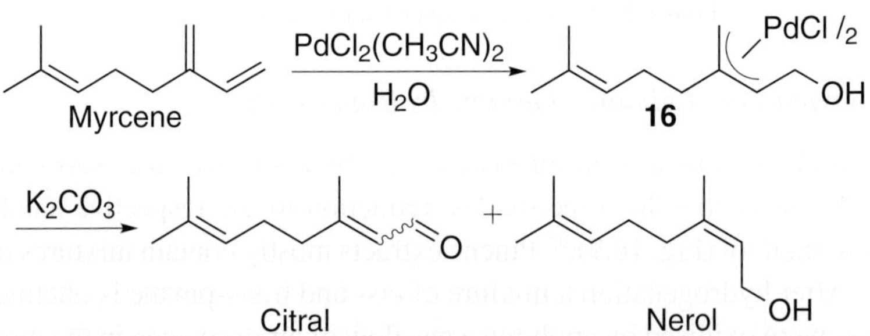


**Figure 16.11.** Two-step procedure from myrcene to citral/nerol.

these free radicals does not result in the formation of side products. Oxygen pressure does not affect the selectivity and thus reaction steps involving oxygen are not rate determining in the oxidation of pinane.[85]

### 16.3.2. *Oxidation of myrcene to citral*

Citral can be obtained from pinene extracts by pyrolysis of β-pinene to myrcene and subsequent oxidation in the presence of palladium(II) complexes and oxoanionic salts (Fig. 16.10).[90–92]

Myrcene was converted into its $\eta^3$-allylpalladium complex by reaction with $PdCl_2(CH_3CN)_2$ in aqueous hexamethylphosphoric triamide or dimethylformamide and a base such as $Li_2CO_3$. The obtained yield was 75% and 33%, respectively.[93] In a second reaction step the complex was treated with a base to give a mixture of citral and nerol (Fig. 16.11). The best system for the second step with regard to selectivity of citral was $K_2CO_3$ as a base (3.8 equiv.) in methanol at room temperature. Citral was obtained in 47% selectivity, nerol in 53%.[93]

In a one-step procedure a catalytic system of $PdCl_2(CH_3CN)_2$ (3 mol%), $Li_2MoO_4$ (15 mol%), $CuCl_2$ (3 mol%) and a phase transfer catalyst was used.[90–92] Water and oxygen were introduced into the reaction vessel and the mixture was heated to 90°C. Citral was isolated in a 66% yield. When milder reaction conditions were applied, the palladium complex **16** (Fig. 16.11) could be isolated and transferred into citral in a second reaction step by adding triphenylphosphine.[91] Interestingly, under similar reaction conditions a different dimeric complex was

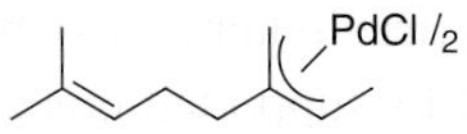


**Figure 16.12.** Isolated complex in the conversion of myrcene with $PdCl_2(CH_3CN)_2$.

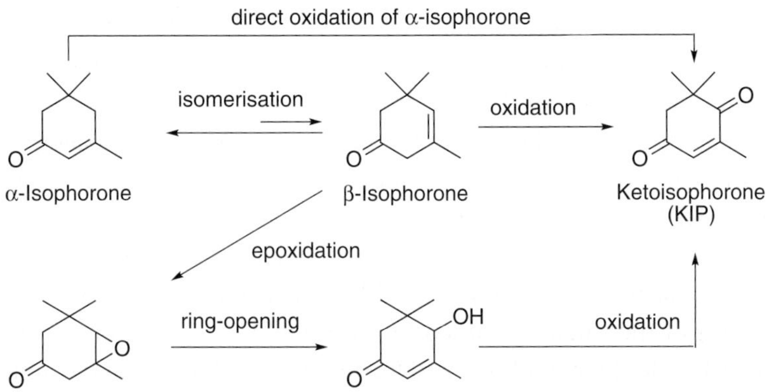


**Figure 16.13.** Syntheses of KIP based on $\alpha$-isophorone.

formed selectively (29% isolated yield, based on Pd) (Fig. 16.12). The authors suggested a 1,2-addition of HCl to myrcene, followed by oxidative addition of Pd(0) to the allylic chloride.[94] Palladium complex **16** was not detected.

### 16.3.3. *Oxidation of isophorone to ketoisophorone*

Ketoisophorone (KIP) is a key intermediate in the production of nutritional products (e.g. vitamins and carotenoids) and in the flavours and fragrances industries. One option for a technical access to KIP is the catalytic oxidation of isophorone (Fig. 16.13). For good selectivity and yield in the oxidation step a thermal isomerisation of $\alpha$-isophorone to $\beta$-isophorone is necessary. However, in order to avoid this additional step and because the isomerisation equilibrium is strongly in favour of the $\alpha$-isomer, a direct oxidation of $\alpha$-isophorone to KIP would clearly be preferred.

### 16.3.4. *Oxidation of $\beta$-isophorone to KIP*

In early vitamin and carotenoid syntheses, the oxidation of $\beta$-isophorone was accomplished by epoxidation of $\beta$-isophorone followed by the ring-opening of the epoxide and oxidation (Fig. 16.13).[95,96] A significant improvement of the three-step synthesis of KIP was made by the one-step liquid phase oxidation of $\beta$-isophorone using oxygen or an oxygen-containing gas in the presence of a transition metal catalyst. In 1975, the aerobic oxidation of $\beta$-isophorone in the presence of Cu(II)-or V(III)-acetylacetonate was published, in which KIP was obtained in a yield of up to 55%.[97]

The addition of a tertiary amine (triethylamine) to the reaction mixture of the oxidation of $\beta$-isophorone, catalysed by supported precious metal catalysts (e.g. Ag/C), allowed a yield increase to 77%.[98] When pyridine was used as the base and solvent in the $V(acac)_3$-catalysed oxidation with molecular oxygen, up to 91% yield was achieved at 70°C.[99,100] A combination of the $Cu(acac)_2$-catalysed oxidation in pyridine as the solvent was filed by Hüls AG in 1988 which allowed a decrease in the required amount of base/solvent by a factor of 10, which thus reduces the volume and facilitates the removal of the base.[101]

The oxidation of $\beta$-isophorone in the presence of Mn- or Co-salen complexes was first investigated by Constantini *et al.* This method significantly improves the selectivity to KIP as well as the space-time yield. In the presence of triethylamine and in an aprotic solvent (e.g. 1,2-dimethoxyethane) KIP was obtained at up to 90% selectivity at full conversion.[102,103] Electron-withdrawing substituents on the aromatic rings of the salen ligand led to lower yields.[102] Variations of the counter ion in [Mn(III)-salen]X complexes and the number of C-atoms in the amine bridge of the salen ligand were described in 1989.[104] By applying additives (such as weak organic acids with a $pK_a$ 2–7 as bidentate chelating ligands, enolisable compounds or buffers for pH 6–8) to the Mn-salen-catalysed oxidation, the selectivity and turnover number (mol product per mol catalyst) could be further improved.[105]

Researchers from BASF developed the Mn(II)-salen- or [Mn(III)-salen]X-catalysed oxidation of $\beta$-isophorone using salen derivatives with electron-withdrawing substituents without[106] or in the presence of acetates as additives.[107] The increase of ignition temperature of the base/solvent mixture (ignition point of triethylamine/diglyme: 0°C) and thus a reduction of the explosion risk, was successfully achieved using tripropylamine in dimethylformamide (DMF) or dimethyl acetamide (DMA). With a chloro-substituted [Mn(III)-salen]Cl catalyst in the presence of lithium acetate and tripropylamine in DMA, KIP was obtained in an 89.4% yield with minor amounts of by-products (1.5% $\alpha$-isophorone, 1.3% hydroxyisophorone) (Fig. 16.14).

In the presence of Cu-salen complexes with electron-withdrawing or electron-donating substituents on the salen ligand (e.g. $SO_3H$-containing substituents), yields of up to 91% were reached in the presence of tripropylamine at 60–80°C.[108] In 2006, the first asymmetric metal–Schiff base catalysts based on arginine were applied to the air oxidation of β-isophorone, affording KIP in an excellent yield of 95%.[109] A continuous process and apparatus for the isomerisation of $\alpha$- to β-isophorone (with an aliphatic polycarboxylic acid catalyst) and a following Schiff base catalysed oxidation of β-isophorone to KIP was claimed by Tomohide *et al.* in 2002.[110]

The heterogenisation of Schiff base complexes was investigated by Halligudi *et al.*[111] Zeolite-encapsulated Co(II)saloph complexes (saloph = salicylaldehyde-*o*-phenylenediimine) oxidised β-isophorone to KIP at ambient temperature and

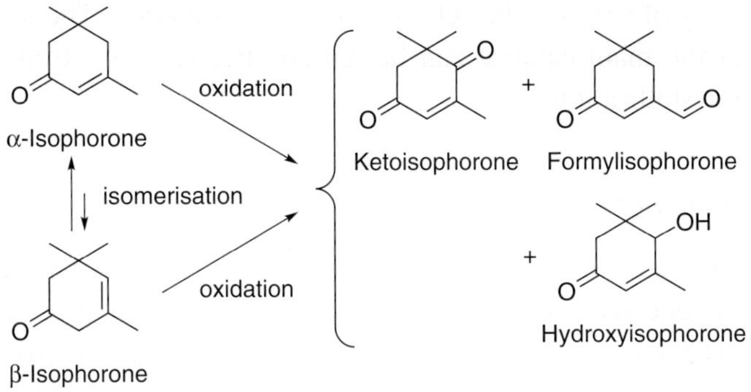


**Figure 16.14.** Possible by-products in the oxidation of isophorone.

pressure in the presence of acetylacetonate and trimethylamine, in methyl ethyl ketone in 60% selectivity at 90% conversion using air as the oxidant. However, higher selectivity (>95%) was obtained only at up to 30% conversion.

In 2005, Fe(III) tetrasulfophthalocyanine (PcS) were grafted onto mesoporous titania nanocrystals in a one-pot sol-gel process by a research group at Hubert-Pfalzgraph and applied as catalysts for the aerobic oxidation of $\beta$-isophorone in the presence of triethylamine as the base. At 99% conversion, 57% KIP was obtained together with 21% hydroxyisophorone.[112] The immobilisation of water-soluble metallo-tetrasulfophthalocyanine complexes on chitosan aerogel micro-spheres affords a class of bifunctional catalysts which combines basic and oxidising sites in one solid material. As chitosan acts as both support of the metal complex and solid organic base, the addition of an amine is not necessary. With 4% of a CoPcS@chitosan catalyst (81.4 $\mu$mol $g^{-1}$ complex loading) in acetonitrile at 80°C under 2 bar oxygen atmosphere, KIP was obtained in 39% yield at 85% conversion (46% selectivity).[113]

In 2008, Mao *et al.* published the preparation of an immobilised Fe(III) catalyst for the oxidation of $\beta$-isophorone: Fe(III) chloride supported on pyridine-modified mesoporous silica.[114] At 66°C in a mixture of pyridine and methyl ethyl ketone up to 88.5% selectivity for KIP was obtained at full conversion using 4 mol% of catalyst. The crude product also contained 4.8% of $\alpha$-isophorone and 2.8% of dimers. Recycling of the catalyst showed a selectivity of 84.1% after the fourth cycle, with an increasing selectivity towards $\alpha$-isophorone of 8.5%. Recently, the same group developed a Fe(III)-acetylacetone-imidazolium catalyst. Here, in pyridine as the solvent with a catalyst loading of 1 mol%, 99% conversion was observed affording KIP in 90% selectivity.[115,116] A metal-free method for the oxidation of $\beta$-isophorone was claimed by Zhejiang NHU Co., Ltd.[117] The catalytic system is composed of N-hydroxyphthalimide as the main catalyst and an organic base as the co-catalyst in an organic solvent. At full conversion up to 74% selectivity to KIP is obtained.

Higher selectivity of 91% can be achieved at 81% conversion. The method has the advantage that the main catalyst can be reused after recovery from the reaction mixture by recrystallisation.

### 16.3.5. *Oxidation of α-isophorone to KIP*

Even though detailed studies of the oxidation of β-isophorone have led to highly selective homogeneous and some heterogeneous oxidation procedures for the synthesis of KIP, all of them still require the isomerisation of α-isophorone to β-isophorone. As α-isophorone is readily available and is the thermodynamically more stable isomer, and to avoid this additional and thermodynamically unfavoured reaction, intense efforts have been made to find a direct oxidation of α-isophorone to KIP.

The non-catalysed direct oxidation of α-isophorone with air is possible, but occurs with low yield. In the presence of transition metal salts or oxides (e.g. of V, Cr, Mn, Fe, Co, Cu, Ni, Rh), KIP can be obtained in moderate yields of 30–40%.[118] Due to competing allylic oxidation of the exocyclic methyl group, the direct oxidation of α-isophorone to KIP is accompanied by the formation of the structural isomer formylisophorone (Fig. 16.14).

When the aerobic oxidation of α-isophorone is performed in the presence of heteropoly acids or salts thereof in combination with additives such as copper sulfate and/or molybdenum oxides, a selectivity of 61% was observed at 83% conversion.[119] Molybdovanadophosphates supported on active carbon also mediate the oxidation of α-isophorone, however in most cases oxidation occurs on the methyl group, affording formylisophorone.[120]

The nature of the solvent was found to play an important role for the catalytic activity and selectivity in the aerobic allylic oxidation of α-isophorone to KIP with phosphomolybdic acid (PMA). With 0.43 mol% PMA and potassium *tert*-butoxide as the additive in dimethyl sulfoxide (DMSO) at 115°C, KIP was obtained in 70% selectivity at 99% conversion.[121] Using a ruthenium–porphyrin complex as the catalyst, the oxidation of α-isophorone with 2,6-dichloropyridine N-oxide in dichloromethane at 40°C afforded KIP at 99% selectivity and 75% conversion.[122] Mn(III) acetate catalyses the allylic oxidation of alkenes to enones with high regio- and chemo-selectivity. Using *tert*-butyl hydroperoxide, α-isophorone was oxidised to KIP in a 74% yield.[123] Selective allylic oxidation of α-isophorone to KIP was carried out over ruthenium grafted onto MgAl-hydrotalcite using *tert*-butyl hydroperoxide. In acetonitrile, KIP was obtained in 100% selectivity at up to 74% conversion.[124,125]

Similar to the oxidation of β-isophorone, Li *et al.* have recently developed a metal- and solvent-free method for the oxidation of α-isophorone to KIP. However, compared to $\beta$-isophorone, good selectivity for the oxidation of $\alpha$-isophorone with N-hydroxyphthalimide (NHPI) was only observed at low conversion (e.g. up to 81.7% selectivity at 10.9% conversion).[126] With variation of solvent and choice of activator, selectivity was improved to up to 92% at 58% conversion.[127]

In 1975, Ohloff *et al.* studied the gas-phase oxidation of $\alpha$-isophorone to KIP over a vanadia/pumice catalyst modified with 1 wt% of lithium phosphate at 230°C.[128] Under these conditions, simultaneous formation of KIP and formylisophorone occurred. More than 20 years later, Baiker *et al.* revisited the catalytic gas-phase oxidation of isophorone.[129] At 200–250°C, 75% combined yields of KIP and formylisophorone were obtained at 17% $\alpha$-isophorone conversion over vanadia/pumice impregnated with lithium phosphate; $\beta$-isophorone was found as a major by-product (18%). Bismuth molybdate or vanadium phosphate showed poor selectivity and rapid deactivation. The Ag/$\gamma$-alumina-catalysed oxidation was unselective and resulted mainly in isomerisation to $\beta$-isophorone. Chromia-based catalysts led to an increased formation of 3,5-xylenol. To efficiently remove coke deposits and to re-oxidise vanadium oxides to vanadia, temperatures higher than 300°C would be needed; however, under these conditions isophorone and KIP are not stable. Thus, highly selective catalysts would be required which are active at lower temperatures.

### 16.3.6. *Manufacture of rose oxide by ene-type allylic oxidation via "dark" singlet oxygenation of β-citronellol*

An important natural fragrance compound with an appreciated green, floral aroma is (−)-cis-rose oxide ((2*S*, 4*R*)-4-methyl-2-(2-methylprop-1-enyl)tetrahydro-2*H*-pyran). Bulgarian roses are a major source of natural rose oxide in the form of fragrant oil. Isolation of 1 kg of rose oil requires 3,000 kg of rose blossoms, and accordingly this oil carries a very high price tag that limits its application to high-end products such as perfumes.[130] Employing ene-type allylic oxidation of $\beta$-citronellol by chemically-generated singlet oxygen as the key step provides a cost-efficient synthetic manufacturing process for the bulk manufacturing of rose oxide as a (−)-cis, (+)-cis, (−)-trans and (+)-trans stereoisomeric mixture.[131] The singlet oxygenation of $\beta$-citronellol generates a 50/50 mixture of the secondary and tertiary hydroperoxides, the latter being the desired intermediates for rose oxide (Fig. 16.15).

In general, singlet oxygen is commonly generated by photosensitisation of triplet oxygen ($^3O_2$). Companies that own suitable photochemical reactors employ photo-oxidation of $\beta$-citronellol on a commercial scale for rose oxide manufacture.[132] Chemical generation of singlet oxygen allows the "dark" singlet oxygenation of organic compounds via the catalytic disproportionation of hydrogen peroxide into water and singlet oxygen. It can be carried out in conventional stirred tank reactors

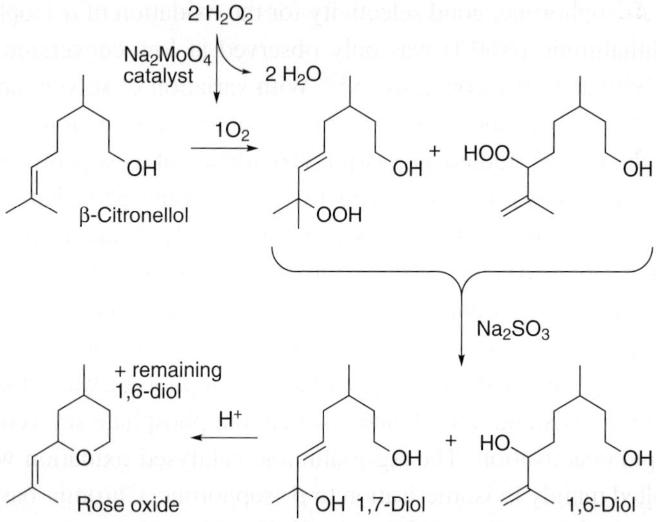


**Figure 16.15.** Rose oxide via "dark" singlet oxygenation of *β*-citronellol.

by simply adding hydrogen peroxide to a solution containing the disproportionation catalyst and the substrate.[133]

Many inorganic compounds catalyse the disproportionation of hydrogen peroxide, with widely varying degrees of singlet versus triplet oxygen efficiency.[134] Sodium molybdate is an inexpensive catalyst and stands out in terms of efficiency provided that a highly polar medium is present. In water, singlet oxygen is generated under molybdate catalysis from hydrogen peroxide in quantitative yield and with high rate, but unfortunately, *β*-citronellol is insoluble in water. A two-phase organic solvent/water system cannot be used in case of "dark" singlet oxygenation catalysed by sodium molybdate since singlet oxygen generated in the aqueous phase is quenched very efficiently by water to triplet oxygen. Therefore, it will not reach the organic layer where *β*-citronellol is located. A variety of water-miscible polar solvents that allow the formation of a single-phase reaction medium in the presence of aqueous hydrogen peroxide were therefore screened for the molybdate catalysed hydrogen peroxide disproportionation to singlet oxygen.[135] It was found that lower alcohols such as methanol are suitable for sufficiently reactive substrates such as *β*-citronellol.

Whereas methanol is a preferred organic solvent for "dark" singlet oxygenations on a laboratory-scale, its oxygen limit concentration (OLC)[136] is very low (8 vol-%) while ethylene glycol has a much more favourable OLC (50 vol-%). The latter solvent was therefore employed on a plant-scale, and by sparging nitrogen from the bottom of the reactor, it was assured that the oxygen concentration in the headspace did not exceed the OLC. In ethylene glycol, the yield of singlet oxygen from hydrogen peroxide disproportionation and the singlet oxygen lifetime are reduced to 70% and 7.5 $\mu$s, respectively. Nevertheless, nearly quantitative conversion of $\beta$-citronellol could be achieved using 4 mol-% $Na_2MoO_4$ as the catalyst with only a 1.5 fold excess of 50% aqueous hydrogen peroxide relative to the disproportionation stoichiometry (2 $H_2O_2$ required for 1 $^1O_2$).[133]

Another critical parameter for safe processing is the hydrogen peroxide addition rate, which in turn depends on the reaction temperature. Hydrogen peroxide should be added at such a rate that the latter equals the rate of its consumption, thus maintaining a low stationary concentration. The rate of hydrogen peroxide consumption via the molybdate catalysed disproportionation reaches its maximum when the predominant peroxomolybdate species in solution equals the triperoxomolybdate $Mo(O_2)_3O^{2-}$.[137,138] Since the prevalent peroxomolybdate species that is present in the reaction mixture depends *inter alia* on the hydrogen peroxide concentration, there is an optimum hydrogen peroxide addition rate. (The pH is another factor that determines the nature of the predominant peroxomolybdate. The "dark" singlet oxygenation of $\beta$-citronellol was carried out under slightly basic conditions at natural pH — i.e., no pH adjustment.) A high space-time yield was obtained without compromising safety by increasing the temperature for the "dark" singlet oxygenation of $\beta$-citronellol to 55°C, at which temperature a very low stationary hydrogen peroxide concentration was maintained.

Multiple runs of the "dark" singlet oxygenation for rose oxide manufacture have been carried out successfully in 10 $m^3$ stirred tank reactors, with easy catalyst recycling via simple extraction steps.[133]

## 16.4. Vitamin B$_5$: Ketopantolactone

Vitamin B$_5$, (*R*)-pantothenic acid (**17**, Fig. 16.16), occurs in nature as a component of coenzyme A. The industrial synthesis of enantiopure **17** and of (*R*)-panthenol (**18**) uses (*R*)-pantolactone ((*R*)-**19**, Fig. 16.16) as a key intermediate.[139] (*R*)-**19** can be obtained by starting from *rac*-**19** via classical optical resolution applying enantiopure amines, or kinetic enzymatic resolution, for example, with a lipase using vinyl acetate, followed by the corresponding recycling loops.[140] Enantioselective syntheses are based on the oxynitrilase-catalysed addition of hydrogen cyanide (HCN) to $\beta$-substituted pivalaldehydes, or the catalytic hydrogenation of prochiral ketopantolactone (2-oxopantolactone, **20**, Fig. 16.16). Beside the corresponding aminolactone, racemic pantolactone ((*R*, *S*)-**19**) is, on the other hand, the starting material for oxidation reactions to yield **20**.[14,139]

Oxidations of pantolactone (**19**) to ketopantolactone (**20**) with stoichiometric amounts of oxidants involve, for example, the reagents DMSO with acetic

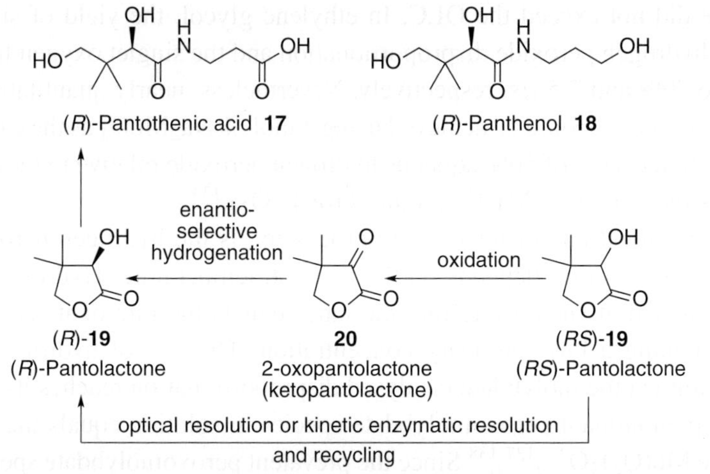

**Figure 16.16.** Ketopantolactone (**20**) as a key intermediate in pantothenate synthesis.

anhydride[141] or oxalyl chloride,[142] manganese dioxide[143] or organic hypohalides.[144] Often practical problems, such as the troublesome isolation of the product, occur with such protocols. Efforts to replace those oxidants with more environmentally benign reagents led to the development of procedures mediated by suitable catalysts.

The use of *tert*-butyl hydroperoxide catalysed by $RuCl_2(PPh_3)_3$ in benzene is claimed in a patent application from the mid 1980s.[145] A system containing NaOBr with catalytic amounts of HCl in a dichloromethane–water mixture delivered **20** in a 93% yield.[146] Trichloroisocyanuric acid in combination with a cheap base such as sodium acetate served as an efficient oxidant of **19** in the presence of either (2,2,6,6-tetramethylpiperidin-1-yl)oxyl (TEMPO)[147] or $RuCl_3$ as a catalyst. In the latter case, non-acidic conditions with 1 mol% of the ruthenium salt and 2 mol% of a phase transfer catalyst gave **20** in over 97% yield in a biphasic system at 20–45°C.[148] The oxidation of **19** by bromine/$H_2SO_4$ in refluxing $CCl_4$ was described as yielding ketopantolactone (**20**) in 93%.[149] Sodium periodate served as an oxidant under catalysis by a ruthenium salt in an aqueous solvent mixture under microwave irradiation.[150]

Air or oxygen as the most attractive reagent was successfully applied in combination with special transition metal catalysts. Nösberger could achieve a selectivity of over 86% for **20** at almost full conversion in a vapour-phase air oxidation at 250–290°C; the catalysts were $MoO_3$ and $V_2O_5$ supported on $\alpha$-$Al_2O_3$.[151] Wet ruthenium dioxide (5 weight%) in *o*-chlorobenzene as a solvent mediated the oxidation of **19** by introducing oxygen gas at 180°C. Ketopantolactone (**20**) was isolated at a 99.5% yield after chromatographic purification.[152] The platinum and palladium catalysed partial oxidation of **19** gave the highest selectivity and conversion

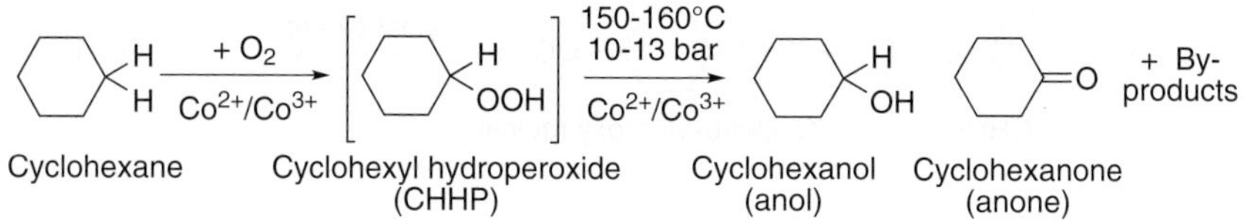


**Figure 16.17.** General reaction scheme of the cobalt-catalysed cyclohexane oxidation.

under water-free conditions.[153] An electrochemical oxidation of pantolactone (**19**) to ketopantolactone (**20**) in an aqueous solution containing inorganic chloride and phosphate salts in the presence of a ruthenium catalyst was claimed by Sato *et al.*[154]

## 16.5. Cyclohexane Oxidation

Caprolactam is the precursor to nylon-6, which has many applications in daily life. One of the key intermediates for the production of caprolactam is cyclohexanone. DSM is the largest merchant caprolactam producer in the world, and the sole company operating and licensing several technologies for the production of cyclohexanone: Hydranone™ (trade name of DSM technology for the production of cyclohexanone via phenol hydrogenation) and Oxanone™ (trade name of DSM technology for the production of cyclohexanone via the uncatalysed oxidation of cyclohexane). Currently, more cyclohexanone is produced worldwide via DSM technology than via technology provided by other companies.

Originally, all cyclohexanone production through the selective oxidation of cyclohexane was carried out via catalysed processes. In most cases, such as the processes of BASF, DSM and Zaklady Azotowe Tarnow (Cyclopol process),[155] soluble cobalt salts were the catalysts of choice. Reactions were carried out at a low per pass conversion to avoid over-oxidation of the primary and secondary reaction products (Fig. 16.17). For the same reason, a reactor system with a plug flow character was used. Per pass conversions of cyclohexane are generally limited to less than 5%.

Cobalt not only catalyses the oxidation reaction, but it also catalyses the decomposition of the first reaction product, i.e. cyclohexyl hydroperoxide (CHHP). However, the exact role, if any, of cobalt in the activation of cyclohexane has always been a subject of dispute.

The main disadvantage of the catalysed oxidation processes is their low selectivity. Often selectivity is not higher than 80% and in many cases even below 77%. One of the reasons for this low selectivity is the decomposition of CHHP to cyclohexylperoxy and cyclohexyloxy radicals. This depends on the oxidation state of cobalt (Fig. 16.18).

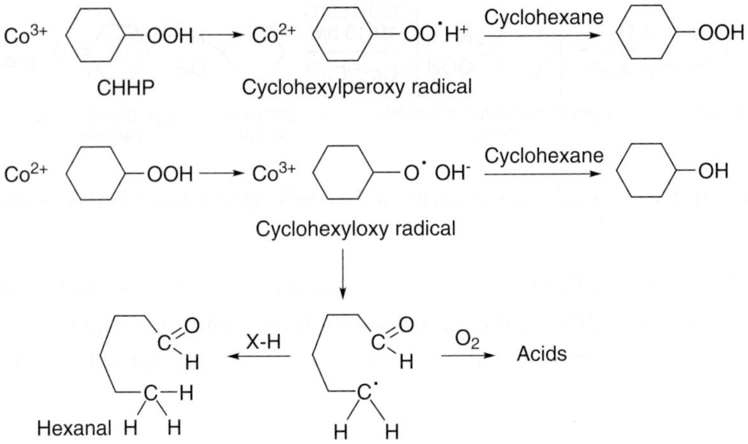


**Figure 16.18.** Cobalt catalysed decomposition of cyclohexyl hydroperoxide under oxidation conditions.

A fraction of the cyclohexyloxy radicals reacts with cyclohexane molecules by abstracting hydrogen atoms and converting to cyclohexanol. Another fraction of the cyclohexyloxy radicals ring opens, thereby resulting in the formation of hexanal, which is further oxidised to organic acids.[156, 157] This undesired ring-opening reaction with its high activation energy is favoured by high temperatures required for the oxidation of cyclohexane to CHHP. An uncatalysed Oxanone™ process has been established by DSM to avoid decomposition of cyclohexyl hydroperoxide under oxidation conditions and to carry out the decomposition with a high selectivity at much lower temperatures.

In DSM's uncatalysed Oxanone™ process, with a high overall selectivity, the formation of CHHP and its decomposition are performed under completely different process conditions. As almost all metals decompose CHHP, although some much more efficiently (especially Co and Cr) than others (such as Fe), cyclohexane oxidation to CHHP has to be carried out in the absence of transition metals. However, a disadvantage of leaving out these transition metals is a severe reduction of the reaction rate to CHHP. This reduced reaction rate could only partly be compensated by performing the uncatalysed oxidation reaction at enhanced reaction temperatures. The affordable increase of oxidation temperature is limited, because at too high a temperature CHHP is thermally decomposed with low selectivity.

In the late 1970s extensive research efforts were made by DSM to solve the problem of low activity during oxidation. An important outcome of these activities was that the oxidation products should be present within a certain concentration range (of between 0.1 and 3 wt%) at the beginning of the reaction or in the cyclohexane feed to the first oxidation reactor, provided that these oxidation products are recycled in sufficient quantities in order to obtain reaction rates that make industrial application feasible.[158,159] In practice, particularly suitable oxidation products are by-products of the cyclohexane oxidation itself, such as lower alcohols and ketones. Berezin *et al.* found that a cyclohexanol concentration below 4.5% retards the reaction rate, whereas cyclohexanone accelerates the reaction.[160] Moreover, it was assumed at that time that, for economical reasons, the quantity of oxidation products at the beginning of the reaction should be minimised, because they would also contain some cyclohexanol and cyclohexanone, which would then be over-oxidised to by-products.[161]

After DSM had developed their novel method of making CHHP in both a technically and commercially superior fashion, it was also important to develop a method to decompose CHHP in a selective manner in order to achieve a synthesis pathway with a high overall selectivity to cyclohexanone. It was evident that lower reaction temperatures were needed to achieve this goal.

One of the routes was the low temperature decomposition of CHHP with the addition of cobalt to the oxidation mixture. The problem of fast deactivation of the cobalt catalyst could partly be solved by introducing a water wash to remove the dibasic acids that were responsible for the fast precipitation of the cobalt catalyst. Nevertheless, even after adding the cobalt catalyst to later stages of the decomposition section, a fraction of the CHHP still remained unconverted. The selectivity losses were caused by radicals obtained from the cobalt-catalysed decomposition of CHHP, which not only reacted with the available cyclohexane, but also with the desired reaction products cyclohexanone and cyclohexanol. Such one-phase decomposition has recently been industrially implemented.[157]

A two-phase decomposition technology of CHHP was developed by DSM.[162] The oxidation mixture is prepared at 60–95°C with a caustic aqueous phase containing a cobalt catalyst. Because CHHP is slightly acidic it dissolves in the aqueous phase, where it is decomposed by the cobalt catalyst to cyclohexanol and cyclohexanone. Because the products formed have low solubility in the aqueous phase, they are transferred to the organic (cyclohexane) phase where they are separated from the reactive radicals. Consequently, the selectivity of the CHHP decomposition was significantly enhanced. The occurrence of this proposed mechanism was checked by performing a reaction in the presence of phenol, which is a water-soluble radical scavenger. The decomposition reaction almost stopped. However, if *p*-methyl-2,6-di-*tert*-butylphenol, an inhibitor which remains in the cyclohexane phase, was used, no effect was observed on activity and selectivity. The two-phase decomposition of CHHP was an improvement compared to the one-phase decomposition. After implementing the two-step technology, the performance of the plant increased as measured by a reduction of over 100 kg in the cyclohexane consumption needed per ton of cyclohexanone product.

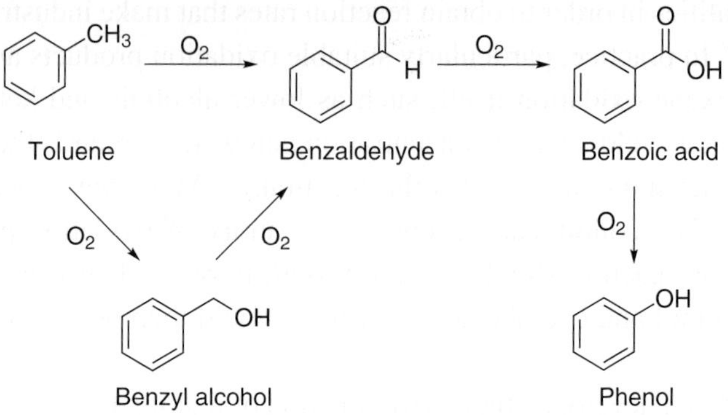


**Figure 16.19.** Reaction pathway to phenol, starting from toluene.

## 16.6. Toluene Side-Chain Oxidation

Until 2005, DSM produced about 130 kt/a of phenol, used as a raw material to produce caprolactam from cyclohexanone. Phenol was produced by a copper-catalysed oxidation of benzoic acid. The raw material, benzoic acid, was produced in the same plant by the cobalt-catalysed oxidation of toluene, which also produced significant amounts of benzaldehyde (Fig. 16.19).

At the end of 2004, phenol production at DSM stopped. The operating process of the existing toluene oxidation plant changed dramatically, producing much less benzoic acid but maintaining the same level of benzaldehyde production.

Figure 16.20 shows the relationship between the conversion and main reaction products. The ratio between benzaldehyde and benzoic acid increases at decreasing conversion. It was therefore decided to decrease the conversion in the plant significantly in order to acheive the desired higher ratio of benzaldehyde and benzoic acid. A drawback of the new operating method was the decreased solubility of the cobalt catalyst due to the lower concentration of benzoic acid. This contributed to scaling problems and inhibition of the oxidation reaction. This disadvantage was omitted by recycling more benzoic acid over the oxidation reactor.

From Fig. 16.20, it is also clear that at low toluene conversion, a significant increase of benzyl alcohol will be produced in the plant. The excess of benzyl alcohol is recovered downstream by distillation, and recycled back to the oxidation where it is oxidised to benzaldehyde. At first sight, this might be seen as a nice spin-off to produce additional benzaldehyde, but the reality was less promising; the production of tar (i.e. benzyl benzoate) increased dramatically, as explained below.

The products benzoic acid, benzaldehyde and benzyl alcohol are separated through distillation, by exposing these products to high temperatures, which normally results in the formation of benzyl benzoate. The common reaction pathway

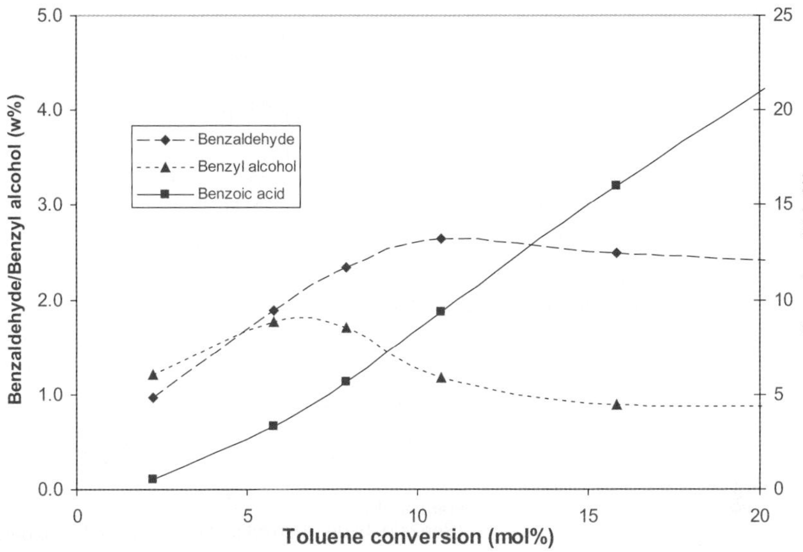


**Figure 16.20.** Relationship between the main products and conversion in the batch-wise oxidation of toluene.

to benzyl benzoate is the esterification reaction between benzoic acid and benzyl alcohol.

However, surprisingly, we have discovered that benzyl benzoate was predominantly formed by a new, unknown route: the formation and decomposition of the acetal derived from benzaldehyde and benzyl alcohol. A proposed mechanism is shown in Fig. 16.21.

In the top section of the distillation, the acetal is formed from the light benzaldehyde and benzyl alcohol. The high boiling acetal falls down to the bottom section and is decomposed by reaction with benzoic acid into benzyl benzoate, benzaldehyde and benzyl alcohol. This is probably also one of the reasons why it is so difficult to remove all traces of benzyl alcohol and benzaldehyde from benzoic acid by distillation.

The large amount of benzyl benzoate formed can also be turned into a new opportunity to produce high-grade benzyl alcohol and benzoic acid. All low boiling by-products are easily removed from the high boiling benzyl benzoate. Benzyl benzoate can thereafter be hydrolysed to benzoic acid and benzyl alcohol under basic conditions (saponification, affording sodium benzoate) or acidic conditions via an equilibrium reaction. Under acidic conditions, large amounts of water and a co-solvent are necessary to reach acceptable high conversions, resulting in high rates of energy consumption due to the evaporation of water.

We have also developed[163] a "neutral" low energy process for the hydrolysis of benzyl benzoate at high temperatures (>240°C) by heating benzyl benzoate with

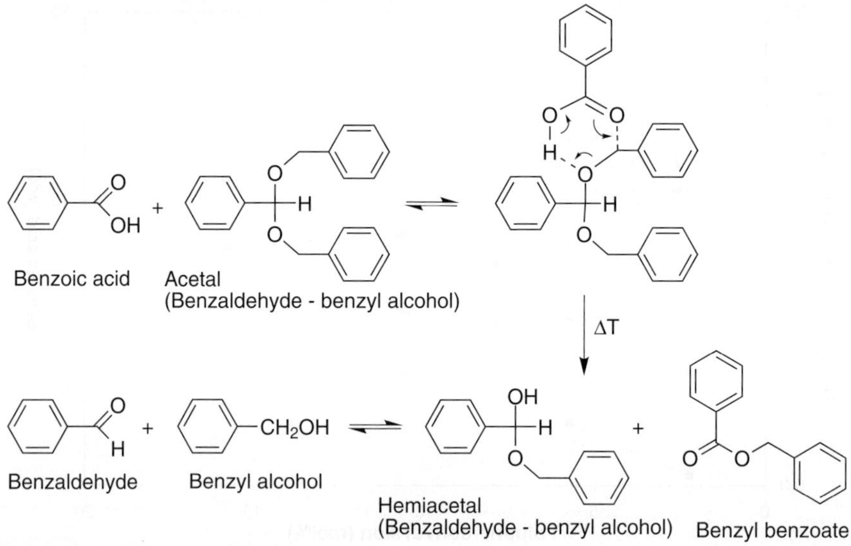


**Figure 16.21.** Proposed mechanism for the formation of benzyl benzoate from benzoic acid and benzaldehyde/benzyl alcohol acetal.

water. Under these conditions, a homogeneous system (one phase) is obtained. The outlet of the reactor is cooled counter-current with the inlet (energy integration) to <140°C, affording a two-phase system of mainly water and an organic phase of mainly benzoic acid, benzyl alcohol and benzyl benzoate. After phase separation, the organic phase is worked up by distillation, affording benzyl alcohol and benzoic acid of high purity. The water phase (and benzyl benzoate) is recycled back to the hydrolysis reactor.

After (partial) implementation of these aspects, the phenol plant in Rotterdam was successfully rebuilt as a plant for the dedicated production of benzaldehyde, benzyl alcohol and benzoic acid. As a result, the economic situation was dramatically shifted from massive yearly losses leading to plant shutdown, to a profitable and attractive business.

## References


1. Szent-Györgyi, A. (1928). Observations on the Function of Peroxidase Systems and the Chemistry of the Adrenal Cortex. Description of a New Carbohydrate Derivative, *Biochem. J.*, **22**, pp. 1387–1409.
2. Haworth, W. and Hirst, E. (1933). Synthesis of Ascorbic Acid. *J. Chem. Soc. Ind.*, **52**, pp. 645–647.
3. Bielski, B. (1982). Chemistry of Ascorbic Acid Radicals, in P. Seib and P. Tolbert (eds), *Ascorbic Acid: Chemistry, Metabolism, and Uses*, American Chemical Society, Washington, DC, pp. 81–100.

4. Bielski, B., Allen, A. and Schwarz, H. (1981). Mechanism of the Disproportionation of Ascorbate Radicals, *J. Am. Chem. Soc.*, **103**, pp. 3516–3518.
5. Martell, A. (1982). Chelates of Ascorbic Acid, in P. Seib and P. Tolbert (eds), *Ascorbic Acid: Chemistry, Metabolism, and Uses*, American Chemical Society, Washington, DC, pp. 153–178.
6. Tolbert, P. and Ward, J. (1982). Dehydroascorbic Acid, in P Seib and P. Tolbert (eds), *Ascorbic Acid: Chemistry, Metabolism, and Uses*, American Chemical Society, Washington, DC, pp. 101–123.
7. Oster, B. and Fechtel, U. (1996). Vitamin C (L-Ascorbic Acid), in *Ullmann's Encyclopedia of Industrial Chemistry*, 5th edition, Vol. A27 Vitamins, Wiley-VCH, Weinheim, Chapt. 10, pp. 547–559.
8. Nakagawa, K., Konaka, R. and Nakata, T. (1962). Oxidation with Nickel Peroxide. I. Oxidation of Alcohols, *J. Org. Chem.*, **27**, pp. 1597–1601.
9. Reichstein, T. and Grüssner, A. (1934). Eine Ergiebige Synthese der L-Ascorbinsäure (Vitamin C). *Helv. Chim. Acta*, **17**, pp. 311–328.
10. Isler, O., Brubacher, G., Ghisla, S., *et al.* (1988). *Vitamine II*, Georg Thieme Verlag, Stuttgart, pp. 396–433.
11. Wittmann, R., Wintermeyer, W. and Butzke, J. (1981). Verfahren zur Herstellung von Diacetonketogulonsäure, *German Patent DE 3019321.*
12. Jaffe, G. and Pleven, E. (1971). Oxidationsverfahren, *German Patent DE 2123621.*
13. Shnaidman, L. and Kushchinskaya, I. (1961). Catalytic Oxidation of Di-O-isopropylidene-L-sorbose to Di-O-isopropylidene-L-xylo-hexulosonic Acid with Atmospheric Oxygen, *Tr. Vses. Nauchn.-Issled.Vitamin. Inst.*, **8** pp. 13–22; (1963). *Chem. Abstr.*, **58**, p. 15064.
14. Bonrath, W. and Netscher, T. (2005). Catalytic Processes in Vitamin Synthesis and Production, *Appl. Catal. A: Gen.*, **280**, pp. 55–73.
15. Heyns, K. and Paulsen, H. (1957). Selective Catalytic Oxidations with Noble-metal Catalysts, *Angew. Chem.*, **69**, pp. 600–608.
16. Broennimann, C., Bodnar, Z., Aeschimann, R., *et al.* (1996). Platinum Catalysts Modified by Adsorbed Amines: A New Method of Enhancing Rate and Selectivity of L-sorbose Oxidation, *J. Catal.*, **161**, pp. 720–729.
17. Sulman, E., Matveeva, V., Bronstein, L., *et al.* (2003). Platinum-containing Polymeric Catalysts in Direct L-sorbose Oxidation, *Green Chem.* **5**, pp. 205–208.
18. Bonrath, W. and Fischesser, J. (2008). Novel Reaction with Gold Catalysts, *International Patent Application WO 2008148549.*
19. Berndt, H., Haji Begli, A., Kowalczyk, J., *et al.* (2009). Procedure for the Selective Carbohydrate Oxidation Using Supported Gold Catalysts, *German Patent DE 10319917.*
20. Yin, G. and Tao, Z. (1980). Studies on the Production of Vitamin C Precursor 2-Keto-L-Gulonic Acid from L-sorbose by Fermentation, *Acta Microbiol. Sin.*, **20**, pp. 246–251.
21. Ning, W., Tao, Z., Wang, C., *et al.* (1988). Fermentation Process for Producing 2-keto-L-gulonic acid, *European Patent EP 278447.*
22. Feng, S., Zhang, Z., Zhang, C., *et al.* (2000). Effect of *Bacillus megaterium* on *Gluconobacter oxydans* in Mixed Culture, *Yingyong Shengtai Xuebao*, **11**, pp. 119–122.
23. Zhang, J., Zhou, J., Liu, J., *et al.* (2011). Development of Chemically Defined Media Supporting High Cell Density Growth of *Ketogulonicigenium vulgare* and *Bacillus megaterium*, *Bioresour. Technol.* **102**, pp. 4807–4814.
24. Sugisawa, T., Hoshino, T., Masuda, S., *et al.* (1990). Microbial Production of 2-keto-L-gulonic Acid from L-sorbose and D-sorbitol by *Gluconobacter melanogenus*, *Agric. Biol. Chem.*, **54**, pp. 1201–1209.
25. Hoshino, T., Ojima, S. and Sugisawa, T. (1994). Process for Producing 2-keto-L-gulonic acid, *US Patent US 5312741.*

26. Takagi, Y., Sugisawa, T. and Hoshino, T. (2009). Continuous 2-keto-L-gulonic Acid Fermentation from L-sorbose by *Ketogulonicigenium vulgare* DSM 4025, *Appl. Microbiol. Biotechnol.*, **82**, pp. 1049–1056.
27. Takagi, Y., Sugisawa, T. and Hoshino, T. (2010). Continuous 2-keto-L-gulonic Acid Fermentation by Mixed Culture of *Ketogulonicigenium vulgare* DSM 4025 and *Bacillus megaterium* or *Xanthomonas maltophila*, *Appl. Microbiol. Biotechnol.*, **86**, pp. 469–480.
28. Running, J., Huss, R. and Olson, P. (1994). Heterotrophic Production of Ascorbic Acid by Microalgae, *J. Appl. Phycol.*, **6**, pp. 99–104.
29. Running, J., Severson, D. and Schneider, K. (2002). Extracellular Production of L-Ascorbic Acid by *Chlorella protothecoides*, *Prototheca species*, and Mutants of *P. moriformis* During Aerobic Culturing at low pH, *J. Ind. Microbiol. Biotechnol.*, **29**, pp. 93–98.
30. Kim, S., Huh, W., Kim, J., *et al.* (1996). D-arabinose Dehydrogenase and Biosynthesis of Erythroascorbic Acid in *Candida albicans*, *Biochim. Biophys. Acta*, **1297**, pp. 1–8.
31. Kim, S., Huh, W., Lee, B. *et al.* (1998). D-arabinose Dehydrogenase and its Gene from *Saccharomyces cerevisiae*, *Biochim. Biophys. Acta*, **1429**, pp. 29–39.
32. Hancock, R., Galpin, J. and Viola, R. (2000). Biosynthesis of L-ascorbic Acid (Vitamin C) by *Saccharomyces cerevisiae*, *FEMS Microbiol. Lett.*, **186**, pp. 245–250.
33. Sauer, M., Branduardi, P., Valli, M., *et al.* (2004). Production of L-ascorbic Acid by Metabolically Engineered *Saccharomyces cerevisiae* and *Zygosaccharomyces bailii*, *Appl. Environ. Microbiol.*, **70**, pp. 6086–6091.
34. Sugisawa, T., Ojima, S., Matzinger, P., *et al.* (1995). Isolation and Characterization of a New Vitamin C Producing Enzyme (L-gulono-$\gamma$-lactone Dehydrogenase) of Bacterial Origin, *Biosci. Biotechnol. Biochem.*, **59**, pp. 190–196.
35. Sugisawa, T., Miyazaki, T. and Hoshino, T. (2005). Microbial Production of L-ascorbic Acid from D-sorbitol, L-sorbose, L-gulose, and L-sorbosone by *Ketogulonicigenium vulgare* DSM 4025, *Biosci. Biotechnol. Biochem.*, **69**, pp. 659–662.
36. Berry, A., Lee, C., Mayer, A. *et al.* (2005). Microbial Production of L-ascorbic Acid, *International Patent Application WO 2005/017172*.
37. Miyazaki, T., Sugisawa, T. and Hoshino, T. (2006). Pyrroloquinoline Quinone-dependent Dehydrogenases from *Ketogulonicigenium vulgare* Catalyze the Direct Conversion of L-sorbosone to L-ascorbic Acid, *Appl. Environ. Microbiol.*, **72**, pp. 1487–1495.
38. Baldenius, K., von dem Bussche-Hünnfeld, L., Hilgemann, E., *et al.* (1996). Vitamin E (Tocopherols, Tocotrienols), in *Ullmann's Encyclopedia of Industrial Chemistry*, 5th edition, Vol. A27 Vitamins, Wiley-VCH, Weinheim, Chapt. 4, pp. 478–488.
39. Ning, Z., Xi, Z., Cao, G., *et al.* (1999). Oxidation of Trimethylphenol Catalyzed by Aqueous Soluble Oxygen Carriers, *Oxidation Commun.*, **22**, pp. 527–531.
40. Dong, Q., Hou-Qun, H. and Kai-Yi, W. (2002). Synthesis of 2,3,5-Trimethylhydroquinone by Direct Oxidation, *Chem. Reag.*, **24**, pp. 231–232.
41. Bonrath, W., Eggersdorfer, M. and Netscher, T. (2007). Catalysis in the Industrial Preparation of Vitamins and Nutraceuticals, *Catal. Today* **121**, pp. 45–57.
42. Diehl, H. and Hach, C. (1950). Bis(N,N′-disalicylalethylenediamine)-?-aquodicobalt(II) *Inorg. Synth. III*, pp. 196–201.
43. Jouffret, M. (1975). Verfahren zur Herstellung von Trimethyl-p-benzochinon, *German Patent DE 2450908*.
44. (1970). Trimethyl-p-benzoquinone by Oxidation of 2,3,6-Trimethylphenol in the Presence of a Cobalt-complexed Salt, *French Patent FR 2015576*.
45. Thoemel, F. and Hoffmann, W. (1983). 2,3,5-Trimethyl-p-benzoquinone, *German Patent DE 3215095*; Bartoldus, D. and Lohri, B. (1981). Trimethyl Benzoquinone, *European Patent EP 35635*.

46. Maassen, R., Krill, S. and Jäger, B. (2001). Verfahren zur Herstellung von 2,3,5-Trimethyl-p-benzochinon, *European Patent EP 1092701*.
47. Brenner, W. (1972). Verfahren zur Herstellung von Chinonen, *German Patent DE 2221624*; Hirose, N., Hamamura, K. and Inai, Y. (1988). Process for preparing 2,3,5-trimethylbenzoquinone, *European Patent EP 0294584*; Hsu, C. and Lyons, J. (1984). Oxidizing a Phenol to a *p*-Benzoquinone, *European Patent EP 107427*.
48. Takehiro, K., Orita, H. and Shimizu, M. (1990). Method for the Preparation of 2,3,5-trimethylbenzoquinone, *European Patent EP 0369823*; Bodnar, Z., Mallat, T. and Baiker, A. (1996). Oxidation of 2,3,6-trimethylphenol to Trimethyl-1,4-benzoquinone with Catalytic Amount of $CuCl_2$, *J. Mol. Catal. A: Chem.*, **110**, pp. 55–63.
49. Sun, H., Harms, K. and Sundermeyer, J. (2004). Aerobic Oxidation of 2,3,6-trimethylphenol to Trimethyl-1,4-benzoquinone with Copper(II) Chloride as Catalyst in Ionic Liquid and Structure of the Active Species, *J. Am. Chem. Soc.*, **126**, pp. 9550–9551.
50. Bockstiegel, B., Hoercher, U. and Laas, H. (1992). Verfahren zur Herstellung von 2,3,5-Trimethyl-p-benzochinon; *German Patent DE 4029198*.
51. Maassen, R., Krill, S. and Huthmacher, K. (2001). Verfahren zur Herstellung von 2,3,5-Trimethyl-p-benzochinon, *European Patent EP 1132367*.
52. Akiyama, A. (2003). Method of Manufacturing 2,3,5-Trimethylhydroquinone, *Japanese Patent JP 96010*.
53. Kholdeeva, O., Golovin, A., Maksimovskaya, R., *et al.* (1992). Oxidation of 2,3,6-Trimethylphenol in the Presence of Molybdovanadophosphoric Heteropoly Acids, *J. Mol. Catal. A: Chem.*, **75**, pp. 235–244.
54. Fujibayashi, S., Nakayama, K. and Hamamoto, M. (1996). An Efficient Aerobic Oxidation of Various Organic Compounds Catalyzed by Mixed Addenda Heteropolyoxometalates Containing Molybdenum and Vanadium, *J. Mol. Catal. A: Chem.*, **110**, pp. 105–117.
55. Vandewalle, M. (1998). Method for preparing trimethylbenzoquinone, *International Patent Application WO 9818746*.
56. Kholdeeva, O., Trubitsina, T. and Maksimov, G. (2005). Synthesis, Characterization, and Reactivity of Ti(IV)-Monosubstituted Keggin Polyoxometalates, *Inorg. Chem.*, **44**, pp. 1635–1642.
57. Adam, W., Herrmann, W., Lin, J., *et al.* (1994). Catalytic Oxidation of Phenols to *p*-Quinones with the Hydrogen Peroxide and Methyltrioxorhenium(VII) System, *J. Org. Chem.*, **59**, pp. 8281–8283; Bernini, R., Mincione, E., Barontini, M., *et al.* (2006). Convenient Oxidation of Alkylated Phenols and Methoxytoluenes to Antifungal 1,4-benzoquinones with Hydrogen Peroxide ($H_2O_2$)/methyltrioxorhenium ($CH_3ReO_3$) Catalytic System in Neutral Ionic Liquid, *Tetrahedron*, **62**, pp. 7733–7737.
58. Bernini, R., Mincione, E., Barontini, M., *et al.* (2007). Dimethyl Carbonate: An Environmentally Friendly Solvent for Hydrogen Peroxide ($H_2O_2$/methyltrioxorhenium ($CH_3ReO_3$, MTO)) Catalytic Oxidations, *Tetrahedron*, **63**, pp. 6895–6900.
59. Murtinho, D., Pineiro, M., Pereira, M. *et al.* (2000). Novel Porphyrins and a Chlorine as Efficient Singlet Oxygen Photosensitizers for Photooxidation of Naphthols or Phenols to Quinones, *J. Chem. Soc. Perkin Trans.*, **2**, pp. 2441–2447; Sorokin, A. and Tuel, A. (1999). Heterogeneous Oxidation of Aromatic Compounds Catalyzed by Metallophthalocyanine Functionalized Silicas, *New J. Chem.*, **23**, pp. 473–476; Sorokin, A., Mangematin, S. and Pergrale, C. (2002). Selective Oxidation of Aromatic Compounds with Dioxygen and Peroxides Catalyzed by Phthalocyanine Supported Catalysts, *J. Mol. Catal. A: Chem.*, **182–183**, pp. 267–281.
60. Hong, Z. and Song-Ying, C. (2003). Catalytic Performance of Magnesium Ferrite for Hydroxylation of 2,3,6-Trimethylphenol, *Chin. J. Catal.*, **24**, pp. 635–638.

61. Li, Y., Liu, W. and Wu, M. (2007). Oxidation of 2,3,5-trimethylphenol to 2,3,5-trimethylbenzoquinone with Aqueous Hydrogen Peroxide in the Presence of Spinel $CuCo_2O_4$, *J. Mol. Catal. A: Chem.* **261**, pp. 73–78.
62. Li, Y., Liu, W., Wu, M., *et al.* (2010). Selective Photoinduced Oxidation of 2,3,5-trimethylphenol Catalyzed by Hypocrellins/$CuCo_2O_4$, *Mendeleev Commun.*, **20**, pp. 218–219.
63. Zhu, H. and Chen, S. (2004). A New Process for the Production of Trimethylhydroquinone, *Ind. J. Chem.* **43B**, pp. 1349–1354; Meng, X., Sun, Z., Lin, S., *et al.* (2002). Catalytic Hydroxylation of 2,3,6-trimethylphenol with Hydrogen Peroxide over Copper Hydroxyphosphate ($Cu_2(OH)PO_4$), *Appl. Catal. A*, **236**, pp. 17–22.
64. Run-Wei, W., Lei, H. and Guang-Shan, Z. (2004). Synthesis, Characteristics and Catalytic Properties of Mesoporous Materials with Loading of TS-1 Precursors, *Chem. J. Chin. University*, **25**, pp. 1485–1488.
65. Cheng, S., Tsai, T., Chou, B., *et al.* (2002). Synthesis of TMBQ with Transition Metal-containing Molecular Sieve as Catalyst, *US Patent US 143198*.
66. Kholdeeva, O., Mel'gunov, M., Shmakov, A., *et al.* (2004). A New Mesoporous Titanium-silicate Ti-MMM-2: A Highly Active and Hydrothermally Stable Catalyst for $H_2O_2$-based Selective Oxidations, *Catal. Today*, **91**, pp. 205–209.
67. Kholdeeva, O., Trukhan, N., Vanina, M., *et al.* (2002). A New Mesoporous Titanium-silicate Ti-MMM-2: A Highly Active and Hydrothermally Stable Catalyst for $H_2O_2$-based Selective Oxidations, *Catal. Today*, **75**, pp. 203–209.
68. Rüttimann, A. (1986). Recent Advances in the Synthesis of K-vitamins, *Chimia*, **40**, pp. 290–306.
69. Isler, O. and Brubacher, G. (1988). *Vitamine I*, Georg Thieme Verlag, Stuttgart, Chapt. 4, pp. 152–171.
70. Weber, F. and Rüttimann, A. (2009). Vitamin K, in *Ullmann's Encyclopedia of Industrial Chemistry*, 7th edition, Vol. 38 Vitamins, Wiley-VCH, Weinheim, Chapt. 5, pp. 47–67.
71. A. Hirschmann, R., Miller, R. and Wendler, N.L. (1954). The Synthesis of Vitamin $K_1$, *J. Am. Chem. Soc.*, **76**, pp. 4592–4594.
72. Lindlar, H. (1957). Verfahren zur Herstellung von Kondensationsprodukten, *Swiss Patent 320,582*.
73. Isler, O. and Doebel, K. (1954). Syntheses in the Vitamin K Series. I. The Total Synthesis of Vitamin $K_1$, *Helv. Chim. Acta*, **37**, pp. 225–233; Isler, O. and Doebel, K. (1954). Synthesis of Vitamin $K_1$ using Boron Trifluoride Catalysts, *US Patent US 2,683,176*; (1956). Novel Naphthalene Compounds and a Process for the Manufacture Thereof, *British Patent GB 752,420*.
74. Puetter, H. and Bewert, W. (1981). 2-Methyl-1,4-naphthoquinones, *German Patent DE 2952709*.
75. Eremin, D. and Petrov, L. (2009). Optimization of Conditions for Preparing Vitamin $K_3$ by Oxidation of 2-Methylnaphthalene with Chromium Trioxide in Acid Solutions, *Russ. J. Appl. Chem.*, **82**, pp. 866–870.
76. Arnold, R. and Larson, R. (1940). Quinones by the Peroxide Oxidation of Aromatic Compounds, *J. Org. Chem.*, **5**, pp 250–252.
77. Narayanan, S., Murthy, K., Reddy, K., *et al.* (2002). A Novel and Environmentally Benign Selective Route for Vitamin K3 Synthesis, *Appl. Catal. A*, **228**, pp. 161–165; Sankarasubbier, N., Murthy, K., Reddy, K., *et al.* (2002). Process for Preparation of 2-methyl-1,4-naphthoquinone, *US Patent US 2002/0188141*; Kholdeeva, O., Zalomaeva, O., Sorokin, A., *et al.* (2007). New Routes to Vitamin $K_3$, *Catal. Today*, **121**, pp. 58–64.
78. Adam, W., Herrmann, W., Lin, J., *et al.* (1994). Homogeneous-catalytic Oxidation of Arenes and a New Synthesis of Vitamin $K_3$, *Angew. Chem. Int. Ed. Engl.*, **33**, pp. 2475–2477; Herrmann, W., Correia, G., Fischer, R., *et al.* (1995). Oxidation of Electron-rich Aromatic Compounds Using Organorhenium Oxide Catalysts, *European Patent EP 665209*; Fischer, R., Haider, J., Herrmann, W., *et al.* (1998). Rhenium Catalysts for Selective Oxidation of

Aromatic Compounds, *International Patent Application WO 9847837*; Herrmann, W., Haider, J. and Fischer, R. (1999). Rhenium-catalyzed Oxidation of Arenes: An Improved Synthesis of Vitamin $K_3$, *J. Mol. Catal. A: Chem.*, **138**, pp. 115–121.

79. Shi, F., Tse, M. and Beller, M. (2007). A Novel and Convenient Process for the Selective Oxidation of Naphthalenes with Hydrogen Peroxide, *Adv. Synth. Catal.*, **349**, pp. 303–308; Shi, F., Tse, M. and Beller, M. (2007). Selective Oxidation of Naphthalene Derivatives with Ruthenium Catalysts Using Hydrogen Peroxide as Terminal Oxidant, *J. Mol. Catal. A: Chem.*, **270**, pp. 68–75; Wienhöfer, G., Schröder, K., Möller, K., *et al.* (2010). A Novel Process for Selective Ruthenium-catalyzed Oxidation of Naphthalenes and Phenols, *Adv. Synth. Cataly.*, **352**, pp. 1615–1620.
80. Kowalski, J., Poszyska, J. and Sobkowiak, A. (2003). Iron(III)-induced Activation of Hydrogen Peroxide for Oxidation of 2-Methylnaphthalene in Glacial Acetic Acid, *Catal. Commun.*, **4**, pp. 603-608; Möller, K., Wienhöfer, G., Schröder, K., *et al.* (2010). Selective Iron-catalyzed Oxidation of Phenols and Arenes with Hydrogen Peroxide: Synthesis of Vitamin E Intermediates and Vitamin $K_3$, *Chem. Eur. J.*, **16**, pp. 10300–10303.
81. Anunziata, O., Beltramone, A. and Cussa, J. (2004). Studies of Vitamin $K_3$ Synthesis over Ti-containing Mesoporous Material, *Appl. Catal. A*, **270**, pp. 77–85; Urus, S., Keles, M. and Serindag, O. (2010). Catalytic Synthesis of 2-Methyl-1,4-Naphthoquinone (Vitamin $K_3$) over Silica-supported Aminomethyl Phosphine-Ru(II), Pd(II), and Co(II) Complexes, *Phosphorus, Sulfur and Silicon and the Related Elements*, **185**, pp. 1416–1424; Selvaraj, M., Kandaswamy, M., Park, D., Ha, C. (2010). Highly Selective Synthesis of Vitamin $K_3$ over Mesostructured Titanium Catalysts, *Catal. Today*, **158**, pp. 377–384; Urus, S., Keles, M. and Serindag, O. (2010). Synthesis of Silica-supported Platinum(II) and Nickel(II) Complexes of Bis(Diphenylphosphinomethyl)Amino Ligand: Applications as Catalysts for the Synthesis of 2-Methyl-1,4-Naphthoquinone (Vitamin $K_3$), *J. Inorg. Organomet. Polymers and Materials*, **20**, pp. 152–160.
82. Machowska, Z., Mieluch, J., Obloj, J., *et al.* (1987). Process for Preparing Vitamin $K_3$ from 2-methylnaphthalene, *Polish Patent PL 140780*; Harrison, S., Fiset, G. and Mahdavi, B. (1999). Preparation of Quinones by Oxidation of Aromatic Compounds Using Electrochemically Prepared Ceric Ion, *European Patent EP 919533*.
83. Monteleone, F., Cavani, F., Felloni, C., *et al.* (2004). A Redox Process for the Production of Menadione and Use of Polyoxometalates as Oxidizing Agents, *International Patent Application WO 2004014832*; Strukul, G., Somma, F., Ballarini, N., *et al.* (2009). The Oxidation of 2-methyl-1-naphthol to Menadione with $H_2O_2$, Catalyzed by Nb-based Heterogeneous Systems, *Appl. Catal. A: Gen.*, **356**, pp. 162–166; Zalomaeva, O., Ivanchikova, I., Kholdeeva, O., *et al.* (2009). Kinetics and Mechanism of the Oxidation of Alkyl Substituted Phenols and Naphthols with TBuOOH in the Presence of Supported Iron Phthalocyanine, *New J. Chem.*, **33**, pp. 1031–1037.
84. John, M. and Hähnlein W. (1996). Vitamin A (Retinoids), in *Ullmann's Encyclopedia of Industrial Chemistry*, 5th edition, Vol. A27 Vitamins, Wiley-VCH, Weinheim, Chapt. 2, pp. 453–469.
85. Semikolenov, V., Ilyna, I. and Simakova, I. (2001). Linalool Synthesis from $\alpha$-Pinene: Kinetic Peculiarities of Catalytic Steps, *Appl. Catal. A: Gen.*, **211**, pp. 9–107.
86. Sercheli, R., Ferreira, A., Baptistella, L., *et al.* (1997). Transition-metal Catalyzed Autoxidation of *cis*- and *trans*-Pinane to a Mixture of Diastereoisomeric Pinanols, *J. Agric. Food Chem.*, **45**, pp. 1361–1364.
87. Yang, G., Xu, L., Huang, K., *et al.* (2004). Synthesis of Pinane Hydroperoxide Used as an Intermediate of Linalool, *Fine Chemical Intermediates*, **35**, pp. 39–41.
88. Fisher, G., Stinson, J., Moore, R., *et al.* (1955). Peroxides from Turpentine, *Eng., Design, and Proc. Dev.*, **47**, pp. 1368–1373.

89. Brose, T, Pritzkow, W. and Thomas, G. (1992). Studies on the Oxidation of *cis*- and *trans*-Pinane with Molecular Oxygen, *J. Prakt. Chem.*, **334**, pp. 403–409.
90. Woell, J. (1990). Processes for the Conversion of Myrcene to Citral, *US Patent US 4, 978,804*.
91. Woell, J. (1991). Processes for the Conversion of Myrcene to Nerol and Citral, *US Patent US 5,017,726*.
92. Woell, J. (1991). Processes for the Conversion of Myrcene to Citral, *European Patent EP 0439368*.
93. Takahashi, M., Urata, H., Suzuki, H., *et al.* (1984). Regioselective Introduction of O-nucleophiles into Myrcene and trans-Ocimene Using Palladium(II) Complexes, *J. Organomet. Chem.*, **226**, pp. 327–336.
94. Netscher, T., Bonrath, W., Bähler, F., *et al.* (2005). Stereoselective Functionalization Reactions of Conjugated Dienes: The Conversion of Myrcene to Citral, *IUPAC Symposium on Organometallic Chemistry Directed Towards Organic Synthesis (OMCOS 13, Geneva)*, Poster.
95. Isler, O., Lindlar, H., Montavon, M., *et al.* (1956). Syntheses in the Carotenoid Series. VII. Total Synthesis of Zeaxanthin and Physaliene, *Helv. Chim. Acta*, **39**, pp. 2041–2053.
96. Isler, O., Montavon, M., Ruegg, R., *et al.* (1959). Process for the Manufacture of Carotenoids, *US Patent US 2917539*.
97. Becker, J., Schulte-Elte, K., Strickler, H., *et al.* (1975). Verfahren zur Herstellung von 2,2,6-Trimethyl-cyclohex-5-en-1,4-dion, *German Patent DE 2457157*.
98. Constantini, M., Dromard, A. and Jouffret, M. (1977). Procédé de Préparation de Triméthyl-3,5,5-cyclohéxène-2-dione-1,4, *French Patent FR 2335486*.
99. Brenner, W. (1978). Verfahren zur Herstellung von Ketoisophoron, *Swiss Patent CH 605536*.
100. Brenner, W. (1985). Verfahren zur Herstellung von 2,6,6-Trimethyl-2-cyclohexen-1,4-dion (Ketoisophoron), *German Patent DE 2515304*.
101. Bellut, H. (1990). Verfahren zur Herstellung von 2,6,6-Trimethyl-2-cyclohex-2-en-1,4-dion, *German Patent DE 3842547*.
102. Zumstein Sr, F., Klingeisen, F. and Zumstein Jr, F. (1985). Verfahren zur Herstellung von 3,5,5-Trimethylcyclohexen-2-dion-1,4, *German Patent DE 2610254*.
103. Constantini, M., Dromard, A., Jouffret, M., *et al.* (1980). Selective Oxidation of 3,5,5-trimethylcyclohexene-3-one ($\beta$-isophorone) to 3,5,5-trimethylcyclohexene-1,4-dione by Oxygen Catalyzed by $Mn^{II}$ or $Co^{II}$ Chelates, *J. Mol. Catal.*, **7**, pp. 89–97.
104. Ito, N., Kinoshita, K., Suzuki, K., *et al.* (1989). Preparation of 3,5,5-Trimethyl-2-cyclohexene-1,4-dione by Manganese Complex-catalyzed Oxidation of 3,5,5-Trimethyl-3-cyclohexen-1-one, *Japanese Patent JP 01090150*.
105. Hahn, R., Gora, U., Huthmacher, K., *et al.* (1997). Verfahren zur Herstellung von Ketoisophoron, *German Patent DE 19619570*.
106. Klatt, M., Müller, T. and Bockstiegel, B. (2000). Verfahren zur Herstellung von Oxoisophoron, *German Patent DE 19929367*.
107. Klatt, M., Müller, T. and Bockstiegel, B. (2000). Verfahren zur Herstellung von Oxoisophoron unter Verwendung von Additiven, *German Patent DE 19929362*.
108. Dong, S. and Huang, Y. (2007). Method for Preparing Keto-isophorone, *Chinese Patent CN 100999453*.
109. Mao, J., Li, N., Li, H., *et al.* (2006). Novel Schiff Base Complexes as Catalysts in Aerobic Selective Oxidation of $\beta$-isophorone, *J. Mol. Catal. A*, **258**, pp. 178–184.
110. Tomohide, I., Noboru, K. and Hiroyuki, M. (2002). Process and Apparatus for Producing Ketoisophorone, *US Patent US 6346651*.
111. Joseph, T., Halligudi, S., Satyanarayan, C., *et al.* (2001). Oxidation by molecular Oxygen Using Zeolite Encapsulated Co(II)saloph Complexes, *J. Mol. Catal. A*, **168**, pp. 87–97.
112. Beyrhouty, M., Sorokin, A., Daniele, S., *et al.* (2005). Combination of Two Catalytic Sites in a Novel Nanocrystalline $TiO_2$–iron Tetrasulfophthalocyanine Material Provides Better Catalytic Properties, *New J. Chem.*, **29**, pp. 1245–1248.

113. Sorokin, A., Quignard, F., Valentin, R., *et al.* (2006). Chitosan Supported Phthalocyanine Complexes: Bifunctional Catalysts with Basic and Oxidation Active Sites, *Appl. Cat. A.*, **309**, pp. 162–168.
114. Mao, J., Hu, X., Li, H., *et al.* (2008). Iron Chloride Supported on Pyridine-modified Mesoporous Silica: An Efficient and Reusable Catalyst for the Allylic Oxidation of Olefins with Molecular Oxygen, *Green Chem.*, **10**, pp. 827–831.
115. Hu, X., Mao, J., Sun, Y., *et al.* (2009). Acetylacetone–Fe Catalyst Modified by Imidazole Ionic Compound and its Application in Aerobic Oxidation of $\beta$-Isophorone, *Catal. Commun.*, **10**, pp. 1908–1912.
116. Li, H., Chen, Z., Hu, X., *et al.* (2007). Method for Preparing Oxoisophorone by Catalytic Oxidation of $\beta$-Isophorone, *Chinese Patent CN 1923782*.
117. Li, H., Chen, Z., Mao, J., *et al.* (2009). Method for Preparing Oxoisophorone by Catalytic Oxidation of $\beta$-Isophorone with Metal-free Catalyst System, *Chinese Patent CN 101417935*.
118. Becker, J., Skorianetz, W. and Hochstrasser, U. (1976). Verfahren zur Herstellung von 2,2,6-Trimethylcyclohex-5-en-1,4-dion, *German Patent DE 2459148*.
119. Widmer, E. and Seuret, M. (1976). Verfahren zur Herstellung von Ketoisophoron, German Patent *DE 2526851*.
120. Hanyu, A., Sakurai, Y., Fujibayashi, S., *et al.* (1997). Selective Aerobic Oxidation of Isophorone Catalyzed by Molybdovanadophosphate Supported on Carbon (NPMoV/C), *Tetrahedron Lett.*, **38**, pp. 5659–5662.
121. Murphy, E., Schneider, M., Mallat, T., *et al.* (2001). Enhanced Catalytic Activity and Selectivity in Oxidation of $\alpha$-Isophorone to Ketoisophorone with Phosphomolybdic Acid, *Synthesis*, **4**, pp. 547–549.
122. Zhang, J. and Che, C. (2005). Dichlororuthenium(IV) Complex of Meso-tetrakis(2,6-dichlorophenyl)porphyrin: Active and Robust Catalyst for Highly Selective Oxidation of Arenes, Unsaturated Steroids, and Electron-deficient Alkenes by Using 2,6-Dichloropyridine N-Oxide, *Chem. Eur. J.*, **11**, pp. 3899–3914.
123. Shing, T., Yeung, Y. and Su, P. (2006). Mild Manganese(III) Acetate Catalyzed Allylic Oxidation: Application to Simple and Complex Alkenes, *Org. Lett.*, **8**, pp. 3149–3151.
124. Kishore, D. and Rodrigues, A. (2007). Catalytic Oxidation of Isophorone to Ketoisophorone over Ruthenium Supported MgAl-hydrotalcite, *Catal. Comm.*, **8**, pp. 1156–1160.
125. Kishore, D. and Rodrigues, A. (2008). Liquid Phase Catalytic Oxidation of Isophorone with *tert.*-butylhydroperoxide over Cu/Co/Fe–MgAl Ternary Hydrotalcites, *Appl. Catal. A: Gen.*, **345**, pp. 104–111.
126. Wang, C., Wang, G., Mao, J., *et al.* (2010). Metal and Solvent-free Oxidation of $\alpha$-Isophorone to Ketoisophorone by Molecular Oxygen, *Catal. Commun.*, **11**, pp. 758–762.
127. Li, H., Chen, Z., Wang, G., *et al.* (2009). Method for Synthesis of Oxoisophorone from $\alpha$-Isophorone with Metal-free Catalyst System, *Chinese Patent CN 101417936*.
128. Strickler, H., Becker, J. and Ohloff, G. (1975). Verfahren zur Herstellung eines acyclischen Diketons, *German Patent DE 2457158*.
129. Murphy, E., Mallat, T., Baiker, A., *et al.* (2000). Catalytic Gas Phase Oxidation of Isophorone to Ketoisophorone, *Appl. Catal. A: Gen.*, **197**, pp. 295–301.
130. Fahlbusch, K., Hammerschmidt, F., Panten, J., *et al.* (2003). Flavors and Fragrances, in *Ullmann's Encyclopedia of Industrial Chemistry (online release)*, Wiley-VCH, Weinheim, DOI: 10.1002/14356007.a11_141.
131. Yamamoto, T., Matsuda, H., Utsumi, Y., *et al.* (2002). Synthesis and Odor of Optically Active Rose Oxide, *Tetrahedron Lett.*, **43**, pp. 9077–9080.
132. Rojahn, W. and Warnecke, H. (1980). *Dragoco Report*, **27**, pp. 159–164.
133. Alsters, P., Jary, W., Nardello-Rataj, V., *et al.* (2010). "Dark" Singlet Oxygenation of $\beta$-Citronellol: A Key Step in the Manufacture of Rose Oxide, *Org. Process Res. Dev.*, **14**, pp. 259–262.

134. Aubry, J. (1985). Search for Singlet Oxygen in the Decomposition of Hydrogen Peroxide by Mineral Compounds in Aqueous Solutions, *J. Am. Chem. Soc.*, **107**, pp. 5844–5849.
135. Nardello, V., Bogaert, S., Alsters, P., *et al.* (2002). Singlet Oxygen Generation from $H_2O_2/MoO_4^{2-}$: Peroxidation of Hydrophobic Substrates in Pure Organic Solvents, *Tetrahedron Lett.*, **43**, pp. 8731–8734.
136. Schmid, A., Kollmer, A., Sonnleitner, B., *et al.* (1999). Development of Equipment and Procedures for the Safe Operation of Aerobic Bacterial Bioprocesses in the Presence of Bulk Amounts of Flammable Organic Solvents, *Bioproc. Eng.*, **20**, pp. 91–100.
137. Nardello, V., Marko, J., Vermeersch, G., *et al.* (1995). Mo-95 NMR and Kinetic-studies of Peroxomolybdic Intermediates Involved in the Catalytic Disproportionation of Hydrogen-peroxide by Molybdate Ions, *Inorg. Chem.*, **34**, pp. 4950–4957.
138. Csányi, L. (2010). On the Peroxomolybdate Complexes as Sources of Singlet Oxygen, *J. Mol. Catal. A*, **322**, pp. 1–6.
139. Kaiser, K. (2009). Pantothenic Acid, in *Ullmann's Encyclopedia of Industrial Chemistry*, 7th edition, Vol. 38 Vitamins, Wiley-VCH, Weinheim, Chapt. 11, pp. 123–130.
140. Bonrath, W., Karge, R. and Netscher, T. (2002). Lipase-catalyzed Transformations as Key-steps in the Large-scale Preparation of Vitamins, *J. Mol. Catal. B: Enzymatic*, **19–20**, pp. 67–72, and references cited therein.
141. Kuroda, N. and Kashiwa, K. (1992). Preparation of Ketopantolactone, *Japanese Patent JP 04095086*.
142. Kin, E. and Tanyama, E. (1993). Preparation of Ketopantolactone, *Japanese Patent JP 05306276*.
143. Kuroda, N. and Kashiwa, K. (1992). Preparation of Ketopantolactone, *Japanese Patent JP 04095087*.
144. Kuroda, N. and Kashiwa, K. (1992). Preparation of Ketopantolactone, *Japanese Patent JP 04095084*.
145. Tanaka, M. and Kobayashi, T. (1985). $\alpha$-Keto Esters, *Japanese Patent JP 60184050*.
146. Chang, H., Woo, J. C., Lee, K., *et al.* (2002). Facile Synthesis of $\alpha$-Ketocarbonyl Compounds from $\alpha$-Hydroxycarbonyl Compounds, *Synth. Commun.*, **32**, pp. 31–35.
147. Jenny, C., Lohri, B. and Schlageter, M. (1997). Oxidation of Primary or Secondary Alcohols Using N-Chloro Compounds and Piperidinyloxyl Derivatives, *European Patent EP 775684*.
148. Yamaoka, H., Moriya, N. and Ikunaka, M. (2004). A Practical RuCl3-catalyzed Oxidation Using Trichloroisocyanuric Acid as a Stoichiometric Oxidant under Mild Nonacidic Conditions, *Org. Process Res. Dev.*, **8**, pp. 931–938.
149. Kaneko, T., Sumya, H. and Kashiwa, K. (1992). Preparation of Ketopantolactone, *Japanese Patent JP 04095085*.
150. Bonrath, W., Karge, R., Nüchter, M., *et al.* (2003). Oxidative Process and Ruthenium Catalysts for the Conversion of Pantolactone into Ketopantolactone in the Presence of Microwave Irradiation, *International Patent Application WO 2003091235*.
151. Nösberger, P. (1983). Ketopantolactone, *European Patent EP 86322*.
152. Matsumoto, M. and Ito, S. (1984). Ruthenium-catalyzed Aerobic Oxidation of Pantoyl Lactone to Ketopantoyl Lactone, *Synth. Commun.*, **14**, pp. 697–700.
153. Mallat, T., Seyler, L., Mir Alai, M., *et al.* (1998). Aerobic Oxidation of $\alpha$-Substituted Alcohols over Promoted Platinum Metal Catalysts, in B. Hodnett, J. Clark and K. Smith (eds), *Supported Reagents and Catalysts in Chemistry*, **216**, Royal Society of Chemistry, Cambridge, pp. 66–71, cit. in: Jenzer, G., Sueur, D., Mallat, T., *et al.* (2000). Partial Oxidation of Alcohols in Supercritical Carbon Dioxide, *Chem. Commun.*, **22**, pp. 2247–2248.
154. Sato, H., Takeda, S. and Yuya, M. (1989). Electrochemical Manufacturing of Ketopantolactone, *Japanese Patent JP 01208488*.
155. Cibrowski, S. (2000). Development of Cyclopol Process, *Polish J. Chem. Technol.*, **2**, pp. 1–4

156. Druliner, J., Krusic, J., Lehr, G., *et al.* (1985). Generation and Chemistry of Cyclohexyloxy Radicals, *J. Org. Chem.*, **50**, pp. 5838–5845.
157. Hermans, I., Peeters, J. and Jacobs, P. (2008). Origin of Byproducts during the Catalytic Autooxidation of Cyclohexane, *J. Phys. Chem. A.*, **112**, pp. 1747–1753.
158. van de Moesdijk, C. (1983). Process for the Preparation of Cyclohexyl Hydroperoxide, *European Patent EP 0579323.*
159. Griszka, M., Krzystoforski, A., Moiuk, W., *et al.* (2005). Cyclopol-Bis- the Second Youth of the Polish Process for Oxidation of Cyclohexane, *Przemysl Chemiczny*, **84**, pp. 493–502, *Chem. Abstr.*, **144**, 173050.
160. Berezin, I., Denisov, E. and Emanuel, N. (1966). *The Oxidation of Cyclohexane*, 1st English edition, Pergamon Press, New York, pp. 82–86.
161. Vasin, A., Chernysheva, L. and Bychof, A. (1988). Improving the Selectivity of the Industrial Process of the Cyclohexane Oxidation, *Khimicheskaya Promyshlenmost*, **20**, pp. 3–6.
162. Bryan, W. (1980). Process for Cycloalkanols and Cycloalkanones, *US Patent US 4238415.*
163. van Soolingen, J., Vrinzen, A., Stoelwinder, C., *et al.* (2001). Process for the Preparation of Benzyl Alcohol, *US Patent US 6326521.*

# Chapter 17

## *In Situ* and *Operando* Raman Spectroscopy of Oxidation Catalysts

Israel E. WACHS* and Miguel BAÑARES†

Over the past two decades, Raman spectroscopy has been extensively applied during catalytic oxidation reactions by mixed-metal oxides and metals under *in situ* and *operando* spectroscopy conditions, which has allowed the direct identification of the catalytic active sites involved in the oxidation reactions. Among the multiple spectroscopic techniques that can provide information about the catalytic active sites under oxidation reaction conditions, Raman spectroscopy is unique because of its ability to directly provide molecular level information that allows discrimination among the different catalytic active sites which may be present in the oxidation catalyst. This chapter provides a snapshot of the types of fundamental information obtainable by Raman spectroscopy, and the different types of catalytic materials and oxidation reactions that have been reported, especially under oxidation reaction conditions.

### 17.1. Introduction

Catalytic oxidation represents the major technology employed by the chemical industry to produce and upgrade chemical intermediates (e.g. the oxidation of methanol to formaldehyde, ethylene to ethylene epoxide, *n*-butane to maleic anhydride, *o*-xylene to phthalic anhydride, etc.), and is also the method of choice for the environmental remediation of toxic emissions (e.g. the oxidation of hydrocarbons, carbon monoxide, $H_2S$ to elemental sulfur and $SO_2$, etc.). These oxidative chemical transformations take place at catalytic active sites present in the oxidation catalysts.

Among the multiple spectroscopic techniques that can provide information about the catalytic active sites under reaction conditions (Raman, IR, UV-vis, X-ray absorption (EXAFS (extended X-ray absorption fine structure)/XANES (X-ray absorption near edge structures)), nuclear magnetic resonance (NMR), electron sprin resonance (ESR), etc.), Raman spectroscopy is the technique of choice because of

**Operando* Molecular Spectroscopy and Catalysis Laboratory, Department of Chemical Engineering, Lehigh University, Bethlehem, PA 18015 USA.

†Catalytic Spectroscopy Laboratory, Institute of Catalysis and Petroleum Chemistry, CSIC, Madrid, Spain.

its ability to directly provide molecular level information that allows discrimination among multiple catalytic active sites that may be present in the oxidation catalyst.

The molecular aspect of the Raman vibrational selection rules allows the determination of the molecular structure of the metal-oxide catalytic active sites in all types of oxidation catalysts (supported metal oxides, zeolites, layered hydroxides, polyoxometalates (POMs), bulk pure metal oxides, bulk mixed oxides and mixed oxide solid solutions, supported and unsupported metals).[1] The ability of Raman spectroscopy to operate in all phases (gas, liquid, solid and their mixtures), over a very wide range of temperatures and pressures, and provide molecular level information about catalysts, especially oxides or partially oxidized metals, makes Raman spectroscopy the most informative characterisation technique for understanding the molecular structure and surface chemistry of the catalytic active sites present in heterogeneous oxidation catalysts.[2–9] This chapter focuses on *in situ* and *operando* Raman spectroscopy investigations of oxidation catalysts. The term *in situ* Raman spectroscopy will refer to Raman measurements under oxidising conditions where either the catalysts are dehydrated and fully oxidised or under relevant oxidation reaction conditions. The recent development of the *operando* Raman spectroscopy methodology (e.g. Raman-Mass Spectrometry (Raman-MS) and Raman-Gas Chromatography (Raman-GC)) allows the establishment of direct structure-activity/selectivity relationships.[10] For detailed discussion of Raman theory,[1–5] instrumentation,[2–5] reaction cells[2–5] and experimental methodology,[2–9] the reader is referred to the indicated references. This chapter will review what is currently known from Raman spectroscopy about heterogeneous oxidation catalysts that are employed for oxidation chemical reactions.

## 17.2. Methanol Oxidation to Formaldehyde

The selective oxidation of $CH_3OH$ to HCHO represents the industrial production of one of the major chemical intermediates. Two different catalytic oxidation industrial processes are in use for methanol oxidation. One process employs unsupported silver catalysts and the other employs bulk iron-molybdate, $Fe_2(MoO_4)_3$, catalysts.[11] The silver catalyst process is used when limited amounts of formaldehyde are desired and the process can be quickly started and shut down.

### 17.2.1. *Ag process*

The methanol oxidation silver process employs an autothermal reactor since the exothermic heat of the reaction maintains the catalyst bed temperature at ~650–725°C. The catalyst bed consists of a thin layer of metallic Ag particles (99.99%) on a gauze substrate. The reaction is conducted with excess methanol, having a ratio of $CH_3OH/O_2$ of ~2.7, and methanol conversion is maintained at 90% or more because unconsumed oxygen will further oxidise the formaldehyde reaction product to $CO_2$.[11,12] The HCHO selectivity at ~90% methanol conversion is ~90%, which gives formaldehyde yields of ~80% or more, with $CO_2$ as the main by-product. Extensive research has been performed over many decades to elucidate the nature of catalytic active sites and reactive oxygen species.[13–29] *In situ* Raman spectroscopy measurements of a commercial-type polycrystalline silver catalyst reveal that there are multiple oxygen species present under oxidising conditions, as shown in Fig. 17.1.

These Raman bands have recently been assigned with the assistance of density functional theory (DFT) calculations.[30] The Raman bands are assigned as follows: 910–960 $cm^{-1}$ to Ag-O-O-Ag vibration on a partially oxidised surface, 780–870 $cm^{-1}$ to the O-O vibration of an adsorbed atomic oxygen coordinated to a lattice atomic oxygen of a partially oxidised silver (referred to as a hybrid Ag-O-O-Ag), and 364 $cm^{-1}$ (not shown in Fig. 17.1) that is a surface atomic oxygen atom

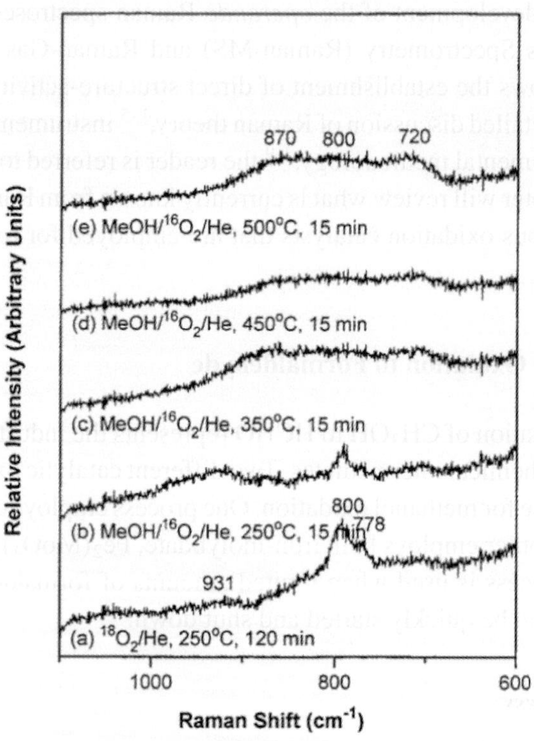


**Figure 17.1.** *In situ* Raman spectra of an unsupported Ag catalyst during the oxidation of methanol to formaldehyde for a catalyst that was pre-oxidised with $^{18}O_2$, which shifted the ~960 $cm^{-1}$ band to 931 $cm^{-1}$. Reproduced with permission from Wang, C., Deo, G. and Wachs, I. (1999), *J. Phys. Chem. B*, **103**, pp. 5645–5656, copyright 1999 American Chemical Society.

coordinated to three Ag atoms ($Ag_3O$). Upon exposure to a flowing $CH_3OH/O_2$ mixture, the strong Raman band at $\sim$800 cm$^{-1}$ from the hybrid oxygen atom on the partially oxidised Ag is selectively consumed indicating that it is the active oxygen species responsible for methanol oxidation to formaldehyde.[12] A recent *operando* Raman-GC study employing microfabricated silver reactors for methanol oxidation confirms the consumption of the Raman band at $\sim$800 cm$^{-1}$ during methanol oxidation to formaldehyde reaction conditions.[31]

### 17.2.2. *Iron molybdate process*

The methanol oxidation iron-molybdate process takes place in a shell-and-tube fixed-bed reactor where the catalysts are packed within the tubes and the circulating oil coolant maintains catalyst temperatures of $\sim$350–450°C, with the higher temperatures occurring in the hot spot region. The process is conducted at 100% methanol conversion with greater than 90% formaldehyde selectivity, which results in formaldehyde yields greater than 90%. The higher formaldehyde yield of the methanol oxidation iron-molybdate process makes it the preferred process for new plants.

The commercial catalyst is in the form of Raschig rings and the catalyst consists of an excess of molybdenum oxide which is denoted as $MoO_3/Fe_2(MoO_4)_3$. This catalyst system was first reported by Adkins and Peterson in 1931[32] and began to be commercially employed in the 1950s. The commercial introduction of the iron-molybdate catalysts for methanol oxidation to formaldehyde has stimulated numerous research studies into the nature of the phases and catalytic active sites responsible for methanol oxidation to formaldehyde.[33–42] The *in situ* Raman spectra of a commercial-type iron-molybdate catalyst is presented in Fig. 17.2 and readily exhibits the bands for crystalline $MoO_3$ (997, 821, 666 and 284 cm$^{-1}$) and $Fe_2(MoO_4)_3$ (970, 784 and 346 cm$^{-1}$) and have previously been reported in the literature.[43] It is not possible to detect the much weaker Raman bands from the surfaces of this bulk mixed oxide catalyst because of the strong features of the crystalline phase and few surface sites due to its low surface area ($<10\ m^2/g$).[44] Recently, combined $CH_3OH$-IR, low energy ion scattering (LEIS), $CH_3OH$-IR and $CH_3OH$-temperature programmed surface reaction (TPSR) spectroscopy characterisation studies have revealed that the surface of the $MoO_3/Fe(MoO_4)_3$ catalyst is covered by a monolayer of a two-dimensional amorphous molybdenum oxide phase that contains the catalytic active sites for methanol oxidation to formaldehyde, and can be directly detected by its terminal Mo = O vibrations at $\sim$1,000 cm$^{-1}$, when molybdenum oxide is dispersed on an $Fe_2O_3$ support. The function of the excess $MoO_3$ is as a reservoir that supplies additional $MoO_x$ to cover exposed $FeO_x$ sites that are responsible for unselective reaction products such as dimethyl ether ($CH_3OCH_3$).

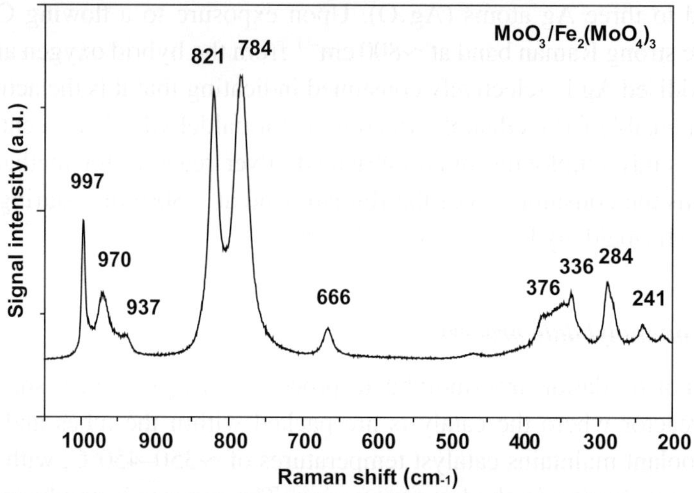


**Figure 17.2.** Raman spectrum of bulk $MoO_3/Fe_2(MoO_4)_3$ catalysts commercially employed for methanol oxidation to formaldehyde. Source: K. Routray and I. Wachs.

Redispersion of $MoO_3$ in $O_2$[45] or methanol[46] environments has been shown to readily take place and can even be employed in the synthesis of bulk iron-molybdate catalysts from physical mixtures of $MoO_3$ and $Fe_2O_3$,[47] as well as the regeneration of spent bulk iron-molybdate catalysts.[48]

## 17.3. Methane Oxidation to Formaldehyde

The direct oxidation of methane to formaldehyde is a highly desired process since it would avoid the production of methanol which currently involves the methane steam reforming and methanol synthesis reaction steps.[11] Unfortunately, a commercially viable catalyst for this oxidation reaction still does not exist because of the uneconomical formaldehyde yields achieved to date, a result of the further oxidation of formaldehyde to carbon oxides at the very high reaction temperature.[49] Nevertheless, many fundamental studies have been undertaken for this oxidation reaction with the supported $MoO_3/SiO_2$ catalyst, which is among one of the most investigated methane oxidation catalysts for formaldehyde production.[50–56] The Raman spectrum of a supported $MoO_3/SiO_2$ catalyst during methane oxidation at ~500°C is depicted in Fig. 17.3. The Raman bands from crystalline $MoO_3$ (997, 821, 666 and 284 $cm^{-1}$) nanoparticles are absent and instead the Raman bands from isolated di-oxo surface $(O=)_2Mo(-O-Si)_2$ species (986 $cm^{-1}$) and the $SiO_2$ support (805, 601 and 487 $cm^{-1}$) are present.

The surface $MoO_4$ species retains its structure and remains fully oxidised under the methane oxidation reaction conditions.[57] Thus, the catalytic active site for

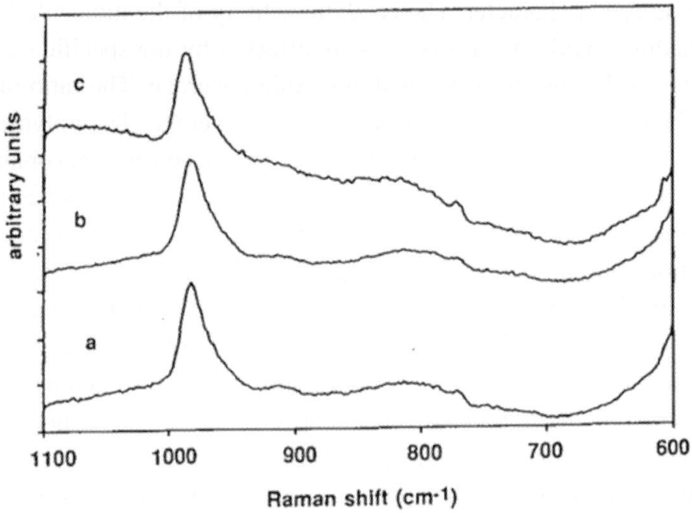


**Figure 17.3.** *In situ* Raman spectra of silica-supported molybdena during (a) methane oxidation at 823 K, (b) dehydrated in flowing $O_2$ at 823 K and (c) dehydrated in flowing $O_2$ at room temperature. Reprinted from Bañares, M., Spencer, N., Jones M. and Wachs, I. (1994). *J. Catal.*, **146**, p. 204, copyright 1994 from Elsevier.

methane oxidation to formaldehyde is the di-oxo surface $(O=)_2Mo(-O-Si)_2$ site. This is further confirmed by assessing the role of sodium poisoning on a silica-supported molybdena catalyst. There is a linear decrease of the Raman band of the dispersed molybdenum oxide species with increasing sodium loading that parallels the decrease of methane turnover frequency (TOF) with sodium doping.[57]

The role of silicomolybdic acid ($H_4SiMo_{12}O_{40}$) Keggins, which form by the reaction between the hydrated surface $MoO_x$ species and the $SiO_2$ support under ambient conditions, for methane oxidation was investigated by *in situ* Raman spectroscopy. It was found that silicomolybdic acid decomposes into dispersed surface $MoO_x$ species at 300°C, well below the temperature range 500–700°C, which is required for activation of methane to formaldehyde.[58] Thus, only dispersed isolated surface molybdenum oxide species are present on $SiO_2$ during the methane oxidation to formaldehyde reaction with supported $MoO_3/SiO_2$ catalysts.

## 17.4. Ethane Oxidative Dehydrogenation (ODH) to Ethylene

Ethylene represents the major feedstock for the petrochemical industry and is generally produced by ethane and propane steam cracking, which also produces a wide range of hydrocarbon products.[11] Currently, there is much interest in developing a selective catalytic route for the oxidative conversion of ethane to ethylene.[59,60] Supported vanadium oxide catalysts are particularly good for this process because of their strong redox character. The catalytic activity of the molecularly-dispersed surface vanadium oxide species is strongly affected by the specific oxide support and less affected by the surface vanadium oxide coverage. The ratio of V-O-V/V-O-Support bonds increases with surface vanadia coverage. The polymeric surface vanadia species are more reducible than isolated species on zirconia or alumina supports;[61,62] this situation appears to reverse on reducible supports such as titania.[61–63] However, combined *in situ* UV-vis and Raman spectroscopy measurements during ethane reaction conditions, reveal that the polymerization degree of the surface vanadium oxide species at high coverage decreases because of preferential reduction of the surface polymeric species,[61] but the interaction of polymeric species with reactants, intermediates or reaction products may also lead to the opening of the V-O-V bonds. Loss of polymeric surface vanadia species has only a minor effect on the ethane oxidation reactions. Although polymerization of the surface vanadia species changes the selectivity, due to the appearance of new Brønsted acid sites, polymerization has little effect on the ethane TOF value.[62] The reducibility of supported vanadium oxide species corresponds with the TOF values, but not with the average oxidation state under a steady-state reaction. The moderate effect of vanadia surface coverage on TOF values to ethylene and the pronounced effect of the support indicate that the bridging V-O-Support bond must be the critical active site for this oxidation reaction. Isotopic labeling with $^{18}O$ demonstrates that the terminal V$=$O bond is not the active site.[64] Such a conclusion is consistent with the fact that the dramatic changes in ethane TOF values with oxide support do not correlate with the changes of the V$=$O bond strength as determined by Raman spectroscopy.[65] Raman spectroscopy during the ethane ODH reaction demonstrates that the catalytic cycle during reaction involves partial reduction of the surface vanadia sites. This is consistent with the relevance of the bridging V-O-Support bond, which would be the site where supported vanadia lattice oxygen participates in the redox cycles and accounts for the strong support effect. DFT calculations demonstrate that this is the site for titania-supported vanadia.[66] An *operando* Raman study of alumina-supported molybdena during ethane oxidative dehydrogenation indicates changes in both the Mo$=$O and Mo-O-Mo bonds during reaction. The ethane ODH rate per Mo shows that it is directly proportional to the presence of Mo-O-Al bonds rather than to the polymerization of supported molybdena species.[67]

Ceria-supported vanadia behave significantly differently than other supported vanadium oxide catalysts. In this system, the active site is associated with the bridging V-O-Support bond ($V^{5+}$-O-$Ce^{3+}$) and the $V^{5+}$ species do not reduce, even under extreme conditions. The redox cycle for oxidative ethane dehydrogenation for supported $V_2O_5/CeO_2$ surprisingly appears to be associated with cerium rather than with the surface vanadium oxide sites.[68] The high affinity between ceria and vanadia also leads to the formation of bulk $CeVO_4$ (with $V^{5+}$ and $Ce^{3+}$ ions). A reducing environment decreases the temperature for this solid-state reaction by 200°C.[69] The Temperature-Programmed Reduction Raman (TPR-Raman) and Temperature-Programmed Oxidation-Raman (TPO-Raman) spectral profiles demonstrate that the redox cycle involves exchange between $Ce^{3+}$ and $Ce^{4+}$, while $V^{5+}$ remains unaltered,[69] which suggests that surface vanadium oxide species promote the reducibility of the ceria support. The stability of $V^{5+}$ ions on ceria is further demonstrated by XANES taken at 1,073 K during the ethane ODH reaction and reducing conditions.[68] Based on these experimental data, a DFT model describing the vanadia-ceria interface provides a rationale for such an interaction, confirming that ceria reduction at the interface is far more favorable than the reduction of surface vanadia species.[70]

An *operando* Raman-GC study during ethane ODH over supported vanadia-ceria catalysts investigated the deactivation due to the formation of bulk $CeVO_4$.[71] The *operando* Raman spectra confirm that the catalyst deactivates above 500°C (Fig. 17.4a) and corresponds to the appearance of sharp Raman bands from crystalline $CeVO_4$ (Fig. 17.4b). The Arrhenius plots measured while the ethane ODH reaction was taking place in the *operando* Raman cell (Fig. 17.4a) show that the apparent activation energy does not change as the catalyst ages (i.e. as the structure changes from surface vanadia species on ceria to crystalline $CeVO_4$ on ceria). Thus, the nature of the active site does not change and the bridging V-O-Ce bonds present

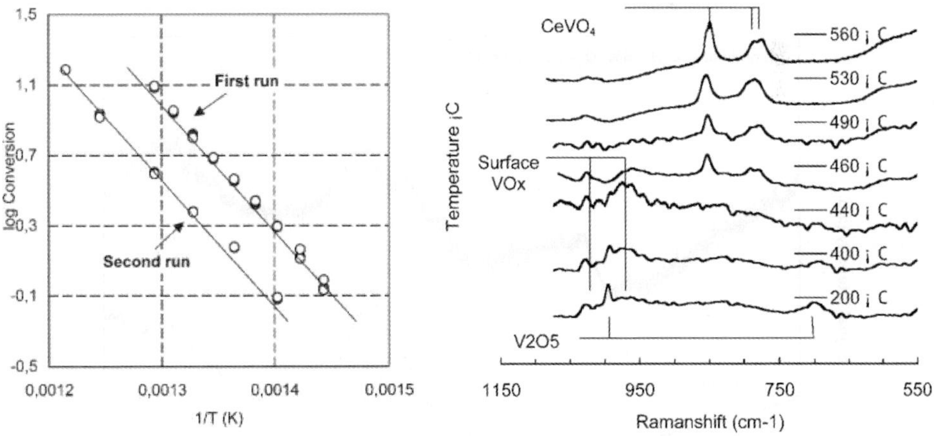


(a) (b)

**Figure 17.4.** (a) Arrhenius plot of ethane conversion vs reaction temperature in the *operando* fixed-bed reaction cell. The data are obtained in the *operando* reactor during the Raman study. (b) *Operando* Raman-GC during ethane oxidative dehydrogenation with 2VCe. The Raman spectra are acquired during the reaction, and the simultaneous on-line activity measurements are presented in the figure. (a) and (b) both reproduced with permission from Martínez-Huerta, M., Deo, G., Fierro, J. and Bañares, M. (2008). *J. Phys. Chem. C.*, **112**, p. 11441, copyright 2008 American Chemical Society.

in both the fresh and aged catalysts are the active sites, which is consistent with a redox cycle occurring on the ceria ions.

## 17.5. Ethylene Oxidation to Ethylene Epoxide

The selective oxidation of ethylene to ethylene epoxide, where an oxygen atom becomes attached to the double bond, is among the major industrial chemical intermediates[11] and has been extensively researched in the catalysis literature.[13–29] This reaction is conducted in a shell-and-tube fixed-bed reactor with the catalysts packed into the tubes. The circulating coolant maintains the reactor at ~200–250°C during this exothermic reaction. The commercial catalysts consist of supported Ag/alpha-$Al_2O_3$ catalysts containing Ag particles in the micron range. The *in situ* Raman spectra for the supported Ag/alpha-$Al_2O_3$ catalyst under oxidising and ethylene oxidation reaction conditions are shown in Fig. 17.5. Unlike the polycrystalline Ag catalysts employed for methanol oxidation to formaldehyde, the supported Ag particles on alpha-$Al_2O_3$ do not possess the Raman band at ~960 $cm^{-1}$ from the Ag-O-O-Ag species. The Raman bands for the hybrid O-O (~800 $cm^{-1}$) and $Ag_3O$ (~350 $cm^{-1}$) oxygen species are present under oxidising conditions. During an ethylene epoxidation reaction at 250°C, the Raman band from both the hybrid O-O oxygen species are significantly diminished relative to the $Ag_3O$ oxygen species and a new weak band at ~1,300 $cm^{-1}$ from surface $C_xH_yO$ species is present. This suggests that the

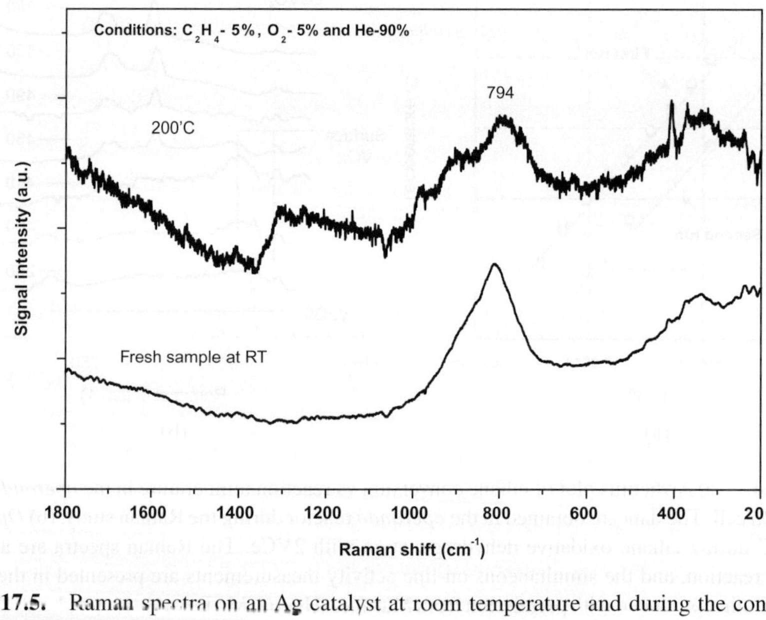


Figure 17.5. Raman spectra on an Ag catalyst at room temperature and during the conversion of ethylene to ethylene oxide at 200°C. Source: K. Routray and I. Wachs.

oxygen from the hybrid O-O sites is the catalytic active site for ethylene oxidation to ethylene epoxide. The selective participation of the hybrid O-O sites in the ethylene epoxidation reaction is even more pronounced with unsupported polycrystalline Ag particles.[12]

## 17.6. Propane Oxidative Dehydrogenation to Propylene

Supported vanadium oxide catalysts have been widely investigated for propane ODH to propylene. The catalytic behavior depends on total surface $VO_x$ coverage, presence of secondary surface metal oxide additives and specific oxide support. Below the monolayer coverage, dispersed metal oxides dominate, while bulk crystalline metal oxides form above the monolayer coverage.[72–76] If more than one component is present (primary metal oxide and secondary metal oxide additives), the two oxides may result in mixed oxide phases, such as Mo-V-O or Sb-V-O systems. If total coverage is below the monolayer coverage, and a bulk phase forms during synthesis, it is likely that it will break and spread during the reaction, as in the case of supported mixed Mo-V oxides on alumina.[77] In the case of propane oxidation, an *operando* Raman-GC study on alumina-supported vanadia was examined at different oxygen-to-propane ratios. During the reaction, isolated vanadium oxide species exhibit a $V=O$ vibrational mode at 1,009 $cm^{-1}$ while polymerized ones, at 1,026 $cm^{-1}$.[78] In line with previous investigations, the polymerization of alumina-supported vanadia decreases as the partial pressure of oxygen decreases.[61] As the partial pressure of oxygen decreases in the *operando* Raman-GC study, the Raman band of the $V=O$ mode of polymeric vanadia species decreases (Fig. 17.6).[78] The simultaneous activity measurement demonstrates that despite the loss of surface polymerized vanadia species due to partial reduction, propane conversion and propylene selectivity remain unchanged, which thus confirms that the polymerization of surface vanadia species is not critical for propane oxidative dehydrogenation.

## 17.7. Propylene Oxidation and Ammoxidation

The oxidation and ammoxidation of propylene ($H_2CHC=CH_3$) to acrolein ($H_2CHC=CHO$) and acrylonitrile ($H_2CHC=CN$) respectively, constitute major chemical intermediates.[11,79] Conversion of propylene to acrolein is conducted in an oxidising environment, while conversion of propylene to acrylonitrile requires both molecular $O_2$ and $NH_3$. The industrial catalyst employed for these oxidation/ammoxidation reactions is a bulk Bi-Mo-O mixed oxide.[11] The alpha-$Bi_2(MoO_4)_3$ phase is one of the most active phases and its Raman spectrum is shown in Fig. 17.7. An *operando* Raman-GC spectroscopy investigation of propylene

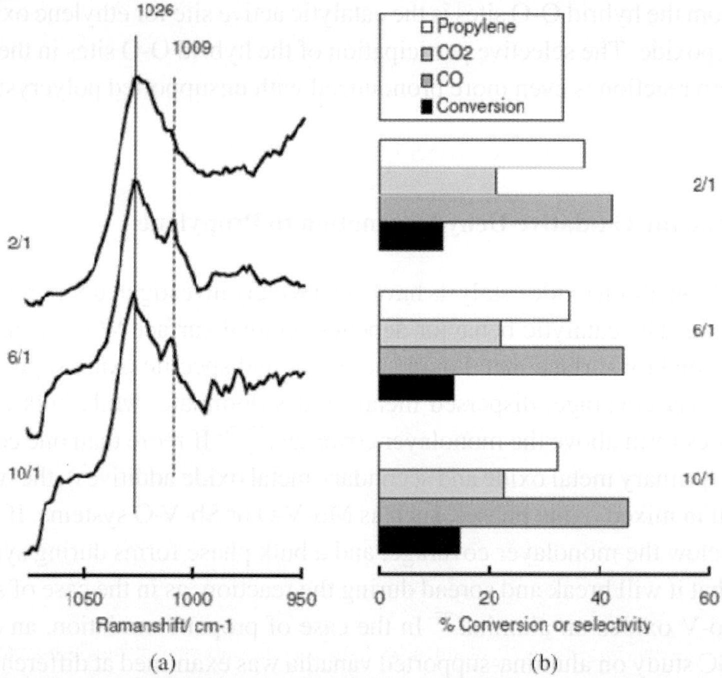


**Figure 17.6.** Raman spectra of an alumina-supported vanadia catalyst during propane ODH vs $O_2/C_3H_8$ molar ratio at 400°C in the *operando* Raman–GC cell during the spectroscopic study of propane ODH (A) and conversion of propane and selectivity to CO, $CO_2$ and propylene analyzed online (B). Reprinted with permission from Cortez, G. and Bañares, M. (2002), *J. Catal.*, **209**, p. 197, copyright 2002 from Elsevier.

oxidation to acrolein at 400°C, over an extended period of reaction time, with propylene conversions of 20–40% and acrolein selectivity of 90–50%, demonstrates that the bulk bismuth-molybdate phase of the catalyst is stable and does not exhibit any significant structural changes under reaction conditions for extended periods of time.[80]

It has generally been assumed that the outermost surfaces of bulk bismuth-molybdate catalysts are just an extension of one of its bulk crystal planes.[81] According to this model, the mechanism has been proposed that molecular $O_2$ dissociatively chemisorbs on the surface bismuth sites, where it becomes incorporated into the bulk lattice and the propylene chemisorbs and reacts on the surface Mo sites where it is oxidised by oxygen supplied from the bismuth-molybdate bulk lattice. The exchange of gas-phase molecular $O_2$ with lattice O in bismuth-molybdate catalysts has been confirmed with Raman studies using isotopically labeled $^{18}O_2$.[82] Recent LEIS spectroscopy analyses of the outermost surface of bismuth-molybdate, however, revealed that its surface is enriched in Mo sites and that instead of Bi, the

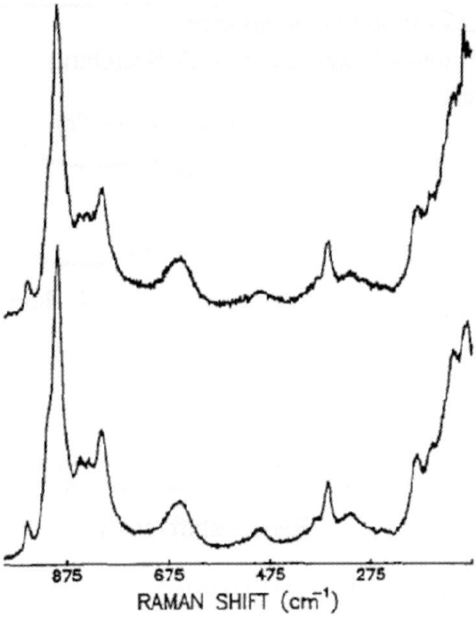


**Figure 17.7.** *In situ* Raman spectra of alpha-$Bi_2(MoO_4)_3$ bismuth-molybdate during partial oxidation of propylene at 400°C (top: 1 h; bottom: 24 h). Reproduced with permission from Snyder T. and Hill, C. (1991). *J. Catal.*, **132**, p. 536, copyright 1991 from Elsevier.

surface is populated with K,[83] which is an impurity present in bismuth oxides. This suggests that the accepted reaction mechanism models need to be re-examined to establish the correct models for propylene oxidation and ammoxidation by bulk bismuth-molybdate catalysts.

The low surface area of bismuth molybdates and the significant solid state reactivity between bismuth and molybdenum oxide complicates surface studies of bismuth-molybdate catalysts.[84] Employment of model-supported $MoO_3$ catalysts, however, did not result in efficient propylene oxidation catalysts.[85] Supported $V_2O_5$ catalysts, however, turn out to be good model catalyst systems for propylene oxidation to acrolein, since they selectively produce acrolein that allows for direct observation of the catalytic active site under reaction conditions.[86] The *operando* Raman spectra of the supported $V_2O_5/Nb_2O_5$ catalysts during propylene oxidation are shown in Fig. 17.8, where the gas-phase oxygen is switched from $^{16}O_2$ to $^{18}O_2$ and back to $^{16}O_2$.[86] The Raman band at 1,036 $cm^{-1}$ is from the V= $^{16}O$ vibration of the surface $VO_4$ species and does not exchange with $^{18}O$ to form V = $^{18}O$ (expected at ~990 $cm^{-1}$) when switching the gas-phase oxygen isotopes. This lack of exchange indicates that the surface $VO_4$ sites are receiving their $^{16}O$ from the lattice of the $Nb_2O_5$ support, which is consistent with the observations made for bismuth-molybdate catalysts. Consequently, the acrolein formed during propylene

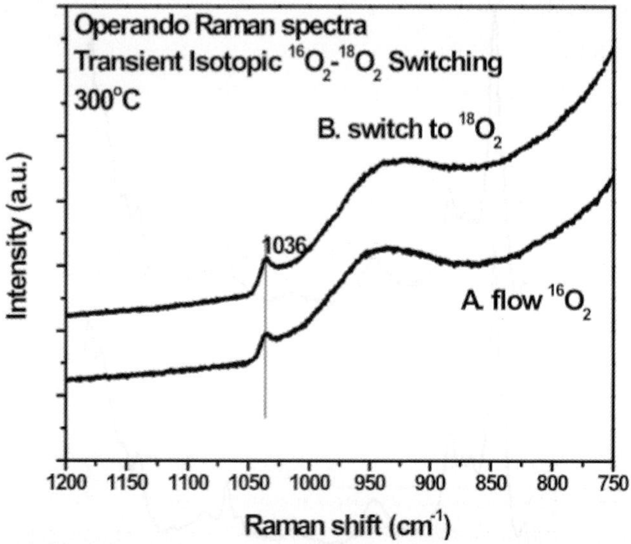


**Figure 17.8.** *In situ* Raman spectra of $^{16}O_2$ to $^{18}O_2$ switch during propylene oxidation by a supported $V_2O_5/Nb_2O_5$ catalyst. Based on Zhao, C. and Wachs, I. (2008). *J. Phys. Chem. C*, **112**, pp. 11363–11372.

oxidation with $^{18}O_2$ is $H_2CHC=CH^{16}O$. Corresponding *operando* IR spectra reveal the surface reaction intermediate is $H_2CHCCH_2$*, surface allyl and $C_3^=$-TPSR; spectroscopy demonstrates that the rate-determining step involves C-H bond breaking of the surface allyl intermediate with the resulting surface $H_2CCHCH$* intermediate rapidly oxidised to the $H_2CHC=CHO$ product. Although the model-supported $V_2O_5/Nb_2O_5$ catalyst system is different to the bulk bismuth-molybdate catalysts, this model catalyst system does capture all the features of the bismuth-molybdate catalyst system and demonstrates that the surface redox sites are catalytic active sites responsible for propylene oxidation to acrolein. It also suggests that the role of Bi may just be to moderate the redox properties of the surface $MoO_x$ sites.

## 17.8. Propane Oxidation and Ammoxidation

The direct ammoxidation of propane into acrylonitrile by its reaction with oxygen and ammonia is an alternative route to the conventional propylene ammoxidation catalytic reaction to acrylonitrile. The use of propane instead of propylene makes the reaction conditions more demanding because of the necessity of activating the propane C-H bond which is much stronger than the propylene methyl C-H bond. Different catalytic systems have been investigated for the ammoxidation of propane into acrylonitrile with vanadium-based catalysts proving particularly efficient, particularly the Sb-V-O mixed oxide catalyst.[87–90]

Vanadium antimonate catalysts used in propane ammoxidation become activated during the initial period on stream. This transient activity is believed to arise from a rearrangement of the mixed Sb-V-O into the active phase, which is not observed during calcination of the precursors. The active catalyst is characterised by a crystalline $SbVO_4$ phase accompanied by some segregated antimony oxide phases. The nature of the catalytic active sites present for the bulk $SbVO_4$ catalyst are difficult to determine since the low surface to volume ratio results in a huge Raman signal from the bulk phase, overwhelming the very small signal from the surface sites. Alumina-support stabilised nanoscaled $SbVO_4$ catalysts were used to investigate the active site of this model system, since it performs like conventional bulk catalysts[91] and the nanosized $SbVO_4$ aggregates facilitate the investigation of the surface structure by minimising the overwhelming signal from the bulk phase. *Operando* Raman-GC has been employed to monitor the states of $SbVO_4$ phases during propane ammoxidation.[92–94] The Raman spectra revealed the restructuring of dispersed vanadium and antimony oxide species into crystalline $SbVO_4$ nanoparticles during ammoxidation reaction conditions,[92] which is concomitant to the formation of acrylonitrile. The presence of surface vanadium oxide species is related to the formation of surface alkoxide species and propane activation.[92] The catalysts that exhibit strong Raman signals of surface alkoxides at 1,060 $cm^{-1}$, which arises from the alkoxide C-O stretch, afford higher acrylonitrile yields.[94] *Operando* Raman-GC studies also demonstrate the relevance of segregated antimony oxide, since it is a pool of $Sb^{5+}$ ions that migrate in and out of the rutile $SbVO_4$ phase during the ammoxidation reaction.[93] During ammoxidation, the redox cycle occurs between $V^{5+}$ dispersed species and reduced $V^{4+}$ lattice species in the nanocrystalline $SbVO_4$. The bulk $VSbO_4$ rutile lattice does not stabilise $V^{5+}$ ions, which are expelled to the surface. The deficit of $V^{4+}$ cations is compensated for by the migration of $Sb^{5+}$ cations from the antimony oxide pool. Computer calculations have demonstrated that the activation barrier for these changes is very low.[95] HR-TEM studies also demonstrate the remarkable structural flexibility of the bulk $SbVO_4$ crystalline phases.[96]

## 17.9. Butane Oxidation to Maleic Anhydride

The selective oxidation of *n*-butane to maleic anhydride is conducted in both shell-and-tube fixed-bed and fluidized-bed reactors. The advantage of the fluid-bed process is the efficient removal of the exothermic heat of reaction, while the advantage of the fixed-bed process is the longer catalyst life. Both processes employ unsupported $(VO)_2P_2O_7$ as the heterogeneous catalyst. The starting material for this catalyst is the hemihydrate precursor $VOHPO_4$ and it must be slowly converted to $(VO)_2P_2O_7$ under reaction conditions for several weeks.[97–100] The *in situ* Raman spectra of the

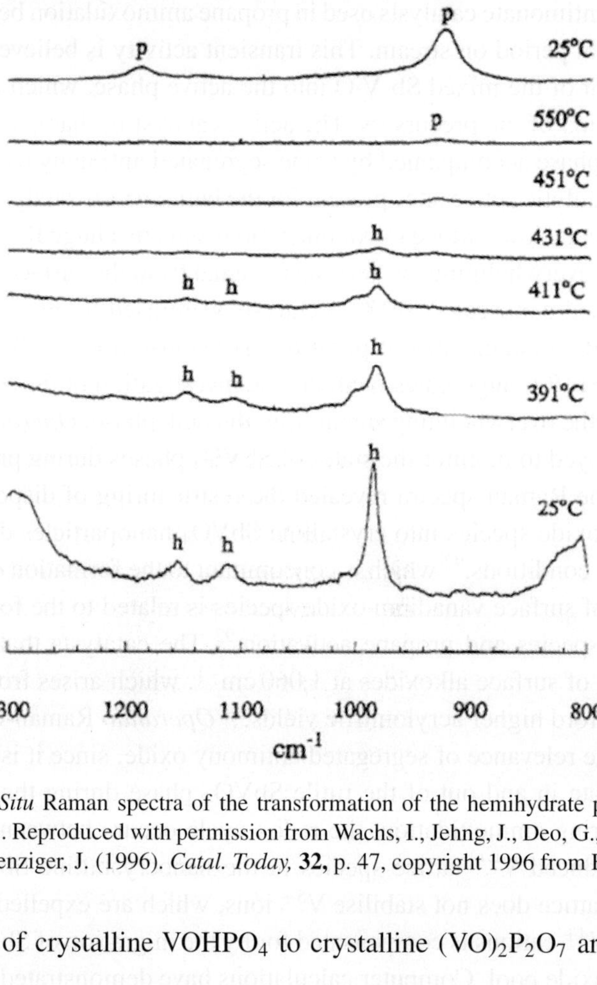


**Figure 17.9.** *In Situ* Raman spectra of the transformation of the hemihydrate precursor to crystalline $(VO)_2P_2O_7$. Reproduced with permission from Wachs, I., Jehng, J., Deo, G., Weckhuysen, B., Guliants, V. and Benziger, J. (1996). *Catal. Today,* **32,** p. 47, copyright 1996 from Elsevier.

transformation of crystalline $VOHPO_4$ to crystalline $(VO)_2P_2O_7$ are presented in Fig. 17.9.[101]

HR-TEM studies have revealed that the conversion to $(VO)_2P_2O_7$ is incomplete during the initial stages because of the presence of an amorphous layer that coats the $(VO)_2P_2O_7$ crystals.[102, 103] The amorphous layer is readily oxidised to $VOPO_4$ nanoparticles under oxidising conditions and, thus, catalyst activation must be performed under the somewhat reducing butane/$O_2$ reaction conditions to minimise the formation of this undesired phase.[104] To examine the catalytic behavior of a model amorphous V-P-O layer, an amorphous V-P-O layer was synthesised by first forming a surface $PO_x$ layer on $TiO_2$ and then depositing surface $VO_x$ species onto the $PO_x/TiO_2$ catalyst.[105] The *in situ* Raman spectra for such an amorphous V-P-O layer on $TiO_2$ is shown in Fig. 17.10 and exhibits the V$=$O vibration of the mono-oxo surface $VO_4$ species that are present in the $V^{5+}$ oxidation state.[104] This amorphous V P O layer on the $TiO_2$ catalyst was indeed able to selectively convert *n*-butane to maleic anhydride.[104] Thus, a catalytic active site present for unsupported $(VO)_2P_2O_7$,

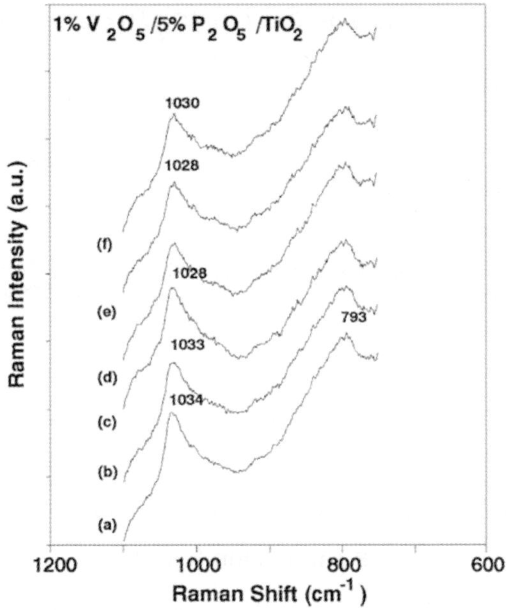


**Figure 17.10.** *In situ* Raman spectra of the supported 1% $V_2O_5$/5% $P_2O_5$/$TiO_2$ catalyst during *n*-butane oxidation. (a) $O_2$, 100 cm$^3$/min, 503 K; (b) $C_4H_{10}/O_2$/He, 100 cm$^3$/min, 503 K; (c) $C_4H_{10}/O_2$/He, 100 cm$^3$/min, 573 K; (d) $C_4H_{10}/O_2$/ He, 100 cm$^3$/min, 623 K; (e) $C_4H_{10}/O_2$/He, 50 cm$^3$/min, 623 K; (f) $O_2$, 100 cm$^3$/min, 623 K. Reprinted with permission from Wachs, I., Jehng, J., Deo, G., Weckhuysen, B., Guliants, V., Benziger, J. and Sundaresan, S. (1997). *J. Catal.*, **170**, p. 75, copyright 1997 from Elsevier.

which possesses $V^{+4}$ sites in its internal crystalline bulk structure, consists of surface redox $V^{+5}$ and acid $P^{+5}$ species in the amorphous outermost layer that are responsible for selective oxidation of *n*-butane to maleic anhydride.

## 17.10. Isobutane Oxidation

Methacrylic acid is a major chemical commodity produced annually via a multistep route that does, unfortunately, produce many environmentally undesirable by-products.[106] Methacrylic acid, however, can also be directly formed by oxidation of isobutane over polyoxometalate Keggin catalysts (e.g. $H_3PMo_{12}O_{40}$ and its vanadium-substituted analogs $H_{3+x}PMo_{12-x}V_xO_{40}$) and such a green process has been investigated by several researchers.[107,108] This possibility has resulted in many oxidation studies with polyoxometalate catalysts over the past two decades.

The *in situ* Raman spectra of the polyoxometalate $H_3PMo_{12}O_{40}$ (PM) and $H_6PMo_9V_3O_{40}$ (PMV) Keggin catalysts, during methanol oxidation to formaldehyde by the redox sites and dehydration to $CH_3OCH_3$ by the acid sites, are shown in Fig. 17.11.

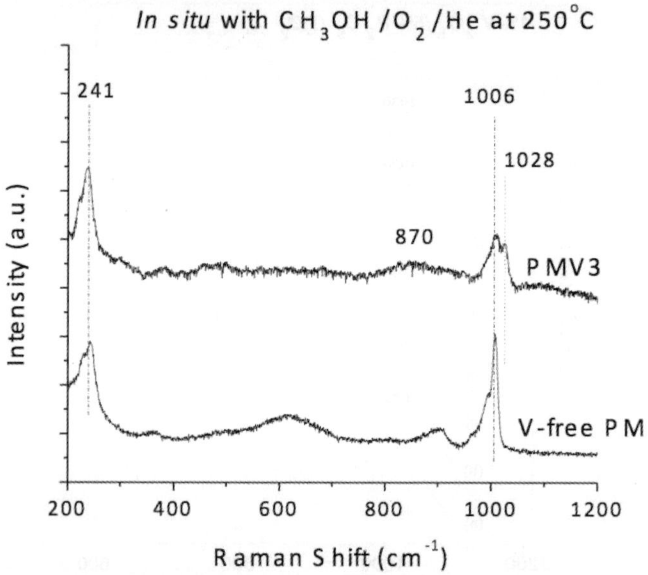


**Figure 17.11.** *In situ* Raman spectra of phosphomolybdic acid Keggin catalysts during methanol oxidation to formaldehyde by the redox sites and dehydration to $CH_3OCH_3$ by the acid sites. *Source*: J. Molinari and I. Wachs.

The V-free $H_3PMo_{12}O_{40}$ Keggin catalyst exhibits a sharp Raman band at 1,007 $cm^{-1}$ that significantly broadens when the primary $MoO_5$ units are replaced with $VO_5$ units. Furthermore, during methanol oxidation/dehydration a new Raman band appears at ~1,028 $cm^{-1}$ from the V = O vibration of surface $VO_4$ species that have been expelled from the Keggin primary structure. The $VO_5$ unit in the Keggin primary structure has its vibrations coupled to those of Mo = O and does not give rise to its own V=O bond vibration. The $VO_x$ sites in the $H_6PMo_9V_3O_{40}$ Keggin primary structure and the secondary surface structure are responsible for redox sites in the $H_6PMo_9V_3O_{40}$ Keggin, since $H_3PMo_{12}O_{40}$ possesses very little redox character and primarily yields dimethyl ether DME as the reaction product. Thus, the $VO_x$ sites in the $H_6PMo_9V_3O_{40}$ Keggin are the catalytic active sites for oxidation catalysis.

## 17.11. *o*-Xylene Oxidation to Phthalic Anhydride

The oxidation of *o*-xylene to phthalic anhydride is commercially conducted with supported $V_2O_5/TiO_2$ catalysts containing several promoters (e.g. $SbO_x$, $KO_x$ and $PO_x$) in a fixed-bed shell-and-tube reactor with a molten salt coolant.[109–113, 115] Although the titania support has the anatase structure, its performance is not dependent on the specific titania phase.[114] The supported vanadium oxide phase under ambient

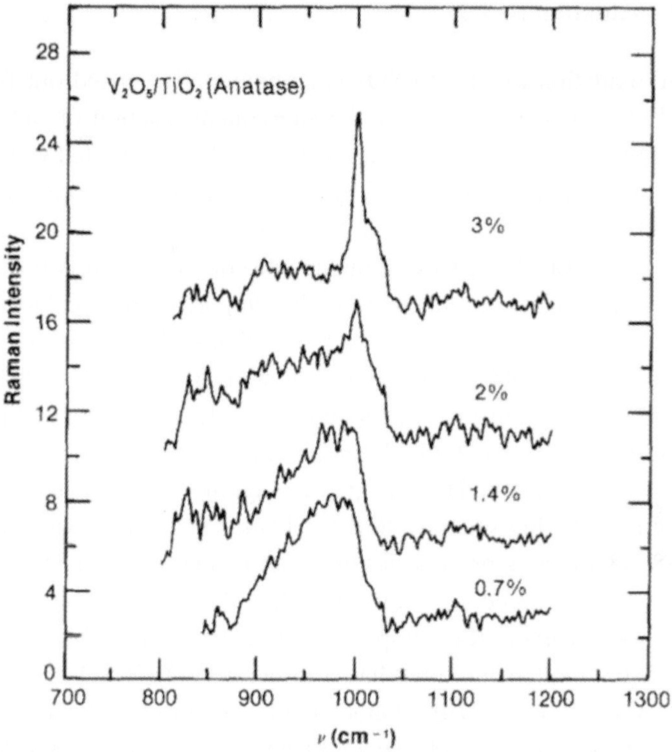


**Figure 17.12.** The supported vanadium oxide phase under ambient conditions in commercial *o*-xylene catalysts is present both as dispersed surface $VO_x$ species (broad Raman band from ~900–1,000 cm$^{-1}$ for 0.7 and 1.4% $V_2O_5/TiO_2$) and crystalline $V_2O_5$ nanoparticles (sharp Raman band at 994 cm$^{-1}$ for 2.0 and 3.0% $V_2O_5/TiO_2$). Reprinted from Wachs, I., Saleh, R., Chan, S., Chersich, C. (1985). *Appl. Catal.*, **15**, p. 339, copyright 1985 from Elsevier.

conditions in commercial *o*-xylene catalysts is present both as dispersed surface $VO_x$ species (broad Raman band from ~900–1,000 cm$^{-1}$ for 0.7 and 1.4% $V_2O_5/TiO_2$) and crystalline $V_2O_5$ nanoparticles (sharp Raman band at 994 cm$^{-1}$ for 2.0 and 3.0% $V_2O_5/TiO_2$) as shown in Fig. 17.12.[115] The corresponding *o*-xylene oxidation performance reveals that high activity and selectivity is only achieved at the monolayer surface $VO_x$ coverage because exposed Ti sites are responsible for side products, and the additional presence of crystalline $V_2O_5$ nanoparticles has only a marginal affect on the catalyst performance. Furthermore, the crystalline $V_2O_5$ nanoparticles are in a reduced state during the *o*-xylene oxidation reaction and, consequently, don't possess the lattice oxygen to supply the reactants.[116] This performance trend clearly demonstrates that the surface $VO_x$ species on the $TiO_2$ support are the catalytic active sites for *o*-xylene oxidation to phthalic anhydride.

## 17.12. $SO_2$ oxidation to $SO_3$

The catalytic oxidation of $SO_2$ to $SO_3$ is commercially carried out for the manufacture of $H_2SO_4$ that is formed by the subsequent reaction of $SO_3$ with $H_2O$. This catalytic reaction is conducted with a unique silica-supported vanadium oxide molten salt.[11] This catalyst is produced by calcination of diatomaceous earth, vanadium pentoxide (or other V salts) and alkali salt promoters (usually in the form of sulfates), with an alkali-to-vanadium molar ratio ranging from two to five. Large quantities of sulfur oxides are taken up by the supported liquid catalyst during the activation process, forming molten alkali pyrosulfates,[117] which dissolve the vanadium salts. A major problem in the $SO_2$ oxidation process is the sudden drop in activity that is experienced in all commercial catalysts at an operating temperature below 420°C. A systematic high-temperature *in situ* Raman investigation by Boghosian *et al.* demonstrated that the phase diagrams of the molten alkali sulfate-vanadia system are highly sensitive to the oxidation state of the vanadium sites.[118] The liquid vanadia phases present in this system show a clear tendency to partial self-reduction, which can be tuned by controlling the partial pressure of oxygen.

High-temperature *in situ* Raman spectra demonstrate that the $V^V$ complex present in $V_2O_5$-$Cs_2S_2O_7$ molten mixtures has a dimeric $(VO)_2O(SO_4)_4^{4-}$ configuration containing a V-O-V bridge with characteristic vibrations at 1,047 (V=O), 996 (S-$O_{terminal}$), 839 (S-$O_{bridge}$) and 765 (V-O-V) $cm^{-1}$.[119] The V-O-V bond breaks when $Cs_2SO_4$ is added, and the vanadium phase converts into $VO_2(SO_4)_3^{2-}$. The most characteristic bands due to the molten $VO_2(SO_4)_3^{2-}$ complex occur at 1038 (ν(V=O)) and 941 (ν(S-$O_{terminal}$)) $cm^{-1}$.

An *operando* ESR study of the deactivation and regeneration in the molten salt-gas model systems $M_2S_2O_7/V_2O_5$-$SO_2/O_2/SO_3/N_2$ was also investigated.[120] This work is, to the best of our knowledge, the first *operando* EPR work and demonstrates that the restoration of deactivated catalysts occurs at 450–500°C, where the low valence compounds decompose. The precipitation of such compounds containing reduced vanadium sites was detected by ESR during reactions below the onset temperature for catalyst deactivation. This pioneering work laid the ground for a multitechnique, multinational approach using *operando* EPR, *in situ* NMR and *operando* Raman.[121] This study shows that the catalytic cycle involves essentially V(V) species, (dimeric or binuclear fragments of larger oligomers) like those described earlier,[122] while a side reaction reduces the active $V^V$ phases to the catalytically inactive $V^{IV}$ phases.

A combined *in situ* Raman and UV-vis spectroscopic study provides details on the interaction of vanadium ($V^{IV}$ and $V^V$) oxosulfato complexes with $SO_2$, which depend on the temperature and on the presence of molecular $O_2$.[123] The UV-vis spectra confirm the presence of $V^V$ sites in the presence of oxygen and the combined

Raman and UV-vis demonstrate that under these conditions, $V^V$ ions do not coordinate with $SO_2$. As the partial pressure of $SO_2$ increases, there is a progressive coordination, concomitant to the reduction of vanadium sites to $V^{IV}$. The first *in situ* studies of molten salts during $SO_2$ oxidation to $SO_3$[124] demonstrates that the $V^V$ complex, $V^VO_2(SO_4)_2^{3-}$, and $Cs_2SO_4$ dominate the catalyst after calcination. The excess sulfate is converted to pyrosulfate under $SO_3/O_2$ feed (disappearance of Raman bands at 960 and 611 $cm^{-1}$ of sulfate species); this dissolves the vanadium oxides leading to the formation of the $V^V$ dimer $(V^VO)_2O(SO_4)_4^{4-}$ complex (770 (V-O-V) and 830 (S-O-V) $cm^{-1}$ Raman bands). Lowering of the temperature to 450–380°C under $SO_2$ and $O_2$ results in the gradual blue shift of the 1,034 $cm^{-1}$ band due to the $V^V = O$ terminal stretch of $VO_2(SO_4)_2^{3-}$ to 1,046 $cm^{-1}$, characteristic of the $V^V = O$ terminal stretch of $(VO)_2O(SO_4)_4^{4-}$. The appearance of bands at 994 and 687 $cm^{-1}$ due to terminal S–O stretches and $\nu 4(SO_4^{2-})$ split component of $(VO)_2O(SO_4)_4^{4-}$ is also apparent. Finally, the growth of the 770 and 830 $cm^{-1}$ features of $VO_2(SO_4)_2^{3-}$ becomes evident. In the absence of oxygen, the reduction of $V^V$ complexes takes place under $SO_2$, leading to the formation of $V^{IV}O(SO_4)_2^{2-}$ and $V^{IV}O(SO_4)_3^{4-}$.

The supported vanadia molten salts exhibit clear changes that can be monitored by Raman spectroscopy.[125] The reduction of vanadia species depends on the alkali ion and on the presence of water vapor.[126] Under oxidising conditions, the interaction between $SO_2$ and vanadia species is minimised. Raman spectra demonstrate that the interaction between $SO_2$ and vanadia occurs upon reduction, and deactivates the catalyst in melt, precipitating $V^{IV}$ phase $K_4(VO)_3(SO_4)$, deactivating the catalyst. The presence of cesium and/or the presence of water vapor in the feed stabilise pentavalent vanadium complexes against their reduction. A very detailed complementary *in situ* Raman and EPR study during reaction was reported confirming these transformations and brings further fundamental insight.[127]

## 17.13. Conclusions and Outlook

Raman spectroscopy has been successfully applied to many different oxidation catalysts (mixed-metal oxides and metals) and oxidation reactions (organic and inorganic) since its introduction into the field of catalysis. The initial Raman studies from 1970–1990 primarily involved measurements under ambient conditions and *in situ* Raman studies under flowing $O_2$ environments began to appear during the 1980–1990 period. *In situ* Raman studies during relevant oxidation reaction conditions were introduced during 1990–2000. In the past decade, *operando* spectroscopy studies have started to appear in the literature that are assisting in the establishment of fundamental molecular structure-activity relationships for different oxidation catalysts and oxidation reactions. It is anticipated that the growth of *in situ* and *operando* Raman spectroscopy studies on oxidation catalysis will continue to significantly grow in the coming years because of the strong Raman signals from oxidation catalysts, as well as the molecular level and time-resolved information afforded by Raman spectroscopy.

## Acknowledgments

The financial support of the US Department of Energy — Basic Energy Sciences (Grant DE-FG02-93ER14350) and US National Science Foundation (Grant 0609018) during the writing of this chapter is gratefully acknowledged. The financial support from the Spanish Ministry of Science and Innovation (Grants CTQ2011-25517/PPQ and CTQ2011-13343E) is gratefully acknowledged.

## References

1. Nakamoto, K. (2009). *Infrared and Raman Spectra of Inorganic and Coordination Compounds, Part A and Part B*, John Wiley & Sons, Hoboken NJ.
2. Bañares, M. (2004). Raman Spectroscopy, in B. Weckhuysen (ed.), *In Situ Spectroscopy of Catalysts*, American Scientific Publishers, Stevenson Ranch CA, pp. 47–92.
3. Bañares, M. and Wachs, I. (2002). Molecular Structures of Supported Metal Oxide Catalysts under Different Environments, *J. Raman Spectrosc.*, **33**, pp. 359–380.
4. Bañares, M. and Mestl, G. (2009). Structural Characterization of Operating Catalysts by Raman Spectroscopy, *Adv. Catal.*, **52**, pp. 43–128.
5. Bañares, M. and Wachs, I. (2010). Raman Spectroscopy of Catalysts, in R. Meyers (ed.), *Encyclopedia of Analytical Chemistry*, Wiley, Hoboken NJ, Supplementary vol. **S1–S3**, pp. 1895–1924.
6. Wachs, I. (1996). Raman and IR Studies of Surface Metal Oxide Species on Oxide Supports: Supported Metal Oxide Catalysts, *Catal. Today*, **27**, pp. 437–455; Wachs, I. (2005). Recent Conceptual Advances in the Catalysis Science of Mixed Metal Oxide Catalytic Materials, *Catal. Today*, **100**, pp. 79–94.
7. Mestl, G. (2000). In Situ Raman Spectroscopy: A Valuable Tool to Understand Operating Catalysts, *J. Mol. Catal. A: Chem.*, **158**, pp. 45–65.
8. Wachs, I. (2001). Raman Spectroscopy of Catalysts, in I. Lewis, and H. Edwards (eds), *Handbook of Raman Spectroscopy*, CRC Press, New York, pp. 799–833.
9. Wachs, I. and Roberts, C. (2010). Monitoring Surface Metal Oxide Catalytic Active Sites with Raman Spectroscopy, *Chem. Soc. Rev.*, **39**, pp. 5002–5017.
10. Bañares, M. (2005). Operando Methodology: Combination of In Situ Spectroscopy and Simultaneous Activity Measurements under Catalytic Reaction Conditions, *Catal. Today*, **100**, pp. 71–77.
11. Bartholomew, C. and Farrauto, R. (2006). Fundamentals of Industrial Catalytic Processes, Wiley, Hoboken NJ.
12. Wang, C., Deo, G. and Wachs, I. (1999). Interaction of Polycrystalline Silver with Oxygen, Water, Carbon Dioxide, Ethylene, and Methanol: In Situ Raman and Catalytic Studies, *J. Phys. Chem. B*, **103**, pp. 5645–5656.

13. Barteau, M. and Madix, R. (1982). The Surface Reactivity of Silver: Oxidation Reactions, in D. King and P. Woodruff (eds), *The Chemical Physics of Solid Surfaces and Heterogeneous Catalysis*, Elsevier, Amsterdam, Vol. 4, pp. 95–120.
14. Wachs, I. and Madix, R. (1978). The Oxidation of Methanol on a Silver (110) Catalyst, *Surf. Sci.*, **76**, pp. 531–558.
15. Sachtler, W., Backx, C. and van Santen, R. (1981). On the Mechanism of Ethylene Epoxidation, *Catal. Rev. Sci. Eng.*, **23**, pp. 127–149.
16. Sajkowski, D. and Boudart, M. (1987). Structure Sensitivity of the Catalytic Oxidation of Ethene by Silver, *Catal. Rev. Sci. Eng.*, **29**, pp. 325–360.
17. van Santen, R. and Kuipers, H. (1987). The Mechanism of Ethylene Epoxidation, *Adv. Catal.*, **35**, pp. 265–321.
18. Sexton, B. and Madix, R. (1980). Vibrational Spectra of Molecular and Atomic Oxygen on Ag(11O), *Chem. Phys. Lett.*, **76**, pp. 294–297.
19. Eickmans, J., Otto, A. and Goldmann, A. (1985). The Transition from Physisorbed to Chemisorbed Oxygen on Silver Films Studied by Photoemission, *Surf. Sci.*, **149**, pp. 293–312.
20. Twigg, G. (1946). The Catalytic Oxidation of Ethylene, *Trans. Faraday Soc.*, **42**, pp. 284–290.
21. Grant, R. and Lambert, R. (1983). Mechanism of the Silver-catalyzed Heterogeneous Epoxidation of Ethylene, *J. Chem. Soc., Chem. Commun.*, **12**, pp. 662–663.
22. Grant, R. and Lambert, R. (1985). A Single Crystal Study of the Silver-catalysed Selective Oxidation and Total Oxidation of Ethylene, *J. Catal.*, **92**, pp. 364–375.
23. Bao, X., Muhler, M., Pettinger, B., *et al.* (1993). On the Nature of the Active State of Silver During Catalytic Oxidation of Methanol, *Catal. Lett.*, **22**, pp. 215–225.
24. Prabhakaran, K. and Rao, C. (1987). A Combined EELS-XPS Study of Molecularly Chemisorbed Oxygen on Silver Surfaces: Evidence for Superoxo and Peroxo Species, *Surf. Sci.*, **186**, pp. L575–L580.
25. Wang, X. and Greenler, R. (1991). Direct Structure Determination of Oxygen Adsorbed on Silver by Reflection-absorption Infrared Spectroscopy with Isotopic Substitution, *Phys. Rev. B*, **43**, pp. 6808–6811.
26. Pettenkofer, C., Pockrand, I. and Otto, A. (1983). Surface Enhanced Raman Spectra of Oxygen Adsorbed on Silver, *Surf. Sci.*, **135**, pp. 52–64.
27. McBreen, P. and Moskovits, M. (1987). A Surface-enhanced Raman Study of Ethylene and Oxygen Interacting with Supported Silver Catalysts, *J. Catal.*, **103**, pp. 188–199.
28. Deng, J., Xu, X., Wang, J., *et al.* (1995). *In Situ* Surface Raman Spectroscopy Studies of Oxygen Adsorbed on Electrolytic Silver, *Catal. Lett.*, **32**, pp. 159–170.
29. Millar, G., Metson, J., Bowmaker, G., *et al.* (1995). In Situ Raman Studies of the Selective Oxidation of Methanol to Formaldehyde and Ethene to Ethylene Oxide on a Polycrystalline Silver Catalyst, *J. Chem. Soc., Faraday Trans.*, **91**, pp. 4149–4159.
30. Chen, T., Pal, A., Jehng, J., *et al.* (2011). Identification of Oxygen Species Active in Ethylene Epoxidation on Silver Catalysts with Raman Spectroscopy and DFT Calculations, Presentation #144 FUEL Division, 241st ACS National Meeting, Anaheim CA.
31. Cao, E., Firth, S., McMillan, P., *et al.* (2007). Application of Microfabricated Reactors for Operando Raman Studies of Catalytic Oxidation of Methanol to Formaldehyde on Silver, *Catal. Today*, **126**, pp. 119–126.
32. Adkins, H. and Peterson, W. (1931). The Oxidation of Methanol with Air Over Iron, Molybdenum, and Iron-Molybdenum Oxides, *J. Am. Chem. Soc.*, **53**, pp. 1512–1520.
33. Routray, K., Zhou, W., Kiely, C., *et al.* (2010). Origin of the Synergistic Interaction Between $MoO_3$ and Iron Molybdate for the Selective Oxidation of Methanol to Formaldehyde, *J. Catal.*, **275**, pp. 84–98.
34. Soares A., Portela M. and Kiennemann A. (2005). Methanol Selective Oxidation to Formaldehyde over Iron-molybdate Catalysts, *Catal. Rev. Sci. Eng.*, **47**, 125–174.

35. Hill, C. and Wilson III, J. (1990). Raman Spectroscopy of Iron Molybdate Catalyst Systems: Part I. Preparation of Unsupported Catalysts, *J. Mol. Catal.*, **63**, pp. 65–94.
36. Burcham, L., Briand, L. and Wachs, I. (2001). Quantification of Active Sites for the Determination of Methanol Oxidation Turnover Frequencies Using Methanol Chemisorption and In Situ Infrared Techniques. 2. Bulk Metal Oxide Catalysts, *Langmuir*, **17**, pp. 6175–6184.
37. House, M., Carley, A. and Bowker, M. (2007). Selective Oxidation of Methanol on Iron Molybdate Catalysts and the Effects of Surface Reduction, *J. Catal.*, **252**, pp. 88–96.
38. House, M., Carley, A., Echeverria-Valda, R., *et al.* (2008). Effect of Varying the Cation Ratio within Iron Molybdate Catalysts for the Selective Oxidation of Methanol, *J. Phys. Chem. C*, **112**, pp. 4333–4341.
39. Farneth, W., Ohuchi, F., Staley, R., *et al.* (1985). Mechanism of Partial Oxidation of Methanol over Molybdenum(VI) Oxide as Studied by Temperature-programmed Desorption, *J. Phys. Chem.*, **89**, pp. 2493–2497.
40. Badlani, M. and Wachs, I. (2001). Methanol: A "Smart" Chemical Probe Molecule, *Catal. Lett.*, **75**, pp. 137–149.
41. Briand, L., Hirt, A. and Wachs, I. (2001). Quantitative Determination of the Number of Surface Active Sites and the Turnover Frequencies for Methanol Oxidation over Metal Oxide Catalysts: Application to Bulk Metal Molybdates and Pure Metal Oxide Catalysts, *J. Catal.*, **202**, pp. 268–278.
42. Holstein, W. and Machiels, C. (1996). Inhibition of Methanol Oxidation by Water Vapor—Effect on Measured Kinetics and Relevance to the Mechanism, *J. Catal.*, **162**, pp. 118–124.
43. Wilson, J., Hill, C. and Dumesic, J. (1990). Raman Spectroscopy of Iron Molybdate Catalyst Systems: Part III. In Situ Studies of Supported and Bulk Catalysts under Reaction and Redox Conditions, *J. Mol. Catal.* **61**, pp. 333–352.
44. Tian, H., Wachs, I. and Briand, L. (2005). Comparison of UV and Visible Raman Spectroscopy of Bulk Metal Molybdate and Metal Vanadate Catalysts, *J. Phys. Chem. B.*, **109**, pp. 23491–23499.
45. Knözinger H. and Taglauer E. (1993). Toward Supported Oxide Catalysts via Solid–Solid Wetting, *Catalysis*, **10**, pp. 1–40.
46. Wang, C., Cai, Y. and Wachs, I. (1999). Reaction-induced Spreading of Metal Oxides onto Surfaces of Oxide Supports during Alcohol Oxidation: Phenomenon, Nature, and Mechanisms, *Langmuir*, **15**, pp. 1223–1235.
47. Wachs, I. and Cai, Y (2001; 2002). Formation of Alcohol Conversion Catalysts, *US Patent No. 6,245,708*, and Alcohol conversion, *US Patent No. 6,350,918*.
48. Wachs, I. and Briand, L. *In-Situ* Formation of Methyl-Molybdate Catalysts, *US Patent No. 6,331,503*.
49. Faraldos, M., Bañares, M., Anderson, J., *et al.* (1996). Comparison of Silica-supported $MoO_3$ and $V_2O_5$ Catalysts in the Selective Partial Oxidation of Methane, *J. Catal.*, **160**, pp. 214–221.
50. Pitchai, R. and Klier, K. (1986). Partial Oxidation of Methane, *Catal. Rev. Sci. Eng.*, **28**, pp. 13–88.
51. Khan, M. and Somorjai, G. (1985). A Kinetic Study of Partial Oxidation of Methane with Nitrous Oxide on a Molybdena-silica Catalyst, *J. Catal.*, **91**, pp. 263–271.
52. Otsuka, K. and Hatano, M. (1987). The Catalysts for the Synthesis of Formaldehyde by Partial Oxidation of Methane, *J. Catal.*, **108**, pp. 252–255.
53. Spencer, N. (1988). Partial Oxidation of Methane to Formaldehyde by Means of Molecular Oxygen, *J. Catal.*, **109**, pp. 187–197.
54. Spencer, N. and Pereira, C. (1989). $V_2O_5$-/$SiO_2$-catalyzed Methane Partial Oxidation with Molecular Oxygen, *J. Catal.*, **116**, pp. 399–406.
55. Bañares, M., Fierro, J. and Moffat, J. (1993). The Partial Oxidation of Methane on $MoO_3/SiO_2$ Catalysts: Influence of the Molybdenum Content and Type of Oxidant, *J. Catal.* **142**, pp. 406–417.

56. Mauti, R. and Mims, C. (1993). Oxygen Pathways in Methane Selective Oxidation over Silica-supported Molybdena, *Catal. Lett.*, **21**, pp. 201–207.
57. Bañares, M., Spencer, N., Jones, M., *et al.* (1994). Effect of Alkali Metal Cations on the Structure of Mo(VI)/$SiO_2$ Catalysts and its Relevance to the Selective Oxidation of Methane and Methanol, *J. Catal.*, **146**, pp. 204–210.
58. Bañares, M., Hu, H. and Wachs, I. (1995). Genesis and Stability of Silicomolybdic Acid on Silica-supported Molybdenum Oxide Catalysts: In Situ Structural-Selectivity Study on Selective Oxidation Reactions, *J. Catal.*, **155**, pp. 249–255.
59. Mamedov, E. and Cortes Corberan, V. (1995). Oxidative Dehydrogenation of Lower Alkanes on Vanadium Oxide-based Catalysts. The Present State of the Art and Outlooks, *Appl. Catal. A: Gen.*, **127**, pp. 1–40.
60. Bañares, M. (1999). Supported Metal Oxide and Other Catalysts for Ethane Conversion: A Review, *Catal. Today*, **51**, pp. 319–348.
61. Bañares, M., Martínez-Huerta, M., Gao, X., *et al.* (2000). Dynamic Behavior of Supported Vanadia Catalysts in the Selective Oxidation of Ethane: *In Situ* Raman, UV–Vis DRS and Reactivity Studies, *Catal. Today*, **61**, pp. 295–301.
62. Martinez-Huerta, M., Gao, X., Tian, H., *et al.* (2006). Oxidative Dehydrogenation of Ethane to Ethylene over Alumina-supported Vanadium Oxide Catalysts: Relationship between Molecular Structures and Chemical Reactivity, *Catal. Today*, **118**, pp. 279–287.
63. Martínez-Huerta M., Fierro J. and Bañares M. (2009). Monitoring the States of Vanadium Oxide during the Transformation of $TiO_2$ Anatase-to-rutile under Reactive Environments: $H_2$ Reduction and Oxidative Dehydrogenation of Ethane, *Catal. Commun.*, **11**, pp. 15–19.
64. Bañares M., Martínez-Huerta M., Gao X., *et al.* (2000). Identification and Roles of the Different Active Sites in Supported Vanadia Catalysts by In Situ Techniques, *Studs. Surf. Sci. Catal.*, **130A**, pp. 3125–3130.
65. Bañares M., Gao X., Fierro J., *et al.* (1997). Partial Oxidation of Ethane over Monolayers of Vanadium Oxide. Effect of the Support and Surface Coverage, *Studs. Surf. Sci. Catal.*, **110**, pp. 295–304.
66. Lewandowska, A., Calatayud, M., Lozano Diz, E., *et al.* (2008). Combining Theoretical Description with Experimental In Situ Studies on the Effect of Alkali Additives on the Structure and Reactivity of Vanadium Oxide Supported Catalysts, *Catal. Today*, **139**, pp. 209–213.
67. Christodoulakis, A., Heracleous, E., Lemonidou, A., *et al.* (2006). An Operando Raman Study of Structure and Reactivity of Alumina-supported Molybdenum Oxide Catalysts for the Oxidative Dehydrogenation of Ethane, *J. Catal.*, **242**, pp. 16–25.
68. Martinez-Huerta, M., Coronado, J., Fernández-García, M., *et al.* (2004). Nature of the Vanadia–ceria Interface In $V^{5+}/CeO_2$ Catalysts and its Relevance for the Solid-state Reaction Toward $CeVO_4$ and Catalytic Properties, *J. Catal.*, **225**, pp. 240–248.
69. Martínez-Huerta, M., Deo, G., Fierro, J., *et al.* (2007). Changes in Ceria-supported Vanadium Oxide Catalysts during the Oxidative Dehydrogenation of Ethane and Temperature-programmed Treatments, *J. Phys. Chem. C*, **111**, pp. 18708–18714.
70. Baron, M., Abbott, H., Bondarchuk, O., *et al.* (2009). Resolving the Atomic Structure of Vanadia Monolayer Catalysts: Monomers, Trimers, and Oligomers on Ceria, *Angew. Chem. Int. Ed.*, **48**, pp. 8006–8009.
71. Martínez-Huerta, M., Deo, G., Fierro, J., *et al.* (2008). Operando Raman-GC Study on the Structure-activity Relationships in $V^{5+}+/CeO_2$ Catalyst for Ethane Oxidative Dehydrogenation: The Formation of $CeVO_4$, *J. Phys. Chem. C*, **112**, pp. 11441–11447.
72. Ermini, V., Finocchio, E., Sechi, S., *et al.* (2000). Propane Oxydehydrogenation over Alumina-supported Vanadia Doped with Manganese and Potassium, *Appl. Catal. A*, **198**, pp. 67–79.
73. Khodakov, A., Olthof, B., Bell, A., *et al.* (1999). Structure and Catalytic Properties of Supported Vanadium Oxides: Support Effects on Oxidative Dehydrogenation Reactions, *J. Catal.*, **181**, pp. 205–216.

74. Wachs, I., Jehng, J., Deo, G., *et al.* (1997). Fundamental Studies of Butane Oxidation over Model-supported Vanadium Oxide Catalysts: Molecular Structure-reactivity Relationships, *J. Catal.*, **170**, pp. 75–88.
75. Eon, J., Olier, R. and Volta, J. (1994). Oxidative Dehydrogenation of Propane on $\gamma$-Al2O3 Supported Vanadium Oxides, *J. Catal.*, **145**, pp. 318–326.
76. Gao, X., Bañares, M. and Wachs, I. (1999). Ethane and *n*-Butane Oxidation over Supported Vanadium Oxide Catalysts: An In Situ UV-Visible Diffuse Reflectance Spectroscopic Investigation, *J. Catal.*, **188**, pp. 325–331.
77. Bañares, M. and Khatib, S. (2004). Structure–activity Relationships in Alumina-supported Molybdena–vanadia Catalysts for Propane Oxidative Dehydrogenation, *Catal. Today*, **96**, pp. 251–257.
78. Garcia Cortez, G. and Bañares, M. (2002). A Raman Spectroscopy Study of Alumina-supported Vanadium Oxide Catalyst during Propane Oxidative Dehydrogenation with Online Activity Measurement, *J. Catal.*, **209**, pp. 197–201.
79. Grasselli, R. (1997). G. Ertl, H. Knoezinger, and J. Weitkamp (eds), *Handbook of Heterogeneous Catalysis*, vol. V, VCH Verlagsgesellschaft mbH, Weinheim, pp. 2303–
80. Snyder, T. and Hill Jr, C. (1991). Stability of Bismuth Molybdate Catalysts at Elevated Temperatures in Air and under Reaction Conditions, *J. Catal.*, **132**, pp. 536–555.
81. Gates, B., Katzer, J. and Schuit, G. (1979). *Chemistry of Catalytic Processes*, McGraw-Hill, New York.
82. Monnier, J. and Keulks, G. (1981). The Catalytic Oxidation of Propylene. IX. The Kinetics and Mechanism over $\beta$-$Bi_2Mo_2O_9$, *J. Catal.*, **68**, pp. 51–66.
83. Merzlikin, S., Tolkachev, N., Briand, L., *et al.* (2010). Anomalous Surface Compositions of Stoichiometric Mixed Oxide Compounds, *Angew. Chem. Intl. Ed.*, **49**, pp. 8037–8041.
84. Arora, N., Deo, G., Wachs, I., *et al.* (1996). Surface Aspects of Bismuth–metal Oxide Catalysts, *J. Catal.*, **159**, pp. 1–13.
85. Oyama, S., Desikan, A., Zhang, W. (1993). Adsorbate Bonding and Selectivity in Partial Oxidation, in S. Oyama and J. Hightower (eds), *Catalytic Selective Oxidation*, *Am. Chem. Soc.*, Washington DC, **523**, pp. 16–30.
86. Zhao, C. and Wachs, I. (2008). An Operando Raman, IR, and TPSR Spectroscopic Investigation of the Selective Oxidation of Propylene to Acrolein over a Model Supported Vanadium Oxide Monolayer Catalyst, *J. Phys. Chem. C*, **112**, pp. 11363–11372.
87. Andersson, A., Hanse, S. and Wickman, A. (2001). The Importance of Site Isolation and Phase Cooperation in Propane Ammoxidation on Rutile-type Vanadia Catalysts, *Top. Catal.*, **15**, pp. 103–110.
88. Guttmann, A., Grasselli, R. and Brazdil, F. (1998). Ammoxidation of Paraffins and Catalysts Thereof, *U.S. Patent* 4,746,64; Cavalcanti, F., Bremer, N. and Brazdil, L. (1995). Method of Improving Oxidation and Ammoxidation Catalysts, *WO Patent* 9505895.
89. Xiong, G., Sullivan, V., Stair, P., *et al.* (2005). Effect of Titanium Substitution on the Structure of $Vsbo_4$ Catalysts For Propane Ammoxidation, *J. Catal.*, **230**, pp. 317–326.
90. Centi, G. and Perathoner, S. (1995). Modification of the Surface Reactivity of Vanadium Antimonate Catalysts during Catalytic Propane Ammoxidation, *App. Catal. A*, **124**, 317–337.
91. Guerrero-Pérez, M., Fierro, J., Vicente, M., *et al.* (2002). Effect of Sb/V Ratio and of Sb + V Coverage on the Molecular Structure and Activity of Alumina-supported Sb–V–O Catalysts for the Ammoxidation of Propane to Acrylonitrile, *J. Catal.*, **206**, pp. 339–348.
92. Guerrero-Pérez, M. and Bañares M. (2002). Operando Raman Study of Alumina-supported Sb–V–O Catalyst during Propane Ammoxidation to Acrylonitrile with On-line Activity Measurement, *Chem. Commun.*, **12**, pp. 1292–1293.
93. Guerrero-Pérez M. and Bañares M. (2007). *Operando* Raman-GC Study of Supported Alumina Sb- and V-Based Catalysts: Effect of Sb/V Molar Ratio and Total Sb+V Coverage in the Structure of Catalysts during Propane Ammoxidation, *J. Phys. Chem. C*, **111**, pp. 1315–1322.

94. Guerrero-Pérez, M. and Bañares, M. (2004). *Operando* Raman–GC Studies of Alumina-supported Sb-V-O Catalysts and Role of the Preparation Method, *Catal. Today*, **96**, pp. 265–272.
95. Irigoyen, B., Juan, A., Larrondo, S., *et al.*, (2005). The Electronic Structure of Vanadium Antimonate: A Theoretical Study, *Catal. Today*, **107–108**, pp. 40–45.
96. Landa Cánovas, A., García-Carcía F. and Hansen, S. (2010). Structural Fexibility in $SbVO_4$, *Catal. Today* **158**, pp. 156–161.
97. Centi, G., Trifiro, F., Ebner, J., *et al.* (1988). Mechanistic Aspects of Maleic Anhydride Synthesis from C4 Hydrocarbons over Phosphorus Vanadium Oxide, *Chem. Rev.*, **88**, pp. 55–80.
98. Centi, G. (1993). Vanadyl Pyrophosphate: A Critical Overview, *Catal. Today*, **16**, pp. 5–26.
99. Ben Abdelouahah, F., Olier, R., Guilhaume, N., *et al.* (1992). A Study By *In Situ* Laser Raman Spectroscopy of VPO Catalysts for *n*-Butane Oxidation to Maleic Anhydride I. Preparation and Characterization of Pure Reference Phases, *J. Catal.*, **134**, pp. 151–167.
100. Igarashi, H., Tsuji, K., Okuhara, T., *et al.* (1993). Effects of Consecutive Oxidation on the Production of Maleic Anhydride in Butane Oxidation over Four Kinds of Well-characterized Vanadyl Pyrophosphates, *J. Phys. Chem.*, **97**, pp. 7065–7071.
101. Wachs, I., Jehng, J., Deo, G., *et al.* (1996). *In Situ* Raman Spectroscopy Studies of Bulk and Surface Metal Oxide Phases during Oxidation Reactions, *Catal. Today*, **32**, pp. 47–55.
102. Guliants, V., Benziger, J., Sundaresan, S., *et al.* (1995). Evolution of the Active Surface of the Vanadyl Pyrophosphate Catalysts, *Catal. Lett.*, **32**, pp. 379–386.
103. Ayub, I., Su, D., Willinger, M., *et al.* (2003). Tribomechanical Modification of Bi Promoted Vanadyl Phosphate Systems 1: An Improved Catalyst and Insight into Structure-function Relationship, *Phys. Chem. Chem. Phys.*, **5**, pp. 970–978.
104. Guliants, V., Benziger, J., Sundaresan, S., *et al.* (1996). The Effect of the Phase Composition of Model VPO Catalysts for Partial Oxidation of *n*-Butane, *Catal. Today*, **28**, pp. 275–295.
105. Wachs, I., Jehng, J., Deo, G., *et al.* (1997). Fundamental Studies of Butane Oxidation over Model-supported Vanadium Oxide Catalysts: Molecular Structure-reactivity Relationships, *J. Catal.*, **170**, pp. 75–88.
106. Bauer Jr, W. (2002). *Ullmann's Encyclopedia of Industrial Chemistry*, Wiley-VCH, Weinheim.
107. Misono, M. (1987). Heterogeneous Catalysis by Heteropoly Compounds of Molybdenum and Tungsten, *Catal. Rev.*, **29**, pp. 269–321.
108. Mizuno, N., Tateishi, M. and Iwamoto, M. (1996). Oxidation of Isobutane Catalyzed by $Cs_xH_{3-x}PMo_{12}O_{40}$-based Heteropoly Compounds, *J. Catal.*, **163**, pp. 87–94.
109. Nikolov, V., Klissurski, D. and Anastasov, A. (1991). Phthalic Anhydride from *o*-Xylene Catalysis: Science and Engineering, *Catal. Rev. Sci. Eng.*, **33**, pp. 319–374.
110. Grabowski, R., Grzybowska, B., Haber, J., *et al.* (1975). Catalytic Activity of $V_2O_5$–$TiO_2$ Systems in the Oxidation of *o*-Xylene, *React. Kinert. Ctal. Lett.*, **2**, pp. 81–87.
111. Bond, G., Sarkany, J. and Parfitt, G. (1979). The Vanadium Pentoxide-titanium Dioxide System: Structural Investigation and Activity for the Oxidation of Butadiene, *J. Catal.*, **57**, pp. 476–493.
112. Gasior, M., Gasior, I. and Grzyboswka, B. (1984). *o*-Xylene Oxidation on the $V_2O_5$-$TiO_2$ Oxide System: I. Dependence of Catalytic Properties on the Modification of $TiO_2$, *Appl. Catal.*, **10**, pp. 87–100.
113. Roozeboom, F., Mittelmeijer-Hazeleger, M., Moulijn, J., *et al.* (1980). Vanadium Oxide Monolayer Catalysts. 3. A Raman Spectroscopic and Temperature-programmed Reduction Study of Monolayer and Crystal-type Vanadia on Various Supports, *J. Phys. Chem.*, **84**, pp. 2783–2791.
114. Saleh, R., Wachs, I., Chan, S., *et al.* (1986). Comparison of Vanadium Pentoxide/Titanium Oxide (anatase) and $V_2O_5$/$TiO_2$ (Rutile): Promoting Effect of the Support, Preprints ACS Div. of Petroleum Chemistry, **31**, pp. 272–276.
115. Wachs, I., Saleh, R., Chan, S., *et al.* (1985). The Interaction of Vanadium Pentoxide with Titania (Anatase): Part I. Effect on *o*-Xylene Oxidation to Phthalic Anhydride, *Appl. Catal.*, **15**, pp. 339–352.

116. Wachs, I., Chan, S. and Saleh, R. (1985). The Interaction of $V_2O_5$ with $TiO_2$(anatase). II. Comparison of Fresh and Used Catalysts for *o*-Xylene Oxidation to Phthalic Anhydride, *J. Catal.*, **91**, pp. 366–369.
117. Frazer, J. and Kirkpatrick, W. (1940). A New Mechanism for the Action of the Vanadium Pentoxide-silica-alkali Pyrosulfate Catalyst for the Oxidation of Sulfur Dioxide, *J. Am. Chem. Soc.*, **62**, pp. 1659–1660.
118. Boghosian, S. (1998). Vibrational Modes and Structure of Vanadium(V) Complexes in $M_2SO_4$–$V_2O_5$ (M=K or Cs) Molten Salt Mixtures, *J. Chem. Soc., Faraday Trans.*, **94**, pp. 3463–3469.
119. Boghosian, S., Borup, F. and Chrissanthopoulos, A. (1997). Vanadium (V) Complexes in Molten Salts of Interest for the Catalytic Oxidation of Sulphur Dioxide, *Catal. Letters*, **48**, pp. 145–150.
120. Eriksen, K., Karydis, D., Boghosian, S., *et al.* (1995). Deactivation and Compound Formation in Sulfuric-acid Catalysts and Model Systems, *J. Catal.*, **155**, pp. 32–42.
121. Lapina, O., Bal'zhinimaev, B., Boghosian, S., *et al.* (1999). Progress on the Mechanistic Understanding of $SO_2$ Oxidation Catalysts, *Catal. Today*, **51**, pp. 469–479.
122. Bal'zhinimaev, B., Ivanov, A., Lapina, O., *et al.* (1989). Mechanism of Sulphur Dioxide Oxidation over Supported Vanadium Catalysts, *Faraday Discuss. Chem. Soc.*, **87**, pp. 133–147.
123. Boghosian, S., Chrissanthopoulos, A. and Fehrmann, R. (2002). Structure of Vanadium Oxosulfato Complexes in $V_2O_5$-$M_2S_2O_7$-$M_2SO_4$ (M = K, Cs) Melts. A High Temperature Spectroscopic Study, *J. Phys. Chem. B*, **106**, pp. 49–56.
124. Giakoumelou, I., Caraba, R., Pârvulescu, V., *et al.* (2002). First *In Situ* Raman Study of Vanadium Oxide Based $SO_2$ Oxidation Supported Molten Salt Catalysts, *Catal. Letters*, **78**, pp. 209–214.
125. Giakoumelu, I., Pârvulescu, V. and Boghosian, S. (2004). Oxidation of Sulfur Dioxide over Supported Solid $V_2O_5/SiO_2$ and Supported Molten Salt $V_2O_5$–$Cs_2SO_4/SiO_2$ Catalysts: Molecular Structure and Reactivity, *J. Catal.*, **225**, pp. 337–349.
126. Christodoulakis, A. and Boghosian, S. (2003). Molecular Structure of Supported Molten Salt Catalysts for $SO_2$ Oxidation, *J. Catal.*, **215**, pp. 139–150.
127. Pârvulescu, V., Paun, C., Pârvulescu, V., *et al.* (2004). Vanadia–silica and Vanadia–cesium–silica Catalysts for Oxidation of $SO_2$, *J. Catal.*, **225**, pp. 24–36.

# Chapter 18

# Infrared Spectroscopy in Oxidation Catalysis

Guido BUSCA*

The fundamentals of the use of infrared spectroscopy in the research on solid oxidation catalysts are briefly summarized. The application of IR to a number of case studies is reported: they include methanol partial and total oxidation and steam reforming, CO oxidation and water gas shift, methylaromatics selective oxidations, oxidative dehydrogenation of butenes to butadiene, total oxidation of halide hydrocarbons and oxygenates, selective catalytic reduction of NOx.

## 18.1. Introduction

Developed as an analytical technique in the early decades of the 20th century,[1] infrared (IR)spectroscopy was first applied to surface chemical phenomena in 1937 by Buswell *et al.*,[2] who reported the spectra of water adsorbed on montmorillonite. In the 1940s and 1950s Terenin and pupils, at Leningrad University, showed that this technique can be widely applied to molecules adsorbed on solid materials, including metal oxides.[3] The use of IR spectroscopy in catalysis and surface chemistry was later developed in the 1950s by Eischens *et al.* at Texaco laboratories (Beacon, New York)[4] and, almost simultaneously, by Sheppard and Yates at Cambridge University in the UK.[5] After more than 50 years IR spectroscopy is still one of the most powerful techniques for the characterisation of solid catalysts and for studies of surface and catalytic phenomena. This is due to the relatively cheap cost of Fourier transform infrared (FTIR) spectrometers, the very rapid measurements, the very deep density of information in an IR spectrum, and the well-established fundamentals of the technique. Several review papers and some books have been published on the use of IR in the field of catalysis. Among these, the last edition of the book of A. A. Davydov, written with N. Sheppard, is particularly related to the present chapter.[6] We recently reviewed the fundamentals of the IR technique as applied to the field of heterogeneous catalysis.[7]

*Università di Genova, *Dipartimento di Ingegneria Chimica e di Processo "G.B. Bonino"*, Laboratorio di Chimica delle Superfici e Catalisi Industriale, P.le J. F. Kennedy, I-16129 Genova, Italy.

In this chapter we will focus on the application of IR spectroscopy methods to oxidation catalysts. We will mainly summarize the work performed in our lab, discussing it in relation to the most relevant papers published in the recent international literature.

## 18.2. Experimental Techniques

Infrared spectroscopy in the catalysis field is mostly performed today with either the transmission/absorption technique or with the diffuse reflectance technique. In the transmission/absorption technique, the IR beam is focused over a sample consisting of a layer which is passed through by part of the radiation (the "transmitted" light), part of the photons being absorbed by the solid causing excitation of its vibrational modes. In the case of the diffuse reflectance technique, part of the light is scattered out from the powder, the rest being absorbed by the solid causing excitation of its vibrational modes. The spectra are the same in the two techniques, whose performances are comparable with modern instruments and optical attachments. Many other experimental set-ups are possible, such as emission, reflection, multiple attenuated reflection, photothermal, and photoacoustic techniques. As discussed elsewhere,[7,8] their application to catalysis is by far less usual, although sometimes very interesting. In this review we will not approach the description of the different experimental techniques which can be found in more specialized publications as well as in refs[7,8] and the papers cited therein.

## 18.3. The Bulk Characterisation of Solid Oxidation Catalysts by IR

### 18.3.1. *IR skeletal spectra of crystalline and amorphous solids*

The application of skeletal IR spectroscopy to the structural characterisation of solids such as heterogeneous catalysts is not a trivial matter. The analysis of the symmetry elements of the crystal structure (i.e. the space group and factor group[9] of the smallest Bravais cell), according to the site symmetry of every atom, allows the determination of the irreducible representation of the total modes and, after the subtraction of the "hindered translational modes" (i.e. the "acoustic modes"), the irreducible representation of the vibrational (or "optical") modes can be obtained. This means that the number of vibrational modes belonging to the symmetry species associated with the molecular or crystal symmetry can be counted. Consequently, the number of active modes can be counted, according to the symmetry selection rules of the different techniques (IR and Raman). In the solid state, the polar phonons (those that are IR-active) split into two components, the transverse mode (TO) and the longitudinal mode (LO). This TO/LO splitting occurs because the electric field associated with the transverse wave is $= 0$, while that associated with the longitudinal wave is $\neq 0$; when coupling of these modes occurs with the electric fields associated with the vibration, this gives rise to $\nu_{LO} > \nu_{TO}$. This factor is relevant in relation to the shape and interpretation of the IR spectra of solid materials. On the bases of the composition of the crystalline solid and of its symmetry elements, the position of the vibrational transitions can be forecast. The vibrational modes, which are in principle coupled movements of all the atoms of the entire, smallest Bravais cell, can be attributed predominantly to stretching and deformation modes of small groups of atoms. The position (measured as transition energy or wavenumber) of the IR-active vibrational transitions depends on the weights of the atoms mainly involved, as well as on the bond strengths. Band intensities depend, as always for IR spectroscopy, on the variation of the dipole moment during vibration. This in practice allows empirical interpretation of the IR spectra and comparison of spectra of different structurally related materials. On the other hand, the complete calculation of IR spectra of solids can also be attempted today with specialized software. The analysis of IR spectra, in particular if coupled with that of the corresponding Raman spectra, allows (at least for some crystal structures) a very precise determination of the structure.

On the other hand, according to the microscopic nature of the absorption of the IR photons, it occurs also with amorphous, disordered, and defective solids. For this reason, vibrational spectroscopies (in particular IR spectroscopy) represent very useful techniques for the structural characterisation of non-metallic solids.

### 18.3.2. *Skeletal IR characterisation of oxidation catalysts: techniques*

Insulating or weakly semiconducting materials, like most inorganic compounds, do not absorb the radiation outside the skeletal region, where all light is essentially reflected, transmitted, refracted, or scattered. The skeletal vibrations give rise not only to absorbed radiation in transmission and diffuse reflectance experiments, but also to reflected radiation in the reflection experiments. Thus the specular reflectance for insulating materials, both in the form of monocrystals and sintered pellets, is frequently the basis for the best determination of the skeletal spectrum, as far as the IR-active modes are concerned.

The most common technique in practice, to obtain a good spectrum of the fundamental vibrations of a powdered insulating material with the transmission/absorption IR technique, is to prepare a layer appropriately diluted and sufficiently thin. To do this the most used technique is that of the KBr pressed discs. KBr in fact is an easily available powdered material which does not absorb in the medium IR region (down to near 400 $cm^{-1}$, i.e. it cuts out the far IR). It can be easily mixed homogeneously with the powder to be investigated, and pressed, thus obtaining "diluted" self-supporting discs very useful for IR transmission. Other materials (such as e.g. CsI or polyethylene for far IR studies) allow the production of similar pressed discs, with cut-off limits at even lower frequencies. Alternatively, the powders can be simply deposited in the form of a thin layer on a disc of a transmitting medium. Skeletal IR spectra can also be obtained in the diffuse reflectance mode by diluting the powder with the powder of non-absorbing materials.

### 18.3.3. *Skeletal IR characterisation of oxidation catalysts: case studies*

IR spectroscopy is largely used for the characterisation of metal oxide-based catalysts in relation to their structural features, with additional possible information on morphology. Several collections of IR, Raman, or both IR and Raman spectra of inorganic materials and minerals have been published and are available electronically. We have previously reported several investigations on this subject.[7,10]

As an example, the joint analysis of IR and Raman spectra provided evidence of the partial ordering of cations in a Fe-Cr corundum-type mixed sesquioxides, which are used industrially as high temperature water-gas shift catalysts, but are also active in olefin oxidative dehydrogenation. X-ray diffraction (XRD) patterns of these solids indicate the corundum-type structure without any superstructure. This implies that iron and chromium ions are randomly distributed. IR and Raman spectra instead definitely show that cations are at least partially ordered in layers such as in the ilmenite-type superstructure.[11] Similarly, XRD analysis shows a cubic (non-ferroelectric) structure of nanometric $BaTiO_3$, while vibrational spectroscopies reveal microscopic asymmetry of this structure.[12] Similarly, IR spectroscopy allowed the determination of the state of vanadium in solid solution in $TiO_2$ anatase catalysts,[13] and the presence of $Ti^{4+}$ in the silicalite framework of TS1 catalysts,[14–16] used for the selective oxidation of phenol and the ammoximation of cyclohexanone with hydrogen peroxide.

Infrared skeletal spectroscopy may reveal the structure of defective situations in complex oxides. An example concerns Ni-Al binary mixed oxides as well as Ni-Mg-Al ternary mixed oxides. These materials find relevant application as the precursors of steam-reforming catalysts as well as total oxidation catalysts of VOCs.

X-ray diffraction patterns of NiMgAl oxides essentially only show the pattern of a rock salt-type structure common to periclase MgO and to NiO,[17] in spite of the significant amount of alumina (33% metal molar ratio in the case of the sample in Fig. 18.1).

The skeletal IR spectra indeed show a well-defined maximum at 430–420 cm$^{-1}$ that can be confidently assigned to $\nu_{TO}$ (the only IR-active transverse optical mode) of the Ni-Mg oxide rock salt-type solid solution. The absence of bands in the far infrared (i.e. below 400 cm$^{-1}$) points to the absence of a spinel phase. However, the additional features, found in the region 900–600 cm$^{-1}$, are typically associated with M-O vibrations of tetrahedrally coordinated cations. Indeed, two absorptions are found in the spectra, near 800 cm$^{-1}$ and 650 cm$^{-1}$. In the case of the Mg-free Ni-Al compound, a similar band was observed at 850 cm$^{-1}$.[18] This suggests that the band at 800–850 cm$^{-1}$ is associated with M-O stretchings of tetrahedrally coordinated

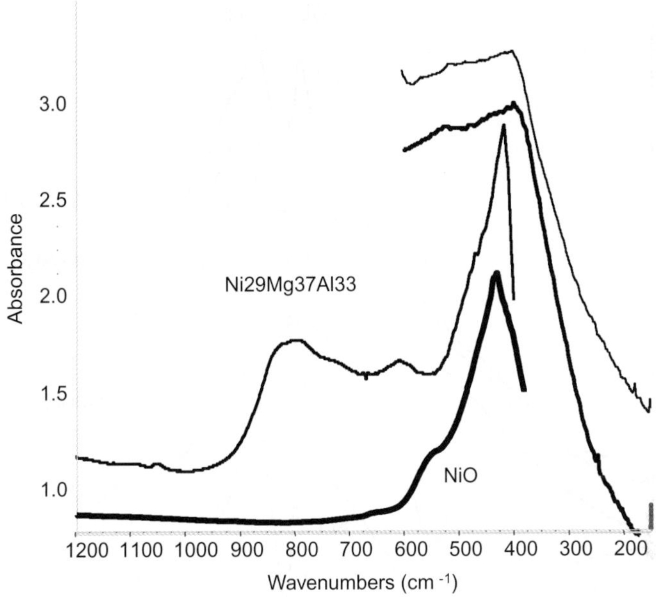


**Figure 18.1.** FTIR skeletal spectra (KBr pressed disc) of NiO and of a Ni-Mg-Al oxide catalyst (Ni : 29; Mg : 37; Al : 33, atomic ratios).

$Al^{3+}$ species, as occurs in the inverse spinel $NiAl_2O_4$, while that at 650 cm$^{-1}$ might be associated with M-O stretchings of tetrahedrally coordinated $Mg^{2+}$ species, as occurs in the normal spinel $MgAl_2O_4$. Thus, IR skeletal spectroscopy reveals the presence of XRD undetectable spinel domains in these solids.

Another example concerns $Co_3O_4/Al_2O_3$ systems also used in the total oxidation of organics as well as precursors of supported Co metal catalysts. In Fig. 18.2 the skeletal IR spectra of the support Siralox 5 (alumina containing 5% $SiO_2$, from Sasol) and of two $Co_3O_4/Al_2O_3$ catalysts are compared.

The two catalysts differ for both Co loading and for preparation procedure. The strong and complex absorption in the region 950–450 cm$^{-1}$ present in the three cases is typical of a spinel transitional alumina (delta phase) partially evolved towards a theta phase.[7] A weak tail towards higher frequencies is associated with the Si-O modes due to the small amount of silica in the support. The spectra of the two catalysts show the same absorption with two bands superimposed. For the lower loading sample these absorptions must be evidenced by subtracting the spectrum of the alumina support. They are quite broad, located at 665 and 530 cm$^{-1}$. Instead the spectrum of the higher loading catalyst shows two quite sharp bands at 665 and 565 cm$^{-1}$. These absorptions are typical of cobalt spinels such as $Co_3O_4$ and $CoAl_2O_4$, whose spectra are also reported in Fig. 18.2. The position of the lower frequency peak in the spectrum of the higher loading sample is actually an intermediate between those of the two spinels, suggesting, in agreement with the literature, that

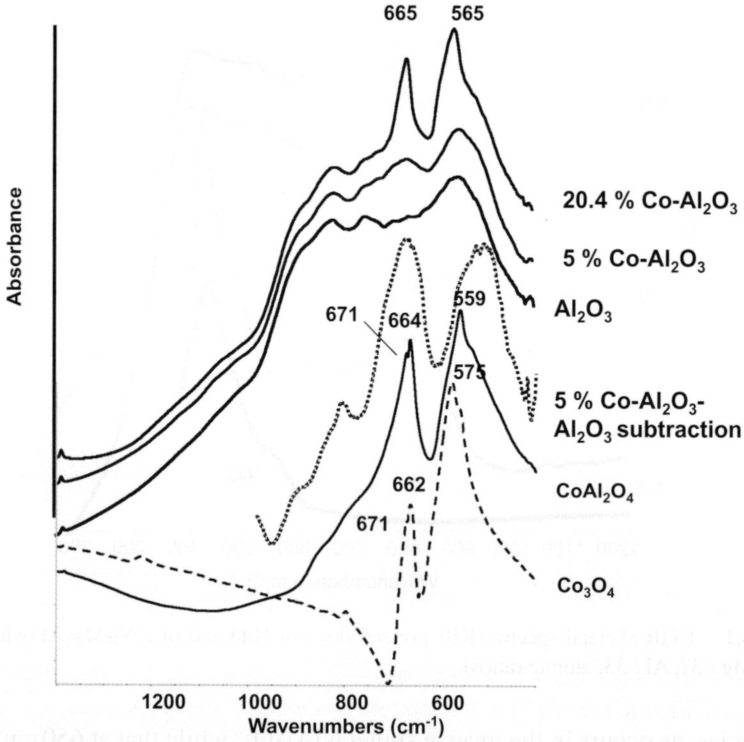


**Figure 18.2.** FTIR skeletal spectra (KBr pressed disc) of Co-Al oxide catalysts.

the spinel phase in our case can be intermediate, i.e. near to $Co_2AlO_4$ (i.e. a solid solution of cobalt aluminate and cobaltite $Co_{1+x}Al_{2-x}O_4$ con $x = 1$) in agreement with literature data.[19] More likely, we have some gradient of Co:Al distribution from the bulk towards the exterior of the catalyst, with more cobalt at the exterior and less at the interior, or the coexistence of more compositions. The spectrum of the low loading sample (definite broader bands and the lower frequency component located at an even lower frequency), suggests that Co entered the transitional alumina support producing a more disordered defective $Co_xAl_{2-x}O_3$ spinel structure. Neither XRD nor UV-vis spectroscopies easily distinguish these situations.

A last example comes from cobalt-exchanged mordenite zeolite (Co-MOR), used as the catalyst for NOx reduction by methane.The FTIR skeletal spectra of both Co-MOR and its precursor $NH_4$-MOR (ammonium exchanged mordenite, Fig. 18.3) show, as usual for materials based on silica networks,[15] a strong band with some complexity in the region 1,300–950 $cm^{-1}$, which is essentially due to the Si-O-(Si,Al) asymmetric stretching mode, a sharper band near 450 $cm^{-1}$, which is due to the Si-O-(Si,Al) rocking mode, and a weak band near 800 $cm^{-1}$, due to the Si-O-(Si,Al) symmetric stretching/in-plane bending mode. The position of the main maximum is clearly shifted upwards when ammonium ions of

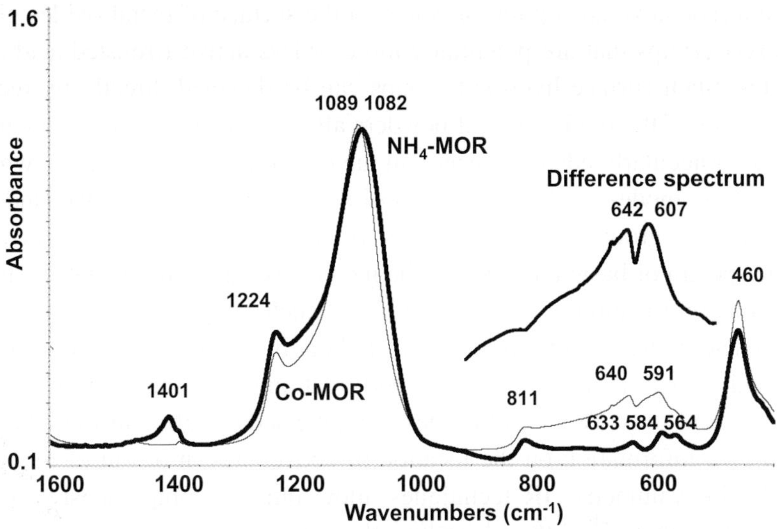


**Figure 18.3.** FTIR skeletal spectra (KBr pressed disc) of ammonium-mordenite and of cobalt-mordenite catalysts, and the subtraction spectrum.

$NH_4$-MOR (1082 $cm^{-1}$) are exchanged by cobalt (1,089 $cm^{-1}$). The subtraction spectrum relative to the exchange with Co of $NH_4$-MOR shows the appearance of a doublet at 642, 607 $cm^{-1}$. This spectrum is roughly similar to that of the spinel type compounds, like the Co oxide $Co_3O_4$ and $CoAl_2O_4$ (see Fig. 18.2). This suggests that massive Co-oxide species are formed, possibly partly arising from the reaction with extra-framework alumina species. These species, neither detected by XRD nor by UV-vis, may have a crucial role in the reaction (see below).

## 18.4. Surface Characterisation of Oxidation Catalysts by IR Spectroscopy

Self-supporting pressed discs of the pure oxide powders are prepared for *in situ* characterisation studies by transmission/absorption IR spectroscopy. These samples are put onto the IR beam, in an appropriate cell allowing heating, cooling, and gas/vapour manipulation. Activation is mostly performed by outgassing at relatively high temperatures. In the case of diffuse reflectance infrared Fourier transform (DRIFT) experiments the pure catalyst powder is deposited on the sample holder, with smooth pressure, and activation is mostly performed by an inert, dry gas flow.

### 18.4.1. *Acid-base characterisation by IR spectroscopy of adsorbed probe molecules*

For many years, IR spectroscopic techniques have found a large application in the characterisation of the acid-base properties of catalysts.[20–22] The fragments arising from the dissociative adsorption of water on the surface of metal oxides give rise to hydroxyl groups that are potentially more or less active Brønsted acid sites or basic sites. Such surface hydroxyl groups can be detected directly by recording the IR spectra of the oxide catalyst powders after treatments,which allow the desorption of molecularly adsorbed water, in the region 3,800–3,000 $cm^{-1}$, where the O-H stretching modes ($\nu$OHs) fall. The position and shape of the $\nu$OH bands of the surface hydroxy groups is informative as to their coordination and surface structure. The spectra of basic adsorbed probe molecules can be investigated, showing, for example, protonation of quite strong bases (such as pyridine or ammonia) over Brønsted acidic OHs, or the formation of H-bondings with weaker bases (such as nitriles and carbon monoxide), and the perturbation of the spectrum of the probes over Lewis acid sites. The use of probes characterised by different steric hindrance can also give information on the location of active sites in or out of micropores. Coupled with valumetry, IR techniques allow quite a comprehensive quantitative/qualitative analysis of the adsorption phenomena.

Similarly, the use of "acidic" molecules allows the probing of surface basicity,[23] perhaps with less success. We have reported a great deal of data concerning the characterisation of acidity and basicity of catalysts in recent review papers.[7,24–26] Acidity and basicity have a complex relationship with catalytic activity in oxidation,[27] depending also upon the nature of the reaction considered.

### 18.4.2. *Detection of surface oxygen species*

The so-called Mars–van Krevelen or redox mechanism[28] is widely accepted to occur in the case of many oxidation reactions over metal-oxide catalysts. In this case the oxidised catalyst surface oxidises the reactant and is reoxidised by gas-phase $O_2$ in a following step. Bulk or subsurface atomic oxide species may be active in this mechanism. As said, skeletal IR spectroscopy provides detailed information on the bulk metal-oxide bonds. During bulk oxidation and reduction, as is obvious, the skeletal spectrum is modified. As we will discuss later on, $Mn_3O_4$ (a random spinel whose mineralogical name is hausmannite) represents an interesting oxidation catalyst. When reduced in bulk, $Mn_3O_4$ converts into MnO, a rocksalt (or periclase) -type oxide called manganosite. The skeletal spectra of these two compounds are reported in Fig. 18.4, right. In the left section of the figure, part of the IR spectrum of a pure powder pressed in a $Mn_3O_4$ disc is shown. It presents a cut-off limit just below 800 $cm^{-1}$ due to the absorption associated with the higher frequency bulk Mn-O band of $Mn_3O_4$, whose maximum is found at 615 $cm^{-1}$ in the skeletal spectrum. This band is associated with the stretchings of the Mn-O bonds involving the Mn ions in tetrahedral coordination sites within the spinel structure. When reduced in hydrogen, the cut-off limit shifts to near 600 $cm^{-1}$. This is likely associated with the

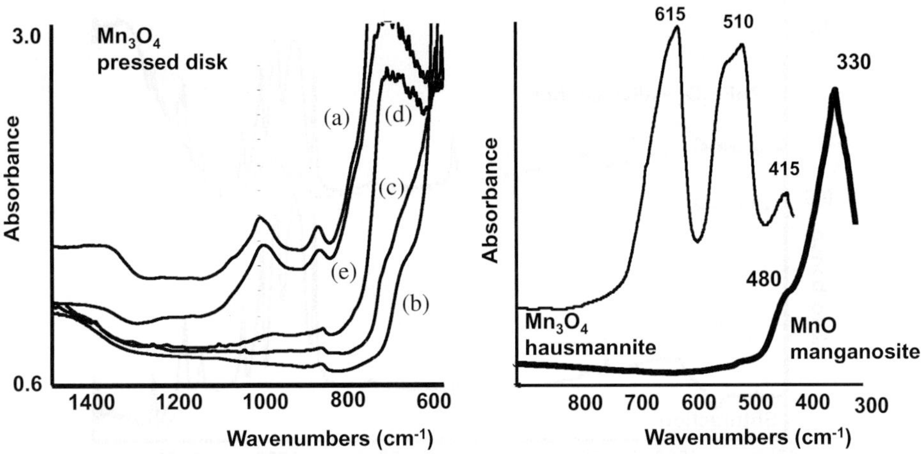


**Figure 18.4.** Left: FTIR pure powder spectra (pure powder pressed disc) of $Mn_3O_4$ (hausmannite) (a) activated in air and vacuum, 450°C; (b) heated in $H_2$ 500 Torr at 450°C for 30 min and outgassed at 450°C; (c) reheated in oxygen 500 Torr at 200°C; (d) reheated in oxygen 500 Torr at 300°C; (e) reheated in oxygen 500 Torr at 450°C. Right: FTIR skeletal spectra (KBr pressed disc) of $Mn_3O_4$ (hausmannite) and MnO (manganosite).

reduction of $Mn_3O_4$ to MnO, that (as do other periclase-type oxides) presents one IR-active mode only, split into the LO (*ca.* 480 $cm^{-1}$) and the TO mode (510 $cm^{-1}$). The much lower position of the skeletal bands of MnO is due to the octahedral-only coordination of the cations in the cubic close packing oxide array of periclase-type oxides.

The adsorption of probe molecules may provide evidence of the existence of metal oxygen bonds at the surface of metal oxides, whose stretching modes are located just at the limit of the cut-off. This is shown in Fig. 18.5 where the spectrum of a disc of $ZnFe_2O_4$ is shown after activation in vacuum (dashed line) and after the adsorption of allyl alcohol.

The spectrum of $ZnFe_2O_4$ shows the weak bands of surface hydroxyl groups (3,735 $cm^{-1}$ weak and inactive in adsorption, and 3,660 $cm^{-1}$, perturbed during adsorption), of trapped $CO_2$ (2,345 $cm^{-1}$), and trapped carbonates, arising from the preparation in an organic medium. Upon adsorption, quite sharp bands appeared, due to the adsorbed species. However, in the region 900–700 $cm^{-1}$ the spectrum after adsorption is located at a lower absorbance than that of the clean catalyst. Correspondingly, a broad "negative" band is observed in the subtraction spectrum (see the dashed line). A similar absorption is found in several other cases, such as on aluminas[29] and aluminates[30] as well as gallia,[7] just above the cut-off due to the skeletal modes, and is attributed to the relaxation of surface metal-oxygen bonds by the adsorption of molecular probes.

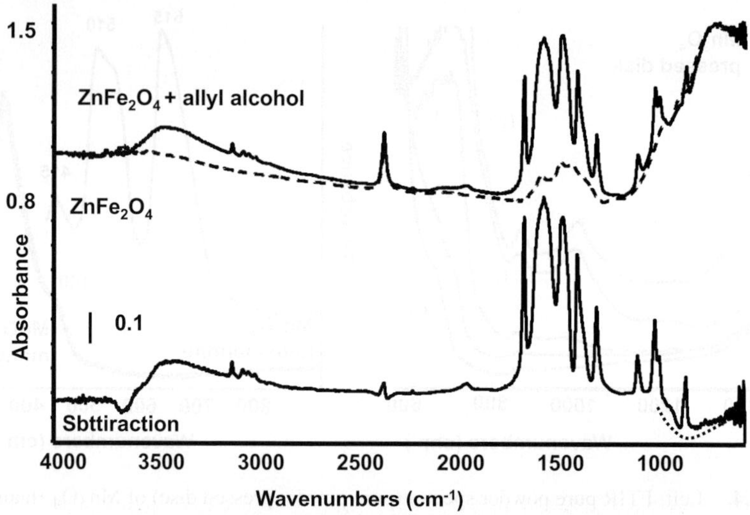


**Figure 18.5.** FTIR pure powder spectra (pure powder pressed disc) of $ZnFe_2O_4$ activated in air and vacuum at 450°C (dashed line) and after adsorption of allyl alcohol (full line).

Metals and other elements in very high oxidation states can give rise to element-oxygen "double bonds" in their oxides, i.e. very short bonds. This is the case in vanadyl, niobyl, molybdenyl, chromyl, wolframyl groups, as well as of P=O bonds present in oxo-compounds of the corresponding elements. The location at the surface of some such bonds has been observed, as in the case of V=O bonds (1,038 $cm^{-1}$) at the surface of bulk $V_2O_5$,[31] Nb=O bonds at the surface of niobic acid ($Nb_2O_5 \cdot n H_2O$, 995 $cm^{-1}$),[32] W=O bonds at the surface of bulk $WO_3$ (1,040 $cm^{-1}$),[33] Cr=O bonds at the surface of oxidised chromia,[34] and metal chromites such as $ZnCr_2O_4$,[35] $CoCr_2O_4$,[36] and $MgCr_2O_4$(1,030–800 $cm^{-1}$).[37]

These features are better observed as negative bands in the subtraction spectra after the adsorption of probe molecules (spectrum recorded after adsorption from which the spectrum of the "clean" sample has been subtracted), showing that such surface M=O species are perturbed upon adsorption. The reactivity of such species in oxidation catalysis has been well evidenced in the case of chromia and metal cromites, whose surface is covered by chromate species. Upon reduction the Cr=O stretching bands disappear, while they reappear upon reoxidation.[37]In Fig. 18.6, part of the spectrum of $CuCr_2O_4$, an important industrial catalyst, is reported in the oxidised state after interaction with methane at increasing temperature. The massive peak at 1,000–800 $cm^{-1}$ is due to the stretching of multiple $Cr^{6+}$=O bonds of chromate species that become reduced to $Cr^{3+}$-O by oxidising methane.

Surface vanadate, molybdate, and tungstate species are also well detected by IR spectroscopy, enhanced when coupled with Raman measurements.

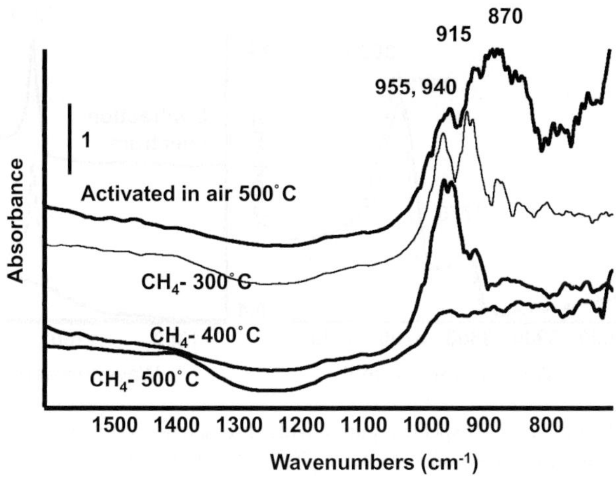


**Figure 18.6.** FTIR pure powder spectra (pure powder pressed disc) of $CuCr_2O_4$ activated in air and vacuum at 450°C and after contact with methane 200 Torr at increasing temperatures.

Supported oxides of vanadium, chromium, molybdenum, and tungsten in their higher oxidation states are used widely as heterogeneous catalysts, for selective oxidation as well as for acid catalysis. Over dispersing oxide carriers, such as alumina, titania, and zirconia, metal oxide surface species are observed (at least at low loadings, i.e. well below the loading corresponding to the "theoretical geometric monolayer") which are spectroscopically characterized, under dry conditions, by:

(i) a very high position of the very intense band observed in IR and Raman, similar to that of gaseous mono-oxo species;
(ii) the coincidence of the position of the band detected in Raman and IR spectra and its non-complexity in both cases;
(iii) the detection in the IR spectrum of a single overtone band;
(iv) the absence of modes assignable to M-O-M asymmetric stretchings;
(v) partial $^{18}O/^{16}O$ exchange experiments which showed a simple splitting found, for example, in the cases of $WO_3/Al_2O_3$[38] and in the case of $V_2O_5/TiO_2$ catalysts.[39]

These data are a strong demonstration that the species are isolated and mono-oxo, being essentially formed by low coordination complexes with a single M=O bond. This topic has been previously discussed by us in detail.[40] In Fig. 18.7 the spectrum of a low loading of $V_2O_5/ZrO_2$ catalyst is shown. The very strong and sharp peak at 1,024 $cm^{-1}$, not present in the spectrum of zirconia, is due to V=O stretching of isolated surface vanadyl species. Note that the spectrum of zirconia in the OH stretching is almost unperturbed. At high loadings, most authors agree that polymeric anions form. As shown by Wachs and co-workers using mostly but not

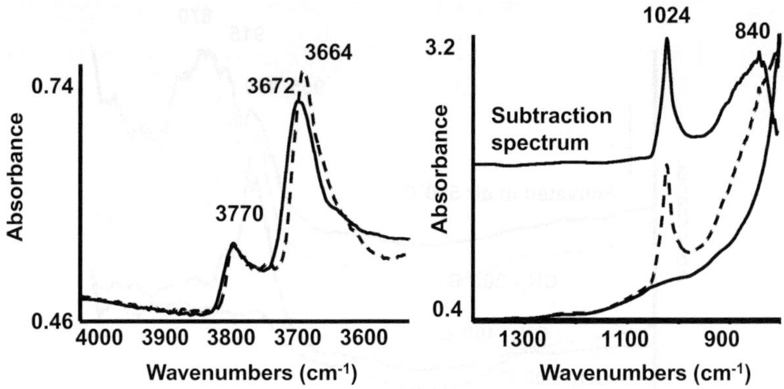


**Figure 18.7.** FTIR pure powder spectra (pure powder pressed discs) of $ZrO_2$ (50 $m^2/g$, 81% tetragonal, 19% monoclinic, full lines) and of 2% $V_2O_5/ZrO_2$ (dashed lines), both activated in air and vacuum at 450°C.

only Raman spectroscopy, in the case of silica-supported catalysts the situation is more complex, with the presence of mono-oxo, di-oxo, and tri-oxo species.[41] Their results concerning vanadia catalysts have been recently reviewed.[42]

Planar model catalysts with similar compositions (such as vanadia-silica, vanadia-alumina on NiAl(110), and vanadia on $CeO_2(111)$ monocrystal faces) have been recently investigated with surface science techniques such as infrared reflection absorption spectroscopy (IRAS).[43,44] Although the authors claim that these studies imply a revisiting of previous interpretations, it appears that conclusions obtained 25 years ago with conventional (and much cheaper) techniques by several research groups are substantially confirmed.

Infrared spectroscopy has also been used to reveal the formation of molecular oxygen species adsorbed on metal oxide surfaces, as discussed in detail by Davydov and Sheppard.[45] Neutral, weakly adsorbed dioxygen species were observed recently on a $TiO_2$ powder at 120 K, characterised by the O–O stretching at 1,550 $cm^{-1}$.[46] The position is the same as the Raman active (and IR-inactive) mode of gas-phase $O_2$, but the interaction with the surface caused its activation in IR. Indeed, several years ago, Davydov and co-workers reported IR bands attributed to dioxygen adsorbed on metal oxides.[45] Over several metal oxides, such as $CeO_2$ and $CeO_2$-$ZrO_2$,[47] and CoO-MgO[48] solid solutions, bands due to dioxygen species in the region 1,000–1,200 $cm^{-1}$ have been well-characterised and attributed to superoxide species $O_2^-$. Peroxide-like species, $O_2^{2-}$, absorbing in the region below 1,000 $cm^{-1}$ have been reported too, for example, on chromia and ceria, but their characterisation by IR is not easy due to the superimposition with bulk vibrations.

Molecular and dissociative adsorption of oxygen has been revealed by vibrational spectroscopies such as IRAS and electron energy loss spectroscopy (EELS) over metal monocrystal faces. As for example, molecular adsorption is observed over Pd(111) at 30 K producing two different species characterised by OO stretchings at 850 and 1,035 $cm^{-1}$.[49] These peaks are assigned to a peroxo-like state and a superoxo-like state, respectively. After saturation of these chemisorbed molecular states, a state of physisorbed oxygen with its vibrational frequency close to the gas-phase value of 1,556 $cm^{-1}$ is populated. Upon warming the sample above 80 K, an additional loss feature at 650 $cm^{-1}$ develops, which is assigned to a second peroxo-like molecular species. The dissociation process is completed at $T \approx 200$ K leaving a layer of atomic oxygen on the surface which is characterised by a peak at 480 $cm^{-1}$ (Pd-O stretching) and by a $2 \times 2$ pattern in LEED.[49] Oxygen atoms are adsorbed on threefold hollow adsorption sites. At 523 K, data show the formation of subsurface oxygen species on Pd(111) characterised by a Pd-O stretching band at 326 $cm^{-1}$.[50] At and above 600 K, several surface oxide phases may form[51] until Pd(111) converts into the bulk PdO(101) face.[52]

PdO crystallises in the so-called "cooperite" (or PdS) structure, with (space group $P4_2/mmc = D^9_{4h}$, $Z = 2$). In this structure $Pd^{2+}$ ions are in square planar coordination, $O^{2-}$ ions being in tetrahedral coordination. The vibrational structure of PdO crystallites gives rise to two IR-active bands (668 $cm^{-1}$, $A_{2u}$, 612 $cm^{-1}$, $E_{2u}$) and two Raman active modes (651 $cm^{-1}$, $B_{1g}$, 445 $cm^{-1}$, $E_g$).[53] Raman studies allow the detection of PdO particles on oxidised supported Pd catalysts such as $Pd/Al_2O_3$[54] while this is less easy using IR, due to the superimposition with bulk vibrations. A band at 740 $cm^{-1}$ has been tentatively assigned to a surface Pd–O mode over a $Pd/CeO_2$-$ZrO_2$ catalyst.[55]

Solid catalysts may also be used for reactions implying oxidants differ by dioxygen. The most popular case is that of titanium-silicalite catalysts for oxidation with hydrogen peroxide. The active species in the presence of water has been characterised to be a side-on peroxo complex characterised by a Raman-detected O–O stretching at 618 $cm^{-1}$.[56] Upon drying, this species converts into a hydroperoxo species characterised using IR by an O–O stretching at 837 $cm^{-1}$ and a broad OH band at 3400 $cm^{-1}$.[57]

In a recent paper the interaction of tert-butyl hydroperoxide with oxide catalysts was investigated.[58] This oxidant is used, with the aid of oxide catalysts, for the oxidation of (benzo) thiophenes to the corresponding sulfones in liquid phase, but also for other liquid-phase oxidation and epoxidation reactions. The adsorption of tert-butyl hydroperoxide on zirconia from the gas-phase is reported to produce an undissociated adsorbed species as well as a terbutylperoxide species whose OO stretching frequency is observed at 867 $cm^{-1}$, i.e. shifted to higher frequencies with respect to the undissociated compound (845 $cm^{-1}$). The presence of vanadium at the surface of the catalyst increases the reactivity of this species.

### 18.4.3. *Determination of the oxidation state of cationic centres by IR spectroscopy of adsorbed probe molecules*

Carbon monoxide is a very weak base largely used for the surface characterisation of cationic centres on metal oxide surfaces.[59] Low temperature CO adsorption experiments allow the observation of weakly acidic but active adsorption sites, including surface hydroxyl groups and alkali cations, whose interaction with CO is not observed in room or higher temperature experiments. The interaction of CO with strongly oxidising cations can also be revealed at lower temperature experiments. A summary of metal sites revealed by low temperature CO adsorption on some alumina-supported catalysts is reported in Table 18.1.

Low temperature adsorption experiments enable a more precise photography of the state of the surface as they avoid misunderstandings due to the reactivity of CO, which may reduce the surface producing carbonate species and $CO_2$, and sometimes react giving rise to undetectable surface carbide species. However, room temperature experiments may provide evidence for some kind of "activated" adsorption phenomena, revealing adsorption modes needing some energy to be allowed. In Fig. 18.8 the spectra of CO adsorbed at low temperature ($-140$°C, full line) and at room temperature (dashed line) over the same pre-reduced Pt-K-$Al_2O_3$ catalyst are compared. At low temperature two species are evident, characterised by CO stretching at 2,143 $cm^{-1}$, sharp and strong, and at 2,073 K, weaker, with an evident tail towards lower frequencies. The former feature fully disappears at room temperature, which is the typical behaviour for CO adsorbed on very weak sites such as $K^+$ ions. On the contrary, the latter feature is greatly increased in intensity at room temperature with the main maximum at 2,056 $cm^{-1}$ and an evident shoulder at 2,008 $cm^{-1}$. These bands are certainly due to CO adsorbed on top of Pt metal particles. Additionally, a further band appears at 1,800 $cm^{-1}$, which is attributable to bridging CO species. It seems likely that the experiment performed at room temperature allows more complex interactions to occur over Pt metal species which are not allowed at low temperatures. It is also possible that the presence of CO at sufficiently high temperatures allows some restructuring of the Pt metal species,[60] as also observed for Pd metal species.[61]

The use of IR spectroscopy of adsorbed CO for the characterisation of supported metal catalysts takes advantage of the many surface science studies performed on metal monocrystal faces. In fact, the technique of low temperature CO adsorption followed by vibrational spectroscopies (such as IRAS, high resolution EELS (HREELS), or, more recently, sum frequency generation spectroscopy (SFG)) represents a largely used technique in metal surface science studies. These techniques give very precise reference data for catalyst characterisation and allow good comparison among the respective results.

**Table 18.1.** Position of CO stretching of surface carbonyls on alumina–based catalysts.

| Catalyst | Adsorbing cation | Species | $\nu$CO, $cm^{-1}$ | Note |
|---|---|---|---|---|
| $Al_2O_3$ | $Al^{3+}$ | | 2,240–2,180 | Different sites depending on activation temperature |
| $Co/Al_2O_3$ | $Co^{3+}$ | On $Co_3O_4$ part. | 2,190 | Oxidation CO to $CO_2$ |
| | $Co^{2+}$ | | 2,170 | |
| | $Co^0$ | On defects | 2,050 | |
| | $Co^0$ | Terminal on faces | 2,030–1,990 | Shift due to vibrat. couplings |
| $Ni/Al_2O_3$ | $Ni^{2+}$ | | 2,167–2,173 | |
| | $Ni^{n+}$ | Polycarbonyls | 2,150–2,000 | Several sharp bands |
| | Ni° | Terminal | 2,080–2,020 | |
| | Ni° | Bridging | 1,930–1,870 | |
| $Cu/Al_2O_3$ | $Cu^{2+}$ | | 2,160–2,140 | Oxidation CO to $CO_2$ |
| | $Cu^+$ | | 2,120–2,100 | Strong, resistant |
| | Cu° | Terminal | 2,080–2,050 | Weak very labile |
| $Rh/Al_2O_3$ | $Rh^+$ | Dicarbonyls on dispersed ions | *ca.* 2,090, *ca.* 2,020 | Strong doublet |
| | Rh° | Terminal | 2,080–2,050 | |
| $Pd/Al_2O_3$ | $Pd^{2+}$ | Dispersed ions | 2,160–2,170 | Oxidation CO to $CO_2$ |
| | $Pd^{2+}$ | PdO particles | 2,150–2,140 | |
| | $Pd^{n+}$ | Partially oxidised metal particles | 2,130 | |
| | Pd° | Terminal on faces | 2,120–2,080 | Shift due to vibrat. couplings |
| | Pd° | Bridging | 1,990–1,980 | Strong |
| | Pd | Triply bridging | 1,820 and 1,778 | |
| $Pt/Al_2O_3$ | $Pt^{4+}$ | Dispersed ions | 2,186 | Oxidation CO to $CO_2$ |
| | $Pt^{2+}$ | | 2,135 | |
| | Pt° | Terminal | 2,100–2,030 | Shift due to vibrat. couplings |
| | Pt° | Bridging | 1,815–1,780 | Weak |

The presence of very strongly oxidising cationic centres is also revealed by the oxidation of CO to $CO_2$ at a temperature as low as 140 K, well evident in the IR spectra. This is the case for $Pd^{4+}$, $Cu^{2+}$, $Pt^{2+}$, and $Co^{3+}$ centres (see Table 18.1). We will further consider this point later on.

Nitrogen monoxide has one electron more than CO, so that the same electron configuration could apply with the additional electron in an antibonding $\pi^*$-type orbital. The bond order is decreased to 2.5 and this is reflected in the lower stretching frequency (1,875 $cm^{-1}$ for NO gas). Thus, the NO molecule is a radical and this makes its reactivity far higher. It easily dimerizes to $N_2O_2$, is easily oxidised to

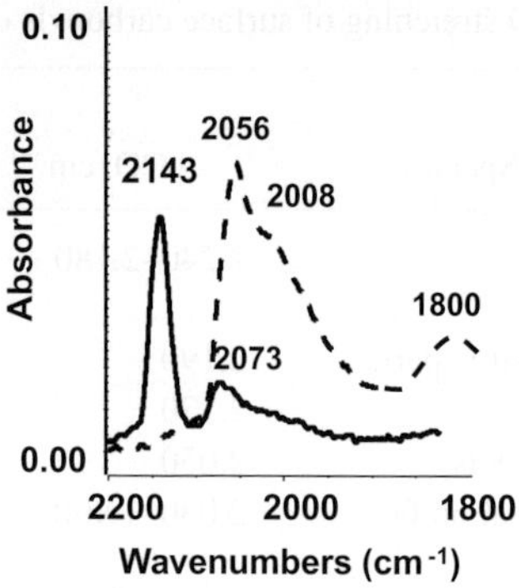


**Figure 18.8.** FTIR spectra of CO adsorbed at on pre-reduced Pt-K/$Al_2O_3$ at $-140°C$ (full line) and at room temperature (dashed line).

$NO^+$, to nitrites $NO_2^-$ or to $NO_2$, and is easily reduced to $NO^-$ and $N_2O$. NO can also easily disproportionate and is thermodynamically unstable towards decomposition to $N_2 + O_2$. This makes this molecule very reactive so that its successful use as a probe for cationic centres is limited.

In addition, NO can also interact with the surface cationic centres on metal oxides, giving rise to surface mono-nitrosyl, gem-dinitrosyl, and tri-nitrosyl species. When the cation or the metal atom contains not only empty orbitals, but also full or partly filled d-type orbitals, they can interact with the $\pi^*$-type orbitals of the NO molecule, via a $\pi$-type electron backdonation. This gives rise to bent nitrosyls, where N is likely to be a $sp^2$ hybrid and the NO stretching frequency is decreased.

Adsorbed NO can be even more sensitive than adsorbed CO to the oxidation state of the cations; for example, the use of NO as a probe allows the determination of the oxidation state of Co ions in zeolites better than CO.[62] On a Co-ZSM5 zeolite, three bands due to adsorbed CO are observed at 2,204, 2,198 and 2,188 $cm^{-1}$: they are due to different cobalt carbonyls but it is difficult to assign them the oxidation state for cobalt. When NO is used as the probe it shows the existence of both $Co^{3+}$ and $Co^{2+}$ species. In fact the bands observed at 1,901, 1,817 $cm^{-1}$ are due to $[Co(NO)_2]^{2+}$gem–dinitrosyl species while that at 1,945 $cm^{-1}$ is assigned to $[Co\text{-}NO]^{3+}$ mononitrosyls.

## 18.5. Studies of Oxidation Reactions Over Solid Catalysts: Methodologies

From the first applications of the IR technique for surface studies and heterogeneous catalysis it was recognized that this technique can be useful, not only for the characterisation of the catalyst surface structure by the detection of the spectra of the adsorbed probe molecules, but also to give almost direct information on the mechanisms of heterogeneous catalysis.

Since the work of Eischens and co-workers,[4] where the reactant adsorption of the catalytic reactions of industrial interest over the relevant catalysts was investigated, a possible indication of the mechanism of the catalytic steps was obtained. This approach was later developed by Kokes and co-workers[63] in the 1960s. The concept of experiments under dynamic conditions was also developed in the 1960s by Tamaru,[64] and applied to IR spectroscopy and other techniques. During his experiments, Tamaru measured the conversion of reactants in parallel with the modification of the spectra of the surface species for several reactions such as (i) the olefin isomerization and hydrogenation over ZnO; (ii) the decomposition of formic acid; (iii) the water-gas shift reaction over ZnO and MgO; (iv) the decomposition of methanol over ZnO; (v) the oxidation of CO on supported Pd.

Starting from the end of the 1970s, appropriate IR cells for experiments performed in flow conditions—with the simultaneous detection of the reaction product—whilst measuring the conversion and selectivities have been developed. These experiments were denoted, in those times, as *in situ* studies. In his extensive review published in 2001, J. Ryczkowski described and discussed most of the designs of IR cells used for these studies.[8]

More recently, this approach evolved into the concept of *operando* spectroscopy,[65] applied to different spectroscopic techniques including IR[66] with the possibility of applying several spectroscopic techniques simultaneously to the same catalyst sample in working conditions.[67] According to the most recent definition, the term *in situ* study should essentially be reserved for studies performed under a controlled atmosphere but without the measurement of gas-phase concentration.[68] Many catalysis laboratories today employ *operando* apparata with IR detection, aided by the commercial availability of cells allowing heating and gas flow. Most of these apparata use diffuse reflectance IR cells and attachments,[69,70] which have the advantage of allowing the use of powdered catalysts without any pressing. This avoids the mass transport limitations which occur with pressed discs used in transmission spectroscopies, thus allowing easier kinetic studies.[68] Other labs prefer home-made transmission/absorption cells for *operando* experiments, which allow an easier subtraction of the gas-phase spectrum.[71,72]

Indeed, it seems obvious that under reaction conditions, when the reactants adsorb onto the surface, and transform into intermediates and products, that they finally desorb rapidly; hence the concentration at the surface of most or all of these species is expected to be very low. In particular, this is expected to be true for species involved in successive steps down to the slowest step. Indeed, the detectability of the active adsorbed species (the surface intermediates) is rare, if even possible, in *operando* conditions (i.e. at relatively high temperatures) with usual

IR measurements. These labile adsorbed species may be detected using unusual measurements such as femtosecond laser excitation followed by nanosecond time resolved FTIR measurements.[73] On the contrary, heavy adsorbed species that are not involved in the main reaction concentrate at the catalyst surface and can be easily detected. It is possible to forecast that, if a species is determined to be highly concentrated and stable in reaction conditions, it is either the true catalyst (e.g. sometimes carbonaceous materials deposited at the surface during the induction period), or a poison (e.g. coke that is a main cause of deactivation of acidic catalysts), or finally, a spectator species. In any case, the information obtained in *operando* conditions is always interesting and sometimes useful for revealing the reaction mechanisms.

In the case of catalysis by metals, useful information arises from comparison with surface science studies on metal faces performed using IRAS, HREELS, and SFG spectroscopy.[74,75] Starting in the 1960s, these studies have been performed over monocrystal metal surfaces, using several molecules as probes. However, with the same techniques, reactant adsorption studies have also been undertaken, allowing the following of the surface chemistry of industrially significant reactions. Among these reactions, several oxidation reactions have been investigated. Surface science studies were of great help towards the understanding of metal catalysis. Recognition of this work is represented by the Nobel Prize in 2007 for Chemistry, awarded to Gerhard Ertl, the most eminent surface scientist. However, most of these investigations present both a materials gap and a conditions gap with respect to industrial catalysis. In fact, they are performed at very low temperatures and pressures (conditions gap) and usually over well-defined and pure monocrystal faces, while industrial catalysis is performed over doped metals and, frequently, small metal nanoparticles deposited over carriers (materials gap). More recent PM-IRAS (polarization-modulation infrared reflection absorption spectroscopy) studies[76] allowed the temperature and pressure gap previously existing between surface science monocrystal studies and studies on real catalytic powder to be at least partially overcome.

The preparation of layered model catalysts in UHV (ultra high vacuum) represents an approach to partially overcome the materials gap. This approach, developed by different authors such as Goodman,[75] and Freund and co-workers,[77] allowed the preparation of materials whose composition is close that of real oxidation catalysts deposited or formed at the surface of conducting monocrystal faces. These authors prepared different kinds of layered model catalysts, such as $Pd/Al_2O_3$ obtained by depositing Pt particles from the vapour phase over an alumina layer formed by the oxidation and annealing of a NiAl(110) single crystal surface,[78] or $V_2O_5/CeO_2$ obtained by depositing and oxidising first Ce and later V over an oxidised Ru(0001) single crystal surface.[79] These solids allow IR reflection/absorption spectroscopy studies as well as IR–vis SFG vibrational spectroscopy. Indeed, these studies do not provide much additional information with respect to *in situ* IR studies than was attained several decades ago with (much cheaper) traditional IR apparata and real catalysts.[61,80]

Supported metals can also be studied using the multilayer enhanced infrared reflection absorption spectroscopy (MEIRAS) developed by Wolff and co-workers,[81] prepared using semi-transparent thin metal films or nanowires over $SiO_2$ or $TiO_2$ dielectric films of varying thicknesses with reflecting gold films underneath.

For over 30 years we have tried in our laboratory to obtain information on the catalytic reaction mechanisms coupling deep surface and bulk catalyst characterisation with surface science studies, i.e. with experiments investigating reactant adsorption at low temperature and following the evolution of the adsorbed species under reaction conditions in different static environments (vacuum, oxygen, hydrogen, ...), depending on the reaction investigated. With this approach, the chemistry of the reactant/catalyst surface chemistry may be understood and intermediates can be frequently "trapped" in conditions where they accumulate at the surface. The dynamic evolution of the species approaching reaction conditions is evidence of the possible role of these species in the reaction.

Hereinafter we will review data concerning oxidation reactions performed in our lab using this approach and, when possible, we will compare these data with those arising from other studies, including studies performed in *operando* conditions.

## 18.6. IR Spectroscopy Studies of Heterogeneously Catalyzed Oxidations: Case Studies

### 18.6.1. *Oxidation and steam reforming of methanol*

The selective oxidation of methanol is an important industrial process for the production of formaldehyde.[82]

$$CH_3OH + 1/2O_2 \rightarrow CH_2O + H_2O$$

The reaction is carried out either over bulk $Fe_2(MoO_4)_3$-$MoO_3$ or over silver catalysts with very high selectivity. Over different catalysts other selective oxidation products or total oxidation can be obtained. Some years ago we focused our attention on vanadia-titania catalysts, where methyl formate can be obtained with high yields.[83–85]

$$2CH_3OH + 2O_2 \rightarrow HCOOCH_3 + 2H_2O$$

Over these catalysts, or modified by the presence of sulfates,[86] dimethoxymethane can also be obtained with high selectivities.

$$3CH_3OH + 1/2O_2 \rightarrow H_2C(OCH_3)_2 + 2H_2O$$

On the other hand, methanol is a toxic and pollutant volatile organic compound (VOC) which can be abated by catalytic total oxidation.

$$CH_3OH + 2O_2 \rightarrow CO_2 + 2H_2O$$

IR spectroscopy has been largely applied to these reactions. Briand[87] recently reviewed part of the literature concerning IR spectroscopy and flow reactor work on methanol oxidation over many oxide catalysts. IR spectra show that methanol adsorption on metal oxides occurs either in a molecular undissociated form or as methoxy groups,[88,89] which can have a more covalent or more ionic character.[90] The typical spectrum of ionic methoxides is shown in Fig. 18.9, after outgassing at 100°C the $\alpha$-$Fe_2O_3$ disc where methanol was adsorbed at room temperature. The C–O stretching is observed at 1,070 $cm^{-1}$, i.e. well above the C–O stretching of methanol (1,032 $cm^{-1}$). A split weak $CH_3$ deformation band (1,461, 1,441 $cm^{-1}$) and two strong bands in the CH stretching region, which are predominantly due to the Fermi resonance of the $CH_3$ asymmetric stretching with an overtone of the deformation mode, are also evident. Covalent methoxides form on very high oxidation state small sized atoms (such as for $Si^{IV}$-$OCH_3$, $Mo^{VI}$-$OCH_3$, and $V^{V}$-$OCH_3$) and are characterised by a higher intensity of the $CH_3$ deformation modes and lower wavenumbers of the $CH_3$ and CO stretchings with respect to ionic methoxides. CO stretching of ionic methoxy groups may be multiple, showing the presence of

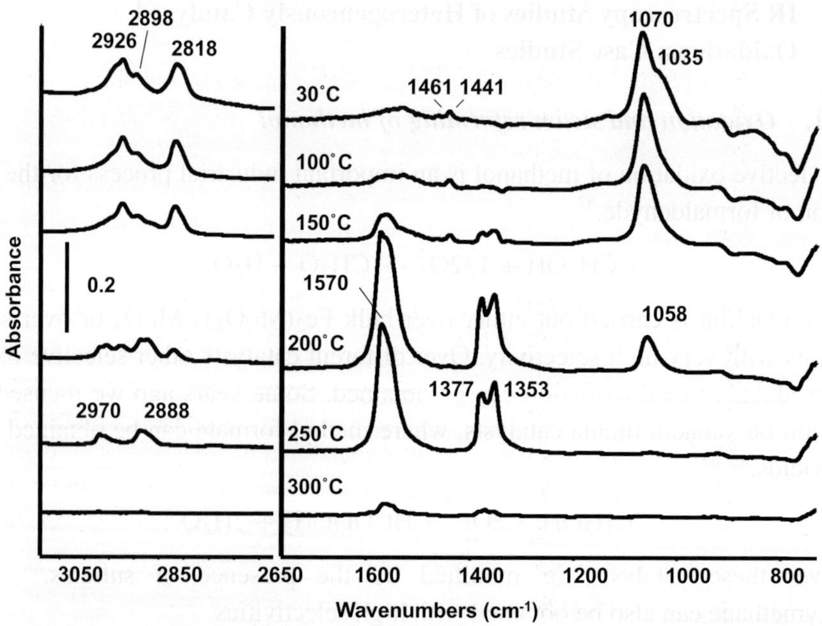


**Figure 18.9.** FTIR spectra of adsorbed species arising from methanol adsorbed on $\alpha$-$Fe_2O_3$ and upon outgassing at increasing temperatures.

terminal, bridging, and/or triply bridging species,[91] characterised by a lower C–O stretching wavenumber the higher the coordination of oxygen.

Infrared spectroscopy has been used to investigate methanol total oxidation over typical transition metal oxide combustion catalysts. On $\alpha$-$Fe_2O_3$ (Fig. 18.9), methoxy groups are totally converted into formate species (well characterised by strong bands at 1,570 $cm^{-1}$, COO asymmetric stretching, and at 1,377; 1,353 $cm^{-1}$, partially superimposed COO symmetric stretching, and CH deformation) at 250°C. Formates totally disappear at 300°C giving rise to gas-phase $CO_2$. All the reaction steps can be performed even at very low temperatures (150–300°C) in the absence of gas-phase oxygen on bulk transition metal oxides such as $\alpha$-$Fe_2O_3$,[92] $Co_3O_4$,[93] CuO,[94]$\alpha$-$Mn_2O_3$, $Mn_3O_4$, and $MnO_2$,[95] as well as on mixed or supported oxides such as $MnO_x/Al_2O_3$,[95] showing the combustion activity of reticular oxygen. Over $MnO_x/Al_2O_3$ combustion catalysts, methoxy groups on $Mn^{3+}$, $Mn^{2+}$, and $Al^{3+}$ can be distinguished, with those interacting with bare alumina being inactive spectator species.

An elegant *operando* IR study has been reported recently by Rousseau *et al.*[96] concerning methanol total oxidation over very active Au/$CeO_2$ catalysts. Methanol is found to adsorb dissociatively on the $CeO_2$ support giving three families of methoxy groups. Only two of them convert into formates by oxidation with the participation of gold, the third acting as a spectator. Formates on ceria decompose into CO which would later be oxidised on gold to $CO_2$, only in the presence of gas-phase oxygen.

The mechanism of the methanol oxidation to methylformate over vanadia-titania was investigated with parallel IR and flow reactor experiments by our groups.[83–85,90] Methanol adsorption over vanadia-titania gives rise to surface methoxygroups that, upon heating, start to be oxidised above 130°C. At this temperature the features of the formate species dominate the spectrum. However, in this case a weak band at 1,645 and 1,300 $cm^{-1}$, likely due to molecularly adsorbed formaldehyde (C=O stretching and $CH_2$ deformation), can be found. Formate species are stable in the adsorbed form up to near 220°C, but later tend to decompose into $CO_x$. The direct adsorption of formaldehyde at room temperature on vanadia-titania reveals the formation of dioxymethylene species, stable up to 100°C; upon heating this species both methoxy and formate groups form via a Cannizzaro-type disproportionation, or by oxidation to formate groups only. This behaviour was interpreted assuming that the desorption of formaldehyde from the dioxymethylene species is slow on vanadia-titania, as on many other oxides, and oxidation or disproportionation may occur. In the presence of methanol, at relatively low temperatures, dioxymethylene reacts with methanol giving rise to dimethoxymethane, which in fact is produced with high selectivity at low conversion. Formate species are also quite stable so that their reaction with methanol gives rise to methylformate with high selectivity over this catalyst.

Methanol oxidation was investigated recently by IRAS (Infrared Reflection Absorption Spectroscopy) and TPD (Temperature Programmed Desorption) over $V_2O_5/CeO_2(111)$ planar model catalysts.[44] The results substantially agree with,but are far less detailed than those obtained with powdered catalysts using transmission/absorption IR.

Similar IR experiments were performed to establish the reaction mechanism for methanol oxidation on unsupported and silica-supported vanadia, which are more selective catalysts for formaldehyde synthesis than vanadia-titania.[97] The formation of methoxy groups from methanol dissociative or "condensative" adsorption was determined while it was established that formaldehyde (directly adsorbed or produced by methoxy group oxidation) mainly adsorbs in the form of dioxymethylene species, stable only at relatively low temperatures. It was concluded that dioxymethylene can react with methanol at low conversion to give rise to dimethoxymethane while it preferentially desorbs as formaldehyde at higher conversions and temperatures. The weakness of the adsorption of formaldehyde was considered to be the key feature of catalysts allowing high selectivity in formaldehyde synthesis.

The IR spectra of methanol and methoxy groups adsorbed on $MoO_3$, ferric molybdate, and other bulk molybdates[98,99] and supported molybdena catalysts,[100] have been reported.

Recently we investigated, by IR and flow reactor studies, the oxidation of methanol over a 25% $MoO_3/Zr_{0.75}Ce_{0.25}O_2$ catalyst. The flow reactor study showed 100% selectivity to formaldehyde at small methanol conversion (below 150°C). At higher temperatures and conversion, selectivity to formaldehyde drops to zero in favour of $CO_2$ before and CO later. The study performed in the IR cell revealed the same products in the same temperature intervals.

The IR spectra (Fig. 18.10) show, at low temperature, the formation of different types of surface methoxy groups. Two well-evident components at 1,145 and 1,063 $cm^{-1}$, which are stable even at 400°C (thus acting as spectator species), likely characterise terminal methoxy groups on $Zr^{4+}$ ($CH_3$ rocking and CO stretching).[101]

The active species (those that disappeared during the reaction in the 100–300°C interval, when formaldehyde first and $CO_2$ later are released in the gas-phase, see the subtraction spectrum) are characterised by bands at 1,159, 1,098, 1,080, and 1,036 $cm^{-1}$. This spectrum is clearly not only due to methoxy groups. The lowest frequency band is confidently assigned to the CO stretching of covalent $CH_3O$-Mo species, while the highest frequency bands may be due to C–O stretching of dioxymethylene species,[102] over which the $CH_3$ rocking mode of methoxy groups is superimposed. Dioxymethylene species are well-characterised adsorbed forms of formaldehyde, as observed on many metal oxides. Formate species form from the evolution of methoxy and dioxymethylene species, and are also clearly multiple. These spectra allow us to distinguish species which are adsorbed on the active sites and are very reactive, with respect to those interacting

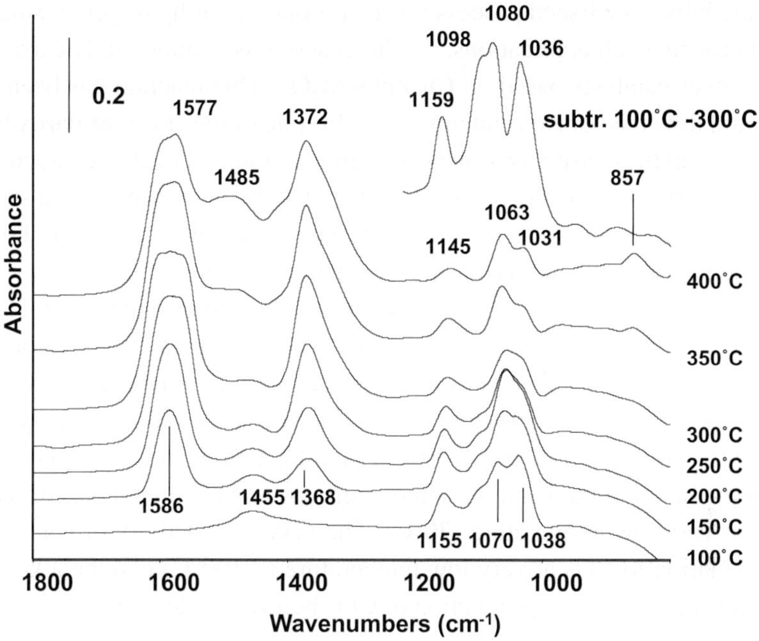


**Figure 18.10.** FTIR spectra of adsorbed species arising from contact of 25% $MoO_3/Zr_{0.75}Ce_{0.25}O_2$ catalyst with methanol vapour at increasing temperatures.

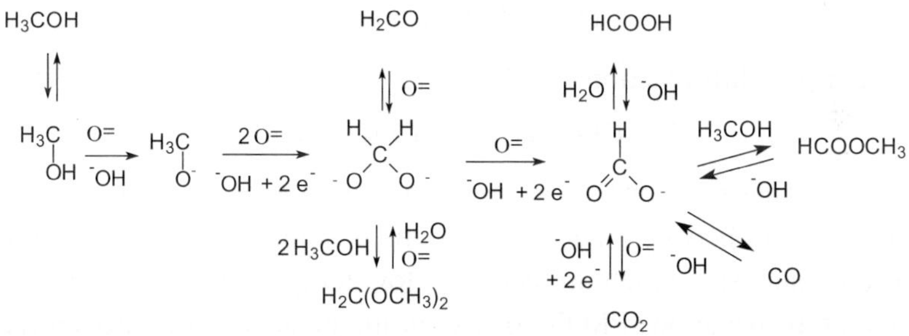


**Figure 18.11.** Reaction mechanism for methanol conversion on oxidising surfaces.

with the support and acting as spectators, as well as the dioxymethylene key intermediate.

The data discussed above allowed us to propose the reaction scheme reported in Fig. 18.11, previously proposed as a generalized mechanism for methanol oxidation over metal oxides.[103]

The steam reforming of methanol

$$CH_3OH + H_2O \rightarrow CO_2 + 3H_2$$

is an industrially established process for the production of hydrogen in small-scale plants and for fuel cell applications.[104] This reaction is commercially carried out at *ca.* 350°C over catalysts based on Cu-ZnO-$Al_2O_3$. This reaction has been studied using IR and flow reactor experiments.[105,106] In spite of the close similarity between the catalysts and those used for water-gas shift reactions, a peculiar characteristic of this process is that $CO_2$ is the primary product, CO being produced as a secondary product by the retro-water-gas shift reaction. The IR spectra show a very high reactivity of methanol on the oxidised catalyst, where the evolution is closely related to that observed over combustion catalysts such as CuO, depicted in Fig. 18.11. On the contrary, the hydrogen reduced catalyst seems to be almost inactive in adsorbing and transforming methanol. This agrees with surface science studies performed on bare and oxidised Cu(110) monocrystal faces.[107] However, the reduced catalyst can be "reoxidised" with water vapour at 250°C. After this treatment it recovers the ability to adsorb methanol in the form of methoxy groups and to convert part of them into formates at very low temperature (30°C). The very high activity of $Cu^{2+}$ sites for the CO oxidation to $CO_2$ at very low temperature (−100°C) may be a key feature of the catalyst allowing the production of $CO_2$ before CO desorption.[105,106]

### 18.6.2. *Low temperature and preferential CO oxidation and water-gas shift*

The oxidation of CO

$$CO + 1/2\,O_2 \rightarrow CO_2$$

and the water-gas shift reaction

$$CO + H_2O \rightarrow CO_2 + H_2$$

are formally similar reactions where the oxidant for CO is either oxygen or air.

The parallelism between the two reactions is somehow supported by the similarity of the catalysts, which are active at low temperature in the two reactions. The preferential oxidation (PROX) of CO in the presence of hydrogen represents an important step in the synthesis of pure hydrogen for application in low temperature fuel cells.[108] Noble metals are active for this reaction at very low temperature (100–200°C). Commercial catalysts are typically based on Pt/$\gamma$-$Al_2O_3$. Supported gold was quite recently discovered to act as an excellent catalyst for PROX as well as for the low temperature (200 K) CO oxidation in waste gases.[109,110] Copper-based catalysts are also reported to be excellent for the PROX reaction.[111]

The low temperature water-gas shift (LTWGS) catalysts commercially used for hydrogen and ammonia syntheses processes are based on Cu-ZnO-$Al_2O_3$. These catalysts work at 185–275°C.[112] Non-pyrophoric and very active LTWGS catalysts, as needed for fuel cell applications, are based on supported noble metals, like Pt/$CeO_2$-$Al_2O_3$,[113] in monolytic structures. Catalysts based on gold, such as

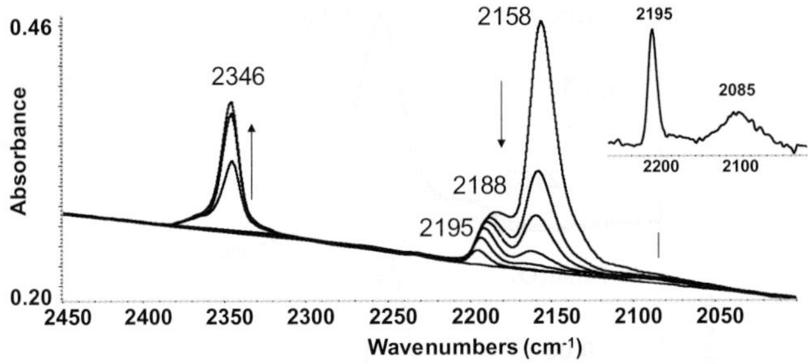


**Figure 18.12.** FTIR spectra of pre-reduced $Pt/Al_2O_3$ at −140°C after admission of CO gas into the cell until saturation and upon outgassing at increasing temperatures from −140 to −90°C (lower spectra). In the insert: the spectrum of residual CO adsorbed after outgassing at −90°C.

$Au\text{-}Fe_2O_3$, $Au\text{-}CeO_2$, and $Au\text{-}TiO_2$, find high activity and commercial interest for LTWGS.[114]

Infrared spectra of CO adsorption at low temperature provides evidence of the very high activity of oxidised noble metal centres in the absence of gaseous oxygen. In Fig. 18.12 the spectra of carbon monoxide adsorbed on an unreduced 2% $Pt/\gamma\text{-}Al_2O_3$ catalyst are reported. The spectra were recorded after adsorption at −140°C (130 K) during outgassing upon warming. At the highest coverage the main maximum is at 2,158 $cm^{-1}$, due to CO interacting with the alumina OH groups. A weak tail at lower frequency is observed on the spectrum recorded during adsorption, which is more clearly evident upon desorption as a single band centred at 2,085 $cm^{-1}$. This absorption is typical for terminal carbonyls on Pt° particles whose average size is of the order of a few nanometers.[115] A band more resistant to outgassing is observed at higher frequencies, shifting from near 2,188 $cm^{-1}$ to 2,195 $cm^{-1}$ upon outgassing. This absorption is in the region of bands typically assigned to CO species interacting with Lewis acidic $Al^{3+}$ cations (Table 18.1). However, upon outgassing and warming at very low temperatures (*ca.* −100°C), a band also grows at 2,346 $cm^{-1}$, due to OCO asymmetric stretching of adsorbed $CO_2$. This shows that unreduced Pt centres exist and are able to oxidise CO to $CO_2$.

In Fig. 18.13 the spectra of CO adsorbed on $Al_2O_3$, unreduced $Pt/Al_2O_3$, and reduced $Pt/Al_2O_3$ (all activated at 350°C) and their deconvolutions are reported. The deconvolution of the spectrum of CO adsorbed on alumina allows only the separation of the component due to CO interacting with $Al^{3+}$, found at 2,183 $cm^{-1}$, from that due to CO H-bonded on OHs, found at 2,152 $cm^{-1}$. In the case of the spectrum of CO adsorbed on unreduced $Pt/Al_2O_3$, the higher frequency band increases in intensity and shifts up to 2,186 $cm^{-1}$, while a new absorption appears clearly centred at 2,135 $cm^{-1}$. In the case of reduced $Pt/Al_2O_3$ the higher frequency band shifts back down to 2,183 $cm^{-1}$ and decreases in intensity, while the lower frequency component increases in intensity and shifts up to 2,139 $cm^{-1}$. These data suggest that highly

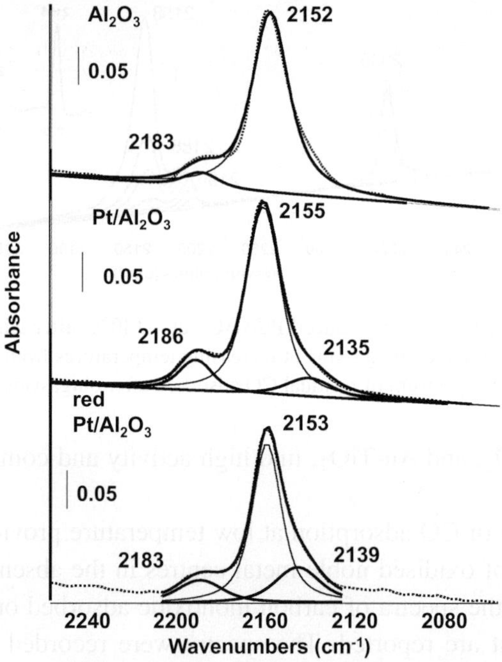


**Figure 18.13.** FTIR spectra of CO adsorbed on $Al_2O_3$, on unreduced and pre-reduced $Pt/Al_2O_3$ at −140°C at saturation (pointed lines), and deconvolution of the spectra.

oxidised and oxidising Pt ions contribute to the band at 2,186 $cm^{-1}$, and are reduced in the reduced sample. We assign the band at 2,186 $cm^{-1}$ to $Pt^{n+}$-CO (likely n = 4). The band at 2,135–2,139 $cm^{-1}$ should be assigned to another form of $Pt^{n+}$-CO, less oxidised and less of an oxidant, likely with n = 2. The amount of the latter species should increase by reduction of the former one. The real existence of such sites is evident upon low temperature CO adsorption experiments,[60,116] but cannot be detected (as is quite obvious indeed) by CO oxidation *operando* studies[117] or room or higher temperature *in situ* adsorption studies.[118]

Similar results can be obtained with supported gold. A low temperature IR study of CO adsorption on partially reduced $Au/Nb_2O_5$[119] (Fig. 18.14) showed CO interacting with the corners of gold metal nanoparticles (CO stretching band at *ca.* 2,134 $cm^{-1}$ and at 2,105–2,120 $cm^{-1}$, terminal species, and 1,938 $cm^{-1}$, bridging species), but also CO interacting with $Au^{3+}$ centres (band shifting from 2,160 to 2,188 $cm^{-1}$) and the oxidation of CO to $CO_2$ at *ca.* −100°C (band at 2,347–2,357 $cm^{-1}$). A similar situation was found for gold on alumina, where CO interacting with gold centres was difficult to identify, but the oxidising effect of cationic gold on CO was evident by the formation of $CO_2$ at *ca.* −100°C.

The detection of very strongly oxidising $Cu^{2+}$ sites has also been obtained using low temperature IR experiments on bulk CuO,[94] on $Cu/Al_2O_3$,[120] $Cu\text{-}ZnO\text{-}Al_2O_3$

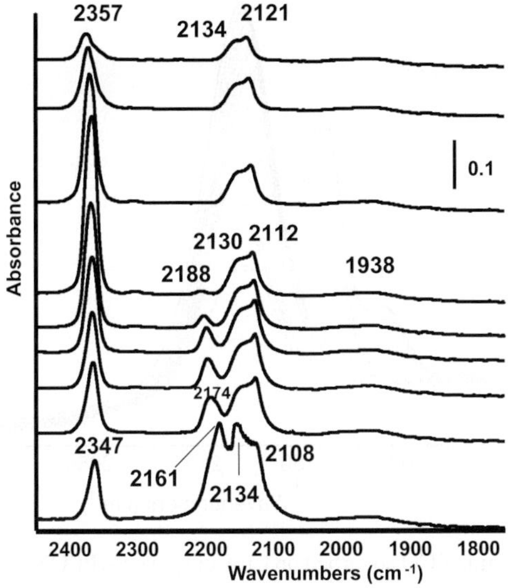


**Figure 18.14.** FTIR spectra of CO adsorbed on $Au/Nb_2O_5$ reduced in hydrogen and outgassed at 500°C and after CO adsorption at −140°C (bottom) and following outgassing upon warming from −140 to 0°C (from bottom to top).

methanol steam reforming and water-gas shift (WGS) catalysts,[106] Cu-ZnO-$TiO_2$,[121] and Cu-$CeO_2$-$Al_2O_3$ PROX catalysts.[111] These sites too are able to oxidise CO to $CO_2$ at *ca.* −100°C. Indeed, the spectra of the surface carbonyls in these systems is very complex, as evident in Fig. 18.15. The low temperature spectrum of CO over an unreduced Cu-$CeO_2$-$Al_2O_3$ PROX catalyst contains features due to CO interacting with $Al^{3+}$ (labile, 2,200–2,170 $cm^{-1}$), $Ce^{4+}$ (more labile, 2,170–2,150 $cm^{-1}$), $Cu^{2+}$ (weak, very labile, 2,160–2,140 $cm^{-1}$), $Cu^+$ (very stable monocarbonyls, 2,120–2,100 $cm^{-1}$; labile dicarbonyls, two bands in the 2,150–2,100 $cm^{-1}$ region), and Cu° (extremely labile, < 2,100 $cm^{-1}$). This produces, during outgassing upon warming at −140–0°C, a very complex behaviour with the main band shifting first up, then down, then up again (Fig. 18.15).

In PROX reaction conditions only a resistant band, although complex, in the region of $Cu^+$ monocarbonyls, is observed.[122–124] However, the intensity of this band correlates with the selective PROX activity. This suggested to the authors that CO and $H_2$ oxidation would occur on the same copper sites and that strong adsorption of CO prevents adsorption and oxidation of $H_2$ on the catalyst.[124] According to our interpretation, however, the possibility of CO oxidation occurring on $Cu^{2+}$,which cannot be revealed in *operando* conditions, should also be considered.

According to the literature, two different mechanisms may justify the WGS reaction over Cu-Zn-Al catalysts: (i) the redox mechanism, where water oxidises the catalyst producing hydrogen and CO re-reduces it producing $CO_2$; and (ii) the

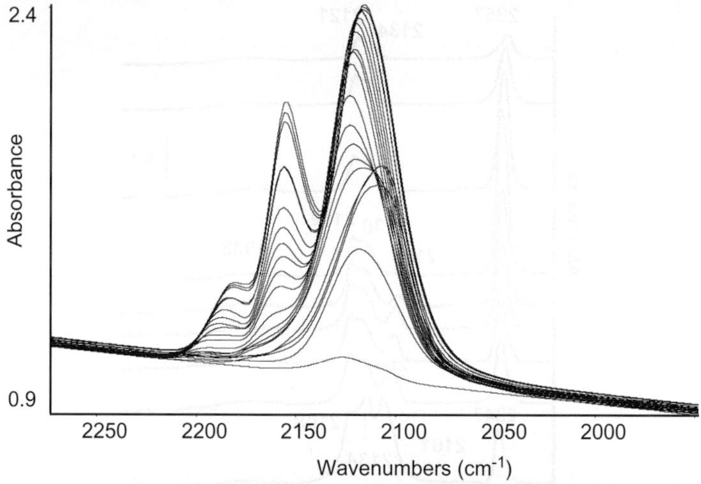


**Figure 18.15.** FTIR spectra of CO adsorbed on Cu-$CeO_2$-$Al_2O_3$PROX catalyst after CO adsorption at −140°C (top) and following outgassing upon warming from −140 to 0°C (from top to bottom).

"associative" mechanism through the formate species, where water hydrates the surface producing active OH species.

The detection of this high CO oxidation activity for Cu-based catalysts activated in water supports the "redox" mechanism for LTWGS over copper catalysts, although the mechanism via formate is also evident and might be active at higher temperatures.[106] *Operando* studies have been performed on the reaction mechanism of LTWGS over supported Pt and Au catalysts by Meunier *et al.*[125,126] Also, Kalamaras *et al.*[127] in their SSITKA-DRIFTS (steady-state isotopic transient kinetic analysis coupled with Diffuse Reflectance Infrared Fourier Transform spectroscopy) experiments concluded that formate are inactive spectators upon WGS reaction on Pt/$TiO_2$ where a redox mechanism would be active. These studies tend to exclude the role of surface formates as intermediates. However, it must be remarked that two kinds of formates may exist over these catalysts. While those on the support surface area are inactive (they are spectators) and may tend to accumulate, thus having strong band intensites, those interacting with the active metal possibly having a role in the reaction should be labile and detectable with much difficulty. *Operando* studies may not allow the detection of the two different formate species, while *in situ* studies may allow the trapping of the active species when they are still not reactive, allowing their distinction from spectator species.[106,128]

### 18.6.3. *Selective oxidation of methylaromatic hydrocarbons*

The synthesis of phthalic anhydride (a precursor of phthalate esters, largely used as lubricants and plasticizers) is performed industrially over vanadia catalysts (4–10%

$V_2O_5$ wt/wt) supported on titania (anatase polymorph) with a surface area of 6–25 $m^2/g$, alkali ions (K, Rb, Cs), with Sb and P playing the role of promoters.[129] The temperature in the bed is 360–450°C. Although this process has been well established for some decades, improvements are needed from several points of view, with the need, in particular, of improving catalyst selectivity.[129–131] Although the content of vanadium in industrial catalysts is generally quite high, it has been shown that sub-monolayer $V_2O_5$ on $TiO_2$ catalysts may also have good activity and selectivity if the support area is sufficiently deactivated.[132] This catalytic system has been thoroughly characterised by IR and Raman spectroscopy.[39,40,42]

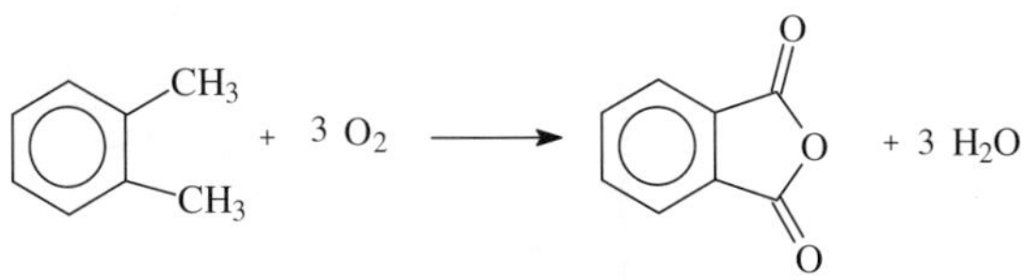


Reactions in similar conditions allow the syntheses of aromatic anhydrides and of aromatic nitriles by oxidation and ammoxidation of toluenes and xylenes over vanadia-based catalysts such as $V_2O_5/TiO_2$ or $V_2O_5/Al_2O_3$.

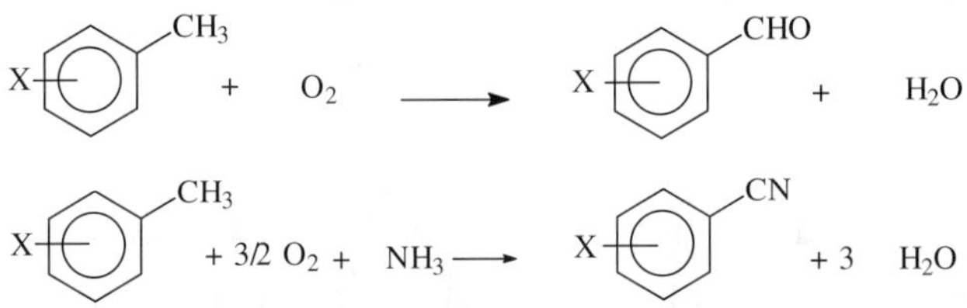


Infrared spectroscopy has been applied to these reactions at the surface of $V_2O_5/TiO_2$ monolayer catalysts. The adsorption of toluene, *ortho*-, *meta*-, and *para*-xylene on vanadia-titania allowed the detection of two different adsorbed species (Fig. 18.16 for toluene and ortho-xylene). The spectrum of one of these species, completely desorbed by outgassing in the temperature range 30–100°C (dashed lines in Fig. 18.16), is that of the reactant species very weakly perturbed. The IR spectra of the other species, resistant to evacuation at 100°C (full lines in Fig. 18.16), are similar in the four cases showing that they are due to closely related entities. They are certainly due to hydrocarbon species, as bands due to metal-oxygen bonds were not found. Further reaction of these species with the catalyst surface, in the presence or in the absence of oxygen, gives rise to spectra indentical to those of adsorbed side-chain oxygenated compounds like benzaldehyde and tolualdehydes (characterised by C=O stretching at *ca.* 1,635 $cm^{-1}$), and adsorbed phthalic anhydride from *o*-xylene (bands at 1,860, 1,790 $cm^{-1}$, C=O stretchings, and at 1,255 $cm^{-1}$, C–O stretching, Fig. 18.17). A detailed analysis of the IR spectra allowed their assignment to benzyl species.[133] As discussed by Gimeno *et al.*,[134] the detail of the

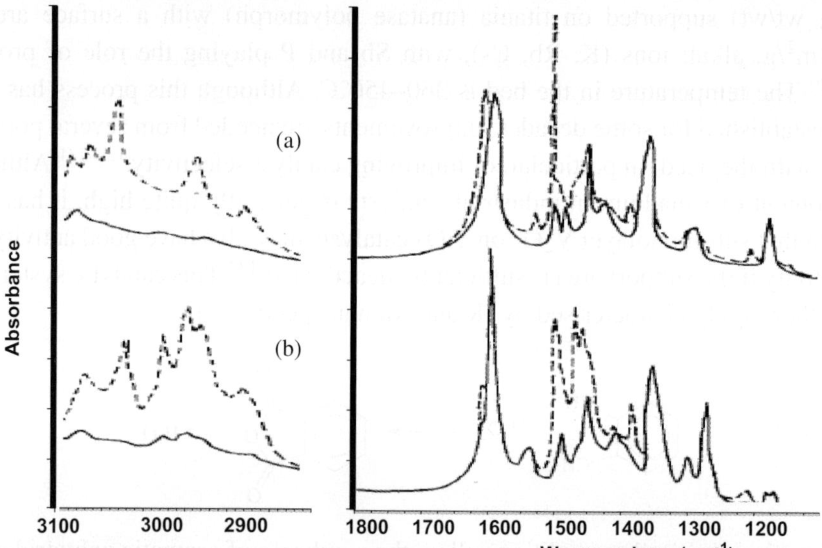


**Figure 18.16.** FTIR spectra of the adsorbed species arising from adsorption of toluene (a) and *o*-xylene (b) over $V_2O_5$-$TiO_2$. Broken lines: in contact with the vapours at room temperature; full lines: after outgassing at 623°C. Reprinted with permission of Elsevier.

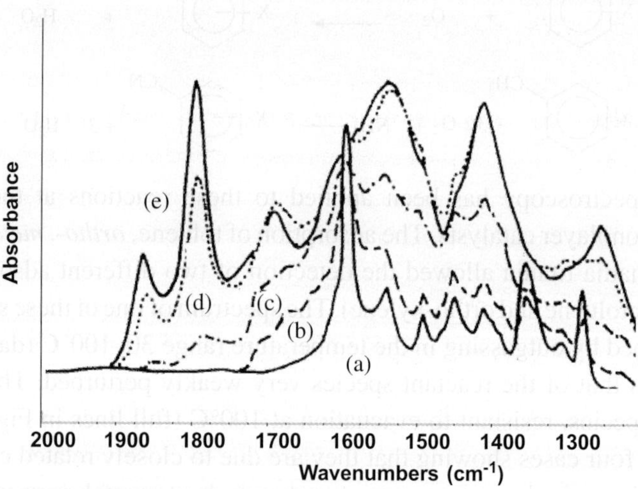


**Figure 18.17.** FTIR spectra of the adsorbed species arising from contact of the $V_2O_5$-$TiO_2$ catalyst with *o*-xylene/oxygen at (a) room temperature, (b) 100°C, (c) 150°C, (d) 200°C, and (e) 250°C.

existence of a previous activation step of *o*-xylene should be introduced in the kinetic model to improve its reliability. IR spectra also show the formation of other species such as phthalide from *o*-xylene, and maleic and citraconic anhydrides, as well as of surface carboxylate species from all alkylaromatics. The data support the idea that the slow step consists of the further reaction of the adsorbed activated hydrocarbon (i.e. the benzyl species), desorption of products (aldehydes and anhydride) also being a slow step. In addition, the*o*-xylene oxidation, the main path producing maleic anhydride, is competitive (although slow) with that giving phthalic anhydride, and the side-chain oxidation products have a minor role in maleic anhydride production. Our data also provides evidence for the competition of the main route to carbon oxides, attributed to decarbonylation of *o*-toluate anions and combustion of the resulting aryl species.[135]

The reaction of the ammoxidation of toluene has also been investigated.[136] Ammonia adsorbs over the $V_2O_5/TiO_2$ monolayer catalysts in two forms, as ammonium ions over Brønsted acidic sites as well as in coordinated molecular form over Lewis acid sites. Benzylamine appears as an intermediate, likely formed through the reaction of amide species, which is a common intermediate in ammonia oxidation over oxide catalysts[137] and benzyl species. Benzylamine is further oxidatively dehydrogenated to benzonitrile, well evident in the IR spectra. A later IR study of the same reaction performed over V-P oxides showed similar intermediates but suggested that, over these catalysts, the main reaction path consists of the formation of benzaldehyde and benzimine.[138]

### 18.6.4. *Selective oxidative dehydrogenation of n-butenes to 1,3-butadiene*

The oxidative dehydrogenation of *n*-butenes to butadiene has been performed industrially for decades (Oxo-D process developed by Petro-Tex[139]). The reaction is carried out at 550–600°C in the presence of oxygen or air and steam, in an adiabatic fixed-bed reactor; 65% conversion and a butadiene selectivity of 93% are obtained. The catalysts are based on ferrite spinels, such as $MgFe_2O_4$ and $ZnFe_2O_4$ sometimes also containing $Cr_2O_3$.[140] Mixed $Fe_2O_3$-$Cr_2O_3$ catalysts have also been reported as active catalysts for this reaction.[141]

IR studies have enabled the revelation of the allylic activation mechanism, which produces the allyloxy species by room temperature adsorption of linear butenes over both $MgFe_2O_4$[142, 143] and mixed $Fe_2O_3$-$Cr_2O_3$ catalysts.[144] Strong bands appear, in fact, in the range 1,200± 1,000 $cm^{-1}$, namely at 1,189, 1,153, 1,089 and 1,056 $cm^{-1}$ over $Fe_2O_3$-$Cr_2O_3$, which closely correspond to those arising from the but-3-en-2-ol adsorbed species, typically due to the C–O and C–C stretchings of alkoxide species. It is straightforward to conclude that but-3-en-2-oxide species (or their isomeric form but-2-en-1-olate) are formed on the catalysts from both the corresponding alcohol and all three *n*-butenes. Bands due to adsorbed methyl-vinyl ketone are also formed. The disappearance of the bands of alkyloxyspecies has been correlated to the appearance of gas-phase butadiene.[145] On the other hand, a typical feature of these catalysts is the weak adsorption of 1,3-butadiene, which is consistent with its desorption as an intact product without significant further over-oxidation. This differentiates these catalysts from those based on vanadia (where the preferential final product is maleic anhydride), which strongly adsorb 1,3-butadiene.[146,147] This suggests that the evolution of an allyl-alkoxide species may represent the selectivity-determining step in the selective oxidation of butenes. The IR study allowed the development of a reaction mechanism which has been used recently for kinetic modelling[148] and the development of a new process based on Mn ferrite catalysts.[149]

### 18.6.5. *Total oxidation of oxygenated VOCs over Mn oxide-based catalysts*

In spite of their lower combustion activity with respect to noble metal-based catalysts, base metal-based catalysts, such as $MnOx/Al_2O_3$,[150] are commercially used for the catalytic combustion of oxygenated VOCs. Manganese-based catalysts, such as unsupported $Mn_3O_4$,[151,152] as well as Mn oxides supported on carriers such as alumina,[153] titania,[154] and zirconia,[155] have been investigated in our lab as combustion catalysts for alcohols. These materials may act as very selective catalysts for alcohol selective dehydrogenation to the corresponding carbonyl compounds, but convert into total combustion catalysts very selective towards $CO_2$ at higher temperatures (*ca.* 300°C) and contact times. *In situ* IR studies strongly suggest that the sequence alkoxide-carbonyl compound-carboxylate is active in catalytic combustion, where C–C bond breaking occurs through previous enolyzation.

In the case of alcohol conversion, the observed behaviour suggests that the evolution of acetates (either formed by direct ethanol oxidation or by C–C breaking in the case of higher alcohol oxidation) may be the rate-determining step of the catalytic combustion reaction. Two types of acetates are instead found over supported catalysts. Some of them essentially reflect the catalytic activity of manganese oxides and are burnt fast, while others behave as those on bare support such as alumina, i.e. inactive in the catalytic oxidation. Part of the acetate species in the case of Mn-alumina catalysts are consequently inactive spectators during the combustion reaction.

### 18.6.6. *Adsorption, conversion, and total oxidation of chlorinated VOCs over oxide catalysts*

Chlorinated organic compounds may be converted and dechlorinated using oxide heterogeneous catalysts. Acidic solids such as silica-alumina are active for dehydrochlorination of chloroalkanes in mild conditions to the corresponding olefins.[156–158] IR spectroscopy shows that this occurs through a previous nucleophylic substitution step giving rise to adsorbed alkoxide species.

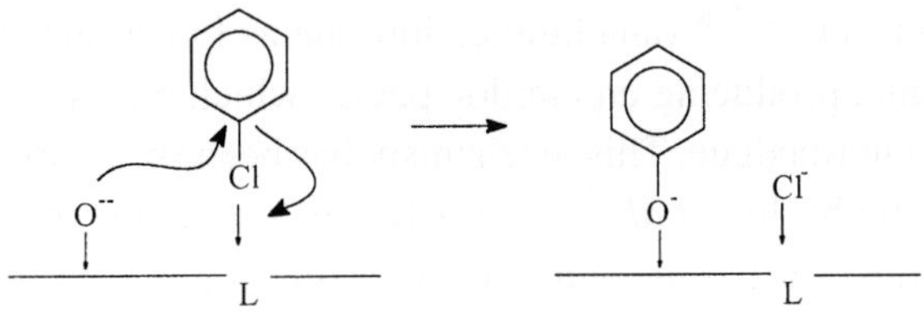


**Figure 18.18.**

Vanadia-titania based catalysts, additionally containing Mo or W oxides, are largely applied for the selective catalytic reduction (SCR) of nitrogen oxides by ammonia to limit emissions of NOx from industrial furnaces as well as diesel engines. According to several authors, such catalysts also allow the simultaneous abatement of dioxins and $NO_x$ in effluents, e.g. from waste incinerators, to limits below those required by EU directives.[159] Among industrial applications, vanadium-rich $V_2O_5$-$WO_3$-$TiO_2$ catalysts (8% vanadium) are reported[160] to represent the catalyst component of Remedia® catalytic filters (W.L. Gore) working at 200°C.

Infrared and catalytic studies[161–163] reveal evidence of a strong adsorption of such compounds associated with fast nucleophilic substitution, giving rise to surface-oxygen-containing species, i.e. phenate species, as the first step of the oxidation reaction. After this step the reaction occurs as on phenol[164] and chlorophenols,[165] as well as benzene itself,[165] with the possibility of the formation of adsorbed intermediates where the aromatic ring is partially broken, such as maleic anhydride and maleate species.

### 18.6.7. *Selective catalytic reduction of NOx by ammonia ($NH_3$-SCR) and selective catalytic oxidation (SCO) of ammonia*

DeNOxing of waste gases from stationary sources can be achieved efficiently by using the so-called SCR process, i.e. the selective catalytic reduction using ammonia as the reductant[166]

$$4NH_3 + 4NO + O_2 \rightarrow 4N_2 + 6H_2O$$

Industrial catalysts are constituted of $V_2O_5$-$WO_3/TiO_2$ or $V_2O_5$-$MoO_3/TiO_2$ monoliths. $TiO_2$ in the anatase form supports a "monolayer" of $V_2O_5$ and $WO_3$ (or $MoO_3$) deposited by impregnation. In general, the overall surface area of the catalysts is 50–100 $m^2/g$, with a $V_2O_5$ virtual content of 0.5–3% w/w and $MoO_3$, or a $WO_3$ content of 8–12% w/w. A typical reaction temperature is around 350°C. The chemical and mechanistic aspects of the ammonia-SCR process were reviewed some years ago.[167] All authors agree that ammonia is strongly adsorbed and activated on the catalysts, later reacting with gas-phase or weakly adsorbed NO. According to the

IR studies of Ramis *et al.*,[168] vanadium cations act as Lewis sites in adsorbing and activating ammonia, producing an amido-species which reacts with NO through a nitrosamide-like intermediate. This mechanism has been successfully verified using computer modelling by Jug *et al.*[169] Alternatively, a study by Topsøe *et al.*,[170] also based on IR studies, shifted the same mechanism proposed by Ramis *et al.* over the Brønsted sites. These authors suggest that ammonia may be protonated over Brønsted acid sites and converted into $NH_3$ radical cations that can later react with NO. This mechanism has also been studied by computational methods and found to be possible.[171,172] The two mechanisms have been compared by Vittadini *et al.*[173] These authors concluded that the reaction may occur on Lewis sites or on Brønsted sites, but without ammonia protonation. Ammonia may first form a hydrogen bond and then give rise to a nitrosamide species. According to these authors, mechanisms involving both Lewis-bonded and hydrogen-bonded $NH_3$ isoenergetic species are viable, which suggests that structural requirements for the active sites could be rather loose. In any case, the ammonia activation step in this high-temperature SCR process is certainly crucial.

Due to its high potential for $NO_x$ reduction under lean conditions, modification of the SCR process (using non-toxic urea instead of ammonia) is presently under investigation for mobile applications.[174] In this case there are limitations to the catalyst volume that can be placed on-board. The operating conditions are also continuously changing in temperature and flow rate. Finally, the temperature window has to be enlarged with respect to stationary applications, in particular towards the low temperature region.

In relation to the last most challenging issue, it has been proposed that the so-called fast SCR reaction be used

$$2NH_3 + NO + NO_2 \rightarrow 2N_2 + 3H_2O$$

in which $NO_2$ essentially substitutes for oxygen in the catalyst reoxidation step, allowing a faster overall reaction rate thus being applicable at much lower temperatures. This reaction is catalyzed efficiently by several materials such as the same $V_2O_5$-$WO_3$/ $TiO_2$ catalyst, iron-exchanged zeolites and also purely acidic zeolites like H-ZSM5.

In Fig. 18.19 the spectra relative to the adsorption of $NO_2$ over an ammonia-covered $V_2O_5$-$WO_3$/$TiO_2$ commercial catalyst are reported. Ammonia adsorption (Fig. 18.19a) occurs in two modes, as ammonium ions on Brønsted acidic sites (broad bands at 3,020, 2,800 $cm^{-1}$, $NH_4$ stretchings, peaks at 1,668 and 1,444 $cm^{-1}$, $NH_4$ asymmetric deformations), and in the coordinated molecular form over Lewis sites ($NH_3$ stretchings at 3,395, 3,320, 3,250 and 3,168 $cm^{-1}$, $NH_3$ deformation at 1,608 $cm^{-1}$). After contact with $NO_2$ gas, the spectrum is significantly modified (Fig. 18.19b): this is evident as the sharp peaks at 3,400–3,200 $cm^{-1}$ disappear,

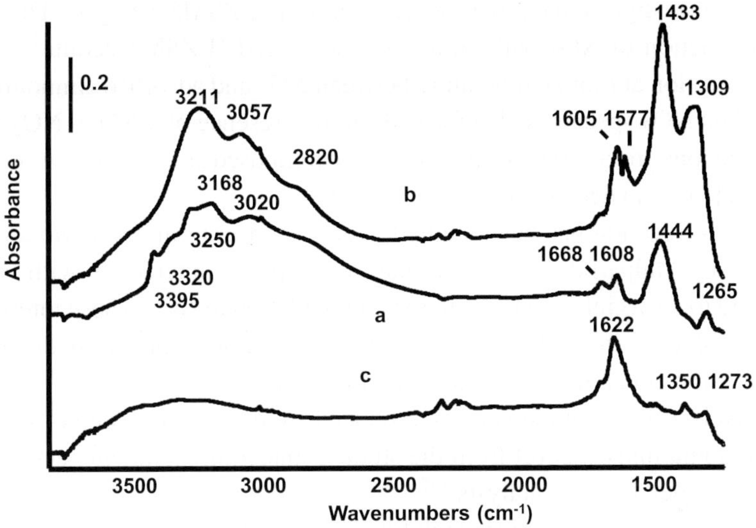


**Figure 18.19.** FTIR spectra of the adsorbed species arising from contact of a commercial automotive $V_2O_5$-$WO_3$-$TiO_2$ catalyst with ammonia (a), and further adsorption of $NO_2$ at (b) room temperature and (c) 100°C.

showing that molecularly-adsorbed ammonia has disappeared, being either desorbed or transformed. However, new bands develop, both in the region 3,300–2,800 $cm^{-1}$ and 1,500–1,200 $cm^{-1}$. The spectrum of the adsorbed species after coadsorption of $NO_2$ and ammonia (Fig. 18.19b) is fully consistent with that of ammonium nitrate $NH_4NO_3$ in its different forms.[175,176] The broad bands at 3,210, 3,049, and 2,820 $cm^{-1}$ are due to the $NH_4$ stretchings, the peak at 1,433 $cm^{-1}$ is due to the $NH_4$ asymmetric deformation, the band at 1,309 $cm^{-1}$ is due to the asymmetric $NO_3$ of trigonal nitrate anions. The weak peaks in the region 1,610–1,500 $cm^{-1}$ are due to other ammonia or coordinated nitrate species.

During heating to 100°C, the spectrum is strongly transformed. Bands typical of ammonia species and of adsorbed NOx species have fully disappeared. The residual species are mainly characterised by a very broad absorption in the region 3,600–3,000 $cm^{-1}$ and a sharp peak at 1,620 $cm^{-1}$, which are assigned to adsorbed water (OH stretchings and $H_2O$ scissoring mode). Weak peaks at 2,288, 2,233, and 1,350 $cm^{-1}$ could be due to N-bonded and O-bonded $N_2O$ adsorbed species.[177] These data show that ammonia reacts completely with $NO_2$ producing gas-phase $N_2$ (not detectable by IR) and partially adsorbed $H_2O$ in the range 25–100°C. The reaction occurs in two steps, i) the formation of ammonium nitrate; and ii) the decomposition of ammonium nitrate. This reveals the completely different mechanism occurring at low temperatures in the presence of $NO_2$ with respect to that occurring at high temperatures where the reactant is NO.

We also investigated the fast SCR reaction over H-ZSM5 catalysts. The IR study of the interaction of $NO_2$ with an ammonia-covered H-ZSM5 zeolite[178] shows a very fast reaction at room temperature between $NO_2$ and adsorbed ammonium ions. An adsorbed $NO_2$ species is also formed. In the presence of a $NO + NO_2$ mixture, ammonium ions are completely and rapidly decomposed at 30–100°C. A nitrate-like intermediate is also detectable.

The same reaction has been studied over Fe-ferrierite in *operando* and *in situ* conditions by Malpartida *et al.*[72] Mononitrosyl species interacting with $Fe^{2+}$ ions were found to act as intermediates in the NO oxidation to $NO_2$. These intermediates are also involved in the SCR reaction, although the authors apparently did not arrive at a conclusion concerning the active ammonia intermediates.

The likely role of amide species produced by oxidation of Lewis-site coordinated ammonia finds support from the studies concerning ammonia activation and oxidation on metal oxide catalysts.[137,179–183]

In recent years, the selective catalytic oxidation (SCO) of ammonia to nitrogen

$$2NH_3 + 3/2\,O_2 \rightarrow N_2 + 3H_2O$$

has become an important reaction to limit the ammonia slip from processes such as the SCR of NOx by ammonia, in the treatment of waste gases from power stations as well as in diesel engine post-treatment. This reaction is of interest also for the combustion of biomass-derived gases. Several catalysts have been studied in the academic research, containing base and noble metal catalysts. Several patents refer to catalysts containing Ni, Cu, Co, or Fe, with small amounts of precious metals, over an oxide or a zeolite support. Among those, catalysts such as $CuO\text{-}TiO_2$ and $MnO_x\text{-}CuO\text{-}TiO_2$[184] have been proposed for the SCO.

We investigated the chemistry of several nitrogen compounds, such as ammonia $NH_3$, hydrazine $NH_2NH_2$, hydroxylamine $NH_2OH$, urea $H_2NCONH_2$, and the oxides $N_2O$, NO, and $NO_2$ over several catalysts. All catalysts able to oxidise ammonia show Lewis acidity but not necessarily Brønsted acidity. A very weak band near $1{,}550\,cm^{-1}$ was observed over $V_2O_5\text{-}TiO_2$ and $Fe_2O_3\text{-}TiO_2$ catalysts and assigned to the $NH_2$ amide species.[168,182] IR spectra show the formation of adsorbed hydrazine as an intermediate species in the ammonia oxidative conversion to nitrogen,[137,181] likely formed by the dimerization of the amide species. Other species tentatively identified as imido species, NH, nitroxyl species, HNO, and nitrogen anions, $N_2^-$, are also observable. Species arising from hydrazine, such as $NH_2O$- and NHO easily convert to NO.

On Cu/Mg/Al mixed oxide catalysts, selective oxidation to $N_2$ predominates, likely via hydrazine and other N≡N triple bond containing species, as surface intermediates. IR experiments show that over Cu/Mg/Al mixed oxide catalysts the reactivity of ammonia to give hydrazine and other compounds is not found in the presence of NO gas. This is explained assuming that the earlier ammonia oxidation intermediate (likely to be amide $NH_2$) reacts faster with NO giving rise to nitrogen.[180]

The role of amide species as a key intermediate in ammonia oxidation found recent support in the surface study on $RuO_2$ 110 using HREELS, where a band at 1,500 $cm^{-1}$ was assigned to the amide species,[185] in agreement with our data. This species appears to be the key intermediate in the conversion of coordinated ammonia into NO over $RuO_2$. A similar mechanism was found for ammonia oxidation over other metal faces.[186]

### 18.6.8. *Selective catalytic reduction of NOx by methane ($CH_4$-SCR)*

The denitrification of waste gases can be obtained using methane as a selective reductant of NOx to nitrogen

$$CH_4 + 2NO + O_2 \rightarrow CO_2 + N_2 + 2H_2O$$

Co-containing zeolites, such as Co-MFI and Co-FER, were found by Armor[187] to be particularly active for this reaction. Using a parallel flow reactor, IR and UV-vis characterisation studies revealed a high complexity in active Co-FER, Co-MFI, and Co-MOR catalysts,[62, 188–191] with the presence of cobalt and protonic sites both in the inner cavity and on the outer surfaces. The joint use of NO as a cationic-sensitive probe and ortho-toluonitrile as a location-sensitive probe[191] allowed the confirmation of the existence of $Co^{3+}$-containing oxide aggregates in the internal surface which act as the active site for the reaction, in agreement with the conclusion of other authors.[192] The role of $NO_2$ gas as an intermediate in the $CH_4$-SCR reaction appears unlikely. The reaction has been investigated by IR *operando* technique.[191] The study shows the existence of trivalent cobalt nitrosyls during reaction conditions. Nitrate-like and isocyanate species were found to be likely sequential intermediates of $CH_4$-SCR. It is concluded that the active sites for $CH_4$-SCR in Co-MFI are trivalent cobalt species located in the cavities but likely in non-classical cation positions, in the form of small aggregates.

### 18.6.9. *Oxidation of benzothiophene by tert-butyl hydroperoxide over vanadia-alumina catalyst*

Oxidative desulfurization (ODS)[193] of benzothiophenes and dibenzothiophenes has proved to be a promising alternative to hydrotreatment used to remove sulfur from oil heavy fractions, due to its moderate process conditions and lack of hydrogen consumption. In this process, organosulfur compounds in fuels are oxidised to form sulfoxides and sulfones. Alumina-supported vanadium-oxide catalysts have been used along with hydrogen peroxide or tert-butyl hydroperoxide, obtaining excellent results in the desulfurization of commercial diesel.[194]

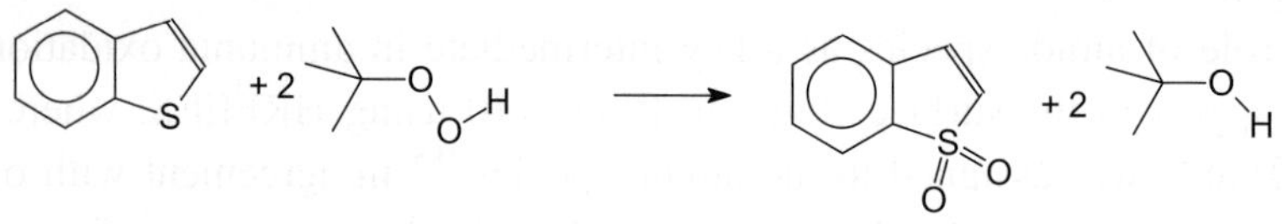


An IR study has been performed on the adsorption, from the vapour phase, and the coadsorption of tert-butyl hydroperoxide (TBHP) and benzothiophene (BT) over the active catalyst, alumina-supported vanadia.[195]

In Fig. 18.20a, the spectrum of the adsorbed species arising from contact of the catalyst with BT vapour (0.2 Torr) is shown. The spectrum is highly consistent with that of pure BT as well as of adsorbed BT which, on metal oxides, bonds through the sulfur atom on Lewis acid sites.[196] Adsorption of BT causes the perturbation of the surface vanadyl species, as evidenced by the negative band at 1,030 $cm^{-1}$ in the subtracted spectrum.

In Fig. 18.20b the spectrum obtained after adsorption of TBHP over the catalyst and outgassing is reported. This spectrum has been assigned to tert-butylperoxy species. Upon further coadsorption of BT, strong bands at 1,295 and 1,152 $cm^{-1}$

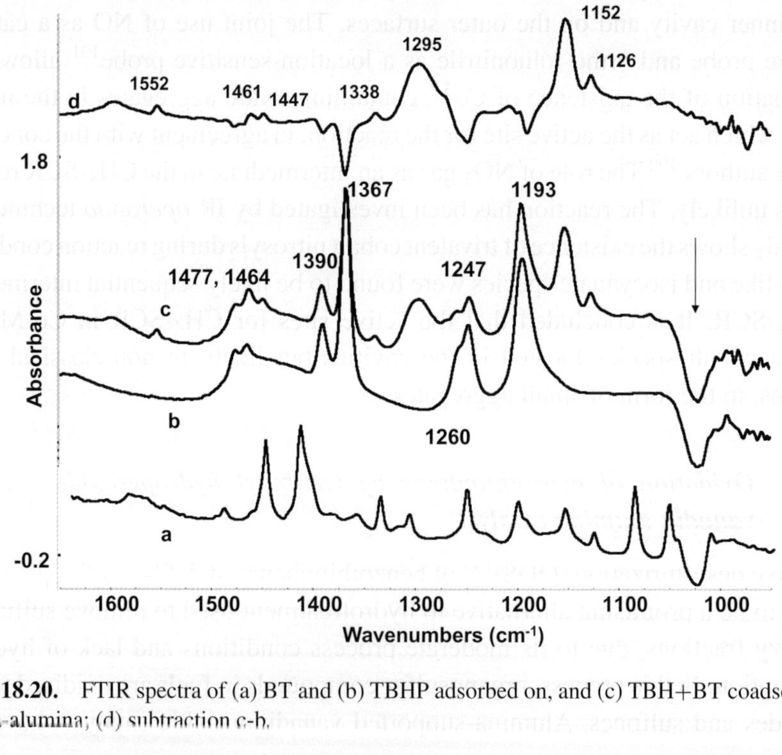


**Figure 18.20.** FTIR spectra of (a) BT and (b) TBHP adsorbed on, and (c) TBH+BT coadsorbed on vanadia-alumina; (d) subtraction c-b.

form. The overall spectrum formed (see the subtraction in Fig. 18.20 d) is clearly due to the sulfone of BT. In the same subtraction, the spectrum of tert-butylperoxide is observed as a negative feature.

These data show that adsorbed tert-butylperoxide species react at room temperature with BT, producing the adsorbed sulfone. No trace of an intermediate role for the sulfoxide species is observed. The sulfone species we observed here are clearly adsorbed on the catalyst surface, in our conditions. This is mostly due to the very low volatility of these heavy molecules near room temperature. However, under real reaction conditions at the catalyst/solution interface the sulfone is expected to desorb and dissolve onto the hydrocarbon mixture.

## Conclusions

Infrared spectroscopy represents one of the most useful techniques for the understanding of the basic chemical phenomena occurring during heterogeneous catalysis, in particular in the field of oxidation reactions. The recent development of spectroscopic techniques in *operando* conditions represents a further improvement, although it does not always fully enable the elucidation of reaction mechanisms. On the other hand, the joint use of several techniques, in parallel conditions, represents a better way for a more comprehensive understanding. In particular, the coupling of deep characterisation studies, of surface science and *operando* experiments, together with careful flow reactor experiments as well as thermal desorption and/or temperature programmed surface reaction techniques represents an optimal approach to mechanistic studies.

## References


1. Jones, R. (1985). Analytical applications of vibrational spectroscopy. An historical review in durig, in J. Durig (ed.), *Chemical, Biological and Industrial Applications of Infrared Spectroscopy*, Wiley, New York, pp. 1–50.
2. Buswell, A., Krebs, K. and Rodebush, W. (1937). Infrared studies. III. Absorption bands of hydrogels between 2.5 and 3.5 micrometers, *J. Am. Chem. Soc.*, **59**, pp. 2603–2605.
3. Terenin, A., Roev, L. (1959). Infra-red spectra of nitric oxide adsorbed on transition metals, their salts and oxides, *Spectrochim. Acta*, **11**, 946–957.
4. Eischens, R. and Pliskin, W. (1958). The infrared spectra of adsorbed molecules, *Advan. Catal. Relat. Subj.*, **10**, pp. 1–56.
5. Sheppard, N. and Yates, D. (1956). Changes in the infra-red spectra of molecules due to physical adsorption, *Proc. Roy Soc.*, **A238**, pp. 69–89.
6. Davydov, A. and Sheppard, N. (2003). *Molecular Spectroscopy of Oxide Catalyst Surfaces*, Wiley, New York.
7. Busca, G. (2009). Use of infrared spectroscopy methods in the field of heterogeneous catalysis by metal oxides, in S. Jackson and J. Hargreaves (eds.), *Metal Oxide Catalysis*, Vol. 1, Wiley-VCH, New York, pp. 95–175.

8. Ryczkowski, J. (2001). IR spectroscopy in catalysis, *Catal. Today*, **68**, pp. 263–381.
9. Fateley, W., Dollish, F., McDevitt, N., *et al.* (1972). *Infarared and Raman Selection Rules for Molecular and Lattice Vibrations*, Wiley, New York.
10. Busca, G. and Resini, C. (2000). Vibrational spectroscopy for the analysis of geological and inorganic materials, in R. Meyers (ed.), *Encyclopedia of Analytical Chemistry*, Wiley, Chichester, pp. 10984–11020.
11. Baraton, M., Busca, G., Prieto, M., *et al.* (1994). On the vibrational spectra and structure of $Fecro_3$ and of the ilmenite-type compounds $CoTiO_3$ and $NiTiO_3$, *J. Solid State Chem.*, **112**, pp. 9–14.
12. Busca, G., Buscaglia, V., Leoni, M., *et al.* (1994). Solid state and surface spectroscopic characterization of $BaTiO_3$ fine powders, *Chem. Mater.*, **6**, pp. 955–961.
13. Busca, G., Tittarelli, P., Forzatti, P., *et al.* (1987). Evidence for the formation of an anatase-type V-Ti oxide solid state solution, *J. Solid State Chem.*, **67**, pp. 91–97.
14. Notari, B. (1987). Synthesis and catalytic properties of titanium containing zeolites, in P. Grobet, J. Martins and P. Jacobs (eds.), *Innovation in Zeolite Materials Science*, Elsevier, Amsterdam, pp. 413–424.
15. Boccuti, M., Rao, K., Zecchina, A., *et al.* (1989). Spectroscopic characterization of silicalite and titanium-silicalite, in C. Morterra, A. Zecchina and G. Costa (eds.), *Structure and Reactivity of Surfaces*, Elsevier, Amsterdam, pp. 133–144.
16. Astorino, E., Peri, J., Willey, R., *et al.* (1995). Spectroscopic characterization of silicalite-1 and titanium silicalite-1, *J. Catal.*, **157**, pp. 482–500.
17. Resini, C., Montanari, T., Barattini, L., *et al.* (2009). Hydrogen production by ethanol steam reforming over Ni catalysts derived from hydrotalcite-like precursors. Catalyst characterization, catalytic activity and reaction path, *Appl.Catal. A: Gen.*, **355**, pp. 83–93.
18. Busca, G., Lorenzelli, V. and Sanchez Escribano, V. (1992). Preparation, solid-state characterization and surface chemistry of high-surface area $Ni_xAl_{2-2x}O_{3-2x}$ mixed oxides, *Chem. Mater.*, **4**, pp. 595–605.
19. Garcia Casado, P. and Rasines, I. (1984). The series of spinels Co3-*s*Al*s*O4 ($0 < s < 2$): Study of $Co_2AlO_4$, *J. Solid State Chem.*, **52**, pp. 187–190.
20. Knözinger, H. (1997). Infrared spectroscopy for the characterization of surface acidity and basicity, in G. Ertl, H. Knözinger and J. Weitkamp (eds.), *Handbook of Heterogeneous Catalysis*, Vol. 2, Wiley-VCH, Weinheim, pp 707–732.
21. Busca, G. (1998). Spectroscopic characterization of the acid properties of metal oxide catalysts, *Catal. Today*, **41**, pp. 191–206.
22. Busca, G. (1999). The surface acidity of solid oxides and its characterization by IR spectroscopic methods. An attempt at systematization, *Phys. Chem. Chem. Phys.*, **1**, pp. 723–736.
23. Lavalley, J. (1997). Infrared spectrometric studies of the surface basicity of metal oxides and zeolites using adsorbed probe molecules, *Catal. Today*, **27**, pp. 377–401.
24. Busca, G. (2005). The surface acidity and basicity of solid oxides and zeolites, in J. Fierro (ed.), *Metal Oxides: Chemistry and Applications*, CRC Press, Boca Raton FL, pp. 247–318.
25. Busca, G. (2007). Acid catalysts in the industrial hydrocarbon chemistry, *Chem. Rev.*, **107**, pp. 5366–5410.
26. Busca, G. (2010). Bases and basic materials in industrial and environmental chemistry. Liquid versus solid basicity, *Chem. Rev.*, **110**, pp. 2217–2249.
27. Busca, G., Finocchio, E., Ramis, G., *et al.* (1996). On the role of acidity in catalytic oxidation, *Catal. Today*, **32**, pp. 133–143.
28. Doornkamp, C. and Ponec, V. (2000). The universal character of the mars and van krebvelen mechanism, *J. Mol. Catal. A: Chem.*, **162**, pp. 19–32.
29. Lavalley, J. and Benaissa, M. (1984). IR spectroscopic evidence for surface vibrational modes upon dehydroxylation of alumina, *J. Chem. Soc. Chem. Commun.*, **14**, pp. 908–909.

30. Busca, G., Lorenzelli, V., Sanchez Escribano, V., *et al.* (1991). FTIR study of the surface properties of the spinels $NiAl_2O_4$ and $CoAl_2O_4$ in relation to those of transitional aluminas, *J. Catal.*, **131**, pp. 167–177.
31. Busca, G., Ramis, G. and Lorenzelli, V. (1989). FTIR study of the surface properties of polycrystalline vanadia, *J. Mol. Catal.*, **50**, pp. 231–240.
32. Armaroli, T., Busca, G., Carlini, C., *et al.* (2000). Acid sites characterization of niobium phosphate catalysts and their activity in fructose dehydration to 5-hydroxymethyl-2-furaldehyde, *J. Mol. Catal.*, **151**, pp. 233–243.
33. Ramis, G., Cristiani, C., Elmi, A., *et al.* (1990). Characterization of the surface properties of polycristalline tungsten trioxide, *J. Mol. Catal.*, **61**, pp. 319–331.
34. Hadjiivanov, K., and Busca, G. (1994). On the surface chemistry of oxidised and reduced chromia: An FTIR study, *Langmuir*, **10**, pp. 4534–4541.
35. Riva, A., Trifirò, F., Vaccari, A., *et al.* (1987). The promoting role of Cr and K in catalysts for high-pressure and high-temperature methanol and higher alcohol synthesis, *J.Chem. Soc., Faraday Trans. I*, **83**, pp. 2213–2225.
36. Busca, G., Trifirò, F. and Vaccari, A. (1990). Characterization and catalytic activity of cobalt-chromium mixed oxides, *Langmuir*, **6**, pp. 1440–1447.
37. Finocchio E., Busca G., Lorenzelli V., *et al.* (1995). The activation of hydrocarbon C–H bonds over transition metal oxide catalysts: A FTIR study of hydrocarbon catalytic combustion over $MgCr_2O_4$, *J. Catal.*, **151**, pp. 204–215.
38. Stencel, J., Makovsky, L., Diehl, J., *et al.* (1984). Structural study of surface tungstate in $WO_3$-$Al_2O_3$: Effects of hydration and $^{18}O$ exchange, *J. Raman Spectrosc.*, **15**, pp. 282–287.
39. Ramis, G., Cristiani, C., Forzatti, P., *et al.* (1990). On the consistency of data obtained from different techniques concerning the surface structure of vanadia-titania catalysts, *J. Catal.*, **124**, pp. 574–575.
40. Busca, G., (2002). Differentiation of mono-oxo and poly-oxo and of monomeric and polymeric vanadate, molybdate and wolframate species in metal oxide catalysts by IR and raman speotroscopies, *J. Raman Spectrosc.*, **33**, pp. 348–358.
41. Lee, E. and Wachs I. (2008). In situ raman spectroscopy of $SiO_2$-supported transition metal oxide catalysts: An isotopic 18O-16O exchange study, *J. Phys. Chem. C*, **112**, pp. 6487–6498.
42. Wachs, I. (2011). The generality of surface vanadium oxide phases in mixed oxide catalysts, *Appl. Catal. A: Gen.*, **391**, pp. 36–42.
43. Magg N., Immaraporn, B., Giorgi, J., *et al.* (2004). Vibrational spectra of alumina- and silica-supported vanadia revisited: An experimental and theoretical model catalyst study, *J. Catal.*, **226**, pp. 88–100.
44. Abbott, H., Uhl, A., Baron, M., *et al.* (2010). Relating methanol oxidation to the structure of ceria-supported vanadia monolayer catalysts, *J. Catal.*, **272**, pp. 82–91.
45. Davydov, A. and Sheppard, N. (2003). *Molecular Spectroscopy of Oxide Catalyst Surfaces*, Wiley, New York, pp. 44–55.
46. Green, I. and Yates Jr, J. (2010). Vibrational spectroscopic observation of weakly bound adsorbed molecular oxygen on powdered titanium dioxide, *J. Phys. Chem. C*, **114**, pp. 11924–11930.
47. Descorme, C, Madier, Y. and Duprez, D. (2000). Infrared study of oxygen adsorption and activation on cerium-zirconium mixed oxides, *J. Catal.*, **196**, pp. 167–173.
48. Giamello, E., Sojka, Z., Che, M., *et al.* (1986). Spectroscopic study of superoxide species formed by low-temperature adsorption of oxygen onto cobalt oxide (CoO)-magnesium oxide solid solutions: An example of synthetic heterogeneous oxygen carriers, *J. Phys. Chem.*, **90**, pp. 6084–6091.
49. Imbihl, R. and Demuth J. (1986). Adsorption of oxygen on a Pd(111) surface studied by high resolution electron energy loss spectroscopy (EELS), *Surf. Sci.*, **173**, pp. 395–410.
50. Leisenberger, F., Koller, G., Sock, M., *et al.* (2000). Surface and subsurface oxygen on Pd(111), *Surf. Sci.*, **445**, pp. 380–393.

51. Klikovits, J., Napetschnig, E., Schmid, M., *et al.* (2007). Surface Oxides on Pd(111): STM and Density Functional Calculations, *Phys. Rev. B*, **76**, 045–405.
52. Kan, H. and Weaver, J. (2009). Mechanism of PdO thin film formation during the oxidation of Pd(111), *Surf. Sci.*, **603**, pp. 2671–2682.
53. McBride J., Hass K. and Weber W., (1991). Resonance-raman and lattice-dynamics studies of single-crystal PdO, *Phys. Rev. B*, **44**, pp. 5016–5028.
54. Otto, K., Hubbard, C., Weber, W., *et al.* (1992). Raman spectroscopy of palladium oxide on $\gamma$-alumina applicable to automotive catalysts, *Appl. Catal. B: Environ.*, **1**, pp. 317–327.
55. Specchia, S., Finocchio, E., Busca, G., *et al.* (2009). Surface chemistry and reactivity of ceria–zirconia-supported palladium oxide catalysts for natural gas combustion, *J. Catal.*, **263**, pp. 134–145.
56. Bonino, F., Damin, A., Ricchiardi, G., *et al.* (2004). Ti-peroxo species in the TS-1/$H_2O_2$/$H_2O$ system, *J. Phys. Chem. B*, **108**, pp. 3573–3583.
57. Un, W. and Frei, H. (2002). Photochemical and FTIR probing of the active site of hydrogen peroxide in Ti silicalite sieve, *J. Am. Chem. Soc.*, **124**, pp. 9292–9298.
58. Gómez Bemal, H., Cedeňo Caero, L., Finocchio, E., *et al.* (2009). An FTIR study ofthe adsorption and reactivity of tert-butyl hydroperoxide over oxide cataiysts, *Appl.Catal. A: Gen.*, **369**, pp. 27–35.
59. Hadjiivanov, K. and Vayssilov, G. (2002). Characterization of oxide surfaces and zeolites by carbon monoxide as an IR probe molecule, *Adv. Catal.*, **47**, pp. 307–511.
60. Montanari, T., Matarrese, R., Artioli, N., *et al.* (2011). FTIR study of the surface redox states on platinum-potassium-alumina catalysts, *Appl. Catal. B: Environ.*, **105**, pp. 15–23.
61. Busca, G. and Finocchio, E. (2011). Surface science and industrial catalysis: Redox states of alumina supported palladium nanoparticles, *La Chimica e l'Industria*, **93**, pp. 85–95.
62. Montanari, T., Marie, O., Daturi, M., *et al.* (2005). Cobalt on and in zeolites and silica–alumina: Spectroscopic characterization and reactivity, *Catal. Today*, **110**, pp. 339–344.
63. Kokes, R. (1973). Characterization of adsorbed intermediates on zinc oxide by infrared, *Acc. Chem. Res.*, **6**, pp. 226–233.
64. Tamaru, K. (1978). *Dynamic Heterogeneous Catalysis*, Academic Press, New York.
65. Bañares, M. (2005). Operando methodology: Combination of in situ spectroscopy and simultaneous activity measurements under catalytic reaction conditions, *Catal. Today*, **100**, pp. 71–77.
66. Vimont, A., Thibault-Starzyk F. and Daturi, M. (2010). Analysing and understanding the active site by IR spectroscopy, *Chem. Soc. Rev.*, **39**, pp. 4928–4950.
67. Tinnemans, S., Mesu, J., Kervinen, K., *et al.* (2006). Combining operando techniques in one spectroscopic-reaction cell: New opportunities for elucidating the active site and related reaction mechanism in catalysis, *Catal. Today*, **113**, pp. 3–15.
68. Meunier, F. (2010). The power of quantitative kinetic studies of adsorbate reactivity by operando FTIR spectroscopy carried out at chemical potential steady-state, *Catal. Today*, **155**, pp. 164–171.
69. Goguet, A., Tibiletti, D., Meunier, F., *et al.* (2004). Spectrokinetic investigation of reverse water-gas shift reaction intermediates over a Pt/$CeO_2$, Catalyst, *J. Phys. Chem. B*, **108**, pp. 20240–20246.
70. Daniel, C., Clarté, M., Provendier, H., *et al.* (2009). Spatially resolved operando infrared analysis of a microstructured catalytic surface for CO oxidation over Pt based catalysts, *C. R. Chimie*, **12**, pp. 647–653.
71. Colinot, F., Meunier, F. and Daturi, M. (2009). Low temperature oxidation catalysis for diesel engine emission control by FTIR, *Spectrosc. Europe*, **21**, pp. 9–12.
72. Malpartida, I., Ivanova, E., Mihaylov, M., *et al.* (2010). CO and NO Adsorption for the IR characterization of $Fe^{2+}$ cations in ferrierite: An efficient catalyst for NOx SCR with $NH_3$ as studied by operando IR spectroscopy, *Catal. Today*, **149**, pp. 295–303.

73. Thibault-Starzyk, F., Seguin, E., Thomas, S., *et al.* (2009). Real-time infrared detection of cyanide flip on silver-alumina Nox removal catalyst, *Science*, **324**, pp. 1048–1051.
74. Somorjai, G. (1994). *Introduction to Surface Chemistry and Catalysis*, Wiley, New York.
75. Goodman, D. (1995). Model studies in catalysis using surface science probes, *Chem. Rev.*, **95**, pp. 523–536.
76. Ozensoy, E., Meier, D. and Goodman, D. (2002). Polarization modulation infrared reflection absorption spectroscopy at elevated pressures: CO adsorption on Pd(111) at atmospheric pressures, *J. Phys. Chem. B*, **106**, pp. 9367–9371.
77. Libuda, J. and Freund, H. (2005). Molecular beam experiments on model catalysts, *Surf. Sci. Reports*, **57**, pp. 157–298.
78. Dellwig, T., Hartmann, J., Libuda, J., *et al.* (2000). Complex model catalysts under UHV and high pressure conditions: CO adsorption and oxidation on alumina-supported Pd particles, *J. Mol. Catal. A: Chem.*, **162**, pp. 51–66.
79. Abbott, H., Uhl, A., Baron, M., *et al.* (2010). Relating methanol oxidation to the structure of ceria-supported vanadia monolayer catalysts, *J. Catal.*, **272**, pp. 82–91.
80. Busca, G., Elmi, A. and Forzatti, P. (1987). Mechanism of selective methanol oxidation over vanadium-titanium oxide catalysts: A FTIR and flow-reactor study, *J. Phys. Chem.*, **91**, pp. 5263–5269.
81. Deshlahra, P., Pfeifer, K., Bernstein, G., *et al.* (2011). CO adsorption and oxidation studies on nanofabricated model catalysts using multilayer enhanced IRAS technique, *Appl. Catal. A: Gen.*, **391**, pp. 2–30.
82. Soares, A., Portela, M. and Kiennemann, A. (2005). Methanol selective oxidation to formaldehyde over iron-molybdate catalysts, *Catal. Rev. Sci. Eng.*, **47**, pp. 125–145.
83. Forzatti, P., Tronconi, E., Busca, G., *et al.* (1987). Oxidation of methanol to methyl formate over V-Ti oxide catalysts, *Catal. Today*, **1**, pp. 209–218.
84. Tronconi, E., Elmi, A., Ferlazzo, N., *et al.* (1987). Methyl formate from methanol oxidation over co-precipitated V-Ti-O catalysts, *Ind. Eng. Chem. Res.*, **26**, pp. 1269–1275.
85. Forzatti P., Tronconi E., Elmi A., *et al.* (1997). Methanol oxidation over vanadia-based catalysts, *Appl. Catal. A: Gen.*, **157**, pp. 387–408.
86. Zhao, H., Bennici, S., Shen, J., *et al.* (2010). Nature of surface sites of $V_2O_5$-$TiO_2$/$SO_4^{2-}$ catalysts and reactivity in selective oxidation of methanol to dimethoxymethane, *J. Catal.*, **272**, pp. 176–189.
87. Briand, L. (2006). Investigation of the nature and number of surface active sites of supported and bulk methanol oxide catalysts, in J. Fierro (ed.), *Metal Oxides: Chemistry and Applications*, CRC Press, Boca Raton, FL, pp. 353–390.
88. Busca, G., Forzatti P., Lavalley J., *et al.* (1985). A TPD, FTIR and catalytic study of the interaction of methanol with pure and KOH doped $TiO_2$ anatase, in B. Imelik, C. Naccache, G. Coudurier, *et al.* (eds.), *Catalysis by Acids and Bases*, Elsevier, Amsterdam, pp. 15–24.
89. Busca, G., Rossi, P., Lorenzelli, V., *et al.* (1985). Microcalorimetric and FTIR spectroscopic studies of methanol adsorption on $Al_2O_3$, *J. Phys. Chem.*, **89**, pp. 5433–5439.
90. Busca, G., Elmi, A. and Forzatti, P. (1987). Mechanism of selective methanol oxidation over vanadium-titanium oxide catalysts: A FTIR and flow-reactor study, *J. Phys. Chem.*, **91**, pp. 5263–5269.
91. Montagne, X., Lynch, J., Freund, E., *et al.* (1987). A study of the adsorption sites on thoria by STEM and FTIR spectroscopy. Adsorption and desorption of water and methanol. *J. Chem. Soc. Faraday Trans. I*, **83**, pp. 1417–1425.
92. Busca, G. and Lorenzelli, V. (1980). Infrared study of methanol, formaldehyde and formic acid adsorbed on hematite, *J. Catal.*, **66**, pp. 155–161.
93. Busca, G., Guidetti, R., and Lorenzelli, V. (1990). FTIR study of the surface properties of cobalt oxides, *J. Chem. Soc. Faraday Trans. I*, **86**, pp. 989–994.

94. Busca, G. (1987). FTIR study of the surface of copper oxide, *J. Mol. Catal.*, **43**, pp. 225–236.
95. Finocchio, E. and Busca, G. (2001). Characterization and hydrocarbon oxidation activity of coprecipitated mixed oxides $Mn_3O_4/Al_2O_3$, *Catal. Today*, **70**, pp. 13–225.
96. Rousseau, S., Marie, O., Bazin, P., *et al.* (2010). Investigation of methanol oxidation over Au/Catalysts using operando IR spectroscopy: Determination of the active sites, intermediate/spectator species and reaction mechanism, *J. Am. Chem. Soc.*, **132**, pp. 10832–10841.
97. Busca, G., (1989). On the mechanism of methanol oxidation over vanadia-based catalysts; A FTIR study of the adsorption of methanol, formaldehyde and formic acid on vanadia and vanadia-silica, *J. Mol. Catal.*, **50**, pp. 241–249.
98. Groff, R. (1984). Infrared study of methanol and ammonia adsorption on molybdenum, *J. Catal.*, **86**, pp. 215–218.
99. Burcham L., Briand L. and Wachs I. (2001). Quantification of active sites for the determination of methanol oxidation turn-over frequencies using methanol chemisorption and in situ infrared techniques. 2. Bulk metal oxide catalysts, *Langmuir*, **17**, pp. 6175–6184.
100. Burcham L., Briand L. and Wachs I. (2001). Quantification of active sites for the determination of methanol oxidation turn-over frequencies using methanol chemisorption and in situ infrared techniques. 1. supported metal oxide catalysts, *Langmuir*, **17**, pp. 6164–6174.
101. Finocchio, E., Daturi, M., Binet, C., *et al.* (1999). Thermal evolution of the adsorbed methoxy species on $Ce_xzr_{1-X}o_2$ solid solution samples: A FTIR study, *Catal. Today*, **52**, pp. 53–63.
102. Busca, G., Lamotte, J., Lavalley, J., *et al.* (1987). FTIR study of the adsorption and transformation of formaldehyde on oxide surfaces, *J. Am. Chem. Soc.*, **109**, pp. 5197–5203.
103. Busca, G. (1996). Infrared studies of the reactive adsorption of organic molecules over metal oxides and of the mechanisms of their heterogeneously-catalyzed oxidation, *Catal. Today*, **27**, pp. 457–496.
104. Busca, G., Resini, C., Montanari, T., *et al.* (2009). Hydrogen from alcohols: IR and flow reactor studies, *Catal. Today*, **143**, pp. 2–8.
105. Turco, M., Bagnasco, G., Costantino, U., *et al.* (2004). Production of hydrogen from oxidative steam reforming of methanol: II. Catalytic activity and reaction mechanism on $Cu/ZnO/Al_2O_3$ hydrotalcite-derived catalysts, *J. Catal.*, **228**, pp. 56–65.
106. Larrubia Vargas, M., Busca, G., Costantino, U., *et al.* (2007). An IR study methanol steam reforming over ex-hydrotalcite Cu-Zn-Al catalysts, *J. Mol. Catal., A: Chem.*, **266**, pp. 188–197.
107. Davies, P. and Bowker, M. (2010). On the nature of the active site in catalysis: The reactivity of surface oxygen on Cu(110), *Catal. Today*, **154**, pp. 31–37.
108. Trimm, D. (2005). Minimisation of carbon monoxide in a hydrogen stream for fuel cell application, *Appl.Catal. A. Gen.*, **296**, pp. 1–11.
109. Louis, C. (2008). Gold nanoparticles: Recent advances in CO oxidation, in astruc, D. (ed.), *Nanoparticles and Catalysis*, Wiley-VCH, Weinheim, pp. 475–503.
110. Bond, G. and Thompson, D. (1999). Catalysis by gold, *Catal. Rev. Sci. Eng.*, **41**, pp. 319–338.
111. Moretti, E., Lenarda, M., Storaro, L., *et al.* (2008). One-step synthesis of a structurally organized mesoporous $CuO\text{-}CeO_2\text{-}Al_2O_3$ system for the preferential CO oxidation, *Appl. Catal. A: Gen.*, **335**, pp. 46–55.
112. Haldor Topsøe, http://www.topsoe.com, $10^{th}$ October 2012.
113. Farrauto, R., Liu, Y., Ruettinger, W. *et al.* (2008). Precious metal catalysts supported on ceramic and metal monolithic structures for the hydrogen economy, *Catal. Rev. Sci. Eng.*, **49**, pp. 141–196.
114. Corti, C., Hollyday, T. and Thompson, D. (2005). Commercial aspects of gold catalysis, *Appl.Catal. A: Gen.*, **291**, pp. 253–261.
115. De Ménorval, L., Chaqroune, A., Coq, B., *et al.* (1997). CO adsorption on pure and alloyed well dispersed platinum catalysts, *J. Chem. Soc. Faraday Trans. I*, **93**, pp. 3715–3720.

116. Malpartida, I., Larrubia Vargas, M., Alemany, L., *et al.* (2008). Pt-Ba-$Al_2O_3$ for NOx storage and reduction: Characterization of the dispersed species, *Appl. Catal. B: Environ.*, **80**, pp. 214–225.
117. Daniel, C., Clarté, M., Teh, S., *et al.* (2010). Spatially resolved catalysis in microstructured reactors by IR spectroscopy: CO oxidation over mono- and bifunctional pt catalysts, *J. Catal.*, **272**, pp. 55–64.
118. Singh, J. and van Bokhoven, J. (2010). Structure of alumina supported platinum catalysts of different particle size during CO oxidation using in situ IR and HERFD XAS, *Catal. Today*, **155**, pp. 199–205.
119. Musialska, K., Finocchio, E., Sobczak, I., *et al.* (2010). Characterization of alumina- and niobia-supported gold catalysts used for oxidation of glycerol, *Appl.Catal. A. Gen.*, **384**, pp. 70–77.
120. Sanchez-Escribano, V., Arrighi, L., Riani, P., *et al.* (2006). Characterization of Pd-Cu alloy nanoparticles on $Al_2O_3$-supported catalysts, *Langmuir*, **22**, pp. 9214–9219.
121. Moretti, E., Storaro, L., Talon, A., *et al.* (2008). Preferential CO oxidation (CO-PROX) over CuO-ZnO/$TiO_2$ catalysts, *Appl. Catal. A: Gen.*, **344**, pp. 165–174.
122. Polster, C., Nair, H. and Baertsch, C. (2009). Study of active sites and mechanism responsible for highly selective CO oxidation in $H_2$ rich atmospheres on a mixed Cu and Ce oxide catalyst, *J. Catal.*, **266,** pp. 308–319.
123. Gamarra, D., Belver, C., Fernández-García, M., *et al.* (2007). Selective CO Oxidation in excess $H_2$ over copper-ceria catalysts: Identification of active entities/species, *J. Am. Chem. Soc.*, **129**, pp. 12064–12065.
124. Kydd, R., Ferri, D., Hug, P., *et al.* (2011). Temperature induce devolution of reaction sites and mechanisms during preferential oxidation of CO, *J. Catal.*, **277**, pp. 64–71.
125. Meunier, F., Goguet, A., Hardacre, C., *et al.* (2007). Quantitative DRIFTS investigation of possible reaction mechanisms for the water–gas shift reaction on high-activity Pt- and Au-based catalysts, *J. Catal.*, **252**, pp. 18–22.
126. Meunier, F., Reid, D., Goguet, A., *et al.* (2007). Quantitative analysis of the reactivity of formate species seen by DRIFTS over a Au/Ce(La)$O_2$ water–gas shift catalyst: First unambiguous evidence of the minority role of formates as reaction intermediates, *J. Catal.*, **247**, pp. 277–287.
127. Kalamaras, C., Panagiotopoulou, P., Kondarides, D., *et al.* (2009). Kinetic and mechanistic studies of the water-gas shift reaction on Pt/$TiO_2$ catalysts, *J. Catal.*, **265**, pp. 117–129.
128. Kiennemann, A., Idriss, H., Hindermann, J., *et al.* (1990). Methanol synthesis on Cu/Zn$Al_2O_4$ and Cu/ZnO-$Al_2O_3$ catalysts: Influence of carbon monoxide pretreatment on the formation and concentration of formate species, *Applied Catalysis*, **59**, pp. 165–184.
129. Rosowski, F., Akltwasser, S., Dobner, C., *et al.* (2010). New silver-and vanadium-containing multimetal oxides for oxidation of aromatic hydrocarbons, *Catal. Today*, **157**, pp. 339–344.
130. Becht, S., Franke, R., Geißelmann, A., *et al.* (2009). An industrial view of process intensification, *Chem. Eng. Process.: Proc. Int.*, **48**, pp. 329–332.
131. Guettel, R. and Turek, T. (2010). Assessment of microstructured fixed-bed reactors for highly exothermic gas-phase reactions, *Chem. Eng. Sci.*, **65**, pp. 1644–1654.
132. Luciani S., Ballarini N., Cavani F., *et al.* (2009). Anatase-supported vanadium oxide catalysts for *o*-Xylene oxidation: From consolidated certainties to unexpected effects, *Catal. Today*, **142**, pp. 132–137.
133. Busca, G. (1993). Infrared characterization of the hydrocarbon species intermediates in the oxidation of toluene and xylenes over vanadia-titania catalysts, *J. Chem. Soc. Faraday Trans. I*, **89,** pp. 753–755.
134. Gimeno, M., Gadvon, J., Tellez, C., *et al.* (2008). Selective oxidation of *o*-Xylene over $V_2O_5$-$TiO_2$: Kinetic study in a fluidized bed reactor, *Chem. Eng. Process.*, **47**, pp. 1844–1852.

135. Busca, G. (1993). Selective and non-selective pathways in the oxidation and ammoxidation of methyl-aromatics over vanadia-titania catalysts: FTIR studies, in S. Oyama and J. Hightower (eds.), *Catalytic Selective Oxidation*, American Chemical Society, Washington, DC, pp. 168–182.
136. Busca, G., Cavani, F. and Trifirò, F. (1987). Oxidation and ammoxidation of toluene over vanadium-titanium oxide catalysts: A fourier transform infrared and flow-reactor study, *J. Catal.*, **106**, pp. 471–482.
137. Gallardo Amores, J., Sanchez Escribano, V., Ramis, G., *et al.* (1997). An FTIR study of ammonia adsorption and oxidation over anatase-supported metal oxides, *Appl. Catal. B: Environ.*, **13**, pp. 45–58.
138. Zhang, Y., Martin, A., Berndt, H., *et al.* (1997). FTIR investigation of surface intermediates formed during ammoxidation of toluene over vanadyl pyrophosphate, *J. Mol. Catal., A: Chem.*, **118**, pp. 205–214.
139. Chauvel, A. and Lefebvre, G. (1989). Petrochemical processes: Synthesis-gas derivatives and major hydrocarbons; *Editions Technip: Paris*, **1**, p. 336.
140. Gibson, M. and Hightower, J. (1976). Oxidative dehydrogenation of butenes over magnesium ferrite kinetic and mechanistic studies, *J. Catal.*, **41**, pp. 420–430.
141. Rennard, R. and Kehl, W. (1971). Oxidative dehydrogenation of butenes over ferrite catalysts, *J. Catal.*, **81**, pp. 282–293.
142. Busca, G., Finocchio, E., Lorenzelli, V., *et al.* (1996). Infrared study of olefin allylic activation on magnesium ferrite and alumina catalysts, *J. Chem. Soc. Faraday Trans. I*, **92**, pp. 4687–4693.
143. Finocchio, E., Busca, G., Ramis, G., *et al.* (1997). On the mechanism of the selective oxy-dehydrogenation of n-butenes to 1,3-butadiene on magnesium ferrite: An FTIR study, in R. Grasselli, T. Oyama, A. Gaffney, *et al.* (eds.), *Oxidation Catalysis*, Elsevier, Amsterdam, pp. 989–998.
144. Busca, G. and Lorenzelli, V. (1992). FTIR study of n-butenes oxidation and oxidative dehydrogenation on the surface of $FeCrO_3$, *J. Chem. Soc. Faraday Trans. I*, **88**, pp. 2783–2789.
145. Busca, G., Finocchio, E., Lorenzelli, V., *et al.* (1999). IR studies on the activation of CH hydrocarbons bonds on oxidation catalysts, *Catal. Today*, **49**, pp. 453–465.
146. Centi, G. and Busca, G. (1989). Surface dynamics of adsorbed species on heterogeneous catalysts: Evidence from the oxidation of $C_4$ and $C_5$ alkanes on vanadyl pyrophosphate, *J. Am. Chem. Soc.*, **111**, pp. 46–54.
147. Busca, G., Ramis, G. and Lorenzelli, V. (1989). On the mechanism of the selective oxidation of $C_4$ linear hydrocarbons to maleic anhydride: A FTIR study of the adsorption and oxidation of 1,3-butadiene on vanadia-titania, *J. Mol. Catal.*, **55**, pp. 1–11.
148. Lee, H., Jung, J. and Song, I. (2009). Oxidative dehydrogenation of n-butene to 1,3-butadieene over sulphated $ZnFe_2O_4$ catalyst, *Catal. Lett.*, **133**, pp. 321–327.
149. Chung, Y., Kwon, Y., Kim, T., *et al.* (2009). Mixed manganese ferrite catalysts, method of preparing thereof and method of preparing 1,3-butadiene using there of, patent WO/2009/075478.
150. Haldor Topsøe, http://www.topsoe.com, 10th October 2012.
151. Baldi, M., Milella, F., Ramis, G., *et al.* (1998). An FTIR and flow reactor study of the selective catalytic oxy-dehydrogenation of C3 alcohols on $Mn_3O_4$, *Appl.Catal. A: Gen.*, **166**, pp. 75–88.
152. Baldi, M., Finocchio, E., Milella, F., *et al.* (1998). Catalytic combustion of C3 hydrocarbons and oxygenates over $Mn_3O_4$, *Appl. Catal. B: Environ.*, **16**, pp. 43–51.
153. Peluso, M., Pronsato, E., Sambeth, J., *et al.* (2008). Catalytic combustion of ethanol on pure and alumina supported K-Mn oxides: An IR and flow reactor study, *Appl. Catal. B: Environ.*, **78**, pp. 73–79.
154. Gallardo Amores, J., Armaroli, T., Ramis, G., *et al.* (1999). A study of anatase-supported Mn oxides and their behaviour as catalysts for 2-propanol total oxidation, *Appl. Catal. B: Environ.*, **22**, pp. 249–259.

155. Fernandez-Lopez, E., Sanchez-Escribano, V., Resini, C., *et al.* (2001). A study of coprecipitated Mn-Zr oxides and their behavior as oxidation catalysts, *Appl. Catal. B: Environ.*, **29**, pp. 251–261.
156. Pistarino, C., Finocchio, E., Romezzano, G., *et al.* (2000). A study of the catalytic dehydrochlorination of 2-chloropropane in oxidizing conditions, *Ind. Eng. Chem. Res.*, **39**, pp. 2752–2760.
157. Pistarino, C., Finocchio, E., Larrubia, M., *et al.* (2001). A study of the dehydrochlorination of 1,2-dichloropropane over silica-alumina catalysts, *Ind. Eng.Chem. Res.*, **40**, pp. 3262–3269.
158. Finocchio, E., Pistarino, C., Dellepiane, S., *et al.* (2002). Studies on the catalytic dechlorination and abatement of chlorided VOC: The cases of 2-chloropropane, 1,2-dichloropropane and trichloroethylene, *Catal. Today*, **75**, pp. 263–267.
159. Finocchio, E., Busca, G. and Notaro, M. (2006). A review of catalytic processes for the destruction of PCDD and PCDF from waste gases, *Appl. Catal. B: Environ.*, **62**, pp. 12–20.
160. Fino, D., Specchia, S., Specchia, V., *et al.* (2006). Catalytic filters for flue gas cleaning, in A. Cybulski and J. Moulijn (eds.), *Structured Catalysts and Reactors*, CRC Press, Boca Raton, FL, pp. 553–578.
161. Lichtenberger J. and Amiridis, M. (2004). Catalytic oxidation of chlorinated benzenes over $V_2O_5$-$TiO_2$ catalysts, *J. Catal.*, **223**, pp. 296–308.
162. Larrubia, M. and Busca, G. (2002). An FTIR study of the conversion of 2-chloropropane, *o*-dichlorobenzene and dibenzofuran on $V_2O_5$-$MoO_3$-$TiO_2$ SCR-DeNO$_x$ catalysts, *Appl. Catal. B: Environ.*, **39**, pp. 343–352.
163. Finocchio, E., Ramis, G. and Busca, G. (2011). A study on catalytic combustion of chlorobenzenes, *Catal. Today*, **16**, pp. 3–9.
164. Busca, G., Ramis, G. and Lorenzelli, V. (1990). FTIR study of selective oxidation intermediates of benzene on the surface of vanadia-titania monolayer catalysts, in G. Centi and F. Trifirò (eds.), *New Developments in Selective Oxidation*, Elsevier, Amsterdam, pp. 825–831.
165. Hetrick, C., Lichtenberger J. and Amiridis, M. (2008). Catalytic oxidation of chlorophenol over $V_2O_5$-$TiO_2$ catalysts, *Appl. Catal. B: Environ.*, **77**, pp. 255–263.
166. Forzatti, P., Lietti L. and Tronconi E. (2009). Catalytic removal of Nox under lean conditions from stationary and mobile sources, in P. Barbaro and C. Bianchini C. (eds.), *Catalysis for Sustainable Energy Production*, J. Wiley-VCH, Weinheim, pp. 393–438.
167. Busca, G., Lietti, L., Ramis, G., *et al.* (1998). Chemical and mechanistic aspects of the selective catalytic reduction of Nox by ammonia over oxide catalysts: A review, *Appl. Catal. B: Environ.*, **18**, pp. 1–36.
168. Ramis, G., Busca, G., Bregani, F., *et al.* (1990). FTIR study of the adsorption and coadsorption of NO, $NO_2$ and ammoniaon vanadia-titania and mechanism of the SCR reaction, *Appl. Catal.*, **64**, pp. 259–278.
169. Jug, K., Homann, T. and Bredow, T. (2004). Reaction mechanism of the selective catalytic reduction of NO with $NH_3$ and $O_2$ to $N_2$ and $H_2O$, *J. Phys. Chem. A*, **108**, pp. 2966–2971.
170. Topsøe, N. and Topsøe, H. (1991). Combined in-situ FTIR and on-line activity studies: Applications to vanadia-titania DeNO$_x$ catalyst, *Catal. Today*, **9**, pp. 77–82.
171. Anstrom, M., Topsøe, N. and Dumesic, J. (2003). Density functional theory studies of mechanistic aspects of the SCR reaction on vanadium oxide catalysts, *J. Catal.*, **213**, pp. 115–125.
172. Kobayashi, Y., Tajima, N., Nakano, H., *et al.* (2004). Selective catalytic reduction of nitric oxide by ammonia: The activation mechanism, *J. Phys. Chem. B*, **108**, pp. 12264–12266.
173. Vittadini, A., Casarin, M. and Selloni, A. (2005). First principles studies of vanadia-titania monolayer catalysts: Mechanisms of NO selective reduction, *J. Phys. Chem. B*, **109**, pp. 1652–1655.
174. Brandenberger, S., Kröcher, O., Tissler, A., *et al.* (2008). The state of the art in selective catalytic reduction of NOx by ammonia using metal-exchanged zeolite catalysts, *Catal. Rev. Sci. Eng.*, **50**, pp. 492–531.

175. Kearley,G., Kettle, S. and Oxton, I. (1980). Ammonium ion symmetry in phases II–V of ammonium nitrate, *Spectrochim. Acta A: Mol. Spectroscopy*, **36**, pp. 419–423.
176. Schlenker, J. and Martin, S. (2005). Crystallization pathways of sulfate-nitrate-ammonium aerosol particles, *J. Phys. Chem. A*, **109**, pp. 9980–9985.
177. Ramis, G., Busca, G. and Bregani, F. (1992). Nitrous oxide adsorption on vanadia-titania and tungsta-titania catalysts for the reduction of nitrogen oxides, *Gazzetta Chim. Ital.*, **122**, pp. 79–84.
178. Sanchez Escribano, V., Montanari, T. and Busca, G. (2005). Low temperature selective catalytic reduction of Nox by ammonia over H-ZSM-5: An IR study, *Appl. Catal. B: Environ.*, **58**, pp. 19–23.
179. Ramis, G., Li, Y. and Busca, G. (1996). Ammonia activation over SCR and SCO catalysts, *Catal. Today*, **28**, pp. 373–380.
180. Trombetta, M., Ramis, G., Busca, G., *et al.* (1997). Ammonia adsorption and oxidation on Cu/Mg/Al mixed oxide catalysts prepared via hydrotalcite-type precursors, *Langmuir*, **13**, pp. 4628–4637.
181. Bagnasco, G., Peluso, G., Russo, G., *et al.* (1997). Ammonia oxidation over $CuO/TiO_2$ catalyst: Selectivity and mechanistic study, in R. Grasselli, T. Oyama, A. Gaffney and J. Lyons (eds.), *Oxidation Catalysis*, Elsevier, Amsterdam, pp. 643–652.
182. Ramis, G., Larrubia, M. and Busca, G. (2000). On the chemistry of ammonia over oxide catalysts: Fourier transform infrared study of ammonia, hydrazine and hydroxylamine adsorption over iron-titania catalyst, *Top. Catal.*, **11/12**, pp. 161–166.
183. Lietti, L., Ramis, G., Busca, G., *et al.* (2000). Characterization and reactivity of $Moo_3/Sio_2$ De-$No_x$ catalysts in the selective catalytic oxidation of ammonia to $N_2$, *Catal. Today*, **61**, pp. 187–195.
184. Wollner, A., Lange, F., Schmelz, H., *et al.* (1993). Characterization of mixed copper-manganese oxides supported on titania catalysts for selective oxidation of ammonia, *Appl. Catal. A: Gen.*, **94**, 181–203.
185. Wang Y., Jacobi K., Schöne W., *et al.* (2005). Catalytic oxidation of ammonia on $RuO_2(110)$ surfaces: mechanism and selectivity, *J. Phys. Chem. B*, **109**, pp. 7883–7893.
186. Carley, A., Davies, P. and Roberts, M. (2011). Oxygen transient states in catalytic oxidation at metal surfaces, *Catal. Today*, **169**, pp. 118–124.
187. Armor, J. (1995). Catalytic reduction of nitrogen oxides with methane in the presence of excess oxygen: A review, *Catal. Today*, **26**, pp. 147–158.
188. Resini, C., Montanari, T., Nappi, L., *et al.* (2003). Selective catalytic reduction of $NO_x$ by methane over o-H-MFI and Co-H-FER zeolite catalysts: Characterisation and catalytic activity, *J. Catal.*, **214**, pp. 179–190.
189. Bagnasco, G., Turco, M., Resini, C., *et al.* (2004). On the role of external co sites in NO oxidation and reduction by methane over Co-H-MFI catalysts, *J. Catal.*, **225**, pp. 536–540.
190. Montanari, T., Bevilacqua, M., Resini, C., *et al.* (2004). UV-vis and FTIR study of the nature and location of the active sites of partially exchanged Co-H-zeolites, *J. Phys. Chem. B*, **108**, pp. 2120–2127.
191. Montanari, T., Marie, O., Daturi, M., *et al.* (2007). Searching for the active sites of Co-H-MFI catalyst for the selective catalytic reduction of NO by methane: A FTIR in situ and operando study, *Appl. Catal. B: Environ.*, **71**, pp. 216–222.
192. Boix, A., Miró, E., Lombardo, E., *et al.* (2003). The nature of cobalt species in co and PtCoZSM5 used for the SCR of $NO_x$ with $CH_4$, *J. Catal.*, **217**, pp. 186–194.
193. Ito, E. and van Veen, J. (2006). On novel processes for removing sulphur from refinery streams, *Catal. Today*, **116**, pp. 446–460.
194. Gómez-Bernal, H., Cedeño-Caero, L. and Gutiérrez-Alejandre, A. (2009). Liquid phase oxidation of dibenzothiophene with alumina-supported vanadium oxide catalysts: An alternative to deep desulfurization of diesel, *Catal. Today.*, **142**, pp. 227–233.

195. Gómez-Bernal, H., Cedeño-Caero, L., Finocchio, E., *et al.* (2009). Oxidation of benzothiophene by tert-butyl hydroperoxide over vanadia–alumina catalyst: An FTIR study at the vapour–solid interface, *Catal, Commun.*, **10**, pp. 1629–1632.
196. Larrubia, M., Gutièrrez-Alejandre, A. Ramìrez J., *et al.* (2002). A FTIR study of the adsorption of indole, carbazole, benzothiophene, dibenzothiophene and 4,6-dibenzo-thiophene over solid adsorbents and catalysts, *Appl. Catal. A Gen.*, **224**, pp. 167–178.

# Chapter 19

# *In Situ* Non-Vibrational Characterization Techniques to Analyse Oxidation Catalysts and Mechanisms

Angelika BRÜCKNER,* Evgenii KONDRATENKO,* Vita KONDRATENKO,*
Jörg RADNIK* and Matthias SCHNEIDER*

Catalytic oxidation processes are usually connected with transfer of electrons and changes of structure and valence state of active catalyst components. This chapter presents methods that are especially suitable for monitoring these kinds of changes (UV-vis-DRS, EPR, X-ray scattering, XPS, XAS, TPO, TPR, TPRS, TAP and SSITKA). After a short section on basic principles and experimental details, the potential of each technique is illustrated by selected application examples that include a wide variety of oxidation catalysts such as mixed metal oxides and oxynitrides, zeolites containing transition metal ions, heteropoly acids and supported noble metals.

## 19.1. Introduction

Heterogeneous catalytic oxidation reactions start with the adsorption of reactants on the catalyst surface, being in many cases molecular oxygen and hydrocarbons, followed by bond breaking/reformation and finally desorption of products. Detailed knowledge about the role of distinct catalyst building blocks in these processes are essential for rational catalyst design beyond empirical trial-and-error approaches, and this can best be obtained when catalyst synthesis, operation and deactivation are studied by *in situ* methods. Tremendous progress has been achieved during the last decade in developing *in situ* characterization methods, including, but not restricted to, oxidation catalysts, which is summarized in several recent books and reviews.[1–4]

While vibrational *in situ* methods such as Fourier transform infrared (FTIR) and Raman spectroscopy are excellent tools for monitoring the formation of organic intermediates and vibrations of molecular moieties in the catalyst lattice and/or surface, these techniques usually fail to detect changes of structure and valence states of transition metal ions (TMI), which are very often part of the active sites

*Leibniz-Institut für Katalyse e. V. an der Universität Rostock, Albert-Einstein-Str. 29a, D-18059 Rostock, Germany.

in oxidation catalysts. Such changes are caused by the ability of the catalyst to reversibly adsorb, dissociate and transfer oxygen to the substrate, i.e. by their redox and coordination properties, which govern their catalytic performance. For this purpose, *in situ* techniques able to visualize transitions of electrons and/or electron spins are needed. The different disciplines of X-ray absorption spectroscopy (XAS),[5–10] X-ray photoelectron spectroscopy (XPS),[11] but also UV-vis[12,13] and electron paramagnetic resonance (EPR)[14–20] spectroscopy belong to this category. When catalysts contain crystalline phases, such redox processes can also alter the phase and/or defect structure as well as the crystallite size. This can be detected by *in situ* X-ray diffraction (XRD), and wide and small angle X-ray scattering (WAXS, SAXS).[21,22] The potential of these methods for analysing oxidation catalysts in the presence of reactants and probe molecules will be discussed in Sections 19.2 and 19.3.

Besides monitoring oxidation catalysts under working conditions in a certain reaction, most of these methods are also helpful for following the assembly of oxide catalyst precursors with a dedicated phase and pore structure during synthesis from parent solutions. Thus, O'Brien *et al.* reviewed the potential of synchrotron-based techniques such as wide and small angle X-ray scattering (WAXS, SAXS) and XAS, sometimes in combination with transmission and atomic force microscopy (TEM, AFM), UV-vis or Raman spectroscopy for analysing the formation of crystallites of zeolites and aluminophosphates, starting from clear precursor solutions or gels.[23] Bentrup first introduced a simultaneous coupling of five *in situ* methods (WAXS/SAXS/Raman/UV-vis/attenuated total reflection (ATR) for online monitoring of the synthesis of mixed-metal molybdates, used as precursors for selective oxidation catalysts. While X-ray scattering reflected the formation, growth and crystallization of particles in the slurry, UV-vis and vibrational spectroscopy provided information about the chemical nature of the polymolybdate anions.[24]

Temperature-programmed reduction, oxidation and desorption (TPR, TPO, TPD), belong probably to the most widely used *in situ* techniques for the characterization of oxidation catalysts and are discussed in more detail in Section 19.4.[25] While TPD (with ammonia as the probe molecule) is frequently used to examine surface acid sites, TPR and TPO (with $H_2$ or $O_2$, respectively) provide information on the redox properties of oxide catalysts being crucial for their performance in catalytic oxidation reactions. Important information on reaction mechanisms can be obtained when the catalysts are heated in the presence of reactants combined with mass spectrometric product analysis. This is called temperature-programmed reaction spectroscopy (TPRS). As far as reaction mechanisms and kinetics are concerned, transient techniques which reflect the response of the catalytic system to a sudden change of reactant are inevitable tools. Two such techniques, namely the temporal analysis of products (TAP)[26] reactor and steady-state isotopic transient kinetic analysis (SSITKA)[27] will be described in more detail in Section 19.5.

New trends are focused on increasing the time and space resolution of *in situ* methods with the final aim of obtaining a comprehensive picture of the working catalyst on an atomic scale. Thus, microscopic and tomographic *in situ* X-ray techniques including time- and space-resolved EXAFS (extended X-ray absorption fine structure) and XANES (X-ray absorption near edge structure) emerged with the development of synchrotron facilities and have recently been comprehensively reviewed.[1,3] Although most of their applications so far are not focused on oxidation catalysis, these methods certainly bear an attractive potential for this research field. One of the still rare examples is the spatio-temporal imaging of the Pt oxidation state in Pt-Rh/$Al_2O_3$ catalysts by X-ray absorption, during the partial oxidation of methane, showing that initially oxidized Pd is reduced after ignition within subseconds and this reduction starts at the reactor outlet and propagates to the inlet.[28]

It is impossible to comprehensively discuss all non-vibrational *in situ* techniques with a potential application to oxidation catalysts within this chapter. Therefore, we have selected only those methods for a more detailed presentation which have seen a widespread application so far and/or offer unique opportunities for understanding the functioning of real catalysts. For more specific *in situ* methods, such as the microscopy techniques mentioned above, Mössbauer spectroscopy which is restricted to the viewing of elements only,[29] or thermo-analytical studies using an oscillating microbalance reactor,[30] the reader is referred to the respective reviews.

## 19.2. Electronic (Resonance) Techniques

Catalytic redox reactions involve the participation of valence electrons of the catalytically active species, which are transition metal ions in the majority of oxidation catalysts. UV-vis spectroscopy (measured in diffuse reflectance mode, UV-vis-DRS) and EPR spectroscopy are unique methods to follow reaction-induced changes of the local coordination as well as the redox behaviour of such ions. In particular, their combination is of high value, since they are, to a certain extent, complementary. EPR can only detect species with unpaired electrons, such as reduced TMI with partially filled d-orbitals. In contrast, UV-vis spectroscopy is particularly sensitive to TMI in their highest oxidation states with empty d-orbitals, since these give rise to intense charge-transfer transitions, while d-d transition bands of reduced TMI are frequently weak and broad. It is the aim of this section to highlight the special benefits arising from a combination of UV-vis-DRS and EPR spectroscopy for the characterization of oxidation catalysts. An in-depth treatise of the theoretical background of both methods is beyond the scope of this section but can be found in the relevant textbooks.[31–33] More comprehensive descriptions of their potential for catalytic *in situ* studies are available in a number of reviews.[12–20]

### 19.2.1. *Basic principle and instrumentation of UV-vis spectroscopy*

#### 19.2.1.1. *Basic principle*

The energetic relationship between UV-vis and EPR spectroscopy is illustrated schematically in Fig. 19.1. As an example, we consider a TMI in an octahedral ligand field, e.g. a $O_5V{=}O$ vanadyl species, which is present in the important class of vanadium-containing oxide catalysts. A simplified molecular orbital (MO) scheme of such a vanadyl species is shown in Fig. 19.1a. Linear combination of the empty s- and p-orbitals of the TMI with six filled ligand orbitals gives rise to fully occupied bonding and empty antibonding orbitals, while the non-bonding d-orbitals are empty in the case of pentavalent$O_5V^{5+}{=}O$, but contain a single electron in the $d_{xy}$ orbital in the case of $O_5V^{4+}{=}O$. Irradiation with light can excite one electron from the highest occupied MO (which bears mainly ligand characteristics) to a $t_2g$ d-orbital at the metal ion. This gives rise to a ligand-to-metal charge transfer (CT) transition. In $O_5V^{5+}{=}O$ this is the only source of UV-vis-DRS bands (Fig. 19.1a, spectrum of $VOPO_4$). In the case of tetravalent $O_5V^{4+}{=}O$, there is already one single electron in the $d_{xy}$-orbital. The CT transition of this species requires higher energy (lower wavelength) due to electron repulsion. Additionally, the single electron in the $t_2g$-orbital can be excited to one of the $e_g$-orbitals. This is called a d-d transition

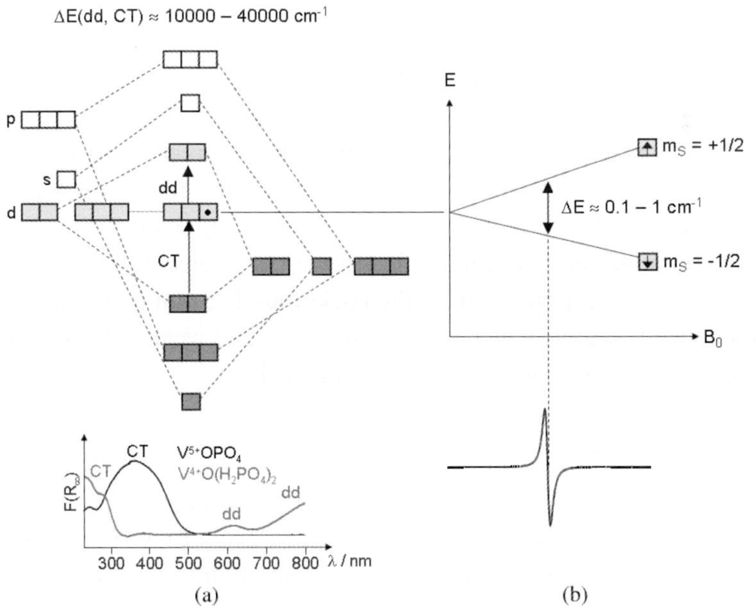


**Figure 19.1.** Energetic relation between UV-vis and EPR spectroscopy: (a) simplified molecular orbital (MO) scheme considering only $\sigma$-bonding of an octahedrally coordinated TMI with bonding (dark grey), non-bonding (light grey) and antibonding MOs and UV-vis spectra of two vanadyl phosphates; (b) simplified scheme of the electron spin level splitting in an external magnetic field giving rise to EPR signals.

(Fig. 19.1a, spectrum of $VO(H_2PO_4)_2$). Because they are symmetry-forbidden, d-d transitions of octahedral TMI are frequently poorly resolved and very weak. As mentioned above, the MO scheme in Fig. 19.1 is simplified. In the real MO scheme of $O_5V{=}O$, metal-O $\pi$-bonding and axial distortion along the O–V=O direction cause additional splitting of the MO levels, so that more than one d-d and/or CT transition occur.[34]

Unlike clear liquid solutions which are measured in transmission mode and the concentration of which is determined using the well-known Lambert–Beer law, UV-vis spectra of solids are recorded in diffuse reflectance mode. An excellent overview highlighting the peculiarities of this technique is given in the review by Jenthoft.[12] When a solid is irradiated, there are three ways of light interaction with the sample: i) specular reflection for which the angle and energy of the incident photon are equal to those of the reflected photon, ii) absorption of photons by the sample and iii) diffuse reflection. In the latter case the incident photons are scattered on particles of the solid sample and are reflected in all directions. During this scattering, they transfer energy to the sample components and excite the transitions, as pictured in Fig. 19.1a. The reflected light is measured as a function of its wavelength and usually converted to the Kubelka–Munk function $F(R_\infty)$ (Eq. 19.1, Fig. 19.1a), in which K and S are the absorption and scattering coefficients, respectively, and $R_\infty$ is the light intensity reflected from the sample, divided by the light intensity reflected from an ideally white standard.

$$F(R_\infty) = \frac{(1 - R_\infty)^2}{2R_\infty} = \frac{K}{S} \quad (19.1)$$

When S does not depend on the wavelength, $F(R_\infty)$ is proportional to the concentration of absorbing species in the sample, when the following conditions are fulfilled: i) diffuse monochromatic irradiation (for strong scatterers, parallel irradiation is also possible since specular reflection can be neglected); ii) infinite sample thickness, to avoid contribution from the background (5 mm is usually sufficient); iii) a homogeneous sample which does not fluoresce; iv) weak light absorption only (otherwise specular reflectance cannot be neglected).

### 19.2.1.2. *Instrumentation*

An instructive overview of the most frequently used instrumental equipment including photographs and schematic drawings has been given by B. M. Weckhuysen,[13] while a more comprehensive summary of the opportunities and limitations of the different experimental variants is available from F. C. Jenthoft.[12] In general, there are three main kinds of set-ups for recording the *in situ* UV-vis spectra of solid oxidation catalysts in reflectance mode:

i) A commercial diffuse reflectance accessory (the Praying Mantis made by Harrick Inc.) which consists of a system of four plane and two concave mirrors

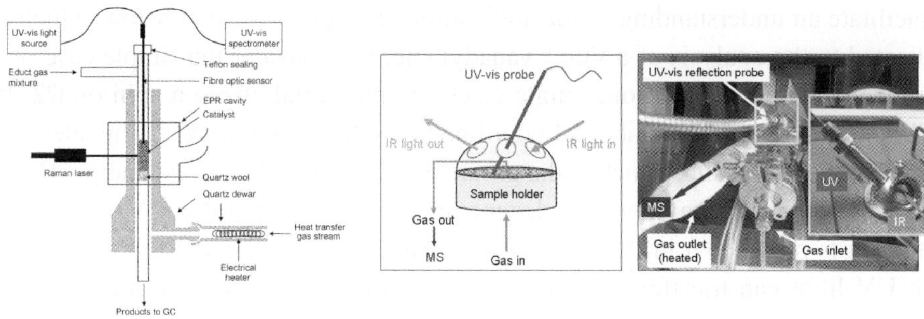


(a) (b)

**Figure 19.2.** Experimental set-ups for (a) simultaneous *operando* EPR/UV-vis/Raman spectroscopy (adopted from Ref. 15) and (b) simultaneous DRIFTS/UV-vis spectroscopy.

surrounding a stainless steel reaction chamber in which the catalyst is placed in a small cup with a sieve at the bottom which can be heated up to about 600°C and through which a gaseous reactant flow can be passed at normal or elevated pressure. With this set-up, only about 10–20% of the diffusely reflected light is collected (which can cause sensitivity problems, particularly for coloured samples, and may require dilution with an ideally white material such as MgO or $BaSO_4$), yet the specular light reflection is suppressed.

ii) An integration sphere which provides a better light yield since more than 95% of the reflected light can be detected. However, no commercial reaction chambers are available in this case and researchers have built a variety of dedicated *in situ* cells compatible with such integration spheres.

iii) Fibre-optic spectrometers equipped with commercial high-temperature reflection probes, which probably offer the highest versatility for catalytic *in situ* studies since they can be easily inserted into or attached onto catalytic fixed-bed tube reactors. There are also several examples in which this fibre-optic technique has been simultaneously coupled with other *in situ* characterization methods such as X-ray scattering, X-ray absorption, nuclear magnetic resonance (NMR) and ATR spectroscopy.[24] Set-ups for the coupling of *in situ* UV-vis spectroscopy with EPR/Raman spectroscopy or DRIFTS using fibre-optic UV-vis probes are shown in Fig. 19.2 below.

## 19.2.2. *Basic principle and instrumentation of EPR spectroscopy*

### 19.2.2.1. *Basic principle*

The theory of EPR spectroscopy is quite complex and cannot be treated in detail in this chapter. Summaries of basic aspects with focus on transition metal ions are given in reviews[14–20] and references therein, and a comprehensive treatise is available in relevant textbooks.[32,33] In this section, only some essential basics shall be mentioned to facilitate an understanding of the application example presented below, which is dedicated to the analysis of a $VO^{2+}$ vanadyl site. We consider the simple case of a paramagnetic species with one single electron and a total electron spin of 1/2. In a TMI with a distorted octahedral coordination sphere such as $V^{4+}$, this electron resides in the $d_{xy}$-orbital which is the ground state and has the lowest energy. (Note that the removal of the energetic degeneration of the d-orbitals of $V^{4+}$ by axial distortion is not shown in the simplified scheme of Fig. 19.1a). While irradiation with UV light can transfer the single electron into an excited state, i.e. into one of the d-orbitals with higher energy (Fig. 19.1a), the EPR spin transition occurs within the ground state. This explains the big energy difference of UV-vis and EPR transitions (Fig. 19.1). When an external magnetic field is applied, the spin vector of the unpaired electron aligns either parallel or antiparallel to the magnetic field vector, and these two orientations correspond to different energies, the difference $\Delta E$ of which is proportional to the applied magnetic field (Fig. 19.1b). When the sample is irradiated with microwaves, a spin flip can be induced when the energy $h\upsilon$ of the microwaves exactly matches the energetic difference $\Delta E$ between the two spin levels and an EPR signal is measured at the resonance field $B_0$ when the resonance condition (Eq. 19.2) is fulfilled, in which h is the Planck constant, $\upsilon$ is the microwave frequency, $\beta$ is the Bohr magneton and $g_e$ is the free electron g factor ($g_e = 2.0023$).

$$\Delta E = h\nu = g_e \beta B_0 \tag{19.2}$$

The energies of the two spin levels in Fig. 19.1b can be calculated using the Schrödinger equation, $H\Psi = E\Psi$, in which H is the so-called spin Hamiltonian (Eq. 19.3) with $S_z$ being the component of the electron spin vector along the direction of the external magnetic field.

$$H = g_e \beta S_z B_0 \tag{19.3}$$

Equation 19.3 is only valid for the hypothetical case of a free electron which in fact does not exist. In real oxidation catalysts, electrons move in atomic orbitals (which couples spin and orbital angular momentum) and atoms are surrounded by ligands, frequently in an anisotropic manner. Both effects, spin-orbit coupling and anisotropy of the spin density, are the reason that g is no longer a simple scalar like $g_e$ but a tensor, i.e. a $3 \times 3$ matrix which reduces to its principal components $g_{xx}$, $g_{yy}$ and $g_{zz}$ reflecting the anisotropic spin density distribution within the paramagnetic centre, with respect to space directions of the external magnetic field. A $VO^{2+}$ ion obeys axial distortion for which $g_{xx} = g_{yy} = g_{\perp}$ and $g_{zz} = g_{||}$. If the single electron belongs to an ion with a non-zero nuclear spin, the electron spin S can couple with the nuclear spin I and split the two electron spin energy levels into $2I + 1$ sublevels. In the EPR spectrum of a $VO^{2+}$ site with a nuclear spin of 7/2, this so-called hyperfine structure (hfs) interaction splits each of the two $g_{||}$ and $g_{\perp}$ components of the electron spin transition into 8 hfs subsignals, so that the EPR powder spectrum of a catalyst containing $VO^{2+}$ single sites is a superposition of 16 hfs lines (see below).[16] When a paramagnetic site contains more than one unpaired electron, dipolar interaction between these electrons can lead to more than one allowed electron spin transition and, thus, to additional fine structure (fs) splitting.[15,16] The general form of the spin Hamiltonian which comprises all these interactions is then given by Eq. 19.4, in which S and I are the electron and nuclear spin vectors, ***g*** is the g tensor, and ***A*** and ***D*** are the hyperfine and fine structure tensors, respectively.

$$H = \beta S\boldsymbol{g}B_0 + S\boldsymbol{A}I + S\boldsymbol{D}S \qquad (19.4)$$

It must be mentioned that signal splitting arising from g anisotropy, hyperfine and/or fine structure interactions are only properly resolved for paramagnetic single sites, i.e. when magnetic dipolar interactions between neighboring sites can be neglected. In catalysts with a high concentration of TMI, such interactions are usually pronounced and average out the above-mentioned signal splitting, which usually leads to more or less broad, isotropic singlets. However, with special modeling procedures based on evaluation of EPR line shapes and intensities, it is possible to derive useful information on the role of such interacting sites in catalytic oxidation reactions. This has been demonstrated for a number of bulk VPO catalysts.[15,16,35]

### 19.2.2.2. *Instrumentation*

For analyzing oxidation catalysts under reaction conditions, continuous-wave EPR spectroscopy is used. Samples are irradiated with microwaves of fixed frequency (usually about 9.5 GHz in so-called X-band) and the magnetic field is swept until the resonance condition is fulfilled. The probehead is a cavity in which a standing wave is established. Therefore, the inner dimension of the cavity must be in the order of the microwave length which is about 3 cm in an X-band. Figure 19.2a shows a homemade tube reactor within a dewar mantle, which has been implemented in a commercial rectangular X-band cavity (Bruker). It can be heated up to about 550°C using $N_2$ as the heat carrier gas. Both the inlet and outlet are connected to a gas dosing device and a gas chromatograph or mass spectrometer for product analysis, respectively.[36] This set-up has been additionally coupled with a fibre-optic UV-vis and Raman spectrometer, thus representing the first three-in-one spectroscopy. Other set-ups for *in situ* EPR measurements under flowing gases at elevated temperatures and pressures used either homemade quartz flow reactors with commercial cavities or a dedicated high-temperature/high-pressure cavity functioning as a reactor.[15,16] Finally, the pioneering work of J. Lunsford should be mentioned, who developed the matrix isolation electron spin resonance (MIERS) technique which enabled him to identify short-lived radicals in selective hydrocarbon oxidation processes.[37] With this set-up it was possible to quickly freeze out a part of the gaseous reaction mixture diluted with argon on a sapphire rod at 12 K in vacuum and transfer this rod into the EPR cavity which was also held at 12 K.

### 19.2.3. *Application examples of in situ UV-vis and in situ EPR spectroscopy in oxidation catalysis*

A wide variety of examples have been published in which these two techniques were used to explore the role of certain TMIs in heterogeneous catalytic oxidation reactions, some of which have been discussed in review papers.[12–20] An exhaustive presentation of the application potential of these methods is impossible within this chapter. Instead, an instructive example highlighting the benefits of combining EPR and UV-vis spectroscopy for elucidating active sites in vanadium-containing oxidation catalysts will be presented in more detail, followed by a summary of selected more recent applications on other oxidation catalysts, for which the reader is referred to the respective literature for a more detailed description.

#### 19.2.3.1. *In situ EPR and UV-vis studies of V-containing oxynitrides in ammoxidation of β-picoline*

Heterogeneous gas-phase ammoxidation of 3-picoline (3-PIC) to 3-cyanopyridine (3-CP), being an important intermediate in fine chemistry, has been mainly performed with vanadium-containing mixed oxides in the past. However, some years ago a new class of amorphous vanadium-aluminium oxynitrides (VAlON) was developed for propane ammoxidation, in which much higher space-time yields (STY) of acrylonitrile could be achieved in comparison with conventional $MoVNbTeO_x$ and $VSbWO_x$ catalysts.[38] These studies have recently been extended to VZrON catalysts, in which the highest space-time yields (STY) ever measured could be achieved. However, while being more active, 3-CP selectivities over VZrON catalysts were lower than over VAlON catalysts.[39] To improve the 3-CP selectivity while maintaining high STY values, the VZrON catalysts have been modified by adding phosphorus, since VPO catalysts are known as good catalysts for 3-PIC ammoxidation as well.[40] Surprisingly, a detrimental effect of phosphorus on catalytic performance was observed. By using *in situ* EPR spectroscopy for certain experiments simultaneously coupled with fibre-optic Raman and UV-vis spectroscopy, and supported by a number of conventional *ex situ* techniques, it was possible to identify structural reasons for the observed differences in the catalytic performance of VMON and VMPON oxynitrides with M = Al or Zr and V/M ratios between 0.25 and 1.

Figure 19.3 shows the *in situ* EPR spectra of the oxynitrides after 1 h exposure to air/$NH_3$ flow at 350°C, followed by switching to the whole feed. A typical EPR

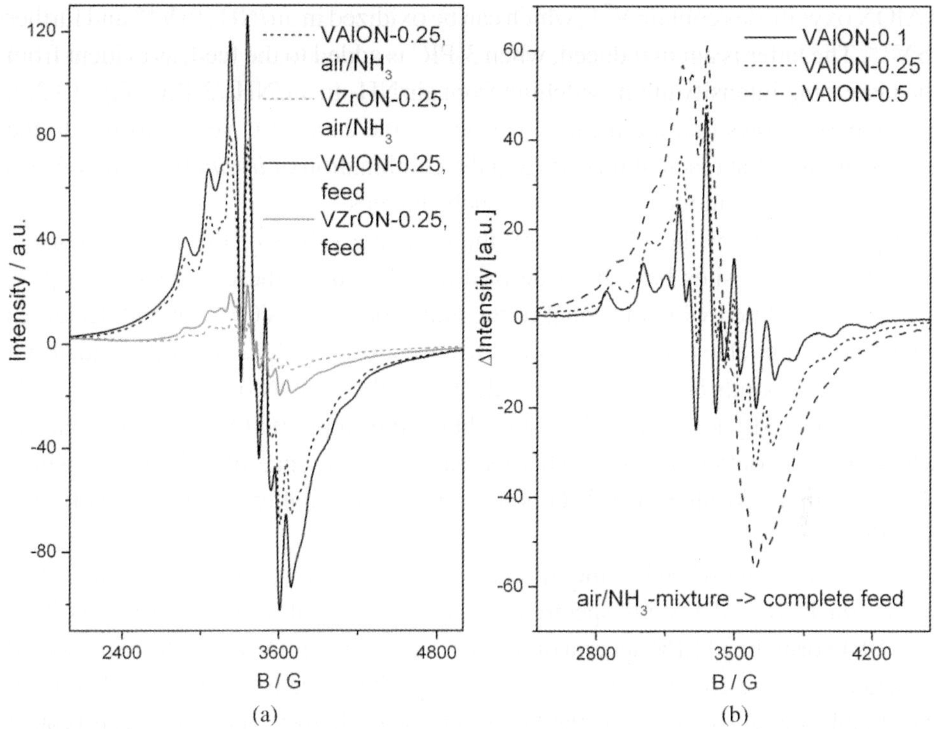


**Figure 19.3.** *In situ* EPR spectra of VMON at 350 °C. Plot (a) shows the original spectra while plot (b) shows difference spectra of VAlON catalysts with different V/Al ratios upon switching from an air/$NH_3$mixture to complete feed.

signal of $VO^{2+}$ sites is seen for both types of catalysts, but its intensity is much higher for M = Al compared with M = Zr (Fig. 19.3a) and, moreover, increases with the total V content (Fig. 19.3b). The higher concentration of reduced $V^{4+}$ in VAlON under working conditions might be one reason for the lower activity and higher selectivity compared to VZrON. Moreover, despite the lower total intensity of the EPR $V^{4+}$ signal, the broad isotropic background signal of magnetically interacting $VO^{2+}$ sites is much more pronounced in comparison with the characteristic hyperfine structure signal of $VO^{2+}$ single sites in VZrON samples (Fig. 19.3a), indicating a higher degree of V polymerization in contrast to the corresponding VAlON catalysts. This feature, which was also confirmed by UV-vis spectra, is regarded as another reason for higher activity but lower selectivity.[39]

The preceding switch from an inert $N_2$ atmosphere to air/$NH_3$ flow has also been studied *in situ*. In the EPR spectra, an increase of the $V^{4+}$ signal intensity was observed, while in the corresponding UV-vis spectra absorbance decreased at around 500 nm (corresponding to a d-d transition of $V^{3+}$ accompanied by an increase at around 750 nm (arising from a $V^{4+}$ d-d transition)).[39] This suggests that the received

VAlON oxynitrides contain $V^{3+}$, which can be oxidized in air/$NH_3$ to $V^{4+}$ and further to $V^{5+}$. The latter is again reduced, when 3-PIC is added to the feed, as evident from the increasing intensity upon switching from air/$NH_3$ to air/$NH_3$/3-PIC (Fig. 19.3a). This clearly shows that the mean V valence of the catalyst is very sensitive to the composition of the reactant mixture and the combination of *in situ* EPR and UV-vis spectroscopy is a unique tool to assess such changes.

By simulation of the EPR spectra, the principle values of the g and A tensors of isolated $VO^{2+}$ sites have been derived and, based on these, a coefficient $\beta_2{}^{*2}$ has been calculated which is equal to the unity for a pure $VO^{2+}$ ion and decreases when the single electron is delocalized towards the four ligands in the equatorial plane, i.e. the bonds become more covalent.[39] In the VAlON oxynitrides, this coefficient is markedly lower than the value obtained in supported $VO_x/Al_2O_3$ catalysts which do not contain nitrogen. This indicates the presence of V-N bonds, which should be more covalent than V-O bonds, due to the smaller difference in electronegativity.

This is different when P is incorporated. The nitridation process has been studied by *in situ* EPR and UV-vis spectroscopy for a series of VZrPON catalysts.[40] In Fig. 19.4 normalized EPR spectra of the best VZrPON with V/Zr = 0.5 are presented to better visualize changes of the spectral shape. The true intensity is evident from the double integrals given on the right side of each EPR spectrum. The typical hyperfine structure signal of isolated $VO^{2+}$ species is much more pronounced when P is added (compare to Fig. 19.3a). Moreover, two different single $VO^{2+}$ sites (S1 and S2) could be discriminated by spectra simulation. It has also been found that with rising V content the reduction of V goes partially to EPR-silent $V^{3+}$, which was also

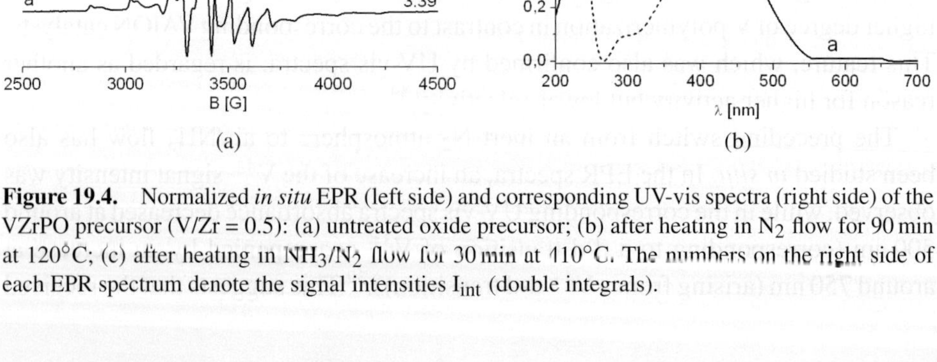


**Figure 19.4.** Normalized *in situ* EPR (left side) and corresponding UV-vis spectra (right side) of the VZrPO precursor (V/Zr = 0.5): (a) untreated oxide precursor; (b) after heating in $N_2$ flow for 90 min at 120°C; (c) after heating in $NH_3/N_2$ flow for 30 min at 410°C. The numbers on the right side of each EPR spectrum denote the signal intensities $I_{int}$ (double integrals).

supported by *in situ* UV-vis spectra that showed a strong increase of d-d transitions of reduced V species detected upon treatment with $NH_3$ (Fig. 19.4b).[40] It can be seen that the strong broad CT band of $V^{5+}$ (Fig. 19.4b, Spectrum a) is replaced upon treatment in $NH_3$ by a CT band of reduced V species at lower wavelength and d-d bands of the latter at wavelength $\geq$500 nm.

Spin Hamiltonian parameters of the isolated $VO^{2+}$ sites S1 after $NH_3$ treatment were found to be very similar to those of single $VO^{2+}$ defects in partially reduced $VOPO_4 \cdot 2H_2O$, suggesting that these V species might be connected via O-bridges to P. In contrast, parameters of the second isolated $VO^{2+}$ species S2 were similar to those derived for the corresponding phosphorus-free VZrON catalyst. From these results it has been concluded that species S2 is most probably connected via O-bridges to Zr rather than to P. The two isolated species are almost equally abundant in the untreated VZrPO (Fig. 19.4a), while the intensity of species S2 decreases in relation to that of species S1 after $NH_3$ treatment (Fig. 19.4c). This suggests that $VO^{2+}$ species connected to Zr (S2) are easier to reduce to EPR-silent $V^{3+}$ than those connected to P (S1).

The coefficient $\beta_2{}^{*2}$, being a measure of the extent to which the unpaired electron of the $V^{4+}$ is delocalized towards the ligands in the basal plane of the $VO^{2+}$ species (i.e. the degree of covalency), amounts to 0.83–0.91 in the VZrPON and is markedly larger compared with those of the VAlON catalysts ($\hat{a}_2^{*2} = 0.69-0.67$), in which V-N bonds are present. This indicates that a partial substitution of V-O by V-N moieties observed for VAlON can be excluded for VZrPON catalysts. In other words, the coordination sphere of V sites in the latter catalysts is most probably almost N-free, and this agrees completely with the results of XPS and FTIR spectroscopy (by which $P\text{-}NH_2$ and -P=N-P vibrations have been identified) which also suggest that N might be preferentially linked to P and not to V. Considering these results, it is probable that the detrimental effect of P on the catalytic performance is due to missing V-N moieties (lower selectivity) and higher V site dispersion compared to VAlON and VZrON (lower activity).

### 19.2.3.2. *Further in situ studies of oxidation catalysts by in situ EPR and UV-vis spectroscopy*

*In situ* EPR and UV-vis studies of oxidation catalysts can be divided in two main categories: 1) monitoring of catalysts at elevated temperature under flowing gases including true reaction conditions, in some cases in connection to product analysis devices such as gas chromatography (GC) or mass spectrometry (MS) (known as *operando* spectroscopy); 2) measurement of catalysts at ambient or lower temperature after pre-treatment in certain gas atmospheres and/or adsorption of probe molecules without intermediate contact to the ambient atmosphere. In addition to

**Table 19.1.** Selected recent examples of *in situ* EPR studies at elevated temperature under flowing gases.

| Catalyst | Gas atmosphere | Main focus | Ref. |
|---|---|---|---|
| Commercial SCR $V_2O_5$,$WO_3/TiO_2$ | Simulated flue gas NO,$O_2$,$H_2O$,$NH_3/N_2$ | Deactivation of active V sites by covering with polyphosphoric acids | 42 |
| Cu-ZSM zeolites | NO/Ar | Activity of $Cu^{2+}$ sites depending on their location within the pore structure | 43 |
| | NO, $NO_2$, $CH_4$, CO, $H_2O$ and/or $O_2$/He | Behaviour of $Cu^{2+}$ dependent on feed composition, identification of active site, reasons for deactivation | 44 45 |
| Fe-ZSM-5 zeolites in SCR of $N_2O$ by CO | $N_2O$ and CO in He | Identification of $O^{\bullet -}$ radical ions as active $\alpha$-oxygen on Fe single sites in combination with *in situ* UV-vis-DRS | 46 |
| Bulk VPO catalysts in selective oxidation of *n*-butane and toluene | 1.5% *n*-butane/air 1.0% toluene/air | Use of spin-spin exchange interactions as monitor for catalytic activity | 47 48 |
| $Bi_xMo_yO_z/SiO_2$ in oxidation of propene to acrolein | $C_3H_6$, $O_2$/He | Impact of $Mo^{5+}$ formed *in situ* on catalytic activity and selectivity | 49 |
| $PtO_2$/polyaniline (PANI) | $CO/O_2$ | Behaviour of conduction electrons in the PANI support as monitor for catalytic activity | 50 |

the systems described in section 19.2.3.1., there are a number of EPR studies which fall into the first of these categories. A representative though not complete selection reflecting the wide variety of catalytic redox processes that can be monitored by *in situ* EPR is listed in Table 19.1.

Besides less frequently applied monitoring of oxidation catalysts under reaction-like conditions at elevated temperature (Table 19.1), there is a wide variety of EPR studies in which catalysts have been pre-treated under model conditions with certain gases at elevated temperature and subsequently quenched to room or even lower temperature for recording the EPR spectra. Prominent examples for this approach are, for example, studies of reduction and oxidation processes in vanadia-molybdena catalysts occurring upon interaction with propene and oxygen and detection of oxygen radical species on the surface of manganese-containing catalysts[20] and studies of the interaction of cobalt centres in Co-ZSM-5 deNOx catalysts with CO, NO and $NO_2$.[41] For further examples, the reader is referred to the relevant reviews.[14–20]

## 19.3. X-ray Techniques

When talking about X-ray techniques, one has to distinguish between two principles, X-ray scattering and X-ray absorption. In the latter techniques, X-ray photons are absorbed by the sample and initiate the emission of electrons followed by secondary relaxation processes such as fluorescence or Auger electron emission. X-ray photoelectron spectroscopy and X-ray absorption spectroscopy belong to this kind of method and are described in Sections 19.3.2 and 19.3.3. In contrast, X-ray scattering is based on the elastic interaction of the X-ray beam with the sample. This means that the energy and the angle of the ingoing and outgoing X-ray beam are identical and there is no energy transfer from the radiation to the sample. One of the most widely used scattering techniques is WAXS (also known as XRD). Nowadays, each laboratory conducting research in the field of oxidation catalysis has access to XRD and different types of commercial cells for *in situ* studies of catalysts under reactive atmospheres are even available and widely used. Therefore, the section on X-ray scattering will focus on a few selected examples only.

### 19.3.1. *Wide and small angle X-ray scattering* (*WAXS and SAXS*)

#### 19.3.1.1. *Basic principle*

Both methods are based on the same physical principle, namely the elastic coherent scattering of the incident X-rays by the electron cloud of the sample, which is described by the Bragg law (Eq. 19.5) in which both differ in the size of the scattering angle θ (Fig. 19.5). In Eq. 19.5, q is the scattering vector, d is the distance between the scattering centres, λ is the wavelength of the X-rays, n is an integer number and θ is the scattering angle.

$$\frac{q}{2\pi} = \frac{n}{d} = \frac{2}{\lambda}\sin\theta \qquad (19.5)$$

This angle depends inversely on the distance between the scattering centres, which means that the incident X-rays are coherently scattered at wide angles when the dimension of the scattering structures is very small, e.g. when scattering occurs from atoms on well-ordered crystal lattice planes which are separated by a distance d being in the same order of magnitude as the wavelength of the X-rays (about 0.1 nm). This is wide angle X-ray scattering (WAXS) or X-ray diffraction (XRD) and is depicted on the left side of Fig. 19.5. Constructive interference of X-rays scattered under the same angle, θ, which leads to the well-known reflections in the XRD pattern, is only observed when the phase shift between the scattered beams is equal to an integer multiple of the wavelength. In a WAXS pattern, the position and

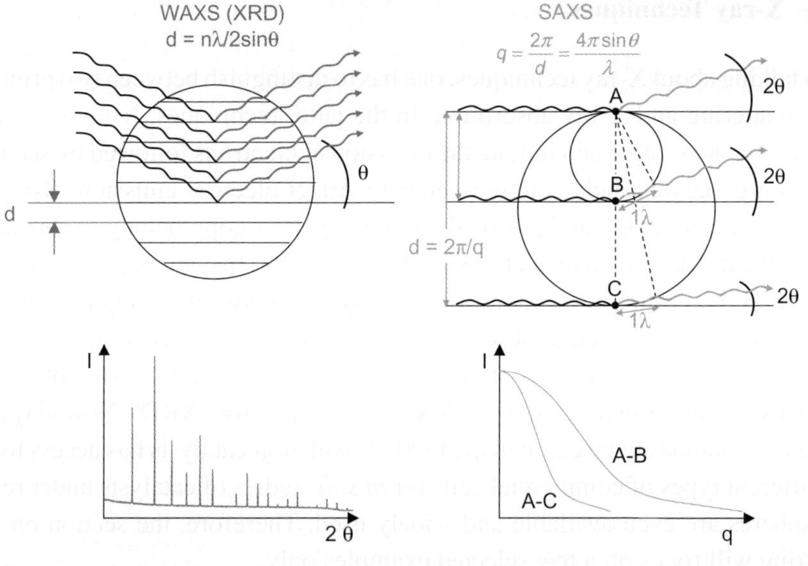


**Figure 19.5.** Simplified scheme illustrating the principle of WAXS (blue) and SAXS (red).

relative intensity of the Bragg reflections plotted as a function of 2θ (Fig. 19.5, left side), is a fingerprint of a crystalline phase and can be identified using the database of the International Centre for Diffraction Data (ICDD).[51] Usually, crystallite sizes of 3–5 nm are required to achieve sufficient signal-to-noise ratios in WAXS. Very small crystallite sizes as well as a high concentration of defects leads to changes of width and shape of the Bragg reflections that can be evaluated by suitable mathematical procedures.[52]

In contrast to WAXS, which comprises scattering from well-ordered lattice planes with very small distances, SAXS is based on diffuse scattering from inhomogeneities of the electron density in a sample, which are especially pronounced at the boundary between the particle and the surrounding medium (Fig. 19.5, right side). In general, the incident beams can be diffusively scattered from all electron clouds (atoms) of the particles at any angles. However, superposition of all scattered waves would then lead to destructive interference and essentially no scattering would be observed. However, for very small scattering angles 2θ (small scattering vectors q), the phase shift of the scattered beams also becomes very small, so that they still reinforce each other by constructive interference. Since the scattering vector q depends inversely on the distance d between the scattering centres, a SAXS pattern can only be obtained for large d values (usually in 1–100 nm range). Therefore, SAXS is ideally suited to determine particle size and shape. On the right side of Fig. 19.5 this is schematically shown for two spherical particles of different size. For the sake of simplification it is assumed that the waves scattered from points A and

B under the same angle 2θ have a phase difference of 1λ and reinforce each other. When the distance between the scattering points increases (A and C for the larger particle), it is clearly seen that the phase difference of 1λ is reached at a markedly smaller scattering angle. In the corresponding SAXS pattern, the scattering curve for the larger particle will be much narrower than for the smaller one. By fitting the experimental scattering curves to those calculated using geometrical models, it is possible to derive the particle dimensions. A comprehensive treatise of the theory of X-ray scattering can be found in textbooks.[52,53]

### 19.3.1.2. *Instrumentation*

For *in situ* XRD measurements, a number of commercial reaction chambers for use in conventional laboratory spectrometers are available, which allow treatment of catalysts at elevated temperatures in different gas atmospheres; among them are the Bühler HTK 1, the Paar XRK 900 and temperature chambers of the type TC and BTS from MRI. Two laboratory accessories used in our own laboratory are depicted in Fig. 19.6. One working in reflection mode contains a flat sample holder heated by a Pt/Rh wire up to 1,500°C. In this chamber, the catalyst is placed as a shallow bed. Although this cell can be connected to gas flows, the disadvantage is the rather high dead volume which does not allow the establishment of flow-through conditions similar to those in catalytic reactors (Fig. 19.6a). This is partially circumvented by the other set-up which operates in transmission (Debye–Scherrer) mode. Here, the reactor is a capillary of 0.7–1.0 mm diameter and a wall thickness of 0.01 mm containing the catalyst as a fixed bed through which the gas flow is directed. The X-rays penetrate through the sample and are collected by a position sensitive detector. Both set-ups can be simultaneously coupled with laser-Raman spectroscopy.[54] Recently, the first set-up has been presented in which the different plug-flow reactors of a high

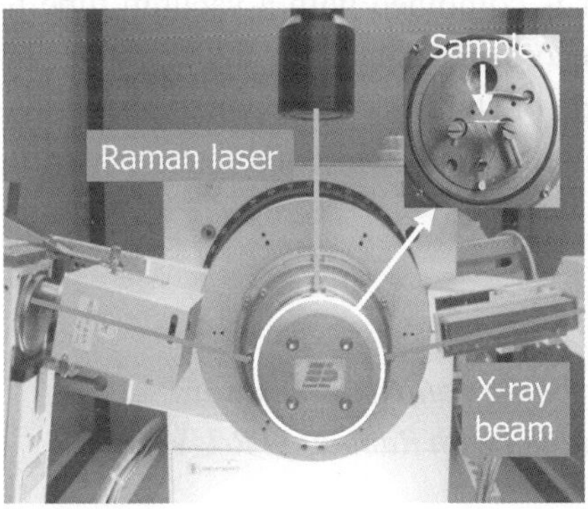


(a)

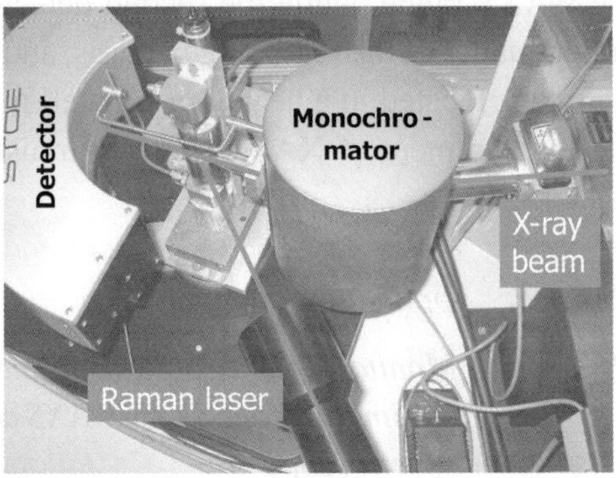


(b)

**Figure 19.6.** *In situ* XRD set-ups for measurements in (a) reflection and (b) Debye–Scherrer mode.

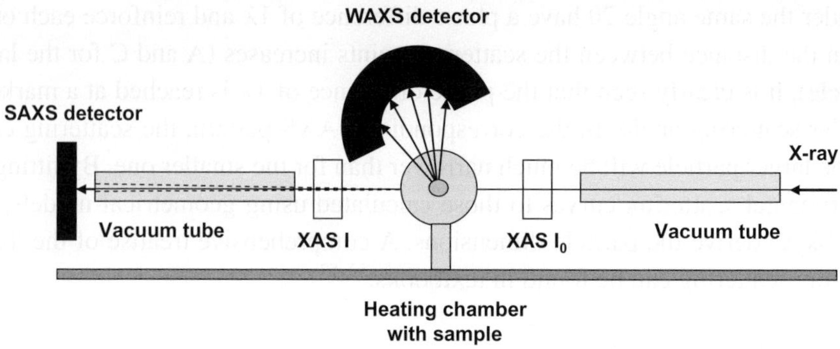


**Figure 19.7.** Scheme of a typical set-up for *in situ* SAXS/WAXS/XAS measurements.

throughput set-up have been equipped with windows through which the oxidative transformation of $V_2O_3$ could be monitored by *in situ* XRD.[55]

Performing *in situ* X-ray scattering experiments is much more advantageous at synchrotron radiation sources due to the high intensity and brilliance of the radiation facilitates, and the use of capillary reactor set-ups in Debye–Scherrer mode with acceptable signal-to-noise ratios which allow a much higher time resolution. One of the first experiments was carried out in 1993 by Clausen with a coupled XRD/EXAFS set-up.[56] Since then, a number of dedicated set-ups have been established, that include, in addition to X-ray scattering, X-ray absorption and spectroscopic techniques such as UV-vis and Raman spectroscopy.[57–60] Coupled SAXS/WAXS set-ups are available nowadays at almost all synchrotron sources. A typical design of an *in situ* SAXS/WAXS set-up (coupled with XAS) is shown in Fig. 19.7. For data collection in the small and wide angle range, two different detectors were used in transmission mode.

The curved INEL position sensitive detector located near the *in situ* cell registers the WAXS range, while the second detector is combined with a vacuum tube for better quality of the scattering patterns, and records the data in the SAXS range. The sample is located in a heated *in situ* cell made of stainless steel, which also allows the measurement of XAS spectra in transmission mode, in combination with the scattering data.

### 19.3.1.3. *Application examples*

#### 19.3.1.3a. *Monitoring the synthesis of molybdate catalyst precursors by simultaneous SAXS/WAXS coupled with Raman/ATR/UV-vis spectroscopy*

Mixed oxides such as molybdates which can also contain additional TMIs, such as iron, vanadium, tungsten or chromium, are widely used as catalysts for a variety of selective oxidation reactions, among them the partial oxidation of propane and propene to acrolein and acrylic acid or of their $C_4$ analogues to methacrolein and methacrylic. Such polyoxometallates are often prepared by co-precipitation procedures which involve a number of subsequent synthesis steps. The composition and structure of the products usually depends on synthesis conditions such as temperature, pH value and the sequence of adding the components. Frequently it is difficult to ensure reproducibility of such synthesis procedures as the influence of those parameters on the phase structure is poorly understood. This calls for the application of *in situ* methods which allow the subsequent preparation steps to be followed in detail. Very recently a set-up has been presented in which, for the first time, SAXS and WAXS have been simultaneously coupled with Raman, FTIR(ATR) and UV-vis spectroscopy at the μ Spot Beamline of the BESSY synchrotron facility in Berlin,to monitor the synthesis of iron-molybdate catalyst precursors, which are used for the selective oxidation of methanol to formaldehyde.[61] For SAXS/WAXS/Raman measurements, the slurry was continuously circulated from the reaction vessel through a glass capillary, on which the X-ray beam of the synchrotron source and the laser beam of a fibre-optic Raman spectrometer were focused. Additional fiber-optic probes for measuring FTIR(ATR) and UV-vis spectra were directly introduced into the reaction vessel. The synthesis procedure has been subdivided into three subsequent steps: 1) mixing of aqueous solutions of $Fe(NO_3)_3$ and ammoniumheptamolybdate $(NH_4)_6Mo_7O_{24} \cdot 4H_2O$ (AHM) at 20°C followed by 1 h stirring, 2) addition of a $(NH_4)_2HPO_4$ solution at 20°C followed by 0.5 h stirring and 3) heating of the slurry to 50°C while stirring for 1 h.

X-ray scattering curves recorded every two minutes show a significant intensity decrease during step 1 in the SAXS range at scattering vectors of $q < 5\,nm^{-1}$ (Fig. 19.8a), which indicates that the formed precipitate contains polydispersed particles that agglomerate, gradually reaching sizes above 5 nm. The sharp reflections which appear immediately after the start of the synthesis point to the formation of a crystalline phase, which has been identified as the so-called Anderson phase of the type $(NH_4)_3H_6MeMo_6O_{24} \cdot 6H_2O$ (Me = Rh, Fe, Ni). It can be seen from Fig. 19.8b that this phase prevails throughout all three synthesis steps. A second crystalline phase appeared after about 30 minutes during step 1, which could be identified as a compound containing $[Mo_8O_{26}]^{4-}$ structural units, based on simultaneously recorded Raman spectra.

Upon addition of $(NH_4)_2HPO_4$ in step 2, the Bragg reflections of the $[Mo_8O_{26}]^{4-}$ containing-phase disappear immediately, indicating dissolution and/or amorphization of this phase (Fig. 19.8b). In the SAXS range of small scattering vectors ($q < 5\,nm^{-1}$), intensity increases again. By modeling the slope of the scattering curve it has been concluded that during step 2, cylindrical particles of about 1 nm diameter and 4 nm length were formed (Fig. 19.8a).

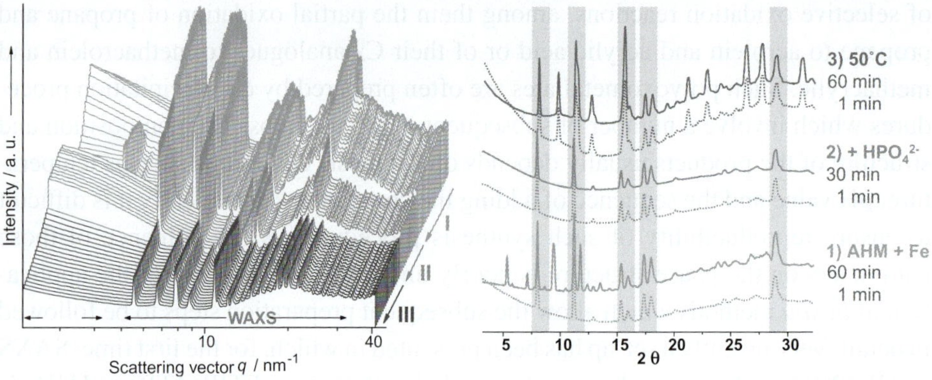


**Figure 19.8.** SAXS/WAXS patterns recorded every two minutes during synthesis steps 1–3 (left) and selected WAXS patterns, in which the reflections of the stable Anderson phase are marked (right).

Upon heating to 50°C (step 3), SAXS intensity decreases again pointing to particle growth. In the corresponding wide angle range, new reflections appear in addition to those of the Anderson phase. Based on FTIR(ATR) and UV-vis results, it was determined that these reflections originated from a phase in which phosphorus was incorporated into the polymolybdate anions. This probably reflects the beginning of the formation of heteropolymolybdate anions, similar to those present in Keggin-type heteropolyacids. While WAXS clearly indicates the presence of crystalline phases, this method cannot identify their chemical nature as long as the respective patterns are not contained in the powder diffraction file database. Yet this information can be derived from other techniques such as UV-vis, FTIR and Raman spectroscopy. This illustrates the added value that derives from simultaneous coupling of X-ray scattering and spectroscopy.

### 19.3.1.3b. *Monitoring phase changes of heteropolyoxometallates (HPOM) during redox reactions*

Heteropolyacids and their corresponding salts, among them those containing anions with Keggin structure, are useful catalysts for a number of selective oxidation reactions, e.g. the oxidation of propane and isobutene to acrolein and methacrolein and the corresponding acids. The crystal structure of such heteropolyoxometallates (HPOMs) is usually very sensitive to temperature and composition of the reactant mixture, which is a major reason for their undesired fast deactivation during catalytic redox reactions. *In situ* XRD is thus a unique tool for analysing reaction-dependent changes of such materials and quite a few such studies have been performed in recent years to investigate structural changes of HPOMs under the influence of temperature and gas flow.[62–64] Ressler studied changes of $H_3[PMo_{12}O_{40}] \cdot 13H_2O$, $Cs_2H[PMo_{12}O_{40}]$ and $Cs_3[PMo_{12}O_{40}]$ by *in situ* XRD and *in situ* XAS during thermal treatment in a reducing atmosphere (5% $H_2$ or 10% propene/He), and under the conditions of partial oxidation (10% $O_2$,10% propene/He).[65] The *in situ* XRD experiments were performed in a Bühler HDK S1 high-temperature chamber under laboratory conditions while the *in situ* XAS transmission experiments were carried out in a flow cell at the Beamline X1,located at the synchrotron radiation source HASYLAB (Hamburg). The outlet of the reaction cells was connected to mass spectrometry for establishing structure–reactivity relationships.

In these experiments, it was observed that the initial compound $H_3[PMo_{12}O_{40}] \cdot 13H_2O$ loses crystal water molecules at around 37°C and is transformed to the octahydrate and subsequently to water-free $H_3[PMo_{12}O_{40}]$, both phases being defined crystalline compounds. Under reducing conditions and above 300°C, Mo ions migrate from the lattice positions of the Keggin anions to interstitial sites outside these anions. Around 325°C another crystalline phase was passed, the structure of which is not yet known. Above 327°C, this phase converted to a cubic HPOM phase with a structure similar to that of the $Cs_2H[PMo_{12}O_{40}]$ salt. Finally, the cubic water-free $H_3[PMo_{12}O_{40}]$ phase decomposed above 500°C, accompanied by the formation of $MoO_3$ and $MoO_2$. A similar decomposition has been observed for the $Cs_2H[PMo_{12}O_{40}]$ salt, while $Cs_3[PMo_{12}O_{40}]$ remained stable under the same conditions.

During propene partial oxidation, migration of Mo ions out of the Keggin ions into interstitial lattice sites has also been observed at around 300°C. Simultaneously, the onset of catalytic activity was detected for $Cs_2H[PMo_{12}O_{40}]$ and $H_3[PMo_{12}O_{40}]$ along with a partial reduction of molybdenum ions. From this observation it has been concluded that defective Keggin anions must be present in the HPOMs in order to become effective heterogeneous catalysts. Similar experiments were also performed with the vanadium-containing analogue $H_4[PVMo_{11}O_{40}] \cdot 13H_2O$.[65] It has been claimed that a cubic $Mo_x[PVMo_{11x}O_{40}]$ phase is formed above 300°C in which Mo is partly expelled from the Keggin anions, while vanadium is supposed to remain inside. This phase was stable up to 347°C and considered to be responsible for the onset of catalytic activity at 300°C. However, it should also be mentioned that investigations of the same system by various spectroscopic *in situ* methods such as EPR/UV-vis/Raman, FTIR, $^1H$ and $^{51}V$-MAS-NMR spectroscopy suggested that a partial extraction of vanadium from the Keggin structure, followed by the stepwise connection of the removed V species via oxygen bridges to the outer surface of the defective Keggin ion occurred during calcination and remained during subsequent isobutene oxidation.[66] This again illustrates the added value of combining the results of different *in situ* techniques.

## 19.3.2. *X-ray absorption spectroscopy (XAS)*

### 19.3.2.1. *Basic principle and instrumentation*

Due to its broad field of application, XAS is one of the more frequently applied analytical methods for deriving structure–reactivity relationships in catalysis. XAS provides information on the average local electronic and geometric structures of elements in a wide variety of working catalysts such as high-surface area-supported catalysts, bulk catalysts and also homogeneous catalysts. The accessible temperature and pressure range is not limited by the principle of the method itself, only the technical parameters of the reaction cells define the limits, as for all synchrotron-based methods.

Upon irradiation with X-rays, an X-ray photon is absorbed by the atom and ejects a photoelectron from a core energy level, when the photon energy $h\nu$ is higher than the binding energy $E_b$ of the electron in the core level. This process leads to a characteristic edge in the absorption spectrum (Fig. 19.9). The photoelectron, which has both particle and wave characteristics, leaves the atom with a kinetic energy $E_{kin} = h\nu - E_b$ and is scattered onto the neighbouring atoms in the sample. The waves of the ejected and backscattered photoelectrons interfere and enhance or destroy each other, depending on their wavelength and phase shift, and the distance of the atoms involved in the scattering event. As a consequence, the X-ray absorption coefficient is modulated which leads to oscillations above the absorption edge. This is called extended X-ray absorption fine structure (EXAFS). In general, an XAS spectrum can be divided into three regions, namely the pre-edge, the X-ray absorption near edge structure (XANES) and the EXAFS region (Fig. 19.9). From the shape of the XANES spectrum the average valence state of the absorbing atoms can be determined, while the pre-edge peaks provide valuable information on their

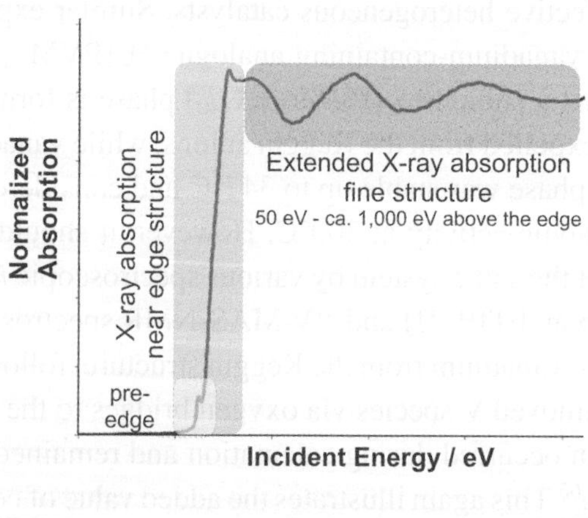


**Figure 19.9.** Schematic X-ray absorption spectrum with the different regions indicated.

coordination geometry. From the EXAFS spectrum, the type, the distance and the number of neighbouring atoms can be determined. These data also indicate the size and shape of the nanoparticles and the Debye–Waller factor.[67]

X-ray absorption spectroscopy is usually performed at synchrotron radiation sources, which provide intense and tunable X-ray beams. The photon energy is tuned to a range where core electrons can be excited (0.1 up to 100 keV), using a crystalline monochromator. A part of the X-ray flux is absorbed during penetration of the sample. For the measurement of this absorption, different methods are used. The simplest way is to measure the photon flux before and after the penetration through the sample with an ionization chamber. This is known as the transmission mode. This method is only applicable when the samples are thin enough to be penetrated and contain a rather high number of the atoms which can be analysed. Other approaches to measure the absorption are the electron yield and the fluorescence mode. In these cases, the hole left in the core level of the excited atom after photoelectron ejection is refilled by an electron from a higher level accompanied by the ejection of another electron from the same level (Auger process, electron yield mode) or by the emission of X-ray photons (fluorescence mode). The energy of the emitted Auger electrons or X-ray photons is characteristic for the absorbing atom. While the electron-yield method is more surface sensitive and often used in vacuum, the fluorescence mode is bulk sensitive and applicable in different atmospheres and for element concentrations down to the ppm range. The mode of measurement influences the design of the sample cell which also serves as a catalytic reactor in order to obtain data of catalysts in the working state. There are a wide variety of different cells designed by different research groups, often for a particular reaction or catalyst. A comprehensive review of the cell designs and application examples is presented in Ref. 10 while a more detailed treatise of the theoretical background can be found in Refs. 6–9.

#### 19.3.2.2. *Application examples*

The importance of XAS spectroscopy in oxidation catalysis research results from its ability to provide information on the local structure of active redox sites under reaction-like conditions, which can often hardly be obtained by other methods, such as XRD which is restricted to crystalline material, EPR spectroscopy which only detects species with unpaired electrons, or UV-vis diffuse reflectance and Raman spectroscopy which are less sensitive to strongly absorbing TMI in reduced valence states. Thus, *in situ* EXAFS has been widely used to determine the local environment and redox behaviour of TMI in nanoporous oxides. A typical example is discussed in 19.3.2.2a for Fe-ZSM-5 zeolites. Another major application is the determination of size and shape of supported metal clusters, especially for particles smaller than

2 nm.[68] Although this has, so far,mainly been applied to metallic catalysts used for hydrogenation reactions, it is a general procedure, which can likewise be used for metal catalysts in oxidation reactions. A detailed description of data evaluation and simulation procedures is given elsewhere.[5,69]

The XANES region has frequently been used to determine the average oxidation state of an element during interaction with reactants, since the energy of the absorption edge and the intensity of the white line above the edge, are sensitively dependent upon the oxidation state. With higher oxidation states, the absorption edge is shifted to higher energies. For electron states with a filled d-band (e.g. for noble metals) a simple step function could be expected at the edge, while a high density of unoccupied d-orbitals (e.g. usually highly oxidized TMI) leads to a pronounced peak (white line) above the absorption edge. Moreover, the XANES region can be used to determine the coordination geometry of the absorbing atom, as illustrated below.

#### 19.3.2.2a. *Behaviour of framework-substituted Fe-ZSM-5 zeolites*

The investigation of zeolites containing transition metal ions is a major application of *in situ* XAS and is used to elucidate the nature of the active sites. Iron-containing zeolites such as Fe-ZSM-5 have attracted particular attention due to their unique properties in selective hydrocarbon oxidation. Thus, Fe-ZSM-5 was found to oxidize benzene to phenol even at room temperature.[70] Usually, such zeolites are activated in air to remove the template from the pores, a process during which water is also liberated. For zeolites in which Fe had been initially incorporated in tetrahedral Si and/or Al framework positions during synthesis, it is possible that these Fe ions are extracted to extra framework positions during calcination and agglomerate to catalytically inactive iron oxide nanoparticles, in which Fe may be octahedrally coordinated. XAS spectroscopy is a suitable method to detect this undesired effect, since it is possible to distinguish between tetrahedral and octahedral Fe coordination by a detailed analysis of the pre-edge region in the XANES spectrum. A pre-edge peak before the adsorption edge is only detected for Fe ions in a non-centrosymmetric environment which is obeyed in tetrahedral coordination. Octahedrally coordinated Fe ions do not show this pre-edge peak. The intensity of the pre-edge peak can be used to determine the ratio between ions in octahedral and tetrahedral coordination. However, a quantitative analysis is only possible with a clear separation between the pre-edge peak and the main edge, which requires a sufficiently high energetic resolution of the spectra.

In a study performed at the Advanced Photon Source of the Argonne National Laboratories (Chicago), a Fe-ZSM-5 zeolite with Fe in framework sites has been treated after the synthesis of the precursor by different heating procedures using

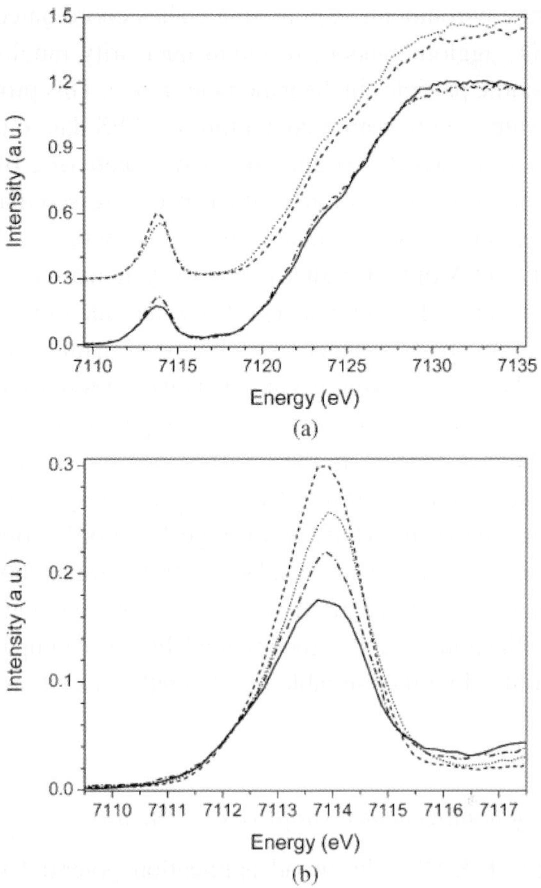


**Figure 19.10.** (a) XANES spectra of framework-substituted Fe-ZSM-5 after template removal (dashed line), after calcination in air at 520°C (dotted line), after mild steaming for 4 h at 550°C with a water partial pressure of 200 mbar in $N_2$(dash-dot line) and after hard steaming for 5h at 600°C with a water partial pressure of 300 mbar (solid line). The spectra were measured in a flow of oxygen (7.7 vol% in He) at 350°C. (b) Zoom of the pre-edge region. Adapted from Ref. 71.

nitrogen and air with and without steam to remove the template.[71] Fig. 19.10 shows the XANES spectra measured in a flow of $O_2$/He at 350°C detecting the Kβ fluorescence radiation in the region of the Fe K edge. It can be seen that the intensity of the pre-edge peak depends on the preceding thermal treatment. In particular, a steam-containing atmosphere leads to a lower pre-edge peak intensity. These results show that the presence of water during thermal treatment favours the extraction of Fe ions from a tetrahedral framework to an octahedral extra framework position. No evidence for the presence of any crystalline $Fe_xO_y$ species was found.

The limitations of EXAFS for analysing real catalysts must also be mentioned and Fe-ZSM 5 is a good example for this purpose. Frequently, those samples contain

Fe ions in different environments, e.g. as single sites coordinated by four or more O atoms and $Fe_xO_y$ agglomerates of different nuclearity ranging from a few Fe atoms only up to oxidic particles in the nanometer range. This problem is discussed in a review,[72] in which the literature containing EXAFS data on Fe-O, Fe-Fe and Fe-Si/Al distances in Fe-ZSM-5 is analysed and a scattering of about 0.4 Å for each of the different distances was found. It has been stressed that such large deviations cannot be explained by statistical error of the method, which is usually in the range of about 0.02 Å only. It rather points to systematic errors in the EXAFS analysis as a reason for such discrepancies. These are due to the fact that parameters used for determining bond distances and coordination numbers are derived from guessed models of the Fe site assuming Gaussian distributions of the bond distances, which may deviate from the real situation in the sample. Another problem concerns the nuclearity of the iron species. In the EXAFS spectrum, contributions from the different species superimpose and it is frequently not easy or may be impossible to differentiate between them. Thus, an average Fe coordination number of one can reflect Fe dimers as main Fe species. However, the same coordination number would also be obtained from a sample containing 50% Fe single sites alongside 50% trimers. This illustrates that it is recommendable to accompany EXAFS investigations of Fe zeolites by other suitable *in situ* methods such as EPR and UV-vis spectroscopy.[45]

#### 19.3.2.2b. *Study of homogeneous oxidative esterification*

Another advantage of XAS is its broad application potential for very different samples. Although the majority of *in situ* investigations in oxidation catalysis are performed on solid catalysts, homogeneous catalytic oxidation reactions can be followed as well. Recently, *in situ* XAS investigations were performed at the Hamburger Synchrotronstrah lungs labor (HASYLAB) to investigate the role of a gold catalyst in the oxidative esterification of aldehydes to the respective esters.[73] The main aim of this study was to clarify whether this reaction is really catalysed homogeneously by mononuclear gold complexes or heterogeneously by small gold nanoparticles formed during the reaction. Benzylaldehyde and *n*-butanol were chosen as the educts, tert-butyl hydroperoxide as the oxidant and $HAuCl_4$ with a pyridine additive as the catalyst. The XANES analysis of the absorption edge points to the existence of gold species in different oxidation states in the reaction mixture (Fig. 19.11a). At the beginning of the reaction the XANES spectra of the reaction mixtures (X, XI, XII) show the white line being typical for threevalent gold (spectrum VIII of an $HAuCl_4$ reference sample). After 24 h (XIII) the spectrum is similar to the absorption edge of AuCl (IX) indicating the reduction of threevalent to monovalent gold. Metallic gold could not be found in the mixture, which thereby excluded

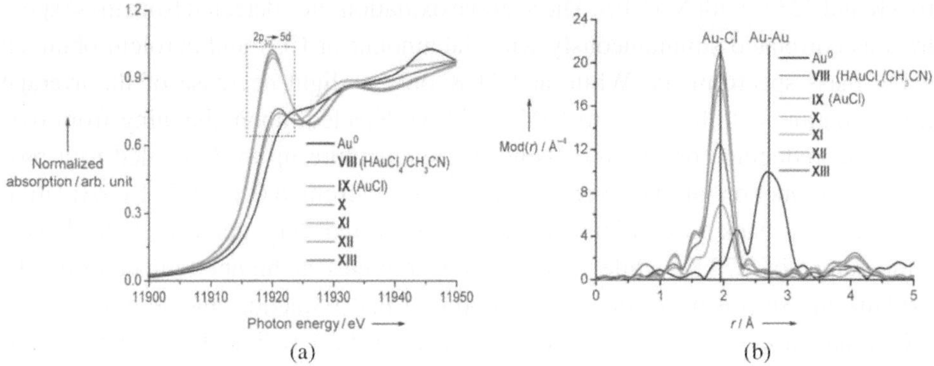


**Figure 19.11.** XANES spectra depending on the oxidation states (a) and Fourier-transformed EXAFS spectra (b) showing Au-Au bond distances in an Au foil ($Au^0$) and Au-Cl distances in $HAuCl_4/CH_3CN$ and AuCl reference samples and different reaction mixtures: X: $HAuCl_4/CH_3CN$ + pyridine derivate; XI: + benzaldehyde +*n*-butanol; XII: + TBHP; XIII: XII after 24 h. Adapted from Ref. 73.

the existence of gold nanoparticles. These observations were supported by EXAFS (Fig. 19.11b) which evidenced only Au-Cl neighbours but no hint of Au-Au bonds. These results proved that the reaction is catalysed homogeneously.

#### 19.3.2.2c. *Time-resolved studies of a Keggin-type partial oxidation catalyst under dynamic conditions*

An important subject in heterogeneous catalysis is the investigation of the reactivity of solids and the kinetics of solid-state reactions, e.g. the response of the catalyst to changing reaction conditions. Such structural changes or phase transitions can only be obtained from time-resolved investigations. For these kinds of investigations special equipment such as quick EXAFS (QEXAFS)[74] were developed, which allowed measurements down to several tenths of a second. An even better time resolution in the millisecond range can be reached with energy-dispersive XAFS.[75] QEXAFS investigations were performed with Keggin-type heteropolyoxomolybdates (e.g. $H_3[PMo_{12}O_{40}]$) which are active catalysts for the partial oxidation of alkanes and alkenes, e.g. of propene to acrolein.[76] In these catalysts, the redox behaviour of the Mo ions, i.e. their ability to switch between different valence states depending on the surrounding atmosphere, has a crucial impact on the catalytic properties.

In this study the average molybdenum valence state of an activated $H_5[PV_2Mo_{10}O_{40}]$ was monitored during isothermal switching of the gas phase from oxidizing (propene/$O_2$) to reducing conditions (propene) in a temperature range between

673 K and 723 K with XANES. The average oxidation state detected from these spectra was correlated simultaneously with the amount of $CO_2$ and acrolein obtained using mass spectrometry. While at 673 K only a slight decrease of the average Mo oxidation state from 6.0 to 5.94 could be detected, upon changing from oxidizing to reducing conditions, raising the temperature up to 723 K led to a pronounced reduction of the average valence state from 6.0 to 5.5. Re-oxidation by switching back to oxidizing conditions occurred very quickly at the lowest temperature of 673 K, while it took a few minutes at higher temperatures. To explain this surprising result it was supposed that reduction and re-oxidation at low temperatures take place only in the near-surface region, but penetrate into the bulk at higher temperatures. Since re-oxidation is a diffusion-limited process, bulk re-oxidation at higher temperatures is slower than surface re-oxidation at low temperatures.

Simultaneously, amounts of acrolein and $CO_2$ were detected. The fast re-oxidation at 673 K was accompanied by a rapid regaining of activity and selectivity for both acrolein and $CO_2$. Re-oxidation at higher temperatures led to catalysts being less selective for acrolein. Based on this result it is clear that the electronic structure governs the selectivity of the catalysts, which is sketched in Fig. 19.12. For V, an analogous exact determination of the valence states was not possible, because the lower photon energy of the V K edge requires longer measuring times and hence does not permit time-resolved measurements. $V^{4+}$ was found as the major oxidation state under both reducing and oxidizing conditions. Only a minor part of V is pentavalent which is proposed to be not active in this reaction due to missing coordination sites for the reactants. From these observations the following model was proposed (Fig. 19.12). At partially reduced $V^{4+}$-$Mo^{5+}$ sites, total oxidation of propene to $CO_2$ dominates while oxidized $V^{4+}$-$Mo^{6+}$ sites favour the partial oxidation of propene to acrolein. Consequently, reaction conditions should be chosen so as to allow a rapid re-oxidation of the $Mo^{5+}$ centres formed during reaction.

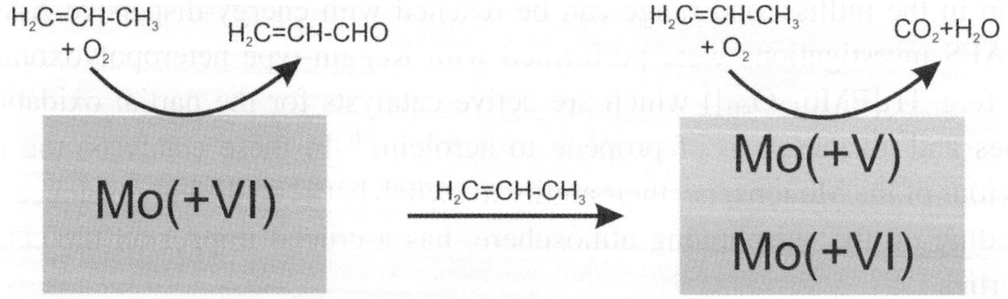


**Figure 19.12.** Sketch of the correlation between Mo oxidation state and oxidation of propene. At Mo(+VI) sites the partial oxidation to acrolein dominates, whereas at Mo(+V) sites the total oxidation occurs.

## 19.3.3. *X-ray photoelectron spectroscopy (XPS)*

### 19.3.3.1. *Basic principle and instrumentation*

The crucial elementary steps in heterogeneous catalysis such as adsorption, dissociation and interaction of reactant molecules, and formation and desorption of products take place at the solid surface of the catalyst. These steps are sensitively influenced by the composition, the electronic and geometrical structure of the surface, which may differ markedly from the bulk properties. The vital role of such surface processes in heterogeneous oxidation catalysis was one of the driving forces in the development of surface analytical techniques.

X-ray photoelectron spectroscopy is generally accepted as one of the best methods for analysing the composition and the electronic structure of solid surfaces and interfaces. In XPS, the sample is irradiated by X-rays from a laboratory source, e.g. from a Mg or an Al anode, or from a Synchrotron storage ring. As in XAS spectroscopy, the X-ray photon absorbed by a surface atom ejects a photoelectron from a core energy level, when the photon energy $h\upsilon$ is higher than the binding energy $E_b$ of the electron in the core level. This electron travels with a kinetic energy $E_{kin}$ to the detector. In an XPS experiment, the number of photoelectrons, which depends on the intensity of the X-ray radiation, and their kinetic energy which is governed by the X-ray wavelength, are simultaneously measured. The binding energy, being a characteristic measure for the valence state and environment of this element in the surface, is calculated according to Eq. 19.6, in which $E_b$ is the binding energy of the ejected photoelectron, $E_{kin}$ is its kinetic energy (experimentally measured), $h\nu$ and $\varphi_{sp}$ are the energy of the X-ray radiation and the work function of the spectrometer, respectively. The latter two parameters are known.

$$E_b = h\upsilon - E_{kin} - \varphi_{sp} \qquad (19.6)$$

Due to collisions of the emitted photoelectrons with atoms in the solid, only electrons from the near-surface region can leave the sample, i.e. the mean free path of the electrons is restricted to a few numbers of atomic layers, which is the reason that this technique is surface sensitive. After leaving the sample, collisions of the electrons and the gas environment must be avoided. Therefore, XPS is routinely used in ultra-high vacuum at a pressure below $10^{-9}$ mbar. Yet, catalytic reactions proceed at ambient or even higher pressure. This pressure gap implies that the surface features analysed by conventional XPS are not necessarily identical with those prevailing under catalytic reaction conditions.

Since the development of the modern surface science techniques in the 1960s, some efforts have been made to bridge this pressure gap. First attempts have been called "quasi *in situ*" experiments, in which the catalyst is pre-treated under reaction conditions at elevated temperature and pressure in a special reaction cell, followed by quenching to room temperature, evacuation and transfer to the ultra high vacuum (UHV) analysis chamber of the spectrometer. This procedure allows the analysis of a "frozen" active state of the catalyst surface and prevents air contact or other contamination during the transfer. It permits the determination of the structure and composition of the catalyst surface before and after the catalytic reaction carried out under realistic conditions.[77] The benefits of this approach have been demonstrated, e.g. in studies of supported Au catalysts which were used for several oxidation reactions. After activation in $H_2$ at 300°C, a negative binding energy shift of the Au 4f electrons was observed depending on the oxidic support, which could not be seen after contact with air.[78] Another example comprises supported Ru oxidation catalysts. The catalytically active $RuO_2$ phase cannot be formed under UHV conditions. Therefore, pre-treatment in an $O_2$ flow at temperatures above 475 K has been performed. Detailed investigations in UHV with XPS, electron diffraction and thermal desorption were undertaken to clarify the structure.[79,80] A "surface oxide" consisting of oxygen incorporated between the first and second atomic layer can only be formed under normal pressure conditions. The disadvantage of this "quasi *in situ*" approach is that the catalyst surface is not investigated under reaction conditions. Composition and structures stable only under these conditions and instable in UHV are not accessible with these investigations. Thus, bridging the above-mentioned pressure gap by designing equipment which allows XPS measurements of the catalytically active surface under a certain reaction pressure is one of the most challenging aims for further technical development.

As described above, the major problem hindering XPS investigations under real conditions is the strong interaction between the emitted electrons and the gas environment. All methods allowing such measurements are based on the idea of capturing the electrons at distances comparable with their mean free path in the gas atmosphere. This means that emitted electrons with a typical kinetic energy of a few hundred eV have to be detected at a distance of a few millimetres from the sample surface when the total pressure is around 1 mbar. This is accomplished by small apertures near the sample and by differential pumping on the other side of the aperture to decrease the pressure and, thus, the collision rate. A typical set-up with an aperture is sketched in Fig. 19.13.[81]

The first system for the investigation of solids in gas atmospheres at pressures up to 1 mbar was reported by Joyner, Roberts and Yates 1979.[82] In this system, the photoelectrons were simply admitted and spatially filtered through the aperture separating the high-pressure environment from the detection chamber. The next generation of systems established at the Synchrotron Storage Ring Advanced Light Source in Berkeley (ALS) and the Berliner Elektronen-Speicherring-Gesellschaftfür Synchrotron Strahlung (BESSY), used an electrostatically controlled and differentially pumped pre-lens system to focus the

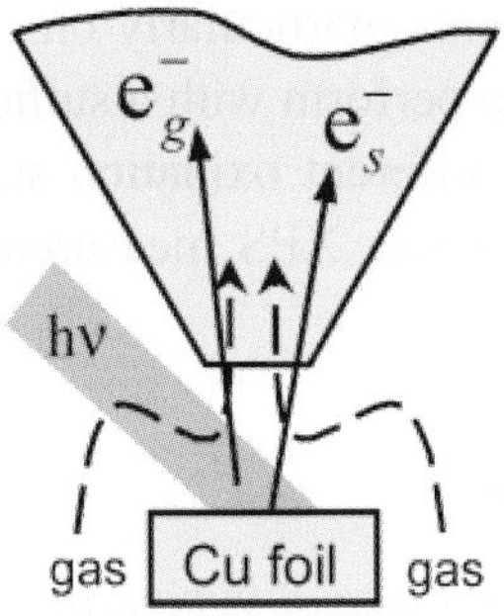


**Figure 19.13.** Sketch of the sample first aperture geometry of an *in situ* XPS experiment. Not only is the surface of the solid sample irradiated by the X-ray, but also the gas molecules. The distance between sample and aperture is 2 mm. Adapted from Ref. 81.

emitted photoelectrons through the aperture.[83,84] These so-called ambient pressure XPS systems allow measurements up to 100 mbar. Recently, a new end station has been designed with a new instrument at the ALS that includes a manifold increase in the photoelectron transmission leading to a better signal-to-noise ratio and a decreased acquisition time.[85] In this system an imaging mode with a spatial resolution of ~16 μm and an angle resolved mode was developed.

Another system using electrostatic focusing and differential pumping was established in both synchrotron and laboratory versions. In contrast to the other equipment, these set-ups are coupled with a molecular beam dosing system, which allows local pressures at the sample to be much higher than the background pressure in the chamber.[86,87] By using two or more molecular beams with different reactants, catalytic reactions on the surface can be directly studied.

Despite the fact that *in situ* XPS is an interesting technical development and significant technical advances have been achieved, especially in the last ten years, its application is still rather limited. This might be due to the scarce availability of dedicated instruments and the limited data quality compared with modern UHV XPS equipments. Another problem is the charging of insulating samples, especially at synchrotron sources. The loss of photoelectrons during the XPS investigations leads to a positive charging of the sample which shifts and broadens the peaks in the spectra, thus making the interpretation of the data sometimes impossible. The traditional way to solve the problem is the use of electron flood guns that send low energy electrons to the surface and neutralize the sample. In recent years, this neutralization was optimized for high-performance XPS equipment using UHV. In general, using ambient pressure XPS, the ionized gas species created by the incident X-rays discharge the sample effectively, similar to flood guns. Unfortunately, especially for strong insulators like alumina, the charge neutralization is not as homogeneous as with modern flood gun systems, resulting in partial charging. For this reason, *in situ* measurements, particularly on catalysts supported on insulating oxides, are very difficult to perform with a sufficiently high energy resolution, as is needed for determining different oxidation states. In the following section, the use and benefit of such *in situ* XPS measurements is illustrated with some examples.

#### 19.3.3.2. *Application examples*

##### 19.3.3.2a. *Evolution of the structure of bimetallic noble metal nanoparticles under reaction conditions*

Bimetallic nanoparticles are used in numerous partial oxidation reactions such as alcohol oxidation[88] and the gas-phase acetoxylation of ethylene to vinyl acetate[89,90] or toluene to benzyl acetate.[91] In many cases, the structure and composition of such bimetallic nanoparticles are unknown under reaction conditions, since most structural and compositional studies have been performed *ex situ*. Recently, a study has been performed to obtain new insights into the structural changes of bimetallic nanoparticles under oxidizing and reducing conditions using XPS at a pressure of 0.1 mbar as the advanced light source in connection with TEM investigations.[92] Three series of bimetallic Rh-Pd, Rh-Pt and Pd-Pt nanoparticles were synthesized using colloidal chemistry and deposited on silicon. Depth profiling was performed by using different X-ray energies leading to different mean free paths of the emitted photoelectrons. These experiments have shown that as-synthesized particles obey a core-shell structure for Rh-Pd (with a Rh-rich shell and a Pd-rich core) and Pd-Pt (with a Pd-enriched shell), a more homogeneous distribution was observed for Rh-Pt. The *in situ* XPS investigations were performed at 300°C under 0.1 mbar $H_2$, CO, NO or $O_2$, or a mixture of 0.1 mbar CO and 0.1 mbar NO. The evolution and concentration of the different elements in the near-surface region under different atmospheres was detected by XPS and is presented in Fig. 19.14.

In the Rh-containing particles, Rh was segregated on the surface under oxidizing conditions in the presence of NO or $O_2$ accompanied by the formation of rhodium oxide. This process was reversible under reducing conditions in the presence of $H_2$ and under catalytic conditions with a mixture of NO and CO. The changes in the compositions were much weaker for Pd-Pt particles. By varying the sample temperature and the pressure, thermodynamic measurements are possible, allowing rationalization of the restructuring behavior under reaction conditions.

##### 19.3.3.2b. *Sulfur oxidation on Pt(335)*

Sulfur is well known to poison catalysts, e.g. active sites are blocked by sulfur atoms. A key step to regain catalytic activity is the removal of sulfur by oxidative treatment.

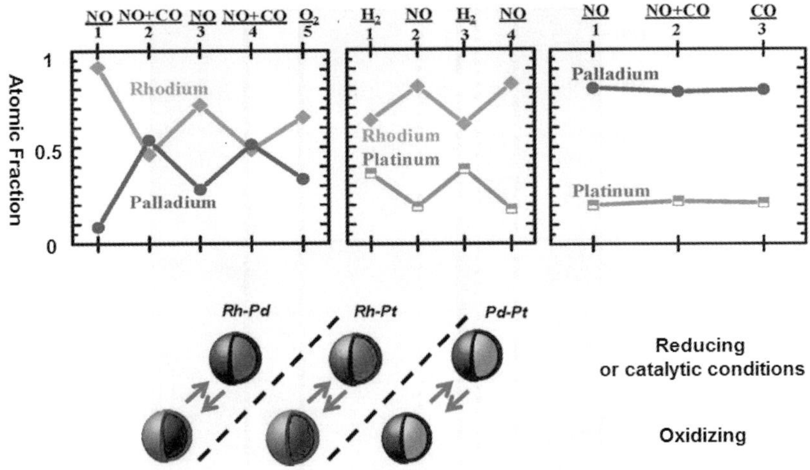


**Figure 19.14.** Evolution of the atomic fractions in bimetallic nanoparticles at 300°C under oxidizing (13.3 Pa NO or $O_2$), reducing (13.3 Pa $H_2$) and catalytic conditions (13.3 Pa NO and 13.3 Pa CO) as denoted on the horizontal axis. Adapted from Ref 92.

In an *in situ* study at BESSY, the kinetics of sulfur oxidation were investigated on a well-defined stepped Pt(335) face to elucidate the role of the steps in the removal of sulfur atoms.[93] A stepped surface was chosen because defects, as steps or kinks, are thought to be the active sites in a catalyst and are omnipresent in highly dispersed catalyst nanoparticles used as real catalysts.

The model catalyst surface was covered with sulfur by adsorbing $H_2S$ at -143°C, followed by heating to 427°C, which leads to decomposition of the $H_2S$. Oxygen was dosed by means of a molecular beam to obtain high local pressure at the sample. The measurements at the stepped Pt(335) surface were compared with the results obtained for a flat Pt(111) surface under similar conditions. For kinetic measurements, temperature and pressure were varied. As reaction intermediates, adsorbed sulfite and sulfate were identified. An increase of the temperature from 27°C to 77°C led to a more pronounced oxidation of sulfide (Fig. 19.15).

At 27°C some sulfide ions are oxidized to sulfite and no hint of sulfate was observed. In contrast, at 77°C all sulfides are oxidized after *ca.* 100 s to sulfite and subsequently to sulfate. After 600 s, sulfate is the dominant species on the surface. Measurements at different temperatures allowed the determination of the activation energy for this oxidation. It was found that this energy on the stepped Pt(335) surface is less than half that on a flat Pt(111) surface, which indicates that sulfur oxidation occurs mainly at steps or other defects of the surface. Such investigations are important for a mechanistic understanding of the catalytic process.

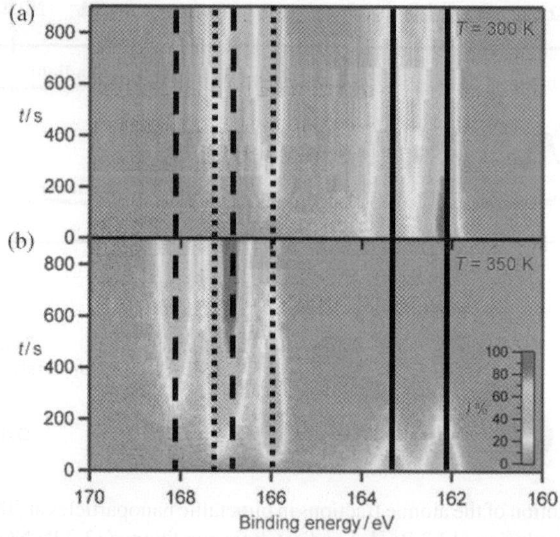


**Figure 19.15.** Colour-coded density plot of the S2p spectra collected at 27°C (300 K) and 77°C (350 K) during the oxidation of sulfur on Pt(335) at an oxygen pressure at $6 \times 10^{-7}$ mbar. The binding energies of sulfide (solid line), sulfite (dotted line) and sulfate (dashed line) are labelled. Adapted from Ref. 93.

#### 19.3.3.2c. *Methanol oxidation catalysed by copper*

Elemental copper is used as a catalyst for the oxidation of primary alcohols to their respective aldehydes. In a study performed at BESSY and ALS, the chemical species near the surface of the catalyst were identified *in situ* with XPS.[81] A polycrystalline Cu foil was investigated at 400°C under oxidizing ($CH_3OH : O_2 = 1{:}2$ and 1:2) and reducing ($CH_3OH : O_2 = 3{:}1$ and 6:1) conditions at a total pressure of 0.6 mbar. The $CH_3OH$ conversion and HCOOH and $CO_2$ yields were calculated from the area of the corresponding gas-phase peaks observed with XPS. The absolute amount of formaldehyde produced in the catalytic reaction is the highest for a $CH_3OH/O_2$ ratio of 3:1. Measurements in the O1s region at an electron binding energy of 530 eV (Fig. 19.16) and near the Fermi edge showed clearly that $Cu_2O$ is the dominating species under the most oxidizing conditions with a $CH_3OH/O_2$ ratio of 1:2. Two other peaks appeared at a binding energy of 529.7 eV and 531.4 eV, which became more abundant when changing the ratio from 1:2 to 1:1. The valence band spectrum near the Fermi edge showed metallic Cu next to $Cu_2O$. Under reducing conditions the valence band was typical for metallic Cu. Two peaks at 530.4 eV and 532.0 eV dominated the O1s region. Depth profiling measurements, by varying the radiation energy, revealed that the oxygen species related to the peak at 532.0 eV is located at the surface, while the other species extended into the subsurface region. It must be noted that this peak vanished after stopping the oxygen flow. The Cu:O stoichiometry

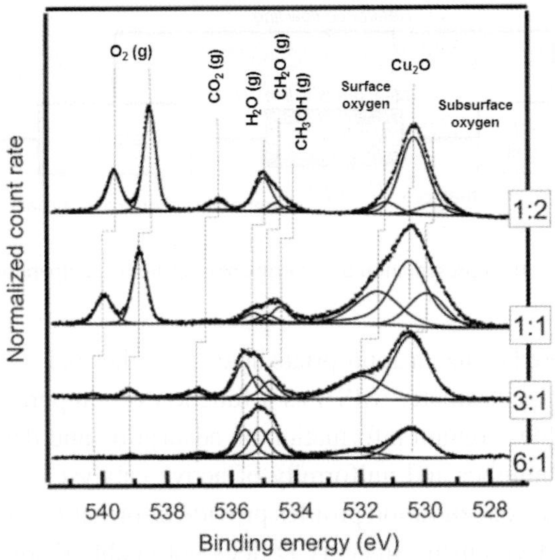


**Figure 19.16.** O1s photoelectron spectra of a copper sample at 400°C at $CH_3OH : O_2$ reactant streams of ratios 1:2, 1:1, 3:1 and 6:1. Gas-phase peaks are found at BE > 534 eV, surface peaks at BE < 534 eV. Adapted from Ref. 81.

for this sample was close to $Cu_2O$, but the valence band measurements and near-edge X-ray absorption fine structure (NEXAFS) investigations performed under similar conditions showed the metallic character of the Cu.[94] The peak abundance of this subsurface oxygen was clearly correlated with the formaldehyde yield which proves the catalytically active role of this species. The authors proposed that this subsurface oxygen species acts as co-catalyst and formed, together with the neighboring Cu species, the active centre in this reaction.

## 19.4. Temperature-programmed Reduction, Oxidation and Reaction Spectroscopy (TPR, TPO and TPRS)

### 19.4.1. *Basic principles and instrumentation*

Nowadays, temperature-programmed (TP) techniques are almost routinely used in many research laboratories.[95] These methods are principally applicable for different types of catalysts, ranging from single crystals to commercial catalysts. The relatively simple and affordable set-up (Fig. 19.17) is another advantage of such methods. The TP experiments are usually performed in continuous flow fixed-bed reactors, in which temperature is programmed to rise typically in a linear fashion. The heating rate is varied between 0.1 and 20 K/min. For quantifying the TP tests, an inert gas is added to reactive components. Gas-phase species leaving the

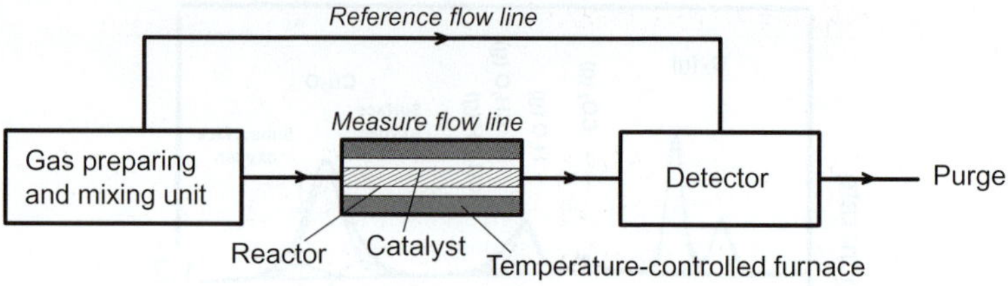


**Figure 19.17.** Schematic representation of experimental set-up for temperature-programmed tests.

reactor are monitored using an appropriate detector (a thermal conductivity detector (TCD) or a mass spectrometer (MS)) as a function of temperature. The resulting concentration profiles contain information on adsorption and desorption processes as well as on the number and uniformity of active catalyst sites. When the reaction takes place, the concentration profiles provide information on the reaction rates and the reaction mechanism. Since TCD does not enable distinguishing between the feed components and the reaction products, MS is a more appropriate analytic tool.

Temperature-programmed reduction (TPR) and oxidation (TPO) are the most frequently applied variants of TP methods. Reaction mixtures used for TPR and TPO usually contain 5 vol% $H_2$ in Ar and 5 vol% $O_2$ in He, respectively. Such feed compositions provide an optimal thermal conductivity difference between the reactants and the carrier gases. TPR is used for elucidating the reduction mechanism of oxide catalytic materials. In these tests, the degree of hydrogen consumption is followed as a function of temperature. From integrating the resulting profile, the amount of oxygen removed from the catalyst is determined providing information on the valence state of the metal ions after the TPR tests. This knowledge is important for identifying optimal conditions for the reductive treatment of catalysts during their preparation. For deriving insights into the re-oxidation of reduced catalysts, TPO can be performed after finishing the TPR runs. Additionally, TPO is useful for the characterization of catalysts deactivated due to coking. In such experiments, the catalysts are treated inan $O_2$-containing feed under temperature-programmed conditions. During coke burning, $H_2O$, CO and $CO_2$ are formed and detected by MS. From the change of their concentrations during TP experiments, optimal temperature range for catalyst regeneration can be determined. Moreover, from the amount of $H_2O$ and $CO_x$ formed, the ratio of H/C can be calculated. It provides insights into the nature of the coke species (aliphatic, aromatic or graphitic).

The above TP approach was also extended to study sulfidation, methanation, hydrogenation, gasification, carburization and other catalytic surface reactions. Applications of these methods for analysing different catalytic systems are comprehensively described in Ref. 96.

### 19.4.2. *Kinetic evaluation of TP tests*

Despite the distinctive mechanism of TP oxidation, reduction or reaction, their kinetic analysis is very similar. It is based on the rate Eq. 19.7, in which á is the degree of conversion, $E_a$ is the activation energy of the reaction and $\beta$ is the heating rate. $k_0$ and R stand for the pre-exponential factor and for the gas constant, respectively. This equation describes the temperature dependence of these processes when the temperature increases linearly with time (Eq. 19.8). f(1-á) relates the reaction rate and the degree of conversion and follows one of the kinetic models which are commonly used for gas-solid reactions.[97] In order to extract the activation energy, the experimental data recorded at different heating rates are plotted on the system of coordinates $\ln(\beta/T^2_{max})$ versus $1/T_{max}$. If this plot results in a straight line, its slope is equal to $-E_a/R$.

$$\beta \cdot \frac{d\alpha}{dT} = k_0 \cdot \exp\left(-\frac{E_a}{RT}\right) \cdot f(1-\alpha) \qquad (19.7)$$

$$T = \beta \cdot t + T_0 \qquad (19.8)$$

The temperature dependence of the reaction is given by an Arrhenius term. The activation energy can be derived after measuring a series of experimental data at different heating rates $\beta$ using a method suggested by Kissinger.[98] In this method a relationship between the temperature ($T_{max}$) of the maximal reaction rate and the activation energy is given by Eq. 19.9. This equation is valid when f(1-á) does not depend on the heating rate.

$$\ln\left(\frac{\beta}{T^2_{max}}\right) = -\frac{E_a}{RT_{max}} + \ln\left(\frac{k_0 R}{E_a}\right) + const \qquad (19.9)$$

### 19.4.3. *Selected application examples*

Reduction of iron oxide is one the most studied topics due to the application of this oxide forammonia synthesis, for the Fischer–Tropsch process and for the production of metallic iron and steel. From a mechanistic point of view, a complete reduction occurs in two or three steps via the formation of intermediate $Fe_3O_4$. Several studies have attempted to determine the apparent activation energies of reduction. Pineau *et al.*[99] listed the obtained energies for reduction of iron oxide with $H_2$. These values vary from 18 to 250 kJ/mol and strongly depend on the starting raw materials, their chemical purity, temperature range and presence of water in the reducing feed.

Recently, TP reduction with $H_2$ ($H_2$-TPR) in combination with isothermal *in situ* time-resolved UV-vis analysis was applied to investigate the effect of $TiO_2$ in $SiO_2$ on the redox behavior of the surface $VO_x$ species.[100] $VO_x/(Ti\text{-}Si)O_2$ materials

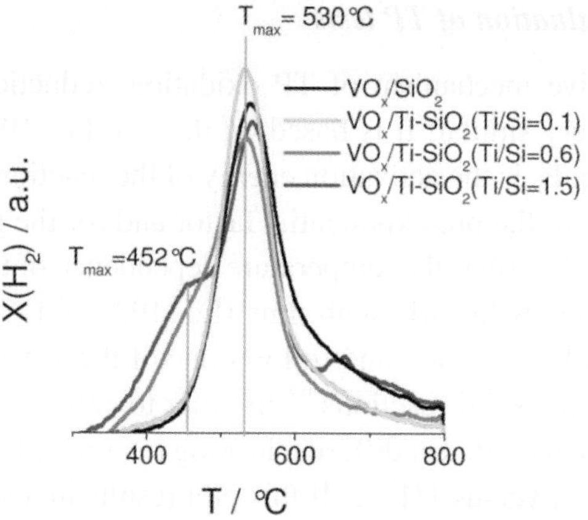


**Figure 19.18.** Profiles of hydrogen conversion during temperature-programmed reduction of $VO_x/Ti\text{-}SiO_2$ catalysts.

containing approximately 3 wt% vanadium catalyse the oxidative dehydrogenation of propane to propene. Figure 19.18 compares the $H_2$-TPR profiles obtained over these materials. No significant $H_2$ consumption was observed over the bare supports. A maximum of $H_2$ consumption ($T_{max}$) at 530–537°C is characteristic for all materials. The single $T_{max}$ obtained for $VO_x/SiO_2$ represents a single-stage reduction of $V^{5+}$ to $V^4/V^{3+}$. An additional shoulder with a maximum shifted to lower temperatures is visible in the $H_2$-TPR profiles of the Ti-containing samples. This was explained by weakening the strength of V-O bonds in the $VO_x$ species connected to $TiO_x$. Further insights into the kinetics of the reduction of oxidized $VO_x$ species by $C_3H_8$ and the oxidation of reduced $VO_x$ species by $O_2$ were obtained from *in situ* transient UV-vis analysis. The corresponding kinetic constants were determined from quantitative analysis of temporal changes in the Kubelka–Munk function at 700 nm, after switching from $O_2/Ne = 20/80$ to $C_3H_8/Ne = 40/60$ and back. It was found that the constant of catalyst reduction by $C_3H_8$ strongly increased with an increase in Ti loading in the support, while the constant of re-oxidation of reduced $VO_x$ species decreased.

## 19.5. Transient Techniques

Generally, transient techniques are used for analysis of catalytic reactions under non-steady state conditions, i.e. one or more reaction parameters, such as temperature, pressure or concentration of reaction components, are temporarily varied. Temporal response of the studied reaction during such changes is monitored as a function of time. Due to its technical characteristics, mass spectroscopy is generally preferred for fast analysis of gas-phase components at the reactor outlet and for discriminating between isotopically-labelled species. In addition, modern *in situ* characterization techniques such as IR, Raman and UV-vis spectroscopy enable the monitoring of the changes in the state of catalytic materials during transient experiments.[100–104] Therefore, kinetic information on surface processes can be derived from analysis of temporal changes in catalyst composition and concentration of active species.

Since one or more reaction parameters are changed during a non-steady state experiment, the reaction is investigated in a broader range of conditions compared to a steady-state experiment. Moreover, transient experiments are performed with a higher time resolution than steady-state tests, which enables the detection of short-lived intermediates. Therefore, these experiments are very helpful for unravelling complex reaction networks in terms of near-to-elementary reaction steps and for determining the kinetic parameters of these steps. The sections below describe the basic principles of individual transient methods and demonstrate their potential for investigating heterogeneous catalytic reactions.

### 19.5.1. *Temporal analysis of products (TAP)*

#### 19.5.1.1. *Basic principles and instrumentation*

The temporal analysis of products (TAP) reactor is a pulse technique operating in vacuum ($p \sim 10^{-5}$ Pa) with very low (0.1–100 nmol/pulse) concentrations of reactants. It was developed by John T. Gleaves and his team at Monsanto in the late 1980s.[105–108] Figure 19.19 schematically shows the main parts of the TAP reactor: high-speed pulse valves, a valve for continuous flow experiments, a micro-reactor and a vacuum system equipped with a quadrupole mass spectrometer as an analytic device. The micro-reactor is a classical fixed-bed tube reactor made of stainless steel, inconel or quartz. Depending on the reactor material, it can operate at up to 1,000°C. Catalytic materials are sandwiched between two layers of particles of an inert material in the isothermal zone of the reactor. Although the set-up design principally allows catalyst treatment at ambient pressure, the reactor is evacuated to a pressure of approximately $10^{-5}$ Pa for transient experiments. However, the peak pressure over the catalyst during pulsing increases up to several Pa for a short time. For a typical UHV pressure of $10^{-7}$ Pa, the ratio of $P_{UHV}/P_{ambient}$ is *ca.* $10^{-12}$, while in the TAP reactor the ratio of $P_{TAP}/P_{ambient}$ is increased up to $10^{-3}$ (for a pulse size of $10^{16}$ molecules). Accordingly, the pressure gap between the TAP and ambient pressure studies is largely reduced with respect to studies in UHV. Moreover, the TAP reactor operates with real catalytic materials. This positively distinguishes it from traditional surface science techniques, which mainly deal with ideal catalytic systems such as single crystals or model catalysts.

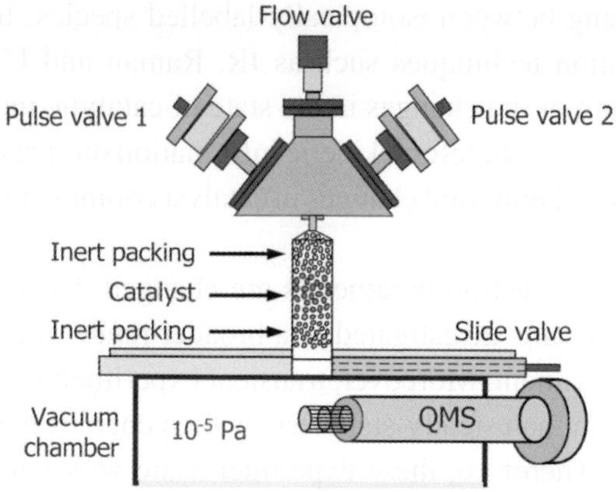


**Figure 19.19.** Schematic representation of the key parts of the TAP reactor system.

The pulse valves permit a release of $10^{13}$ to $10^{18}$ molecules per pulse. Since this amount is usually $10^2–10^5$ times smaller than the number of surface atoms,[108, 109] the changes in the state of the catalyst surface are negligible. Furthermore, isothermal reactor operation is ensured for even highly exothermic or endothermic reactions. For example, oxidation of 10 nmol of $NH_3$ to $N_2$ over a Pt gauze (100% conversion and selectivity) at 800°C results in an adiabatic temperature increase of only 2.8°C.[110] This value is even overestimated since no heat conduction away from the gauze was considered.

Another very important feature of transient experiments in the TAP-2 reactor is that they can be performed in different diffusion regimes, which are determined by the reactor geometry and the number of molecules pulsed. When the pulse size is below $10^{15}$ molecules, gas transport in the micro-reactor occurs via Knudsen diffusion. This means that any collisions between gas-phase molecules are strongly minimized. Therefore, pure heterogeneously-catalysed reactions can be investigated. Transient experiments with higher pulse sizes (molecular diffusion) provide important information about the contribution of gas-phase processes to the overall reaction studied.

The concentration of feed components and reaction products is monitored directly at the reactor outlet using a quadrupole mass spectrometer (QMS) with a time resolution below 100 μs. Since QMS requires at least 10 μs to switch between different atomic mass units (AMUs), only one AMU per pulse can be monitored. In order to detect all feed components and possible reaction products during pulse experiments, QMS switches from the measured AMU to the next one after the data collection interval is completed. Usually 10 pulses for each AMU are recorded and averaged to improve the signal-to-noise ratio. Recently, Gleaves and co-workers[109] equipped their TAP reactor with a time-of-flight (TOF) mass spectrometer. It allows rapid detection of gas-phase components at a rate which is several orders of magnitude faster than using QMS. The obtained concentration profiles of feed components and reaction products at the reactor outlet are called transient responses. Their shape and position relative to each other contain information on the mechanism and kinetics of studied reactions. As exemplified in Section 19.5.1.3, this information can be extracted from a kinetic evaluation of the transient experiments.

#### 19.5.1.2. *Types of TAP experiments*

Three different types of transient experiments can be performed using the TAP reactor. In single pulse experiments, which are often used, several (*ca.* 100) low-sized (*ca.* $10^{13}-10^{14}$ molecules) pulses of a certain composition are injected into the micro-reactor. The pulsed mixture usually contains reactants and an inert gas, which serves as an internal standard. Due to the low amount of pulsed gases, the catalyst's state is not changed during these experiments. Therefore, they provide intrinsic kinetic information.

In multi-pulse experiments, a series (1,000–10,000) of low-sized pulses is introduced into the reactor. These experiments alter the surface state of the catalyst. They are helpful for deriving information on the ability of the catalyst to adsorb the pulsed gases (concentration of active sites can be estimated) or to provide its active surface/bulk species for the reaction. For example, such experiments were demonstrated to be very useful for elucidating mechanistic aspects of selective catalytic reduction (SCR) of NO by $NH_3$ over an $Ag/Al_2O_3$ catalyst.[111] The authors reported an increase in the conversion of nitric oxide in the presence of gas-phase oxygen. The accelerating effect of oxygen was related to the activation of gas-phase $O_2$ over reduced Ag species, yielding reactive oxygen species. These oxygen species dehydrogenated ammonia and enabled efficient liberation of molecular nitrogen with one N coming from $NH_3$ and one from NO. In addition, the $O_2$ multi-pulse experiments showed that $^{15}N_2O$, $^{15}N^{14}NO$, $^{14}N_2O$, $^{15}N^{14}N$, $^{14}N_2$, $^{15}N_2$ and $^{15}NO$ were formed when $O_2$ was pulsed over $Ag/Al_2O_3$ treated in a $^{15}NH_3$-$^{14}NO$-$O_2$ mixture at 450°C. A typical profile of the $O_2$ outlet concentration as a function of the number of oxygen pulses is given in Fig. 19.20. One can see that the $O_2$ concentration in the initial $O_2$ pulses is very low but increases with the amount of $O_2$ pulsed. In contrast to oxygen, the outlet concentration of nitrogen-containing products is highest in the first $O_2$ pulse and decreases with the number of $O_2$ pulses. These concentration profiles were explained by the fact that gas-phase $O_2$ converted adsorbed nitrogen-containing species to gas-phase products. Thus, in addition to accelerating $NH_3$ dehydrogenation, gas-phase $O_2$ is responsible for removing the strongly adsorbed species from the catalyst surface.

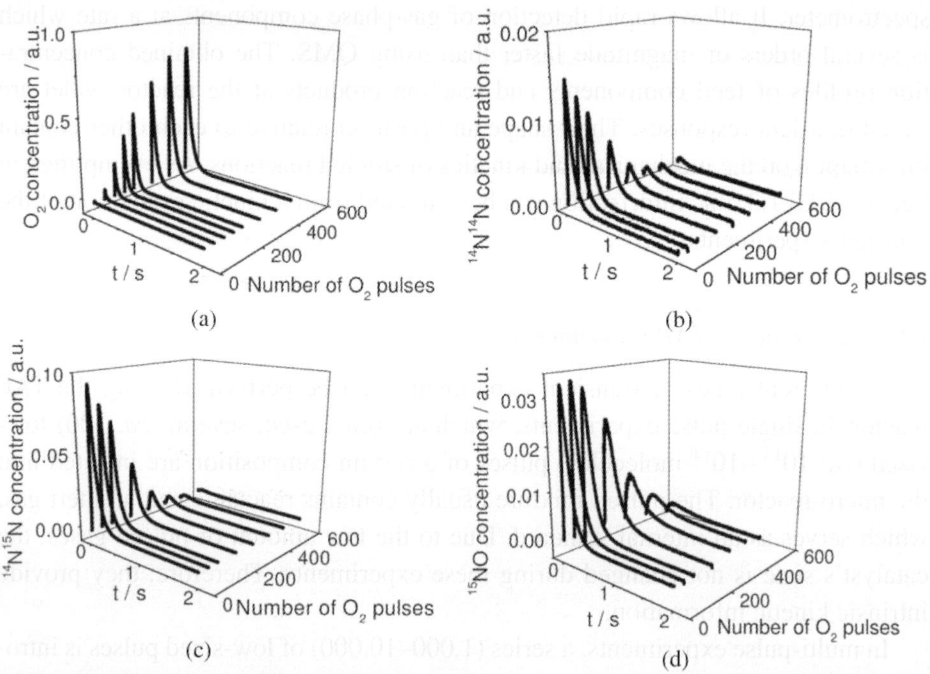


**Figure 19.20.** Concentrations of (a) $O_2$, (b) $^{14}N^{14}N$, (c) $^{14}N^{15}N$ and (d) $^{15}NO$ measured at the reactor outlet upon $O_2$ pulsing at 450°C over $2Ag/Al_2O_3$ having been pre-pulsed by an $^{15}NH_3$-$^{14}NO$-$O_2$ mixture. $O_2$ pulse size was $5 \cdot 10^{15}$ molecules.

Sequential pulse experiments, also known as pump–probe experiments, are very useful for elucidating the reactivity and selectivity of adsorbed species as well as their surface residence time. In these experiments two different reaction mixtures are pulsed alternately from two separate pulse valves with different time delays ($\Delta t$) between the pulses. The time delay can be varied from 0 to several seconds. In these experiments, adsorbed species originating from one feed mixture are probed for the reaction with the components from the second mixture. By plotting yields of products formed in the second pulse versus the time delay between the pulses. For example, oxygen species formed from $N_2O$ upon decomposition over Fe-containing zeolites lost their activity for oxidation of $CH_4$ and $C_3H_8$ when the pulses of $N_2O$ and the reducing agent were separated by more than 0.1 s.[112] This is due to the rapid transformation of highly reactive oxygen species into non-reactive ones.

This type of transient experiment helped, for the first time, to derive mechanistic insights into the origin of $N_2O$ in the high-temperature ammonia oxidation (Ostwald process) over commercial Pt-Rh gauzes. The experiments with labelled ammonia ($^{15}NH_3$) showed that nitrous oxide originated mainly from coupling the $^{14}NO$ and $^{15}NH_x$ ($x = 0$–$2$) fragments.[113] The heterogeneous nature of the reaction steps resulting in $^{15}N^{14}NO$ was conclusively proven by comparing its formation at different time delays between the $^{15}NH_3$ and $^{14}NO$ pulses. It was found that the amount of nitrous oxide formed decreased with increases in the time delay between the ammonia and nitric oxide pulses. This is due to the fact that the formation of nitrous oxide strongly depends on surface coverage by reactive $NH_x$ intermediates.

#### 19.5.1.3. *Kinetic evaluation of TAP experiments*

Kinetic evaluation of transient experiments provides incisive insights into reaction processes occurring inside the TAP micro-reactor. The evaluation procedure of transient experiments is based on fitting the experimental data to different kinetic models and discriminating between them. When transient experiments are performed in the Knudsen diffusion regime, the description of the gas transport is simplified by the absence of temperature gradients and radial concentration in the catalyst zone. The TAP micro-reactor is described as a one-dimensional pseudo-homogeneous system divided into three different zones, which are represented by the catalyst and the two layers of inert material on top and below the catalyst.[114] Outside the catalyst layer the mass balance for each gas-phase species is defined by Eq. 19.10. For reacting gas-phase and surface species in the catalyst zone, reaction terms are considered in the mass balance in Eqs. 19.11 and 19.12, respectively.

$$\frac{\partial C_i}{\partial t} = D_{Knudsen,i} \cdot \frac{\partial^2 C_i}{\partial x^2} \tag{19.10}$$

$$\frac{\partial C_i}{\partial t} = D_{Knudsen,i} \cdot \frac{\partial^2 C_i}{\partial x^2} + \sum_{j=1}^{n} \rho_{cat} \nu_{ij} r_j \tag{19.11}$$

$$\frac{\partial \Theta_i}{\partial t} = \sum_{j=1}^{n} \frac{\nu_{ij} r_j}{c^x} \tag{19.12}$$

In these equations, $r_j = k_j \Pi_l C_l^{n_l} \Pi_m \theta_m^{n_m}$, $C_i$ is the concentration of gas-phase species i (mol/$m^3$), $c^x$ is the concentration of adsorption centres (mol/$m^3$), $D_{Knudsen,i}$ is the effective Knudsen diffusion coefficient of species i ($m^2$/s), $\Theta_i$ is the fractional surface coverage of species i (dimensionless), $k_j$ is the rate coefficient of reaction j, $\nu_{ij}$ is the stoichiometric coefficient of species in reaction j (dimensionless) and $\rho_{cat}$ is the catalyst density (kg/$m^3$).

The detailed information on initial and boundary conditions can be found in Ref. 115 and references therein. Partial differential equations (PDEs) describing processes of diffusion transport, adsorption/desorption, and reaction in the TAP micro-reactor can be solved either analytically[109,116,117] or numerically.[115,118–121] Although the analytical solution allows fast estimation of kinetic parameters, the choice of kinetic models is strongly restricted. Only simplified models including single reaction steps described by first-order rate equations with respect to reactants can be applied. The numerical solution approaches are much slower in comparison to the analytic one. However, they do not have any restrictions with respect to the complexity of the kinetic models evaluated.

Recently, the TAP reactor was successfully applied for deriving the micro-kinetics of direct $N_2O$ decomposition over Rh-ZSM-5 and Fe-ZSM-5/silicalite differing in the nature of $FeO_x$ species (isolated, oligomerized and nanoparticles).[122–124] The kinetics was validated by extrapolating to the conditions of steady-state ambient pressure catalytic tests (approximately over three orders of magnitude of partial $N_2O$ pressure). Both for Fe- and Rh-containing zeolites, the respective micro-kinetic models confirmed first-order dependence of $N_2O$ decomposition on the partial pressure of $N_2O$ and the absence of an inhibiting effect of $O_2$ under steady-state conditions. Moreover, they qualitatively predicted the differences in the steady-state performance of differently structured $FeO_x$ species as well as in the de-$N_2O$ activity of $FeO_x$ and $RhO_x$ species in a wide range of temperatures and partial $N_2O$ pressures. However, the quantitative description of the steady-state activity failed. Therefore, further thorough studies are required to identify the reasons limiting the quantitative relevance of transient micro-kinetics.

### 19.5.2. *Steady-state isotopic transient kinetic analysis* (*SSITKA*)

#### 19.5.2.1. *Basic principles*

Steady-state isotopic transient kinetic analysis (SSITKA) is one of the powerful techniques for kinetic and mechanistic analysis of reactions on the catalyst surface under real reaction conditions.[125,126] The SSITKA method is based upon the inclusion of one or more stable isotopic labels in a reactant flow and can operate in a broad range of pressures (from vacuum to elevated pressures). Usually in a SSITKA experiment, two reaction feeds differing only in their isotopic composition are used. One feed is directed to the reactor filled with a catalyst, while another one is not fed to the reactor. Practically, the feeds are separated by a four-port valve. Due to costs, steady-state catalytic operation is usually achieved using the feed with a common isotopic label. Thereafter, this feed is replaced by another one containing a different isotopically labelled feed component. In order to determine the gas-phase hold-up for the applied reactor, an inert tracer is introduced in a small concentration into the new feed. It is important to stress that reaction parameters (temperature, pressure and contact time) are not changed after the switching. As a result, surface reaction intermediates also remain undisturbed. As the new feed progresses through the reactor and reacts on the catalyst surface to form products, the new

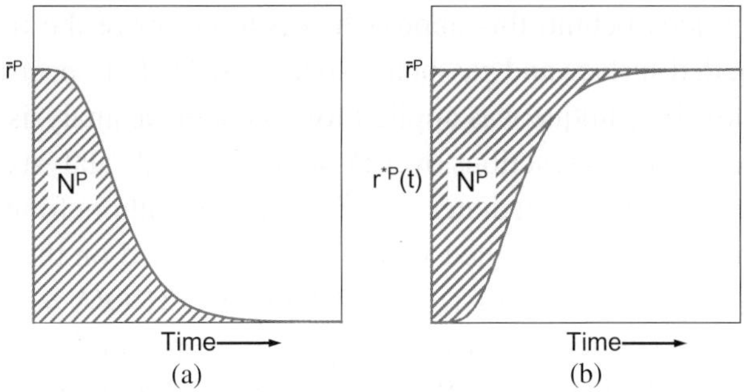


**Figure 19.21.** Transient rates of (a) old and (b) new isotopically labelled products. The filled area corresponds to the number of surface intermediates leading to these products.

isotopic label is distributed between reaction products (gas phase and surface) and unreacted reactant. Temporal concentration of both the initial (before switching) and new (after switching) isotopic labels in the reaction products is monitored at the reactor outlet by a mass spectrometer. The resulting transients are called step-decay and step-input responses, respectively. The temporal rates of reaction products can be calculated from these signals. Figure 19.21 schematically shows these rates after isotopic switching ($t = 0$ in this case); the rate of the new isotopic label increases with time while that of the old one decreases. From integration of these rates, the total number ($\bar{N}^P$) of steady-state intermediate surface species leading to these products can be obtained. Compared with other techniques, this value determined in SSITKA does not depend on the reaction mechanism and kinetic model. Another important parameter derived from SSITKA is surface lifetime ($\bar{\tau}^P$) of adsorbed intermediates. It is obtained from integration of the temporal rates of a certain reaction product normalized by its steady-state rate (Eq. 19.13). However, in contrast to the number of reactive intermediates, this parameter can be misinterpreted if the reaction pathways are reversible.[125]

$$F^P(t) = \frac{r^P(t)}{\bar{r}^P} \quad or \quad F^{*P}(t) = \frac{r^{*P}(t)}{\bar{r}^P} \tag{19.13}$$

In this equation $r^P(t)$ and $r^{*P}(t)$ are the temporal rates of new and old isotopically labelled reaction product, respectively, and $\bar{r}^P = r^P(t) + r^{*P}(t)$.

### 19.5.2.2. *Coupling of SSITKA with in situ catalyst characterisation*

The two examples below demonstrate further developments in SSITKA for investigating heterogeneous reactions. The group of Burch and Meunier at Queen's University Belfast developed a set-up combining *in situ* diffuse reflectance infrared Fourier transform (DRIFT) spectroscopy and on-line mass spectrometry (MS) with

SSITKA.[127] The idea behind this approach was to compare the time response of isotopically labelled surface and gas-phase species by DRIFT spectroscopy and MS analysis. This novel technique was applied for mechanistic analysis water-gas shift (WGS) and reversed WGS reactions over Pt- and Au-based catalysts with the aim of identifying true surface intermediates.[102,128,129] These authors found that both formate and carbonate species labelled with $^{13}C$ were formed on the catalyst surface after switching from $^{12}CO/H_2O/Ar = 2/7/91$ to $^{13}CO/H_2O/A = 2/7/91$. In order to determine which surface species actively participated in $CO_2$ formation, they compared temporal changes in the IR bands of these surface species with those of gas-phase $^{13}CO_2$. Since the rate of $CO_2$ formation was *ca.* 60 times higher than the rate of the exchange of formate species it was concluded that the formates detected by DRIFT spectroscopy could not be the main surface intermediates of gas-phase $CO_2$. However, the role of surface formates in $CO_2$ production may change with rising temperature as demonstrated in Ref. 129 where the formate species were spectators at 160°C but became main reaction intermediates at 220°C.

For simultaneous monitoring of events taking place over catalytic materials and in the gas phase under real reaction conditions, Kondratenko and co-workers[105] combined the SSITKA technique with *in situ* time- and spatially-resolved UV-vis spectroscopy (Fig. 19.22). In contrast to the separate single technique applications, this novel coupling provides more comprehensive and more relevant information about the working catalytic system, since problems arising from differences in reaction conditions and cell designs are avoided and a broader range of catalyst properties can be assessed.

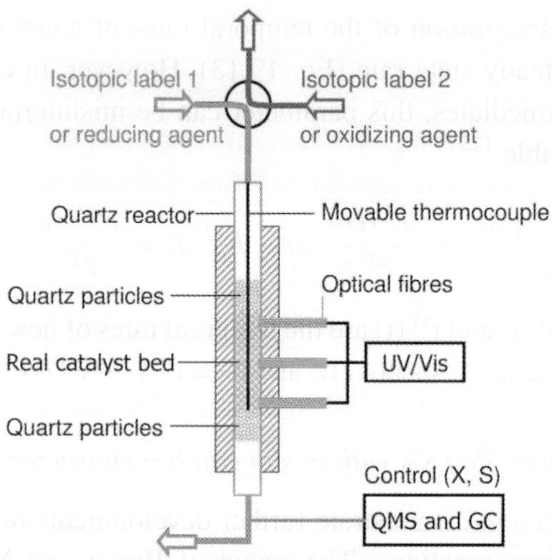


**Figure 19.22.** Schematic representation of a set-up developed in-house for simultaneous SSITKA experiments and catalyst characterization by *in situ* time- and spatially-resolved UV/Vis spectroscopy.

The set-up is fully automated and pre-designed kinetic and characterization runs can be carried out. It was applied for identifying selectivity-determining factors in the oxidative dehydrogenation of propane (ODP) over vanadium-based catalysts.[130,131] From a kinetic analysis of redox behaviour of differently structured $VO_x$ species (from isolated ones to bulk $V_2O_3$, $VO_2$ and $V_2O_5$) by time-resolved UV-vis spectroscopy, it was concluded that the steady-state reduction degree of $VO_x$ species under ODP conditions determines the propene selectivity; the higher the reduction degree, the higher the selectivity that can be achieved. In addition, SSIKTA experiments of the ODP reaction with $^{16}O_2$ and $^{18}O_2$ provided insights into the role of oxygen species from $VO_x$ species and from gas-phase $O_2$ in $CO_2$ formation. Analysis of the statistical distribution of $^{16}O$ and $^{18}O$ in differently labelled carbon dioxides led to the suggestion that differently active oxygen species participate in $CO_2$ formation. Further insights into the nature of these species were derived from density functional theory (DFT) calculations of $O_2$ interaction with reduced $VO_x$ species.[132] These calculations predicted the formation of peroxovanadates as precursors of vanadyl species. The peroxovanadates species were highly reactive for consecutive propene oxidation to $CO_x$.

## 19.6. Concluding Remarks

Tremendous progress has been achieved in the last decade in tailoring the setups of characterization techniques for monitoring catalytic systems under conditions as close as possible to those of real catalytic processes. In particular, X-ray-based techniques have seen a rapid improvement with the development of modern synchrotron facilities that offer highly brilliant radiation, which is the basis for performing such experiments, with high time and space resolution reaching the atomic scale.[3,28] Beyond the examples presented in this work, this will lead to further advanced opportunities for analysing structure–reactivity relationships in oxidation catalysis, when access to such modern facilities becomes increasingly available for a broader field of users. However, the examples discussed in this chapter have also shown that it is almost always necessary to use several *in situ* techniques to solve a special problem. Combinations and simultaneous couplings of non-vibrational and vibrational methods are therefore particularly helpful for extending the variety of information that can be obtained on the same catalytic system to support the reliability of the derived conclusion. An important issue, which is rarely tackled in spectroscopic studies, is kinetic evaluation on a level near the elementary reaction steps. To this end, there is still a need for developing numerical tools for kinetic analysis of gas-phase and surface species.

## References


1. Weckhuysen, B. (ed.) (2010). *Chem. Soc. Rev.*, **39**, In-Situ Characterization of Heterogeneous Catalysts Themed Issue, pp. 4541–5072.
2. Gates, B. and Knözinger, H. (eds) (2006). *Adv. Catal.*, **50**, pp. 1–291; Gates, B. and Knözinger, H. (eds) (2007). *Adv. Catal.*, **51**, pp. 1–389; Gates, B., Knözinger, H. and Jentoft, F. (eds) (2009). *Adv. Catal.*, **52**, pp. 1–474.
3. Weckhuysen, B. (ed.) (2004). *In-Situ Spectroscopy of Catalysts*, American Scientific Publishers, Stevenson Ranch, CA.
4. Haw, J. (ed.) (2002). *In-Situ Spectroscopy in Heterogeneous Catalysts*, Wiley-VCH, Weinheim.
5. Meitzner, G. (2002). In Situ XAS Characterization of Heterogeneous Catalysts, in J. Haw (ed.), *In-Situ Spectroscopy in Heterogeneous Catalysts*, Wiley-VCH, Weinheim, pp. 179–194.
6. de Groot, F., Knop-Gericke, A. and Ressler, T. (2004). X-Ray Absorption Near Edge Spectroscopy, in B. Weckhuysen (ed.), *In-Situ Spectroscopy of Catalysts*, American Scientific Publishers, Stevenson Ranch, CA, pp. 107–122.
7. van Bokhoven, J., Ressler, T., de Groot, F., *et al.* (2004). Extended X-Ray Absorption Fine Structure Spectroscopy, in B. Weckhuysen (ed.), *In-Situ Spectroscopy of Catalysts*, American Scientific Publishers, Stevenson Ranch, CA, pp. 123–144.
8. Knop-Gericke, A., de Groot, F., van Bokhoven, J., *et al.* (2004). Soft X-Ray Absorption Methods, in B. Weckhuysen (ed.), *In-Situ Spectroscopy of Catalysts*, American Scientific Publishers, Stevenson Ranch, CA, pp. 145–160.
9. Ressler, T., van Bokhoven, J., Knop-Gericke, A., *et al.* (2004), Time-Resolved X-ray Absorption Spectroscopy Methods, in B. Weckhuysen, B. M. (ed.), *In-Situ Spectroscopy of Catalysts*, American Scientific Publishers, Stevenson Ranch, CA, pp. 161–174.
10. Bare, S. and Ressler, T. (2009). Characterization of Catalysts in Reactive Atmospheres by X-Ray Absorption Spectroscopy, *Adv. Catal.*, **52**, pp. 339–465.
11. Knop-Gericke, A., Kleimenov, E., Hävecker, M., *et al.* (2009). X-Ray Photoelectron Spectroscopy for Investigation of Heterogeneous Catalytic Processes, *Adv. Catal.*, **52**, pp. 213–272.
12. Jenthoft, F. (2009). Ultraviolet-Visible-Near Infrared Spectroscopy in Catalysis: Theory, Experiment, Analysis, and Application under Reaction Conditions, *Adv. Catal.*, **52**, pp. 129–211.
13. Weckhuysen, B. (2004). Ultraviolet-Visible Spectroscopy, in B. Weckhuysen (ed.), *In-Situ Spectroscopy of Catalysts*, American Scientific Publishers, Stevenson Ranch, CA, pp. 255–270.
14. Brückner, A. (2010). In Situ Electron Paramagnetic Resonance: A Unique Tool for Analyzing Structure-Reactivity Relationships in Heterogeneous Catalysis, *Chem. Soc. Rev.*, **39**, pp. 4673–4684.
15. Brückner, A. (2009). Electron Paramagnetic Resonance: A Powerful Tool for Monitoring Working Catalysts, *Adv. Catal.*, **52**, pp. 265–308.
16. Brückner, A. (2004). Electron Paramagnetic Resonance, in B. Weckhuysen (ed.), *In-Situ Spectroscopy of Catalysts*, American Scientific Publishers, Stevenson Ranch, CA, pp. 219–252.
17. Louis, C., Lepetit, C. and Che, M. (1994). EPR Characterization of Oxide Supported Transition Metal Ions: Relevance to Catalysis, *Molecular Engineering*, **4**, pp. 3–38.
18. Dyrek, K. and Che, M. (1997). EPR as a Tool to Investigate the Transition Metal Chemistry on Oxide Surfaces, *Chem. Rev.*, **97**, pp. 305–331.
19. Sojka, Z. (1995). Molecular Aspects of Catalytic Reactivity. Application of EPR Spectroscopy to Studies of the Mechanism of Heterogeneous Catalytic Reactions, *Catal. Rev. Sci. Eng.*, **37**, pp. 461–512.
20. Labanowska, M. (2001). EPR Monitoring of Redox Processes in Transition Metal Oxide Catalysts, *Chem Phys Chem*, **2**, pp. 712–731.
21. Schlögl, R. (2009). X-ray Diffraction: A Basic Tool for Characterization of Solid Catalysts in the Working State, *Adv. Catal.*, **52**, pp. 273–338.

22. Canton, P., Riello, P., Meneghini, C., *et al.* (2004). X-ray Diffraction and Scattering, in B. Weckhuysen (ed.), *In-Situ Spectroscopy of Catalysts*, American Scientific Publishers, Stevenson Ranch, CA, pp. 293–322.
23. O'Brien, M., Beale, A. and Weckhuysen, B. (2010). The Role of Synchroton Radiation in Examining the Self–Assembly of Crystalline Nanoporous Framework Materials: From Zeolites and Aluminophosphates to Metal Organic Hybrids, *Chem. Soc. Rev.*, **39**, pp. 4767–4782.
24. Bentrup, U. (2010). Combining In Situ Characterization Methods in One Set-Up: Looking with More Eyes into the Intricate Chemistry of the Synthesis and Working of Heterogeneous Catalysts, *Chem. Soc. Rev.*, **39**, pp. 4718–4730.
25. Niemantsverdriet, J. (2007). Temperature-Programmed Techniques, in J. Niematsverdriet, *Spectroscopy in Catalysis*, Wiley-VCH, Weinheim, pp. 11–38.
26. Hinrichsen, O., van Veen, A., Zanthoff, H., *et al.* (2002). TAP Reactor Studies, in J. Haw (ed.), *In-Situ Spectroscopy in Heterogeneous Catalysts*, Wiley-VCH, Weinheim, pp. 237–269.
27. Bal'zhinimaev, B., Sadovskaya, E. and Suknev, A. (2010). Transient Isotopic Kinetics Study to Investigate Reaction Mechanisms, *Chem. Eng. J.*, **154**, pp. 2–8.
28. Grunwaldt, J. and Schroer, C. (2010). Hard and Soft X-ray Microscopy and Tomography in Catalysis: Bridging the Different Time and Length Scales, *Chem. Soc. Rev.*, **39**, pp. 4741–4753.
29. Millet, J. (2007). Mössbauer Spectroscopy in Heterogeneous Catalysis, *Adv. Catal.*, **51**, pp. 309–350.
30. Chen, D., Bjorgum, E., Christensen, K., *et al.* (2007). Characterization of Catalysts under Working Conditions with an Oscillating Microbalance Reactor, *Adv. Catal.*, **51**, pp. 351–382.
31. Kortüm, G. (1969). *Reflectance Spectroscopy*, Springer Verlag, Berlin, Heidelberg, New York.
32. Weil, J., Bolton, J. and Wertz, J. (1994). *Electron Paramagnetic Resonance, Elementary Theory and Practical Applications*, Wiley, New York.
33. Pilbrow, J. (1990). *Transition Ion Electron Paramagnetic Resonance*, Clarendon Press, Oxford.
34. Ballhausen, C. and Gray, H. (1962). The Electronic Structure of the Vanadyl Ion, *Inorg. Chem.*, **1**, pp. 111–122.
35. Brückner, A. (2006). Spin–Spin Exchange in Vanadium-Containing Catalysts Studied by In Situ EPR: A Sensitive Monitor for Disorder-Related Activity, *Top. Catal.*, **38**, pp. 133–139.
36. Brückner, A. (2005). Killing Three Birds with One Stone – Simultaneous Operando EPR/UV-Vis/Raman Spectroscopy for Monitoring Catalytic Reactions, *Chem. Commun.*,**13**, pp. 1761–1763.
37. Matir, W. and Lunsford, J. (1981). The Formation of Gas-Phase $\pi$-Allyl Radicals from Propylene over Bismuth Oxide and $\gamma$-Bismuth Molybdate Catalysts, *J. Am. Chem. Soc.*, **103**, pp. 3728–3732.
38. Prada Silvy, R., Florea, M., Blangenois, N., *et al.* (2003). Propane Ammoxidation Catalyst Based on Vanadium-Aluminum Oxynitride, *AIChE Journal*, **49**, pp. 2228–2231.
39. Janke, C., Radnik, J., Bentrup, U., *et al.* (2009). Vanadium Containing Oxynitrides: Effective Catalysts for the Ammoxidation of 3-Picoline, *Chem. Cat. Chem*, **1**, pp. 485–491.
40. Janke, C., Schneider, M., Bentrup, U., *et al.* (2011). Impact of Phosphorus and Nitrogen on Structure and Catalytic Performance of Vzrpon Oxynitrides in the Ammoxidation of 3-Picoline, *J. Catal.*, **277**, pp. 196–207.
41. Pietrzyk, P. and Sojky, Z. (2007). $Co^{2+}/Co^0$ Redox Couple Revealed by EPR Spectroscopy Triggers Preferential Coordination of Reactants during SCR Of $No_x$ with Propene over Cobalt-Exchanged Zeolites, *Chem. Commun.*, **19**, pp. 1930–1932.
42. Castellino, F., Rasmussen, S., Jensen, A., *et al.* (2008). Deactivation of Vanadia-Based Commercial SCR Catalysts by Polyphosphoric Acids, *Appl. Catal. B*: *Environ.*, **83**, pp. 110–122.
43. Kustova, M., Kustov, A., Christiansen, S., *et al.* (2006), Cu-ZSM-5, Cu-ZSM-11, and Cu-ZSM-12 Catalysts for Direct NO Decomposition, *Catal. Commun.*, **7**, pp. 705–708.

44. Kucherov, A., Gerlock, J., Jen, H., *et al.* (1995). In Situ E. S. R. Monitoring of the Coordination and Oxidation States of Copper in Cu-ZSM-5 up to 500°C in Flowing Gas Mixtures: 2. Interaction with $CH_4$ and CO, *Zeolites*, **15**, pp. 15–120.
45. Kucherov, A., Gerlock, J., Jen, H., *et al.* (1995). In Situ E. S. R. Monitoring of the Coordination and Oxidation States of Copper in Cu-Zsm-5 Up To 500°C in Flowing Gas Mixtures: 1. Interaction with He, $O_2$, NO, $NO_2$, and $H_2O$, *Zeolites*, **15**, pp. 9–14.
46. Berrier, E., Ovsitser, O., Kondratenko, E., *et al.* (2007). Temperature-Dependent $N_2O$ Decomposition over Fe-ZSM-5: Identification of Sites with Different Activity, *J. Catal.*, **249**, pp. 67–78.
47. Brückner, A., Kubias, B., Lücke, B., *et al.* (1996). In Situ — ESR Study of Vanadium Phosphate Catalysts (VPO) during the Selective Oxidation of *n*-Butane to Maleic Anhydride (MA), *Colloids Surf. A*, **115**, pp.179–186.
48. Brückner, A. (2000). A New Approach to Study the Gas-Phase Oxidation of Toluene: Probing Active Sites in Vanadia Based Catalysts Under Working Conditions, *Appl. Catal. A*: *Gen.*, **200**, pp. 287–297.
49. Sancier, K., Dozono, T. and Wise, H. (1971). ESR Spectra of Metal Oxide Catalysts during Propylene Oxidation, *J. Catal.*, **23**, pp. 270–280.
50. Klasovsky, F., Hohmeyer, J., Brückner, A., *et al.* (2008). Catalytic and Mechanistic Investigation of Polyaniline Supported $PtO_2$ Nanoparticles: A Combined In Situ/Operando EPR, DRIFTS, and EXAFS Study, *J. Phys. Chem. C*, **112**, pp. 19555–19559.
51. International Centre for Diffraction Data (ICDD), 12 Campus Boulevard, Newtown Square, PA, 19073, USA, http://www.icdd.com/.
52. Glatter, O. and Kratky, O. (eds.) (1982). *Small Angle X-Ray Scattering*, Academic Press London, available free of charge at: http//physchem.kfunigraz.ac.at/sm/Software.htm.
53. Glegg, W., Blake, A., Gould, R., *et al.* (2001). *Crystal Structure Analysis Principles and Practise*, Oxford University Press, Oxford.
54. Schneider, M., Winkler, S. and Brückner, A. (2008). The Influence of Calcination Conditions on Phase Formation in Movtenbox Catalysts : A Simultaneous In Situ XRD/Raman Study, SNBL Workshop on Simultaneous Raman-X-Ray Diffraction/Absorption Studies for the In Situ Investigations of Solid State Transformations and Reactions at Non Ambient Conditions, ESRF Grenoble, 18–19 June 2008.
55. Fait, M., Schneider, M., Kontratenko, E., *et al.* (2009). Combination of a Multi-Channel Reactor System with UV/vis Spectroscopy or XRD: Proof of Principle, Third International Congress on Operando Spectroscopy, 19–23 April 2009, Rostock-Warnemünde, Germany.
56. Clausen, B., Grabek, L., Steffensen, G., *et al.* (1993). A Combined QEXAFS/XRD Method for On-Line In Situ Studies of Catalysts: Examples of Dynamic Measurements of Cu-Based Methanol Catalysts, *Catal. Lett.*, **20**, pp. 23–36.
57. Newton, M. and van Beek, W. (2010). Combining Synchrotron-Based X-Ray Techniques with Vibrational Spectroscopy for the In Situ Study of Heterogeneous Catalysts: A View from a Bridge, *Chem. Soc. Rev.*, **39**, pp. 4845–4863.
58. Grunwaldt, J. and Clausen, B. (2002). Combining XRD and EXAFS with On-Line Catalytic Studies for In Situ Characterization of Catalysts, *Top. Catal.*, **18**, pp. 37–41.
59. Beale, A. and Sankar, G. (2003). In Situ Characterization of Iron Phosphate and Bismuth Molybdate Prepared by Hydrothermal Methods: An EDXRD and Combined XRD/XAS Study, *Nucl. Instr. Meth. Phys. Res.*, **B 199**, pp. 504–508.
60. Beale, A., van der Erden, M., Jaques, S., *et al.* (2006). A Combined SAXS/WAXS/XAFS Setup Capable of Observing Concurrent Changes Across the Nano-to-Micrometer Size Range in Inorganic SolidCrystallization Processes, *J. Am. Chem. Soc.*, **128**, pp. 12386–12387.
61. Bentrup, U., Radnik, J., Armbruster, U., *et al.* (2009). Linking Simultaneous In Situ WAXS/SAXS/Raman/ATR/UV-vis Spectroscopy: Comprehensive Insight into the Synthesis of Molybdate Catalyst Precursors, *Top. Catal.*, **52**, pp. 1350–1359.

62. Marosi, L., Cox, G., Tenten, A., *et al.* (2000). In Situ XRD Investigations of Heteropolyacid Catalysts in the Methacroleinto Methacrylic Acid Oxidation Reaction: Structural Changes during the Activation/Deactivation Process, *J. Catal.*, **194**, pp. 140–145.
63. Jentoft, F., Klokishner, S., Kröhnert, J., *et al.* (2003). The Structure of Molybdenum Heteropoly Acids under Conditions of Gas Phase Selective Oxidation Catalysis: A Multi Method In Situ Study, *Appl. Catal. A*: *Gen.*, **256,** pp. 291–317.
64. Wienhold, J., Timpe, O. and Ressler, Th. (2003). In Situ Investigations of Structure-Activity Relationships in Heteropolyoxomolybdates as Partial Oxidation Catalysts, *Chem. Eur. J.*, **9**, pp. 6007–6017.
65. Ressler, T., Timpe, O., Girgsdies, F., *et al.* (2005). In Situ Investigations of the Bulk Structural Evolution of Vanadium-Containing Heteropolyoxomolybdate Catalysts During Thermal Activation, *J. Catal.*, **231**, pp. 279–291.
66. Brückner, A., Scholz, G., Heidemann, D., *et al.* (2007). Structural Evolution of $H_4PVMo_{11}O_{40}xH_2O$ during Calcination and Isobutane Oxidation: New Insights into Vanadium Sites by a Comprehensive In Situ Approach, *J. Catal.*, **245**, pp. 369–380.
67. Stern, E., Sayers, D. and Lytle, F. (1975). Extended X-Ray-Absorption Fine-Structure Technique: III. Determination of Physical Parameters, *Phys. Rev. B*, **11**, pp. 4836–4846.
68. Clausen, B., Topsoe, H., Hansen L., *et al.* (1994). Determination of Metal-Particle Sizes from EXAFS, *Catal. Today*, **21**, pp. 49–55.
69. Koningsberger, D., Mojet, B., van Dorssen, G., *et al.* (2000). XAFS Spectroscopy: Fundamental Principles and Data Analysis, *Topics Catal.*, **10**, pp. 143–155.
70. Panov, G., Sheveleva, G., Kharitonov, A., *et al.* (1992). Oxidation of Benzene to Phenol by Nitrous Oxide over Fe-ZSM-5 Zeolites, *Appl. Catal. A*: *Gen.*, **82**, pp. 31–36.
71. Heijboer, W., Glatzel P., Sawant K., *et al.* (2004). K$\beta$-Detected XANES of Framework-Substituted FeZSM-5 Zeolites, *J. Phys. Chem. B*, **108**, pp. 10002–10011.
72. Zecchina, A., Rivallan, M., Berlier, G., *et al.* (2007). Structure and Nuclearityof Active Sites in Fe-Zeolites: Comparison with Iron Sites in Enzymes and Homogeneous Catalysts, *Phys. Chem. Chem. Phys.*, **9**, pp. 3483–3499.
73. Hashmi, A., Lothschütz, C., Ackermann, M., *et al.* (2010). Gold Catalysis: In Situ EXAFS Study of Homogeneous Oxidative Esterification, *Chem. Eur. J.*, **16**, pp. 8012–8019.
74. Frahm, R. (1989). New Method For Time-Dependent X-Ray Absorption Studies, *Rev. Sci. Instrum.*, **60**, pp. 2515–2518.
75. Hagelstein, M., San Miguel A., Fontaine, A., *et al.* (1997), The Beamline ID24 at ESRF for Energy-Dispersive X-Ray Absorption Spectroscopy, *J. Phys. IV France*, **7**, pp. C2–303–C2–308.
76. Ressler, T., and Timpe, O. (2007). Time-Resolved Studies on Correlations Between Dynamic Electronic Structure and Selectivity of a $H_5[PV_2Mo_{10}O_{40}]$ Partial Oxidation Catalyst, *J. Catal.*, **247**, pp. 231–237.
77. Rodriguez, J. and Goodman, D. (1991). High-Pressure Catalytic Reactions over Single-Crystal Metal Surfaces, *Surf. Sci. Rep.*, **14**, pp. 1–107.
78. Radnik, J., Mohr. C. and Claus P. (2003). On the Origin of Binding Energy Shifts of Core Levels of Supported Gold Nanoparticles and Dependence of Pretreatment and Material Synthesis, *Phys. Chem. Chem. Phys.*, **5**, pp. 172–177.
79. Blume, R., Niehus, H., Conrad, H., *et al.* (2005). Identification of Subsurface Oxygen Species Created during Oxidation of Ru(0001), *J. Phys. Chem. B*, **109**, pp. 14052–14058.
80. Over, H., Kim, Y., Seitsonen, A., *et al.* (2000), Atomic-Scale Structure and Catalytic Reactivity of the $Ru_2O$ (110) Surface, *Science*, **287**, pp. 1474–1476.
81. Bluhm, H., Hävecker, M., Knop-Gericke, A., *et al.* (2004). Methanol Oxidation on a Copper Catalyst Investigated Using In Situ X-Ray Photoelectron Spectroscopy, *J. Phys. Chem. B*, **108**, pp. 14340–14347.

82. Joyner, R., Roberts, M., and Yates, K. (1979). A "High-Pressure" Electron Spectrometer for Surface Studies, *Surf. Sci.*, **87**, pp. 501–509.
83. Ogletree, D., Bluhm H., Lebedev G., *et al.* (2002). A Differentially Pumped Electrostatic Lens System for Photoemission Studies in the Millibar Range, *Rev. Sci. Instrum.*, **73**, pp. 3872–3877.
84. Salmeron M. and Schlögl R. (2008), Ambient Pressure Photoelectron Spectroscopy: A New Tool for Surface and Nanotechnology, *Surf. Sci. Rep.*, **63**, pp. 169–199.
85. Grass, M., Karlsson, P., Aksoy, F., *et al.* (2010), New Ambient Pressure Photoemission Endstationat Advanced Light Source Beamline 9.3.2, *Rev. Sci. Instrum.*, **81**, 053106.
86. Dencke, R., Kinne, M., Whelan, C., *et al.* (2002). In Situ Core Level Photoelectron Spectroscopy of Adsorbates on Surfaces Involving a Molecular Beam — General Setup and First Experiments, *Surf. Rev. Lett.*, **9**, pp. 797–801.
87. Pantförder J., Pöllmann S., Zhu J., *et al.* (2005). New Setup for In Situ X-Ray Photoelectron Spectroscopy from Ultrahigh Vacuum to 1 mbar, *Rev. Sci. Instrum.*, **76**, 014102.
88. Enache, D., Edwards, J., Landon, P., *et al.* (2006). Solvent-Free Oxidation of Primary Alcohols to Aldehydes Using Au-Pd/$TiO_2$Catalysts, *Science*, **311**, pp. 362–365.
89. Chen, M., Kumar D., Yi, C., *et al.* (2005). The Promotional Effect of Gold in Catalysis by Palladium-Gold, *Science*, **310**, pp. 291–293.
90. Pohl, M., Radnik, J., Schneider, M., *et al.* (2009). Bimetallic PdAu-KOac/$SiO_2$ Catalysts for Vinyl Acetate Monomer (VAM) Synthesis: Insights into Deactivation under Industrial Conditions, *J. Catal.*, **262**, pp. 314–323.
91. Benhmid, A., Narayana, K., Martin, A., *et al.* (2004). Highly Efficient Pd-Sb-$TiO_2$ Catalysts for the Gas Phase Acetoxylation of Toluene to Benzyl Acetate, *J. Catal.*, **230**, pp. 420–435.
92. Tao, F., Grass, M., Zhang, Y., *et al.* (2010). Evolution of Structure and Chemistry of Bimetallic Nanoparticle Catalysts under Reaction Conditions, *J. Am. Chem. Soc.*, **132**, pp. 8697–8703.
93. Streber, R., Papp, C., Lorenz, M., *et al.* (2009). SulfurOxidation on Pt(335): It is the Steps!, *Angew. Chem. Int. Ed.* **48**, pp. 9743–9746.
94. Hävecker, M., Knop-Gericke, A., Schedel-Niedrig, T., *et al.* (1998). High-PressureSoft X-Ray Absorption Spectroscopy: A Contribution to Overcoming the "Pressure Gap" in the Study of Heterogeneous Catalytic Processes, *Angew. Chem. Int. Ed.*, **37**, pp. 1939–1942.
95. Ertl, G., Knözinger, H., Schüth, F., *et al.* (eds) (2008). *Handbook of Heterogeneous Catalysis*, Wiley-VCH, Weinheim.
96. Bhatia, S., Beltramini, J. and Do, D. (1990). Temperature Programmed Analysis and its Applications in Catalytic Systems, *Catal. Today*, **7**, pp. 309–438.
97. Brown, M., Dollimore, D. and Galwey, A. (1980). *Reactions in the Solid State,* Elsevier, Amsterdam.
98. Kissinger, H. (1957). Reaction Kinetics in Differential Thermal Analysis, *Anal. Chem.*, **29**, pp. 1702–1706.
99. Pineau, A., Kanari, N. and Gaballah, I. (2006). Kinetics of Reduction of Iron Oxides by $H_2$. Part I: Low Temperature Reduction of Hematite, *Thermochim. Acta*, **447**, pp. 89–100.
100. Ovsitser, O., Cherian, M., Brückner, A., *et al.* (2009). Dynamics of Redox Behavior of Nano-Sized $VO_x$ Species over Ti–Si-MCM-41 from Time-Resolved In Situ UV/Vis Analysis, *J. Catal.*, **265**, pp. 8–18.
101. Tibiletti, D., Goguet, A., Meunier, F., *et al.* (2004). On the Importance of Steady-State Isotopic Techniques for the Investigation of the Mechanism of the Reverse Water-Gas-Shift Reaction, *Chem. Commun.*, **14**, pp. 1636–1637.
102. Meunier, F., Reid, D., Goguet, A., *et al.* (2007). Quantitative Analysis of the Reactivity of Formate Species Seen by DRIFTS over an Au/Ce(La)$O_2$ Water-Gas Shift Catalyst: First Unambiguous Evidence of the Minority Role of Formatesas Reaction Intermediates, *J. Catal.*, **247**, pp. 277–287.

103. Bravo-Suarez, J., Lu, J., Dallos, C., *et al.* (2007). Kinetic Study of Propylene Epoxidation with $H_2$ and $O_2$ over a Gold/Mesoporous Titanosilicate Catalyst, *J. Phys. Chem. C*, **111**, pp. 17427–17436.
104. Lu, J., Zhang, X., Bravo-Suarez, J., *et al.* (2007). Direct Propylene Epoxidation over Barium-Promoted Au/Ti-TUD Catalysts with $H_2$ and $O_2$: Effect of Au Particle Size, *J. Catal.*, **250**, pp. 350–359.
105. Ebner, J. and Gleaves, J. (1986) Method and Apparatus for Carring Out Catalyzed Chemical Reactions and for Studing Catalysts, *US Patent 4626412*.
106. Gleaves, J. (1996) Method for Study and Analysis of Products of Catalytic Reaction, *US Patent 5500371*.
107. Gleaves, J., Ebner, J. and Kuechler, T. (1988). Temporal Analysis of Products (TAP): A Unique Catalyst Evaluation System with Submillisecond Time Resolution, *Catal. Rev. Sci. Eng.*, **30**, pp. 49–116.
108. Gleaves, J., Yablonsky, G., Phanawadee, P., *et al.* (1997). TAP-2: An Interogative Kinetics Approach, *Appl. Catal. A*, **160**, pp. 55–88.
109. Gleaves, J., Yablonsky, G., Zheng, X., *et al.* (2010). Temporal Analysis of Products (TAP): Recent Advances in Technology for Kinetic Analysis of Multi-Component Catalysts, *J. Molec. Catal. A*, **315,** pp. 108–134.
110. Perez-Ramirez, J., Kondratenko, E., Kondratenko, V., *et al.* (2004). Selectivity-Directing Factors of Ammonia Oxidation over PGM Gauzes in the Temporal Analysis of Products Reactor: Primary Interactions of $NH_3$ and $O_2$, *J. Catal.*, **227**, pp. 90–100.
111. Kondratenko, E., Kondratenko, V., Richter, M., *et al.* (2006). Influence of $O_2$ and $H_2$ on NO Reduction by $NH_3$ over $Ag/Al_2O_3$: A Transient Isotopic Approach, *J. Catal.*, **239**, pp. 23–33.
112. Kondratenko, E., Pérez-Ramírez, J. (2006). Importance of the Lifetime of Oxygen Species Generated by $N_2O$ Decomposition for Hydrocarbon Activation over Fe-Silicalite, *Appl. Catal. B*, **64**, pp. 35–41.
113. Perez-Ramirez, J. and Kondratenko, E. (2004). Evidences of the Origin of $N_2O$ in the High-Temperature $NH_3$ Oxidation over Pt-Rh Gauze, *Chem. Commun.*, **4**, pp. 376–377.
114. Soick, M., Wolf, D. and Baerns, M. (2000). Determination of Kinetic Parameters for Complex Heterogeneous Catalytic Reactions by Numerical Evaluation of TAP Experiments, *Chem. Eng. Sci.*, **55**, pp. 2875–2882.
115. Constales, D., Shekhtman, S., Yablonsky, G., *et al.* (2006). Multi-Zone TAP-Reactors Theory and Application IV. Ideal and Non-Ideal Boundary Conditions, *Chem. Eng. Sci.* **61**, pp. 1878–1891.
116. Svoboda, G., Gleaves, J., and Mills, P. (1992). New Method for Studying the Pyrilysis of VPE/CVD Precursors under Vacuum Conditions. Application to Trimethylantimony and Tetramethyltin, *Ind. Eng. Chem. Res.*, **31**, pp. 19–29.
117. Creten, G., Lafyatis, D. and Froment, G. (1995). Transient Kinetics from the TAP Reactor System: Application to the Oxidation of Propylene to Acrolein, *J. Catal.*, **154**, pp. 151–162.
118. Rothaemel, M. and Baerns, M. (1996). Modeling and Simulation of Transient Adsorption and Reaction in Vacuum Using the Temporal Analysis of Products Reactor, *Ind. Eng. Chem. Res.*, **35**, pp. 1556–1565.
119. van der Linde, S., Nijhuis, T., Dekker, F., *et al.* (1997). Mathematical Treatment of Transient Kinetic Data: Combination of Parameter Estimation with Solving the Related Partial Differential Equations, *Appl. Catal. A*, **151**, pp. 27–57.
120. Kondratenko, V. and Baerns, M. (2004). Mechanistic and Kinetic Insights into $N_2O$ Decomposition over Pt Gauze, *J. Catal.*, **225**, pp. 37–44.
121. Kondratenko, E., Kondratenko, V., Santiago, M., *et al.* (2010). Mechanism and Micro-Kinetics of Direct $N_2O$ Decomposition over $BaFeAl_{11}O_{19}$ Hexaaluminate and Comparison with Fe-MFI Zeolites, *Appl. Catal. B*, **99**, pp. 66–73.

122. Kondratenko, E. and Pérez-Ramírez, J. (2006). Mechanism and Kinetics of Direct $N_2O$ Decomposition over Fe-MFI Zeolites with Different Iron Speciation from Temporal Analysis of Products, *J. Phys. Chem. B*, **110**, pp. 22586–22595.
123. Kondratenko, E. and Pérez-Ramírez, J. (2007). Micro-Kinetic Analysis of Direct $N_2O$ Decomposition over Steam-Activated Fe-Silicalite from Transient Experiments in the TAP Reactor, *Catal. Today*, **121**, pp. 197–203.
124. Kondratenko, E., Kondratenko, V., Santiago M., *et al.* (2008). Mechanistic Origin of the Different Activity of Rh-ZSM-5 and Fe-ZSM-5 in $N_2O$ Decomposition, *J. Catal.*, **256**, pp. 248–258.
125. Shannon, S. and Goodwin Jr, J. (1995). Characterization of Catalytic Surfaces by Isotopic-Transient Kinetics during Steady-State Reaction, *Chem. Rev.*, **95**, pp. 677–695.
126. Kondratenko, E. (2010). Using Time-Resolved Methods to Monitor and Understand Catalytic Oxidation Reactions, *Catal. Today*, **157**, pp. 16–23.
127. Tibiletti, D., Goguet, A., Meunier, F., *et al.* (2004). On the Importance of Steady-State Isotopic Techniques for the Investigation of the Mechanism of the Reversed Water-Gas-Shift Reaction, *Chem. Commun.*, **14**, pp. 1636–1637.
128. Meunier, F., Goguet, A., Hardacre, C., *et al.* (2007). Quantitative DRIFTS Investigation of Possible Reaction Mechanisms for the Water-Gas Shift Reaction on High-Activity Pt- and Au-based Catalysts, *J. Catal.*, **252**, pp. 18–22.
129. Meunier, F., Tibiletti, D., Goguet, A., *et al.* (2007). On the Complexity of the Water-Gas Shift Reaction Mechanism over a $Pt/CeO_2$ Catalyst: Effect of the Temperature on the Reactivity of Formate Surface Species Studied By Operando DRIFT during Isotopic Transient at Chemical Steady-State, *Catal. Today*, **126**, pp. 143–147.
130. Kondratenko, E., Ovsitser, O., Radnik, J., *et al.* (2007). Influence of Reaction Conditions on Catalyst Composition and Selective/Non-Selective Reaction Pathways of the ODP Reaction over $V_2O_3$, $VO_2$ and $V_2O_5$ with $O_2$ and $N_2O$, *Appl. Catal. A*, **319**, pp. 98–110.
131. Ovsitser, O. and Kondratenko, E. (2009). Similarity and Differences in the Oxidative Dehydrogenation of $C_2$-$C_4$ Alkanes over Nano-Sized $VO_x$ Species Using $N_2O$ and $O_2$, *Catal. Today*, **142**, pp. 138–142.
132. Rozanska, X., Kondratenko, E. and Sauer, J. (2008). Oxidative Dehydrogenation of Propane: Differences between $N_2O$ and $O_2$ in the Reoxidation of Reduced Vanadia Sites and Consequences for Selectivity, *J. Catal.*, **256**, pp. 84–94.

# Chapter 20

# Vanadium-Phosphorus Oxide Catalyst for *n*-Butane Selective Oxidation: From Catalyst Synthesis to the Industrial Process

Elisabeth BORDES-RICHARDS*, Ali SHEKARI†
and Gregory S. PATIENCE†

The knowledge accrued over 40 years of using vanadium-phosphorus oxide(VPO) catalysts in the oxidation of *n*-butane to maleic anhydride is reviewed from both fundamental and applied points of view. The importance of the $VOHPO_4 \cdot 0.5H_2O$ precursor and its decomposition to vanadium pyrophosphate is highlighted. The phenomena occurring during the transient state are better understood thanks to recent *in situ* investigations. The depth and surface of the (eventually pre-treated) precursor are modified and restructured while adapting to the $C_4/O_2$ redox mixture. The atomic phosphorus-to-vanadium ratio plays a critical role because it regulates the redox behavior. To maximize commercial reactor performance, the VPO catalyst should be formulated to be adapted to the oxidizing environment typical of a fixed-bed reactor, or to a more reducing environment as in a circulating fluidized bed (CFB) reactor. Transient redox conditions in which the oxidation step is timely (pulse or periodic feed reactor) or physically separated (by membrane reactor, CFB reactor) from the reduction step are also described. Results show that the catalyst activity may be moderated by catalyst oxidation time, $C_4/O_2$ ratio, temperature and pressure. The advantages and drawbacks of several types of reactors are also discussed.

## 20.1. Introduction

Selective catalysts for the partial oxidation of hydrocarbons, like those leading to bulk chemicals intermediates (aldehydes, ketones, acids, anhydrides) contain transition metaloxides. Typically, these reactions involve the exchange of several electrons ($e^-$) that come from the valence band of the catalyst: two electrons in the case of the oxidative dehydrogenation of alkanes, up to 14 electrons in the case of the partial

*Unité de Catalyse et de Chimie du Solide, UMR CNRS 8181, Ecole Nationale Supérieure de Chimie de Lille — Université Lille 1, Cité scientifique, 59655 Villeneuve d'Ascq, France.
†Department of Chemical Engineering, Ecole Polytechnique de Montréal, C.P. 6079, Succ. "CV", Montréal, QC, Canada H3C 3A7.

oxidation of $n$-butane to maleic anhydride. Obviously, the more electrons involved, the more demanding is the reaction and consequently the more specific is the catalyst. As an example, more than three systems (Sn-Sb-O, U-Sb-O, bismuthmolybdates, though only the latter is commercialized), are known to oxidize the 4 $e^-$ oxidation of propylene to acrolein, but only V-P-O phases or $VO_x/TiO_2$ are selective in the 14 $e^-$ or 12 $e^-$ oxidation of $n$-butane or $o$-xylene to the respective anhydrides. The specificity of the selective catalyst is determined by its structure and reactivity, and not only its superficial structure. The selectivity to the targeted molecule is due to the participation of surface lattice oxide ($O^{2-}$) ions, which are, most often, linked to vanadium and/or molybdenum cations which undergo reduction. The contribution of lattice $O^{2-}$species during the oxidation of propylene to acrolein on bismuth-molybdate was demonstrated a long time ago by means of $^{18}O$ isotopic exchange (e.g. Ref.[1]) which showed that labeled oxygen was incorporated into the reaction products. Such experiments were conceived to demonstrate the redox mechanism,[2] but many others showed that it applies to most selective oxidation reactions.

For the sake of simplificity, the oxidic material $MeO_y$ (Me transition metal) may be seen as an oxide ion reservoir to be (partly) depleted depending on the reducibility of the associated cation(s). During reaction, the given catalytic $Me^{n+}/Me^{(n-p)+}$ red/ox couple is faced with the reactant feed (e.g. hydrocarbon and air), which itself is a red/ox mixture. While the first oxidized molecules are formed (transient state), the catalytic solid (both surface and bulk) restructures according to the composition of the reducing (hydrocarbon, HC) and oxidizing mixture($O_2$, air, $N_2O$...). At the steady state, the stability of the catalytic bulk must be achieved, and on the surface the rates of reduction (due to oxidation of the reactant) and of reoxidation (by co-fed oxygen) are equal. A picture of the phenomena that may occur during these transient or stationary regimes will first be proposed, before application to vanadium-phosphorus oxide (VPO) catalysts for $n$-butane oxidation to maleic anhydride (MA). VPO catalysts are unique in providing the 14 electrons during the $C_4H_{10}$-MA reaction, and up to now they have not been significantly challenged by any other composition. The state of the art will be discussed as far as the catalyst properties are concerned. More particularly, the role of oxidized ($VOPO_4$) and reduced $(VO)_2P_2O_7$ phases will be examined in the light of recent experiments published in the literature, including their reactivity (equilibration, ageing), and the possible transformations depending on the catalyst composition (atomic phosphorus-to-vanadium ratio, P/V and the operating conditions (mainly temperature, and $C_4H_{10}/O_2$ ratio).

Obviously, the manner in which the gas phase is put in contact with the catalyst surface is important as the reactor type influences the catalytic performance through heat and mass transfer (as well as hydrodynamics in special cases). This is true for any catalyst/reaction couple, but alternative designs exploiting the redox mechanism have been proposed in the case of selective oxidation reactions. Redox decoupling by separating the oxidation of the HC reactant (reduction of the catalyst) from the oxidation of the catalyst is a means to tentatively optimize each step that may require different operating conditions than in a co-feeding regime. The separation may be achieved "in time" by carrying out transient experiments (alternately feeding HC and oxygen), as first applied to $n$-butane oxidation by Emig *et al.*[3] It can be achieved "spatially" by using different reactor designs, like dense membrane reactors,[4] porous membrane reactors,[5,6] two-zone fludizedbed[7] or circulating fluidized-bed reactors (CFB).[8] DuPont developed a commercial process in which $n$-butane was oxidized to maleic anhydride in a CFB, the maleic acid being subsequently hydrogenated to tetrahydrofuran.[9] In a continuous process, the "oxidized" form of the catalyst was reacted with $n$-butane in a riser/transport bed; it was stripped of hydrocarbons and most products and the reduced catalyst was then oxidized in a fluidized-bed reactor before returning to the riser/transport bed. The morphological changes occurring as the VPO catalyst cycles between these regimes will be reviewed.

Research into new technologies and catalysts for butane oxidation to maleic anhydride continue, but at the moment it would appear that fixed beds are the preferred commercial reactor technology. Providers of fixed-bed technology include Huntsmann (who purchased Monsanto's maleic anhydride business), Scientific Design and Technobell Limited. Lummus/Polynt license fluidized-bed technology, while Ineos (formally part of BP and Amoco) operates major fluidized-bed reactors. Both DuPont's CFB and membrane reactor will be discussed. Forced concentration cycling experiments — operation between a net reducing zone and an oxidizing zone — at a wide range of feed concentrations and elevated pressure will also be reviewed.

## 20.2. Portrait of a Selective Oxidation Catalyst

A portrait of the underlying phenomena transpiring in a small crystalline particle ($<$100 nm) of a $MeO_y$ oxide (most often Me = V or Mo) when exposed to a HC and oxygen mixture at a given temperature during the transient state or "equilibration" period, is proposed in Fig. 20.1. A "multiplet site" (constituted by at least one $Me^{n+}$–$O^{2-}$ group, but as large as required by the reaction) on the surface reacts with the reactant R, which becomes the targeted product P (rate of reaction $r_R$) by picking up the oxygen linked to $Me^{n+}$, which in turn becomes $Me^{(n-p)+}$ (rate of reduction $r_{red}$). The product desorbs and the $Me^{n+}$–o□ (where □ represents an oxygen vacancy) site is re-oxidized to $Me^{n+}$–$O^{2-}$ (rate of oxidation $r_{ox}$). There are two possibilities to refill the oxygen vacancies. One is by means of dioxygen co-fed with R. The reduction of $O_2$ to $2O^{2-}$ requires four electrons which are provided by the catalytic material. This step depends on the type and value of semi-conductivity ($\sigma_e$ by electrons or

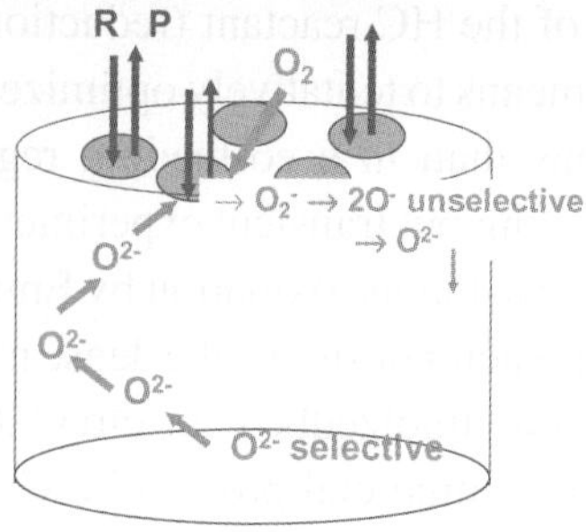


**Figure 20.1.** Schematics of reactant (R) to product (P) reaction on multiplet sites in a <100 nm-thick crystal of an early transition metal oxide. Re-oxidation of the catalyst surface after desorption of P, by $O^{2-}$ diffusion from bulk or after reduction of dioxygen $O_2$ to $O^{2-}$.

$\sigma_p$ by holes). Oxygen species (other than $O^{2-}$) which are said to be non-selective can form on the surface if the material is not able to provide enough electrons,[10] as shown by the following successive reactions:

$$O_2 + e = O_2^-; O_2^- + e = 2O^-; 2O^- + 2e = 2O^{2-} \quad (20.1)$$

Alternatively, the sub-surface or bulk $O^{2-}$ species may diffuse to the surface to refill the vacancy. Note that only reactors operating in a redox decoupling mode, like CFBs, would derive oxygen from the bulk because the oxidation step is carried out in a distinct fluidized-bed reactor. Obviously this diffusion process is slower than the surface re-oxidation by molecular (gaseous) $O_2$, but depending on the cation reducibility during catalyst "equilibration", part of the bulk becomes reduced. At the steady state ($r_R = r_{red} = r_{ox}$) there is a limited amount of $O^{2-}$ species able to participate in the reaction at each cycle (turn-over).

The solid-state chemistry of vanadium and molybdenum oxides is related to the ability of $Mo^{6+}$ and $V^{5+}$ (formal valences) to bind one oxygen (or more) as a short $\sigma-\pi Me = O$ bond, and to reversibly undergo coordination changes from six to four, this change being accompanied or not by reduction. However, the difference between Mo and V is that molybdates ($MoO_4^{2-}$) are more easily formed than the corresponding vanadate salts, and many molybdates with Bi, Fe, Co, etc., cations are known. On the contrary, both $V^{5+}{=}O$ and $V^{4+}{=}O$ bonds in distorted octahedra are quite stable, and they can also be stabilized by anionic ligands, like phosphates. The other main difference is their reactivity. Typically, the rate of oxidation, $r_{ox}$, of molybdenum-based oxides is very high, and at steady state the oxidation state of Mo is very close to 6+. It has been necessary to perform electron spin resonance spectroscopy (ESR) or *in situ* experiments to detect the presence of some reduced $Mo^{5+}$ species in molybdate catalysts.[11] On the contrary, the oxidation rate of vanadium-based oxides is generally lower than in Mo-based oxides, so that mixtures of $V^{5+}$ or $V^{4+}$

species are generally present at steady state. UV-vis spectroscopy of used catalysts is a means to check the formation of reduced vanadium, and often this is deduced just by looking at the change of color of the catalyst after use.[12]

## 20.3. Application to VPO Catalysts in *n*-butane Oxidation to Maleic Anhydride

Originally, maleic anhydride was manufactured from butenes (in 1962 in the US) and was eventually converted tobenzene over a vanadium pentoxide catalyst. Most plants have converted to *n*-butane as a feedstock (although Mitsubishi operates a plant based on mixed butenes in a fluidized bed) while global production outside the US of MA from benzene remains around 15–20%. In the late 1960s and early 1970s, major corporations patenting catalysts and processes included BASF, ICI, Halcon and Japan Catalyst Chem. Ind. In 1971 Mitsubishi published a patent for butenes while one of Monsanto's first patents was issued in 1973. Few studies of VPO catalysts for the oxidation of butene to MA have been published.[13–15] $VOPO_4$ phases were identified as capable of activating butenes with MA molar yields in the vicinity of 50%.[13] The drawback of this process was the wide range of by-products, which made the separation too expensive. At the same time significant advances in activating *n*-butane — a cheaper feedstock — were being achieved.

### 20.3.1. *Preparation of active and selective VPO catalyst*

A peculiar method of preparation had to be developed to get an active and selective VPO material.[13–15] Indeed, it was shown early on that selectivity to MA was related to the platy morphology of vanadyl pyrophosphate $(VO)_2P_2O_7$ crystals, supposedly the active phase. The method consisted of refluxing a mixture of vanadium oxide and phosphoric acid in an organic medium, which is now practiced commercially. The $VOHPO_4 \cdot 0.5H_2O$ precursor precipitates and is then further dehydrated to produce an active and selective catalyst. There have been many attempts to improve the method, e.g. by distilling water during reflux to eliminate other hydrates. More particularly, resistance to attrition for use in fluidized-bed reactors or CFBs was achieved by developed spray drying technology.[13] The effects of promotors have also been studied, as well as several other preparation methods. Herein, we focus on undoped VPO materials and the effect of the P/V (atomic ratio) which affects the catalyst reactivity.

The crystal structure of the $VOHPO_4 \cdot 0.5H_2O$ precursor is described as pairs of face-sharing $[VO_6]$ octahedra that are linked to hydrogenophosphate groups. The layered structure is due to the presence of water between $[VOHPO_4]$ sheets[14] (Fig. 20.2a). The crystal structure of $(VO)_2P_2O_7$ (VPP) was solved by Gorbunova

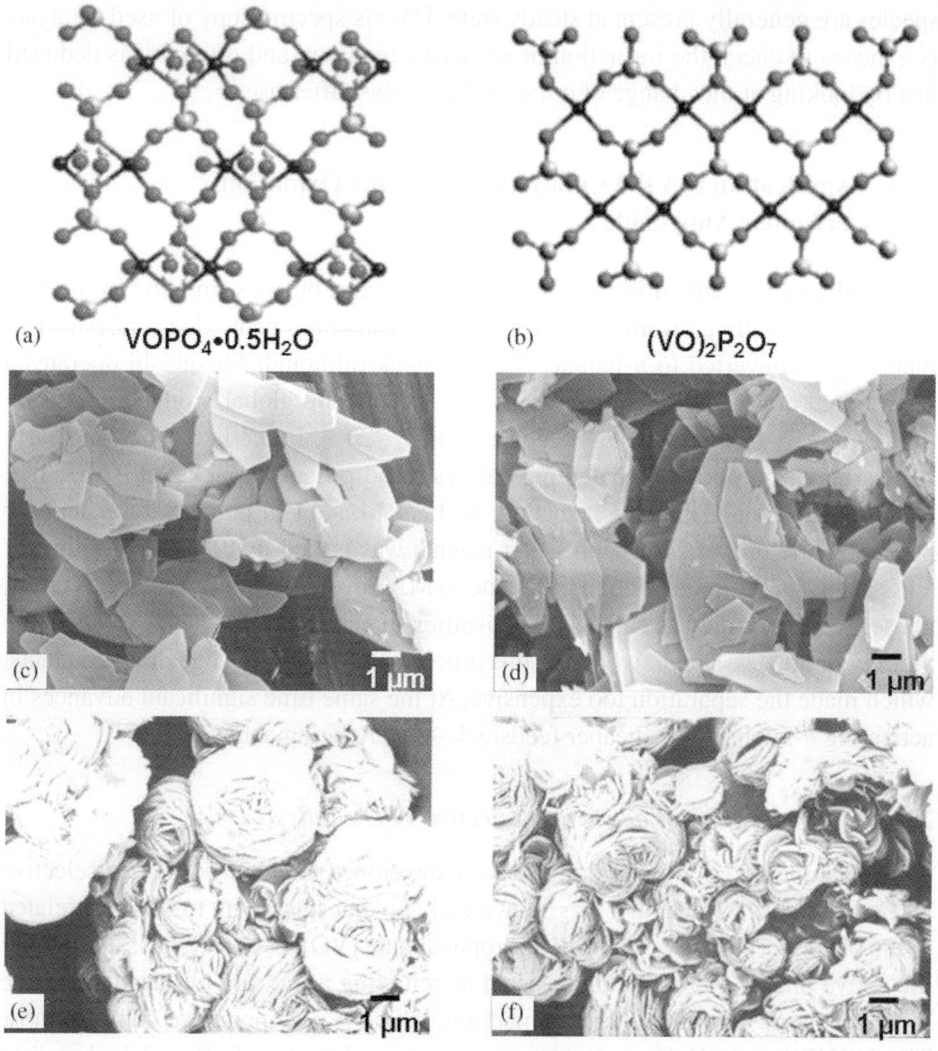


**Figure 20.2.** Dehydration of the $VOHPO_4 \cdot 0.5H_2O$ precursor (a, c, e) to $(VO)_2P_2O_7$ (b, d, f). Crystal structure of {001} face (a) and of {100} face (b); black: V; violin: P; red: O; white: H (after Petit *et al.*[22]). SEM pictures of primary particles (c and d); SEM pictures of secondary particles (e and f).

and Linde,[15] and later refined by Nguyen *et al.*[16] The crystal framework of VPP is made of pairs of edge-sharing $[VO_6]$ octahedra linked to pyrophosphate groups (Fig. 20.2b).

The topotactic character of the thermal dehydration of $VOHPO_4 \cdot 0.5H_2O$ to $(VO)_2P_2O_7$ was studied by Torardi and Calabrese[17] and by Bordes *et al.*[18–20] The pseudomorphicity between the precursor and VPP was demonstrated at two scales, for primary (crystalline plates, nano to microscale) (Figure 20.2c, d) as well as for secondary (rosets, microscale) particles (Figure 20.2e, f). DuPont samples were prepared at the lab scale and industrially using the same P/V ratio (1/1) and alcohol mixture, but for a longer time in the laboratory (delivering bigger crystals) than for future commercial use. The topotactic mechanism was indirectly confirmed by *in situ* X-ray diffraction (XRD), scanning electron microscopy (SEM), transmission electron microscopy (TEM) and X-ray photoelectron spectroscopy (XPS) experiments.[18] $VOPO_4 \cdot 2H_2O$ was synthesized in water when using $V_2O_5$ and $H_3PO_4$ with a high excess of P (P/V = 7/1).[13] It was also used as a precursor to make well-defined crystals of $VOHPO_4 \cdot 0.5H_2O$.[19,24] The possible control of VPP crystal morphology by controlling the precursor morphology makes the alcohol refluxing method very attractive. The alcohol is a reducing agent of $V^{5+}$ to $V^{4+}$ while in water chlorhydric acid is necessary to avoid $VOPO_4 \cdot 2H_2O$ ($V^{5+}$).[20] The $VO^{2+}$ cation is stabilized by the $HPO_4^{2-}$ ligand and this favors the precipitation of the vanadyl hydrogenophosphate precursor. $HPO_4^{2-}$ anion is preferred over other phosphate anions because phosphoric acid is a milder acid in alcoholic solvent than in water, and it becomes milder as the hydrocarbon chain of the alcohol becomes longer. Kestemann *et al.*[21] showed that the crystal size and shape of the precursor depends on the chemical nature (formula, secondary/primary, etc.) of alcohols, and on the boiling point temperature, and, consequently, on the rate of Reaction 20.2:

$$V_2O_5 + 2H_3PO_4 \rightarrow 2VOHPO_4 \cdot 0.5H_2O + H_2O + 0.5O_2 \qquad (20.2)$$

A distillation column might be added when phosphoric acid containing water ($H_3PO_4$85%) is used, to avoid precipitating higher hydrates ($VOHPO_4 \cdot nH_2O$, $n \leq 4$). Isobutanol/benzyl alcohol mixtures seem to be the most appropriate from all these points of view, and they yield platy crystals of precursor with well-developed {001} face areas. The roset shape of the secondary particles of the precursor is due to strains during the synthesis in alcohol (non-Newtonian behavior). After appropriate dehydration this microstructure is maintained for the catalyst, and consequently it allows a better contact with reacting molecules than if it were made of piled plates.

### 20.3.2. *Occurrence of $VOPO_4$ phases and ageing*

The presence of $O_2$ at sufficiently high concentrations during calcination and/or *in situ* heat treatment by $C_4/O_2$ leads to the formation of $VOPO_4$ phases, which may or may not coexist with VPP, depending on the temperature. Though the value of the P/V ratio influences the amount and nature of $VOPO_4$ phases, here we shall discuss their presence in general. Up to now, seven forms of $VOPO_4(\alpha_I, \alpha_I, \beta, \delta, \varepsilon, \gamma, \omega)$ have been identified, the formation of which depends on the preparation methods. According to literature data, the frequency of their occurrence in catalysts, before or after use, decreases along the $\alpha_{II}, \beta > \delta > \gamma, \alpha_I > \omega \gg \varepsilon$ series. The $\delta$ and $\gamma$

polymorphs were first identified when $VOHPO_4 \cdot 0.5H_2O$ was oxidatively dehydrated in air or $O_2$,[23] but they were also found in catalysts after use.[24] Recently, the structures of $\delta$, $\varepsilon$, $\gamma$ and $\omega$ have been described.[25–28] All $VOPO_4$ frameworks are constituted by single $[VO_6]$ octahedra linked to phosphates in various ways. The $\varepsilon$ phase has been identified as an excellent candidate for electrochemical applications. It was pointed out[25] that the same building blocks appear in both $\varepsilon$-$VOPO_4$ and $\beta$-$VOPO_4$ polymorphs, but that they are stacked in a different way. This different connectivity pattern was also found in the case of $\alpha_I$ compared with $\alpha_{II}$ and similarly due to the easy inversion of $V=O$ bonds vs. phosphate groups.[26] It is important to note that it affects the reactivity since $\alpha_{II}$ is quite stable whereas $\alpha_I$ is readily and topotactically hydrated to $VOPO_4 \cdot nH_2O$ ($n \leq 2$) in the presence of steam or moisture. The $\omega$ polymorph was prepared from $VOHPO_4 \cdot nH_2O$ ($n \leq 4$).[36] Compounds with XRD patterns close to that of the $\omega$ form have been often incorrectly assigned. For example, $\omega$-$VOPO_4$ could be detected during the dehydration in air of the $VOHPO_4 \cdot 0.5H_2O$ precursor in a fluidized bed, and the XRD lines of $\delta$ and $\omega$ were noticed to be very close.[27] A great value was added by Volta[28,29] and Hutchings *et al.*[30] They carried out several experiments on VPO pure phases and catalysts by innovative techniques such as[31] P nuclear magnetic resonancemagic angle spinning (NMR MAS) and spin-echo mapping, extensive high resolution electron microscopy (HREM) and transmission electron microscopy (TEM), and *in situ* laser Raman spectroscopy (first pioneered by Schrader *et al.*)[31,32] Taking into account the ease in which $\delta$-$VOPO_4$ is formed from the precursor as compared with the formation of $\alpha_I$, $\alpha_{II}$, $\beta$, Bordes *et al.*[25,31] proposed that, similarly to VPP, pairs of edge-sharing octahedra should be present in the structure of both $\delta$ and $\gamma$ polymorphs. Today we know that pairs of $[V^{5+}O_6]$ octahedra in any $VOPO_4$ framework are non-existent. An important outcome is that the reduction of any form of $VOPO_4$ to VPP requires the pairing of the single $[V^{5+}O_6]$ octahedra. The nucleation of crystallographic shear planes was proposed to account for the reduction of $\alpha_I$, $\alpha_{II}$, $\beta$–$VOPO_4$,[25] similarly for the case of $V_2O_5$ to its suboxides. *A priori* it can also be considered as a plausible mechanism for the reduction of other $\delta$, $\gamma$, $\omega$ phases. A consequence is that, once $VOPO_4$ phases have crystallized, it is more difficult for them to be reduced to platy VPP crystals during the catalytic reaction, which proceeds at low temperature (350–450°C).

The ageing of catalysts is related to the above considerations, as far as the reactivity of the phases is concerned. Due to its crystal structure, the equilibrium shape of VPP is not platy but prismatic. As a consequence, platy VPP crystals age by thickening more and more with time and tend to become prismatic. It results in the decrease of the whole surface area but more particularly of the area of selective {100} faces, and consequently the selectivity to MA decreases. As in the case of $\beta$-$VOPO_4$, the reduction of $\alpha_I$-$VOPO_4$ was shown to lead to prismatic-shaped crystals of VPP. On the contrary, crystals of $\delta$ could be reduced to platy VPP crystals exhibiting

{100} faces.[31] The counterpart of over-oxidation is over-reduction leading to the local formation of $V^{3+}$. However, these species may help to activate dioxygen. In the worst cases, $V^{3+}$-containing phosphates such as $VPO_4$, $V(PO_3)_3$ and $V_4(P_2O_7)_3$ can be formed, which are hardly converted to VPP by oxidation. A slight excess of phosphorus vs stoichiometry is said to be favorable to maintain the VPP structure by avoiding the crystallization of $VOPO_4$. However too much P leads to the formation of $VO(PO_3)_2$ which itself is poorly active in $n$-butane oxidation.[33] Some loss of P could also occur upon the re-oxidation of VPP, ($[P_2O_7]$ giving rise to 2 $[PO_4]$ orthophosphate groups), with the production of $[VO_x]$ species which would not be stable. Such a process will be discussed further (see below). Other ways of ageing are related to the catalyst being affected by reactants and products during reaction. Catalyst deactivation due to carbonaceous residue will not be discussed here. This is common to CFB processes when there is insufficient oxygen co-fed with $n$-butane, or when the residence time in the reducing zone is too long.[9] The formation of 4 $H_2O$ per maleic anhydride molecule (not counting those accompanying the formation of acetic acid or CO, $CO_2$) leads to a significant hydration of the surface. When crystals of $VOPO_4$ phases are already present, they tend to transform to $VOPO_4 \cdot nH_2O$ ($n \leq 2$), as seen by XRD performed at room temperature. At reaction temperature the formation of $\alpha_I$-$VOPO_4$ is observed. Too much water on the surface is also responsible for the segregation of phosphorus, which may lead to sublimation of phosphoric anhydride, but also to the formation of strong phosphoric acid sites, accompanying the formation of $[VO_x]$ species as already mentioned.

### 20.3.3. *Reactivity and structure sensitivity*

*Ex situ* analyses of catalysts after reaction have shown the presence of both VPP and $VOPO_4$ phases, and a number of interpretations have been proposed. According to the literature, most forms of $VOPO_4$ have been identified, apart from $\varepsilon$-$VOPO_4$. With the exception of early papers, the $\beta$ form of $VOPO_4$ was unequivocally associated with poor performance, while high selectivity to MA was generally related to the presence of the $\delta$ form, and to a lesser extent of $\alpha_{II}$-$VOPO_4$, besides VPP. The presence of a number of $V^{5+}$ species on the surface of VPP has been related to high activity in $C_4H_{10}$ oxidation. So, various combinations of partial oxidation (calcination in air) and/or *in situ* heat treatment in $n$-butane/oxygen mixtures have been proposed in the open literature to activate the catalyst. The dehydration of the precursor in an inert gas, or in an $O_2$-containing gas mixture (air, $C_4/O_2$) leads to VPP and/or $VOPO_4$ phases by Reactions 20.3 and 20.4, respectively:

$$2VOHPO_4 \cdot 0.5H_2O \rightarrow (VO)_2P_2O_7 + H_2O \tag{20.3}$$

$$VOHPO_4 \cdot 0.5H_2O + 0.25O_2 \rightarrow (\delta) - VOPO_4 + H_2O \tag{20.4}$$

The conditions (heating rate, temperature and duration, atmosphere) in which this step is achieved are very important as they further determine the activity and the selectivity to MA.

It was soon recognized that selectivity to MA was high when crystals of VPP exhibited mostly {100} faces.[15,34] As mentioned above, the formation of such {100} layered crystals is favored because of the pseudomorphism with {001} faces of the precursor, provided the atmosphere is inert and the temperature of dehydration is kept below 500°C. Duvauchelle *et al.*[35] examined the relationship between crystal morphology and catalytic properties of VPP particles (P/V = 1.0). The precursor was heated in $N_2$ at T = 450, 750 and 870°C to get thin plates, thicker plates and prismatic crystals of VPP, respectively (Table 20.1). TEM experiments showed that the surface of platy {100} crystals (T450) was constituted of mosaic crystals of VPP.[26] As shown by the oxygen uptake followed by temporal analysis of products (TAP) method, the oxygen capacity of platy mosaic crystals T450 was higher than that of thicker T750 plates or T870prisms. The high specific activity in *n*-butane conversion observed with platy crystals dropped for prismatic crystals, and accordingly the yield of MA decreased along T450 > T750 > T870 (Table 20.1). Thus, as expected, the decrease of the {100} face area when crystals are thicker led to lower MA productivity. However, the catalytic performance of T450 was poorer compared with those reported in the literature. The low surface area inherited from the precursor strongly limited the butane conversion, and the heat treatment under a nitrogen atmosphere minimized the formation of $V^{5+}$ species, which is related to selectivity. The main point was the relatively high selectivity to carbon oxides (68–81%). As the mean oxidation state of vanadium was the highest (4.15 vs 4.00) in platy crystals of T450, it was assumed that the $V^{5+}$ species were mostly present in the defect areas located between the {100} mosaic facets, and were not "associated"with the VPP phase. This means that to achieve high MA selectivity the $(O\text{-}V)_{defect}$ species must be linked to the VPP surface, and not stored in defects areas in which case combustion would be facilitated.

**Table 20.1.** Relationship between crystal morphology and catalytic properties of VPP particles (P/V) = 1.0) in *n*-butane oxidation to maleic anhydride. $T_{HT}$: temperature of heat treatment; SSA: specific surface area; n: oxidation state of vanadium; Th: thickness of particles along [020]; $X_{C4}$: conversion of *n*-butane; $S_{MA}$, $Y_{MA}$: selectivity to MA.

| $T_{HT}$ (°C) | SSA ($m^2$/g) | n in $V^{n+}$ | Th(Å) | $X_{C4}$ (mol%) | $S_{MA}$ (mol%) | $Y_{MA}$ (mol%) | Specific Activity [a] |
|---|---|---|---|---|---|---|---|
| 450 | 5.8 | 4.15 | 437 | 30.2 | 31.9 | 9.6 | 8.0 |
| 750 | 6.1 | 4.01 | 995 | 27.6 | 23.9 | 6.6 | 7.2 |
| 870 | 1.0 | 4.00 | ≫1000 | 21.7 | 19.2 | 4.1 | 2.5 |

[a] $mol_{C4}\ s^{-1}\ g^{-1} \times 10^7$; 420°C; contact time 1.5 s; $C_4H_{10}/O_2/N_2$ = 1.5/19.7/78.8.

All XPS results in the literature suggest that the VPP surface is enriched in phosphorus. In a paper devoted to XPS, Coulston *et al.*[36] presented a calibration method to determine the P/V ratio, and showed that during the reaction $\delta$, $\gamma$ and $\beta$-$VOPO_4$ transformed to VPP (or more exactly to the presence of pyrophosphate groups). Interpretations differ about the way $P^{5+}$ species are displayed on the surface, but such excess phosphorus could stabilize $(O\text{-}V)_{defect}$ species. The question remains regarding the actual oxidation state of vanadium.

One of the major controversies concerns whether only VPP (with some surface $V^{5+}$ species) is required in a selective catalyst, or whether $V^{5+}$ in $VOPO_4$ phases also play an important role.[42] Microdomains of [$VOPO_4$] structures on the surface of VPP were proposed to address this question.[25] The interfaces between [$VOPO_4$] and VPP were assumed to be coherent because the lattice mismatch between the very similar frameworks should be small. Both contain [$VO_6$] octahedra (even though they are deformed, whether paired or single) and phosphate groups. Coherent interfaces means that the potential barrier for exchange of electrons or $H^+$, $OH^-$, $O^{2-}$, etc. between them is very low. This is a condition to conceive that both [$VOPO_4$] and VPP participate in the reaction ($V^{5+}$ in the former and $V^{4+}$ in the latter), to keep a very active/selective mixed oxide catalyst.[37] However, the type of [$VOPO_4$] is also important. Undoubtedly the decomposition of the $VOHPO_4 \cdot 0.5H_2O$ precursor brings about the formation of $\delta$ (or $\gamma$ at high temperature) instead of $\beta$, which is detrimental to performance.

HRTEM experiments on precursor samples heated in $C_4/O_2$ at given reaction times[42] showed that the precursor can be directly converted to VPP (route A) or that VPP is formed by the reduction of $\delta$-$VOPO_4$ (route B). One or the other process seems to be related to the thickness of the platy crystals. It was observed that VPP developed at the rims of the $VOHPO_4 \cdot 0.5H_2O$ crystals (route A), while inside the plates the transformation between the two phases occurred indirectly via an intermediate $\delta$-$VOPO_4$ phase (route B). The topotaxicity of the $VOHPO_4 \cdot 0.5H_2O$ to VPP reaction was thus fully demonstrated by the epitaxial orientation relationship between $[001)_{prec}//[100)_{VPP}$ and $[010)_{prec}//[010)_{VPP}$. Interestingly the topotaxicity of the $VOHPO_4 \cdot 0.5H_2O$ to $\delta$-$VOPO_4$ reaction proposed by Bordes *et al.*[31] was also confirmed by a similar relationship between $[001)_{prec}//[100)_{delta}$ and $[010)_{prec}//[001)_{delta}$. Duvauchelle and Bordes[47] addressed the role of crystal morphology and showed that, compared to plates, prismatic VPP were more easily re-oxidized (although the size of particles was greater). Moreover the formation of $\beta$-$VOPO_4$, and not of $\delta$, $\gamma$-$VOPO_4$, was observed. As demonstrated in the literature,[18,38–40] fast oxygen exchange occurs only in the near-surface region, and below this region the diffusion of lattice oxygen is quite slow. Apart from the fact that different methods were utilized for such studies, the difference in depth of this near-surface region (two to ten layers) is probably due to differences in crystallinity and in the actual surface P/V ratios. Values reported for the activation energy of reaction also varied from 40 to 150 kJ/mol, and for reaction order of the gas-phase oxygen from 0 to 4. Centi *et al.*[15] suggested that such discrepancies were found because the re-oxidation reaction is sensitive to the structure. We could add that it is also sensitive to the exhibited crystal faces.[41,42]

### 20.3.4. *Reaching the steady state in model VPO Systems*

Several teams have shown that the oxidation state of vanadium varies as a function of time when the catalyst was facing the reaction gas flow. Bordes and Contractor[43] studied the behavior of $\beta$-$VOPO_4$ ($VPO_{5.0}$) and of VPP ($VPO_{4.5}$) in 1% butene/air feed in a thermobalance. After 8 h at 450°C, they found that the final stoichiometry was $VPO_{4.75}$ and $VPO_{4.68}$ respectively, the oxidation of VPP being faster than the reduction of $VPO_{5.0}$. On the contrary, feeding 1.5% butane/air over VPP at 380°C resulted in little change, in accordance with the oxidation state 4.02–4.04 of vanadium.[50] This experiment illustrates again the influence of the redox mixture on solid-state transient reactivity, the reduction rate of the catalyst being greater than its re-oxidation rate. Centi *et al.*[15] noticed that, owing to the presence of crystal defects in fresh or non-equilibrated VPO, the catalyst lattice oxygen was more active (order 0–0.5 in oxygen). Once equilibrated, the order in oxygen was close to one. Wang and Barteau[44] studied the reaction kinetics during *in situ* activation of the catalyst precursor, using both steady state and transient methods to permit their direct comparison. They showed that the activation energy of the reduction process on an equilibrated catalyst was 85 kJ/mol, which is very close to the apparent energy of the steady state oxidation of $n$-butane (88 kJ/mol). The reduction rate was of the 4th order in the available lattice oxygen concentration. They assumed that, four oxygens at a time were involved in the activation of $n$-butane (rate-determining step), in accordance with the model presented in Ziolkowski *et al.*[53]

As far as the MA production is concerned, it depends on the pre-treatment of the precursor before loading into the reactor. Two types of evolution of the conversion (selectivity) vs time to reach the steady state are found in the literature. Either the initial conversion of $n$-butane is low at first and then it increases (selectivity to MA decreases) or it is high and then decreases (selectivity to MA increases) with time. The performance stabilization period up to the actual steady state may vary. The first case generally holds when the precursor is activated *in situ* by a $C_4/O_2$ mixture, as schematized in Fig.20.3. In such cases, the composition on the surface is determined by the operating conditions (atmosphere, heating ramp, temperature and duration) in which the precursor is decomposed when exposed to $HC/O_2$ red/ox mixtures. In the best situation the resulting composition is generally VPP and $\delta$-$VOPO_4$ (and/or $\alpha_{II}$-$VOPO_4$). However, many other compositions were found in the literature, thereby

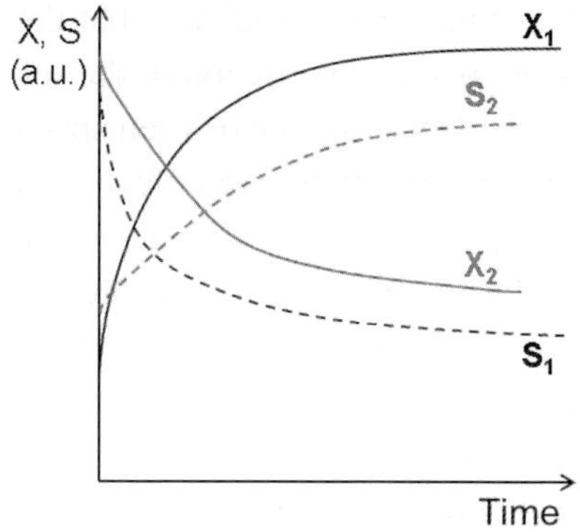


**Figure 20.3.** Two typical evolutions of conversion of reactant (X) and selectivity to product (S) during transient state (arbitrary drawings). $X_1$, $S_1$ for VPO precursor *in situ* activated in the reactor by $C_4/O_2$; $X_2$, $S_2$ when the precursor is first heat treated before loading in the reactor.

contributing to a blurred picture of the active and selective catalyst. As explained, the formation of only VPP with $V^{5+}$ species is desired. For this reason, the second method is practiced commercially because it ensures a higher MA yield.[8,45,46] From an industrial perspective, the preferred process is indeed one in which a stable, active and selective VPO is injected directly into the reactor. Most experimental studies rely on an *ex situ* procedure for calcination, followed by a variable prolonged "activation" with butane (see below). The formation of VPP is easily observed, but a delicate balance of operating conditions during decomposition of the hemihydrate precursor has to be found to favor the platy morphology of primary particles of VPP, as well as the existence of $V^{5+}$ species on its surface, but without crystallization of $VOPO_4$ forms.

Before coming to the results of *in situ* experiments, it is necessary to briefly recall that the specificity of VPP is also related to its acidic properties, studied by Védrine *et al.*[47] They are mainly owed to hydrogenophosphate groups, as shown by theoretical works using density functional theory (DFT) calculations. Bulk VPP, together with stoichiometric and phosphorus-enriched {100} surfaces, were analyzed using periodic DFT.[48] The most nucleophilic (basic) surface oxygen atoms in both stoichiometric (with bulk and optimized surface geometry) and phosphorus-enriched configurations, are the terminal P–O oxygen species, as expected. They would be involved in the post-activation nucleophilic attack on substrate C–H bonds, allowing selective substrate oxidation to non-combustion products. The surface enrichment of phosphorus, commonly observed, would be offered as an explanation for promoting selectivity to MA. Calculations also indicate that, by means of surface ionic relaxation, the P–O–V oxygen would be as nucleophilic as terminal P–O, which may explain its role in the rupture of reactant C–H bonds. Interestingly, calculations dealing with the hydration of the surface indicate that the dissociative chemisorption of water plays a key role in the perpetuation of the selective oxidation cycle by contributing to the supply of selective lattice oxygen species. Haras *et al.*[49] stressed that in most practical cases oxide surfaces are defective and irregular. They used a cluster model to show how the formation of oxygen vacancies could modify the catalytic properties of the (100) VPP surface. For example, if the oxygen of the vanadyl ($V=O$) bond is lacking, the acidic character of the vanadium atom is enhanced and the desorption of an acidic product such as maleic anhydride from the surface to avoid deeper oxidation is facilitated.

Another characteristic of VPP is related to its *p*-type semi-conductivity associated with $V^{5+}$–$O^{2-}$ defects by Eq. 20.5.[41] *n*-Butane would be activated by means of electrophilic oxygen linked to $V^{4+}$ according to Eq. 20.6, as proposed by Herrmann *et al.*,[50] the $C_4H_9^{\bullet}$ radicals being the starting reacting entities for further oxygen insertion:

$$V^{5+}\text{–}O^{2-} \rightarrow V^{4+}\text{–}O^{-} \quad (20.5)$$

$$C_4H_{10} + O^{-} \rightarrow C_4H_9^{\bullet} + O^{2-} + H^{+} \quad (20.6)$$

Recently the picture of the catalytically active VPP has dramatically changed. Hävecker *et al.*[51] and Bluhm *et al.*[52] performed *in situ* X-ray absorption spectroscopy and XPS. The attack of the methyl groups of butane proposed by Ziolkowski *et al.*[53,54] on the basis of thermodynamic considerations was ruled out, but oxygen-terminated sites were indeed considered as the reaction centers controlled by the underlying metal-oxygen bonds. Edges, as well as basal planes of equilibrated catalysts, were found to be covered by a *ca.* 1–3 nm-thick adlayer of a non-crystalline material, as observed by other authors.[42] This non-crystalline material, shown to occur during the first minutes of the reaction, was thought to be representative of the active structure. An important point is that near edge X-ray absorption fine structure spectroscopy NEXAFS experiments suggest that it is not one single site, but an ensemble of sites which is involved in the *n*-butane oxidation reaction. Such a "multiplet" site is by no means contradictory with the concept of "site isolation".[53] The results show that dynamic interactions proceed between either different phases on the active surface onto a core of VPP, as already postulated,[31,45] or on a distorted surface modification of VPP. In their model, the authors claim that the active and selective oxygen do not belong to the core VPP phase, arguing that a stoichiometric compound such as $(VO)_2P_2O_7$ vanadyl pyrophosphate could not deliver 14 $e^-$ and 7 oxygens without collapsing. This is to forget the versatility of vanadium in its coordination (VI to IV) and oxidation (V to III) states, as well as the large covalency of most V–O and P–O bonds that allows an easy pathway for electrons.[18] Furthermore, the observed "enormous excess" of oxygen was assigned to a mixture of hydrated phosphoric acid and [$VO_x$] resulting from hydrolysis of the pyrophosphate adlayer, as suggested above. Such an excess of oxygen could as well be related to the oxygen stored in the defective zones that we observed in between the mosaic crystals of VPP.[26,47]

Cavani's team[54,55,57] brought significant insight into the working structure of VPP in relation to the influence of the P/V ratio (0.99–1.15), the hydration state of the surface as well as the temperature of reaction. After the precursor was dried, calcination (air) was followed by heat treatment ($N_2$), and the catalysts were maintained for 150 h at 400°C in the reaction stream. These equilibrated catalysts were investigated under steady and unsteady, forced conditions. The dynamic character of surfaces when submitted to forced conditions (oxidizing or hydrating atmosphere and variation of temperature) was demonstrated. When $P/V > 1.0$, an active layer of vanadium oxide and polyphosphoric acid was generated by hydrolysis at temperatures higher than 340–350°C. When $P/V \approx 1.0$, VPP was oxidized to $VOPO_4$, with the development of a very active but unselective surface layer at 380°C. However, if heated to higher temperatures an active layer was generated by the hydrolysis of $VOPO_4$, with the result of a similar performance to the $P/V > 1.0$ catalyst. *In situ* LRS and XPS[56] confirmed that the major reversible variations of catalytic performance were observed after hydration and dehydration treatments. They were related to the distribution of vanadium phosphates ($\delta$-$VOPO_4$ or $\alpha_I$-$VOPO_4$) and/or $[(VO)_x + (PO_4)_n]$ or $VOPO_4 \cdot 2H_2O$ on the catalyst surface. These transient states made it possible to infer that the nature of the true active surface is a function of the VPO characteristics and of the reaction conditions. However, it is necessary to recall that after 1000 h of equilibration, only well-crystallized VPP with a homogeneous distribution of surface centers was detected. As mentioned above, the recrystallization is a normal trend for any amorphous particle, and such recrystallization was observed in many cases in used VPO commercial catalysts. Since this crystalline state at steady state is not related to ageing (otherwise the production of MA would be too low), the old picture of a VPP based on which $V^{5+}$ sites "keep the memory" of the underlying VPP seems to be more favorable to explain the fact that a unique catalyst fulfills all the requirements of the $n$-butane oxidation reaction.

### 20.3.5. *Partial conclusion*

In fixed-bed reactors the P/V ratio is higher than 1.0 because the catalyst has to endure the highly oxidizing $C_4/O_2$ gas mixture (typically 1.8–2.0% butane in air). An important point is that a VPO formulation designed to work in a fixed-bed reactor may not be optimal for other operating conditions.[55,57] In studies mimicking the behavior of the catalyst in a membrane reactor, Mota *et al.*[58] showed that the surface was rapidly reduced, but that it could be restored by re-oxidation after 2 h.

Mallada *et al.*[59] found that both selectivity and yield rose when 10% $C_4$ was fed in a membrane reactor while increasing the inlet oxygen concentration from 10% to 26%. In these studies the P/V ratio ($> 1.0$) was not modified and it cannot be ascertained that such performance could be sustained for a long time. On the contrary, the P/V ratio was deliberately fixed to 1.0 for use in a CFB because the atmosphere is more reducing in the riser more reducing. Thus there is still room for improvement of VPP catalyst performance.

It is now well established that, in the most active and selective catalysts used in fixed- or fluidized-bed reactors, only a few layers of VPO participate in the reaction. One may think that the Mars–van Krevelen redox mechanism is ruled out, or at least that it is confined to the outermost layers. The question arises as to how this picture holds when working deliberately in the transient state using alternate pulses or periodic feed,[3] with the idea of separately optimizing the two steps of the redox mechanism, or when alternative reactors in which the redox system is physically separated by decoupling. The next paragraphs address this question, as well as show some important differences in using the same DuPont catalyst in different reactor configurations.

## 20.4. Transient Regimes

### 20.4.1. *Redox operational conditions*

The VPO catalyst reactivity has been extensively examined under redox regimes. The experimental procedures that have mostly been used are TAP,[60] fixed bed reactors, fluidized bed reactors, thermogravimetric measurements[61] and *in situ* catalyst characterization, namely by Fourier transform infrared (FTIR) and Raman spectroscopies.[68] Two main objectives have been to investigate the transient kinetics and reaction mechanism by identifying the reaction intermediates,and to characterize the catalyst active phases during reaction. Most studies are generally conducted over a limited range of operating conditions. The specific experimental configurations during transient regimes have limited the direct correlation of these results with realistic commercial operations. However, forced concentration cycling between oxidizing and reducing environments in fluidized beds have been used to evaluate the redox kinetics of CFB reactors.

As mentioned above, the performance of the VPP catalyst is largely dependent on the transient operating conditions. Some of the redox parameters studied in the literature are feed compositions, solids residence time and temperature, or in rare cases pressure. The feed compositions for fixed-bed or fluidized-bed reactors vary between 1.8–4.0% *n*-butane, while in an industrial CFB reactor the feed may contain up to 20% *n*-butane. Catalyst regeneration times may vary from 40 seconds to 1 minute in industrial operations and up to several minutes in the laboratory scale reactors. MA yield improvements have been reported by increasing the catalyst oxidation time or by supplying adequate oxygen during fuel-rich reaction conditions.[3,62,72] Also, a higher $n$-butane concentration in the feed has been reported to be favorable for MA yield. However, catalyst deactivation could occur under high $n$-butane concentrations in the feed due to carbon build-up on the surface, and excessive $V^{4+}$ to $V^{3+}$ transformation. These studies confirmed that the key for an active and selective catalyst is to keep the catalyst surface at an optimized oxidation state, either by prolonged regeneration, or by supplying adequate oxygen during reaction. Temperature has been reported to improve either the catalyst re-oxidation rate[63] or to enhance the catalyst active phase transformations ($V^{4+}$ to $V^{5+}$).[68] However, only a few studies have mentioned the influence of pressure on VPP transient reactivity.[58,74] Despite current knowledge, it seems there is still a need to comprehensively study the effects of different redox regimes on VPP transient reactivity. In the following sections, the effects of feed composition and catalyst regeneration time on the transient reactivity of DuPont's VPO catalyst — as measured in a laboratory scale fluidized bed reactor operating under forced concentration cycling conditions — are discussed. These conditions adequately cover the operating range for all commercial reactors — fixed bed, turbulent fluidzed bed and CFBs. In addition, the effect of pressure (up to 4 bar) on catalytic performance will also be presented. All experiments were conducted with DuPont's commercial VPO catalyst (particles encapsulated in porous silica shell) that was calcined in the Asturias plant CFB.

### 20.4.2. *VPO redox reactivity*

#### 20.4.2.1. *Feed composition and regeneration time*

Transient data show that $n$-butane conversion is strongly affected by the feed composition or catalyst regeneration time (Fig. 20.4). Irrespective of the oxygen to $n$-butane ratio in the feed ($O_2/C_4H_{10}$), $n$-butane conversion increases considerably with catalyst regeneration time. Up to 45% increase in conversion is observed after 10 min of catalyst regeneration in an oxidizing environment, and a feed composition of 1.4% $n$-butane and 18.2% $O_2$ in the reducing environment. As the oxygen content in the feed decreases, the $n$-butane conversion drops significantly. This drop is larger when the feed is under fuel-rich conditions ($O_2/C_4H_{10} < 1.1$). A sudden drop is observed in conversion as the oxygen is depleted (from 2.6% $O_2$ in the feed to the pure redox mode). Under these conditions, the $n$-butane conversion drops from a range of 6–10% to below 2%. The drop in $n$-butane conversion may be related to the fact that the $n$-butane activation over the VPP catalyst might proceed through reaction with the nucleophilic surface lattice ($O^{2-}$) or adsorbed electrophilic

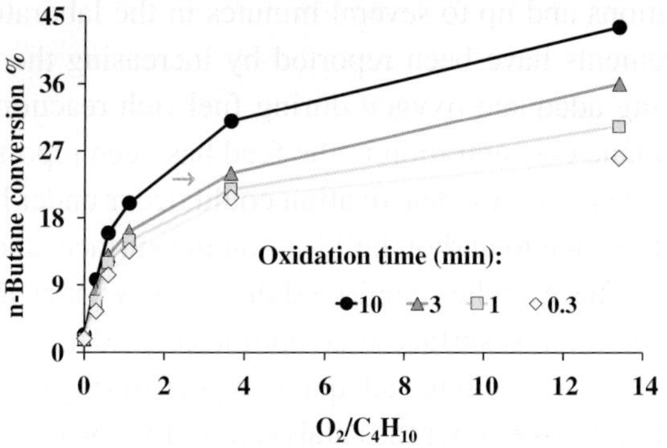


**Figure 20.4.** n-Butane conversion vs reduction feed composition. T = 380°C, Flow rate = 40 mL/min (STP), P = 1 bar, 500 mg calcined/activated VPO.

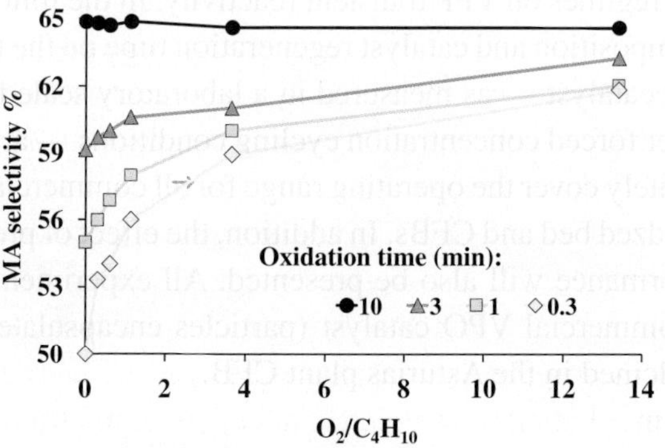


**Figure 20.5.** Maleic anhydride selectivity vs feed composition. T = 380°C, Flow rate = 40 mL/min (STP), P = 1 bar, 500 mg calcined/activated VPO.

oxygen (particularly $O_2^-$). The role of such species in $n$-butane activation has been reported to be significant.[52,72] As the oxygen partial pressure in the feed decreases, the amount of available surface $O^{2-}$ species,which are supplied through reduction of gas-phase oxygen and incorporation into the catalyst surface layers,decrease, and hence $n$-butane conversion is decreased. These results may also show that the contribution of lattice oxygen in the surface reaction is very limited as the loss of surface lattice oxygen cannot be compensated when the oxygen in the gas phase is limiting.

Similar to $n$-butane conversion, MA selectivity is also strongly dependent on the reduction feed composition (Fig. 20.5). Depending on the catalyst regeneration time, as the feed becomes more reducing (lower $O_2/C_4H_{10}$ratio), a drop in MA selectivity is observed from a steady value of about 65% to about 50% at a catalyst regeneration time of 0.3 min. MA selectivity tends to decrease as the feed composition moves towards more reducing conditions. Interestingly, this effect was not observed when the catalyst was pre-oxidized for 10 min. This indicates the important role of the catalyst regeneration stage in a redox operation. When the catalyst is adequately oxidized, MA selectivity tends to stay constant even at the highest concentration of *n*-butane (9.9%) in the feed, which corresponds to a pure redox mode with no oxygen. The adverse effect of *n*-butane concentration on MA selectivity is more evident when the regeneration step is short (<3 min). The drop in selectivity when the feed composition is changed from the most oxidizing ($O_2/C_4H_{10} = 13.4$) to the most reducing ($O_2/C_4H_{10} = 0.0$) conditions is around 4% (from 63 to 59%) when the catalyst oxidation time is 3 min, compared to a drop of about 12% (from 62 to 50%) when the oxidation time is 0.3 min. This drop in MA selectivity obviously shows the favorable effect of catalyst regeneration on reaction yield, especially at the industrial range of catalyst regeneration times, which could be less than a minute.

Figure 20.5 also shows that the MA selectivity converges to a close range of 62–65% when the feed composition becomes oxidizing. This indicates that the effect of catalyst oxidation time on MA selectivity might be less critical in more oxidizing feed conditions. Under such conditions (typically when $O_2/C_4H_{10} \geq 3.7$), the improvement observed in the reaction yield by increasing the catalyst oxidation time could mainly be attributed to the increase in *n*-butane conversion (see Fig. 20.4) rather than to the increase in MA selectivity.

Data presented in Figs 20.4 and 20.5 show that when operating under highly oxidizing feed conditions, ($O_2/C_4H_{10} > 3.7$), the reaction yield is only improved by the increase observed in *n*-butane conversion. Under these conditions, the MA selectivity increases only slightly — from 62 to 65%. On the other hand, while operating under fuel-rich conditions (typically $O_2/C_4H_{10} \leq 1.1$), the catalyst regeneration time becomes critically important in any improvement observed in the reaction yield. Under such conditions, extending the catalyst regeneration time improves both the MA selectivity and *n*-butane conversion. It could be concluded that the regeneration step plays a critical role in reactor performance while operating in the fuel-rich conditions typical in industry.

#### 20.4.2.2. *Pressure*

The data in Fig. 20.6 show the effect of reactor pressure on *n*-butane conversion. The effect is more evident when the oxygen concentration in the feed is increased. The conversion increases by about 40% for a feed containing 7.3% $O_2$ while this increase is about 55% when the feed is strongly oxidizing. These data show that the partial pressure of oxygen plays an important role in increasing the *n*-butane conversion. Probably, when the pressure is higher, the kinetics of the catalyst oxidation moves forward to generate more surface oxygen species on the catalyst.

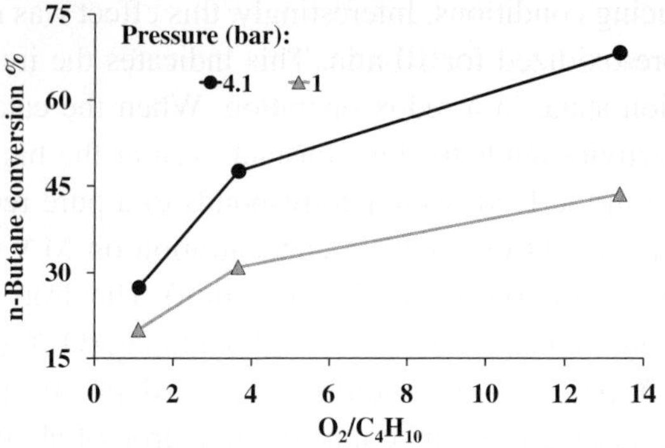


**Figure 20.6.** Effect of pressure on $n$-butane conversion. T = 380°C, Oxidation time = 10 minutes, Flow rate = 40 mL/min (STP), 500 mg calcined/activated VPP.

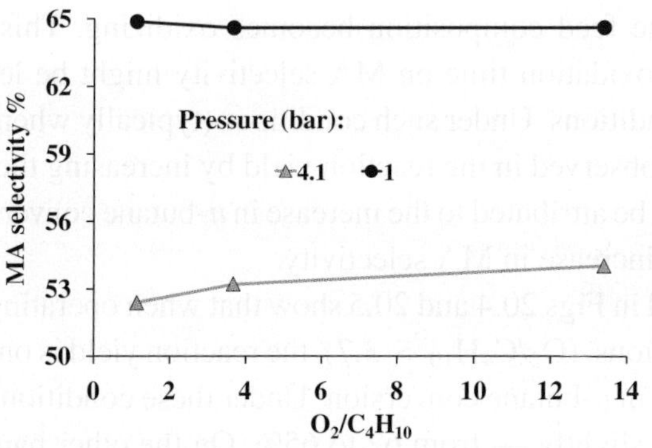


**Figure 20.7.** Effect of pressure on maleic anhydride selectivity. T = 380°C, Oxidation time = 10 minutes, Flow rate = 40 mL/min (STP), 500 mg calcined/activated VPP.

The MA selectivity showed an opposite trend vs the reactor pressure. Data shows an average decrease of about 20% when the reactor pressure increases to 4 bar (Fig. 20.7). The drop in MA selectivity could be attributed to higher concentrations of non-selective (electrophilic) surface oxygen species as a result of increased oxygen partial pressure. In other words, when the pressure increases, the surface population of oxygen species responsible for $n$-butane activation increases, but these species are not necessarily responsible for a selective MA formation. Other reasons for MA selectivity drop could be the promotion of the gas-phase combustion of $n$-butane which results in a lower maleic anhydride concentration at the reactor outlet. Generally speaking, despite the drop in MA selectivity with increased pressure, as the pressure increases, the increase in $n$-butane conversion (37–56%) results in an overall increase in MA productivity and yield. Actually, the MA yield increases from 10 to 30% when the $O_2/C_4H_{10}$ ratio in the feed changed from 1.1 to 13.4 and the pressure increased from ambient to 4 bar.

### 20.4.3. *Partial conclusions*

There is a strong relationship between VPP catalyst reactivity with reduction feed composition and catalyst regeneration time, as well as with the pressure and temperature. The MA yield could be improved by increasing the catalyst oxidation time and also by feeding more oxygen. Both temperature and pressure improve reactor performance. These effects were more important when the concentration of oxygen in the feed was higher. A strong dependency of maleic anhydride selectivity is observed on the feed composition at relatively short catalyst oxidation times ($<1$ min), corresponding to operations in industrial reactors.

Generally, while the concentration ratio of the feed to the reactor is oxidizing ($O_2/C_4H_{10} \geq 3.7$) any improvement in the reaction yield can only be attributed to the increase in $n$-butane conversion with catalyst oxidation time. The surface adsorbed or lattice oxygen species are probably the main source for $n$-butane activation under such conditions. However, when operating under fuel-rich feed conditions and short regeneration times, both $n$-butane conversion and MA selectivity contribute to MA yield improvement. Under these conditions, the effect of catalyst oxidation time on catalytic performance is critical. The contribution of catalyst sub-surface lattice oxygen should be higher under fuel-rich conditions, but data show that the catalyst over-reduction cannot be effectively compensated, even after extensive catalyst regeneration. The presence of gas-phase oxygen is critical to maintain a high catalytic activity.

The $n$-butane conversion increases with increasing pressure in the reactor. The MA selectivity drops, but the increase in $n$-butane conversion results in an overall improvement in the MA yield of up to 30%. Data show that there should be a certain combination of operating temperature and pressure at which the catalyst activity could be enhanced. The maximum catalytic performance should be achievable at a higher pressure. However, the negative effect of temperature on MA selectivity and also on the useful life of the catalyst has to be minimized.

## 20.5. Experiments in Alternative Reactors

Capital investment and operating costs are the drivers when developing new technologies for MA. Fixed beds operate at low butane concentrations and achieve high throughput capacity (space-time yield, STY) by operating at high conversion while maintaining selectivity. Fluidized beds operate with higher butane concentrations, thus requiring lower reactor volumes for the same levels of conversion and selectivity

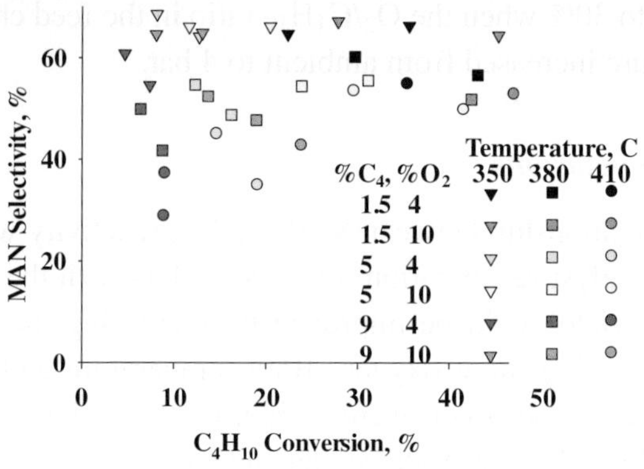


**Figure 20.8.** Conversion of *n*-butane vs selectivity to MA of DuPont's commercial VPO.

as a fixed bed. In the CFB process technology, butane concentrations double that of fluidized beds are possible and thus — coupled with higher pressures, catalyst activity and gas superficial velocities — even lower catalyst inventories are required. Figure 20.8 illustrates the relationship between conversion and selectivity as a function of the gas-phase composition and temperature as measured in a 4.1 cm diameter fluidized-bed reactor.[64]

With most gas-phase compositions, the conversion increases with increasing temperature. The selectivity is relatively insensitive to gas-phase composition at 350°C and lies above 60%. The only exception is the most reducing condition of 9 vol% butane and 4 vol% oxygen. At higher temperatures, differences between conversion and selectivity become more apparent: the selectivity decreases sharply for the conditions that are highly reducing, i.e. 9 vol% butane and 4 vol% $O_2$. Under this condition, as the temperature is increased, conversion barely changes but the selectivity drops from 60% to below 30%.

Figure 20.9 replots the data in terms of yield and productivity (STY). The yield is expressed as the conversion multiplied by the selectivity while the STY is a product of the yield and the vol% butane in the feed. Since the yield is a function of conversion, the plot increases linearly, but it demonstrates that yield is highest at lowest inlet butane vol%. The second plot shows that the productivity is highest at 9 vol% butane and 10 vol% $O_2$ followed by 5 vol% butane and 10 vol% $O_2$. Under these conditions, butane conversion is low, and to make an economically viable process a significant recycle of stream would be required. However, this would appear to be justified based on the fact there is a threefold increase in production.

### 20.5.1. *Fixed bed*

Fixed-bed process research has been devoted to catalyst development — selectivity and longevity — increasing butane concentration and quenching free radicals. As

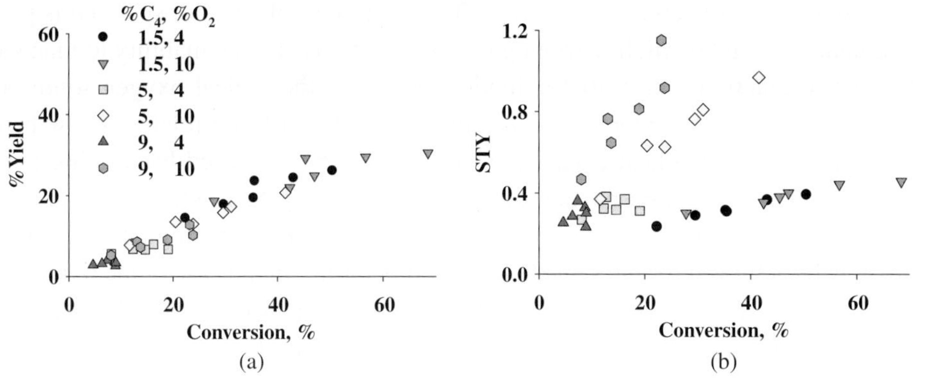


**Figure 20.9.** (a) Yield of MA vs conversion of *n*-butane; (b) STY MA vs conversion (STY: $\%C_4H_{10}*X*S$) of DuPont's commercial VPO.

demonstrated in Fig. 20.9, the productivity can be improved by increasing the butane feed concentration. The reactors may have as many as 35,000 tubes and weigh 740 kt. The tube diameters may be as low as 19 mm in order to reduce the increase in the surface-to-volume ratio to maximize heat transfer. Catalyst activity is not necessarily the primary interest because of the difficulty in withdrawing the heat generated. Rather, by increasing the selectivity, the total heat load is reduced thereby increasing the yield and perhaps allowing for the possibility of increasing gas throughput.

Clearly, heat transfer, flammability and catalyst stability are the major concerns of fixed-bed technology. In the early 1990s, tests were conducted at a commercial plant to increase MA productivity by feeding butane concentrations beyond the flammability limit (which is nominally 1.8 vol% butane in air). Apparently, the butane feed concentration in the vessel approached 2.3 vol% before the reactor head lifted off the body of the vessel (the reactor was designed to allow for such an event). Many patents have been issued that claim the interior vessel walls can be passivated to reduce the frequency of free radical formation (or quench free radicals). Free radicals initiate thermal events that lead to combustion or deflagration and even to detonation when the flame front exceeds the speed of sound.

Combustion is enhanced by free radicals that build up in stagnant zones. Stagnant zones can be as large as the corners of vessels, or as small as the tips of the spargers that introduce hydrocarbons into the air stream (or the inverse) in the reactor. When the nozzles are cut square at the tip, eddies form and free radicals can build up in this region which might measure a couple of millimeters. New nozzle technology has been proposed to reduce this area and involves beveling the edges to a sharp point.[65]

Another way to manage the heat generation is to increase the heat capacity of the gases.[66] Exothermic reactions may cause regions in which the local temperature may rise over 100°C above the average. This region is called a hot spot and is predominantly caused by high reactant gas concentrations and the inability to manage the heat of reaction. For partial oxidation reactions, the typical oxygen source is air. Nitrogen in the air represents 78% by volume but it has a relatively low heat capacity. The average heat capacity of the gas can be increased by diluting pure oxygen with a compound having a higher heat capacity than nitrogen but that is inert. In this way, the hot spot may be reduced significantly or a higher production rate may be achieved for the same hot spot temperature. In the case of the oxidation of methanol to formaldehyde, substituting nitrogen for propane was theoretically demonstrated to increase production rates by as much as 50%.

### 20.5.2. *Membrane reactors*

Membrane reactors are ideal for minimizing the risk of deflagration or detonation and at the same time maximize the feed rate of oxygen and butane, which would hypothetically result in higher production rates.[67–71] As discussed above, the feed gas compositions are limited by the flammability limits of the reactants at the entrance as well as hot spots. Formed near the entrance of the reactor, hot spots not only reduce selectivity, they may also have a detrimental effect on the catalyst life. The reaction rates are the highest and concentrations of the reactants are the greatest at the entrance, and concentrations decrease along the reactor length as the hydrocarbon and oxygen are consumed. One strategy to attenuate the hot spot is to dilute the catalyst with an inert solid. Since reaction rates are proportional to the mass of the catalyst, the reaction rate will drop in proportion to the dilution factor. Another strategy would be to dose the reactants (particularly oxygen) along the entire length, hence the risks of deflagration and detonation are eliminated because the flammability region is avoided, and the hot spots are more easily managed because the average concentration is lowered.The membrane reactor has been studied extensively for butane oxidation, particularly with the aim of increasing the butane concentration.[67,68,72] The VPO catalyst was either deposited as a thin layer on mesoporous MFI membrane or packed in a tube of porous alumina. The conventional fixed bed was used as a novel reactor configuration consisting of a porous metallic membrane immersed in a gas-solid fluid bed[69,70] A pilot plant was built with a tube diameter of 32 mm. Figure 20.10 illustrates this configuration that would solve the problem of flammability. Most of the tubes are filled with commercial VPO catalyst and a mixture of butane and inert gas is introduced to the bottom. In order to minimize the risk of over-reduction, a sufficient quantity of oxygen should also be co-fed with the butane. The air (or air/oxygen mixture) which fluidizes an inert bed of solid particles on the shell sidepasses across the membrane — in this case, a porous metal filter. The flow rate of fluidization gas from the shell side to the tube

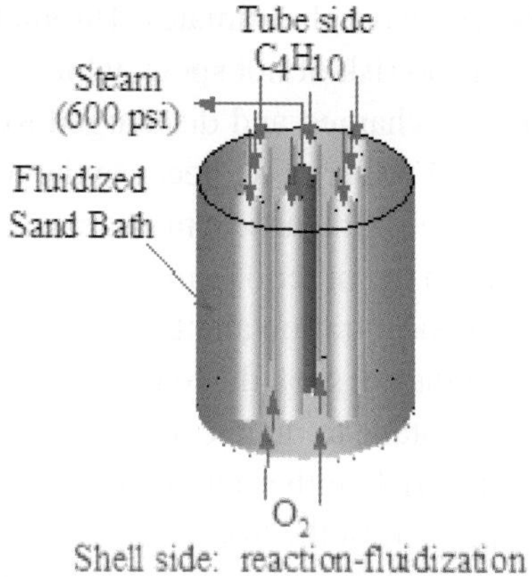


**Figure 20.10.** Fluidized-bed membrane reactor for $n$-butane oxidation.

is easily controlled by the pressure differential between the two. In both cases — fluidized bed and fixed bed — the pressure is higher at the entrance compared with the exit. To maintain a constant flow across the length, the particle size and velocity in the tube must be selected to match the pressure drop across the fluidized bed. To control the reaction temperature, cooling coils are distributed throughout the bed. Fluidized beds have extremely high heat transfer rates — as much as 10 times higher than the heat transfer coefficients between the inner wall and the fluid of the fixed beds.

Despite the advantages of excellent heat transfer, control of oxygen concentration as well as lowered safety hazard risks, experimental data have not demonstrated an increase in performance — higher STY — compared to a fixed bed operating in the same conditions. The local heat transfer coefficient between the fluidized bed and the outer wall is very high, however, the local heat transfer coefficient between the interior wall and the gas is much lower than that in a standard fixed bed. Thus, the overall heat transfer coefficient is lower and the membrane reactor operates at higher temperatures compared to a standard fixed bed. It would appear that the gas crossing the membrane forms a "film" and may preferentially channel up the wall, which results in lower radial mixing and lower heat transfer rates because the average gas temperature at the wall can be lower than the cross-sectional average.

### 20.5.3. *Turbulent fluidized beds*

In contrast to either the membrane reactor or the fixed-bed reactor, turbulent fluidized beds have excellent heat transfer characteristics. In addition, they have good mass transfer rates and the ability to operate at extremely large scales. While a typical fixed-bed reactor may weigh up to 740 kt, a fluidized bed with a similar capacity may weigh less than 200 kt (very rough estimate). Turbulent fluidized beds operate isothermally and thus there is no risk of hot spots, which results in a reduced deactivation rate. Catalysts may be charged and discharged to the vessel easily, which ensures continuous operation. Furthermore, because the diameter of catalyst particles are in the order of 70 $\mu$m, the intraparticle mass transfer resistance is negligible. Pellets in fixed beds are often in the order of 3 mm (hollow cylinders are commonly used) and this may result in concentration and temperature gradients for extremely fast reactions. In the region of the hot spot, the reaction rates may be fast in fixedbeds for the partial oxidation of $n$-butane to maleic anhydride.

To reduce the flammability risk, either butane or oxygen may be introduced at different points within the reactor. In the Sohio process for production of acrylonitrile from propylene, the fuel gases — propylene and ammonia — are fed through spargers positioned over each orifice in the grid through which the air is introduced. A similar configuration is employed for the oxychlorination of ethylene by HCl and oxygen: fuel spargers to feed ethylene and HCl are placed directly in the grid orifices through which the air is fed.

The catalyst mechanical resistance is a critical property for the success of fluidized beds. Particle–particle collisions at the gas spargers and abrasion in the cyclones tend to attrit the powder and thus, to maintain constant production, make-up catalyst must be added at regular time intervals. Particle size management is another key issue: fines (powder with a diameter between 20 and 44 $\mu$m) are carried upwards with the gas to cyclones and must be returned to the bed through diplegs. Under certain conditions (e.g. start-up), the powder may become cohesive and can block the diplegs. As a consequence, the fines are carried through the cyclone and accumulate in the filters. This will eventually result in an unscheduled shutdown.

Although fines represent operational complexity, they are required to maintain good mass transfer rates between the gas and solids. In many cases, the bed should be maintained with as much as 30–35% fines. As the particle terminal velocity of the fines is so low, the elutriation rate to the cyclones can be high. Special care must be devoted to design efficient cyclones to return the powder to the bed while minimizing attrition. Arnold *et al.*[71] reported improved reactor performance after they modified the fluidized-bed cyclones to maintain a higher fraction of fine particles.

Fixed-bed reactors are much simpler to operate compared to fluidized beds. After charging the catalyst pellets to the tubes and reaching steady state conditions, the reactor may run with little intervention for extended periods of time — even several years. Charging the catalyst and ramping up to steady state are the two most critical stages. It is imperative that all the tubes — in the case of maleic anhydride they may be as many as 35,000 — be loaded with equal amounts of catalyst in order to ensure an even distribution of gases. Several years ago, one tube was left empty in a plant in Europe. As the plant was approaching steady state, the operators realized that the cooling fluid level was dropping: the fluid was draining through a hole in the empty tube that was created by the combustion of butane in the absence of the catalyst.

Both fixedbed and fluidizedbed reactors are generally operated with a high single pass conversion, probably exceeding 90 to 95%, to maximize butane utility. Figure 20.8 shows that the MA selectivity remains relatively flat with low butane concentrations. However, as the conversion increases, the selectivity begins to decrease. Low butane concentrations are adopted to avoid the flammability region, either at the entrance or even at the exit. Higher reactant concentrations are possible in fluidized beds because the butane can be introduced at multiple points in the bed. Thus, not only the flammability risk is lower, but also there is the possibility to operate at a higher STY. However, if the butane conversion is insufficient, there is a risk of combustion in the freeboard region of the fluidized bed. Flammability at the entrance is controlled by multiple spargers, but the risk of combustion at the exit must be minimized by ensuring high conversion, which, in the case of MA production, might also result in lower selectivity.

### 20.5.4. *Circulating fluidized beds*

As an adaption of turbulent fluidized beds, circulating fluidized beds are the next logical step in reactor technology. Multiple vessels are required at the expense of the simplicity of turbulent fluidized beds, but by operating the reactor with high butane concentrations at the entrance and exit of the bed, higher production rates (STY) are possible. Coincidentally, single pass $n$-butane conversions are kept low (on the order of 20–40%) to maintain high MA selectivity.

The concept is based on spatial redox decoupling as already explained (*vide supra*). It relies on circulating the VPO catalyst between a region rich in butane and deficient in gas-phase oxygen and a region absent of butane and rich in oxygen.[72–75] Figure 20.11 is a schematic diagram of DuPont's commercial reactor. The solid particles enter the "fast bed"(also called a transport bed orriser in the present case), from the side through a standpipe. Butane is added to the recycled gas, which contains as much as 3 vol% oxygen and as much as 4 vol% butane. Ideally the riser could be operated without gas-phase oxygen since MA selectivity is linked to catalyst lattice oxygen. In an experimental reactor operated at atmospheric pressure and with 1 kg of catalyst inventory, as much as 50% of the oxygen came from the catalyst lattice. However to avoid over-reduction in the commercial reactor, oxygen diluted to about 90 vol% by nitrogen is fed through three levels of spargers within the bed. The gas velocity in this part of the bed was slightly greater than found in turbulent fluidized beds. The gas from the fast bed carries the catalyst upwards through the riser where the gas velocity exceeds 6 m/s. The gas–solids suspension enters the stripper/rough cut cyclone. Most of the particles descend into the bed

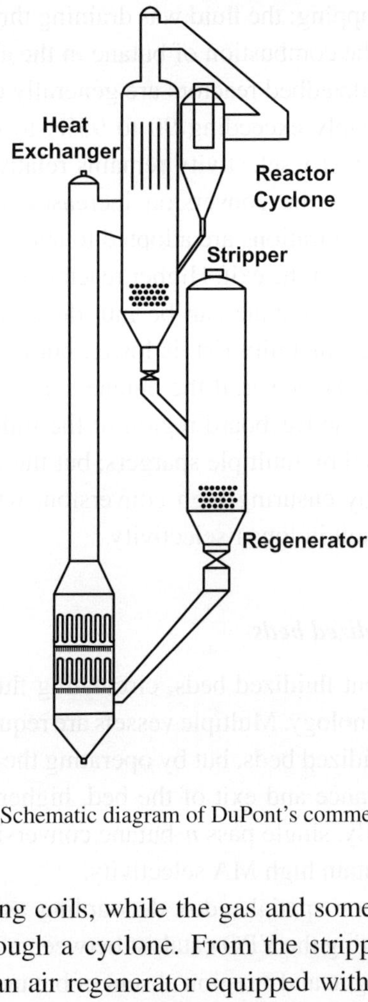


**Figure 20.11.** Schematic diagram of DuPont's commercial CFB reactor

and pass through stripping coils, while the gas and some fines rise through a heat exchanger and then through a cyclone. From the stripper, the catalyst descends through a standpipe to an air regenerator equipped with horizontal cooling coils. After the regenerator in which the reduced catalyst is reoxidized to an appropriate extent, the solids enter the standpipe leading to the fast bed.

The reactor was operated at a temperature of 400°C at the exit and the catalyst was circulated at rates as high as 7 kt/h. Basic data was collected in a pilot plant for almost two years and showed that as much as 20% of the total oxygen requirement was supplied by the catalyst lattice (*vs* 50% in the experimental reactor). In the commercial reactor, oxygen contribution from the lattice was *ca.*50% of the pilot plant. The consequences of the lower lattice oxygen contribution were lower activity, selectivity and a greater requirement to feed oxygen through the spargers in the fast bed.

The design of CFB reactors is substantially more complicated than that of turbulent fluidized beds because of the external circulation loop. Together with multiple fluidized beds, solids must flow through standpipes, slide valves (for control), cones, cooling coils, etc. However, the design principles and concerns are similar: particle size management, cyclone operation, diplegs, attrition and agglomeration. Catalyst mechanical resistance must be at least as good as in turbulent fluidized beds. To handle the powder, filters, hoppers and silos must be included, which adds to the overall investment costs.

The flammability region is avoided at the entrance of the reactor by introducing a large portion of the required oxygen through spargers within the bed. However, as the oxygen from the sparger mixes with the gas in the bed, the composition necessarily passes through the flammable region. In addition, the gas composition at the cyclone — 4% vol% butane and 3% vol% $O_2$ — is also flammable. The flammable mixture at the cyclone limits the feed butane concentrations at the entrance and consequently the production rate. The heat exchanger, shown at the top of the riser stripper, was designed and mounted after the plant construction, to drop the temperature below the autoignition temperature. In the original design, the solids went directly into the cyclone after the stripper.

In a recent economic study, fixed beds, fluidized beds and circulating fluidized beds were compared for the production of acrylic acid from propane.[76] The catalyst (MoVSb) oxygen capacity was assumed to equal that of an active VPO at 110 mmol $O_2$/kg. Capital costs were the lowest for turbulent beds and highest for fixed beds. Although the economies of scale are greater for the CFB, the turbulent fluidized bed had lower investment costs even at a scale over 200 kt/y, which is at least four times as large as the largest scale MA plant. In order to match the economics of the fluidized bed, circulating the catalyst from the regenerator must increase the selectivity (or conversion) by at least 20%. However, the major uncertainty with respect to the turbulent fluidized-bed process is the possibility that the flammable mixture of propane and acrylic acid might form in the freeboard region. If it approaches a flammable composition then either the temperature in the freeboard must be reduced (to increase the induction time) or the propane and oxygen concentration must be reduced, which would result in lost production.

## 20.6. Conclusions

VPO catalysts are uniquely able to transform *n*-butane to MA with high yield. The *ca.*40 years of work on that matter have shed considerable light on the complexity of the system. Thanks to recent *in situ* experiments, an efficient catalyst is today described as made of $[V^{5+}O_x]$ species supported onto a vanadyl pyrophosphate core. This picture satisfies the principle of least change (during the reaction) leading to higher turnover frequency. At variance, everytime a $VOPO_4$ form crystallizes it leads to an irreversible loss of surface area and consequently a loss of performance.

Several studies showed that the $\delta$ form was the least detrimental, probably because its reduction to VPP is easier. As described, VPO resembles the case of $VO_x$ species when supported on $TiO_2$-anatase, that are active and selective in the oxidation of *o*-xylene to phthalic anhydride. The latter is another highly specific ($12e^-$) reaction requiring the insertion of 3 + 3 oxygen molecules. Polyvanadates are claimed to be the "active phase", but the properties of V-O-Ti (and not only V-O) bonds are determinant for the catalytic properties. However, in this case the occurrence of crystalline $V_2O_5$ is responsible for the loss of performance, whereas in VPO catalysts the presence of VPP is a prerequisite to attain high activity and selectivity.

*In situ* experiments also pinpointed the existence of "amorphous" layers on a working catalyst surface that were associated with high selectivity, whereas long term, and/or commercial experiments showed a well-crystallized VPP phase. One has to note that *n*-butane does not transform to any $C_4$-oxidized molecule (e.g. crotonaldehyde) but to the highly symmetric MA molecule. In our opinion, to cope with the disorder shown during transient experiments, as well as with the presence of crystalline VPP in commercial catalysts, pairs of edge-sharing $[VO_6]$ octahedra surrounded by phosphate groups could be assembled as nanodomains in a disordered manner. Such V-O-P multiplet sites are necessarily linked to the VPP core, to ensure both acidity (P-O bonds) and redox ($V^{5+}/V^{4+}$), but also oxygen lability requirements. It is conceivable that with time on stream such multiplet sites would resemble more and more VPP itself. One may remember that simple calculations of the bond strength by Brown and Wu's method[77] for the simplest VPP structure[21] showed that in each pair of edge-sharing octahedra the formal charge of two neighboring vanadium is 3.69–3.79 and 4.19–4.44, that is something like "redox couples" inside the crystalline VPP matrix itself.[25]

If today $VOHPO_4 \cdot 0.5H_2O$ is universally considered as the precursor to be synthesized to obtain an efficient catalyst, the conditions of its decomposition to optimize catalytic performance at steady state depend on the P/V ratio as well as on the type of reactor in which the catalyst is loaded. In our opinion there is no universal recipe valid for any reactor configuration. Indeed to maximize the commercial reactor performance, the VPO catalyst should be formulated to be adapted to the operating conditions (temperature, butane and oxygen partial pressures). The P/V ratio plays a critical role in catalyst activity. P/V ratios should be maintained slightly higher than unity in an oxidizing environment (1.8–2 vol% *n*-butane in air), which is typical of a fixedbed reactor. Higher MA production rates are achievable by increasing the *n*-butane concentration beyond 4 vol%, as long as the oxygen concentration is kept high enough to avoid carbon build-up or over-reduction. The optimum P/V ratio should be slightly lower for these conditions.

The catalyst activity may be moderated by working in transient redox conditions in which the oxidation step is physically separated from the reduction step. MA yields improve with increased catalyst oxidation time, feeding more oxygen together with the $n$-butane in the reduction step, temperature and pressure. MA selectivity is strongly dependent on the feed composition at relatively short catalyst oxidation times ($<1$ min) corresponding to operations in industrial reactors. However, this effect is less sensitive when the catalyst is adequately oxidized. On the contrary, $n$-butane conversion improves by extending the catalyst oxidation time even under highly oxidizing feed conditions. Surface adsorbed or lattice oxygen species are the main source for $n$-butane activation under such conditions. Data show that under fuel-rich conditions, the catalyst over-reduction cannot be effectively compensated even after excessive catalyst regeneration, and the presence of gas-phase oxygen is critical to maintain a high catalytic activity.

From an industrial point of view, choosing the best reactor configuration for MA production depends largely on the advantages and shortcomings of the underlying technology. Obviously, the best technology would finally be decided by their operating costs and the required capital investment. Capital costs could be the lowest for turbulent beds and highest for fixed beds. Although the economies of scale are greater for the CFB, the turbulent fluidized bed might have lower investment costs.

Fixedbed reactors operate at relatively low $n$-butane concentrations due to flammability limits. Clearly, heat transfer, flammability and catalyst stability are the major concerns of fixed-bed technology. Fluidized-bed and membrane reactors could operate with higher $n$-butane concentrations and thus are more compact for the same production level. Natural advantages of membrane reactors are to minimize the hot spot formation and to suppress the runaway combustion risk. However, an increase in catalytic performance has not been proven for membrane reactors compared with fixed beds. Turbulent fluidized beds show excellent heat and mass transfer capability which eliminates the risk of hot spots. The mechanical stability of the catalyst, management of fines and the risk of combustion in the freeboard region could be the major challenges for fluidized-bed reactors. In CFB technology, higher $n$-butane concentrations could be introduced, while at the same time managing the flammability risk, thus lowering the required catalyst inventories. However, the design leads to highly complicated operation. Other disadvantages are related to the catalyst (low VPP oxygen transfer, particle size management, attrition and agglomeration), and to necessary equiment (cyclone, diplegs).

## References

1. Keulks G.W. (1970). The mechanism of oxygen atom incorporation into the products of propylene oxidation over bismuth molybdate, *J. Catal.*, **19**, pp. 232–235.
2. Mars P. and van Krevelen D.W. (1954). Oxidations carried out by means of vanadium oxide catalysts, *Chem. Eng. Sci.*, Suppl.1, **3**, pp. 41–49.

3. Emig G., Uihlein K. and Hacker C.-J. (1994). 'Separation of Catalyst Oxidation and Reduction — An Alternative to the Conventional Oxidation of n-Butane to Maleic Anhydride?' in Cortés-Corberan V. and Vic Bellon S. (eds), *New Developments in Selective Oxidation II*, Elsevier, Amsterdam, *Stud. Surf. Sci. Catal.*, **82**, pp. 243–251.
4. Löfberg A., Bodet H., Pirovano C., Vannier R.N., Steil M.C. and Bordes-Richard E. (2006). Transient behaviour of dense catalytic membranes based on Cu- and Co-doped $Bi_4V_2O_{11}$ (BIMEVOX) in the oxidation of propene and propane, *Catal. Today*, **112**, pp. 8–15.
5. Mota S., Miachon S., Volta J.-C. and Dalmon J.-A. (2001). Membrane reactor for selective oxidation of butane to maleic anhydride, *Catal. Today*, **67**, pp. 169–176.
6. Téllez C., Menéndez M. and Santamaría J. (1997). Oxidative dehydrogenation of butane using membrane reactors, *AIChE J.*, **43**, pp. 777–784.
7. Soler J., López Nieto J.M., Herguido J., Menéndez M. and Santamaría J. (1998). Oxidative dehydrogenation of n-butane on V/MgO catalysts. Influence of the type of contactor, *Catal. Letters*, **50**, pp. 25–30.
8. Contractor R.M., Bergna H.E., Horowitz H.S., Blackstone C.M., Malone B., Torardi C.C., Griffiths B., Chowdhry U. and Sleight A.W. (1987). Butane oxidation to maleic anhydride over vanadium phosphate catalysts, *Catal. Today*, **1**, pp. 49–58; Contractor R.M., Garnett D.I., Horowitz H.S., Bergna H.E., Patience G.S., Schwartz J.T. and Sisler G.M. (1994). 'A New Commercial Scale Process for n-Butane Oxidation to Maleic Anhydride Using a Circulating Fluidized Bed Reactor', in *New Developments in Selective Oxidation II*, Cortés Corberán V. and Vic Bellón S. (eds), *Stud. Surf. Sci. Catal.*, **82**, pp. 233–242.
9. Patience G.S. and Bockrath R.E. (2010). Butane oxidation process development in a circulating fluidized bed, *Appl. Catal. A: General*, **376**, pp. 4–12.
10. Haber J. (1994). In 'Perspectives in catalysis', Thomas J.M. and Zamaraev K.I. (eds), Blackwell, pp. 371–384.
11. Hanuza J., Jeżowska-Trzebiatowska B. and Oganowski W. (1978). The oxygen bond in Mo(V)/Mo(VI) active centres and its role in the oxidizing dehydrogenation process, *J. Molec. Catal.*, **4**, pp. 271–287.
12. Gao X., Bañares M.A. and Wachs I.E. (1999). Ethane and n-Butane Oxidation over Supported Vanadium Oxide Catalysts: An in Situ UV–Visible Diffuse Reflectance Spectroscopic Investigation, *J. Catal.*, **188**, pp. 325–331.
13. Bordes E. and Courtine P. (1979). Some selectivity criteria in mild oxidation catalysis : V-P-O phases in butene oxidation to maleic anhydride, *J. Catal.*, **57**, pp. 237–252.
14. Poli G., Resta I., Ruggeri O. and Trifirò F. (1981). The chemistry of catalysts based on vanadium-phosphorous oxides: Note II. The role of the method of preparation, *Appl. Catal.*, **1** pp. 393–404.
15. Centi G., Manenti I., Riva A. and Trifirò F. (1984). The chemistry of catalysts based on vanadium-phosphorus oxides: Note III: Catalytic behaviour of different phases in 1-butene oxidation to maleic anhydride, *Appl. Catal.*, **9**, pp. 177–190; Centi G., Trifiró F., Ebner J.R. and Franchetti V.M. (1988). Mechanistic aspects of maleic anhydride synthesis from C4 hydrocarbons over phosphorus vanadium oxide, *Chem. Rev.*, **88**, pp. 55–80.
16. Hodnett B.K. and Delmon B. (1984). Influence of calcination conditions on the phase composition of vanadium-phosphorus oxide catalysts, *Appl. Catal.*, **9**, 203–211. Hodnett B.K. (1985). Vanadium-Phosphorus Oxide Catalysts for the Selective Oxidation of $C_4$ Hydrocarbons to Maleic Anhydride *Catal. Rev.-Sci. Eng.*, **27**, pp. 373–424.
17. Centi G., Trifirò F., Busca G., Ebner J. and Gleaves J. (1989). Nature of active species of $(VO)_2P_2O_7$ for selective oxidation of n-butane to maleic anhydride, *Faraday Discuss.*, **87**, pp. 215–225.
18. Pepera M.A., Callahan J.L., Desmond M.J., Milberger E.C., Blum P.R. and Bremer N.J. (1985). Fundamental study of the oxidation of butane over vanadyl pyrophosphate*J. Am. Chem. Soc.*, **107**, pp. 4883–4892.

19. Bergna H.E. (1988). Method of making maleic anhydride, DuPont, U.S. Patent 4,769,477.
20. Leonowicz M.E., Johnson J.W., Brody J.F., Shannon H.F. and Newsam J.M. (1985). Vanadyl hydrogenphosphate hydrates: $VO(HPO_4)\cdot 4H_2O$ and $VO(HPO_4)\cdot 0.5H_2O$, *J. Solid State Chem.*, **56**, pp. 370–378.
21. Linde S.A. and Gorbunova Yu.E. (1979). Crystal structure of vanadyl pyrophosphate, *Dokl. Akad. Nauk SSSR*, **245**, pp. 584–591.
22. Nguyen P.T., Hoffman R.D. and Sleight A.W. (1995). Structure of $(VO)_2P_2O_7$ *Mater. Res. Bull.*, **30**, pp. 1055–1063.
23. Torardi C.C. and Calabrese J.C. (1984). Ambient- and low-temperature crystal structure of vanadyl hydrogen phosphate $(VO)_2H_4P_2O_9$, *Inorg. Chem.*, **23**, pp. 1308–1310.
24. Bordes E., Courtine P. and Johnson J.W. (1984). On the topotactic dehydration of $VOHPO_4.0.5H_2O$ into vanadyl pyrophosphate, *J. Solid State Chem.*, **55**, pp. 270–279.
25. Bordes E. (1987a). Crystallochemistry of V-P-O phases and application to catalysis, *Catal. Today*, **1**, 499–526; Bordes E. (1987b). Reactivity and crystal chemistry of V-P-O phases related to $C_4$-hydrocarbon catalytic oxidation, *Catal. Today*, **3**, pp. 163–174.
26. Duvauchelle N., Kesteman E., Oudet F. and Bordes E. (1998). Mosaic Crystals of Vanadyl Pyrophosphate Obtained by Oriented Nucleation and Growth, *J. Solid State Chem.*, **137**, pp. 311–324.
27. O'Mahony L., Henry J., Sutton D., Curtin T. and Hodnett B.K. (2003). Crystallisation of $VOHPO_4 \cdot 0.5H_2O$, *Appl. Catal. A: General*, **253**, pp. 409–416.
28. Hutchings G.J., Olier R., Sananès M.T. and Volta J.-C. (1994). 'Vanadium Phosphate Catalysts Prepared by the Reduction of $VOPO_4$, $2H_2O$', in Cortés Corberán V. and Vic Bellón S. (eds), *New Developments in Selective Oxidation II, Stud. Surf. Sci. Catal.*, **82**, 213–220.
29. Centi G., Fornasari G. and Trifirò F. (1984). On the mechanism of n-butane oxidation to maleic anhydride: Oxidation in oxygen-stoichiometry-controlled conditions, *J. Catal.*, **89**, pp. 44–51.
30. Kestemann E., Merzouki M., Taouk B., Bordes E. and Contractor R.M. (1995). 'Systematic control of crystal morphology during preparation of selective vanadyl pyrophosphate', in Poncelet G., Martens J., Delmon B., Jacobs P.A. and Grange P. (eds), *Preparation of catalysts VI,Stud. Surf. Sci. Catal.*, **91**, pp. 707–716.
31. Petit, S., Borshch, S.A. and Robert, V. (2003). *Ab initio* Simulation of Paramagnetic NMR Spectra: the $^{31}P$ NMR in Oxovanadium Phosphates, *J. Amer. Chem. Soc.* **125**, pp. 3959–3966.
32. Bordes E. and Courtine P. (1985a). New phases in V–P–O catalysts and their role in oxidation of butane in maleic anhydride, *J. Chem. Soc., Chem. Comm.*, **1985**, pp. 294–296; Bordes E., Johnson J.W., Raminosona A. and Courtine P. (1985b). Topotactic reactions yielding new phases in V-P-O system, *Mater. Sci. Monograph.*, **28B**, pp. 887–892.
33. Ben Abdelouahab F.B., Olier R., Guilhaume N., Lefebvre F. and Volta J.-C. (1992). A study by in situ laser Raman spectroscopy of VPO catalysts for n-butane oxidation to maleic anhydride. I. Preparation and characterization of pure reference phases, *J. Catal.*, **134**, 151–167; Benabdelouahab F.B., Volta J.-C. and Olier R. (1994). New Insights into $VOPO_4$ Phases Through Their Hydration, *J. Catal.*, **148**, pp. 334–340.
34. Girgsdies F., Schneider M., Brückner A., Ressler T., Schlögl R. (2009). The Crystal Structure of $\delta$-$VOPO_4$ and its Relationship to $\omega$-$VOPO_4$, *Solid State Sci.*, **11**, pp. 1258–1264.
35. Lim S.C., Vaughey J.T., Harrison W.T.A., Dussac L.L., Jacobson A.J. and Johnson J.W. (1996). Redox transformations of simple vanadium phosphates: the synthesis of $\varepsilon$-$VOPO_4$, *Solid State Ionics*, **84**, pp. 219–226.
36. Harlow R.L., Li Z.G., Herron N., Horowitz H.S., McCarron E.M., Richardson Jr. J.W. and Toby B.H. (2005). Private communication to the ICSD, ICSD collection code 415213.
37. Amorós P., Marcos M.D., Roca M., Alamo J., Beltrán-Porter A. and Beltrán-Porter D. (2001). Crystal structure of a new polytype in the V–P–O system: Is $\omega$-$VOPO_4$ a dynamically stabilised metastable network? *J. Phys. Chem. Solids*, **62**, pp. 1393–1399.

38. Girgsdies F., Dong W.-S., Bartley J.K., Hutchings G.J., Schlögl R. and Ressler T. (2006). The crystal structure of $\varepsilon$-$VOPO_4$, *Solid State Sci.*, **8**, pp. 807–812.
39. Tachez M., Theobald F. and Bordes E. (1981). A structural explanation for the polymorphism of the alpha form of anhydrous vanadyl phosphate, *J. Solid State Chem.*, **40**, pp. 280–284.
40. Hannour K., Lamari A., Kesteman E., Duvauchelle N. and Bordes E. (1996). 11th Int. Congress on Catalysis, Baltimore, USA, July 1–7, 1996, abstract p. 247.
41. Volta J.C., Olier R., Roullet M., Abdelouahab F.B. and Béré K. (1993a). *Stud. Surf. Sci. Catal.*, **75**, pp. 1531–1534; Volta J.-C., Béré K., Zhang Y.J. and Olier R. (1993b). V–P–O Catalysts in n-Butane Oxidation to Maleic Anhydride Study Using an *In Situ* Raman Cell in Oyama T.S., Hightower J.W. (eds), *Catalytic selective oxidation, ACS Symposium Series* **523**, Washington DC, pp. 217–230.
42. Volta J.-C. (2000). Vanadium phosphorus oxides, a reference catalyst for mild oxidation of light alkanes: A review, *C. R. Acad. Sci. Paris, IIc, Chemistry*, **3**, pp. 717–723.
43. Kiely C.J., Burrows A., Sajip S., Hutchings G.J., Sananès M.T., Tuel A. and Volta J.-C. (1996a). Characterisation of Variations in Vanadium Phosphate Catalyst Microstructure with Preparation Route, *J. Catal.*, **162**, pp. 31–47; Kiely C.J., Burrows A., Hutchings G.J., Béré K.E., Volta J.-C., Tuel A. and Abon M. (1996b). Structural transformation sequences occurring during the activation of vanadium phosphorus oxide catalysts, *J. Chem. Soc., Faraday Disc.*, **105**, pp. 103–118.
44. Moser T.P. and Schrader G.L. (1987). Stability of model V–P–O catalysts for maleic anhydride synthesis, *J. Catal.*, **104**, 99–108.
45. Xue Z.-Y. and Schrader G.L. (1992). *In Situ* Laser Raman Spectroscopy Studies of VPO Catalyst Transformations, *J. Phys. Chem. B*, **103**, pp. 9459–9467.
46. Centi G. (1993). Vanadyl Pyrophosphate — A Critical Overview, *Catal. Today*, **16**, pp. 5–26; Centi G. (Ed.) (1993). *Catal. Today*, **16**, papers therein.
47. Inumaru K., Okuhara T. and Misono M. (1992). Active Crystal Face of Vanadyl Pyrophosphate for Catalytic Oxidation of n-Butane to Maleic Anhydride, *Chem. Lett.*, **10**, pp. 1955–1958.
48. Duvauchelle N. and Bordes E. (1999). Influence of the Nanostructure and Morphology of $(VO)_2P_2O_7$ on its Catalytic Reactivity, *Catal. Lett.*, **57**, pp. 81–88.
49. Coulston G.W., Thompson E.A. and Herron N. (1996). Characterization of VPO Catalysts by X-Ray Photoelectron Spectroscopy, *J. Catal.*, **163**, pp. 122–129.
50. Courtine P. and Bordes E. (1997a). 'Synergistic Effects in Multicomponent Catalysts for Selective Oxidation', in Grasselli R.K., Oyama S.T., Gaffney A.M. and Lyons J.E. (eds), *Third World Congress on Oxidation Catalysis, Stud. Surf. Sci. Catal.*, **110**, pp. 177–184; Courtine P. and Bordes E. (1997b). Mode of arrangement of components in mixed vanadia catalysts and its bearing for oxidation catalysis, *Appl. Catal. A: Gen.*, **157**, pp. 45–65.
51. Ebner J.R. and Thompson M.R. (1993). An active site hypothesis for well-crystallized vanadium phosphorus oxide catalyst systems, *Catal. Today*, **16**, pp. 51–60.
52. Wang D., Kung H.H. and Barteau M.A. (2000). Identification of vanadium species involved in sequential redox operation of VPO catalysts, *Appl. Catal. A: General*, **201**, pp. 203–213.
53. Gascón J., Valenciano R., Téllez C., Herguido J. and Menéndez M. (2006). A generalized kinetic model for the partial oxidation of n-butane to maleic anhydride under aerobic and anaerobic conditions, *Chem. Eng. Sci.*, **61**, pp. 6385–6394.
54. Ziolkowski J., Bordes E. and Courtine P. (1990a). Butane and butene on the (100) face of $(VO)_2P_2O_7$: a dynamic view in terms of the crystallochemical model of active sites, *J. Catal.*, **122**, pp. 126–150; ibid. (1990b). A dynamic model of the oxidation of n-butane and 1-butene on various crystalline faces of $(VO)_2P_2O_7$, New Developments in Selective Oxidation, G. Centi and F. Trifirò Eds., *Stud. Surf. Sci. Catal.*, **55**, 625–633.
55. Ziolkowski J. and Bordes E. (1993). Dynamic description of the oxidation of n-butane on various faces of $(VO)_2P_2O_7$ in terms of the crystallochemical model of active sites, *J. Molec. Catal.*, **84**, pp. 307–326.

56. Bordes E. and Contractor R.M. (1996). Adaptation of the solid state properties of redox catalysts to the type of gas-solid reactor, *Topics Catal.*, **3**, 365–375.
57. Wang D. and Barteau M. (2001). Kinetics of Butane Oxidation by a Vanadyl Pyrophosphate Catalyst, *J. Catal.*, **197**, pp. 17–25.
58. Albonetti S., Cavani F., Trifirò F., Venturoli P., Calestani G., López Granados M. and Fierro J.L.G. (1996). A Comparison of the Reactivity of 'Nonequilibrated' and 'Equilibrated' V–P–O Catalysts: Structural Evolution, Surface Characterization, and Reactivity in the Selective Oxidation of *n*-Butane and *n*-Pentane, *J. Catal.*, **160**, pp. 52–64.
59. Patience G.S., Bockrath R.E., Sullivan J.D. and Horowitz H.S. (2007). Pressure Calcination of VPO Catalyst, *Ind. Eng. Chem. Res.*, **46**, pp. 4374–4381.
60. Vedrine J.C., Millet J.M.M., Volta J.-C. (1996). Molecular description of active sites in oxidation reactions: Acid-base and redox properties, and role of water, *Catal. Today*, **32**, pp. 115–123.
61. Thompson D.J., Ciobîcă I.M., Hodnett B.K., van Santen R.A. and Fanning M.O. (2003). A DFT periodic study of the vanadyl pyrophosphate (1 0 0) surface, *Surf. Sci.*, **547**, pp. 438–451.
62. Haras A., Duarte H.A., Salahub D.R. and Witko M. (2002). Changes of local electronic structure of perfect $(VO)_2P_2O_7$ (100) surface in response to oxygen vacancy formation: effect of electron trapping, *Surf. Sci.*, **513**, pp. 367–380.
63. Herrmann J.-M., Vernoux P., Béré K.E. and Abon M. (1997). *In Situ* Study of Redox and of p-Type Semiconducting Properties of Vanadyl Pyrophosphate and of V–P–O Catalysts during the Partial Oxidation of n-Butane to Maleic Anhydride, *J. Catal.*, **167**, pp. 106–117.
64. Hävecker M., Mayer R.W., Knop-Gericke A., Bluhm H., Kleimenov E., Liskowski A., Su D.S., Follath R., Requejo F.G., Ogletree D.F., Salmeron M., Lopez-Sanchez J.A., Bartley J.K., Hutchings G.J. and Schlögl R. (2003). In Situ Investigation of the Nature of the Active Surface of a Vanadyl Pyrophosphate Catalyst during n-Butane Oxidation to Maleic Anhydride, *J. Phys. Chem. B*, **107**, pp. 4587–4596.
65. Bluhm H., Hävecker M., Kleimenov E., Knop-Gericke A., Liskowski A., Schlögl R. and Su D.S. (2003). In Situ Surface Analysis in Selective Oxidation Catalysis: n-Butane Conversion Over VPP, *Topics Catal.*, **23**, pp. 99–107.
66. Grasselli R.K. (2005). Selectivity issues in (amm) oxidation catalysis, *Catal. Today*, **99**, pp. 23–31.
67. Ballarini N., Cavani F., Cortelli C., Ligi S., Pierelli F., Trifirò F., Fumagalli C., Mazzoni G. and Monti T. (2006a). VPO catalyst for n-butane oxidation to maleic anhydride: A goal achieved, or a still open challenge?, *Topics Catal.*, **38**, pp. 147–156.
68. Ballarini N., Cavani F., Cortelli C., Ricotta M., Rodeghiero F., Trifirò F., Fumagalli C. and Mazzoni G. (2006). Non-steady catalytic performance as a tool for the identification of the active surface in VPO, catalyst for n-butane oxidation to maleic anhydride, *Catal. Today*, **117**, pp. 174–179.
69. Cavani F., Luciani S., Esposti E. D., Cortelli C. and Leanza R. (2010a). Surface Dynamics of A Vanadyl Pyrophosphate Catalyst for n-Butane Oxidation to Maleic Anhydride: An In Situ Raman and Reactivity Study of the Effect of the P/V Atomic Ratio, *Chem. Eur. J.*, **16**, pp. 1646–1655; Cavani F., De Santi D., Luciani S., Löfberg A., Bordes-Richard E., Cortelli C. and Leanza R. (2010b). Transient reactivity of vanadyl pyrophosphate, the catalyst for n-butane oxidation to maleic anhydride, in response to in-situ treatments, *Appl. Catal. A: General*, **376**, pp. 66–75.
70. Bordes E. (2000). Influence of structural properties of catalysts at various stages of selective oxidation: from catalyst preparation to catalytic reactors, *Topics Catal.*, **11/12**, pp. 61–65.
71. Mota S., Abon M., Volta J.-C. and Dalmon J.A. (2000). Selective Oxidation of n-Butane on a V–P–O Catalyst: Study under Fuel-Rich Conditions, *J. Catal.*, **193**, pp. 308–318.
72. Mallada R., Menendez M. and Santamaria J. (2000). Use of membrane reactors for the oxidation of butane to maleic anhydride under high butane concentrations, *Catal. Today*, **56**, pp. 191–197.
73. Schuurman Y. and Gleaves J. T. (1997). A comparison of steady-state and unsteady-state reaction kinetics of n-butane oxidation over VPO catalysts using a TAP-2 reactor system, *Catal. Today*, **33**, pp. 25–37.

74. Wang D. and Barteau M. A. (2003). Differentiation of active oxygen *specie* for butane oxidation on vanadyl pyrophosphate, *Catal. Letters*, **90**, pp. 7–11.
75. Lorences M. J., Patience G. S., Cenni R., Díez F. and Coca J. (2006). VPO transient lattice oxygen contribution, *Catal. Today*, **112**, pp. 45–48.
76. Huang X.-F., Chen B.-H., Liu B.-J., Silveston P. L. and Li C.-Y. (2002). Re-oxidation kinetics of a VPO catalyst, *Catal. Today*, **74**, pp. 121–130.
77. Patience, G.S. and Lorences, M.J. (2006). VPO Transient Oxidation Kinetics, *Int. J. Chem. Reactor Eng.*, **4**, pp. 1–16.
78. Shekari A., Patience G.S. and Bockrath R.E. (2010). Effect of feed nozzle configuration on n-butane to maleic anhydride yield: From lab scale to commercial, *Appl. Catal. A: General*, **376**, pp. 83–90.
79. Patience, G.S. and Cenni R. (2007). Formaldehyde process intensification through gas heat capacity, *Chem. Eng. Sci.*, **62**, pp. 5609–5612.
80. Coronas J. and Santamaría J. (1999). Catalytic reactors based on porous ceramic membranes, *Catal. Today*, **51**, pp. 377–389.
81. Julbe A., Farrusseng D., Guizard C. (2001). Porous ceramic membranes for catalytic reactors–overview and new ideas, *J. Membr. Sci.*, **181**, pp. 3–20.
82. Macano G.S., Tsotsis T.T. (2002). '*Catalytic Membranes and Membrane Reactors*', Wiley/VCH, Weinheim.
83. Saracco G., Neomagus H.WJ.P., Veersteeg G.F., van Swaaij W.P.M. (1999). High-temperature membrane reactors: potential and problems, *Chem. Eng. Sci.*, **54**, pp. 1997–2017.
84. Sirkar K.K., Shanbhag P.V.and A.S. Kovvali. (1999). Membrane in a reactor: A functional perspective, *Ind. Eng. Chem. Res.*, **38**, pp. 3715–3737.
85. Xue, E. and Ross, J. (2000). The Use of Membrane Reactors for Catalytic *n*-Butane Oxidation to Maleic Anhydride with a Butane-Rich feed, *Catal. Today*, **61**, pp. 3–8.
86. Cruz-López. A, Guihaume, N., Miachon, S. *et al.* (2005). Selective Oxidation of Butane to Maleic Anhydride in a Catalytic Membrane Reactor Adapted to Rich Butane Feed, *Catal. Today*, **107–108**, pp. 949–956.
87. Alonso, M., Lorences, M.J., Pina, M.P. *et al.* (2001)., *Catal. Today*, **67**, pp. 151–157.
88. Alonso, M., Lorences, M.J., Patience, G.S. *et al.* (2005). Membrane Pilot Reactor Applied to Selective Oxidation Reactions, *Catal. Today*, **104**, pp. 177–184; Alonso, M., Patience, G., Fernandez, J., *et al.* (2006). Heat Transfer Studies in an Inorganic Membrane Reactor at Pilot Plant Scale, *Catal. Today*, **118**, pp. 32–38.
89. Alonso, M., G.S. Patience, J.R. Fernandez, M.J. Lorences, F. Díez, A. Vega, R. Cenni, R. (2006). Heat transfer studies in an inorganic membrane reactor at pilot plant scale, *Catal. Today*, **118**, pp. 32–38.
90. Arnold, S.C., Cecchini, M., Fenati, G. (2001). Revitalization of fluidized-bed reactor technology for maleic anhydride (Alma process) by enhancing hydrodynamics via cyclone improvements, *Fluidization X*, Engineering Foundation, New York, pp. 181–188.
91. Shekari A. and Patience G.S. (2010). Maleic anhydride yield during cyclic n-butane/oxygen operation, *Catal. Today*, **157**, pp. 334–338.
92. Hutchenson K.W., La Marca C., Patience G.S., Laviolette J.-P. and Bockrath R.E. (2010). Parametric study of n-butane oxidation in a circulating fluidized bed reactor, *Appl. Catal. A: General*, **376**, pp. 91–103.
93. Dummer N.F., Weng W., Kiely C., Carley A.F., Bartley J.K., Kiely C.J. and Hutchings G.J. (2010). Structural evolution and catalytic performance of DuPont V-P-O/$SiO_2$ materials designed for fluidized bed applications, *Appl. Catal. A: General*, **376**, pp. 47–55.
94. Fernández J. R., Vega A. and Díez F. V. (2010). Partial oxidation of n-butane to maleic anhydride over VPO in a simulated circulating fluidized bed reactor, *Appl. Catal. A: General*, **376**, pp. 76–82.

95. Godefroy, A., G.S. Patience, T. Tzakova, D. Garrait and J.-L. Dubois (2009). Reactor technologies for propane partial oxidation to acrylic acid, *Chem. Eng. Technol.*, **32**, pp. 1–8.
96. Brown I.D. and Wu K.K. (1976). Empirical parameters for calculating cation–oxygen bond valences, *Acta Cryst. B*, **32**, pp. 1957–1959.

# Chapter 21

# Polyoxometalates Catalysts for Sustainable Oxidations and Energy Applications

Mauro CARRARO*, Giulia FIORANI*, Andrea SARTOREL* and Marcella BONCHIO*

The aim of this capter is to present an overview on the use of polyoxometalates (POMs) as oxidation catalysts. POMs display several interesting features which make them attractive for condidates for these kinds of applications: (i) their composition mainly includes oxidatively stable $d^0$ early series transition metals; (ii) they can be used as robust inorganic ligands for the coordination of redox active transition metals; (iii) they represent useful platforms for the development of hybrid organic-inorganic compounds, where the organic domain can synergistically contribute to implement the catalytic activity; (iv) their polyanionic nature can be exploited to foster the association with solubilizing organic-inorganic cations or to promote their immobilization onto different solid supports. The chapter will thus be focused on both homogeneous and heterogeneous POM-based oxidations of organic substrates by sustainable oxidants, as dioxygen and hydrogen peroxide, employing different activation techniques, including UV and microwave irradiation. Finally, the structural and mechanistic analogy of some POMs with natural metallo-enzymes has been underlined, as in the case of oxygenase activity and water oxidation catalysis.

## 21.1. Polyoxometalates

Polyoxometalates (POMs) are oxo-clusters of early transition metals in their highest oxidation state, namely Mo(VI), W(VI), V(V), Nb(V) and Ta(V). A simple classification is based on their chemical composition, with two types of general formula:

(i) $[M_mO_y]^{p-}$, isopolyanions; and
(ii) $[X_xM_mO_y]^{q-}$, heteropolyanions

*ITM-CNR and Department of Chemical Sciences, University of Padova, via Marzolo 135131 Padova, Italy.

where M is the main transition metal constituent of the polyoxometalate (addenda atom), O is the oxygen atom and X can be a non-metal (such as P, Si, As, Sb), another element of the p block or a different transition metal.[1–4]

Polyoxometalates are generally formed only by transition metals with dimensions (cationic radius) compatible with an octahedral coordination. In addition, empty low energy d-orbitals on the metal ion are required to form at least one terminal M=O double bond.[5] The aggregation of octahedral units $MO_6$ occurs upon condensation between two central metal ions through μ-oxo bridged bonds, while the presence of less basic, terminal oxygen atoms prevents the linear polymerization (as in common metal oxides: silicates, germanates, tellurates) and promotes the formation of discrete molecular units.[2–6]

These discrete, soluble multimetal oxides are characterized by a formidable structural variety that can be achieved by tuning reaction parameters (such as concentrations and/or stoichiometric ratio of the reagents, temperature and pH), thus yielding nanosized complexes with a different shape, charge density and surface reactivity. When considering isostructural POMs, in particular, such properties can be controlled at the molecular level by changing their constituent elements (other transition metals or the heteroatom X).

In addition, the choice of suitable counterions gives access to their solubilization and use in a wide range of solvents, from water, with alkaline cations or protons, to apolar ones (toluene, dichloromethane), with lipophilic cations such as tetraoctylammonium.

Polyoxometalates are able to act as ligands for different redox-active transition metals (M') such as iron, manganese, cobalt and ruthenium or $d^0$ metals (such as zirconium, hafnium, titanium). The coordination of a heterometal by a polyoxometalate framework takes place either through the superficial coordination of a metal cation by electrostatic interaction, or by incorporation of the transition metal within the POM structure resulting in the formation of transition metals substituted polyoxometalates (TMSPs), which generally exhibit a higher stability.

Transition metals substituted polyoxometalates are obtained starting from vacant or "lacunary" POMs. The latter are derived from the saturated parent POM, through formal loss of one or more $MO_6$ octahedral units. Such complexes present several coordinatively unsaturated oxygen atoms, which surround the defect and form a "polydentate" site, able to coordinate a multitude of transition metals, M'. Fundamental modes of coordination of the transition metals are "in pocket" and "out of pocket", the latter including "sandwich-like" structures and metal oxide cluster stabilization, depending both on the dimensions of M' and on the stoichiometic ratio between M' and the vacant units.[7–13]

The nucleophilicity of coordinatively unsaturated oxygen atoms can also be exploited to synthesize organic–inorganic hybrid complexes, by reaction with electrophilic organic derivatives. Organo-trichlorosilane and trialkoxysilanes, or organophosphonic acids and phosphonyl chlorides are generally used as electrophilic reagents.[14–17] The derivatization of POMs is useful to: i) stabilize molecular architectures which can otherwise give isomerization or decomposition;[18] ii) exploit the robust polyanionic scaffold to support organic molecules, including chiral moieties,[19] organometallic catalysts,[20] and tune their solubility in the reaction media; and iii) introduce polyfunctional groups, to link different POMs or for the preparation of dendrimeric and polymeric hybrid materials.[21,22]

Hybrid POMs have been successfully employed as building blocks to prepare self-assembled supramolecular aggregates and nanostructured systems.[23,24] The aggregation of amphiphilic polyoxoanions may lead to vesicles and tubular channels, thus providing a microenvironment of interest for catalytic application.[25,26]

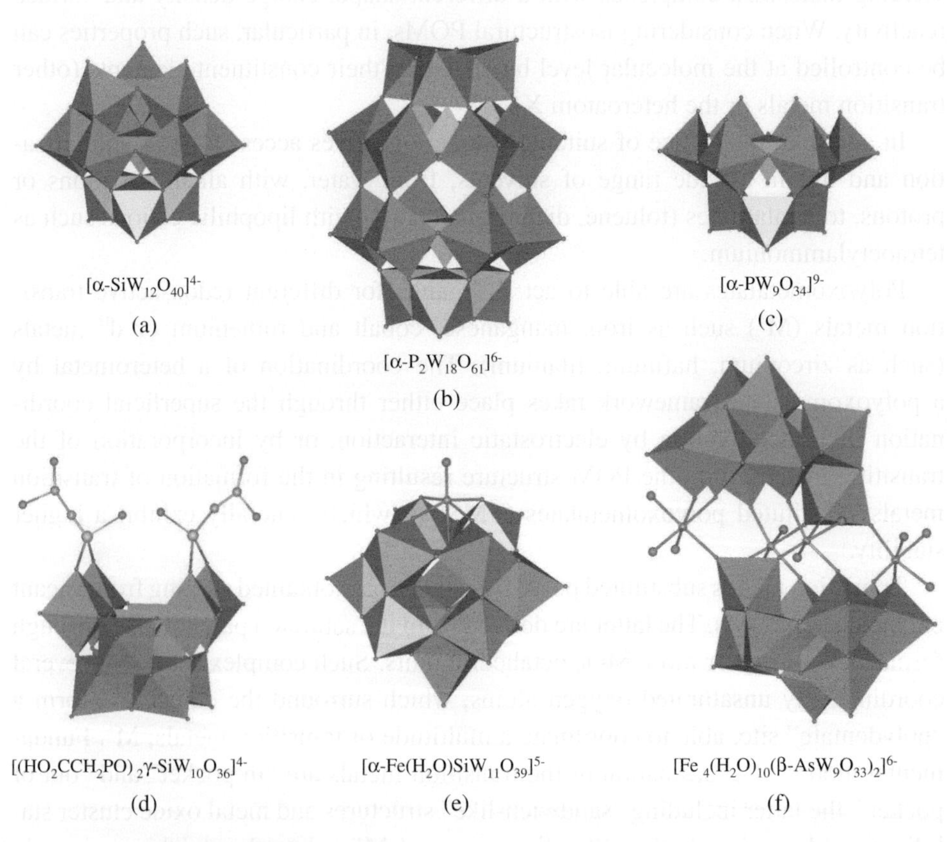


**Figure 21.1.** Polyhedral representation of some POMs (cations and protons are omitted): examples of saturated Keggin (a) and Dawson (b), vacant (c), hybrid (d) and TMSP (e and f) complexes are represented.

The easy accessibility to a variety of different structures and, consequently, to molecules with different chemical and physical properties, justifies the potential application of POMs in different fields, from material science, to medicinal chemistry and homogeneous/heterogeneous catalysis. Catalytic applications of POMs have been systematically studied since the late 1970s.[27] POMs have been known mainly as Brønsted acids, but their structural diversity and the merging with different catalytic domains has expanded their application to include hydrogenations,[28] click chemistry,[29] and cross-couplings reactions.[20,30] However, due to their thermal and oxidative stabilities, POMs still retain a major appeal as oxidation catalysts.[31]

Relevant structures of some POMs that will be described in this chapter are summarized in Fig. 21.1.

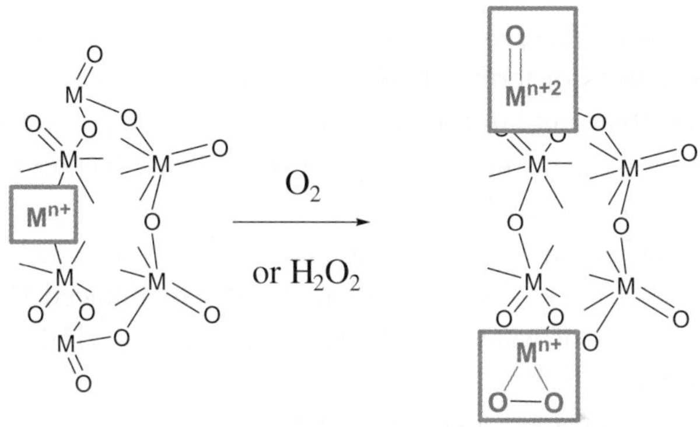


**Figure 21.2.** Activation of $O_2$ or $H_2O_2$ by POMs, through the formation of oxene intermediates (upper right) or metal-peroxo intermediates (lower right).

## 21.2. Oxidation Catalysis by POMs

The growing need for sustainable chemical processes requires reactions with high atom economy that use environmentally friendly solvents. Catalytic metal-based oxidation, using molecular oxygen and hydrogen peroxide as intrinsically non-waste producing terminal oxidants, is an important alternative to classic stoichiometric procedures. Research published since the 1980s has established the potential of POMs as homogeneous oxidation catalysts, especially in the context of the use of such oxidants. The occurrence of different transition metals within the POM structure can be exploited to activate these oxidants through the formation of peroxo/oxo-active species (Fig. 21.2). The totally inorganic ligand system derived from POMs represents a distinct advantage over the commonly employed coordination complexes based on organic moieties, because of their remarkable stability in harsh oxidative reaction conditions.[27,31]

Some relevant examples of the use of POMs in different oxidation processes will be presented in the next section, including a brief discussion of possible heterogenization/recovery strategies of the POM-based catalysts.

### 21.2.1. *POM catalysts for aerobic oxidations*

Dioxygen is the most attractive terminal oxidant, since it is cheap and abundant in the atmosphere. Moreover, it presents the highest active oxygen content with no harmful by-products. Several POMs have thus been used as catalysts to activate dioxygen by using different approaches.[32]

Keggin-type mixed-addenda heteropolyanions such as $H_{3+n}[PMo_{12-n}V_nO_{40}]\cdot nH_2O$, (HPA-n, with n = 1, 2, 3, etc.), have found many applications in catalysis.[33,34] In the area of liquid-phase oxidation, the earliest use was as co-catalysts in Wacker-type oxidations of terminal alkenes.[35] From a mechanistic point of view there is now quite abundant information indicating that the polyoxomolybdates act as electron-transfer oxidants in the catalytic cycle. HPA-2 has been employed as a catalyst for the oxidative dehydrogenation of α-terpinene to *p*-cymene. Ultraviolet-visible (UV-vis) and infrared (IR) kinetic measurements have led to the proposal that one molecule of $O_2$ reacts with two molecules of reduced $H_7PV^2_{IV}Mo_{10}O_{40}$ to yield two molecules of water and the reoxidized catalyst $H_5PV^2_VMo_{10}O_{40}$ in a four-electron oxidation.[32,36]

*vic*-Diols like 1,2-cyclohexanediol have been smoothly cleaved by a catalytic amount of HPA-2 with dioxygen, at 75°C, to yield carboxylic acids with 90% selectivity and up to 92% yield.[37] The cleavage of representative substituted and unsubstituted ketones has been obtained with HPA-3 and HPA-2,[38,39] and relevant results are summarized in Table 21.1. For the oxidation of cyclohexanone to adipic acid, the use of acetic acid as a co-solvent has led to a strongly enhanced conversion with respect to the reaction carried out solely in water. A recent kinetic investigation has suggested that a high POM-to-cyclohexanone molar ratio favors a POM-mediated redox-type mechanism, which is also preferred in the absence of acetic acid. Conversely, when the POM:cyclohexanone ratio is low, the prevailing mechanism is a radicalic chain autoxidation. Interestingly, the latter is more selective toward adipic acid, due to the lower contribution of cyclohexanone oxidative degradation, leading to lighter acids and $CO_2$.[40]

Regioselective cleavage of 2-hydroxycyclohexanone by HPA-2 has given adipic acid or its dimethyl ester as the major products (yields 80–90%) at 65°C in aqueous acetic acid (in 3.5 h) or methanol (in 7 h), respectively.

HPA-3 has been used with benzoyl derivatives at room temperature with completion of dioxygen uptake within 5 h, and methyl benzoate and/or benzoic acid have been formed with 90–100% selectivity.[39]

**Table 21.1.** Catalytic oxidation of ketones by $H_{3+n}[PMo_{12-n}V_nO_{40}]\cdot mH_2O$ (n = 2 or 3) with $O_2$.

| Entry | Substrate | Product | POM | T(° C); solvent | t (h) | Yield (%) |
|---|---|---|---|---|---|---|
| 1[38] | 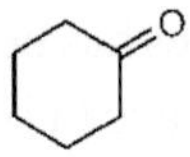 | 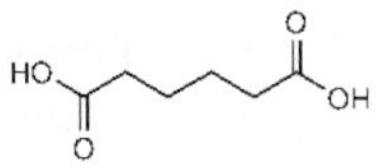 | HPA-2[a] | 70°C; AcOH/$H_2O$ 9:1 | 7 | 51 |
| 2[38] | 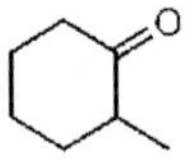 | 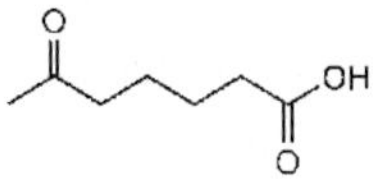 | HPA-2[a] | 60°C; $H_2O$ | 8 | 97 |
| 3[39] | 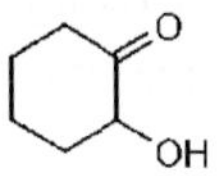 | 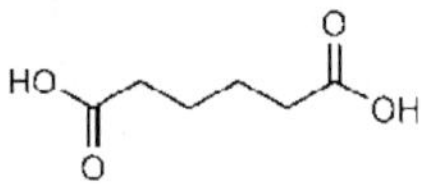 | HPA-3[b] | 65°C; AcOH/$H_2O$ 9:1 | 3.5 | 80 |
| 4[39] | 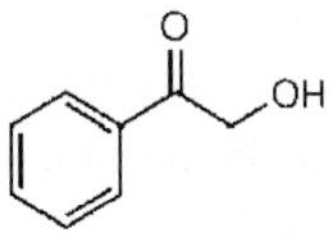 | 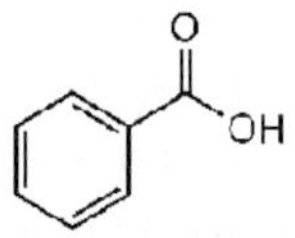 | HPA-3[b] | r.t.; AcOH/$H_2O$ 9:1 | <5 | >99 |

a = $H_5[PMo_{10}V_2O_{40}]\cdot 30H_2O$; b = $H_6[PMo_9V_3O_{40}]\cdot xH_2O$.

Pyridine-modified HPA-1 ha also been applied for the liquid-phase oxidation of benzene to phenol with 2.0 MPa of dioxygen at 110°C, in acetic acid. When using ascorbic as reducing reagent, phenol was obtained with 9% yield in 10 h.[41] A 13% yield was observed in the presence of β-cyclodextrin as phase transfer agent, under analogue conditions.[42]

As already mentioned, vacant POMs can be considered as inorganic ligands featuring rigid polydentate binding sites. They can thus coordinate redox-active transition metals, such as Ru, Fe, Co, Mn and Cu, leading to the formation of TMSPs.[43,44]

In some cases, the coordination geometry of a representative structural type of TMSP finds a close correspondence with the active site of natural oxygenase enzymes.

Iron-substituted POMs have indeed attracted a great deal of interest, because of the well-known catalytic properties of iron both in biological and in synthetic systems. In particular, the Keggin-type complexes $[\alpha\text{-Fe}(H_2O)SiW_{11}O_{39}]^{5-}$,[8] $[\gamma\text{-Fe}_2(H_2O)_2SiW_{10}O_{38}]^{6-}$,[45] and the Krebs-type derivative $[\beta\text{-Fe}_4(H_2O)_{10}(XW_9O_{33})_2]^{n-}$ ($Fe_4X_2W_{18}$, X = $Se^{IV}$, $Te^{IV}$, $As^{III}$, $Sb^{III}$; n = 4, 6)[46,47] have inspired a POM-based inorganic mimicry of heme-cytochrome P450, the $Fe_2(\mu\text{-O})_2$ diamond core of methane mono-oxygenase (MMO) and the non-heme dioxygenase iron site, respectively.[48–52]

Iron-substituted Krebs POMs $[\beta\text{-Fe}_4(H_2O)_{10}(XW_9O_{33})_2]^{n-}$ have been employed to catalyze the oxidative cleavage of 3,5-di-tert-butylcatechol (DTBC) in the presence of molecular oxygen. Oxidation of DTBC by $Fe_4X_2W_{18}$ (X = $As^{III}$, $Sb^{III}$,

$Se^{IV}$, $Te^{IV}$) has been examined in 1,2-dichloroethane (DCE) under an oxygen atmosphere at 60°C: 80–90% substrate conversion occurred in 3 h.[52] In all cases, the two-electron oxidation product 3,5-di-tert-butyl-1,2-benzoquinone has been formed in 50–60% yield. The mimicry of the catechol dioxygenase enzymes has been established by studying DTBC cleavage selectivity in wet tetrahydrofurane (THF), in the presence of 2,6-di-tert-butylcresol (BHT) as the radical scavenger, to inhibit free radical oxidation and polymerization pathways. Indeed, a favorable enhancement of the cleavage selectivity (up to 40%) has been observed, yielding 3,5-di-tert-butyl-1-oxacyclohepta-3,5-diene-2,7-dione and 4,6-di-tert-butyl-1-oxacyclohepta-4,6-diene-2,3-dione, arising from intra- and extra-diol cleavage mechanisms, respectively.[49]

Iron-substituted complexes have been employed in a number of oxidations with $O_2$, including high synthetic value transformations, such as hydrocarbon hydroxylations and epoxidations.[53–58] The relevant results observed for the aerobic epoxidation of cyclic and terminal olefins are summarized in Table 21.2.

Cyclic olefins have been oxidized to the corresponding epoxides in 12 days at 75°C, with 30–78% yield and 59–66% selectivity, by using tetrahexylammonium (THA) POM salts in 1,2-dichloroethane.

Higher selectivity has been registered at lower conversion and with a low catalyst loading. A turnover number (TON) of up to 8,000 has been achieved with $(THA)_4$ $[\beta\text{-}Fe_4(H_2O)_{10}(SbW_9O_{33})_2]$ (Fig. 21.3).

For the epoxidation reactions, an induction time (3–7 days) for the formation of the oxidation products has been observed. The reaction rate is directly proportional to Fe concentration, whereas alkene consumption is characterized by a zero order regime. The induction time is affected by the nature of the catalyst, being shorter (< 100 h) for stronger oxidants (Sb and As derivatives), while turnover frequencies (TOFs), between 43 and 50 TON $h^{-1}$, are basically unaffected by these electronic modulations. This behavior is consistent with a Haber–Weiss mechanism, which is also supported by the shortening of the induction time when a radical initiator such as benzoyl peroxide is added to the reaction mixture.[57] The stability of the Fe-substituted POMs used has been assessed by UV-vis and Fourier transform infrared (FTIR) spectroscopy: for $[\alpha\text{-}Fe(H_2O)SiW_{11}O_{39}]^{5-}$, FTIR analysis, performed after precipitation of the catalyst at different time intervals, has confirmed that POM structural features are maintained even after a prolonged heating. Analogous conclusions have been drawn for the Krebs-type complexes, where only a modest shift of the W-O stretching absorptions at higher frequencies has been observed.[57] The tetrabutylammonium (TBA) salt of $[\gamma\text{-}Fe_2(H_2O)_2SiW_{10}O_{38}]^{6-}$ has been reported to epoxidize cyclic olefins at 83°C in 16 days with 74–82% yield and 64–98% selectivity, in a mixture of 1,2-dichloroethane and acetonitrile.[58] Both terminal and internal linear olefins have been oxidized with 11–24% yield and 79–94% selectivity. However, both FTIR analysis and the solution speciation of this

**Table 21.2.** Aerobic epoxidation of alkenes with Fe-substituted POMs.

| Alkene | TMS Fe-POM catalyst (% mol) | t (h) | C (%) | S (%) | Induction time (h) | TON | Ref. |
|---|---|---|---|---|---|---|---|
| 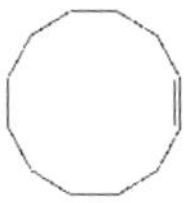 | $(TBA)_6[\gamma\text{-}Fe_2(H_2O)_2SiW_{10}O_{38}]$ (0.008) | 385 | 74 | 64 | n.d. | 9,600 | 58 |
| 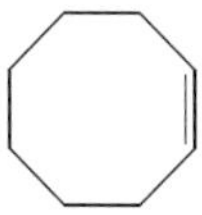 | $(TBA)_6[\gamma\text{-}Fe_2(H_2O)_2SiW_{10}O_{38}]$ (0.008) | 385 | 82 | 98 | n.d | 10,000 | 58 |
| | $(THA)_6[\gamma\text{-}Fe_2(H_2O)_2SiW_{10}O_{38}]$ (0.041) | 300 | 32 | 65 | 120 | 2,660 | 57 |
| | $(THA)_5[\alpha\text{-}Fe(H_2O)SiW_{11}O_{39}]$ (0.041) | 300 | 68 | 57 | 100 | 936 | 57 |
| | $(THA)_6[\beta\text{-}Fe_4(H_2O)_{10}(AsW_9O_{33})_2]$ (0.005) | 300 | 51 | 59 | 95 | 5,684 | 57 |
| | $(THA)_6[\beta\text{-}Fe_4(H_2O)_{10}(SbW_9O_{33})_2]$ (0.005) | 300 | 68 | 62 | 75 | 7,958 | 57 |
| | $(THA)_4[\beta\text{-}Fe_4(H_2O)_{10}(SeW_9O_{33})_2]$ (0.005) | 300 | 40 | 63 | 170 | 7,200 | 57 |
| | $(THA)_4[\beta\text{-}Fe_4(H_2O)_{10}(TeW_9O_{33})_2]$ (0.005) | 300 | 58 | 66 | 105 | 4,737 | 57 |
|   | $(TBA)_6[\gamma\text{-}Fe_2(H_2O)_2SiW_{10}O_{38}]$ (0.008) | 385 | 17 | 93 | n.d. | 2,200 | 58 |
| cis/trans = 8:2 | $(TBA)_6[\gamma\text{-}Fe_2(H_2O)_2SiW_{10}O_{38}]$ (0.008) | 385 | 20 | 94 | n.d. | 2,500 | 58 |
|  cis/trans = 8:2 | $(TBA)_6[\gamma\text{-}Fe_2(H_2O)_2SiW_{10}O_{38}]$ (0.008) | 385 | 24 | 91 | n.d. | 2,900 | 58 |
|  cis/trans = 4:6 | $(TBA)_6[\gamma\text{-}Fe_2(H_2O)_2SiW_{10}O_{38}]$ (0.008) | 385 | 11 | 79 | n.d. | 1,300 | 58 |

C: conversion; S: epoxide selectivity based on substrate conversion.
Reaction conditions for ref.[57]: solvent: 1.2-dichloroethane V = 0.8 ml; substrate: 3.6 mmol; $O_2$ = 1 atm; T = 75°C.
Reaction conditions for ref.[58]: solvent: 1,2-dichloroethane/$CH_3CN$ 15:1; $V_{tot}$ = 1.6 ml; catalyst: 1.5 μ mol; substrate: 18.5 mmol; $O_2$ = 1 atm, T = 83°C.

complex have revealed that it undergoes a conversion to another active, undefined species.[56]

Microwave (MW) induced dielectric heating has provided a rapid reaction protocol for screening the activity and selectivity of Fe-substituted POMs. Under the conditions explored (summarized in Fig. 21.4), cyclohexane oxygenation occurs with a conversion of up to 3% in 250 min, yielding cyclohexylhydroperoxide (HP), cyclohexanol (A) and cyclohexanone (K) with 90–95% selectivity.[54]

Again, several experimental pieces of evidence have confirmed a radical chain mechanism. For all the oxidation reactions studied, both conversion and selectivity

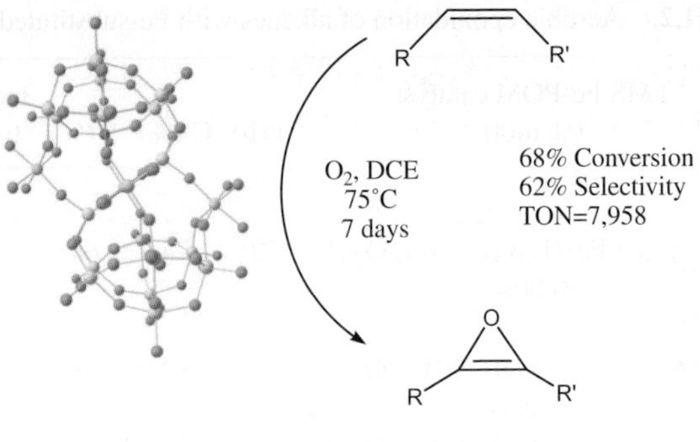


**Figure 21.3.** Aerobic alkene epoxidation catalyzed by $(THA)_6[\beta\text{-}Fe_4(H_2O)_{10}(SbW_9O_{33})_2]$.

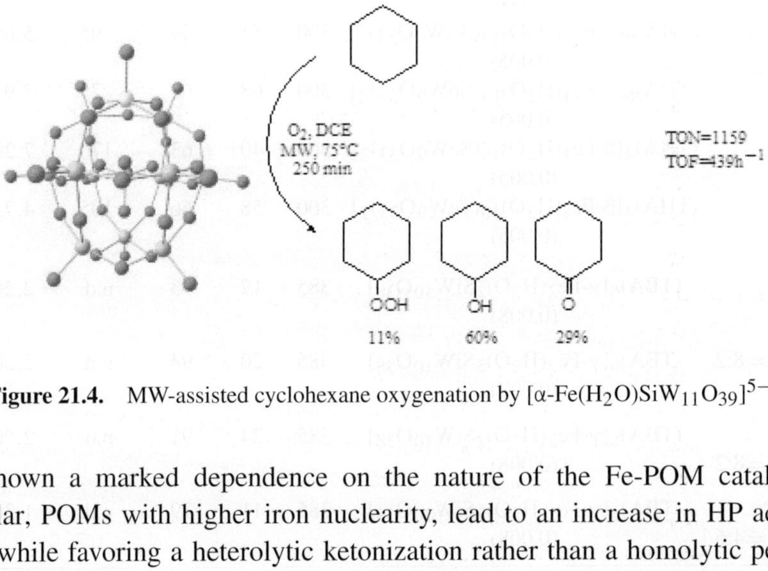


**Figure 21.4.** MW-assisted cyclohexane oxygenation by $[\alpha\text{-}Fe(H_2O)SiW_{11}O_{39}]^{5-}$.

have shown a marked dependence on the nature of the Fe-POM catalyst. In particular, POMs with higher iron nuclearity, lead to an increase in HP accumulation, while favoring a heterolytic ketonization rather than a homolytic peroxide decomposition.[59]

Transition metals substituted polyoxometalate embedding different transition metals have been employed too: for example, the ruthenium substituted POM $\{[WZnRu_2^{III}(OH)(H_2O)](ZnW_9O_{34})_2\}^{6-}$ displays dioxygenase reactivity and can be activated (10 h at 80°C) under a dioxygen atmosphere to form a $Ru^{IV}$-oxo species able to give selective adamantane hydroxylation on the tertiary carbon position as well as olefin epoxidation, with selectivity >99%, following a non-free radical oxidation mechanism.[60] Another Ru-based TMSP, $(TBA)_4[HRu(H_2O)SiW_{11}O_{39}]$, has shown high catalytic activity towards aerobic oxidation of alcohols and hydrocarbons, with yields of up to 90%, selectivities up to 88%, and TONs up to 2,000 in 48–120 h, at 110°C.[61] $Li_5[Ru^{II}(DMSO)PW_{11}O_{39}]$ (DMSO = dimethylsulfoxide) has been used in water to convert DMSO to the corresponding sulfone under an oxygen atmosphere: the reaction has been carried out in an acetate buffer solution (pH = 4.8), under MW-induced dielectric heating ($T_{bulk} = 200$°C).[55]

### 21.2.1.1. *Photocatalytic oxygenation by POMs*

Photocatalytic oxygenation has been successfully employed for the development of photo-assisted synthetic procedures or advanced oxidation processes (AOPs) applied to environmental remediation, exploiting the selective oxidation of organic substrates or their full mineralization, respectively.[62–67] In this context, POMs have been referred to as "molecular fragments" or as the homogeneous analogues of photoactive semiconductor metal oxides.[2,4,27,66,68] In addition, the possibility to undergo photoinduced multi-electron transfers without changing their structure makes POMs very attractive catalysts for the oxidation of organic substrates in the presence of $O_2$.[68–72] The choice of POMs to be employed in photocatalytic processes is affected by their electronic properties, namely the ease of re-oxidation by dioxygen; this step is favored with polyoxotungstates with respect to polyoxomolybdates and polyoxovanadates.[69,73,74]

The most extensively studied photoactive polyanion is the decatungstate $[W_{10}O_{32}]^{4-}$.[72,75,76,77] The general reaction mechanism involves: (i) irradiation with wavelengths of <350–400 nm, to promote a ligand to metal charge transfer (LMCT) transition; (ii) oxidation of the organic substrate through hydrogen (or electron) abstraction, generating radical species which can react with $O_2$ to yield organic peroxides; (iii) reaction between the reduced POM (heteropolyblue complex) and dioxygen, to restore the initial form of the catalyst while forming reactive oxygenated species (ROS).[78,79]

The chemical environment assisting the photo-oxidation events may affect some fundamental aspects, such as the lifetime of the photoactive transient and dioxygen availability. When using POMs as photocatalysts, the photo-assisted reaction can easily be performed in a wide range of different media, including aqueous phases, tuning POMs solubility by convenient counterion metathesis. Acetonitrile is mostly used, but acetone has also been employed.[80] The main parameter for the solvent choice is its stability towards radicals. The impact of solvent-derived reactive radicals can also elicit the autoxidation cycles. In water, oxidation occurs through the production of highly reactive hydroxyl radicals $OH^{\cdot}$ from the solvent activation routine.[77,81] Relative values of the rate constants for the reaction of the photocatalyst with propan-2-ol in acetone, acetonitrile and water are 1/1.8/23.[82] This enhanced reactivity can override more selective pathways, involving a direct interaction of the substrate with the POM photocatalyst.[64,82,83] The main drawback of POM-based photocatalysis lies in their low quantum yields ($\Phi$, defined as moles of products per mole of adsorbed photons, in the range 0.1–0.5), associated with a limited absorption in the most desirable visible range.[84,85] High photocatalyst/substrate ratios are thus generally required to obtain fast and efficient oxygenation cycles.

Heterogeneous systems have been successfully designed to recycle photoactive POMs. $(TBA)_4W_{10}O_{32}$ has indeed been supported on amorphous silica or on mesoporous aluminosilicate MCM-41, and used in an organic dispersing medium. The positive tetra-alkylammonium cations act as a bridge between the negative surface of silica and the decatungstate anion. Photoexcitation ($\lambda > 300$ nm) of a powdered dispersion of $(TBA)_4W_{10}O_{32}/SiO_2$ system can promote the oxygenation of cyclohexane and cyclohexene at 20°C and 1 atm of $O_2$. Cyclohexane is oxidized to an equimolar mixture of cyclohexanol and cyclohexanone, while cyclohexene is converted to the corresponding cyclohexenyl hydroperoxide (about 90% of the overall oxidized substrate) and to cyclohex-2-en-1-one (about 6%).[86,87]

To improve the catalyst performance and recovery, a different heterogenization strategy consists of the preparation of hybrid polymeric films or membranes with embedded POMs.[88,89] The application of catalytic membrane reactors (CMRs) in catalysis combines advanced molecular separation with selective transport properties, while allowing catalyst compartmentalization.[90,91] Moreover, by varying the nature of the polymer and of the resulting hybrid membrane, the hydrophobic/hydrophilic surface properties can be tailored to optimize the affinity towards target reagents. To obtain heterogeneous photocatalysts with tunable composition and structural/surface properties, decatungstate has been incorporated within the polymeric membranes of polyvinylidene difluoride (PVDF) and polydimethylsiloxane (PDMS). The surfaces and cross-section morphology (thickness range: from 15 to 600 μm) of the resulting membranes have been examined by scanning electron microscopy (SEM). These catalytic systems have been applied for the aerobic photo-oxidation of alcohols in aqueous media (Fig. 21.5).[88,89]

The performances of both heterogeneous photocatalysts (PDMS-$(TBA)_4W_{10}O_{32}$ and PVDF-$(TBA)_4W_{10}O_{32}$) have been compared, in terms of the initial reaction rates and product selectivity, with the homogeneous reaction. Despite the faster reactivity of the homogeneous photocatalyst, both membrane-based systems retain a good reactivity when compared to the homogeneous system. Concerning the heterogeneous catalysis, higher product selectivity towards the formation of carbonyl products is observed for substrate conversions lower than 30%,

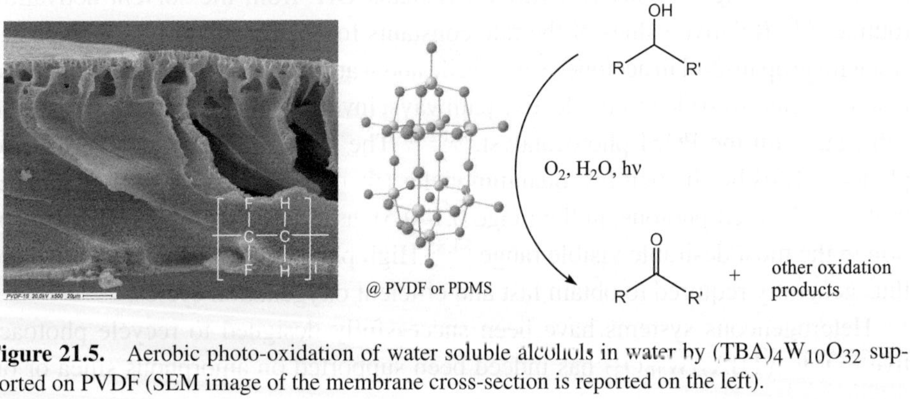


**Figure 21.5.** Aerobic photo-oxidation of water soluble alcohols in water by $(TBA)_4W_{10}O_{32}$ supported on PVDF (SEM image of the membrane cross-section is reported on the left).

before undergoing a consecutive over-oxidation. Interestingly, a structure-dependent reactivity induced by the nature of the polymeric membrane has been observed, affecting reaction rate and selectivity. Both steric hindrance and hydrophobicity seems to play an important role for membrane-induced substrate discrimination, modifying the behavior observed under homogeneous conditions.[88,89]

The possibility of recycling the heterogeneous photocatalyst has been evaluated. A moderate temperature increase (up to 40°C) is needed to improve the extraction of residual products responsible of membrane fouling. The PDMS-$W_{10}$ photocatalyst appears to be the most stable in multiple reaction runs, reaching an activity plateau for catalyst loading of 6 wt%.

To optimize membrane preparation, a fluorous-tagged decatungstate $\{[CF_3(CF_2)_7CH_2CH_2CH_2]_3NCH_3]\}_4W_{10}O_{32}$, $(RfN)_4W_{10}O_{32}$ has been embedded within the fluoropolymeric membrane films. Due to the peculiar nature of C-F bonds, perfluoropolymers exhibit outstanding thermal and oxidative resistance,[92–94] moreover, the preferential permeability of dioxygen in fluorinated membranes can be conveniently used to perform aerobic oxidation reactions.[95]

$(R_fN)_4W_{10}O_{32}$ has been prepared starting from the sodium salt, by counterion metathesis with a trifluoroalkylmethyl ammonium cation obtaining the complex, soluble in neat 1,1,1,3,3,3-hexafluoro-2-propanol (HFIP) or in a mixture with perfluorocarbons. $(R_fN)_4W_{10}O_{32}$ has thus been dispersed in a mixture of HFIP-Galden together with Hyflon®, and embedded in flat hybrid membranes ($(R_fN)_4W_{10}O_{32}$@Hyflon®) by using phase inversion techniques.[96,97]

The structural and spectroscopic features of the photoactive complex are preserved inside the polymeric film, as confirmed by UV-vis spectra. In addition, SEM analysis of both the film surface and cross-section highlight a well-dispersed, homogeneous distribution of the catalyst domains, which appear as spherical particles of approximately 1–2 μm in diameter. The structural similarity of $(R_fN)_4W_{10}O_{32}$ with other surfactant encapsulated polyoxometalates (SEPs) suggests the formation of organized domains, similar to spherical onion-like vesicles.[98–100]

Catalytic tests have been performed on different benzylic substrates, to evaluate the aerobic photo-oxygenation of the reactive C-H bond under atmospheric pressure of dioxygen (Table 21.3). Photo-oxygenation of ethylbenzene, both in homogeneous and heterogeneous conditions, by $(R_fN)_4W_{10}O_{32}$ yields the autooxidation products: ethylbenzene hydroperoxide (EBHP), 1-phenylethanol (PE) and acetophenone (AP), as reported in Fig. 21.6.

The reactivity of $(R_fN)_4W_{10}O_{32}$ has been compared with that of $(TBA)_4W_{10}O_{32}$ in $CH_3CN$. Under homogeneous conditions, both in $CH_3CN$ or in HFIP, the kinetic presents a bell-shaped profile for the hydroperoxide, which is initially formed and then converts to 1-phenylethanol and acetophenone.[101–103] In HFIP, the reaction is more efficient: after 4 h of irradiation ($\lambda > 345$ nm at 500 W) 580 TON has been obtained with a product distribution of 56:23:21 for EBHP, PE and AP, respectively.

**Table 21.3.** Photocatalytic oxygenation of ethylbenzene by decatungstate both in homogeneous and heterogeneous conditions with $(R_fN)_4W_{10}O_{32}$@Hyflon® membrane.

| Entry[a] | Photocatalyst; thickness (μm) | Solvent | Catalyst[b] (μmol) | Products[c] (mM) (% EBHP:PE:AP) | TON[d] |
|---|---|---|---|---|---|
| 1[e] | $(TBA)_4W_{10}O_{32}$ | $CH_3CN$ | 0.20 | 64 (36:32:32) | 351 |
| 2[e] | $(R_fN)_4W_{10}O_{32}$ | HFIP | 0.18 | 95 (56:23:21) | 581 |
| 3 | $(TBA)_4W_{10}O_{32}$@Hyflon; 76[f] | neat | 0.20 | 81 (14:66:20) | 443 |
| 4 | $(R_fN)_4W_{10}O_{32}$@Hyflon; 50 | neat | 0.18 | 196 (25: 41:34) | 1,198 |
| 5 | $(R_fN)_4W_{10}O_{32}$@Hyflon; 7 | neat | 0.03 | 94 (16:46:38) | 3,447 |
| 6 | $(R_fN)_4W_{10}O_{32}$@Hyflon; 94 | neat | 0.70 | 270 (15:48:37) | 424 |

[a]Reaction conditions: 1.1 ml ethylbenzene, $O_2 = 1$ atm, $\lambda > 345$ nm, T = 20°C, irradiation time: 4 h.
[b]Photocatalyst content provided as homogeneous complex or embedded in membranes with 20–25 wt. % loading.
[c]Total oxidation products and % distribution determined by GC-FID and GC-MS analyses.
[d]TON: Turnover number calculated as products (mol)/catalyst (mol).
[e]"Pseudo neat" conditions obtained by addition of 20 μl of solvent.
[f]Porous membrane.

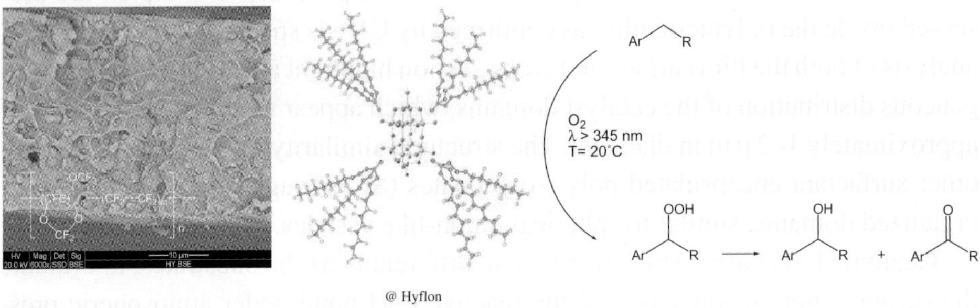


**Figure 21.6.** SEM-BSE image of the cross-section of the catalytic membrane $(R_fN)_4W_{10}O_{32}$@Hyflon®, employed for the aerobic photo-oxidation of alkylbenzene derivatives.

The competitive oxidation of 1-phenylethanol to acetophenone can be decreased by working with excess substrate. The polymeric membrane has been used in solvent-free conditions.[96] For membranes with 25% loading and a cross-section in the range of 7–94 μm, the correlation between film thickness and observed TON has been studied. A steady increase in the total oxidation products has been observed as a function of the overall photocatalyst content, and an inverse correlation between the turnover and the membrane thickness has been observed, revealing a preferential catalytic activity on the surface of the membranes. For membranes with a 7 μm cross-section, 3,447 TON has been obtained after 4 h of irradiation, with a product distribution of 16:46:38 for EBHP, PE and AP, respectively.

The reactivity studies have been extended to other substituted aryls such as tetralin, indane and cumene, that have been oxidized with TONs up to 6,100 and a remarkable 48–53% alcohol selectivity, providing an interesting alternative to other radical-based oxygenation systems (Table 21.4).

Swelling measurements have emphasized the decreased affinity of Hyflon® towards these alcohols, which leads to a decreased formation of over-oxidation products. The use of these hybrid membranes in continuous flow systems with tangential and transmembrane flows is expected to further increase catalytic performances.

A promising strategy to modulate the photoreactivity of polyoxotungstates is the use of multicomponent systems. For example, yields and selectivity for cyclohexene oxidation by irradiated $(TBA)_4W_{10}O_{32}$ have been improved in the presence of $Fe^{III}$(Cl)TDCPP (TDCPP = tetra(2,6-dichlorophenyl)porphyrin). The presence of iron porphyrin induces an increase of quantum yield to give 1.6 ketone to alcohol ratio (instead of 4.1 with the decatungstate alone), which has been ascribed to its

**Table 21.4.** Photocatalytic oxygenation of substituted aryls in the presence of $(R_fN)_4W_{10}O_{32}$@Hyflon® membrane.

| Substrate[a] | Products (mM)[b] (% HP:A:K)[c] | TON |
|---|---|---|
| 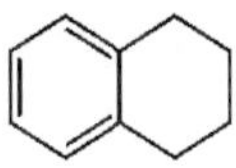 | 167<br>14: 58: 6[d] | 6,124 |
| 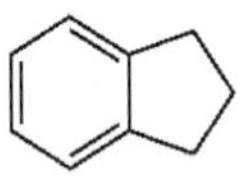 | 165<br>31: 48: 10[d] | 6,051 |
| 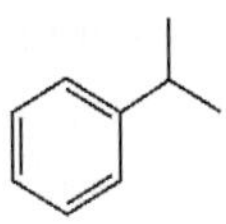 | 91<br>47:-: 53 | 3,337 |
| 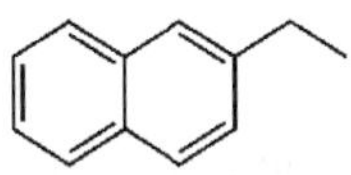 | 30<br>21:26: 53 | 1,100 |

[a]Reaction conditions: substrate 1.1 ml, $pO_2 = 1$ atm, $\lambda > 345$ nm, T = 20°C, irradiation time: 4 h, heterogeneous catalyst: $(R_fN)_4W_{10}O_{32}$@Hyflon® (7 μm, 25 wt. % catalytic loading).
[b]mM concentration of total oxidation products.
[c]% distribution of benzylic oxidation products: hydroperoxide (HP), alcohol (A), ketone (K), as determined by GC-FID.
[d]10–20% of dimerization products have also been observed.

ability to decompose allylic hydroperoxides to give the corresponding alcohols. As far as porphyrin stability is concerned, no appreciable bleaching of its UV-vis spectrum has been observed.[104]

A combined catalytic system POM/$TiO_2$ has been used for the direct oxidation of benzene to phenol. For this reaction, selectivity is a crucial parameter: phenol is much more reactive than benzene, leading to the formation of several over-oxidation products (such as cathecol, hydroquinone, benzoquinone isomers and tars). When using a $H_3PW_{12}O_{40}/TiO_2$ in $H_2O/CH_3CN$ 0.02:1 (v/v), phenol was obtained in 4 h with 11% yield and the minor formation of cathecol and hydroquinone.[105]

Polyoxometalates have been used as a means to heterogenize cationic organic sensitizers, such as methylene blue and tris(2,2'-bipyridine)ruthenium(II) ($[Ru(bpy)_3]^{2+}$), by means of electrostatic interactions.[106] The activity of such a hybrid heterogeneous photocatalyst has been assessed in water, using visible light ($\lambda > 375$ nm) and dioxygen (1 atm). With the complex $[Ru(bpy)_3]_2W_{10}O_{32}$, 84% conversion of phenol has been obtained in 150 minutes with TON = 45. The extensive degradation of the substrate has been highlighted by a loss of chemical oxygen demand (COD) of 29%.[107]

Another interesting class of photoactive POM-based hybrids has been obtained by covalent grafting of fulleropyrrolidine derivatives on the divacant decatungstosilicate $[\gamma\text{-}SiW_{10}O_{36}]^{8-}$. The hybrid combination of fullerene derivatives with a POM moiety provides a rigid platform to tune the physico-chemical behavior as a function of the inorganic component and of its counterion, improving the complexes dispersibility in aqueous solutions while avoiding hydrophobic aggregation of the fullerenic moieties. N-[3-(triethoxysilyl)propyl]-2-carbomethoxy-3,4-fulleropyrrolidine,[108] and N-methyl-2-[10-(triethoxysilyl)decyl]-3,4-fulleropyrrolidine,[109] bearing a silicon-alkoxide moiety, have been used to graft the [60]-fullerene sensitizer on a $[\gamma\text{-}SiW_{10}O_{36}]^{8-}$ surface. The activity of the hybrid photocatalysts has been assessed in water at 25°C, upon irradiation with $\lambda > 375$ nm, in the presence of dioxygen (1 atm), performing photo-oxygenation reactions of aqueous phenol solution at pH = 10.5.[107] Phenol oxidation occurs with a conversion of 90% in 150 min, and a 30% decrease in the COD content. 1,4-Benzoquinone is the only detectable intermediate observed during photo-oxidation, mainly involving $^1O_2$ as the active oxidant.[62,110] The photocatalytic behavior of the hybrid complexes has also been evaluated in the selective photo-oxygenation of the L-methionine methyl ester to the corresponding sulfoxide (Fig. 21.7). The reaction occurs quantitatively in 90 min with TONs up to 200.[107]

### 21.2.1.2. *Hydrogen peroxide activation by polyoxometalates*

As dioxygen, hydrogen peroxide can be considered a sustainable oxidant because of its high atom efficiency, oxygen content and the formation of water as the only

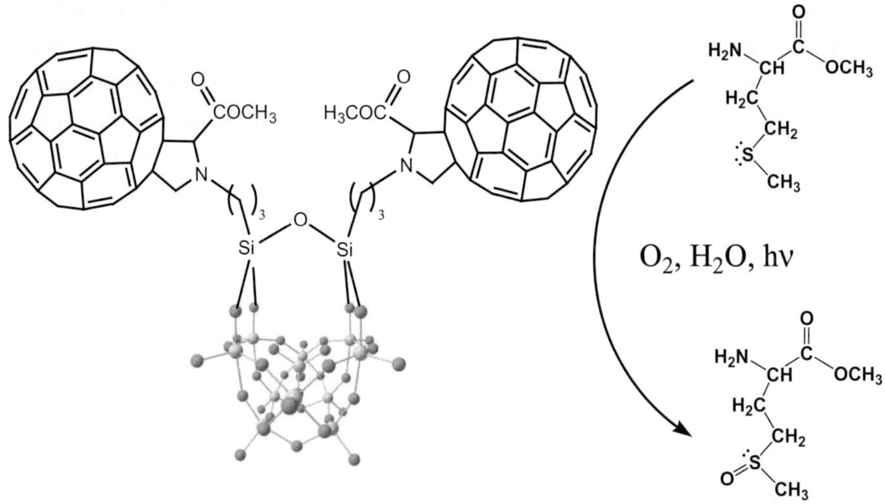


**Figure 21.7.** Aerobic photo-oxygenation of L-methionine methyl ester in $H_2O$ by the fulleropyrrolidine-POM conjugate.

by-product. Among the oxidative processes with hydrogen peroxide, those catalyzed by high valent $d^0$ transition metals are between the most important and selective.[111] In this respect, POMs are suitable candidates for the activation of hydrogen peroxide aimed at performing heterolytic oxidations. The activation of $H_2O_2$ by transition metals occurs through the generation of metal-peroxo complexes with different coordination modes ($\eta^2$–$O_2$, $\mu$–$\eta^1$: $\eta^1$–$O_2$, $\mu$–$\eta^1$: $\eta^2$–$O_2$, $\mu$–$\eta^2$: $\eta^2$–$O_2$, OOH, etc.). All these active oxygenated species may play an important role in the various oxidative transformations of organic substrates.

Several research groups have studied the interaction between vacant, TMSP or their parent Keggin polyanions and hydrogen peroxide.[112–119] In some cases, alkyhydroperoxides have also been used.[120]

$[PM_{12}O_{40}]^{3-}$ (M=Mo(VI), W(VI)), associated with suitable ammonium salts such as cetylpyridinium chloride have been used, in biphasic conditions using chloroform as the solvent, to obtain the epoxidation of terminal/internal olefins and enols in 2–24 h at 60°C with yields up to 98%.

The isolated tris(cetylpyridinium) salt of $[PW_{12}O_{40}]^{3-}$ efficiently catalyzes the ketonization of the secondary hydroxy group of alcohols and diols with $H_2O_2$ under homogeneous conditions, using tert-butyl alcohol as the solvent under refluxing conditions, in 24 h, with yields up to 98%. In the same experimental conditions, oxidative cleavage of *vic*-diols and carbon-carbon double bonds gives carboxylic acids with good yields.[121] The oxidant species in solution are dimeric peroxotungstate complexes such as $\{PO_4[MO(O_2)_2]_4\}^{3-}$.[122–125]

In contrast to monomeric or dimeric peroxo species, stable polynuclear peroxo species are expected to show specific reactivity and selectivity because of their electronic and structural features. Activation of hydrogen peroxide with $(R_4N)_4W_{10}O_{32}$, where R=n-$C_{16}H_{33}(CH_3)_3$-, has led to the selective oxidation of aliphatic alcohols to the corresponding carboxylic acids and ketones. In refluxing aqueous $H_2O_2$ solution, cyclohexanol has been converted to cyclohexanone in 2 h with 89% yield and 97% selectivity, while 1-hexanol and 1-octanol have been oxidized to the corresponding carboxylic acids in 12–13 h with yields of 87–89% and selectivities >93%.[126]

Promising oxidation catalysts belong to the vacant polyoxotungstates family: the monovacant Keggin polyanion $[CoW_{11}O_{39}]^{9-}$, grafted with four peroxo moieties to give $[\beta_3\text{-}CoW_{11}O_{35}(O_2)_4]^{10-}$, has been isolated, characterized in the solid state by X-ray crystallography and used to epoxidize 2-cyclohexenol with hydrogen peroxide in a biphasic system.[127]

It has been reported that $Na_9[SbW_9O_{33}]$, in conjunction with methyltricaprylammonium chloride as the phase transfer agent, is a highly efficient catalytic system for the selective epoxidation of alkenes with aqueous $H_2O_2$ in solvent-free conditions: yields up to 99% were obtained in 4–9 h, at 35 or 60°C.[128]

$K_8[HBW_{11}O_{39}]\cdot 13H_2O$ has been used as a precatalyst for the oxidation of alcohols in water at 90°C with $H_2O_2$ as the oxidant. Secondary alcohols such as cyclohexanol have been converted (99%) in 3 h to the corresponding ketones and separated from the aqueous phase by extraction. Due to the occurrence of an alcoholic layered phase, benzyl alcohol has been oxidized in 98% conversion and 83% selectivity, without extensive over-oxidation to benzoic acid.[129]

$[(C_{18}H_{37})_2(CH_3)_2N]_{10}[SiW_9O_{34}]$ has been used in water at 65°C as an effective and recyclable catalyst for alcohol oxidation with hydrogen peroxide. Secondary alcohols have been oxidized to ketones in 6 h with quantitative yields. For the oxidation of 2-octanol, the catalyst has been isolated by extraction with diethyl ether and recycled up to four times with no appreciable loss of activity and selectivity towards ketone formation.[130]

The tetraprotonated divacant silicodecatungstate, $(TBA)_4[\gamma\text{-}SiW_{10}O_{34}(H_2O)_2]$, has been employed to catalyze oxygen-transfer reactions on various substrates including olefins, allylic alcohols and sulfides in the presence of 30% aqueous $H_2O_2$. As shown in Table 21.5, terminal olefins have been oxidized with 90% yield (calculated on the limiting reagent $H_2O_2$) in 10 h, while cis and cyclic internal olefins have been oxidized in <6 h with yields up to 99%.[131–133]

For the epoxidation of cis-and trans-2-octenes, the configuration of the double bond has been retained in the corresponding epoxides. In competitive reactions, cis-2-octene has been oxygenated much faster than the trans isomer, with a cis/trans ratio of 11.5:1. This value is higher than those obtained for the other tungstate-$H_2O_2$

**Table 21.5.** Olefin epoxidation catalyzed by $(TBA)_4[\gamma\text{-}SiW_{10}O_{34}(H_2O)_2]^{4-}$ with $H_2O_2$ as the oxidant.

| Entry[a] | Substrate | Product | Selectivity (%) | T (h) | Yield (%) | $H_2O_2$ conversion (%) |
|---|---|---|---|---|---|---|
| 1 | 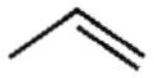 | 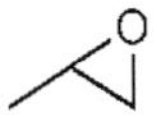 | > 99 | 8 | 90 | >99 |
| 2 | 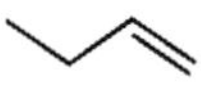 | 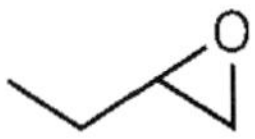 | 99 | 8 | 88 | >99 |
| 3 | 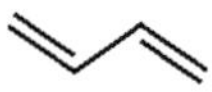 | 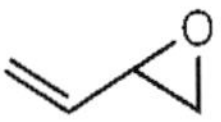 | 99 | 9 | 91 | 99 |
| 4 |  | 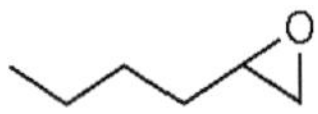 | 99 | 10 | 90 | >99 |
| 5 | 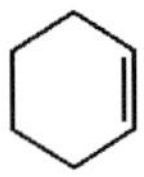 | 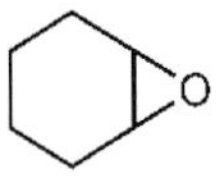 | >99 | 6 | 84 | > 99 |

[a]Reaction conditions: substrate (5 mmol); $(TBA)_4[\gamma\text{-}SiW_{10}O_{34}(H_2O)_2]^{4-}$ (8 μmol); $H_2O_2$ (1 mmol) in 6 ml of $CH_3CN$; T = 32°C.

systems and for the stoichiometric epoxidation with organic oxidants,[134,135] and suggests the contribution of a structurally rigid, non-radical oxidant generated on $(TBA)_4[\gamma\text{-}SiW_{10}O_{34}(H_2O)_2]$.

Non-conjugated dienes, such as 1-methyl-1,4-cyclohexadiene, were epoxidized at the less hindered, even if less nucleophilic, double bond with a high regioselectivity (up to 89%)(Table 21.6). Unsaturated alcohols, such as geraniol, have been mainly epoxidized at the allylic position, giving only small amounts of α, β-unsaturated aldehydes.

Mechanistic and theoretical studies have been performed to address the role of protons on the vacant sites of $(TBA)_4[\gamma\text{-}SiW_{10}O_{34}(H_2O)_2]$, which seem to foster the formation and the reactivity of peroxotungstate groups as active species.[136,137,138] Convergent results suggest the occurrence of two terminal W-$(OH_2)$ (aquo ligand) fragments on the POM surface.[139] The formation of peroxo species, where the two aquo ligands W-$(OH_2)$ are replaced by two W$(O_2)$ groups, has been observed by positive-ion cold spray ionization mass spectrometry (CSI-MS) of the isolated intermediate.[133]

The structural stability of $(TBA)_4[\gamma\text{-}SiW_{10}O_{34}(H_2O)_2]$ has been assessed by *in situ* FTIR spectroscopy. Moreover, the kinetic studies have revealed a first-order dependence of the reaction rate on the concentration of the catalyst, ruling out the occurrence of smaller tungstate fragments.

**Table 21.6.** Olefin epoxidation catalyzed by $(TBA)_4[\gamma\text{-}SiW_{10}O_{34}(H_2O)_2]^{4-}$ with $H_2O_2$ as the oxidant.

| Entry[a] | Substrate | Products | t (h) | Yield (%) | $H_2O_2$ conversion (%) |
|---|---|---|---|---|---|
| 1 | 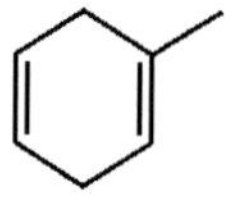 | 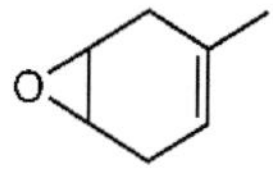 | 4 | 83 | >99 |
| | | 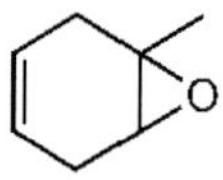 | | 10 | |
| 2 | 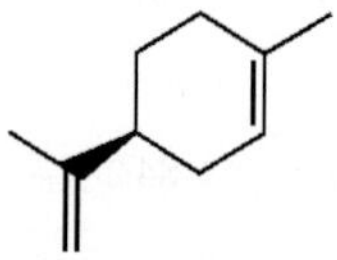 | 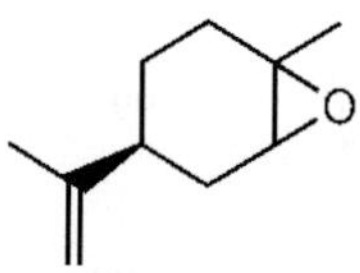 | 7 | 34 | 93 |
| | | 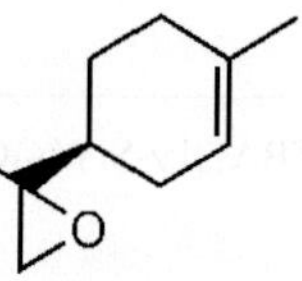 | | 55 | |
| 3 | 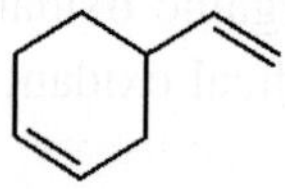 | 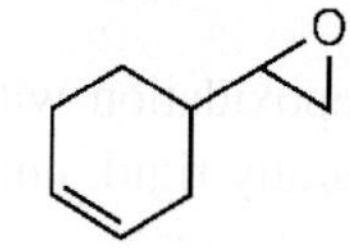 | 4 | 4 | 81 |
| | | 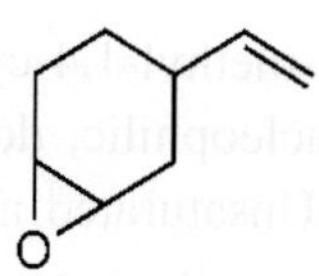 | | 94 | |
| 4 | 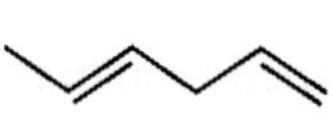 | 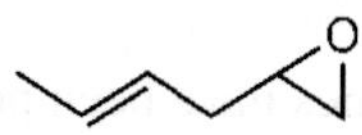 | 17 | 43 | 98 |
| | | 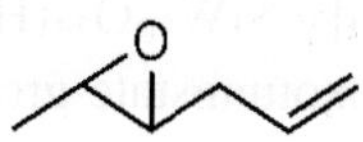 | | 27 | |

[a]Reaction conditions: substrate (5 mmol); $(TBA)_4[\gamma\text{-}SiW_{10}O_{34}(H_2O)_2]^{4-}$ (8 μmol); $H_2O_2$ (1 mmol) in 6 ml of $CH_3CN$; T = 32°C.

Another important approach in obtaining POM-based catalysts is the complementary assembly of organic and inorganic molecular components: the merging of different domains may lead to synergistic effects with the ultimate scope to improve the catalytic performance.[16,17] The hybrids can be prepared using straightforward literature protocols, exploiting the nucleophilicity of the coordinatively unsaturated oxygen atoms on vacant POMs which are able to react with electrophilic organic moieties, such as organosilanes or organophosphonic reagents.[140–142]

Among the catalysts employed, the hybrid polyoxoanion $[(PhPO)_2SiW_{10}O_{36}]^{4-}$ has been used to perform the oxidation of several classes of substrates in acetonitrile (Fig. 21.8 and Table 21.7).[143] The functionalization of the vacant site prevents

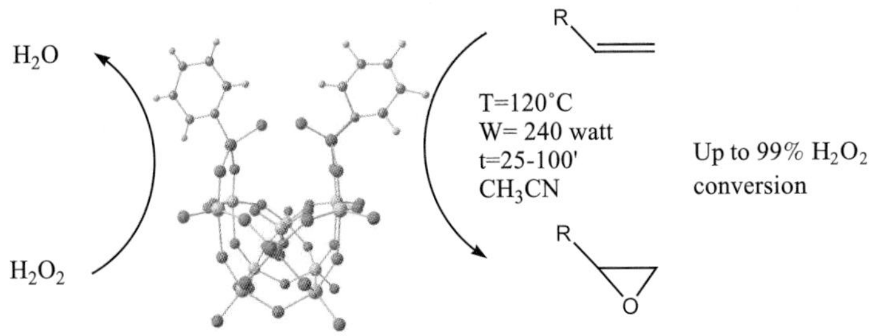


**Figure 21.8.** Catalytic epoxidation of olefins by $(TBA)_4[(PhPO)_2SiW_{10}O_{36}]$ with $H_2O_2$ in $CH_3CN$.

**Table 21.7.** MW-assisted oxidation with $H_2O_2$ catalyzed by $(TBA)_4[(PhPO)_2SiW_{10}O_{36}]$.[a]

| Entry[a] | Substrate | Product | t (min) | Yield (%)[b] |
|---|---|---|---|---|
| 1 | 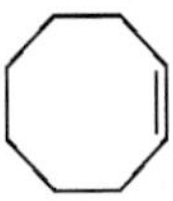 | 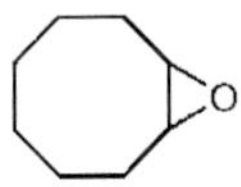 | 50 | 97 |
| 2[c] |  | 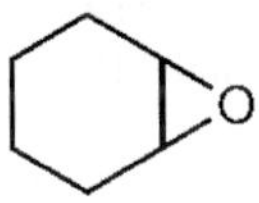 | 25 | 99 |
| 3 | 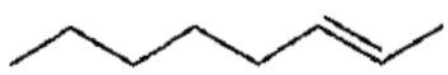 | 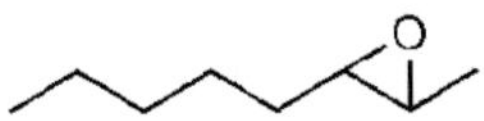 | 25 | 99 |
| 4[d] | 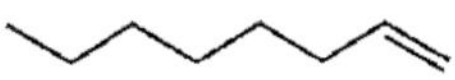 | 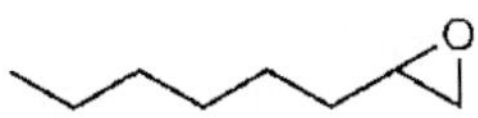 | 50 | 79[e] |
| 5[d] | 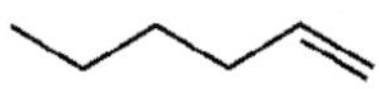 | 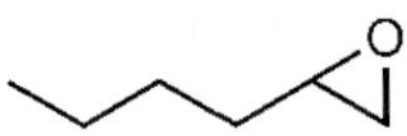 | 50 | 50[f] |
| 6 | 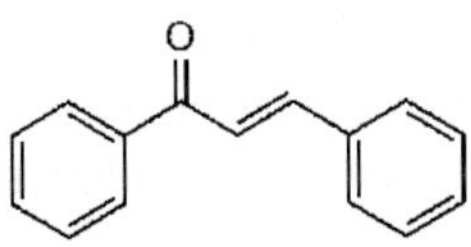 | 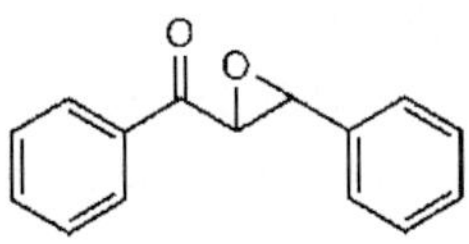 | 50 | 29[g] |
| 7 | 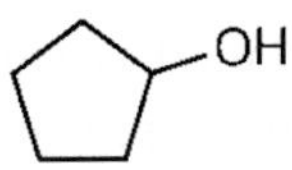 | 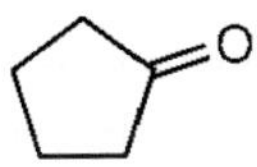 | 50 | >99 |
| 8 | | 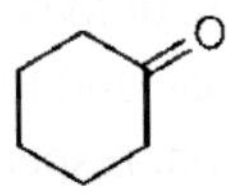 | 50 | 95 |

(*Continued*)

**Table 21.7.** (*Continued*)

| Entry[a] | Substrate | Product | t (min) | Yield (%)[b] |
|---|---|---|---|---|
| 9 |  | 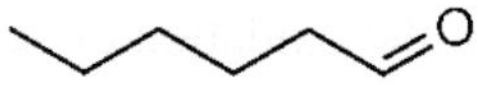 | 50 | 59[h] |
| 10 | 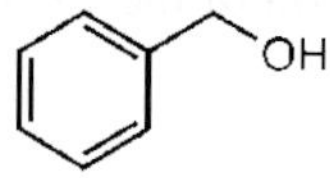 | 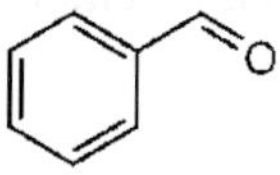 | 10 | >99 |
| 11[d] | 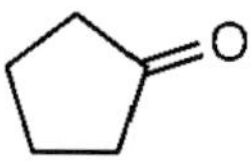 | 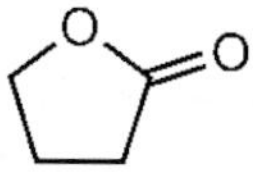 | 50 | 62[g,i] |
| 12 | 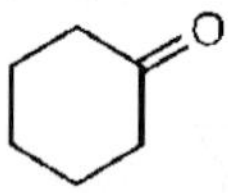 | 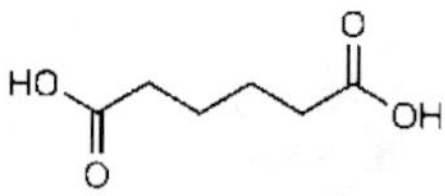 | 100 | 75[g,j] |
| 13 | 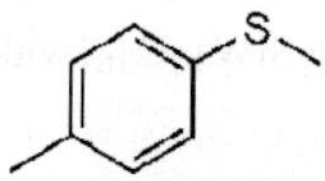 | 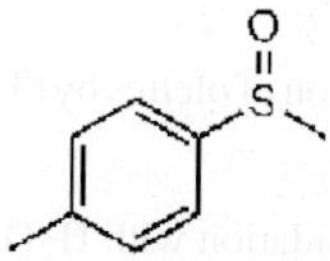 | 25 | >99 |
| 14 | 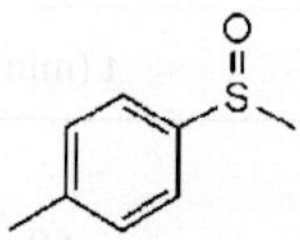 | 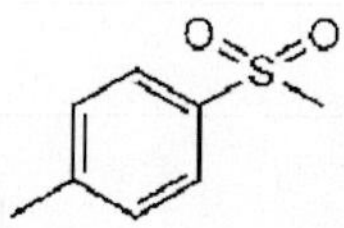 | 25 | >99 |

[a]$(TBA)_4[(PhPO)_2SiW_{10}O_{36}]$ (0.8 μmol), substrate (0.5 mmol), 35% $H_2O_2$ (0.1 mmol), $CH_3CN$ (0.6 mL), $T_{bulk}$ = 90–120°C; MW irradiation at 180–240 W.
[b]Yields calculated with respect to initial $H_2O_2$.
[c]70% $H_2O_2$ (0.1 mmol).
[d]Catalyst (2.4 μmol), substrate (1.5 mmol), 70% $H_2O_2$ (0.3 mmol).
[e]Heptanoic acid formed as by-product.
[f]α-Hydroxy hexanoic and pentanoic acids formed as by-products.
[g]$H_2O_2$ partially decomposed.
[h]Pentanoic acid formed as by-product.
[i]The uncatalyzed reaction proceeds with 27% yield.
[j]Catalyst (2.4 μmol), substrate (0.5 mmol), 70% $H_2O_2$ (1.5 mmol).

the rearrangement of the POM structure: stability studies by means of heteronuclear magnetic resonance ($^{31}$P NMR) and electro spray ionization mass spectrometry (ESI-MS) analyses have revealed that the complex is stable in oxidative conditions, even under MW irradiation.

The best catalytic performances have been observed for the oxidation of internal olefins, secondary benzylic alcohols and organic sulfur compounds, with >95% $H_2O_2$ conversion after 10–50 min of MW irradiation. Oxygen transfer to the electron deficient double bond of chalcone and to cyclic ketones (Baeyer–Villiger reaction) has given 29 and 62% conversion, respectively, after 50 min.

A mechanistic study has revealed a weak electrophilic character for the competent oxidant, turning to a biphilic nature with respect to sulfoxide oxidation.[144] This behavior is likely to be stimulated by the polynuclear $W^{VI}$ framework, which may foster a dual activation of both the oxidant and the sulfoxide by coordination to proximal Lewis acid sites on the polyoxotungstate surface. The postulated asymmetric binding of the peroxo-ligand, not evolving to a bidentate $\eta^2$-coordination mode, could explain the atypical selectivity behavior, with respect to classical $d^0\eta^2$-peroxometal complexes.[145,146]

The hybrid organic-inorganic catalytic complex $[(PhPO)_2SiW_{10}O_{36}]^{4-}$ has been used in ionic liquids (ILs), as an alternative reaction media to replace hazardous volatile organic solvents (VOCs).[147] The IL acts as a solvating/immobilization medium for the hybrid polyanion. The selective epoxidation of cis-cyclooctene, with quantitative conversion of $H_2O_2$ (up to >99%) is achieved at 50°C, in hydrophobic ILs as $[bmim^+][(CF_3SO_2)_2N^-]$ and $[bmim^+][PF_6^-]$, in 45 and 90 min respectively. A good recycling performance has been obtained by extracting the spent catalytic phase (IL + POM) with hexane and water, followed by vacuum dehydration over $P_2O_5$. Upon this treatment the epoxidation product has been recovered in quantitative yields for at least four consecutive runs (total TON = 500). The polyelectrolytic nature of the catalytic phase (hybrid-POM+IL) guarantees negligible vapor pressure, as well as fast and selective MW-induced heating by ionic conduction mechanism, even at low irradiation power within a monomodal MW oven (4–10 watt). Continuous MW-irradiation has been applied with simultaneous cooling to prevent bulk overheating ($T_{bulk}$ < 80°C). Under the conditions explored, quantitative epoxidation of cis–cyclooctene has occured in 1 min, with TOF > 200 $min^{-1}$, i.e. 35 times higher than the reaction with conventional heating.

Other internal and terminal olefins have been efficiently epoxidized with 99% selectivity (Table 21.8). Substituted olefins have been converted into the corresponding epoxides (>97% yield, calculated on the limiting reagent $H_2O_2$) in 15–60 min. High epoxide yields have also been achieved for less reactive terminal olefins (up to 99% in 2–3 h). The combined use of POMs, ionic liquids and MW irradiation thus represents a convenient strategy for the immobilization, recycling and activation of the catalytic phase within a solvent-free reaction set-up.

The divacant Keggin-type polyoxotungstate $[\gamma\text{-}SiW_{10}O_{36}]^{8-}$ has also been decorated with chiral phosphonic groups, yielding the bis-functionalized hybrid $[\gamma\text{-}SiW_{10}O_{36}(R^*PO)_2]^{4-}$ with R* = N-protected aminoalkyl groups or O-protected amino acid derivatives. The chirality transfer at the POM level has been highlighted by the chiroptical solution behavior of the systems, which shows distinct Cotton effects up to 400 nm.[19] Oxidation of pro-chiral methyl *p*-tolylsulfide with $H_2O_2$ has been used to screen the potential of the chiral POM-phosphonates as catalysts for enantioselective oxygen transfer. The reactions have been performed between 0°C

**Table 21.8.** Catalytic epoxidation by $(TBA)_4[(PhPO)_2SiW_{10}O_{36}]$ with aqueous $H_2O_2$ in $[bmim^+][(CF_3SO_2)_2N^-]$ under MW irradiation.[a]

| Entry | Substrate | t (h) | Epoxide yield (%)[b] |
|---|---|---|---|
| 1 | 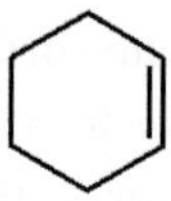 | 0.25 | >99 |
| 2 |  | 0.75 | >99 |
| 3 | 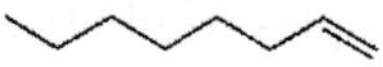 | 3 | 54 |
| 4 |  | 2 | 99 |
| 5 | 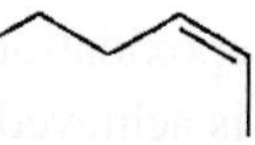 | 0.5 | 97 |
| 6 |  | 1 | 99 |
| 7 | 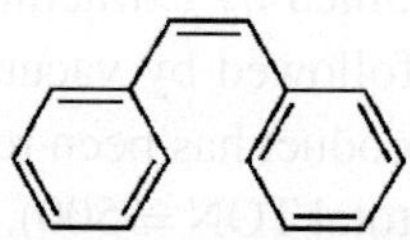 | 0.5[c] | 44 |

[a]Reactions performed in the presence of a layered hydrocarbon phase (2.0 mmol), $H_2O_2$ (40 μmol), provided by 24 M aqueous solution, $(TBA)_4[(PhPO)_2SiW_{10}O_{36}]$ (3.2 μmol), 200 μL of $[bmim^+][(CF_3SO_2)_2N^-]$, MW irradiation at 5 W, under stirring and simultaneous cooling, $T_{bulk} = 57$–65°C.
[b]Yields calculated with respect to $H_2O_2$.
[c]MW irradiation at 4 W, $T_{bulk} = 58$–73°C.

and −10°C, in the presence of excess sulfide. Oxidation has occurred in 24–72 h, with yields of up to 75%. Even if the enantioselectivity is very scarce, the oxidative activity and stability of the chiral complexes offer good premises for the development of novel nanodimensional stereoselective catalysts. Interesting developments in this field are expected upon association with chiral cations and/or polycations.[148,149]

The TMSPs can be exploited to obtain $H_2O_2$ activation, with M' playing different roles, depending on its nature. Isostructural TMSP $(TBA)_n[PXW_{11}O_{39}]^{n-}$ (X = Ni, Co, Cu, n = 5; X = Fe, n = 4 have been tested observing either extensive decomposition of the oxidant or formation of unidentified products.[150] There are also several examples of TMSPs featuring an interesting selectivity. Sandwich-like complexes containing Fe(III),[119] Zn(II),[151] Pt(II), Pd(II), Ru(III),[152] Rh(III),[153] Mn(II),[154] have been employed for the epoxidation of olefins and allyl alcohols, showing little dependence on the nature of the transition metal, thus suggesting the occurrence of W-peroxo groups. Formation of peroxotungstic groups have indeed been observed by FTIR and heteronuclear magnetic resonance ($^{183}$W-NMR).

$[ZnWM_2(H_2O)_2(ZnW_9O_{34})_2]^{n-}$ with M=Mn(II), Ru(III), Fe(III), Pd(II), Pt(II), Zn(II), in particular, have shown to be very efficient for the diastereoselective epoxidation of chiral allyl alcohols in a biphasic system DCE/$H_2O$ under mild conditions.[154] The Zn(II) derivative, which can be self-assembled in aqueous solution starting from simple precursors, has been used to oxidize different primary and secondary alcohols in water, in 7 h at 85°C, with yields and selectivities of up to 99%.[155] Upon cation exchange with $[(CF_3(CF_2)_7CH_2CH_2CH_2)_3NCH_3]^+$, the Mn(II) and Zn(II) substituted POMs have also been used in biphasic systems containing fluorinated solvents for olefins epoxidation and alcohol/enol oxidation to the corresponding ketones, with yields and selectivities up to 99% in 8–13 h at 22–88°C.[156]

$Na_4H_3[SiW_9Al_3(H_2O)_3O_{37}]\cdot 12(H_2O)$ has been used for the selective oxidation of secondary and benzylic alcohols to carbonylic products (up to 99% yields in 7–12 h at 90°C) using $H_2O_2$as the oxidant in acetonitrile or solvent-free conditions.[157]

The reaction between $(TBA)_4[\gamma\text{-}SiW_{10}O_{34}(H_2O)_2]$ and $Zn(acac)_2$ has given $TBA_8[\{Zn(OH_2)(\mu_3\text{-}OH)\}_2\{Zn\text{-}OH_2)_2\}_2\{\gamma\text{-}HSiW_{10}O_{36}\}_2]\cdot 9H_2O$. This complex selectively oxidizes secondary alcohols in the presence of primary hydroxyl groups and allylic double bonds, to give the corresponding ketones, in 1–18 h at 56°C in acetone, with > 78% yield. In analogue conditions, the parent POM and other tungsten-based catalysts have preferentially given the epoxidation of cyclohexenol.[158] $Na_6[H_2ZnSiW_{11}O_{40}]\cdot 12H_2O$ has been used to quantitatively oxidize alcohols and enols to ketones in alcohol/water biphasic systems, at 90°C in 2–9 h.[159]

A polyfluorooxometalate complex, $[Ni(H_2O)NaH_2W_{17}O_{55}F_6]^{9-}$ has been employed in the epoxidation reactions with hydrogen peroxide. This Ni-substituted TMSP has been used as a catalyst for the oxidation of cis-cyclohexene in a two-phase system of DCE/$H_2O$ at 60°C obtaining the corresponding epoxide with >99% selectivity and quantitative yield. The stability of the complex has been assessed by FTIR, atomic absorption and heteronuclear magnetic resonance ($^{19}$F-NMR) measurements. Formation of the intermediate oxo/peroxo species has been observed by FTIR analysis.[160]

The use of fourth or fifth group $d^0$ transition metals may be very useful to drive the formation of reactive peroxo ligands on such metals. This could be particularly convenient to increase selectivity through well-defined mechanistic pathways.

The complex $(TBA)_4[\gamma\text{-}H_2SiV_2W_{10}O_{40}]$, containing a $\{VO\text{-}(\mu\text{-}OH)_2\text{-}VO\}$ fragment, has been used in the presence of a stoichiometric amount of $H_2O_2$ to epoxidize non-activated aliphatic terminal olefins including propylene, that could be transformed into the corresponding epoxide with 99% selectivity and 87% efficiency of $H_2O_2$ utilization in $CH_3CN$/t-BuOH at 20°C, in 24 h. More accessible, but less nucleophilic double bonds in non-conjugated dienes have been regioselectively epoxidized in high yields.[161,162]

Zr- and Ti-based POMs have been suggested as molecular models for the well-known heterogeneous titanium silicalite catalyst, TS-1, with major industrial appeal in the field of $H_2O_2$ activation and selective oxidation.[163,164,165] In some cases, POMs bearing stable peroxometal groups have been isolated and characterized.[166,167] $(TBA)_4[PTi(OH)W_{11}O_{39}]$ presents homolytic reactivity towards olefins, sulfides and alkylphenols, while $[Ti_2(OH)_2As_2W_{19}O_{67}(H_2O)]^{8-}$, containing two square-pyramidal $[Ti(OH)]^{3+}$groups, exhibits heterolytic reactivity. Furthermore, for the oxidation of alkenes, the reaction selectivity is strongly affected by the protonation of the POM, which can be controlled by changing pH during the precipitation of the TBA salt. Cyclohexene has been epoxidized with 80% conversion and 81% selectivity, while diols, α-hydroxyketones, and C-C bond-cleavage products can be obtained by using the diprotonated dititanium 19-tungstodiarsenate.[163–165]

$Zr^{IV}$ and $Hf^{IV}$ ions present a larger ionic radius, and may provide accessible and adjacent sites for both substrate coordination and peroxide activation at the metal center.[168] Heteropolyoxotungstates incorporating butterfly-type $Zr^{IV}$ and $Hf^{IV}$ peroxides with a mono- or bis-$\mu - \eta^2$:$\eta^2$ arrangement of the peroxo ligand for each Zr/Hf substituent have been isolated.[169,170] Such a structural feature is shared by a family of three isostructural dimers, $[\{M(O_2)(\alpha\text{-}XW_{11}O_{39})\}_2]^{12-}$ (M = Zr, X = Si, Ge; M = Hf, X = Si). Different to the generally found inertness of related classical coordination compounds, POM ligands boost the reactivity of the coordinated Zr/Hf peroxides.[171] The reactivity of these TMSPs has been assessed for oxygen transfer to L-methionine in aqueous solution, yielding the corresponding sulfoxide and sulfone (Fig. 21.9).

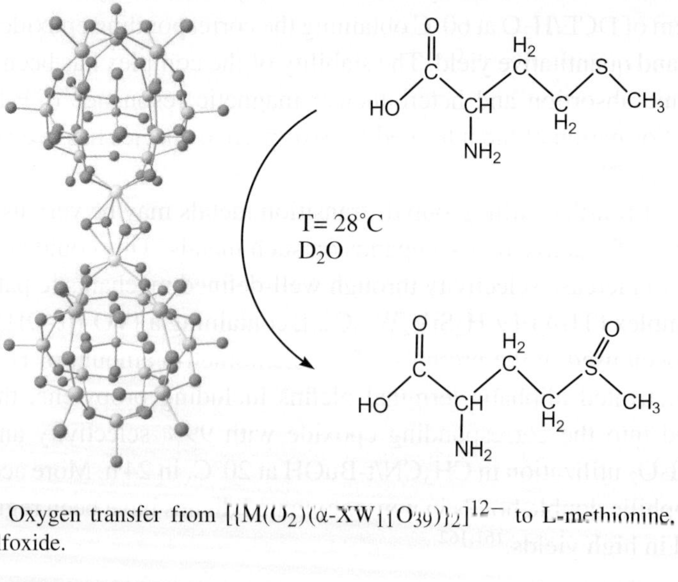


**Figure 21.9.** Oxygen transfer from $[\{M(O_2)(\alpha\text{-}XW_{11}O_{39})\}_2]^{12-}$ to L-methionine, to give DL-methionine sulfoxide.

At neutral pH, in stoichiometric conditions, both Hf and Zr complexes display a remarkable activity towards sulfide oxidation, yielding quantitative oxygen transfer to the electron-rich sulfur atom, with quantitative formation of sulfoxide within a few seconds. The stability of the spent peroxo-POMs was confirmed by FTIR, UV-vis and Raman analysis after completion of the reaction. The access to catalytic conditions was evaluated upon recharging the spent reaction mixtures with a supplementary stoichiometric amount of both $H_2O_2$ and substrate. An immediate restoring of the reactivity has been observed, with analogue kinetic behavior, confirming the nature of the catalytic site.

Catalytic oxidation of L-methionine and DL-methionine sulfoxide has thus been successfully achieved in the presence of excess $H_2O_2$, 70–99% yields have been obtained over 20–48 h. For the TMSP $\{[Zr(O_2)(\alpha\text{-}GeW_{11}O_{39})]_2\}^{12-}$, the reaction scope has been extended to alcohol oxidation in aqueous solutions through a MW-assisted protocol (Table 21.9).[172]

**Table 21.9.** Catalytic oxidation of alcohols by $\{[Zr(O_2)(\alpha\text{-}GeW_{11}O_{39})]_2\}^{12}$ and $H_2O_2$ in aqueous solution.[a]

| Entry | Substrate | Product | Conversion (%)[b] |
|---|---|---|---|
| 1 | 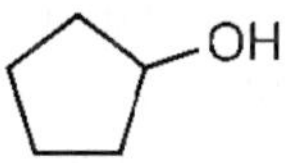 | 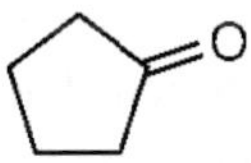 | 99 |
| 2 | 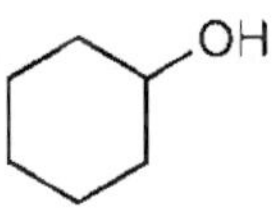 | 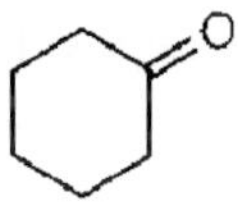 | 99 |
| 3 | 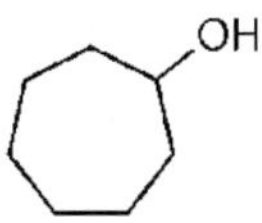 | 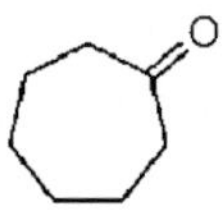 | 99 |
| 4 | 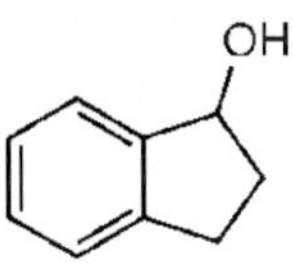 | 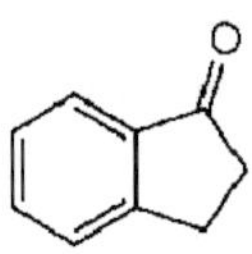 | 99 |
| 5 | 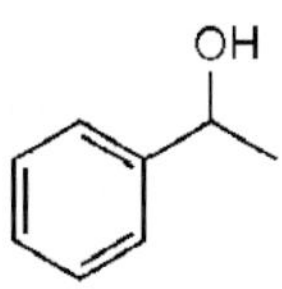 | 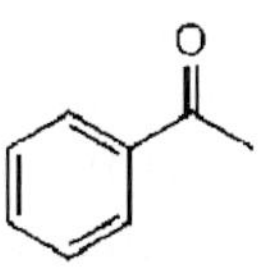 | 80 |

[a]Reaction conditions: alcohol 0.8–1.0 mmol, POM 1.6 μmol, $H_2O_2$ 0.08–0.1 mmol in $H_2O$ (500 μl); irradiation power 70 W, time = 50 minutes, $T_{bulk}$ = 90°C, with applied compressed air.
[b]% conversion calculated with respect to the limiting reagent $H_2O_2$.

Selective oxidation of benzylic and cyclic substrates (99% $H_2O_2$ conversion and 80–99% selectivity) has been achieved upon an irradiation power of 70 W. The present protocol represents a further advancement with respect to on-water catalysis, and environmentally sustainable oxidations.[173]

The impact of the geometry/structure of the peroxide bridge on oxygen transfer catalysis, has been addressed by density functional theory (DFT) computations, including relativistic and solvent effects, highlighting a high electron density localized on the peroxidic moieties and on W-O-M bridging oxygens,[138,139] providing insights on the role of the highest occupied molecular orbital (HOMO) energy to control proton affinity of the peroxo group, which is likely responsible for triggering the electrophilic reactivity of peroxides.[167,174]

### 21.2.2. *Miscellaneous oxygen donors for POM catalyzed oxidations*

Polyoxometalates have been used in the presence of different oxidants, confirming their versatility in terms of reactivity and the possibility of accessing diverse activation mechanisms.

Nitrous oxide ($N_2O$) is an interesting oxidant for the selective oxidation of organic substrates since it contains 36 wt. % oxygen, and the by-product of an oxidation reaction is molecular nitrogen. These advantages are difficult to realize since $N_2O$ is generally considered to be scarcely reactive. Nevertheless, $[Mn_2^{III}ZnW(ZnW_9O_{34})_2]^{10-}$ catalyzed the selective epoxidation of alkenes under 1 atm $N_2O$ in fluorobenzene at 150°C, with up to 25% yield.[175]

Primary and secondary alcohols have been selectively oxidized to carbonyl products with a yield of 42–89% by $(TBA)_5[PV_2Mo_{10}O_{40}]$ with $N_2O$ (1 atm), in benzonitrile, at 150°C in 15 h, while alkyl aromatics have been oxygenated at the benzylic C-H bonds with up to 89% yield in 48 h.[176]

Alkylarenes have been oxidized to the corresponding benzylic acetates and carbonyl products by nitrate salts in acetic acid in the presence of HPAs, with $H_5PV_2Mo_{10}O_{40}$ being the most effective. The conversion to the aldehyde/ketone has been increased by the addition of water to the reaction mixture. In an electron transfer reaction, the proposed $N^VO_2[H_4PV_2Mo_{10}O_{40}]$ complex reacts with the alkylarene substrate to yield a radical cation-based donor acceptor intermediate, $N^{IV}O_2[H_4PV_2Mo_{10}O_{40}]$-$ArCH_2R^{+\cdot}$. Concurrent proton transfer yields an alkylarene radical, $ArCHR^{\cdot}$, and $NO_2$, which undergo heterocoupling to yield a benzylic nitrite, which is hydrolyzed or acetolyzed, finally yielding benzylic acetates and the corresponding aldehydes or ketones.[177] Contrary to what has been observed for nitric acid as an oxidant, over-oxidation of methylarenes to the corresponding carboxylic acids has not been observed.

Sulfoxides have been exploited as oxygen donors. Dimethylsulfoxide has been used in the presence of $[PMo_{12}O_{40}]^{3-}$, under anaerobic conditions, for the oxidation of alkylarenes such as xanthene and diphenylmethane to xanthen-9-one and benzophenone, respectively, with up to 96% yield, in 15 h at 170°C in 1,2-dichlorobenzene.[178]

The 12-tungstocobaltate $[Co^{II}W_{12}O_{40}]^{6-}$ has been used as a catalyst for the oxidation of aliphatic and benzylic alcohols to the corresponding carbonyl compounds with up to 92% yield in 2.5–3 h at 50°C, in aqueous acetonitrile. The reduced POM, formed in the oxidation of alcohols, can be re-oxidized by potassium monopersulfate ($KHSO_5$).[179]

The reactivity of $[Ru^{II}(DMSO)PW_{11}O_{39}]^{5-}$ has been examined in the presence of sodium periodate ($NaIO_4$) and potassium monopersulfate ($KHSO_5$), as tested in the oxidation of two model hydrocarbons, cis-cyclooctene and adamantane, respectively. The cleavage of the olefinic double bond by $NaIO_4$ has been performed in water, as depicted in Fig. 21.10, leading to quantitative conversion of the substrate in 4 h at 50°C, with the formation of suberic acid as the major product (84% yield). In analogous conditions, cyclohexene yields adipic acid (90% yield).[55]

### 21.2.2.1. *Use of POMs as water splitting oxidants*

Polyoxometalates have been shown to present biomimicking activity of the Photosystem II (PSII) enzyme, thus emerging as a promising class of catalysts to perform water oxidation.[180,181] In green plants and some bacteria, water oxidation occurs at the heart of the PSII enzyme, a homodimer protein of 650 KDa, where $O_2$ evolution is catalyzed by a polynuclear metal-oxo-cluster with four manganese and one calcium atom held together by oxygen bridges ($CaMn_4O_x$).[182] The adoption of such a catalytic core, featuring adjacent multi-transition metal centers and multiple-μ-hydroxo/oxo bridging units, is the key strategy to master a four-electron/four-proton

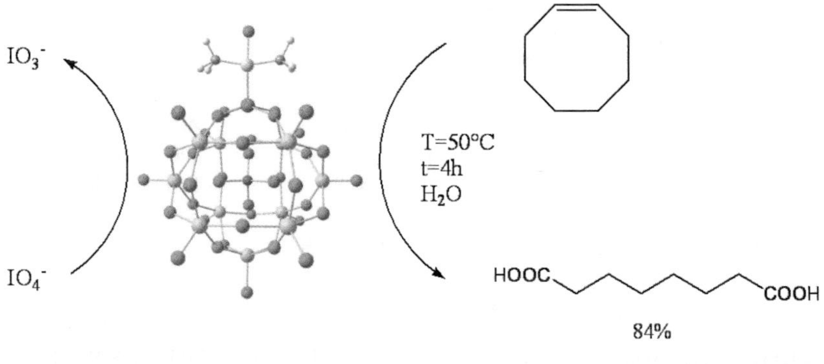


**Figure 21.10.** $[Ru^{II}(DMSO)PW_{11}O_{39}]^{5-}$oxidation of cyclooctene with $NaIO_4$ as the oxidant.

mechanism through sequential redox steps with high efficiency and minimal energy cost. In this process, $H_2O$ is the multi-electron source, providing a total of four electrons/mol by the oxidative half reaction ($2H_2O \rightarrow O_2 + 4H^+ + 4\,e^-$; $E^0 = 1.23$ V vs NHE), required to accomplish the photosynthetic process.

The possibility of accessing artificial photosynthesis is a Holy Grail of modern science.[183,184] Indeed, efficient light-driven catalytic splitting of water into high-energy content chemical species, that are molecular oxygen and hydrogen ($2H_2O \rightarrow O_2 + 2H_2$), represents one of the most ambitious research goals for the development of systems able to convert a ubiquitous energy source like solar energy into a cheap, abundant and readily available fuel.[185]

The tetraruthenium polyoxometalate $\{Ru_4^{IV}(\mu\text{-}OH)_2(\mu\text{-}O)_4(H_2O)_4[\gamma\text{-}SiW_{10}O_{36}]\}^{10-}$, $Ru_4(POM)$ is an efficient water oxidation catalyst.[186,187] The structure of $Ru_4(POM)$, reported in Fig. 21.11, is constituted by two staggered $[\gamma\text{-}SiW_{10}O_{36}]^{8-}$ units which coordinate an adamantine-like tetraruthenium-oxo core: as for the tetra-manganese oxygen evolving site of PSII, four redox-active transition metals are connected through μ-oxo or μ-hydroxo bridges, with the metal centers coordinating a water molecule as the terminal ligand.

The activity of $Ru_4(POM)$ in water oxidation catalysis has been studied in the presence of sacrificial oxidants as cerium ammonium nitrate and $[Ru(bpy)_3]^{3+}$ (bpy = 2,2′-bipyridine).[186,187] In the presence of an excess of Ce(IV), up to 500 catalytic cycles for oxygen evolution have been observed (calculated as moles of oxygen produced per moles of catalyst employed), with an initial TOF of $0.125\,s^{-1}$.

When $[Ru(bpy)_3]^{3+}$ is used as the sacrificial oxidant, up to 18 turnovers are obtained in 30–40 seconds of reaction, corresponding to a TOF of $0.45–0.60\,s^{-1}$. This catalyst seems to undergo several consecutive proton-coupled electron transfers, finally yielding a high valent intermediate responsible for oxygen production, in a single four-electron step.[188,189]

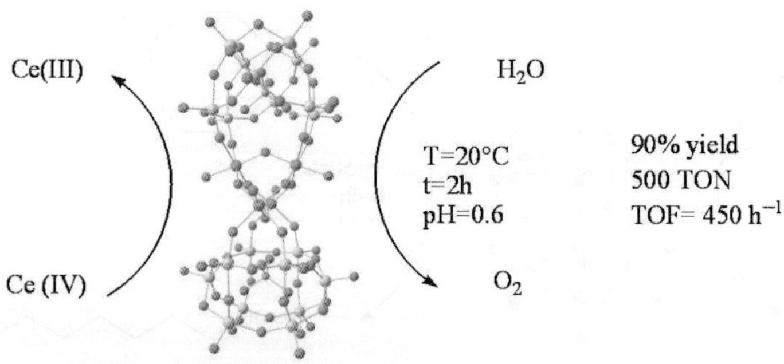


**Figure 21.11.** Water oxidation catalyzed by $\{Ru_4^{IV}(\mu\text{-}OH)_2(\mu\text{-}O)_4(H_2O)_4[\gamma\text{-}SiW_{10}O_{36}]\}^{10-}$ in the presence of $Ce^{IV}$ as a sacrificial oxidant.

The species $Ru_4(POM)$ has also been studied in light-driven water oxidation.[190–192] $[Ru(bpy)_3]^{2+}$ has been used as the photosensitizer with persulfate as the sacrificial acceptor in a phosphate buffer.[191] With this system, up to 350 turnovers and an initial turnover frequency of $8{\cdot}10^{-2}\ s^{-1}$ have been observed for the tetraruthenium-substituted POM, while the quantum yield has been found to be 0.045. A lower efficiency has been observed for the isostructural catalyst $\{Ru_4(\mu\text{-}O)_5(\mu\text{-}OH)(H_2O)_4(\gamma\text{-}PW_{10}O_{36})_2\}^{9-}$, where phosphorous instead of silicon is present as the central heteroatom of the polyoxometalate units.[193]

The reaction kinetics of the electron transfer from the catalyst to the photogenerated oxidant $[Ru(bpy)_3]^{3+}$ has been studied by nanosecond laser flash photolysis experiment. The kinetics of the hole scavenging, are pseudo first order when $[[Ru(bpy)_3]^{3+}] \ll [Ru_4(POM)]$, and are characterized by a bimolecular rate constant of $(2.1 \pm 0.4){\cdot}10^9\ M^{-1}\ s^{-1}$, close to the diffusion-controlled limiting rate, likely due to the efficient pairing between the positively charged Ru-polypyridine sensitizers and the negatively charged $Ru_4(POM)$ catalyst.[191]

Coupling $[Ru\{(\mu\text{-}dpp)Ru(bpy)_2\}_3]^{8+}$ (dpp = 2,3-bis(2′-pyridyl)pyrazine) with $Ru_4(POM)$ has allowed the expansion to the useful wavelength region, thus maximizing the overlap with solar emission. An outstanding photoreaction quantum yield for oxygen production of 0.3 has indeed been calculated by irradiating at 550 nm.[192]

The tetracobalt-substituted POM $[Co_4(H_2O)_2(\alpha\text{-}PW_9O_{34})_2]^{10-}$ has also been reported as a water oxidation catalyst. This species features a $Co_4O_4$ core, again resembling the active site of the natural enzyme PSII. Water oxidation has been studied in a phosphate buffer (pH = 7.5–8.0), using $[Ru(bpy)_3]^{3+}$ as the sacrificial oxidant, and its activity has been compared to that of $Co(NO_3)_2$ to confirm its stability and reactivity.[194,195]

## 21.3. Heterogeneous Polyoxometalate-Based Systems

To reduce the environmental impact of POM-based catalysts, many different techniques of immobilization/heterogenization of these compounds have been developed.[196] This final section outlines some relevant examples of the preparation of hybrid POM-based materials that could be useful to further expand the catalytic applications of POMs.

### 21.3.1. *Physical entrapment*

Hybrid composite films enabling physical entrapment can be obtained by spin coating or dipping, starting from a suitable homogeneous solution. Water-soluble polymers, such as polyvinyl alcohol (PVA), polyacrylamide (PAA) and polyvinyl pyrrolidone (PVP), are generally employed for this type of immobilization.

Examples of physical entrapment are the uniform dispersion of $[PMo_{12}O_{40}]^{3-}$ in polyethylene glycol (PEG),[197] obtaining a material with distinct new physical properties, that retains the photochromic properties of the reduced POM, or the synthesis of transparent and flexible films of agarose able to tune the luminescence of the doping $[Eu(SiW_{10}MoO_{39})_2]^{13-}$.[198] As reported above, similar approaches have been extended to hydrophobic polymers, by using cationic surfactants bearing long alkyl chains to prepare surfactant encapsulated POMs (SEPs), thus obtaining nanostructures with improved dispersion in organic mixtures.[199]

The limit in the preparation of hybrid materials by physical entrapment is the lack of an effective attachment between the organic and the inorganic domain, which affects the stability of the resulting hybrid material and can lead to phase separations between POM domains and the polymer matrix, severely limiting their applications.

### 21.3.2. *Electrostatic interactions*

A different approach to increasing the stability of the hybrid material is to anchor the POM to the polymeric support via electrostatic interactions. Due to the cooperative effect of multiple ionic interactions, polycationic species allow an efficient interaction between the two domains.

A heterogeneous catalytic system was prepared upon grafting a cationic dihydroimidazolium-tagged silane on solid $SiO_2$. The resulting supported ionic liquid phase (SILP) has been used to immobilize $[\gamma\text{-}1,2\text{-}H_2SiV_2W_{10}O_{40}]$ and was employed in a mixture of acetonitrile/t-butyl alcohol at 20°C for the oxidation of different substrates: terminal olefins (66–82% yield), non-hindered internal olefins (> 70% yield) in 24 h, and sulfides (81–95% yield) in 4–10 h.[200]

Adducts between POMs and hydrosoluble polyelectrolytes have been formed through ionic exchange, by mixing a polymer matrix of positively charged, cross-linked polyethylenamine (PEA) and a suitable POM. A hybrid material with enhanced affinity for liphophilic substrates, such as for secondary alcohol has thus been obtained.[201]

Cationic polyaniline (PANI) has also been associated with POMs to prepare thin films of hybrid conductive material.[202] Cationic PANI and polydiallyldimethyammonium (PDDA) have been used to support $[Ru_4(\mu\text{-}OH)_2(\mu\text{-}O)_4(H_2O)_4(\gamma\text{-}SiW_{10}O_{36})_2]^{10-}$. This POM presents catalase activity and leads to hydrogen peroxide decomposition to water with the simultaneous generation of oxygen: the latter process has been used to promote the motion of the composite material itself.[203]

Polyoxometalates can be assembled with cationic polyelectrolytes using layer-by-layer (LBL) techniques.[204,205] With this methodology, mono- and multilayer polyvinylpyridine/POM films have been cast on gold electrodes, with good control of the film thickness and of the resulting physico-chemical properties. The materials thus obtained can be used for applications including electrochromism, photo-electrochemistry, sensing and catalysis.[206] This technique can be used for LBL assembly on spherical inert templates, to be removed in order to obtain spherical microcapsules containing POM within the shell layers.[207]

Chitosan derivatives have been exploited to prepare nanoparticles by using ionic gelification with alkali metal cations,[208] or upon LBL assembly.[209] 12-Phosphotungstic acid, $H_3PW_{12}O_{40}$, has been heterogenized by conjugation with a chitosan matrix and applied to the epoxidation of allylic alcohols.[210]

The assembly with dendrimeric polycations has been taken into account to prepare bulky catalytic species, easy to remove from the solution.[211]

Multiwalled carbon nanotubes (MWCNTs) have been decorated with polyamidoamine (PAMAM) ammonium dendrimers and associated through ionic interactions with $\{[Ru_4(\mu\text{-}OH)_2(\mu\text{-}O)_4(H_2O)_4(\gamma\text{-}SiW_{10}O_{36})_2]^{10-}$, reported above. In this case, the resulting material has been cast on an indium tin oxide (ITO) electrode to prepare an oxygen-evolving anode, operating at low overpotential.[212]

### 21.3.3. *Covalent interactions*

As reported above, many POMs can be covalently modified with functional organic groups. By using unsaturated groups, hybrid POM derivatives may be copolymerized with different monomers to give stable hybrid materials.

The polyanion $[Mo_6O_{18}(NC_6H_4CHNCH_2)]2-$ has been obtained through a reaction between the Lindqvist POM $[Mo_6O_{19}]^{2-}$ and $Ph_3P{=}NC_6H_4CH{=}CH_2$; the hybrid complex has then been linked to side chains of a polystyrene matrix via radical copolymerization.[213] Alternatively, POMs as $\{[CH_2{=}C(CH_3)C(O)O(CH_2)_3Si]_2O(\gamma\text{-}SiW_{10}O_{36})\}^{4-}$ and $\{[(CH_2{=}CH)(C_6H_4)Si]_2O(\alpha\text{-}SiW_{11}O_{39})\}^{4-}$ have been grafted into polymeric methacrylate matrices via covalent bonds. In this case, the hybrid derivatives act as cross-linking monomers, increasing the stability of the resulting material.[21]

In addition to extensive polymer matrices, POMs can be covalently linked to polymeric nanoparticles. The vacant complex $[P_2W_{17}O_{61}]^{10-}$ has been functionalized with mercaptosilanes and linked to polystyrene nanoparticles functionalized with benzylchloride, via nucleophilic substitution.[214]

Despite the good results obtained using the covalent approach for the synthesis of hybrid polymers, this strategy has an inherent limit, since only certain classes of POM, usually polyoxomolybdates or vacant polyoxotungstates are suitable for the preparation of hybrid monomers. An alternative and more generally applicable method is the encapsulation of POMs with cationic surfactants bearing unsaturated groups, in order to obtain polymerizable SEPs. These complexes are soluble in solutions of non-polar monomers and can be copolymerized *in situ*, without the addition of any solvent. With this strategy, the lacunary fluorescent POM $[EuW_{10}O_{36}]^{9-}$ has been encapsulated with dodecyl(11-methacryloyloxyundecyl)dimethylammonium bromide (DMDA), to give a uniform dispersion of the POM in a polymethylmethacrilate matrix.[215] In addition, uniformly SEP-doped nanoparticles have been obtained by microemulsion polymerization.[216]

Finally, the complex $[PW_{12}O_{40}]^{3-}$ has been paired with di(11-hydroxyundecyl) dimethylammonium (DOHA) and bound onto a silica matrix through sol-gel condensation with tetraethylorthosilicate (TEOS). The resulting catalyst has been shown to be a recyclable oxidation catalyst.[217]

## 21.4. Conclusions

Due to their remarkable oxidative stability, oxidation catalysis by POMs continues to be a timely field of investigation. The possibility to tune their composition at the molecular level offers the advantage of controlling fundamental properties of interest for selective oxidations. Steric hindrance, redox potential and single-site activation via coordination/Lewis acid catalysis are instrumental to direct the chemo- and regio-selectivity required for the processing of polyfunctional substrates. In addition, the easy access to a large structural/compositional variety, combined with their unique molecular nature and nanosized dimensions, offers straightforward tools for mechanistic investigation under a turnover regime, with the possibility to trace fundamental structure–activity descriptors, validated by computational tools, and representing a functional model of extended metal-oxides heterogeneous surfaces.

Indeed, the next frontier is represented by the exploitation of POM chemistry under heterogeneous conditions. Their polyanionic charge and dimensions offer a distinctive advantage for immobilizations/confinement strategies into either solid or liquid phases. The design of tailored POM counterions affords new species with increased affinity towards diverse materials spanning from solid surfaces, polymers, membranes and also carbon nanostructures. The interplay of POMs with the surrounding nanostructured environment has shown to be a promising strategy for the design of innovative and functional hybrid materials with enhanced catalytic activity and selectivity.

In conclusion, the use of POM catalysts, along with benign oxidants represents a true possibility for the development of novel and selective synthetic procedures. In this scenario, water splitting for energy applications represents a challenging frontier in TMSP-based catalysis, nowadays attracting great attention.

## Acknowledgments

Financial support from University of Padova (Progetto Strategico 2008, HELIOS, Prot. STPD08RCX and PRAT CPDA084893/08), MIUR (PRIN Contract No. 20085M27SS) and Fondazione Cariparo, (Nano-Mode, progetti di eccellenza 2010) is gratefully acknowledged.


## References


1. Kepert, D. (1972). *The Early Transition Metals*, Academic Press, London.
2. Pope, M. (1983). *Heteropoly and Isopoly Oxometalates*, Springer Verlag, Berlin.
3. Pope, M. and Müller, A. (1991). Polyoxometalate Chemistry: An Old Field with New Dimensions in Several Disciplines, *Angew. Chem. Int. Ed.*, **30**, pp. 34–48.
4. Pope, M. and Müller, A. (eds) (2001). *Polyoxometalate Chemistry: From Topology via Self-Assembly to Applications,* Kluwer Academic Publishers, Dordrecht.
5. Lipscomb, W. (1965). Paratungstate Ion, *Inorg. Chem.*, **4**, pp. 132–134.
6. Chen, Q. and Zubieta, J. (1992). Coordination Chemistry of Soluble Metal Oxides of Molybdenum and Vanadium, *Coord. Chem. Rev.*, **114**, pp. 107–167.
7. Pope, M. (1987). Main Group and Early Transition Elements, in G. Wilkinson, R. Gillard and J. McCleverty (eds), *Comprehensive Coordination Chemistry*, Pergamon Press, Oxford, pp. 1023–1058.
8. Zonnevijlle, F., Tourne, C. and Tourne, G. (1982). Preparation and Characterization of Iron(III)- and Rhodium(III)-Containing Heteropolytungstates. Identification of Novel Oxo-Bridged Iron(III) Dimers, *Inorg. Chem.*, **21**, pp. 2751–2757.
9. Finke, R. and Droege, M. (1983). Trivacant Heteropolytungstate Derivatives. 2. Synthesis, Characterization, and Tungsten-183 NMR of $P_4W_{30}M_4(H_2O)_2O_{112}^{16-}$ (M = Co, Cu, Zn), *Inorg. Chem.*, **22**, pp. 1006–1008.
10. Lyon, D., Miller, W., Novet, T., *et al.* (1991). Highly Oxidation Resistant Inorganic-Porphyrin Analog Polyoxometalate Oxidation Catalysts. 1. The Synthesis and Characterization of Aqueous-Soluble Potassium Salts of $\alpha$-2-$P_2W_{17}O_{61}(M^{n+}{\cdot}OH_2)^{(n-10)}$ and Organic Solvent Soluble tetra-*n*-butylammonium Salts of $\alpha$-2-$P_2W_{17}O_{61}(M^{n+}{\cdot}Br)^{(n-11)}$ ($M = Mn^{3+}$, $Fe^{3+}$, $Co^{2+}$, $Ni^{2+}$, $Cu^{2+}$), *J. Am. Chem. Soc.*, **113**, pp. 7209–7221.
11. Müller, A. and Kögerler, P. (1999). From Simple Building Blocks to Structures with Increasing Size and Complexity, *Coord. Chem. Rev.*, **182**, pp. 3–17.
12. Kortz, U., Nellutla, S., Stowe, A., *et al.* (2004). Sandwich-Type Germanotungstates: Structure and Magnetic Properties of the Dimeric Polyoxoanions $[M_4(H_2O)_2(GeW_9O_{34})_2]^{12-}$ ($M = Mn^{2+}$, $Cu^{2+}$, $Zn^{2+}$, $Cd^{2+}$), *Inorg. Chem.*, **43**, pp. 2308–2317.
13. Botar, B., Kögerler, P. and Hill, C. (2007). Tetrairon and Hexairon Hydroxo/Acetato Clusters Stabilized by Multiple Polyoxometalate Scaffolds. Structures, Magnetic Properties, and Chemistry of a Dimer and a Trimer, *Inorg. Chem.*, **46**, pp. 5398–5403.
14. Gouzerh, P. and Proust, A. (1998). Main-Group Element, Organic, and Organometallic Derivatives of Polyoxometalates, *Chem. Rev.*, **98**, pp. 77–112.
15. Proust, A., Thouvenot, R. and Gouzerh, P. (2008). Functionalization of Polyoxometalates: Towards Advanced Applications in Catalysis and Materials Science, *Chem. Commun.*, pp. 1837–1852.

16. Dolbecq, A., Dumas, E., Mayer, C., *et al.* (2010). Hybrid Organic-Inorganic Polyoxometalate Compounds: From Structural Diversity to Applications, *Chem. Rev.*, **110**, pp. 6009–6048.
17. Berardi, S., Carraro, M., Sartorel, A., *et al.* (2011). Hybrid Polyoxometalates: Merging Organic and Inorganic Domains for Enhanced Catalysis and Energy Applications, *Isr. J. Chem.*, **51**, pp. 259–274.
18. Proust, A., Gouzerh, P. and Robert, F. (1993). Molybdenum Oxo Nitrosyl Complexes. 1. Defect Lindqvist Compounds of the Type $[Mo_5O_{13}(OR)_4(NO)]^{3-}$ ($R = CH_3, C_2H_5$). Solid-State Interactions with Alkali-Metal Cations, *Inorg. Chem.*, **32**, pp. 5291–5298.
19. Carraro, M., Modugno, G., Sartorel, A., *et al.* (2009). Optically Active Polyoxotungstates Bearing Chiral Organophosphonate Substituents, *Eur. J. Inorg. Chem.*, pp. 5164–5174.
20. Berardi, S., Carraro, M., Iglesias, M., *et al.* (2010). Polyoxometalate-Based *N*-Heterocyclic Carbene (NHC) Complexes for Palladium-Mediated C-C Coupling and Chloroaryl Dehalogenation Catalysis, *Chem. Eur. J.*, **16**, pp. 10662–10666.
21. Mayer, C., Thouvenot, R. and Lalot, T. (2000). New Hybrid Covalent Networks Based on Polyoxometalates: Part 1. Hybrid Networks Based on Poly(ethyl methacrylate) Chains Covalently Cross-Linked by Heteropolyanions: Synthesis and Swelling Properties, *Chem. Mater.*, **12**, pp. 257–260.
22. Zeng, H., Newkome, G. and Hill, C. (2000). Poly(polyoxometalate) Dendrimers: Molecular Prototypes of New Catalytic Materials, *Angew. Chem. Int. Ed.*, **39**, pp. 1771–1774.
23. Carraro, M., Sartorel, A., Scorrano, G., *et al.* (2008). Chiral Strandberg-Type Molybdates $[(RPO_3)_2Mo_5O_{15}]^{2-}$ as Molecular Gelators: Self-Assembled Fibrillar Nanostructures with Enhanced Optical Activity, *Angew. Chem. Int. Ed.*, **47**, pp. 7275–7279.
24. Long, D., Tsunashima, R. and Cronin, L. (2010). Polyoxometalates: Building Blocks for Functional Nanoscale Systems, *Angew. Chem. Int. Ed.*, **49**, pp. 1736–1758.
25. Bu, W., Li, H., Sun, H., *et al.* (2005). Polyoxometalate-Based Vesicle and its Honeycomb Architectures on Solid Surfaces, *J. Am. Chem. Soc.*, **127**, pp. 8016–8017.
26. Landsmann, S., Lizandara-Pueyo, C. and Polarz, S. (2010). A New Class of Surfactants with Multinuclear, Inorganic Head Groups, *J. Am. Chem. Soc.*, **132**, pp. 5315–5321.
27. Hill, C. (ed.) (1998). *Chem. Rev.*, **98**, pp. 1–390.
28. Bar-Nahum, I. and Neumann, R. (2003). Synthesis, Characterization and Catalytic Activity of a Wilkinson's Type Metal-Organic-Polyoxometalate Hybrid Compound, *Chem. Commun.*, pp. 2690–2691.
29. Yamaguchi, K., Kotani, M., Kamata, K., *et al.* (2008). An Efficient One-Pot Three-Component Reaction to Produce 1,4-Disubstituted-1,2,3-triazoles Catalyzed by a Dicopper-Substituted Silicotungstate, *Chem. Lett.*, **37**, pp. 1258–1259.
30. Kogan, V., Aizenshtat, Z., Popovitz-Biro, R., *et al.* (2002). Carbon-Carbon and Carbon-Nitrogen Coupling Reactions Catalyzed by Palladium Nanoparticles Derived from a Palladium Substituted Keggin-Type Polyoxometalate, *Org. Lett.*, **4**, pp. 3529–3532.
31. Vazylyev, M., Sloboda-Rozner, D., Haimov, A., *et al.* (2005). Strategies for Oxidation Catalyzed by Polyoxometalates at the Interface of Homogeneous and Heterogeneous Catalysis, *Top. Catal.*, **34**, pp. 93–99.
32. Neumann, R. (2010). Activation of Molecular Oxygen, Polyoxometalates, and Liquid-Phase Catalytic Oxidation, *Inorg. Chem.*, **49**, pp. 3594–3601.
33. Kozhevnikov, I. (1998). Catalysis by Heteropoly Acids and Multicomponent Polyoxometalates in Liquid-Phase Reactions, *Chem. Rev.*, **98**, pp. 171–198.
34. Brégeault, J., Launay, F. and Atlamsani, A. (2001). Catalytic Oxidative Carbon-Carbon Bond Cleavage of Ketones with Dioxygen: Assessment of Some Metal Complexes. Some Alternatives for Preparing $\alpha, \varpi$-dicarboxylic Acids, *C. R. Acad. Sci. Paris, Série IIc, Chimie/Chemistry*, **4**, pp. 11–26.

35. Kozhevnikov, I. and Matveev, K. (1983). Homogeneous Catalysts Based on Heteropoly Acids, *Appl. Cat.*, **5**, pp. 135–150.
36. Neumann, R. and Levin, M. (1992). Aerobic Oxidative Dehydrogenations Catalyzed by the Mixed-Addenda Heteropolyanion $PV_2Mo_{10}O_{40}^{5-}$: A Kinetic And Mechanistic Study, *J. Am. Chem. Soc.*, **114**, pp. 7278–7286.
37. Brégeault, J., El Ali, B., Mercier, J., *et al.* (1989). Novel Catalytic Carbon-Carbon Bond Cleavage of α-Diols in the Presence of Vanadium Precursors, *C. R. Acad. Sci. Paris, Série II*, **309**, pp. 459–462.
38. Atlamsani, A., Bregeault, J. and Ziyad, M. (1993). Oxidation of 2-Methylcyclohexanone and Cyclohexanone by Dioxygen Catalyzed by Vanadium-Containing Heteropolyanions, *J. Org. Chem.*, **58**, pp. 5663–5665.
39. El Aakel, L., Launay, F., Atlamsani, A., *et al.* (2001). Efficient and Selective Catalytic Oxidative Cleavage of a-Hydroxy Ketones Using Vanadium-Based HPA and Dioxygen, *Chem. Commun.*, pp. 2218–2219.
40. Cavani, F., Ferroni, L., Frattini, A., *et al.* (2011). Evidence for the Presence of Alternative Mechanisms in the Oxidation of Cyclohexanone to Adipic Acid with Oxygen, Catalysed by Keggin Polyoxometalates, *Appl. Cat. A: Gen.*, **391**, pp. 118–124.
41. H. Ge, H., Leng, Y., Zhang, F., *et al.* (2008). Direct Hydroxylation of Benzene to Phenol with Molecular Oxygen over Pyridine-modified Vanadium-substituted Heteropoly Acids, *Catal. Lett.*, **124**: 250–255.
42. Ge, H., Leng, Y., Zhou, C., *et al.* (2008). Direct Hydroxylation of Benzene to Phenol with Molecular Oxygen over Phase Transfer Catalysts: Cyclodextrins Complexes with Vanadium-Substituted Heteropoly Acids, *Catal. Lett.*, **124**, pp. 324–329.
43. Hill, C. and Prosser-McCartha, C. (1995). Homogeneous Catalysis by Transition Metal Oxygen Anion Clusters, *Coord. Chem. Rev.*, **143**, pp. 407–455.
44. Putaj, P. and Lefebvre, F. (2011). Polyoxometalates Containing Late Transition and Noble Metal Atoms, *Coord. Chem. Rev.*, **255**, pp. 1642–1685.
45. Nozaki, C., Kiyoto, I., Minai, Y., *et al.* (1999). Synthesis and Characterization of Diiron(III)-Substituted Silicotungstate, $[\gamma\text{-}(1,2)\text{-}SiW_{10}\{Fe(OH_2)\}_2O_{38}]^{6-}$, *Inorg. Chem.*, **38**, pp. 5724–5729.
46. Liu, J., Ortega, F., Sethuraman, P., *et al.* (1992). Trimetallo Derivatives of Lacunary 9-Tungstosilicate Heteropolyanions. Part 1. Synthesis and Characterization, *J. Chem. Soc., Dalton Trans.*, pp. 1901–1906.
47. Kortz, U., Savelieff, M., Bassil, B., *et al.* (2002). Synthesis and Characterization of Iron(III)-Substituted, Dimeric Polyoxotungstates, $[Fe_4(H_2O)_{10}(\beta\text{-}XW_9O_{33})_2]^{n-}$ ($n = 6$, $X = As^{III}$, $Sb^{III}$; $n = 4$, $X = Se^{IV}$, $Te^{IV}$), *Inorg. Chem.*, **41**, pp. 783–789.
48. Wallar, B. and Lipscomb, J. (1996). Dioxygen Activation by Enzymes Containing Binuclear Non-Heme Iron Clusters, *Chem. Rev.*, **96**, pp. 2625–2658.
49. Costas, M., Mehn, M., Jensen, M., *et al.* (2004). Dioxygen Activation at Mononuclear Nonheme Iron Active Sites: Enzymes, Models, and Intermediates, *Chem. Rev.*, **104**, pp. 939–986.
50. de Visser, S., Kumar, D., Neumann, R., *et al.* (2004). Computer-Generated High-Valent Iron–Oxo and Manganese–Oxo Species with Polyoxometalate Ligands: How do they Compare with the Iron–Oxo Active Species of Heme Enzymes?, *Angew. Chem. Int. Ed.*, **43**, pp. 5661–5665.
51. Groves, J. (2005). Models and Mechanisms of Cytochrome P450 Action, in P. Ortiz de Montellano (ed.), *Cytochrome P450: Structure, Mechanism, and Biochemistry,* 3rd ed., Kluwer Academics / Plenum Publishers, New York.
52. Sartorel, A., Carraro, M., Scorrano, G., *et al.* (2009). Iron-Substituted Polyoxotungstates as Inorganic Synzymes: Evidence for a Biomimetic Pathway in the Catalytic Oxygenation of Catechols, *Chem. Eur. J.*, **15**, pp. 7854–7858.

53. Okun, N., Anderson, T., Hill, C. (2003). $[(Fe^{III}(OH_2)_2)_3(A\text{-}\alpha\text{-}PW_9O_{34})_2]^{9-}$ on Cationic Silica Nanoparticles, a New Type of Material and Efficient Heterogeneous Catalyst for Aerobic Oxidations, **125**, pp. 3194–3195.
54. Bonchio, M., Carraro, M., Scorrano, G., *et al.* (2005). Microwave-Assisted Fast Cyclohexane Oxygenation Catalyzed by Iron-Substituted Polyoxotungstates, *Adv. Synth. Catal.*, **347**, pp. 1909–1912.
55. Bonchio, M., Carraro, M., Sartorel, A., *et al.* (2006). Bio-Inspired Oxidations with Polyoxometalate Catalysts, *J. Mol. Catal. A: Chem.*, **251**, pp. 93–99.
56. Botar, B., Geletii, Y., Kögerler, P., *et al.* (2006). The True Nature of the Di-iron(III) $\gamma$-Keggin Structure in Water: Catalytic Aerobic Oxidation and Chemistry of an Unsymmetrical Trimer, *J. Am. Chem. Soc.*, **128**, pp. 11268–11277.
57. Bonchio, M., Carraro, M., Farinazzo, A., *et al.* (2007). Aerobic Oxidation of cis-Cyclooctene by Iron-Substituted Polyoxotungstates: Evidence for a Metal Initiated Auto-Oxidation Mechanism, *J. Mol. Catal. A: Chem.*, **262**, pp. 36–40.
58. Nishiyama, Y., Nakagawa, Y. and Mizuno, N. (2001). High Turnover Numbers for the Catalytic Selective Epoxidation of Alkenes with 1 atm of Molecular Oxygen, *Angew. Chem. Int. Ed.*, **40**, pp. 3639–3641.
59. Chen, J., Dakka, J. and Sheldon, R. (1994). Selective Decomposition of Cyclohexyl Hydroperoxide to Cyclohexanone Catalyzed by Chromium Aluminophosphate-5, *Appl. Cat. A: Gen.*, **108**, pp. L1–L6.
60. Neumann, R. and Dahan, M. (1997). A Ruthenium-Substituted Polyoxometalate as an Inorganic Dioxygenase for Activation of Molecular Oxygen, *Nature*, **388**, pp. 353–355.
61. Yamaguchi, K. and Mizuno, N. (2002). Heterogeneously Catalyzed Liquid-Phase Oxidation of Alkanes And Alcohols with Molecular Oxygen, *New J. Chem.*, **26**, pp. 972–974.
62. Bonchio, M., Carofiglio, T., Carraro, M., *et al.* (2002). Efficient Sensitized Photooxygenation in Water by a Porphyrin-Cyclodextrin Supramolecular Complex, *Org. Lett.*, **4**, pp. 4635–4637.
63. Legrini, O., Oliveros, E. and Braun, A. (1993). Photochemical Processes for Water Treatment, *Chem. Rev.*, **93**, pp. 671–698.
64. Mylonas, A., Papaconstantinou, E. and Roussis, V. (1996). Photocatalytic Degradation of Phenol and *p*-Cresol by Polyoxotungstates. Mechanistic Implications, *Polyhedron*, **15**, pp. 3211–3217.
65. Meunier, B. and Sorokin, A. (1997). Oxidation of Pollutants Catalyzed by Metallophthalocyanines, *Acc. Chem. Res.*, **30**, pp. 470–476.
66. Maldotti, A., Molinari, A. and Amadelli, R. (2002). Photocatalysis with Organized Systems for the Oxofunctionalization of Hydrocarbons by $O_2$, *Chem. Rev.*, **102**, pp. 3811–3836.
67. Szaci lowski, K., Macyk, W., Drzewiecka-Matuszek, A., *et al.* (2005). Bioinorganic Photochemistry: Frontiers and Mechanisms, *Chem. Rev.*, **105**, pp. 2647–2694.
68. Renneke, R., Kadkhodayan, M., Pasquali, M., *et al.* (1991). Roles of Surface Protonation on the Photodynamic, Catalytic, and Other Properties of Polyoxometalates Probed by the Photochemical Functionalization of Alkanes. Implications for Irradiated Semiconductor Metal Oxides, *J. Am. Chem. Soc.*, **113**, pp. 8357–8367.
69. Papaconstantinou, E. (1989). Photochemistry of Polyoxometallates of Molybdenum and Tungsten and/or Vanadium, *Chem. Soc. Rev.*, **18**, pp. 1–31.
70. Hill, C. and Prosser-McCartha, C. (1993). Photocatalytic and Photoredox Properties of Polyoxometalates Systems, in M. Grätzel and K. Kalyanasundaram (eds), *Photosensitization and Photocatalysis Using Inorganic and Organometallic Compounds*, Kluwer Academic Publishers, Dordrecht, pp. 307–330.
71. Maldotti, A., Amadelli, R., Varani, G., *et al.* (1994). Photocatalytic Processes with Polyoxotungstates: Oxidation of Cyclohexylamine, *Inorg. Chem.*, **33**, pp. 2968–2973.

72. Ermolenko, L. and Giannotti, C. (1996). Aerobic Photocatalysed Oxidation of Alkanes in the Presence of Decatungstates: Products and Effects of Solvent and Counter-Ion of the Catalyst, *J. Chem. Soc., Perkin Trans. 2*, pp. 1205–1210.
73. Papaconstantinou, E. (1982). Photocatalytic Oxidation of Organic Compounds Using Heteropoly Electrolytes of Molybdenum and Tungsten, *J. Chem. Soc., Chem. Commun.*, pp. 12–13.
74. Mizuno, N., Watanabe, T. and Misono, M. (1985). Catalysis by Heteropoly Compounds. VIII. Reduction-Oxidation and Catalytic Properties of 12-Molybdophosphoric Acid and its Alkali Salts. The Role of Redox Carriers in the Bulk, *J. Phys. Chem.*, **89**, pp. 80–85.
75. Duncan, D., Netzel, T. and Hill, C. (1995). Early-Time Dynamics and Reactivity of Polyoxometalate Excited States. Identification of a Short-Lived LMCT Excited State and a Reactive Long-Lived Charge-Transfer Intermediate following Picosecond Flash Excitation of $[W_{10}O_{32}]^{4-}$ in Acetonitrile, *Inorg. Chem.*, **34**, pp. 4640–4646.
76. Molinari, A., Varani, G., Polo, E., *et al.* (2007). Photocatalytic and Catalytic Activity of Heterogenized $W_{10}O_{32}^{4-}$ in the Bromide-Assisted Bromination of Arenes and Alkenes in the Presence of Oxygen, *J. Mol. Catal. A: Chem.*, **262**, pp. 156–163.
77. Mylonas, A., Hiskia, A., Androulaki, E., *et al.* (1999). New Aspect of the Mechanism of Photocatalytic Oxidation of Organic Compounds by Polyoxometalates in Aqueous Solutions. The Selective Photooxidation of Propan-2-ol to Propanone: The Role of OH Radicals, *Phys. Chem. Chem. Phys.*, **1**, pp. 437–440.
78. So, H. and Pope, M. (1972). Origin of Some Charge-Transfer Spectra. Oxo Compounds of Vanadium, Molybdenum, Tungsten, and Niobium Including Heteropoly Anions and Heteropoly Blues, *Inorg. Chem.*, **11**, pp. 1441–1443.
79. Hiskia, A. and Papaconstantinou, E. (1992). Photocatalytic Oxidation of Organic Compounds by Polyoxometalates of Molybdenum and Tungsten. Catalyst Regeneration by Dioxygen, *Inorg. Chem.*, **31**, pp. 163–167.
80. Tanielian, C., Cougnon, F. and Seghrouchni, R. (2007). Acetone, a Substrate and a New Solvent in Decatungstate Photocatalysis, *J. Mol. Catal. A: Chem.*, **262**, pp. 164–169.
81. Bonchio, M., Carraro, M., Conte, V., *et al.* (2005). Aerobic Photooxidation in Water by Polyoxotungstates: The Case of Uracil, *Eur. J. Org. Chem.*, **2005**, pp. 4897–4903.
82. Fox, M., Cardona, R. and Gaillard, E. (1987). Photoactivation of Metal Oxide Surfaces: Photocatalyzed Oxidation of Alcohols by Heteropolytungstates, *J. Am. Chem. Soc.*, **109**, pp. 6347–6354.
83. Hill, C. and Bouchard, D. (1985). Catalytic Photochemical Dehydrogenation of Organic Substrates by Polyoxometalates, *J. Am. Chem. Soc.*, **107**, pp. 5148–5157.
84. Tanielian, C., Duffy, K. and Jones, A. (1997). Kinetic and Mechanistic Aspects of Photocatalysis by Polyoxotungstates: A Laser Flash Photolysis, Pulse Radiolysis, and Continuous Photolysis Study, *J. Phys. Chem. B*, **101**, pp. 4276–4282.
85. Tanielian, C. (1998). Decatungstate Photocatalysis, *Coord. Chem. Rev.*, **178–180**, pp. 1165–1181.
86. Maldotti, A., Molinari, A., Varani, G., *et al.* (2002). Immobilization of $(n\text{-}Bu_4N)_4W_{10}O_{32}$ on Mesoporous MCM-41 and Amorphous Silicas for Photocatalytic Oxidation of Cycloalkanes with Molecular Oxygen, *J. Catal.*, **209**, pp. 210–216.
87. Molinari, A., Amadelli, R., Mazzacani, A., *et al.* (2002). Tetralkylammonium and Sodium Decatungstate Heterogenized on Silica: Effects of the Nature of Cations on the Photocatalytic Oxidation of Organic Substrates, *Langmuir*, **18**, pp. 5400–5405.
88. Bonchio, M., Carraro, M., Scorrano, G., *et al.* (2003). Heterogeneous Photooxidation of Alcohols in Water by Photocatalytic Membranes Incorporating Decatungstate, *Adv. Synth. Catal.*, **345**, pp. 1119–1126.
89. Bonchio, M., Carraro, M., Gardan, M., *et al.* (2006). Hybrid Photocatalytic Membranes Embedding Decatungstate for Heterogeneous Photooxygenation, *Top. Catal.*, **40**, pp. 133–140.

90. Vankelecom, I. (2002) Polymeric Membranes in Catalytic Reactors, *Chem. Rev.*, **102**, pp. 3779–3810.
91. Ozdemir, S., Buonomenna, M. and Drioli, E. (2006). Catalytic Polymeric Membranes: Preparation and Application, *Appl. Cat. A*: *Chem.*, **307**, pp. 167–183.
92. Maldotti, A. and Molinari, A. (1998). Novel Reactivity of Photoexcited Iron Porphyrins Caged into a Polyfluoro Sulfonated Membrane in Catalytic Hydrocarbon Oxygenation, *Chem. Commun.*, pp. 507–508.
93. Arcella, V., Colaianna, P., Maccone, P., *et al.* (1999). A Study on a Perfluoropolymer Purification and its Application To Membrane Formation, *J. Membrane Sci.*, **163**, pp. 203–209.
94. Arcella, V., Troglia, C. and Ghielmi, A. (2005). Hyflon Ion Membranes for Fuel Cells, *Ind. Eng. Chem. Res.*, **44**, pp. 7646–7651.
95. Prabhakar, R., Freeman, B. and Roman, I. (2004). Gas and Vapor Sorption and Permeation in Poly(2,2,4-Trifluoro-5-Trifluoromethoxy-1,3-Dioxole-Co-Tetrafluoroethylene), *Macromolecules*, **37**, pp. 7688–7697.
96. Carraro, M., Gardan, M., Scorrano, G., *et al.* (2006). Solvent-Free, Heterogeneous Photooxygenation Of Hydrocarbons by Hyflon® Membranes Embedding a Fluorous-Tagged Decatungstate, *Chem. Commun.*, pp. 4533–4535.
97. Fontananova, E., Drioli, E., Donato, L., *et al.* (2006). Hybrid Photocatalytic Membranes Embedding Decatungstate for Heterogeneous Photooxydation, *Desalination*, **200**, pp. 705–707.
98. Sun, H., Li, H., Bu, W., *et al.* (2006). Self-Organized Microporous Structures Based on Surfactant-Encapsulated Polyoxometalate Complexes, *J. Phys. Chem. B*, **110**, pp. 24847–24854.
99. Li, H., Sun, H., Qi, W., *et al.* (2007). Onionlike Hybrid Assemblies Based on Surfactant-Encapsulated Polyoxometalates, *Angew. Chem. Int. Ed.*, **46**, pp. 1300–1303.
100. Qi, W., Li, H. and Wu, L. (2008). Stable Photochromism and Controllable Reduction Properties of Surfactant-Encapsulated Polyoxometalate/Silica Hybrid Films, *J. Phys. Chem. B*, **112**, pp. 8257–8263.
101. Attanasio, D., Suber, L. and Thorslund, K. (1991). Aerobic Photooxidation of Substituted Benzenes Catalyzed by the Tungsten Isopolyanion $[W_{10}O_{32}]^{4-}$, *Inorg. Chem.*, **30**, pp. 590–592.
102. Lykakis, I., Tanielian, C. and Orfanopoulos, M. (2003) Decatungstate Photocatalyzed Oxidation of Aryl Alkanols. Electron Transfer or Hydrogen Abstraction Mechanism?, *Org. Lett.*, **5**, pp. 2875–2878.
103. Lykakis, I. and Orfanopoulos, M. (2004). Photooxidation of Aryl Alkanes by a Decatungstate/Triethylsilane System in the Presence of Molecular Oxygen, *Tetrahedron Lett.*, **45**, pp. 7645–7649.
104. Maldotti, A., Molinari, A., Bergamini, P., *et al.* (1996). Photocatalytic Oxidation of Cyclohexane by $(nBu_4N)_4W_{10}O_{32}$/Fe(III) Porphyrins Integrated Systems, *J. Mol. Catal. A*: *Chem.*, **113**, pp. 147–157.
105. Park, H. and Choi, W. (2005). Photocatalytic Conversion of Benzene to Phenol Using Modified $TiO_2$ and Polyoxometalates, *Catal. Today*, **101**, pp. 291–297.
106. Han, Z., Wang, E., Luan, G., *et al.* (2001). Synthesis and Crystal Structure of a Novel Compound Constructed from Tris-(2,2'-bipy)ruthenium(II) and Decatungstate, *Inorg. Chem. Commun.*, **4**, pp. 427–429.
107. Bonchio, M., Carraro, M., Scorrano, G., *et al.* (2004). Photooxidation in Water by New Hybrid Molecular Photocatalysts Integrating an Organic Sensitizer with a Polyoxometalate Core, *Adv. Synth. Catal.*, **346**, pp. 648–654.
108. Bianco, A., Gasparrini, F., Maggini, M., *et al.* (1997). Molecular Recognition by a Silica-Bound Fullerene Derivative, *J. Am. Chem. Soc.*, **119**, pp. 7550–7554.
109. Kordatos, K., Prato, M., Menna, E., *et al.* (2001). Synthesis of Fullerene Derivatives for Incorporation in Sol-Gel Glasses, *J. Sol-Gel Sci. Techn.*, **22**, pp. 237–244.

110. Yamakoshi, Y., Umezawa, N., Ryu, A., *et al.* (2003). Active Oxygen Species Generated from Photoexcited Fullerene (C60) as Potential Medicines: $O_2^{-\cdot}$ Versus $^1O_2$, *J. Am. Chem. Soc.*, **125**, pp. 12803–12809.
111. Lane, B. and Burgess, K. (2003). Metal-Catalyzed Epoxidations of Alkenes with Hydrogen Peroxide, *Chem. Rev.*, **103**, pp. 2457–2474.
112. Khenkin, A. and Hill, C. (1993). Selective Homogeneous Catalytic Epoxidation of Alkenes by Hydrogen Peroxide Catalysed by Oxidatively- and Solvolytically-Resistant Polyoxometalate Complexes, *Mendeleev Commun.*, **3**, pp. 140–141.
113. Zhang, X., Chen, Q., Duncan, D., *et al.* (1997). Multiiron Polyoxoanions. Synthesis, Characterization, X-Ray Crystal Structure, and Catalytic $H_2O_2$-Based Alkene Oxidation by $[(n\text{-}C_4H_9)_4N]_6[Fe_4^{III}(H_2O)_2(PW_9O_{34})_2]$, *Inorg. Chem.*, **36**, pp. 4381–4386.
114. Zhang, J., Tang, Y., Li, G., *et al.* (2005). Room Temperature Direct Oxidation of Benzene to Phenol Using Hydrogen Peroxide in the Presence of Vanadium-Substituted Heteropolymolybdates, *Appl. Cat. A*: *Gen.*, 278, pp. 251–261.
115. Mizuno, N., Nozaki, C., Kiyoto, I., *et al.* (1998). Highly Efficient Utilization of Hydrogen Peroxide for Selective Oxygenation of Alkanes Catalyzed by Diiron-Substituted Polyoxometalate Precursor, *J. Am. Chem. Soc.*, **120**, pp. 9267–9272.
116. Mizuno, N., Seki, Y., Nishiyama, Y., *et al.* (1999). Aqueous Phase Oxidation of Methane with Hydrogen Peroxide Catalyzed by Di-Iron-Substituted Silicotungstate, *J. Catal.*, **184**, pp. 550–552.
117. Mizuno, N., Kiyoto, I., Nozaki, C., *et al.* (1999). Remarkable Structure Dependence of Intrinsic Catalytic Activity for Selective Oxidation of Hydrocarbons with Hydrogen Peroxide Catalyzed by Iron-Substituted Silicotungstates, *J. Catal.*, **181**, pp. 171–174.
118. Seki, Y., Min, J., Misono, M., *et al.* (2000). Reaction Mechanism of Oxidation of Methane with Hydrogen Peroxide Catalyzed by 11-Molybdo-1-Vanadophosphoric Acid Catalyst Precursor, *J. Phys. Chem. B*, **104**, pp. 5940–5944.
119. Zhang, X., Anderson, T., Chen, Q., *et al.* (2001). A Baker–Figgis Isomer of Conventional Sandwich Polyoxometalates. $H_2Na_{14}[Fe_2^{III}(NaOH_2)_2(P_2W_{15}O_{56})_2]$, a Diiron Catalyst for Catalytic $H_2O_2$-Based Epoxidation, *Inorg. Chem.*, **40**, pp. 418–419.
120. Khenkin, A. and Neumann, R. (2001). Redirection of Oxidation Reactions by a Polyoxomolybdate: Oxydehydrogenation Instead of Oxygenation of Alkanes with tert-Butylhydroperoxide in Acetic Acid, *J. Am. Chem. Soc.*, **123**, pp. 6437–6438.
121. Ishii, Y., Yamawaki, K., Ura, T., *et al.* (1988). Hydrogen Peroxide Oxidation Catalyzed by Heteropoly Acids Combined with Cetylpyridinium Chloride. Epoxidation of Olefins and Allylic Alcohols, Ketonization of Alcohols and Diols, and Oxidative Cleavage of 1,2-Diols and Olefins, *J. Org. Chem.*, **53**, pp. 3587–3593.
122. Prandi, J., Kagan, H. and Mimoun, H. (1986). Epoxidation of Isolated Double Bonds with 30% Hydrogen Peroxide Catalyzed by Pertungstate Salts, *Tetrahedron Lett.*, **27**, pp. 2617–2620.
123. Ishii, Y., Tanaka, H. and Nishiyama, Y. (1994). Selectivity in Oxidation of Sulfides with Hydrogen Peroxide by $[p\text{-}C_5H_5N^+(CH_2)_{15}CH_3]_3PM_{12}O_{40}^{3-}$ and $[p\text{-}C_5H_5N^+(CH_2)_{15}CH_3]_3$ $\{PO_4[M(O)(O_2)_2]_4\}^{3-}$ (M = Mo or W), *Chem. Lett.*, **23**, pp. 1–4.
124. Venturello, C. and D'Aloisio, R. (1988). Quaternary Ammonium Tetrakis(diperoxotungsto)phosphates$^{(3-)}$ as a New Class of Catalysts for Efficient Alkene Epoxidation with Hydrogen Peroxide, *J. Org. Chem.*, **53**, pp. 1553–1557.
125. Sato, K., Aoki, M., Ogawa, M., *et al.* (1997). A Halide-Free Method for Olefin Epoxidation with 30% Hydrogen Peroxide, *Bull. Chem. Soc. Jpn.*, **70**, pp. 905–915.
126. Guo, M. (2004). Quaternary Ammonium Decatungstate Catalyst for Oxidation of Alcohols, *Green Chem.*, **6**, pp. 271–273.
127. Server-Carrió, J., Bas-Serra, J., González-Núñez, M., *et al.* (1999). Synthesis, Characterization, and Catalysis of $\beta_3\text{-}[(Co^{II}O_4)W_{11}O_{31}(O_2)_4]^{10-}$, the First Keggin-Based True Heteropoly

Dioxygen (Peroxo) Anion. Spectroscopic (ESR, IR) Evidence for the Formation of Superoxo Polytungstates, *J. Am. Chem. Soc.*, **121**, pp. 977–984.
128. Ingle, R. and Raj, N. (2008). Lacunary Keggin Type Polyoxotungstates in Conjunction with a Phase Transfer Catalyst: An Effective Catalyst System for Epoxidation of Alkenes with Aqueous $H_2O_2$, *J. Mol. Catal. A: Chem.*, **294**, pp. 8–13.
129. Zhao, W., Zhang, Y., Ma, B., *et al.* (2010). Oxidation of Alcohols with Hydrogen Peroxide in Water Catalyzed by Recyclable Keggin-Type Tungstoborate Catalyst, *Catal. Commun.*, **11**, pp. 527–531.
130. Ding, Y., Zhao, W., Zhang, Y., *et al.* (2011). An Effective and Recyclable Catalytic System for Alcohol Oxidation in Water Based on a Temperature-Responsive Catalyst, *Reac. Kinet. Mech. Cat.*, **102**, pp. 85–92.
131. Kamata, K., Yonehara, K., Sumida, Y., *et al.* (2003). Efficient Epoxidation of Olefins with $>$ 99% Selectivity and Use of Hydrogen Peroxide, *Science*, **300**, pp. 964–966.
132. Kamata, K., Nakagawa, Y., Yamaguchi, K., *et al.* (2004). Efficient, Regioselective Epoxidation of Dienes with Hydrogen Peroxide Catalyzed by $[\gamma\text{-SiW}_{10}O_{34}(H_2O)_2]^{4-}$, *J. Catal.*, **224**, pp. 224–228.
133. Kamata, K., Kotani, M., Yamaguchi, K., *et al.* (2007). Olefin Epoxidation with Hydrogen Peroxide Catalyzed by Lacunary Polyoxometalate $[\gamma\text{-SiW}_{10}O_{34}(H_2O)_2]^{4-}$, *Chem. Eur. J.*, **13**, pp. 639–648.
134. Baumstark, A. and Vasquez, P. (1988). Epoxidation by Dimethyldioxirane. Electronic and Steric Effects, *J. Org. Chem.*, **53**, pp. 3437–3439.
135. Ueno, S., Yoshida, K., Ebitani, K., *et al.* (1998). Hydrotalcite Catalysis: Heterogeneous Epoxidation of Olefins Using Hydrogen Peroxide in the Presence of Nitriles, *Chem. Commun.*, **3**, pp. 295–296.
136. Musaev, D., Morokuma, K., Geletii, Y., *et al.* (2004). Computational Modeling of Di-Transition-Metal-Substituted g-Keggin Polyoxometalate Anions. Structural Refinement of the Protonated Divacant Lacunary Silicodecatungstate, *Inorg. Chem.*, **43**, pp. 7702–7708.
137. Prabhakar, R., Morokuma, K., Hill, C., *et al.* (2006). Insights into the Mechanism of Selective Olefin Epoxidation Catalyzed by $[\gamma\text{-}(SiO_4)W_{10}O_{32}H_4]^{4-}$. A Computational Study, *Inorg. Chem.*, **45**, pp. 5703–5709.
138. Sartorel, A., Carraro, M., Bagno, A., *et al.* (2007). Asymmetric Tetraprotonation of $\gamma\text{-}[(SiO_4)W_{10}O_{32}]^{8-}$ Triggers a Catalytic Epoxidation Reaction: Perspectives in the Assignment of the Active Catalyst, *Angew. Chem. Int. Ed.*, **46**, pp. 3255–3258.
139. Sartorel, A., Carraro, M., Bagno, A., *et al.* (2008). $H_2O_2$ Activation by Heteropolyacids with Defect Structures: The Case of $\gamma\text{-}[(XO_4)W_{10}O_{32}]^{n-}$ (X=Si, Ge, n=8; X=P, n=7), *J. Phys.Org. Chem.*, **21**, pp. 596–602.
140. Knoth, W. (1979) Derivatives of Jeteropolyanions. 1. Organic Derivatives of $W_{12}SiO_{40}^{4-}$, $W_{12}PO_{40}^{3-}$, and $Mo_{12}SiO_{40}^{4-}$, *J. Am. Chem. Soc.*, **101**, pp. 759–760.
141. Kim, G., Hagen, K. and Hill, C. (1992). Synthesis, Structure, Spectroscopic Properties, and Hydrolytic Chemistry of Organophosphonoyl Polyoxotungstates of Formula $[C_6H_5P(O)]_2X^{n+}W_{11}O_{39}^{(8-n)-}$ ($X^{n+}$=$P^{5+}$, $Si^{4+}$), *Inorg. Chem.*, **31**, pp. 5316–5324.
142. Mayer, C., Herson, P. and Thouvenot, R. (1999). Organic-Inorganic Hybrids Based on Polyoxometalates. 5.1 Synthesis and Structural Characterization of Bis(organophosphoryl) decatungstosilicates $[\gamma\text{-SiW}_{10}O_{36}((RPO)_2]^{4-}$, *Inorg. Chem.*, **38**, pp. 6152–6158.
143. Carraro, M., Sandei, L., Sartorel, A., *et al.* (2006). Hybrid Polyoxotungstates as Second-Generation POM-Based Catalysts for Microwave-Assisted $H_2O_2$ Activation, *Org. Lett.*, **8**, pp. 3671–3674.
144. Bonchio, M., Calloni, S., Di Furia, F., *et al.* (1997). Titanium(IV)-(*R*,*R*,*R*)-Tris(2-Phenylethoxy)amine-Alkylperoxo Complex Mediated Oxidations: The Biphilic Nature of the Oxygen Transfer to Organic Sulfur Compounds, *J. Am. Chem. Soc.*, **119**, pp. 6935–6936.

145. Conte, V., Di Furia, F. and Modena, G. (1992). Transition Metal Catalyzed Oxidation. The Role of Peroxometal Complexes, in Andō, W. (ed.), *Organic Peroxides*. John Wiley & Sons, Chichester, pp. 559–598.
146. Bonchio, M., Campestrini, S., Conte, V., *et al.* (1995). A Theoretical and Experimental Investigation of the Electrophilic Oxidation of Thioethers and Sulfoxides by Peroxides, *Tetrahedron*, **51**, pp. 12363–12372.
147. Berardi, S., Bonchio, M., Carraro, M., *et al.* (2007). Fast Catalytic Epoxidation with $H_2O_2$ and $[g\text{-}SiW_{10}O_{36}(PhPO)_2]^{4-}$ in Ionic Liquids under Microwave Irradiation, *J. Org. Chem.*, **72**, pp. 8954–8957.
148. Jahier, C., Cantuel, M., McClenaghan, N., *et al.* (2009). Enantiopure Dendritic Polyoxometalates: Chirality Transfer from Dendritic Wedges to a POM Cluster for Asymmetric Sulfide Oxidation, *Chem. Eur. J.*, **15**, pp. 8703–8708.
149. Jahier, C., Coustou, M., Cantuel, M., *et al.* (2011). Optically Active Tripodal Dendritic Polyoxometalates: Synthesis, Characterization and their Use in Asymmetric Sulfide Oxidation with Hydrogen Peroxide, *Eur. J. Inorg. Chem.*, pp. 727–738.
150. Schwegler, M., Floor, M. and van Bekkum, H. (1988). Heteropolyanions as Oxidation Catalysts in a 2-Phases System, *Tetrahedron Lett.*, **29**, pp. 823–826.
151. Witte, P., Alsters, P., Jary, W., *et al.* (2004). Self-Assembled $Na_{12}[WZn_3(ZnW_9O_{34})_2]$ as an Industrially Attractive Multi-Purpose Catalyst for Oxidations with Aqueous Hydrogen Peroxide, *Org. Proc. Develop.*, **8**, pp. 524–531.
152. Neumann, R. and Khenkin, A. (1995). Noble Metal ($Ru^{III}$, $Pd^{II}$, $Pt^{II}$) Substituted "Sandwich" Type Polyoxometalates: Preparation, Characterization, and Catalytic Activity in Oxidations of Alkanes and Alkenes by Peroxides, *Inorg. Chem.*, **34**, pp. 5753–5760.
153. Neumann, R. and Khenkin, A. (1996). A New Dinuclear Rhodium(III) "Sandwich" Polyoxometalate, $[(WZnRh^{III2})(ZnW_9O_{34})_2]^{10-}$. Synthesis, Characterization and Catalytic Activity, *J. Mol. Catal. A: Chem.*, **114**, pp. 169–180.
154. Adam, W., Alsters, P., Neumann, R., *et al.* (2003). A Highly Chemoselective, Diastereoselective, and Regioselective Epoxidation of Chiral Allylic Alcohols with Hydrogen Peroxide, Catalyzed by Sandwich-Type Polyoxometalates: Enhancement of Reactivity and Control of Selectivity by the Hydroxy Group through Metal-Alcoholate Bonding, *J. Org. Chem.*, **68**, pp. 1721–1728.
155. Sloboda-Rozner, D., Alsters, P. and Neumann, R. (2003). A Water-Soluble and "Self-Assembled" Polyoxometalate as a Recyclable Catalyst for Oxidation of Alcohols in Water with Hydrogen Peroxide, *J. Am. Chem. Soc.*, **125**, pp. 5280–5281.
156. Maayan, G., Fish, R. and Neumann, R. (2003) Polyfluorinated Quaternary Ammonium Salts of Polyoxometalate Anions: Fluorous Biphasic Oxidation Catalysis with and without Fluorous Solvents, *Org. Lett.*, **5**, pp. 3547–3550.
157. Wang, J., Yan, L., Qian, G., *et al.* (2007) $Na_4H_3[SiW_9Al_3(H_2O)_3O_{37}]\cdot 12H_2O/H_2O$: A New System for Selective Oxidation of Alcohols with $H_2O_2$ as Oxidant, *Tetrahedron*, **63**, pp. 1826–1832.
158. Kikukawa, Y., Yamaguchi, K. and Mizuno, N. (2010). Zinc(II) Containing γ-Keggin Sandwich-Type Silicotungstate: Synthesis in Organic Media and Oxidation Catalysis, *Angew. Chem. Int. Ed.*, **49**, pp. 6096–6100.
159. Wang, J., Yan, L., Li, G., *et al.* (2005). Mono-Substituted Keggin-Polyoxometalate Complexes as Effective and Recyclable Catalyst for the Oxidation of Alcohols with Hydrogen Peroxide in Biphasic System, *Tetrahedron Lett.*, **46**, pp. 7023–7027.
160. Ben-Daniel, R., Khenkin, A. and Neumann, R. (2000). The Nickel-Substituted Quasi-Wells–Dawson-Type Polyfluoroxometalate, $[Ni^{II}(H_2O)H_2F_6NaW_{17}O_{55}]^{9-}$, as a Uniquely Active Nickel-Based Catalyst for the Activation of Hydrogen Peroxide and the Epoxidation of Alkenes and Alkenols, *Chem. Eur. J.*, **6**, pp. 3722–3728.

161. Nakagawa, Y., Kamata, K., Kotani, M., *et al.* (2005). Polyoxovanadometalate-Catalyzed Selective Epoxidation of Alkenes with Hydrogen Peroxide, *Angew. Chem. Int. Ed.*, **44**, pp. 5136–5141.
162. Nakagawa, Y. and Mizuno, N. (2007) Mechanism of [g-$H_2SiV_2W_{10}O_{40}]^{4-}$-Catalyzed Epoxidation of Alkenes with Hydrogen Peroxide, *Inorg. Chem.*, **46**, pp. 1727–1736.
163. Villanneau, R., Carabineiro, H., Carrier, X., *et al.* (2004). Synthesis and Characterization of Zr(IV) Polyoxotungstates as Molecular Analogues of Zirconia-Supported Tungsten Catalysts, *J. Phys. Chem. B*, **108**, pp. 12465–12471.
164. Kholdeeva, O. and Maksimovskaya, R. (2007). Titanium- and Zirconium-Monosubstituted Polyoxometalates as Molecular Models for Studying Mechanisms of Oxidation Catalysis, *J. Mol. Catal. A: Chem.*, **262**, pp. 7–24.
165. Kholdeeva, O., Donoeva, B., Trubitsina, T., *et al.* (2009). Unique Catalytic Performance of the Polyoxometalate $[Ti_2(OH)_2As_2W_{19}O_{67}(H_2O)]^{8-}$: The Role of 5-Coordinated Titanium in $H_2O_2$ Activation, *Eur. J. Inorg. Chem.*, **2009**, pp. 5134–5141.
166. Yamase, T., Ishikawa, E., Asai, Y., *et al.* (1996). Alkene Epoxidation by Hydrogen Peroxide in the Presence of Titanium-Substituted Keggin-Type Polyoxotungstates $[PTi_xW_{12-x}O_{40}]^{(3+2x)-}$ and $[PTi_xW_{12-x}O_{40-x}(O_2)_x]^{(3+2x)-}$ ($x = 1$ and 2), *J. Mol. Catal. A: Chem.*, **114**, pp. 237–245.
167. Kholdeeva, O., Trubitsina, T., Maksimovskaya, R., *et al.* (2004). First Isolated Active Titanium Peroxo Complex: Characterization and Theoretical Study, *Inorg. Chem.*, **43**, pp. 2284–2292.
168. Nomiya, K., Sakai, Y. and Matsunaga, S. (2011). Chemistry of Group IV Metal Ion-Containing Polyoxometalates, *Eur. J. Inorg. Chem.*, **2011**, pp. 179–196.
169. Bassil, B., Mal, S., Dickman, M., *et al.* (2008). 6-Peroxo-6-Zirconium Crown and Its Hafnium Analogue Embedded in a Triangular Polyanion: $[M_6(O_2)_6(OH)_6(\gamma\text{-}SiW_{10}O_{36})_3]^{18-}$ (M = Zr, Hf), *J. Am. Chem. Soc.*, **130**, pp. 6696–6697.
170. Mal, S., Nsouli, N., Carraro, M., *et al.* (2009). Peroxo-Zr/Hf-Containing Undecatungstosilicates and Germanates, *Inorg. Chem.*, **49**, pp. 7–9.
171. Agarwal, D., Jain, R., Bhatnagar, R., *et al.* (1990). Synthesis and Characterization of Some Peroxo Complexes of Zirconium, *Polyhedron*, **9**, pp. 1405–1409.
172. Carraro, M., Nsouli, N., Oelrich, H., *et al.* (2011). Reactive $Zr^{IV}$ and $Hf^{IV}$ Butterfly Peroxides on Polyoxometalate Surfaces: Bridging the Gap between Homogeneous and Heterogenous Catalysis, *Chem. Eur. J.*, **17**, pp. 8371–8378.
173. Carraro, M., Sartorel, A., Scorrano, G., *et al.* (2008). Catalytic Strategies for Sustainable Oxidations in Water, *Synthesis*, **2008**, pp. 1971–1978.
174. Antonova, N., Carbó, J., Kortz, U., *et al.* (2010). Mechanistic Insights into Alkene Epoxidation with $H_2O_2$ by Ti- and other TM-Containing Polyoxometalates: Role of the Metal Nature and Coordination Environment, *J. Am. Chem. Soc.*, **132**, pp. 7488–7497.
175. Ben-Daniel, R., Weiner, L. and Neumann, R. (2002). Activation of Nitrous Oxide and Selective Epoxidation of Alkenes Catalyzed by the Manganese-Substituted Polyoxometalate, $[Mn^{III}{}_2ZnW(Zn_2W_9O_{34})_2]_{10}$, *J. Am. Chem. Soc.*, **124**, pp. 8788–8789.
176. Ben-Daniel, R. and Neumann, R. (2003). Activation of Nitrous Oxide and Selective Oxidation of Alcohols and Alkylarenes Catalyzed by the $[PV_2Mo_{10}O_{40}]^{5-}$ Polyoxometalate Ion, *Angew. Chem. Int. Ed.*, **42**, pp. 92–95.
177. Khenkin, A. and Neumann, R. (2004). Oxidation of Alkylarenes by Nitrate Catalyzed by Polyoxophosphomolybdates: Synthetic Applications and Mechanistic Insights, *J. Am. Chem. Soc.*, **126**, pp. 6356–6362.
178. Khenkin, A. and Neumann, R. (2002). Oxygen Transfer from Sulfoxides: Oxidation of Alkylarenes Catalyzed by a Polyoxomolybdate, $[PMo_{12}O_{40}]^{3-}$, *J. Am. Chem. Soc.*, **124**, pp. 4198–4199.
179. Maradur, S., Halligudi, S. and Gokavi, G. (2004). Oxidation of Aliphatic and Benzylic Alcohols by Oxone®, Catalysed by 12-Tungstocobaltate (II), *Catal. Lett.*, **96**, pp. 165–167.

180. Carraro, M., Sartorel, A., Toma, F., *et al.* (2011) Artificial Photosynthesis Challenges: Water Oxidation at Nanostructured Interfaces *Top. Curr. Chem.* pp. 1–30.
181. Geletii, Y., Yin, Q., Hou, Y., *et al.* (2011). Polyoxometalates in the Design of Effective and Tunable Water Oxidation Catalysts, *J. Isr. Chem.* **51**, pp. 238–246.
182. Yano, J., Kern, J., Sauer, K., *et al.* (2006). Where Water Is Oxidized to Dioxygen: Structure of the Photosynthetic $Mn_4Ca$ Cluster, *Science*, **314**, pp. 821–825.
183. Gray, H. (2009) Powering the Planet with Solar Fuel, *Nature Chem.*, **1**, pp. 7.
184. Gust, D., Moore, T. and Moore, A. (2009). Solar Fuels via Artificial Photosynthesis, *Acc. Chem. Res.*, **42**, pp. 1890–1898.
185. Balzani, V., Credi, A. and Venturi, M. (2008). Photochemical Conversion of Solar Energy, *ChemSusChem*, **1**, pp. 26–58.
186. Sartorel, A., Carraro, M., Scorrano, G., *et al.* (2008). Polyoxometalate Embedding of a Tetraruthenium(IV)-Oxo-Core by Template-Directed Metalation of $[\gamma\text{-}SiW_{10}O_{36}]^{8-}$: A Totally Inorganic Oxygen-Evolving Catalyst, *J. Am. Chem. Soc.*, **130**, pp. 5006–5007.
187. Geletii, Y., Botar, B., Kögerler, P., *et al.* (2008). An All-Inorganic, Stable, and Highly Active Tetraruthenium Homogeneous Catalyst for Water Oxidation, *Angew. Chem. Int. Ed.*, **47**, pp. 3896–3899.
188. Sartorel, A., Mirò, P., Salvadori, E., *et al.* (2009). Water Oxidation at a Tetraruthenate Core Stabilized by Polyoxometalate Ligands: Experimental and Computational Evidence To Trace the Competent Intermediates, *J. Am. Chem. Soc.*, **131**, pp. 16051–16053.
189. Geletii, Y., Besson, C., Hou, Y., *et al.* (2009). Structural, Physicochemical, and Reactivity Properties of an All-Inorganic, Highly Active Tetraruthenium Homogeneous Catalyst for Water Oxidation, *J. Am. Chem. Soc.*, **131**, pp 17360–17370.
190. Puntoriero, F., Sartorel, A., Orlandi, M., *et al.* (2011). Photoinduced Water Oxidation Using Dendrimeric Ru(II) Complexes as Photosensitizers, *Coord. Chem. Rev.*, **255**, pp. 2594–2601.
191. 191. Orlandi, M., Argazzi, R., Sartorel, A., *et al.* (2010). Ruthenium Polyoxometalate Water Splitting Catalyst: Very Fast Hole Scavenging from Photogenerated Oxidants, *Chem. Commun.*, **46**, pp. 3152–3154.
192. Puntoriero, F., La Ganga, G., Sartorel, A., *et al.* (2010). Photo-Induced Water Oxidation with Tetra-Nuclear Ruthenium Sensitizer and Catalyst: A Unique 4 X 4 Ruthenium Interplay Triggering High Efficiency with Low-Energy Visible Light, *Chem. Commun.*, **46**, pp. 4725–4727.
193. Besson, C., Huang, Z., Geletii, Y., *et al.* (2010). $Cs_9[(\gamma\text{-}PW_{10}O_{36})_2Ru_4O_5(OH)(H_2O)_4]$, a New All-Inorganic, Soluble Catalyst for the Efficient Visible-Light-Driven Oxidation of Water, *Chem. Commun.*, **46**, pp. 2784–2786.
194. Yin, Q., Tan, J., Besson, C., *et al.* (2010). A Fast Soluble Carbon-Free Molecular Water Oxidation Catalyst Based on Abundant Metals, *Science,* **328**, pp. 342–345.
195. Huang, Z., Luo, Z., Geletii, Y., *et al.* (2011). Efficient Light-Driven Carbon-Free Cobalt-Based Molecular Catalyst for Water Oxidation, *J. Am. Chem. Soc.*, **133**, pp 2068–2071.
196. Qi, W. and Wu, L. (2009). Polyoxometalate/Polymer Hybrid Materials: Fabrication and Properties, *Polym. Int.*, **58**, pp. 1217–1225.
197. Chen, J., Ai, L., Feng, W., *et al.* (2007). Preparation and Photochromism of Nanocomposite Thin Film Based on Polyoxometalate and Polyethyleneglycol, *Mater. Lett.*, **61**, pp. 5247–5249.
198. Wang, Z., Ma, Y., Zhang, R., *et al.* (2009). Reversible Luminescent Switching in a $[Eu(SiW_{10}MoO_{39})_2]^{13-}$-Agarose Composite Film by Photosensitive Intramolecular Energy Transfer, *Adv. Mater.*, **21**, pp. 1737–1741.
199. Kurth, D., Lehmann, P., Volkmer, D., *et al.* (2000). Surfactant-Encapsulated Clusters (SECs): $(DODA)_{20}(NH_4)[H_3Mo_{57}V_6(NO)_6O_{183}(H_2O)_{18}]$, a Case Study, *Chem. Eur. J.*, **6**, pp. 385–393.
200. Kasai, J., Nakagawa, Y., Uchida, S., *et al.* (2006). $[\gamma\text{-}1,2\text{-}H_2SiV_2W_{10}O_{40}]$ Immobilized on Surface-Modified $SiO_2$ as a Heterogeneous Catalyst for Liquid-Phase Oxidation with $H_2O_2$, *Chem. Eur. J.*, **12**, pp. 4176–4184.

201. Haimov, A. and Neumann, R. (2006). An Example of Lipophiloselectivity: The Preferred Oxidation, in Water, of Hydrophobic 2-Alkanols Catalyzed by a Cross-Linked Polyethyleneimine-Polyoxometalate Catalyst Assembly, *J. Am. Chem. Soc.*, **128**, pp. 15697–15700.
202. Fukaya, K., Srifa, A., Isikawa, E., *et al.* (2010). Synthesis and Structural Characterization of Polyoxometalates Incorporating with Anilinium Cations and Facile Preparation of Hybrid Film, *J. Mol. Sci.*, **979**, pp. 221–226.
203. Sartorel, A., Truccolo, M., Berardi, S., *et al.* (2011). Oxygenic Polyoxometalates: A New Class of Molecular Propellers, *Chem. Commun.*, **47**, pp. 1716–1718.
204. Wang, B., Vyas, R. and Shaik, S. (2007) Preparation Parameter Development for Layer-by-Layer Assembly of Keggin-Type Polyoxometalates, *Langmuir*, **23**, pp. 11120–11126.
205. Nagaoka, Y., Shiratori, S. and Einaga, Y. (2008). Photo-Control of Adhesion Properties by Detachment of the Outermost Layer in Layer-by-Layer Assembled Multilayer Films of Preyssler-Type Polyoxometalate and Polyethyleneimine, *Chem. Mater.*, **20**, pp. 4004–4010.
206. Cheng, L., Niu, L., Gong, J., *et al.* (1999). Electrochemical Growth and Characterization of Polyoxometalate-Containing Monolayers and Multilayers on Alkanethiol Monolayers Self-Assembled on Gold Electrodes, *Chem. Mater.*, **11**, pp. 1465–1475.
207. Gao, L., Wang, E., Kang, Z., *et al.* (2005). Layer-by-Layer Assembly of Polyoxometalates into Microcapsules, *J. Phys. Chem. B*, **109**, pp. 16587–16592.
208. Geisberger, G., Paulus, S., Carraro, M., *et al.* (2011). Synthesis, Characterisation and Cytotoxicity of Polyoxometalate/Carboxymethyl Chitosan Nanocomposites, *Chem. Eur. J.*, **17**, pp. 4619–4625.
209. Feng, Y., Han, Z., Peng, J., *et al.* (2006). Fabrication and Characterization of Multilayer Films Based on Keggin-Type Polyoxometalate and Chitosan, *Mater. Lett.*, **60**, pp. 1588–1593.
210. Yamada, M. and Maeda, A. (2009). Heteropolyacid-Conjugated Chitosan Matrix for Triphase Catalyst, *Polymer*, **50**, pp. 6076–6082.
211. Plault, L., Hauseler, A., Nlate, S., *et al.* (2004). Synthesis of Dendritic Polyoxometalate Complexes Assembled by Ionic Bonding and Their Function as Recoverable and Reusable Oxidation Catalysts, *Angew. Chem. Int. Ed.*, **43**, pp. 2924–2928.
212. Toma, F., Sartorel, A., Iurlo, M., *et al.* (2010). Efficient Water Oxidation at Carbon Nanotube-Polyoxometalate Electrocatalytic Interfaces, *Nature Chem.*, **2**, pp. 826–831.
213. Moore, A., Kwen, H., Beatty, A., *et al.* (2000). Organoimido-Polyoxometalates as Polymer Pendants, *Chem. Commun.*, pp. 1793–1794.
214. Cannizzo, C., Mayer, C., Sécheresse, F., *et al.* (2005). Covalent Hybrid Materials Based on Nanolatex Particles and Dawson Polyoxometalates, *Adv. Mater.*, **17**, pp. 2888–2892.
215. Li, H., Qi, W., Li, W., *et al.* (2005). A Highly Transparent and Luminescent Hybrid Based on the Copolymerization of Surfactant-Encapsulated Polyoxometalate and Methyl Methacrylate, *Adv. Mater.*, **17**, pp. 2688–2692.
216. Li, H., Li, P., Yang, Y., *et al.* (2008). Incorporation of Polyoxometalates into Polystyrene Latex by Supramolecular Encapsulation and Miniemulsion Polymerization, *Macromol. Rapid Commun.*, **29**, pp. 431–436.
217. Qi, W., Wang, Y., Li, W., *et al.* (2010). Surfactant-Encapsulated Polyoxometalates as Immobilized Supramolecular Catalysts for Highly Efficient and Selective Oxidation Reactions, *Chem. Eur. J.*, **16**, pp. 1068–1078.

# Chapter 22

# Supported Metal Nanoparticles in Liquid-Phase Oxidation Reactions

Nikolaos DIMITRATOS,* Jose A. LOPEZ-SANCHEZ**
and Graham J. HUTCHINGS*

This chapter provides an updated overview of the recent developments in the selective transformation of organic compounds such as alkanes, alkenes, alcohols, polyols and aldehydes using green catalytic processes and metal supported nanoparticles in liquid phase oxidation processes. We focused on the recent developments and especially the design of novel metal supported nanoparticles, where the control of particle size and morphology of the fine metal supported nanoparticle is essential for producing a new generation of catalysts that can be highly active, sustainable and selective at moderate temperatures and pressures under green catalytic conditions. In this view we report the recent achievements, future challenges and directions for the synthesis of supported metal nanoparticles and the successful utilization for a large variety of chemical processes.

## 22.1. Introduction

The selective oxidation of organic compounds, such as alkanes, alcohols, polyols and aldehydes for the production of fine and specialty chemicals using supported metal nanoparticles as heterogeneous catalysts has been investigated thoroughly in recent years in academia as well as within industry. The main problem with current industrial processes for the synthesis of chemicals is the utilisation of typically stoichiometric inorganic reagents, which creates a massive environmental concern due to the associated toxicity and corrosion problems. An alternative is the development of heterogeneous catalysts that can easily be recovered, recycled and have a prolonged lifetime, and the use of green oxidants such as molecular oxygen or hydrogen peroxide to minimise chemical waste.[1–4] Several excellent reviews have addressed

*Cardiff Catalysis Institute, School of Chemistry, Cardiff University, Cardiff, UK, CF10 3AT. Present address: Department of Chemistry, University College London, 20 Gordon Street, WC1H 0AJ London, UK; Research Complex at Harwell, Rutherford Appleton Laboratory, Harwell Oxford, Didcot OX11 0FA, UK. **Stephenson Institute for Renewable Energy, Chemistry Department, The University of Liverpool, Crown Street, L69 7ZD, Liverpool, UK. ***Cardiff Catalysis Institute, School of Chemistry, Cardiff University, Cardiff, UK, CF10 3AT.

the catalytic application of solid materials in liquid-phase oxidation and the subject is very extense.[5–9] The aim of this chapter is to provide an updated overview of metal-supported nanoparticles that have been applied as catalysts in liquid-phase oxidation reactions, with emphasis on the more recent developments.

## 22.2. Oxidation of Alcohols and Aldehydes using Molecular Oxygen

### 22.2.1. *Ru-based catalysts*

Ruthenium-supported nanoparticles have been used for the aerobic oxidation of alcohols to carbonyl compounds (see Table 22.1). Kaneda and co-workers synthesised Ru-supported catalysts, where the monomeric Ru cation species were uniformly fixed on the surface of calcium hydroxyapatite, and it was found that the synthesised $Ru^{3+}$-hydroxyapatite acted as an effective heterogeneous catalyst for the oxidation of various alcohols using molecular oxygen (Eq. 22.1).[10] The oxidation of various alcohols such as benzylic and allylic alcohols was performed at 80°C using toluene as the solvent, with yields in the range of 92–99%. The catalyst was reusable without any detectable leaching. The proposed mechanism is based on the initiation of the oxidation of the alcohol by a ligand exchange between an alcohol and a chlorine species of the Ru-hydroxyapatite to form a Ru-alcoholate species, which undergoes a $\beta$-hydride elimination to produce the corresponding carbonyl compound and a Ru-hydride species. Further reaction of the hydride species with oxygen affords a Ru-hydroperoxide species, followed by a ligand exchange to regenerate the Ru-alcoholate species with the formation of oxygen and water.

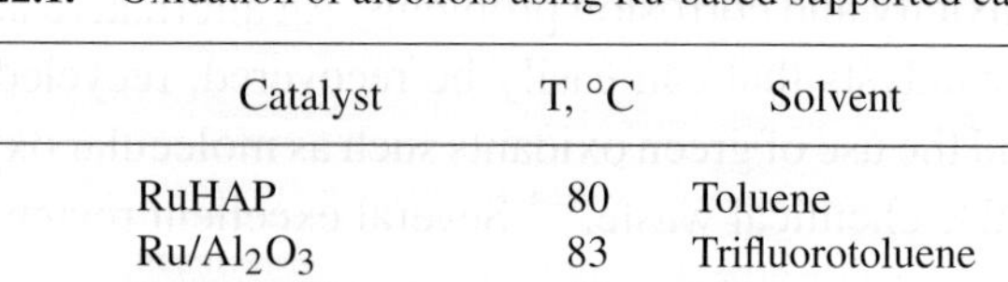


(22.1)

**Table 22.1.** Oxidation of alcohols using Ru-based supported catalysts.*

| Substrate | Catalyst | T, °C | Solvent | Conv. % | Ref. |
|---|---|---|---|---|---|
| Benzyl alcohol | RuHAP | 80 | Toluene | 100 | 10 |
| 2-Octanol | $Ru/Al_2O_3$ | 83 | Trifluorotoluene | 91 | 11 |
| Cyclohexanol | $RuO_2$-FAU | 80 | Toluene | 17 | 13 |
| 2-Thiophenmethanol | Ru-$Co(OH)_2$-$CeO_2$ | 60 | Benzotrifluoride | 100 | 14 |
| Benzyl alcohol | RuCoHAP | 90 | Toluene | 100 | 15 |
| Benzyl alcohol | Ru(III)/HAP-Bacid | 60 | Toluene | 100 | 17 |
| Benzyl alcohol | Ru-CHNAP-MgO | 80 | Toluene | 100 | 18 |
| Benzyl alcohol | Ru/C | 50 | Toluene | 100 | 19 |
| Benzyl alcohol | $Ru(OH)_x/TiO_2$ | 80 | Toluene | 100 | 20 |

*Note that selectivity to the aldehydes/ketone was ca. 99% in all entries.

One of the first examples of the solvent-free oxidation of alcohols was reported by Mizuno *et al.*[11] It was demonstrated that the effective aerobic heterogeneous oxidation of alcohols, which can possess a sulfur atom, a nitrogen atom or a carbon-carbon double bond, could be achieved by using a $Ru/Al_2O_3$ catalyst. The reusability of the catalyst was demonstrated as it was reused several times without loss of activity and without any leaching. The rate of the reactions was of zero order dependence on the pressure of molecular oxygen, and mechanistic investigation showed that the oxidation of alcohols proceeded by the formation of a Ru-alcoholate species via the formation of a ligand exchange between Ru-hydroxide and alcohol.[12] Subsequently, alcohol undergoes $\beta$ elimination to form the corresponding carbonyl compound and Ru-H species, which was finally re-oxidised by molecular oxygen. White and co-workers synthesised a zeolite-confined nanometre-sized $RuO_2$ with a mean particle size of 1.3 nm as determined using a one-step hydrothermal method.[13] It was reported that the $RuO_2$ nanoclusters exhibit high conversion and selectivity in the aerobic oxidation of various activated (benzylic and allylic) and unactivated (saturated) alcohols under mild conditions (80°C, toluene or chlorobenzene as the solvent and atmospheric pressure) with turnover number (TON) values of 5–15 (Eq. 22.2). It was demonstrated that the zeolitic framework displayed substrate shape-selectivity and the high activity of the encapsulated $RuO_2$ nanoparticles was attributed to the much higher density of active sites in nano$RuO_2$.

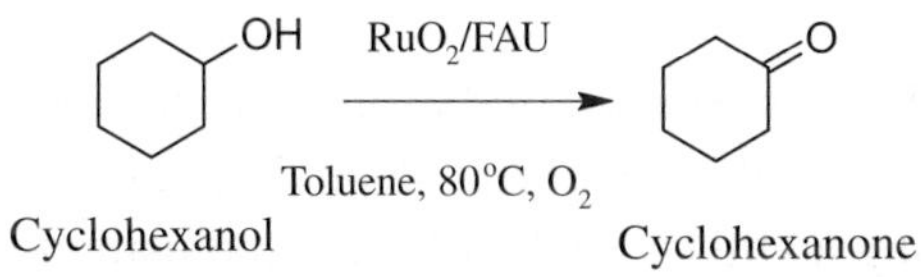


(22.2)

A Ru cation combined with cobalt hydroxide and cerium oxide ($RuCo(OH)_2$-$CeO_2$ was developed by Kaneda and co-workers using a co-precipitation method.[14] The obtained catalyst exhibited high catalytic activity for the oxidation of alcohols in the presence of molecular oxygen. It was found that a combination of Ru with Co and Ce elements is necessary to achieve high conversion levels. The addition of water enhanced the formation of carboxylic acids and the possible role of Co was to maintain the high oxidation state of Ru.

Baiker and co-workers developed a new methodology based on the hypothesis that the isolated Ru species should be close to the surface of the hydroxyapatite particles to improve the catalytic activity.[15] In order to achieve this goal, metal promoters were added, the amount of ruthenium used was reduced and the pre-treatment conditions were altered. It was claimed that by following this approach the efficient oxidation of aromatic and aliphatic alcohols was possible at 90°C using toluene as the solvent at 1 bar of $O_2$ pressure with turnover frequency (TOF) values in the range 15–80 $h^{-1}$. The higher activity of the promoted Ru catalysts was attributed to the steric effect of the promoter and the presence of $Ru(OH)_2$ species. Mechanistic studies confirm that the alcohol oxidation proceeds via the formation of Ru-alcoholate species, which undergo $\beta$-hydride elimination to produce the carbonyl compound and a hydrido-ruthenium species, which is then re-oxidised by molecular oxygen to close the catalytic cycle.[16] In subsequent studies Baiker and co-workers developed Ru organically-modified hydroxyapatite catalysts.[17] The enhanced catalytic activity they observed was due to the higher intrinsic activity the Ru species owed to their improved location and coordination in the organically modified hydroxyapatite. Ru species were located mainly on the outer surface and anchored to phosphate and hydroxyl groups, therefore these sites were more accessible to the alcohol substrate and as a consequence the catalytic activity was enhanced. The organic modifiers acted as templating agents for the controlled location and coordination of the Ru species.

An aerogel-prepared nanocrystalline MgO (NAP-MgO) modified by the incorporation of choline hydroxide was used for the deposition and stabilisation of the ruthenium species.[18] By using a combination of characterisation techniques, it was claimed that the development of $Ru^{3+}$ took place during the preparation method and the $Ru^{3+}$ species were distributed on the outer surface by a combination of strong electrostatic interaction and coordination between the surface functionalised MgO and $Ru^{3+}$. The catalyst showed high conversion of aromatic alcohols with yields of approximately 95% to ketones, using toluene as the solvent at 80°C and molecular oxygen as the oxidant. The catalyst was reused several times without significant loss of catalytic activity.

A Ru/C commercial catalyst was used as a heterogeneous catalyst for the oxidation of primary and secondary benzylic alcohols to the corresponding carbonyl compounds, without any additives such as bases, and using toluene as the solvent, molecular oxygen and temperatures in the range 50–90°C.[19] The authors also carried out the oxidation of allylic and aliphatic alcohols with good yields although reusability studies were not reported. Mizuno *et al.*[20] supported ruthenium hydroxide $Ru(OH)_x$ on three different $TiO_2$ supports and an $Al_2O_3$ support and showed high catalytic activity for the oxidation of alcohols including benzylic, allylic, aliphatic and heteroatom-containing ones into the corresponding carbonyl compounds with molecular oxygen as the oxidant. It was reported that the catalytic activity was dependent on the coordination number of the nearest neighbour Ru atoms (the size of $Ru(OH)_x$ species) and the $Ru(OH)_x/TiO_2$ catalyst could be reused several times without an appreciable loss of this catalytic activity. Subsequent theoretical studies[21] confirmed the proposed reaction mechanism via alcoholate formation and $\beta$-hydride elimination steps. It was found that the alcoholate formation was promoted by the "concerted activation" of an alcohol by the Lewis acid and Brønsted bases on a $Ru(OH)_x$/support, and the formed Ru-H species reacted with molecular oxygen to form Ru-OOH, which could further react with an alcohol or water regenerating the alcoholate or hydroxide species with the simultaneous formation of hydrogen peroxide.

**Table 22.2.** Oxidation of alcohols using Pd-based supported catalysts.

| Substrate | Catalyst | T, °C | Solvent | Conv. % | Sel. % (Ald/ket) | Ref. |
|---|---|---|---|---|---|---|
| Cinnamyl alcohol | $Pd_{561}phen_{60}(OAc)_{180}/TiO_2$ | 60 | Ac. acid | 100 | 99 | 23 |
| Benzyl alcohol | Pd(II)/HT | 65 | Toluene | 98 | 100 | 24 |
| 1-Phenylethanol | Pd/HAP-0 | 90 | TFT | 99 | 99 | 25 |
| Cinnamyl alcohol | Pd-OMC | 80 | SC $CO_2$ | 82.5 | 99 | 26 |
| Cinnamyl alcohol | $Pd/PEG/SiO_2$ | 80 | SC $CO_2$ | 96.8 | 98.5 | 27 |
| Benzyl alchohol | Pd/MS | 60 | SC $CO_2$ | 92.5 | 98.4 | 28 |
| 1-Octanol | Pd/NiZn | 80 | TFT | 99 | 99 | 29 |
| Benzyl alcohol | Pd-G/SBA-16-G | 80 | $SF/K_2CO_3$ | 99 | 99 | 30 |

TFT: trifluorotoluene; HT: hydrotalcite; SC: supercritical; SF: solvent-free.

### 22.2.2. *Pd-based catalysts*

Pd-based catalysts have been shown to catalyse many oxidation reactions, including hydroxylation of benzenes, oxidation of alkanes, oxidative coupling reactions, epoxidation of alkenes and oxidation of alcohols/aldehydes (see Table 22.2).[22] Different synthetic approaches have been used for the synthesis of active Pd-based catalysts.

Kaneda *et al.* immobilised giant Pd clusters with five shells, $Pd_{561}phen_{60}(OAc)_{180}$ on the surface of metal oxides, i.e. $TiO_2$ and studied their catalytic application.[23] The authors reported that the immobilised giant Pd clusters (2–3 nm mean particle size) efficiently catalysed the oxidation of primary allylic alcohols in the presence of molecular oxygen with conversion levels higher than 90%. The active sites are Pd-Pd paired sites with an oxidation state smaller than +2, where the oxidations took place via multiple interactions of these sites with the allylic alcohols.

A heterogenised Pd catalyst was developed by Kakiuchi *et al.*,[24] where a Pd(II)-hydrotalcite catalyst was synthesised by supporting a palladium(II) acetate-pyridine complex on commercially available hydrotalcite. Using a non-polar solvent (toluene) and pyridine they reported the oxidation of benzyl alcohol under mild conditions (65°C) to benzaldehyde with 98% yield, and the general applicability of the Pd(II)-hydrotalcite in the oxidation of aliphatic and alkenic alcohols as well as of diols. Moreover, the authors reported that the addition of pyridine is crucial for efficient conversion of the alcohol and that the temperature affects the final oxidation state of Pd. At high temperature (80°C) the reduction of Pd(II) to Pd(0) that took place resulted in an inactive catalyst (Eq. 22.3). Reusability tests showed that there is a progressive decrease of catalytic activity due to Pd leaching.

Kaneda and co-workers[25] extended their initial work with palladium nanoclusters[23] and demonstrated the synthesis of palladium-grafted hydroxyapatite, where Pd-supported nanoclusters can be synthesised in the presence of alcohol. The nanoclustered Pd(0) species can effectively oxidise a variety of aromatic alcohols, such as 1-phenylethanol, using molecular oxygen with a TOF of up to 9,800 $h^{-1}$ and TON of up to 236,000. They demonstrated that the catalyst is recyclable and that there is a dependence of catalytic activity on particle size (Eq. 22.4). They proposed that the oxidation of alcohols occurs primarily on low-coordinated Pd atoms so that the alcohol oxidation is considered "structure sensitive".

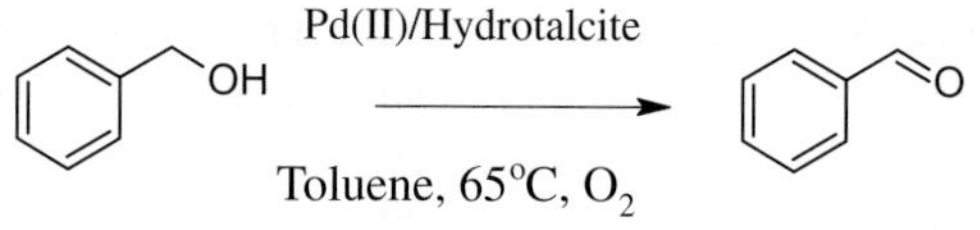


(22.3)

Benzyl alcohol Benzaldehyde

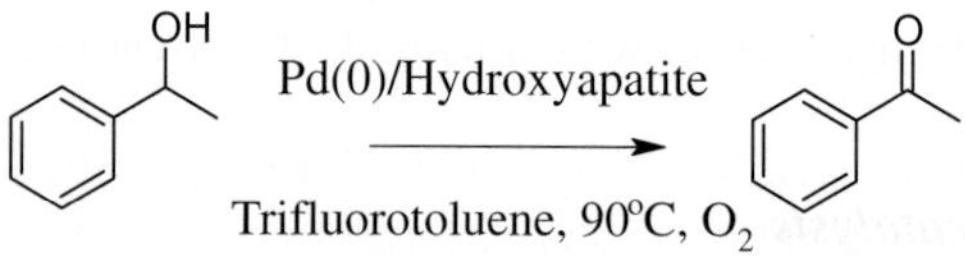


(22.4)

1-Phenylethanol Acetophenone

Schüth and co-workers[26] reported the synthesis of well-dispersed metallic nanoparticles on ordered mesoporous carbon (Pd-OMC), where highly temperature-stable dispersed Pd clusters, below 1 nm, were uniformly embedded in the carbon walls. They reported high catalytic activity under mild conditions (80–100°C) for the oxidation of aromatic alcohols, such as benzyl alcohol, cinnamyl alcohol and 1-phenylethanol, with selectivity around 99%, using supercritical $CO_2$ as the reaction medium (Eq. 22.5).

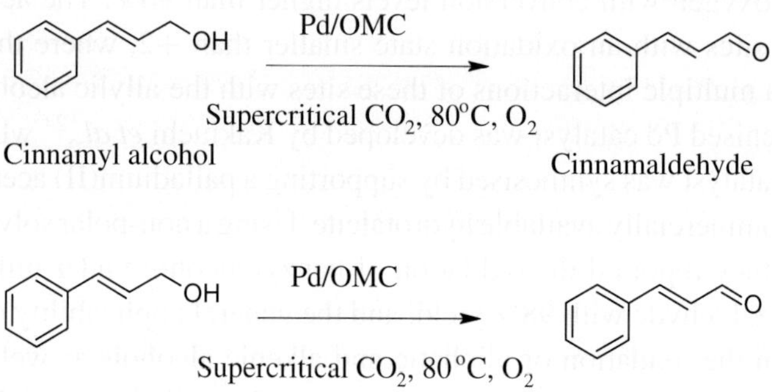


(22.5)

Leitner and co-workers also studied the oxidation activity of palladium using supercritical $CO_2$ as the solvent.[27] They reported efficient and stable catalysts for the selective aerobic oxidation of benzylic and allylic alcohols to aldehydes and ketones with selectivities over 98% and TON values in the range of 22–47, using supercritical $CO_2$ as the mobile phase in a batch as well as in continuous-flow process. The palladium nanoparticles were stabilised by polyethylene glycol (PEG)-modified silica and deposited on the surface of modified silica, and the authors claim that this procedure results in a decrease in agglomeration of the metal nanoparticles. In following studies[28] they reported the immobilisation of palladium nanoparticles on mesoporous silica using 2,2′-dipyridylamine as a linker, which facilitates the immobilisation and stabilisation of the nanoparticles. They observed that the method of reduction, either using benzyl alcohol under reflux conditions or molecular hydrogen, affected the particle size as well as the oxidation state. They concluded that the formation of very small palladium particles with a metallic oxidation state was achieved using the benzyl alcohol treatment and it was beneficial in terms of catalytic activity. The hydrogen treatment resulted in an average particle size of approximately 6 nm, which was larger than the 2 nm particles obtained by the alcohol reduction approach.

Shimazu and co-workers[29] used a new approach for the synthesis of Pd(II) catalysts supported on Ni-Zn mixed basic salt. Specifically, their strategy was based on the intercalation of an anionic $[Pd(OH)_4]^{2-}$ species into an interlayer of NiZn. The intercalated anionic Pd hydroxide complex was rigidly fixed by the strong electrostatic interactions characteristic of the NiZn. The Pd/NiZn could act as an efficient heterogeneous catalyst for the aerobic oxidation of benzylic, aliphatic and allylic alcohols to the corresponding aldehydes/ketones without the need of any additives, and give a yield of around 90%. Reactions were performed in trifluorotoluene at 80°C using air flow conditions. The proposed reaction mechanism was based on the initial binding of the alcohol to the Pd(II) hydroxide species to form a Pd(II)-alkoxide species and the carbonyl product, and then the Pd(II)-hydride species could reductively eliminate water to form Pd(0), which could then be re-oxidised by molecular oxygen to form a Pd-peroxide species.

Ji and co-workers[30] reported the synthesis of supported Pd catalysts using the grafting method. Using SBA-16 as the chosen support, a mixture of a palladium-guanidine complex and guanidine was grafted onto SBA-16 via a one-pot silylation. They claimed that the Pd-quanidine complex was introduced into the interior of SBA-16 and the mesoporous material structure of SBA-16 remained intact during the preparation method. The oxidation of alcohols using toluene as a solvent and $K_2CO_3$ as a base at 80–100°C was efficient, with selectivities to aldehydes/ketones of 99%. The catalyst was recyclable, and after 10 cycles only a minimum loss of activity was observed with no change of selectivity. Studies with a transmission electron microscope (TEM) show the formation of Pd nanoparticles that were stabilised by the nanocages of SBA-16, therefore preventing aggregation and agglomeration into larger particles.

Mechanistic and structural studies on palladium systems have been extensively investigated by Baiker and co-workers[31] in the selective oxidation of benzyl alcohol, cinnamyl alcohol and 1-phenylethanol. The oxidation of the aforementioned substrates using site-selective blocking by CO and isotope labelling with *in situ* attenuated total reflection infrared spectroscopy (ATR-IR) proved to be highly informative. The discrimination of the active sites was studied using a commercial $Pd/Al_2O_3$ catalyst with a mean metal particle size of 3–4 nm. High-resolution transmission electron microscopic images (HR-TEM) of the commercial $Pd/Al_2O_3$ showed that the palladium particles are round-shaped particles with defined (111) and (100) crystalline planes, that can be approximated by the cub-octahedron shape. During the liquid-phase oxidation of benzyl alcohol, the sites active for the desired dehydrogenation reaction are all planes exposed by the Pd particles, whereas those catalysing the undesired decarbonylation of the product are (111) planes. Moreover, Baiker and co-workers investigated the possible role of the oxidation state of Pd and the desired oxidation state that will favour aerobic alcohol oxidation.[32] Using a combination of *in situ* X-ray adsorption spectroscopy with on-line catalytic measurements and Fourier transform infrared (FTIR) spectroscopy, the authors identified that metallic palladium is the more active phase compared to palladium oxide. The increased catalytic activity observed with the metallic palladium was a factor of 50 times higher than the palladium oxide. The authors stated that for optimising conditions for the oxidation of alcohols, the Pd constituent should be kept in the metallic state but at the same time be allowed an optimal availability of oxygen to remove the formed hydrogen and degradation products from the catalyst surface.

### 22.2.3. *Au-based catalysts*

Au-based catalysts have been explored extensively over the last 15 years in the selective oxidation of alcohols, polyols, aldehydes and sugars.[33,34] Rossi and Prati demonstrated the efficient transformation of vicinal diols to $\alpha$-hydroxy carboxylates, with molecular oxygen in alkaline solution using gold-based catalysts.[35] The high activity and especially high chemoselectivity of gold towards the primary alcoholic function with comparison to more commonly used Pd- and Pt-supported catalysts was demonstrated. Gold on carbon showed the best stability and neither deactivation nor metal leaching were observed after recycling experiments. Prati and Martra[36] investigated the effect of the preparation method using deposition-precipitation (DP) versus the sol-immobilisation method in the liquid-phase oxidation of ethane-1,2-diol (see Table 22.3). It was demonstrated that the activity of Au/C catalysts prepared by the sol method showed an increase of catalytic activity by a factor of two, compared to a similar catalyst prepared by the deposition-precipitation method. The difference in catalytic activity was ascribed to the smaller particle size and higher dispersion of gold particles on the support using the sol-immobilisation method. Studies by Rossi and co workers[37] demonstrated that catalytic activity depends not only on the gold particle size but also its surface concentration. Moreover, different

**Table 22.3.** Oxidation of alcohols using Au-based supported catalysts.

| Substrate | Catalyst | T, °C | Solvent | Conv. % | Sel. % (Ald/ket) | Ref. |
|---|---|---|---|---|---|---|
| 1,2-Propanediol | Au/C | 90 | Water/NaOH | 78 | 100(Acid) | 35 |
| Ethane-1,2-diol | Au/C | 70 | Water/NaOH | 100 | 96(Acid) | 36 |
| D-Glucose | Au/C | 50 | Water/NaOH | 99 | 99(Acid) | 39 |
| Glycerol | Au/G | 60 | Water/NaOH | 56 | 100(Acid) | 50 |
| Glycerol | Au/C | 60 | Water/NaOH | 90 | 83(Acid) | 52 |
| Glycerol | Au/$CeO_2$ | 60 | Water/NaOH | 33 | 44(Acid) | 54 |
| Salicylic alcohol | Au/$Fe_2O_3$ | 60 | Water | 90 | 90 | 58 |
| 3-Octanol | Au/$CeO_2$ | 80 | Solvent free | 97 | 99 | 59 |
| Ethanol | Au/$MgAl_2O_4$ | 150 | Water | 97 | 86 | 60 |
| Benzyl alcohol | Au/GMS | 130 | Toluene | 99 | 79 | 61 |
| 4-MBA | Au/$Cu_5Mg_1Al_2O_x$ | 90 | Mesitylene | 98 | 99 | 62 |
| 1-Phenyethanol | Au-Microgel | 60 | Water | 75 | 100 | 64 |
| Benzyl alcohol | Au/$MnO_2$NR | 120 | Solvent free | 41 | 99 | 65 |
| Benzyl alcohol | Au/$\gamma$-Ga2O3 | 130 | Solvent free | 40 | 98 | 66 |
| Benzyl alcohol | Au/Ga3Al3O9 | 80 | Toluene | 98 | 99 | 67 |
| Benzyl alcohol | Au/PoPD | 25 | Water | 99 | 99(Acid) | 68 |
| 1-Phenylethanol | Au/Zn | 90 | Solvent free | 6.5 | 99 | 69 |
| Benzyl alcohol | Au/SBA-15 | 80 | Water | 100 | 91(Acid) | 70 |
| 1-Phenylethanol | Au/PNIPAM | 80 | Water | 99 | 99 | 71 |
| Cyclohexanol | Au/$Fe_3O_4$@$SiO_2$ | 100 | Toluene | 42 | 100 | 72 |
| Benzyl alcohol | Au/$ZrO_2$ | 94 | Solvent free | 59.5 | 81.5 | 79 |
| 1-Phenylethanol | Au/TiO2 | 90 | Water | 99 | 100 | 81 |
| Benzyl alcohol | Au/SBA-15 | 60 | Water/$K_2CO_3$ | 99 | 87(Acid) | 82 |

MBA: methylbenzylalcohol; NR: nanorods.

trends in activity were found when using gold-supported oxides as compared to gold-supported on carbon. In the former case, an increase of gold particle size led to a decrease of catalytic activity, whereas in the latter case a maximum catalytic activity was observed with gold particles with a mean diameter of around 7–8 nm. The authors ascribed this difference to the possibility that small gold particles can lie deeper in the carbon than larger particles; there is, therefore, a limitation to the reagent accessibility to the deeper gold particles. Rossi *et al.* also studied several parameters in the synthesis of metal sols and their effect on the liquid-phase oxidation of polyols.[38] It was concluded that the choice of stabiliser, concentration of metal and choice of support affected metal particle size. Moreover, the particular nature of the support drastically influenced the choice of the sol for maintaining, once supported, the gold particle size observed in the solution. It was concluded that the effect of the stabilizer on catalyst activity was negligible, since similar-sized particles showed the same activity although differently generated. The selective aerobic oxidation of D-glucose to D-gluconate was achieved with the highest selectivity using gold-based catalysts under mild reaction conditions. It was shown that Au/C could be an alternative catalyst to most of the multimetallic catalysts based on palladium and platinum metals. In addition, a gold-supported catalyst was the least sensitive to chemical poisoning, and metal leaching was dependent upon the reaction conditions.[39] In subsequent studies Rossi and co-workers synthesised an Au/C catalyst with a mean particle size of 3.6 nm and showed that the role of the support is to stabilise the gold nanoparticles.[40] Furthermore, by optimising the reaction conditions and comparing with the enzymatic process, it was shown that TOF values of 150,000 $h^{-1}$ at 50°C could be achieved with the inorganic catalyst, whereas with the enzymatic catalyst TOF values of 550,000 $h^{-1}$ were reported,[41] showing the potential of gold-based catalysts to achieve high plant productivity.

Claus and co-workers investigated the effect of gold particle size in glucose oxidation with Au/C catalysts, with a mean gold particle size in the range of 3–6 nm also prepared using a colloidal method. It was observed that decreasing the particle size resulted in an increase in the specific gold surface area and therefore the rate of glucose oxidation increased. A Langmuir–Hinshelwood model was proposed where the overall reaction rate is limited by the surface oxidation reaction. The proposed mechanism was based on a dehydrogenation pathway, where D-glucose transforms to an aqueous solution in the hydrated form which, after adsorption on the catalyst surface, is dehydrogenated and finally desorbed.[42]

Prüße and co-workers used a DP methodology for the synthesis of gold catalysts for glucose oxidation.[43] When specifically using NaOH or urea as precipitation agents and alumina as the support, they reported highly active and selective catalysts which showed an excellent long-term stability. DP-urea was found to be a better method due to the fact that no loss of gold occurred during the preparation.

Haruta and co-workers[44] reported the effect of support and size of the gold particles on glucose oxidation using a new preparation method, which is based on the solid grinding, by using a volatile organogold complex [$Me_2Au(acac)$] (acac = acetylacetonate). Following this methodology it is possible to deposit gold clusters smaller than 2 nm in diameter onto porous coordination polymers, several kinds of metal oxides and carbon supports. From their studies they concluded that the most active catalyst was gold on $ZrO_2$ and the control of particle size is critical in the liquid-phase oxidation.

Glycerol is a by-product of biodiesel production which has recently attracted significant research interest since it is a highly functionalised molecule and a large number of products can be formed from glycerol oxidation.[45–49] Hutchings and co-workers demonstrated for the first time the utilization of gold-based catalysts in the selective oxidation of glycerol to glycerate using alkaline conditions (Eq. 22.6). By synthesising gold-supported nanoparticles of around 25 nm mean particle size and varying parameters such as pressure, sodium hydroxide and catalyst amount, they optimised the conversion of glycerol and yield to glycerate.[50] It was proposed that the role of sodium hydroxide was essential for the initial dehydrogenation pathway, since in the presence of the base, hydrogen is readily abstracted from one of the primary hydroxyl groups of glycerol and becomes the rate-determining step. In subsequent studies they demonstrated the superior performance of the gold catalyst with respect to Pd- and Pt-based catalysts in terms of high selectivity to glycerate and higher catalytic activity.[51]

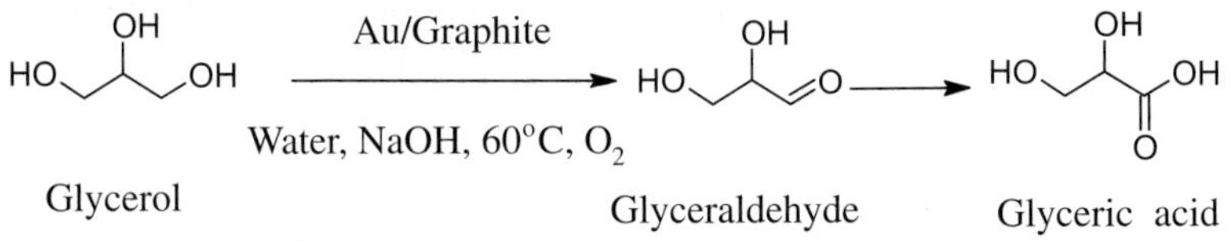


(22.6)

Prati and Porta investigated the effect of different preparation methods where a variation of particle size could be achieved and from their studies concluded that small gold nanoparticles, of 6 nm mean diameter and well dispersed, were responsible for the high activity to glycerate.[52] However, the initial selectivity could not be maintained due to the consecutive oxidation of glycerate to tartronate. Large nanoparticles of around 20 nm were responsible for high selectivity to glycerate without over-oxidation of glyceric acid.[52] From mechanistic studies they concluded that the overall selectivity of the reaction is affected by a combination of factors such as initial selectivity of the catalyst, base-catalysed interconversion and stability of the products.

Claus and co-workers investigated the effect of the support as well as gold particle size and concluded that under the same reaction conditions and with similar particle size, the carbon-supported gold catalysts are more active than the oxide-supported gold catalysts.[53] From the investigation of the gold particle size they concluded that the reaction is structure sensitive in agreement with the previous observations by Hutchings and Prati. In subsequent studies they investigated the effect of ceria as a support; although the ceria-supported catalysts were active, there was a considerable decrease in catalytic activity during recycling tests due to gold leaching.[54] The effect of the gold particle was also studied by Davis and co-workers who confirmed that small gold nanoparticles are more active than large gold nanoparticles and the selectivity to glycerate dropped as the gold particle size decreased. Most importantly, Davis *et al.* observed the formation of hydrogen peroxide over all gold catalysts and that the concentration of hydrogen peroxide had a direct relationship with the selectivity to glycerate; lower amounts of hydrogen peroxide were associated with higher selectivity to glycerate and less formation of glycolate. They concluded that the formation of glycolic acid, which is due to C-C cleavage, may be unavoidable over monometallic gold catalysts.[55,56]

The effect of the preparation method was studied by Prati and co-workers and it was found that a low temperature chemical reduction of the gold-supported catalyst enhanced the activity due to the formation of gold particles in the range of 2–5 nm, in a narrower particle size distribution. The use of a higher pre-treatment temperature (400°C) resulted in lower activity due to the increase of gold particle size and also higher selectivity to glycerate due to the suppression of the over-oxidation.[57]

The oxidation of alcohols with gold-based catalysts has been extensively investigated over the last decade (see Table 22.3). Galvagno and co-workers investigated the liquid-phase oxidation of *o*-hydroxybenzylalcohol under mild conditions (50°C, $P_{O2} = 1$ atm), using an $Au/Fe_2O_3$ catalyst synthesised by the co-precipitation method, and it was found that the catalytic activity increased with gold loading, while the selectivity to aldehyde at similar levels of conversion was higher with lower metal loading. Moreover, it was reported that the reaction is of the first order with respect to the organic substrate and zero order with respect to oxygen partial pressure.[58]

Corma and co-workers demonstrated that the combination of small gold particles (2–5 nm) and nanocrystalline ceria (5 nm) can produce a highly active, selective and recyclable catalyst for the oxidation of alcohols into aldehydes and ketones with high TON values under solvent-free conditions (Eq. 22.7). Based on mechanistic studies it was demonstrated that the deposition of gold nanoparticles transform nanocrystalline cerium oxide from a stoichiometric oxidant into a catalytic material.[59]

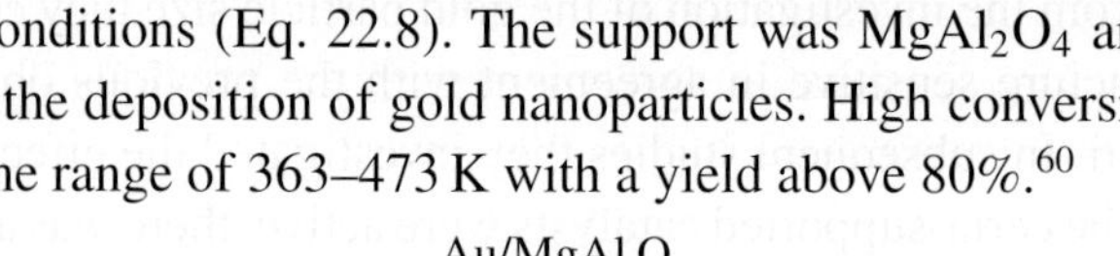


(22.7)

Christensen and co-workers demonstrated the feasible formation of acetic acid by aqueous-phase oxidation of ethanol with air, using gold-supported nanoparticles under mild conditions (Eq. 22.8). The support was $MgAl_2O_4$ and the DP method was used for the deposition of gold nanoparticles. High conversion of ethanol was achieved in the range of 363–473 K with a yield above 80%.[60]

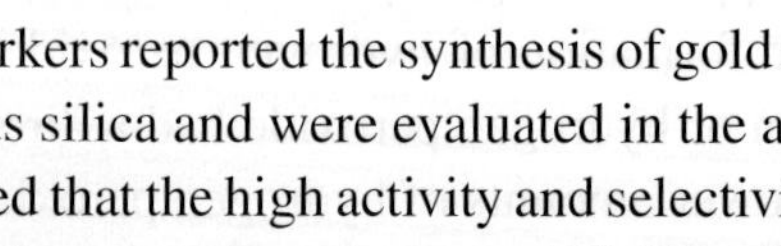


(22.8)

Richards and co-workers reported the synthesis of gold nanoparticles confined in the walls of mesoporous silica and were evaluated in the aerobic oxidation of alcohols. The authors showed that the high activity and selectivity to aldehyde (79–94%) was achievable using toluene as the solvent and $K_2CO_3$. The catalyst was reusable showing that the gold nanoparticles will not sinter when inside the pore channels of the mesoporous silica.[61]

Baiker and co-workers deposited gold nanoparticles (6–9 nm range) on Cu-Mg-Al mixed oxides using a DP method and the synthesised catalysts were tested in the aerobic liquid-phase oxidation of alcohols.[62] The catalytic tests showed that the activity strongly depended on the composition of the support, especially of the Cu/Mg molar ratio. X-ray absorption near edge structure (XANES) analysis revealed the presence of reduced and oxidized gold species on the ternary mixed-oxide support before and after the reaction. The best catalytic performance for a ternary mixed-oxide-supported gold catalyst was achieved using the following composition $Au/Cu_5Mg_1Al_2O_x$. The oxidation of several alcohols was achieved at full conversion and at high selectivity (98%). In subsequent studies, Baiker and co-workers studied in depth the oxidation state of gold during aerobic alcohol oxidation using XANES and they observed that the catalytic activity increased with the increase in the reduction of the gold component.[63] Therefore, they concluded that the main active species in the catalytic aerobic oxidation is metallic gold.

The use of microgels for stabilising gold nanoclusters and their use as "guasi-homogeneous" catalysts for the aerobic oxidation of primary and secondary alcohols in water was demonstrated by Prati and co-workers.[64] It was shown that the microgels containing gold can exhibit high and comparable activity with known and well-used gold-supported catalysts.

Cao and co-workers deposited gold nanoparticles (4–5 nm) on $MnO_2$ nanorods and demonstrated high catalytic activity in the liquid-phase oxidation of benzyl alcohol and 1-phenylethanol.[65] Comparison with commercial $Au/MnO_2$ showed that the $Au/MnO_2$ nanorods exhibited a much higher catalytic performance which was attributed to the collaborative effects resulting from the interaction of the gold nanoparticles and the well-defined reactive surface of the $MnO_2$ nanorods.

Gold nanoparticles supported on different polymorphs of Gallia ($\alpha$-$\beta$- and $\gamma$-$Ga_2O_3$) were synthesised and tested in the solvent-free liquid-phase oxidation of benzyl alcohol with molecular oxygen.[66] Different activity was obtained among the gold supported on different polymorphs of Gallia, suggesting the influence of the support for determining catalytic activity. The most active gold-supported catalyst was obtained when $\gamma$-$Ga_2O_3$ was the chosen support. The general applicability and the reusability of the Au/$\gamma$-$Ga_2O_3$ as an efficient catalyst for oxidizing a range of aromatic and aliphatic alcohols, with selectivities up to 98% to their corresponding ketones and aldehydes, has been demonstrated. Enlightenment for the high activity of the gold-supported nanoparticles on Gallia was based on spectroscopic investigation and it was found that a significantly higher dehydrogenation capability, using $\gamma$-$Ga_2O_3$ in comparison with other supports, exists. In subsequent studies, a series of binary mesostructured Ga-Al mixed-oxide supports ($Ga_xAl_{6-x}O_9$, $x = 2, 3, 4$) were synthesised and gold was deposited.[67] The $Au/Ga_xAl_{6-x}O_9$ materials showed high catalytic performance in the aerobic oxidation of alcohols under mild conditions

(80°C) and high selectivity to the aldehyde/ketone formation (Eq. 22.9). Comparison with other conventional oxide-supported systems showed that Au/$Ga_xAl_{6-x}O_9$ was much more active and the authors concluded that the collaborative interaction between gold and the mixed-oxide support is crucial for alcohol oxidation. Reusability tests confirmed that the catalyst is reusable and the spinel structure of the oxide was retained during the oxidation process.

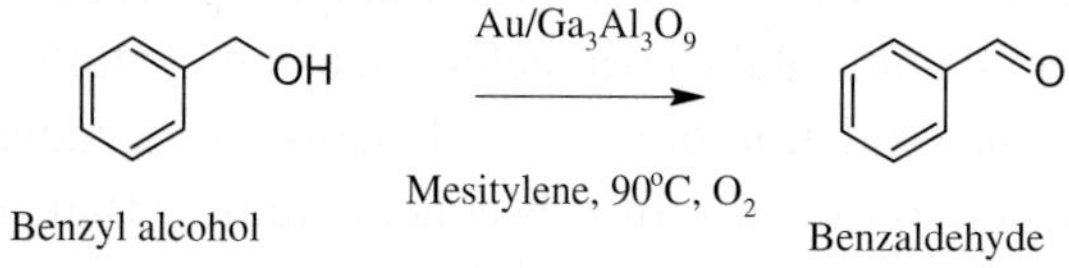


(22.9)

A novel method for the synthesis of gold nanoparticles with different particle size (3–15 nm) and shape, supported on both the inner and outer surfaces of poly(*o*-phenylenediamine) (PoPD) hollow microspheres was developed by Guo and co-workers.[68] The supported gold nanoparticles were active in the liquid-phase aerobic oxidation of alcohols using water as a solvent and $K_2CO_3$, under very mild conditions (room temperature) with yields of 91–95%.

Hensen and co-workers demonstrated that gold nanoparticles (4–6 nm range) supported by basic hydrozincite or bismuth carbonate are active catalysts for liquid-phase aerobic alcohol oxidation.[69] The catalytic performance of a series of metal (Zn, Bi, Ce, La, Zr) carbonate-supported gold catalysts shows a strong dependence on the basicity of the supporting materials. The high catalytic activity was related to the presence of accessible strong basic sites, where the initial O-H bond cleavage is possible on the support basic sites. The authors claim that this approach is promising for the development of heterogeneous catalysts possessing strong base sites for alcohol oxidation because in this way there is no need for the utilisation of soluble bases.

Immobilisation of around 1 nm Au clusters within mesoporous silicas (SBA-15, MCF, HMS) using triphenylphosphione-protected Au clusters (Au11:TPP) as precursors was presented by Tsukuda and co-workers. It was shown that the Au11:TPP clusters were homogeneously dispersed and by controlled calcinations the removal of the protecting ligands was achieved without aggregation of the resulting Au clusters. The catalytic properties of the gold-supported clusters were studied in water with the addition of $K_2CO_3$ and using benzyl alcohol as a model reaction with a yield of 91% to benzoic acid (Eq. 22.10).[70]

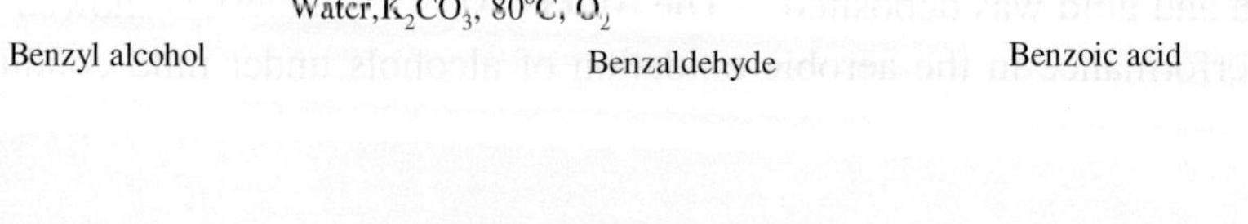


(22.10)

Zhang and co-workers reported the synthesis of size-controlled Au nanoparticles (2.6–6.3 nm) within a porous chelating hydrogel of poly(N-isopropylacrylamide)-co-poly[2-methacrylic acid 3-(bis-carboxymethylamino)-2-hydroxypropyl ester] referred to as (PNIPAM-co-PMACHE].[71] The encapsulated Au nanoparticles were tested for aerobic alcohol oxidation (oxidation of 1-phenyethanol) in the presence of water and KOH. It was demonstrated that the catalytic activity depends on the Au particle size. The authors claimed that the Au-encapsulated nanoparticles in the hydrogel were highly efficient and reusable catalysts.

Rossi and co-workers synthesised Au nanoparticles and immobilised them on a magnetically recoverable support (core-shell $Fe_3O_4@SiO_2$). The synthesised Au-supported catalysts were tested in the liquid-phase oxidation of benzyl alcohol, 1-phenylethanol and cyclohexanol using toluene as the solvent, $K_2CO_3$ at 100°C with conversions of 100%.[72] The reusability of the Au-supported catalyst was poor and the authors claimed this was due to particle size growth and a change in the morphology of the support. The authors claim that a magnetically-recoverable support offers advantages for the recovery of the catalyst.

Methyl esters are important products in the chemical industry, e.g. in the synthesis of flavouring agents, solvent extractants and intermediates as well as in the fragrance industry. Therefore, the synthesis of esters, by avoiding the traditional route (which involves the reaction of carboxylic acid and methanol) and using strong acid catalysts, such as sulfuric, sulfonic and *p*-toluenesulfonic acid or base alkoxides, will be desirable due to environmental concerns.[73,74] The selective oxidation of alcohols in methanol for the synthesis of the corresponding methyl esters has been reported by several groups (see Table 22.4). Christensen and co-workers have synthesised potassium titanate nanowires and deposited gold nanoparticles, and have shown that at ambient temperature the aerobic oxidation of benzyl alcohol in methanol, to give methyl benzoate using a catalytic amount of base, can proceed with high efficiency (over 99% conversion and a yield of 93% to methyl benzoate). They demonstrated the significance of adding a base to the solution, since in the

**Table 22.4.** Oxidative esterification of alcohols/aldehydes using Au-based supported catalysts.

| Substrate | Catalyst | T, °C | Solvent | Conv. % | Sel. % | Ref. |
|---|---|---|---|---|---|---|
| Benzyl alcohol | $Au/K_2Ti_6O_{13}$ | 20 | Methanol/KOH | 99 | 93(MB) | 75 |
| Propane-1,2-diol | $Au/Fe_2O_3$ | | Methanol/NaOH | 99 | 72(ML) | 77 |
| 5-HMF | $Au/TiO_2$ | 130 | Methanol/NaOH | 100 | 98(DMF) | 83 |
| 5-HMF | $Au/TiO_2$ | 30 | Water/NaOH | 100 | 71(FDCA) | 84 |
| 5-HMF | $Au/CeO_2$ | 130 | Methanol/NaOH3 | 100 | 100(DMF) | 85 |
| 5-HMF | $Au/TiO_2$ | 130 | Water/NaOH | 100 | 95(FDCA) | 86 |

HMF: hydroxymethylfurfural; MB: methylbenzoate; ML: methyllactate.

presence of a base deactivation of the catalyst was prevented.[75] In subsequent work, they demonstrated the production of methyl esters by selectively oxidising aldehydes in a primary alcohol environment and using gold-supported catalysts.[76] They reported the formation of methyl benzoate at temperatures below 0°C. The general applicability was shown by synthesising acrylate esters at room temperature using air as the oxidant, with selectivities of more than 85% at 97% conversion. The proposed mechanism indicates that the alcohol is first oxidised to an aldehyde, which then forms a hemiacetal with methanol, and is further oxidised to the corresponding ester; the rate-determining step is the aerobic oxidation of alcohols to the corresponding aldehyde. In subsequent studies Christensen *et al.* reported the synthesis of methyl esters using polyols such as glycerol, 1,2-propanediol and 1,3-propanediol. Using gold-supported nanoparticles in the range of 3–7 nm it was shown that the synthesis of important chemicals such as methyl lactate from 1,2-propanediol, methyl acrylate from 1,3-propanediol and dimethyl mesoxalate from glycerol can be obtained with yields in the range of 40–90%.[77]

The oxidative esterification of alcohols was also demonstrated by the group of Rossi *et al.*; using the Brust method they synthesised gold-supported nanoparticles using $SiO_2$ as the support with a mean gold particle size of 5.7 nm.[78] They demonstrated the general applicability of the oxidation of alcohols to the corresponding methyl esters and the reusability of the synthesised catalyst obtaining high yields to the corresponding methyl esters in the range of 70–95%.

The utilisation of tertbutylhydroperoxide (TBHP) or hydrogen peroxide ($H_2O_2$) as alternative oxidants has been investigated using gold-supported nanoparticles. Choudhary and co-workers have shown the successful liquid-phase oxidation of benzyl alcohol to benzaldehyde using TBHP under mild conditions (95°C), with gold-supported nanoparticles synthesised by using a homogeneous precipitation method.[79,80] They studied the effect of the support and preparation method and concluded that the most active catalysts consisted of gold-supported nanoparticles with a high surface ratio of $Au^{3+}/Au^0$, whereas gold particle size seemed not to be a crucial parameter in the range 3–8 nm. In addition, the most effective supports were $TiO_2$, $ZrO_2$ and MgO.

The utilisation of $H_2O_2$ as an oxidant for the efficient oxidation of alcohols by gold-supported nanoparticles using solvent-free conditions was reported by Cao and co-workers. 1-Phenyethanol was oxidised at 90°C using gold-supported catalysts supplied by the World Gold Council and the most effective support was ceria. The general applicability of the methodology was demonstrated by oxidising a range of non-activated alcohols and the catalysts were reused several times.[81] Tsukuda and co-workers reported the oxidation of benzyl alcohol using hydrogen peroxide and microwave irradiation, and studied the effect of particle size by synthesising gold-supported nanoparticles in the range of 0.8–1.9 nm. They concluded that there is a size dependence of the activity of gold clusters on SBA-15, with smaller gold particles showing higher catalytic activity.[82]

The scientific community has recently focused on the transformations of 5-hydroxymethyl-furfural (HMF), which is an important chemical for the development of several compounds which finds applications in pharmaceuticals, antifungals and polymer precursors.[47] In particular, the most important future application could be in the synthesis of a polymer precursor and the synthesis of 2,5-furandicarboxylic acid (FDCA) from furfural, which can replace the current industrial production of terephthalic acid. Another approach is via the oxidative esterification of HMF to synthesise dimethylfuroate (DMF), which can be used as a monomer for the replacement of terephthalic acid in plastics and has the advantage of being soluble in many solvents. Christensen and co-workers reported the oxidative esterification of HMF to DMF using methanol as the solvent at 130°C, and a catalytic amount of sodium methoxide ($NaOCH_3$) in the presence of an $Au/TiO_2$ catalyst with a yield of 98% (see Table 22.4).[83] In subsequent studies they reported the catalytic oxidation of HMF to FDCA using gold-supported nanoparticles under mild conditions (30°C) and water as the solvent, with a 71% yield.[84] Corma and co-workers reported a similar method to synthesise DMF and FDCA. By using an $Au/CeO_2$ catalyst composed of gold nanoparticles and nanoparticulated ceria they demonstrated the efficient conversion of HMF to DMF, using methanol as the solvent and molecular oxygen as the oxidant.

By tuning reaction parameters such as temperature, pressure and substrate to catalyst ratio, 100% yield of DMF was obtained (Eq. 22.11). The gold-supported catalyst was reusable taking into account that a special regeneration procedure had to be followed.[85] In the case of FDCA they used the same material and by tuning the reaction conditions, (base amount, temperature and pressure) a yield of 95% was achieved.[86]

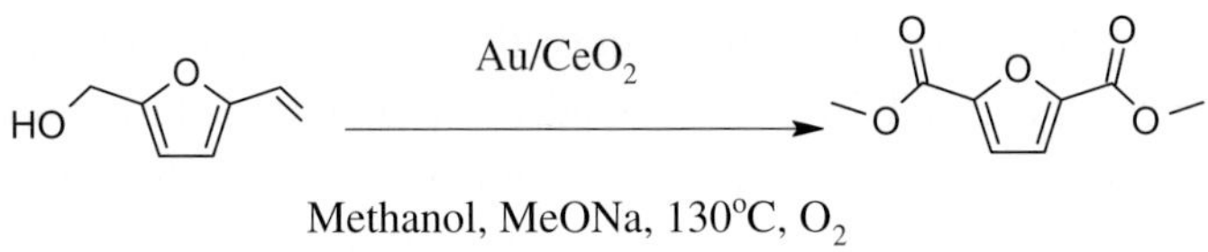


(22.11)

5-Hydroxymethyl-2-furfural 2,5-Dimethylfuroate

### 22.2.4. *Pt-based catalysts*

The utilisation of Pt-based catalysts for the liquid-phase oxidation of alcohols and polyols to highly valuable products has been extensively explored in academia in recent years (Table 22.5).[5,6] Inaya and co-workers reported the selective aerobic

**Table 22.5.** Oxidation of alcohols using Pt-based supported catalysts.

| Substrate | Catalyst | T, °C | Solvent | Conv. % | Sel. % (Ald/ket) | Ref. |
|---|---|---|---|---|---|---|
| Glycerol | Pt-Bi/C | 50 | Water | 95 | 80(DHA) | 87 |
| Cinnamyl alcohol | Pt-Bi/$Al_2O_3$ | 40 | Water/$Li_2CO_3$/DBSA | 99 | 97 | 89 |
| Cyclohexanol | Pt-Bi/C | 50 | Water-dioxane | 99 | 83 | 92 |
| Cinnamyl alcohol | Pt-Bi/Graphite | 60 | Ethanol | 65 | 95 | 93 |
| 1-Octanol | Pt-Bi/C | 60 | Toluene | 60 | 81 | 94 |
| Cinnamyl alcohol | Pt-Bi/$Al_2O_3$ | 65 | Toluene | 38 | 99 | 95 |
| 1-Phenylethanol | Pt/AC | 60 | Water | 81 | 100 | 96 |

DBSA: dodecylbenzensulfonic acid sodium salt; DHA: dihydroxyacetone.

oxidation of glycerol to dihydroxyacetone using Pt supported on charcoal under mild reaction conditions (50°C).[87] By incorporating bismuth in platinum, the selectivity to dihydroxyacetone increased remarkably from 10% to 80%. It was proposed that the formation of a bismuth submonolayer on Pt is responsible for the high selectivity towards dihydroxyacetone. Bismuth adatoms block the sites, which are responsible for unselective oxidation and control the glycerol orientation towards dihydroxyacetone formation. Comparison with the conventional fermentation process showed that the catalytic method they proposed showed higher productivity.

Gallezot and co-workers, in similar studies, confirmed that the deposition of bismuth on platinum particles using a low Bi/Pt molar ratio, orientates the selectivity towards the oxidation of the secondary hydroxyl group to form dihydroxyacetone with selectivities of 50% at 70% conversion. No leaching of bismuth into the solution was reported.[88] Baiker and co-workers synthesised Bi-Pt-supported catalysts by the selective deposition of Bi onto supported Pt particles of 3–4 nm. The synthesised Bi-Pt catalysts were used for the selective aerobic oxidation of cinnamyl alcohol to cinnamaldehyde as the desired product and using water as the solvent. From their studies it was concluded that an increase in Bi/Pt surface ratio resulted in a higher Bi coverage, which suppressed the hydrogen adsorption on Pt and therefore improved not only the activity but also the selectivity to cinnamaldehyde. The increased activity of the Bi-Pt-supported catalyst in comparison with the monometallic Pt catalyst was attributed to the presence of Bi particles that block the Pt sites (geometric blocking) responsible for the over-oxidation of cinnamaldehyde, reducing the self-poisoning effect of the substrate and by-products, and suppressing deactivation overall.[89]

Gallezot and co-workers investigated the liquid-phase oxidation of cyclohexanol to adipic acid using Pt/C catalysts and demonstrated that high activity can be obtained using water as the solvent at moderate temperature and pressure (150°C, 5 MPA air). The production of adipic acid using a green process and a heterogeneous catalyst is highly desirable due to the highly industrial importance of adipic acid as a crucial intermediate for the manufacture of nylon, but also as adipic acid is used as a plasticiser and food additive. During their investigation they showed that high conversion of cyclohexanol is achievable (90%) with 50% selectivity to adipic acid and the major by-products comprising glutaric and succinic acids. The main drawback of this method is the low solubility of cyclohexanol in water, however the use of water as a green solvent and air as an oxidant makes this process environmentally cleaner.[90]

The synthesis of hydroxypyruvic acid, a starting material for the synthesis of L-tyrosine, and its use as a flavour component has been reported by Bekkum and co-workers, using a Pt/C catalyst modified by bismuth.[91] In addition, they reported selectivity of 93% with 95% conversion of sodium glycerate. Exact control of the pH was necessary to regulate the selectivity towards the desired product. Gallezot and co-workers showed the efficient liquid-aerobic oxidation of unsaturated alcohols (9-decen-1ol) using Pt-supported catalysts under mild conditions.[92] The formation of 9-decenoic acid, which has been used in the preparation of flavour and fragrance ingredients, was targeted. The oxidation was performed at 50°C using molecular oxygen, NaOH as the base and water/dioxane as the solvent. They reported high conversion (99%) with selectivity of 83% by adding Bi on the Pt catalyst, minimising the deactivation and improving the activity and resistance of the catalyst. The oxidation of allylic alcohols was also studied by Lee and co-workers using Pt-Bi/graphite catalysts.[93] They performed systematic studies for understanding the influence of the double bond position within the chain in terms of activity and selectivity. It was concluded that the carbon double bond plays a role in the control of oxidation and depending on the position of the double bond within the chain, it can facilitate the anchoring of the hydroxyl group to the catalyst surface and therefore influence the oxidative dehydrogenation step to the aldehyde. Moreover, the effect of air pressure, catalyst mass and stirrer speed were investigated and it was concluded that aldehyde formation was zero order in dioxygen and the main role of oxygen was in the removal of carbonaceous species from the catalyst surface, and that the reaction was not mass transport limited.

Griffin and co-workers used a high throughput screening technique to identify trends in catalyst activity and product selectivity using different compositions of Pt-Bi-supported catalysts.[94] Using water as the solvent and air as an oxidant, they identified that the most efficient catalyst for the transformation of a variety of alcohols was 5%Pt-1%Bi/C. In addition, they demonstrated the utilisation of hydrogen peroxide as an oxidant. Baiker and co-workers investigated the effect of a promoter by using a simple approach.[95] The catalytic performance of various promoted (Bi, Pb) and unpromoted Pt-group catalysts in the oxidation of aliphatic, aromatic and allylic alcohols in the presence and absence of molecular oxygen was compared. From their studies it was concluded that the role of the promoter depends on the substrate used; it influences either the reaction rate and selectivity of the alcohol dehydrogenation reaction or the adsorption and transfer of oxygen. The latter ultimately involves the oxidation of the co-product, hydrogen, to water and the oxidative removal of surface impurities which results in the improved resistance of Pt metal against over-oxidation.

A methodology that was reported by Ikeda and co-workers, facilitates the incorporation of Pt nanoparticles of 3 nm particle size inside the framework of porous carbon and in this way prevents particle aggregation, movement and leaching, and has been shown to synthesise active and reusable catalysts for the oxidation of a range of alcohols.[96] The synthesised Pt on carbon catalyst was selectively transformed at high yield aromatic alcohols such as benzyl and 1-phenylethanol under mild conditions, using atmospheric oxygen, at 60°C and water as the solvent. The authors claim that the high activity observed is due to the incorporation of Pt nanoparticles inside the pores of the carbon, which consist of three-dimensional hydrophobic channels where efficient mass transfer and preferential adsorption of the reaction substrate can occur.

### 22.2.5. *Au-Pd, Au-Pt based supported catalysts*

There has been an explosion of interest in the synthesis of Au-Pd-supported catalysts and their utilisation in the aerobic liquid-phase oxidation of alcohols (Table 22.6). The main reasons for the academic and industrial interest rely on the fact that gold and palladium can form solid solutions over the whole range of gold/palladium atomic ratios and the addition of a second metal can alter the electronic and geometrical

**Table 22.6.** Oxidation of alcohols using Au-Pd-based supported catalysts.

| Substrate | Catalyst | T, °C | Solvent | Conv. % | Sel. % (Ald/ket) | Ref. |
|---|---|---|---|---|---|---|
| D-Sorbitol | Au-Pd/C | 50 | Water/NaOH | 90 | 90 (Acid) | 97,98 |
| Glycerol | Au-Pd/C | 50 | Water/NaOH | 90 | 90 (Acid) | 99–102 |
| Cinnamyl alcohol | Au-Pd/C | 60 | Water | 95 | 83 | 105 |
| Benzyl alcohol | Au80-Pd20/C | 60 | Water/NaOH | 90 | 99 | 106 |
| Benzyl alcohol | Au-Pd/$TiO_2$ | 100 | Solvent free | 90 | 95 | 107 |
| Benzyl alcohol | Au-Pd/C | 120 | Solvent free | 80 | 65 | 110 |
| 1,2-Propanediol | Au-Pd/$TiO_2$ | 60 | Water/NaOH | 94 | 96 (Acid) | 112 |
| Benzyl alcohol | Au-Pd/PI | 100 | Toluene | 99 | 98 | 114 |
| Benzyl alcohol | Au-Pd/SBA-15 | 80 | Water/$Na_2CO_3$ | 40 | 99 | 115 |
| Glycerol | Au-Pt/C | 50 | Water/NaOH | 70 | 60(GA) | 117 |
| Glycerol | Au-Pt/Mordenite | 100 | Water | 70 | 83(GA) | 118 |
| 1-Phenyethanol | Au-Pt/PI | 25 | Water/BTF | 99 | 99 | 119 |

BTF: benzotrifluoride; GA: glyceric acid.

properties of the synthesised particle, which will therefore affect catalytic activity, selectivity to the desired product and catalyst stability.

Prati and co-workers studied the effect of gold with palladium or platinum by synthesising bimetallic colloids and immobilising the synthesised bimetallic colloids on supports such as carbon and graphite.[97,98] The chosen model reactions were the selective oxidation of polyols (sorbitol and glycerol) and alcohols (aliphatic and benzylic alcohols) using mild reaction conditions (30–60°C, $P_{O2}$ lower than 4 atm). In the case of polyols oxidation (sorbitol and glycerol) bimetallic catalysts showed a remarkable, enhanced catalytic activity as well as selectivity to the desired product (gluconic acid) with respect to the monometallic catalysts and, in addition, exhibited enhanced stability due to the resistance of poisoning by dioxygen. Moreover, the effect of the Au-Pd atomic ratio was studied and a typical volcano-type catalytic behaviour was found with the most active bimetallic catalyst possessing an Au/Pd atomic ratio of 6:4. From their results the authors concluded that to enhance the catalytic activity of a bimetallic Au-Pd system, a small amount of the one metal in the presence of the other seems to be enough.

In the case of glycerol, the choice of the bimetallic system (Au-Pd, or Au-Pt), preparation method and support (carbon versus graphite) not only influenced catalytic activity, but also played a role in the distribution of products.[99,100] The utilisation of Au-Pd supported nanoparticles was shown to improve the selectivity towards the oxidation products of the terminal hydroxyl groups (glyceric acid and tartronic acid) with selectivities over 90% at conversion levels of 90% (Eq. 22.12). In the case of Au-Pt-supported catalysts, similar TOF values were obtained with respect to the Au-Pd-supported catalysts, and enhancement in the formation of glycolic acid indicates the facilitation of the oxidation of the secondary hydroxyl group of glycerol. The effect of the Au/Pd molar ratio was studied by the same authors and they observed that the catalytic activity was improved by increasing the Au/Pd molar ratio, with the highest activity corresponding to a rich Au system (Au/Pd = 9/1).[101,102] From their studies they concluded that the surface Pd monomer in contact with Au has a prominent promoting effect on activity and stability.

Synthesis of single-phase Au-Pd catalysts exhibited higher activity and reusability than random Au-Pd catalysts in the selective oxidation of glycerol.[103,104] By comparison, the catalytic performance of Au-Pt and Au-Pd single-phase catalysts in the selective oxidation of various primary alcohols (benzyl alcohol, cinnamyl alcohol and 1-octanol), showed that Au-Pd catalysts were far more active than Au-Pt catalysts (Eq. 22.13). A significant improvement in catalytic activity was found when water, instead of toluene, was used as a solvent, with TOF values increasing by a factor of 1.5–6.[105] Moreover, using the same preparation method, they demonstrated the general applicability of the single-phase alloy Au-Pd/C catalysts and systematically studied the effect of the $Au_xPd_y$ molar ratio on a range of alcohols. They concluded that the most efficient Au/Pd composition was $Au_{80}/Pd_{20}$, which showed the highest catalyst activity.[106]

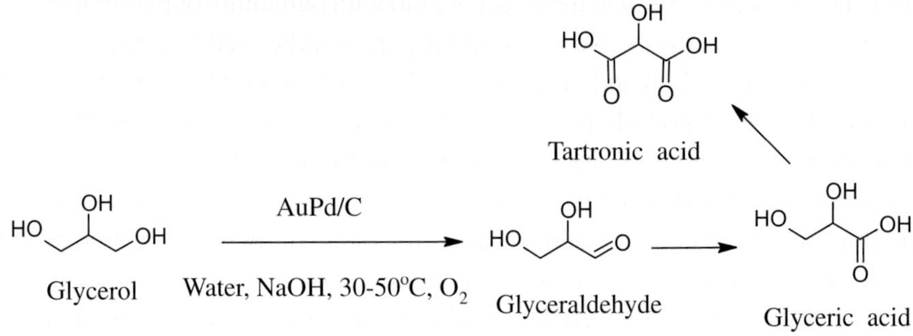


(22.12)

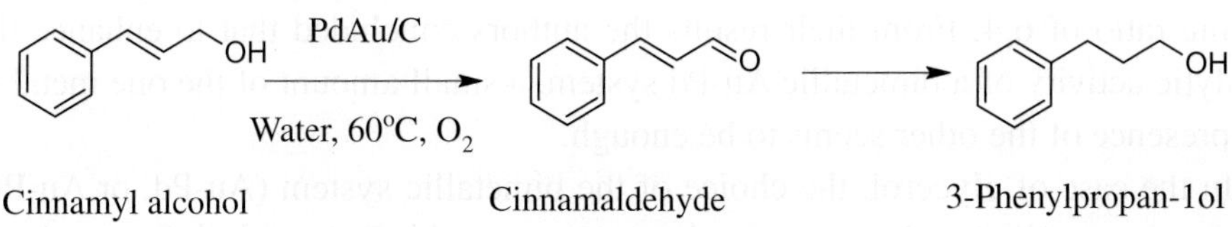


(22.13)

The most active bimetallic Au-Pd-supported catalysts were reported by Hutchings and co-workers, where the solvent-free liquid-phase oxidation of a variety of alcohols was possible at 160°C, employing molecular oxygen without the use of initiators or solvents, and using an Au-Pd/$TiO_2$ catalyst synthesised by the impregnation method (Scheme 22.1).[107] The TOF values reached up to 269,000 $h^{-1}$ and the high activity of the Au-Pd-supported catalyst was due to the gold core-palladium shell structure on the support surface and the electronic promotion of Au for Pd. Moreover, the Au/Pd weight ratio was studied and it was found that the most active catalyst comprised an Au/Pd weight ratio of 1:1, whereas the highest selectivity to benzaldehyde was observed with Au-rich catalysts.[108] In subsequent studies, Hutchings and co-workers investigated the synthesis of Au-Pd-supported catalysts using a colloidal method and they demonstrated the efficient aerobic oxidation of benzyl alcohol with very high TOF values under milder reaction conditions.[109] Scanning transmission electron microscopy-energy-dispersive X-ray spectroscopy (STEM-EDX) and X-ray photoelectron spectroscopy (XPS) analysis showed the presence of random homogeneous alloys with a metallic oxidation state for Au and Pd. The higher activity of the Au-Pd-supported catalysts synthesised by the colloidal method instead of the impregnation method was attributed to the smaller particle size, narrow particle size distribution and metallic oxidation state.

A preparation strategy for the synthesis of bimetallic hydrosols with the formation core-shell structures involving the sequential addition and reduction of the metal and

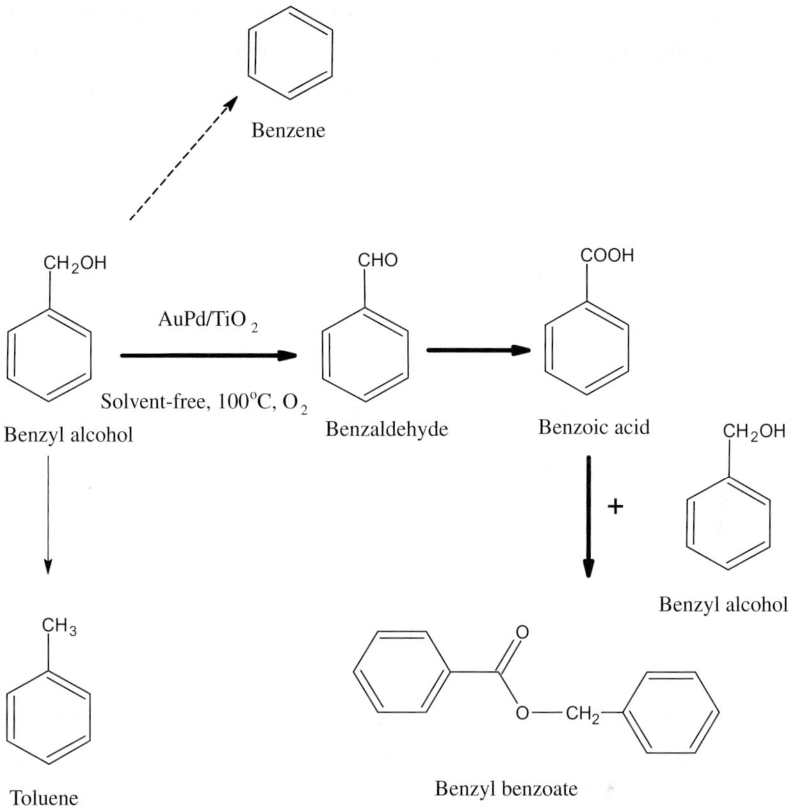


**Scheme 22.1.** Solvent-free oxidation of benzyl alcohol using Au-Pd-supported nanoparticles.

deposition of the bimetallic sols on carbon and titania was used for the synthesis of bimetallic-supported catalysts.[110] It was found that the catalytic activity of the aerobic oxidation can be achieved under mild conditions (120°C, $P_{O2}$ =10 bar) and the order of metal addition has a marked effect on activity. The choice of support (carbon versus titania) was shown to significantly affect the activity and distribution of products, with carbon-supported materials giving an increased activity by a factor of 2, and a lower selectivity to benzaldehyde at iso-conversion level compared to the titania-supported catalysts. It was found that the reaction is zero order in oxygen and the oxidation of benzaldehyde is dependent on the concentration of oxygen at the surface.

Comparison of two preparation methods (impregnation and sol-immobilisation methods) for the synthesis of Au-Pd-supported catalysts in the liquid-phase oxidation of glycerol led to the conclusion that high activity coupled with high selectivity to the desired product (glycerate) can be achieved by gold-rich surface bimetallic nanoparticles with a mean particle size of 3–5 nm and metallic oxidation state, whereas larger particles (over 6 nm) led to significantly lower catalytic activity, as observed with the impregnation method.[111] In subsequent studies Hutchings and co-workers demonstrated the excellent catalytic performance of Au-Pd-supported nanoparticles and the strong synergistic effect of the addition of gold into the palladium metal during the liquid-phase oxidation of 1,2-propanediol to the sodium salt of lactic acid, which is an important monomer for the synthesis of biodegradable polymers (Eq. 22.14).[112] It was shown that high selectivity to lactate was possible (96%) at 94% conversion and the usage of oxidants such as molecular oxygen as well as hydrogen peroxide was achievable under mild conditions (60°C and $P_{O2}$ =10 bar or atmospheric pressure).

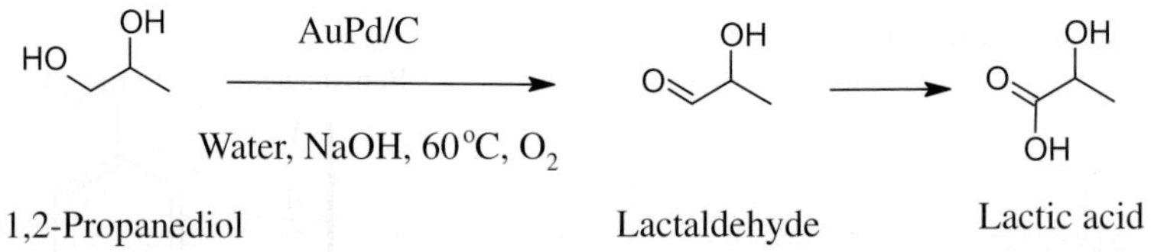


(22.14)

Mechanistic studies investigating the oxidation of benzyl alcohol using an Au-Pd-alloy catalyst synthesised by the sol-immobilisation method found that in the absence of oxygen, benzyl alcohol was transformed into benzaldehyde and toluene at initial equal rates.[113] Introduction of oxygen significantly increased the rate of benzyl alcohol disappearance and the appearance of benzaldehyde at the expense of the formation of toluene. It was found that at low partial pressures (below 3 bar of oxygen) rates are dependent on oxygen, suggesting that oxygen can participate in the reaction pathway as an adsorbed species. The beneficial interaction of gold and palladium in bimetallic catalysts was also demonstrated by Baiker and Marx by synthesising bimetallic catalysts using a colloidal route, where the admixing of Pd to Au resulted in the synthesis of bimetallic nanoparticles in a narrow range (2.4–3.7 nm) and an Au-rich core and a Pd-rich shell.[114] The synthesised Au-Pd-supported nanoparticles on polyaniline showed high activity in the liquid-phase oxidation of benzyl alcohol and high selectivity to benzaldehyde (98%) using toluene as the solvent and an aqueous solution of NaOH.

Qiao and co-workers reported the synthesis of Au-Pd/SBA-15 material by impregnation and grafting methods.[115] It was found that by using the grafting method, metal nanoparticles (5 nm mean particle size) were well-dispersed in mesoporous channels of SBA-15, showing high catalytic performance in the selective oxidation of benzyl alcohol to benzaldehyde under mild reaction conditions (80°C, water as the solvent and $Na_2CO_3$). It was demonstrated that agglomeration and leaching of metal nanoparticles was avoided by restricting the nanoparticles inside the mesopores of SBA-15, therefore leading to enhanced stability and reusability of Au-Pd/SBA-15. The effective confinement of Au-Pd nanoparticles was also reported by Yang and co-workers using SBA-16 as the support. Au-Pd nanoparticles were supported on SBA-16 using an adsorption method. The authors demonstrated the synergistic effect of the bimetallic Au-Pd-supported nanoparticles by showing the enhanced catalytic performance of the Au-Pd catalyst compared to the Au and Pd monometallic catalysts. STEM and EDX analysis revealed the alloyed structure of the bimetallic nanoparticles with a Pd-rich shell and an Au core structure.[116]

The utilisation of Au-Pt catalysts has been reported for the efficient aerobic liquid-phase oxidation of alcohols and polyols.[117] Au-Pt-supported nanoparticles were immobilised on carbon using colloidal methodology with a mean particle size in the range of 3–6 nm and the catalytic performance of the aerobic liquid-phase oxidation of glycerol was investigated. It was found that the effect of the reducing agent and the nature of the Pt precursor were affecting activity and selectivity. Au-Pt-supported nanoparticles showed a higher activity and resistance to poisoning in comparison to the monometallic Pt catalysts, and a substantial increase of activity with respect to the monometallic Au- or Pt-supported nanoparticles. It was also demonstrated that the oxidation of polyols (glycerol, 1,2-propanediol) is possible without the use of a base under mild conditions, with high selectivity to the carboxylic acid using Au-Pt-supported nanoparticles on mordenite, whereas monometallic Au could not be activated without the use of a base, and Pt exhibited high conversion but with low selectivity to glyceric acid.[118] A colloidal method was used for the synthesis of the Au-Pt nanoparticles and from STEM analysis the mean particle size was in the range of 3.7–4 nm.

The utilisation of Au-Pt-supported nanoparticles for the liquid-phase oxidation of a variety of alcohols has been presented with high yields (90–99%) of the corresponding aldehydes or ketones at room temperature, using atmospheric pressure and benzotrifluoride/water as the solvent (Eq. 22.15).[119] A colloidal method was used for the synthesis of the metal nanoparticles. X-ray spectroscopy confirmed the presence of both metals in each nanoparticle and the particle size distribution was in the range of 1.5–5 nm. Catalysts were reused several times without loss of activity.

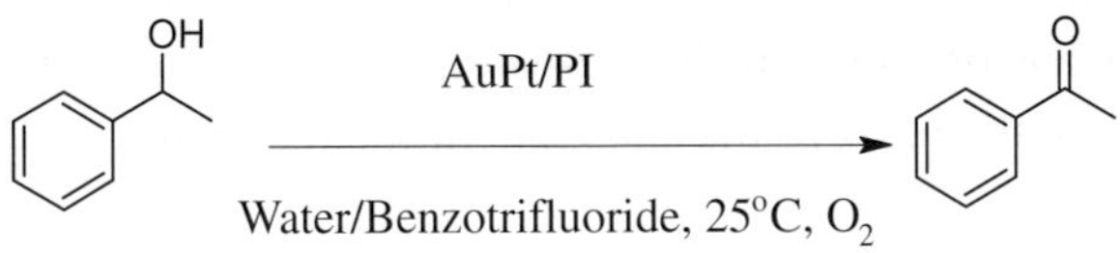


(22.15)

1-Phenylethanol Acetophenone

## 22.2.6. *Ag-based supported catalysts*

Recently, it has been reported that the utilisation of an Ag metal in combination with other metals or supports can provide active catalysts for a variety of reactions in liquid-phase oxidations (see Table 22.7).

**Table 22.7.** Oxidation of alcohols using Ag-based supported catalysts.

| Substrate | Catalyst | T, °C | Solvent | Conv. % (Ald/ket) | Sel. % | Ref. |
|---|---|---|---|---|---|---|
| Benzyl alcohol | Pd-Ag/Pumice | 60 | Acetonitrile | 100 | 100(BZ) | 120 |
| Benzyl alcohol | $CeO_2$+Ag/$SiO_2$ | Reflux | Xylene | 98 | 95(BZ) | 121 |

BZ: benzaldehyde.

Liotta and co-workers have reported the synthesis of Pd-Ag-supported nanoparticles and their utilisation in the liquid-phase oxidation of benzyl alcohol to benzaldehyde using molecular oxygen under mild conditions.[120] They proposed that the role of Pd is mainly the activation of the substrate and the role of Ag is the activation of oxygen, which then migrates via a "spill over" or "hopping" process to the near palladium site. Grunwaldt and co-workers reported that a physical mixture of ceria nanoparticles and Ag/$SiO_2$ performed similarly or even better, compared to palladium and gold catalysts, in the catalytic oxidation of benzyl alcohol to benzaldehyde with a conversion of 98% and selectivity up to 95%.[121] They proposed that Ag is activating the substrate and $CeO_2$ activates molecular oxygen.

## 22.3. Selective Oxidation of Hydrocarbons

The liquid-phase selective oxidation of hydrocarbons is a wide area of research, and relevant heterogeneously-catalysed industrial processes are mostly carried out with mixed oxides or ordered porous catalytic systems, which escape the scope of this chapter. For a more comprehensive survey of the industrial and technological developments within industrial liquid-phase oxidation reactions, we recommend generic literature, in particular the *Handbook of Heterogenous Catalysis*,[122] and recent reviews on oxidation with with transition-metal complexes by Bregeault *et al.*[123] and Punniyamurthy.[22] However, from a more academic perspective, there has been significant progress in this research area which we describe below, in sections divided on the basis of the substrate of the reaction.

### 22.3.1. *Alkene epoxidation*

Epoxidation continues to be an essential part of many chemical syntheses and is often used to functionalise less reactive starting materials to ensure they can be used in downstream processes. In particular, alkene epoxidation is of great relevance for the chemical industry. Molecular oxygen is the preferred oxidant for economic and environmental motivations, however, many molecules and catalysts at present are unreactive with molecular oxygen and consequently more reactive forms of oxygen are employed. These include chromium and manganese compounds that are environmentally non-benign.[124] As a consequence, recent research has been devoted to identifying environmentally friendly designs for new oxidation catalysts that can operate with molecular oxygen. The application of supported metal catalysts in the epoxidation of ethylene by silver is a successful example of this approach. Industrially, large amounts of ethene are oxidised with molecular oxygen to ethylene oxide using supported silver catalysts.[125] However, to achieve high selectivities, non-green additives have to be added to prevent unselective side reactions. Unfortunately, higher alkenes such as propene[126] are not so easily oxidised with molecular oxygen, and ethene appears to be a unique case. In particular, there is great interest in the production of propylene oxide, and propylene can be epoxidised with hydrogen peroxide on a commercial scale. Haruta and co-workers[127,128] have shown that propene can be epoxidised using gold supported on titanium oxide supports, using oxygen in the presence of hydrogen. In this process, it is considered that the $O_2/H_2$ forms a surface hydroperoxy species that is responsible for the selective oxidation. Initially, only low selectivities based on propene were observed, but by using mesoporous titanium silicate supports, high selectivities have been achieved. However, the utilisation of $H_2$ remains very low, and this remains a challenge.[129]

More progress has been achieved in the oxidation of higher alkenes in the liquid phase. The direct route to adipic acid via the direct oxidation of cyclohexenes with hydrogen peroxide is a good example and a major highlight from Noyori *et al.*[130] in the late 1990s. Although the most recent literature addresses these reactions using Ti, porous and framework solids, polyoxometalates and single site catalysts, some research has been dedicated to the epoxidation of cyclohexene with supported metal nanoparticles. Recent work by Hutching and co-workers has shown that gold nanoparticles supported on graphite can epoxidise cyclic alkenes using molecular oxygen as long as a radical initiator is present in catalytic amounts.[131] In these studies it was demonstrated that cis-cyclooctene could be epoxidised with a selectivity of $>80\%$ under mild solvent-free conditions, whereas the epoxidation of cyclohexene, styrene and cis-stilbene resulted in selectivities of ca.60%. In the absence of the catalyst, some oxidation was observed but this was minor and non-selective giving very low selectivities to the epoxide. It was also shown that it was possible to achieve epoxidation in the absence of the radical initiator, but typically low selectivities ensued. Hence, it was concluded that for high selectivities, an appropriate combination of molecular oxygen and a radical initiator were required. More recently,[132] the reaction was explored in more detail and it was found that the catalyst preparation method played an important role. Both activity and selectivity were enhanced using a colloidal method instead of impregnation or deposition-precipitation, and this is probably due to a smaller particle size distribution. Hutchings *et al.* have recently extended their earlier studies[131] to explore whether alternative supports or radical initiators can improve the performance.[133] TBHP has been found to be the most effective initiator for this catalytic oxidation. The catalyst is found to be inhibited by the epoxide product but it was demonstrated that the effect of this is negligible for reused catalysts over a long reaction time. It was also found that the catalyst was fully reusable and this is a key aspect of green chemistry that had not been explored previously with these catalysts. On a similar note, Patil *et al.*[134] have reported that gold nanoparticles deposited on MgO, CaO, SrO and BaO by deposition-precipitation using urea are very active and selective in the epoxidation of styrene to styrene oxide with TBHP. The authors highlighted the importance of the use of urea as a precipitating agent in the catalyst preparation to obtain superior activity than that observed using NaOH. This is attributed to higher gold loading and small particle sizes obtained using urea, resulting in styrene conversion above 53%, and selectivities towards styrene oxide between 45 and 60%. The same authors improved these results using the same preparation method to deposit gold on $Yb_2O_3$, $Sm_2O_3$, $Eu_2O_3$ and $Tb_2O_3$, and the novel catalysts are highly active, selective and reusable.[135] Interestingly, there seemed to be a promotional effect when using aqueous TBHP rather than the anhydrous form. They reported styrene conversions and selectivities to styrene oxide of over 70%, with reusable catalysts containing 6.6 wt% of gold and an average particle size of 11 nm. Additional research is, however, required to understand the promotional role of water as well as the effect of particle size.

Lignier *et al.*[136] recently described that gold catalysts promote the liquid-phase epoxidation of trans-stilbene in methylcyclohexane by taking part in a chain reaction which involves a radical from the solvent. The reaction was performed at atmospheric pressure using catalytic amounts of TBHP and the mechanism was fully investigated.[136–138] Lignier *et al.* used $Au/TiO_2$ reference gold catalysts provided by the World Gold Council and proposed that the combination of the oxidation catalytic activity of gold and titania activity makes a bifunctional catalyst where both sites contribute to the catalytic activity.[137] They concluded that TBHP acts as a radical initiator, that oxygen from air is the oxidant and that the solvent plays a major role as it is also oxidised and acts as a propagating radical. The authors suggest that both titania and gold play roles in the production of the active radicals as well as trapping unselective radicals and stabilising the reaction intermediate. They expanded their studies to include gold on other supports including carbon, alumina or iron oxide, and titania.[137] The authors also studied the effect of the radical initiator in a subsequent study[138] where they described that the nature of the radical initiator has a critical influence on the reaction selectivity. In particular, TBHP leads to high yields of epoxide, whereas hydrogen peroxide and di-tert-butylperoxide cause the undesired degradation of trans-stilbene. They studied a range of solvents, but only alkyl-substituted cyclohexanes led to high yields of epoxide, whereas more polar solvents or cyclohexane resulted in poorer results, both in terms of activity and epoxide yield. A recent report by Mendez *et al.*[139] also carried out the methylcyclohexane/stilbene co-oxidation reaction at lower temperature with gold nanoparticles, but this time supported on Gd-doped titania, which the authors show to be superior to the World Gold Council $Au/TiO_2$ reference catalyst. Mendez *et al.* found that the support has little influence on the intrinsic activity of gold, but affects the apparent reaction rates, which are a combination of catalytic activity and diffusion limitations. Gadolinium-doped titania nanocrystallites, obtained by mild hydrolysis of a new $Gd_4TiO(O^iPr)_{14}$ bimetallic oxoalkoxide, was used as a support and the authors claim that this leads to enhanced wettability of gold particles in the TBHP-initiated epoxidation of stilbene in methylcyclohexane. The gold-catalysed homolytic decomposition of TBHP generating radicals was described as the rate-determining step. It is remarkable that the reaction can also be catalysed by naked gold colloids of about 2 nm under the same reaction conditions with superior activity than the reference catalyst supported on titania.[140] Interestingly, the authors claim that the protecting ligand turns into a siloxane polymer which "might supply the reaction with active intermediates" and that the colloids studied are expected to be highly efficient catalysts for radical oxidations in general.

The use of molecular oxygen instead of peroxides is highly desirable, and Lambert and co-workers[141] have shown that very small $Au_{55}$ nanocrystals (*ca.* 1.4 nm) within inert supports are very active catalysts for styrene oxidation with oxygen under conditions where catalysts prepared by impregnation or sol methods were inactive. They used toluene as the solvent and found minor selectivity to the epoxide whereas the major product was benzaldehyde. The authors highlight the importance of the particle morphology and, in particular, find a size threshold in catalytic activity, in that particles larger than 2 nm are completely inactive which suggests that the origin of catalytic activity lies in the altered electronic structure intrinsic to small gold nanoparticles. A very recent report by Cai *et al.*[142] also describes the catalytic activity of gold for the selective oxidation of an alkene without the need for an initiator. This is achieved with Au/OMS-2 and Au/La-OMS-2 systems in the aerobic liquid-phase oxidation of cyclohexene. La was exchanged in K-OMS-2 and gold deposited by deposition-precipitation with NaOH yielding gold loadings between 0.1 and 1.5 wt%. The reactions were performed at 80°C without solvents for 24 h. High cyclohexene conversions of up to 48% were obtained with Au/La-OMS-2 (0.24) and over 85% selectivity towards $C_6$ oxidation products (2-cyclohexene-1-ol and 2-cyclohexene-1-one made-up >80% selectivity).

### 22.3.2. *Oxidative alkene carbonylation*

Several reports in the recent literature propose reactions by combining two functionalities or reaction steps in a one-pot synthesis with very promising results. This is clearly exemplified by the direct oxidative carbonylation of olefins to produce cyclic carbonates. This comes as a result of the olefin reacting in the presence of $CO_2$ under conditions where the epoxide is rapidly carbonylated, thereby avoiding the need for isolation of the epoxide, which would normally be needed to produce the organic carbonate. Although the oxidative carbonylation of olefins has been known for 50 years, moderate yields, high temperature, long reaction time and toxic organic solvents are still major drawbacks that appear to be overcome with the use of supported metal catalysts. A one-pot synthesis for the oxidative carbonylation of styrene to styrene carbonate catalysed with supported metals has been reported by Sun *et al.*[143] The authors report the preparation of styrene carbonate from styrene and $CO_2$ using TBHP with a multifunctional catalytic system consisting of $Au/SiO_2$, zinc bromide, and tetrabutylammonium bromide ($Bu_4NBr$). As previously reported, $Au/SiO_2$ is active for the epoxidation of styrene, whereas zinc bromide and $Bu_4NBr$ are responsible for the cycloaddition of $CO_2$ to the *in situ* formed epoxide. In follow-up work[144] the same authors describe an improved gold catalyst supported on a resine (R201) that allows the carbonate yield to reach 51%. $CO_2$ acts as the solvent and reagent and the Au/R201 catalyst was reusable. They later reported that gold supported in ferric hydroxide is more efficient for this reaction due to a synergistic effect between gold and the iron; styrene carbonate yields of 53% were obtained at 80°C at 4 MPa of $CO_2$ after 10 h reaction.[145]

### 22.3.3. *Selective oxidation of alkylaromatics*

One of the most important selective oxidation processes in the production of chemicals is the side chain oxidation of alkyl aromatics, which are then further reacted to higher value products that end up in a variety of polymer compositions and specialised chemicals.[122] The largest oxyfunctionalised aromatic products with regard to world production are terephthalic acid, phthalic anhydride and benzoic acid which are produced worldwide with capacities of >30,000, 5,000 and 500 kt $a^{-1}$ respectively.[122] Worldwide production capacities for benzaldehyde and pyromellitic dianhydride do not rise above 50 kt $a^{-1}$. In smaller capacities as specialty chemicals for the pharmaceutical industry, halo-substituted oxyfunctionalized aromatics are also produced. Of the aforementioned products, phthalic anhydride and pyromellitic dianhydride are produced from gas-phase processes using $V_2O_5$-$TiO_2$ catalysts[122] and a liquid-phase alternative process does not appear immediately desirable. On the other hand, terephtalic and benzoic acids, and benzaldehyde are produced catalytically in the liquid phase with Co-Mn-Br homogeneous catalysts for which competitive heterogeneous catalysts have not been found. Homogeneously catalysed processes have a variety of disadvantages, including corrosive reaction conditions that require the use of titanium or Hastelloy in the reactor, as well as the high cost for product and catalyst separation. Commercially, benzaldehyde is produced by the chlorination of toluene followed by saponification,[146] whereas benzoic acid is produced by the liquid-phase cobalt-catalysed oxidation of toluene using oxygen at 165°C with acetic acid as the solvent, but the conversion has to be limited to <15% to retain high selectivities.[147] The use of halogens and acidic solvents makes these processes environmentally unfriendly, which is a motivation for developing new catalytic systems and/or finding alternative processes and reaction conditions. Interestingly, it is also possible to produce benzoic acid and benzaldehyde via heterogeneously catalysed gas-phase oxidation with vanadium-containing catalysts, although the selectivities achieved in such processes are significantly lower,[148] and conversion must be limited to avoid over-oxidation to $CO_2$ and other by-products.[149] Heterogeneous catalysts could be readily used in flow reactors, facilitating the efficient production of materials using continuous processes. For the oxidation of toluene, there have been many attempts to find a suitable liquid-phase oxidation catalyst, and to date these have used oxides of copper and manganese,[150–152] cobalt in SBA-15,[153] or chromium in chromium silicalite-1 (CrS-1)[154] catalysts, but all perform very poorly with TON values of less than 100, even at temperatures in excess of 190°C (Table 22.8). In view of this, there is clearly a need to develop heterogeneous catalysts for toluene oxidation that have greatly improved activity while retaining selectivity and, in particular, some advances have been made utilising supported precious metals.

Hutching and co-workers have shown that Au-Pd alloy nanoparticles are very effective for the direct synthesis of hydrogen peroxide[129] and the oxidation of primary alcohols using oxygen.[107] This catalyst operates by establishing a reactive hydroperoxy intermediate. Because these intermediates are known to be involved

**Table 22.8.** Comparison of toluene oxidation activity of some supported catalysts.

| | | | Selectivity % | | | | | |
|---|---|---|---|---|---|---|---|---|
| Catalyst | T/P/oxidant | Conv. % | BAL | BAD | BAC | BB | TON | Ref. |
| Cu-Mn (1/1) | 190°C/1MPa/$O_2$ | 21.6 | 1.6 | 9.2 | 73.7 | 13.6 | 8 | 150 |
| Cu-Fe/$\gamma$-$Al_2O_3$ | 190°C/1MPa/$O_2$ | 25.4 | 1.0 | 27.4 | 71.6 | n.d. | 74 | 151 |
| $MnCO_3$ | 190°C/1MPa/$O_2$ | 25.0 | 5.3 | 9.7 | 80.8 | n.d. | 50 | 152 |
| CoSBA-15 | 80°C/1 atm/TBHP | 8.0 | n.d. | 64.0 | n.d. | n.d. | 103 | 153 |
| Cr/Silicalite | 80°C/1 atm/TBHP | 18.4 | 5.2 | 23.3 | 25.7 | n.d. | n.d. | 154 |

BAC: benzoic acid; BAD: benzaldehyde; BAL: benzyl alcohol; BB: benzyl benzoate.

in the enzymatic oxidation of primary C-H bonds,[155] it was reasoned that it should be feasible for Au-Pd nanoparticles to be active for the oxidation of the primary C-H bonds in toluene. Additionally, it was found that a sol-immobilisation method produces more active Au-Pd catalysts for a range of selective oxidation reactions with a good control of morphology and particle composition.[109,110,112,156–158] Consequently, Hutchings and co-workers prepared Au-Pd alloyed nanoparticles by a sol-immobilisation technique which can give significantly improved activity for the oxidation of toluene under mild solvent-free conditions.[159] These catalysts have TON values that are a factor of ~30 greater than those of previous heterogeneous catalysts for this reaction and also display a remarkably high selectivity to benzyl benzoate. Table 22.9 summarises the results obtained.

**Table 22.9.** Comparison of catalytic activity for the oxidation of toluene in the absence of solvent with $O_2$ (10 bar), catalyst mass varied between 0.2 and 0.6 g to give a substrate/metal molar ratio of 6,500, toluene 10–20 ml, stirring rate 1,500 rpm, TON calculated on the basis of the total metal. All catalysts contain 1 wt% of metal and were prepared using the sol-immobilisation method.

| | | | | Selectivity % | | | | |
|---|---|---|---|---|---|---|---|---|
| Catalyst Au:Pd | T, °C | Time, h | Conv. % | BAL | BAD | BAC | BB | TON |
| Au-Pd/C*7:1 | 160 | 7 | 0.3 | 28.4 | 57.6 | 6.2 | 7.8 | 17 |
| Au-Pd/C*3:1 | 160 | 7 | 1.5 | 1.8 | 63.4 | 3.1 | 31.4 | 95 |
| Au-Pd/C*1:1.85 | 160 | 7 | 4.8 | 0.9 | 12.7 | 10.3 | 76.1 | 310 |
| Au-Pd/C*1:2 | 160 | 7 | 5.3 | 1.2 | 8.3 | 11.1 | 79.3 | 350 |
| Au-Pd/C*1:3 | 160 | 7 | 5.2 | 1.9 | 8.5 | 10.3 | 79.3 | 340 |
| Au-Pd/C*1:7 | 160 | 7 | 4.3 | 9.6 | 13.6 | 7.3 | 69.5 | 280 |
| Pd/C*0:1 | 160 | 7 | 1.6 | 3.9 | 56.4 | 3.3 | 36.4 | 105 |
| Au/C+Pd/C† 1:1.85 | 160 | 7 | 1.6 | 0.8 | 26.2 | 11.4 | 61.6 | 105 |
| Au-Pd/TiO2*1:1.85 | 160 | 7 | 2.1 | 2.9 | 6.6 | 1.0 | 89.5 | 135 |
| Au-Pd/TiO2*1:1.85 | 160 | 7 | 2.2 | 2.2 | 6.5 | 2.3 | 89.0 | 141 |
| Au-Pd/C‡ 1:1.85 | 160 | 48 | 50.8 | 0.1 | 1.1 | 4.5 | 94.3 | 3,300 |
| Au-Pd/TiO2‡ 1:1.85 | 160 | 48 | 24.1 | 0.5 | 1.2 | 2.8 | 95.5 | 1,570 |
| Au-Pd/C‡ 1:1.85 | 120 | 48 | 10.6 | 0.2 | 7.1 | 13.1 | 79.7 | 690 |
| Au-Pd/TiO2‡ 1:1.85 | 120 | 48 | 4.0 | 1.1 | 6.0 | 4.8 | 88.1 | 260 |
| Au-Pd/C‡ 1:1.85 | 80 | 48 | 0.9 | 8.6 | 34.2 | 0.1 | 57.2 | 60 |
| Au-Pd/C§ 1:1.85 | 160 | 0.5 | 1.8 | 2.1 | 11.6 | 0.8 | 85.5 | 29 |
| Au-Pd/C§ 1:1.85 | 160 | 7 | 21.5 | 0.3 | 1.8 | 3.1 | 94.8 | 350 |
| Au-Pd/C§ 1:1.85 | 160 | 24 | 82.9 | 0.1 | 0.8 | 7.3 | 91.8 | 1,347 |
| Au-Pd/C§ 1:1.85 | 160 | 27 | 94.4 | 0.2 | 1.0 | 13.3 | 85.5 | 1,534 |

*Reaction conditions: toluene = 20 ml, substrate/metal mol ratio = 6,500, mass of catalyst variable.
†Reaction of a physical mixture comprising 1 wt% Au/C and 1 wt% Pd/C, toluene 20 ml, substrate/metal mol ratio 6,500.
‡Reaction conditions: toluene = 10 ml, substrate/metal = 6,500, mass of catalyst = 0.2 g.
§Reaction conditions: toluene = 20 ml, substrate/metal mol ratio = 1,625, mass of catalyst = 1.6 g.

Hutchings *et al.* initially investigated Au-Pd/$TiO_2$ catalysts, prepared by impregnation, however, these catalysts were not found to be particularly active for toluene oxidation, although they did not produce any $CO_2$.[159] The Au-Pd nanoparticles synthesised by the impregnation method can be relatively large (typically > 6 nm) and have substantial compositional variations, which limits their reactivity. Catalysts prepared by sol-immobilisation were then tested and the only products were benzyl alcohol, benzaldehyde, benzoic acid and benzyl benzoate. By itself, Au was not active, but the addition of Pd significantly enhanced the conversion, demonstrating a clear synergistic effect with an optimum catalyst composition of a 1:2 Au:Pd molar ratio. For this catalyst, the TOF after 7 hours of reaction was ~50. The oxidation of 2-, 3- and 4- methoxytoluene, and 2-, 3- and 4-nitrotoluene was demonstrated showing the general applicability of the AuPd/C catalyst and the reactivity trend found (4-methoxy- ~ 2-methoxy- > 3-methoxy ~ toluene > 2-nitro- > 3-nitro ~ 4-nitro) is indicative of the involvement of electron-deficient intermediate(s). Additional products formed were identified as a family of esters and C-C coupling products. In the selective oxidation of xylenes the catalyst formed the aldehyde, acid and esters, with the relative amounts being dependent on the conversion. The reaction profile over a longer time scale was also investigated using a lower substrate/metal molar ratio and the conversion continued to increase steadily, fully depleting the toluene after 110 hours, while the selectivity to benzyl benzoate also increased (Fig. 22.1).

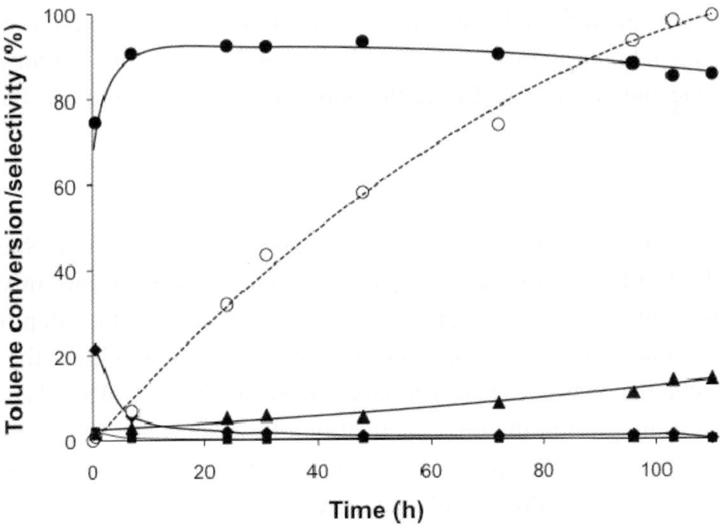


**Figure 22.1.** Toluene conversion and selectivity to partial oxidation products. Reaction conditions: 160°C, 0.1 MPa $pO_2$, 20 ml toluene, 0.8 g of catalyst (1 wt% AuPd/C prepared by sol-immobilisation with 1:1.85 Au/Pd ratio), toluene/metal molar ratio of 3,250 and reaction time: 110 h. Key: ○ conversion, ■ selectivity to benzyl alcohol, ♦ selectivity to benzaldehyde, selectivity to benzoic acid, • selectivity to benzyl benzoate.

It was considered that the high selectivity to benzyl benzoate could result from the coupling of the aldehyde and the alcohol to give the hemiacetal, followed by oxidation to the ester. Characterisation and reactivity studies confirmed that any sintering or structural modification of these highly active catalysts is minimal, and the catalysts are stable and reusable. Finally, at 80°C and using TBHP as a co-oxidant it was found that the conversion increased appreciably up to TON values of 850 to 1,200, although the reaction generated a different product profile. At longer reaction times, benzoic acid became the main product, together with appreciable amounts of benzaldehyde and benzyl alcohol.

Moreover, carboxylic acids can be added as initiators for selective oxidation under solvent-free conditions to oxidise alkyl aromatics in the presence of silver nanoparticles. Beier *et al.* recently reported that silver-supported silica catalyses the solvent-free aerobic side-chain oxidation of alkyl aromatics under solvent-free conditions even at atmospheric pressure.[160] Benzoic acid or *p*-toluic acid are added (3% molar) as promoters whereas additional $CeO_2$ showed both promoting and inhibiting effects depending on substrate and reaction conditions. Toluene, *p*-xylene, ethylbenzene and cumene were investigated. In general, the authors highlight that addition of a Ce precursor to the flame spray pyrolysis (FSP) catalyst results in significantly smaller silver particles that exhibit a superior catalytic performance with TON values of up to 2,000 and higher stability compared to impregnated catalysts. In the specific case of toluene oxidation, 3 mol% benzoic acid was added as the promoter in addition to ceria and biphenyl, and the reaction was carried out at 10 bar air for 1 hour to give a combined yield of 2.6% at 170°C. Benzyl alcohol, benzaldehyde and benzoic acid were the only products detected in similar yields.

### 22.3.4. *Selective oxidation of cycloalkanes*

The oxidation of C–H bonds in cycloalkanes and, in particular, cyclohexane activation under mild conditions is highly desired. This process is currently operated industrially to produce cyclohexanol and cyclohexanone as a first step in the production of adipic acid and caprolactam. Recent reports highlighting the ability of supported gold nanoparticles to selectively oxidise alkenes and alcohols have drawn attention to using gold in the direct activation of cyclohexane. In particular, Zhao *et al.*[161,162] have shown that gold can activate cyclohexane at 150°C. Gold nanoparticles on MCM-41 and ZSM-5 displayed selectivities in excess of 90% which gradually changed from cyclohexanone to cyclohexanol as the catalysts deactivated. More recently, gold nanoparticles of sizes in the range of 2–4 nm, dispersed on ordered mesoporous silica, were prepared via a one-pot synthesis route and the resulting catalysts exhibited high catalytic activity and selectivity using molecular oxygen.[163] The catalytic performance of catalyst Au/xMPTMS-cal was dependent on the porous structure, the amount of Au loading and the Au particle size, which depended on the amount of mercapto-propyl-trimethoxysilane (MPTMS) utilised. The catalysts appear to be very stable as a consequence of avoiding the sintering of gold nanoparticles during high temperature calcination and reaction due to the protective effect of the silica. Hutchings and co-workers[164] investigated the oxidation of cyclohexane with gold catalysts at temperatures well below 100°C, since, at this temperature, higher selectivities might be expected. In this study, Au/C catalysts were contrasted with supported Pt and Pd catalysts and a reaction inhibitor was also investigated (1,4-difluorobenzene). The selectivity to cyclohexanone and cyclohexanol was very high at low conversion, but this declined rapidly with enhanced conversion at longer reaction times, and the gold catalysts were found to give similar performance to the Pt and Pd catalysts. Xu *et al.*[165] observed a promotional effect of titania on the activity of gold on silica which affords a very stable catalyst and selectivity of 90% for *ca.* 10% conversion. Some remaining questions regarding the role of the metal motivated the recent work of Weckhuysen *et al.*,[166] who aimed to clarify the effect of gold in the reaction mechanism and carried out reaction studies over Au/$Al_2O_3$, Au/$TiO_2$ and Au/SBA-15 and compared their reactivity with the industrial autoxidation process. The authors challenged the previous literature reports and indicated that gold-based catalysts do not exhibit excellent catalytic performance, and that the process is not catalytic but rather, on the contrary, that the oxidation follows a radical-chain mechanism instead. They observed product distributions and evolution on-line typical of the autoxidation process, although they acknowledged a significant increase in adipic acid and $CO_2$ formation. They performed additional testing using hydroquinone as the radical scavenger as final proof. They indicated that their observations explain the low selectivity observed at increasing conversion and that the discrepancies with published literature can be understood on the basis of the complicated analysis of products required.

### 22.3.5. *Benzene hydroxylation with in situ generated* $H_2O_2$

The direct hydroxylation of aromatics to phenols has attracted much attention in recent years based on the direct catalytic formation of hydrogen peroxide on Pd and Pt from hydrogen and oxygen *in situ*. The concept has been extrapolated from the enzymatic activity of monooxygenases in the oxygenation of aromatics, where two protons and two electrons are supplied by the NAD(P)H (nicotinamide adenine dinucleotide (phosphate)), to convert molecular oxygen into the active oxygen species.[167] Although hydrogen, carbon monoxide or aldehydes can be used as reductants, good results have only been achieved with hydrogen as the reductant using heterogenous catalysts. Developments in this area are particularly desirable from an economic and industrial point of view because hydrogen peroxide is five times more expensive than hydrogen or carbon monoxide, but there would also be other benefits from the *in situ* production of peroxide species such as improved safety as transportation could be avoided, and the handling of $H_2O_2$ and the use of $H_2O$ in the system could both be minimised.

Some early studies by Miyake *et al.*[168] indicated that $Pt/V_2O_5$ supported on silica was able to afford high phenol productivities of 1,230 g $kg_{catalyst}^{-1}$ $h^{-1}$ at 60°C and $5 \times 10^5$ Pa, with acetic acid as the solvent. The vanadium oxide acted as the promoter, accelerating oxygen transfer to the noble metal and preventing the loss of hydrogen peroxide by over-hydrogenation. Later reports enhanced catalytic activity by combining Pd as the active site for the intermediate hydrogen peroxide formation, with another metal as the oxidation function, such as Ti in TS-1[169] or molecular vanadium sites.[170] Remias *et al.* reported that both benzene and cyclohexane can be oxidized.[170] More recently, acidic zeolites and resins were successfully used as supports for noble metals[171] in order to eliminate the need for liquid acids in the system. Mizukami and co-workers have carried out several studies on the application of Pd membrane reactors for the *in situ* formation of hydrogen peroxide during catalytic oxidation.[172–175] They report that methyl benzoate can be directly hydroxylated to methyl salicylate[173] with a yield of 4.7% at 423 K; whereas the same reactor achieved benzene conversion of 15% and a phenol selectivity of 95%. An increase in reaction temperature, however, caused simultaneous hydrogenation.[174] Vulpescu *et al.* performed a technical and economic feasibility study on the application of Pd-based catalytic membrane reactors for the formation of phenol and found that $CO_2$ was also produced; they then discussed the limitations of the available technology for commercial application.[176] In particular, the total oxidation of hydrogen to water and the total oxidation of benzene to carbon dioxide are the major drawbacks. In a more recent work by Mizukami *et al.*[175] various active metals were loaded on the alpha-$Al_2O_3$ porous tube which was the substrate of the thin Pd membrane. This resulted in a number of side reactions, such as complete oxidation and hydrogenation, and a decrease in the hydroxylation activity. However, the loading of Cu suppressed complete oxidation and enhanced the hydroxylation activity. Centi and Perathoner[177] have recently highlighted some perspectives for this reaction in a context for sustainable chemical production.

## 22.4. Other Selective Oxidation Reactions

### 22.4.1. *Selective oxidation of amines*

The selective oxidation of amines using oxygen is an important goal for green chemistry. Recently, several authors have reported that gold, either as supported nanoparticles[178–181] or in powder form,[178] can perform the selective oxidation of amines to imines. Zhu *et al.* carried out an extensive piece of research which demonstrates that gold does not need to be finely derived to be an effective catalyst for the aerobic oxidative dehydrogenation of amines (CH-NH) to imines (C=N) under the mild conditions of 1 atm $O_2$ and 100°C.[178] A 5% $Au/Al_2O_3$ catalyst was particularly efficient for the practical oxidation of secondary and primary amines to imines. The catalyst contained large gold particles (50–150 nm) as the sample was prepared by impregnation, but this was no deterrent for activity in a reaction that gold can catalyse even in powder form. The higher dispersion of gold in the supported catalyst does, however, increase activity and the catalytic activity of 5 mg of gold in the $Au/Al_2O_3$ catalyst is greater than that of 1 g of gold powder. Aschwanden *et al.*[179] found that this reaction can be carried out by both homogeneous and heterogeneous gold catalysts in toluene at mild temperatures. Initially, $Au(OAc)_3$ was used for the selective oxidation of dibenzylamine to dibenzylimine using molecular oxygen, but the TOFs were lower than unity. However, when $Au(OAc)_3$ was pre-adsorbed onto $CeO_2$, the resulting catalyst was more active, yielding 100% conversion and 91% selectivity at 108°C (TOF = 7.2 $h^{-1}$; related to the total amount of gold). It was then found that $Au(OAc)_3$ is reduced by the amine and metallic gold is formed on the support. The authors claim that the *in situ*-formed gold nanoparticles are the real active species of the reaction, which leads to a simple procedure whereby, starting from a gold salt and upon interaction with the amine, highly active and selective gold catalysts are formed.

So *et al.* further expanded previous catalytic studies in the selective oxidation of cyclic and acyclic benzylic amines to imines with graphite-supported gold catalysts and found that this activity could be translated to substituted quinolines.[180] Conversions ranging from 43 to 100% and product yields between 66 and 99% were obtained and the catalyst was fully reusable. The authors proposed a hydrogen transfer from the amine to the metal and the oxidation of M-H as steps in the mechanism based on radical trap experiments, a Hammett plot and kinetic experiments. The same authors recently reported a gold catalyst supported on silica able to react anilines with aldehydes to form quinolines in a one-pot reaction with yields up to 95%.[182] Corma *et al.*[181] reported that the oxidation of benzylamines to N-benzylidene benzylamines with gold catalysts is a general process, and para-substituted benzylamines as well as heterocyclic methanamines undergo oxidative condensation. They also reported the activity of less active Pd and Pt catalysts with much higher metal loading (5 wt%). Despite the ability of gold to catalyse this reaction in powder form, as described by Zhu *et al.*,[178] Corma *et al.* underline that the efficiency of gold increases exponentially as the average particle size is reduced, and the TOF increases accordingly for a gold catalyst supported on titania. Additionally, they also found that oxidative condensation occurs selectively with sulfur-containing heterocyclic amines and that the selective formation of secondary benzylamines is catalysed by a gold catalyst through a one-pot, two-step reaction which involves oxidation and hydrogenation steps. In a recent publication,[183] the same authors reveal that gold supported on titania is also able to selectively oxidize aromatic anilines to azo compounds with yields above 98%.

### 22.4.2. *Selective oxidation of silanes to silanols*

Silanols are utilised for silicon-based polymeric materials and also find use as nucleophilic coupling partners in organic synthesis. Traditional synthetic methods utilise toxic reagents and are non-environmentally friendly, and other recently reported synthetic methods, in the absence of organic solvents, suffer the main drawback of the production of disiloxanes. Recent results by Kaneda *et al.*[184] overcome this by using water as the solvent, with silver supported in hydroxyapatite with little condensation to the disiloxanes. They show that the reaction can also be catalysed by homogeneous silver, although the supported nanoparticles were superior and reusable without any loss of activity or selectivity.

### 22.4.3. *Selective oxidation of oximes*

The industrially-relevant oxidative transformation of carvone oxime into carvone with gold-, palladium- and platinum-supported nanoparticles has been reported by Corma *et al.*[185] This reaction represents an example of the general applicability of highly active and selective nanoparticles for the aerobic oxidation of oximes to the corresponding carbonylic compounds. Carvone is an essential oil utilised in the fragrance industry, but its synthesis is difficult due to the reluctance of its oxime to undergo hydrolysis. Supported precious metal nanoparticles afford an alternative route via the oxidation of the C=N bond. The reaction was carried out in several solvents, but aqueous solutions of ethanol were most efficient. The reaction is carried out in the presence of oxygen, but most importantly, in the absence of corrosive Brønsted acids and without significant waste as in the current synthetic methods. Ceria was a better support than titania and carbon. Gold again shows superior catalytic activity than Pd, but very similar to Pt. The authors claim this process to be a suitable replacement to current methods that require high amounts of acids and that are environmentally problematic.

## 22.5. Conclusions and Final Remarks

The utilisation of metal-supported catalysts as an alternative approach for the synthesis of fine chemicals has shown significant development in recent years and can now contribute to the sustainable development of chemical processes. The use of supported metal nanoparticles instead of stoichiometric oxidising agents and homogenous catalytic methods remains an appealing challenge for both industry and academia. Generally, greater effort is needed for developing new technologies, which can compete with the traditional selective oxidation industrial processes, and at the same time remain economically feasible. However, much progress has been made in this field in recent years.

The incorporation of two metals on the surface of the support seems to be a key parameter for enhancing catalytic activity and resistance to poisoning, thereby lengthening the lifetime of the catalyst. In most selective oxidations, the synthesis of nanoparticles of small particle size (2–6 nm) seems to be the "key" for obtaining high catalytic activity. Moreover, it has been demonstrated that control of the metal nanoparticle size is also key with respect to influencing selectivity. Some progress has been exemplified by using milder conditions and lower amounts of catalyst for maintaining the process under safe operating conditions. The utilisation of gold-based catalysts seems to improve the existing chemical processes as the high conversion, improved selectivity and longer lifetime of the catalytic systems has been demonstrated. However, there are still challenges to be solved, such as the improvement in the stability of the supported nanoparticles. One promising option could be confinement into mesoporous materials. Also, the high cost and volatility in the price of gold and other precious metals still raises concerns when the commercialisation of these catalysts is discussed. Some future research directions to address this issue include the utilisation of lower metal loading, prolonged catalyst lifetime and the utilisation of transition metals in the synthesis of active gold-based materials in the form of alloys.

There are still challenges in the synthesis of supported metal nanoparticles with precise particle size and morphology and, in particular, the case of alloyed nanoparticles. Another important challenge is the understanding of the mechanism by which bimetallic-supported nanoparticles generate enhanced catalytic performance when compared with their monometallic counterparts. Therefore, better understanding of the active sites responsible for catalysis will be beneficial in designing and controlling the properties of metal nanoparticles. Finally, there is still room for the discovery of the activity of supported metal nanoparticles in a larger variety of chemical processes, especially considering the great advances in recent years regarding the control in catalyst preparation and the discovery of the reactivity of gold and its alloys.

## References

1. Muhler, M. (1997). *Handbook of Heterogeneous Catalysis*, Wiley-VCH, Weinheim.
2. Sheldon, R., Arends, I. and Dijksman, A. (2000). New Developments in Catalytic Alcohol Oxidations for Fine Chemicals Synthesis, *Catal. Today*, **57**, pp. 157–166.

3. Sheldon, R., Arends, I. and Hanefeld, U. (2007). *Green Chemistry and Catalysis*, Wiley-VCH, Weinheim.
4. Cavani, F. (2010). Catalytic Selective Oxidation Faces the Sustainability Challenge: Turning Points, Objectives Reached, Old Approaches Revisited and Solutions Still Requiring Further Investigation, *J. Chem. Tech. Biot.*, **85**, pp. 1175–1183.
5. Mallat, T. and Baiker, A. (1994). Oxidation of Alcohols with Molecular-Oxygen on Platinum Metal-Catalysts in Aqueous-Solutions, *Catal. Today*, **19**, pp. 247–283.
6. Besson, M. and Gallezot, P. (2000). Selective Oxidation of Alcohols and Aldehydes on Metal Catalysts, *Catal. Today*, **57**, pp. 127–141.
7. Matsumoto, T., Ueno, M., Wang, N., *et al.* (2008). Recent Advances in Immobilized Metal Catalysts for Environmentally Benign Oxidation of Alcohols, *Chemistry – An Asian Journal*, **3**, pp. 196–214.
8. Mallat, T. and Baiker, A. (2004). Oxidation of Alcohols with Molecular Oxygen on Solid Catalysts, *Chem. Rev.*, **104**, pp. 3037–3058.
9. Vinod, C., Wilson, K. and Lee, A. (2011). Recent Advances in the Heterogeneously Catalysed Aerobic Selective Oxidation of Alcohols, *J. Chem. Tech. Biot.*, **86**, pp. 161–171.
10. Yamaguchi, K., Mori, K., Mizugaki, T., *et al.* (2000). Creation of a Monomeric Ru Species on the Surface of Hydroxyapatite as an Efficient Heterogeneous Catalyst for Aerobic Alcohol Oxidation, *J. Am. Chem. Soc.*, **122**, pp. 7144–7145.
11. Yamaguchi, K. and Mizuno, N. (2002). Supported Ruthenium Catalyst for the Heterogeneous Oxidation of Alcohols with Molecular Oxygen, *Angew. Chem. Int. Edit.*, **41**, 4538–4542.
12. Yamaguchi, K. and Mizuno, N. (2003). Scope, Kinetics, and Mechanistic Aspects of Aerobic Oxidations Catalyzed by Ruthenium Supported on Alumina, *Chemistry – A European Journal*, **9**, pp. 4353–4361.
13. Zhan, B., White, M., Sham, T., *et al.* (2003). Zeolite-Confined Nano-RuO2: A Green, Selective, and Efficient Catalyst for Aerobic Alcohol Oxidation, *J. Am. Chem. Soc.*, **125**, pp. 2195–2199.
14. Ji, H., Mizugaki, T., Ebitani, K., *et al.* (2002). Highly Efficient Oxidation of Alcohols to Carbonyl Compounds in the Presence of Molecular Oxygen Using a Novel Heterogeneous Ruthenium Catalyst, *Tetrahedron Lett.*, **43**, pp. 7179–7183.
15. Opre, Z., Grunwaldt, J., Maciejewski, M., *et al.* (2005). Promoted Ru-Hydroxyapatite: Designed Structure for the Fast and Highly Selective Oxidation of Alcohols with Oxygen. *J. Catal.*, **230**, pp. 406–419.
16. Opre, Z., Grunwaldt, J., Mallat, T., *et al.* (2005). Selective Oxidation of Alcohols with Oxygen on Ru-Co-Hydroxyapatite: A Mechanistic Study, *J. Mol. Catal. A: Chem.*, **242**, pp. 224–232.
17. Opre, Z., Ferri, D., Krumeich, F., *et al.* (2006). Aerobic Oxidation of Alcohols by Organically Modified Ruthenium Hydroxyapatite, *J. Catal.*, **241**, pp. 287–295.
18. Kantam, M., Pal, U., Sreedhar, B., *et al.* (2008). Aerobic Alcohol Oxidation by Ruthenium Species Stabilized on Nanocrystalline Magnesium Oxide by Basic Ionic Liquids, *Adv. Synth. Catal.*, **350**, pp. 1225–1229.
19. Mori, S., Takubo, M., Makida, K., *et al.*, (2009). A Simple and Efficient Oxidation of Alcohols with Ruthenium on Carbon, *Chem. Commun.*, **34**, pp. 5159–5161.
20. Yamaguchi, K., Kim, J. W., He, J., *et al.* (2009). Aerobic Alcohol Oxidation Catalyzed by Supported Ruthenium Hydroxides, *J. Catal.*, **268**, pp. 343–349.
21. Nikaidou, F., Ushiyama, H., Yamaguchi, K., *et al.* (2010). Theoretical and Experimental Studies on Reaction Mechanism for Aerobic Alcohol Oxidation by Supported Ruthenium Hydroxide Catalysts, *J. Phys. Chem. C*, **114**, pp. 10873–10880.
22. Punniyamurthy, T., Velusamy, S. and Iqbal, J. (2005). Recent Advances in Transition Metal Catalyzed Oxidation of Organic Substrates with Molecular Oxygen, *Chem. Rev.*, **105**, pp. 2329–2363.

23. Ebitani, K., Fujie, Y. and Kaneda, K. (1999). Immobilization of a Ligand-Preserved Giant Palladium Cluster on a Metal Oxide Surface and its Nobel Heterogeneous Catalysis for Oxidation of Allylic Alcohols in the Presence of Molecular Oxygen, *Langmuir*, **15**, pp. 3557–3562.
24. Kakiuchi, N., Maeda, Y., Nishimura, T., *et al.* (2001). Pd(II)-Hydrotalcite-Catalyzed Oxidation of Alcohols to Aldehydes and Ketones Using Atmospheric Pressure Of Air, *J. Org. Chem.*, **66**, pp. 6620–6625.
25. Mori, K., Hara, T., Mizugaki, T., *et al.* (2004). Hydroxyapatite-Supported Palladium Nanoclusters: A Highly Active Heterogeneous Catalyst for Selective Oxidation of Alcohols by Use of Molecular Oxygen, *J. Am. Chem. Soc.*, **126**, pp. 10657–10666.
26. Lu, A., Li, W., Hou, Z., *et al.* (2007). Molecular Level Dispersed Pd Clusters in the Carbon Walls of Ordered Mesoporous Carbon as a Highly Selective Alcohol Oxidation Catalyst, *Chem. Commun.*, **10**, pp. 1038–1040.
27. Hou, Z., Theyssen, N. and Leitner, W. (2007). Palladium Nanoparticles Stabilised on PEG-Modified Silica as Catalysts for the Aerobic Alcohol Oxidation in Supercritical Carbon Dioxide, *Green Chem.*, **9**, pp. 127–132.
28. Hou, Z., Theyssen, N., Brinkmann, A., *et al.* (2008). Supported Palladium Nanoparticles on Hybrid Mesoporous Silica: Structure/Activity-Relationship in the Aerobic Alcohol Oxidation Using Supercritical Carbon Dioxide, *J. Catal.*, **258**, pp. 315–323.
29. Hara, T., Ishikawa, M., Sawada, J., *et al.* (2009). Creation of Highly Stable Monomeric Pd(II) Species in an Anion-Exchangeable Hydroxy Double Salt Interlayer: Application to Aerobic Alcohol Oxidation under an Air Atmosphere, *Green Chem.*, **11**, pp. 2034–2040.
30. Yang, H., Han, X., Ma, Z., *et al.* (2010). Palladium-Guanidine Complex Immobilized on SBA-16: A Highly Active and Recyclable Catalyst for Suzuki Coupling and Alcohol Oxidation, *Green Chem.*, **12**, pp. 441–451.
31. Ferri, D., Mondelli, C., Krumeich, F., *et al.* (2006). Discrimination of Active Palladium Sites in Catalytic Liquid-Phase Oxidation of Benzyl Alcohol, *J. Phys. Chem. B*, **110**, pp. 22982–22876.
32. Grunwaldt, J., Caravati, M. and Baiker, A. (2006). Oxidic or Metallic Palladium: Which is the Active Phase in Pd-Catalyzed Aerobic Alcohol Oxidation? *J. Phys. Chem. B*, **110**, pp. 25586–25589.
33. Hashmi, A. and Hutchings, G. (2006). Gold Catalysis, *Angew. Chem. Int. Edit.*, **45**, pp. 7896–7936.
34. Della Pina, C., Falletta, E., Prati, L., *et al.* (2008). Selective Oxidation Using Gold, *Chem. Soc. Rev.*, **37**, pp. 2077–2095.
35. Prati, L. and Rossi, M. (1998). Gold on Carbon as a New Catalyst for Selective Liquid Phase Oxidation of Diols, *J. Catal.*, **176**, pp. 552–560.
36. Prati, L. and Martra, G. (1999). New Gold Catalysts for Liquid Phase Oxidation, *Gold Bull.*, **32**, pp. 96–101.
37. Bianchi, C., Porta, F., Prati, L., *et al.* (2000). Selective Liquid Phase Oxidation Using Gold Catalysts, *Top. Catal.*, **13**, pp. 231–236.
38. Porta, F., Prati, L., Rossi, M., *et al.*, (2000). Metal Sols as a Useful Tool for Heterogeneous Gold Catalyst Preparation: Reinvestigation of a Liquid Phase Oxidation, *Catal. Today*, **61**, pp. 165–172.
39. Biella, S., Prati, L. and Rossi, M. (2002). Selective Oxidation of D-Glucose on Gold Catalyst, *J. Catal.*, **206**, pp. 242–247.
40. Comotti, M., Della Pina, C., Matarrese, R., *et al.* (2004). The Catalytic Activity of "Naked" Gold Particles, *Angew. Chem. Int. Edit.*, **43**, pp. 5812–5815.
41. Comotti, M., Della Pina, C., Falletta, E., *et al.* (2006). Is the Biochemical Route Always Advantageous? The Case of Glucose Oxidation, *J. Catal.*, **244**, pp. 122–125.
42. Onal, Y., Schimpf, S. and Claus, P. (2004). Structure Sensitivity and Kinetics of D-Glucose Oxidation to D-Gluconic Acid over Carbon-Supported Gold Catalysts, *J. Catal.*, **223**, pp. 122–133.

43. Baatz, C., Thielecke, N. and Prusse, U. (2007). Influence of the Preparation Conditions on the Properties of Gold Catalysts for the Oxidation of Glucose, *Appl. Catal. B: Environ.*, **70**, pp. 653–660.
44. Ishida, T., Kinoshita, N., Okatsu, H., *et al.* (2008). Influence of the Support and the Size of Gold Clusters on Catalytic Activity for Glucose Oxidation, *Angew. Chem. Int. Edit.*, **47**, pp. 9265–9268.
45. Gallezot, P. (2007). Catalytic Routes from Renewables to Fine Chemicals, *Catal. Today*, **121**, pp. 76–91.
46. Behr, A., Eilting, J., Irawadi, K., *et al.* (2008). Improved Utilisation of Renewable Resources: New Important Derivatives of Glycerol, *Green Chem.*, **10**, pp. 13–30.
47. Corma, A., Iborra, S. and Velty, A. (2007). Chemical Routes for the Transformation of Biomass into Chemicals, *Chem. Rev.*, **107**, pp. 2411–2502.
48. Zhou, C., Beltramini, J., Fan, Y., *et al.* (2008). Chemoselective Catalytic Conversion of Glycerol as a Biorenewable Source to Valuable Commodity Chemicals, *Chem. Soc. Rev.*, **37**, pp. 527–549.
49. Pagliaro, M., Ciriminna, R., Kimura, H., *et al.* (2007). From Glycerol to Value-Added Products, *Angew. Chem. Int. Edit.*, **46**, pp. 4434–4440.
50. Carrettin, S., McMorn, P., Johnston, P., *et al.*, (2002). Selective Oxidation of Glycerol to Glyceric Acid using a Gold Catalyst in Aqueous Sodium Hydroxide, *Chemical Communications*, **7**, pp. 696–697.
51. Carrettin, S., McMorn, P., Johnston, P., *et al.* (2003). Oxidation of Glycerol using Supported Pt, Pd and Au Catalysts, *Physical Chemistry Chemical Physics*, **5**(6), pp. 1329–1336.
52. Porta, F. and Prati, L. (2004). Selective Oxidation of Glycerol to Sodium Glycerate with Gold-On-Carbon Catalyst: An Insight into Reaction Selectivity, *Journal of Catalysis*, **224**(2), pp. 397–403.
53. Demirel-Gulen, S., Lucas, M. and Claus, P. (2005). Liquid Phase Oxidation of Glycerol over Carbon Supported Gold Catalysts, *Catalysis Today*, **102**, pp. 166–172.
54. Demirel, S., Kern, P., Lucas, M., *et al.* (2007). Oxidation of Mono- and Polyalcohols with Gold: Comparison of Carbon and Ceria Supported Catalysts, *Catalysis Today*, **122**(3–4), pp. 292–300.
55. Ketchie, W., Fang, Y., Wong, M., *et al.* (2007). Influence of Gold Particle Size on the Aqueous-Phase Oxidation of Carbon Monoxide and Glycerol, *J. Catal.*, **250**, pp. 94–101.
56. Ketchie, W., Murayama, M. and Davis, R. (2007). Promotional Effect of Hydroxyl on the Aqueous Phase Oxidation of Carbon Monoxide and Glycerol over Supported Au Catalysts, *Top. Catal.*, **44**, pp. 307–317.
57. Dimitratos, N., Villa, A., Bianchi, C., *et al.* (2006). Gold on Titania: Effect of Preparation Method in the Liquid Phase Oxidation, *Appl. Catal. A: Gen.*, **311**, pp. 185–192.
58. Milone, C., Ingoglia, R., Neri, G., *et al.* (2001). Gold Catalysts for the Liquid Phase Oxidation of *o*-Hydroxybenzyl Alcohol, *Appl. Catal. A: Gen.*, **211**, pp. 251–257.
59. Abad, A., Concepcion, P., Corma, A., *et al.* (2005). A Collaborative Effect between Gold and a Support Induces the Selective Oxidation of Alcohols, *Angew. Chem. Int. Edit.*, **44**, pp. 4066–4069.
60. Christensen, C., Jorgensen, B., Rass-Hansen, J., *et al.* (2006). Formation of Acetic Acid by Aqueous-Phase Oxidation of Ethanol with Air in the Presence of a Heterogeneous Gold Catalyst, *Angew. Chem. Int. Edit.*, **45**, pp. 4648–4651.
61. Hu, J., Chen, L., Zhu, K., *et al.* (2007). Aerobic Oxidation of Alcohols Catalyzed by Gold Nano-Particles Confined in the Walls of Mesoporous Silica, *Catal. Today*, **122**, pp. 277–283.
62. Haider, P. and Baiker, A. (2007). Gold Supported on Cu-Mg-Al-Mixed Oxides: Strong Enhancement of Activity in Aerobic Alcohol Oxidation by Concerted Effect of Copper and Magnesium, *J. Catal.*, **248**, pp. 175–187.
63. Haider, P., Grunwaldt, J., Seidel, R., *et al.* (2007). Gold Supported on Cu-Mg-Al and Cu-Ce Mixed Oxides: An *In Situ* XANES Study on the State of Au during Aerobic Alcohol Oxidation, *J. Catal.*, **250**, pp. 313–323.

64. Biffis, A., Cunial, S., Spontoni, P., *et al.* (2007). Microgel-Stabilized Gold Nanoclusters: Powerful "Quasi-Homogeneous" Catalysts for the Aerobic Oxidation of Alcohols in Water, *J. Catal.*, **251**, pp. 1–6.
65. Wang, L., Liu, Y., Chen, M., *et al.* (2008). MnO2 Nanorod Supported Gold Nanoparticles with Enhanced Activity for Solvent-Free Aerobic Alcohol Oxidation, *J. Phys. Chem. C*, **112**, pp. 6981–6987.
66. Su, F., Chen, M., Wang, L., *et al.* (2008). Aerobic Oxidation of Alcohols Catalyzed by Gold Nanoparticles Supported on Gallia Polymorphs, *Catal. Commun.*, **9**, pp. 1027–1032.
67. Su, F., Liu, Y., Wang, L., *et al.* (2008). Ga-Al Mixed-Oxide-Supported Gold Nanoparticles with Enhanced Activity for Aerobic Alcohol Oxidation, *Angew. Chem. Int. Edit.*, **47**, pp. 334–337.
68. Han, J., Liu, Y. and Guo, R. (2009). Reactive Template Method to Synthesize Gold Nanoparticles with Controllable Size and Morphology Supported on Shells of Polymer Hollow Microspheres and Their Application for Aerobic Alcohol Oxidation in Water, *Adv. Funct. Mater.*, **19**, pp. 1112–1117.
69. Yang, J., Guan, Y., Verhoeven, T., *et al.* (2009). Basic Metal Carbonate Supported Gold Nanoparticles: Enhanced Performance in Aerobic Alcohol Oxidation, *Green Chem.*, **11**, pp. 322–325.
70. Liu, Y., Tsunoyama, H., Akita, T., *et al.* (2009). Preparation of Similar to 1 nm Gold Clusters Confined within Mesoporous Silica and Microwave-Assisted Catalytic Application for Alcohol Oxidation, *J. Phys. Chem. C*, **113**, pp. 13457–13461.
71. Wang, Y., Yan, R., Zhang, J., *et al.* (2010). Synthesis of Efficient and Reusable Catalyst of Size-Controlled Au Nanoparticles within a Porous, Chelating and Intelligent Hydrogel for Aerobic Alcohol Oxidation, *J. Mol. Catal. A: Chem.*, **317**, pp. 81–88.
72. Oliveira, R., Kiyohara, P. and Rossi, L. (2010). High Performance Magnetic Separation of Gold Nanoparticles for Catalytic Oxidation of Alcohols, *Green Chem.*, **12**, pp. 144–149.
73. Hudlicky, M. (1990). *Oxidations in Organic Chemistry*: American Chemical Society, Washington, DC.
74. Taft, R., Newman, M. and Verhoek, F. (1950). The Kinetics of the Base-Catalyzed Methanolysis of Ortho-Substituted, Meta-Substituted and Para-Substituted L-Menthyl Benzoates, *J. Am. Chem. Soc.*, **72**, pp. 4511–4519.
75. Klitgaard, S., DeLa Riva, A., Helveg, S., *et al.* (2008). Aerobic Oxidation of Alcohols over Gold Catalysts: Role of Acid and Base, *Catal. Lett.*, **126**, pp. 213–217.
76. Marsden, C., Taarning, E., Hansen, D., *et al.* (2008). Aerobic Oxidation of Aldehydes under Ambient Conditions Using Supported Gold Nanoparticle Catalysts, *Green Chem.*, **10**, pp. 168–170.
77. Taarning, E., Madsen, A., Marchetti, J., *et al.* (2008). Oxidation of Glycerol and Propanediols in Methanol over Heterogeneous Gold Catalysts, *Green Chem.*, **10**, pp. 408–414.
78. Oliveira, R., Kiyohara, P. and Rossi, L. (2009). Clean Preparation of Methyl Esters in One-Step Oxidative Esterification of Primary Alcohols Catalyzed by Supported Gold Nanoparticles, *Green Chem.*, **11**, pp. 1366–1370.
79. Choudhary, V., Dumbre, D. and Bhargava, S. (2009). Oxidation of Benzyl Alcohol to Benzaldehyde by Tert-Butyl Hydroperoxide over Nanogold Supported on TiO2 and Other Transition and Rare-Earth Metal Oxides, *Ind. Eng. Chem. Res.*, **48**, pp. 9471–9478.
80. Choudhary, V. and Dumbre, D. (2009). Magnesium Oxide Supported Nano-Gold: A Highly Active Catalyst for Solvent-Free Oxidation of Benzyl Alcohol to Benzaldehyde by TBHP, *Catal. Commun.*, **10**, pp. 1738–1742.
81. Ni, J., Yu, W., He, L., *et al.* (2009). A Green and Efficient Oxidation of Alcohols by Supported Gold Catalysts Using Aqueous $H_2O_2$ under Organic Solvent-Free Conditions, *Green Chem.*, **11**, pp. 756–759.
82. Liu, Y., Tsunoyama, H., Akita, T., *et al.* (2010). Size Effect of Silica-Supported Gold Clusters in the Microwave-Assisted Oxidation of Benzyl Alcohol with $H_2O_2$, *Chem. Lett.*, **39**, pp. 159–161.

83. Taarning, E., Nielsen, I., Egeblad, K., *et al.* (2008). Chemicals from Renewables: Aerobic Oxidation of Furfural and Hydroxymethylfurfural over Gold Catalysts, *Chemsuschem*, **1**, pp. 75–78.
84. Gorbanev, Y., Klitgaard, S., Woodley, J., *et al.* (2009). Gold-Catalyzed Aerobic Oxidation of 5-Hydroxymethylfurfural in Water at Ambient Temperature, *Chemsuschem*, **2**, pp. 672–675.
85. Casanova, O., Iborra, S. and Corma, A. (2009). Biomass into Chemicals: One Pot-Base Free Oxidative Esterification of 5-Hydroxymethyl-2-Furfural into 2,5-Dimethylfuroate with Gold on Nanoparticulated Ceria, *J. Catal.*, **265**, pp. 109–116.
86. Casanova, O., Iborra, S. and Corma, A. (2009). Biomass into Chemicals: Aerobic Oxidation of 5-Hydroxymethyl-2-Furfural into 2,5-Furandicarboxylic Acid with Gold Nanoparticle Catalysts, *Chemsuschem*, **2**, pp. 1138–1144.
87. Kimura, H., Tsuto, K., Wakisaka, T., *et al.* (1993). Selective Oxidation of Glycerol on a Platinum Bismuth Catalyst, *Appl. Catal. A: Gen.*, **96**, pp. 217–228.
88. Garcia, R., Besson, M. and Gallezot, P. (1995). Chemoselective Catalytic-Oxidation of Glycerol with Air on Platinum Metals, *Appl. Catal. A: Gen.*, **127**, pp. 165–176.
89. Mallat, T., Bodnar, Z., Hug, P., *et al.* (1995). Selective Oxidation of Cinnamyl Alcohol to Cinnamaldehyde with Air over Bi-Pt/Alumina Catalysts, *J. Catal.*, **153**, pp. 131–143.
90. Beziat, J., Besson, M. and Gallezot, P. (1996). Liquid Phase Oxidation of Cyclohexanol to Adipic Acid with Molecular Oxygen on Metal Catalysts, *Appl. Catal. A: Gen.*, **135**, L7–L11.
91. Abbadi, A. and vanBekkum, H. (1996). Selective Chemo-Catalytic Routes for the Preparation of Beta-Hydroxypyruvic Acid, *Appl. Catal. A: Gen.*, **148**, pp. 113–122.
92. Crozon, A., Besson, M. and Gallezot, P. (1998). Oxidation of 9-Decen-1-Ol (Rosalva) by Air in Aqueous Media on Platinum Catalysts, *New J. Chem.*, **22**, pp. 269–273.
93. Lee, A., Gee, J. and Theyers, H. (2000). Aspects of Allylic Alcohol Oxidation – A Bimetallic Heterogeneous Selective Oxidation Catalyst, *Green Chem.*, **2**, pp. 279–282.
94. Anderson, R., Griffin, K., Johnston, P., *et al.* (2003). Selective Oxidation of Alcohols to Carbonyl Compounds and Carboxylic Acids with Platinum Group Metal Catalysts, *Adv. Synth. Catal.*, **345**, pp. 517–523.
95. Keresszegi, C., Mallat, T., Grunwaldt, J., *et al.* (2004). A Simple Discrimination of the Promoter Effect in Alcohol Oxidation and Dehydrogenation over Platinum and Palladium, *J. Catal.*, **225**, pp. 138–146.
96. Ng, Y., Ikeda, S., Harada, T., *et al.* (2008). An Efficient and Reusable Carbon-Supported Platinum Catalyst for Aerobic Oxidation of Alcohols In Water, *Chem. Commun.*, **27**, pp. 3181–3183.
97. Dimitratos, N. and Prati, L. (2005). Gold Based Bimetallic Catalysts for Liquid Phase Applications, *Gold Bull.*, **38**, pp. 73–77.
98. Dimitratos, N., Porta, F., Prati, L., *et al.* (2005). Synergetic Effect of Platinum or Palladium on Gold Catalyst in the Selective Oxidation of D-Sorbitol, *Catal. Lett.*, **99**, pp. 181–185.
99. Bianchi, C., Canton, P., Dimitratos, N., *et al.* (2005). Selective Oxidation of Glycerol with Oxygen Using Mono and Bimetallic Catalysts based on Au, Pd and Pt Metals, *Catal. Today*, **102**, pp. 203–212.
100. Dimitratos, N., Porta, F. and Prati, L. (2005). Au, Pd (Mono and Bimetallic) Catalysts Supported on Graphite Using the Immobilisation Method – Synthesis and Catalytic Testing for Liquid Phase Oxidation of Glycerol, *Appl. Catal. A: Gen.*, **291**, pp. 210–214.
101. Villa, A., Campione, C. and Prati, L. (2007). Bimetallic Gold/Palladium Catalysts for the Selective Liquid Phase Oxidation of Glycerol, *Catal. Lett.*, **115**, pp. 133–136.
102. Wang, D., Villa, A., Porta, F., *et al.* (2008). Bimetallic Gold/Palladium Catalysts: Correlation between Nanostructure and Synergistic Effects, *J. Phys. Chem. C*, **112**, pp. 8617–8622.
103. Wang, D., Villa, A., Porta, F., *et al.* (2006). Single-Phase Bimetallic System for the Selective Oxidation of Glycerol to Glycerate, *Chem. Commun.*, **18**, pp. 1956–1958.
104. Prati, L., Villa, A., Porta, F., *et al.* (2007). Single-Phase Gold/Palladium Catalyst: The Nature of Synergistic Effect, *Catal. Today*, **122**, pp. 386–390.

105. Dimitratos, N., Villa, A., Wang, D., *et al.* (2006). Pd and Pt Catalysts Modified by Alloying with Au in the Selective Oxidation of Alcohols, *J. Catal.*, **244**, pp. 113–121.
106. Villa, A., Janjic, N., Spontoni, P., *et al.* (2009). Au-Pd/AC as Catalysts for Alcohol Oxidation: Effect of Reaction Parameters on Catalytic Activity and Selectivity, *Appl. Catal. A*: *Gen.*, **364**, pp. 221–228.
107. Enache, D., Edwards, J., Landon, P., *et al.* (2006). Solvent-Free Oxidation of Primary Alcohols to Aldehydes Using Au-Pd/$TiO_2$ Catalysts, *Science*, **311**, pp. 362–365.
108. Enache, D., Barker, D., Edwards, J., *et al.* (2007). Solvent-Free Oxidation of Benzyl Alcohol Using Titanic-Supported Gold-Palladium Catalysts: Effect of Au-Pd Ratio on Catalytic Performance, *Catal. Today*, **122**, pp. 407–411.
109. Lopez-Sanchez, J., Dimitratos, N., Miedziak, P., *et al.* (2008). Au-Pd Supported Nanocrystals Prepared by a Sol Immobilisation Technique as Catalysts for Selective Chemical Synthesis, *Phys. Chem. Chem. Phys.*, **10**, pp. 1921–1930.
110. Dimitratos, N., Lopez-Sanchez, J., Morgan, D., *et al.* (2009). Solvent-Free Oxidation of Benzyl Alcohol Using Au-Pd Catalysts Prepared by Sol Immobilisation, *Phys. Chem. Chem. Phys.*, **11**, pp. 5142–5153.
111. Dimitratos, N., Lopez-Sanchez, J., Anthonykutty, J., *et al.* (2009). Oxidation of Glycerol Using Gold-Palladium Alloy-Supported Nanocrystals, *Phys. Chem. Chem. Phys.*, **11**, pp. 4952–4961.
112. Dimitratos, N., Lopez-Sanchez, J., Meenakshisundaram, S., *et al.* (2009). Selective Formation of Lactate by Oxidation of 1,2-Propanediol Using Gold Palladium Alloy Supported Nanocrystals, *Green Chem.*, **11**, pp. 1209–1216.
113. Meenakshisundaram, S., Nowicka, E., Miedziak, P., *et al.* (2010). Oxidation of Alcohols Using Supported Gold and Gold-Palladium Nanoparticles, *Faraday Discuss.*, **145**, pp. 341–356.
114. Marx, S. and Baiker, A. (2009). Beneficial Interaction of Gold and Palladium in Bimetallic Catalysts for the Selective Oxidation of Benzyl Alcohol, *J. Phys. Chem. C*, **113**, pp. 6191–6201.
115. Ma, C., Dou, B., Li, J., *et al.* (2009). Catalytic Oxidation of Benzyl Alcohol on Au or Au-Pd Nanoparticles Confined in Mesoporous Silica, *Appl. Catal. B*: *Environ.*, **92**, pp. 202–208.
116. Chen, Y., Lim, H., Tang, Q., *et al.* (2010). Solvent-Free Aerobic Oxidation of Benzyl Alcohol over Pd Monometallic and Au-Pd Bimetallic Catalysts Supported on SBA-16 Mesoporous Molecular Sieves, *Appl. Catal. A*: *Gen.*, **380**, pp. 55–65.
117. Dimitratos, N., Messi, C., Porta, F., *et al.* (2006). Investigation on the Behaviour of Pt(0)/Carbon and Pt(0),Au(0)/Carbon Catalysts Employed in the Oxidation of Glycerol with Molecular Oxygen in Water, *J. Mol. Catal. A*: *Chem.*, **256**, pp. 21–28.
118. Villa, A., Veith, G. and Prati, L. (2010). Selective Oxidation of Glycerol under Acidic Conditions Using Gold Catalysts, *Angew. Chem. Int. Edit.*, **49**, pp. 4499–4502.
119. Miyamura, H., Matsubara, R. and Kobayashi, S. (2008). Gold-Platinum Bimetallic Clusters for Aerobic Oxidation of Alcohols under Ambient Conditions, *Chem. Commun.*, **17**, pp. 2031–2033.
120. Liotta, L., Venezia, A., Deganello, G., *et al.* (2001). Liquid Phase Selective Oxidation of Benzyl Alcohol over Pd-Ag Catalysts Supported on Pumice, *Catal. Today*, **66**, pp. 271–276.
121. Beier, M., Hansen, T. and Grunwaldt, J. (2009). Selective Liquid-Phase Oxidation of Alcohols Catalyzed by a Silver-Based Catalyst Promoted by the Presence of Ceria, *J. Catal.*, **266**, pp. 320–330.
122. Rosowski, F., Storck, S. and Zühlke, J. (2008). Oxyfunctionalization of Alkyl Aromatics. Handbook of Heterogeneous Catalysis, Wiley-VCH, Weinheim.
123. Bregeault, J. (2003). Transition-Metal Complexes for Liquid-Phase Catalytic Oxidation: Some Aspects of Industrial Reactions and of Emerging Technologies, *Dalton T.*, **17**, pp. 3289–3302.
124. Sheldon, R. (1991). Heterogeneous Catalytic-Oxidation and Fine Chemicals, in M. Guisnet, J. Barrault, C. Bouchoule, *et al.* (eds), *Heterogeneous Catalysis and Fine Chemicals*, Elsevier Science Publ B V, Amsterdam, pp. 33–54.
125. Lambert, R., Williams, F., Cropley, R., *et al.* (2005). Heterogeneous Alkene Epoxidation: Past, Present and Future, *J. Mol. Catal. A*: *Chem.*, **228**, pp. 27–33.

126. Klemm, E., Dietzsch, E., Schwarz, T., *et al.* (2008). Direct Gas-Phase Epoxidation of Propene with Hydrogen Peroxide on TS-1 Zeolite in a Microstructured Reactor, *Ind. Eng. Chem. Res.*, **47**, pp. 2086–2090.
127. Hayashi, T., Tanaka, K. and Haruta, M. (1998). Selective Vapor-Phase Epoxidation of Propylene over Au/$TiO_2$ Catalysts in the Presence of Oxygen and Hydrogen, *J. Catal.*, 178, pp. 566–575.
128. Haruta, M. and Date, M. (2001). Advances in the Catalysis of Au Nanoparticles, *Appl. Catal. A: Gen.*, **222**, pp. 427–437.
129. Edwards, J. and Hutchings, G. (2008). Palladium and Gold-Palladium Catalysts for the Direct Synthesis of Hydrogen Peroxide, *Angew. Chem. Int. Edit.*, **47**, pp. 9192–9198.
130. Sato, K., Aoki, M. and Noyori, R. (1998). A "Green" Route to Adipic Acid: Direct Oxidation of Cyclohexenes with 30 Percent Hydrogen Peroxide, *Science*, **281**, pp. 1646–1647.
131. Hughes, M., Xu, Y., Jenkins, P., *et al.* (2005). Tunable Gold Catalysts for Selective Hydrocarbon Oxidation under Mild Conditions, *Nature*, **437**, pp. 1132–1135.
132. Bawaked, S., Dummer, N., Dimitratos, N., *et al.* (2009). Solvent-Free Selective Epoxidation of Cyclooctene Using Supported Gold Catalysts, *Green Chem.*, **11**, pp. 1037–1044.
133. Bawaked, S., Dummer, N., Bethell, D., *et al.* (2011). Solvent-Free Selective Epoxidation of Cyclooctene Using Supported Gold Catalysts: An Investigation of Catalyst Re-Use, *Green Chem.*, **13**, pp. 127–134.
134. Patil, N., Uphade, B., Jana, P., *et al.* (2004). Epoxidation of Styrene by Anhydrous T-Butyl Hydroperoxide over Reusable Gold Supported on MgO and Other Alkaline Earth Oxides, *J. Catal.*, **223**, pp. 236–239.
135. Patil, N., Uphade, B., Jana, P., *et al.* (2004). Epoxidation of Styrene by T-Butyl Hydroperoxide over Gold Supported on $Yb_2O_3$ and Other Rare Earth Oxides, *Chem. Lett.*, **33**, pp. 400–401.
136. Lignier, P., Morfin, F., Mangematin, S., *et al.* (2007). Stereoselective Stilbene Epoxidation over Supported Gold-Based Catalysts, *Chem. Commun.*, **2**, pp. 186–188.
137. Lignier, P., Morfin, F., Piccolo, L., *et al.* (2007). Insight into the Free-Radical Chain Mechanism of Gold-Catalyzed Hydrocarbon Oxidation Reactions in the Liquid Phase, *Catal. Today*, **122**, pp. 284–291.
138. Lignier, P., Mangematin, S., Morfin, F., *et al.* (2008). Solvent and Oxidant Effects on the Au/$TiO_2$-Catalyzed Aerobic Epoxidation of Stilbene, *Catal. Today*, **138**, pp. 50–54.
139. Mendez, V., Guillois, K., Daniele, S., *et al.* (2010). Aerobic Methylcyclohexane-Promoted Epoxidation of Stilbene over Gold Nanoparticles Supported on Gd-Doped Titania, *Dalton Transactions*, **39**, pp. 8457–8463.
140. Boualleg, M., Guillois, K., Istria, B., *et al.* (2010). Highly Efficient Aerobic Oxidation of Alkenes over Unsupported Nanogold, *Chem. Commun.*, **46**, pp. 5361–5363.
141. Turner, M., Golovko, V., Vaughan, O., *et al.* (2008). Selective Oxidation with Dioxygen by Gold Nanoparticle Catalysts Derived from 55-atom Clusters, *Nature*, **454**, pp. 981–983.
142. Cai, Z., Zhu, M., Chen, J., *et al.* (2010). Solvent-Free Oxidation of Cyclohexene over Catalysts with Molecular Oxygen, *Catal. Commun.*, **12**, pp. 197–201.
143. Sun, J., Fujita, S., Zhao, F., *et al.* (2005). A Direct Synthesis of Styrene Carbonate from Styrene with the Au/$SiO_2$-$ZnBr_2$/$Bu_4NBr$ Catalyst System, *J. Catal.*, **230**, pp. 398–405.
144. Xiang, D., Liu, X., Sun, J., *et al.* (2009). A Novel Route for Synthesis of Styrene Carbonate Using Styrene and $CO_2$ as Substrates over Basic Resin R201 Supported Au Catalyst, *Catal. Today*, **148**, pp. 383–388.
145. Wang, Y., Sun, J., Xiang, D., *et al.* (2009). A Facile, Direct Synthesis of Styrene Carbonate from Styrene and $CO_2$ Catalyzed by Au/Fe(OH)(3)-$ZnBr_2$/$Bu_4NBr$ System, *Catal. Lett.*, **129**, pp. 437–443.
146. Partenheimer, W. (1995). Methodology and Scope of Metal Bromide Autoxidation of Hydrocarbons, *Catal. Today*, **23**, pp. 69–158.
147. Ishii, Y., Sakaguchi, S. and Iwahama, T. (2001). Innovation of Hydrocarbon Oxidation with Molecular Oxygen and Related Reactions, *Adv. Synth. Catal.*, **343**, pp. 393–427.

148. Miki, J., Osada, Y., Konoshi, T., *et al.* (1996). Selective Oxidation of Toluene to Benzoic Acid Catalyzed by Modified Vanadium Oxides, *Appl. Catal. A: Gen.*, **137**, pp. 93–104.
149. Konietzni, F., Kolb, U., Dingerdissen, U., *et al.* (1998). AMM-MnxSi-Catalyzed Selective Oxidation of Toluene, *J. Catal.*, **176**, pp. 527–535.
150. Li, X., Xu, J., Zhou, L., *et al.* (2006). Liquid-Phase Oxidation of Toluene by Molecular Oxygen over Copper Manganese Oxides, *Catal. Lett.*, **110**, pp. 149–154.
151. Wang, F., Xu, J., Li, X., *et al.* (2005). Liquid Phase Oxidation of Toluene to Benzaldehyde with Molecular Oxygen over Copper-Based Heterogeneous Catalysts, *Adv. Synth. Catal.*, **347**, pp. 1987–1992.
152. Gao, J., Tong, X., Li, X., *et al.*, (2007). The Efficient Liquid-Phase Oxidation of Aromatic Hydrocarbons by Molecular Oxygen in the Presence of $MnCO_3$, *J. Chem. Tech. Biot.*, **82**, pp. 620–625.
153. Brutchey, R., Drake, I., Bell, A., *et al.* (2005). Liquid-Phase Oxidation of Alkylaromatics by a H-Atom Transfer Mechanism with a New Heterogeneous CoSBA-15 Catalyst, *Chem. Commun.*, **29**, pp. 3736–3738.
154. Singh, A. and Selvam, T. (1996). Liquid Phase Oxidation Reactions over Chromium Silicalite-1 (CrS-1) Molecular Sieves, *J. Mol. Catal. A: Chem.*, **113**, pp. 489–497.
155. Colby, J., Stirling, D. and Dalton, H. (1977). Soluble Methane Mono-Oxygenase of Methylococcus-Capsulatus-(Bath): Ability to Oxygenate Normal-Alkanes, Normal-Alkenes, Ethers, and Alicyclic, Aromatic and Heterocyclic-Compounds, *Biochem. J.*, **165**, pp. 395–402.
156. Dimitratos, N., Lopez-Sanchez, J., Morgan, D., *et al.* (2007). Solvent Free Liquid Phase Oxidation of Benzyl Alcohol Using Au Supported Catalysts Prepared Using a Sol Immobilization Technique, *Catal. Today*, **122**, pp. 317–324.
157. Pritchard, J., Kesavan, L., Piccinini, M., *et al.* (2010). Direct Synthesis of Hydrogen Peroxide and Benzyl Alcohol Oxidation Using Au-Pd Catalysts Prepared by Sol Immobilization, *Langmuir*, **26**, pp. 16568–16577.
158. Lopez-Sanchez, J., Dimitratos, N., Glanville, N., *et al.* (2011). Reactivity Studies of Au-Pd Supported Nanoparticles for Catalytic Applications, *Appl. Catal. A: Gen.*, **391**, pp. 400–406.
159. Kesavan, L., Tiruvalam, R., Ab Rahim, M., *et al.* (2011). Solvent-Free Oxidation of Primary Carbon-Hydrogen Bonds in Toluene Using Au-Pd Alloy Nanoparticles, *Science*, **331**, pp. 195–199.
160. Beier, M., Schimmoeller, B., Hansen, T., *et al.* (2010). Selective Side-Chain Oxidation of Alkyl Aromatic Compounds Catalyzed by Cerium Modified Silver Catalysts, *J. Mol. Catal. A: Chem.*, **331**, pp. 40–49.
161. Zhao, R., Ji, D., Lv, G., *et al.*, (2004). A Highly Efficient Oxidation of Cyclohexane over Au/ZSM-5 Molecular Sieve Catalyst with Oxygen as Oxidant, *Chem. Commun.*, **7**, pp. 904–905.
162. Lu, G., Zhao, R., Qian, G., *et al.* (2004). A Highly Efficient Catalyst Au/MCM-41 for Selective Oxidation Cyclohexane Using Oxygen, *Catal. Lett.*, **97**, pp. 115–118.
163. Wu, P., Bai, P., Lei, Z., *et al.* (2011). Gold Nanoparticles Supported on Functionalized Mesoporous Silica for Selective Oxidation of Cyclohexane, *Micropor. Mesopor. Mat.*, **141**, pp. 222–230.
164. Xu, Y., Landon, P., Enache, D., *et al.* (2005). Selective Conversion of Cyclohexane to Cyclohexanol and Cyclohexanone Using a Gold Catalyst under Mild Conditions, *Catal. Lett.*, **101**, pp. 175–179.
165. Xu, L., He, C., Zhu, M., *et al.* (2007). Silica-Supported Gold Catalyst Modified by Doping with Titania for Cyclohexane Oxidation, *Catal. Lett.*, **118**, pp. 248–253.
166. Hereijgers, B. and Weckhuysen, B. (2010). Aerobic Oxidation of Cyclohexane by Gold-Based Catalysts: New Mechanistic Insight by Thorough Product Analysis, *J. Catal.*, **270**, pp. 16–25.
167. Moro-oka, Y. and Akita, M. (1998). Bio-Inorganic Approach to Hydrocarbon Oxidation, *Catal. Today*, **41**, pp. 327–338.

168. Miyake, T., Hamada, M., Sasaki, Y., *et al.* (1995). Direct Synthesis of Phenol by Hydroxylation of Benzene with Oxygen and Hydrogen, *Appl. Catal. A: Gen.*, **131**, pp. 33–42.
169. Tatsumi, T., Yuasa, K. and Tominaga, H. (1992). Hydroxylation of Benzene and Hexane by Oxygen and Hydrogen over Palladium-Containing Titanium Silicalites, *J. Chem. Soc. Chem. Comm.*, **19**, pp. 1446–1447.
170. Remias, J., Pavlosky, T. and Sen, A. (2003). Catalytic Hydroxylation of Benzene and Cyclohexane Using In Situ Generated Hydrogen Peroxide: New Mechanistic Insights and Comparison with Hydrogen Peroxide Added Directly, *J. Mol. Catal. A: Chem.*, **203**, pp. 179–192.
171. Laufer, W., Niederer, J. and Hoelderich, W. (2002). New Direct Hydroxylation of Benzene with Oxygen in the Presence of Hydrogen over Bifunctional Palladium/Platinum Catalysts, *Adv. Synth. Catal.*, **344**, pp. 1084–1089.
172. Ye, S., Hamakawa, S., Tanaka, S., *et al.* (2009). A One-Step Conversion of Benzene to Phenol Using MEMS-Based Pd Membrane Microreactors, *Chem. Eng. J.*, **155**, pp. 829–837.
173. Sato, K., Niwa, S., Hanaoka, T., *et al.* (2004). Direct Hydroxylation of Methyl Benzoate to Methyl Salicylate by Using New Pd Membrane Reactor, *Catal. Lett.*, **96**, pp. 107–112.
174. Sato, K., Hanaoka, T., Niwa, S., *et al.* (2005). Direct Hydroxylation of Aromatic Compounds by a Palladium Membrane Reactor, *Catal. Today*, **104**, pp. 260–266.
175. Sato, K., Hamakawa, S., Natsui, M., *et al.* (2010). Palladium-Based Bifunctional Membrane Reactor for One-Step Conversion of Benzene to Phenol and Cyclohexanone, *Catal. Today*, **156**, pp. 276–281.
176. Vulpescu, G., Ruitenbeek, M., van Lieshout, L., *et al.* (2004). One-Step Selective Oxidation over a Pd-Based Catalytic Membrane: Evaluation of the Oxidation of Benzene to Phenol as a Model Reaction, *Catal. Commun.*, **5**, pp. 347–351.
177. Centi, G. and Perathoner, S. (2009). One-Step $H_2O_2$ and Phenol Syntheses: Examples of Challenges for New Sustainable Selective Oxidation Processes, *Catal. Today*, **143**, pp. 145–150.
178. Zhu, B., Lazar, M., Trewyn, B., *et al.* (2008). Aerobic Oxidation of Amines to Imines Catalyzed by Bulk Gold Powder and by Alumina-Supported Gold, *J. Catal.*, **260**, pp. 1–6.
179. Aschwanden, L., Mallat, T., Grunwaldt, J., *et al.* (2009). Gold-Catalyzed Aerobic Oxidation of Dibenzylamine: Homogeneous or Heterogeneous Catalysis? *J. Mol. Catal. A: Chem.*, **300**, pp. 111–115.
180. So, M., Liu, Y., Ho, C., *et al.* (2009). Graphite-Supported Gold Nanoparticles as Efficient Catalyst for Aerobic Oxidation of Benzylic Amines to Imines and N-Substituted 1,2,3,4-Tetrahydroisoquinolines to Amides: Synthetic Applications and Mechanistic Study, *Chemistry — An Asian Journal*, **4**, pp. 1551–1561.
181. Grirrane, A., Corma, A. and Garcia, H. (2009). Highly Active and Selective Gold Catalysts for the Aerobic Oxidative Condensation of Benzylamines to Imines and One-Pot, Two-Step Synthesis of Secondary Benzylamines, *J. Catal.*, **264**, pp. 138–144.
182. So, M., Liu, Y., Ho, C., *et al.* (2011). Silica-Supported Gold Nanoparticles Catalyzed One-Pot, Tandem Aerobic Oxidative Cyclization Reaction for Nitrogen-Containing Polyheterocyclic Compounds, *Chemcatchem*, **3**, pp. 386–393.
183. Grirrane, A., Corma, A. and Garcia, H. (2010). Preparation of Symmetric and Asymmetric Aromatic Azo Compounds from Aromatic Amines or Nitro Compounds Using Supported Gold Catalysts, *Nature Protocols*, **5**, pp. 429–438.
184. Mitsudome, T., Arita, S., Mori, H., *et al.* (2008). Supported Silver-Nanoparticle-Catalyzed Highly Efficient Aqueous Oxidation of Phenylsilanes to Silanols, *Angew. Chem. Int. Edit.*, **47**, pp. 7938–7940.
185. Grirrane, A., Corma, A. and Garcia, H. (2009). Gold Nanoparticles Supported on Ceria Promote the Selective Oxidation of Oximes into the Corresponding Carbonylic Compounds, *J. Catal.*, **268**(2), pp. 350–355.

# Chapter 23

# Sustainability Trends in Homogeneous Catalytic Oxidations

Alessandro SCARSO* and Giorgio STRUKUL*

Green chemistry is strongly influencing all aspects of chemical research and oxidation reactions in particular are witnessing a shift from the use of toxic and less atom efficient oxidants to the preferred employment of hydrogen peroxide and molecular oxygen. Recent developments in the field span from the discovery of more efficient metal catalysts characterized by higher turnovers or selectivities (especially enantioselectivity), to the advent of more economic and less toxic metal catalysts, to the replacement of organic solvents with water as reaction medium and to the implementation of efficient techniques to ensure simple catalyst recycling. The present contribution adequately covers all the above subjects, enriching the discussion with recent examples of timely and cutting edge catalytic systems characterized by high degree of novelty and possible future developments.

## 23.1. Introduction

The field of catalytic oxidation involving soluble transition metal complexes has experienced a boom in research over the past three decades. In this period a huge variety of new catalysts have been discovered and tested in reactions such as alkene epoxidation, sulfoxidation, alkane oxidation, the Baeyer–Villiger oxidation of ketones, N-oxidation, etc. The synthesis and mechanistic operating principles of prominent categories of catalysts, such as biomimetic systems based on synthetic metalloporphyrins or polyoxometalates, have been established and a variety of mono-oxygen donors from alkyl hydroperoxides, to dioxygen, to bleach, etc. have been successfully tested, contributing to an impressive wealth of results. However, because of the complexity of oxygen transfer with respect to other catalytic processes, the field has not witnessed the discovery of "the" catalyst, similar to the situation found in e.g. the hydrogenation or hydroformylation with Rh-based systems or in C-C forming reactions with Pd-based systems. Still, quite distinct from these examples is the degree of efficiency in terms of activity and selectivity observed in oxidation reactions.

*Dipartimento di Scienze Molecolari e Nanosistemi, Università Ca' Foscari, Venezia, Italy.

**Table 23.1.** Papers focused on oxidation presented at different ISHC symposia.

| Symposium | Year | Total contributions | Total on oxidation | Invited lectures (on oxidation) | Oral presentations (on oxidation) |
|---|---|---|---|---|---|
| ISHC 8, Amsterdam | 1992 | 252 | 46 | 6 | 3 |
| ISHC 12, Stockholm | 2000 | 295 | 45 | 4 | 2 |
| ISHC 17, Poznan | 2010 | 312 | 24 | — | 2 |

In more recent times, the interest in the fundamental aspects of homogeneous catalytic oxidations has been declining as witnessed by e.g. the number of contributions presented at the International Symposium on Homogeneous Catalysis (ISHC). A comparison between the Amsterdam (1992), Stockholm (2000) and the recent Poznan (2010) meetings is reported in Table 23.1 and shows that of approximately 250–300 contributions presented at each symposium, over roughly 20 years, the number of those devoted to oxidations has decreased from 46 in Amsterdam, including 6 invited lectures and 3 oral presentations, to 24 in Poznan with just 2 oral communications and no invited lectures.

At the same time an analysis of the recent literature shows that the focus is shifting towards aspects related to making homogeneous oxidation catalytic systems more sustainable and/or more compatible with practical applications, looking for higher turnovers or selectivities (especially enantioselectivity), trying to get rid of organic solvents and eventually attempting to overcome the major problem with homogeneous systems, i.e. the separation of the catalyst from the reaction mixture and possible recycling. In other words, even this area is being strongly influenced by the implementation of the now binding twelve principles of green chemistry.

In this respect, in order to give an overview of the area, we have envisaged some trends towards the sustainability of homogeneous oxidation processes, namely:

- the use of environmentally acceptable oxidants such as $O_2$ and $H_2O_2$;
- enantioselective oxidations with $O_2$ and $H_2O_2$;
- the use of water as reaction medium;
- the use of less toxic metals as catalysts;
- the heterogenization of the most successful homogeneous systems.

These will be the subject of the sections to come. Because of the limitations of space, subjects such as enzymatic oxidation, the use of polyoxometalates in oxidation reactions, oxidations mediated by metal nanoparticles, oxidation of bioavailable feedstocks, oxidation of water to dioxygen, organocatalytic asymmetric oxidations and asymmetric phase transfer oxidations are not covered in this chapter.

## 23.2. Use of Oxygen and Hydrogen Peroxide

In accordance with the twelve principles of green chemistry,[1] oxidation reactions should evolve toward the exclusive employment of more benign oxidants for the benefit of both the environment and living organisms. Oxidants are also characterized by different kinetic stability which means very different handling conditions, also implying the need for correct storage and the limited lifetime for the most active ones, while the most stable require activation via catalysis in order to express their oxidizing power. After reaction, the reduced form of the oxidant is a by-product which needs to be removed and properly treated. In this respect, low molecular weight oxidants are preferred because the remaining mass of the by-product is the lowest possible. Atom efficiency[2] is a common criterion in green chemistry and is usually also applied to oxidants. In this case the value is calculated by dividing the atomic weight of one oxygen atom transferred to the product by the molecular weight of the oxidant. The meaning is determined by establishing how much of the mass of the oxidant is wasted in order to transfer one oxygen atom to the substrate. Table 23.2 shows the reported values for a series of common oxidants. In this respect it is evident that $O_2$ and $H_2O_2$ are characterized by the highest possible atom efficiency (50% and 48%, respectively) that is extremely important when considering large-scale industrial oxidation processes. More importantly, the common by-product formed by these oxidants is water, which is absolutely compatible with release into the environment. Just as a comparison, other common and widely employed oxidizing species such as hypochlorite, alkyl hydroperoxides and percarboxylic acids are characterized by much lower atom efficiencies (21, 22–12 and 27–29%, respectively) and the related by-products are usually difficult to remove from the crude product and require extraction or distillation steps. General reviews on the use of dioxygen[3] and hydrogen peroxide[4] have been published.

This is why in the following sections oxidations employing dioxygen and hydrogen peroxide will be considered whenever possible.

## 23.3. Enantioselective Oxidations

The importance of enantiomerically pure drugs is now common knowledge, but it is a relatively recent discovery. In fact, it was only in 1992 that it became mandatory for the chemical industry to meet the new criteria from the American Food and Drug Administration (FDA) that established a policy towards single enantiomer drugs.[5]

In this section landmark examples of homogeneous asymmetric catalytic oxidations using chiral organometallic complexes will be considered; the discussion will be limited to systems based on the use of $O_2$ and $H_2O_2$ as terminal oxidants because of the improvement in their environmental impact. However, when the catalytic system

**Table 23.2.** Comparison between different terminal oxidants in terms of green character.

| Oxidant | By-product | Atom efficiency (%)[a] | Catalysis required | Cost (€/Kg) or (€/l)[b] | Title, appearance, packaging [b] |
|---|---|---|---|---|---|
| $O_2$ | $H_2O$ | 50 | ✓ | 3.5 | 99.6%, gas, cylinder |
| $H_2O_2$ | $H_2O$ | 47 | ✓ | 90 | <30%, solution, bottle |
| Urea · $H_2O_2$ | Urea + $H_2O$ | 17 | ✓ | 400 | 97%, solid, bottle |
| $O_3$ | $O_2$ | 33 | ✓ | c | c |
| $CH_3COOOH$ | $CH_3COOH$ | 27 | — | 380 | 40%, solution, bottle |
| t-BuOOH | t-BuOH | 22 | ✓ | 45 | 70%, solution, bottle |
| NaClO | NaCl | 21 | — | 13 | 10%, solution, bottle |
| Pyridine-N-oxide | Pyridine | 17 | ✓ | 440 | 95%, solid, bottle |
| N-methylmorpholine N-oxide | N-methylmorpholine | 14 | ✓ | 1500 | 97%, solid, bottle |
| CumOOH | CumOH | 12 | ✓ | 106 | 80%, solution, bottle |
| $KMnO_4$ | $MnO_2 + H_2O$ | 10 | — | 85 | >98%, solid, bottle |
| Caro's acid or Oxone® $2KHSO_5 + KHSO_4 + K_2SO_4$ | $3\ KHSO_4 + K_2SO_4$ | 10[d] | — | 53 | ≥47% $KHSO_5$, solid, bottle |
| 4-Ph-pyridine-N-oxide | 4-Ph-pyridine | 9 | — | 8500 | 98%, solid, bottle |
| m-CPBA | m-CBA | 9 | — | 450 | <77%, solid, bottle |
| $NaIO_4$ | $NaI + H_2O$ | 8 | — | 350 | >99%, solid, bottle |
| PhIO | PhI | 7 | ✓ | c | c |
| $OsO_4$ | $OsO_2(OH)_2$ | 6 | — | 300000 | <98%, solid, bottle |
| $PhI(OAc)_2$ | PhI + 2HOAc | 5 | — | 1500 | >98%, solid, bottle |

[a]Calculated considering transfer of only one O atom to the product.
[b]Data from www.sigmaaldrich.com, June 2011.
[c]Not sold.
[d]Two O atoms are transferred to products.

is highly enantioselective (enantioselectivity is measured by the enantiomeric excess or ee), examples of oxidations with urea · $H_2O_2$ will also be discussed. The most recent contributions are divided by reaction category and the most efficient and stereoselective catalysts are reported. Special emphasis is placed on those catalysts that show potential applications to large-scale production, synthesis of natural products or drugs and those that are suitable for catalyst recycling under homogeneous conditions.

### 23.3.1. *Asymmetric epoxidation*

This is the most studied reaction. For simplicity of presentation the different catalytic systems are divided by metal center.

#### 23.3.1.1. *Manganese*

Given its well-known role in redox processes such as peroxidases, catalases and photosystem II, manganese is a well-established metal center for oxidation reactions. Since the landmark discovery of Jacobsen and Katsuki in the 1990s concerning asymmetric epoxidation with salen Mn(III) species using iodosylbenzene as the oxidant, more than 40 papers have appeared in the literature concerning asymmetric epoxidation with hydrogen peroxide as the terminal oxidant, and many more if we consider other related oxidants such as urea $\cdot$ $H_2O_2$ adduct or peracetic acid. While for a detailed and exhaustive description of all systems we refer the reader to a recent publication,[6] we here describe selected contributions characterized by high asymmetric induction and original ligand design.

Salen-type ligands have been extensively investigated for Mn(III)-based epoxidation catalysts, the first example concerning the use of hydrogen peroxide coming from the group of Katsuki.[7] The catalytic system was based on chiral salen ligands with stereogenic centers, both on the diimine backbone as well as in the ortho position of the phenolic groups. After a careful optimization of the experimental conditions involving optimization of the solvent and of the concentrations of the species in solution, good yields (55–98%) and good enantioselectivities (up to 95%) were obtained for a series of chromene-based substrates (Scheme 23.1).

Allyl alcohol derivatives proved to be suitable substrates for the asymmetric epoxidation with classical chiral salen Mn(III) complexes, showing both good regioselectivity in the case of geraniol derivatives where only the allylic double bond was converted into the epoxide,[8] as well as moderate enantioselectivities and yields, the latter being obtained with a large excess of oxidant because of the concomitant decomposition of $H_2O_2$ induced by the chiral catalysts itself, as commonly observed with catalase (Scheme 23.2).

A step forward in the reaction was made when Katsuki disclosed the asymmetric epoxidation reaction with achiral salen Mn(III) complexes in the presence of chiral additives, whose role was to coordinate as fifth ligand to the metal center, steering the existing equilibrium between two enantiomeric conformations of the salen ligand preferentially towards only one. Subsequently, this led to the direct covalent connection of a nitrogen-based ligand to the salen scaffold leading to pentadentate

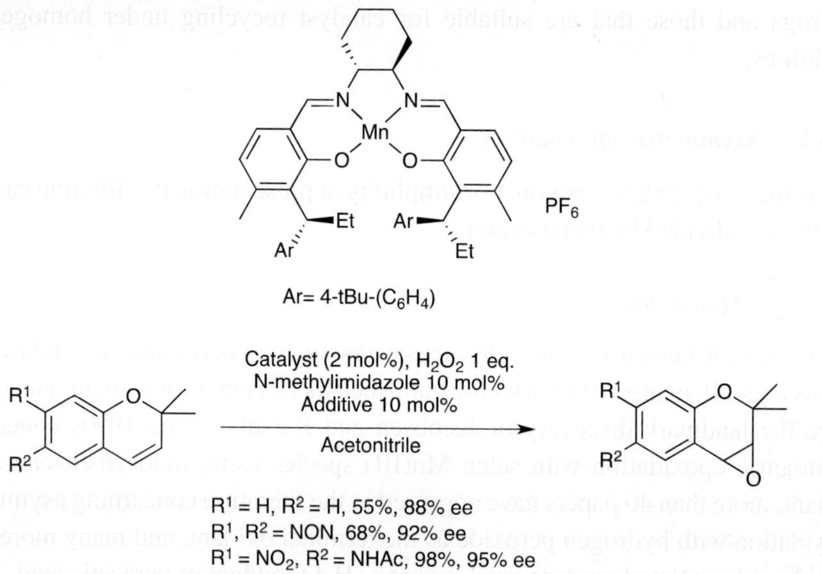


**Scheme 23.1.** Asymmetric epoxidation of chromene derivatives mediated by a Mn(salen) complex with $H_2O_2$ as the terminal oxidant.

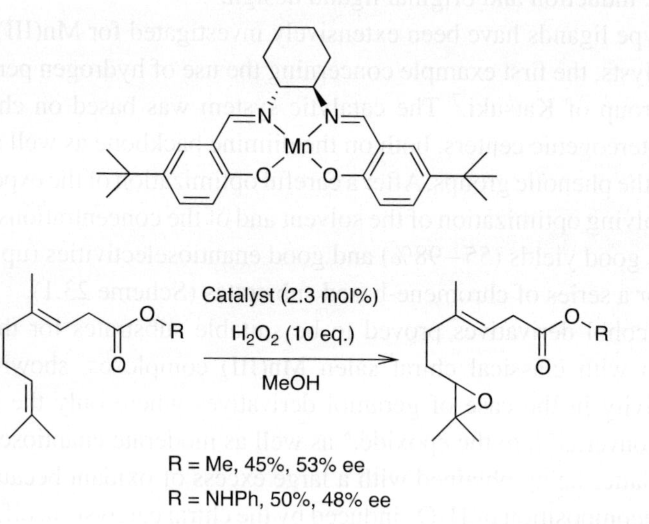


**Scheme 23.2.** Epoxidation of geraniol derivatives with a Mn(salen) complex and $H_2O_2$ as the oxidant.

ligands bearing atropoisomeric binaphthyl moieties (Scheme 23.3).[9] In particular, the complex containing a methylimidazole residue as a fifth coordinating moiety turned out to be the best catalyst towards electron-rich alkene substrates belonging to the chromene or styrene families, with excellent yields and extremely high enantiomeric excesses.

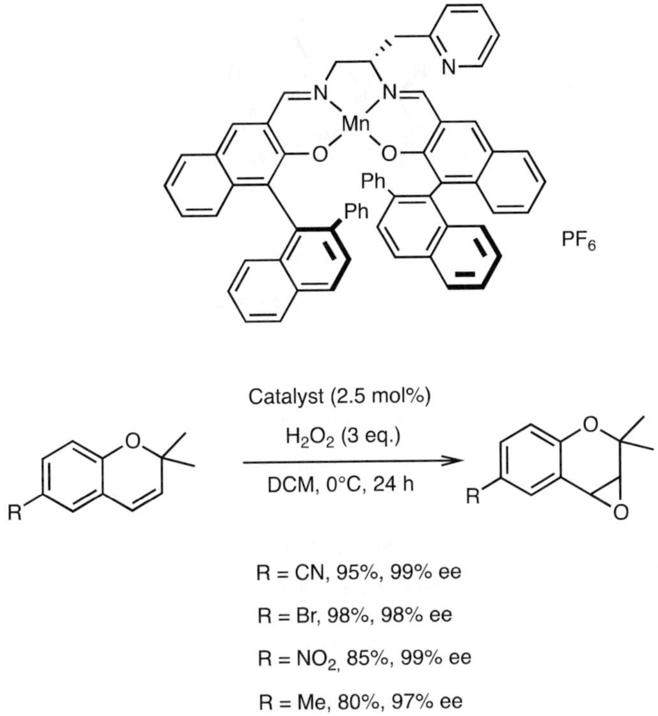


**Scheme 23.3.** Asymmetric epoxidation of chromene derivatives with $H_2O_2$ mediated by a Mn(salen) complex bearing a pentadentate ligand.

More recently, other nitrogen-based chiral ligands started to be employed for the development of Mn(II) epoxidation catalysts. In particular, pyridine and bipyridine-based ligands obtained more encouraging results. In most cases the oxidant employed was peracetic acid or combinations of $H_2O_2$ and acetic acid that provide the percarboxylic oxidant *in situ*. One representative system is reported in Scheme 23.4 based on tetradentate nitrogen ligands bearing two pyridines and two tri-substituted amines.[10] The catalytic system was active towards styrenes as well as chalcone derivatives with good catalytic activities and moderate enantioselectivities.

### 23.3.1.2. *Titanium*

Chiral Ti complexes with tetradentate NOON ligands of the salan family were intensively investigated, in particular by the group of Katsuki. As far as non-atropoisomeric ligands are concerned, the effects of the substituents on the aromatic rings of the phenol residues were investigated in detail, and it was observed that the presence in the ortho position of an aromatic residue,[11] ranging from phenyl to 9-anthracenyl, ensured high ees in the range 78–89%, with the best value with an *o*-OMe-phenyl group, indicating that the enantioselectivity was only slightly affected by the steric hindrance present on the ligand (Scheme 23.5).

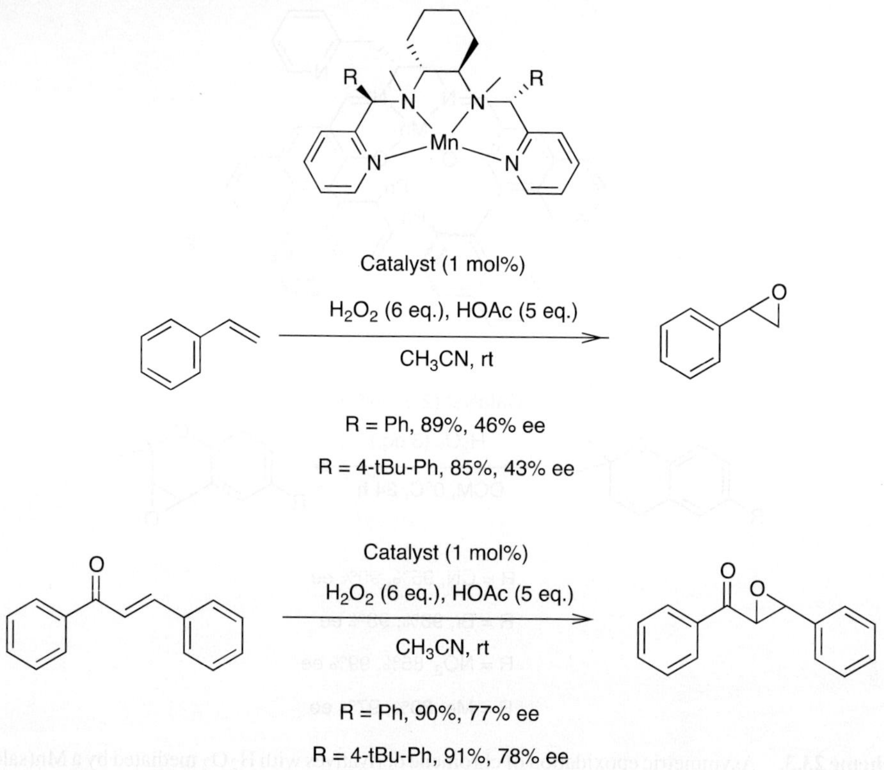


**Scheme 23.4.** Asymmetric epoxidation of styrene and chalcone with $H_2O_2$ mediated by a Mn catalysts bearing tetradentate pyridine ligands.

Berkessel further extended the study tailoring the substituents in the *ortho* and *para* positions of the phenol group position for half-reduced Salalen ligands. The results observed were in agreement with the findings of Katsuki, once again stressing the importance of aromatic substituents in the three position and confirming that even electronic effects are not very important in determining the enantioselectivity of the reaction (Scheme 23.6).[12]

Atropoisomeric salalen ligands showed even better results forming robust dimeric $\mu$-oxo species that ensured high catalytic activity and enantioselectivity with catalyst loading as low as 0.02%, keeping the amount of oxidant stoichiometric with respect to the substrate.[13] This catalytic system proved to be very general in terms of the structure of the substrate. In fact, not only rigid cis aromatic alkenes such as indene, tetrahydronaphthane or styrene derivatives were efficiently and selectively oxidized with ee in the range of 93–99%, but also poorly reactive terminal (72–95% ee) and internal (71–97% ee) aliphatic alkenes provided the enantioenriched corresponding epoxide, with good enantioselectivity and moderate to good activity.

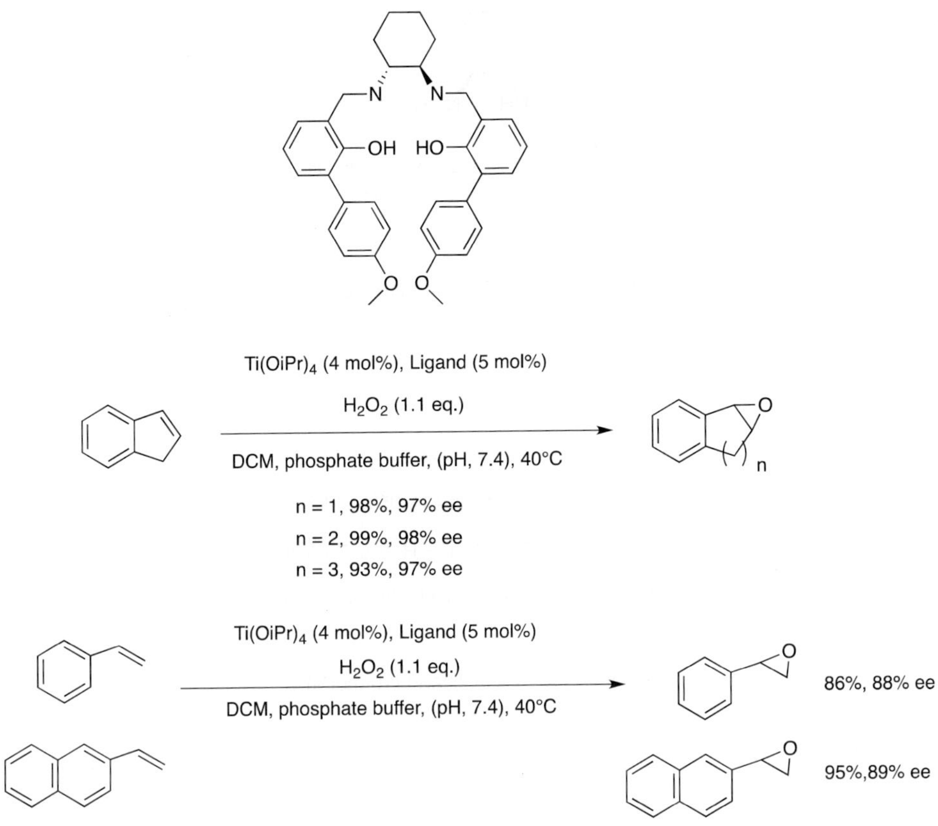


**Scheme 23.5.** Asymmetric epoxidation of vinyl arenes with $H_2O_2$ mediated by Ti(IV) complex bearing salan ligand.

The preference for unsubstituted C=C double bonds enabled the development of highly region- and stereoselective epoxidation of dienes bearing both an electron-rich substituted and an electron-poor terminal double bond (Scheme 23.7),[14] which is an uncommon behavior observed for the first time in the asymmetric epoxidation with chiral Pt(II) complexes[15] and described in the following paragraphs. Though the intrinsic mechanism remains to be solved, Ti salalen catalysts bearing atropoisomeric ligands represent one of the most versatile, efficient and selective catalysts for asymmetric epoxidation with hydrogen peroxide.

### 23.3.1.3. *Niobium*

While Ti(salan) complexes showed good activity and enantioselectivity towards unfunctionalized olefins, the corresponding Nb(salan) allowed asymmetric epoxidation of allyl alcohols via coordination of the heteroatom to the metal center, which is

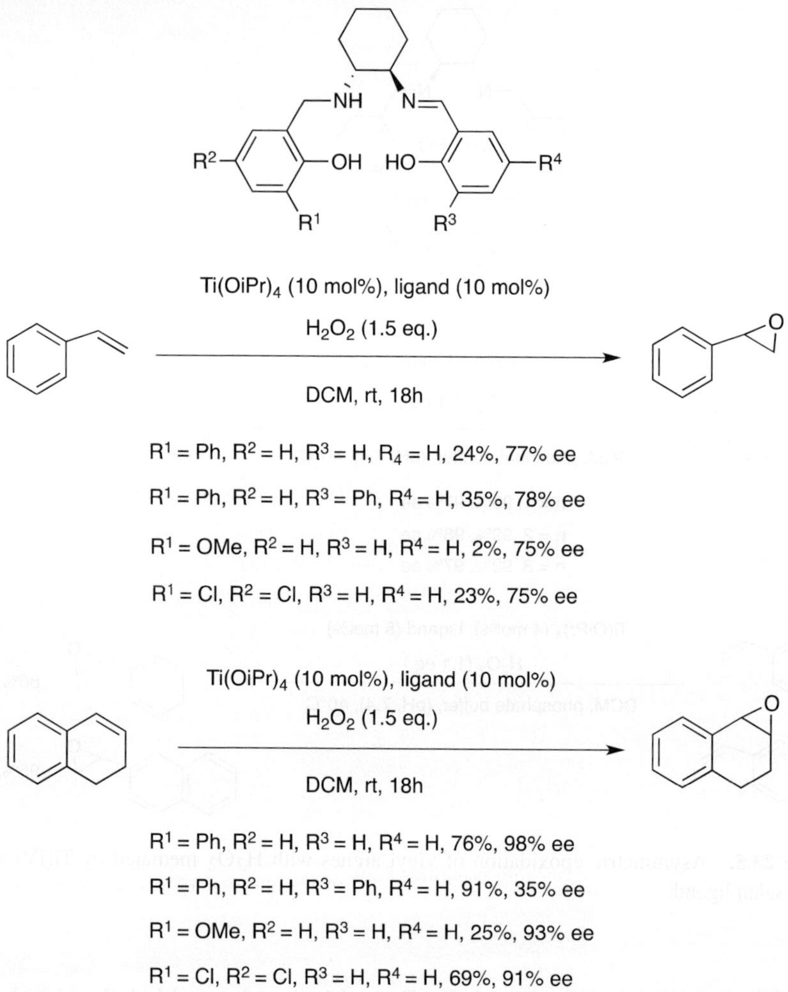


**Scheme 23.6.** Asymmetric epoxidation of viny arenes with $H_2O_2$ mediated by Ti(IV) complex bearing salalen ligands.

a stringent requirement for this oxidation reaction (Scheme 23.8a).[16,17] The catalytic system employs urea-$H_2O_2$ as the terminal oxidant under mild conditions, which is atypical for asymmetric epoxidation of allylic alcohols that are usually obtained with alkyl hydroperoxides. Second generation salen ligands turned out to be better performing, and the correct combination of chirality of the binaphthalene residues and diamine backbone was investigated, observing that the complex reported in Scheme 23.8b enabled high catalytic activity and high enantioselectivity towards allyl alcohols, with higher selectivity for three substituted substrates and lower for geminal substituted substrates.

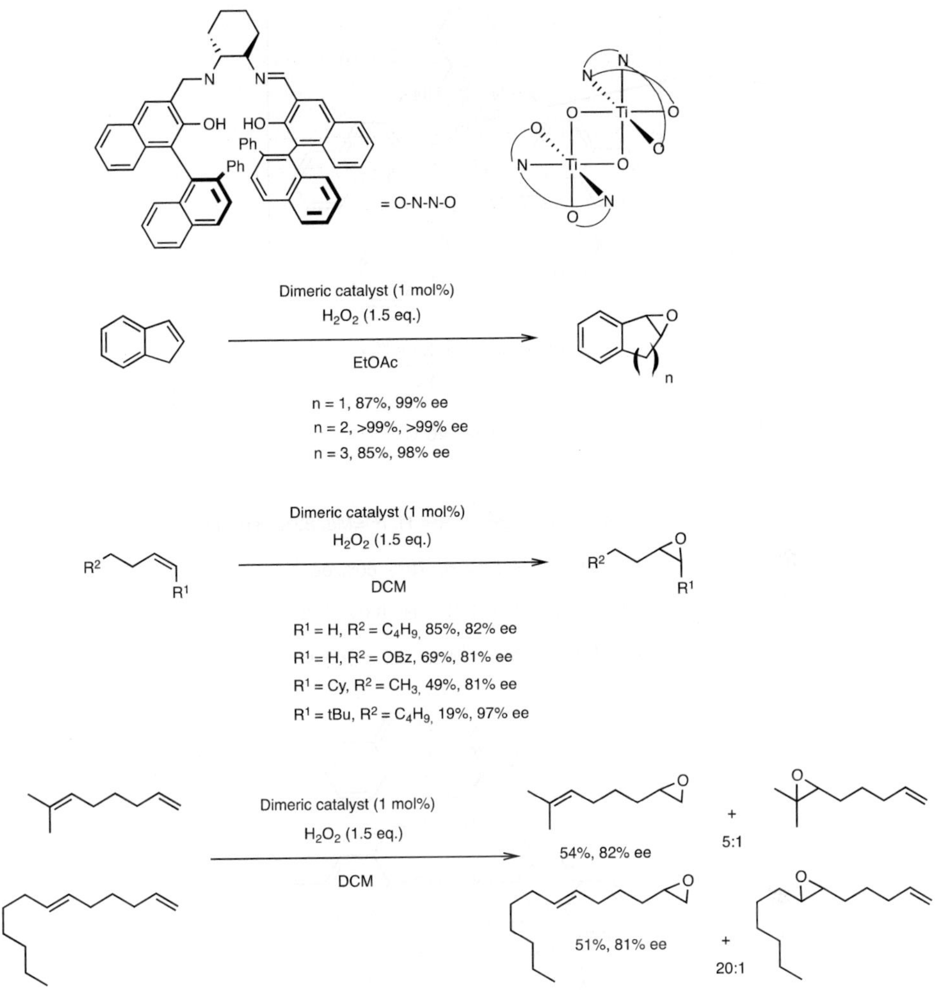


**Scheme 23.7.** Preferential asymmetric epoxidation of terminal over internal alkenes with $H_2O_2$ mediated by a dimeric Ti(IV) catalyst bearing atropoisomeric salalen ligand.

### 23.3.1.4. *Iron*

The widespread presence in nature of Fe containing enzymes for oxidative degradation of xenobiotics, together with the recent attention paid to the development of iron-based homogeneous catalysis because of the much lower cost of this metal compared to other transition metals, prompted the development of several iron-based asymmetric epoxidation systems. Moreover, the stringent requirements concerning trace metals in marketed pharmaceutical products represent an important push for the use of iron in homogeneous catalysis. Most of the known catalytic systems rely on oxidants which do not include $O_2$ or $H_2O_2$. It was only

(a)

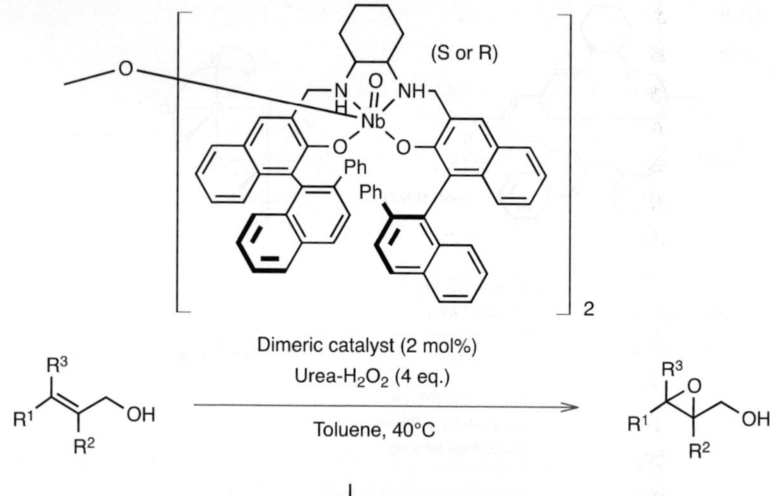


R[1] = (3-methylbut-2-enyl, drawn), R[2]= H, R[3]=Me, 83%, 81% ee

R[1] = Et, R[2] = H, R[3] = H, 78%, 80% ee

R[1] = $PhCH_2$, R[2] = H, R[3] = H, 83%, 79% ee

R[1] = Ph, R[2] = Me, R[3] = H, 86%, 72% ee

(b)

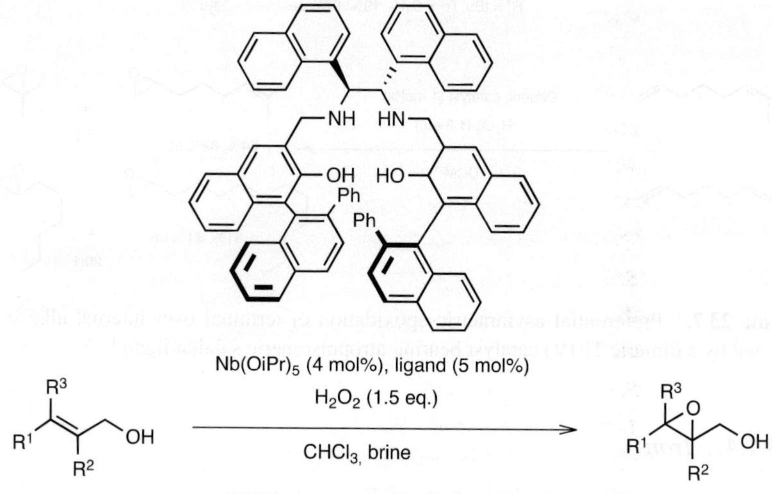


R[1] = n-Pr, R[2] = H, R[3] = H, 57%, 91% ee

R[1] = $C_5H_{11}$, R[2] = H, R[3] = H, 79%, 93% ee

R[1] = Cy, R[2] = H, R[3] = H, 82%, 93% ee

R[1] = tBu, R[2] = H, R[3] = H, 52%, 95% ee

R[1] = Ph, R[2] = H, R[3] = H, 61%, 74% ee

**Scheme 23.8.** Examples of asymmetric epoxidation of allyl alcohols with urea · $H_2O_2$ and $H_2O_2$ mediated by Nb(V) salan complexes.

after the year 2000 that the first asymmetric epoxidation methods based on iron catalysts with non-heme ligands came on the scene. Beller introduced the use of chiral N-arenesulfonyl-N′-benzyl-substituted ethylene diamine ligands in combination with tridentate pyridine-2,6-dicarboxylic acid and $FeCl_3$, observing that with 12 mol% catalyst loading under mild experimental conditions, it was possible to obtain good conversions and moderate to good yields in epoxides, together with enantioselectivity that was very sensitive to the steric hindrance of the substrate. Best results in terms of asymmetric induction were observed with *p*-disubstituted trans-stilbenes as reported in Scheme 23.9.[18,19] Detailed mechanistic investigations on the reaction and solution speciation with different techniques established that the oxidation proceeds via radical intermediates, the relative concentrations and reactivities of which determine the observed ee value. This is confirmed by the observation of a positive nonlinear effect (NLE)[20] on enantioselectivity, suggesting the participation of more than one iron center in the rate-determining step of the reaction.

A recent example of an Fe asymmetric epoxidation catalyst is based on the employment of a polypyridine ligand bearing two rigid chiral bicyclic structures at

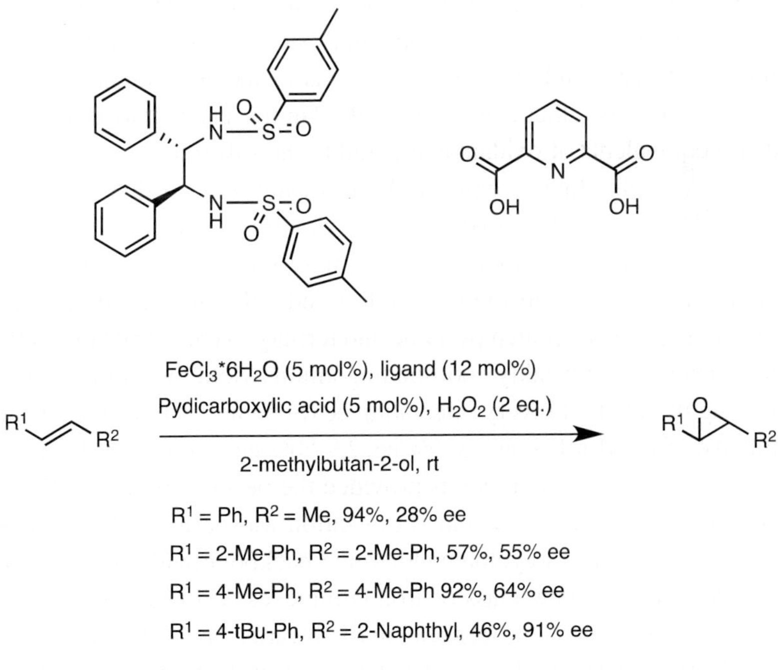


**Scheme 23.9.** Asymmetric epoxidation of stilbenes with $H_2O_2$ catalyzed by Fe(III) complexes bearing the chiral bidentate sulfonamide ligand and pyridine dicarboxylate as the achiral tridentate ligand.

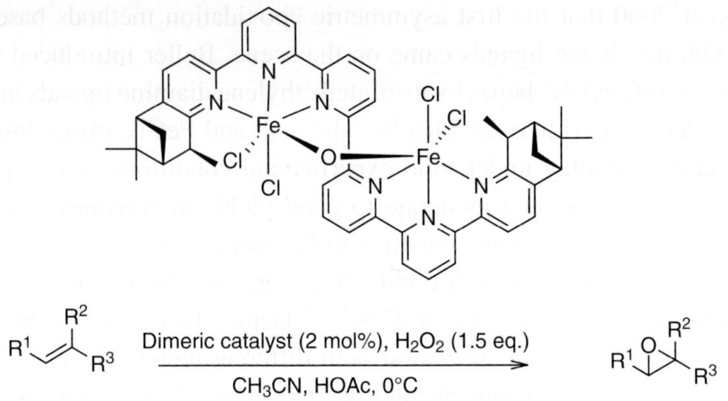


$R^1$ = Ph, $R^2$ = H, $R^3$ = H, 95%, 43% ee

$R^1$ = 4-Cl-Ph, $R^2$ = H, $R^3$ = H, 90%, 42% ee

$R^1$ = Ph, $R^2$ = H, $R^3$ = Me, 96%, 37% ee

$R^1$ = Ph, $R^2$ = Me, $R^3$ = H, 100%, 40% ee

**Scheme 23.10.** Asymmetric epoxidation of vinylarenes with $H_2O_2$ mediated by a dinuclear Fe(III) catalyst bearing a chiral poly-pyridine ligand.

the extremities, that fold like a helix coordinating two iron centers (Scheme 23.10.)[21] The dimeric complex is active and moderately stereoselective in the asymmetric epoxidation of styrenes and cis- and trans-aromatic alkenes with an ee up to 43%. The catalytic system is rather active with only 2 mol% catalyst loading working at 0°C with 1.5 equivalents of oxidant compared to the substrate.

Depending on the ligand employed, iron-catalyzed oxidation reactions of alkenes with $H_2O_2$ provide, in some cases, cis-diols rather than epoxides. Que and co-workers, inspired by natural dioxygenase enzymes that catalyze the cis-dihydroxylation of arene and olefin double bonds, developed tetradentate chiral ligands bearing 2,6-disubstituted pyridine and tertiary asymmetric amines that with Fe(II) metal centers and weakly coordinating anions yield low conversion into the corresponding diol, but with very interesting ee (Scheme 23.11a). Specifically, cis-disubstituted olefins afforded only 3–9% ee, 23–60% ee was achieved with terminal olefins, and trans-disubstituted olefins provided the best results with 82% ee in the oxidation of trans-2-octene.[22] Ligand optimization was performed and the replacement of the chiral 1,2-trans-cyclohexyldiamine backbone with bis-pyrrolidine led to the development of a new class of iron complexes still bearing $\alpha$-methyl pyridine residues that are crucial to steer selectivity towards the formation of the cis-diol products rather than the epoxides. The level of enantioselectivity observed is the best so far reported for iron catalysts, with values up to 96% and 97% for trans-4-octene and trans-2-heptene, respectively (Scheme 23.11b).[23]

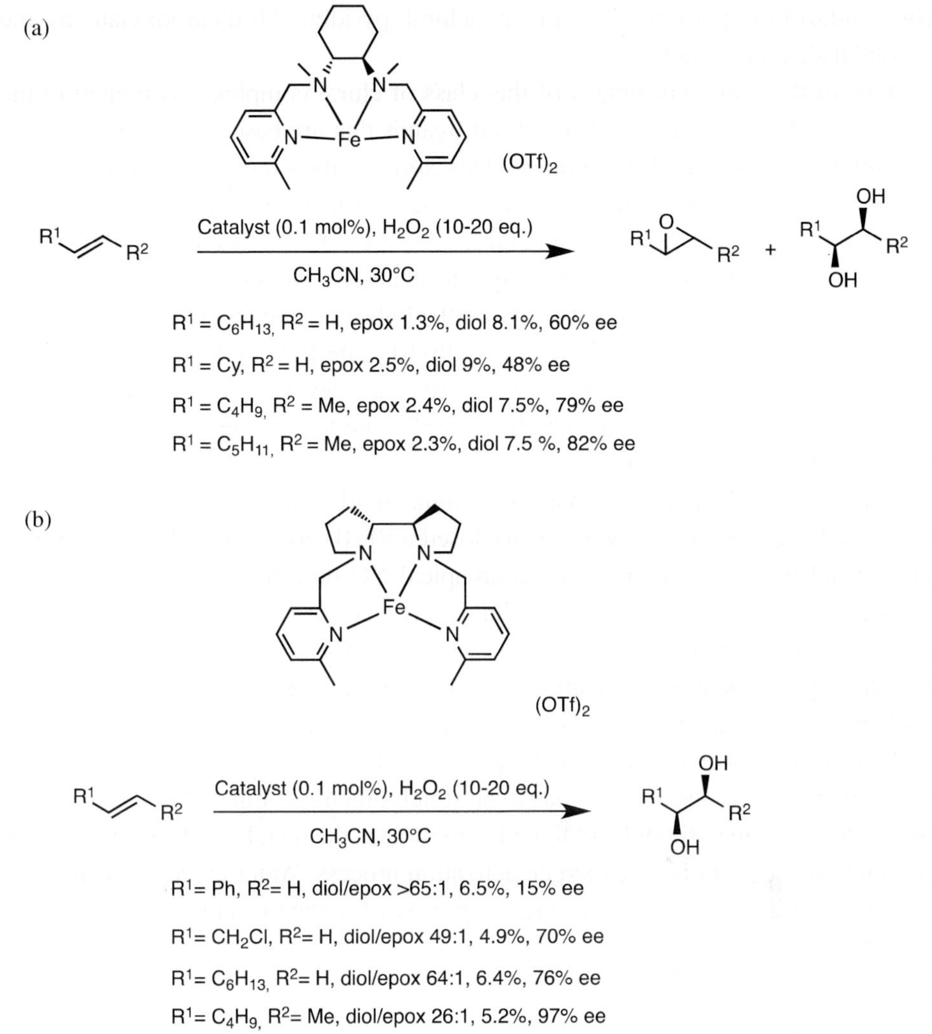


**Scheme 23.11.** Asymmetric epoxidation of various alkenes with $H_2O_2$ mediated by Fe(II) catalyst bearing tetradentate nitrogen ligands.

### 23.3.1.5. *Ruthenium*

While asymmetric epoxidation reactions with Ru species bearing porphyrin, Schiff base and poly-pyridyl ligands constitute a well-documented field of research,[23] examples concerning the employment of $H_2O_2$ as the oxidant are less frequent. In particular, the field was opened by Nishiyama[24] and further developed by Beller[25] employing chiral tridentate nitrogen ligands generally characterized by $C_2$ symmetry such as Pyboxazine (2,2′-pyridine-2,6-dily-bi(5,6-dihydron-4h-1,3-oxazine), Pybox (bis(oxazolinyl)pyridine) and Pybims (bis(imidazolinyl)pyridine)[26] with an achiral pyridine-2,6-dicarboxylate as the second tridentate ligand.

One of the main advantages of this class of chiral complexes is related to the relative synthetic versatility of the chiral ligands that are based on available chiral 1,2-aminoalcohols and 1,2-diamines. This allowed the preparation of a series of complexes generally showing good catalytic activity in the oxidation of styrenes and disubstituted aromatic alkenes using an excess of $H_2O_2$ slowly added to the system in tert-amyl alcohol. Enantioselectivity was generally moderate, observing higher values for Pyboxazine compared to the smaller Pybox analogues on trans-substituted aromatic alkenes. For example, $\beta$,$\beta$-dimethylstyrene provided the corresponding epoxide in 84% ee (Scheme 23.12). A slightly lower level of enantioselectivity was possible using Pybims ligands which allowed ee up to 71% and quantitative yield for trans-stilbene oxidation.

The use of molecular oxygen as the terminal oxidant in asymmetric epoxidation is unusual. Katsuki and co-workers disclosed a Ru(II) complex with a second generation Salen tetradentate ligand and an apical NO molecule that enables oxygen transfer to disubstituted trans-alkenes in the presence of water as a hydrogen donor molecule and, more importantly, under visible light irradiation. The catalytic system displayed good to high enantioselectivity (76–92%) and medium to good yields (Scheme 22.13). The peculiarity of the system consists in requiring neither a proton and electron transfer system, nor the presence of a sacrificial reductant. A pivotal role is played by water which needs to be stoichiometric compared to the substrate. Water displaces the original NO ligand, coordinates the metal center and serves as a proton transfer agent for the oxygen activation process. Water is recycled and used as a proton transfer mediator during the process, favoring turnover and increasing the rate of the reaction.[27]

### 23.3.1.6. *Platinum*

Pt(II) complexes bearing diphosphine ligands are able to activate hydrogen peroxide and, in particular, complexes bearing strong electron-withdrawing ligands such as trifluoromethyl or pentafluorophenyl residues. They are catalytically active in the epoxidation of intrinsically poorly reactive terminal alkenes. This is based on a peculiar feature of such a catalytic system, which promotes the nucleophilic oxidation of the alkene[28] rather than a classical electrophilic pathway. While the oxidation of electron-rich alkenes with hydrogen peroxide as the terminal oxidant is performed by the vast majority of chiral organometallic complexes, the most selective towards terminal alkenes are monomeric bis-cationic Pt(II) complexes containing chiral diphosphines.[15] A set of equilibrium reactions involving the

(a)

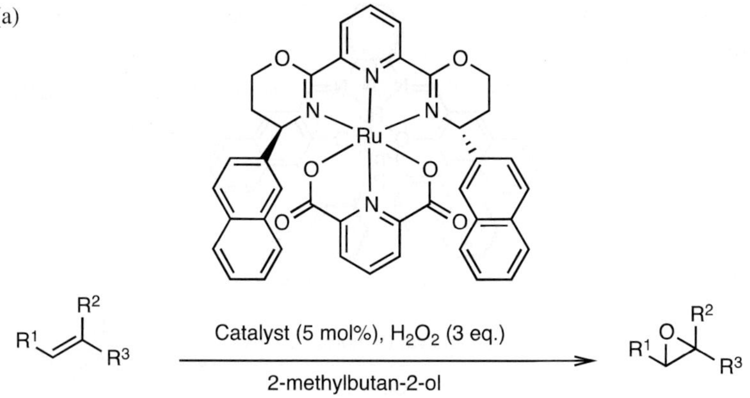


$R^1$ = Ph, $R^2$ = H, $R^3$ = H, 85%, 59% ee

$R^1$ = 2-Me-Ph, $R^2$ = H, $R^3$ = H, >99%, 64% ee

$R^1$ = Ph, $R^2$ = Me, $R^3$ = Me, >91%, 84% ee

$R^1$ = Ph, $R^2$ = H, $R^3$ = Ph, 100%, 54% ee

(b)

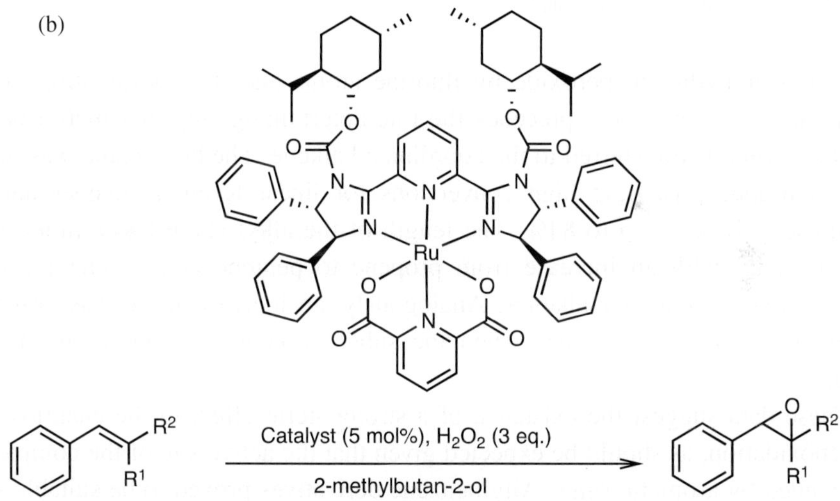


$R^1$ = H, $R^2$ = H, 76%, 42% ee

$R^1$,$R^2$ = -$(CH_2)_5$-, >99%, 54% ee

$R^1$ = Me, $R^2$ = Me, 82%, 68% ee

**Scheme 23.12.** Asymmetric epoxidation of vinyl arenes with $H_2O_2$ mediated by Ru(II) catalysts bearing Pyboxazine (a) and Pybims (b) as chiral tridentate nitrogen ligands.

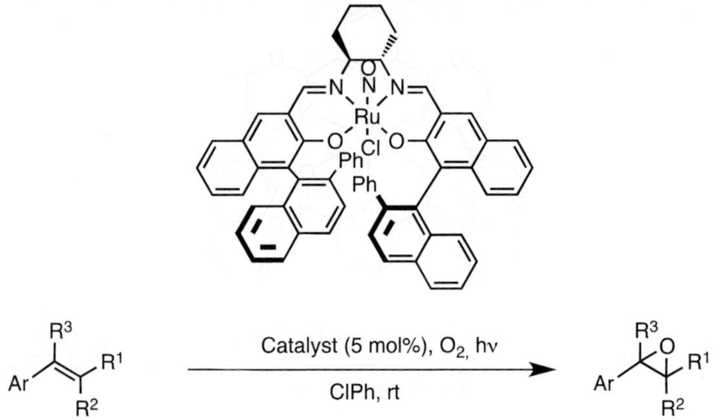


Ar = 2-Naphthyl, $R^1$ = Me, $R^2$ = H, $R^3$ = H, 75%, 92% ee

Ar = 3-Me-Ph, $R^1$ = Me, $R^2$ = H, $R^3$ = H, 67%, 90% ee

Ar = Ph, $R^1$ = Me, $R^2$ = H, $R^3$ = H, 34%, 90% ee

**Scheme 23.13.** Asymmetric epoxidation of vinylarenes with dioxygen mediated by a Ru(II) catalyst bearing an atropoisomeric salen ligand.

activation of hydrogen peroxide by fluorine atoms and the coordination of the alkene to the metal center precedes the rate-determining step in which a nucleophilic oxygen is transferred to the coordinated alkene. The best ligand was found to be chiraphos, yielding high conversions for linear terminal alkenes and an enantioselectivity of up to 84%. The length of the alkyl chain had a mild effect on selectivity with an increase from propene to pentene and hexene followed by a decrease for longer alkenes. Analogously, an increase in ee was observed for branched terminal alkenes when the substitution was closer to the double bond.

These data suggest the existence of a strong steric effect in the enantioselective epoxidation, as should be expected given that the active site of the complex is surrounded by aromatic rings. Allylbenzene derivatives proved to be suitable substrates, with the electron density of the aromatic ring causing a marked negative effect on the activity, but positive effect on the enantioselectivity. A maximum 87% ee was observed for safrole. The nucleophilic character of the epoxidation allowed the selective regio- and enantioselective asymmetric oxidation of substrates bearing both an internal and a terminal carbon-carbon double bond. The results reported in Scheme 23.14 are impressive: for all the three dienes investigated, the epoxidation occurred exclusively at the terminal double bond with complete regioselectivity and ee up to 98%. A greener version of the reaction was developed using water as the solvent in the presence of surfactants to mediate the solubility of both substrates and catalyst in the polar solvent (see Section 23.4.2).[29]

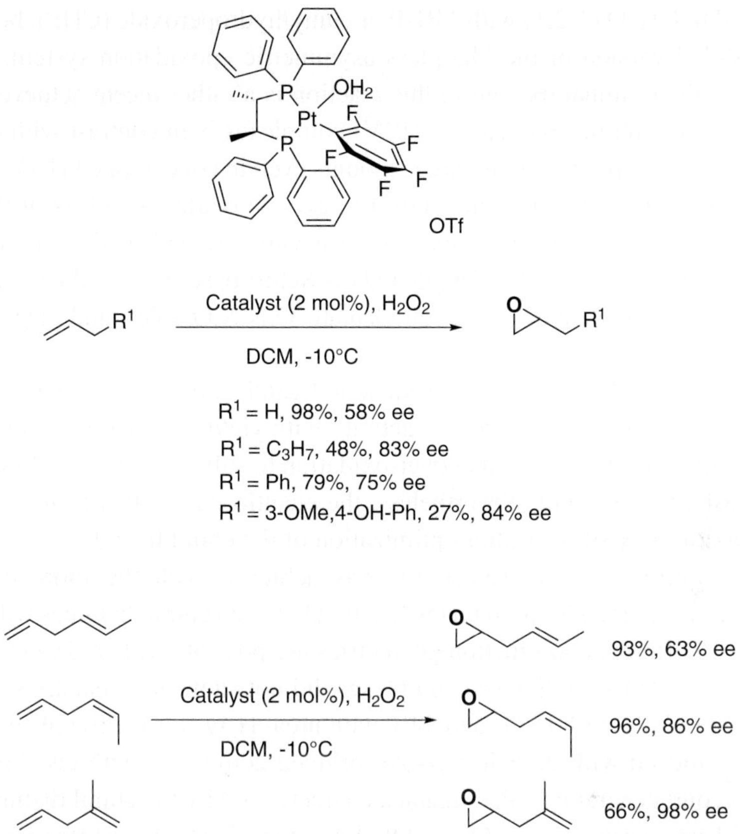


**Scheme 23.14.** Asymmetric epoxidation of terminal alkenes with $H_2O_2$ catalyzed by a Pt(II) chiral complex.

### 23.3.2. *Asymmetric sulfoxidation*

Oxidation of thioethers to the corresponding sulfides is a rather facile reaction because of the electron-rich character of the sulfur atom. Sulfoxidation can be performed with a wide range of oxidants and $H_2O_2$ can be easily employed exploiting simple activation with Brønsted acids or a plethora of Lewis acids. Chiral enantiopure sulfoxides are extremely interesting molecules[30] and one of the ten most sold drugs in 2003 was NexiumTM, the (S) enantiomer of omeprazole, a known anti-ulcer agent.[31] The asymmetric sulfoxidation is a benchmark test to analyze the asymmetric induction properties of newly prepared complexes.

#### 23.3.2.1. *Titanium*

In the past, concomitant pioneering contributions came from Modena[32] using $Ti(OiPr)_4$/DET (1:4) with tert-butylhydroperoxide (TBHP) and Kagan[33] employing

$Ti(OiPr)_4/DET/H_2O$ (1:2:1) with TBHP or cumylhydroperoxide (CHP), both based on a modified version of the Sharpless asymmetric epoxidation system. The use of $H_2O_2$ as the terminal oxidant in this reaction is a rather recent achievement. In fact the high oxophilic character of Ti(IV) complexes is in contrast with the presence of water, usually present in large amounts even in concentrated $H_2O_2$, because of decomposition of the original catalyst to give Ti oxides with loss of the chiral ligand. The robustness of Ti complexes for asymmetric sulfoxidation with $H_2O_2$ increased with the advent of tridentate O,O,N Schiff bases prepared from aminoalcohols and salicylaldehydes, and tetradentate O,N,N,O salen and salalen chiral ligands.

Asymmetric sulfoxidation with tridentate ligand Ti complexes and 30% $H_2O_2$ is quite efficient in terms of catalytic activity with yields of up to 96%, high selectivity to sulfoxide (90%), minimal over-oxidation to sulfone and rather low catalyst loading, usually <1% mol. Nevertheless, the enantioselectivity was not very high with a maximum ee of 60% after optimization of the chiral ligand.[34]

Better control on stereoselectivity was achieved with the more congested tetradentate salen ligands based on the reaction between chiral diamines with salicylaldehydes. Different coordination geometries are possible with Ti(IV) precursors, in particular, octahedral dichloro complexes did not show good enantioselectivity, while ee up to 76% with $H_2O_2$ and 94% with urea·$H_2O_2$ for methyl phenyl sulfide could be achieved with dinuclear $\mu$-oxo bridged complexes with cis-$\beta$ geometry bearing second generation salen ligands, characterized by the chiral diamine backbone and large atropoisomeric binaphthyl residues.[35] The lower stereoselectivity afforded by $H_2O_2$ is probably caused by the presence of several Ti-hydroperoxo equilibrium species formed by the addition of water to an original Ti-peroxo species. The asymmetric induction was high, regardless of the electronic nature of the sulfide and, surprisingly, more electron-rich substrates provided lower conversions compared to more electron-poor substrates. Detailed speciation analysis in solution confirmed that in methanol and in the presence of the oxidant, the original dimeric catalyst dissociates into monomeric species (Scheme 23.15).[36]

The same catalytic system proved to be effective and highly enantioselective towards a much less investigated class of sulfides like dithioacetals. In particular, 1,3-dithiolanes and 1,3-dithianes were converted smoothly into a single diastereoisomer with ee >90% using urea·$H_2O_2$ as the oxidant (Scheme 23.15). The catalytic activity and enantioselectivity were not affected by the electronic nature of the substituent in position two, but were more sensitive to steric requirements like 2-t-Bu-dithiane and dithiolane which showed lower ee.[37] The diastereoselectivity was always high and decreased slightly with alkyl or alkynyl substituents, favoring in all cases the trans isomer that was produced with the highest ee compared to the corresponding cis isomer.

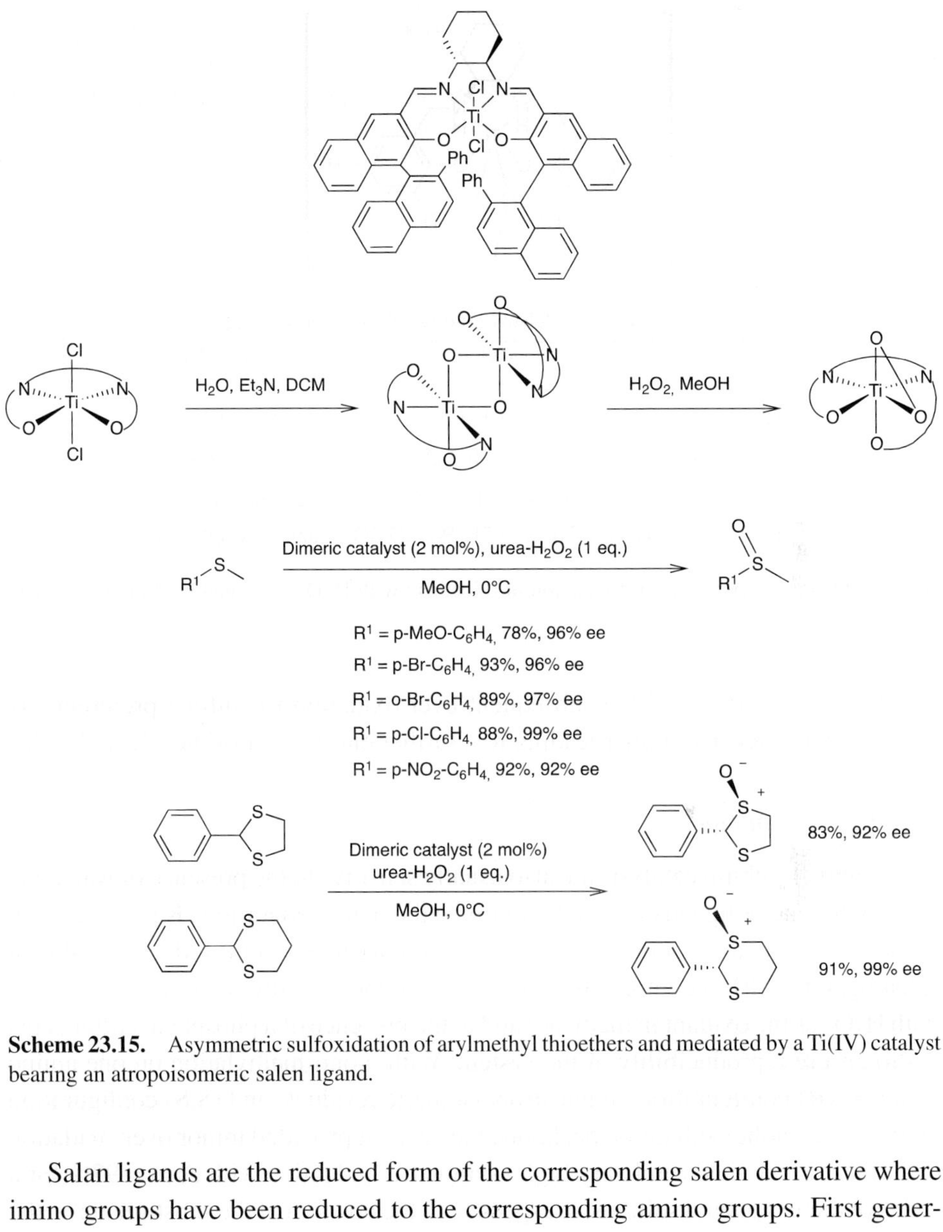


**Scheme 23.15.** Asymmetric sulfoxidation of arylmethyl thioethers and mediated by a Ti(IV) catalyst bearing an atropoisomeric salen ligand.

Salan ligands are the reduced form of the corresponding salen derivative where imino groups have been reduced to the corresponding amino groups. First generation salan ligands bearing substituted salicylaldehyde residues provided dimeric $\mu$-oxo-Ti(IV) complexes that allowed the asymmetric sulfide oxidation with $H_2O_2$ in dichloromethane.[38] The reaction, performed in the presence of 1.6 equivalents of oxidant with respect to the substrate and with as low as 0.2% mol of catalyst, allowed high conversion and ee up to 97% thanks to a tandem enantioselective oxidation of the sulfide to the sulfoxide followed by kinetic resolution of the sulfoxide to sulfone (Scheme 23.16). With the (S,S) catalyst, sulfide was preferentially formed with

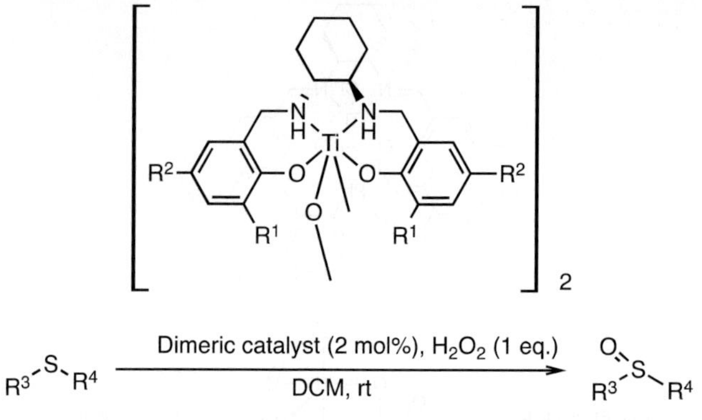


$R^1$ = H, $R^2$ = H, $R^3$ = Ph, $R^4$ = $CH_2Ph$, $SO/SO_2$ = 1.07, 51%, 86% ee

$R^1$ = H, $R^2$ = H, $R^3$ = $CH_3$, $R^4$ = Ph, $SO/SO_2$ = 6.5, 78%, 45% ee

$R^1$ = Ph, $R^2$ = H, $R^3$ = iPr, $R^4$ = Ph, $SO/SO_2$ = 1.8, 48%, 69% ee

$R^1$ = Ph, $R^2$ = H, $R^3$ = Ph, $R^4$ = $CH_2Ph$, $SO/SO_2$ = 3.9, 75%, 88% ee

**Scheme 23.16.** Asymmetric sulfoxidation of thioethers with $H_2O_2$ mediated by dimeric Ti(salan) complexes.

the (S) configuration and the subsequent over-oxidation to sulfone preferentially consumed the (R) enantiomer leading to a further enrichment of the (S) sulfoxide.

### 23.3.2.2. *Aluminum*

Aluminum(III) chiral catalysts are also usually sensitive to the presence of water, but with multidentate ligands and under optimized reaction conditions Katsuki and co-workers developed a monomeric second generation salalen (semi-reduced version of salen ligands) Al(III) complex that enabled asymmetric sulfoxidation of thioethers with $H_2O_2$ as the oxidant in methanol and in the presence of a phosphate buffer at pH 7.4 to ensure reproducibility of the system. With mono-methylation on one amino N donor, (R) configuration on the atropoisomeric residues, and (S,S) configuration on the 1,2-cyclohexyldiamino backbone the catalyst provided minor over-oxidation to sulfone, good yields and very high asymmetric induction with ee at >97% for a series of methyl phenyl sulfides, irrespective of the position of the substituents on the aromatic ring and the electronic nature of the aryl substituents (Scheme 23.17).[39,40]

The same catalytic system enabled the kinetic resolution of sufoxides to sulfones confirming that the high ee observed is again the result of concomitant asymmetric oxidation of the sulfide, followed by kinetic resolution of the sulfoxide, both favoring the (S) enantiomer. An extensive investigation of the asymmetric oxidation of two substituted 1,3-dithianes and dithiolanes bearing alkyl, alkenyl, alkynyl and aryl groups showed that the same catalytic system was able to provide the corresponding

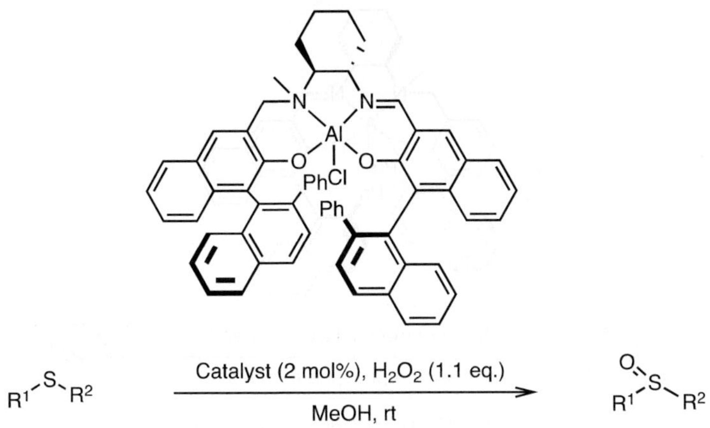


$R^1$ = Ph, $R^2$ = Me, $SO/SO_2$ = 9.6, 86%, 98% ee

$R^1$ = 4-MeO-$C_6H_4$, $R^2$ = Me, $SO/SO_2$ = 9.1, 82%, 97% ee

$R^1$ = 2-$NO_2$-$C_6H_4$, $R^2$ = Me, $SO/SO_2$ = 42, 84%, 99% ee

$R^1$ = Ph, $R^2$ = Et, $SO/SO_2$ = 100, 80%, 91% ee

**Scheme 23.17.** Asymmetric sulfoxidation of thioethers with $H_2O_2$ mediated by an Al(salalen) complex.

trans-monoxides in high yields with >20:1 diastereomeric ratio (dr) and 98–99% ee with $H_2O_2$ in ethyl acetate. Non-substituted dithianes were preferentially oxidized to the monoxide with high ee (Scheme 23.18).[41] The catalytic system is extremely selective since starting from the cyclic dithioacetals potentially four stereoisomers for the monoxide, six for the dioxide and four and one for the tri and tetraoxide are possible, respectively. Dithianes substituted in position five and two, and substituted dithiepane (seven-membered ring thioacetals) afforded high yields and ee for the corresponding monoxides, while only the sulfoxidation of acyclic thioacetals was sluggish and poorly enantioselective (Scheme 23.18).

### 23.3.2.3. *Vanadium*

Asymmetric sulfoxidation with chiral vanadium complexes is much older than Ti and Al because this metal provides more robust catalysts which are not deactivated by the presence of water. The first contribution to the field was made by Bolm and co-workers in the second half of 1990s. These authors developed chiral catalysts formed *in situ* by the reaction of $VO(acac)_2$ with chiral enantiopure Schiff ligands bearing one stereocenter based on a t-leucinol scaffold.[42] The maximum ee achieved was 85% using low catalyst loadings (<1% mol) and without any precautions to avoid moisture or oxygen, which was unusual at that time as hydroperoxides

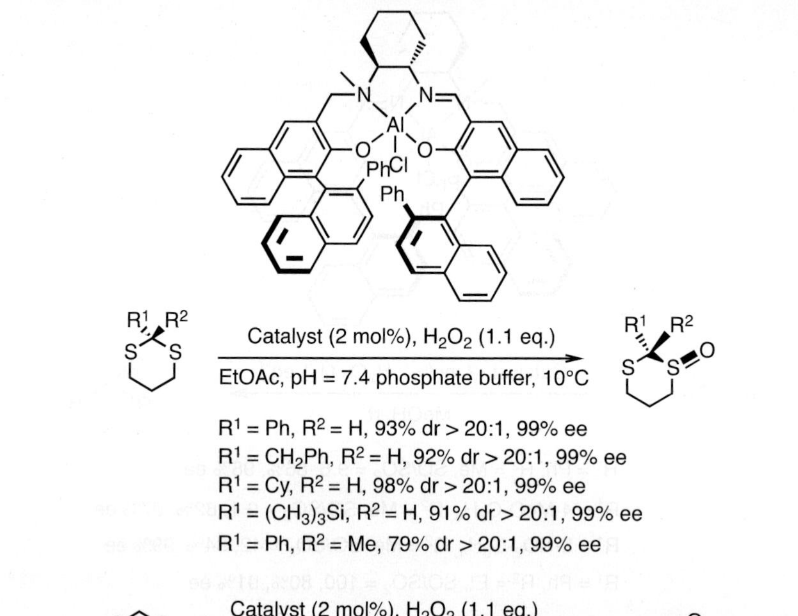


**Scheme 23.18.** Asymmetric sulfoxidation of 1,3-dithianes with $H_2O_2$ mediated by an Al(salalen) complex.

where the most common oxidants and strict anhydrous conditions were mandatory (Scheme 23.19).

An optimized version of Bolm's catalytic system was developed after careful variation of the substituents on the aromatic ring, observing that the di-iodo derivative reported in Scheme 23.20 allowed formation of aryl benzyl sulfoxides with ee

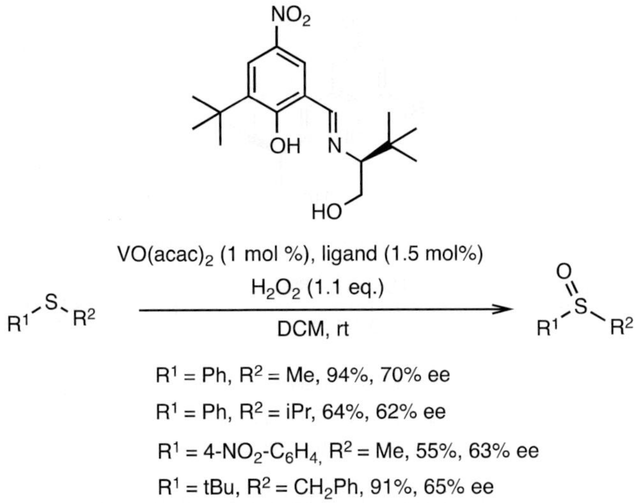


**Scheme 23.19.** Asymmetric sulfoxidation of thioethers with $H_2O_2$ mediated by a V(IV) Schiff ligand complex.

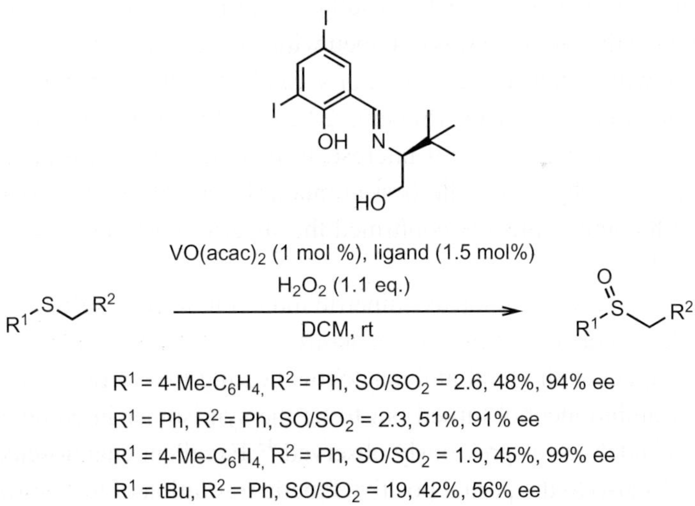


**Scheme 23.20.** Asymmetric sulfoxidation of thioethers with $H_2O_2$ mediated by a V(IV) catalyst bearing a halogenated Schiff ligand.

typically >90% in the presence of a tandem cooperative asymmetric oxidation and kinetic resolution.[43] The kinetic resolution of sulfoxides was optimized using the *in situ* generated catalyst with 0.8 equivalents of $H_2O_2$ and the slow addition of the oxidant, thus obtaining good enantioselectivity with rather low conversions and selectivity factors up to 17.6 between the two enantiomers of the sulfoxide.

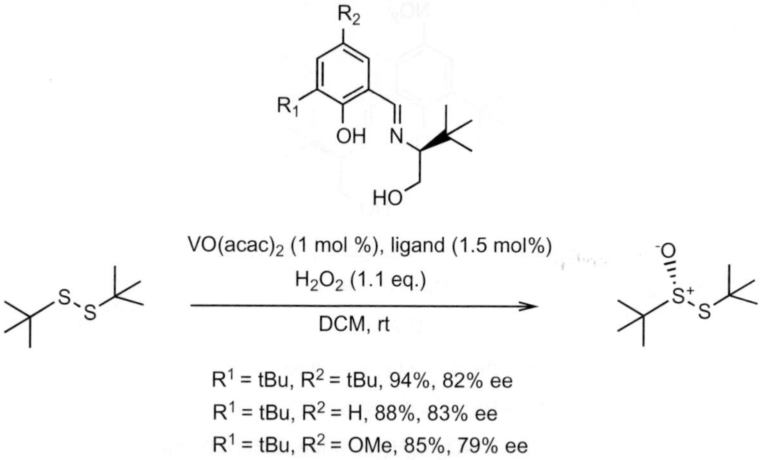


**Scheme 23.21.** Asymmetric sulfoxidation of alkyl disulfide with $H_2O_2$ mediated by a V(IV) Schiff base ligand.

The same catalytic system disclosed by Bolm was further extended by Ellman to the asymmetric oxidation of t-Bu-disulfide with $H_2O_2$. The corresponding tert-butyl tert-butane thiosulfinate was obtained with 91% ee and with 92% on scales as large as 1 mol with as low as 0.25% mol of V catalyst, with a Schiff base ligand and stoichiometric amount of $H_2O_2$ (Scheme 23.21).[44] Extensive ligand modification was necessary for the substrate of interest. In particular, it turned out that 3,5-di-tert-butyl salicylaldehyde was the best aromatic fragment, while variation on the properties of the amino alcohol confirmed the superior enantioselectivity possible with t-leucinol.

Further improvements in the asymmetric induction for the sulfoxidation of aryl methyl sulfides was achieved employing Schiff base ligands derived by condensation of salicylaldehydes with chiral aminoalcohols bearing two stereocenters.[45] In this case the best amino alcohol turned out to be the syn isomer in combination with di-iodo salicylaldehyde as previously observed.[43] Excellent enantioselectivity was possible thanks also to the partial positive kinetic resolution of the sulfoxide which was optimized using a slightly larger amount of oxidant (1.35 eq. compared to the sulfide). The catalytic system was also active towards substrates bearing longer alkyl residues without detrimental effects on either the yield and enantioselectivity (Scheme 23.22).

On the basis of the high versatility of salen and salan ligands for asymmetric oxidation with other metal centers, Zhu investigated the employment of such ligands for the *in situ* preparation of V catalysts. In particular, the first generation of salen ligands, having stereogenic centers on the 1,2-diamino backbone, showed poor asymmetric induction, while the reduced salan version surprisingly

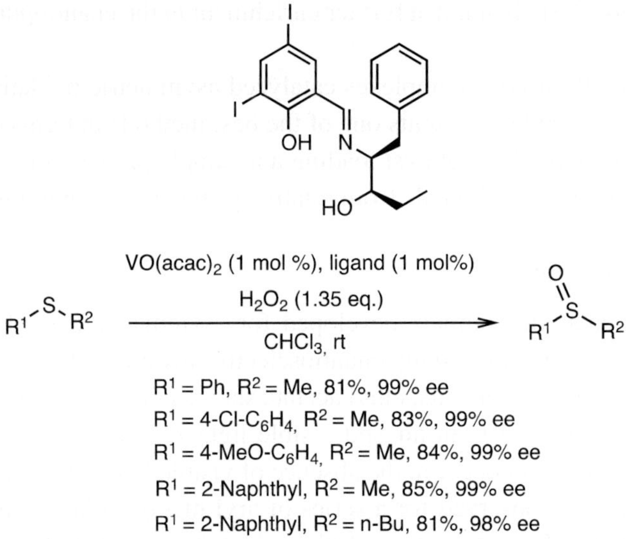


**Scheme 23.22.** Asymmetric sulfoxidation of thioethers with $H_2O_2$ mediated by a V(IV) complex bearing a Schiff base ligand comprising two stereocenters.

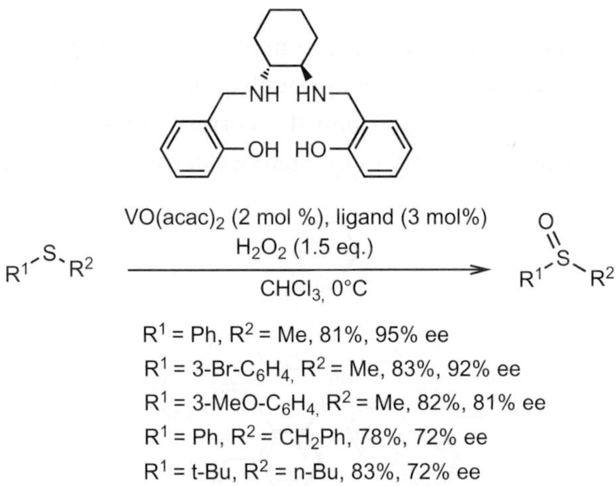


**Scheme 23.23.** Asymmetric sulfoxidation of thioethers with $H_2O_2$ mediated by a V(IV) complexes with a salan ligand.

allowed good enantioselectivity in the oxidation of simple aryl methyl thioethers (Scheme 23.23).[46] Methylation of the amino groups of the ligand and the use of aromatic residues bearing t-Bu moieties, which are common strategies to enhance enantioselectivity in other reactions, did not improve the asymmetric induction. Better results were obtained by exploiting the partial over-oxidation of the enantioenriched sulfoxide to the corresponding sulfone (selectivity factor for methyl phenyl sulfoxide 7.3) obtaining a further enrichment in the enantiopurity of the sulfoxide.

Overall, Schiff-based V complexes catalyzed asymmetric oxidation of sulfides with hydrogen peroxide represents one of the best methods in terms of high yields and enantioselectivity, low catalyst loading and simple preparation of the catalyst, as well as employment of ligands from relatively cheap chiral building blocks.

### 23.3.2.4. *Ruthenium*

The same Ru catalytic complex developed for asymmetric epoxidation with air proved to be active and also highly enantioselective towards asymmetric sulfoxidation.[27] Among different solvents, ethyl acetate showed the best results with 5% mol of catalyst loading at 25°C in air and visible light irradiation over 24–48 h. The oxygen transfer does not occur in the absence of visible light and no over-oxidation of the sulfoxide was observed for a series of aryl methyl sulfides and two substituted 1,3-dithianes that were successfully oxidized with ee in the range 72–98% (Scheme 23.24).

### 23.3.2.5. *Iron*

One of the first attempts to perform asymmetric sulfoxidation with Fe catalysts was reported by Fontecave and Ménage with mononuclear $[Fe(pb)_2\text{-}(CH_3CN)_2]$ (pb=(−)4,5-pinene-2,2′-bipyridine) and the corresponding $\mu$-oxo dinuclear system, observing that the latter was more stereoselective and active.[47] The level of

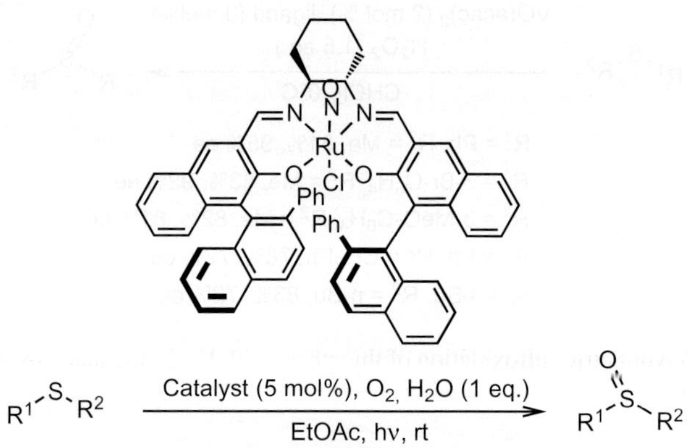


$R^1$ = Ph, $R^2$ = Me, 74%, 94% ee
$R^1$ = 4-Cl-$C_6H_4$, $R^2$ = Me, 69%, 84% ee
$R^1$ = 2-MeO-$C_6H_4$, $R^2$ = Me, 86%, 96% ee
$R^1$ = $CH_2Ph$, $R^2$ = Me, 26%, 75% ee

**Scheme 23.24.** Asymmetric sulfoxidation of thioethers with $O_2$ mediated by a Ru(III)salen complex.

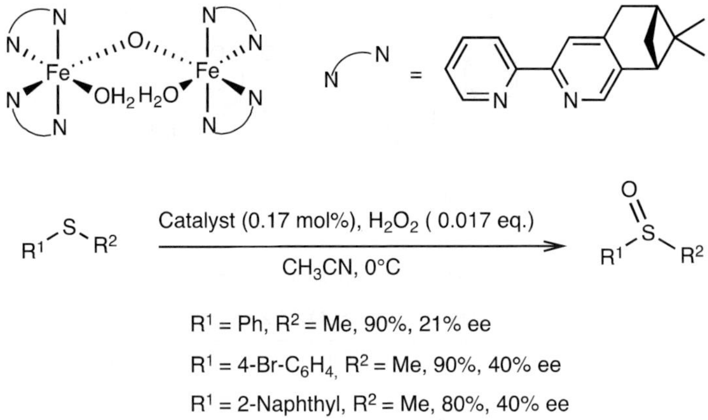


**Scheme 23.25.** Asymmetric sulfoxidation of thioethers with substoichiometric $H_2O_2$ mediated by a dimeric Fe(III) catalyst bearing a bipyridine ligand.

asymmetric induction and the turnover ability were not high, with yields up to 90% and ee up to 40% (Scheme 23.25). It is worth noting that the catalytic system did not withstand the presence of large amounts of oxidant, in fact $H_2O_2$ was the limiting reactant of the reaction.

This system inspired the application of wellknown Schiff base ligands for Fecatalyzed asymmetric sulfoxidation. Bolm and coworkers investigated this area in detail, initially observing that di-iodo-substituted ligands ensured high enantioselectivity.[48] The major disadvantage was the rather low conversion of the substrate. In order to improve this aspect, a series of aromatic carboxylic acids and salts were investigated as promoters observing a marked increase in the yield of sulfoxide when employing electron-rich benzoic acids, with a further increase in the stereoselectivity. As an example, in the oxidation of phenyl methyl sulfide, the addition of 4-methoxy-benzoic acid increased the yield from 27% to 63% and ee from 26% to 90%.[49] All other substrates behaved similarly.

The effect of the acid was probably to promote the dimerization of the original complex, as confirmed by the observation of the non-linear effect[20] between the ee of the sulfoxide product and the ee of the ligand.[50] The catalytic system showed a good chemoselectivity with only marginal over-oxidation of the sulfoxide to the corresponding sulfone, with negligible effect on the stereocontrol of the reaction, which remains driven by the first oxidation step (Scheme 23.26). Overall the optimized catalytic system allowed high enantioselective oxidations (up to 96% ee) of prochiral sulfides obtained in moderate to good yields (up to 78%), under simple reaction conditions using a readily available *in situ* iron catalyst (<4 mol%) and aqueous hydrogen peroxide, using small quantities of a carboxylic acid or a carboxylate salt (1 mol%).

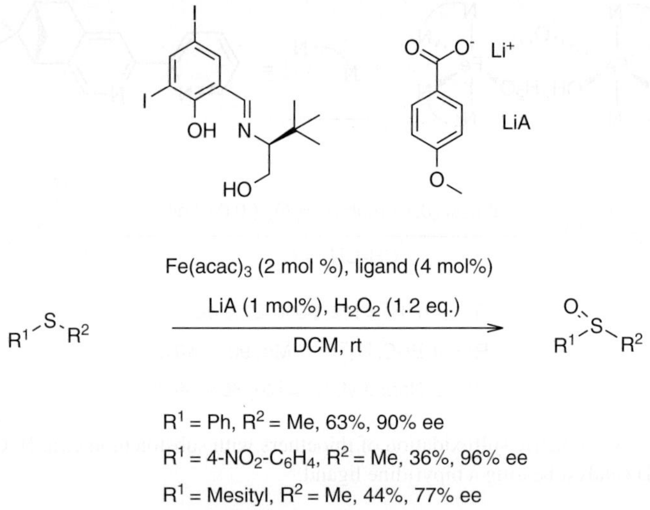


**Scheme 23.26.** Asymmetric sulfoxidation of thioethers with substoichiometric $H_2O_2$ mediated by a dimeric Fe(III) catalyst bearing a bipyridine ligand.

The use of Fe as metal center instead of the more oxophilic Ti and Al, associated with a more sterically demanding ligand such as salalen bearing atropoisomeric binaphthyl moieties, allowed the development of an asymmetric oxidation of sulfides with $H_2O_2$ in pure water.[51] The system was studied by Katsuki observing good conversion of the reagent with partial formation of sulfone and good enantioselectivity, which is only induced in the first oxidation step as kinetic resolution was poorly selective. The catalytic system was active and selective towards aryl alkyl as well as dialkyl sulfides with both high yields and ee (Scheme 23.27). Using water as the solvent, the system is biphasic, but both catalytic activity and selectivity are higher compared to the reaction in methanol, where the system is monophasic, and is an indication that hydrophobic effects[52] and "on-water conditions"[53] probably play a role.

Chiral iron porphyrin systems have been studied for asymmetric sulfoxidation usually with iodosylbenzene as the terminal oxidant. The employment of $H_2O_2$ is hampered by the strong tendency of the iron species to promote spontaneous decomposition of the latter oxidant, the so-called catalase reaction, with consequent catalyst destruction by hydroxyl radicals released by homolytic $H_2O_2$ decomposition. Very recently, it has been demonstrated that using chiral tetra-sulfonated electron-rich porphyrin Fe(II) complexes, it is possible to favor heterolytic $H_2O_2$ splitting when using polar solvents such as methanol. This led to the development of an efficient and enantioselective sulfoxidation reaction with very high yields and enantioselectivities in the range 76–84%, regardless of the substituents on the aromatic residue of the substrate (Scheme 23.28).[54] The addition of 10 eq. of 2-methylimidazole

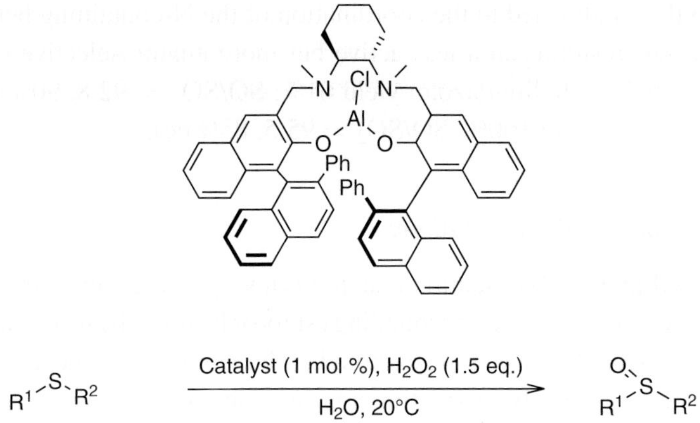


$R^1$ = 4-Me-Ph, $R^2$ = Me, 91%, 96% ee
$R^1$ = 2-Cl-$C_6H_4$, $R^2$ = Me, 97%, 96% ee
$R^1$ = $CH_2Ph$, $R^2$ = Me, 93%, 87% ee
$R^1$ = $C_{12}H_{25}$, $R^2$ = Me, 82%, 94% ee

**Scheme 23.27.** Asymmetric sulfoxidation with $H_2O_2$ mediated by an Al(III) salan ligand.

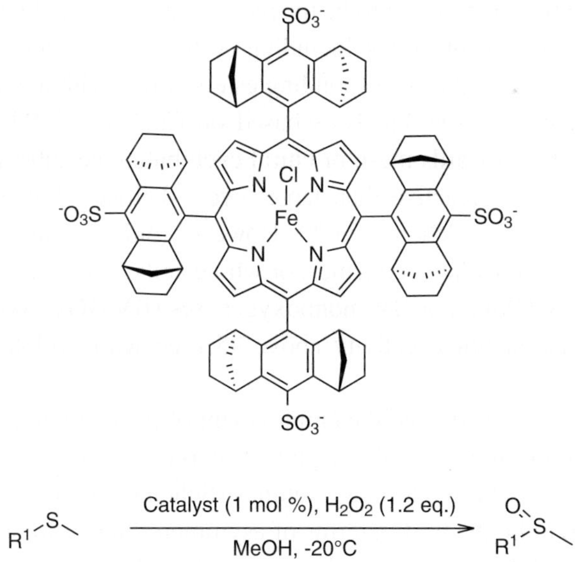


$R^1$ = Ph, $R^2$ = Me, 100%, SO/$SO_2$ = 95:5, 85% ee
$R^1$ = 4-$NO_2$-$C_6H_4$, $R^2$ = Me, 90%, SO/$SO_2$ = 95:5, 85% ee
$R^1$ = 2-Bu-$C_6H_4$, $R^2$ = Me, 98%, SO/$SO_2$ = 98:2, 87% ee

**Scheme 23.28.** Asymmetric sulfoxidation with $H_2O_2$ mediated by an Fe(III) porphyrin ligand.

compared to the catalyst led to the coordination of the N-containing heterocycle to the Fe(II) center, resulting in a less active but more enantioselective catalyst (for thioanisole with 2-methylimidazole: yield 61%, $SO/SO_2 = 92:8$, 90% ee; without 2-methylimidazole: yield 100%, $SO/SO_2 = 95:5$, 85% ee).

### 23.3.3. *Asymmetric Baeyer–Villiger*

The Baeyer–Villiger (BV) oxidation reaction is a very old chemical transformation that converts ketones into the corresponding esters or lactones by inserting an oxygen atom from the oxidant between the carbonyl carbon and the one next to it. The first report on the asymmetric BV oxidation appeared about fifteen years ago thanks to the pioneering work of Bolm[55] and Strukul.[56] These authors independently demonstrated, with two different catalysts and oxidant combinations, that a certain degree of stereocontrol of the process was possible, with ee up to 92% on chiral cyclobutanones and 58% in the kinetic resolution of chiral-substituted cyclohexanones, respectively. After these seminal contributions, a series of different catalysts have been proposed.

Most catalytic systems are active towards four-membered ring ketones leading to the corresponding substituted γ-butyrolactones. Pt(II) complexes modified with chelating diphosphines turned out to be the most efficient in catalyzing the BV oxidation of a wide range of substrates, including cyclohexanones and acyclic ketones that can be converted into the corresponding esters, albeit with low turnover.

Worthy of mention is that catalysts based on Co,[57] Cu,[58] Pd,[59] Pt,[20,60] Zr[61] and Al[62] are all active towards meso or chiral cyclobutanone substrates, which are intrinsically much more reactive than larger cyclic ketones. Some representative examples are reported in Scheme 23.29. However, only biocatalytic BV oxidations[63] performed with isolated enzymes, or whole cells containing cyclohexanone monooxygenases (CHMOs) or BV monooxygenases (BVMOs) showed good conversions as well as enantioselectivity above 95% ee with cyclohexanone as the substrate.[64]

Recently, Strukul investigated the employment of bis-cationic Pt(II) complexes bearing chiral diphosphines for the asymmetric BV oxidation of six- and four-membered ring cyclic chiral and meso ketones (Scheme 23.30).[65] Atropoisomeric complexes proved to be the best in terms of asymmetric induction in the oxidation of 2,6-dimethyl cyclohexanones which exists in two diastereoisomers, one meso and one chiral compound. The reaction occurs in a stereoselective fashion, with complete retention of configuration of the migrating carbon atom, which is a common observation for the BV reaction. All catalysts are much more active towards the chiral trans-isomer than the meso cis-isomer, in particular, binaphthyl ligands imparted both higher activity and higher enantioselectivity.

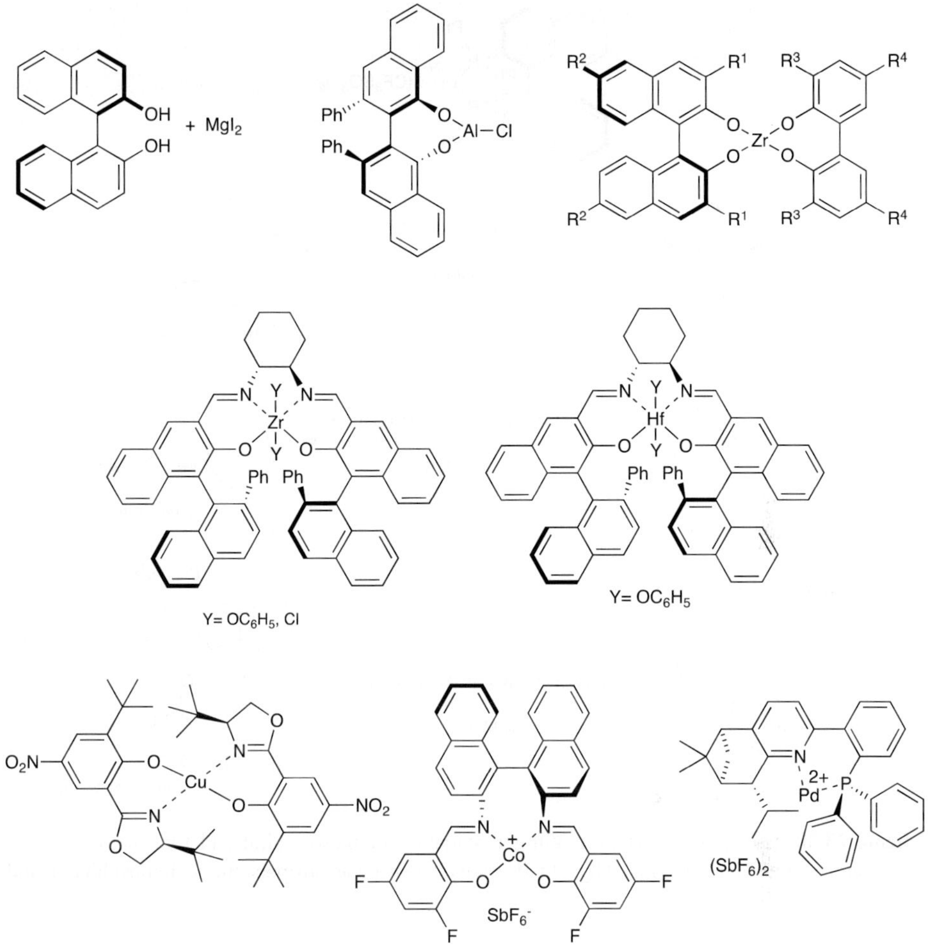


**Scheme 23.29.** Examples of chiral orgsnometallic catalysts developed for the asymmetric BV oxidation cyclic ketones.

When an intrinsically more reactive substrate such as the chiral cyclobutanone reported in Scheme 23.30 was investigated, the interaction between the substrate and the chiral enantiopure catalysts led to substrate kinetic resolution, while the insertion of the oxygen atom between the carbonyl moiety and the adjacent C atom provides two possible diastereoisomeric five-membered ring lactones. Overall, the reaction is a double parallel kinetic resolution and high ee of products in combination with high conversions are possible. The normal lactone (NL) was the predominant product of the reaction in agreement with the migratory aptitude for BV reactions. In particular, catalysts bearing a biaryl ligand reacted with the cyclobutanone-substrate more reactive enantiomer to give both NL and the abnormal lactone (AL) at a greater

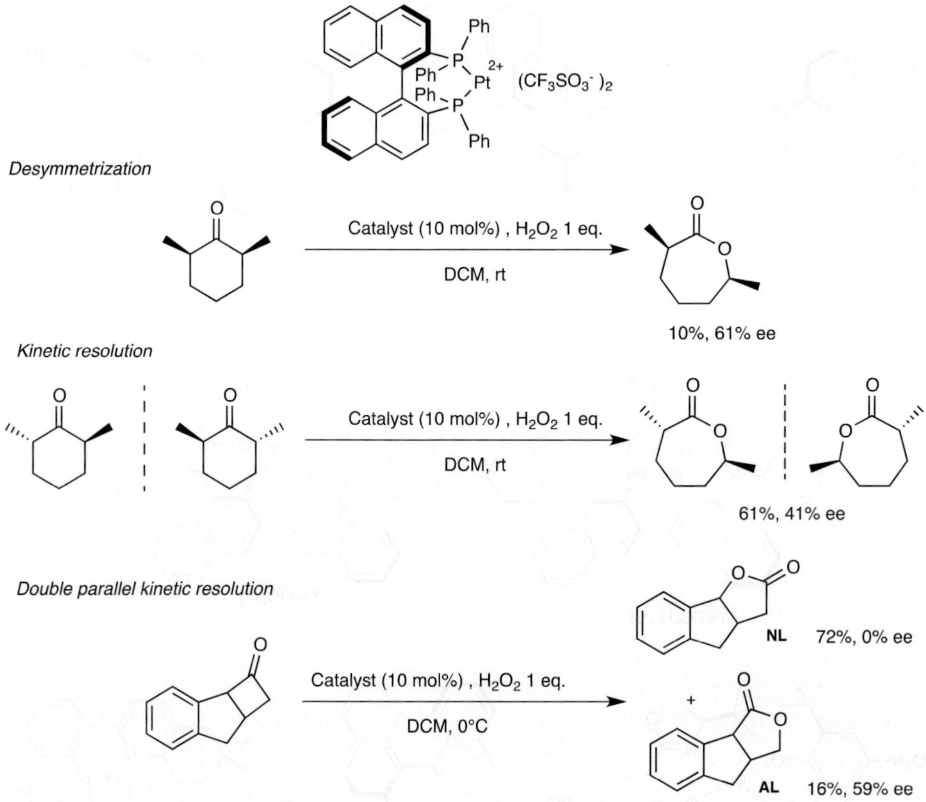


**Scheme 23.30.** Asymmetric Baeyer–Villiger oxidation of meso, chiral and four-membered ring cyclic ketones with $H_2O_2$ mediated by a Pt(II) catalyst bearing an atropoisomeric diphosphine ligand.

rate than the corresponding reaction on the less reactive enantiomer. On the contrary, the other catalysts characterized by larger bite angles showed a small degree of regiodivergent parallel kinetic resolution.

At present, chiral Pt(II) complexes represent the most efficient catalytic system in terms of activity and enantioselectivity for poorly reactive meso and chiral cyclohexanones, with important improvements observed when working in water, as reported in the following sections.

### 23.3.4. *Other asymmetric oxidations*

The atropoisomerism of binaphthol ligands has been extensively exploited for the preparation of enantiopure bidentate ligands. These are usually produced in racemic mixtures and subsequently treated with a resolving agent. Alternative pathways for the direct synthesis of enantioenriched binaphthols are based on the oxidative dimerization of 2-naphthol derivatives which directly provides the chiral ligand. This asymmetric oxidation can be performed with different methods. In the following, two procedures will be discussed based on the use of molecular oxygen as the terminal oxidant.

We have already discussed the use of iron-based chiral complexes in asymmetric oxidation reactions with $H_2O_2$. In order to employ $O_2$ as the terminal oxidant it is necessary to choose the correct ligand to suitably reduce its oxidation potential. Salan ligands, bearing two phenolic and two amino donor atoms, provide the necessary requirements to promote the aerobic oxidation of 2-naphthols to the corresponding bi-naphthols, which are invaluable chiral precursors and ligands. Katsuki observed that first generation chiral dimeric $\mu$-oxo-Fe(salan) complexes were poorly active, while second generation Fe(salan) with a chiral 1,2-diphenyl backbone and atropoisomeric bi-naphthalene substituents allowed substantial asymmetric induction. Once optimized, the catalytic system allowed the dimerization of 2-naphthols with 4% mol catalyst at 60°C in toluene under air as the terminal oxidant, with yields in the range 77–94% and 77–96% ee.[66] In particular, 2-naphthols substituted in position six provided lower asymmetric induction compared to substrates substituted in position three (Scheme 23.31).

An alternative approach to the asymmetric synthesis of binaphthols via biaryl coupling in the presence of molecular oxygen as the terminal oxidant was disclosed by Kozlowski using chiral diamine ligands, in particular, using (S,S)-1,5-diaza-cis-decalin with Cu(II).[67] The peculiarity of the system is to work efficiently (2.5–10% mol of catalyst) with high yields and high enantioselectivity (>90%) exclusively for 2-naphthols bearing ester, ketone, phosphonyl and sulfonyl derivatives in the $R^1$ position (Scheme 23.32). The role of such functional groups in that position is manifold: they play a pivotal role in regulating the oxidation potential of the substrate, they provide a coordination site for the Cu(II) metal center, in association with the phenolic OH, and they enable turnover ability of the catalyst. Electron-rich 2-naphthols lacking the substituents in position three provide good catalytic activity but with much lower enantioselectivity because of the less rigid monodentate coordination to the metal center. The mechanism of the reaction was investigated in detail[68] observing that the aerobic oxidation of the catalyst is the turnover limiting step of the reaction.

This catalytic system operating on less electron-rich 2-naphthols complements well more traditional systems working on highly electron-rich naphthols, which have been employed in the synthesis of chiral perylenequinones-based derivatives that are potent protein kinase C inhibitors and are promising agents for photodynamic cancer therapy.[69]

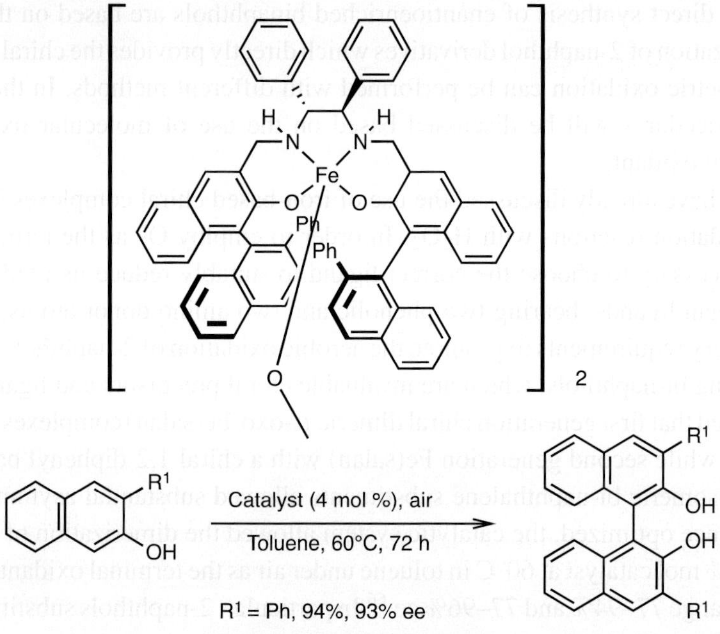


$R^1$ = Ph, 94%, 93% ee

$R^1$ = 2-naphthyl, 91%, 94% ee

$R^1$ = Cl, 82%, 94% ee

$R^1$ = I, 77%, 96% ee

$R^1$ = Me, 89%, 77% ee

**Scheme 23.31.** Asymmetric oxidation of naphthols with air to provide atropoisomeric binaphthols mediated by a dimeric Fe(III) salan catalyst.

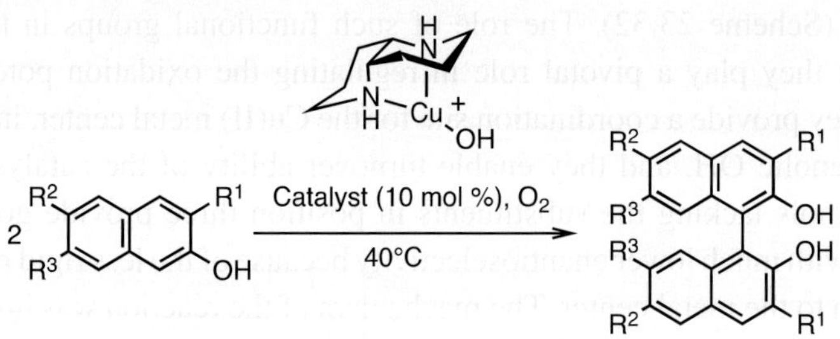


$R^1$ = COOMe, $R^2$ = Br, $R^3$ = H, 27%, 92% ee

$R^1$ = $CONEt_2$, $R^2$ = H, $R^3$ = H, 48%, 72% ee

$R^1$ = $CO(C_6H_4)$-p-$NMe_2$, $R^2$ = H, $R^3$ = H, 84%, 94% ee

$R^1$ = $CO(C_6H_4)$-p-OMe, $R^2$ = H, $R^3$ = H, 93%, 90% ee

**Scheme 23.32.** Asymmetric oxidation of naphthols to atropoisomeric binaphthols with $O_2$ mediated by a dimeric Cu(II) cornplex bearing a diamine ligand.

## 23.3.5. *Resolution of chiral secondary alcohols via asymmetric oxidation*

### 23.3.5.1. *Palladium*

Despite its high value in the synthesis of chiral natural products and pharmaceutical drugs, the asymmetric kinetic resolution of secondary alcohols via oxidation to the corresponding ketones is a much younger reaction than other asymmetric oxidations, just a little more than ten years old. The major problem that hampered its development was the choice of oxidant[70] which can participate in the general mechanism based on the substrate oxidation by Pd(II) species followed by re-oxidation of the Pd(0) species by the terminal oxidant (Wacker-type oxidation). The seminal work of Uemura[71] on the achiral version of the reaction based on the employment of $Pd(OAc)_2$ with pyridine and $O_2$ as the oxidant under anhydrous conditions was independently extended to the use of chiral ligands by Sigman[72] and Stoltz[73] when in 2001 they selected (−)-sparteine as the best ligand for the reaction. The two groups identified similar catalytic systems based on different Pd precursors, solvents and bases. In particular, Sigman's system is based on the employment of 5% mol of Pd[(−)-sparteine]$Cl_2$ with an extra 20% mol of ligand that acts as a base in t-butanol as the solvent at 65°C, for 20–24 h with oxygen, and it is active and selective with $K_{rel}$ ($K_R/K_S$) between 10 and 20 for a wide range of secondary benzylic and aliphatic alcohols, as well as for meso 1,3-diols (Scheme 23.33) with possible extension up to 10 mmol scale.[74] The catalytic system optimized by Stoltz uses 5% mol of [Pd(nbd)$Cl_2$] (nbd = norbornadiene) with 12% of (−)-sparteine in chloroform,

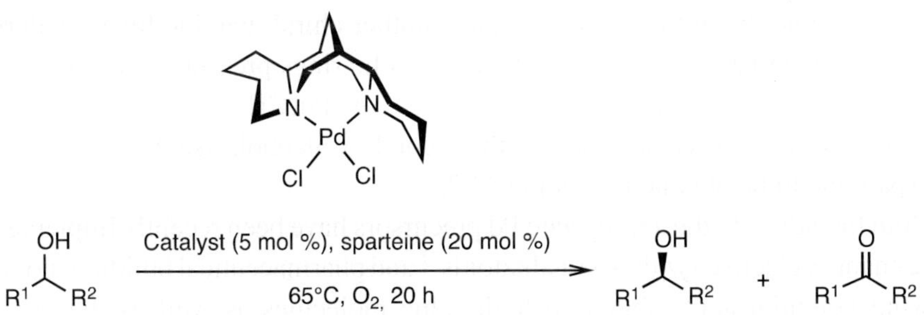


$R^1$ = 4-Me-Ph, $R^2$ = Me, 61.6%, 98.5% ee, $K_{rel}$ = 19

$R^1$ = 2-naphthyl, $R^2$ = Me, 55.9%, 90.2% ee, $K_{rel}$ = 17.7

$R^1$ = 2-furyl, $R^2$ = Me, 65.2%, 99.3% ee, $K_{rel}$ = 16.3

$R^1$ = t-Bu, $R^2$ = Me, 60.9%, 97.2% ee, $K_{rel}$ = 17.3

$R^1$ = Cy, $R^2$ = Me, 53%, 61% ee, $K_{rel}$ = 6.1

**Scheme 23.33.** Kinetic resolution of secondary alcohols via oxidation with $O_2$ mediated by a Pd(II) catalyst bearing a bidentate amine ligand.

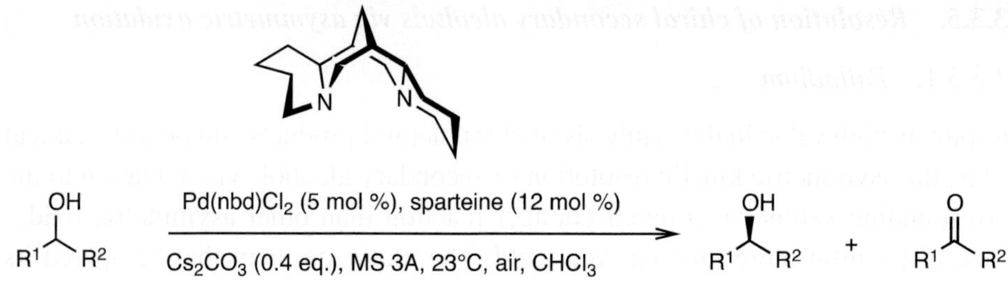


$R^1$ = 4-Me-Ph, $R^2$ = Me, 62.3%, 99.8% ee, $K_{rel}$ = 25.7

$R^1$ = 2-naphthyl, $R^2$ = Me, 55.5%, 98.0% ee, $K_{rel}$ = 37.3

$R^1$ = Ph, $R^2$ = Et, 56.8%, 94.9% ee, $K_{rel}$ = 21.7

$R^1$ = 4-F-$C_6H_4$, $R^2$ = Me, 56.7%, 93.0% ee, $K_{rel}$ = 19.5

**Scheme 23.34.** Kinetic resolution of secondary alcohols via oxidation with air mediated by a Pd(II) bearing (−)-sparteine ligand.

with $Cs_2CO_3$ as the base under anhydrous conditions at room temperature and with air as the oxidant.[75] This system works under milder experimental conditions and allows $K_{rel}$ between 10 and 30, in particular, for benzylic alcohols (Scheme 23.34).

In spite of its high selectivity, the intrinsic limitation of the Pd(−)-sparteine catalytic system is related to the natural origin of the chiral ligand which is only available in one enantiomeric form. This restricts the stereochemistry of the enantioenriched secondary alcohol recovered after the kinetic resolution to one enantiomeric series. Stoltz overcame the problem by selecting another chiral diamine ligand[76] derived by (−)-cytisine that shares three of the four cyclic rings present in the structure of (−)-sparteine. The new ligand, in the presence of $[Pd(CH_3CN)_2Br_2]$ as the metal precursor, allowed the enantiomers of the secondary alcohols usually achieved with (−)-sparteine to be obtained (Scheme 23.35).

Both ligands with the appropriate Pd precursors have been recently implemented in the enantioselective synthesis of alkaloids[77] and pharmaceutical building blocks,[78] showing potential applicability to high-value molecules as well as the possible scale-up of the reaction.

### 23.3.5.2. *Vanadium*

Asymmetric secondary alcohol oxidation can also be performed with other metal complexes, in particular Toste showed that traditional Schiff based V catalysts prepared *in situ* from the corresponding ligand and VO(O-*i*Pr)$_3$ allowed the kinetic resolution of α-hydroxy esters in acetone under mild experimental conditions and with 1 atm of $O_2$ (Scheme 23.36).[79] The reaction works well for both benzylic and

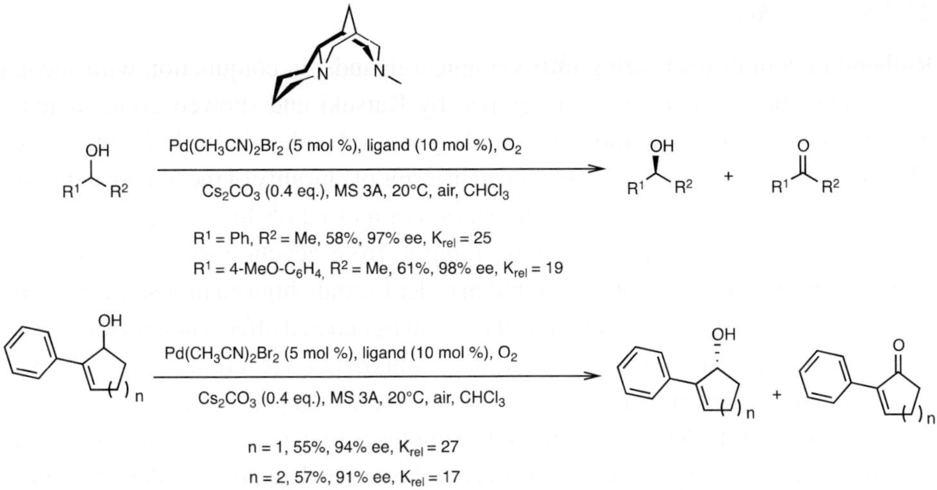


**Scheme 23.35.** Kinetic resolution of secondary alcohols with a Pd(II) catalyst bearing a bidentate N ligand that provides the opposite enantiomers compared to (−)-sparteine.

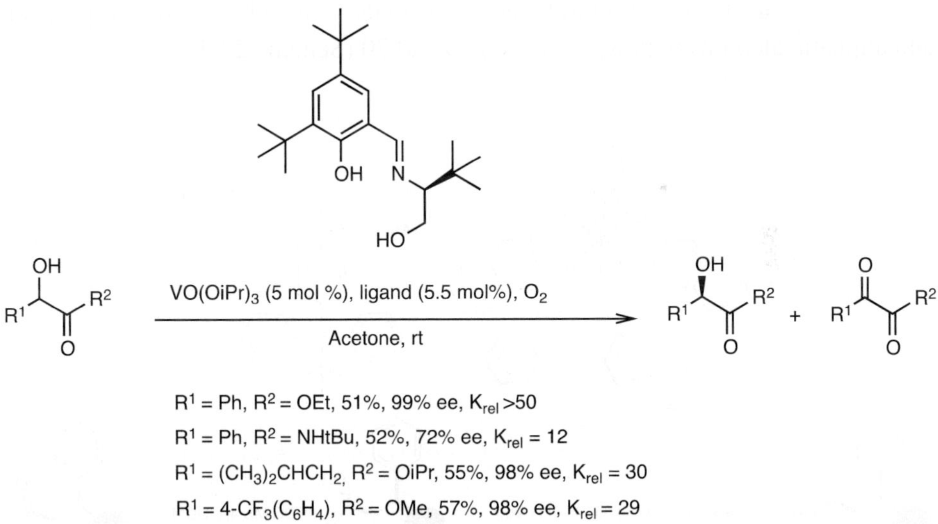


**Scheme 23.36.** Kinetic resolution of secondary α-keto alcohols with $O_2$ mediated by a V(V) with a Schiff base ligand.

non-benzylic substrates, in the latter case longer reaction times being necessary. Although V complexes are known to efficiently promote the epoxidation of alkenes, the present catalytic system is highly chemoselective and also provides the carbonyl product in the presence of alkene residues which remain unchanged. The selectivity factor, which correlates the ee of the reagent and conversion, is in most cases >10 and up to more than 50 for ethyl mandelate.

### 23.3.5.3. *Ruthenium*

Ruthenium complexes bearing nitroxyl apical ligands, in conjunction with second generation salen ligands, were developed by Katsuki and showed good catalytic activity in the aerobic oxidation of secondary alcohol under visible light irradiation. The catalytic system was accurately tuned in terms of chirality of the ligand and it was subsequently applied to the desymmetrization of meso diols to give optically active lactols.[80] In particular, each substrate, ranging from acyclic diols to monocyclic diols, needs the catalytic system tailored in order to attain high enantioselectivity (up to 93%) with a peculiar role played by the apical ligand that affects not only the enantioselectivity, but also the kinetics of the desymmetrization reaction (Scheme 23.37).

Subsequent implementation of the knowledge developed in this work enabled the application of the Ru(salen)(nitrosyl) complexes to the oxidative aerobic kinetic resolution of secondary alcohols, modifying the original catalyst by the addition of 1,3-bis(p-bromophenyl)propane-1,3-dione as a bidentate ligand whose effect was to steer the coordination geometry of the Ru(III) center to a cis-$\beta$ configuration.[81] This allowed a highly enantioselective catalyst which, under mild conditions (air at room temperature), converts chiral racemic secondary benzylic, allylic, propargylic and aliphatic alcohols with $K_{rel}$ between 14 and 30 (Scheme 23.38).

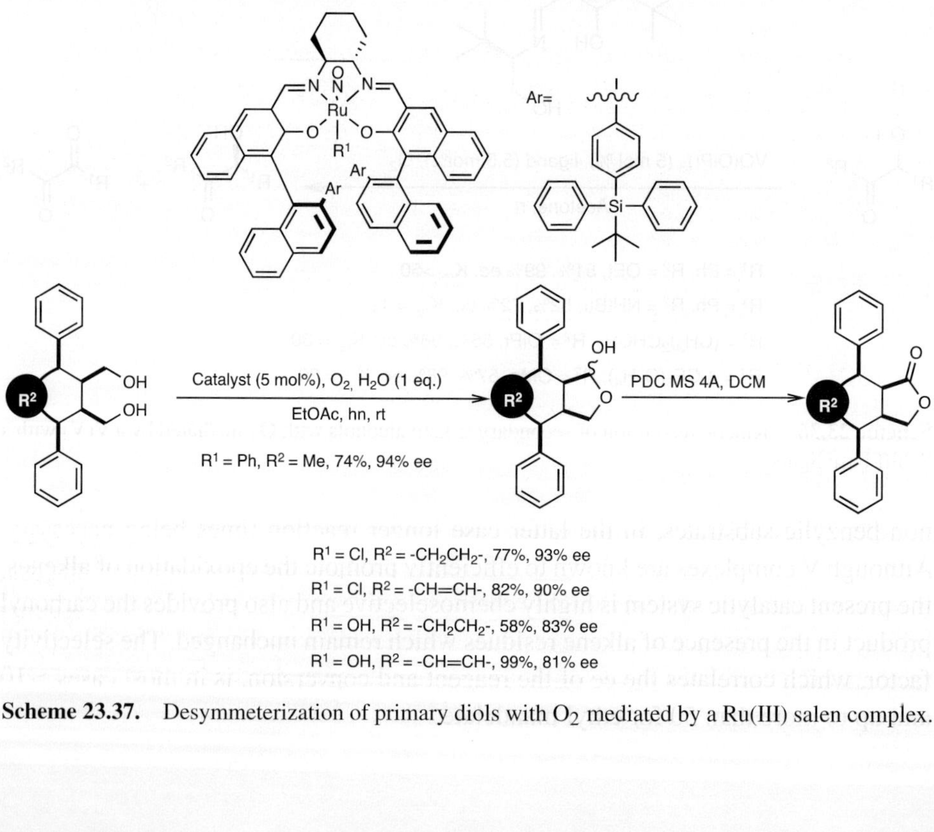


**Scheme 23.37.** Desymmeterization of primary diols with $O_2$ mediated by a Ru(III) salen complex.

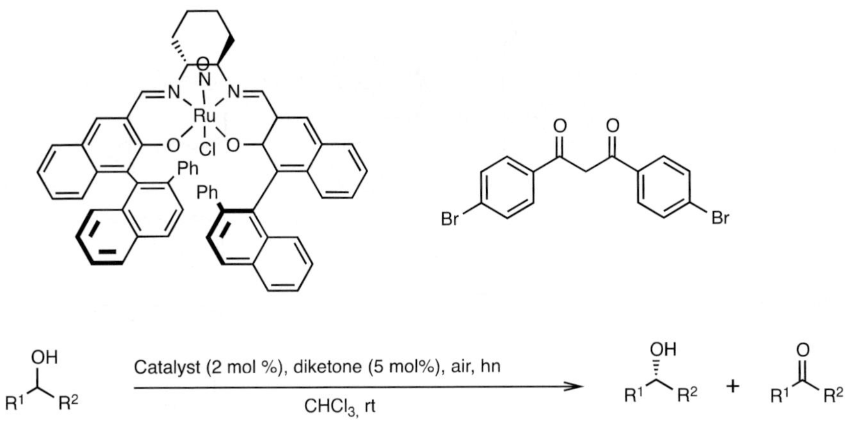


$R^1$ = Ph, $R^2$ = Me, 61%, 98% ee, $K_{rel}$ = 19

$R^1$ = PhCC, $R^2$ = Me, 58%, 99% ee, $K_{rel}$ = 30

$R^1$ = 4-Cl($C_6H_4$), $R^2$ = Me, 57%, 95% ee, $K_{rel}$ = 23

$R^1$ = $PhCH_2$, $R^2$ = Me, 41%, 55% ee, $K_{rel}$ = 15

**Scheme 23.38.** Kinetic resolution of secondary alcohols with air mediated by Ru(III) salen complex.

## 23.4. Water as the Reaction Medium

Water is the medium where all biological reactions take place, including oxidation reactions, but it is a rather unfamiliar solvent for chemists who tend to avoid it, often in an over-prudent approach. When $H_2O_2$ and $O_2$ are used as oxidants, water is present as a by-product and this prompted the investigation of catalytic asymmetric oxidation reactions in water. The hydrophobic effect,[52] which consists of the tendency for organic species to self-assemble in water, is the most peculiar effect of this solvent and operates both on apolar catalysts and organic substrates. This overall "squeeze out" effect produces, in several cases, positive effects on both the catalytic activity and the enantioselectivity of asymmetric reactions, as described in the following examples of stereoselective oxidation.

### 23.4.1. *Water as solvent*

The vanadium-catalyzed epoxidation reaction of allylic alcohols with bidentate chiral hydroxamic acids is, similar to several other oxidation reactions, an example of ligand-decelerated catalysis when comparing the original V-alkoxide precursor and the V-ligand complex for reactions performed in organic media. In water the effect is reversed, with the chiral ligand accelerating the reaction if compared to the original V precursor for a ratio of up to 1:1 ligand to metal. The epoxidation reaction with

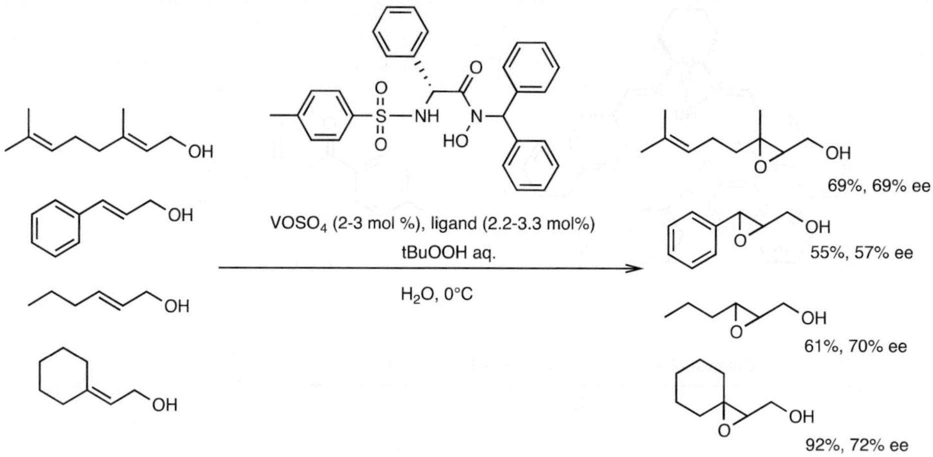


**Scheme 23.39.** Asymmetric epoxidation of allyl alcohols with tBuOOh in water mediated by a V(IV) with a NO bidentate ligand.

t-BuOOH as the oxidant occurs in the organic phase formed by the poorly soluble substrate where the catalyst is confined by the hydrophobic effect. This catalytic system does not suffer from epoxide ring-opening caused by the presence of water, and a series of allyl alcohols can be efficiently oxidized with 41–92% yield and 57–72% ee (Scheme 23.39).[82]

More impressive results were observed in two other asymmetric reactions both based on the use of well-known catalysts developed for other transformations that turned out to perform well in new reactions, thanks to the beneficial use of water as solvent. One case is based on the oxidative kinetic resolution of secondary alcohols with chiral Mn(salen) complexes using $PhI(OAc)_2$ as the oxidant (Scheme 23.40). The reaction is poorly enantioselective in dichloromethane (2% ee $K_{rel} < 1.1$), while in water, in the presence of tetraethylammonium bromide as the phase transfer agent, the reaction is fast (63.4% conversion in 2 h) and highly enantioselective (85.2% ee with $K_{rel}$ 23.7).[83]

### 23.4.2. *Micellar catalysis*

As exemplified above, the use of water as the solvent for organometallic catalysis is hampered by the low general solubility of both substrates and catalysts in this medium. Possible approaches to solving this are based on i) the use of polar co-solvents that often decrease the green character of the catalytic system or ii) the employment of surfactants that form micelles in water as apolar nano-environments where substrates and catalyst get together and react, often with enhanced activity and selectivity.[84] Micelles are self-assembled devices arising from neutral, anionic

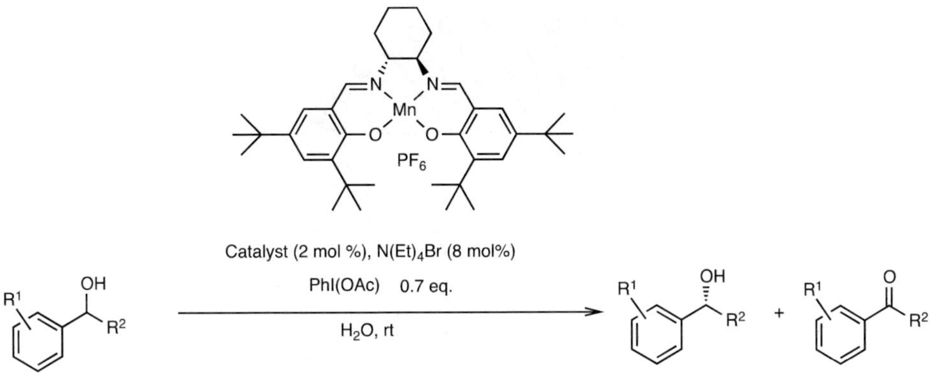


$R^1$ = H, $R^2$ = Me, 51.7%, 85.2% ee, $K_{rel}$ = 23.7

$R^1$ = 4-Cl, $R^2$ = Me, 61.1%, 97.8% ee, $K_{rel}$ = 18.2

$R^1$ = 4-Br, $R^2$ = Me, 65.9%, 96.8% ee, $K_{rel}$ = 11.3

$R^1$ = 4-Me, $R^2$ = Et, 64.8%, 5.4% ee, $K_{rel}$ = 1.1

**Scheme 23.40.** Kinetic resolution of secondary alcohols with Phl(OAc)$_2$ in water mediated by the Mn(II)salen complex.

or cationic amphiphilic molecules exposing the polar heads to water and aligning the hydrophobic tails in the core. Because of this structure, ongoing from the surface to the core of a micelle, a polarity gradient is present. Asymmetric oxidation reactions, such as Baeyer–Villiger epoxidation and sulfoxidation reactions, all showed better performances when carried out in water under micellar conditions rather than with the use of organic chlorinated solvents.

One remarkable example is based on a Co(salen) complex that was employed as the catalyst for the Baeyer–Villiger oxidation of cyclic meso ketones with $H_2O_2$ as the oxidant. The complex was not active in organic media with poor yields and no enantioselectivity, but when tested in micellar media the system became active and selective (Scheme 23.41).[85] After optimizing the experimental conditions, the Co(salen) catalyst in the aqueous micellar medium allowed high yields (up to 98%) in normal lactone, with high diastereoselectivity (up to 86%) and high ee (up to 90%) in the double parallel kinetic resolution of chiral cyclobutanones via asymmetric Baeyer–Villiger oxidation. It is worth noting that when this catalytic system operates in water the exceptional increase in activity and selectivity is solely due to the micelles where both catalyst and substrate are dissolved.

Using water as the solvent also increased the selectivity in the asymmetric Baeyer–Villiger oxidation of cyclic six- and four-membered ring ketones. In particular, the use of surfactants under micellar conditions allowed i) the direct solubilization of otherwise water insoluble complexes and ii) the increase of

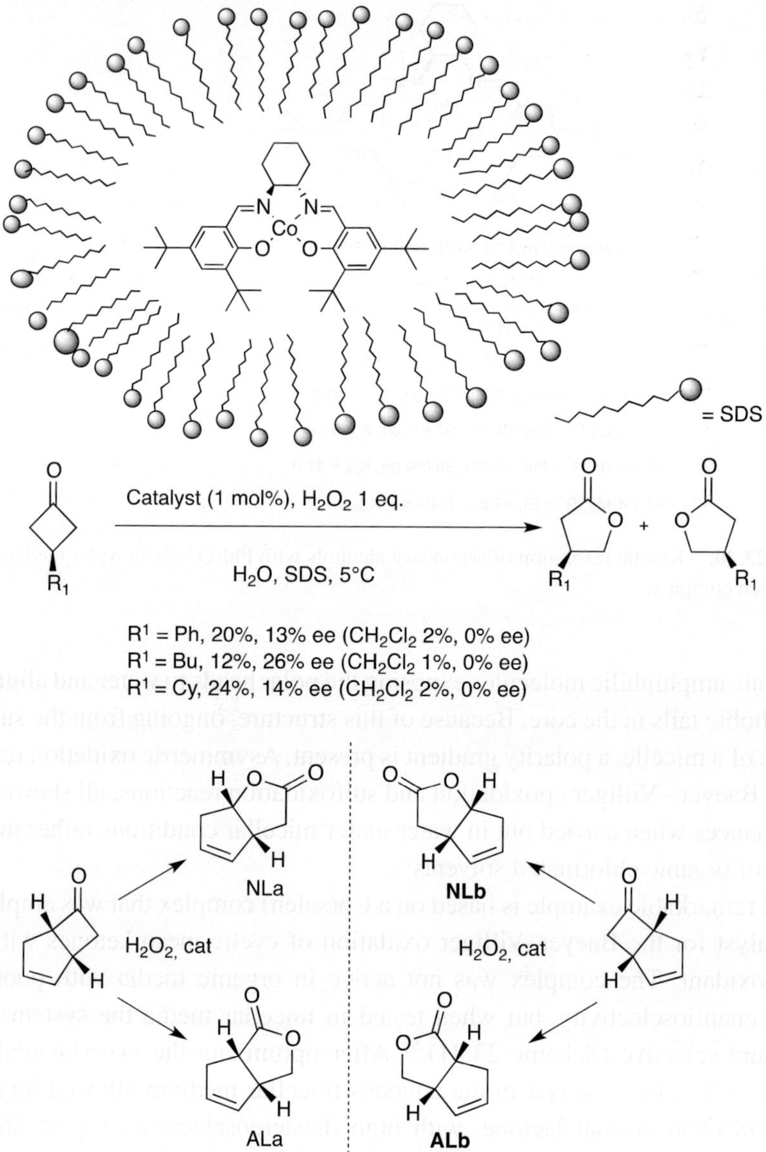


| Solvent | TON NL+AL | NL/AL | ee (%) NL |
|---|---|---|---|
| $CH_2Cl_2$ | 15 | 34 | 0 |
| $H_2O$/TritonX100 | 48 | 74 | 51 |
| $H_2O$/TritonX114 | 60 | 70 | 90 |

**Scheme 23.41.** The asymmetric Baeyer–Villiger oxidation of 3-substituted cyclobutanones chiral cyclobutanones with $H_2O_2$ mediated by a Co(II)salen catalyst benefits greatly from the employment of micellar media instead of common organic media.

enantioselectivity and green character compared to the same reaction carried out in common chlorinated organic solvents.[86]

Scarso and Strukul observed that anionic micelles based on sodium dodecylsulfate (SDS) efficiently solubilized Pt bis-cationic complexes and ketones in the apolar core of the micelles. This favored the contact between substrate and catalyst and, more importantly, led to a higher steric control of the asymmetric reaction thanks to the more ordered nano-environment present in the micelles compared to bulk organic solvents. Each substrate required the dedicated optimization of the catalyst, the surfactant and the experimental conditions since the distribution of the substrates and catalysts is greatly affected by the kind of surfactant and aggregate considered (Scheme 23.42). Overall, in all the cases tested, an increase in enantioselectivity was observed for the asymmetric BV oxidation of meso-4-substituted cyclohexanones with bis-diphenylphoshinobinaphthyl (BINAP) as the ligand and SDS as the surfactant. A different scenario was present in the kinetic resolution of chiral cyclobutanones or with meso cyclobutanones where an increase of enantioselectivity was observed with the neutral polyoxyethanyl-α-tocopheryl sebacate (PTS) surfactant (Scheme 23.42).

Overall, the use of surfactants in water for the studied BV reactions implies the partition of all reaction partners (substrate, oxidant and catalyst) between the micelle, bulk water and the interphase between the two. As a consequence, the lipophilicity of all species is crucial to rationalize their positioning in the micellar system, and as a general observation, more hydrophilic substrates worked well in neutral surfactants while more lipophilic ones worked well in anionic micelles.

Asymmetric epoxidation of terminal alkenes with hydrogen peroxide was optimized with electron-poor chiral Pt(II) complexes bearing a pentafluorophenyl residue, as described in Section 23.3.1.6. The same catalytic system was made more sustainable by the employment of water as the solvent under micellar conditions. Surfactant optimization revealed the preferential use of neutral species like Triton-X100 to solubilize both the catalyst and substrates. In several cases an increase of the asymmetric induction was observed (Scheme 23.43).[29] The use of an aqueous phase and the strong affinity of the catalyst for the micelle allowed the recycling of the catalytic system by means of phase separation and extraction of the reaction products using an apolar solvent (hexane). The aqueous phase containing the catalyst was reused for up to three cycles with no loss of activity or selectivity.

In addition, asymmetric sulfoxidation could benefit from the use of aqueous micellar systems. In the reaction catalyzed by dimeric chiral Pt(II) species, both the chemoselectivity (chiral sulfoxide vs achiral sulfone) and the enantioselectivity of the former product increased, moving from dichloromethane as the organic solvent to water/SDS forming anionic micelles where the bis-cationic catalyst interacts via ion pairing (Scheme 23.44).[87]

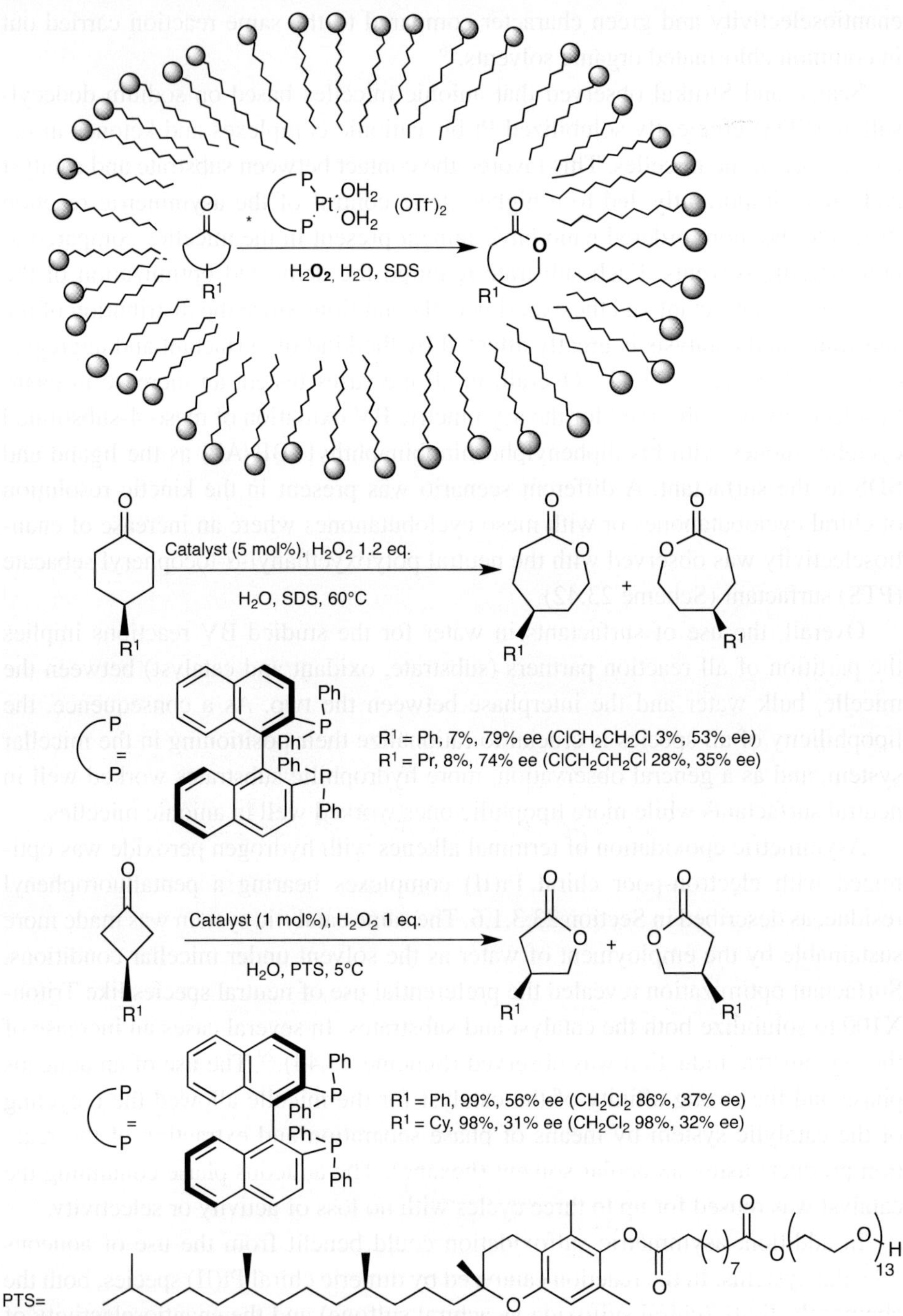


**Scheme 23.42.** The asymmetric Baeyer–Villiger oxidation of meso cyclohexanones and cyclobutanones with $H_2O_2$ mediated by a Pt(II) diphosphine catalyst in water with surfactant addition.

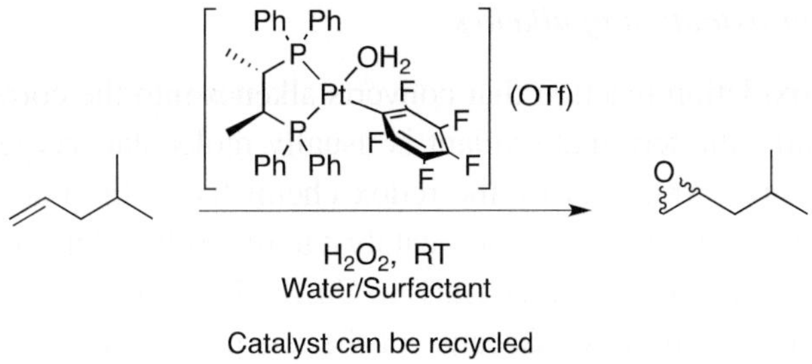


| Yield (%) | EE (%) | Medium |
|---|---|---|
| 56 | 58 | DCE |
| 51 | 84 | $H_2O$/Triton-X100 |

| Cycle | Yield (%) | EE (%) |
|---|---|---|
| 1 | 26 | 75 |
| 2 | 26 | 72 |
| 3 | 33 | 77 |

**Scheme 23.43.** Asymmetric epoxidation of terminal alkenes with $H_2O_2$ mediated by electron-poor Pt(II) catalysts bearing diphosphine ligands.

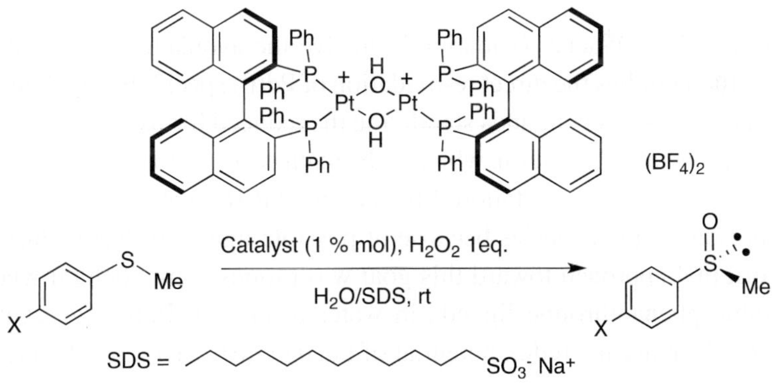


$R^1$ = H, 98%, 40% ee (CH$_2$Cl$_2$, 99%, 16% ee)
$R^1$ = CN, 68%, 63% ee
$R^1$ = NO$_2$, 63%, 88% ee

**Scheme 23.44.** Asymmetric sulfoxidation of thioethers with $H_2O_2$ mediated by a dimeric Pt(II) catalyst in micellar media.

### 23.4.3. *Wacker oxidation of alkenes*

In the Wacker oxidation reaction that converts alkenes into the corresponding carbonyl compounds, the terminal oxidant is usually molecular oxygen, but the real oxidation steps are carried out by the redox chemistry of the Pd(II) centers. The classical system requires Cu(II) as a co-catalyst to re-oxidize Pd(0) to Pd(II) and $O_2$ as the terminal oxidant to re-oxidize Cu(I) to Cu(II). The oxygen atom transferred to the substrate comes from water. Because of the role played by $H_2O$, recent updates on Wacker-type oxidations are described below. The mechanism of the general oxidation with $PdCl_2$, $CuCl_2$ and $O_2$ has also been reviewed recently, and different pathways are possible as a function of the different concentrations and other experimental conditions.[88] The oxidation reaction usually follows the Markovnikov rule, but important exceptions are known. In particular, heteroatoms in allylic positions tend to mitigate the preference for the formation of methyl ketones. This effect can be exploited in order to selectively form the aldehyde when using allylamines as substrates protected with a phthalimide functional group that ensures quantitative oxidation and >99:1 preference for the aldehyde (Scheme 23.45).[89] The catalyst used in this case is $PdCl_2$ with stoichiometric CuCl in DMF/$H_2O$ as the solvent, using $O_2$ as the oxidant; the reaction is compatible with the presence of stereocenters on the substrate. In fact, enantioenriched allylic phthalimides maintain their chirality and provide chiral aldehydes which are precursors of $\beta$-aminoacids.

Recent trends in Wacker oxidation[90] involve the avoidance of Cu salts as co-catalysts, which implies the direct re-oxidation of Pd(0) species by $O_2$. This requires the presence of the correct ligands stabilizing the reduced form of the catalyst avoiding zerovalent metal aggregation, which is the main deactivation pathway hampering catalyst turnover. The use of tailored ligands for Pd(II) centers is also in contrast with the presence of Cu species because of possible ligand exchange during catalysis. An original approach toward this goal was proposed by Sheldon when using water-soluble phenanthroline ligands in water to prevent Pd(0) aggregation and contextually decrease the redox potential of Pd(II)/Pd(0), thereby reducing catalyst

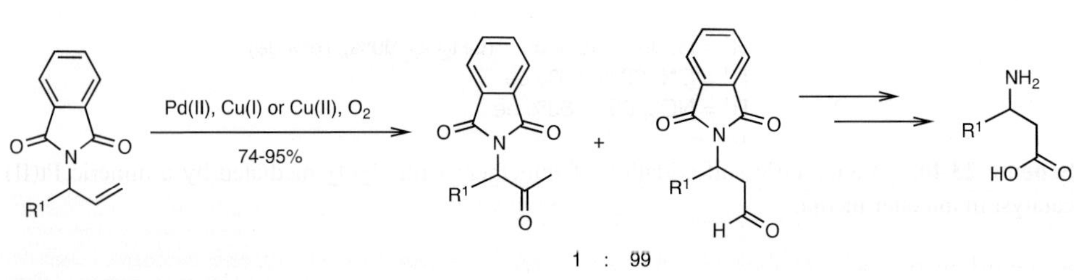


**Scheme 23.45.** Aldehyde selective Wacker oxidation of terminal alkenes bearing a phthalimide substituent with $O_2$.

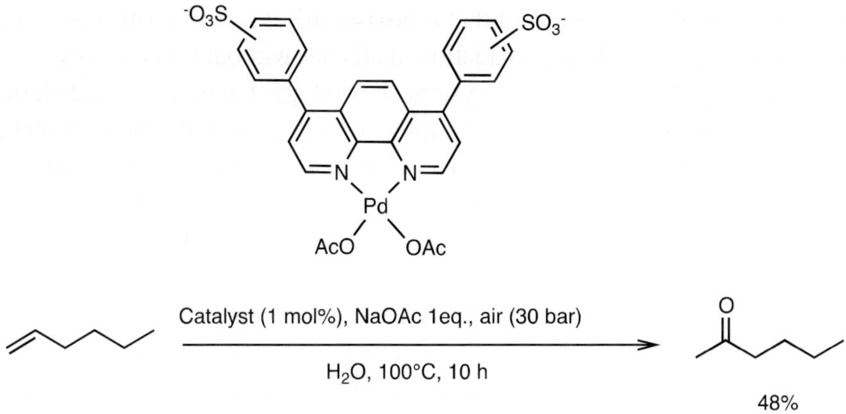


**Scheme 23.46.** Wacker oxidation of terminal alkenes with air in water with a water-soluble Pd(II) catalyst bearing a sulfonated phenanthroline ligand.

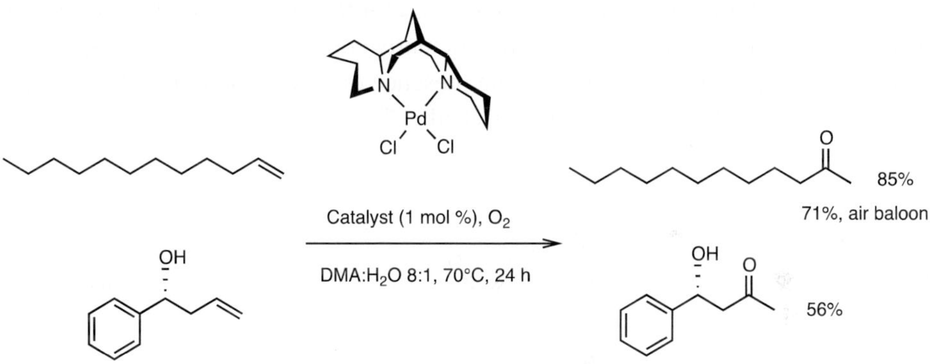


**Scheme 23.47.** Wacker oxidation of alkenes with $O_2$ mediated by a (−)-sparteine Pd(II) catalyst.

decomposition (Scheme 23.46).[91] Both terminal and internal alkenes were oxidized to ketones at 100°C and 30 bar of air. Using water as the solvent, catalyst recycling was achieved by means of a simple extraction of the organic product and reuse of the aqueous phase.

A more active system working under less harsh conditions was obtained using (−)-sparteine as a bidentate ligand for Pd(II).[92] A lower concentration of $O_2$ can be used and the possible isomerization of terminal olefins is suppressed with the exclusive formation of terminal methyl ketones (Scheme 23.47). Enantiomerically-enriched protected alcohols bearing terminal alkenes provided the corresponding chiral methyl ketones without racemization, further emphasizing the potential synthetic utility of the oxidation method.

Very recently, Reiser[93] showed that a bis(isonitrile) ligand forms robust Pd(II) complexes for the direct Wacker oxidation of alkenes without Cu co-catalysts under 1 atm of $O_2$ at 70°C. The catalytic system showed good activity towards terminal aliphatic alkenes, but also styrene substrates, which are usually more challenging substrates for this kind of oxidation because of the competitive double-bond cleavage under oxidative conditions reacting readily, favoring the acetophenone products and concomitant formation of benzaldehydes as side products in 4–20% yield (Scheme 23.48).

An extremely simple solution for the direct use of $O_2$ in the Wacker reaction was achieved by switching to N,N-dimethylacetamide (DMA) as the water co-solvent because of its intrinsic properties that promote the re-oxidation of Pd(0) species. Efficient and regioselective oxidation on $C_2$ of terminal alkenes was possible using $PdCl_2$ as the catalyst at 60°C with 6 atm of $O_2$.[94] If the reaction is carried out under anhydrous conditions in the presence of NaOAc and acetic acid with DMA as the solvent, the reaction provides selective oxidation of the terminal alkene on $C_1$ leading to the corresponding allyl acetates (Scheme 23.49). The catalytic system is also active towards internal alkenes that are efficiently converted into the corresponding ketones, showing good tolerance for the presence of other functional groups in the substrate such as alcohol, nitrile and allyl ether.[95]

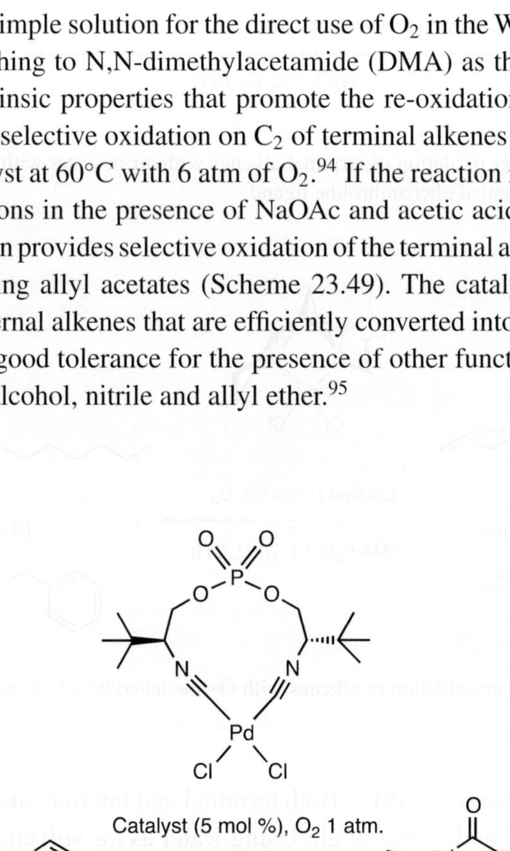


$R^1 = C_6H_{13}$, 98%
$R^1 = PhCH_2$, 75%
$R^1 =$ Cy, 84%
$R^1 =$ 4-MeO-Ph, 99%, ket./ald. = 21
$R^1 =$ Ph, 84%, ket./ald. = 17

**Scheme 23.48.** Wacker oxidation of terminal alkenes with $O_2$ mediated by a Pd(II) catalyst bearing a bidentate isonitrile ligand without a Cu(I) co-catalyst.

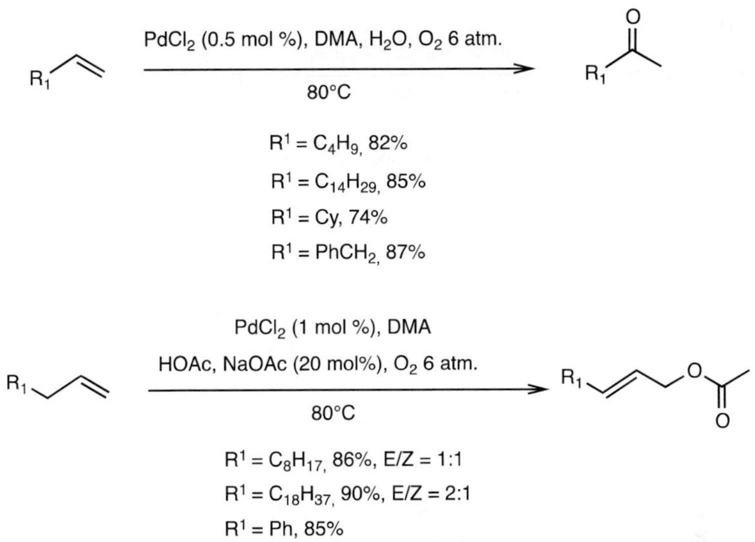


**Scheme 23.49.** Wacker oxidation of terminal alkenes with $O_2$ in DMA with a $PdCl_2$ without Cu(I) co-catalyst provides the corresponding methyl ketones or, in the presence of HOAc and NaOAc, allyl acetates.

An alternative approach to the Wacker reaction, excluding the use of Cu(II), is based on the use of oxidants alternative to $O_2$. Early examples dating back to the 1980s for the reaction using $H_2O_2$ instead of $O_2$ were reported by Mimoun using five equivalents of oxidant in acetic acid.[96] The reaction is highly selective, forming the corresponding methyl ketone for a series of terminal alkenes from $C_8$ to $C_{12}$, while it does not work on internal double bonds or alkenes with heteroatoms in the allylic position (Scheme 23.50a). A more recent example is based on the *in situ* formation of $H_2O_2$ from $O_2$ using $Pd(OAc)_2$ as the catalyst, with pyridine as the ligand in the presence of i-PrOH. This system efficiently provides methyl ketones from terminal alkenes but it was inactive towards internal ones. A plausible mechanism of the reaction is based on alcohol oxidation that forms the Pd-hydride species that further react with molecular oxygen leading to Pd-hydroperoxo species that eventually transfer the oxygen of the activated oxidant directly to the substrate (Scheme 23.50b).[97]

In 2005 Sigman disclosed that the dimeric $[Pd\text{-}(I\text{-}i\text{-}Pr)Cl_2]_2$ complex bearing a carbene ligand with AgOTf to replace the chlorine ligand, was able to activate t-butyl hydroperoxide towards the oxidation of styrene to acetophenone in methanol under mild conditions.[98] Conversion was high with a wide range of styrene substrates regardless of the electronic nature of the substituents in the aromatic ring, with all cases exhibiting an extremely high regioselective Markovnikov oxidation of the double bond (Scheme 23.51). Mechanistic investigation and isotopic

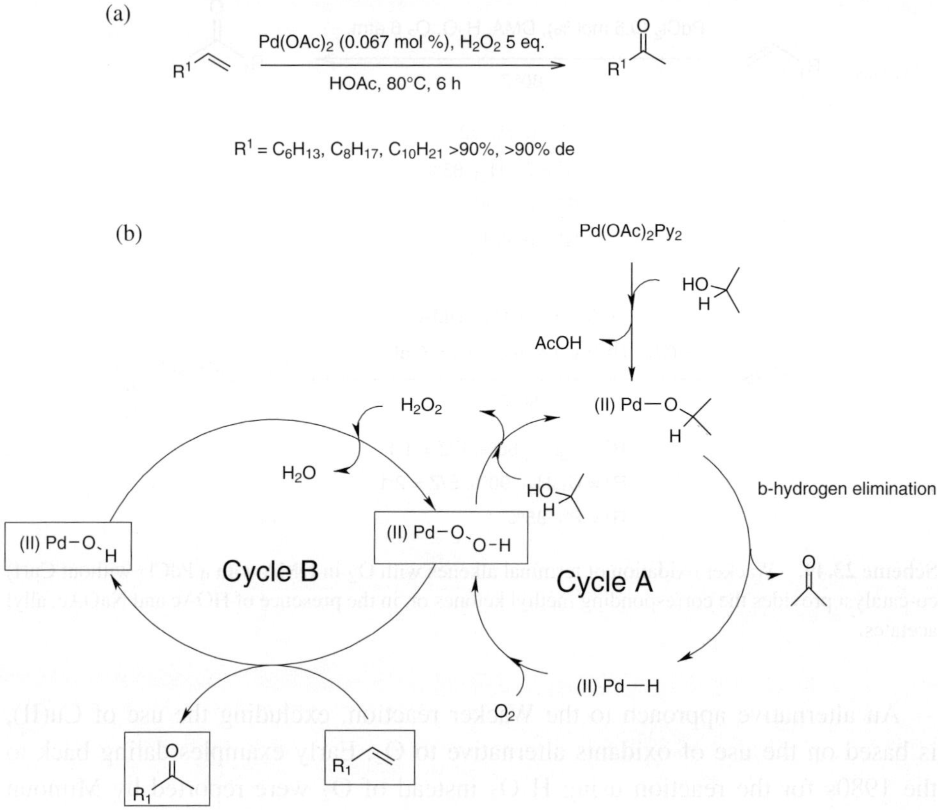


**Scheme 23.50.** (a) Wacker oxidation of terminal alkenes to methyl ketones with $H_2O_2$ mediated by Pd(II) in acetic acid; (b) Mechanistic hypothesis.

labeling experiments ensured that the oxidant provides the oxygen atom transferred to the substrate and the hydrogen atoms present on styrene are maintained in the product in agreement with a 1,2-hydride shift mechanism wherein an enol is formed.

Very recently, Sigman extended the substrate scope of the oxidation with hydroperoxides. With a bidentate ligand belonging to the Quinox family comprising a quinoline and an oxazoline N donor, it was possible to achieve extreme regioselective oxidation of protected allyl alcohol derivatives with t-butyl hydroperoxide to the corresponding methyl ketones (Scheme 23.52), while the same substrates under classical Wacker conditions provide the aldehyde as the preferred product.[99] The chelating nature of the ligand plays a key role by allowing the simple and efficient conversion of styrene and other terminal alkenes exclusively towards the corresponding methyl ketones (Scheme 23.52). Chiral protected allyl alcohols can be transformed into the methyl ketone with retention of configuration and enantiomeric

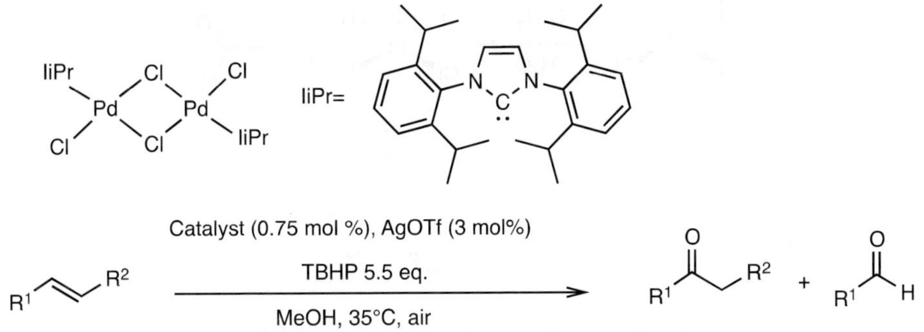


$R^1$ = Ph, $R^2$ = H, 75%, ket./ald. > 130:1

$R^1$ = 4-Me($C_6H_4$), $R^2$ = H, 86%, ket./ald. = 22:1

$R^1$ = Ph, $R^2$ = Ph, 42%, ket./ald. = 42:35

$R^1$ = Ph, $R^2$ = Me, 97%, ket./ald. = 2.3:1

**Scheme 23.51.** Wacker oxidation of alkenes with t-BuOOH mediated by a Pd(II) catalyst bearing a NHC (N heterocyclic carbene ligand) leading to ketones and aldehydes.

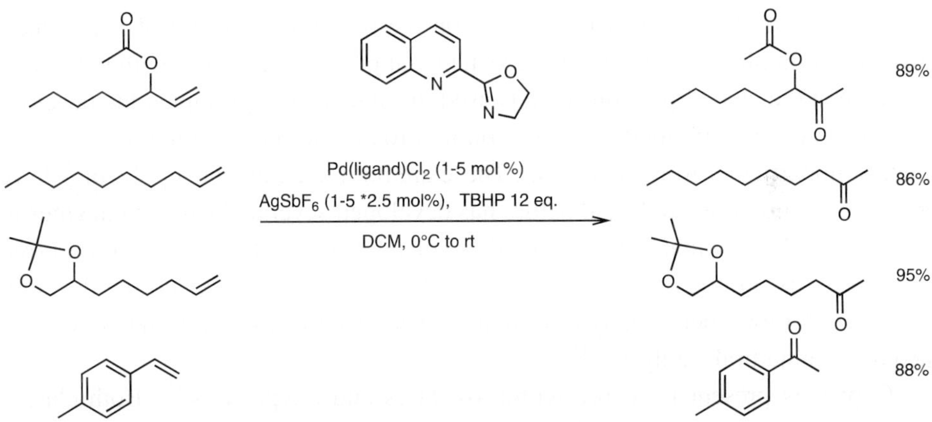


**Scheme 23.52.** Wacker oxidation of terminal alkenes with t-BuOOH mediated by a Pd(II) catalyst bearing a bidentate N,N ligand.

excess of the molecule. Even highly aldehyde orienteering substrates such as allylic phthalimides led to the Markovnikov oxidation, and the reaction showed analogous regioselectivity for other protected chiral and achiral allylamines.[100]

Wacker oxidation is not an exclusive characteristic of Pd(II) complexes. In 2004, Atwood reported an example of the conversion of ethylene to acetaldehyde mediated by a water-soluble Pt(II) complex of general formula cis-Pt$(Cl)_2(TPPTS)_2$ bearing intrinsically water-soluble monophosphines (TPPTS = triphenylphosphine

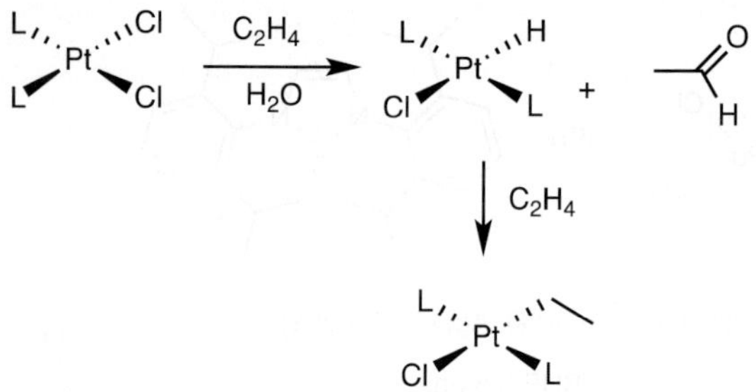

**Scheme 23.53.** Wacker oxidation of alkenes and $O_2$ mediated by a Pt(II) catalyst.

trisulfonate).[101] The reaction is stoichiometric at low temperature but up to 90 turnovers are possible when the reaction temperature is raised to 95°C and Cu(II) salts as co-catalysts remain absent. This behavior is probably due to the higher stability imparted by the phosphine ligands to the Pt(II) metal center (Scheme 23.53).

## 23.5. The Use of Less Toxic Metals as Active Ingredients

In recent years the stringent regulations on the maximum content of heavy metals allowed in pharmaceutical and agrochemical products, together with the steady increase in the cost of precious metals used in catalysis has prompted the development of new catalytic methods based on non-toxic and more economic transition metals, in particular iron and copper. These are both in abundant supply, they have accessible redox potentials and nature has developed several classes of enzymes to perform catalytic oxidations based on these metal centers. This observation spurred investigations on both the mechanism of action of the enzymes and the development of new biomimetic catalytic systems that are inspired by increasing knowledge acquired on natural catalysts.[102]

Copper is present in nature within oxidases and oxygenases.[103] Both classes of enzymes are characterized by the use of $O_2$ as the terminal oxidant but only in the latter class is it directly incorporated into the product. Oxidases are active mainly towards alcohols and amines, and oxygenases towards aromatic residues and ketones, thus requiring a wide variety of active sites bearing different structures and numbers of metal atoms. Notable examples are galactose oxidase which contains a copper-tyrosyl radical unit to perform two-electron redox chemistry, and tyrosinase and catechol oxidase which contain a dicopper(I) active site. As far as iron is concerned, enzymes can be classified in two main families based on either the heme ligand, as in the well-known cytochromes P450 family that lead to the hydroxylation of aliphatic C H bonds and epoxidation of C=C double bonds, or non-heme ligands. Examples of enzymes belonging to the latter class are the di-iron methane monoxygenases which convert methane to methanol, and Rieske dioxygenases that provide cis-dihydroxylation on aromatic rings and contribute to the biodegradation of such substrates in the soil. All these enzymes use dioxygen as the terminal oxidant and all synthetic bio-inspired complexes reported below make use of $O_2$ as a green oxidant, making the overall oxidation systems sustainable, at least in principle.

### 23.5.1. *Bio-inspired systems*

Among the recently developed wide range of copper-based oxidation methods, several are related to alcohol oxidation to the corresponding aldehydes and ketones using $O_2$ as the oxidant and releasing $H_2O_2$ as a by-product, mimicking galactose oxidases and sharing with the latter the presence of ligands containing phenol residues via the intermediacy of Cu(II)-phenoxyl radical species.[102] A conceptually related system was proposed by Sheldon based on copper salts with nitrogen containing ligands such as bipyridine and 2,2,6,6-tetramethyl-1-piperidinyloxyl radical (TEMPO) as the radical intermediate species.[104] The catalytic system converts primary alcohols efficiently under mild conditions, but it is completely inactive towards secondary substrates (Scheme 23.54).[105]

A similar system was proposed by Markó based on CuCl, phenanthroline, di-tert-butyl azodicarboxylate (DBAD) and N-methylimidazole (NMI) as additives showing efficient oxidation of both primary and secondary alcohols to carbonyl compounds, with molecular oxygen or air as the oxidant releasing water as the sole by-product.[106] The catalytic system exhibits good tolerance of the presence of electron-rich alkenes, thioethers and pyridine moieties, which could be involved in other oxidation reactions but remain unaffected by the catalyst. In the case of

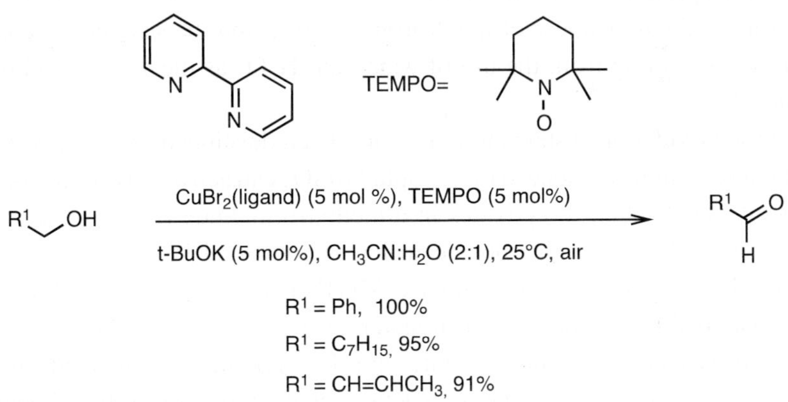


**Scheme 23.54.** Oxidation of primary alcohols with air and TEMPO radical mediated by a Cu(II) complex bearing a phenanthroline ligand.

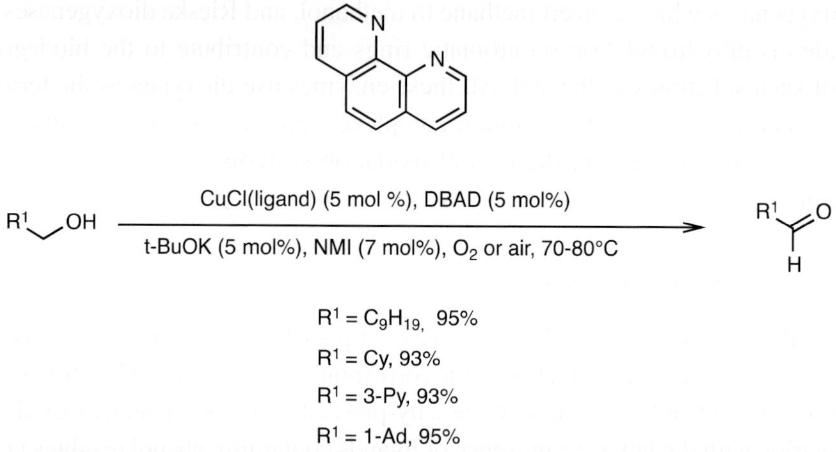


**Scheme 23.55.** Oxidation of primary alcohols to the corresponding aldehydes with $O_2$ mediated by a Cu(I) catalyst in the presence of N-methylimidazole and di-tert-butyl azodicarboxylate.

primary alcohols as substrates the reaction is extremely chemoselective and no over-oxidation of the aldehyde to the corresponding carboxylic acid is observed (Scheme 23.55).

As mentioned before, methane monoxygenases are enzymes capable of the direct oxidation of methane. Only recently, a synthetic catalyst reminiscent of the heme structure of that enzyme was introduced by Sorokin. It is based on a $\mu$-nitrido bridged di-iron phthalocyanine structure showing methane oxidation activity with $H_2O_2$ as the terminal oxidant.[107] It operates in acetonitrile with partial oxidation of the latter, but it can also work in water as a green inert medium by means of heterogenization on silica. The optimized catalytic system shows both formaldehyde and formic acid as oxygenated products, 15 turnover at room temperature and more than 430 at 60°C, demonstrating the robustness of the di-iron structure and the sustainable character of the process coming from the use of water and $H_2O_2$ as the solvent and oxidant (Scheme 23.56).

Mn-based oxidation systems have recently been developed. Burgess, via a high throughput screening, evidenced that simple $MnSO_4$ can activate hydrogen peroxide towards alkenes leading to epoxides in dimethylformamide (DMF). The key role in this case is played by bicarbonate which is converted into the corresponding percarboxylic acid which forms either a Mn(II)-peroxocarbonato or Mn(IV)=O species responsible for the oxygen transfer to the substrate.[108] The presence of simple additives such as sodium acetate (in t–BuOH) and salicylic acid (in DMF) enhanced the rate of the epoxidation reaction, enabled the use of bicarbonate in a catalytic amount, required less hydrogen peroxide and shorter addition times, and favored the epoxidation of less electron-rich alkenes (Scheme 23.57).

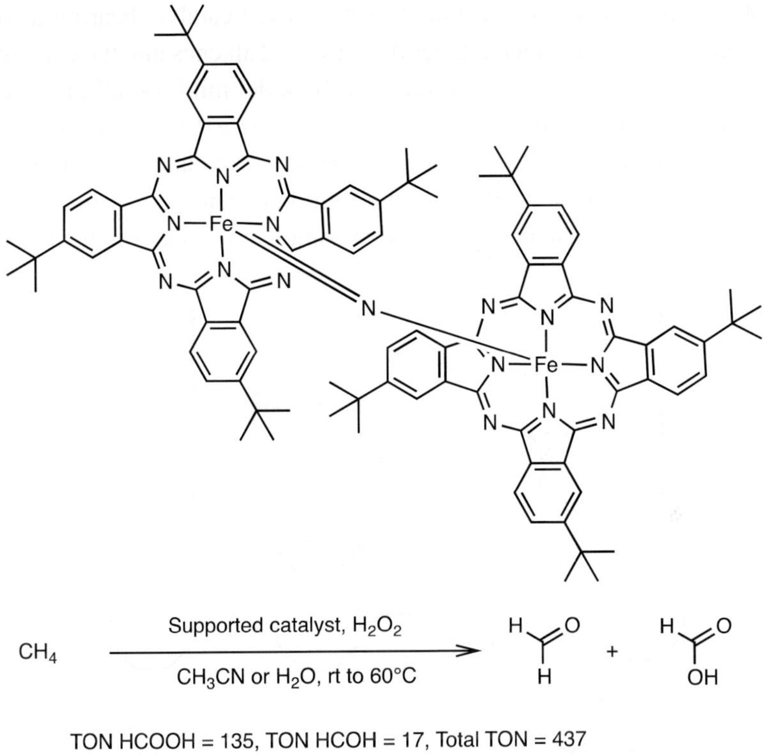


**Scheme 23.56.** Oxidation of methane to formaldehyde and formic acid with $H_2O_2$ mediated by a dimeric Fe(III) catalyst.

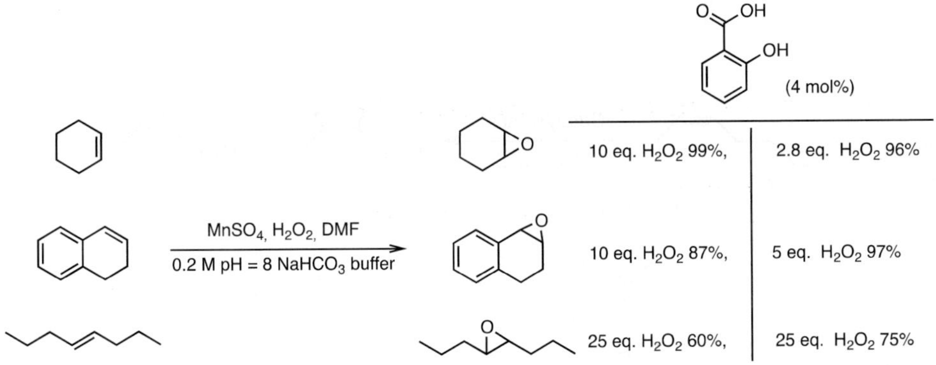


**Scheme 23.57.** Epoxidation of alkenes with $H_2O_2$ mediated by a Mn(II) catalyst in the presence of $NaHCO_3$.

In 2005, Feringa disclosed a dinuclear Mn-based catalyst bearing a trimethyl triazacyclononane ligand which efficiently converted alkenes into the corresponding epoxides and cis-diols, with $H_2O_2$ as the oxidant under mild conditions. In this case too, the presence of carboxylic acids as additives is extremely important. Their role is *in primis* to suppress the decomposition of the oxidant, a common problem with Mn species, but also to influence the ratio between the possible oxygenated products, as shown in Scheme 23.58, where complete reversal of chemoselectivity

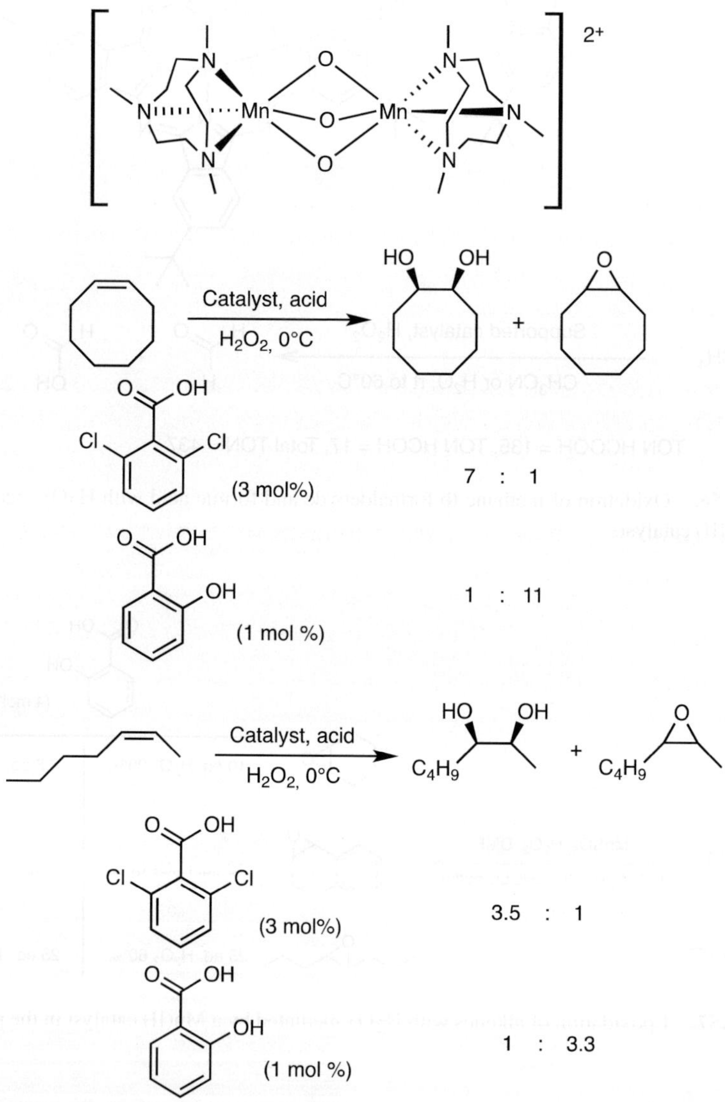


**Scheme 23.58.** Oxidation of alkenes to cis-diols and epoxides with $H_2O_2$ mediated by a dimeric Mn(III) catalyst bearing a tridentate cyclic ligand.

is possible using either 2,6-dichlorobenzoic acid which promotes diol formation or salicylic acid that favors the epoxide.[109] Overall the catalytic system is very active providing up to 2,000 turnover numbers for cis-cycloctane diol representing a real, green non-toxic alternative to Os-based oxidation systems.

### 23.5.2. *Iron complexes*

The well-known Fenton chemistry involves the oxidation process activated by Fe(II) salts in the presence of $H_2O_2$, which generates radical species in solution and oxidizes a wide range of organic substrates with high activity but generally poor selectivity. The main drawbacks of these systems are the tendency to decompose the oxidant, which implies the need for over-stoichiometric amounts of $H_2O_2$, the difficult control of the coordination geometry and the oxidation state of the metal leading to the coexistence of several parallel oxidation processes, all of which have limited the development of iron-based oxidation methods for years. On the contrary, the very low cost of this metal compared to others, in combination with its poor toxicity, support its employment in homogeneous catalysis,[110] and in oxidation processes in particular.

One landmark example was developed by Collins and Lenoir based on a Fe(III) metal center coordinated to a tetra-amido macrocyle which coordinates the metal with four anionic amide groups. The complex is highly active in the activation of $H_2O_2$ as the terminal oxidant towards the complete degradation of different species in aqueous solution such as dyes, water effluents from paper industries and resistant chlorinated phenol derivatives. The catalysts operate under $\mu$M concentrations and mild experimental conditions allowing hundreds to thousands of cycles within minutes before coming to a stop (Scheme 23.59).[111] More importantly, among the decomposition products of pentachlorophenol or trichlorophenol, dioxines were not

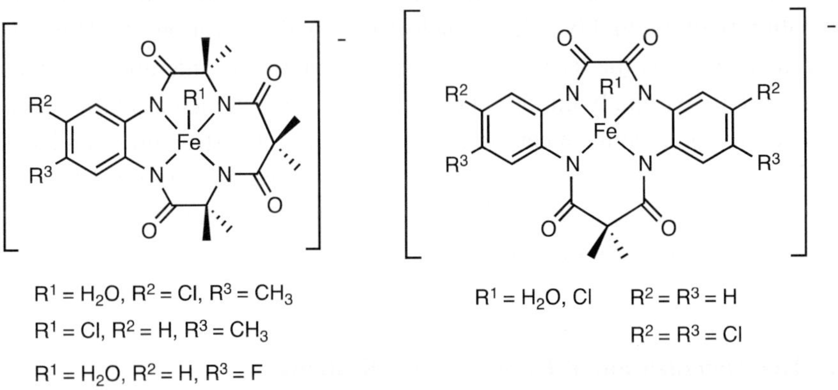


**Scheme 23.59.** Structure of Fe(III) catalysts developed for waste water treatment.

detected; instead a mixture of small biodegradable organic acids and CO, $CO_2$ and HCl were present when working under basic conditions, clearly showing the high environmental compatibility of this catalytic system which soon found practical applications. Further optimization of the ligand structure led to the development of a second generation of tetramide systems showing high activity at pH 5–8, which makes the new class of catalysts even more suitable for purifying environmental waters. The endocrine disrupting activity of the new Fe(III) tetramide complexes was checked by observing their absolute compatibility with aquatic life[112] as a further confirmation of the green character of this oxidation method.

Among the possible oxidation reactions, the direct CH selective functionalization of alkanes to oxygenated compounds is both a highly valuable and challenging transformation. White recently reported two important contributions concerning Fe(II)-mediated oxidation of tertiary and secondary CH bonds to tertiary alcohols[113] or carbonyl compounds,[114] respectively, by using $H_2O_2$ as the oxidant under mild experimental conditions. A pivotal role in this catalytic system is played by the tetradentate ligand employed bearing two pyridines and two pyrrolidines as donor atoms, whose effect is to greatly increase both the catalytic activity and the selectivity of the iron catalyst. In particular, the catalyst is more active towards tertiary CH bonds if these are sterically accessible and if no electron-withdrawing groups are in close proximity. This allows extremely regioselective functionalization of elaborated organic structures such as in the cases reported in Scheme 23.60.

In the absence of tertiary CH groups or if these do not meet the above described criteria, the catalytic system operates on secondary methylene residues with high selectivity, again with preference for $CH_2$ groups that are sterically accessible and with greater conversion if electron-activating groups such as cyclopropyl or ether moieties are present (Scheme 23.60).

Overall, despite the high catalyst loading required and the generally low turnover number of this Fe(II) catalytic system, the use of $H_2O_2$ under mild conditions, in combination with the highly valuable selective functionalization of CH bonds and high predictability on the basis of electronic and steric considerations, makes this catalytic system extremely important and valuable. In fact, the possibility to specifically insert oxygen atoms in certain positions in complex structures at a late stage of synthesis represents an alternative streamline approach compared to classical synthesis, thereby reducing unproductive chemical manipulations associated with carrying them through a sequence.

## 23.6. Heterogenization of Homogeneous Systems

The quest for separation and recycling has always been a key issue in homogeneous catalysis in general. This is why methods have been devised to anchor the

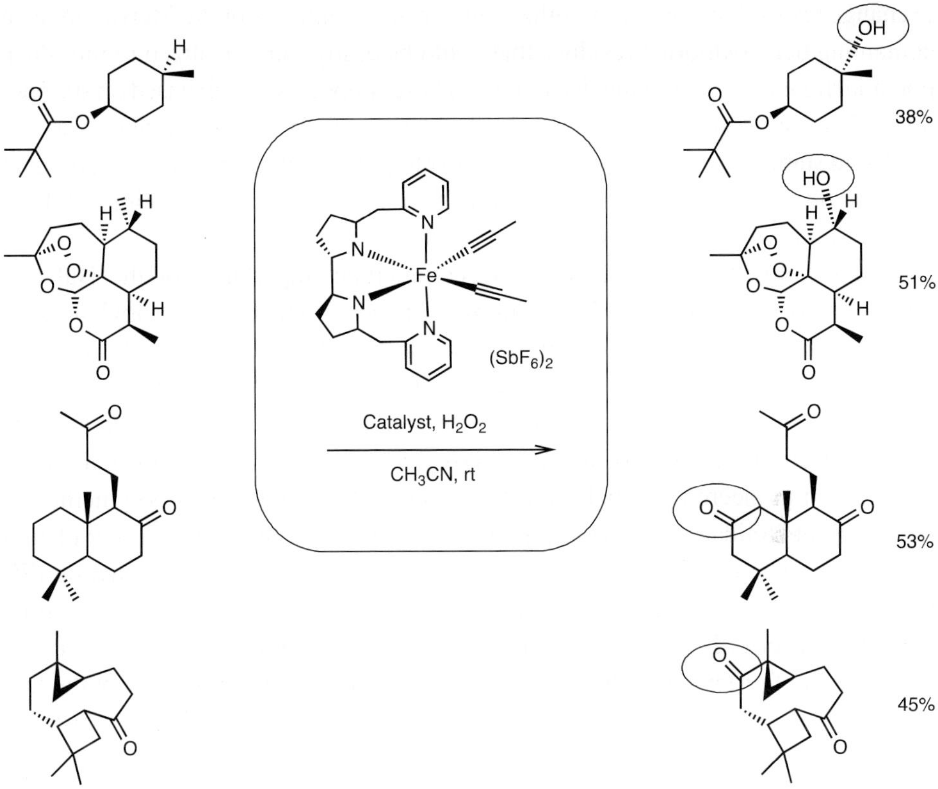


**Scheme 23.60.** Selective CH oxidation of complex molecules with $H_2O_2$ mediated by a Fe(II) catalyst bearing a tetradentate N ligand.

most interesting soluble catalysts to suitable solid phases. The possibility of easily separating the catalyst from the reaction mixture without harming its catalytic properties (activity, selectivity) is an important step forward towards possible commercialization, increasing the overall exploitable turnover number, minimizing metal contamination of the products, simplifying the process engineering, saving energy, etc., and generally moving towards a more sensible and sustainable use of costly materials and primary resources. Over the years the topic has been reviewed several times, so throughout this section, reference to these reviews will be provided, while only some selected examples taken from those among the most interesting and/or recent papers will be overviewed, emphasizing those aspects that are most relevant to sustainability.

### 23.6.1. *Polymer-based systems*

The use of commercial polymers to immobilize homogeneous catalytic species dates back to the early 1970s. Initial supports were in most cases commercial ion exchange resins that could immobilize ionic metal complexes or the Merrifield resin containing benzylchloride residues that could be easily functionalized to introduce potential ligands to anchor metal catalysts. These supports were all based on styrene-divinylbenzene copolymers. The major limitation of these early systems was the moderate stability of the catalysts upon recycling due to the leaching of significant amounts of metal into solution in the majority of cases. Critical evaluations of these early systems were reported several times.[115]

In the field of oxidation, given the commercial importance of the Halcon–Arco process using soluble Mo complexes for the epoxidation of propylene with t-BuOOH, epoxidation has been by far the most studied reaction. Early attempts in this area focused on the immobilization of soluble $Mo^{VI}$ catalysts on polymer supports by anion or cation exchange, or by modifying the polymer with suitable functional groups capable of acting as possible ligands towards molybdenum.[116] Sherrington pioneered this field developing aminated polystyrene, polymethacrylate, polybenzimidazole (PBI) and polysiloxane to immobilize $Mo^{VI}$ epoxidation catalysts.[117] The most successful catalyst (PBI.Mo) was obtained by reacting PBI with $MoO_2(acac)_2$ (Scheme 23.61) and proved to be highly active in generating the active species. It showed a remarkable aging behavior over 10 recycles in the epoxidation of propylene with the yield of epoxide increasing from 59% at the first

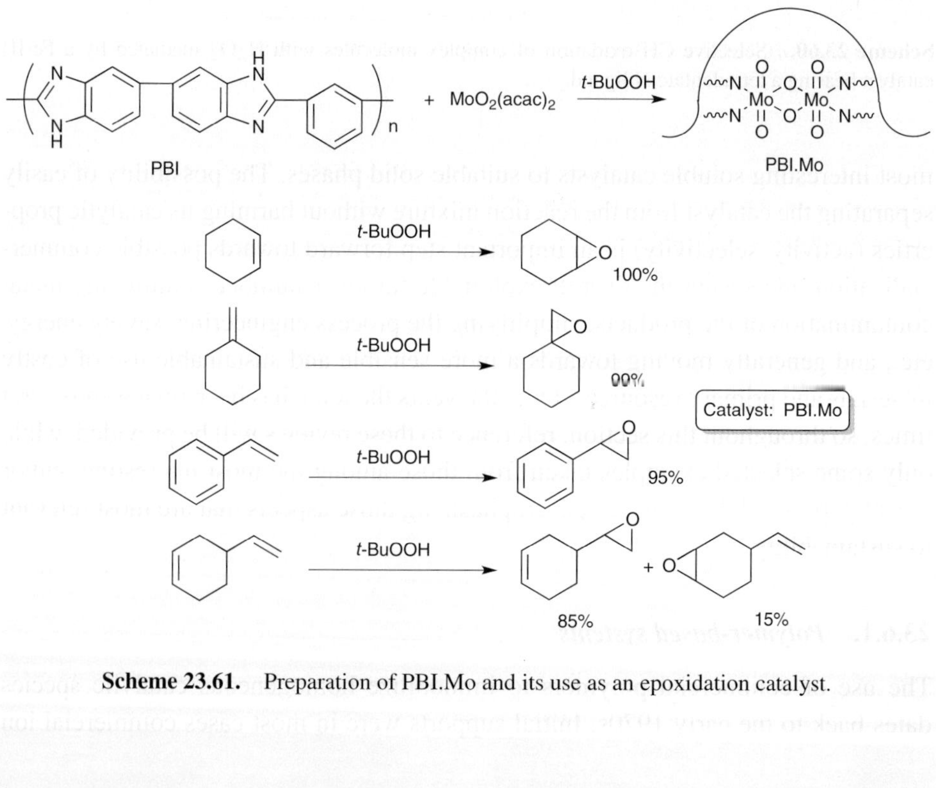


**Scheme 23.61.** Preparation of PBI.Mo and its use as an epoxidation catalyst.

cycle to 99.8% at the tenth. Furthermore, apart from the first catalytic run, there was negligible leaching of Mo from PBI.Mo. It also proved to be very useful in the epoxidation of a range of alkenes (Scheme 23.61).

Using cyclohexene as a model substrate and TBHP as the oxidant, Cu, Mn, Fe, Ru and Ti have all been shown to be catalytically active when supported on PBI. PBI.Cu, PBI.Mn and PBI.Fe are all potentially useful allylic oxidation catalysts; PBI.Ru offers an opportunity in dihydroxylation, and perhaps not surprisingly, PBI.Ti shows significant selectivity towards epoxidation.[118]

Attempts to switch to more appealing oxidants, such as hydrogen peroxide, by immobilizing simple tungstate anions on poly(methacrylate)-based aminophosphorilated resins to carry out epoxidation were only moderately successful because of significant metal leaching into solution with consequent loss of activity.[119]

Remarkably stable Mo catalysts have been obtained with the use of epoxy resins for their immobilization.[120] The epoxy monomers are shown in Scheme 23.62. Tetraglycidyl-4,4′-methylenedianiline (TGMDA) was treated with either a series of metal acetylacetonates ($TiO(acac)_2$, $VO(acac)_2$, $MoO_2(acac)_2$) or metal alkoxides ($Ti(OiPr)_4$, $VO(OiPr)_3$, $Mo(OEt)_5$). Metal contents of 1.5 wt% were adjusted by dispersing the corresponding amount of metal complex in the resin. The resins were cured in aluminum molds at temperatures between 120 and 230°C and thin plates with about 1 mm thickness were obtained. The thermosets thus obtained were tested as epoxidation catalysts. Outstanding long-term catalyst activities and selectivities were observed for Mo-containing resins in the epoxidation of cyclohexene with tert-butyl hydroperoxide. Mechanistic investigations indicate the true heterogeneity of the system with an extremely low metal leaching that depends on the catalyst preparation mode. The catalysts were employed batchwise in up to 60 consecutive runs exceeding an application period of 50 days. They were applied without any reconditioning or loss of activity, thus offering promising perspectives for industrial applications in continuously operating processes.

Methyltrioxorhenium (MTO) has perhaps been the most successful homogeneous catalyst for the epoxidation of simple olefins with hydrogen peroxide,[121] however, attempts to modify the catalyst to achieve enantioselective epoxidation or anchoring to a solid support have met with only moderate success, probably because

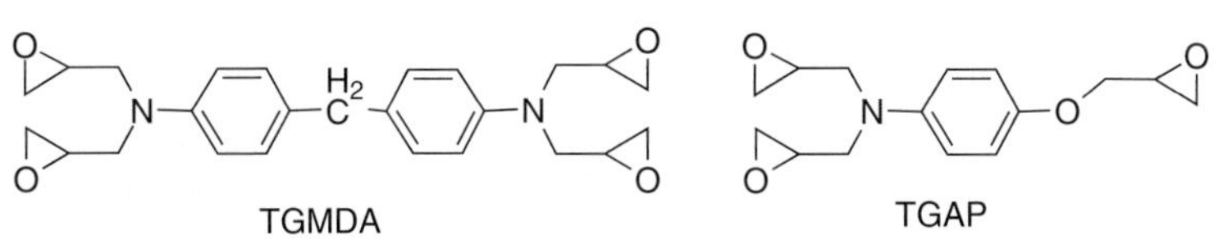


**Scheme 23.62.** Monomers used for the preparation of epoxy resins to be used as catalyst supports.

these modifications change the electronic properties of the metal center generally decreasing its catalytic activity.

Heterogenization could be achieved using either simple poly-4-vinylpyridine polymers,[122] exploiting the donor porperties of pyridine to create a six-coordinate surface species or simple encapsulation in polystyrene. Both heterogenized catalysts could be used for the conversion of monoterpenes such as carene, limonene, geraniol and nerol to their corresponding epoxides using $H_2O_2$ or its urea adduct (UHP) as the oxygen atom donors.[122] Heterogeneous catalysts were stable systems for at least four recycling experiments. The oxidation of geraniol and nerol proceeded selectively at the more electron-rich 6,7-double bond in accordance with the electrophilic character of the oxidant. A comparison between molecular solvents and ionic liquids as reaction media showed that the latter enhance the reaction rates and improve the regioselectivity of epoxidation.

Outside epoxidation a significant contribution was reported on the Baeyer–Villiger oxidation of ketones with hydrogen peroxide catalyzed by $[(P\text{-}P)Pt(\mu\text{-}OH)]_2^{2+}$ cationic complexes immobilized on ordinary sulfonated styrene-DVB anionic resins (Scheme 23.63) via simple electrostatic interaction.[123]

Polymer swelling was found to be a critical factor and EtOH was the best solvent to ensure a better mixing of reactants (ketone and hydrogen peroxide) in proximity to the active sites. In any case the activity observed was lower than the same system used in solution.

The latter example emphasizes a major problem encountered when using organic polymers to heterogenize homogeneous systems. While easy to modify chemically, in order to construct the proper bonding interaction between solid phase and soluble

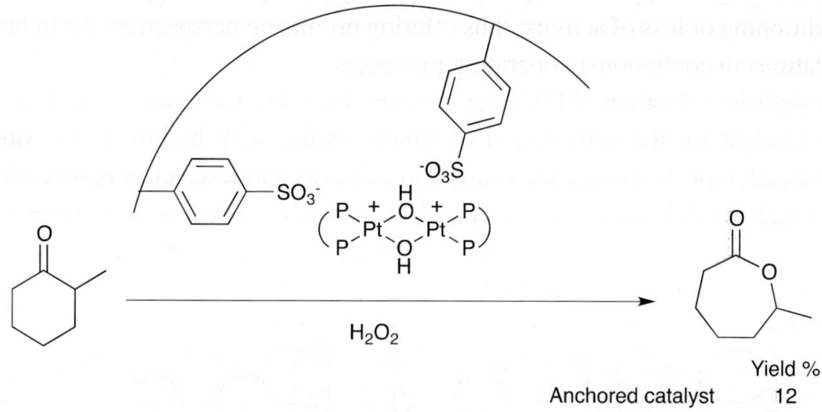


| | Yield % |
|---|---|
| Anchored catalyst | 12 |
| Free catalyst | 65 |

**Scheme 23.63.** Platinum complex anchored on an ion exchange resin and its use as a BV oxidation catalyst.

catalyst, insoluble organic polymers can introduce severe diffusion problems related to the swelling of the organic matrix that hamper the reaction under study, because a good solvent for the polymer may be a bad one for the reaction and *vice versa.*

### 23.6.2. *Silica-based systems*

The problems associated with organic polymers account for why, in the past 10–15 years, major attention has been paid to mesoporous inorganic oxides (especially ordered silicas) that do not suffer from such limitations, and plenty of work has been devoted to organic-inorganic nanocomposites obtained by surface modification. Grafting has been the method of choice to introduce into physically inert inorganic oxides with suitable porosity, additional properties such as (i) the functional groups necessary to bind the complex catalyst component; (ii) the desired philicity (inorganic oxides are generally hydrophilic) to facilitate reactant diffusion and (iii) the ability to maintain single-site catalysts preventing undesired dimerization reactions that may lead to deactivation. As an alternative method for heterogenizing homogeneous catalysts, encapsulation of large soluble complexes into the cages of microporous oxides e.g. zeolites has met with considerable success. In general, if the bonding interaction between the solid support and catalyst is sufficiently strong and no ligand dissociation is involved, or if the size of the support pores is smaller than the diameter of the catalyst (ship-in-the-bottle concept), then the major problem of catalyst leaching can be avoided.[124]

Again epoxidation remains the most investigated reaction in this area with a dominant role for Mo- and W-based catalysts. The use of heterogenized complexes in epoxidation has been recently reviewed.[125] Because of the molecular simplicity of the Mo species involved in Halcon chemistry, early studies exploited the ligand ability of surface hydroxyls present on inorganic oxides such as MCM-41 to bind simple precursors such as $MoO_2Cl_2(solvent)_2$ to produce single-site catalysts capable of epoxidizing alkenes such as cyclooctene, 1-octene, norbornene, styrene, etc., with t-BuOOH obtaining moderate to good yields.[126] However, the grafted species are essentially atomically dispersed Mo oxides with a pronounced tendency to coalesce into clusters during operation, thus leading to loss of activity and metal leaching into solution.

The creation of a robust metal ligand interaction has been the preferred strategy to prevent metal leaching. As a typical example worthy of note, recent work by Masteri-Farahani reported the synthetic approach shown in Scheme 23.64 for the synthesis of Mo complexes on functionalized MCM-41 materials.[127]

As can be seen, stepwise synthesis of robust chelating ligands allows a strong anchoring of the metal center on the silica surface. Studies on the epoxidation of different olefins with t-BuOOH as the oxygen source indicated the following reactivity

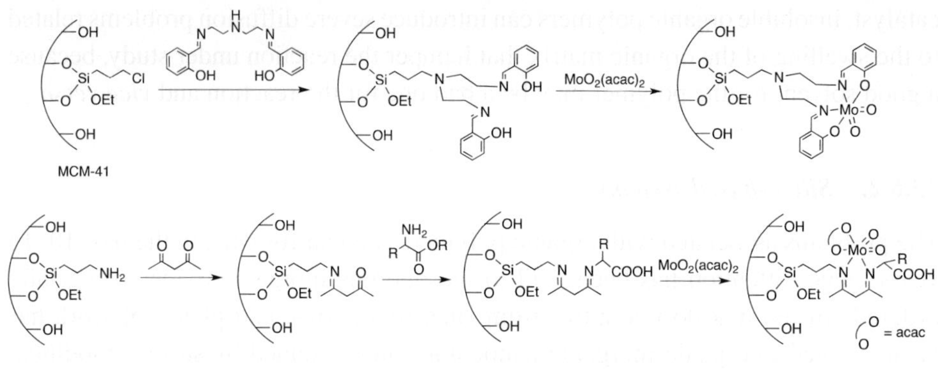


**Scheme 23.64.** Synthesis of Mo(VI) complexes anchored on MCM-41 to be used as catalysts for the epoxidation of alkenes.

order: cyclooctene > cyclohexene > 1-hexene > 1-octene and the catalyst could be recycled without loss of activity and selectivity after hot filtration,[128] indicating the heterogeneous nature of the reaction.

Cyclopentadienyl molybdenum carbonyl complexes are stable precursors to achieve (Cp)Mo(O)$_2$Cl by *in situ* oxidation with t-BuOOH. The latter complexes are perhaps the most active catalysts for the epoxidation of alkenes with turnover frequencies (TOF) as high as 21,000 h$^{-1}$ in the case of cyclooctene.[129] Kühn and co-workers reported the direct grafting of (Cp)Mo(CO)$_3$Cl on mesoporous MCM-41, MCM-48 and their aluminum-substituted analogues by simple exchange of the chloro ligand with the surface hydroxyls. These materials were excellent catalyst precursors for the selective epoxidation of cyclooctene with t-BuOOH, with TOFs exceeding 4,000 h$^{-1}$, similar to the homogeneous system under identical conditions, thus demonstrating the absence of diffusional limitations induced by the mesoporous material.[130] The observed higher yield and activity in the case of Al-containing mesoporous molecular sieves was attributed either to the higher Lewis acidity of Mo in these systems or to the activation of t-BuOOH by the Al sites.[131] After reaction, used catalysts were washed several times with dichloromethane to remove the physisorbed molecules (coke) and reused. Catalysts were found to be quite active even after four catalytic runs. However, the catalytic activities decreased in all examined cases to about two thirds of the original activities after four runs. The observed activity decrease was attributed — at least partially — to an increasing number of chemisorbed organic molecules on the surface of the materials. Leaching was proved to be negligible demonstrating the true heterogeneity of the system.

The same authors grafted (Cp)Mo(CO)$_3$ moieties onto MCM-41/48, starting from complexes functionalized with trialkoxysilane groups connected to the Mo center via a hydrocarbon chain (Scheme 23.65).[132] The new materials retained the original mesoporosity of the siliceous supports and proved to be active and robust

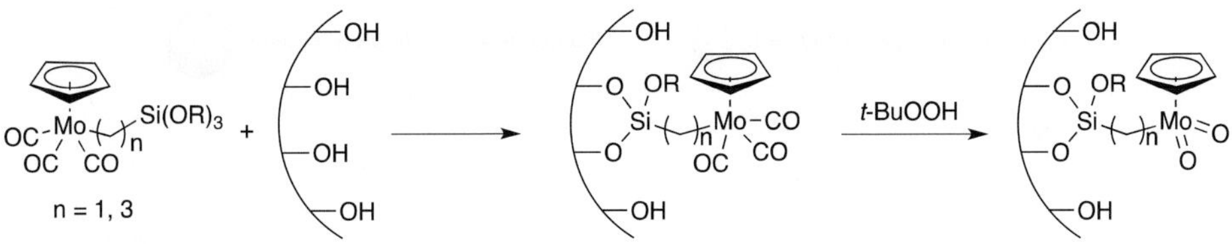

**Scheme 23.65.** Grafting and activation of catalytically active Mo complexes.

catalysts for the selective epoxidation of cyclooctene with t-BuOOH (TOF: 2,900–10,200 $h^{-1}$, selectivity: 73–100%). The formation of the Mo=O species upon treatment with t-BuOOH was confirmed by Fourier transform infrared spectroscopy (FTIR) analysis. It was found that the material containing the longer chain between molybdenum and the silica surface displayed higher catalytic activity, because the Lewis acidity of the Mo center is less influenced by the donor ability of the surface. The catalyst can be successfully recycled four times, although again some loss of activity was experienced and attributed to coke formation.

Polyoxometalates (POMs) constitute a large family of anionic metal-oxygen clusters of early transition metals and have stimulated many current research activities in many fields of science such as catalysis, materials and medicine, because their chemical properties such as redox potentials, acidities and solubilities in various media can be finely tuned by choosing constituent elements and counter cations. In addition, POMs are thermally and oxidatively stable in comparison to organometallic complexes and enzymes. The use of POMs in oxidation catalysis has received much attention and has been reviewed several times.[133]

Recently, Mizuno and co-workers have successfully developed highly atom efficient, catalytic green oxidation systems with $H_2O_2$ or $O_2$ as the oxidants by precise design and synthesis of novel POM-based molecular catalysts.[133] Hybrid supports have been synthesized by covalently anchoring N-octyldihydroimidazolium cation fragments onto $SiO_2$ and the catalytically active peroxotungstate and POM anions, such as A and B, can be immobilized on the modified support via simple anion exchange (Scheme 23.66).[134]

The structure of polyanions was preserved after the anion exchange process. The supported catalysts A/Im–$SiO_2$ and B/Im–$SiO_2$ showed high catalytic performance for epoxidation and sulfoxidation with $H_2O_2$.[134] The catalyst A/Im–$SiO_2$ showed high yields (>90%) in the conversion of a variety of alkenes and allylic alcohols to the corresponding epoxides, and its activity could be compared with that of the corresponding homogeneous n-dodecyltrimethylammonium salt of A under the same conditions.[135] Similarly, B/Im–$SiO_2$ also proved very active towards the epoxidation of different alkenes and in the sulfoxidation of a variety of substituted thioanisoles.[135] The catalyzed epoxidation of 3-methyl-1-cyclohexene was highly

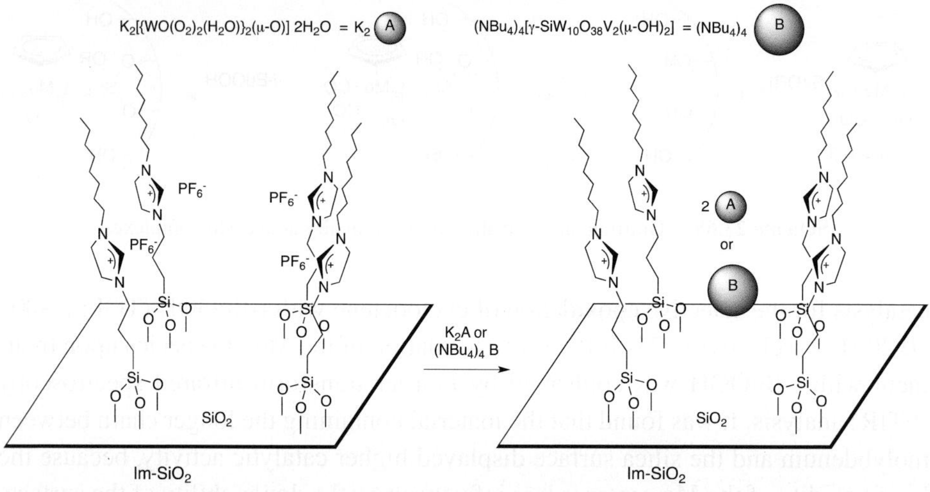


**Scheme 23.66.** Immobilization of POM catalysts onto hybrid organically-modified silica.

diastereoselective and gave the corresponding epoxide with the oxirane ring trans to the substituent (anti-configuration). Furthermore, the epoxidation of trans-1,4-hexadiene gave only 1,2-epoxide. These unusual diastereo- and regioselectivities were very close to those for the homogeneous epoxidation.[135] These facts indicate that POM-based homogeneous molecular catalysts can be heterogenized with retention of their intrinsic catalytic performance (reaction rates, chemo-, regio- and diastereoselectivities). After filtration, no POM species could be found in the filtrate ruling out any contribution from leached species. In addition, the catalysts could be reused several times without loss of their catalytic performance.

A still unanswered problem with POMs is related to the inherent acidity of these catalysts and the use of aqueous $H_2O_2$ as the oxidant (the commercial solution is acid stabilized), as they may cause hydrolysis of the epoxide. Many approaches such as the addition of basic pH-adjusting additives,[136] insufficient amount of aqueous $H_2O_2$,[136] the use of phase transfer catalysts[136,137] and organic solvents[136,137,138] have been applied to overcome this problem. Given these limitations, the development of heterogeneous POM catalysts that can suppress the acid-catalyzed hydrolysis of epoxide without the need for any additives has long been a challenge.

Recently Liu *et al.* reported[139] on a series of self-assembled polyoxotungstates modified with heteroatoms such as Zn, Co, Mn, that were directly immobilized into hydrotalcite-like layered double hydroxides (LDH) by a selective ion exchange method. Sandwich-type POM species were found to be more favorable than Keggin-type POM for the direct immobilization in LDH, because strict pH control was not needed and the LDH hosts can be kept intact. The resulting LDH–POM catalysts

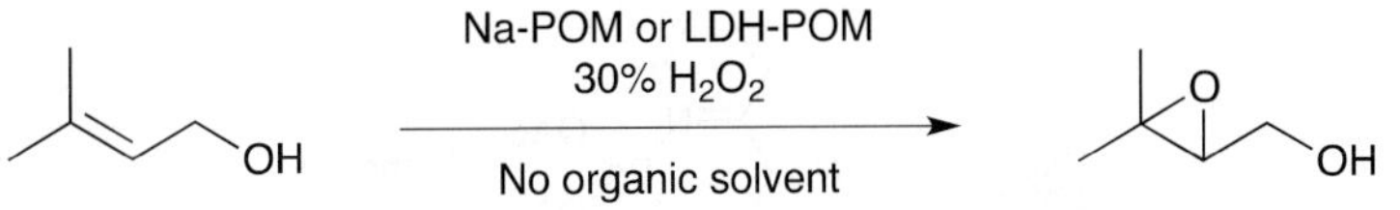


| | Conv.(%) | Sel. (%) |
|---|---|---|
| Na-POM | 92–98 | 15–27 |
| LDH-POM | 85–95 | 81–98 |

**Scheme 23.67.** Epoxidation of allylic alcohols with free and hydrotalcite-immobilized POM catalysts.

were evaluated in the epoxidation of allylic alcohols with aqueous $H_2O_2$ as the oxidant without using any organic solvents. The heterogeneous LDH–POM catalysts show much higher epoxide selectivity than the corresponding homogeneous Na–POM catalysts (Scheme 23.67), which was attributed to the beneficial effect of the basic LDH host suppressing the acid-catalyzed epoxide hydrolysis. The cooperation between the POM guest and the LDH host can lead to 98% epoxide selectivity, 95% $H_2O_2$ efficiency and 37,200 $h^{-1}$ TOF without the need of base additives and pH control, and the host–guest catalysts can be readily recycled with no apparent loss of catalytic performance.

The selective oxidation of alcohols to aldehydes and ketones is a highly desirable and much sought after transformation, in both industrial chemistry and organic synthesis, due to the wide-ranging utility of these products as important precursors and intermediates for many drugs, vitamins and fragrances.[140] A palladium catalyst covalently anchored onto the surface of silica gel for the selective oxidation of alcohols to carbonyl compounds has been reported by the group of Clark.[141] This is a quite efficient and reusable supported palladium catalyst for the selective oxidation of alcohols to carbonyl compounds using molecular oxygen (Scheme 23.68). This catalyst can be used for the oxidation of primary benzylic alcohols using atmospheric air, but is less effective for aliphatic and secondary alcohols. Although it requires relatively long reaction times (4–14 h), it was quite stable and could be recycled at least three times without appreciable loss of activity.

An improved series of catalysts bearing N–N, N–S and N–O chelating ligands has been reported by the same group.[142] The most active catalyst (Scheme 23.69) was used for the oxidation of a series of primary and secondary alcohols to carbonyl compounds. The reactivity was improved with respect to the previous example and no significant differences were observed between primary and secondary alcohols. The catalyst is highly selective and no over-oxidation products were detected. Catalyst 1 in Scheme 23.69 can also be used for the oxidation of primary

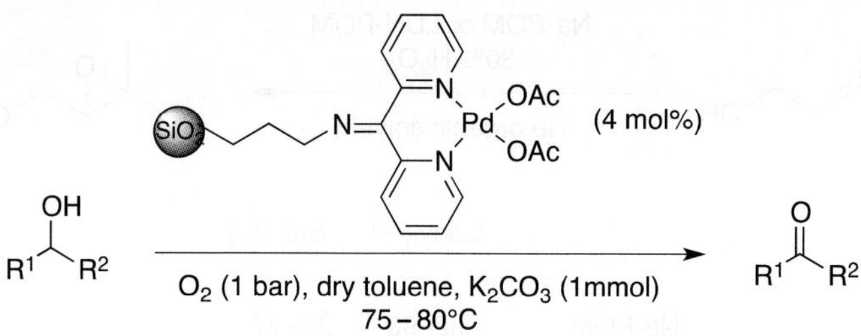


| $R^1$ | $R^2$ | Time (hr) | Yield (%) |
|---|---|---|---|
| Ph | H | 8 | 100 |
| Ph | Et | 12 | 100 |
| 4-Me-$C_6H_4$ | H | 5 | 100 |

**Scheme 23.68.** Oxidation of alcohols with a silica-anchored Pd catalyst.

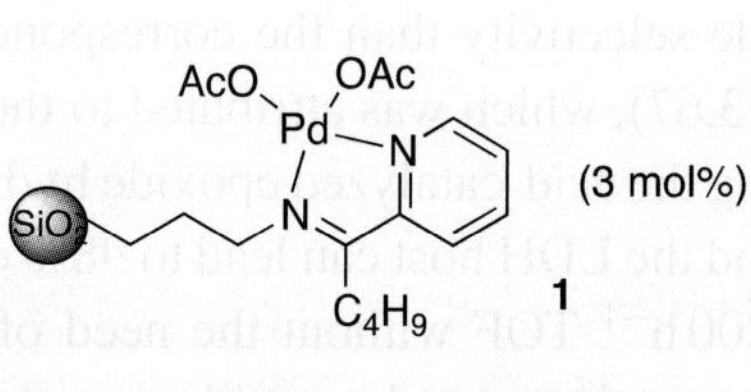


| Alcohol (1 mmol) | Oxidant (1 bar) | Time (hr) | Yield (%) |
|---|---|---|---|
| Benzyl alcohol | $O_2$ | 2.5 | 95 |
| 2-Butanol | $O_2$ | 5 | 82 |
| Benzyl alcohol | Air | 7 | 100 |

**Scheme 23.69.** Evolution of the catalyst reported in Scheme 23.68.

benzyl alcohols to carbonyl compounds with comparable yields in the presence of atmospheric air.

The selective oxidation of alcohols to carbonyl compounds has recently been achieved through the use of the stable nitroxyl radical TEMPO, in combination with transition metals such as Ru[143] and Cu[144] using molecular oxygen or air under mild reaction conditions. Toy and co-workers have developed an attractive multipolymer reaction system for the aerobic oxidation of alcohols using poly(ethylene glycol) monomethyl ether (MPEG)-supported TEMPO and MPEG-supported bipyridine to bind $Cu^{II}$. Both TEMPO and the $Cu^{II}$ complex could be recovered at the same time and recycled.[145] However, this method is only effective for the aerobic oxidation of highly active primary benzylic alcohols after prolonged heating (18 h) in $CH_3CN$-$H_2O$ solvent mixtures.

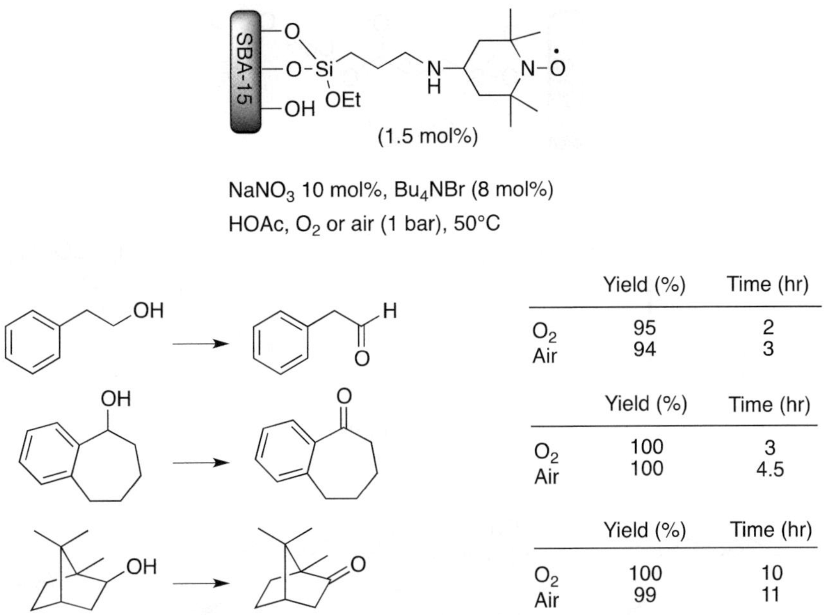


| | Yield (%) | Time (hr) |
|---|---|---|
| $O_2$ | 95 | 2 |
| Air | 94 | 3 |

| | Yield (%) | Time (hr) |
|---|---|---|
| $O_2$ | 100 | 3 |
| Air | 100 | 4.5 |

| | Yield (%) | Time (hr) |
|---|---|---|
| $O_2$ | 100 | 10 |
| Air | 99 | 11 |

**Scheme 23.70.** Oxidation of alcohols with air using a SBA-15-supported TEMPO catalyst.

More recently Clark and co-workers have reported the use of a stable version of supported TEMPO grafted on SBA-15 as an efficient and selective heterogeneous catalyst for the aerobic oxidation of a wide range of primary, secondary, and even sterically-hindered alcohols (Scheme 23.70) in the absence of any transition-metal-containing co-catalyst.[146] Air can be conveniently used instead of pure oxygen without affecting the efficiency of the reaction. Benzyl alcohol was tested in a catalyst recycling experiment and, over 14 subsequent runs, a total turnover number (TON) of over 1,00 was observed. Detailed spectroscopic analyses of the catalyst, prior to and after, the 14th reaction cycle showed that neither the anchored organic group nor the nanometer-scale voids and channels of the parent SBA-15 are significantly affected by prolonged use in the reaction system.

The same group also reported the selective allylic oxidation of unsaturated steroids and valencene by a series of metal complexes based on Co(II), Cu(II), Mn(II) and V(II) immobilized on mesoporous silica using a synthetic strategy similar to that reported in Scheme 23.71. The catalysts could be recycled with only minor loss of activity.[147]

With a similar synthetic approach, a cobalt (II) salen complex catalyst has been supported on silica and successfully employed in the aerobic oxidation of alkyl aromatics at atmospheric pressure in the presence of N-hydroxyphthalimide (NHPI). The reaction is particularly selective for the oxidation of the benzylic $CH_2$ group and the major product obtained was ketone (Scheme 23.72). The immobilized catalyst

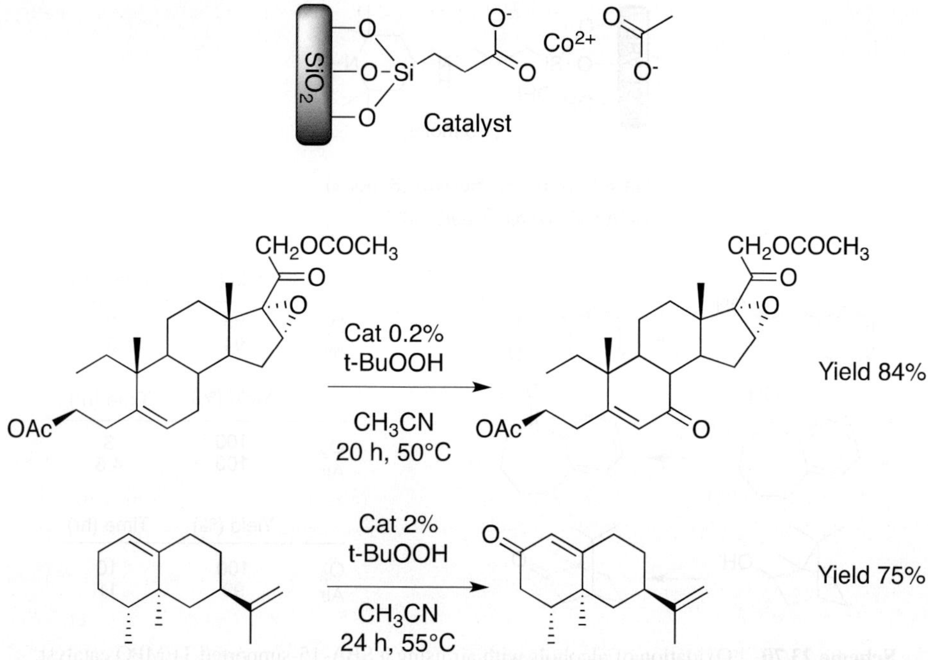


**Scheme 23.71.** Use of a silica-anchored Co catalyst for the allylic oxidation of an unsaturated steroid and valencene.

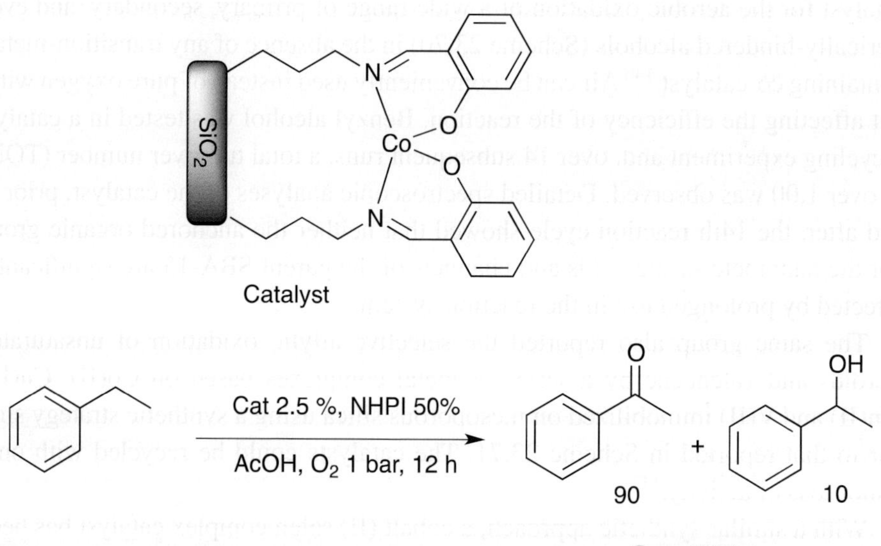


**Scheme 23.72.** A silica-immobilized Co(salen) complex and its use as a catalyst for the oxidation of ethylbenzene.

can be easily recovered and reused for at least four reaction cycles without any significant loss of its catalytic activity.[148]

### 23.6.3. *Ship-in-the-bottle systems*

Koldeeva and co-workers recently reported the encapsulation of phosphotungstates $[PW_4O_{24}]^{3-}$ $[PW_{12}O_{40}]^{3-}$ and their titanium- and cobalt-monosubstituted heteropolyanions $[PW_{11}CoO_{39}]^{5-}$ and $[PW_{11}TiO_{40}]^{5-}$ into the cages of the chromium terephthalate polymer matrix MIL-101 via electrostatic binding.[149] MIL-101 is a metal organic framework, zeolite-type innovative material with a large pore size (14.5 x 16 Å hexagonal windows) and very high surface area (4,500–5,500 $m^2/g$) obtained from terephthalic acid and chromium nitrate.[150] The MIL-supported polyoxometalate catalysts (Scheme 23.73) were characterized by elemental analysis, x-ray diffraction XRD, $N_2$ adsorption and FTIR spectroscopy. The catalytic performance of both MIL-101 and the novel composite materials M-POM/MIL-101 (M = Ti, Co) was assessed in the oxidation of three representative alkenes —$\alpha$-pinene, caryophyllene and cyclohexene — using molecular oxygen and aqueous hydrogen peroxide as oxidants, observing poor to moderate selectivities in allylic oxidation and epoxidation. On the other hand, operating under mild conditions ($[H_2O_2]$ = 0.1–0.2 M, 50°C, MeCN), the unmodified phosphotungstates ($PW_4$, $PW_{12}$) demonstrated fairly good catalytic activities and selectivities in the epoxidation of various alkenes (3-carene, limonene, R-pinene, cyclohexene, cyclooctene,

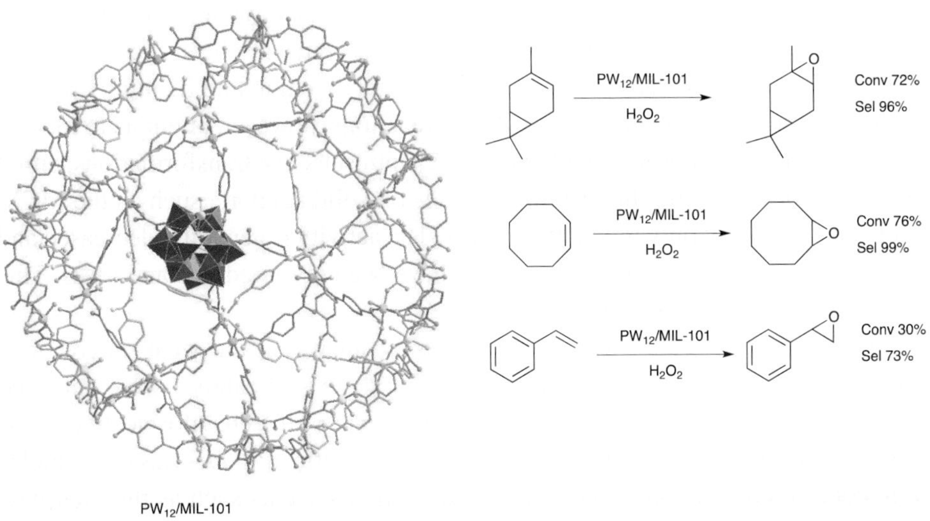


**Scheme 23.73.** MIL-101-encapsulated POM catalyst and its use in the epoxidation of alkenes.

1-octene), with conversions close to the corresponding ones achieved with homogeneous $PW_x$ (Scheme 23.73).

For the oxidation of substrates with aromatic groups (styrene, cis- and trans-stilbenes), a higher level of olefin conversion was attained using $PW_{12}$/MIL-101. Moreover, confinement of $PW_{12}$ within MIL-101 nanocages allowed the realization of higher epoxide selectivities at higher alkene conversions. The hybrid $PW_x$/MIL-101 materials were stable to leaching, behaved as true heterogeneous catalysts, were easily recovered by filtration and were reused several times without sustaining a loss of activity or selectivity.

The ship-in-the-bottle approach to heterogenization just exemplified has also been frequently investigated in oxidation. Two recent papers reported the use of complexes of Fe(III), Mn(III), V(VI), Cu(II), Ni(II) modified with salen-type ligands entrapped into the cages of zeolite Y.[151,152] Typically, the complexes were formed *in situ* by stepwise introduction of the different components (a suitable transition metal precursor, followed by the ligand or *vice versa*) so that the assembly of the small components into a complex larger than the zeolite pores can take place in the relatively large cages of the support. Catalysts were tested in the epoxidation of styrene, but only with t-BuOOH as the oxidant; moderate to good selectivities were observed (60–71% in the case of the V and Cu derivative).[151] In the oxidation of cyclohexane, cyclohexanol and cyclohexanone were observed with excellent combined yields (>90%), but high TOFs (220 $h^{-1}$) were detected only in the case of the Cu derivative with hydrogen peroxide as the oxidant. The oxidation of thioanisol with hydrogen peroxide yielded 93% conversion with 96% selectivity to sulfoxide in the presence of the VO(salen)/Y catalyst.[151]

The implementation of enantioselectivity issues in heterogenized oxidation catalysts has been a key strategy for a number of years which has been successful only with certain classes of catalysts. For example, molybdenum and rhenium catalysts are among the most efficient in epoxidation reactions, however, attempts to modify them with chiral ligands in order to achieve enantioselective transformations, either in solution or anchoring the resulting species on solid supports such as the MCM family, has met with only moderate success. The activity is generally decreased and only poor to moderate enantioselectivities have been observed in some cases. This topic has been recently reviewed.[153]

In 2004 Can Li published a seminal review in which the weak interactions between the surface of the support and the heterogenized catalyst were analyzed with respect to their influence on the transition state of the enantioselective transformation, exploiting the concept of the confinement effect.[154] A series of examples were considered including a variety of epoxidation systems such as the Sharpless Ti catalyst for the epoxidation of allylic alcohols or the Jacobsen family of catalysts for the epoxidation of simple alkenes, all prepared by grafting, encapsulation and intercalation. The confinement effect has been initially proposed by Thomas *et al.*[155] to explain some surface effects on enantioselectivity in heterogeneous asymmetric hydrogenation, and subsequently by Hutchings *et al.*[156] in the asymmetric aziridination of styrene and the hydrogenation of carbonyl- and imino-ene compounds. The confinement effect is a consequence of the change in transition states induced by weak interaction in pores or on surfaces, such as hydrogen bonding, physical adsorption and electronic interaction, in addition to the chemical bonding between the catalysts and the surfaces. It may increase the enantioselectivity (positive effect) or decrease the enantioselectivity (negative effect) depending on how the interaction changes the transition states of the chiral products.

Li and co-workers reported that chiral Mn(salen) complexes can be successfully immobilized onto inorganic mesoporous materials such as activated silica (pore size 9.7 nm with sharp pore distribution), SBA-15 (pore size 7.6 nm) and two MCM-41 (pore sizes 2.7 and 1.6 nm) via a phenyl sulfonic group, and that these catalysts have higher ee values than the corresponding homogeneous ones for the asymmetric epoxidation of some olefins.[157] The anchoring phenyl sulfonic group is bound to silica through chains of different length and the influence of the latter on the reaction performance was investigated comparing the behavior of immobilized chiral Mn(salen) complexes either into the nanopores or onto the external surface of mesoporous materials (Scheme 23.74).

Increasing the linkage length inside the nanopores caused an increase in the ee values (e.g. ee from 14% to 64%) but had practically no effect on the external

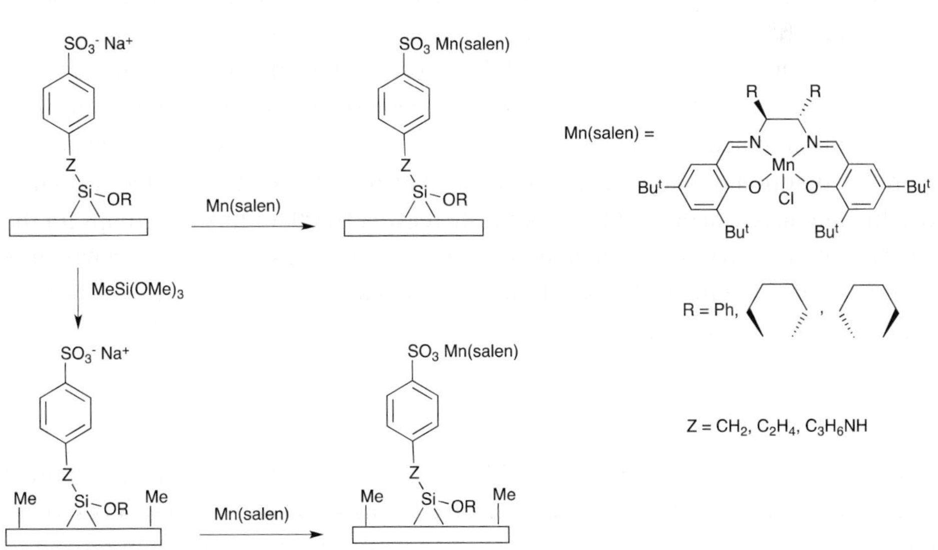


**Scheme 23.74.** Anchoring/encapsulation of a chiral Mn(salen) complex on a mesoporous silica support.

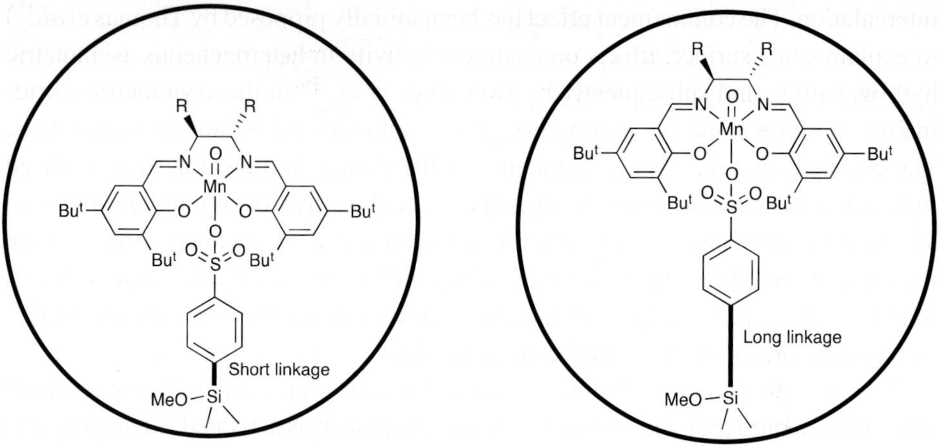


**Scheme 23.75.** Steric effect on a Mn(salen) complex covalently immobilized on a porous support as a function of the length of the tether.

surface (e.g. 47% to 44%). When comparing the same catalyst inside the pores or anchored on the external surface ee values also decreased (e.g. 58% vs 40%). The modification of the nanopore surface by methyl groups further improves the reaction performance for the asymmetric epoxidation (e.g. ee 62% non-methylated vs 70% methylated). In the epoxidation of 1-phenyl-cyclohexene the effect of pore size is to decrease the conversion (86% to 71% on going from activated silica (9.7 nm) to MCM-41 (2.7 nm)) while increasing the enantioselectivity (15% to 25% for the same supports).

As illustrated in Scheme 23.75, all these results show that because of the space available on top of the catalyst where the oxo moiety to be released to the substrate resides, the confinement effect originating from nanopores can not only enhance the chiral recognition of chiral catalysts, but also restrict the rotation of intermediates, resulting in increasing ee values with the increasing length of linkage groups for the catalysts immobilized in the nanopores. For the same reason, a similar behavior is observed changing the pore size of the support and no effect is observed when the catalyst resides on the external surface.

### 23.6.4. *Conclusion*

In this overview we have tried to highlight some of the major advances reported in the heterogenization of oxidation catalysts emphasizing those aspects (where present) that can add extra properties to the heterogeneous system with respect to the homogeneous counterpart. The examples cited have tried to cover the major strategies that have been developed to achieve heterogenization in a stable manner preserving the catalysts' original activity and selectivity properties. The two major issues, i.e. recyclability and absence of metal leaching, have been addressed by all researchers involved in this area, although they have been fulfilled in a satisfactory manner only in a minority of cases. In all the examples cited in this overview, according to the authors, catalysts have been recycled at least four times without appreciable loss of metal. In very few examples has the stability been proven for longer times. In general, while this may not be sufficient to replace existing technologies on a large-scale, and based on low-cost simple homogeneous species, it can otherwise constitute a sustainable alternative for small-scale processes based on costly catalysts (e.g. in enantioselective transformations) where an easy catalyst recovery and/or prolonging catalyst lifetime can be important issues to decrease process costs.

## References

1. Anastas, P. and Warner, J. (1998). *Green Chemistry: Theory and Practice.* Oxford University Press, New York.
2. Trost, B. (1991). The Atom Economy: A Search for Synthetic Efficiency, *Science*, **254**, pp. 1471–1477; Trost, B. (1995). Atom Economy: a Challenge for Organic Synthesis: homogeneous Catalysis Leads the Way, *Angew. Chem. Int. Ed.*, **34**, pp. 259–281.
3. Gligorich, K. and Sigman, M. (2009). Recent Advancements and Challenges of Palladium$^{ii}$-Catalyzed Oxidation Reactions with Molecular Oxygen as the Sole Oxidant, *Chem. Commun.*, **26**, pp. 3854–3867; Punniyamurthy, T., Velusamy, S. and Iqbal, J. (2005). Recent Advances in Transition Metal Catalyzed Oxidation of Organic Substrates with Molecular Oxygen, *Chem. Rev.*, **105**, pp. 2329–2364; Ishii, Y., Sakaguchi, S. and Iwahama, T. (2001). Innovation of Hydrocarbon Oxidation with Molecular Oxygen and Related Reactions, *Adv. Synth. Catal.*, **343**, pp. 393–427.
4. Noyori, R., Aoki, M. and Sato, K. (2003). Green Oxidation with Aqueous Hydrogen Peroxide. *Chem. Commun.*, **16**, pp. 1977–1986; Lane B. and Burgess K. (2003). Metal-Catalyzed Epoxidations of Alkenes with Hydrogen Peroxide, *Chem. Rev.*, **103**, pp. 2457–2473.
5. http://www.fda.gov/Drugs/GuidanceComplianceRegulatoryInformation/Guidances/ucm122883.htm (June 8, 2011); Agranat, I. and Caner, H. (1999). Intellectual Property and Chirality of Drugs, *Drug Discovery Today*, **4**, pp. 313–321.
6. De Faveri, G., Ilyashenko, G. and Watkinson, M. (2011). Recent Advances in Catalytic Asymmetric Epoxidation Using the Environmentally Benign Oxidant Hydrogen Peroxide and its Derivatives, *Chem. Soc. Rev.*, **40**, pp. 1722–1760.
7. Irie, R., Hosoya, N. and Katsuki, T. (1994). Enantioselective Epoxidation of Chromene Derivatives Using Hydrogen Peroxide as a Terminal Oxidant, *Synlett*, **4**, pp. 255–256.
8. Garcia, M., Meou, A. and Brun, P. (1996). (Salen) MnIII-Catalyzed Asymmetric Epoxidation of Geraniol Derivatives, *Synlett*, **12**, pp. 1049–1050; Meou A., Garcia M. and Brun, P. (1999). Oxygen Transfer Mechanism in the Mn-Salen Catalyzed Epoxidation of Olefins, *J. Mol. Catal. A: Chem.*, **138**, pp. 221–226.
9. Shitama, H. and Katsuki, T. (2006). Asymmetric Epoxidation Using Aqueous Hydrogen Peroxide as Oxidant: Bio-Inspired Construction of Pentacoordinated Mn–Salen Complexes and their Catalysis, *Tetrahedron Lett.*, **47**, pp. 3203–3207.

10. Wu, M., Wang, B., Wang, S., *et al.* (2009). Asymmetric Epoxidation of Olefins with Chiral Bioinspired Manganese Complexes, *Org. Lett.*, **11**, pp. 3622–3625.
11. Sawada, Y., Matsumoto, K., Kondo, S., *et al.* (2006). Titanium–Salan-Catalyzed Asymmetric Epoxidation with Aqueous Hydrogen Peroxide as the Oxidant, *Angew. Chem. Int. Ed.*, **45**, pp. 3478–3480; Matsumoto, K., Sawada, Y. and Katsuki, T. (2006). Catalytic Enantioselective Epoxidation of Unfunctionalized Olefins: Utility of a Ti(Oi-Pr)$_4$-salan-$H_2O_2$ System, *Synlett*, **20**, pp. 3545–3547.
12. Berkessel, A., Brandenburg, M., Leitterstorf, E., *et al.* (2007). A Practical and Versatile Access to Dihydrosalen (Salalen) Ligands: highly Enantioselective Titanium In Situ Catalysts for Asymmetric Epoxidation with Aqueous Hydrogen Peroxide, *Adv. Synth. Catal.*, **349**, pp. 2385–2391.
13. Matsumoto, K., Sawada, Y., Saito, B., *et al.* (2005). Construction of Pseudo-Heterochiral and Homochiral di-$\mu$-Oxotitanium(Schiff Base) Dimers and Enantioselective Epoxidation Using Aqueous Hydrogen Peroxide, *Angew. Chem. Int. Ed.*, **44**, pp. 4935–4939.
14. Sawada, Y., Matsumoto, K. and Katsuki, T. (2007). Titanium-Catalyzed Asymmetric Epoxidation of Non-Activated Olefins with Hydrogen Peroxide, *Angew. Chem. Int. Ed.*, **46**, pp. 4559–4561.
15. Colladon, M., Scarso, A., Sgarbossa P., *et al.* (2006). Asymmetric Epoxidation of Terminal Alkenes with Hydrogen Peroxide Catalyzed by Pentafluorophenyl Pt(II) Complexes, *J. Am. Chem. Soc.,* **128**, pp. 14006–14007.
16. Egami, H., Oguma, T. and Katsuki, T. (2010). Oxidation Catalysis of Nb(Salan) Complexes: Asymmetric Epoxidation of Allylic Alcohols Using Aqueous Hydrogen Peroxide as an Oxidant, *J. Am. Chem. Soc.*, **132**, pp. 5886–5895.
17. Egami, H. and Katsuki, T. (2008). Nb(salan)-Catalyzed Asymmetric Epoxidation of Allylic Alcohols with Hydrogen Peroxide, *Angew. Chem., Int. Ed.*, **47**, pp. 5171–5174.
18. Gadissa Gelalcha, F., Bitterlich, B., Anilkumar, G., *et al.* (2007). Iron-Catalyzed Asymmetric Epoxidation of Aromatic Alkenes Using Hydrogen Peroxide, *Angew. Chem. Int. Ed.*, **46**, pp. 7293–7296.
19. Gadissa Gelalcha, F., Anilkumar, G., Tse, M., *et al.* (2008). Biomimetic Iron-Catalyzed Asymmetric Epoxidation of Aromatic Alkenes by Using Hydrogen Peroxide, *Chem. Eur. J.*, **14**, pp. 7687–7698.
20. Satyanarayana, T., Abraham, S. and Kagan, H. (2009). Nonlinear Effects in Asymmetric Catalysis, *Angew. Chem. Int. Ed.*, **48**, pp. 456–494.
21. Yeung, H., Sham, K., Tsang, C., *et al.*, (2008). A Chiral Iron-Sexipyridine Complex as a Catalyst for Alkene Epoxidation with Hydrogen Peroxide, *Chem. Commun.*, **32**, pp. 3801–3803.
22. Costas, M., Tipton, A., Chen, K., *et al.*, (2001). Modeling Rieske Dioxygenases: The First Example of Iron-Catalyzed Asymmetric Cis-Dihydroxylation of Olefins, *J. Am. Chem. Soc.*, **123**, pp. 6722–6723.
23. Chatterjee, D. (2008). Asymmetric Epoxidation of Unsaturated Hydrocarbons Catalyzed by Ruthenium Complexes, *Coord. Chem. Rev.*, **252**, pp. 176–198.
24. Nishiyama, H., Shimada, T., Itoh, H., *et al.* (1997). Novel Ruthenium–Pyridinedicarboxylate Complexes of Terpyridine and Chiral Bis(Oxazolinyl)Pyridine: a New Catalytic System for Alkene Epoxidation with [Bis(Acetoxy)Iodo.Benzene as an Oxygen Donor, *Chem. Commun.*, **19**, pp. 1863–1864.
25. Tse, M., Döbler, C., Bhor, S., *et al.* (2004). Development of a Ruthenium-Catalyzed Asymmetric Epoxidation Procedure with Hydrogen Peroxide as the Oxidant, *Angew. Chem. Int. Ed.*, **43**, pp. 5255–5260; Tse, M., Bhor, S., Klawonn, M., *et al.*, (2006). Ruthenium-Catalyzed Asymmetric Epoxidation of Olefins Using $H_2O_2$. Part II: catalytic Activities and Mechanism, *Chem. Eur. J.*, **12**, pp. 1875–1888.
26. Bhor, S., Anilkumar, G., Tse, M., *et al.* (2005). Synthesis of a New Chiral N,N,N-Tridentate Pyridinebisimidazoline Ligand Library and its Application in Ru-Catalyzed Asymmetric Epoxidation, *Org. Lett.*, **7**, pp. 3393–3396.

27. Tanaka, H., Nishikawa, H., Uchida, T., *et al.* (2010). Photopromoted Ru-Catalyzed Asymmetric Aerobic Sulfide Oxidation and Epoxidation Using Water as a Proton Transfer Mediator, *J. Am. Chem. Soc.*, **132**, pp. 12034–12041.
28. Strukul G. and Michelin R. (1984). Selective Epoxidation of Terminal Alkenes with Diluted Hydrogen Peroxide Catalysed by Pt–OH Species, *J. Chem. Soc. Chem. Commun.*, **22**, pp. 1538–1539; Strukul G., Michelin R. (1985). Catalytic Epoxidation of 1-Octene with Diluted Hydrogen Peroxide. On the Basic Role of Hydroxo Complexes of Platinum(II) and Related Species, *J. Am. Chem. Soc.*, **107**, pp. 7563–7569; Zanardo, A., Pinna, F., Michelin, R., *et al.* (1988). Kinetic Study of the Epoxidation of 1-Octene with Hydrogen Peroxide Catalyzed by Platinum(II) Complexes. Evidence of the Involvement of Two Metal Species in the Oxygen-Transfer Step, *Inorg. Chem.*, **27**, pp. 1966–1973; Colladon M., Scarso A., Sgarbossa P., *et al.* (2007). Regiospecificity and Diasteroselectivity of Pt(II) Mediated "Green" Catalytic Epoxidation of Terminal Alkenes with Hydrogen Peroxide: mechanistic Insight into its Peculiar Substrate Selectivity, *J. Am. Chem. Soc.*, **129**, pp. 7680–7689.
29. Colladon, M., Scarso, A. and Strukul, G. (2007). Towards a Greener Epoxidation Method: Use of Water-Surfactant Media and Catalyst Recycling in the Platinum-Catalyzed Asymmetric Epoxidation of Terminal Alkenes with Hydrogen Peroxide, *Adv. Synth. Catal.*, **349**, pp. 797–801.
30. Legros, J., Dehli, J. and Bolm, C. (2005). Applications of Catalytic Asymmetric Sulfide Oxidations to the Syntheses of Biologically Active Sulfoxides, *Adv. Synth. Catal.*, **347**, pp. 19–31.
31. Rouhi, A. (2004). Chiral Chemistry, *Chem. Eng. News*, **82**, pp. 47–62.
32. Di Furia, F., Modena, G. and Seraglia, R. (1984). Synthesis of Chiral Sulfoxides by Metal-Catalyzed Oxidation with T-Butyl Hydroperoxide, *Synthesis*, **4**, pp. 325–326.
33. Pitchen, P., Duñac, E., Desmukh, M., *et al.* (1984). An Efficient Asymmetric Oxidation of Sulfides to Sulfoxides, *J. Am. Chem. Soc.*, **106**, pp. 8188–8193.
34. Bryliakov, K. and Talsi, E. (2007). Asymmetric Oxidation of Sulfides with $H_2O_2$ Catalyzed by Titanium Complexes with Aminoalcohol Derived Schiff Bases, *J. Mol. Catal. A: Chem.*, **264**, pp. 280–287.
35. Saito, B. and Katsuki, T. (2001). Ti(Salen)-Catalyzed Enantioselective Sulfoxidation Using Hydrogen Peroxide as a Terminal Oxidant, *Tetrahedron Lett.*, **42**, pp. 3873–3876.
36. Saito, B. and Katsuki, T. (2001). Mechanistic Consideration of Ti(Salen)-Catalyzed Asymmetric Sulfoxidation, *Tetrahedron Lett.*, **42**, pp. 8333–8336.
37. Tanaka, T., Saito, B. and, Katsuki, T. (2002). Highly Enantioselective Oxidation of Cyclic Dithioacetals by Using a Ti(Salen) and Urea · Hydrogen Peroxide System, *Tetrahedron Lett.*, **43**, pp. 3259–3262.
38. Bryliakov, K. and Talsi, E. (2008). Titanium-Salan-Catalyzed Asymmetric Oxidation of Sulfides and Kinetic Resolution of Sulfoxides with $H_2O_2$ as the Oxidant, *Eur. J. Org. Chem.*, **19**, pp. 3369–3376.
39. Matsumoto, K., Yamaguchi, T., Fujisaki, J., *et al.* (2008). Aluminum Oxidation Catalysis under Aqueous Conditions: highly Enantioselective Sulfur Oxidation Catalyzed by Al(salalen) Complexes, *Chem. Asian J.*, **3**, pp. 351–358.
40. Yamaguchi, T., Matsumoto, K., Saito, B., *et al.* (2007). Asymmetric Oxidation Catalysis by a Chiral Al(Salalen) Complex: highly Enantioselective Oxidation of Sulfides with Aqueous Hydrogen Peroxide, *Angew. Chem. Int. Ed.*, **46**, pp. 4729–4731.
41. Fujisaki, J., Matsumoto, K., Matsumoto, K., *et al.* (2011). Catalytic Asymmetric Oxidation of Cyclic Dithioacetals: highly Diastereo- and Enantioselective Synthesis of the S-oxides by a Chiral Aluminum(salalen) Complex, *J. Am. Chem. Soc.*, **133**, pp. 56–61.
42. Bolm, C. and Bienewald, F. (1995). Asymmetric Sulfide Oxidation with Vanadium Catalysts and $H_2O_2$, *Angew. Chem. Int. Ed.*, **34**, pp. 2640–2642.
43. Kelly, P., Lawrence, S. and Maguire, A. (2006). Asymmetric Synthesis of Aryl Benzyl Sulfoxides by Vanadium-Catalysed Oxidation: a Combination of Enantioselective Sulfide Oxidation and Kinetic Resolution in Sulfoxide Oxidation, *Eur. J. Org. Chem.*, **19**, pp. 4500–4509.

44. Cogan, D., Liu, G., Kim, K., *et al.* (1998). Catalytic Asymmetric Oxidation of Tert-Butyl Disulfide. Synthesis of Tert-Butanesulfinamides, Tert-Butyl Sulfoxides, and Tert-Butanesulfinimines, *J. Am. Chem. Soc.*, **120**, pp. 8011–8019.
45. Wu, Y., Liu, J., Li, X., *et al.* (2009). Vanadium-Catalyzed Asymmetric Oxidation of Sulfides Using Schiff Base Ligands Derived from B-Amino Alcohols with Two Stereogenic Centers, *Eur. J. Org. Chem.*, **16**, pp. 2607–2610.
46. Sun, J., Zhu, C., Dai, Z., *et al.* (2004). Efficient Asymmetric Oxidation of Sulfides and Kinetic Resolution of Sulfoxides Catalyzed By A Vanadium-Salan System, *J. Org. Chem.*, **69**, pp. 8500–8503.
47. Mekmouche, Y., Hummel, H., Ho, R., *et al.* (2002). Sulfide Oxidation by Hydrogen Peroxide Catalyzed by Iron Complexes: Two Metal Centers Are Better Than One, *Chem. Eur. J.*, **8**, pp. 1196–1204.
48. Legros, J. and Bolm, C. (2003). Iron-Catalyzed Asymmetric Sulfide Oxidation with Aqueous Hydrogen Peroxide, *Angew. Chem. Int. Ed.*, **42**, pp. 5487–5489.
49. Legros, J. and Bolm, C. (2004). Highly Enantioselective Iron-Catalyzed Sulfide Oxidation with Aqueous Hydrogen Peroxide under Ssimple Reaction Conditions, *Angew. Chem. Int. Ed.*, **43**, pp. 4225–4228.
50. Legros, J. and Bolm, C. (2005). Investigations on the Iron-Catalyzed Asymmetric Sulfide Oxidation, *Chem. Eur. J.*, **11**, pp. 1086–1092.
51. Egami, H. and Katsuki, T. (2007). Fe(salan)-Catalyzed Asymmetric Oxidation of Sulfides with Hydrogen Peroxide in Water, *J. Am. Chem. Soc.*, **129**, pp. 8940–8941.
52. Chandler, D. (2005). Interfaces and the Driving Force of Hydrophobic Assembly, Nature, **437**, pp. 640–647; Breslow, R. (2006). The Hydrophobic Effect in Reaction Mechanism Studies and in Catalysis by Artificial Enzymes, *J. Phys. Org. Chem.*, **19**, pp. 813–822; Pratt, L. and Pohorille, A. (2002). Hydrophobic Effects and Modeling of Biophysical Aqueous Solution Interfaces, *Chem. Rev.*, **102**, pp. 2671–2692.
53. Chanda, A. and Fokin, V. (2009). Organic Synthesis "On Water," *Chem. Rev.*, **109**, pp. 725–748.
54. Le Maux, P. and Simonneaux, G. (2011). First Enantioselective Iron-Porphyrin-Catalyzed Sulfide Oxidation with Aqueous Hydrogen Peroxide. *Chem. Commun.*, **47**, pp. 6957–6959.
55. Bolm, C., Schlingloff, G. and Weickardt, K. (1994). Optically Active Lactones from a Baeyer–Villiger-Type Metal-Catalyzed Oxidation with Molecular Oxygen, *Angew. Chem. Int. Ed.*, **33**, pp. 1848–1849.
56. Gusso, A., Baccin, C., Pinna, F., *et al.* (1994). Platinum-Catalyzed Oxidations with Hydrogen Peroxide: Enantiospecific Baeyer-Villiger Oxidation of Cyclic Ketones, *Organometallics*, **13**, pp. 3442–3451.
57. Uchida, T. and Katsuki, T. (2001). Cationic Co(III)(Salen)-Catalyzed Enantioselective Baeyer–Villiger Oxidation of 3-Arylcyclobutanones Using Hydrogen Peroxide as a Terminal Oxidant, *Tetrahedron Lett.*, **42**, pp. 6911–6914.
58. Bolm, C., Khanh Luong, T. and Schlingloff, G. (1997). Enantioselective Metal-Catalyzed Baeyer-Villiger Oxidation of Cyclobutanones, *Synlett*, **10**, pp. 1151–1152.
59. Ito, K., Ishii, A., Kuroda, T., *et al.* (2003). Asymmetric Baeyer–Villiger Oxidation of Prochiral Cyclobutanones Using a Chiral Cationic Palladium(II) 2-(Phosphinophenyl)Pyridine Complex as Catalyst, *Synlett*, **5**, pp. 643–646; Malkov, A., Friscourt, F., Bell, M., *et al.* (2008). Enantioselective Baeyer–Villiger Oxidation Catalyzed by Palladium(II) Complexes with Chiral P,N-Ligands, *J. Org. Chem.*, **73**, pp. 3996–4003.
60. Paneghetti, C., Gavagnin, R., Pinna, F., *et al.* (1999). New Chiral Complexes of Platinum(II) as Catalysts for the Enantioselective Baeyer-Villiger Oxidation of Ketones with Hydrogen Peroxide: Dissymmetrization of Meso-Cyclohexanones, *Organometallics*, **18**, pp. 5057–5065.
61. Watanabe, A., Uchida, T., Irie, R., *et al.* (2004). Zr[bis(salicylidene)ethylenediaminato]-Mediated Baeyer–Villiger Oxidation: stereospecific Synthesis of Abnormal and Normal Lactones, *Proc. Natl. Acad. Sci. USA*, **101**, pp. 5737–5742; Watanabe, A., Uchida, T., Ito, K.,

*et al.* (2002). Highly Enantioselective Baeyer–Villiger Oxidation Using Zr(Salen) Complex as Catalyst, *Tetrahedron Lett.*, **43**, pp. 4481–4485; Bolm, C. and Beckmann, O. (2000). Zirconium-Mediated Asymmetric Baeyer-Villiger Oxidation, *Chirality*, **12**, pp. 523–525.

62. Bolm, C., Frison, J., Zhang, Y., *et al.* (2004). Vaulted Biaryls: Efficient Ligands for the Aluminum-Catalyzed Asymmetric Baeyer-Villiger Reaction, *Synlett*, **9**, pp. 1619–1621; Frison, J., Palazzi, C. and Bolm, C. (2006). Ligand Effects in Aluminium-Catalyzed Asymmetric Baeyer–Villiger Reactions, *Tetrahedron*, **62**, pp. 6700–6706; Bolm, C., Beckmann, O., Kühn, T., *et al.* (2001). Influence of Hydroperoxides on the Enantioselectivity of Metal-Catalyzed Asymmetric Baeyer–Villiger Oxidation and Epoxidation with Chiral Ligands, *Tetrahedron Asymm.*, **12**, pp. 2441–2446.
63. Stewart, J. (1998). Cyclohexanone Monooxygenase: a Useful Reagent for Asymmetric Baeyer-Villiger Reactions, *Curr. Org. Chem.*, **2**, pp. 195–216.
64. Colonna, S., Gaggero, N., Pasta, P., *et al.* (1996). Enantioselective Oxidation of Sulfides to Sulfoxides Catalysed by Bacterial Cyclohexanone Monooxygenases, *Chem. Commun.*, **20**, pp. 2303–2307; Roberts, S. and Wan, P. (1998). Enzyme-Catalysed Baeyer–Villiger Oxidations, *J. Mol. Catal. B: Enzym.*, **4**, pp. 111–136; Stewart, J., Reed, K., Martinez, C., *et al.* (1998). Recombinant Baker's Yeast as a Whole-Cell Catalyst for Asymmetric Baeyer-Villiger Oxidations, *J. Am. Chem. Soc.*, **120**, pp. 3541–3548; Ottolina, G., Bianchi, S., Belloni, B., *et al.* (1999). First Asymmetric Oxidation of Tertiary Amines by Cyclohexanone Monooxygenase, *Tetrahedron Lett.*, **40**, pp. 8483–8486; Mihovilovic, M., Chen, G., Wang, S., *et al.*, (2001). Asymmetric Baeyer-Villiger Oxidations of 4-Mono- and 4,4-Disubstituted Cyclohexanones by Whole Cells of Engineered *Escherichia coli*, *J. Org. Chem.*, **66**, pp. 733–738; Mihovilovic, M., Müller, B. and Stanetty, P. (2002). Monooxygenase-Mediated Baeyer-Villiger Oxidations, *Eur. J. Org. Chem.*, **22**, pp. 3711–3730; Kamerbeek, N., Janssen D., van Berkel W., *et al.* (2003). Baeyer–Villiger Monooxygenases, an Emerging Family of Flavin-Dependent Biocatalysts, *Adv. Synth. Catal.*, **345**, pp. 667–678; Ottolina, G., de Ponzalo, G., Carrea, G., *et al.* (2005). Enzymatic Baeyer–Villiger Oxidation of Bicyclic Diketones, *Adv. Synth. Catal.*, **347**, pp. 1035–1040; Kyte, B., Rouvière, P., Cheng, Q., *et al.* (2004). Assessing the Substrate Selectivities and Enantioselectivities of Eight Novel Baeyer-Villiger Monoxygenases Toward Alkyl Substituted Cyclohexanones, *J. Org. Chem.*, **69**, pp. 12–17; Snajdrova, R., Braun, I., Bach, T., *et al.* (2007). Biooxidation of Bridged Cycloketones Using Baeyer-Villiger Monooxygenases of Various Bacterial Origin, *J. Org. Chem.*, **72**, pp. 9597–9603; Mihovilovic, M., Rudroff, F., Winninger, A., *et al.* (2006). Microbial Baeyer-Villiger Oxidation: stereopreference and Substrate Acceptance of Cyclohexanone Monooxygenase Mutants Prepared by Directed Evolution, *Org. Lett.*, **8**, pp. 1221–1224; Reetz, M., Brunner, B., Schneider, T., *et al.* (2004). Directed Evolution as a Method to Create Enantioselective Cyclohexanone Monooxygenases for Catalysis in Baeyer–Villiger Reactions, *Angew. Chem. Int. Ed.*, **43**, pp. 4075–4078; Ríos, M., Salazar, E. and Olivo, H. (2007). Baeyer–Villiger Oxidation of Substituted Cyclohexanones via Lipase-Mediated Perhydrolysis Utilizing Urea–Hydrogen Peroxide in Ethyl Acetate, *Green Chem.*, **9**, pp. 459–462.
65. Colladon, M., Scarso, A. and Strukul, G. (2006). Tailoring Pt(II) Chiral Catalyst Design for Asymmetric Baeyer-Villiger Oxidation of Cyclic Ketones with Hydrogen Peroxide, *Synlett*, **20**, pp. 3515–3520.
66. Egami, H. and Katsuki, T. (2009). Iron-Catalyzed Asymmetric Aerobic Oxidation: oxidative Coupling of 2-Naphthols, *J. Am. Chem. Soc.*, **131**, pp. 6082–6083.
67. Li, X., Hewgley, J., Mulrooney, C., *et al.* (2003). Enantioselective Oxidative Biaryl Coupling Reactions Catalyzed by 1,5-Diazadecalin Metal Complexes: efficient Formation of Chiral Functionalized BINOL Derivatives, *J. Org. Chem.*, **68**, pp. 5500–5511.
68. Hewgley, B., Stahl S. and Kozlowski, M. (2008). Mechanistic Study of Asymmetric Oxidative Biaryl Coupling: evidence for Self-Processing of the Copper Catalyst to Achieve Control of Oxidase vs Oxygenase Activity, *J. Am. Chem. Soc.*, **130**, pp. 12232–12233.

69. Mulrooney, C., Li, X., Di Virgilio, E., *et al.* (2003). General Approach for the Synthesis of Chiral Perylenequinones via Catalytic Enantioselective Oxidative Biaryl Coupling, *J. Am. Chem. Soc.*, **125**, pp. 6856–6857.
70. Sigman, M. and Jensen, D. (2006). Ligand-Modulated Palladium-Catalyzed Aerobic Alcohol Oxidations, *Acc. Chem. Res*, **39**, pp. 221–229.
71. Nishimura, T., Onoue, T., Ohe, K., *et al.* (1999). Palladium(II)-Catalyzed Oxidation of Alcohols to Aldehydes and Ketones by Molecular Oxygen, *J. Org. Chem.*, **64**, pp. 6750–6755.
72. Jensen, D., Pugsley, J. and Sigman, M. (2001). Palladium-Catalyzed Enantioselective Oxidations of Alcohols Using Molecular Oxygen, *J. Am. Chem. Soc.*, **123**, pp. 7475–7476.
73. Ferreira, E. and Stoltz, B. (2001). The Palladium-Catalyzed Oxidative Kinetic Resolution of Secondary Alcohols with Molecular Oxygen. *J. Am. Chem. Soc.*, **123**, pp. 7725–7726.
74. Mandal, S., Jensen, D., Pugsley, J., *et al.* (2003). Scope of Enantioselective Palladium(II)-Catalyzed Aerobic Alcohol Oxidations with (−)-Sparteine, *J. Org. Chem.*, **68**, pp. 4600–4603.
75. Bagdanoff, J. and Stoltz, B. (2004). Palladium-Catalyzed Oxidative Kinetic Resolution with Ambient Air as the Etoichiometric Oxidation Gas, *Angew. Chem. Int. Ed.*, **43**, pp. 353–357.
76. Ebner D., Trend R., Genet C., *et al.* (2008). Palladium-Catalyzed Enantioselective Oxidation of Chiral Secondary Alcohols: access to Both Enantiomeric Series, *Angew. Chem. Int. Ed.*, **47**, pp. 6367–6370.
77. Krishnan, S., Bagdanoff, J., Ebner, D., *et al.* (2008). Pd-Catalyzed Enantioselective Aerobic Oxidation of Secondary Alcohols: applications to the Total Synthesis of Alkaloids, *J. Am. Chem. Soc*, **130**, pp. 13745–13754.
78. Caspi, D., Ebner, D., Bagdanoff, J., *et al.* (2004). The Resolution of Important Pharmaceutical Building Blocks by Palladium-Catalyzed Aerobic Oxidation of Secondary Alcohols, *Adv. Synth. Catal.,* **346**, pp. 185–189.
79. Radosevich, A., Musich, C. and Toste, F. (2005). Vanadium-Catalyzed Asymmetric Oxidation of A-Hydroxy Esters Using Molecular Oxygen as Stoichiometric Oxidant, *J. Am. Chem. Soc.*, **127**, pp. 1090–1091.
80. Shimizu, H., Onitsuka, S., Egami, H., *et al.* (2005). Ruthenium(Salen)-Catalyzed Aerobic Oxidative Desymmetrization of Meso-Diols and its Kinetics, *J. Am. Chem. Soc.*, **127**, pp. 5396–5413.
81. Nakamura, Y., Egami, H., Matsumoto, K., *et al.* (2007). Aerobic Oxidative Kinetic Resolution of Racemic Alcohols with Bidentate Ligand-Binding Ru(Salen) Complex as Catalyst, *Tetrahedron*, **63**, pp. 6383–6387.
82. Bourhani, Z. and Malkov, A., (2005). Ligand-Accelerated Vanadium-Catalyzed Epoxidation in Water, *Chem. Commun.*, **36**, pp. 4592–4594.
83. Sun, W., Wang, H., Xia, C., *et al.* (2003). Chiral-Mn(Salen)-Complex-Catalyzed Kinetic Resolution of Secondary Alcohols in Water, *Angew. Chem. Int. Ed.*, **42**, pp. 1042–1044.
84. Lindström, U. (2006). Hydrophobically Directed Organic Synthesis, *Angew. Chem. Int. Ed.*, **45**, pp. 548–551; Pirrung, M. (2006). Acceleration of Organic Reactions through Aqueous Solvent Effects, *Chem. Eur. J.*, **12**, pp. 1312–1317; Blokzijl, W. and Engberts, J. (1993). Hydrophobic Effects. Opinions and Facts, *Angew. Chem. Int. Ed.*, **32**, pp. 1545–1579; Dwars, T., Paetzold, E. and Oehme, G. (2005). Reactions in Micellar Systems, *Angew. Chem. Int. Ed.*, **44**, pp. 7174–7199.
85. Bianchini, G., Cavarzan, A., Scarso, A., *et al.* (2009). Asymmetric Baeyer–Villiger Oxidation with Co(Salen) and $H_2O_2$ in Water: striking Supramolecular Micelles Effect on Catalysis, *Green Chem.*, **11**, pp. 1517–1520.
86. Cavarzan, A., Bianchini, G., Sgarbossa, P., *et al.* (2009). Catalytic Asymmetric Baeyer–Villiger Oxidation in Water by Using Ptii Catalysts and Hydrogen Peroxide: supramolecular Control of Enantioselectivity, *Chem. Eur. J.*, **15**, pp. 7930–7939.

87. Scarso, A. and Strukul, G. (2005). Asymmetric Sulfoxidation of Thioethers with Hydrogen Peroxide in Water Mediated by Platinum Chiral Catalyst, *Adv. Synth. Catal.*, **347**, pp. 1227–1234.
88. Keith, J. and Henry, P. (2009). The Mechanism of the Wacker Reaction: a Tale of Two Hydroxypalladations, *Angew. Chem. Int. Ed.*, **48**, pp. 9038–9049.
89. Weine, B., Baeza, A., Jerphagnon, T., *et al.* (2009). Aldehyde Selective Wacker Oxidations of Phthalimide Protected Allylic Amines: a New Catalytic Route to $\beta$3-Amino Acids, *J. Am. Chem. Soc.*, **131**, pp. 9473–9474.
90. Cornell, C. and Sigman, M. (2007). Recent Progress in Wacker Oxidations: moving Toward Molecular Oxygen as the Sole Oxidant, *Inorg. Chem.*, **46**, pp. 1903–1909.
91. ten Brink, G., Arends, I., Papadogianakis, G., *et al.* (1998). Catalytic Conversions in Water. Part 10. Aerobic Oxidation of Terminal Olefins to Methyl Ketones Catalysed by Water Soluble Palladium Complexes, *Chem. Commun.*, **21**, pp. 2359–2360.
92. Cornell, C. and Sigman, M. (2006). Discovery of a Practical Direct $O_2$-Coupled Wacker Oxidation with Pd[(−)-Sparteine.$Cl_2$, *Org. Lett.*, **8**, pp. 4117–4120.
93. Naik, A., Meina, L., Zabel, M., *et al.* (2010). Efficient Aerobic Wacker Oxidation of Styrenes Using Palladium Bis(isonitrile) Catalysts, *Chem. Eur. J.*, **16**, pp. 1624–1628.
94. Mitsudome, T., Umetani, T., Nosaka, N., *et al.* (2006). Convenient and Efficient Pd-Catalyzed Regioselective Oxyfunctionalization of Terminal Olefins by Using Molecular Oxygen as Sole Reoxidant, *Angew. Chem. Int. Ed.*, **45**, pp. 481–485.
95. Mitsudome, T., Mizumoto, K., Mizugaki, T., *et al.* (2010). Wacker-Type Oxidation of Internal Olefins Using a $PdCl_2$/N,N-Dimethylacetamide Catalyst System under Copper-Free Reaction Conditions, *Angew. Chem. Int. Ed.*, **49**, pp. 1238–1240.
96. Roussel, M. and Mimoun, H., (1980). Palladium-Catalyzed Oxidation of Terminal Olefins to Methyl Ketones by Hydrogen Peroxide, *J. Org. Chem.*, **45**, pp. 5387–5390.
97. Nishimura, T., Kakiuchi, N., Onoue, T., *et al.* (2000). Palladium(II)-Catalyzed Oxidation of Terminal Alkenes to Methyl Ketones Using Molecular Oxygen, *J. Chem. Soc., Perkin Trans.*, **1**, pp. 1915–1918.
98. Cornell, C. and Sigman M. (2005). Discovery of and Mechanistic Insight into a Ligand-Modulated Palladium-Catalyzed Wacker Oxidation of Styrenes Using TBHP, *J. Am. Chem. Soc.*, **127**, pp. 2796–2797.
99. Michel, B., Camelio, A., Cornell C., *et al.* (2009). A General and Efficient Catalyst System for a Wacker-Type Oxidation Using TBHP as the Terminal Oxidant: application to Classically Challenging Substrates, *J. Am. Chem. Soc.*, **131**, pp. 6076–6077.
100. Michel, B., McCombs, J., Winkler, A., *et al.* (2010). Catalyst-Controlled Wacker-Type Oxidation of Protected Allylic Amines, *Angew. Chem. Int. Ed.*, **49**, pp. 7312–7315.
101. Helfer, D. and Atwood, J. (2004). Platinum-Catalyzed Wacker Oxidation of Alkenes Utilizing cis-$Pt(Cl)_2(TPPTS)_2$ (TPPTS=$P(m\text{-}C_6H_4SO_3Na)_3$), *Organometallics*, **23**, pp. 2412–2420.
102. Que Jr, L. and Tolman, W. (2008). Biologically Inspired Oxidation Catalysis, *Nature*, **455**, pp. 333–340.
103. Monti, D., Ottolina, G., Carrea, G., *et al.* (2011). Redox Reactions Catalyzed by Isolated Enzymes, *Chem. Rev.*, **111**, pp. 4111–4140.
104. Sheldon, R. and Arends I. (2006). Catalytic Oxidations Mediated by Metal Ions and Nitroxyl Radicals, *J. Mol. Cat. A: Chem.*, **251**, pp. 200–214.
105. Gamez, P., Arends, I., Reedijk, J., *et al.* (2003). Copper(II)-Catalysed Aerobic Oxidation of Primary Alcohols to Aldehydes, *Chem. Commun.*, **19**, pp. 2414–2415.
106. Markó, I., Gautier, A., Dumeunier, R., *et al.* (2004). Efficient, Copper-Catalyzed, Aerobic Oxidation of Primary Alcohols, *Angew. Chem. Int. Ed.*, **43**, pp. 1588–1591.
107. Sorokin, A., Kudrik, E. and Bouchu, D. (2008). Bio-Inspired Oxidation of Methane in Water Catalyzed by N-Bridged Diiron Phthalocyanine Complex, *Chem. Commun.*, **22**, pp. 2562–2564.

108. Lane, B., Vogt, M., DeRose, V., *et al.* (2002). Manganese-Catalyzed Epoxidations of Alkenes in Bicarbonate Solutions, *J. Am. Chem. Soc.*, **124**, pp. 11946–11954.
109. de Boer, J., Brinksma, J., Browne, W., *et al.* (2005). cis-Dihydroxylation and Epoxidation of Alkenes by $Mn_2O(RCO_2)_2(tmtacn)_2$.: Tailoring the Selectivity of a Highly $H_2O_2$-Efficient Catalyst, *J. Am. Chem. Soc.*, **127**, pp. 7990–7991.
110. Bolm, C., Legros, J., Le Paih, J., *et al.* (2004). Iron-Catalyzed Reactions in Organic Synthesis, *Chem. Rev.*, **104**, pp. 6217–6254.
111. Sen Gupta, S., Stadler, M., Noser, C., *et al.* (2002). Rapid Total Destruction of Chlorophenols by Activated Hydrogen Peroxide, *Science*, **296**, pp. 326–328.
112. Chadwick Ellis, W., Tran C., Roy R., *et al.* (2010). Designing Green Oxidation Catalysts for Purifying Environmental Waters, *J. Am. Chem. Soc.*, **132**, pp. 9774–9781.
113. Chen, M. and White, M. (2007). A Predictably Selective Aliphatic C–H Oxidation Reaction for Complex Molecule Synthesis, *Science*, **318**, pp. 783–787.
114. Chen, M. and White, M. (2010). Combined Effects on Selectivity in Fe-Catalyzed Methylene Oxidation, *Science*, **327**, pp. 566–571.
115. Akelah, A. and Sherrington, D. (1981). Application of Functionalized Polymers in Organic Synthesis, *Chem. Rev.*, **81**, pp. 557–587; Hartley, F. (1985). *Supported Metal Complexes: A New Generation of Catalysts*, Reidel, Dordrecht.
116. Sherrington, D. (1988). Polymer-Supported Metal Complex Oxidation Catalysts, *Pure Appl. Chem.*, **60**, pp. 401–414; Linden, G. and Farona, M. (1977). A Resin-Bound Vanadyl Catalyst for the Epoxidation of Olefins, *Inorg. Chem.*, **16**, pp. 3170–3173; Ivanov, S., Boeva, R. and Tanielyan, S. (1979). Catalytic Epoxidation of Propylene with Tert-Butyl Hydroperoxide in the Presence of Modified Carboxy Cation-Exchange Resin "Amberlite" IRC-50, *J. Catal.*, **56**, pp. 150–159; Sobczak, J. and Ziolkowski, J. (1977/78). The Molybdenum(V) Complexes as the Homogeneous and Heterogenized Catalysts in Epoxidation Reactions of Olefins with the Organic Hydroperoxides, *J. Mol. Catal.*, **3**, pp. 165–172; Yokoyama, T., Nishizawa, M., Kimura, T., *et al.* (1985). Catalytic Epoxidation of Olefins with t-Butyl Hydroperoxide in the Presence of Polymer-Supported Vanadium(V) and Molybdenum(VI) Complexes, *Bull. Chem. Soc. Jpn.*, **58**, pp. 3271–3276; Kuruso, Y., Masuyama, Y., Saito, M., *et al.* (1986). Epoxidation with t-Butyl Hydroperoxide in the Presence of Molybdenum Peroxide and Polymer-Immobilized Molybdenum Peroxide, *J. Mol. Catal.*, **37**, pp. 235–241; Linden, G. and Farona, M. (1977). Polymer-Anchored Vanadyl Catalysts for the Oxidation Of Cyclohexene, *J. Catal.*, **48**, pp. 284–291; Bhaduri, S., Ghosh, A. and Khwaja, H. (1981). Polymer-Supported Oxobis(Pentane-2,4-Dionato)vanadium(IV) Catalyst for Reactions Involving t-Butyl Hydroperoxide, *J. Chem. Soc. Dalton Trans,* **2**, pp. 447–451.
117. Sherrington, D. and Simpson, S. (1991). Polymer-Supported Mo and V Cyclohexene Epoxidation Catalysts: activation, Activity, and Stability, *J. Catal.*, **131**, pp. 115–126; Sherrington, D. and Simpson, S. (1993). Polymer-Supported Mo Alkene Epoxidation Catalysts, *Reactive Polym.*, **19**, pp. 13–25; Miller, M., Sherrington, D. and Simpson, S. (1994). Alkene Epoxidations Catalysed by Molybdenum(VI) Supported on Imidazole-Containing Polymers. Part 3. Epoxidation Of Oct-1-Ene And Propene, *J. Chem. Soc.*, *Perkin Trans.*, **2**, pp. 2091–2096; Miller, M. and Sherrington, D. (1995). Alkene Epoxidations Catalyzed by Mo(VI) Supported on Imidazole-Containing Polymers: I. Synthesis, Characterization, and Activity of Catalysts in the Epoxidation of Cyclohexene, *J. Catal.*, **152**, pp. 368–376; Miller, M. and Sherrington, D. (1995). Alkene Epoxidations Catalyzed by Mo(VI) Supported on Imidazole-Containing Polymers: II. Recycling of Polybenzimidazole-Supported Mo(VI) in the Epoxidation of Cyclohexene, *J. Catal.*, **152**, pp. 377–383; Sherrington, D. (2000). Polymer-Supported Metal Complex Alkene Epoxidation Catalysts, *Catal. Today*, **57**, pp. 87–104.
118. Olason, G. and Sherrington, D. (1999). Oxidation of Cyclohexene by T-Butylhydroperoxide and Dioxygen Catalysed by Polybenzimidazole-Supported Cu, Mn, Fe, Ru and Ti Complexes, *React. Funct. Polymers*, **42**, pp. 163–172.

119. Gelbard, G., Breton, F., Quenard, M., *et al.* (2000). Epoxidation of Cyclohexene with Polymethacrylate-Based Peroxotungstic Catalysts, *J. Mol. Catal. A: Chem.*, **153**, pp. 7–18; Gelbard, G., Gauducheau, T., Vidal, E., *et al.* (2002). Epoxidation with Peroxotungstic Acid Immobilised onto Silica-Grafted Phosphoramides, *J. Mol. Catal. A: Chem.*, **182–183**, pp. 257–266.
120. Arnold, U., Habicht, W. and Döring, M. (2006). Metal-Doped Epoxy Resins — New Catalysts for the Epoxidation of Alkenes with High Long-Term Activities, *Adv. Synth. Catal.*, **348**, pp. 142–150.
121. Herrmann, W., Fischer, R. and Marz, D. (1991). Methyltrioxorhenium as Catalyst for Olefin Oxidation, *Angew. Chem. Int. Ed.*, **30**, pp. 1638–1641; Herrmann W., Fischer, R., Scherer, W., *et al.* (1993). Methyltrioxorhenium(VII) as Catalyst for Epoxidations: structure of the Active Species and Mechanism of Catalysis, *Angew. Chem. Int. Ed.*, **32**, pp. 1157–1160; Herrmann, W., Fischer, R., Rauch, M., *et al.* (1994). Alkylrhenium Oxides as Homogeneous Epoxidation Catalysts: activity, Selectivity, Stability, Deactivation, *J. Mol. Catal.*, **86**, pp. 243–266; Romão, C., Kühn, F. and Herrmann, W. (1997). Rhenium(VII) Oxo and Imido Complexes: synthesis, Structures, and Applications, *Chem. Rev.*, **97**, pp. 3197–3246.
122. Saladino, R., Neri, V., Pelliccia, A., *et al.* (2002). Preparation and Structural Characterization of Polymer-Supported Methylrhenium Trioxide Systems as Efficient and Selective Catalysts for the Epoxidation of Olefins, *J. Org. Chem.*, **67**, pp. 1323–1332; Saladino, R., Neri, V., Pelliccia, A., *et al.* (2003). Selective Epoxidation of Monoterpenes with $H_2O_2$ and Polymer-Supported Methylrheniumtrioxide Systems, *Tetrahedron*, **59**, pp. 7403–7408; Saladino, R., Bernini, R., Neri, V., *et al.* (2009). A Novel and Efficient Catalytic Epoxidation of Monoterpenes by Homogeneous and Heterogeneous Methyltrioxorhenium in Ionic Liquids, *Appl. Catal. A: Gen.*, **360**, pp. 171–176.
123. Palazzi, C., Pinna, F. and Strukul, G. (2000). Polymer-Anchored Platinum Complexes as Catalysts for the Baeyer–Villiger Oxidation of Ketones: preparation and Catalytic Properties, *J. Mol. Catal. A*: *Chem.*, **151**, pp. 245–252.
124. Jones, M., Raja, R., Thomas, J., *et al.* (2003). Enhancing the Enantioselectivity of Novel Homogeneous Organometallic Hydrogenation Catalysts, *Angew. Chem. Int. Ed.*, **42**, pp. 4326–4331; Raja, R., Thomas, J., Johnson B., *et al.* (2003). Constraining Asymmetric Organometallic Catalysts within Mesoporous Supports Boosts Their Enantioselectivity, *J. Am. Chem. Soc.*, **125**, pp. 14982–14983; Schögl, R. and Abd Hamid, S. (2004). Nanocatalysis: mature Science Revisited or Something Really New? *Angew. Chem. Int. Ed.,* **43**, pp. 1628–1637; Grunes, J., Zhu, J. and Somorjai, G. (2003). Catalysis and Nanoscience, *Chem. Commun.*, **18**, pp. 2257–2260.
125. Shylesh, S., Jia, M. and Thiel, W. (2010). Recent Progress in the Heterogenization of Complexes for Single-Site Epoxidation Catalysis, *Eur. J. Inorg. Chem.*, **28**, pp. 4395–4410.
126. Ferreira, P., Gonçalves, I., Kühn, F., *et al.* (2000). Mesoporous Silicas Modified with Dioxomolybdenum(VI) Complexes: synthesis and Catalysis., *Eur. J. Inorg. Chem.*, **10**, pp. 2263–2270; Nunes, C., Valente, A., Pillinger, M., *et al.* (2003). Molecular Structure–Activity Relationships for the Oxidation of Organic Compounds Using Mesoporous Silica Catalysts Derivatised with Bis(halogeno)dioxomolybdenum(VI) Complexes, *Chem. Eur. J.*, **9**, pp. 4380–4390; Monteiro, B., Balula, S., Gago, S., *et al.* (2009). Comparison of Liquid-Phase Olefin Epoxidation Catalysed by Dichlorobis-(Dimethylformamide) Dioxomolybdenum(VI) in Homogeneous Phase and Grafted onto MCM-41, *J. Mol. Catal. A*: *Chem.*, **297**, pp. 110–117; Yang, Q., Copéret, C., Li, C., *et al.* (2003). Molybdenum Containing Surface Complex for Olefin Epoxidation, *New J. Chem.*, **27**, pp. 319–323.
127. Masteri-Farahani, M., Farzeneh, F. and Ghandi, M. (2006). Synthesis and Characterization of a New Epoxidation Catalyst by Grafting cis-$MoO_2$(salpr) Complex to Functionalized MCM-41, *J. Mol. Catal. A*: *Chem.*, **243**, pp. 170–175; Masteri-Farahani, M. (2010). Investigation of Catalytic Activities of New Heterogeneous Molybdenum Catalysts in Epoxidation of Olefins, *J. Mol. Catal. A*: *Chem.*, **316**, pp. 45–51.

128. Sheldon, R., Wallau, M., Arends, I., *et al.* (1998). Heterogeneous Catalysts for Liquid-Phase Oxidations: Philosophers' Stones or Trojan Horses?, *Acc. Chem. Res.*, **31**, pp. 485–493.
129. Trost, M. and Bergman, R. (1991). Cp*$MoO_2Cl$-Catalyzed Epoxidation of Olefins by Alkyl Hydroperoxides, *Organometallics*, **10**, pp. 1172–1178; Abrantes, M., Santos, M., Mink, J., *et al.* (2003). A Simple Entry to ($\eta$5-$C_5R_5$)Chlorodioxomolybdenum(VI) Complexes (R = H, $CH_3$, $CH_2Ph$) and Their Use as Olefin Epoxidation Catalysts, *Organometallics*, **22**, pp. 2112–2118.
130. Sakthivel, A., Zhao, J., Hanzlik, M., *et al.* (2004). Heterogenisation of $CpMo(CO)_3Cl$ on Mesoporous Materials and its Application as Olefin Epoxidation Catalyst, *Dalton Trans.*, **20**, pp. 3338–3341.
131. Jia, M. and Thiel, W., (2002). Oxodiperoxo Molybdenum Modified Mesoporous MCM-41 Materials for the Catalytic Epoxidation of Cyclooctene, *Chem. Commun*, **20**, pp. 2392–2393; Jia, M., Seifert, A. and Thiel, W., (2003). Mesoporous MCM-41 Materials Modified with Oxodiperoxo Molybdenum Complexes: Efficient Catalysts for the Epoxidation of Cyclooctene, *Chem. Mater.*, **15**, pp. 2174–2180.
132. Sakthivel, A., Zhao, J., Hanzlik, M., *et al.* (2005). Heterogenization of Organometallic Molybdenum Complexes with Siloxane Functional Groups and their Catalytic Application, *Adv. Synth. Catal.*, **347**, pp. 473–483.
133. Hill, C. and Prosser-McCartha C. (1995). Homogeneous Catalysis by Transition Metal Oxygen Anion Clusters, *Coord. Chem. Rev.*, **143**, pp. 407–455; Neumann, R. (1998). Polyoxometallate Complexes in Organic Oxidation Chemistry, *Prog. Inorg. Chem.*, **47**, pp. 317–345; Kozhevnikov, I. (2002). *Catalysis by Polyoxometalates*. Wiley, Chichester; Mizuno, N., Kamata, K. and Yamaguchi, K. (2010). Green Oxidation Reactions by Polyoxometalate-Based Catalysts: From Molecular to Solid Catalysts, *Top. Catal.*, **53**, pp. 876–893.
134. Yamaguchi, K., Yoshida, C., Uchida, S., *et al.* (2005). Peroxotungstate Immobilized on Ionic Liquid-Modified Silica as a Heterogeneous Epoxidation Catalyst with Hydrogen Peroxide, *J. Am. Chem. Soc.*, **127**, pp. 530–531; Kasai J., Nakagawa Y., Uchida S., *et al.* (2006). $\gamma$-1,2-$H_2SiV_2W_{10}O_{40}$. Immobilized on Surface-Modified $SiO_2$ as a Heterogeneous Catalyst for Liquid-Phase Oxidation with $H_2O_2$, *Chem. Eur. J.*, **12**, pp. 4176–4184.
135. Nakagawa, Y., Kamata, K., Kotani, M., *et al.* (2005). Polyoxovanadometalate-Catalyzed Selective Epoxidation of Alkenes with Hydrogen Peroxide, *Angew. Chem. Int. Ed.*, **44**, pp. 5136–5141; Nakagawa, Y. and Mizuno, N. (2007). Mechanism of [$\gamma$-$H_2SiV_2W_{10}O_{40}$]4-Catalyzed Epoxidation of Alkenes with Hydrogen Peroxide, *Inorg. Chem.*, **46**, pp. 1727–1736.
136. Villa de P., A., Sels, B., De Vos, D., *et al.* (1999). A Heterogeneous Tungsten Catalyst for Epoxidation of Terpenes and Tungsten-Catalyzed Synthesis of Acid-Sensitive Terpene Epoxides, *J. Org. Chem.*, **64**, pp. 7267–7170; Hoegaerts, D., Sels, B., De Vos, D., *et al.* (2000). Heterogeneous Tungsten-Based Catalysts for the Epoxidation of Bulky Olefins, *Catal. Today*, **60**, pp. 209–218; Yamada, Y., Ichinohe, M., Takahashi, H., *et al.* (2001). Development of a New Triphase Catalyst and Its Application to the Epoxidation of Allylic Alcohols, *Org. Lett.*, **3**, pp. 1837–1840; Yamada, Y., Tabata, H., Ichinohe, M., *et al.* (2004). Oxidation of Allylic Alcohols, Amines, and Sulfides Mediated by Assembled Triphase Catalyst of Phosphotungstate and Non-Cross-Linked Amphiphilic Copolymer, *Tetrahedron*, **60**, pp. 4087–4096; Venturello, C. and D'Aloisio, R. (1988). Quaternary Ammonium Tetrakis(diperoxotungsto) Phosphates(3-) as a New Class of Catalysts for Efficient Alkene Epoxidation with Hydrogen Peroxide, *J. Org. Chem.*, **53**, pp. 1553–1557.
137. Ishii, Y., Yamawaki, K., Ura, T., *et al.* (1988). Hydrogen Peroxide Oxidation Catalyzed by Heteropoly Acids Combined with Cetylpyridinium Chloride. Epoxidation of Olefins And Allylic Alcohols, Ketonization of Alcohols and Diols, and Oxidative Cleavage of 1,2-Diols and Olefins, *J. Org. Chem.*, **53**, pp. 3587–3593; Sato, K., Aoki, M., Ogawa, M., *et al.* (1997). A Halide-Free Method for Olefin Epoxidation with 30% Hydrogen Peroxide, *Bull. Chem. Soc. Jpn.*, **70**, pp. 905–915; Xi, Z. W., Zhou, N., Sun, Y., *et al.* (2001). Reaction-Controlled Phase-Transfer Catalysis for Propylene Epoxidation to Propylene Oxide, *Science*, **292**, pp. 1139–1141; Neumann, R.

and Gara, M. (1994). Highly Active Manganese-Containing Polyoxometalate as Catalyst for Epoxidation of Alkenes with Hydrogen Peroxide, *J. Am. Chem. Soc.*, **116**, pp. 5509–5510; Neumann, R. and Gara, M. (1995). The Manganese-Containing Polyoxometalate, $[WZnMn_2^{II}(ZnW_9O_{34})_2]^{12-}$, as a Remarkably Effective Catalyst for Hydrogen Peroxide Mediated Oxidations, *J. Am. Chem. Soc.*, **117**, pp. 5066–5074; Neumann, R. and Juwiler, D. (1996). Oxidations with Hydrogen Peroxide Catalysed by the $[WZnMn(II)_2(ZnW_9O_{34})_2]^{12-}$ Polyoxometalate, *Tetrahedron*, **52**, pp. 8781–8785; Adam, W., Alsters, P., Neumann, R., *et al.* (2002). A New Highly Selective Method for the Catalytic Epoxidation of Chiral Allylic Alcohols by Sandwich-Type Polyoxometalates with Hydrogen Peroxide, *Synlett*, **12**, pp. 2011–2014; Adam, W., Alsters, P., Neumann, R., *et al.* (2003). A Highly Chemoselective, Diastereoselective, and Regioselective Epoxidation of Chiral Allylic Alcohols with Hydrogen Peroxide, Catalyzed by Sandwich-Type Polyoxometalates: enhancement of Reactivity and Control of Selectivity by the Hydroxy Group through Metal-Alcoholate Bonding, *J. Org. Chem.*, **68**, pp. 1721–1728; Maayan, G., Fish, R. and Neumann, R. (2003). Polyfluorinated Quaternary Ammonium Salts of Polyoxometalate Anions: fluorous Biphasic Oxidation Catalysis with and without Fluorous Solvents, *Org. Lett.*, **5**, pp. 3547–3550; Witte, P., Alsters, P., Jary, W., *et al.* (2004). Self-Assembled $Na_{12}[WZn_3(ZnW_9O_{34})_2]$ as an Industrially Attractive Multi-Purpose Catalyst for Oxidations with Aqueous Hydrogen Peroxide, *Org. Proc. Res. Dev.*, **8**, pp. 524–531; Grigoropoulou, G. and Clark, J. (2006). A Catalytic, Environmentally Benign Method for the Epoxidation of Unsaturated Terpenes with Hydrogen Peroxide, *Tetrahedron Lett.*, **47**, pp. 4461–4463.

138. Vasylyev, M. and Neumann, R. (2004). New Heterogeneous Polyoxometalate Based Mesoporous Catalysts for Hydrogen Peroxide Mediated Oxidation Reaction. *J. Am. Chem. Soc.*, **126**, pp. 884–890.
139. Liu, P., Wang, C. and Li, C. (2009). Epoxidation of Allylic Alcohols on Self-Assembled Polyoxometalates Hosted in Layered Double Hydroxides with Aqueous $H_2O_2$ as Oxidant, *J. Catal.*, **262**, pp. 159–168.
140. Sheldon, R. and Kochi, J. (1981). Metal-Catalyzed Oxidation of Organic Compounds, Academic Press, New York; Hudlicky, M. (1990). *Oxidations in Organic Chemistry*, American Chemical Society, Washington, DC.
141. Karimi, B., Zamani, A. and Clark, J. (2005). A Bipyridyl Palladium Complex Covalently Anchored onto Silica as an Effective and Recoverable Interphase Catalyst for the Aerobic Oxidation of Alcohols, *Organometallics*, **24**, pp. 4695–4698.
142. Choudhary, D., Paul, S., Gupta, R., *et al.* (2006). Catalytic Properties of Several Palladium Complexes Covalently Anchored onto Silica for the Aerobic Oxidation of Alcohols, *Green Chem.*, **8**, pp. 479–482.
143. Dijksman, A., Arends, I. and Sheldon, R. (1999). Efficient Ruthenium–TEMPO-Catalysed Aerobic Oxidation of Aliphatic Alcohols into Aldehydes and Ketones, *Chem. Commun.*, **16**, pp. 1591–1592; Dijksman, A., Marino-Gonzalez, A., Mairati i Payeras, A., *et al.* (2001). Efficient and Selective Aerobic Oxidation of Alcohols into Aldehydes and Ketones Using Ruthenium/TEMPO as the Catalytic System, *J. Am. Chem. Soc.*, **123**, pp. 6826–6833.
144. Semmelhack, M., Schmid, C., Cortes, D., *et al.* (1984). Oxidation of Alcohols to Aldehydes with Oxygen and Cupric Ion, Mediated by Nitrosonium Ion, *J. Am. Chem. Soc.*, **106**, pp. 3374–3376; Gamez, P., Arends, I., Reedijk, J., *et al.* (2003). Copper(II)-Catalysed Aerobic Oxidation of Primary Alcohols to Aldehydes, *Chem. Commun.*, **19**, pp. 2414–2415.
145. Chung, C. and Toy, P. (2007). Multipolymer Reaction System for Selective Aerobic Alcohol Oxidation: simultaneous Use of Multiple Different Polymer-Supported Ligands, *J. Comb. Chem.*, **9**, pp. 115–120.
146. Karimi, B., Biglari, A., Clark, J., *et al.* (2007). Green, Transition-Metal-Free Aerobic Oxidation of Alcohols Using a Highly Durable Supported Organocatalyst, *Angew. Chem. Int. Ed.*, **46**, pp. 7210–7213.

147. Salvador, J. and Clark, J. (2002). The Allylic Oxidation of Unsaturated Steroids by Tert-Butyl Hydroperoxide Using Surface Functionalised Silica Supported Metal Catalysts, *Green Chem.*, **4**, pp. 352–356.
148. Rajabi, F., Luque, R., Clark, J., *et al.* (2011). A Silica Supported Cobalt (II) Salen Complex as Efficient and Reusable Catalyst for the Selective Aerobic Oxidation of Ethyl Benzene Derivatives, *Catal. Commun.*, **12**, pp. 510–513.
149. Maksimchuk, N., Timofeeva, M., Melgunov, M., *et al.* (2008). Heterogeneous Selective Oxidation Catalysts Based on Coordination Polymer MIL-101 and Transition Metal-Substituted Polyoxometalates, *J. Catal.*, **257**, pp. 315–323; Maksimchuk, N., Kovalenko, K., Arzumanov, S., *et al.* (2010). Hybrid Polyoxotungstate/MIL-101 Materials: synthesis, Characterization, and Catalysis of $H_2O_2$-Based Alkene Epoxidation, *Inorg. Chem.*, **49**, pp. 2920–2930.
150. Ferey, G., Mellot-Draznieks, C., Serre, C., *et al.* (2005). A Catalytic, Environmentally Benign Method for the Epoxidation of Unsaturated Terpenes with Hydrogen Peroxide, *Science*, **309**, pp. 2040–2042.
151. Maurya, M., Chandrakar, A. and Chand, S. (2007). Zeolite-Y Encapsulated Metal Complexes of Oxovanadium(VI), Copper(II) and Nickel(II) as Catalyst for the Oxidation of Styrene, Cyclohexane and Methyl Phenyl Sulfide, *J. Mol. Catal. A: Chem.*, **274**, pp. 192–201.
152. Correa, R., Salomao, G., Olsen M., *et al.* (2008). Catalytic Activity of $Mn^{III}$(Salen) and $Fe^{III}$(Salen) Complexes Encapsulated in Zeolite Y, *Appl. Catal. A: Gen.*, **336**, pp. 35–39.
153. Jain, K., Herrmann, W. and Kühn, F. (2008). Synthesis and Catalytic Applications of Chiral Monomeric Organomolybdenum(VI) and Organorhenium(VII) Oxides in Homogeneous and Heterogeneous Phase, *Coord. Chem. Rev.*, **252**, pp. 556–568.
154. Li, C. (2004). Chiral Synthesis on Catalysts Immobilized in Microporous and Mesoporous Materials, *Catal. Rev.*, **46**, pp. 419–492.
155. Thomas, J., Maschmeyer, T., Johnson, B., *et al.* (1999). Constrained Chiral Catalysts, *J. Mol. Catal. A: Chem.*, **141**, pp. 139–144; Jones, M., Raja, R., Thomas, J., *et al.* (2003). A New Approach to the Design of Heterogeneous Single-Site Enantioselective Catalysts, *Top. Catal.*, **25**, pp. 71–79.
156. Taylor, S., Gullick, J., McMorn, P., *et al.* (2001). Catalytic Asymmetric Heterogeneous Aziridination of Styrene Using Cuhy: effect of Nitrene Donor on Enantioselectivity, *J. Chem. Soc., Perkin Trans.*, **2**, pp. 1714–1723; Caplan, N., Hancock, F., Page, P., *et al.* (2004). Heterogeneous Enantioselective Catalyzed Carbonyl- and Imino-Ene Reactions Using Copper Bis(Oxazoline) Zeolite Y, *Angew. Chem. Int. Ed.*, **43**, pp. 1685–1688.
157. Zhang, H., Xiang, S. and Li, C. (2005). Enantioselective Epoxidation of Unfunctionalised Olefins Catalyzed by Mn(Salen) Complexes Immobilized in Porous Materials via Phenyl Sulfonic Group, *Chem. Commun.*, **9**, pp. 1209–1211; Zhang, H., Zhang, Y. and Li, C. (2006). Enantioselective Epoxidation of Unfunctionalized Olefins Catalyzed by the Mn(Salen) Catalysts Immobilized in the Nanopores of Mesoporous Materials, *J. Catal.*, **238**, pp. 369–381.

# Chapter 24

# Light Alkanes Oxidation: Targets Reached and Current Challenges

Francisco IVARS* and José M. LÓPEZ NIETO*

The gas-phase selective oxidative transformation of light alkanes is an important challenge as it could reduce the number of process steps, decreasing both the energy required and $CO_2$ emissions, and improve the atom economy. In this chapter a summary of both the oxidative dehydrogenation and the O-insertion for $C_2$-$C_4$ alkanes is presented. In addition, an alternative method for a better selective oxidative transformation of methane and a description of the best catalytic systems are discussed.

## 24.1. Introduction

Selective oxidation by heterogeneous catalysis is one of the most important technologies currently used in the chemical industry and almost 25% of the most relevant organic compounds utilized in the production of consumer goods are obtained from them.[1] The functionalization of organic feeds into monomers of high interest in the polymer industry has been one of the most outstanding applications of the last six decades. However, over the last two decades, only a few reviews and books have been published on this issue.[1–22]

Most of the selective oxidation chemicals are produced from olefins and to a lesser extent from aromatic hydrocarbons. Thus, ethylene oxide, acrylonitrile, acrylic acid, acrolein, methacrolein, methacrylic acid, and 1,2-dichloroethane can be achieved from the corresponding olefins, while phthalic anhydride, benzonitrile, or benzoic acid are produced from aromatic hydrocarbons. The relatively easy production from petroleum, together with their high reactivity (moderate reaction temperatures can be employed, in the range of 300–450°C) have made olefins

*Instituto de Tecnología Química (UPV-CSIC), Campus de la Universidad Politécnica de Valencia,Avenida de los Naranjos s/n, 46022 Valencia, Spain.

and aromatic hydrocarbons one of the basic raw materials in the petrochemical industry.

Since the nature of raw materials has an important influence on the operating cost, substantial efforts have recently been made to develop alternative and cheaper raw materials for the production of several of the aforementioned chemicals. In this way, maleic anhydride (MA) production from *n*-butane represents an interesting model to be followed in other partial oxidation processes. Historically, MA was only obtained from benzene, then $C_4$-olefins were also employed, but for several decades and still today it is mainly produced (*ca.* 70%) from *n*-butane.[23]

Nowadays, the use of cleaner and more efficient feedstocks for the production of chemicals is one of the main technological concerns, and it is directly related to the search for solutions to global problems such as global warming, sustainability, the depletion of natural resources, energy efficiency, and $CO_2$ reduction. In this way, the gas-phase selective oxidative transformation of light alkanes is an important challenge because it could reduce the number of process steps, decreasing both the energy required and $CO_2$ emissions, and improving the atom economy.[2,3,8–16,20–22,24–42] Moreover, this could also permit the development of processes based on natural gases, not only methane but also ethane or propane.

On the other hand, a great interest in the development of selective process involving alkanes as feedstock, in both oxidative dehydrogenation or oxygen-insertion reactions, can be concluded from the evolution of the number and type of publications (in patents and open literature) presented in the last few years (Fig. 24.1).

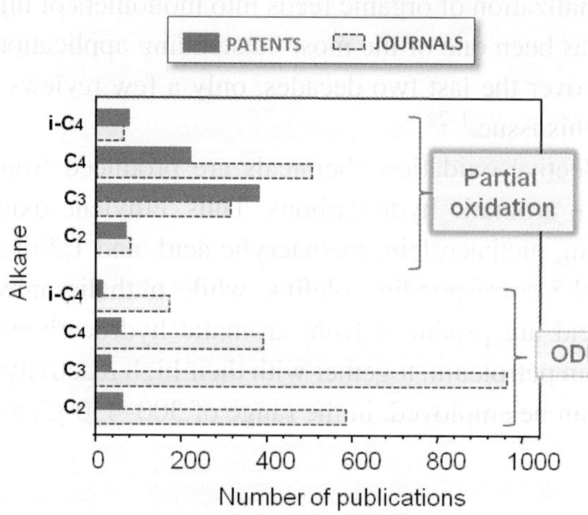


**Figure 24.1.** Number of publications in the last two decades (1989–2009) focused on oxidative dehydrogenation (ODH) of short chain alkanes and partial oxidation products (considering acetic acid, acrylic acid, methacrolein/methacrylic acid, maleic anhydride).

### 24.1.1. *Alkanes rather than olefins as raw materials for partial oxidation processes*

One of the most important applications of olefins is their functionalization by selective oxidation to obtain partial oxidation products of added value which are employed in the chemical industry to produce consumer goods. Nonetheless, the use of olefins involves several drawbacks related to their production processes. Currently, olefins are mainly obtained from petroleum or natural gas by steam cracking (around 70% of the global worldwide production) and by catalytic cracking (*ca.* 28%). The remaining 2% of olefins is obtained by catalytic dehydrogenation of alkanes and platforming of aromatics.[43] Ethylene and propylene are the most used olefins; in fact, they are the most used raw materials in petrochemistry as reflected by their high worldwide production (*ca.* 127 million tons of ethylene[44] and 65 million tons of propylene in 2009). Thus, ethylene is mainly used for the production of polyethylene, ethylene oxide, dichloroethylene, vinyl chloride, ethylbenzene (styrene), and acetic acid, whereas propylene is mainly employed for producing polypropylene, propylene oxide, acrylonitrile, acrylic acid, acrolein, isopropylbenzene (cumene), and isopropylic alcohol.

The specific raw material used in cracking processes depends on the natural resources from each geographic region, and this determines the final distribution of olefins.[43,45] Thus, in Europe and Asia, ethylene is produced from naphtha, gas oil, and natural gas condensates (co-producing propylene, $C_4$-olefins and aromatic compounds), whereas in the US, Canada, and the Middle East, steam cracking of ethane and propane is mainly employed. One aspect to be considered is the fact that for the production of 1 ton of ethylene in a typical naphtha cracker, 3.3 tons of naphtha are required,[45] while when using ethane only 1.7 tons of feed are needed.

To date, some of the best results reported for steam cracking of ethane are[46] i) selectivity to ethylene of 84% for ethane conversion of 54% (at 800°C, residence time of 0.79 s, ethane partial pressure of 154 kPa, and 0.3 $kg_{water}/kg_{raw}$); ii) selectivity to ethylene of 78% for ethane conversion of 69% (at 833°C, residence time of 0.75 s, ethane partial pressure of 154 kPa, and 0.3 $kg_{water}/kg_{raw}$). Therefore, neither the energy efficiency (reaction temperatures of 800°C and above in an endothermic reaction) nor the productivity are optimal in the steam cracking processes.

Accordingly, from an economical point of view, these processes have some important drawbacks; they need high reaction temperatures, especially for steam cracking which can reach up to 850°C, and many by-products are formed which must be separated from the desired reaction product through processes that are complicated and expensive. Coke is included among these by-products as it has a negative affect not only on the selectivity, but also on the process itself.

In addition, all these aspects require huge energy consumption. In fact, steam cracking is the chemical process with the highest energy consumption (934 kW h/$t_{C2H4}$) if compared with reported data for atmospheric crude oil distillation (133 kW h/t) or liquefied petroleum gas (LPG) separation (298 kW h/t).[47] Thus, steam cracking implies the heating of hydrocarbon in reactors with the presence of steam. The heat is transferred by radiation, and the heat transfer coils are made of alloys which are capable of withstanding high temperatures (around 800°C inside the reactor). The reaction mixture must be quickly cooled to avoid undesired reactions which can reduce the selectivity of olefins in the process. This cooling step is carried out on-line in heat exchangers placed immediately after the cracking zone. Hence, the reaction products, olefins and by-products, are cooled and separated. The heavier the feed, the greater the amount of by-products is. This means that ethane mainly produces ethylene with hardly any by-products, while gas oil gives less ethylene and more by-products which must be separated from the desired product, mainly employing distillation columns by complicated and expensive processes.

In steam cracking processes, the amount of coke produced and deposited on the heated pipeline walls depends on the type of fuel employed, operation conditions and the metallurgic nature of pipelines. In addition, coke is also produced in heat exchangers (where temperatures can be between 400 and 700°C).[48] Coke deposits with a thickness of some millimeters/centimeters make heat transfer difficult, so the temperature in the reactor must be increased which in turn leads to higher coke formation. Moreover, coke accumulation favors a pressure drop which results in reduced production of olefins. Over time, production must be frequently stopped to remove coke (decoking) from the reaction system. Decoking is carried out with a mixture of water and air to burn the coke. This process is undesirable as it results in a drop in the production of olefins, is expensive to maintain and reduces the longevity of the pipelines.

On the other hand, from an environmental point of view, very high $CO_2$ emissions are associated with the production processes of olefins,[45,47,49] either the ones inherent to chemical reactions themselves or those derived from the high energy consumption. In this way, $CO_2$ emissions associated with ethylene production in a steam cracking process are of *ca.* 1,200 $g_{CO2}/kg_{olefin}$.[47] This means that in 2004, global $CO_2$ emissions associated with ethylene production was 180–200 million tons,[49] and it is estimated to have been about 250 million tons in 2008.[45]

Finally, an increase in demand for olefins is expected (mainly for ethylene and propylene, of around 3.5% and 4.5–5.0%, respectively), while it seems that this will not be accompanied by a proportional increase in olefin production from the current processes.

So far, the use of alkanes as raw materials versus the current use of olefins in industrial processes has three important advantages: i) higher availability; ii) lower cost; and iii) lower environmental impact. Accordingly, it seems justified and necessary to develop new methods to produce olefins from alkanes, as these raw materials are more available, cheaper, and eco-friendly (achieved through oxidative dehydrogenation (ODH), being especially interesting for ethylene production). Alternatively (or additionally), alkanes could be directly employed in the selective oxidation or ammoxidation processes to obtain the desired oxygenated partial oxidation products, replacing the olefin in the current industrial process (especially interesting are propane and isobutane).

Industrial chemical processes actually developed are in general well optimized in terms of yield, productivity, and efficiency, which give very limited improvement options with respect to reducing both costs and $CO_2$ emissions. Therefore, when a replacement of a well-established current commercial process is proposed, several considerations must be taken into account. Overall the most important are the economic benefits that justify both the high capital investments required and operational risks assumed for any new process. The major costs of production are related to the feedstock, and in this sense light alkanes could be a cheaper alternative.

Light alkanes, mainly propane and *n*-butane, are an overplus from LPG fractions. On the other hand, liquefied natural gas (LNG), mainly composed of methane, also contains other hydrocarbons such as ethane (3–10%) or propane (0.5–2%). However, to date, LPG and LNG fractions (except in the case of ethane) have mainly been employed as domestic/industrial fuel to generate heat. The ready availability of LPG fractions, as an overplus in refineries, and the existence of huge natural gas deposits imply the low merit of their components, which has increased interest in profiting from them by their employment as raw materials in the petrochemical industry.

Alkanes can be obtained directly from natural gas, which means very low associated $CO_2$ emissions. For example, $CO_2$ emissions related to obtaining propane from natural gas is 230 $g_{CO2}/Kg_{propane}$, while $CO_2$ emissions associated with the steam cracking process for obtaining propylene is around five times higher (i.e. 1,200 $g_{CO2}/kg_{C3H6}$).[47] Thus, employing alkanes would mean a drastic reduction of $CO_2$ emissions with respect to olefin processes.

The higher reaction heats of alkanes, which could be an *a priori* disadvantage since it involves a major investment in heat exchangers and reactors, means the possibility of a higher energy saving once the initial investment in the appropriate infrastructure had been made.

Unfortunately, the use of light alkanes also presents some drawbacks. Indeed, they are less reactive, and more reaction steps are intrinsically involved in their reactions than for olefins, so that it is necessary to design catalysts which are more active, multifunctional, and selective enough to avoid over-oxidation of the desired products, since the latter are usually more reactive than the starting alkane.

At the moment, selective oxidation of $n$-butane to MA is the only catalytic oxidation process employing a light alkane as the feedstock which has been fully established at an industrial level. It must be indicated that surprisingly the results from the $n$-butane process (MA yield around 80%, with selectivity of 60%) are better than those obtained from butenes.[23,50–52] This fact has encouraged the scientific community to study similar catalytic reactions with other alkanes. Thus, other successful processes could be developed by using the appropriate catalyst, optimal reaction conditions, and an improved reactor technology.

### 24.1.2. *Key aspects of the selective oxidation of light alkanes*

At the moment, catalysts based on transition metal oxides with redox properties present the most promising results in the gas-phase partial oxidation reactions of light alkanes (Table 24.1). Except in the case of methane, which will be discussed later, alkanes can be transformed into the corresponding olefins and/or partial oxidation

**Table 24.1.** Summary of the best results in the most interesting partial oxidation reactions of $C_2$-$C_3$ alkanes.

| | Temp. (°C) | Conv.[a] (%) | S[b] (%) | Y[b] (%) | Ref. |
|---|---|---|---|---|---|
| **Ethane → Ethylene** | | | | | |
| MoVTeNbO | 400 | 85 | 88 | 75 | 56 |
| $NiNbO_x$ | 400 | 51 | 90 | 46 | 57 |
| VCoAPO-18 | 600 | 57 | 60 | 34 | 58 |
| **Ethane → Acetonitrile** | | | | | |
| $NbSbO/Al_2O_3$ | 540 | 40 | 50 | 20 | 59 |
| Co-Beta | 475 | 47 | 57 | 26 | 60 |
| **Ethane → Acetic acid** | | | | | |
| MoVSbNbReCaO | 277 | 14 | 78 | 11 | 61, 62 |
| $MoVNbPdO_x$ | 300 | 5 | 82 | 4.1 | 62 |
| **Propane → Propylene** | | | | | |
| VMgO | 540 | 62 | 38 | 24 | 39, 41 |
| V-Silicalite | 550 | 30 | 70 | 21 | 63 |
| **Propane → Acrylonitrile** | | | | | |
| MoVTeNbO | 420 | 86 | 72 | 62 | 64 |
| $VSbWO_x/SiO_2$-$Al_2O_3$ | 500 | 67 | 60 | 40 | 65, 66 |
| **Propane → Acrylic acid** | | | | | |
| MoVTeNbO | 432 | 69 | 67 | 46 | 67 |
| MoVTeNbO | 380 | 80 | 61 | 49 | 68 |

[a]Alkane conversion.
[b]Selectivity or yield to desired partial oxidation product.

O- or N-containing products in a single step at relatively moderate temperatures (i.e. lower than 450°C). However, to date, the selectivity to the partial oxidation products is the key factor to be improved in order to achieve competitive processes.[2,3,8–16,24–42]

However, several aspects should be considered in the development of new catalytic systems. Thus, in addition to an extensive knowledge of the nature of active sites, the multifunctionality (redox, acid-base, etc.) of catalysts should be finely tuned. Moreover, the stability of well-defined crystalline structures and the effect of promoters will also be important factors to be considered. But, in addition to these, the alkane feed, the reactivity of olefinic intermediates, and the stability of partial oxidation products are key aspects to take into account. In this way, modification of the catalyst surface reactivity by chemisorbed species in the oxidation of alkanes is lower than that observed for the oxidation of olefins, which can facilitate a lower volume of undesired products in the case of alkane-based processes.[53–55]

It is well accepted that, in all cases the oxidation reaction of a saturated hydrocarbon starts with a hydrogen abstraction from the alkane molecule, which has been proposed to be the rate-limiting step.[2,23–42,50–55,69–73] Alkane activation involving H-abstraction on surface $O^-$ species has been generally proposed,[69–75] while it is also accepted that the oxygen feed does not participate directly in the reaction. In any case, the nature of oxygen species involved in partial oxidation reactions will depend on the nature of active sites and the reaction conditions. Thus, peroxo and superoxo species could also be present on the catalyst surface generating different active sites.[69,71–73] The $O^-$ species formation would be related to structural defects such as excess oxygen anions or cationic vacancies, revealed by the p-type semi-conduction properties of the catalysts, or to charge transfers $[V^{5+}O^{2-} \rightarrow V^{4+} - O^-]$ thermally activated and independent of the conduction properties.[69] Moreover, site isolation seems to play an important role in the stabilization of the active species intermediates between lattice-oxygen and adsorbed-oxygen species.[2,3,69–72] The possible role of $O^-$ species in oxidation reactions is mainly considered for alkane activation processes and not in other reaction steps. In fact, it is generally accepted that the formation of oxygenated molecules from olefins in consecutive reactions is carried out following a Mars–van Krevelen-based mechanism involving nucleophilic $O^{2-}$ species.[1–22] This aspect should also be considered when starting from alkanes, although it is probable that the importance of the concentration of reactants will be different to that proposed in olefins, due to the high reactivity of olefins in respect to alkanes. On the other hand, the role of the adsorbed oxygen species is not completely clear. Although they have been related to deep oxidation reactions during ODH,[69–72] some additional role should be considered in the case of partial oxidation reactions. This is the case when considering *n*-butane oxidation to MA,[50,52,74,75] but is probably also the case in propane oxidation to acrylic acid, in which several oxygen atoms are involved.

Additional studies are required in order to understand the role of defective sites over the catalyst surface on alkane activation. For example, promoted NiO catalysts, such as Ni-Nb-O[57,76,77] or supported NiO,[78–81] also seem to be interesting catalysts for the ODH of ethane, while NiO is very active but unselective. The role of different oxygen species related to non-stoichiometric $Ni^{n+}$ sites also seems to be an important factor to be discussed.

After the first hydrogen abstraction from the alkane (or in parallel to the first H-abstraction), a second hydrogen must be removed to form the corresponding adsorbed olefin. This could simply be desorbed as olefin, but in the presence of a second active site the desorbed olefin could be re-adsorbed and transformed into other oxidation products. Alternatively, the adsorbed olefin could be directly transformed by successive steps, without intermediate desorption, to form final partial oxidation products which should be finally desorbed. Accordingly, the consecutive reactions could be carried out in the same active site or in different active sites (after successive adsorption of the olefin) favoring subsequent reaction steps as indicated in Scheme 24.1. The number of possible reactions is relatively high, since apparently olefins, aldehydes, acids, or anhydrides could be obtained. However, as will be discussed later, the predominance of each reaction product will depend on the characteristics of the catalysts and/or the nature and stability of the reactant and reaction products.

Firstly, the capacity of the catalysts to carry out the catalytic redox cycle at reaction conditions will depend on the redox abilities of metal sites on the catalyst surface. Additionally, the contribution of bulk in the catalyst re-oxidation must be considered in many cases.[1–42] In this way, active and selective catalysts will be those based on transition metal oxides presenting:[4,5] i) double bonds Me=O (Me: metal); ii) capacity to show different oxidation states; iii) non-stoichiometric oxide composition in which oxidation states could be varied without major changes in

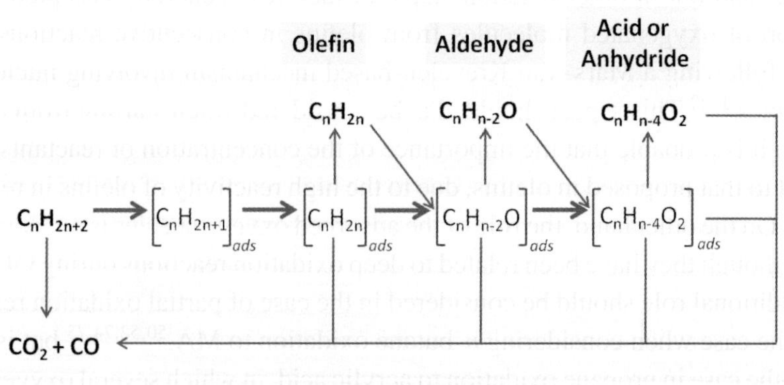


**Scheme 24.1.** Possible reaction network for the partial oxidation of short chain alkanes.

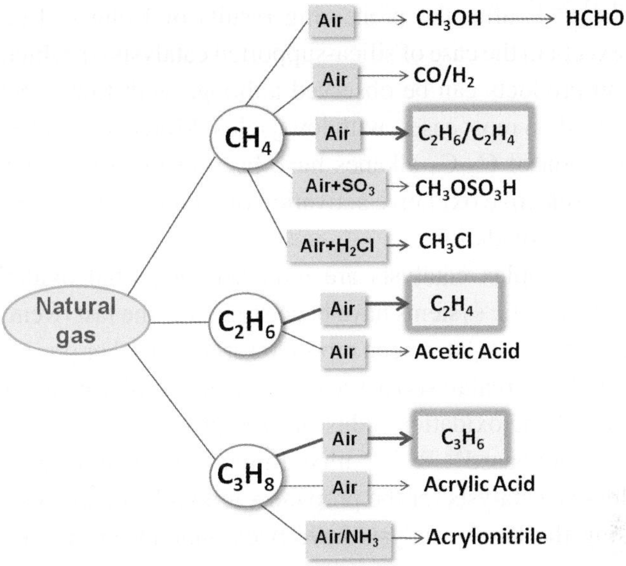


**Scheme 24.2.** Possible reaction network for the partial oxidation of short chain alkanes.

crystalline structure; and iv) some acid characteristics. It is clear that the presence of other metal oxides as promoters could modify the characteristics of the catalyst, changing the Me=O double bond character, the acid-base, and/or redox properties, but could also modify the crystalline structure and stability. These aspects have been discussed in general for selective oxidation reactions for both olefins[1–22] and paraffins.[13,17,18,20–22,24–42]

With respect to the type of catalyst, it is evident that just as molybdenum and antimony catalytic systems have been extensively used for the partial oxidation of olefins,[1,2] V-containing catalysts seem to show more adequate properties for the partial oxidation processes for light alkanes.[13,17,20–22,24–42]

The capability to form a large range of non-stoichiometric solids for molybdenum, antimony, and vanadium oxides (as, for example, $MoO_3$, $Mo_8O_{23}$, $Mo_5O_{14}$, $Mo_{17}O_{47}$, $Mo_4O_{11}$, $MoO_2$, $Mo_9O_{16}$; $Sb_2O_5$, α-$Sb_2O_4$, $Sb_6O_{13}$, $Sb_6O_{13}OH$, $Sb_2O_3$; and $V_2O_5$, $V_3O_7$, $V_4O_9$, $V_6O_{13}$, $VO_2$, $V_4O_7$, $V_3O_5$, $V_2O_3$, respectively) can explain the importance of these elements in the development of active and selective catalysts in oxidation reactions by redox mechanisms.

### 24.1.2.1. *Importance of the catalyst structure*

For the oxidative dehydrogenation of alkanes, vanadium supported on metal oxides or incorporated in molecular sieves have been the most studied catalyst systems,[13,17,20–22,24–42] after the pioneering results of Kung and co-workers.[39,41] However, and except in the case of silica-supported catalysts (in which O-containing partial oxidation products can be observed although with low selectivity), olefins were only produced, in most cases, with low yields. Moreover, studies on the oxidative dehydrogenation of $C_2$-$C_4$ alkanes have been of great importance for understanding both the selective oxidative activation of alkanes and the deep oxidation of olefins during alkane oxidation.

Indeed, more complex catalysts are required for partial oxidation reactions. Although several catalytic systems have been studied in the last twenty years, a very limited number of catalysts have been reported for industrial or pre-industrial use. In fact, in addition to V-P-O catalysts (based on vanadyl pyrophosphate), the unique catalyst used for an alkane oxidation industrialized process,[23,50–55] only V-Sb-[65,82–86] and MoVTe(Sb)NbO-based[31,68,87–93] mixed-metal oxides have been proposed as sufficiently effective catalysts for the propane ammoxidation process. In both cases pilot plants using the latter catalysts have been announced on the bases of their catalytic results.

It is clear that in the case of selective catalysts, the presence of a second element involved in the formation of a specific crystalline phase improves the catalytic behavior of V-based materials. This is the case for V-P-O catalysts, in which the presence of phosphorous permits the synthesis of several crystalline compounds with different compositions, in which vanadium atoms can exhibit different oxidation states.[23,50–55] However, the dominant presence of the vanadyl pyrophosphate (Fig. 24.2a) phase is the key factor in the development of active and selective sites.

Also in the case of V-Sb and Mo-V based catalysts, the presence of Sb or Mo facilitates the synthesis of specific crystalline phases in which vanadium atoms can be easily accommodated, changing the behavior with respect to pure oxides, although, as will be discussed later, Sb sites and Mo sites will also be directly involved in both alkane and olefin transformation.

One important characteristic of V-Sb-O based catalysts, initially developed for the partial ammoxidation of propane to acrylonitrile in the 1980s by BP,[65] is the presence of cation-deficient rutile-type $VSbO_4$ (Fig. 24.2b) and $\alpha$-$Sb_2O_4$ structures. Antimony is mainly present in a pentavalent state favoring a large amount of partially reduced vanadium species, with the formation of $V^{III}SbO_4$ and the possible presence of $V^{IV}$ in substitutional or interstitial solid solution coordinations.[82–86] The incorporation of other metal elements (such as Mo, Al, W, etc.)[65,82–86] as promoters or by the synthesis of ternary V-Sb-Me mixed oxide crystalline phases[94,95] can lead to a significant improvement in catalytic behavior.

In the case of MoVTe(Sb)NbO-based catalysts, initially developed by Mitsubishi for the partial oxidation and ammoxidation of propane,[68,87–89] it is now

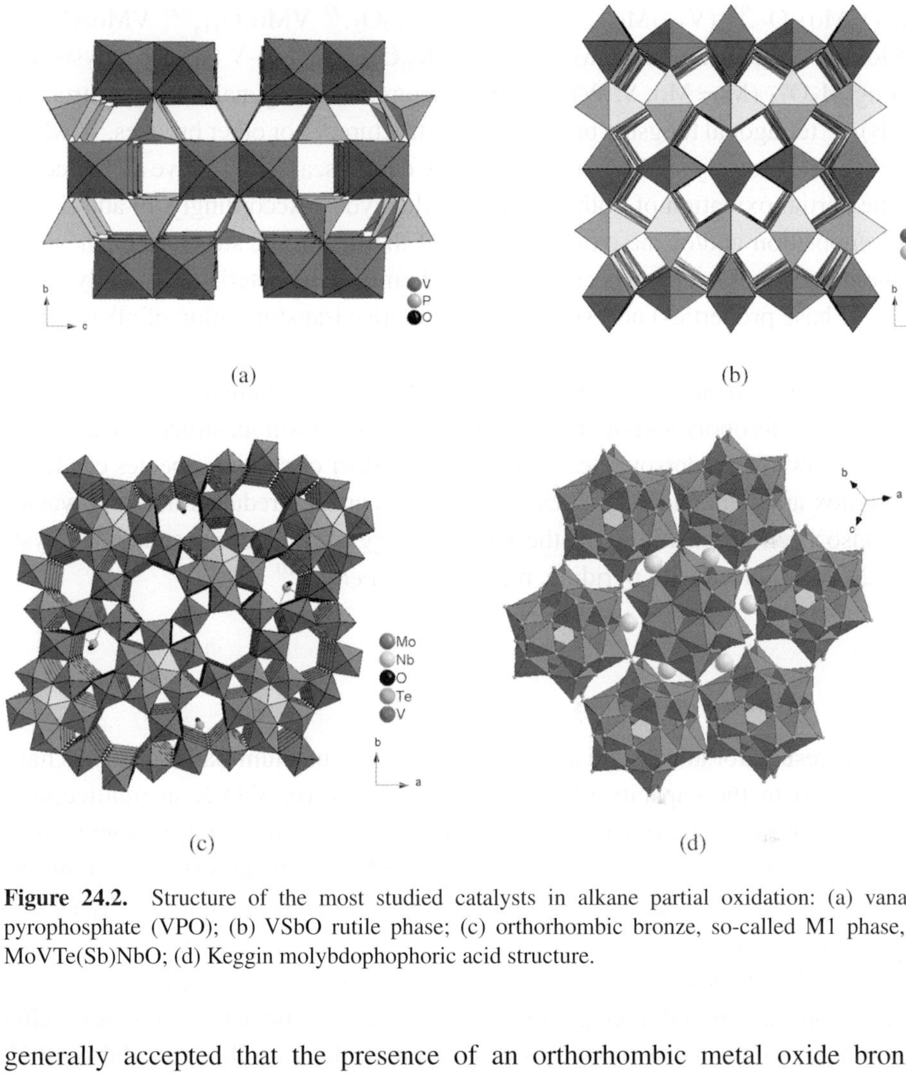


**Figure 24.2.** Structure of the most studied catalysts in alkane partial oxidation: (a) vanadyl pyrophosphate (VPO); (b) VSbO rutile phase; (c) orthorhombic bronze, so-called M1 phase, i.e MoVTe(Sb)NbO; (d) Keggin molybdophophoric acid structure.

generally accepted that the presence of an orthorhombic metal oxide bronze, i.e. $(TeO)_2M_{20}O_{56}$ or $(SbO)_2M_{20}O_{56}$ (M = Mo, V, Nb) phase, the so-called M1 phase,[31,89–93] is essential in obtaining optimal catalytic performance. This crystalline phase, isostructural with $Cs_x(Nb,W)_5O_{14}$,[96] exhibits a quite complex network of $MO_6$ octahedrons forming pentagonal, hexagonal, and heptagonal channels (Fig. 24.2c).[93,94] The pentagonal channels are occupied by M cations, forming $MO_7$ pentagonal bipyramidal groups, while Te or Sb occupy the hexagonal channels, and the heptagonal channels usually remain empty. The importance of the nature of the crystalline structure in this case can be easily deduced from the comparison with the catalytic behavior of the M1-type orthorhombic metal oxide bronze, which is very effective in propane and propylene oxidation,[31,68,87–93] to other mixed oxides formed with the same elements. Thus, it has been reported that $(V,Mo)_2O_5$,[97] $(V_{1.23}Mo_{0.66}O_5)$,[98] $V_{1.1}Mo_{0.9}O_5$,[99] $VMo_3O_{11}$,[100] $VMo_4O_{14}$,[101] $V_2MoO_8$,[102] $V_6Mo_4O_{25}$, or $V_2MoO_8$,[103] $V_9Mo_6O_{40}$,[98] or Mo-V based catalysts presenting $M_5O_{14}$ (M = Mo, V, W) bronze structure,[104] hexagonal tungsten structure (HTB),[105] tetragonal tungsten bronze (TTB) structure,[106] or other bronzes,[107] which have been extensively studied for the last three decades, are only active and selective for the partial oxidation of both olefins and aldehydes. Accordingly, in addition to the composition and/or the oxidation state of the metal, the structure of the crystalline phase can facilitate specific physico-chemical characteristics (mainly redox and acid-base properties) necessary for an oxidative transformation of alkanes into chemicals.

On the other hand, a fourth type of catalytic system studied over the last few years is the heteropolyacids and their salts.[108,109] Composition, structure (including Keggin, Dawson, Anderson, etc.), nature, and number of cationic species can tailor both redox and acid characteristics. In addition, partially reduced heteropolyacids have also been employed as in the case of Keggin molybdophosphate catalysts (Fig. 24.2d) reduced with pyridine, reported by Li *et al.*[110]

#### 24.1.2.2. *Nature of active sites*

Catalytic results for alkane activation obtained over vanadium-based catalysts have been related to the capacity of V=O and/or V-O-V (or V-O-X in multicomponent catalysts) species to activate the C–H bonds of alkanes and, in some cases, to also insert oxygen atoms in activated hydrocarbon molecules (transforming the reaction intermediates in a partially oxygenated compound such as aldehyde, carboxylic acid, anhydride, etc.). In general, the catalytic behavior of V-based catalysts for ODH reactions, can be explained by considering three aspects[2,3,8–16,24–42] (i) the coordination and aggregation of $V^{5+}O$ species (the active sites in paraffin activation); (ii) the redox properties of catalysts (related to the reducibility of V-sites), which can be changed by the incorporation of promoters or interaction with the support or matrix; and (iii) the acid-base character of the catalysts (including the support or matrix in which the active sites are incorporated). In this way, not only the V-sites but also the V-environment, strongly influence the adsorption of reactants and desorption of partial oxidation products (facilitating — or not facilitiating — consecutive reactions, which finally determines the selectivity to olefins).

One of the key aspects in the catalytic behavior for alkane partial-oxidation catalysts seems to be related to the coordination number (Fig. 24.3), the aggregation degree, and the oxidation state of the vanadium species as active sites.[13,14,40] It fact, it has been shown that the V-environment plays an important role in catalytic behavior,

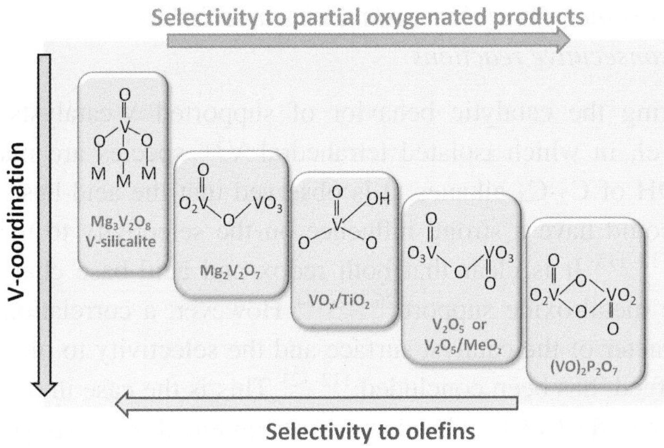


**Figure 24.3.** Importance of the V-structure in V-catalysts for both H-abstraction and O-insertion reactions. From Ref. 36.

as demonstrated by comparing the catalytic performance of different V-based catalysts. In this way, it is generally accepted that catalysts presenting isolated vanadium species with tetrahedral coordination (as in $Mg_3V_2O_8$, supported-$VO_x$ at the sub-monolayer level, V-silicalite or VAPO-5) are some of the most effective catalysts for the ODH of $C_2$-$C_4$ alkanes, especially for propane and *n*-butane.[2,3,8–16,24–42] In fact, these catalysts mainly favor the H-abstraction from alkane, facilitating the formation of olefins without the apparent formation of other partial oxidation products.[13,14,40] Thus, vanadium species with low coordination numbers favor ODH reactions.[26,27,36] On the other hand, the higher the coordination number the higher the capacity of the catalyst for O-insertion, thus facilitating the appearance of O-containing partial-oxidation products (Fig. 24.3). This is also mainly favored by the presence of double bonds V=O in V-O-V pairs,[36,111] as in the case of $VO_2P_2O_7$, $TiO_2$-supported vanadium oxides (at monolayer level), and multilayered supported vanadium oxide (presenting $V_2O_5$ crystals) in which the main partial oxidation products are oxygenated. In these cases, the V-sites are directly involved in alkane activation, but also in the selective attack to the olefin intermediate facilitating not only the formation of olefins but also the insertion of lattice oxygen species into the hydrocarbon intermediate.

A slightly different behavior to that observed in the latter catalysts has been shown by VSb or MoV multicomponent catalysts, in which vanadium is mainly present as dimeric V-O-X pairs, with octahedral $V^{n+}$ species (n = 4 and/or 5). However, these multicomponent catalytic systems present at least two different active sites, i.e. V-sites involved in the oxidative activation of alkanes, and Sb- or Mo-sites involved in olefin partial oxidation.[31,65,68,82–93]

### 24.1.2.3. *Importance of catalyst surface characteristics in consecutive reactions*

When comparing the catalytic behavior of supported V-catalysts (at the sub-monolayer level, in which isolated tetrahedral $V^{5+}$ species are mainly present) during the ODH of $C_2$-$C_4$ alkanes, it is observed that the acid-base properties of the catalysts could have a strong influence on the selectivity to the corresponding olefins.[36,112–115] It is clear that both redox and acid-base character strongly depend on the metal-oxide support.[26,112–124] However, a correlation between the acid-base character of the catalyst surface and the selectivity to olefin, depending on the alkane feed, has been concluded.[112–121] This is the case in the selectivity to olefin during the ODH of $C_2$-$C_4$ alkanes over undoped and K-doped $VO_x/Al_2O_3$ catalysts.[112] Thus, the selectivity to olefins over the undoped catalyst decreases when increasing the number of carbon atoms of the alkane feed, while for the K-doped catalyst the opposite trend is observed. Since only the elimination of acid sites (mainly related to the alumina support) was observed after K-addition (while no modification of the V-species reducibility in both catalysts was detected), it was concluded that the presence of acid sites in these catalysts could have a different influence in the formation (or in the deep oxidation) of olefins depending on the alkane feed.

Similar results were observed in different V-containing catalysts, presenting isolated $V^{5+}$ species, which confirmed the importance of the presence/absence of acid sites on the catalyst surface on the selectivity to propene or butenes.[113] A similar performance was observed in V-containing molecular sieves (VAPO-5 or MgVAPO-5), in which the presence of acid sites ($Mg^{2+}$-containing aluminophosphate materials, i.e. MgVAPO-5) seemed to favor a higher catalytic activity during the ODH of ethane than that achieved over siliceous materials (V-silicalite).[113] Accordingly, the selectivity to ethylene decreases as VMgAPO-5 > $VO_x/Al_2O_3$ > $VO_x/Al_2O_3$(K-doped) > $V/SiO_2$ = VAPO-5 = V-silicalite > $VO_x/MgO$, while the selectivity to butenes (during ODH of *n*-butane) or propene (for propane ODH) shows the opposite trend.[113] Thus, the presence of acid sites has a positive effect during the ODH of ethane but is negative in the ODH of *n*-butane or propane.

Changes in the selectivity to butenes during the ODH of *n*-butane can be more easily explained by considering the *in situ* infrared (IR) study on the adsorption of $C_3$-$C_4$ olefins over supported vanadium oxide catalysts, when using metal oxide supports with different acid-base properties.[114,115] On catalysts presenting acid sites, a fast isomerization of $C_4$-olefins with the selective formation of 2-butene (more easily transformable into carbon oxides) is observed. In these cases, the presence of different adsorbed O-containing species (carbonyl and alkoxide species), which can be considered as precursors in the formation of carbon oxides, was also observed.

However, adsorbed butadiene species were mainly observed on catalysts presenting basic sites (mainly when using MgO as the support) or in those in which acid sites were previously eliminated (as in the case of a K-doped alumina-supported catalyst).[115] Accordingly, the presence of acid sites in the catalyst surface favors a strong interaction between the olefin and the catalyst surface facilitating the deep oxidation.[114] The same conclusion has also been proposed for propane ODH, indicating that the different selectivity to olefins achieved at high propane conversions could be explained by considering the higher/lower adsorption of the olefinic intermediate in the presence/absence of acid sites.[125] In this way, the formation of adsorbed alkoxide species, which are the precursors of carbon oxides, is favored in catalysts presenting acid sites such as $Al_2O_3$-supported vanadium oxide. However, adsorbed alkoxide species are not observed in K-doped $Al_2O_3$-supported vanadium catalysts in which the acid sites were selectively eliminated.

In the case of the ODH of ethane, the catalytic behavior of supported V-catalysts is completely different, and a positive role of acid sites in catalysts on the selectivity to ethylene is observed as a consequence of i) the presence of acid sites which do not have a negative effect on the olefin stability, and ii) the positive effect of the support surface acid sites on the catalytic activity of V-atoms.[114,115] In this way, the catalytic activity of supported vanadium oxide catalysts, when using metal oxide supports with an acid character (such as $Al_2O_3$), is higher than when supported on metal oxides with a basic character (such as MgO).[126]

However, although these types of catalysts present relatively high selectivity to olefins at low alkane conversions in ODH reactions, their main drawback is related to their high catalytic activity for deep oxidation of the corresponding olefins. Thus, in most cases, the olefin oxidation is 5 to 10 times faster than the alkane oxydehydrogenation,[36] limiting the yield to $C_3$ or $C_4$ olefins below *ca.* 40%[34–41] (Fig. 24.4).

Accordingly, we need to design new catalysts in which the active sites for the alkane oxidative activation are less effective in the deep oxidation of olefins or, in the case of O-insertion reactions, to favor a fast selective transformation of olefin to O-containing partial oxidation products. At the moment there are only two examples in which the catalyst shows an extremely low activity for olefin oxidation with respect to that for alkane oxidation, i.e. MoVTeNbO[152] and Ni-modified catalysts[57,76–81] for the ODH of ethane to ethylene (which will be further discussed in Section 24.1.4). This can be explained by the absence of allyl hydrogen in ethylene and the low activity of the catalyst in the abstraction of the vinyl hydrogen. Conversely, the interaction between longer olefins (such as propylenes and butenes) and catalysts with allyl abstraction affinity leads to further oxidized products which will depend on the nature and reactivity of the corresponding olefin.

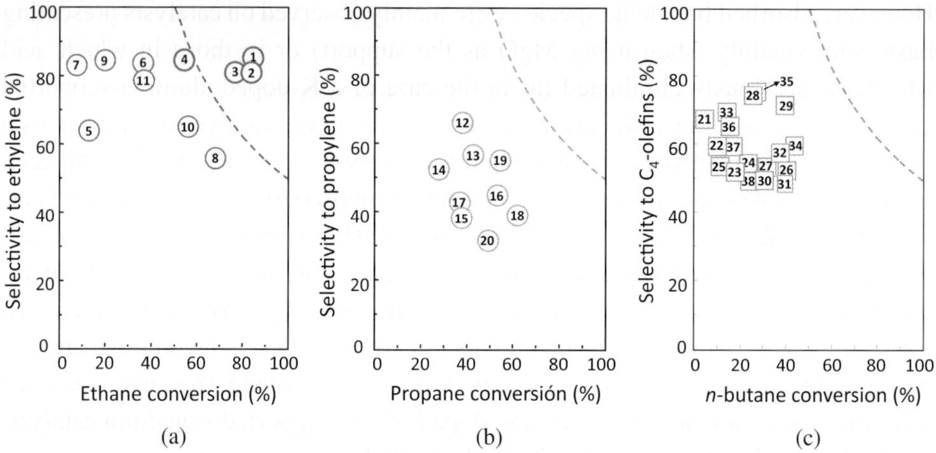


**Figure 24.4.** Catalytic results representative of catalysts used for oxidative dehydrogenation: **(a) Ethane ODH:** 1) MoVTeNbO,[127] 2) MoVTeNbO,[128] 3) MoVTeNb,[129] 4) MoVSbO,[130] 5) MoVNbO,[131] 6) Ni-Nb-O,[57] 7) Ni/$Al_2O_3$,[78] 8) NiO/MgO,[132] 9) NiTaNbO,[133] 10) NiO/$Al_2O_3$,[134] 11) MoVTaTeO.[135] **(b) Propane ODH:** 12) V-MCM,[136] 13) V-SBA-15,[137] 14) V-MCM-48,[138] 15) V-MCM-41,[139] 16) Mo-Cl-$SiO_2$/$TiO_2$,[140] 17) V-Mg-O/$SiO_2$,[141] 18) V-Mg-O,[142] 19) V-Mg-Ga-Mo-O,[143] 20) V-Mg-O/$Al_2O_3$.[144] **(c) *n*-Butane ODH:** 21–27) V-Mg-O,[145,146] 28,29) NiMoPO,[147] 30,31) MgNiSnO,[148] 32) Mg-Ni-$SO_4$,[148] 33,34) Ni-P-O,[149] 35) V-K-$SO_4$,[150] 36–38) V/$SiO_2$.[151]

### 24.1.2.4. *Multifunctionality in partial oxidation reactions*

In general, mixed-metal oxides are at present the most important catalytic system for the partial oxidation of $C_2$-$C_5$ alkanes into O-containing partial oxidation products. However, different characteristics should be considered among the several catalysts within this group. For instance, VPO- and MoVO-based catalysts are the most effective systems for *n*-butane (or *n*-butene) oxidation to MA and propane (or propene) oxidation to acrylic acid, respectively. It must be indicated that, although some similarities are observed for the ammoxidation[153,154] over MoVO-[155–162] or SbVO-based[65,82–86,163] catalyts, the latter show very low selectivity during the partial oxidation of propane and no results with other alkanes have been reported.[54,83] Therefore, SbVO-based catalysts will not be explicitly included in the following discussion.

It is clear that strong differences are observed between VPO and MoVTeNbO catalysts, which are responsible for their particular catalytic behavior. Thus, VPO catalysts with a vanadyl pyrophosphate structure present[23] i) one electron redox couple ($V^{5+}/V^{4+}$, $V^{4+}/V^{3+}$); ii) two electron redox couples, $V^{5+}/V^{3+}$; iii) bridging oxygen, V–O–V, V–O–P, or VO(P)V; iv) terminal oxygen ($V^{5+} = O$ and $V^{4+} = O$) and activated molecular oxygen, $\eta^1$-peroxo and $\eta^2$-superoxo species; and v) Lewis (probably $V^{5+}$ and $V^{4+}$) and Brønsted (probably —POH groups) acid sites. On the other hand, selective MoVTe(Sb)NbO catalysts are characterized by[31,89–93] i)

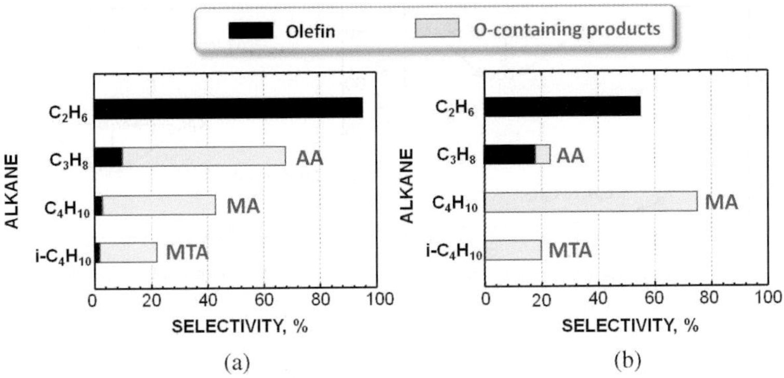


**Figure 24.5.** Selectivity to the main partial oxidation products achieved during the partial oxidation of $C_2$-$C_4$ alkanes over MoVTeNbO (a) and VPO (b) catalysts. AA = Acrylic acid; MA = maleic anhydride; MTA = methacrolein. From Ref. 170.

Mo–O–V–O–X (X = Mo or Nb) clusters in which the activation of propane occurs, specifically localized at V-sites; ii) Te–O–Mo–O–Y (Y = Te or V) such as sites where the oxidation of propylene to acrolein (or the ammoxidation to ACN) takes place; (iii) Nb–O–Mo–O–Z (Z = Nb or V) such as sites for the oxidation of acrolein to AA; and iv) Lewis acid sites (in general, the absence of Brønsted sites favors higher selectivity to acrylic acid).[164–169]

Figure 24.5 shows the selectivity to the main reaction products achieved during the partial oxidation of $C_2$-$C_4$ alkanes over VPO and MoVTeNbO catalysts.[170] The VPO catalyst is active and selective in the *n*-butane oxidation to MA and also in the partial oxidation of *n*-pentane to maleic/phthalic anhydrides.[171] However, the VPO catalyst shows a poor selectivity to partial oxidation products during propane[172–174] or ethane oxidation.[175] On the other hand, the multicomponent MoVTeNbO catalyst shows high activity and selectivity in the partial oxidation of propane to acrylic acid, and extremely high selectivity to ethylene in ethane ODH (even at high ethane conversion).[152] In the partial oxidation of *n*-butane to MA,[176] an interesting behavior has been observed over MoVTeNbO-based catalysts, presenting a reasonable selectivity to MA and a catalytic activity which can be even higher than that obtained by commercial VPO catalysts. In all cases the catalyst was studied in the 380–420°C temperature range.

In either VPO or MoVTeNbO catalysts, V-sites are directly involved in the selective oxidative activation of paraffins, while the presence of a second element is generally required in order to facilitate the formation of a defined crystalline phase (i.e. vanadyl pyrophosphate or orthorhombic metal oxide bronze, respectively) (Figs. 24.2a and 24.2c), but also for facilitating the multifuctionality of the catalysts. However, although both catalytic systems are active for the oxidative activation in alkanes, different selectivities are achieved depending on the alkane feed. Thus, a

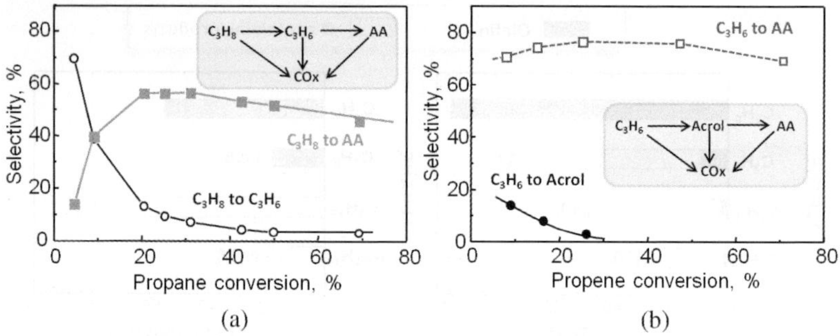


**Figure 24.6.** Selective oxidation of propane (a) and propylene (b) over a MoVTeNbO catalyst at 380°C and a $C_3/O_2/H_2O/He$ molar ratio of 4/8/30/58. Symbols:•, • = acrylic acid, • = propylene,• = acrolein. From Ref. 92.

comparative study on the catalytic results as well as on the nature of adsorption intermediates (from the study of the adsorbed olefins using IR) over these catalysts can help to understand the evolution of olefinic intermediates in the different alkane oxidation reactions.

Figure 24.6 shows the variation of the main partial oxidation products during the oxidation of propane (Fig. 24.6a) and propylene (Fig. 24.6b) over a MoVTeNbO catalyst.[92] It can be observed that propane is initially transformed into propylene (unstable primary product), which is then selectively oxidized into acrylic acid (a secondary reaction product). The formation of the olefin is clear, since the initial selectivity to propylene of *ca.* 90% is observed in effective catalysts for the partial oxidation of propane.[31,54,125,177] A similar reaction network has been proposed for propane oxidation with modified VPO catalysts,[178–180] but also in the propane ammoxidation over SbVO-[14,26–29,82–86] or MoVO-based catalysts[27,30,31,89–92,181] to give acrylonitrile. On the other hand, we must indicate that the selectivity to acrylic acid from propylene is higher than that observed from propane (Fig. 24.6), which could be related to the higher olefin adsorption during propylene oxidation. This is also observed for other catalytic systems (i.e. VPO or VSbO-based catalysts). May be the higher olefin adsorption during propylene oxidation could favor the best results with propylene.

At this point it is interesting to compare the evolution of propylene adsorption over catalysts with different surface acid characteristics, i.e. a MoVTeNbO catalyst (active and selective in the partial oxidation of propane to acrylic acid), an alumina-supported vanadium oxide (active in the ODH of propane to propylene), or a MoVNbO mixed oxide (active in the oxidative transformation of propane to propylene and acetic acid).[125] The final products observed in each case were related to the characteristics of the adsorbed intermediates (Fig. 24.7): (i) a π-allylic compound, interacting with a redox site intermediate in the selective oxidation of

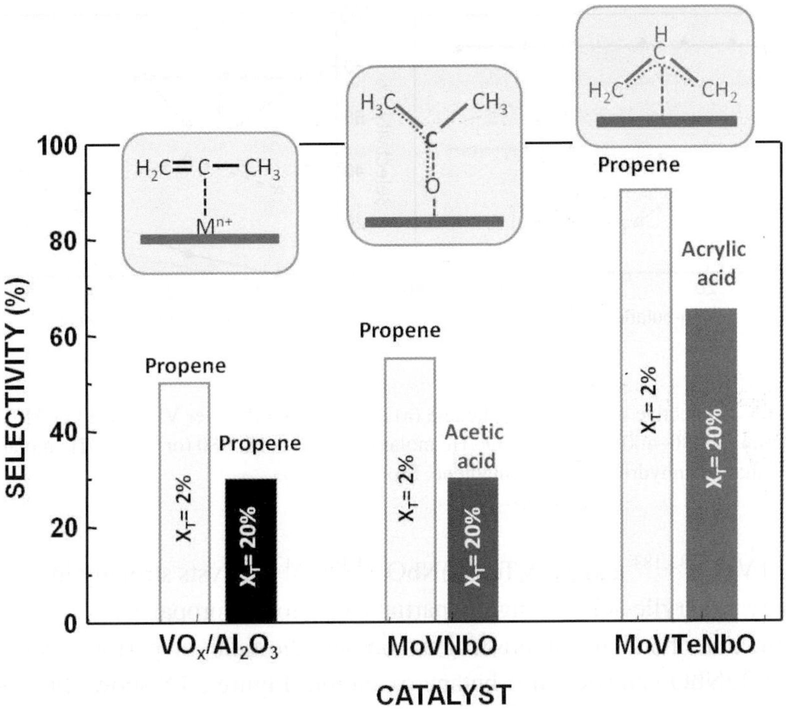


**Figure 24.7.** Selectivity to the main partial oxidation product achieved at low conversion ($X_T = 2\%$) and high conversion ($X_T = 20\%$) during propane oxidation at 400°C over: (i) $Al_2O_3$-supported vanadium oxide ($VOx/Al_2O_3$); (ii) Mo-V-Nb mixed oxides calcined in air at 450°C, presenting an amorphous material (MoVNbO); and (iii) Mo-V-Te-Nb mixed oxides heat-treated at 600°C in $N_2$ presenting M1 phase (MoVTeNbO) catalysts. For comparison, the main adsorbed species observed during the adsorption of propylene on each catalyst it is also presented. From Ref. 125.

propylene to acrylic acid (Fig. 24.7a); (ii) an enolic-type compound, formed in the presence of Brønsted acid sites, intermediate in the hydration/oxidation of the olefin to form acetone and acetic acid (Fig. 24.7b); and iii) $\pi$-bonded propylene species, interacting with Lewis acid sites, a precursor in the deep oxidation of propylene (Fig. 24.7c).

We must note that the acid characteristics of the latter three catalysts are different, and both the number and strength of Brønsted acid sites decreases as follows: MoVNbO (mainly Brønsted acid sites) $>VO_x/Al_2O_3$ (mainly Lewis acid sites) $>$ MoVTeNbO. Accordingly, and in addition to the presence of an alkane activation site ($V^{5+}$ sites), an $\alpha$-hydrogen abstracting site ($Te^{4+}$ or $Sb^{3+}$) and an O-insertion site ($Mo^{6+}$) are also present in a well-defined host structure, i.e. orthorhombic $Te_2M_{20}O_{57}$ (M1 phase, with M = Mo, V, Nb). Therefore, the selective catalysts for propane oxidation to acrylic acid should have a low number of Brønsted acid sites. Indeed, it has been reported that elimination of Brønsted acid sites in both

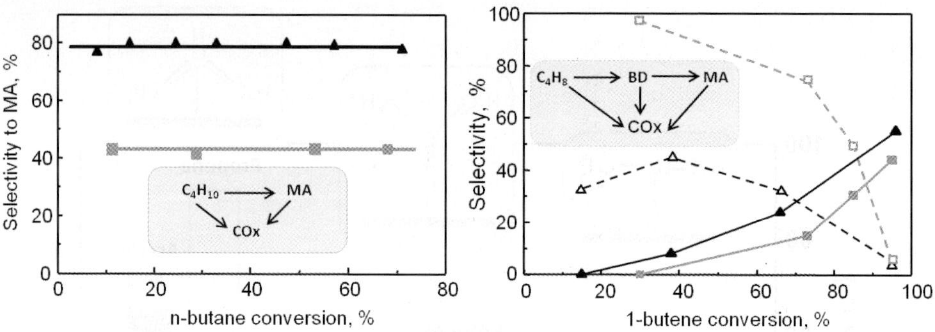


(a) (b)

**Figure 24.8.** Selective oxidation of n-butane (a) and 1-butene (b) over VPO (•,Δ) or MoVTeNbO catalysts (•,•) at 380–400°C and a $C_4/O_2$/He molar ratio of 1.5/18.5/80 (or $C_4^=/O_2$/He molar ratio of 4/8/88). = maleic anhydride; Δ, • = butadiene. From Ref. 176.

modified VPO[182,183] and MoVTe(Sb)NbO[164,169,184] catalysts strongly increases the selectivity to acrylic acid during the partial oxidation of propane.

On the other hand, it is interesting to compare the catalytic performance of VPO and MoVTeNbO catalysts in *n*-butane oxidation. Figure 24.8 shows the variation of the selectivity to MA with the conversion of *n*-butane (Fig. 24.8a) or 1-butene (Fig. 24.8b)[176] over these catalysts. From these results, it can be concluded for both catalysts that MA is a primary unstable product in *n*-butane oxidation, while carbon oxides are formed by both parallel (from *n*-butane) and consecutive (from MA) reactions, in agreement with those previously reported for VPO[23,50–53] (Fig. 24.8a). In this way, it has been proposed that the olefinic intermediates adsorbed on the V-sites of VPO catalysts are transformed (by consecutive reactions) without the presence of butenes in the gas output stream.[23,50–53] Although butenes are observed as primary products in experiments using low $O_2$/butane ratios, the catalytic results for *n*-butane oxidation confirm that they are formed, but rapidly transformed (without desorbing) to form MA. However, some difference is observed between both catalytic systems. Thus, whereas the selectivity to MA during *n*-butane oxidation on the VPO catalyst (*ca.* 70%) was higher than that achieved during 1-butene oxidation (*ca.* 50%), the opposite was observed over MoV-based catalysts for which the selectivity to MA during the oxidation of 1-butene (*ca.* 40%) was higher than that from *n*-butane (*ca.* 35%). Accordingly, the selectivity to MA could be affected by the characteristics of the catalyst. In this way, butadiene has been observed as a primary partial oxidation product during the 1-butene oxidation on a MoVTeNbO catalyst (Fig. 24.8b), while it is a secondary product on VPO.[176] In fact, butadiene is formed on VPO by consecutive reactions from 2-butene (which can be considered as the primary product during 1-butene oxidation on this catalyst). This behavior is related to the acid character of the catalyst. Thus, acid sites favor the isomerization of 1-butene to 2-butene, while redox sites favor the oxidative dehydrogenation of olefin to butadiene (allyl mechanism), which has been confirmed by Fourier transform infrared (FTIR) spectroscopy of the adsorbed 1-butene on these catalysts.[176] In both cases, MA is the final reaction product and its selectivity increases with 1-butene conversion.

Accordingly, a direct formation of MA from $n$-butane can be concluded for both VPO and MoVTeNbO catalysts (Fig. 24.8a), while acrylic acid is mainly formed from propane by consecutive reactions (in which propylene is initially formed) (Fig. 24.6a). The olefinic intermediate is also initially formed in $n$-butane oxidation over VPO, although presenting a different reactivity than that proposed in the case of propene. Thus, although the nature and strength of acid sites between the VPO or MoVTe(Sb)NbO (Mo-based MMO) catalysts are considerably different, the difference in the reaction network between propane oxidation (Fig. 24.6) and $n$-butane oxidation (Fig. 24.8) is not directly related to the catalytic system,[170,176] but to the different reactivity of the corresponding olefin intermediate (propene and butenes).[176]

Finally, it is interesting to address that both VPO and Mo-based MMO catalysts show low yields in the partial oxidation of isobutane. In addition, methacrolein rather than methacrylic acid is formed as the main partial oxidation product.[25,168,183,184] Moreover, the highest yields to methacrylic acid can be achieved during isobutane oxidation over Keggin-type heteropolyacid based catalysts.[110,187–192]

In general, isobutene is practically unobserved even at low isobutane conversion (Fig. 24.5). This suggests that isobutene, if formed, should be rapidly transformed into further oxidized reaction products (mainly methacrolein in the case of VPO and MoV-based catalysts). Indeed, in a recent study, no IR bands were observed after isobutene adsorption over a MoVTeNb catalyst, which could tentatively be explained by the formation of carbocation-type intermediate species weakly adsorbed on the catalyst surface as a consequence of the high reactivity of isobutene.[193] The absence of IR bands associated with adsorbed methacrolein could be related to a fast surface desorption, hence avoiding the secondary oxidation towards methacrylic acid and favoring the deep oxidation (Fig. 24.9).

Taking into account the catalytic results during olefin oxidation, it is observed that methacrolein is again mainly obtained over MoV-based[170] or VPO catalysts,[23,194] while MA is also observed with the latter catalyst. In this way, it has been suggested that a skeletal rearrangement of a carbocation or radical intermediate could occur due to the presence of strong Brønsted acid sites on the VPO catalyst surface, which could explain the formation of maleic anhydride.

Apparently, small differences between propane and isobutene are expected, which should be related to the presence of a methyl group in the second carbon atom in isobutane. In fact, it is well known that similar catalysts can be used for the partial

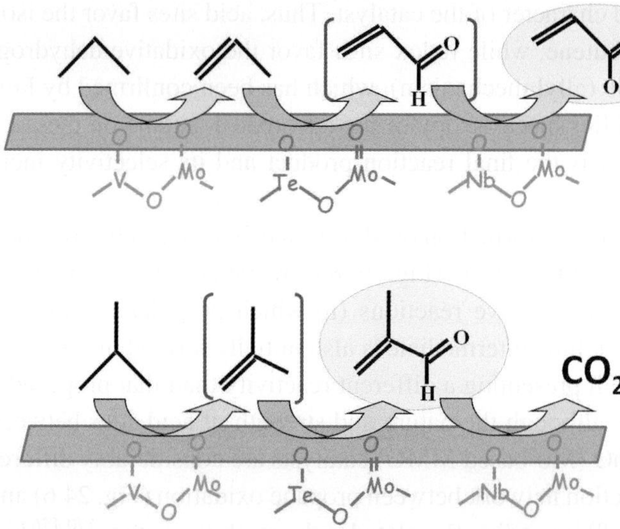


**Figure 24.9.** Reaction network for propane and isobutene oxidation over a M1-phase containing MoVTeNbO catalyst. From Ref. 170.

oxidation of propene or isobutene to the corresponding unsaturated aldehyde.[170] Nevertheless, different catalysts are used industrially for the oxidation of acrolein ($(MoVW)_5O_{14}$-type oxides)[104] or methacrolein (heteropolyacid-based catalysts)[195] to the corresponding unsaturated carboxylic acid.

Since $(MoVW)_5O_{14}$-type oxides[104] and the M1 phase in MoVTe(Nb)O catalysts have a relatively similar framework, they are built from center-occupied pentagonal rings that are linked together forming corner-sharing octahedral sites,[31,89–93] similar behavior in the oxidation of acrolein to acrylic acid could be expected. Indeed, catalytic results of acrolein oxidation over an orthorhombic M1 phase MoVTeO-based catalyst showed 99.0% conversion of acrolein with 97.3% selectivity to acrylic acid at 320°C.[196] This could explain the results presented in Fig. 24.5, which confirm the relatively low consecutive reactions of acrylic acid to carbons oxides in MoVTeO-based catalysts, while methacrolein degradation to $CO_x$ is the main reaction at high isobutene conversion over these catalysts.[170] Therefore, the effective transformation of isobutane/isobutene to methacrylic acid over heteropolyacid-based catalysts should be related to both their specific redox and acid sites, as well as to their structural phases (though difficult to quantify in these cases),[197] favoring a reaction network different to that proposed for propane oxidation to acrylic acid.

Obviously, the typical characteristics which define the catalytic behavior of the catalysts (redox and acid-base properties, distance between active sites, etc.) have an important effect on the selectivity and on the nature of the main reaction products obtained in these catalytic reactions. However, we will also have to consider the characteristics of the reactants, reaction products, and even those of the intermediates. Thus, ethylene, acrylic acid, and maleic anhydride/butadiene are the most favored reaction products during the oxidation of ethane, propane, and *n*-butane, respectively. This behavior can be explained in terms of the different stability between reactants and reaction products, in agreement with a previously proposed explanation by Hodnett and co-workers.[183,198] This behavior has been explained in terms of the different stability between reactants and reaction products. In this way, Hodnett related the selectivity to the partial oxidation product with the difference between the dissociation enthalpy of the weakest C–H bond from the reactant and that of the weakest C–H or C–C bond from the partially oxidized product. Experimentally, it has been shown that if the weakest bond from the reactant has a bond energy higher than 40 kJ $mol^{-1}$ above that of the weakest bond from the desired product, a pronounced decrease in the selectivity to this partial oxidation product will be observed.[198]

## 24.2. Oxidative Dehydrogenation of Light Alkanes to Olefins

From a theoretical point of view, the most interesting catalytic process to obtain olefins from their corresponding alkanes would be catalytic dehydrogenation (CDH), since in this process hydrogen (with a high value as an energy source) is also produced: Alkane $\rightarrow$ Olefin + $H_2$.

Currently, this process is employed to produce propylene (lower than 2% of world production), butenes, butadiene, isobutene, and styrene, operating at high temperatures (550–600°C). Nevertheless, this reaction carries the drawback of being an endothermic process (positive formation enthalpy, $\Delta G$, of 132, 119, and 116 kJ/mol, for ethane, propane, and *n*-butane, respectively). Moreover, undesired reactions occur such as cracking of both paraffins and olefins and especially coke formation which blocks the catalyst's active sites, causing deactivation.

An alternative to the catalytic dehydrogenation process is the ODH of the alkane in order to produce the corresponding olefin: Alkane + $O_2$ $\rightarrow$ Olefin + $H_2O$.

Oxidative dehydrogenation of $C_2$-$C_4$ alkanes has been extensively studied for the last fifteen years as an alternative to the industrial production of olefin[34–42] because of its advantages with respect to steam cracking or catalytic dehydrogenation, which are i) no thermodynamic equilibrium limitation (this is an exothermic reaction with formation energies, $\Delta G$, of –102, –115, and –118 kJ/mol, for ethane, propane, and *n*-butane, respectively); ii) relatively low reaction temperatures required; iii) no need for a catalyst regeneration step (oxygen presence allows the possibility of *in situ* catalyst regeneration); and iv) smaller quantities of by-products with respect to cracking reactions (coke is not formed but instead carbon oxides are obtained).

Therefore, ODH processes offer the possibility of reducing costs due to energy savings, as the operation takes place at lower temperatures, and also due to the absence of the expensive separation steps of by-products. In fact, the main studies of alkane oxidation in the open literature have been focused on ODH reactions, especially for propane ODH (Fig. 24.1).

Moreover, and although the same mechanism has been proposed for all light alkanes, ethane shows a different behavior to that observed for $C_3$–$C_4$ paraffins because its activation usually occurs at higher temperatures (due to the low reactivity of ethane), while the selectivity to ethylene is usually higher (due to its higher stability against deep oxidation). This is also related to the different type of hydrogen in the corresponding olefins. Thus, the presence of allyl hydrogen in propene or butenes can facilitate consecutive reactions, while ethylene (which contains only vinyl hydrogen, which is less reactive than allyl hydrogen) shows a higher stability.

Unfortunately, some doubts have been raised regarding the competitiveness of the ODH processes as the yields to olefin obtained to date are not high enough to replace the conventional processes.[46] Moreover, it has been suggested that yields to olefin of around 65% could be competitive and would be interesting for the establishment of an industrial ODH process.[24] However, we must emphasize that not all the hydrocarbons studied behave in the same way. Thus, as will be commented on in the following section, the reported yields to ethylene have nothing to do with the yields to propylene and isobutene.

The main drawback of ODH is the low olefin yields obtained (meaning low olefin selectivity at high alkane conversion). Indeed, in most catalytic systems an increase in alkane conversion causes a considerable enhancement in the formation of carbon oxides, decreasing the selectivity to olefin. This fact is due to i) the combustion rate of olefin being higher than the formation rate of olefin, and ii) increasing alkane conversion usually impling an increased oxygen concentration in the feed, favoring combustion processes. In addition, there are some other factors to take into account for an industrial application, such as the difficulty of removing CO (formed during the reaction), especially in the case of ethane, and the formation of small amounts of by-products with corrosive behavior.[13]

As indicated previously, several common features are found in the ODH reactions of $C_2$–$C_4$ alkanes, but significant differences are also found regarding reactivity and, in general, catalytic behavior. As aforementioned, reactivity can be related to the energy of C–H bonds of each hydrocarbon; the C–H bonds of tertiary carbons have a lower energy than secondary C–H bonds, and in turn the latter are weaker than primary C–H bonds (Table 24.2). So, *n*-butane (with two secondary carbons) is more reactive than propane (with only one), and the latter is more reactive than ethane which contains two primary carbons. Therefore, ethane will need higher temperatures to reach reaction rates similar to the former ones, or more active sites than those required in the ODH of $C_3$–$C_4$ alkanes.[39,41]

**Table 24.2.** Energy bonds of C–C and C–H bonds.

| Bond | C–C | C–H Primary | C–H Secondary | C–H Tertiary | C–H Allyl | C–H Vinyl |
|---|---|---|---|---|---|---|
| Energy (kJ/mol) | 376 | 420 | 401 | 390 | 361 | 445 |

On the other hand, selectivity to olefins during ODH processes is also related to their own C–H bonds. Thus, ethylene has only vinyl bonds while propylene and butene also have allyl C–H bonds which are more reactive than the former. Indeed, olefin degradation is easier in the cases of propylenes and butenes than ethylenes. In fact, the best results currently achieved in ODH reactions have been reported for ethane oxydehydrogenation.[24–42] Examples of the most effective catalytic systems for the ODH of light alkanes are comparatively shown in Fig. 24.4. Indeed, olefin yields higher than 50% are only achieved in the case of ethane oxidative dehydrogenation on specific catalytic systems. It must be indicated that although the first studies on the ODH of light alkanes appeared in the early 1980s,[39,41] relatively selective catalytic systems were only obtained from the mid 1990s.

The use of Mo-V-containing catalysts for ethane ODH was studied[199–205] after the important catalytic results on MoVNbO catalysts by Union Carbide researchers in 1978.[206] However, no improvement was observed on this type of catalyst. More recently, ethylene yields of around 75% (at reaction temperatures in the 350–450°C range) have been reported employing Mo-V-Te-Nb mixed-oxide-based catalysts. The catalytic behavior in this type of catalyst is directly related to the presence of the so-called M1 phase (Fig. 24.2c).[127–130,207,208] The replacement of Te by Sb improves catalytic activity but slightly decreases selectivity to ethylene.[130] No acetic acid is observed in either case. Thus, oxydehydrogenation of ethane using M1-phase-based catalysts has become the ODH process with higher expectations for an industrial application.[127–130,152,207,208] However, certain aspects must be improved in this type of catalyst in order to optimize productivity at this time.[207,208]

On the other hand, an interesting alternative could be the use of Ni-based catalysts, especially after the motivating results achieved by Lemonidou's group on ethane ODH over Nb-promoted NiO catalysts,[57,76–79] with yields to ethylene of around 50%. More recently it has been reported that the incorporation of a small amount of WOx,[80,81] or $CeO_x$[209] to NiO improves the selectivity during the ODH of ethane. The presence of electrophilic oxygen species, which have been proposed to be responsible for the deep oxidation of ethane, is partially eliminated by the incorporation of Nb.[57,76] On the other hand, Symyx Tech. has also reported the catalytic properties of Ni-based catalysts,[210] presenting a yield to ethylene of *ca.* 26% at 300°C over a $Ni_{0.63}Nb_{0.19}Ta_{0.18}O_x$ catalyst. In all of these cases, the presence of a second element during the preparation of NiO-based catalysts has a strong influence on the crystal size of NiO particles and/or in the lattice parameter of NiO (which means a certain incorporation of the second element in NiO crystals). Both of these aspects should be considered in order to improve the catalytic performance of this type of catalyst.

As previously indicated, the major challenge of the ODH reactions in the majority of catalyst systems reported is to avoid olefin over-oxidation. In this way, the use of catalytic membranes could be a good choice for studying these reactions because of its excellent selectivity to the initial product. Recent reported works are focused on the incorporation of oxygen.[211,212] The use of hollow-fibre perovskite membranes,[213] packed-bed membrane reactors,[214] fluidized-bed membrane reactors (FLBMR),[215] or electrochemical packed-bed membrane reactors have been studied as alternatives to conventional reactors.[216,217] Moreover, metal membranes have also been proposed due to their excellent mechanical stability, relatively low costs, and the possibility to control mass transfer, if this is compatible with the rate of reaction, which can be adjusted by the transmembrane pressure and the catalyst activity, respectively.[218] Other alternatives arising from industry research involve reactors comprising dense membranes of multiphasic materials.[219] In this way, an interesting study has been published by Rodríguez *et al.* on an industrial-scale ethylene production using a novel membrane multi-tubular reactor for the ethane ODH process employing a Ni–Nb–O based catalyst.[220] They concluded that it is possible to reach high selectivity to ethylene, at relatively high ethane conversions, if the reactor is operated at conditions where the reaction is controlled by the permeation flow of $O_2$ through the membrane. More recently, a comparison between a conventional liquid-cooled multi-tubular reactor and a multi-tubular membrane reactor, both using a Ni–Nb–O catalyst washcoated over raschig rings inside the tubes, was reported.[221] It was observed that the variation of the bed density (different thickness of the catalytic washcoated over the pellets) demonstrated opposite effects in both reactor designs: i) for the conventional reactor, the increase in bed density leads to more pronounced hot spots (and undesired oxygen depletion inside the tubes), and ii) for the membrane reactor, higher bed densities prevent oxygen accumulation along the tube length (favoring selectivity). Thus, it was concluded that the heat-generation rate can be efficiently controlled by the permeation flow of oxygen through the membrane.

## 24.3. Partial Oxidation of $C_2$–$C_4$ Alkanes

### 24.3.1. *Partial oxidation of n-butane*

Maleic anhydride is currently used in the production of unsaturated polyesters and butanediol. World consumption over the last few years has exceeded $1.3 \times 10^6$ metric tons per year. Currently, *ca.* 70% of MA is produced from *n*-butane partial oxidation.[52] In this way, partial oxidation of *n*-butane to MA is the only light gas-phase alkane selective oxidation process fully established as an industrial application,[14,15,50–52] which has, for some time now, replaced the old processes from benzene or butenes.

Replacement of the latter by *n*-butane considerably reduces the number of by-products, particularly in comparison to the heavy by-products obtained (i.e. phthalic anhydride and benzoquinone) from the benzene process.[53] Moreover, the *n*-butane process avoids the toxicity associated with the employment of benzene as the raw material (carcinogen) and becomes important in saving raw material costs (around 64%). It is also important to emphasize that the *n*-butane process drastically decreases the formation of carbon oxides (atomic economy). Thus, whereas benzene has six carbon atoms, MA has four, and as a result, two $CO_2$ molecules for each MA molecule are obtained. Accordingly, under optimal conditions, 100 kg of benzene generate 129 kg of maleic anhydride and 113 kg of $CO_2$, whereas 100 kg of *n*-butane generate 170 kg of maleic anhydride.

Currently, the industrial single-step process from *n*-butane is carried out with an active phase based on vanadyl pyrophosphate, $(VO)_2P_2O_7$, commonly named as a VPO catalyst. Yields to MA of around 60% obtained from alkane[14,15,23,50–52] are lower than those from benzene (near 73%), but important cost savings on the final product (higher than 40%) are achieved through the *n*-butane process.[14]

On the other hand, yields to MA similar to those from *n*-butane can be reached from processes using butenes,[222,223] but the number of by-products obtained is also higher than from the *n*-butane process. Indeed, the formation of small amounts of furan, acetaldehyde, crotonaldehyde, or methyl-vinyl-ketone are observed from *n*-butene, which makes the process more expensive due to the need for a purification step.[224] During the selective oxidation of *n*-butane (a cheaper raw material) these products are not formed, but the main by-products are CO and $CO_2$. Indeed, both parallel and consecutive reactions of total combustion must be taken into account in the process from *n*-butane since they are mainly responsible for the limited yield to MA. In this way, *n*-butane conversion lower than 75–80% has been reported[225] to be adequate in order to avoid further oxidation of MA, an important factor in the decrease of the process selectivity.

In order to break the yield borderline, research on both improved catalysts and alternative reactor technologies have been developed.[52] Fixed bed, fluidized bed, and transport bed (circulating fluidized bed) are the main reactor technologies proposed by the different companies for the commercial production of MA from *n*-butane.[13] Within the range of possibilities and depending on the reactor technology employed, companies have been improving their processes over the years, including the way in which MA is recovered, which differs depending on the processes used (i.e. employing organic solvents, anhydrous process; or water, aqueous process) and is a key factor in the process efficiency.

VPO catalysts are the most active and selective for *n*-butane oxidation to MA. The most selective and active equilibrated VPO catalysts preferentially expose the [100] planes of $(VO)_2P_2O_7$.[50–52] In fact, vanadyl pyrophosphate has been identified as the main crystalline phase, although various crystalline and disordered $V^{5+}/V^{4+}$ phosphate phases are also observed depending on synthesis conditions, such as the redox properties of reactants, the time on stream, and the reaction temperature; as a consequence of the facile interconversion between $V^{5+}$ and $V^{4+}$ oxidation states no important changes in crystalline structure occur.[1–12] On the other hand, different composition, preparation, activation, or post-synthesis treatments and regeneration procedures, as well as the use of supports and promoters for VPO, have been continuously proposed by researchers in order to improve activity, selectivity, and/or attrition resistance of the VPO-based catalysts.[52,226–229] A brief review presenting a summary of the most important advances on some representative parameters has been published by Ballarini *et al.*[52] The use of promoters, which has been widely investigated, is suggested as the best approach for the improvement of VPO catalytic properties,[52] although further efforts should be made to better understand the mechanisms taking place upon the addition of these promoters. In this way, Nb has been recently proposed as a novel promoter of the catalytic activity of VPO, apparently due to the formation of Lewis acid sites likely related to surface defects generated by its addition.[230]

On the other hand, fixed-bed reactors are the conventional and the most simple for heterogeneous catalyzed gas-phase reactions. However, as the reaction requires molecular oxygen in the feed, the *n*-butane partial pressure in the feed is restricted in fixed-bed reactors by flammability limits. Therefore, productivity is limited and reactors of a greater size are required to compensate for this. As an example of fixed-bed reactor technology for *n*-butane selective oxidation, the Monsanto process stands out, employing bulk VPO as the catalyst in a tubular reactor to transform *n*-butane into MA (which is then recovered with an organic solvent and finally purified by batch distillation). Maleic anhydride yields of around 60% (alkane conversion of 79% with selectivity of 76%) and productivities of around 130 kg $m^{-3}$ $h^{-1}$ have been reported by Monsanto at a reaction temperature of around 400°C.[231,232] It must be indicated that in order to maximize the MA productivity, the reaction conditions employed by Monsanto (alkane concentrations up to 2.4% in *ca.* 20% of oxygen) are within the range of flammability, exceeding the lower flammability limit established at 1.22% of *n*-butane at 400°C.[14] However, fluidized-bed technology allows the employment of richer *n*-butane conditions (up to 5% in feed) so that higher productivity can be achieved, but with the drawback of selectivity loss. As a noteworthy example we can mention the ALMA process which has been jointly developed by Lummus Technology and Polynt SpA (Lonza group), reporting a yield of 58% to MA of high purity (the highest currently available), with 82% alkane conversion and a selectivity of 71%.[233,234] The most important contributions to the ALMA process are the efficient use of the reaction heat and the effective recovery system (a patented non-aqueous system that uses a non-aromatic organic solvent).

However, the restricted yield encountered when utilizating the latter reactor technology has not been overcome. It is probably due in part to limitations of the VPO catalyst, which tends to be oxidized and hence deactivated under the strong oxidizing conditions necessary to reach high conversion levels during the alkane oxidation reaction. In this way, another technological approach to overcome the aforementioned limitations for the yield to MA was originally undertaken by DuPont, with a project based on circulating fluidized-bed (CFB) technology, in which the catalyst is shuttled between an oxidizing environment and a reducing (butane-rich) environment,[235–239] thus applying the concept of individual steps for oxidation and reduction coupled in two separated reactors:[235] i) alkane oxidation with the corresponding catalyst reduction, and ii) re-oxidation of the oxygen-depleted catalyst in a physically detached reactor. Since oxygen for alkane selective oxidation comes from the catalyst lattice, this technology allows the oxygen concentration into the *n*-butane oxidation reactor to be limited so that high alkane/$O_2$ ratios can be provided, thus avoiding catalyst deactivation by the strong oxidizing conditions during the reaction, and minimizing MA over-oxidation and flammability risks. In addition, DuPont's researchers developed a highly attrition-resistant VPO-based catalyst prepared by coating the active phase with a very strong porous silica shell, giving high mechanical resistance without loss in selectivity under the reaction conditions employing this CFB technology.[238,240,241] Moreover, results suggesting the possibility of industrial CFB reactor utilization of this process, to be improved by efficient regeneration of the catalyst and optimization of the reducing feed-gas composition, have recently been reported by the Patience group.[242]

With the aim of minimizing investment and operating costs, DuPont's approach included an integration of the process for MA from *n*-butane with a process employing the MA obtained to directly produce tetrahydrofuran (used for the manufacture of lycra fibers) in the same production plant, in order to replace the Reppe method with a new process and with minimized $CO_2$ emissions.[13] The DuPont process was initially commercialized in 1996 in a plant on Gijón, Spain. However, due to several operational problems, the plant was closed in 2004.[236]

Considering the latter, the application of membrane reactors with an external fluidized bed has been successfully applied in the partial oxidation of *n*-butane as a way to improve the yield to MA of the process.[243,244] The use of inert membranes could provide adequate doses of molecular oxygen to a fixed bed of a VPO catalyst in a reactor containing porous walls, allowing high concentrations of *n*-butane in the feed which could safely be supplied into the reactor. Re-oxidizing the catalyst would be performed in parallel by the controlled addition of oxygen along the porous reactor walls, which is an advantage over a circulating fluidized process, where the catalyst is shuttled between oxidizing and reducing vessels. Accordingly, results published indicate that yields to MA of 50% and higher, compared with using a conventional fixed-bed reactor, can potentially be reached by employing this technology combining the advantages of both fixed and fluidized beds, minimizing hot spots, and hence providing both isothermal conditions in the reactor and an efficient means to generate steam.[243]

Thus, as a summary, 30 years of the commercial technology for producing maleic anhydride from *n*-butane have revolved around the VPO catalyst. However, no great advances have been made regarding the improvement of the catalytic behavior of the latter material; in general, yields higher than 60% are hard to find in the literature. Ten years ago, only yields to MA higher than 70% were reported, employing a recycling process with a global *n*-butane conversion higher than 95%.[228] Therefore, without detracting from the large number of technical improvements reported, from both equipment and active phase, in order to make the process stages more economically efficient, there is still a wide range for optimizing the yield of the process. The VPO catalyst itself has been the real breakthrough, which has made the successful innovative and sustainable process from *n*-butane possible, but has also encouraged research around alkanes as an alternative sustainable raw material over the years. So, it is likely that VPO will be remembered as the beginning of change.

### 24.3.2. *Selective (amm)oxidation of propane*

Extensive efforts have been undertaken for propane selective oxidation and ammoxidation to acrylic acid and acrylonitrile, respectively. In fact, as can be seen in Fig. 24.1, and according to the literature of patents, the major interest in alkane heterogeneous catalysis over the last few years has covered the selective oxidation of propane.

#### 24.3.2.1. *Selective ammoxidation of propane to acrylonitrile*

Acrylonitrile is currently the second largest outlet for propylene (after polypropylene). It is used as a monomer for synthetic fibers and acrylic plastics (thermoplastics and food packaging mainly), AS (acrylonitrile-styrene) resins, and ABS (acrylonitrile-butadiene-styrene) thermoplastics, as well as in the synthesis of acrylamide, adiponitrile, and nitrile elastomers. The manufacture of acrylonitrile is exclusively based on the one-step propylene ammoxidation process. Originally developed by Sohio, Standard Oil Company (now part of BP America), the conventional method used since 1957 employs a fluidized-bed reactor and multicomponent catalysts based on Mo-containing mixed-metal oxides. Over the years, the industrial process has been improved and now acrylonitrile yields are higher than 80%.[1–22,153] Optimization of this process has also been in terms of maximum profit from the by-products obtained (cyanhydric acid, acetonitrile, unreacted ammonia), most of them with important applications.[245] Development of this process from propylene was an important innovation and advance in terms of sustainability and energy efficiency, since acrylonitrile was originally produced from ethylene oxide or acetylene, and hydrogen cyanide. In addition, the concept of employing multicomponent catalysts based on mixed-metal oxides in the selective ammoxidation of propylene was a key factor for the succesful development of other processes for selective (amm)oxidation of hydrocarbons.[1–23]

In 2008, global production of acrylonitrile was around 4.531 million tons,[245,246] which is currently forecast to increase to an annual demand of *ca.* 4%. The current global overcapacity of production is high enough to satisfy the acrylonitrile demand in the near future.[245,246] However, since propylene accounts for roughly 60% of the cost of making acrylonitrile in current production plants, solutions for replacing the conventional process by an even more sustainable and cost-efficient process from propane have been investigated; this has recently attracted major attention, with the selective (amm)oxidation reaction of light alkanes being the process at the closest stage to a commercial application.

Typical propane ammoxidation catalysts are essentially constituted by a combination of metallic mixed oxides. To date, there are two catalytic systems: i) vanadium-antimonates with a rutile-type structure, represented by the $VSb_xM_yO_z$ formula, where "M" are elements used as the promoter such as W, Al, Te, Nb, Sn, Bi, Cu, or Ti,[12,65,82–86,247] and ii) vanadium-molybdates with a bronze structure, represented by $MoV_xM'_yM''_zO_d$, where M' is most often Te or Sb, and M" is generally Nb.[31,68,87–93]

British Petroleum (BP) reported, at the end of the 1980s, a heterogeneous gas-phase oxidation process to obtain acrylonitrile from propane with a multicomponent catalytic system based on V-antimoniates with a rutile-type structure, among which the composition containing V-Sb-W-Mo-Al corresponds to the most effective catalyst reported (showing acrylonitrile yield of *ca.* 39%).[65,82–85] The structure and composition of the crystalline phases forming the latter catalyst, a non-stoichiometric rutile $VSbO_4$-like phase (Fig. 24.2b) and $\alpha$-$Sb_2O_4$, strongly depend on the catalyst preparation procedure, the catalyst activation (temperature and atmosphere), and the V/Sb ratio, although they can also vary during the catalytic tests.[12,65,82–86,247] In this process a fluidized-bed reactor was used with feed rich in propane (around 50% of the total flow) and reaction temperatures relatively high, in the range of 450–500°C, due to the low activity of VSb-based catalysts. Even so, low propane conversion is obtained and unconverted propane must be recycled to the reactor, increasing both investment and cost of the process.

Furthermore, in the early 1990s Mitsubishi reported a new catalyst system based on MoVTe(Sb)NbO mixed oxides,[31,68,87–93] which achieved a high propane conversion (>90%) with a 60% yield to acrylonitrile[67,89] at reaction temperatures close to 420°C in a fluidized-bed reactor. It is well accepted that the catalytic behavior of this kind of catalyst is strongly related to the presence of the so-called M1-phase,[248,249] an orthorhombic bronze with stoichiometry $(Te_2O)\,M_{20}O_{56}$ or $(SbO)_2M_{20}O_{56}$ (M = Mo, V, Nb) and characterized by the arrangement of molybdenum atoms in an octahedral coordination (partially substituted by other elements such as V or Nb), forming pentagonal (locating Mo, V, or Nb with pentagonal bipyramid coordination), hexagonal (where Te or Sb are located forming oxide chains), and heptagonal channels (occupied or not by Te or Sb) (Fig. 24.2c).[154,196,250–252] The results obtained with these types of catalysts represent a major improvement upon the 40% yield reported in BP patents (with VSb-based catalysts), and operate at even lower reaction temperatures.[154] Currently, the main patents in propane ammoxiation have been assigned to Asahi Chemical Industry, Mitsubishi Chemical, BP, BASF, Mitsui Toatsu Chemicals, Nippon Catalytic Chemical, Rohm and Haas, Atofina (now Arkema Inc.), Rhone-Poulenc, and Air Products.

Actually, the major success in propane ammoxidation into acrylonitrile has been to discover MoVTeNb mixed oxide materials presenting a crystalline phase with a so-called M1 structure,[155] since the results obtained with this type of catalyst are remarkably better than those from the other catalytic systems reported for propane ammoxidation.[83–85,94,95,153,163,253–256] In fact, the complex structure of the M1 phase[64,257,258] represents the clearest example of a multifunctional catalyst in which each element, in close geometrical and electronic synergy with the surrounding elements, plays a specific role in turn, as an isolated active site, in every reaction step for the alkane transformation into the partial oxidation product desired. The flexibility of the structure allows modification of the catalyst composition and hence its catalytic behavior. Moreover, this type of mixed-metal oxide catalyst has the ability to catalyze other different oxidation reactions starting from alkanes,[170] such as propane oxidation to acrylic acid,[259–264] oxidative dehydrogenation of ethane to ethylene,[152] and *n*-butane selective oxidation.[176]

Reaction network analyses for propane ammoxidation over a M1-containing catalyst revealed that i) Mo and V in a framework structure are responsible for the oxidative activation of propane to propene, which is the rate-determining step; ii) Te or Sb clearly promotes the conversion of the formed propene to acrylonitrile; iii) the absence of Te or Sb clearly promotes the destructive conversion of propene to $CO_x$; and iv) the presence of Nb suppresses the further reaction of acrylonitrile, improving selectivity.[31,64,181,251,253,265–269]

Taking into account the product yield necessary for an alkane-based process to achieve variable production costs equivalent to a conventional process from propylene, and on the basis of the current price differential between both feedstocks,[270] MoVTeNb mixed oxide catalysts could be competitive enough to allow replacement of the propylene feedstock process with propane as the raw material. In this sense, a once-through process would be the best option since it would allow the possibility of retrofitting existing conventional plants to move from propylene by simply replacing the catalyst and feedstock. Indeed, the development of a selective ammoxidation process from propane has recently been reported.[245] In fact, Asahi Kasei Corporation announced in 2007 that Asahi Kasei Chemicals Corp. had begun the validation of a propane process line at the Ulsan Plant of its Korean subsidiary Tongsuh Petrochemical Corp. In this way, an existing 70,000 ton/year acrylonitrile plant was modified to use the propane process.[271,272] The catalyst employed is a multicomponent Mo-V-containing mixed oxide diluted with silica, which yields 59% with a propane conversion of 90% reported. Likewise, Asahi and Thailand's PTT Chemical are building a 200,000 ton/year propane-based plant for acrylonitrile production in Malaysia, employing the same technology. Saudi Basic Industries Corp. (SABIC), Asahi Kasei, and Mitsubishi Corp. as a joint venture have just advertised their intention to build an acrylonitrile plant with a capacity of 200,000 ton/year in Jubail (Saudi Arabia) using propane and also producing 40,000 ton/year of sodium cyanide as a co-product.[273]

On the other hand, another approach to optimize the selectivity to acrylonitrile has been to operate at the optimum low level of propane conversion. However, leading the unreacted feedstock back to the reactor would be necessary, suggesting a disadvantage in comparison to the once-through process due to the additional equipment required.[270] Moreover, this approach has an additional interest in recovering and recycling the propylene formed (a reaction intermediate prior to acrylonitrile) and carbon oxides, the latter probably capable of partially increasing the yield to acrylonitrile by avoiding over-oxidation in a similar way to that recently proposed for the MA from *n*-butane process.[274]

An advanced variant of the latter process has been proposed involving multireaction steps, operated under different reaction conditions, to give sequential process steps, each one specifically operated with increasing conversion of propane. In this case, the product is separated after each step and the unreacted feedstock is led to the next reactor with the fresh feed mixture. It has been reported that an overall propane conversion of 95% with overall acrylonitrile yield of 63% has been achieved in these three steps.[245,270]

Finally, an interesting variant of the once-through process has been proposed to produce multiple products, i.e. by varying propane/ammonia in the feed, different acrylonitrile/acrylic acid ratios can be achieved,[275] which provides an interesting opportunity to optimize the operation of the plant, based on changes in feedstock, product price, and demand. In this sense, it is illustrative of the parallel between

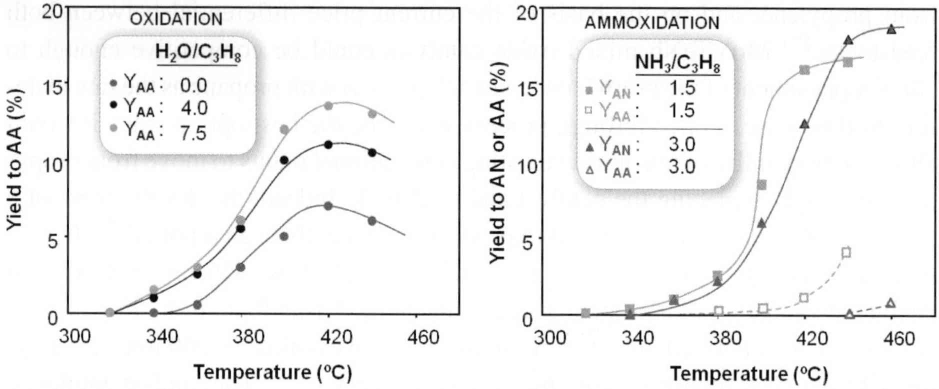


**Figure 24.10.** Variation of the yield of acrylic acid (oxidation) and acrilonitrile and acrylic acid (ammoxidation) with the reaction temperature, with the propane/steam and propane/ammonia ratio, respectively. $Y_{AA}$ = yield to acrylic acid; $Y_{AN}$ = yield to acrilonitrile. From Ref. 250.

the variation of yield to acrylic acid and acrylonitrile with reaction temperature and the $NH_3/C_3H_8$ and $H_2O/C_3H_8$ ratios fed into a fixed-bed reactor loaded with a MoVTeNbO catalyst (Fig. 24.10).[250]

### 24.3.2.2. *Partial oxidation of propane to acrylic acid*

Acrylic acid has traditionally been used as the raw material for acrylic esters, polyacrylates, cross-linked polyacrylates, and copolymers. The global acrylic acid capacity was *ca.* 4.7 million tons in 2006, with an estimated average growth of 4%.[276] Nowadays, acrylic acid is industrially obtained from a physically separated two-step process with propylene as the starting raw material.[276,277] Firstly, propylene is selectively oxidized to acrolein at 300–350°C employing multicomponent catalysts based on metallic mixed oxides, i.e. MoBiO, FeSbO, or SnSbO. Then, the acrylic acid is obtained in a second step from acrolein oxidation at 200–260°C using multicomponent catalysts based on Mo-V-W mixed oxides.[27,30,104,105,277] Thus, an overall acrylic acid yield of 85–90% is reached.

The substitution of propylene by propane as the feedstock in the current industrial process for acrylic acid manufacture would lead to many advantages:[47] i) propane is cheaper than propylene; ii) only one step would be required in contrast to the current two-step process used in industry; and iii) the $CO_2$ emissions from the global process would be lower if propane were used. Thus, taking into account the former advantages of the process from alkane, it is indeed interesting to consider substituting the current acrylic acid process in the near future.

In order to be successful in the former task it is necessary to develop catalysts capable of i) activating propane, breaking the C–H bond; ii) inserting oxygen selectively; iii) blocking the oxidation of intermediates preventing the formation of undesired final products; iv) avoiding the split of C–C bonds; and v) favoring the acrylic acid desorption in order to avoid further oxidation.

The propane reaction is very exothermic and its enthalpy is even higher than both enthalpy steps added from the propylene process, since alkane dehydrogenation must be included in the former process. The propane reaction takes place via an eight-electron transfer, requiring a specific catalyst structure on which an adequate isolated active site distribution exists, in order to carry out the coordinated steps. Moreover, an appropriate element redox balance must be present to complete the catalytic oxidation-reduction cycle, including *in situ* catalyst regeneration.

In this way, catalysts based on heteropolyacids and their salts (HPC),[278–281] vanadyl pyrophosphate (VPO),[178,184,282] or Ni-molybdates[283] have been reported as active and relatively selective catalysts for propane partial oxidation. However, multicomponent catalysts based on mixed-metal oxides (MMO), especially those based on mixed oxides of Mo-V-Te-Nb, are presently considered as the most promising catalysts for propane oxidation to acrylic acid, with reported yields of *ca.* 50%.[259] In addition, MMO catalysts show excellent thermal stability at reaction temperatures of 360–420°C, an important aspect to be considered in order to obtain a feasible industrial implementation. Thus, the partial oxidation of propane to acrylic acid is the most outstanding example for the application of M1-phase-based MoVTe(Sb)NbO catalysts, being one of the most important and attractive issues in the gas-phase heterogeneous catalysis oxidation of alkanes. However, the yields to acrylic acid obtained are slightly lower than those achieved for acrylonitrile employing the same type of catalyst, due to the decreased stability of the former product which further favors consecutive oxidation reactions.

It must also be indicated that the presence of water in the reaction gas mixture is essential to obtain high acrylic acid yields over these mixed-metal oxide catalysts.[250] An IR spectra of propylene on these catalysts suggest that $\pi$-allyl intermediate (key for the acrylic acid formation) is not formed in the absence of water, while it is clearly observed when both propylene and water are present.[125]

Substitution of propylene by propane in the commercial production of acrylic acid would lead to an important feedstock cost saving (the price differential varies over time and with location, but is always positive and substantial), but also in power consumption (directly related to $CO_2$ emissions) due to both the feedstock obtaining process and the more negative reaction heat to obtain acrylic acid from propane ($-171$ kcal/mol$^{-1}$) than from propylene (–142.1 kcal/mol$^{-1}$). An estimated cost saving of over 300 $/Tm, based on feedstock prices (in 2006), which would be achieved with the replacement of propylene by propane has been reported.[270] In this sense, the prices of propane and propylene depend on many factors and have been increasing over the last few years, though the price differential has remained essentially constant.

From an environmental point of view, a substantial decrease in $CO_2$ emissions would follow the replacement of propylene with a propane feedstock, considering the overall process (*ca.* five times higher in the propylene process than for propane).[47] In this way, a recent study on the evaluation of the environmental impact by a systematic analytical method comparing the current commercial process from propylene with a hypothetical propane process (assuming in both a yield to acrylic acid close to 90%, currently obtained with the propene process) concluded that the propane process implied a decrease of 20% in the environmental impact of the industrial process.[284] Moreover, a yield to acrylic acid higher than 61% was calculated to be enough for the propane process to be more environmentally benign.[284]

On the other hand, the higher reaction heat of propane to acrylic acid implies some drawbacks in fixed costs related to the investment needed, as larger heat exchangers and reactors would be required.[270] Therefore, sufficient acrylic acid yield from the propane feedstock is necessary to achieve equivalent production costs from the current process, based on propylene, and to compensate for the required capital investment and other additional variable and fixed costs. In this sense, Vitry *et al.*[285] estimated that setting up a new alkane feedstock-based process would start to be profitable with acrylic selectivities between 60% and 75%. Brazdil[270] is more specific and gives a product yield value of around only 50% as necessary to achieve equivalent production costs to those of the current propylene-based process. Nevertheless, yields to acrylic acid higher than 50% have not yet been reported, despite the great efforts made by the large number of companies and research groups. In this sense, a recent overview advocates,[286] on the bases of kinetics studies, that selectivity to acrylic acid has an intrinsic limit of 83% for the catalysts based on MoVTeNbO mixed oxides. This has been associated with the low relative stability of acrylic acid under the process conditions, which decreases the selectivity to acrylic acid as propane conversion increases. With acrylic acid yields around 50%, unreacted alkane should be recycled to reach the required target of higher overall yields.

Regarding the development of reactor technologies and strategies in order to improve catalytic results and the efficiency of the process for propane oxidation to acrylic acid, the same approaches discussed above for propane ammoxidation have also been conceptualized.[270] In this way, detailed evaluation of advantages and disadvantages for the main three types of reactor i.e. fixed bed, fluidized bed and circulating fluidized bed (CFB), and the translation into economic factors for estimating either investment or operating cost has been recently published.[287] This has enabled the determination, in each case, of the requirements in terms of alkane conversion and selectivity to acrylic acid in order for it to be a competitive process versus conventional technology from propylene. Thus, selectivity to acrylic acid of 65% from propane using a fluidized-bed reactor is reported to be necessary to achieve operating costs similar to the commercial process from propylene, versus the selectivity of 85% estimated for the case of fixed-bed reactors. On the other hand, a CFB process (in a way similar to how the aforementioned process produces MA from *n*-butane which was eventually commercialized by Dupont) would imply the most complex option but with additional advantages, among which include lower flammability risks due to the fact that a lower molecular oxygen ratio in the feed is necessary since the reactive oxygen species come from the lattice catalyst and the latter is re-oxidized (regenerated) in a separate step.[287] A CFB propane process to produce acrylic acid would be even more interesting than for *n*-butane oxidation to MA on the basis of the intrinsic characteristics of both reactions (the heat of the reaction and the oxygen demand are lower for the propane to acrylic acid than for *n*-butane to MA). Moreover, MoV-based catalysts (for propane oxidation to acrylic acid) were recently determined, in a comparative study, to have a higher capacity of oxygen lattice and faster re-oxidation rate than VPO catalysts (for *n*-butane selective oxidation), with carbon deposition decreasing as the reaction temperature increases, a trend opposite that for VPO.[288]

On the other hand, it can be mentioned as a curious novelty that an electrochemical membrane reactor using $Bi_4Cu_{0.2}V_{1.8}O_{11-\delta}$ as a solid electrolyte membrane has been recently reported to enhance the catalytic behavior of a $MoV_{0.3}Te_{0.17}Nb_{0.12}O_x$ catalyst with respect to the use of a conventional fixed-bed reactor, achieving 42% propane conversion with selectivity to acrylic acid of 80% at 380°C.[289]

Therefore, it seems that at the moment, the process to produce acrylic acid from propane is a promising reality which could be at the limit of being a profitable alternative under the current outlook. However, the economic motivation for replacing the current acrylic acid production by the alternative route of propane oxidation is now a little uncertain and strongly dependent on many economic interests, among which the most important seems to be the price difference between propylene and propane.

Finally, the results for the partial oxidation of propane are rather frustrating, as after two decades of intense study, the yields to acrylic acid have not been enhanced. In fact, the highest yield to acrylic acid in both the open and patent literature indicates a limit to the maximum yield which can be obtained, as specifically reported by Muller.[286] This is a consequence of the limitations of the reaction network in consecutive steps, with propylene as the primary reaction product, as indicated in Fig. 24.6. Accordingly, maybe we should design catalysts by considering the model of *n*-butane selective oxidation over VPO catalysts (Fig. 24.8), in which the olefinic intermediate was not desorbed and acrylic acid was directly formed as the primary reaction product. In this sense, new crystalline structures could be required in which active sites for propane activation and those for propylene oxidation were near enough to directly transform the propylene intermediate into acrylic acid. Furthermore, knowledge of the synthesis of bronzes could facilitate the synthesis of other multifunctional catalysts that also embody these advanced concepts.

### 24.3.3. *Partial oxidation of isobutane*

The partial oxidation of isobutene into methacrolein and methacrylic acid would be the best option among all approaches to replace the technology employed in the acetone-cyanohydrin process.[290]

The methyl ester of methacrylic acid is a highly valuable product primarily used to manufacture polymethyl methacrylate, a clear plastic known under the trade name of Plexiglas or Perspex. Although there are a wide number of different technologies currently applied for methyl methacrylate production, the main one is the acetone-cyanohydrin (ACH) process, especially in Europe and North America, which accounts for about 85% of worldwide production.[290] However, from an environmental and safety point of view, this process presents important drawbacks as it uses and produces highly toxic compounds, i.e. hydrogen cyanide (extremely poisonous), which react with acetone to form ACH as an intermediate (extremely hazardous substance), which is treated with sulfuric acid to give the sulfate ester of the methacrylamide. With this process, a mixture of methyl methacrylate and methacrylic acid is obtained, but in addition, large quantities of contaminating $(NH_4)HSO_4$ are co-produced, which has to be pyrolyzed or recovered. All this involves important economic aspects which, combined with a supply shortage of hydrogen cyanide, makes this process susceptible to replacement.[290] In this way, several environmentally conscientious alternatives have been developed in the last few years, among which is a new and improved ACH process developed by Mitsubishi Gas Chem. Co. Inc. The new process avoids the co-production of $(NH_4)HSO_4$ and HCN acid waste. However, excessive steps and high energy consumption are involved. The other coexisting commercial processes for methyl methacrylate production are i) direct catalytic oxidation of isobutene or tert-butanol (TBA); ii) use of the methacrylonitrile method via the isobutene ammoxidation (Asahi Kasei Co. Ltd.); iii) isobutene or TBA oxidation followed by oxidative esterification of methacrolein (Asahi Chemical Co. Ltd.); and iv) CO, $H_2$, and formaldehyde as raw materials from ethane (BASF).

On the other hand, new processes are being researched, such as i) the named Alpha process via ethylene carbonylation to methyl propionate, very similar to the aforementioned BASF route; ii) the one-step method of propyne catalytic carbonylation; and iii) direct oxidation of isobutane to methacrolein or methacrylic acid. The use of isobutyraldehyde[291] and isobutyric acid[292] as feedstocks has also been explored. Among all these methods, the one employing the alkane in a one-step reaction to form methacrylic acid is the most simple and interesting process from both an economic and environmental point of view, due to the lower cost of isobutane, the lower formation of co-products, and the safety and ecofriendly aspects involved compared with the other processes. From the point of view of an interesting industrial application, the complete process should imply a coupled step for methacrylic acid to be esterified with methanol to finally form methyl methacrylate.

Systems based on Keggin-type polyoxometalates (POMs) (phosphomolybdic acid derivatives) have been the most effective and widely used catalysts in gas-phase isobutane selective oxidation research. This Keggin POM catalyst application was first claimed in 1981 by Rohm and Haas,[293] and after several catalyst modifications has been patented by other companies such as Asahi Kasei, Sumitomo, Mitshubishi, etc. However, in addition to a Keggin phosphomolybdate, vanadium must also be present in the anion structure in order to obtain a catalyst effective in the selective oxidation of isobutane. Moreover, the addition of +3 metal cations to the phosphomolybdate anion increases the thermal stability of the catalyst, significantly minimizing its deactivation.[294]

Processes based on selective isobutane oxidation catalyzed with these kinds of catalysts, reported by the different companies, have several common technical aspects, i.e. all of them use isobutane-rich conditions (between 2 and 0.5 alkane/$O_2$ ratios) which implies a reduced reaction environment. In this way, it has been proposed that these conditions seem to be determinant in obtaining a high selectivity to methacrylic acid (MAA),[295] since a partially reduced Keggin-type catalyst presents less of a trend to total oxidation, hence improving selectivity to MAA.[110,295] However, reduction can be also be achieved during catalyst synthesis by the incorporation of transition metal cations in proper oxidation states and suitable redox features to reduce $Mo^{6+}$ to $Mo^{5+}$,[295] and also employ organic cations as in the case of Keggin molybdophosphate catalysts reduced with pyridine reported by Li *et al.*[110] It must also be indicated that only traces of methacrolein were obtained with these Keggin molybdophosphate catalysts reduced with pyridine, while presenting high selectivity to MAA and acetic acid.[110]

Recently, it has been reported that mechanical-phase mixtures of a phosphomolybdic cesium salt containing V and Te as counter-cations[191] and $\alpha$-$La_2Mo_2O_9$ (LM) or $\beta$-$La_2Mo_{1.9}V_{0.1}O_{8.95}$ (LMV) have been studied as catalysts for the partial oxidation of isobutane into MAA and methacrolein. The catalysts with 50 wt.% of LMV in the phase mixture appeared to be the most efficient, with selectivity and yield into MAA of 71% and 12%, respectively, at 360°C and atmospheric pressure.[192]

On the other hand, another common aspect in most of the published results is the presence of steam in the process which seems to be essential in the feed mixture, either from an engineering or catalytic point of view, as it i) favors no flammability conditions by decreasing the concentration of isobutane and $O_2$, and ii) improves catalytic performance by favoring both surface reconstruction of the Keggin catalyst structure, and MAA desorption from the catalyst, avoiding its further oxidation by consecutive reactions.

However, no great advances have been achieved to date since major difficulties have been found in improving activity and selectivity in this reaction; there are also problems with the structural stability of the catalyst in reaction conditions employed to increase the alkane conversion. In this way, it has been specifically observed that the catalyst surface Keggin structure decomposes during reaction at high temperature. As a consequence, low isobutane conversions (less than 40%) have usually been reported (Fig. 24.11), which has led to the approach of processes with a recycle step for unconverted isobutane. It would be doubly necessary, from a commercial point of view, to not discard the methacrolein (preceding reaction intermediate of MAA), which is almost always obtained in considerabe quantities during this process. In this respect, an interesting advance assigned to Asahi Chem. Industry has been the use of an organic solvent (a mixture of decane, undecane, and dodecane) which can absorb isobutane and methacrolein from the off-gas (to be recirculated or recycled) with 99.5% recovery efficiency.[296] Another example has been proposed by Sumitomo[297] in a process where methacrolein is separated and recycled to the oxidation reactor, achieving an overall yield of 52% to MAA (global recycle conversion of 96%, and a per pass isobutane conversion of 10%). It is interesting to comment on the use of $CO_2$ as a ballast component in the latter process, where $CO_2$ is obtained by integrating a CO combustion reactor whose heat of reaction is also beneficial as steam to supply the reaction mixture feed to the oxidation reactor.

Therefore, most attention in the last few years has been focused on improving isobutane conversion and selectivity to MAA, by developing modified POMs

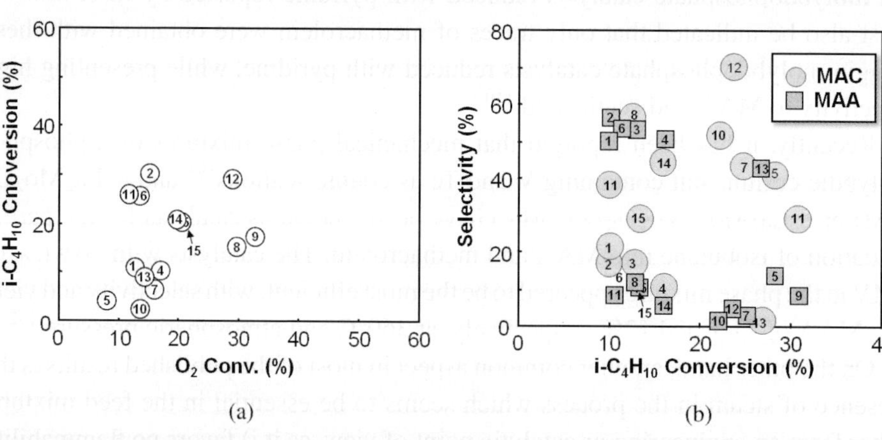


**Figure 24.11.** The most representative results in the selective oxidation of isobutane: (a) feed composition; (b) conversion of isobutane vs selectivity to methacrolein (MAC) and/or methacrylic acid (MAA). Refs. 110, 298–311.

containing different transition metal cations as well as promoters which also improve thermal stability. In addition, several studies have been developed that investigate the regeneration procedures after reaction with these Keggin-type POMs catalysts, which are not usually stable above 380°C. However, it seems that much remains to be done to overcome these key aspects which limit the industrialization of the process.[312] In general, a solution is required for the strong decrease in selectivity to methacrolein and methacrylic acid when increasing the alkane conversion, which decreases productivity, as the increase of conversion does not compensate for the drop of selectivity. Thus, the selectivity falls in a fairly linear manner because the consecutive oxidation rate of the produced MAA is much larger than the rate of isobutane conversion.

### 24.3.4. *Partial oxidation of ethane to acetic acid*

Acetic acid is used for the production of acetic anhydride (18% of total global acetic acid consumption), which is mainly used for cellulose flake production; terephthalic acid (TPA, 17% of global consumption), which is mainly used as an intermediate to produce resins, fibers, and films for a wide range of applications, including polyester fibers, and bottles for water and soft drinks; and acetate ester (17% of total global acetic acid), which is mainly used as a solvent for inks, paints, and coatings. The remainding acetic acid (15%) is consumed for other purposes such as monochloroacetic acid (MCA), ketene and diketene derivatives, pharmaceuticals, dyestuffs, and in other industries.[313]

Currently, acetic acid is mainly obtained (around 75% of world production) by the Monsanto process of methanol carbonylation, in which carbon monoxide reacts with methanol at 180°C and pressures of 30–40 atm in a homogeneous catalysis process employing a catalyst based on a rhodium complex.[314] For to economic reasons, this process replaced the original 1950 process based on acetaldehyde oxidation.[313] It must be noted that in most countries, for legal reasons, acetic acid for vinegar production is usually provided by the bacterial aerobic fermentation of ethanol (around 10% of world production of acetic acid). On the other hand, different processes have been commercially introduced to a lesser extent over the years, such as acetylene oxidation and the uncatalyzed oxidative scission of $C_4$–$C_8$ hydrocarbons (mainly *n*-butane[315] and naphtha).[316] The former was rejected because of the toxicity of the catalysts employed (which were mercury based), and the latter produced significant quantities of by-products which are very complex and expensive to remove.[313] Palaniappan *et al.*[317] have described a systematic methodology for the inherent safety analysis during the route selection stage and have illustrated it using different processes for producing acetic acid, while Smejkal *et al.*[318] have presented an energetic and economic evaluation of the production of acetic acid via ethane oxidation including a discussion about investment and production costs.

Direct oxidation of ethane to acetic acid is an attractive alternative to conventional processes for obtaining acetic acid. In the 1980s, Union Carbide researchers developed a process for the production of ethylene via the oxidative dehydrogenation of ethane with the co-production of acetic acid.[206,319] Using MoVNbO mixed oxides as catalyst systems, different amounts of ethylene and acetic acid were obtained from ethane oxidation depending on the reaction conditions. In this way, selectivity to acetic acid of 26%, at ethane conversion of 5% was reported. After this, several patents were reported by Union Carbide (now known as Dow Chemical).

Interest in ethane selective oxidation to acetic acid has been rising in the last few years due to the low cost of ethane which is now easily available from the large sources of natural gas which have recently been discovered. Although different catalytic systems have been proposed,[313] Pd-doped MoVNbO mixed oxides are one of the more promising catalysts.[62,320–324]

In an interesting study Li and Iglesia[62,323] show how the precipitation of the active phase, i.e. $Mo_{0.61}V_{0.31}Nb_{0.08}O_x$, in the presence of colloidal $TiO_2$ led to a tenfold increase in all rate constants suggesting a higher dispersion of the active phase. In addition to this, the presence of $PdO_x$as the co–catalyst markedly increased the formation of acetic acid (acetic acid rate formation of 829 g $kg_{cat}^{-1}h^{-1}$).[62] On the other hand, they also observed that the presence of water increased acetic acid selectivity by promoting desorption of the adsorbed acetate species. However, increasing the selectivity to acetic acid by increasing the water concentration could have some disadvantages in an industrial process since the energy consumption in the separation stage increases greatly with increasing water concentration. In this way, simulation of the whole process can facilitate the discovery of optimal conditions.[325]

Many companies, such as Standard Oil Company, Mitsubishi Chemical Corporation, BP Chemicals, Celanese, Hoechst Company, and Saudi Basic Industries Corp., have reported catalysts and processes. In 2000, SABIC developed a new process for acetic acid production through ethane oxidation (the Sabox process), and the construction of a 30,000 ton/year plant to produce acetic acid by ethane oxidation at Yanbu has been announced.[320,321]

It is clear that the substitution of the conventional route of acetic acid production by a single-step heterogeneous catalyst process from ethane would mean an extreme reduction in energy consumption and associated $CO_2$ emissions, but it would also see the advantage of working at low pressures instead of the high pressure required in the conventional methanol carbonylation process.

On the other hand, other related technology involving ethane has been proposed over the last few years. This is the production of vinyl acetate monomer (VAM, used in the polymer manufacture of adhesives and coatings), which is the largest end use of acetic acid. Therefore, integrated processes in an ethane oxidation plant for the manufacture of vinyl acetate and ethyl acetate by the reaction of ethylene and acetic acid intermediates would be the most efficient industrial exploitation, allowing less investment and saving on operating costs, as recently proposed.[326]

Finally, there is a great interest in the development of ethane-based technology for the production of vinyl chloride monomer (VCM). Indeed, several processes have been patented by several companies, but as yet none have been commercialized. Ineos (formerly EVC) has operated a 1,000 ton/year pilot plant at Wilhelmshaven, Germany, and has plans to commercialize the technology at some stage.[205]

## 24.4. Selective Oxidative Activation of Methane

Because of the high percentage of methane in the large reserves of natural gas throughout the world, the oxidative selective transformation of methane is becoming a real technical and economic challenge.[327–329] Thus, numerous studies have been devoted to the oxidative transformation of methane (OTM), either toward the partial oxidation of methane (POM) or the oxidative coupling of methane (OCM), but also to the transformation of $CH_4$ into higher hydrocarbons.[48,330,331] In addition, halogenation and oxihydrohalogenation of methane to methyl chloride (or higher chlorines) and sulfonation of methane to methylbisulfate have also been of great interest for future industrial application. However, no industrial application is under consideration at the moment for the aforementioned processes. In general, the low conversion/selectivity and technical problems are the main drawbacks to their applicability at the moment.[328] As the catalytic systems and the reaction conditions are very different, they will be presented separately in the following section.

### 24.4.1. *Methane oxidation to syngas*

Methane oxidation to syngas has recently been presented as the most interesting method for methane valorization.[47,330] In fact, the catalytic partial oxidation (CPO) of methane to syngas has widely been studied as a potential alternative to methane steam reforming since it is mildly exothermic and a $H_2/CO$ ratio of 2 is obtained in the ideal reaction: $CH_4 + {}^1\!/\!_2O_2 \rightarrow CO + 2H_2$ ($\Delta H^\circ_{298} = -36\,\text{kJ/mol}^{-1}$), which is comparable to Fischer–Tropsch synthesis.[332] This has been reviewed recently.[328,333]

However, the selectivities towards $H_2$ and CO can be strongly influenced by the simultaneous occurrence of methane total combustion and/or secondary oxidation reactions of CO and $H_2$, such as the formation of water and carbon dioxide, which can be favored at high reaction temperatures.

Ni-based catalysts were initially reported,[330,334] with further investigations of different group VIII metals (Rh, Pt, Pd, Ir, Ni, Fe, Co, Re, and Ru).[335,336] The highest conversions and selectivity were observed in the case of Ni-[333,337–339] and Rh-based[340] catalysts. In the case of Ni-based catalysts, a non-acid metal oxide support or promoted alumina are preferred in order to decrease the carbon deposit formation.[336] However, a relatively fast deactivation is generally observed. Recently, it has been proposed that the use of sol-gel or the microemulsion method for preparing supported Ni catalysts could facilitate a high resistance of carbon deposition by favoring the synthesis of materials in which the growth of Ni particles can be inhibited during the reaction.[338,339]

Most of these studies have dealt with catalyst development and testing in fixed-bed reactors at atmospheric pressure.[341,342] Moreover, different reactor types have been studied in the last few years: fluidized-bed reactors,[343] monolith reactors,[337,344] membrane reactors,[345] or short contact time reactors,[340,346,347] but also reactors at elevated pressures.[341,342] Recent work has treated the modeling of short contact time reactors.[340,342,347,348] The mechanism of methane conversion to syngas by catalytic partial oxidation is still under discussion. Thus, although the reaction was explained by considering a direct route, there are many other studies suggesting that the reaction is carried out by an indirect route (i.e. total combustion and reforming reactions).[347]

Since catalytic partial oxidation seems to be appropriate for Fischer–Tropsch syntheses, it has also been proposed that the heat required for methane steam reforming could be balanced by the exothermic oxidation of methane.[349] In this way, a quasi-homogeneous, one-dimensional model has been developed from the study of partial oxidation of methane in the presence of steam over a 5% Ru-supported on $\gamma$-$Al_2O_3$, which according to authors can help in pilot reactor design, materials, and further scale-up.

### 24.4.2. *Methane to ethane/ethene by coupling*

The OCM towards ethylene received much attention in the 1980s and 1990s, after Keller and Bhasin published the first results on OCM.[350] Over the years, the higher yields of ethane/ethylene were observed over modified alkali-earth[351,352] and $Mn/Na_2WO_4/SiO_2$ catalysts.[353] However, yields lower than 20% were reported. In order to reach the yield required, as proposed by the industry, the catalyst should reach 90% selectivity at 40% methane conversion. Nevertheless, recent studies suggest that this may be overcome by producing not only ethylene but also electricity, using the heat from the very exothermic coupling reaction.[354] More recently, the effect of promoters[355,356] as well as mechanistic aspects using kinetic and isotopic methods[357] over $Na_2WO_4$–$MnO_x/SiO_2$ catalysts have been studied.

On the other hand, it has been reported that a better OCM yield, which is 23% (70% $C_2$-selectivity at 32% methane conversion, using a $CH_4$/air molar ratio of 1), is obtained using air as the oxidant instead of oxygen (16% yield, with 80%

$C_2$-selectivity at 20% methane conversion, using a $CH_4/O_2$ molar ratio of 7.5) over a $Na_2WO_4$–$MnO_x/SiO_2$ catalyst.[356] Moreover, by changing the chemical composition of the catalysts, both ethylene and syngas could be obtained. Thus, Wu *et al.*[358] report $CH_4$ conversion of 38% and a yield of 21% for ($C_2H_4$ + CO), with a $C_2H_4$/CO/$H_2$ ratio of 1/0.7/0.7 when the reaction was carried out over a $Na_2WO_4$/Co–Mn/$SiO_2$ catalyst. An interesting alternative is the oxidative coupling via bromine activation, which could also be used in partial oxygenation reactions.[359,360]

Takanabe and Iglesia[357] suggested that intermediate $O_2$ pressures and the presence of $H_2O$ improve the $C_2$-yields during the OCM over a $Na_2WO_4$–$MnO_x/SiO_2$ catalyst. This behavior is explained by considering the mechanistic aspects and the reaction pathways on the studied catalyst. More recently, it has also been proposed that more steam should be added to the reactants when working in a $BaCo_xFe_yZr_zO_{3-\delta}$ (BCFZ, $x + y + z = 1$) perovskite hollow fiber membrane reactor, to prevent hydrocarbons from deep oxidation in the gas phase, because of the better removal of excess heat from the catalyst bed.[361]

Finally, most of catalytic tests have been carried out using fixed-bed reactors. However, other reactor types should be considered. Kao *et al.*[362] present an analysis of methane oxidative coupling on a Li/MgO packed porous-membrane reactor (PMR) and by a fixed-bed reactor (FBR). They conclude a maximal 30% yield at 53% selectivity in PMR, while the maximal yield achieved in the FBR of identical dimensions and temperature was 20.7%, with 52.5% selectivity.

Oxidative coupling of methane using a catalytic-membrane reactor (CMR), catalyst packed-bed reactor (PBR), and catalyst packed-bed membrane reactor (PBMR), have been also compared.[363] The authors conclude that the catalytic activity of PBR and PBMR (using Na-WMn/$SiO_2$) were lower than that observed for CMR (with a yield of 34.7%).

A model-based performance analysis of fixed-bed, fluidized-bed, and porous-membrane reactors has been reported,[364] concluding that fluidized-bed reactors can improve the yield up to 26% (which is still below the industrial requirements), while fixed-bed reactors cannot be used industrially. However, membrane reactors offer the possibility of increasing the yield by fine oxygen distribution through the membrane. In this way, it has been suggested that it is possible to optimize the reaction conditions in all investigated reactor concepts. This alternative operation mode using a PBMR, is proposed in order to increase the efficiency of the OCM with a performance of 23.2% yield, 53.9% selectivity, and 42.7% methane conversion.[365]

The particle/monolithic has been also tested for OCM. With the feed gas passed from the particle bed down to the monolithic bed, the dual catalyst bed reactor with $Na_2WO_4$(5%)-Mn(2%)/$SiO_2$ particle catalyst of 15 mm in height and Ce(3%)-$Na_2WO_4$(5%)-Mn(2%)/SBA-15/$Al_2O_3$/FeCrAl monolithic catalyst of 50 mm in height showed the best catalytic performances for OCM. The conversion of $CH_4$

and the selectivity to $C_2H_4$ were 38.4% and 41.5%, respectively. Compared with the single-bed reactor, the dual-catalyst-bed reactor exhibited much better catalytic performances for OCM.

Recently, it has been reported that Siluria Technologies has developed a nanowire-based catalyst that can convert methane to ethylene more efficiently and at a significantly lower temperature than anything reported previously.[366,367] This method has not yet been demonstrated beyond the laboratory.

### 24.4.3. *Methane oxidation to oxygenate compounds*

Some reviews in the last few years have reported on the selective oxidation of methane to oxygenated products,[329,368] methane partial oxidation to methanol and/or formaldehyde being the biggest scientific challenge and of most interest to industry.[368–370] Over the last two decades many studies have been published on the latter issue, although at the moment the yields achieved are too low. The main difficulty arises from the fact that the target product ($CH_3OH$ or HCHO) is far more reactive than the $CH_4$ molecule[371] and would easily undergo consecutive oxidation to CO and $CO_2$ under the reaction conditions, with considerable $CH_4$ conversion. In general, highly dispersed transition metal oxides supported on a metal oxide or incorporated into microporous/mesoporous materials, seem to be the most effective catalysts. However, the nature of the transition metal, the amount of metal, and the characteristics of support for microporous/mesoporous materials as well as the reaction conditions, determine the nature of partial oxidation products. On the other hand, formaldehyde is mainly formed when the reaction is carried out at high reaction temperature, while methanol could only be obtained at low reaction temperature. This is so because methanol quickly reacts to formaldehyde at temperatures higher than 500°C.

Initially, supported $MoO_x$ and $VO_x$ species were proposed as active and relatively selective in the partial oxidation of methane to formaldehyde.[372] The nature of metal oxide support and the V- or Mo-content strongly influence the dispersion of molybdenum and vanadium species, which is a key factor for obtaining better selectivity to formaldehyde. At the moment, the highest productivity to formaldehyde, in terms of space-time yields, seems to be obtained over a vanadium oxide supported on mesoporous silica synthesized by a novel method.[373] In this case, the best results have been achieved with a water pressure of 8 kPa, suggesting that water mainly influences the number of accessible active sites and only weakly affects the catalytic properties of these sites.

Moreover, other transition metal oxides have been investigated in the last ten years. Supported iron oxide catalysts have also been reported as active and relatively selective for partial oxidation of methane to formaldehyde. In general, the nature of the support and the quantity of active metal oxides are key factors for preparing selective catalytic materials, with a silica-based support appearing to be the preferred option. However, the catalyst preparation procedure is also a key factor. In this way, the efficiency in the use of a preparation method based on adsorption/precipitation of $Fe^{n+}$ ions on the silica surface for tailoring highly effective $FeO_x/SiO_2$ catalysts has been reported.[372] More recently, it has been shown that the use of mesoporous materials can favor a better yield to oxygenated products. Thus, tetrahedral Fe(III) sites incorporated inside the framework of SBA-15 showed higher selectivity to formaldehyde than the oligomeric $FeO_x$ clusters located in the mesopores.[213,374]

Cu-containing zeolites are also relatively effective in partial oxidation of methane. However, methanol rather than formaldehyde is mainly formed.[375,376] Initially, it was reported that the oxygen chemisorbed over a Cu-ZSM-5 zeolite is active for the partial oxidation of methane to methanol at temperatures lower than 175°C,[375] the$[Cu_2\text{-}(\mu\text{-}O)_2]^{2+}$ species being responsible for this catalytic behavior, as suggested by catalyst characterization. However, we must indicate that the reaction is not carried out in a catalytic manner;[376] the adsorbed methanol, formed during the reaction, must be extracted from the surface using a solvent. Recently, it has been suggested that the amount of methanol produced decreases according to the solvent used: ethanol > acetonitrile/water > acetonitrile > hexane.[377]

However, the latter catalytic behavior is not exclusive of Cu-species. In fact, Co-containing ZSM-5 is also an active and relatively selective catalyst in the partial oxidation of methane to methanol/formaldehyde.[378] However, as occurs in the case of Cu-ZSMA-5, oxygenated products must be extracted from the surface of catalyst.

A different approach was reported by the Wang group. They observed that, when comparing the catalytic behavior of transition metal oxides deposited over mesoporous materials, the Cu-containing catalyst was the most effective.[379] After this preliminary result, they studied the influence of the catalyst preparation on catalytic behavior, reporting a formaldehyde selectivity of *ca.* 60–70% at methane conversion of 2% when working at 500–650°C.[380,381] They proposed a relatively different mechanism with the stabilization of $Cu^{n+}$ species during the catalytic tests.[381]

Olah and Prakash have reported a method of producing methanol from methane in a three-step system, in which formaldehyde is obtained initially in a quartz-tube continuous-flow reactor. Formaldehyde is then transformed into formic acid on alkaline-earth oxides and finally methanol/formic acid is transformed into methanol methyl formate employing a $WO_3/Al_2O_3$ catalyst.[382] However, at the moment, only low yield to oxygenated products has been obtained.

On the other hand, an alternative to gas-phase heterogeneous catalysis has been the use of metal complexes in the presence of sulfuric acid. However, this method currently presents some important drawbacks. Schüth and co-workers have developed a solid catalyst for the direct oxidation of methane to methanol, over several recycling steps, using $H_2SO_4$ and $SO_3$ as oxidants.[383] In other approaches, partial success in the conversion of methane to methanol has been achieved by Periana *et al.*,[384] who used a bipyrimidal platinum (II) complex in concentrated $H_2SO_4$. At 220°C, 90% of the methane reacted with $H_2SO_4$ in a catalytic manner to form methyl bisulfate at 81% selectivity. It is expected that the methyl bisulfate could be hydrolyzed to methanol, although this was not demonstrated. A complete cycle would require the regeneration of the concentrated $H_2SO_4$.

Finally, there are also some alternatives reported which raise the possibility of preparing oxygenated products by an initial oxidative halogenation using chloride[385] or bromide.[386–388] Moreover, a promoted $Ru/SiO_2$ catalyst has been proposed as active in the synthesis of acetic acid from methane via oxidative bromination, carbonylation, and hydrolysis.[389]

## 24.5. Conclusions

Natural gas and LPG are interesting feedstocks for petrochemical processes since their components (light alkanes as methane, ethane, propane, and butanes) could be alternative raw materials in well-known industrial processes.

As indicated in Fig. 24.12, the development of catalysts during the first quarter of the last century permitted the expansion of processes for obtaining monomers which were crucial for the expansion of the petrochemical industry. However, these processes included the use of non-desirable raw materials and the formation of many

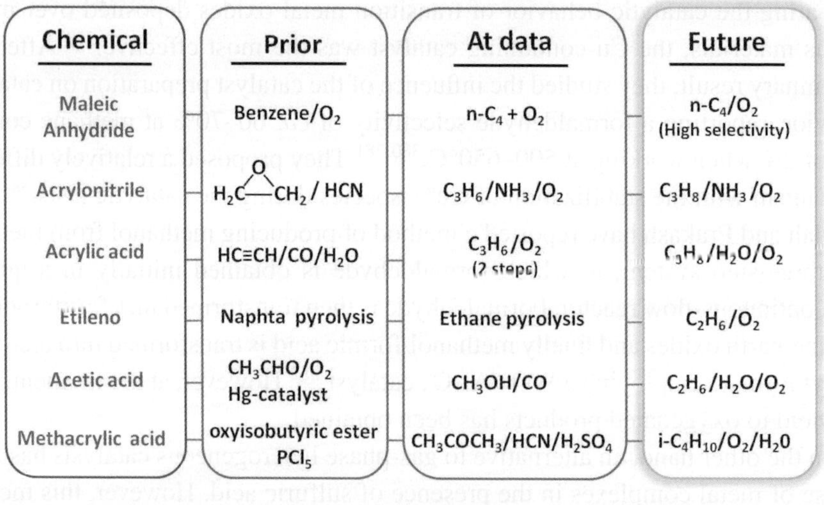


**Figure 24.12.** Evolution of the main industrial processes of possible compounds to be achieved by partial oxidation of alkanes.

by-products. The majority of these industrial processes were improved (from both economical and environmental points of view) during the last 50 years, achieving important industrial advances which are working at the moment.

In this chapter, a summary of alternative industrial partial oxidation processes using light alkanes has been presented, which could offer an important breakthrough to the current industrial processes using olefins and aromatic hydrocarbons as raw materials. In this way, light hydrocarbons would favor a clear reduction in the number steps in global processes, decreasing both the energy required and $CO_2$ emissions, and improving the atom economy. Some of these are summarized in Fig. 24.12.

In this way, the more investigated reactions for selective transformation of alkanes are both the oxidative dehydrogenation (in order to achieve the corresponding olefins) and the O- and N-insertion (achieving aldehydes, acids, anhydrides, and nitriles α,β-unsaturated). In addition, an alternative method for an improved transformation of methane through selective oxidation is also presented.

With respect to the catalysts, i.e. mixed-metal oxides and/or heteropolyacids, those presenting well-known structures — as is the case for vanadyl pyrophosphate, orthorhombic bronzes, or Keggin-type heteropolyacids — have been developed in well-known catalytic processes. However, the roles of amorphous and/or quasi-crystalline phases present in the more active and selective catalysts are still unclear and therefore should also be considered.

With respect to the catalytic reactions, there are well-established industrial reactions (as occurs in the case of *n*-butane to maleic anhydride), reactions in the pre-industrial stage (such as the transformation of propane to acrylonitrile), very promising reactions (such as ethane oxidative dehydrogenation to ethylene), and potential reactions whose economical viability will depend on the prices of crude and natural gas in the future (such as propane selective oxidation to acrylic acid or methane transformation).

## References

1. Grasselli, R. and Burrington, J. (1981). Selective Oxidation and Ammoxidation of Propylene by Heterogeneous Catalysis, *Adv. Catal.*, **30**, pp. 133–163.
2. Grasselli, R. (2001). Genesis of Site Isolation and Phase Cooperation in Selective Oxidation Catalysis, *Top. Catal.*, **15**, pp. 93–101.
3. Grasselli, R. (2002). Fundamental Principles of Selective Heterogeneous Oxidation Catalysis, *Top. Catal.*, **21**, pp. 79–88.
4. Haber, J. (1992). Catalytic Oxidation. State-of-the-Art and Prospects, *Stud. Surf. Sci. Catal.*, **72**, pp. 279–304.
5. Dadyburjor, D., Jewur, S. and Ruckenstein, E. (1979). Selective Oxidation of Hydrocarbons on Composite Oxides, *Catal. Rev.*, **19**, pp. 293–350.
6. Vedrine, J., Coudurier, G. and Millet, J. (1997). Molecular Design of Active Sites in Partial Oxidation Reactions on Metallic Oxides, *Catal. Today*, **33**, pp. 3–13.

7. Vedrine, J., Millet, J. and Volta, J. (1996). Molecular Description of Active Sites in Oxidation Reactions: Acid-Base and Redox Properties, and Role of Water, *Catal. Today*, **32**, pp. 115–123.
8. Oyama, S. (1996). Factors Affecting Selectivity in Catalytic Partial Oxidation and Combustion Reactions, *Heterogeneous Hydrocarbon Oxidation*, **638**, pp. 2–19.
9. Bordes-Richard, E. (2008). Multicomponent Oxides in Selective Oxidation of Alkanes Theoretical Acidity versus Selectivity, *Top. Catal.*, **50**, pp. 82–89.
10. Grzybowska-Swierkosz, B. (2000). Thirty Years in Selective Oxidation in Oxides: What have We Learned? *Top. Catal.*, **11**, pp. 23–42.
11. Volta, J. (2001). Site Isolation for Light Hydrocarbons Oxidation, *Top. Catal.*, **15**, pp. 121–129.
12. Grasselli, R. (1999). Advances and Future Trends in Selective Oxidation and Ammoxidation Catalysis, *Catal. Today*, **49**, pp. 141–153.
13. Arpentinier, P., Cavani, F. and Trifirò, F. (2001). *The Technology of Catalytic Oxidations. Chemical, Catalytic and Engineering Aspects*, Vol. 1, Editions Technip, Paris.
14. Centi, G., Cavani, F. and Trifiró, F. (2001). *Selective Oxidation by Heterogeneous Catalysis*, Kluwer Academic/Plenum Publishers, New York.
15. Hodnett, B. (2000). *Heterogeneous Catalytic Oxidation: Fundamental and Technological Aspects of the Selective and Total Oxidation of Organic Compounds*, John Wiley & Sons, New York.
16. Schlögl, R. (2009). Concepts in Selective Oxidation of Small Alkane Molecules, in M. Noritaka (ed.), *Modern Heterogeneous Oxidation Catalysis: Design, Reactions and characterization*, Wiley-VCH Verlag GmbH & Co. KGaA, Weinheim, Germany, pp. 1–35.
17. Centi, G. and Trifiro, F. (1984). Some Aspects of the Control of Selectivity in Catalytic-Oxidation on Mixed Oxides: A Review, *Appl. Catal.*, **12**, pp. 1–21.
18. Grasselli, R., Centi, G. and Trifiro, F. (1990). Selective Oxidation of Hydrocarbons Employing Tellurium Containing Heterogeneous Catalysts, *Appl. Catal.*, **57**, pp. 149–166.
19. Kung, H. (1988). *Transition Metal Oxides: Surface Chemistry and Catalysis*, (Stud. Surf. Sci. Catal., series, vol. 45), Elsevier Sci. Publishers B.V., Amsterdam.
20. Vedrine, J. (2002). The Role of Redox, Acid-Base and Collective Properties and of Crystalline State of Heterogeneous Catalysts in the Selective Oxidation of Hydrocarbons, *Top. Catal.*, **21**, pp. 97–106.
21. Novakova, E. and Vedrine, J. (2006). Propane Selective Oxidation to Propene and Oxygenates on Metal Oxides, in J. Fierro (ed.), *Metal Oxides, Chemistry and Applications*, CRC Press, New York, pp. 413–462.
22. Grzybowska-Swierkosz, B. (2002). Effect of Additives on the Physicochemical and Catalytic Properties of Oxide Catalysts in Selective Oxidation Reactions, *Top. Catal.*, **21**, pp. 35–46.
23. Guliants, V. and Carreon, M. (2005). *Vanadium-Phosphorus-Oxides: From Fundamentals of n-Butane Oxidation to Synthesis of New Phases*, in J.J. Spivey (ed.), *Catalysis* (vol. 18), The Royal Society of Chemistry, London, pp. 1–45.
24. Buyevskaya, O. and Baerns, M. (2002). Oxidative Functionalization of Ethane and Propane, in J.J. Spivey (ed.), Catalysis (vol. 16), The Royal Society of Chemistry, London, pp. 155–197.
25. Ushikubo, T. (2003). Activation of Propane and Butanes over Niobium- and Tantalum-Based Oxide Catalysts, *Catal. Today*, **78**, pp. 79–84.
26. Albonetti, S., Cavani, F. and Trifiro, F. (1996). Key Aspects of Catalyst Design for the Selective Oxidation of Paraffins, *Catal. Rev.*, **38**, pp. 413–438.
27. Bettahar, M., Costentin, G., Savary, L., *et al.* (1996). On the Partial Oxidation of Propane and Propylene on Mixed Metal Oxide Catalysts, *App. Catal. A: Gen.*, **145**, pp. 1–48.
28. Cavani, F. and Trifiro, F. (1997). Some Aspects that Affect the Selective Oxidation of Paraffins, *Catal. Today*, **36**, pp. 431–439.
29. Cavani, F. and Trifiro, F. (1999). Selective Oxidation of Light Alkanes: Interaction between the Catalyst and the Gas Phase on Different Classes of Catalytic Materials, *Catal. Today*, **51**, pp. 561–580.

30. a) Lin, M. (2001). Selective Oxidation of Propane to Acrylic Acid with Molecular Oxygen, *Appl. Catal. A: Gen.*, **207**, pp. 1–16; b) Lin, M., Desai, T., Kaiser, F., *et al.* (2000). Reaction Pathways in the Selective Oxidation of Propane over a Mixed Metal Oxide Catalyst, *Catal. Today*, **61**, pp. 223–229.
31. Grasselli, R., Burrington, J., Buttrey, D., *et al.* (2003). Multifunctionality of Active Centers in (Amm)oxidation Catalysts: From Bi–Mo–Ox to Mo–V–Nb–(Te, Sb)–Ox, *Top. Catal.*, **23**, pp. 5–22.
32. López Nieto, J. (2006). The Selective Oxidative Activation of Light Alkanes. From Supported Vanadia to Multicomponent Bulk V-Containing Catalysts, *Top. Catal.*, **41**, pp. 3–15.
33. Cavani, F. and Teles, J. (2009). Sustainability in Catalytic Oxidation: an Alternative Approach or a Structural Evolution? *Chemsuschem*, **2**, pp. 508–534.
34. Cavani, F., Ballarini, N. and Cericola, A. (2007). Oxidative Dehydrogenation of Ethane and Propane: How Far from Commercial Implementation? *Catal. Today*, **127**, pp. 113–131.
35. Cavani, F. and Trifiro, F. (1995). The Oxidative Dehydrogenation of Ethane and Propane as an Alternative Way for the Production of Light Olefins, *Catal. Today*, **24**, pp. 307–313.
36. Blasco, T. and López Nieto, J. (1997). Oxidative Dehydrogenation of Short Chain Alkanes on Supported Vanadium Oxide Catalysts, *Appl. Catal. A: Gen.*, **157**, pp. 117–142.
37. Bañares, M. (1999). Supported Metal Oxide and Other Catalysts for Ethane Conversion: A Review, *Catal. Today*, **51**, pp. 319–348.
38. Dai, H. and Au, C. (2002). The Oxidative Dehydrogenation of Ethane to Ethene, *Curr. Top. Catal.*, **3**, pp. 33–80.
39. Kung, H. (1994). Oxidative Dehydrogenation of Light (C-2 to C-4) Alkanes, *Adv. Catal.*, **40**, pp. 1–38.
40. Mamedov, E. and Corberan, V. (1995). Oxidative Dehydrogenation of Lower Alkanes on Vanadium Oxide-Based Catalysts: The Present State-of-the-Art and Outlooks, *Appl. Catal. A: Gen.*, **127**, pp. 1–40.
41. Kung, H. and Kung, M. (1997). Oxidative Dehydrogenation of Alkanes over Vanadium-Magnesium-Oxides, *Appl. Catal. A: Gen.*, **157**, pp. 105–116.
42. Grabowski, R. (2006). Kinetics of Oxidative Dehydrogenation of C-2-C-3 Alkanes on Oxide Catalysts, *Catal. Rev.*, **48**, pp. 199–268.
43. Ren, T., Patel, M. and Blok, K. (2006). Olefins from Conventional and Heavy Feedstocks: Energy Use in Steam Cracking and Alternative Processes, *Energy*, **31**, pp. 425–451.
44. Nakamura, D. (2009). Special Report: Global Ethylene Production Rises 7 Million Tones per Year in 2008, *Oil Gas J.*, **107**, pp. 43–54.
45. Seddon, D. (2010). *Petrochemical Economics. Technology Selection in a Carbon Constrained World*, Imperial Colleges Press, London.
46. Bhasin, M. (2003). Is True Ethane Oxydehydrogenation Feasible? *Top. Catal.*, **23**, pp. 145–149.
47. Dubois, J. (2005). Selective Oxidation of Hydrocarbons and the Global Warming Problem, *Catal. Today*, **99**, pp. 5–14.
48. Cai, H., Krzywicki, A. and Oballa, M. (2002). Coke Formation in Steam Crackers for Ethylene Production, *Chem. Eng. Process.*, **41**, pp. 199–214.
49. Ren, T., Patel, M. and Blok, K. (2008). Steam Cracking and Methane to Olefins: Energy Use, $CO_2$ Emissions and Production Costs, *Energy*, **33**, pp. 817–833.
50. Centi, G., Trifiro, F., Ebner, J., *et al.* (1988). Mechanistic Aspects of Maleic-Anhydride Synthesis from C4-Hydrocarbons over Phosphorus Vanadium-Oxide, *Chem. Rev.*, **88**, pp. 55–80.
51. Thompson, D. and Hodnett, B. (2008). Hydrocarbon Selective Oxidation on Vanadium Phosphorus Oxide Catalysts: Insights from Electronic Structure Calculations, *Top. Catal.*, **50**, pp. 116–123.
52. Ballarini, N., Cavani, F., Cortelli, C., *et al.* (2006). VPO Catalyst for *n*-Butane Oxidation to Maleic Anhydride: A Goal Achieved, or a Still Open Challenge? *Top. Catal.*, **38**, pp. 147–156.

53. Cavani, F. and Trifiro, F. (1994). Selective Oxidations of C4 Paraffins, in *Catalysis*, Vol. 11, Royal Society of Chemistry, Cambridge, Chapter 7, pp. 246–317.
54. Centi, G. and Perathoner, S. (2001). Reaction Mechanism and Control of Selectivity in Catalysis by Oxides: Some Challenges and Open Questions, *Int. J. Mol. Sci.*, **2**, pp. 183–196.
55. Centi, G. (1993). Vanadyl Pyrophosphate: A Critical Overview, *Catal. Today*, **16**, pp. 5–26.
56. López Nieto, J., Botella Asunción, P., Vázquez Navarro, M., *et al.* (2003). *Worldwide Patent 2003064035 A1, Method for the Oxidative Dehydrogenation of Ethane* (CSIC-UPV).
57. Heracleous, E. and Lemonidou, A. (2006). Ni-Nb-O Mixed Oxides as Highly Active and Selective Catalysts for Ethene Production via Ethane Oxidative Dehydrogenation. Part I: Characterization and Catalytic Performance, *J. Catal.*, **237**, pp. 162–174.
58. Concepción, P., Blasco, T., López Nieto, J., *et al.* (2004). Preparation, Characterization and Reactivity of V- and/or Co-Containing AlPO-18 Materials (VCoAPO-18) in the Oxidative Dehydrogenation of Ethane, *Micropor. Mesopor. Mat.*, **67**, pp. 215–227.
59. Catani, R. and Centi, G. (1991). Selective Ethane Ammoxidation to Acetonitrile on Alumina-Supported Niobium Antimony Oxides, *Chem. Commun.*, **16**, pp. 1081–1083.
60. Li, Y. and Armor, J. (1997). Ammoxidation of Ethane to Acetonitrile over Co-Beta Zeolite, *Chem. Commun.*, **20**, pp. 2013–2014.
61. Kitson, M. (1991). *European Patent 407091 A1* (BP Chemicals Ltd., UK).
62. Li, X. and Iglesia, E. (2008). Support and Promoter Effects in the Selective Oxidation of Ethane to Acetic Acid Catalyzed by Mo-V-Nb Oxides, *Appl. Catal. A: Gen.*, **334**, pp. 339–347.
63. Centi, G. and Trifiro, F. (1996). Catalytic Behavior of V-Containing Zeolites in the Transformation of Propane in the Presence of Oxygen, *Appl. Catal. A: Gen.*, **143**, pp. 3–16.
64. Grasselli, R., Buttrey, D., DeSanto, P., *et al.* (2004). Active Centers in Mo-V-Nb-Te-$O_x$ (Amm)oxidation Catalysts, *Catal. Today*, **91–92**, pp. 251–258.
65. Guttmann, A., Grasselli, R. and Brazdil, J. (1988). *US Patent 4746641*, Ammoxidation of paraffins and catalysts therefor (Standard Oil Co Ohio).
66. Guttmann, A., Grasselli, R. and Brazdil, J. (1988). *US Patent 4788317*, Ammoxidation of paraffins and catalysts therefor (Standard Oil Co., USA).
67. Tsuji, H. and Koyasu, Y. (2002). Synthesis of MoVNbTe(Sb)$O_x$ Composite Oxide Catalysts via Reduction of Polyoxometales in an Aqueous Medium, *J. Am. Chem. Soc.*, **124**, pp. 5608–5609.
68. Ushikubo, T., Nakamura, H., Koyasu, Y., *et al.* (1995). *US Patent 5,380,933*, Method for producing an unsaturated carboxylic acid (Mitsubushi Kasei Corporation).
69. Millet, J. (2006). Mechanism of First Hydrogen Abstraction from Light Alkanes on Oxide Catalysts, *Top. Catal.*, **38**, pp. 83–92.
70. Dinse, A., Schomacker, R. and Bell, A. (2009). The Role of Lattice Oxygen in the Oxidative Dehydrogenation of Ethane on Alumina-Supported Vanadium Oxide, *Phys. Chem. Chem. Phys.*, **11**, pp. 6119–6124.
71. Ovsitser, O. and Kondratenko, E. (2009). Similarity and Differences in the Oxidative Dehydrogenation of $C_2$-$C_4$ Alkanes over Nano-Sized $VO_x$ Species Using $N_2O$ and $O_2$, *Catal. Today*, **142**, pp. 138–142.
72. Kondratenko, E. and Bruckner, A. (2010). On the Nature and Reactivity of Active Oxygen Species Formed from $O_2$ and $N_2O$ on $VO_x$/MCM-41 Used for Oxidative Dehydrogenation of Propane, *J. Catal.*, **274**, pp. 111–116.
73. Rozanska, X., Kondratenko, E. and Sauer, J. (2008). Oxidative Dehydrogenation of Propane: Differences between $N_2O$ and $O_2$ in the Reoxidation of Reduced Vanadia Sites and Consequences for Selectivity, *J. Catal.*, **256**, pp. 84–94.
74. Patience, G. and Lorences, M. (2006). VPO Transient Oxidation Kinetics, *Int. J. Chem. React. Eng.*, **4**, pp. 1–18.
75. Gascon, J., Valenciano, R., Tellez, C., *et al.* (2006). A Generalized Kinetic Model for the Partial Oxidation of *n*-Butane to Maleic Anhydride under Aerobic and Anaerobic Conditions, *Chem. Eng. Sci.*, **61**, pp. 6385–6394.

76. Heracleous, E. and Lemonidou, A. (2006). Ni-Nb-O Mixed Oxides as Highly Active and Selective Catalysts for Ethene Production via Ethane Oxidative Dehydrogenation. Part II: Mechanistic Aspects and Kinetic Modeling, *J. Catal.*, **237**, pp. 175–189.
77. Savova, B., Loridant, S., Filkova, D., *et al.* (2010). Ni-Nb-O Catalysts for Ethane Oxidative Dehydrogenation, *Appl. Catal. A*: *Gen.*, **390**, pp. 148–157.
78. Heracleous, E., Lee, A., Wilson, K., *et al.* (2005). Investigation of Ni-Based Alumina-Supported Catalysts for the Oxidative Dehydrogenation of Ethane to Ethylene: Structural Characterization and Reactivity Studies, *J. Catal.*, **231**, pp. 159–171.
79. Heracleous, E. and Lemonidou, A. (2010). Ni-Me-O Mixed Metal Oxides for the Effective Oxidative Dehydrogenation of Ethane to Ethylene: Effect of Promoting Metal Me, *J. Catal.*, **270**, pp. 67–75.
80. Solsona, B., Ivars, F., Dejoz, A., *et al.* (2009). Supported Ni-W-O Mixed Oxides as Selective Catalysts for the Oxidative Dehydrogenation of Ethane, *Top. Catal.*, **52**, pp. 751–757.
81. Solsona, B., López Nieto, J., Concepción, P., *et al.* (2011). Oxidative Dehydrogenation of Ethane over Ni-W-O Mixed Metal Oxides Catalysts, *J. Catal.*, **280**, pp. 28–39.
82. Catani, R., Centi, G., Trifiro, F., *et al.* (1992). Kinetics and Reaction Network in Propane Ammoxidation to Acrylonitrile on V-Sb-Al Based Mixed Oxides, *Ind. Eng. Chem. Res.*, **31**, pp. 107–119.
83. Centi, G., Perathoner, S. and Trifiro, F. (1997). V-Sb-Oxide Catalysts for the Ammoxidation of Propane, *Appl. Catal. A*: *Gen.*, **157**, pp. 143–172.
84. Nilsson, J., Landa-Canovas, A. Hansen, S., *et al.* (1999). An Investigation of the Al-Sb-V-W-Oxide System For Propane Ammoxidation, *J. Catal.*, **186**, pp. 442–457.
85. Zanthoff, H., Grunert, W., Buchholz, S. *et al.* (2000). Bulk and Surface Structure and Composition of V-Sb Mixed-Oxide Catalysts for the Ammoxidation of Propane, *J. Mol. Catal. A*: *Chem.*, **162**, pp. 443–462.
86. Guerrero-Pérez, M. and Bañares, M. (2004). Operando Raman-GC Studies of Alumina-Supported Sb-VO Catalysts and Role of the Preparation Method, *Catal. Today*, **96**, pp. 265–272.
87. Hatano, M. and Kayo, A. (1992). *European Patent 0318295*, Process for producing nitriles (Mitsubishi Kasei Corporation).
88. Ushikubo, T., Kazunori, K., Atsushi, U., *et al.* (1993). *European Patent 529853 A2*, Process for producing nitriles (Mitsubishi Kasei Corp., Japan).
89. Tsuji, H., Oshima, K. and Koyasu, Y. (2003). Synthesis of Molybdenum And Vanadium-Based Mixed Oxide Catalysts with Metastable Structure: Easy Access to the MoVNbTe(Sb) O-x Catalytically Active Structure Using Reductant And Oxoacid, *Chem. Mater.*, **15**, pp. 2112–2114.
90. Millet, J., Roussel, H., Pigamo, A., *et al.* (2002). Characterization of Tellurium in MoVTeNbO Catalysts for Propane Oxidation or Ammoxidation, *Appl. Catal. A*: *Gen.*, **232**, pp. 77–92.
91. DeSanto, P., Buttrey, D., Grasselli, R., *et al.* (2003). Structural Characterization of the Orthorhombic Phase M1 in MoVNbTeO Propane Ammoxidation Catalyst, *Top. Catal.*, **23**, pp. 23–38.
92. Botella, P., López Nieto, J., Solsona, B., *et al.* (2002). The Preparation, Characterization, and Catalytic Behavior of Movtenbo Catalysts Prepared by Hydrothermal Synthesis, *J. Catal.*, **209**, pp. 445–455.
93. Sadakane, M., Yamagata, K., Kodato, K., *et al.* (2009). Synthesis of Orthorhombic Mo-V-Sb Oxide Species by Assembly of Pentagonal Mo6O21 Polyoxometalate Building Blocks, *Angew. Chem. Int. Edit.*, **48**, pp. 3782–3786.
94. Albonetti, S., Blanchard, G., Burattin, P., *et al.* (1998). A New Ternary Mixed Oxide Catalyst for Ammoxidation of Propane: Sn/V/Sb, *Catal. Lett.*, **50**, pp. 17–23.
95. Ballarini, N., Cavani, F., Marion, P., *et al.* (2009). The Role of V in Rutile-Type Sn/V/Nb/Sb Mixed Oxides, Catalysts for Propane Ammoxidation to Acrylonitrile, *Catal. Today*, **142**, pp. 170–174.

96. Lundberg, M. and Sundberg, M. (1993). New Complex Structures in the Cesium-Niobium-Tungsten-Oxide System Revealed by Hrem, *Ultramicroscopy*, **52**, pp. 429–435.
97. Schimanke, G., Martin, M., Kunert, J., *et al.* (2005). Characterization of Mo-V-W Mixed Oxide Catalysts by Ex Situ and In Situ X-Ray Absorption Spectroscopy, *Z. Anorg. Allg. Chem.*, **631**, pp. 1289–1296.
98. Tichy, J. (1997). Oxidation of Acrolein to Acrylic Acid over Vanadium-Molybdenum Oxide Catalysts, *Appl. Catal. A: Gen.*, **157**, pp. 363–385.
99. Duc, F., Gonthier, S., Brunelli, M., *et al.* (2006). Hydrothermal Synthesis and Structure Determination of the New Vanadium Molybdenum Mixed Oxide V1.1Mo0.9O5 from Synchrotron X-Ray Powder Diffraction Data, *J. Solid State Chem.*, **179**, pp. 3591–3598.
100. Tichy, J., Svachula, J., Farbotko, J., *et al.* (1991). Active Component of Molybdenum Vanadium Oxide Catalysts in Acrolein Oxidation, *Zesz. Nauk. Politech. Lodzka, Chem.*, **616**, pp. 95–106.
101. Werner, H., Timpe, O., Herein, D., *et al.* (1997). Relevance of a Glassy Nanocrystalline State of Mo4VO14 for its Action as Selective Oxidation Catalyst, *Catal. Lett.*, **44**, pp. 153–163.
102. Giebeler, L., Kampe, P., Wirth, A., *et al.* (2006). Structural Changes of Vanadium-Molybdenum-Tungsten Mixed Oxide Catalysts during the Selective Oxidation of Acrolein to Acrylic Acid, *J. Mol. Catal. A: Chem.*, **259**, pp. 309–318.
103. Andrushkevich, T. (1993). Heterogeneous Catalytic-Oxidation of Acrolein to Acrylic-Acid: Mechanism and Catalysts, *Catal. Rev.*, **35**, pp. 213–259.
104. Mestl, G. (2006). MoVW Mixed Metal Oxides Catalysts for Acrylic Acid Production: From Industrial Catalysts to Model Studies, *Top. Catal.*, **38**, pp. 69–82.
105. Botella, P., López Nieto, J. and Solsona, B. (2002). Preparation, Characterisation and Catalytic Behaviour of a New TeVMoO Crystalline Phase, *Catal. Lett.*, **78**, pp. 383–387.
106. Botella, P., Garcia-Gonzalez, E., Solsona, B. *et al.* (2009). Mo-Containing Tetragonal Tungsten Bronzes. The Influence of Tellurium on Catalytic Behaviour in Selective Oxidation of Propene, *J. Catal.*, **265**, pp. 43–53.
107. Wang, F. and Ueda, W. (2008). Steric Effect on the Catalytic Performance of the Selective Oxidation of Alcohols over Novel Crystalline Mo-V-O Oxide, *Top. Catal.*, **50**, pp. 90–97.
108. Mizuno, N. and Misono, M. (1997). Heteropolyacid Catalysts, *Curr. Opin. Solid St. M.*, **2**, pp. 84–89.
109. Davis, M., Dillon, C., Holles, J., *et al.* (2002). A New Catalyst for the Selective Oxidation of Butane and Propane, *Angew. Chem. Int. Edit.*, **41**, pp. 858–860.
110. Li, W. and Ueda, W. (1997). Catalytic Oxidation of Isobutane to Methacrylic Acid with Molecular Oxygen over Activated Pyridinium 12-Molybdophosphate, *Catal. Lett.*, **46**, pp. 261–265.
111. Oyama, S. (1991). Adsorbate Bonding and the Selection of Partial and Total Oxidation Pathways, *J. Catal.*, **128**, pp. 210–217.
112. Galli, A., Nieto, J., Dejoz, A., *et al.* (1995). The Effect of Potassium on the Selective Oxidation of N-Butane and Ethane over $Al_2O_3$-Supported Vanadia Catalysts, *Catal. Lett.*, **34**, pp. 51–58.
113. Concepcion, P., Galli, A., Nieto, J., *et al.* (1996). On the Influence of the Acid Base Character of Catalysts on the Oxidative Dehydrogenation of Alkanes, *Top. Catal.*, **3**, pp. 451–460.
114. López Nieto, J., Concepcion, P., Dejoz, A., *et al.* (2000). Selective Oxidation of *n*-Butane and Butenes over Vanadium-Containing Catalysts, *J. Catal.*, **189**, pp. 147–157.
115. López Nieto, J., Concepcion, P., Dejoz, A., *et al.* (2000). Oxidative Dehydrogenation of *n*-Butane and 1-Butene on Undoped and K-Doped $VO_x/Al_2O_3$ Catalysts, *Catal. Today*, **61**, pp. 361–367.
116. Courcot, D., Grzybowska, B., Barbaux, Y., *et al.* (1996). Effect of Potassium Addition to the $TiO_2$ Support on the Structure of $V_2O_5/TiO_2$ and its Catalytic Properties in the Oxidative Dehydrogenation of Propane, *J. Chem. Soc. Faraday T.*, **92**, pp. 1609–1617.
117. Grabowski, R., Grzybowska, B., Kozlowska, A., *et al.* (1996). Effect of Alkali Metals Additives to $V_2O_5/TiO_2$ Catalyst on Physicochemical Properties and Catalytic Performance in Oxidative Dehydrogenation of Propane, *Top. Catal.*, **3**, pp. 277–288.

118. Cortez, G., Fierro, J. and Banares, M. (2003). Role of Potassium on the Structure and Activity of Alumina-Supported Vanadium Oxide Catalysts for Propane Oxidative Dehydrogenation, *Catal. Today*, **78**, pp. 219–228.
119. Santamaria-Gonzalez, J., Luque-Zambrana, J., Merida-Robles, J., *et al.* (2000). Catalytic Behavior of Vanadium-Containing Mesoporous Silicas in the Oxidative Dehydrogenation of Propane, *Catal. Lett.*, **68**, pp. 67–73.
120. Capek, L., Adam, J., Grygar, T., *et al.* (2008). Oxidative Dehydrogenation of Ethane over Vanadium Supported on Mesoporous Materials of M41S Family, *Appl. Catal. A*: *Gen.*, **342**, pp. 99–106.
121. Cherian, M., Deo, G. and Rao, T. (2006). Metal Oxides. Chemistry and Applications, in J. Fierro (ed.), *Oxidative Dehydrogenation (ODH) of Alkanes over Metal Oxide Catalysts*, CRC Press, Boca Raton, pp. 491–516.
122. Bond, G. and Tahir, S. (1991). Vanadium Oxide Monolayer Catalysts Preparation, Characterization and Catalytic Activity, *Appl. Catal. A*: *Gen.*, **71**, pp. 1–31.
123. Lemonidou, A., Nalbandian, L. and Vasalos, I. (2000). Oxidative Dehydrogenation of Propane over Vanadium Oxide Based Catalysts: Effect of Support and Alkali Promoter, *Catal. Today*, **61**, pp. 333–341.
124. Chen, K., Bell, A. and Iglesia, E. (2000). Kinetics and Mechanism of Oxidative Dehydrogenation of Propane on Vanadium, Molybdenum, and Tungsten Oxides, *J. Phys. Chem. B*, **104**, pp. 1292–1299.
125. Concepción, P., Botella, P. and López Nieto, J. (2004). Catalytic and FT-IR Study on the Reaction Pathway for Oxidation of Propane and Propylene on V- or Mo-V-Based Catalysts, *Appl. Catal. A*: *Gen.*, **278**, pp. 45–56.
126. López Nieto, J., Soler, J., Concepción, P., *et al.* (1999). Oxidative Dehydrogenation of Alkanes over V-Based Catalysts: Influence of Redox Properties on Catalytic Performance, *J. Catal.*, **185**, pp. 324–332.
127. Botella, P., García-González, E., Dejoz, A., *et al.* (2004). Selective Oxidative Dehydrogenation of Ethane on MoVTeNbO Mixed Metal Oxide Catalysts, *J. Catal.*, **225**, pp. 428–438.
128. Xie, Q., Chen, L., Weng, W., *et al.* (2005). Preparation of MoVTe(Sb)Nb Mixed Oxide Catalysts Using a Slurry Method for Selective Oxidative Dehydrogenation of Ethane, *J. Mol. Catal. A*: *Chem.*, **240**, pp. 191–196.
129. López Nieto, J., Botella, P., Concepción, P., *et al.* (2004). Oxidative Dehydrogenation of Ethane on Te-Containing MoVNbO Catalysts, *Catal. Today*, **91–92**, pp. 241–245.
130. Botella, P., Dejoz, A., López Nieto, J., *et al.* (2006). Selective Oxidative Dehydrogenation of Ethane over MoVSbO Mixed Oxide Catalysts, *Appl. Catal. A*: *Gen.*, **298**, pp. 16–23.
131. Roussel, M., Bouchard, M., Karim, K., *et al.* (2006). MoVO-Based Catalysts for the Oxidation of Ethane to Ethylene and Acetic Acid: Influence of Niobium and/or Palladium on Physico-Chemical and Catalytic Properties, *Appl. Catal. A*: *Gen.*, **308**, pp. 62–74.
132. Nakamura, K., Miyake, T., Konishi, T., *et al.* (2006). Oxidative Dehydrogenation of Ethane to Ethylene over Nio Loaded on High Surface Area MgO, *J. Mol. Catal. A*: *Chem.*, **260**, pp. 144–151.
133. Liu, Y., Cong, P., Doolen, R., *et al.* (2003). Discovery from Combinatorial Heterogeneous Catalysis: A New Class of Catalyst for Ethane Oxidative Dehydrogenation at Low Temperatures, *Appl. Catal. A*: *Gen.*, **254**, pp. 59–66.
134. Zhang, X., Liu, J., Jing, Y., *et al.* (2003). Support Effects on the Catalytic Behavior of Nio/$Al_2O_3$ for Oxidative Dehydrogenation of Ethane to Ethylene, *Appl. Catal. A*: *Gen.*, **240**, pp. 143–150.
135. Ryan, D. and Ryan, D. (2006). *Worldwide Patent 2006/130288 A1*, Method for Selectively Oxidizing Ethane to Ethylene (Celanese Int. Corp.).
136. Liu, Y., Feng, W., Li, T., *et al.* (2006). Structure and Catalytic Properties of Vanadium Oxide Supported on Mesocellulous Silica Foams (MCF) for the Oxidative Dehydrogenation of Propane to Propylene, *J. Catal.*, **239**, pp. 125–136.

137. Liu, Y., Cao, Y., Yi, N., *et al.* (2004). Vanadium Oxide Supported on Mesoporous SBA-15 as Highly Selective Catalysts in the Oxidative Dehydrogenation of Propane, *J. Catal.*, **224**, pp. 417–428.
138. Buyevskaya, O., Bruckner, A., Kondratenko, E., *et al.* (2001). Fundamental and Combinatorial Approaches in the Search for and Optimisation of Catalytic Materials for the Oxidative Dehydrogenation of Propane to Propene, *Catal. Today*, **67**, pp. 369–378.
139. Solsona, B., Blasco, T., López Nieto, J., *et al.* (2001). Vanadium Oxide Supported on Mesoporous MCM-41 as Selective Catalysts in the Oxidative Dehydrogenation of Alkanes, *J. Catal.*, 203, pp. 443–452.
140. Liu, C., Watson, R. and Ozkan, U. (2006). Spectroscopic Characterization of Cl-Modified Mo/Si: Ti Catalysts for Oxidative Dehydrogenation of Propane, *Top. Catal.*, **41**, pp. 63–72.
141. Solsona, B., Dejoz, A., Vázquez, M., *et al.* (2001). $SiO_2$-Supported Vanadium Magnesium Mixed Oxides as Selective Catalysts for the Oxydehydrogenation of Short Chain Alkanes, *Appl. Catal. A: Gen.*, **208**, pp. 99–110.
142. Sam, D., Soenen, V. and Volta, J. (1990). Oxidative Dehydrogenation of Propane over V-Mg-O Catalysts, *J. Catal.*, **123**, pp. 417–435.
143. Kondratenko, E., Buyevskaya, O. and Baerns, M. (2001). Characterisation of Vanadium-Oxide-Based Catalysts for the Oxidative Dehydrogenation of Propane to Propene, *Top. Catal.*, **15**, pp. 175–180.
144. Gao, X., Xin, Q. and Guo, X. (1994). Support Effects on Magnesium-Vanadium Mixed Oxides in the Oxidative Dehydrogenation of Propane, *Appl. Catal. A: Gen.*, **114**, pp. 197–205.
145. Chaar, M., Patel, D., Kung, M., *et al.* (1987). Selective Oxidative Dehydrogenation of Butane over V-Mg-O Catalysts, *J. Catal.*, **105**, pp. 483–498.
146. Patel, D.; Andersen, P.J.; Kung, H.H. (1990). Oxidative Dehydrogenation of Butane over Orthovanadates, *J. Catal.*, **125**, 132–142.
147. Bertus, B. (1978). *US Patent 4094819*, Catalyst and process for oxidative dehydrogenation (Philips Petroleum Co.).
148. Bertus, B. (1975). *US Patent 388609*, Nickel and cobalt catalysts including a group IIA and VIA component (Philips Petroleum Co.).
149. Ripley, D. (1977). *US Patent 4044066*, Nickel-phosphorus oxidative dehydrogenation catalyst (Philips Petroleum Co.).
150. Walker, D., Hogan, R. and Farha, F. (1980). *US Patent 4218343*, Dehydrogenation of organic compounds (Philips Petroleum Co.).
151. Owens, L. and Kung, H. (1992). Effects of Loading and Cesium Modifier on Silica-Supported Vanadia in Oxidative Dehydrogenation of Butane, Preprints-American Chemical Society, Division of Petroleum Chemistry, 37(4), pp. 1194–1200.
152. López Nieto, J., Botella, P., Vázquez, M., *et al.* (2002). The Selective Oxidative Dehydrogenation of Ethane over Hydrothermally Synthesized MoVTeNb Catalysts, *Chem. Commun.*, **17**, pp. 1906–1907.
153. Centi, G., Grasselli, R. and Trifiro, F. (1992). Propane Ammoxidation to Acrylonitrile: An Overview, *Catal. Today*, **13**, pp. 661–666.
154. Vaarkamp, M. and Ushikubo, T. (1998). Limitations of V-Sb-W and Mo-V-Nb-Te Mixed Oxides in Catalyzing Propane Ammoxidation, *Appl. Catal. A: Gen.*, **174**, pp. 99–107.
155. Hatano, M. and Kayo, A. (1989). *European Patent 318295 A1*, Process for Producing Nitriles (Mitsubishi Kasei Corp., Japan).
156. Ushikubo, T., Oshima, K., Umezawa, T., *et al.* (1992). *European Patent 512846 A1*, Process for Producing Nitriles (Mitsubishi Kasei Corp., Japan).
157. Ushikubo, T., Sawaki, I., Oshima, K., *et al.* (1994). *European Patent 603836 A1*, Process for preparing a catalyst useful for producing a nitrile (Mitsubishi Kasei Polytec Co., Japan).

158. Ushikubo, T., Oshima, K., Kayo, A., *et al.* (1995). *US Patent 5472925 A*, Catalyst for the production of nitriles (Mitsubishi Chemical Corporation, Japan).
159. Ushikubo, T., Oshima, K., Kayou, A., *et al.* (1997). Ammoxidation of Propane over Mo-V-Nb-Te Mixed Oxide Catalysts, *Stud. Surf. Sci. Catal.*, **112**, pp. 473–480.
160. Tsuji, H., Oshima, K. and Koyasu Y. (2003). Synthesis of Molybdenum and Vanadium-Based Mixed Oxide Catalysts with Metastable Structure: Easy Access to the MoVNbTe(Sb)O$x$Catalytically Active Structure Using Reductant and Oxoacid, *Chem. Mater.*, **15**, pp. 2112–2114.
161. Ushikubo, T., Oshima, K., Kayou, A., *et al.* (1997). Ammoxidation of Propane over Catalysts Comprising Mixed Oxides of Mo and V, *J. Catal.*, **169**, pp. 394–396.
162. Ueda, W., Vitry, D., Kato, T., *et al.* (2006). Key Aspects of Crystalline Mo-V-O-Based Catalysts Active in the Selective Oxidation of Propane, *Res. Chem. Intermediat.*, **32**, pp. 217–233.
163. Centi, G., Marchi, F. and Perathoner, S. (1997). Effect of Ammonia Chemisorption on the Surface Reactivity of V-Sb-Oxide Catalysts for Propane Ammoxidation, *Appl. Catal. A: Gen.*, **149**, pp. 225–244.
164. Blasco, T., Botella, P., Concepción, P., *et al.* (2004). Selective Oxidation of Propane to Acrylic Acid on K-Doped MoVSbO Catalysts: Catalyst Characterization and Catalytic Performance, *J. Catal.*, **228**, pp. 362–373.
165. Botella, P., Concepcion, P., López Nieto, J., *et al.* (2003). Effect of Potassium Doping on the Catalytic Behavior of Mo-V-Sb Mixed Oxide Catalysts in the Oxidation of Propane to Acrylic Acid, *Catal. Lett.*, **89**, pp. 249–253.
166. Ueda, W., Endo, Y. and Watanabe, N. (2006). K-Doped Mo-V-Sb-O Crystalline Catalysts for Propane Selective Oxidation to Acrylic Acid, *Top. Catal.*, **38**, pp. 261–268.
167. Ivars, F., Solsona, B., Rodríguez-Castellón, E., *et al.* (2009). Selective Propane Oxidation over MoVSbO Catalysts. On the Preparation, Characterization and Catalytic Behavior of M1 Phase, *J. Catal.*, **262**, pp. 35–43.
168. Ivars, F., Solsona, B., Botella, P., *et al.* (2009). Selective Oxidation of Propane over Alkali-Doped Mo-V-Sb-O Catalysts, *Catal. Today*, **141**, pp. 294–299.
169. Baca, M., Pigamo, A., Dubois, J., *et al.* (2005). Fourier Transform Infrared Spectroscopic Study of Surface Acidity by Pyridine Adsorption on the M1 Active Phase of the MoVTe(Sb)NbO Catalysts Used in Propane Oxidation, *Catal. Commun.*, **6**, pp. 215–220.
170. López Nieto, J., Solsona, B., Concepcion, P., *et al.* (2010). Reaction Products and Pathways in the Selective Oxidation of $C_2$-$C_4$ Alkanes on MoVTeNb Mixed Oxide Catalysts, *Catal. Today*, **157**, pp. 291–296.
171. Centi, G., López Nieto, J., Pinelli, D., *et al.* (1989). Synthesis of Phthalic and Maleic Anhydrides from Normal-Pentane. 1. Kinetic-Analysis of the Reaction Network, *Ind. Eng. Chem. Res.*, **28**, pp. 400–406.
172. Ieda, S., Phiyanalinmat, S., Komai, S., *et al.* (2005). Involvement of Active Sites of Promoted Vanadyl Pyrophosphate in Selective Oxidation of Propane, *J. Catal.*, **236**, pp. 304–312.
173. Ai, M. (1996). Comparison of Catalytic Properties for Partial Oxidation between Heteropoly-acids and Phosphates of Vanadium and Iron, *J. Mol. Catal. A: Chem.*, **114**, pp. 3–13.
174. Volta, J. (1996). Dynamic Processes on Vanadium Phosphorous Oxides for Selective Alkane Oxidation, *Catal. Today*, **32**, pp. 29–36.
175. López Nieto, J., Zazhigalov, V., Solsona, B., *et al.* (2000). Oxidative Dehydrogenation of Ethane on Anadium-Phosphorus Oxide Catalysts, *Stud. Surf. Sci. Catal.*, **130B**, pp. 1853–1858.
176. Solsona, B., Ivars, F., Concepción, P., *et al.* (2007). Selective Oxidation of $n$-Butane over Mov-Containing Oxidic Bronze Catalysts, *J. Catal.*, **250**, pp. 128–138.
177. Védrine, J., Novakova, E. and Derouane, E. (2003). Recent Developments in the Selective Oxidation of Propane to Acrylic and Acetic Acids, *Catal. Today*, **81**, pp. 247–262.

178. Landi, G., Lisi, L. and Volta, J. (2003). Effect of Water on the Catalytic Behaviour of VPO in the Selective Oxidation of Propane to Acrylic Acid, *Chem. Commun.*, **4**, pp. 492–493.
179. Taufiq-Yap, Y., Saw, C. and Irmawati, R. (2005). Activation of $VOHPO_4$ Center Dot 0.5H(2)O in Propane/Air Mixture: Effect on Structural, Morphological, Oxidant's Behaviour and Catalytic Property of (VO)(2)$P_2O_7$ Catalysts for Propane Oxidation, *Catal. Lett.*, **105**, pp. 103–110.
180. Li, X., Ji, W., Zhao, J., *et al.* (2006). A Comparison Study on the Partial Oxidation of *n*-Butane and Propane over VPO Catalysts Supported on SBA-15, MCM-41, and Fumed $SiO_2$, *Appl. Catal. A: Gen.*, **306**, pp. 8–16.
181. Shiju, N., Kale, R., Iyer, S., *et al.* (2007). C-13 Isotope Labeling Study of Propane Ammoxidation over M1 Phase Mo-V-Te-Nb-O Mixed Oxide Catalyst, *J. Phys. Chem. C*, **111**, pp. 18001–18003.
182. Ai, M. (1992). Oxidation of Propane to Acrylic Acid, *Catal. Today*, **13**, pp. 679–684.
183. Cassidy, F. and Hodnett, B. (1998). Selective Oxidation Catalysts: An Evaluation of the Discriminating Capacity of Active Sites on Oxide Catalysts with Molecular Oxygen as Oxidant, *CATTECH*, **2**, pp. 173–180.
184. Ueda, W., Vitry, D. and Katou, T. (2005). Crystalline Mo-V-O Based Complex Oxides as Selective Oxidation Catalysts of Propane, *Catal. Today*, **99**, pp. 43–49.
185. Guan, J., Wu, S., Wang, H., *et al.* (2007). Synthesis and Characterization of MoVTeCeO Catalysts and Their Catalytic Performance for Selective Oxidation of Isobutane and Isobutylene, *J. Catal.*, **251**, pp. 354–362.
186. Shishido, T., Inoue, A., Konishi, T., *et al.* (2000). Oxidation of Isobutane over Mo-V-Sb Mixed Oxide Catalyst, *Catal. Lett.*, **68**, pp. 215–221.
187. Cavani, F., Mezzogori, R., Pigamo, A., *et al.* (2001). Main Aspects of the Selective Oxidation of Isobutane to Methacrylic Acid Catalyzed by Keggin-Type Polyoxometalates, *Catal. Today*, **71**, pp. 97–110.
188. Mizuno, N., Suh, D., Han, W., *et al.* (1996). Catalytic Performance of $Cs_{2.5}Fe_{0.08}H_{1.26}PV Mo_{11}O_{40}$ for Direct Oxidation of Lower Alkanes, *J. Mol. Catal. A: Chem.*, **114**, pp. 309–317.
189. Sultan, M., Paul, S., Fournier, M., *et al.* (2004). Evaluation and Design of Heteropolycompound Catalysts for the Selective Oxidation of Isobutane into Methacrylic Acid, *Appl. Catal. A: Gen.*, **259**, pp. 141–152.
190. Ballarini, N., Cavani, F., *et al.* (2007). The Oxidation of Isobutane to Methacrylic Acid: An Alternative Technology for MMA Production, in P. Tundo, A. Perosa and F Zecchini (Eds), *Methods and Reagents for Green Chemistry: An Introduction*, Wiley-Interscience, New Jersey, Ch. 14, pp. 265–279.
191. Huynh, Q., Schuurman, Y., Delichere, R., *et al.* (2009). Study of Te and V as Counter-Cations in Keggin Type Phosphomolybdic Polyoxometalate Catalysts for Isobutane Oxidation, *J. Catal.*, **261**, pp. 166–176.
192. Huynh, Q., Selmi, A., Corbel, G., *et al.* (2009). Atypical Synergetic Effect between Te- And V-Substituted Phosphomolybdic Cesium Salt and LAMOX-Type Phases for the Oxidation of Isobutane into Methacrylic Acid, *J. Catal.*, **266**, pp. 64–70.
193. Ledo, B., Rives, V., Sanchezescribano, V., *et al.* (1993). A Ft-Ir Assessment of Iso-C4h8 Reactivity with $V_2O_5/TiO_2$ Catalysts, *Catal. Lett.*, **18**, pp. 329–335.
194. Guliants, V. and Holmes, S. (2001). Probing Polyfunctional Nature of Vanadyl Pyrophosphate Catalysts: Oxidation of 16 C-4 Molecules, *J. Mol. Catal. A: Chem.*, **175**, pp. 227–239.
195. Misono, M. and Nojiri, N. (1990). Recent Progress in Catalytic Technology in Japan, *Appl. Catal. A: Gen.*, **64**, pp. 1–30.
196. Vitry, D., Morikawa, Y., Dubois, J., *et al.* (2003). Propane Selective Oxidation over Monophasic Mo-V-Te-O Catalysts Prepared by Hydrothermal Synthesis, *Top. Catal.*, **23**, pp. 47–53.
197. Kendell, S., Alston, A. and Brown, T. (2009). Kinetic Simulation of Methacrolein and Lactone Production from the Catalytic Oxidation of Isobutane over Lanthanide Phosphomolybdates, *Chemical Product and Process Modeling*, **4**, pp. No pp given.

198. Batiot, C. and Hodnett, B. (1996). The Role of Reactant and Product Bond Energies in Determining Limitations to Selective Catalytic Oxidations, *Appl. Catal. A: Gen.*, **137**, pp. 179–191.
199. Ruth, K., Burch, R. and Kieffer, R. (1998). Mo-V-Nb Oxide Catalysts for the Partial Oxidation Of Ethane. II. Chemical and Catalytic Properties and Structure Function Relationships, *J. Catal.*, **175**, pp. 27–39.
200. Ruth, K., Kieffer, R. and Burch, R. (1998). Mo-V-Nb Oxide Catalysts for the Partial Oxidation of Ethane. I. Preparation and Structural Characterisation, *J. Catal.*, **175**, pp. 16–26.
201. Merzouki, M., Taouk, B., Tessier, L., *et al.* (1993). Correlation between Catalytic and Structural Properties of Modified Molybdenum and Vanadium Oxides in the Oxidation of Ethane in Acetic Acid or Ethylene, *Stud. Surf. Sci. Catal.*, **75**, pp. 753–764.
202. Botella, P., López Nieto, J., Dejoz, A., *et al.* (2003). Mo-V-Nb Mixed Oxides as Catalysts in the Selective Oxidation of Ethane, *Catal. Today*, **78**, pp. 507–512.
203. Ueda, W., Chen, N. and Oshihara, K. (1999). Hydrothermal Synthesis of Mo-V-M-O Complex Metal Oxide Catalysts Active for Partial Oxidation of Ethane, *Chem. Commun.*, **6**, pp. 517–518.
204. Ueda, W., Oshihara, K., Vitry, D., *et al.* (2002). Hydrothermal Synthesis of Mo-Based Oxide Catalysts and Selective Oxidation of Alkanes, *Catal. Surv. Jpn*, **6**, pp. 33–44.
205. Grubert, G., Kondratenko, E., Kolf, S., *et al.* (2003). Fundamental Insights into the Oxidative Dehydrogenation of Ethane to Ethylene over Catalytic Materials Discovered by an Evolutionary Approach, *Catal. Today*, **81**, pp. 337–345.
206. Thorsteinson, E., Wilson, T., Young, F., *et al.* (1978). Oxidative Dehydrogenation of Ethane over Catalysts Containing Mixed Oxides of Molybdenum and Vanadium, *J. Catal.*, **52**, pp. 116–132.
207. López Nieto, J. (2006). Deshidrogenacion Oxidativa de Etano, ¿Proceso Alternativo en la Producción Industrial Española?, *OILGAS, Petróleo, Petroquímica y Gas*, **445**, pp. 88–100.
208. Botella, P., Dejoz, A., Abello, M., *et al.*, (2009). Selective Oxidation of Ethane: Developing an Orthorhombic Phase in Mo-V-X (X = Nb, Sb, Te) Mixed Oxides, *Catal. Today*, **142**, pp. 272–277.
209. Solsona, B., Concepción, P., Hernández, S., *et al.* (2011). Oxidative Dehydrogenation of Ethane over NiO-$CeO_2$ Mixed Oxides Catalysts, *Catal. Today*, **180**, pp. 51–58.
210. Liu, Y. (2007). *US Patent 7227049 B2*, Ni catalysts and methods for alkane dehydrogenation (Symyx Technologies, Inc.).
211. Yang, W., Wang, H., Zhu, X., *et al.* (2005). Development and Application of Oxygen Permeable Membrane in Selective Oxidation of Light Alkanes, *Top. Catal.*, **35**, pp. 155–167.
212. Håkonsen, S. and Holmen, A. (eds) (2008). Handbook of Heterogeneous Catalysis, in *Oxidative Dehydrogenation of Alkanes*, Wiley, Weinheim, pp. 3384–3400.
213. Wang, H., Tablet, C., Schiestel, T., *et al.* (2006). Hollow Fiber Membrane Reactors for the Oxidative Activation of Ethane, *Catal. Today*, **118**, pp. 98–103.
214. Deshmukh, S., Heinrich, S., Morl, L., *et al.* (2007). Membrane Assisted Fluidized Bed Reactors: Potentials and Hurdles, *Chem. Eng. Sci.*, **62**, pp. 416–436.
215. Ahchieva, D., Peglow, M., Heinrich, S., *et al.* (2005). Oxidative Dehydrogenation of Ethane in a Fluidized Bed Membrane Reactor, *Appl. Catal. A: Gen.*, **296**, pp. 176–185.
216. Chalakov, L., Rihko-Struckmann, L., Munder, B., *et al.* (2007). Feasibility Study of the Oxidative Dehydrogenation of Ethane in an Electrochemical Packed-Bed Membrane Reactor, *Ind. Eng. Chem. Res.*, **46**, pp. 8665–8673.
217. Chalakov, L., Rihko-Struckmann, L., Munder, B., *et al.* (2009). Oxidative Dehydrogenation of Ethane in an Electrochemical Packed-Bed Membrane Reactor: Model and Experimental Validation, *Chem. Eng. J.*, **145**, pp. 385–392.
218. Hamel, C., Wolff, T. and Seidel-Morgenstern, A. (2011). Compatibility of Transport and Reaction in Membrane Reactors Used for the Oxidative Dehydrogenation of Short-Chain Hydrocarbons, *International Journal of Chemical Reactor Engineering*, **9**, Article Number: A12.
219. Besecker, C., Kleefisch, M. and Zhang, J. (2007). *US Patent 20070245897 A1*, Electron, hydrogen and oxygen conveying membranes (Innovene USA, USA).

220. Rodriguez, M., Ardissone, D., Lemonidou, A., *et al.* (2009). Simulation of a Membrane Reactor for the Catalytic Oxidehydrogenation of Ethane, *Ind. Eng. Chem. Res.*, **48**, pp. 1090–1095.
221. Rodriguez, M., Ardissone, D., Lopez, E., *et al.* (2011). Reactor Designs for Ethylene Production via Ethane Oxidative Dehydrogenation: Comparison of Performance, *Ind. Eng. Chem. Res.*, **50**, pp. 2690–2697.
222. Cavani, F., Centi, G., Manenti, I., *et al.* (1983). Oxidation of 1-Butene and Butadiene to Maleic-Anhydride. 1. Role of Oxygen Partial-Pressure, *Ind. Eng. Chem. Prod. Res. Develop.*, **22**, pp. 565–570.
223. Cavani, F., Centi, G. and Trifiro, F. (1983). Oxidation of 1-Butene and Butadiene to Maleic-Anhydride. 2. Kinetics and Mechanism, *Ind. Eng. Chem. Prod. Res. Develop.*, **22**, pp. 570–577.
224. Brkic, D. and Trifiro, F. (1979). Kinetic-Analysis of 1-Butene Oxidation to Maleic-Anhydride with a Polyfunctional Catalyst, *Ind. Eng. Chem. Prod. Res. Develop.*, **18**, pp. 333–339.
225. Cavani, F., Centi, G., Riva, A., *et al.* (1987). Functionalization of Paraffinic Hydrocarbons by Heterogeneous Oxidation. I. Control of Selectivity in *n*-Butane Conversion to Maleic Anhydride, *Catal. Today*, **1**, pp. 17–26.
226. Dobner, C., Duda, M., Raichle, A., *et al.* (2007). *WO Patent 2007012620 A1*, Catalyst and Methods for Producing Maleic Anhydride (BASF Aktiengesellschaft, Germany).
227. Datta, A., Dasgupta, S. and Agarwal, M. (2004). *US Patent 6774081 B1*, Process for preparing vanadyl pyrophosphate catalyst (Council of Scientific and Industrial Research, India).
228. Bertola, A., Cassarino, S. and Nsunda, V. (2001). *US Patent 6174833 B1*, Process for the preparation of improved vanadium-phosphorus catalysts and use thereof for the production of maleic anhydride (Pantochim S.A., Feluy (BE)).
229. Caldarelli, A., Cavani, F., Folco, F., *et al.* (2010). The Design of a New $ZrO_2$-supported V/P/O Catalyst for *n*-Butane Oxidation to Maleic Anhydride the Build-up of the Active Phase during Thermal Treatment, *Catal. Today*, **157**, pp. 204–210.
230. Duarte de Farias, A., Gonzalez, W., Pries de Oliveira, P., *et al.* (2002). Vanadium Phosphorus Oxide Catalyst Modified by Niobium Doping for Mild Oxidation of *n*-Butane to Maleic Anhydride, *J. Catal.*, **208**, pp. 238–246.
231. Ebner, J. (1993). *US Patent 5185455*, Method for improving the performance of VPO catalysts (Monsanto Co., USA).
232. Ebner, J., Keppel, R. and Mummey, M. (1993). *Worldwide Patent 9301155 A1*, High Productivity Process for the Production of Maleic Anhydride (Monsanto Co., USA).
233. Suciu, G., Giancarlo, S. and Fumagalli, C. (1985). *US Patent 4511670*, Catalysts containing mixed oxides of vanadium, phosphorus, and aluminum and/or boron (The Lummus Company, Bloomfield, N.J; Allusuisse Italia S.p.A., Milan, Italy).
234. Suciu, G., Giancarlo, S. and Fumagalli, C. (1986). *US Patent 4594433*, Production of Maleic Anhydride (Lummus Crest, Inc., Bloomfield, NJ; Alusuisse Italia S.p.A., Milan, Italy).
235. Contractor, R., Garnett, D., Horowitz, H., *et al.* (1994). A New Commercial-Scale Process for N-Butane Oxidation to Maleic-Anhydride Using a Circulating Fluidized-Bed Reactor, *Stud. Surf. Sci. Catal.*, **82**, pp. 233–242.
236. Patience, G.S. and López Nieto, J.M., Eds. (2010). International VPO Workshop: Butane Oxidation Technology and Catalyst Characterization, *Appl. Catal. A: Gen.*, **376**, pp. 1–103.
237. Hutchenson, K., La Marca, C., Patience, G., *et al.* (2010). Parametric Study of *n*-Butane Oxidation in a Circulating Fluidized Bed Reactor, *Appl. Catal. A: Gen.*, **376**, pp. 91–103.
238. Schwartz, J. and Cline, D. (2005). *US Patent 6878668-B1*, Process for manufacture of an attrition resistant catalyst (Du Pont De Nemours & Co. E. I.).
239. Patience, G. and Bockrath, R. (2010). Butane Oxidation Process Development in a Circulating Fluidized Bed, *Appl. Catal. A: Gen.*, **376**, pp. 4–12.
240. Dummer, N., Weng, W., Kiely, C., *et al.* (2010). Structural Evolution and Catalytic Performance of DuPont V-P-O/$SiO_2$ Materials Designed for Fluidized Bed Applications, *Appl. Catal. A: Gen.*, **376**, pp. 47–55.

241. Thon, A. and Werther, J. (2010). Attrition Resistance of a VPO Catalyst, *Appl. Catal. A: Gen.*, **376**, pp. 56–65.
242. Shekari, A. and Patience, G. (2010). Maleic Anhydride Yield during Cyclic *n*-Butane/Oxygen Operation, *Catal. Today*, **157**, pp. 334–338.
243. Alonso, M., Lorences, M., Pina, M., *et al.* (2001). Butane Partial Oxidation in an Externally Fluidized Bed-Membrane Reactor, *Catal. Today*, **67**, pp. 151–157.
244. Marin, P., Hamel, C., Ordonez, S., *et al.* (2010). Analysis of a Fluidized Bed Membrane Reactor for Butane Partial Oxidation to Maleic Anhydride: 2D Modelling, *Chem. Eng. Sci.*, **65**, pp. 3538–3548.
245. Cavani, F., Centi, G. and Marion, P. (2009). Catalytic Ammoxidation of Hydrocarbons on Mixed Oxides, in S. Jackson and J. Hargreaves (eds), *Metal Oxide Catalysis*, Vol. 2, Wiley-VCH Verlag GmbH & Co., Weinheim, pp. 771–818.
246. Tullo, A. (2008). A Solvent Dries up. Acetonitrile is in Short Supply, and Chemists are Concerned, *Chem. Eng. News*, **86**, p. 27.
247. Guerrero-Perez, M. and Bañares, M. (2002). Operando Raman Study of Alumina-Supported Sb-V-O Catalyst during Propane Ammoxidation to Acrylonitrile with On-Line Activity Measurement, *Chem. Commun.*, **12**, pp. 1292–1293.
248. Shiju, N., Guliants, V., Overbury, S., *et al.* (2008). Toward Environmentally Benign Oxidations: Bulk Mixed Mo-V-(Te-Nb)-O M1 Phase Catalysts for the Selective Ammoxidation of Propane, *Chemsuschem*, **1**, pp. 519–523.
249. Shiju, N., Liang, X., Weimer, A., *et al.* (2008). The Role of Surface Basal Planes of Layered Mixed Metal Oxides in Selective Transformation of Lower Alkanes: Propane Ammoxidation over Surface AB Planes of Mo-V-Te-Nb-O M1 Phase, *J. Am. Chem. Soc.*, **130**, pp. 5850–5851.
250. Oliver, J., López Nieto, J. and Botella, P. (2004). Selective Oxidation and Ammoxidation of Propane on a Mo-V-Te-Nb-O Mixed Metal Oxide Catalyst: A Comparative Study, *Catal. Today*, **96**, pp. 241–249.
251. Watanabe, N. and Ueda, W. (2006). Comparative Study on the Catalytic Performance of Single-Phase Mo-V-O-Based Metal Oxide Catalysts in Propane Ammoxidation to Acrylonitrile, *Ind. Eng. Chem. Res.*, **45**, pp. 607–614.
252. Baca, M., Aouine, M., Dubois, J., *et al.* (2005). Synergetic Effect between Phases in MoVTe(Sb)NbO Catalysts Used for the Oxidation of Propane into Acrylic Acid, *J. Catal.*, **233**, pp. 234–241.
253. Grasselli, R. (2005). Selectivity Issues in (Amm)oxidation Catalysis, *Catal. Today*, **99**, pp. 23–31.
254. Shishido, T., Konishi, T., Matsuura, I., *et al.* (2001). Oxidation and Ammoxidation of Propane over Mo-V-Sb Mixed Oxide Catalysts, *Catal. Today*, **71**, pp. 77–82.
255. Mimura, Y., Ohyachi, K., Matsuura, I., *et al.* (1999).*Selective Ammoxidation of Propane over Sb-Nb-V Oxide Catalysts with α-$Sb_2O_4$ and Rutile Type Structure*, *Stud. Surf. Sci. Catal.*, **121**, pp. 69–74.
256. Centi, G., Tosarelli, T. and Trifiro, F. (1993). Acrylonitrile from Propane on $(VO)_2P_2O_7$ with Preadsorbed Ammonia. 1. Role of Competitive Adsorption Phenomena in Determining Selectivity, *J. Catal.*, **142**, pp. 70–83.
257. DeSanto, P., Buttrey, D., Grasselli, R., *et al.* (2004). Structural Aspects of the M1 and M2 Phases in MoVNbTeO Propane Ammomidation Catalysts, *Z. Kristallogr.*, **219**, pp. 152–165.
258. Pyrz, W., Blom, D., Shiju, N., *et al.* (2008). Using Aberration-Corrected STEM Imaging to Explore Chemical and Structural Variations in the M1 Phase of the MoVNbTeO Oxidation Catalyst, *J. Phys. Chem. C*, **112**, pp. 10043–10049.
259. Ushikubo, T., Nakamura, H., Koyasu, Y., *et al.* (1994). *European Patent 608838 A2*, Method for producing an unsaturated carboxylic acid (Mitsubishi Kasei Corp., Japan).
260. Maher, J., Warren, B., Etzkorn, W., *et al.* (1997). *Worldwide Patent 9736849 A1*, Processes for the Oxidation of Alkanes (Union Carbide Chemicals and Plastics Technology Corp., USA).

261. Takahashi, M., To, S. and Hirose, T. (1998). *Japanese Patent 10120617*, Production of Acrylic Acid (Toa Gosei Chemical Industry Co., Ltd., Japan).
262. Ushikubo, T. (1998). *Japanese Patent 10045643*, Production of Acrolein and Acrylic Acid (Mitsubishi Chemical Industries Ltd., Japan).
263. Ushikubo, T., Kinoshita, H. and Watanabe, A. (1998). *Japanese Patent 10057813*, Manufacture of Mixed Metal Oxide Catalyst and Acrylic Acid Production Using Thereof (Mitsubishi Chemical Industries Ltd., Japan.).
264. Ushikubo, T., Koyasu, Y., Nakamura, K., *et al.* (1998). *Japanese Patent 10045664*, Production of alpha beta-Unsaturated Carboxylic Acid (Mitsubishi Chemical Industries Ltd., Japan).
265. Holmberg, J., Grasselli, R. and Andersson, A. (2003). A Study of the Functionalities of the Phases in Mo–V–Nb–Te Oxides for Propane Ammoxidation, *Top. Catal.*, **23**, pp. 55–63.
266. Holmberg, J., Grasselli, R. and Andersson, A. (2004). Catalytic Behaviour of M1, M2, and M1/M2 Physical Mixtures of the Mo-V-Nb-Te-Oxide System in Propane and Propene Ammoxidation, *Appl. Catal. A: Gen.*, **270**, pp. 121–134.
267. Guerrero-Pérez, M., Al-Saeedi, J., Guliants, V., *et al.* (2004). Catalytic Properties of Mixed Mo-V-Sb-Nb-O Oxides Catalysts for the Ammoxidation of Propane to Acrylonitrile, *Appl. Catal. A: Gen.*, **260**, pp. 93–99.
268. Kubo, J., Watanabe, N. and Ueda, W. (2008). Propane Ammoxidation with Lattice Oxygen of Mo-V-O-Based Complex Metal Oxide Catalysts, *Chem. Eng. Sci.*, **63**, pp. 1648–1653.
269. López Nieto, J., Botella Asuncion, P. and Solsona Espriu, B. (2003). Worldwide *Patent 2003008096 A1*, Catalyst for the Selective Oxidation and Ammoxidation of Alkanes and/or Alkenes, Particularly in Processes for Obtaining Acrylic Acid, Acrylonitrile and the Derivatives Thereof (CSIC-UPV, Spain).
270. Brazdil, J. (2006). Strategies for the Selective Catalytic Oxidation of Alkanes, *Top. Catal.*, **38**, pp. 289–294.
271. Hinago, H. and Komada, S. (1999). *German Patent 19835247 A1*, Catalyst for Ammoxidation of Propane or Isobutane to (Meth)Acrylonitrile (Asahi Kasei Kogyo K.K., Japan).
272. Hinago Hidenori, K. and Satoru Komada, Y. (2000). *US Patent 6143916*, Ammoxidation catalyst for use in producing acrylonitrile or methacrylonitrile from propane or isobutane by ammoxidation (Asahi Kasei Kogyo Kabushiki Kaisha, Japan).
273. (2011). SABIC and Partners Slate Acrylonitrile, *Chemical & Engineering News*, **89**, pp. 14–15.
274. Padia, A., Sachs, H. and Millman, M. (2007). *US Patent 2007/0249848 A1*, Yield improvement in the production of maleic anhydride (SD Lizenzverwertungsgesellschaft mbH & Co. KG).
275. Kayou, A. and Ihara, T. (2000). *US Patent 616624*, Process for the simultaneous preparation of acrylonitrile and arcylic acid (Mitsubishi Chemical Corporation, Tokyo, Japan).
276. http://www.ides.com/news/2007/0718_sric.asp. Accessed 12 December 2013.
277. Ogawa, M. (1982). *Japanese Patent 57009737 A*, Preparation of Acrylic Acid and its Catalyst (Nippon Kayaku Co., Ltd., Japan.).
278. Mizuno, N., Tateishi, M. and Iwamoto, M. (1995). Pronounced Catalytic Activity of $Fe_{0.08}Cs_{2.5}H_{1.26}PVMo_{11}O_{40}$ for Direct Oxidation of Propane into Acrylic Acid, *Appl. Catal. A: Gen.*, **128**, pp. L165–L170.
279. Li, W., Oshihara, K. and Ueda, W. (1999). Catalytic Performance for Propane Selective Oxidation and Surface Properties of 12-Molybdophosphoric Acid Treated with Pyridine, *Appl. Catal. A: Gen.*, **182**, pp. 357–363.
280. Dillon, C., Holles, J., Davis, R., *et al.* (2003). A Substrate-Versatile Catalyst for The Selective Oxidation of Light Alkanes. II. Catalyst Characterization, *J. Catal.*, **218**, pp. 54–66.
281. Holles, J., Dillon, C., Labinger, J., *et al.* (2003). A Substrate-Versatile Catalyst for the Selective Oxidation of Light Alkanes. I. Reactivity, *J. Catal.*, **218**, pp. 42–53.
282. Landi, G., Lisi, L. and Volta, J. (2004). Role of Water in the Partial Oxidation of Propane to Acrylic Acid, *Catal. Today*, **91–92**, pp. 275–279.

283. Kaddouri, A., Mazzocchia, C. and Tempesti, E. (1999). The Synthesis of Acrolein and Acrylic Acid by Direct Propane Oxidation with Ni-Mo-Te-P-O Catalysts, *Appl. Catal. A: Gen.*, **180**, pp. 271–275.
284. Holman, P., Shonnard, D. and Holles, J. (2009). Using Life Cycle Assessment to Guide Catalysis Research, *Ind. Eng. Chem. Res.*, **48**, pp. 6668–6674.
285. Vitry, D., Dubois, J. and Ueda, W. (2004). Strategy in Achieving Propane Selective Oxidation over Multi-Functional Mo-Based Oxide Catalysts, *J. Mol. Catal. A: Chem.*, **220**, pp. 67–76.
286. Lintz, H. and Muller, S. (2009). The Partial Oxidation of Propane on Mixed Metal Oxides: A Short Overview, *Appl. Catal. A: Gen.*, **357**, pp. 178–183.
287. Godefroy, A., Patience, G., Tzakova, T., *et al.* (2009). Reactor Technologies for Propane Partial Oxidation to Acrylic Acid, *Chem. Eng. Technol.*, **32**, pp. 373–379.
288. Godefroy, A., Patience, G., Cenni, R., *et al.* (2010). Regeneration Studies of Redox Catalysts, *Chem. Eng. Sci.*, **65**, pp. 261–266.
289. Wang, J., Ji, B., Chu, W., *et al.* (2010). $Bi_4Cu_{0.2}V_{1.8}O_{11}-\delta$ based Electrolyte Membrane Reactor for Selective Oxidation of Propane to Acrylic Acid, *Catal. Today*, **149**, pp. 157–162.
290. Kendell, S. and Brown, T. (2010). Detailed Product and Kinetic Analysis for the Low-Pressure Selective Oxidation of Isobutane over Phosphomolybdic Acid, *React. Kinet. Mech. Catal.*, **99**, pp. 251–268.
291. Hu, J., Burns, R. and Guerbois, J. (2000). The Solid-State Thermal Rearrangement of the Dawson Anion $P_2Mo_{18}O_{62}.^{6-}$ into a Keggin-type $PMo_{12}O_{40}.^{3-}$-Containing Phase and their Reactivity in the Oxidative Dehydrogenation of Isobutyraldehyde, *J. Mol. Catal. A: Chem.*, **152**, pp. 141–155.
292. Marchal-Roch, C., Laronze, N., Villanneau, R., *et al.* (2000). Effects of $NH^{4+}$, $Cs^+$, and $H^+$ Counterions of the Molybdophosphate Anion in the Oxidative Dehydrogenation of Isobutyric Acid, *J. Catal.*, **190**, pp. 173–181.
293. Krieger, H. and Kirch, L. (1980). *European Patent 10902*, Process for the Production of (Meth)acrylic Acid by the Catalytic Vapour Phase Oxidation of Isobutane or Propane (Rohm and Haas Co., USA).
294. Kendell, S., Alston, A., Ballam, N., *et al.* (2011). Structural and Activity Investigation into $Al^{3+}$, $La^{3+}$ and $Ce^{3+}$ Addition to the Phosphomolybdate Heteropolyanion for Isobutane Selective Oxidation, *Catal. Lett.*, **141**, pp. 374–390.
295. Cavani, F., Mezzogori, R., Pigamo, A., *et al.* (2001). Improved Catalytic Performance of Keggin-Type Polyoxometalates in the Oxidation of Isobutane to Methacrylic Acid under Hydrocarbon-Lean Conditions Using Antimony-Doped Catalysts, *Chem. Eng. J.*, **82**, pp. 33–42.
296. Kiyoshi, K., Setsuo, Y. and Tatsuo, Y. (1991). *Japanese Patent 03-176438 A*, Recovering of Isobutane and Methacrolein (Asahi Chem. Ind. Co. Ltd.).
297. Nagai, K. and Ui, T. (2004). Trends and Future of Monomer-MMA Technologies, *Sumitomo Kagaku*, 2004 (2), pp. 4–14.
298. Krieger, H. and Kirch, L.S. (1981). *US Patent 4260822*, Process for the Production of Unsaturated Acids (Rohm and Haas Co.).
299. Yamamatsu, S. and Yamaguchi, T. (1990). *Worldwide Patent 9014325 A1*, Process for Producing Methacrylic Acid and Methacrolein (Asahi Chemical Industry Co., Ltd., Japan).
300. Yamamatsu, S. and Yamaguchi, T. (1993). *US Patent 5191116*, Process for Producing Methacrylic Acid and Methacrolein (Asahi Kasei Kogyo Kabushiki Kaisha).
301. Kuroda, T. and Okita, M. (1992). *Japanese Patent 4128247 A*, Production of Methacrolein and Methacrylic Acid (Mitsubishi Rayon Co., Ltd., Japan).
302. Kawakami, K., Yamamatsu, S. and Yamaguchi, T. (1991). *Japanese Patent 03176438 A*, Recovering of Isobutane and Methacrolein (Asahi Chemical Industry Co., Ltd., Japan).

303. Koichi, N., Yoshihiko, N., Hiroshi, S., *et al.* (1991). *European Patent 0418657 A2*, Process for producing methacrylic acid and methacrolein by catalytic oxidation of isobutene (Sumitomo Chemical Co., Japan).
304. Imai, H., Nakatsuka, M. and Aoshima, A. (1987). *Japanese Patent 62132832 A*, Production of Methacrolein and Methacrylic Acid *A* (Asahi Chemical Industry Co., Ltd., Japan).
305. Bielmeier, E., Haeberle, T. and Gruber, W. (1993). *US Patent 5380932*, Process for producing methacrylic acid and methacrolein by oxidation of isobutane with molybdenum heteropoly acid catalyst (Roehm GmbH Chemische Fabrik).
306. Mizuno, N., Tateishi, M. and Iwamoto, M. (1994). Enhancement of Catalytic Activity of $Cs_{2.5}Ni_{0.08}H_{0.34}PMo_{12}O_{40}$ by $V^{5+}$ Substitution for Oxidation of Isobutane into Methacrylic Acid, *Appl. Catal. A: Gen.*, **118**, pp. L1–L4.
307. Jalowiecki-Duhamel, L., Monnier, A., Barbaux, Y., *et al.* (1996). Oxidation of Isobutane on a Heteropolycompound Hydrogen Reservoir, *Catal. Today*, **32**, pp. 237–241.
308. Deng, Q., Jiang, S., Cai, T., *et al.* (2005). Selective Oxidation of Isobutane over $H_xFe_{0.12}Mo_{11}VPAs_{0.3}O_y$ Heteropoly Compound Catalyst, *J. Mol. Catal. A: Chem.*, **229**, pp. 165–170.
309. Motoyama, A. and Nakamura, I. (2004). *US Patent 6747172 B1*, Method for preparing methacrylic acid (Nippon Shokubal Co. Ltd.).
310. Guan, J., Xu, C., Liu, B., *et al.* (2008). Partial Oxidation of Isobutane over Hydrothermally Synthesized Mo-V-Te-O Mixed Oxide Catalysts, *Catal. Lett.*, **126**, pp. 301–307.
311. Guan, J., Wang, H., Yang, Y., *et al.* (2009). Effect of pH on the Catalytic Properties of Mo-V-Te-P-O Catalysts for Selective Oxidation of Isobutane, *Catal. Lett.*, **131**, pp. 512–516.
312. Schindler, G., Knapp, C., Ui, T., *et al.* (2003). Enhancing the Productivity of Isobutane Selective Oxidation over a Mo-V-P-As-Cs-Cu-O Heteropoly Acid Catalyst, *Top. Catal.*, **22**, pp. 117–121.
313. Yoneda, N., Kusano, S., Yasui, M., *et al.* (2001). Recent Advances in Processes and Catalysts for the Production of Acetic Acid, *Appl. Catal. A: Gen.*, **221**, pp. 253–265.
314. Haynes, A. (2010). Catalytic Methanol Carbonylation, in B. Gates, H. Knoezinger, F. Jentoft (eds), *Advances in Catalysis*, Vol. 53, Elsevier Academic Press Inc., San Diego, pp. 1–45.
315. Hobbs, C. (1998). The Liquid-Phase Oxidation of *n*-Butane. A Search for Plausible Mechanisms, *DGMK Tagungsbericht*, **9803**, pp. 53–63.
316. Lloyd, D., Eve, P. and Gammer, D. (1992). A Comparison of Naphtha Oxidation and Catalyzed Low-Pressure Methanol Carbonylation for the Production of Acetic Acid, *Ber. Dtsch. Wiss. Ges. Erdoel, Erdgas Kohle, Tagungsber.*, **9204**, pp. 1–16.
317. Palaniappan, C., Srinivasan, R. and Tan, R. (2004). Selection of Inherently Safer Process Routes: A Case Study, *Chem. Eng. Process.*, **43**, pp. 641–647.
318. Smejkal, Q., Linke, D. and Baerns, M. (2005). Energetic and Economic Evaluation of the Production of Acetic Acid via Ethane Oxidation, *Chem. Eng. Process.*, **44**, pp. 421–428.
319. McCain, J. (1985). *US Patent 4524236*, Process for oxydehydrogenation of ethane to ethylene (Union Carbide Corp., USA).
320. Karim, K., Al-Hazmi, M. and Khan, A. (2000). *US Patent 6060421*, Catalysts for the oxidation of ethane to acetic acid, methods of making and using the same (Saudi Basic Industries Corporation, Saudi Arabia.).
321. Karim, K., Al-Hazmi, M., Khan, A., *et al.* (2003). *US Patent 6531631 B*, Oxidation of ethane to acetic acid and ethylene using molybdenum and vanadium based catalysts (Saudi Basic Industries Corporation).
322. Linke, D., Wolf, D., Baerns, M., *et al.* (2002). Catalytic Partial Oxidation of Ethane to Acetic Acid over $Mo_1V_{0.25}Nb_{0.12}Pd_{0.0005}O_x$. Catalyst Performance and Reaction Mechanism, *J. Catal.*, **205**, pp. 16–31.

323. Li, X. and Iglesia, E. (2008). Kinetics and Mechanism of Ethane Oxidation to Acetic Acid on Catalysts Based on Mo-V-Nb Oxides, *J. Phys. Chem. C*, **112**, pp. 15001–15008.
324. Roussel, M., Barama, S., Lofberg, A., *et al.* (2009). MoV-Based Catalysts in Ethane Oxidation to Acetic Acid: Influence of Additives on Redox Chemistry, *Catal. Today*, **141**, pp. 288–293.
325. Montolio-Rodriguez, D., Linke, D. and Linke, P. (2007). Systematic Identification of Optimal Process Designs for the Production of Acetic Acid via Ethane Oxidation, *Chem. Eng. Sci.*, **62**, pp. 5602–5608.
326. Johnston, V., Chen, L., Zink, J., *et al.* (2009). *Worldwide Patent 2010056299 A1*, Integrated Process for the Production of Vinyl Acetate from Acetic Acid via Ethylene (Celanese International Corporation, USA.).
327. Tabata, K., Teng, Y., Takemoto, T., *et al.* (2002). Activation of Methane by Oxygen and Nitrogen Oxides, *Catal. Rev.*, **44**, pp. 1–58.
328. Holmen, A. (2009). Direct Conversion of Methane to Fuels and Chemicals, *Catal. Today*, **142**, pp. 2–8.
329. Wang, Y., An, D. and Zhang, Q. (2010). Catalytic Selective Oxidation or Oxidative Functionalization of Methane and Ethane to Organic Oxygenates, *Sci. China Chem.*, **53**, pp. 337–350.
330. Goula, M., Lemonidou, A., Grunert, W., *et al.* (1996). Methane Partial Oxidation to Synthesis Gas Using Nickel on Calcium Aluminate Catalysts, *Catal. Today*, **32**, pp. 149–156.
331. Lunsford, J. (2000). Catalytic Conversion of Methane to More Useful Chemicals and Fuels: A Challenge for the 21st Century, *Catal. Today*, **63**, pp. 165–174.
332. Bharadwaj, S. and Schmidt, L. (1995). Catalytic Partial Oxidation of Natural-Gas to Syngas, *Fuel Process. Technol.*, **42**, pp. 109–127.
333. Choudhary, T. and Choudhary, V. (2008). Energy-Efficient Syngas Production through Catalytic Oxy-Methane Reforming Reactions, *Angew. Chem. Int. Edit.*, **47**, pp. 1828–1847.
334. Prettre, M., Eichner, C. and Perrin, M. (1946). The Catalytic Oxidation of Methane to Carbon Monoxide and Hydrogen, *T. Faraday Soc.*, **42**, pp. 335–340.
335. Torniainen, P., Chu, X. and Schmidt, L. (1994). Comparison of Monolith-Supported Metals for the Direct Oxidation of Methane to Syngas, *J. Catal.*, **146**, pp. 1–10.
336. Silva, F., Martinez, D., Ruiz, J., *et al.* (2008). The Effect of the Use of Cerium-Doped Alumina on the Performance of $Pt/CeO_2/Al_2O_3$, and $Pt/CeZrO_2/Al_2O_3$ Catalysts on the Partial Oxidation of Methane, *Appl. Catal. A: Gen.*, **335**, pp. 145–152.
337. Li, Y., Wang, Y., Hong, X., *et al.* (2006). Partial Oxidation of Methane to Syngas over Nickel Monolithic Catalysts, *AIChE J.*, **52**, pp. 4276–4279.
338. Luna, A. and Iriarte, M. (2008). Carbon Dioxide Reforming of Methane over a Metal Modified $Ni\text{-}Al_2O_3$ Catalyst, *Appl. Catal. A: Gen.*, **343**, pp. 10–15.
339. Song, Y., Liu, H., Liu, S., *et al.* (2009). Partial Oxidation of Methane to Syngas over $Ni/Al_2O_3$ Catalysts Prepared by a Modified Sol-Gel Method, *Energy & Fuels*, **23**, pp. 1925–1930.
340. Beretta, A., Groppi, G., Lualdi, M., *et al.* (2009). Experimental and Modeling Analysis of Methane Partial Oxidation: Transient and Steady-State Behavior of Rh-Coated Honeycomb Monoliths, *Ind. Eng. Chem. Res.*, **48**, pp. 3825–3836.
341. Burke, N. and Trimm, D. (2005). Coke Formation during High Pressure Catalytic Partial Oxidation of Methane to Syngas, *React. Kinet. Catal. Lett.*, **84**, pp. 137–142.
342. Basini, L., Aasberg-Petersen, K., Guarinoni, A., *et al.* (2001). Catalytic Partial Oxidation of Natural Gas at Elevated Pressure and Low Residence Time, *Catal. Today*, **64**, pp. 9–20.
343. Olsbye, U., Moen, O., Slagtern, A., *et al.* (2002). An Investigation of the Coking Properties of Fixed and Fluid Bed Reactors during Methane-to-Synthesis Gas Reactions, *Appl. Catal. A: Gen.*, **228**, pp. 289–303.
344. Stutz, M. and Poulikakos, D. (2008). Optimum Washcoat Thickness of a Monolith Reactor for Syngas Production by Partial Oxidation of Methane, *Chem. Eng. Sci.*, **63**, pp. 1761–1770.

345. Zhu, X., Wang, H., Cong, Y., *et al.* (2006). Partial Oxidation of Methane to Syngas in $BaCe_{0.15}Fe_{0.85}O_{3-\delta}$ Membrane Reactors, *Catal. Lett.*, **111**, pp. 179–185.
346. Hickman, D. and Schmidt, L. (1993). Production of Syngas by Direct Catalytic-Oxidation of Methane, *Science*, **259**, pp. 343–346.
347. Ramaswamy, R., Ramachandran, P. and Dudukovic, M. (2007). Modeling Catalytic Partial Oxidation of Methane to Syngas in Short-Contact-Time Packed Bed Reactors, *Ind. Eng. Chem. Res.*, **46**, pp. 8638–8651.
348. Nogare, D., Degenstein, N., Horn, R., *et al.* (2011). Modeling Spatially Resolved Data of Methane Catalytic Partial Oxidation on Rh Foam Catalyst at Different Inlet Compositions and Flowrates, *J. Catal.*, **277**, pp. 134–148.
349. Barrio, V., Schaub, G., Rohde, M., *et al.* (2007). Reactor Modeling to Simulate Catalytic Partial Oxidation and Steam Reforming of Methane. Comparison of Temperature Profiles and Strategies for Hot Spot Minimization, *Int. J. Hydrogen Energ.*, **32**, pp. 1421–1428.
350. Keller, G. and Bhasin, M. (1982). Synthesis of Ethylene via Oxidative Coupling of Methane. 1. Determination of Active Catalysts, *J. Catal.*, **73**, pp. 9–19.
351. Ito, T. and Lunsford, J. (1985). Synthesis of Ethylene and Ethane by Partial Oxidation of Methane over Lithium-Doped Magnesium-Oxide, *Nature*, **314**, pp. 721–722.
352. Deboy, J. and Hicks, R. (1988). The Oxidative Coupling of Methane over Alkali, Alkaline-Earth, and Rare-Earth Oxides, *Ind. Eng. Chem. Res.*, **27**, pp. 1577–1582.
353. Wang, D., Rosynek, M. and Lunsford, J. (1995). Oxidative Coupling of Methane over Oxide-Supported Sodium-Manganese Catalysts, *J. Catal.*, **155**, pp. 390–402.
354. Hugill, J., Tillemans, F., Dijkstra, J., *et al.* (2005). Feasibility Study on the Co-Generation of Ethylene and Electricity through Oxidative Coupling of Methane, *Appl. Therm. Eng.*, **25**, pp. 1259–1271.
355. Malekzadeh, A., Dalai, A., Khodadadi, A., *et al.* (2008). Structural Features of $Na_2WO_4$-$MO_x/SiO_2$ Catalysts in Oxidative Coupling of Methane Reaction, *Catal. Commun.*, **9**, pp. 960–965.
356. Wu, J., Zhang, H., Qin, S., *et al.* (2007). La-Promoted $Na_2WO_4/Mn/SiO_2$ Catalysts for the Oxidative Conversion of Methane Simultaneously to Ethylene and Carbon Monoxide, *Appl. Catal. A: Gen.*, **323**, pp. 126–134.
357. Takanabe, K. and Iglesia, E. (2009). Mechanistic Aspects and Reaction Pathways for Oxidative Coupling of Methane on $Mn/Na_2WO_4/SiO_2$ Catalysts, *J. Phys. Chem. C*, **113**, pp. 10131–10145.
358. Wu, J., Qin, S. and Hu, C. (2007). $Na_2WO_4$/Co-Mn/$SiO_2$ Catalyst for the Simultaneous Production of Ethylene and Syngas from $CH_4$, *Catal. Lett.*, **118**, pp. 285–289.
359. Lorkovic, I., Yilmaz, A., Yilmaz, G., *et al.* (2004). A Novel Integrated Process for the Functionalization of Methane and Ethane: Bromine as Mediator, *Catal. Today*, **98**, pp. 317–322.
360. Lorkovic, I., Noy, M., Schenck, W., *et al.* (2004). C-1 Oxidative Coupling via Bromine Activation and Tandem Catalytic Condensation and Neutralization over CaO/Zeolite Composites. II. Product Distribution Variation and Full Bromine Confinement, *Catal. Today*, **98**, pp. 589–594.
361. Czuprat, O., Schiestel, T., Voss, H., *et al.* (2010). Oxidative Coupling of Methane in a BCFZ Perovskite Hollow Fiber Membrane Reactor, *Ind. Eng. Chem. Res.*, **49**, pp. 10230–10236.
362. Kao, Y., Lei, L. and Lin, Y. (2003). Optimum Operation of Oxidative Coupling of Methane in Porous Ceramic Membrane Reactors, *Catal. Today*, **82**, pp. 255–273.
363. Bhatia, S., Thien, C. and Mohamed, A. (2009). Oxidative Coupling of Methane (OCM) in a Catalytic Membrane Reactor and Comparison of its Performance with Other Catalytic Reactors, *Chem. Eng. J.*, **148**, pp. 525–532.
364. Jaso, S., Godini, H., Arellano-Garcia, H., *et al.* (2010). Oxidative Coupling of Methane: Reactor Performance and Operating Conditions, *20th European Symposium on Computer Aided Process Engineering*, **28**, pp. 781–786.

365. Godini, H., Arellano-Garcia, H., Omidkhah, M., *et al.* (2010). Model-Based Analysis of Reactor Feeding Policies for Methane Oxidative Coupling, *Ind. Eng. Chem. Res.*, **49**, pp. 3544–3552.
366. Tullo, A. (2011). Ethylene from Methane, *Chem. Eng. News*, **89**, pp. 20–21.
367. Coons, R. (2010). Start-Up Claims Ethylene Conversion Breakthrough, *Chem. Week*, **12**, p. 15.
368. Otsuka, K. and Wang, Y. (2001). Direct Conversion of Methane into Oxygenates, *Appl. Catal. A: Gen.*, **222**, pp. 145–161.
369. Arena, F. and Parmaliana, A. (2003). Scientific Basis for Process and Catalyst Design in the Selective Oxidation of Methane to Formaldehyde, *Accounts Chem. Res.*, **36**, pp. 867–875.
370. He, J., Li, Y., An, D., *et al.* (2009). Selective Oxidation of Methane to Formaldehyde by Oxygen over Silica-Supported Iron Catalysts, *J. Nat. Gas Chem.*, **18**, pp. 288–294.
371. Labinger, J. and Bercaw, J. (2002). Understanding and Exploiting C–H Bond Activation, *Nature*, **417**, pp. 507–514.
372. Parmaliana, A., Arena, F., Frusteri, F., *et al.* (2002). Effect of Fe-Addition on the Catalytic Activity of Silicas in the Partial Oxidation of Methane to Formaldehyde, *Appl. Catal. A: Gen.*, **226**, pp. 163–174.
373. Nguyen, L., Loridant, S., Launay, H., *et al.* (2006). Study of New Catalysts Based on Vanadium Oxide Supported on Mesoporous Silica for the Partial Oxidation of Methane to Formaldehyde: Catalytic Properties and Reaction Mechanism, *J. Catal.*, **237**, pp. 38–48.
374. Zhang, Q., Li, Y., An, D., *et al.* (2009). Catalytic Behavior and Kinetic Features of Feox/SBA-15 Catalyst for Selective Oxidation of Methane by Oxygen, *Appl. Catal. A: Gen.*, **356**, pp. 103–111.
375. Groothaert, M., Smeets, P., Sels, B., *et al.* (2005). Selective Oxidation of Methane by the bis($\mu$-oxo)Dicopper Core Stabilized on ZSM-5 and Mordenite Zeolites, *J. Am. Chem. Soc.*, **127**, pp. 1394–1395.
376. Woertink, J., Smeets, P., Groothaert, M., *et al.* (2009). A $Cu_2O^{2+}$ Core in Cu-ZSM-5, the Active Site in the Oxidation of Methane to Methanol, *Proc. Nat. Acad. Sci.*, **106**, pp. 18908–18913.
377. Beznis, N., Weckhuysen, B. and Bitter, J. (2010). Cu-ZSM-5 Zeolites for the Formation of Methanol from Methane and Oxygen: Probing the Active Sites and Spectator Species, *Catal. Lett.*, **138**, pp. 14–22.
378. Beznis, N., van Laak, A., Weckhuysen, B., *et al.* (2011). Oxidation of Methane to Methanol and Formaldehyde over Co-ZSM-5 Molecular Sieves: Tuning the Reactivity and Selectivity by Alkaline and Acid Treatments of the Zeolite ZSM-5 Agglomerates, *Micropor. Mesopor. Mat.*, **138**, pp. 176–183.
379. Li, Y., Chen, S., Zhang, Q., *et al.* (2006). Copper-Catalyzed Selective Oxidation of Methane to Formaldehyde by Oxygen, *Chem. Lett.*, **35**, pp. 572–573.
380. An, D., Zhang, Q. and Wang, Y. (2010). Copper Grafted on SBA-15 as Efficient Catalyst for the Selective Oxidation of Methane by Oxygen, *Catal. Today*, **157**, pp. 143–148.
381. Li, Y., An, D., Zhang, Q., *et al.* (2008). Copper-Catalyzed Selective Oxidation of Methane by Oxygen: Studies on Catalytic Behavior and Functioning Mechanism of $CuO_x$/SBA-15, *J. Phys. Chem. C*, **112**, pp. 13700–13708.
382. Olah, G. and Prakash, G. (2010). *US Patent 2010152474 A1*, Selective oxidative conversion of methane to methanol, dimethyl ether and derived compounds (Univ Southern California).
383. Palkovits, R., Antonietti, M., Kuhn, P., *et al.* (2009). Solid Catalysts for the Selective Low-Temperature Oxidation of Methane to Methanol, *Angew. Chem. Int. Edit.*, **48**, pp. 6909–6912.
384. Periana, R., Taube, D., Gamble, S., *et al.* (1998). Platinum Catalysts for the High-Yield Oxidation of Methane to a Methanol Derivative, *Science*, **280**, pp. 560–564.
385. Bowman, R., Stangland, E., Jones, M., *et al.* (2010). *Worldwide Patent 2010062427 A2*, Oxidative Mono-Halogenation of Methane (Dow Global Technologies Inc., USA.).

386. Liu, Z., Li, W. and Zhou, X. (2010). Product Oriented Oxidative Bromination of Methane over $Rh/SiO_2$ Catalysts, *J. Nat. Gas Chem.*, **19**, pp. 522–529.
387. Yang, F., Liu, Z., Li, W., *et al.* (2008). The Oxidative Bromination of Methane over Rh/SiO2 Catalyst, *Catal. Lett.*, **124**, pp. 226–232.
388. Lin, R., Ding, Y., Gong, L., *et al.* (2009). Oxidative Bromination of Methane on Silica-Supported Non-Noble Metal Oxide Catalysts, *Appl. Catal. A: Gen.*, 353, pp. 87–92.
389. Wang, K., Xu, H., Li, W., *et al.* (2006). The Synthesis of Acetic Acid from Methane via Oxidative Bromination, Carbonylation, and Hydrolysis, *Appl. Catal. A: Gen.*, **304**, pp. 168–177.

# Chapter 25

# Opportunities for Oxidation Reactions under Supercritical Conditions

Udo ARMBRUSTER* and Andreas MARTIN*

This chapter discusses the feasibility and potential of selective and total oxidation reactions using heterogeneous catalysts under supercritical conditions. After a short introduction on physicochemical properties, the advantages and drawbacks of supercritical fluids (SCF) as reaction media, thermodynamics and kinetics of oxidation reactions in SCF will be discussed. In addition, some comments are made on tools for the direct observation of reactions run in SCFs as well as suitable oxidants. Furthermore, the chapter will cover aspects of heterogeneously catalysed oxidation reactions in carbon dioxide, water, alcohols and some other media. In addition, some questions on corrosion effects, heterogeneous catalysts stability and reactor design are illuminated. The chapter will provide a comprehensive review of the state of scientific and technical knowledge. Research using $scCO_2$ as the reaction medium mainly deals with the synthesis of value-added products, whereas the research using supercritical water splits into work on waste treatment as well as synthesis reactions.

## 25.1. Introduction

On a large scale, hydrocarbons are oxidised either in continuous or in batch processes using heterogeneous or homogeneous catalysts, preferably using air or oxygen. In contrast, small quantities in the pharmaceutical industry are sometimes still oxidised stoichiometrically by various organic or inorganic oxidants. Though homogeneously catalysed oxidations are more selective, from an industrial point of view, a continuous process has advantages over batch operation (no catalyst separation necessary, no downtime, etc.).

Industrial liquid-phase oxidations of hydrocarbons are usually multiphase processes with a gas phase containing oxygen, and a liquid comprising feed, solvent and a catalyst at temperatures up to 300°C. Such reactions may run in usual batch reactors, bubble columns, slurry-phase reactors or trickle-bed reactors. Limitations

*Leibniz Institute for Catalysis, Albert-Einstein-Str. 29a, D-18059 Rostock, Germany.

often arise due to the poor solubility and diffusivity of oxygen in the liquid phase. This results in low effective concentrations of oxygen at the catalytically active sites, mass transfer limitation effects and low turnover frequencies. In some cases, these effects also decrease the selectivity.[1] Consequently, many of these reactions run at elevated oxygen partial pressure to raise the concentration of dissolved oxygen and to increase reaction rates. On the other hand, this operational mode bears a high risk due to the possible formation of explosive mixtures, and needs careful operation and additional safety measures. Another aspect that limits the attractivity of liquid-phase oxidation rises from corrosion problems caused by the combination of high temperature, aggressive solvents (such as acetic acid) and corrosive catalyst compounds (such as halides).[1] There is a persistent need for techniques to increase the efficient oxygen concentration in the liquid phase and for the replacement of corrosive or toxic solvents.

A typical heterogeneous catalyst in gas-phase oxidation operates at a higher temperature than homogeneous catalysts and this inevitably decreases selectivities for oxygenated products. An intrinsic problem in heterogeneous catalysis is the deposition of reactants, intermediates and/or products on the catalyst surface, which may lead to catalyst deactivation or selectivity problems. If the desired oxygenated products are intermediates, the residence time on the catalyst surface and adsorption effects determine product yields.

Supercritical fluids exist beyond their critical loci (temperature, pressure and density) in pure compounds or in a multicomponent mixture. Due to their outstanding liquid- and gas-like properties, they may potentially solve some problems of catalysed liquid- or gas-phase reactions.

### 25.1.1. *Properties of supercritical fluids*

When increasing the temperature and pressure of a material in a closed system, the pressure rises according to the vapour pressure line in the p-T phase diagram up to the critical point. Further heating establishes the supercritical state where the liquid and gas phase have the same properties and the fluid phase becomes homogeneous. The system comprises one triple point, one liquid-gas line, one liquid-solid line, one gas-solid line and one critical point (Figure 25.1, Table 25.1).

When the system contains more than one chemical compound, the phase behaviour becomes more complex; several critical points may exist and critical lines describe the transition from the subcritical to the supercritical region. Some authors point out that the supercritical state is not only defined by critical temperature and pressure, but also by a minimum density.[2] A widely accepted classification describes and distinguishes the phase behaviour according to the nature of the investigated mixtures.[3]

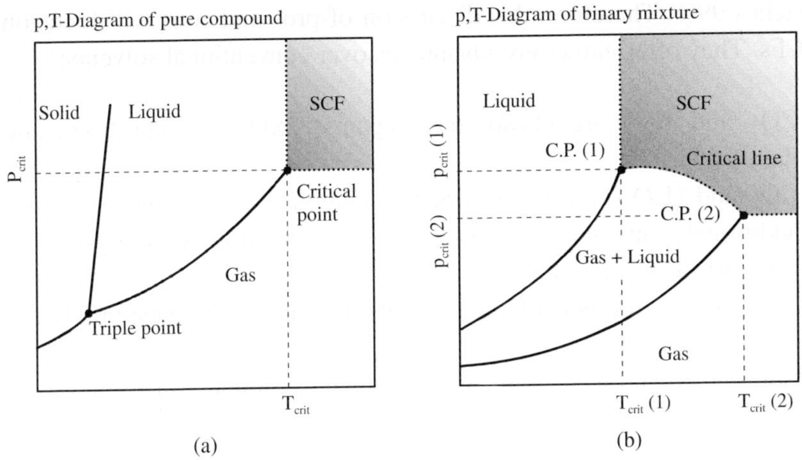


**Figure 25.1.** p,T-diagram for (a) pure compound and (b) binary mixture.

**Table 25.1.** Critical data for carbon dioxide and water.

| Compound | $T_{crit}$/°C | $P_{crit}$/bar | $\rho_{crit}$/g/cm$^3$ |
|---|---|---|---|
| $CO_2$ | 31.1 | 73.8 | 0.468 |
| $H_2O$ | 374 | 221 | 0.322 |

**Table 25.2.** Typical properties of gases, liquids and supercritical fluids.

| Parameter | Gas | SCF | Liquid |
|---|---|---|---|
| Density (g/cm$^3$) | $10^{-3}$ | 0.3 | $\approx 1$ |
| Viscosity (mPa·s) | $10^{-2}$ | 0.1 | $\approx 1$ |
| Diffusion coefficient (cm$^2$/s) | $10^{-1}$ | 0.001 | $5 \times 10^{-6}$ |

The physico-chemical properties of supercritical fluids (SCFs) are often described as a mixture of gas and liquid properties.[2] Densities and solvent power are more liquid-like, whereas transfer properties such as diffusivity are more similar to the gas phase (Table 25.2). Thus, SCFs offer features that might be beneficial for overcoming intrinsic problems of liquid-phase processes.

From these and other physico-chemical properties, selected effects may promote chemical reactions and in particular, heterogeneously catalysed oxidations. A large number of compounds could act as the supercritical reaction medium as their critical data match the temperature range of typical oxidation processes, but only a few provide high oxidation stability, which makes them suitable candidates. Supercritical carbon dioxide ($scCO_2$) and supercritical water ($scH_2O$) account for more than 90% of the literature and patents regarding SCFs, and for oxidation reactions this reaches 99%. Therefore, this discussion of pros and cons will focus on these two SCFs. They offer numerous advantages over conventional solvents:

(i) $CO_2$ and $H_2O$ are cheap, inert against oxidation, non-toxic and non-flammable.
(ii) $CO_2$ and $H_2O$ often form during oxidation whereas other solvents act as additional contaminants. This is beneficial for downstream processing and the work-up of products.
(iii) They are environmentally benign solvents and can replace problematic solvents.
(iv) Both $scCO_2$ and $scH_2O$ can dissolve organic compounds (in contrast to ambient conditions $scH_2O$ is non-polar; on the other hand, polar salts precipitate from $scH_2O$).
(v) The complete miscibility of SCFs with gases such as $O_2$ or air eliminates mass transfer limitations known in liquid-phase oxidation.
(vi) Solubilisation of homogeneous metal complex catalysts in $scCO_2$ is possible by adding surfactants, mixing with water or modification of ligands with fluorine.
(vii) Use of SCFs can simplify multiphase reaction systems to homogeneous mixtures. Better control of oxygen concentration (gradientless operation) is possible.
(viii) Correct setting of pressure and temperature allows tailoring of density and polarity. Reaction and separation can run in the same unit (process intensification).
(ix) They provide inherent safety due to the inert nature of the reaction mixture. In particular, $CO_2$ eliminates the risk of phase separation and formation of explosive gas mixtures.
(x) They may affect elementary reaction steps or transition states (e.g. via ion product or polarity) and have a positive impact on conversion and selectivity.
(xi) Due to good transfer properties, SCFs can effectively dissipate the heat of reaction in exothermic oxidation in $scCO_2$ as well as in $scH_2O$.
(xii) Their solvent power can affect surface coverage on heterogeneous catalysts (removal of valuable oxygenated intermediates or high boiling point deposits that deactivate the catalyst).

Compared to other compounds including $CO_2$, $H_2O$ offers unique features as a solvent and reactant.[4] Heating affects $H_2O$ density as well as the number and strength of hydrogen bonds. Their number decreases when switching from ambient to supercritical conditions.[5] Nuclear magnetic resonance (NMR) measurements have shown that at 400°C and 400 bar, 29% of the initial hydrogen bonds still exist.[6] Raman investigations revealed that at 500°C, only 5–8% of the initial hydrogen bonds are

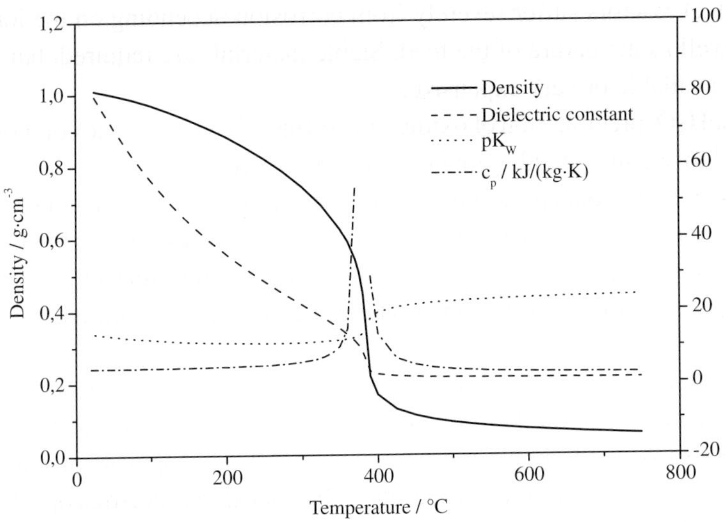


**Figure 25.2.** Physico-chemical properties of water depending on temperature at a given pressure of 250 bar.

detectable.[7] The ion product shows a maximum of $10^{-11}$ at around 250–300°C and water is more acidic than in ambient conditions.[8] On the other hand, its corrosivity is higher than in the supercritical state. In sc$H_2O$, the ion product drops to $10^{-20}$ and less. This effect, together with low polarity, explains the unusual solvent properties of sc$H_2O$.

Many density-dependent properties of $H_2O$, such as viscosity, polarity (dielectric constant $\varepsilon$ changes from 74 to 2), heat capacity at constant pressure (which is infinite at the critical point), ion product and solvent power can be tuned for specific requirements by setting the correct temperature and pressure, and they show significant changes near the critical point (Figure 25.2). Several studies have demonstrated that the transition from sub- to supercritical conditions also affects the elementary steps in reaction mechanisms, and radical intermediates are favoured over ionic species.[9] Another consequence is that subcritical water shows potential for acid catalysis. Reactions can be run either under non-polar/aprotic or polar/pH controlled conditions (water can take part in these reactions). Consequently, non-polar compounds like aromatics become soluble whereas inorganic salts precipitate. Therefore, the properties of water as a solvent are tunable over much wider parameter ranges than for most other compounds.

Besides these advantages, one has to be aware of some drawbacks that limit the economy and applicability of SCF processes:

(i) Reactors and peripherals are more expensive due to required pressure limits.
(ii) High pressure causes high compression costs.
(iii) Investment in additional active and passive safety measures is necessary.

(iv) $scH_2O$ reactors suffer severely from corrosion depending on the ion product as well as the nature of the feed. Stable materials are required, but are either not available or very expensive.
(v) In $scH_2O$, precipitation of oxidic salts formed from inorganic compounds may lead to scaling and plugging of tubes and valves.
(vi) New reactor concepts are necessary to overcome corrosion and scaling.
(vii) Few materials that may act as heterogeneous catalysts are stable in $scH_2O$.
(viii) Stability of reactants and catalysts may be limited in $scH_2O$.
(ix) Knowledge of explosive ranges at elevated pressure is limited.

Some early constraints for the progress in SCF reaction engineering have been overcome. Knowledge on thermodynamic properties of supercritical mixtures has been extended beyond simple measurement to the development of mathematical tools to predict thermodynamic properties.[10] Furthermore, spectroscopic techniques are now available to investigate supercritical mixtures *in situ* and provide new insights into SCF features.[11]

The first industrial SCF applications utilised $scCO_2$ for the extraction of natural compounds (caffeine, hops) and were successfully established in the early 1970 s. In the following decades, research focus also shifted towards reactions in $scCO_2$ and $scH_2O$ (however, it is noteworthy that ammonia and methanol syntheses were sometimes considered as supercritical processes).[2] From all these processes, fundamental thermodynamic data and practical experience in high-pressure reaction engineering are available and promote the development of supercritical oxidation processes.

Though $scCO_2$ and $scH_2O$ share many of the above features, their fields of application are different. The SCF density limits the optimum application regime for practical reasons. If density and therefore solvent power is low, a process cannot run at high feed concentration and the space-time yield is poor. On the other hand, if high density at a given reaction temperature can only be achieved by setting extreme pressure, the process becomes too expensive. Accordingly $scCO_2$ offers mild conditions for thermolabile compounds like homogeneous catalysts and fine chemicals in selective reactions (up to 150°C and 200 bar), whereas the harsh reaction conditions in $scH_2O$ (typically >400°C) favour rapid degradation processes and the cracking of C-C bonds that are not affected by selectivity issues. This latter application field established supercritical water oxidation (SCWO) for the destruction of waste materials.

### 25.1.2. *Scope*

This chapter will cover aspects of heterogeneously catalysed oxidation reactions in $scCO_2$, $scH_2O$ and a few other SCFs. Research using $scCO_2$ as the reaction medium mainly deals with the synthesis of value-added products, whereas the literature on supercritical water splits into work on waste treatment as well as synthesis reactions.

Numerous books and reviews regarding the chemistry in SCF with substantial contributions to the field of oxidation are available from the last decade (e.g. Refs. 12–14) and from the years before 2000 (e.g. Refs. 2, 12–14). Specialised reviews focus on reactions either in $scCO_2$ (general issues,[12,13] homogeneous catalysis,[14] oxidation,[15] catalytic oxidations[16]) or in $scH_2O$ (thermodynamic data,[17] fundamentals,[18,19] basic work on catalyst stability,[20] heterogeneous catalysis,[21,22] oxidations[23]).

### 25.1.2.1. *Thermodynamics for oxidation reactions in SCF*

Pressure and temperature control the chemical equilibrium and solubilities of chemical compounds in solvents. Pressure impact comes to the foreground when the mole number and therefore the volume in a system changes during reaction (Le Chatelier's principle). For the successful realisation of supercritical fluid reaction engineering, it is essential to investigate the phase properties of the reaction mixtures. This ensures that the reaction mixtures are truly homogeneous to take advantage of the aforementioned SCF features. It is also helpful to study the solubility of reactants and products in detail as interesting effects may occur. Sometimes slight changes in composition may shift the mixture properties away from SCF behaviour. *Vice versa*, tuning the solvent properties by slight pressure (or density) changes allows the separation of selected compounds from the mixture and the simplification of product recovery. Rapid depressurisation can quench short-life intermediates.[24] A large amount of literature is available on SCF properties (mostly $scCO_2$) and the solubilities of inorganic and organic compounds,[25–28] which often stem from the investigation of extraction processes. Such data serve to verify the applicability of common equations of state, mixing rules or other thermodynamic models and allow them to be refined. In accordance, various databases and software tools have been developed in recent decades (DETherm®,[25] UNIFAC,[26] Aspen Plus®[27]) as tools for modelling customized reaction mixtures and their properties with sufficient quality for industrial purposes.

Two examples may demonstrate the importance of profound knowledge of the phase behaviour of multicomponent mixtures, which undergo changes during catalytic reaction. In cyclohexane oxidation with molecular oxygen in $scCO_2$, the water solubility is suppressed in the presence of other organic products and water would be the first to condense out of the reaction mixture though it is not the compound in the mixture with the highest boiling point.[28] In the oxidation of benzyl alcohol by molecular oxygen, the ternary mixture of products ($CO_2$-benzaldehyde-water) and the intermediate multicomponent mixtures containing both products and reactants require lower pressure than the corresponding feed mixture of the reactants ($CO_2$-benzyl alcohol-$O_2$) in order to form a single phase.[29]

A very important feature in oxidation is the effective removal of high exothermic heat of reaction from solid catalysts and reactors to avoid hot spots and catalyst destruction. Supercritical fluids are superior to gases with regard to heat dissipation and they allow pressure tuning of heat capacities. $scCO_2$ shows a maximum heat capacity in the near-critical region which improves temperature control e.g. during the total oxidation of $H_2$ in $scCO_2$ compared to $N_2$ at the same pressure.[30] The specific heat capacity $c_p$ of liquid $H_2O$ shows an extraordinary increase in temperature from $4.19\,kJ\cdot kg^{-1}\cdot K^{-1}$ at ambient conditions to $13.02\,kJ\cdot kg^{-1}\cdot K^{-1}$ at 400°C and 250 bar[31] and makes $scH_2O$ an excellent heat carrier.

### 25.1.2.2. *Kinetics of oxidation reactions in SCF*

Changing temperature affects the kinetics of chemical reactions more than pressure. Typically, the Arrhenius exponential equation describes temperature dependency. Some more advanced and specialised theories exist to describe the temperature and pressure impact in general. Among them, the theory of activation state, which is based upon statistical mechanics, is widely used.

For a reaction $A + B \leftrightarrows C^{\#} \rightarrow D$ where $C^{\#}$ represents the transition state, a fundamental equation can be derived that comprises several terms:[2]

$$k_x = \frac{k_B T}{h} \cdot \overline{K_p}^{\#0} \cdot \frac{p}{p_0} \cdot \frac{f_A f_B}{f_c}$$

The reaction rate depends on a temperature term (with Boltzmann and Planck constants), the pseudo-equilibrium constant for formation of the transition state, a pressure term and finally fugacities f of reactants and activation state. The last term also represents SCF properties, but determination or calculation is difficult. However, with further modifications, the rate constants of chemical reactions relate to the activation volume of the transition state. The logarithmic nature of the equation implies that large pressure changes are necessary to affect the reaction rate significantly:

$$RT\left(\frac{\partial \ln k_x}{\partial p}\right)_{T,X} = -\Delta V^{\#}$$

The global reaction rates for heterogeneously catalysed reactions comprise several elementary processes such as diffusion, adsorption/desorption and surface reaction. Thus, the situation becomes more complex, as described above. Due to mass transfer and phase boundaries, the overall reaction rate is completely different from the intrinsic reaction rate on active sites on the catalyst surface. The residence time behaviour in a reactor under SCF conditions has to be considered (for modelling in $scCO_2$ and $scH_2O$, see Refs. 32, 33). Supercritical fluids have the potential to affect some of these elementary steps, in particular mass transfer and adsorption/desorption. Some studies on the theoretical and experimental validation of mass transfer effects on heterogeneously catalysed reactions in SCF media can be found in the literature for $scCO_2$[34] and $scH_2O$.[35,36]

### 25.1.2.3. *Tools for direct investigation of catalytic reactions in SCF*

The development of *in situ* or *operando* techniques (infrared (IR), Raman, UV-vis) to investigate working heterogeneous catalysts in gas-phase reactions has flourished over the last decade. It is state-of-the-art to couple several of these methods at once.[37]

Direct optical observation of SCF with high-pressure view cells is a basic tool to determine solubilities. Meanwhile many spectroscopic techniques have developed far enough to allow the study of high-pressure systems *in situ*.[11] The UV-vis method was among the first and helps in the study of solvatochromic effects and solubilities in SCF.[38] X-ray diffraction is applicable at high pressures for the investigation of hydrogen bonds[39] or solid characterisation (during particle formation in $scH_2O$).[40] Neutron scattering helps to investigate hydrogen bonds in $scH_2O$ and the structure of $scH_2O$.[41] Hydrogen bonds can also be studied with Raman spectroscopy ($scH_2O$,[42,43] sc methanol,[44] sc ethanol[45]). The Raman technique is also suited for investigation of solute-solvent interactions in $scCO_2$ systems.[46,47] Complementary to that, IR has also been used for SCF characterisation.[48] Currently, Raman and IR cells can operate successfully with SCF up to 1,500 bar and 550°C.[49] These techniques allow the detection of reaction intermediates in the bulk of SCFs and determination of the kinetic parameters of oxidation reactions,[50] but also provide information about solid catalysts under high-pressure conditions. The attenuated total reflectance infrared spectroscopy (ATR-IR) technique helps to determine phase behaviour during the heterogeneously catalysed oxidation of benzyl alcohol or geraniol in $scCO_2$.[51] Surface properties and species during the adsorption/desorption of benzaldehyde on a vanadyl pyrophosphate catalyst have been detected by means of IR measurements.[52] Figure 25.3 shows a heated IR cell that can be operated up to 125 bar in dense and $scCO_2$ phase.

Beyond vibrational spectroscopy, additional information about hydrogen bonds, structure and kinetics is accessible with high-pressure NMR spectroscopy for many SCFs ($CO_2$ up to 250°C, 300 bar; $H_2O$ up to 600°C, 400 bar;[53] methanol;[54,55] ethanol[55]). Data are mostly based on $^1H$ at high time resolution, but also on other isotopes such as $^{13}C$ or $^{17}O$.[56] Monitoring of reactions with NMR in $scCO_2$[57] and in $scH_2O$[58,59] has been reported. Furthermore, NMR provides data for the calculation of diffusion coefficients in SCF, e.g. for the self-diffusion of $scH_2O$[53,60] or of $H_2O$ in $scCO_2$[61] that are crucial for kinetic modelling of catalytic processes. Observations of diffusion phenomena inside mesoporous materials with NMR pulse techniques using pentane as the model compound have recently been published.[62] In relation

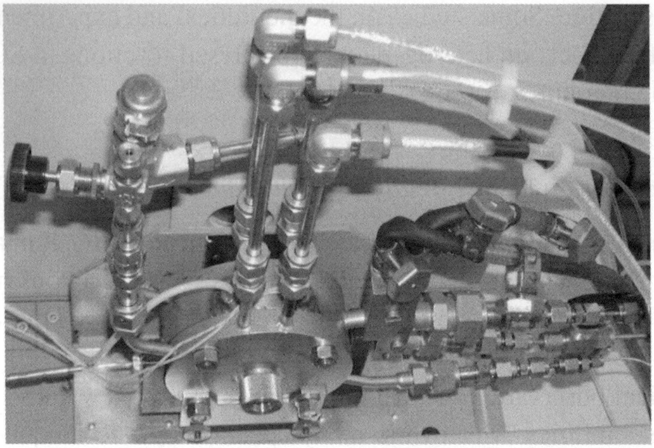

**Figure 25.3.** Heatable IR cell for *in situ* IR studies in dense and $scCO_2$.[52]

to mass transfer issues, SCF chromatography is another mighty tool for the determination of binary diffusion coefficients of supercritical mixtures.[63]

Direct investigation of solid catalysts under high-pressure conditions benefits, in particular, from X-ray techniques that provide valuable information about the solid structure and transformations. Successful X-ray absorption spectroscopy (XAS) studies have been made to monitor the oxidation state of metal catalysts in benzyl alcohol oxidation in $scCO_2$.[64,65] High-pressure *in situ* X-ray absorption near-edge structure (XANES) and extended X-ray absorption fine structure (EXAFS) data are available from the continuous selective oxidation of benzyl alcohol or cinnamic alcohol to the corresponding aldehydes in $scCO_2$ over a $Pd/Al_2O_3$ catalyst.[66,67]

### 25.1.2.4. *Oxidants*

In principle, all known oxidants can be used in SCF media. However, for several reasons some of them are less attractive under such reaction conditions. As discussed above, SCF features may strongly promote processes with gaseous oxidants. Regarding this, transferring an oxidation process with liquid organic peroxides or solid oxidants (persulfates or oxides) to SCF conditions is uneconomical, unless other features will generate sufficient benefit for the process. Peroxides are oxidants with explosion risk, and knowledge on their high-pressure behaviour and explosive ranges is scant.[68] In the case of using air at high pressure, the large excess of $N_2$ ballast increases compression costs. Molecular $O_2$ tends to be the first choice for oxidation in SCF media, similar to current trends in the ongoing development of liquid-phase processes. Nevertheless, many lab-scale studies on oxidation in $scH_2O$ use hydrogen peroxide instead of $O_2$ because handling and feeding are more convenient.

## 25.2. Oxidation in Supercritical Carbon Dioxide

Supercritical $CO_2$ is predestined as the solvent for homogeneous catalysis due to its high density at mild temperatures. The reaction conditions can even be set to realise enzyme catalysis.[69] Hydrogenation is predominant, followed by C-C coupling reactions and various oxidation reactions. Hydrogenation in particular, benefits from complete miscibility of $H_2$ and $scCO_2$, which enhances reaction rates significantly. In oxidation reactions, selectivity and atom efficiency are worse than in hydrogenation. Hydrogen at mild conditions will only attack unsaturated C atoms, whereas oxidation typically runs at a higher temperature to activate oxygen, which afterwards may attack each C atom. Consequently, heterogeneously catalysed hydrogenation in $scCO_2$ has been widely studied and commercial processes are known.[70]

Supercritical $CO_2$ is a non-polar, aprotic solvent and promotes radical mechanisms in oxidation reactions, similar to liquid-phase oxidation.[1] Thus, wall effects might occur as known, e.g. from olefin epoxidation with $O_2$[71] or $H_2O_2$[72] which may decrease epoxide selectivities. The literature covers the synthesis of fine chemicals by oxidation either without catalysts (alkene epoxidation,[73] cycloalkane oxidation,[74] Baeyer–Villiger oxidation of aldehydes and ketones to esters[75]), or with homogeneous metal complex catalysts (epoxidation with porphyrins,[76,77] salenes[78] or carbonyls[79,80]). Also, the homogeneously catalysed oxidation of typical bulk chemicals like cyclohexane (with acetaldehyde as the sacrificial agent[81]), toluene (with $O_2$, $Co^{2+}$/NaBr[82]) or the Wacker oxidation of 1-octene or styrene[83] has been demonstrated.

Heterogeneously catalysed oxidations, which are mostly carried out between 250 and 400°C, are difficult to realise under typical $scCO_2$ conditions because pressures above 500 bar are needed to reach the critical density of $CO_2$ and to take advantage of its liquid-like properties. Operation at lower, more viable pressures is inevitably linked with a feed dilution which lowers the space-time yield and makes the process less attractive. In this case, benefits may occur mostly from its gas-like transfer properties and only reactions that are limited by mass transfer are a target for possible improvement. Consequently, the full potential of $scCO_2$ as a reaction medium is utilised only with catalysts that are active at moderate temperatures up to 150°C.

### 25.2.1. *Partial oxidation of alkanes*

The benefits of SCF reaction engineering are supposed to be highest for liquid-phase oxidation processes with mass transfer limitations for molecular oxygen. The reported results show some inconsistency with regard to a possible improvement by setting the SCF conditions.

Partial oxidation of propane with air in $CO_2$ up to 400°C and 113 bar, over supported metal oxide catalysts in a flow reactor, revealed an increase in total yield of oxygenated intermediates such as acetic acid, methanol, acrolein or acetone with increasing pressure.[84–88] Adsorption studies by means of SCF chromatography suppose that the growing solvent power of $scCO_2$ removed those compounds from the catalyst surface prior to deep oxidation.[84] Similar results obtained in a stirred autoclave indicate that the operational mode has only a weak influence on the reaction.[85,86]

Cyclohexane oxidation with air to cyclohexanol/cyclohexanone was studied in batch and continuous mode in $scCO_2$ at 80–400°C and 80–160 bar, with oxides or salene complexes of the same metals (Fe, Co, Mn) on microporous and mesoporous supports (mordenite, ALPO-31, MCM-41). Selectivities at low conversion were comparable with the commercial process.[87] A comparative study of the same reaction with MnAPO-5 in $scCO_2$ and other solvents of similar polarity such as benzene or $CCl_4$[88] showed the best results at 125°C in the presence of $scCO_2$. The transition of the reaction mixture cyclohexane/$O_2$ from a sub- to supercritical state had a strong impact on the catalyst performance. With a similar CoAPO-5 catalyst, an increase in density of compressed $CO_2$ led to a lower conversion but higher total selectivity for the desired products.[89] When using tert-but OOH or $H_2O_2$ at 40°C and up to 120 bar, in a batch reactor with Fe porphyrine catalysts (also immobilised on zeolite Y), higher cyclohexanol yields (up to 55%) were obtained than in acetonitrile.[90]

### 25.2.2. *Partial oxidation of alkenes*

The continuous flow oxidation of propylene with $O_2$ to propylene glycol in $scCO_2$ at 138 bar on a $Cu/Cu_2O/MnO_2$ catalyst showed strong pressure dependence with a maximum selectivity of 95%. Catalyst dissolution and deactivation did not occur over a run time of 50 hours. However, the space-time yields were still one order of magnitude too low for scale-up.[91]

Recently, the partial oxidation of propylene with $H_2O_2$ into propylene oxide, which is currently mostly produced with chlorohydrine or organic peroxides as the oxidants with large amounts of by-products, has been commercialised.[92] Production of cheap hydrogen peroxide was a prerequisite for this development. Supercritical $CO_2$ was tested as the reaction medium for both $H_2O_2$ synthesis as well as propylene epoxidation. Generation of $H_2O_2$ from $O_2$ and $H_2$ is possible in emulsions of compressed $CO_2$ and water using modified anthraquinones.[93] It is possible to use reasonable concentrations of $H_2$ and $O_2$ without danger of explosion.

Feasibility of propylene epoxidation has been proved with a Pd-Pt/TS-1 or TS-1 catalyst in a continuous fixed-bed reactor at high pressure.[93,94] Initial selectivity can reach 99%,[94] but the catalyst deactivated rapidly and selectivity shifted towards methyl formate. Operation in $scCO_2$ had a beneficial effect on propylene oxide formation compared to $N_2$ and other solvents, due to the improved removal of deposits from the catalyst surface.

The partial oxidation of cyclohexene with molecular oxygen in $scCO_2$ using supported noble metal catalysts up to 150°C and 200 bar resulted in a mixture of dehydrogenation and oxidation products such as ketone and alcohol, and only small amounts of epoxides.[95]

Closely related to this, other olefins were tested in epoxidation reactions. The heterogeneous epoxidation of cyclooctene with $H_2O_2$ in $scCO_2$[72] using a manganese porphyrinate catalyst in the presence of hexafluoroacetone hydrate as the co-catalyst at 40°C and 200 bar, leads to complete transformation into cyclooctene oxide. The co-catalyst forms perhydrates and thereby helps to stabilise the porphyrine catalyst and to solubilise $H_2O_2$ in $scCO_2$.

### 25.2.3. *Partial oxidation of aromatics*

Toluene oxidation with air in $scCO_2$ with classical catalysts such as CoO, $MoO_3$, CoO/$MoO_3$, W, Ni, zeolites or $Al_2O_3$ forms benzaldehyde, benzyl alcohol and cresols. The rate of deactivation by coking is lower than in the gas phase.[24] In addition, dense $CO_2$ removes benzaldehyde from the catalyst surface during toluene oxidation and thus suppresses total oxidation.[52]

Oxidation of cumene with air in $scCO_2$ over different metal catalysts at 110°C did not outmatch the commercial process, which was explained by inhibition effects.[96]

### 25.2.4. *Deep oxidation of hydrocarbons and oxygenated compounds*

A few studies have focused on the deep oxidation of hydrocarbons in $scCO_2$, aimed at a post-treatment of recycle streams in $scCO_2$-based extraction units without intermediate depressurisation and expensive recompression. Model waste compounds such as alkanes and aromatic hydrocarbons (0.6% by weight of hexane, octane, decane, benzene, toluene, *p*-xylene and cumene) were treated with air in dense $CO_2$ at 80–140 bar and 180–280°C over various alumina-supported catalysts (Pt, CoO, NiO). The reported rates were by one order of magnitude higher than for the comparable gas-phase oxidation at the same temperature.[34] Oxidation of feeds with approximately 2,000 ppm of toluene and tetralin with $O_2$ in $scCO_2$ over a Pt/$Al_2O_3$ catalyst required higher temperatures, up to 390°C at 107 bar, to reach more than 90% conversion.[97] Similarly, deep oxidation of aromatic volatile organic compounds (VOC) such as benzene, toluene and *m*-xylene with $O_2$ in $scCO_2$ over a Pt/$Al_2O_3$ catalyst up to 350°C and 105 bar has been reported.[98] The $scCO_2$ based reaction showed higher conversion than the gas-phase reaction. However, as the feed concentration in these studies was rather low, the high pressure and large $CO_2$ ballast would make such a treatment rather inefficient if used as a stand-alone process.

For the same purpose, a Pt/$TiO_2$ catalyst was used in the deep oxidation of ethanol and acetaldehyde with molecular oxygen in $scCO_2$ at 150 to 300°C, but no significant improvement compared to the gas-phase process was achieved.[99]

### 25.2.5. *Oxidation of alcohols*

Alcohol oxidation is theoretically the simplest route towards the formation of valuable aldehydes and ketones for fine chemistry. Unfortunately, these intermediates rapidly react towards the corresponding carboxylic acids. Better control of oxygen availability, as well as the adsorption and desorption on catalyst surfaces in SCF media, might strongly affect product distribution.

Selectivity of methanol partial oxidation with $O_2$ in $scCO_2$ at 200–300°C, using iron-based aerogels can be shifted towards either dimethyl ether (70%), formaldehyde or methyl formate as the predominant products depending on the reaction conditions and catalyst.[100] Though methanol conversion was lower than in the gas-phase process, the production of desired ether, aldehyde and ester in the SCF process is higher.

Partial oxidation of benzyl alcohol to benzaldehyde has been studied frequently either in gas or liquid phase as a model reaction to characterise potential oxidation catalysts, but there is no real economic focus. The oxidation with air runs well in $scCO_2$ with a $H_5PV_2Mo_{10}O_{40}$ heteropolyacid catalyst in batch mode.[101] No catalyst leaching was detected and recycling was possible. It was demonstrated in an IR cell up to 40°C, that the benzaldehyde desorption rate from the vanadyl pyrophosphate $(VO)_2P_2O_7$ surface strongly depends on the $CO_2$ density.[52] An extensive investigation using mainly the commercial $Pd/Al_2O_3$ catalyst (see Ref. 67), but also $Pt/Al_2O_3$ and $Ru/Al_2O_3$, in a fixed-bed reactor and $O_2$ as the oxidant in $scCO_2$ at 80°C and 150 bar, resulted in turnover frequencies (TOF) up to 1,800 $h^{-1}$ and a constant selectivity of 95%.[74–76] Accompanying phase behaviour studies revealed that raising the pressure from just 140 to 150 bar doubles the activity because single-phase SCF conditions were established that enhance the external and internal mass transfer of reactants. *In situ* ATR-IR spectroscopy validated these results. In a continuation of these tests, toluene was investigated as either the co-solvent or as the $CO_2$ expanded solvent.[102] A small concentration of toluene in the feed had already boosted the TOFs from 1,500 $h^{-1}$ to 2,500 $h^{-1}$ under the same conditions. The partial oxidation kinetics of benzyl alcohol and *m*-hydroxy-benzyl alcohol to the corresponding aldehydes with $O_2$ over a Pd/charcoal catalyst in $scCO_2$ were measured up to 140°C and 200 bar (the maximum $O_2$ partial pressure is 20 bar).[103] Above 120°C and high $O_2$ partial pressure, deep oxidation was observed. It was assumed that $scCO_2$ improved the heat dissipation. Conversion of benzyl alcohol and aldehyde selectivity was higher than in the case of the partial oxidation of *m*-hydroxybenzyl alcohol.

Gold nanoparticles are currently a hot topic in catalysis and consequently first reports on their application as oxidation catalysts in the oxidation of benzyl alcohol towards benzaldehyde as well as of other alcohols in $scCO_2$ have been published.[104]

With $Au/TiO_2$ and $O_2$ as the oxidant at 70°C, high conversion of 97% and selectivity of 95% were obtained.

Tetrapropylammonium perruthenate (TPAP) organically modified silicas (Ormosils) are catalysts that have been tested in the oxidation of benzyl, allyl and aliphatic alcohols with $O_2$ or $H_2O_2$ in $scCO_2$ at 75°C and 220 bar. Active Ru species in the pores were protected from aggregation because they are insoluble in $scCO_2$.[105,106] Similarly, Pd nanoparticles stabilised either with polyethylene glycol (PEG)[107] or 2,2′-dipyridylamine[108] on organically modified $SiO_2$ are suitable catalysts for the aerobic oxidation of benzaldehyde and other alcohols in $scCO_2$. These catalysts show high activity and excellent stability under continuous flow operation. This concept in particular, benefits from the poor solubility of metals in $scCO_2$ that suppresses leaching effects and loss of valuable noble metals.

Oxidation of cinnamyl alcohol to cinnamaldehyde in a continuous fixed-bed reactor with $O_2$ in $scCO_2$ and co-solvent toluene with a commercial $Pd/Al_2O_3$ catalyst[51,67] showed that the pressure and phase behaviour had a strong impact on reaction rates. At 80°C and 120 bar the aldehyde selectivity reached 60% (TOF $400\,h^{-1}$). Toluene concentrations which were too high led to biphasic systems and a large drop in activity (TOF $130\,h^{-1}$). Again, pressure and phase behaviour have a strong impact on the reaction rate. ATR-IR spectroscopy proved that the reaction surprisingly performed best in the biphasic region. Concentration profiles in bulk and catalyst pores differed significantly, indicating that $scCO_2$ was less effective in eliminating the mass transfer limitations than in tests with benzaldehyde on the same catalyst (see above).

Similar observations were made for the continuous oxidation of geraniol with $O_2$ to citral in $scCO_2$ using the same catalyst at 80°C and 150 bar.[51] Reaction rates were higher in the biphasic region, and the catalyst deactivated rapidly. This was explained by the adsorption of water and geranic acid on the catalyst surface. Another study with mesoporous MCM-41 as the support for Cr, Co, Pt and Pd compounds under similar conditions (80°C, 6 h) also reported excellent citral selectivities up to 98% at 52% conversion.[109] Catalyst recycling was possible and the SCF regime gave better results than biphasic systems.

In some of these examples, the higher molecular weight of the reactants may result in their strong adsorption on the solid surface and their low solubility in $scCO_2$ might hamper an effective removal of intermediates from the catalyst. The studies on geraniol oxidation demonstrate that the moderate acidity of MCM-41 compared to $Al_2O_3$ eases desorption of the product citral. Superposition of such effects might lead to misinterpretation of phase behaviour effects.

It can be seen that slight changes in feed nature may significantly affect phase behaviour and catalytic performance. A comparative study of the oxidation of 1-octanol and 2-octanol with $O_2$ to carbonyl compounds in $scCO_2$ with $Pd/Al_2O_3$

in a fixed-bed reactor up to 140°C and 125 bar demonstrated that oxidation rates were 2–4 times higher than the rates in $N_2$ at similar pressure.[110] Oxidation of the secondary alcohol runs with excellent selectivities (>99.5%) and without deactivation, whereas the selectivity with 1-octanol was low. This highlights the importance of tuning the phase behaviour for successful SCF reaction engineering.

The rather unusual combination of photocatalysis and $scCO_2$ medium was tested in the oxidation of 1-octanol to octanal with $O_2$ over a $TiO_2$ catalyst up to 46°C and 100 bar.[111] It was demonstrated that temperature, but also $CO_2$ pressure, affect the reaction rate. The latter was explained by better mass transfer of aldehyde product near the catalytically active sites.

With non-polar, water-insoluble alcohols, another effect occurs: they are soluble in $scCO_2$, but adsorption onto the solid catalyst surface with acid sites is poor. An interesting solution might be surface modification with Teflon.[112] Thereby, alcohols like 9-(hydroxymethyl)anthracene or *m*-hydrobenzoin were oxidised on Teflon-modified Pt/graphite catalysts in $scCO_2$ and the impact of $CO_2$ density on conversion and product distribution was strong. In contrast, water-insoluble alcohols were oxidised into the corresponding aldehydes and ketones with oxygen in $scCO_2$ using a Pd-Pt-Bi/C catalyst in a continuous fixed-bed reactor, and high rates and yields up to 98% were reported.[113]

From the numerous studies on alcohol oxidation, it is evident that not every substrate is suited for a $scCO_2$-based oxidation process. Therefore, results from different researchers may lead to different conclusions. Mass, polarity and solubility, as well as the interaction with a solid catalyst surface, determine the distribution of reactants in the environment of a catalyst particle, and proper tuning of solvent properties is essential.

### 25.2.6. *New concepts for $scCO_2$-based oxidation processes*

Against this background, researchers have introduced new concepts to SCF reaction engineering. A possible strategy to overcome the problem with heterogeneous catalysts is the combination of $scCO_2$ with other solvents to create new types of reaction media that preserve or improve SCF properties under more moderate reaction conditions, e.g. $CO_2$ expanded liquids, where up to 80% of a conventional solvent is replaced by $scCO_2$.[114] The reaction pressure can thereby be significantly lower without losing solvent power and, at the same time, the $O_2$ solubility can increase by two orders of magnitude compared to neat solvent. This leads to significant improvement of reaction rates and selectivities in several homogeneously catalysed oxidation reactions. First attempts with solid catalysts were made in the oxidation of cyclohexene with iron porphyrinates immobilised on MCM-41 up to 50°C and 127 bar with $O_2$ and iodosylbenzene.[115] Yields were double in comparison to neat acetonitrile as the solvent and leaching of the iron catalyst from the support was decreased. The same group also reported on the oxidation of 2,6-di-tert-butylphenol with $O_2$ in $scCO_2$ and $CO_2$ expanded acetonitrile with immobilised Co-complexes.[116] Conversion was highest (60%) with neat $scCO_2$ due to improved $O_2$ and reactant mass transfer inside the pores. A study of benzyl alcohol oxidation to benzaldehyde in $scCO_2$ with $Pd/Al_2O_3$[102] revealed that low amounts of toluene (co-solvent) or large amounts of toluene (expanded liquid) led to single-phase conditions and increased reaction rates.

Mixing of $scCO_2$ with polyethylene glycol (and polypropylene glycol) leads to swelling of the polymer and significantly changes the solvent properties of the mixture.[117] Suitable linkers disperse and stabilise solid catalysts that are otherwise insoluble in neat $scCO_2$. Examples were reported on alcohol oxidation with Pd nanoparticle catalysts[118,119] or styrene oxidation to acetophenone with a $PdCl_2/CuCl$ catalyst.[120]

Multiphase catalysis occurs in $CO_2$/water by adding surfactants, which form micelles that disperse catalysts in pressurised $CO_2$/water mixtures.[121,122] This is known for homogeneous toluene oxidation,[123] toluene oxidation on immobilised micelles,[124] oxybromination of phenol and aniline derivatives,[125] or enzyme catalysis.[126]

Ionic liquids are miscible with $scCO_2$, and this offers the opportunity to immobilise catalysts in the ionic liquid whereas reactants stay dissolved in $scCO_2$. It is also possible to tune the solubilities for the easy separation of products or catalysts. Literature reports describe the oxidation of 1-hexene with $O_2$[127] or electro-oxidation of benzyl alcohol.[128]

Some of these concepts might be transferable to heterogeneous catalysis. Realisation of oxidation with heterogeneous catalysts in $scCO_2$ on a technical scale can only be successful if intelligent usage of the aforementioned advantages leads to significant progress either in space-time yields or to the replacement of solvents.

### 25.2.7. *Patents on oxidation in supercritical $CO_2$*

Only a limited number of patents are granted on the application of $scCO_2$ as a reaction medium for heterogeneously catalysed oxidation reactions (Table 25.3). However, none of these patents is known to be realised in a commercial process.

## 25.3. Oxidation in Supercritical Water

Supercritical water as a reaction medium has been investigated since the early 1980s, in particular in studies by the research groups of Modell,[129] Antal[130] and Tester.[131] Due to the severe conditions, focus was mainly set on degradation reactions. Supercritical water oxidation of waste compounds can be seen as a further

**Table 25.3.** Patents on heterogeneously catalysed oxidation in $scCO_2$.

| Reactant, oxidant | Product | Catalyst | T [°C] | p [bar] | Remarks | Ref. |
|---|---|---|---|---|---|---|
| Methane | Methanol | Oxides of Mn, Co, Pd, Mo, Cr, V, Cu | 25–250 | 1–69 | Continuous, homogeneous or heterogeneous catalyst $CO_2$ mixed with perfluorinated compounds (expanded liquid) | 139 |
| Propylene $O_2$ | Propylene oxide acrolein | TlO | 50–120 | 80–350 | | 140 |
| Propylene $O_2$ | Propylene glycol | CuI-$Cu_2O$-$MnO_2$ with supports | 100–250 | 100–500 | Continuous reactor $H_2O$ in feed S = 95% | 141 |
| Propylene $O_2/H_2$ | Propylene oxide | Pd/TS1 | 20–100 | 100–300 | S = 91% | 142 |
| Methanol $O_2$ | Dimethyl ether, Formaldehyde methyl formate | $Fe_2O_3$, $Fe_2O_3/SiO_2$, $Fe_2O_3$-$MoO_3$ | 200–330 | 83–124 | Continuous reactor $Y_{max}$ = 81% (DME) $Y_{max}$ = 95% (FA) $Y_{max}$ = 69% (MF) | 143 |
| Benzene $O_2/H_2/N_2O$ | Phenol | Noble metal on $TiO_2$ or V-silicalite, Pd/TS-1 | 20–100 | 10–300 | Batch reactor Y = 0.11% | 144 |
| Propylene and other alkenes $H_2O_2$ | Epoxides | Mole sieves with Ti (TS-1, TS-2, TS-3), Zr, Hf, V, Nb, Ta, Cr, Mo and W | 0–100 | 1–700 | Use of co-solvents S = 100% (PO) | 145 |

development of wet air oxidation technology starting with the Zimpro process in the late 1950s for wastewater treatment at elevated temperatures and pressures (150–325°C, 2–20 bar).[132] Maximum SCWO temperatures currently reach around 750°C (at pressures up to 300 bar). This technology quickly proved its capability to destroy recalcitrant compounds such as acetic acid, methanol, phenol or ammonia (which are hard to degrade by wet air oxidation), and has been developed to pilot scale for almost 20 years. A summary of commercially designed SCWO facilities, which were in operation in 2007, is given in Ref. 133.

In SCWO, hydrocarbons are almost completely converted into $CO_2$ and $H_2O$, and heteroatoms (sulfur or halides) are transformed into minerals or corresponding acids. Nitrogen-containing compounds including ammonia form mainly $N_2$ or $N_2O$,[134] but traces of $NO_X$ were also found.[135] Very hazardous wastes such as dioxins (degradation 99.9%),[136] polychlorinated biphenyls (degradation 99.999%)[137] or toxic warfare agents (degradation 99.99%)[138] have been successfully treated in SCWO plants. Beyond model compounds, the deep oxidation of many different real wastewater streams has been demonstrated (Fig. 25.4).

Supercritical $H_2O$ as a solvent allows the running of ionic as well as radical reactions.[146] A special feature of $scH_2O$ compared to other SCFs is its ability to act as a reaction partner, e.g. for hydrolysis or gasification. This has led to numerous studies on gasification of (wet) biomass for syngas production. The water-gas shift reaction plays an important role in SCWO, and the presence of $H_2$ allows reductive steps to take part in the reaction mechanisms.[147,148] The contribution of all these

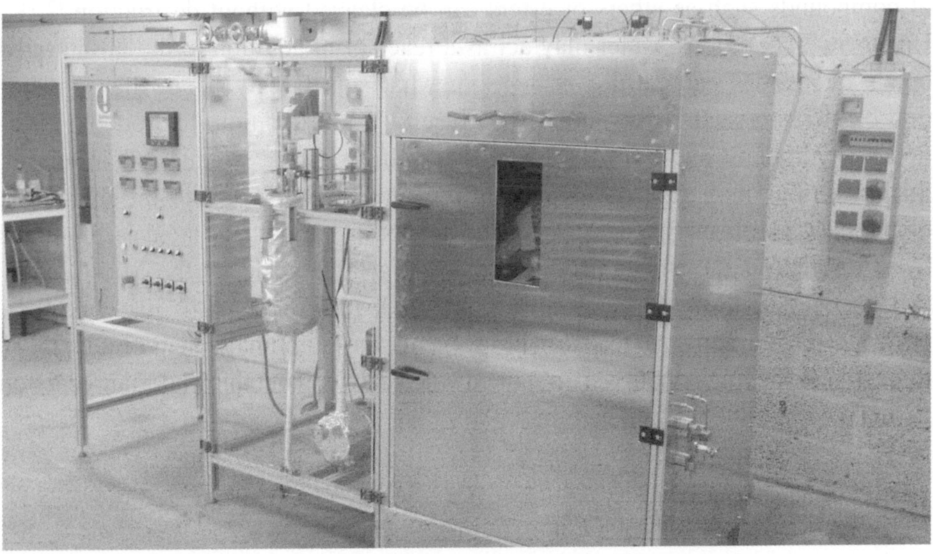

**Figure 25.4.** Test plant for oxidation in near-critical water (courtesy of HDT GmbH, Berlin).

side reactions to the overall process often outnumbers a quantifiable effect of the catalysts.

Processes at such extreme conditions require a large energy input to heat the large amount of water (with the heat capacity of sc$H_2O$ being very high) and effective heat exchangers. It is necessary to compress oxidant air or $O_2$, which contributes significantly to running costs. In addition, expensive, high-temperature, high-pressure and corrosion-resistant construction materials are required. As inorganic compounds like salts are almost insoluble in supercritical water,[149] they precipitate in reactors and downstream processing units and lead to severe scaling and plugging; fouling might also become a problem.[150] New reactor concepts have to be developed to overcome this particular problem (see below). To date, the main problems in SCWO reactors are limited corrosion resistance and the handling of formed solids in the reactor (and in downstream parts like backpressure valves).

It is evident that such a demanding and expensive technology will be limited to special applications where efficiency is favoured over economic issues. Supercritical water oxidation as a waste treatment technology will most likely be used for waste streams that are hard to dispose of in other ways. A possible solution is the use of catalysts to lower the process temperature and to soften the requirements for construction materials and energy consumption. Consequently, attempts to apply hydrothermally stable catalysts in SCWO plants have been reported (see below).

Although sc$H_2O$ conditions are severe and even presumably stable inorganic materials fail to withstand, it is possible to run organic reactions other than deep oxidation.[9] Hydrolysis suggests itself as the dominant reaction and various types of compounds such as ethers or esters can be cleaved without destruction.[151,152] Under SCF conditions, water may act as a reactant in addition to the unsaturated bonds of alkenes and nitriles or — somewhat surprisingly — in the dehydration of alcohols such as ethanol[153] or glycerol.[154] In addition, hydrodeoxygenation,[155] hydrodenitrogenation[156] and hydrodesulfurisation[157] are feasible in sc$H_2O$. Such reactions may gain importance in the future with regard to the upgrading of biomass pyrolysis products. Last but not least, terephthalic acid is produced by the liquid-phase oxidation of *p*-xylene on a large scale by the Mid-Century (MC) process (Mn-Co-Br catalyst in acetic acid, $T \approx 280°C$, $p \approx 30$ bar). Supercritical $H_2O$ is among the few suitable solvents for terephthalic acid (Table 25.4). Recently, promising results were obtained when carrying out this important reaction in sc$H_2O$.[158]

### 25.3.1. *Corrosion suppression*

Typical corrosion effects on construction materials are intergranular corrosion, pitting, stress corrosion cracking and area corrosion.[160] Though sc$H_2O$ is considered

**Table 25.4.** Solubility of terephthalic acid in water (g/g of solvent). From Ref. 159.

| | 25°C | 150°C | 200°C | 250°C |
|---|---|---|---|---|
| Water | 0.0017 | 0.24 | 1.7 | 12.6 |
| Acetic acid | 0.013 | 0.38 | 1.5 | 5.7 |

less corrosive due to its low ion product, instationary operation may change the situation. Depending on temperature and $H_2O$ density, the ion product changes by orders of magnitude and reaches a maximum around 250°C. During heating, an aqueous mixture inevitably passes this subcritical stage with accelerated corrosion,[161] which first slows down above the critical point.[162] Common stainless steel, e.g. SS316, is resistant to weakly acidic media due to the formation of passivated oxide layers such as $(Fe,Cr)_2O_3$, $(Fe,Cr)_3O_4$ or $Cr_2O_3$,[163] and can therefore often satisfactorily withstand pure $scH_2O$.

The presence of other compounds in the feed, however, will often increase the corrosion stress dramatically. When running oxidation reactions, $CO_2$ often forms as a by-product and can be another source of corrosion in aqueous mixtures.

Real wastewaters often contain heteroatoms that transform into the corresponding salts or mineral acids, such as HCl, HBr and $H_2SO_4$, during SCWO. Corrosion studies often use HCl or chlorinated hydrocarbons in long-term tests, up to many thousands of hours.[164, 165] Under such conditions, Ni-based alloys (Hastelloy C-276, Inconel-625, Incoloy 800) perform better than stainless steel, but still significant corrosion occurs in the presence of HCl, $H_2SO_4$ and $HNO_3$,[166] or HF, HBr or HI.[161] Oxidants such as $O_2$ and $H_2O_2$ enhance the corrosion of many alloys (stainless steel,[167] Ni-base alloys[168]). Some improvement is possible by adding Ti,[167] Zr[169] or Al/Nb/Ti[170] to steels. Also, iron-free but very expensive metals like pure titanium or zirconium,[171] or alloys like Monel (Ni+Cu)[165] and Ti60[164, 172] have been tested.

To avoid classical corrosion mechanisms, ceramics and composite materials have also been tested (some of these materials are also common components in heterogeneous catalysts). An alumina reactor for SCWO was proposed;[173] among ceramic materials only a few aluminas and zirconias did not corrode severely, whereas SiC or BN lost up to 90% by weight under SCWO conditions in the presence of HCl.[174] The combination of steel and ceramic coatings should theoretically provide high-pressure stability and improved corrosion resistance, but only slight improvements were reported for stainless steel SS316 coated with sol-gel-prepared Ti, Zr or Hf oxides,[175] stainless steel SUS-304 with TiN[176] or Ni alloys and ceramics.[177] Often the adhesion of the ceramic layer on the steel surface is not sufficient.

### 25.3.2. *Reactor design*

Besides improved construction materials, new reactor types may help to suppress corrosion. They must ensure that proper temperature control is possible in deep oxidation reactions (it is favoured to run the process at least autothermally). Mixing zones of proper size are necessary to allow the reactants to dissolve in $scH_2O$. Finally, and most importantly, the precipitation of inorganic materials and plugging has to be circumvented. The first known reactors contained an oxidation zone and a solid separator (settling) zone.[178] Another concept for downstream solid removal is the use of zyclones.[179] Many newer concepts are based on fluid dynamics and try to avoid contact between the supercritical reaction zone and the reactor wall by a protective zone of fresh sub- or supercritical water acting as a buffer.[180] The most common types of reactors currently being investigated are:

(i) Transpiring wall reactor: an inner porous concentric steel, titanium[181] or ceramic tube[182] in the reactor directs an additional stream of pure water permanently and equally into the reaction zone and establishes a cool water layer on the inner wall. The inner tube is not exposed to high-pressure gradients. The enveloping $H_2O$ layer has three effects: it removes solids by convective flow, dissolves inorganic salts and scale, and suppresses corrosion of the outer pressure vessel. This is currently the most-tested configuration for SCWO plants.
(ii) Hydrothermal flame reactor: $O_2$ is injected into a mixture of fuel and $scH_2O$ and ignites at high temperature.[183–185] A hydrothermal flame forms that creates temperatures above 1,000°C in a confined volume usable for the oxidative destruction of organic materials. The outer water zone is much cooler and acts as a protective layer for the reactor. A combination of transpiring wall reactor and hydrothermal flame was investigated.[183]
(iii) Reactors with intermitting flushing or mixing/diluting: alternately feed or flushing stream are supplied in regular intervals. Thus, several SCWO reactors have to be connected (switch operation) to maintain continuous operation.[184]
(iv) Cooled-wall reactor:[185] an example for the realisation of such a concept is depicted in Fig. 25.5.

Other proposals describe reverse flow tanks filled with brine pools, reverse flow tubular reactors, centrifuge reactors, high velocity flow conditions, mechanical brushing or rotating scrapers.[180] Additives help to lower the corrosivity of the aqueous mixtures, e.g. neutralising agents such as $KHCO_3$, $K_2CO_3$ or KOH.[186]

Obviously, corrosion, plugging and other demanding challenges from a technical point of view, and precise control of temperatures, flows and heat transfer at such pressures hamper the safe operation of SCWO plants. Reactor designs differ from classical tube reactors in industrial catalysis, and further research on engineering

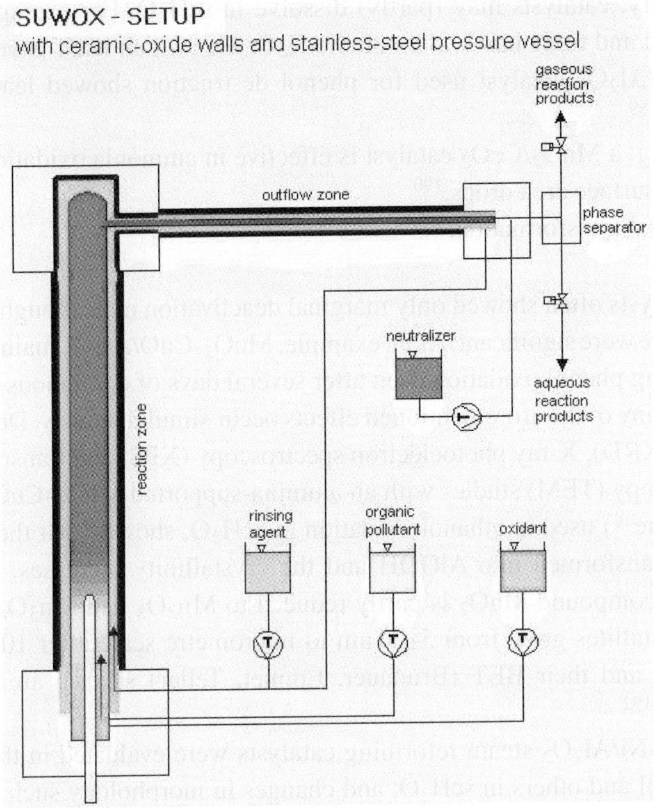


**Figure 25.5.** Set-up for supercritical water oxidation (SUWOX) using water as the rinsing agent and $H_2O_2$ as the oxidant. From Ref. 191.

and catalysis at these extreme gradients in flow, concentration and temperature is necessary.

### 25.3.3. *Stability of heterogeneous catalysts*

Catalysts are used to try to lower SCWO temperatures from 700°C to approximately 400°C. In a survey focused on metal oxides known to be stable at high temperatures and to promote deep oxidation or gasification of organics in $scH_2O$, only a few materials proved to be stable, e.g. zirconia.[20] Many reported catalysts comprise oxides of Mn, Cu and Ce as active compounds. Different effects seem to be responsible for poor catalyst stability:

(i) Change of oxidation state: the presence of $O_2$ changes the metal oxidation state and forms less stable materials. $MnO_2$ transforms into $Mn_2O_3$ during phenol oxidation in $scH_2O$.[187]

(ii) Solubility: catalysts may (partly) dissolve in $scH_2O$. For example, $Cr_2O_3$ is oxidised and turns into water-soluble $H_2Cr_2O_4$ (which itself is very toxic).[188] A $CuO/Al_2O_3$ catalyst used for phenol destruction showed leaching of Cu and Al.[189]
(iii) Sintering: a $MnO_2/CeO_2$ catalyst is effective in ammonia oxidation in $scH_2O$, but the surface area drops.[190]
(iv) Structural transformation.[190]

The catalysts often showed only marginal deactivation rates though the changes in solid nature were significant. As an example, $MnO_2$-$CuO/Al_2O_3$ maintained initial activity during phenol oxidation even after several days of continuous operation.[35]

Often, many of the aforementioned effects occur simultaneously. Detailed X-ray diffraction (XRD), X-ray photoelectron spectroscopy (XPS) and transmission electron microscopy (TEM) studies with an alumina-supported $MnO_2$-$CuO/Al_2O_3$ catalyst (Carulite®) used in ethanol oxidation in $scH_2O$, showed that the amorphous support is transformed into AlOOH and the crystallinity increases, whereas the main active compound $MnO_2$ is partly reduced to $Mn_2O_3$ and $Mn_3O_4$. The initial primary crystallites grow from 5–20 nm to micrometre scale after 100–200 hours of operation and their BET (Brunauer, Emmet, Teller) surface area also drops significantly.[192]

Classical $Ni/Al_2O_3$ steam reforming catalysts were evaluated in the SCWO of phenol, cresol and others in $scH_2O$, and changes in morphology such as softening and swelling were reported.[193] These were at first assigned to the support material. In general, metal catalysts are expected to be more stable than metal oxide catalysts in hydrothermal conditions as they tend to form thermodynamically stable oxide layers under an oxidative atmosphere, e.g. CuO and $Cu_2O$ from Cu.[194] If such protective layers cannot form, metal catalysts are highly sensitive to corrosion. Ag and Ru are less stable, as their oxides are less stable. In contrast, noble metals like Pd that do not form oxide layers are stable under SCWO conditions.

### 25.3.4. *Heterogeneously catalysed deep oxidation of model compounds*

Known recalcitrant compounds like ammonia or phenol are frequently used as model compounds as their destruction rate limits the overall kinetics in SCWO, and also in the presence of solid catalysts.

Ammonia is stable under SCWO conditions without a catalyst up to 600°C. Degradation leads to the formation of $N_2$ and $N_2O$. The destruction efficiency at 530–700°C and 246 bar with a catalyst reaches more than 90%.[195] Although the available solid surface increases by a factor of 30, the reaction rate increases only by a factor of 4. Thus, the reaction is mainly homogeneous. Application of a $MnO_2/CeO$

catalyst under milder conditions (450°C, 272 bar) and a residence time of 1 s led to significant ammonia conversion of 40%.[190]

Phenol is the most intensively studied model compound for SCWO. In addition to its high activity, a suitable catalyst must be able to shift selectivity towards $CO_2$. Bulk $MnO_2$ enhanced phenol conversion at 380–420°C and 222–304 bar,[196] but did not affect $CO_2$ selectivity at a given phenol concentration. This catalyst seemingly accelerated the formation of phenoxy radicals, which lead to similar reaction mechanisms as in non-catalysed SCWO. Studies with bulk $MnO_2$ at 425°C, 227–272 bar and residence times of 0.02–1.0 s revealed severe internal mass transfer limitations.[197] Reaction orders are 1 for phenol, 0.7 for oxygen and −2 for water, pointing to inhibition by water adsorption. $MnO_2$ itself transforms into $Mn_2O_3$ at SCWO conditions[187] and its specific surface area is reduced.[196] Bulk $TiO_2$ accelerates phenol destruction at 380–440°C and 222–304 bar compared to non-catalytic runs, but selectivity for phenol dimers increases.[198] After 120 h, the structure changed from anatase to rutile. The BET surface area dropped from 12.1 to 3.2 $m^2/g$ whereas metal leaching was not observed. Both bulk $MnO_2$ and $TiO_2$ accelerate the phenol degradation rate compared to homogeneous SCWO at the same temperature by factors of 10 and 4, respectively.

A supported $CuO/Al_2O_3$ catalyst that was tested under the same conditions as $MnO_2$ and $TiO_2$[196,198] at 380–450°C and 222–304 bar, resulted in a similar enhancement of phenol destruction but without the formation of phenol dimers.[189] The catalyst showed a slight initial deactivation due to the transformation of CuO into $Cu_2O$ and of $Al_2O_3$ into AlOOH, but subsequently maintained activity over 100 h though the BET surface dropped by a factor of 20, and Cu and Al were present in the reactor effluent. This Cu-based catalyst showed the highest activity related to mass, but the lowest related to surface area among these three materials.[199] $MnO_2$ performed the best related to specific surface area. All three catalysts maintained activity for more than 100 h and $MnO_2$ was considered the best choice for SCWO processes.

A supported binary metal oxide catalyst $MnO_2$-$CuO/Al_2O_3$ (Carulite 150®) at 380–430°C and 253 bar gave a phenol conversion above 70% compared to 6% in runs without a catalyst; in addition $CO_2$ selectivity was higher.[35] Under these conditions, internal mass transfer limitation was evident, as reported for bulk $MnO_2$ (see above). Catalysts such as $MnO_2/CeO_2$ and $V_2O_5$ enhanced phenol conversion (>99%) and $CO_2$ selectivity at 390°C and 14.3 s residence time relative to the homogeneous case,[200] and for $MnO_2/CeO_2$ no metal ions were detected in the effluent. In the case of low $CO_2$ selectivity, many by-products such as low molecular organics (formic acid, acetic acid), aromatics (benzyl alcohol, benzoic acid), condensation products (dibenzofuran, xanthone) and dimerisation products (bibenzyl, dibenzyl ether and biphenyl) were found.

A CuO-ZnO-$Co_2O_3$ catalyst (Süd-Chemie AG) was able to destroy phenol by more than 90% at 400–440°C and 240 bar.[201] The spectrum of products was smaller than in the homogeneous reaction and their nature indicates a free-radical mechanism. Use of a cement-supported CuO-ZnO catalyst (Süd-Chemie AG) at 380–390°C and 230–235 bar at a residence time of 15 s led to complete phenol conversion and strong total organic carbon (TOC) reduction.[202] The latter indicates an increase in $CO_2$ selectivity. The catalyst showed Cu leaching but less than in subcritical wet oxidation.

Besides metal oxide catalysts, somewhat surprisingly, activated carbon was also used as a cheap and clean catalyst for the SCWO of phenol at 400°C and 250 bar. The overall reaction is composed of three reactions: homogeneous and heterogeneous phenol oxidation and the combustion of active carbon.[203,204] Although the carbon catalyst was oxidised, its destruction was sufficiently low to observe a catalytic effect on phenol oxidation. The catalyst increased $O_2$ consumption for phenol degradation (from 39 to 65%) as well as gas yield and decreased tar formation; internal mass transfer limited the reaction.[205] This unusual concept was successful with other carbonaceous materials such as graphite, coke or carbon fibres, which were all catalytically active in phenol oxidation in $scH_2O$.[206]

Like terephthalic acid, benzoic acid is stable in $scH_2O$ and withstands non-catalysed oxidation up to 400°C (conversion $\approx 1\%$).[202] With a supported CuO-ZnO catalyst (Süd-Chemie AG), conversion is complete at 380–390°C and 230–235 bar, and TOC destruction reaches 73%.

Heteroaromatic compounds are often only partially oxidisable in non-catalysed SCWO and by-products like ammonia are more stable than the original feed. Pyridine is known to be stable in $scH_2O$[207] and below 500°C conversion is negligible without a catalyst. More than 20 intermediate products (carboxylic acids, amines and amides) are known.[208] Catalytic degradation in $scH_2O$ with 0.5%$Pt/Al_2O_3$ at 365–400°C and 242 bar resulted in conversion levels above 99%,[36] leading mainly to the formation of $CO_2$, $N_2O$ and $NO_3^-$, whereas $NH_3$ and $NO_X$ were not detected. The impact of external and internal mass transfer on the catalytic performance was experimentally validated and modelled. External mass transfer limitation was not relevant, but internal diffusion was significant when increasing the catalyst particle size from 0.09 to 0.7 mm; the catalyst effectiveness factor then decreased from 0.96 to 0.2. Other catalysts $MnO_2/Al_2O_3$ and $MnO_2/CeO_2$ (that have also been used in phenol degradation)[200] were sufficiently stable in pyridine oxidation, however, only $Pt/Al_2O_3$ achieved complete conversion at a subcritical temperature of 370°C.[209] The $Pt/Al_2O_3$ catalyst favoured the formation of $NO_3^-$ and $N_2O$ whereas Mn-containing catalysts mostly formed $N_2$ and $NO_3^-$.

Quinoline, which shows similar stability in $scH_2O$ to pyridine, can be almost completely converted over a commercial $MnO_2/CuO$ mixed oxide catalyst

(Carulite 300®) in $scH_2O$ at 400°C, 230–300 bar and a residence time of approximately 25 s.[210] However, TOC analysis of the liquid effluent indicated that deep oxidation only reached 90%. Neither external nor internal mass transfer was experimentally measurable. Catalyst stability changed during the first two hours of operation as the conversion dropped by approximately 20% and was then stable for another 6 hours at 75% quinoline conversion and 65% TOC removal.

Similarly, 1-methyl-2-pyrrolidone is very difficult to decompose in non-catalysed SCWO (conversion ≈37% at 380–390°C and 230–235 bar) and forms amines and carboxylic acids.[202] A cement-supported CuO-ZnO catalyst increases conversion to 59% and $CO_2$ selectivity to 86% and essentially reduces the number of intermediate products.

Halides and, in particular, chlorinated compounds are often present in toxic wastes and some of them, like polychlorinated biphenyls or dioxins, are extremely stable under oxidation conditions. Various chlorinated model compounds serve as model feeds in SCWO investigations.

A study of the deep oxidation of 1,4-dichlorobenzene with air, either in batch or continuous mode, provided data on the catalytic performance of $V_2O_5/Al_2O_3$, $MnO_2$ and $Cr_2O_3$, and corresponding adsorption equilibria. Independent of catalyst presence and nature, the same product spectrum was always observed. Temperatures ranged from 343–412°C and pressures of 100-670 bar were applied.[211] In comparison to gas-phase reactions, the reaction rate in $scH_2O$ was lower.

CuO supported on zeolites (Y, ZSM-5, ZSM-48) was tested in the deep oxidation of 2-chlorophenol with $H_2O_2$ at 400°C. Condensation to polyaromatics and the formation of higher chlorinated phenols was lower and thus shape selectivity was evident. All catalysts increased the conversion (>80%) compared to the homogeneous case (43%). The enhancement was explained by the fast precipitation of copper chloride during the SCWO reaction.[212] The additional dosage of $Na_2C_2O_4$ led to almost complete conversion.[213] It was shown that $Na^+$ alone led to a remarkable increase in the rate of degradation. Further tests with $Li^+$ alone or together with zeolite-supported CuO confirmed this result.[214] However, Cu-O and Cu-Cu species were proposed as the main active species. Characterisation of the spent catalysts with EXAFS, XANES and electron paramagnetic resonance (EPR) methods demonstrated that the active CuO and $Cu_2O$ species in the zeolite channels were transformed by $H_2O_2$ into $Cu_3O_2$ clusters and subsequently into $Cu_3O_4$.[215]

Model pollutants (2-propanol, tert-butanol, acetic acid) were tested with a cement-supported CuO-ZnO catalyst.[202] All these compounds were stable in the absence of a catalyst, but the catalyst managed to convert them almost completely at 380–390°C and 230–235 bar ($X = 87$–$98\%$). In particular, the alcohols react with $CO_2$ selectivities close to 100%. The same group investigated the performance of a CuO-ZnO-$Co_2O_3$ catalyst (Süd-Chemie AG, also tested in phenol destruction)[201] in

**Table 25.5.** Treatment of real wastewater from chemical plants with heterogeneous catalysts in $scH_2O$.

| Source | Catalyst | Conditions | Efficiency | Ref. |
|---|---|---|---|---|
| Terephthalic acid manufacturing plant COD 15000 mg/l* | $\gamma$-$Al_2O_3$ | 418–513°C 220–300 bar flow reactor | 98% COD | 218 |
| Cellulose, coconut oil, brewery and dairy effluents | $TiO_2$ | 400°C 276 bar batch reaction time 5 min | 95% TOC formation of carboxylic acids was investigated | 219 |
| Benzothiophene, vacuum residual oil | Co-Mo/$\gamma$-$Al_2O_3$ | 400–500°C 300–400 bar batch time 60 min | Desulfurisation in presence of $O_2$ up to 67% | 220 |
| Municipal sludge | Active carbon | | Partial oxidation for $H_2$ generation | 221 |

*COD = chemical oxygen demand.

acetic acid oxidation, and CO, $CO_2$ and $H_2O$ were the products.[216] With bulk $MnO_2$ as the catalyst, the $CO_2$ selectivity is almost quantitative at 380°C and 277 bar.[217] Conversion was doubled compared to the non-catalysed reaction.

### 25.3.5. *Heterogeneously catalysed deep oxidation of real wastewater*

In contrast to the large amount of data on non-catalysed SCWO of real wastewaters, very few studies are available on heterogeneously catalysed SCWO (Table 25.5).

### 25.3.6. *Heterogeneously catalysed partial oxidation*

Methane reacts to methanol and CO at 450°C over $Cr_2O_3$ in batch reactors and reaction times up to 40 min.[222] The reproducibility was limited due to the small reactor size (1.26 ml). At 10% methane conversion, methanol selectivity reached 40%. Compared to a gas-phase reaction, conversion was less but the yield was higher. Continuous partial oxidation of methane with catalysts $Cr_2O_3/Al_2O_3$ and $MnO_2/CeO_2$ at 400–475°C led to the formation of methanol, formic acid and other partial oxidation products.[223] Metals Ag, Cu and Au/Ag as catalysts are also able to convert methane (375–500°C, 220–350 bar, residence time 0.5–60 s) into methanol and formaldehyde with 50–80% selectivity at conversion below 1%.[224–226]

Propane oxidation in $scH_2O$ up to 420°C and 167–280 bar on mixed-metal oxide catalysts (Carulite 300®, $MnO_2$, $Co_2O_3$, $MnO_2$-$Co_2O_3$, $MoO_3$, supported on $\gamma$-$Al_2O_3$) gives oxygenated products with yields up to 15% at 90% conversion, and methanol is the predominant product.[224] This is similar to that reported for the reaction in $scCO_2$ (see above). Transition from the sub- to supercritical regime showed the influence on propane and $O_2$ conversion due to homogenisation of the reaction mixture. The overall influence of these heterogeneous catalysts is rather small.

The industrially important partial oxidation of cyclohexane was investigated at 350–420°C and 250–300 bar using Cu, Ag and Pt as active metal catalysts. Cu was inert, whereas Ag and Pt catalysed deep oxidation. From these experiments, a maximum yield of 30% was obtained for value-added products.[225,226]

Oxidation of ethanol and ethyl acetate[192] over Carulite $MnO_2$-$CuO/Al_2O_3$ was studied at 400°C and 240 bar. Although the catalyst suffered from structural changes, neither activity nor selectivity changed significantly over 200 h, pointing to a rather small catalyst effect.

Aromatic carboxylic acids are stable in $scH_2O$ in the absence of catalysts. Conversion in the temperature range below 500°C is only possible with catalysts. The partial oxidation of benzoic acid to phenol (Dow phenol process) in a continuous flow reactor was realised with several commercial catalysts such as NiO (Fisher Scientific), CuO (Aldrich), $MnO_2/CuO$ (Carulite 300®), Carulite 110®, $MnO_2$ (Aldrich) and $Al_2O_3$ (Acros).[227] Temperature ranged from 200 to 400°C and pressure was set to 140–250 bar; feed concentration was 2,500 ppm. The best performing catalyst was Carulite 300®, which was also sufficiently stable. Phenol yield in $scH_2O$ reached 11%, however, in the subcritical regime the results were even better.

## 25.4. Heterogeneously Catalysed Oxidation in Other Supercritical Fluids

Due to the absence of stable alternatives to $scCO_2$ and $scH_2O$, beyond these solvents the oxidation of hydrocarbons under SCF conditions is possible almost exclusively in neat substance, comparable to solventless technical oxidation processes like cyclohexane oxidation. This implies that the starting material itself is in a supercritical state.

The most prominent example is the partial oxidation of sc isobutane ($T_c$ = 134.7°C, $p_c$ = 36.3 bar, $\rho_c = 0.225$ g/cm$^3$) towards tert-butanol via hydroperoxides. The reaction is autocatalytic and follows a similar mechanism as in cyclohexane oxidation. Isobutane oxidation has gained importance because of the applications of the oxidation products, tert-butyl hydroperoxide and tert-butyl alcohol, in the manufacture of important chemicals like propylene oxide and methyl tert-butyl ether (MTBE). A comparative study between supercritical oxidation and liquid-phase oxidation with air, but without a catalyst, provides thermodynamic and kinetic data on isobutane oxidation.[228] Supercritical conditions provided higher rates and selectivities than in liquid phase, because a liquid phase-like mechanism runs at temperatures beyond liquid-phase conditions. In both, subcritical and supercritical oxidations, conversion and temperature have an adverse effect on selectivity towards tert-butyl hydroperoxide.[229] Attempts with heterogeneous catalysts like $SiO_2/TiO_2$ or Pd/C resulted in a remarkably increased conversion in the supercritical regime, and tert-butyl alcohol was the predominant oxidation product.[230]

Another example is known for the partial oxidation of sc propene with $O_2$ ($T_{crit}$ = 92.4°C, $p_{crit}$ = 46.6 bar, $\rho_{crit}$ = 0.223 g/cm$^3$) at mole fractions of propene from 20–90% over Ag catalysts.[231] In the optimum temperature range of 180–220°C, conversion reaches 3.6% and propylene oxide selectivity is 36%.

One example is reported in which the oxidant $N_2O$ was in a supercritical state ($T_{crit}$ = 36.4°C, $p_{crit}$ = 71.5 bar, $\rho_{crit}$ = 0.452 g/cm$^3$). Its critical data are similar to those for $CO_2$ and $N_2O$ might be an attractive oxidant as $N_2$ is the only by-product. It was possible to oxidise phosphines to phosphine oxide without a catalyst below 100°C. Compounds with other functional groups remain unoxidised while in sc$N_2O$ solution. With a Pt/C catalyst, it became feasible to oxidise secondary alcohols to ketones, e.g. isoborneol to campher with 100% yield.[232]

## 25.5. Summary and Outlook

Both, sc$CO_2$ and sc$H_2O$ are applicable as reaction media in special fields of oxidation. Some of the advantages, postulated from their physico-chemical properties, are truly beneficial for chemistry and, in particular, oxidation. Homogenisation of reaction mixtures can lead to significant enhancement of reaction rates. In the case of heterogeneous catalysis, additional surface effects of a (highly porous) catalyst have to be considered. Depending on the reaction regime, severe limitations in mass transfer may occur and the overall benefit is marginal. This is a minor obstacle in case of sc$CO_2$, however, when working with sc$H_2O$, the catalytic processes are often controlled by the internal diffusion rate. Supercritical fluid application is ruled by some general constraints such as the necessity of a minimum density to maintain the high solvent power, or by the thermal and chemical stability of reactants, catalysts and materials.

An intrinsic drawback of SCF in general is set by the high investment costs for high-pressure equipment and high operational costs, for gas compression in particular. In sc$CO_2$-based processes, temperatures are low and heating costs play a minor role. In many cases, it may also be possible to separate $CO_2$ easily from the reactants. Partial oxidation products such as alcohols, aldehydes or acids are often water soluble, and in a water-based process, additional energy is required for downstream work-up. In SCWO processes for the destruction of organics, the high overall energy demand can be equalised by the formed heat of reaction when a certain feed concentration/heating value is set, otherwise auxiliary fuel is necessary.[233] Autothermal operation for an overall energy balance in the SCWO process should therefore be a prerequisite.

Exemplary calculations for applications in supercritical fluid extraction[234] or enzyme catalysis[2] in $scCO_2$ and partial oxidation in $scH_2O$[235] tried to show that a SCF-based process leads to costs, which are comparable to such processes using conventional solvents. Heterogeneously catalysed hydrogenation has been realised successfully at the industrial scale.[236] Regarding oxidation reactions, the picture is different: compared to conventional technologies, oxidation in SCF is still not competitive. Reported activities and selectivities may be higher in some cases, but the gain does not justify a technology change. This is true for the partial oxidation in $scCO_2$ and deep oxidation in $scH_2O$. Several authors have postulated that the combination of SCWO technology with a heterogeneous catalyst will significantly lower the investment and operating costs, but such a process has not been realised.[237]

Summarising all these issues, SCF technology will obviously not replace established gas-phase oxidation processes to manufacture value-added chemicals, least of all bulk chemicals or commodities. To date, no such process development is known. While $scCO_2$ might be an alternative for liquid-phase reactions, known advances are yet not significant enough to justify a change in technology. Beyond these issues, for SCF processes the same rules are valid as for the implementation of any other technology. The replacement of current technologies is only attractive when the jump in efficiency and return of investment is high enough.

The most advanced development has been made in SCWO technology beyond chemistry and thermodynamics, also covering the identification of suitable materials for construction and innovative reactor design. However, heterogeneously catalysed SCWO processes are not in operation due to the small benefit they offer compared to the homogeneous reaction and insufficient catalyst stability. Wastewater treatment is a market that is driven by legislation rather than by customers and economic benefit is poor. Therefore, the introduction of the expensive SCWO technology will be limited to niches where the costs are ranked lower than efficiency.

## References


1. Suresh, A., Sharma, M. and Sridhar, T. (2000). Engineering aspects of industrial liquid-phase air oxidation of hydrocarbons, *Ind. Eng. Chem. Res.*, **39**, pp. 3958–3997.
2. Jessop, P. and Leitner, W. (eds) (1999). *Chemical Synthesis Using Supercritical Fluids*, Wiley-VCH, Weinheim.
3. Van Konynenburg, P. and Scott, R. (1980). Critical lines and phase equilibria in binary van der waals mixtures, *Philos. Trans. R. Soc. Lond. A*, **298**, pp. 495–540.

4. Kruse, A. and Dinjus, E. (2006). Hot compressed water as reaction medium and reactant: Properties and synthesis reactions, *J. Supercrit. Fluid*, **39**, pp. 362–380.
5. Cochran, H., Cummings, P. and Karabomi, S. (1992). Solvation in supercritical Water, *Fluid Phase Equilibr.*, **71**, pp. 1–16.
6. Hoffmann, M. and Conradi, M. (1997). Are there hydrogen bonds in supercritical water?, *J. Am. Chem. Soc.*, **119**, pp. 3811–3817.
7. Franck, E. (1976). Properties of water, in *High Temperature, High Pressure Electrochemistry in Aqueous Solutions*, D. Jones and R. Staehle (eds); National Association of Corrosion Engineers, Houston, pp. 109.
8. Lesutis, H., Gläser, R. and Liotta, C. (1999). Acid/base-catalyzed ester hydrolysis in near-critical water, *Chem. Commun.*, **20**, pp. 2063–2064.
9. Bröll, D. Kaul, C. Krämer, A., *et al.* (1999). Chemistry in supercritical water, *Angew. Chem., Int. Ed.*, **38**, pp. 2998–3014.
10. Aim, K. and Fermeglia, M. (2003). Solubility of solids and liquids in supercritical fluids, in G. Hefter, R. Tomkins (eds), *The Experimental Determination of Solubilities*, John Wiley & Sons, Chichester.
11. Grunwaldt, J., Wandeler, R. and Baiker, A. (2003). Supercritical fluids in catalysis: Opportunities of in situ spectroscopic studies and monitoring phase behavior, *Catal. Rev.*, **45**, pp. 1–96.
12. Subramaniam, B., Lyon, C. and Arunajatesan, V. (2002). Environmentally benign multiphase catalysis with dense phase carbon dioxide, *Appl. Catal. B: Environ.*, **37**, pp. 279–292.
13. Rayner, C., Oakes, R., Sakakura, T., *et al.* (2005). Supercritical carbon dioxide, in K. Mikami (ed.), *Green Reaction Media in Organic Synthesis*, Blackwell Publishing Ltd, Oxford, pp. 125–180.
14. Leitner, W. (2002). Supercritical carbon dioxide as a green reaction medium for catalysis, *Acc. Chem. Res.*, **35**, pp. 746–756.
15. Beckman, E. (2003). Oxidation reactions in $CO_2$: Academic exercise or future green processes?, *Environ. Sci. Technol.*, **37**, pp. 5289–5296.
16. Seki, T. and Baiker, A. (2009). Catalytic oxidations in dense carbon dioxide, *Chem. Rev.*, **109**, pp. 2409–2454.
17. Tremaine, P., Hill, P., Irish, D., *et al.* (eds) (2000). *Steam, Water and Hydrothermal Systems: Physics and Chemistry Meeting the Needs of Industry*, NRC Press, Ottawa.
18. Savage, P. (1999). Organic chemical reactions in supercritical water, *Chem. Rev.*, **99**, pp. 603–621.
19. Watanabe, M., Sato, T., Inomata, H., *et al.* (2004). Chemical reactions of C1 compounds in near-critical and supercritical water, *Chem. Rev.*, **104**, pp. 5803–5822.
20. Ding, Z., Frisch, M., Li, L., *et al.* (1996). Catalytic oxidation in supercritical water, *Ind. Eng. Chem. Res.*, **35**, pp. 3257–3279.
21. Kruse, A., Vogel, H. (2008). Heterogeneous catalysis in supercritical media. 2. Near-critical and supercritical water, *Chem. Eng. Technol.*, **31**, pp. 1241–1245.
22. Savage, P. (2000). Heterogeneous catalysis in supercritical water, *Catal. Today*, **62**, pp. 167–173.
23. Brunner, G. (2009). Near and supercritical water. Part II: Oxidative processes, *J. Supercrit. Fluid*, **47**, pp. 382–390.
24. Dooley, K. and Knopf, F. (1987). Oxidation catalysis in a supercritical fluid medium, *Ind. Eng. Chem. Res.*, **26**, pp. 1910–1916.
25. Westhaus, U., Dröge, T. and Sass, R. (1999). DETHERM® — a thermophysical property database, *Fluid Phase Equilibr.*, **158**, pp. 429–435.
26. Wittig, R., Lohmann, J. and Gmehling, J. (2003). Vapor-liquid equilibria by UNIFAC group contribution. 6th revision and extension, *Ind. Eng. Chem. Res.*, **42**, pp. 183–188.
27. www.aspentech.com. Accessed on 09.10.2012.

28. Mukhopadhyay, M. and Srinivas, P. (1996). Multicomponent solubilities of reactants and products of cyclohexane oxidation in supercritical carbon dioxide, *Ind. Eng. Chem. Res.*, **35**, pp. 4713–4717.
29. Beier, M., Tsivintzelis, I., Grunwaldt, J., *et al.* (2011). Experimental determination and modeling of the phase behavior for the selective oxidation of benzyl alcohol in supercritical $CO_2$, *Fluid Phase Equilibr.*, **302**, pp. 83–92.
30. Jin, H. and Subramaniam, B. (2003). Exothermic oxidations in supercritical $CO_2$: effects of pressure-tunable heat capacity on adiabatic temperature rise and parametric sensitivity, *Chem. Eng. Sci.*, **58**, pp. 1897–1901.
31. VDI Waermeatlas, Verein deutscher Ingenieure (ed.), VDI-Verlag Düsseldorf, 8. Auflage, 1998, Db3.
32. García-Serna, J., García-Verdugo, E., Hyde, J., *et al.* (2007). Modelling residence time distribution in chemical reactors: A novel generalised n-laminar model: Application to supercritical $CO_2$ and subcritical water tubular reactors, *J. Supercrit. Fluid*, **41**, pp. 82–91.
33. Plugatyr, A. and Svishchev, I. (2008). Residence time distribution measurements and flow modeling in a supercritical water oxidation reactor: Application of transfer function concept, *J. Supercrit. Fluids*, **44**, pp. 31–39.
34. Pang, T., Ye, M., Knopf, F., *et al.* (1991). Catalytic oxidation of model waste aromatic hydrocarbons in a dense fluid, *Chem. Eng. Commun.*, **110**, pp. 85–97.
35. Zhang, X. and Savage, P. (1998). Fast catalytic oxidation of phenol in supercritical water, *Catal. Today*, **40**, pp. 333–342.
36. Aki, S. and Abraham, M. (1999). Catalytic supercritical water oxidation of pyridine: Kinetics and mass transfer, *Chem. Eng. Sci.*, **54**, pp. 3533–3542.
37. Weckhuysen, B. (2003). Determining the active site in a catalytic process: Operando spectroscopy is more than a buzzword, *Phys. Chem. Chem. Phys.*, **5**, pp. 4351–4360.
38. Reichardt, C. (1994). Solvatochromic dyes as solvent polarity indicators, *Chem. Rev.*, **94**, pp. 2319–2358.
39. Ohtaki, H. (2003). Effects of temperature and pressure on hydrogen bonds in water and in formamide, *J. Mol. Liq.*, **103–104**, pp. 3–13.
40. Bremholm, M., Jensen, H., Brummerstedt Iversen, S., *et al.* (2008). Reactor design for in situ X-ray scattering studies of nanoparticle formation in supercritical water syntheses, *J. Supercrit. Fluid*, **44**, pp. 385–390.
41. Bellissent-Funel, M. (2001). Structure of supercritical water, *J. Mol. Liq.*, **90**, pp. 313–322.
42. Kohl, W., Linder, H. and Franck, E. (1991). Raman spectra of water to 400°C and 3000 bar, *Ber. Bunsen. Phys. Chem.*, **95**, pp. 1586–1593.
43. Masten, D., Foy, B., Harradine, D., *et al.* (1993). In situ raman spectroscopy of reactions in supercritical water, *J. Phys. Chem.*, **97**, pp. 8557–8559.
44. Ebukuro, T., Takami, A., Oshima, Y., *et al.* (1999). Raman spectroscopy studies on hydrogen bonding in methanol and methanol/water mixtures under high temperature and pressure, *J. Supercrit. Fluid*, **15**, pp. 73–78.
45. Lalanne, P., Tassaing, T., Danten, Y., *et al.* (2002). Raman and infrared studies of hydrogen-bonding in supercritical ethanol, *J. Mol. Liq.*, **98–99**, pp. 203–214.
46. Hegarty, J., McGarvey, J., Bell, S., *et al.* (1996). Time-resolved resonance raman scattering of triplet state anthracene in supercritical $CO_2$, *J. Phys. Chem.*, **100**, pp. 15704–15707.
47. Kachi, Y., Tsukahara, T., Kayaki, Y., *et al.* (2007). Raman spectral shifts of $CO_2$ as measure of $CO_2$-philicity of solutes in supercritical carbon dioxide, *J. Supercrit. Fluid*, **40**, pp. 20–26.
48. Kazarin, S. (1997). Applications of FTIR spectroscopy to supercritical fluid drying, extraction and impregnation, *Appl. Spectrosc. Rev.*, **32**, pp. 301–348.

49. Poliakoff, M., Howdle, S. and Kazarin, S. (1995). Vibrational spectroscopy in supercritical fluids: From analysis and hydrogen bonding to polymers and synthesis, *Angew. Chem., Int. Ed.*, **34**, pp. 1275–1295.
50. Rice, S. and Croiset, E. (2001). Oxidation of simple alcohols in supercritical water III. Formation of intermediates from ethanol, *Ind. Eng. Chem. Res.*, **40**, pp. 86–93.
51. Burgener, M., Tyszewski, T., Ferri, D., *et al.* (2006). Palladium-catalyzed oxidation of geraniol in dense carbon dioxide, *Appl. Catal. A: Gen.*, **299**, pp. 66–72.
52. Müller, B., Martin, A. and Lücke, B. (2002). Effect of dense $CO_2$ on the removal of strongly adsorbed benzaldehyde from $(VO)_2P_2O_7$ surfaces: a transmission FTIR study, *J. Supercrit. Fluid*, **23,** pp. 243–250.
53. Marcus, Y. (2000). Supercritical water: relationships of certain measured properties to the extent of hydrogen bonding obtained from a semi-empirical model, *Phys. Chem. Chem. Phys.*, **4**, pp. 1465–1472.
54. Schnabel, T., Srivastava, A., Vrabec, J., *et al.* (2007). Hydrogen bonding of methanol in supercritical CO2: Comparison between 1 H NMR spectroscopic data and molecular simulation results, *J. Phys. Chem. B*, **111**, pp. 9871–9878.
55. Hoffmann, M. and Conradi, M. (1998). Are there hydrogen bonds in supercritical methanol and ethanol?, *J. Phys. Chem. B*, **102**, pp. 263–271.
56. Yonker, C. and Linehan, J. (2005). The use of supercritical fluids as solvents for NMR spectroscopy, *Prog. Nucl. Mag. Res. Sp.*, **47**, pp. 95–109.
57. Thurecht, K., Hill, D. and Whittaker, A. (2006). Investigation of spontaneous microemulsion formation in supercritical carbon dioxide using high-pressure NMR, *J. Supercrit. Fluid*, **38**, pp. 111–118.
58. Nagai, Y., Matubayasi, N. and Nakahara, M. (2005). Mechanisms and kinetics of noncatalytic ether reaction in supercritical water. 1. Proton-transferred fragmentation of diethyl ether to acetaldehyde in competition with hydrolysis, *J. Phys. Chem. A*, **109**, pp. 3550–3557.
59. Nagai, Y., Matubayasi, N. and Nakahara, M. (2005). Mechanisms and kinetics of noncatalytic ether reaction in supercritical water. 2. Proton-transferred fragmentation of dimethyl ether to formaldehyde in competition with hydrolysis, *J. Phys. Chem. A*, **109**, pp. 3558–3564.
60. Lamb, W., Hoffman, G. and Jonas, J. (1981). Self-diffusion in compressed supercritical water, *J. Chem. Phys.*, **74**, pp. 6875–6880.
61. Xu, B., Nagashima, K., DeSimone, J., *et al.* (2003). Diffusion of water in liquid and supercritical carbon dioxide: An NMR study, *J. Phys. Chem. A*, **107**, pp. 1–3.
62. Dvoyashkin, M., Valiullin, R., Kärger, J., *et al.* (2007). Direct assessment of transport properties of supercritical fluids confined to nanopores, *J. Am. Chem. Soc.*, **129**, pp. 10344–10345.
63. Lin, R. and Tavlarides, L. (2010). Determination of diffusion coefficients by supercritical fluid chromatography: Effects of mobile phase mean velocity and column orientation, *J. Chromatogr. A*, **1217**, pp. 4454–4462.
64. Caravati, M., Grunwaldt, J. and Baiker, A. (2004). Selective oxidation of benzyl alcohol to benzaldehyde in "supercritical" carbon dioxide, *Catal. Today*, **91–92**, pp. 1–5.
65. Caravati, M., Grunwaldt, J. and Baiker, A. (2007). Comparative in situ XAS investigations during aerobic oxidation of alcohols over ruthenium, platinum and palladium catalysts in supercritical $CO_2$, *Catal. Today*, **126**, pp. 27–36.
66. Grunwaldt, J., Caravati, M. and Baiker, A. (2006). In situ extended X-ray absorption fine structure study during selective alcohol oxidation over $Pd/Al_2O_3$ in supercritical carbon dioxide, *J. Phys. Chem. B*, **110**, pp. 9916–9922.
67. Caravati, M., Meier, D., Grunwaldt, J., *et al.*, (2006). Continuous catalytic oxidation of solid alcohols in supercritical $CO_2$: A parametric and spectroscopic study of the transformation of cinnamyl alcohol over $Pd/Al_2O_3$, *J. Catal.*, **240**, pp. 126–136.

68. Piqueras, C., García-Serna, J. and Cocero, M. (2011). Estimation of lower flammability limits in high-pressure systems. Application to the direct synthesis of hydrogen peroxide using supercritical and near-critical $CO_2$ and air as diluents, *J. Supercrit. Fluid*, **56**, pp. 33–40.
69. Randolph, T., Clark, D., Blanch, H., *et al.* (1988). Enzymatic oxidation of cholesterol aggregates in supercritical carbon dioxide, *Science*, **239**, pp. 387–390.
70. Ciriminna, R., Carraro, M., Campestrini, S., *et al.* (2008). Heterogeneous catalysis for fine chemicals in dense phase carbon dioxide, *Adv. Synth. Catal.*, **350**, pp. 221–226.
71. Loeker, F. and W. Leitner, W. (2000). Steel-promoted oxidation of olefines in supercritical carbon dioxide using dioxygen in the presence of aldehydes, *Chem. Eur. J.*, **6**, pp. 2011–2015.
72. Campestrini, S. and Tonellato, U. (2002). Catalytic olefin epoxidation with $H_2O_2$ in supercritical $CO_2$. Synergic effect by hexafluoroacetone and manganese-porphyrins, *Adv. Synth. Catal.*, **343**, pp. 819–825.
73. Nolen, S., Lu, J., Brown, J., *et al.* (2001). Olefin epoxidations using supercritical carbon dioxide and hydrogen peroxide without added metallic catalysts or peroxy acids, *Ind. Eng. Chem. Res.*, **41**, pp. 316–323.
74. Theyssen, N. and Leitner, W. (2002). Selective oxidation of cyclooctane to cyclootanone with molecular oxygen in the presence of compressed carbon dioxide, *Chem. Commun.*, **5**, pp. 410–411.
75. Bolm, C., Palazzi, C., Franciò, G., *et al.*, (2002). Baeyer-villiger oxidation in compressed $CO_2$, *Chem. Commun.*, **15**, pp. 1588–1589.
76. Birnbaum, E., Le Lacheur, R., Horton, A., *et al.* (1999). Metalloporphyrin-catalyzed homogeneous oxidation in supercritical carbon dioxide, *J. Mol. Catal. A*, **139**, pp. 11–24.
77. Kokubo, Y., Wu, X., Oshima, Y., *et al.* (2004). Aerobic oxidation of cyclohexene catalyzed by Fe(III)(5,10,15,20-tetrakis(pentafluorophenyl)porphyrin)Cl in supercritical $CO_2$, *J. Supercrit. Fluid*, **30**, pp. 225–235.
78. Haas, G. and Kolis, J. (1998). The diastereoselective epoxidation of olefins in supercritical carbon dioxide, *Tetrahedron Lett.*, **39**, pp. 5923–5926.
79. Haas, G. and Kolis, J. (1998). Oxidation of alkenes in supercritical carbon dioxide catalyzed by molybdenum hexacarbonyl, *Organometallics*, **17**, pp. 4454–4460.
80. Kreher, U., Schebesta, S., Walther, D., *et al.* (1998). Übergangsmetall-organoverbindungen in superkritischem kohlendioxid: Löslichkeiten, reaktionen, katalyse, *Z. Anorg. Allg. Chem.*, **624**, pp. 602–612.
81. Wu, X., Oshima, Y. and Koda, S. (1997). Aerobic oxidation of cyclohexane catalyzed by Fe(III)(5,10,15,20)-tetrakis(pentafluorophenyl)porphyrin) Cl in Sub- and Supercritical $CO_2$, *Chem. Lett.*, **26**, pp. 1045–1046.
82. Zhu, J., Robertson, A. and Tsang, S. (2002). Aqueous emulsion containing fluorous cobalt species in supercritical $CO_2$ for catalytic air oxidation of toluene, *Chem. Commun.*, **18**, pp. 2044–2045.
83. Jiang, H., Jia, L. and Li, J. (2000). Wacker reaction in supercritical carbon dioxide, *Green Chem.*, **2**, pp. 161–164.
84. Kerler, B., Martin, A., Pohl, M., *et al.* (2002). $(VO)_2P_2O_7$ catalysed partial oxidation of propane in dense carbon dioxide, *Catal. Lett.*, **78**, pp. 259–265.
85. Kerler, B. and Martin, A. (2001). Partial oxidation of propane using dense carbon dioxide, *Chem. Eng. Technol.*, **24**, pp. 41–44.
86. Kerler, B. and Martin, A. (2000). Partial oxidation of alkanes to oxygenates in supercritical carbon dioxide, *Catal. Today*, **61**, pp. 9–17.
87. Armbruster, U., Martin, A., Smejkal, Q., *et al.* (2004). Heterogeneously catalysed partial oxidation of cyclohexane in supercritical carbon dioxide, *Appl. Catal. A: Gen.*, **265**, pp. 237–246.

88. Hou, Z., Han, B., Gao, L., *et al.* (2002). Selective oxidation of cyclohexane in compressed $CO_2$ and in liquid solvents over MnAPO-5 molecular sieve, *Green Chem.*, **4**, pp. 426–430.
89. Zhang, R., Qin, Z., Dong, M., *et al.* (2005). Selective oxidation of cyclohexane in supercritical carbon dioxide over CoAPO-5 molecular sieves, *Catal. Today*, **110**, pp. 351–356.
90. Olsen, M., Salomão, G., Drago, V., *et al.* (2005). Oxidation of cyclohexane in supercritical carbon dioxide catalyzed by iron tetraphenylporphyrin, *J. Supercrit. Fluid*, **34**, pp. 119–124.
91. Gaffney, A. and Sofranko, J. (1992). Selective oxidation of propylene glycol in supercritical media, *ACS Div. Fuel Chem. Prep.*, **37**, pp. 1273–1279.
92. http://www.chemicals-technology.com/projects/BASF-HPPO/. Accessed on 09.10.2012.
93. Beckman, E. (2003). Production of $H_2O_2$ in $CO_2$ and its use in the direct synthesis of propylene oxide, *Green Chem.*, **5**, pp. 332–336.
94. Jenzer, G., Mallat, T., Maciejewski, M., *et al.* (2001). Continuous epoxidation of propylene with oxygen and hydrogen on a Pd-Pt/TS-1 catalyst, *Appl. Catal. A: Gen.*, **208**, pp. 125–133.
95. Sahle-Demessie, E., Gonzalez, M., Enriquez, J., *et al.* (2000). Selective oxidation in supercritical carbon dioxide using clean oxidants, *Ind. Eng. Chem. Res.*, **39**, pp. 4858–4864.
96. Suppes, G., Occhiogrosso, R. and McHugh, M. (1989). Oxidation of cumene in supercritical reaction media, *Ind. Eng. Chem. Res.*, **28**, pp. 1152–1156.
97. Zhou, L., Erkey, C. and Akgerman, A. (1995). Catalytic oxidation of toluene and tetralin in supercritical carbon dioxide, *AIChE J.*, **41**, pp. 2122–2130.
98. Lee, S. and Hong, I. (2003). Catalytic oxidation of VOC mixtures in supercritical fluid media, *J. Ind. Eng. Chem. (Seoul, Republic of Korea)*, **9**, pp. 590–594.
99. Zhou, L. and Akgerman, A. (1995). Catalytic oxidation of ethanol and acetaldehyde in supercritical carbon dioxide, *Ind. Eng. Chem. Res.*, **34**, pp. 1588–1595.
100. Wang, C. and Willey R. (1999). Fine particle iron oxide based aerogels for the partial oxidation of methanol, *Catal. Today*, **52**, pp. 83–89.
101. Maayan, G., Ganchegui, B., Leitner, W., *et al.* (2006). Selective aerobic oxidation in supercritical carbon dioxide catalyzed by the $H_5PV_2Mo_{10}O_{40}$ polyoxometalate, *Chem. Commun.*, **21**, pp. 2230–2232.
102. Caravati, M., Grunwaldt, J. and Baiker, A. (2006). Solvent-modified supercritical $CO_2$: A beneficial medium for heterogeneously catalyzed oxidation reactions, *Appl. Catal. A: Gen.*, **298**, pp. 50–56.
103. Sato, T., Watanabe, A., Hiyoshi, N., *et al.* (2007). Partial oxidation kinetics of m-hydroxybenzyl alcohol with noble metal catalysts in supercritical carbon dioxide, *J. Supercrit. Fluid*, **43**, pp. 295–302.
104. Wang, X., Kawanami, H., Dapurkar, S., *et al.* (2008). Selective oxidation of alcohols to aldehydes and ketones over $TiO_2$-supported gold nanoparticles in supercritical carbon dioxide with molecular oxygen, *Appl. Catal. A: Gen.*, **349**, pp. 86–90.
105. Campestrini, S., Carraro, M., Ciriminna, R., *et al.* (2005). A mechanistic study on alcohol oxidations with oxygen catalysed by TPAP-doped ormosils in supercritical carbon dioxide, *Adv. Synth. Catal.*, **347**, pp. 825–832.
106. Ciriminna, R., Carraro, M., Campestrini, S., *et al.* (2008). Sol-gel entrapped TPAP: an off-the-shelf catalyst set for the clean oxidation of alcohols, *Curr. Org. Chem.*, **12**, pp. 257–261.
107. Hou, Z., Theyssen, N. and Leitner, W. (2007). Palladium nanoparticles stabilised on PEG-modified silica as catalysts for the aerobic alcohol oxidation in supercritical carbon dioxide, *Green Chem.*, **9**, pp. 127–132.
108. Hou, Z., Theyssen, N., Brinkmann, A., *et al.* (2008). Supported palladium nanoparticles on hybrid mesoporous silica: Structure/activity-relationship in the aerobic alcohol oxidation using supercritical carbon dioxide, *J. Catal.*, **258**, pp. 315–323.
109. Dapurkar, S., Kawanami, H., Chatterjee, M., *et al.* (2011). Selective catalytic oxidation of geraniol to citral with molecular oxygen in supercritical carbon dioxide, *Appl. Catal. A: Gen.*, **394**, pp. 209–214.

110. Jenzer, G., Schneider, M., Wandeler, R., *et al.* (2001). Palladium-catalyzed oxidation of octyl alcohols in "supercritical" carbon dioxide, *J. Catal.*, **199**, pp. 141–148.
111. Hirakawa, T., Whitesell, J. and Fox, M. (2004). Effect of temperature and pressure in the photocatalytic oxidation of n-octanol on partially desilanized hydrophobic $TiO_2$ Suspended in aerated supercritical $CO_2$, *J. Phys. Chem. B*, **108**, pp. 10213–10218.
112. Tsang, S., Zhu, J., Steele, A., *et al.* (2004). Partial aerial oxidation of nonpolar alcohols over Teflon-modified noble metal catalysts in supercritical carbon dioxide, *J. Catal.*, **226**, pp. 435–442.
113. Jenzer, G., Sueur, D., Mallat, T., *et al.* (2000). Partial oxidation of alcohols in supercritical carbon dioxide, *Chem. Commun.*, **22**, pp. 2247–2248.
114. Subramaniam, B. (2010). Gas-expanded liquids for sustainable catalysis and novel materials: Recent advances, *Coordin. Chem. Rev.*, **254**, pp. 1843–1853.
115. Kerler, B., Robinson, R., Borovik, A., *et al.* (2004). Application of $CO_2$-expanded solvents in heterogeneous catalysis: a case study, *Appl. Catal. B: Environ.*, **49**, pp. 91–98.
116. Sharma, S., Kerler, B., Subramaniam, B., *et al.* (2006). Immobilized metal complexes in porous hosts: catalytic oxidation of substituted phenols in $CO_2$ media, *Green Chem.*, **8**, pp. 972–977.
117. Guadagno, T. and Kazarian, S. (2004). High-pressure $CO_2$-expanded solvents: simultaneous measurement of $CO_2$ Sorption and swelling of liquid polymers with in-situ near-IR spectroscopy, *J. Phys. Chem. B*, **108**, pp. 13995–13999.
118. Hou, Z., Theyssen, N., Brinkmann, A., *et al.* (2005). Biphasic aerobic oxidation of alcohols catalyzed by poly(ethylene glycol)-stabilized palladium nanoparticles in supercritical carbon dioxide, *Angew. Chem. Int. Ed.*, **44**, pp. 1346–1349.
119. Wang, X., Yang, H., Feng, B., *et al.* (2009). Functionalized poly(ethylene glycol)-stabilized palladium nanoparticles as an efficient catalyst for aerobic oxidation of alcohols in supercritical carbon dioxide/poly(ethylene glycol) biphasic solvent system, *Catal. Lett.*, **132**, pp. 34–40.
120. Wang, J., Cai, F., Wang, E., *et al.* (2007). Supercritical carbon dioxide and poly(ethylene glycol): an environmentally benign biphasic solvent system for aerobic oxidation of styrene, *Green Chem.*, **9**, pp. 882–887.
121. Desset, S. and Cole-Hamilton, D. (2009). Biphasic catalysis: catalysis in supercritical $CO_2$ and in water, in M. Benaglia (ed.), *Recoverable and Recyclable Catalysts*, John Wiley & Sons, Ltd, Chichester, pp. 199–257.
122. Morgenstern, D., LeLacheur, R., Morita, D., *et al.* (1996). Supercritical carbon dioxide as a substitute solvent for chemical synthesis and catalysis, *ACS Symp. Ser.*, **626**, pp. 132–151.
123. Zhu, J. and Tsang, S. (2003). Micellar catalysis for partial oxidation of toluene to benzoic acid in supercritical $CO_2$: effects of fluorinated surfactants, *Catal. Today*, **81**, pp. 673–679.
124. Tsang, S., Yu, K., Steele, A., *et al.* (2003). Solid supported micellar catalysis: some syntheses and characterisations, *Catal. Today*, **81**, pp. 573–581.
125. Ganchegui, B. and Leitner, W. (2007). Oxybromination of phenol and aniline derivatives in $H_2O/scCO_2$ biphasic media, *Green Chem.*, **9**, pp. 26–29.
126. Kane, M., Baker, G., Pandey, S., *et al.* (2000). Performance of cholesterol oxidase sequestered within reverse micelles formed in supercritical carbon dioxide, *Langmuir*, **16**, pp. 4901–4905.
127. Hou, Z., Han, B., Gao, L., *et al.* (2002). Wacker oxidation of 1-hexene in 1-n-butyl-3-methylimidazolium hexafluorophosphate ([bmim][PF6]), supercritical (SC) $CO_2$, and SC $CO_2$/[bmim][PF6] mixed solvent, *New. J. Chem.*, **26**, pp. 1246–1248.
128. Zhao, G., Jiang, T., Wu, W., *et al.* (2004). Electro-oxidation of Benzyl Alcohol in a Biphasic System Consisting of Supercritical $CO_2$ and Ionic Liquids, *J. Phys. Chem. B*, **108**, pp. 13052–13057.
129. Modell, M. (1982). Processing methods for the oxidation of organics in supercritical water, *US Patent 4338199*.
130. Antal, M., Mok, W., Roy, J., *et al.* (1985). Pyrolytic sources of hydrocarbons from biomass, *J. Anal. Appl. Pyrol.*, **8**, pp. 291–3203.

131. Modell, M., Gaudet, G., Simson, M., *et al.* (1982). Supercritical water: Testing reveals new process holds promise, *Solid Wastes Management*, **25**, p. 26.
132. Zimmermann, F., (1961). Wet air combustion, *Ind. Water and Waste*, **102**.
133. Verinasyah, B. and Kim, J. (2007). Supercritical water oxidation for the destruction of toxic organic wastewaters: A review, J. Environ. Sci., **19**, pp. 513–522.
134. Gloyna, E. and Li, L. (1993). Supercritical water oxidation: An engineering update, *Waste Management*, **13**, pp. 379–394.
135. Oe, T., Suzugaki, H., Naruse, I., *et al.* (2007). Role of methanol in supercritical water oxidation of ammonia, *Ind. Eng. Chem. Res.*, **46**, pp. 3566–3573.
136. Serikawa, R., Usui, T., Nishimura, T., *et al.* (2002). Hydrothermal flames in supercritical water oxidation: investigation in a pilot scale continuous reactor, *Fuel*, **81**, pp. 1147–1159.
137. Hatakeda, K., Ikushima, Y., Sato, O., *et al.* (1999). Supercritical water oxidation of polychlorinated biphenyls using hydrogen peroxide, *Chem. Eng. Sci.*, **54**, pp. 3079–3084.
138. Marrone, P., Cantwell, S. and Dalton, D. (2005). SCWO system designs for waste treatment: Application to chemical weapons destruction, *Ind. Eng. Chem. Res.*, **44**, pp. 9030–9039.
139. Brandvold, T. and Kocal, J. (2007). Process for the direct production of methanol from methane, *US Patent 7288684* (UOP LLC).
140. Jacobson, S. E. (1984). Process for production of olefin oxides and ketones, *US Patent 4483996*, (The Halcon SD Group, Inc).
141. Gaffney, A. and Sofranko, J. (1993). Oxidation of olefin to glycol, *US Patent 5210336*, (ARCO Chemical Technology L.P.).
142. Hancu, E., Beckman, E. and Danciu, T. (2004). Dense phase epoxidation, *US Patent 6710192*, (ARCO Chemical Technology L.P.).
143. Wang, C. and Willey, R. (1998). Catalytic process for making ethers, aldehydes esters and acids from alcohols using a supercritical fluid mobile, *US Patent 5831116* (Northeastern University, Boston, MA).
144. Cochran, R., Miller, J., Beckman, E., *et al.* (2005). Dense phase oxidation of benzene, *US Patent 6936740*, (ARCO Chemical Technology, L.P.).
145. Dakka, J., Goris, H. and Mathys, G. (2010). Oxidation process in the presence of carbon dioxide, *US Patent 7649100*, (ExxonMobil Chemical Patents Inc).
146. Gopalan, S. and Savage, P. (1995). A reaction network model for phenol oxidation in supercritical water: A comprehensive quantitative model, *AIChE J.*, **41**, pp. 1864–1873.
147. Arai, K., Adschiri, T. and Watanabe, M. (2000). Hydrogenation of hydrocarbons through partial oxidation in supercritical water, *Ind. Eng. Chem. Res.*, **39**, pp. 4697–4701.
148. Watanabe, M., Mochiduki, M., Sawamoto, S., *et al.* (2001). Partial oxidation of n-hexadecane and polyethylene in supercritical water, *J. Supercrit. Fluid*, **20**, pp. 257–266.
149. Rogak, S. and Teshima, P. (1999). Deposition of sodium sulfate in a heated flow of supercritical water, *AIChE J.*, **45**, pp. 240–247.
150. Bermejo, M., Fernandez-Polanco, F. and Cocero, M. (2006). Effect of the transpiring wall on the behavior of a supercritical water oxidation reactor: Modeling and experimental results, *Ind. Eng. Chem. Res.*, **45**, pp. 3438–3446,
151. Penninger, J., Kersten, R. and Baur, H. (1999). Reactions of diphenylether in supercritical water — mechanism and kinetics, *J. Supercrit. Fluid*, **16**, pp. 119–132.
152. Penninger, J. (1988). Reactions in di-n-butylphthalate at near-critical temperature and pressure, *Fuel*, **67**, pp. 490–496.
153. Xu, X., DeAlmeida, C. and Antal, M. (1990). Mechanism and kinetics of the acid-catalyzed dehydration of ethanol in supercritical water, *J. Supercrit., Fluid*, **3**, pp. 228–232.
154. Ramayya, S., Brittain, A., DeAlmeida, C., *et al.* (1987). Acid-catalyzed dehydration of alcohols in supercritical water, *Fuel*, **66**, pp. 1364–1371.

155. de Miguel Mercader, M., Groeneveld, M., Kersten, S., *et al.* (2010). Pyrolysis oil upgrading by high pressure thermal treatment, *Fuel*, **29**, pp. 2829–2837.
156. Yuan, P., Cheng, Z., Zhang, X., *et al.* (2005). Catalytic denitrogenation of hydrocarbons through partial oxidation in supercritical water, *Fuel*, **85**, pp. 367–373.
157. Yuan, P., Cheng, Z., Jiang, W., *et al.* (2005). Catalytic desulfurization of residual oil through partial oxidation in supercritical water, *J. Supercrit. Fluid*, **35**, pp. 70–75.
158. Hamley, P., Ilkenhans, T., Webster, J., *et al.* (2002). Selective partial oxidation in supercritical water: the continuous generation of terephthalic acid from para-xylene in high yield, *Green Chem.*, **4**, pp. 235–238.
159. Hoffmann, G. and Mayer, D. Terephthalsäure mit Isophthalsäure, in Ernst Bartholomé (ed) (1972), *Ullmanns Enzyklopädie der Technischen Chemie, 4. Auflage*, VCH, Weinheim, Band 22, p. 519.
160. Kritzer, P. (2004). Corrosion in high-temperature and supercritical water and aqueous solutions: a review, *J. Supercrit. Fluid*, **29**, pp. 1–29.
161. Kritzer, P., Schacht, M. and Dinjus, E. (1999). The corrosion behaviour of nickel-base alloy 625 (NiCr22Mo9Nb; 2.4856) and ceria stabilized tetragonal zirconia polycrystal (Ce-TZP) against oxidizing aqueous solutions of HF, HBr, and HJ at sub- and supercritical temperatures, *Mater. Corros.*, **50**, pp. 505–516.
162. Huang, S., Daehling, K., Carleson, T., *et al.* (eds). (1989). Thermodynamic analysis of corrosion of iron alloys in supercritical water, *ACS Symp. Ser.*, **406**, 276–286.
163. Sun, M., Wu, X., Han, E., *et al.*, (2009). Microstructural characteristics of oxide scales grown on stainless steel exposed to supercritical water, *Scripta Mater.*, **61**, pp. 996–999.
164. Delville, M., Botella, P., Jaszay, T., *et al.* (2002). Electrochemical study of corrosion in aqueous high pressure, high temperature media and measurements of materials corrosion rates: applications to the hydrothermal treatments of organic wastes by SCWO, *J. Supercrit. Fluid*, **26**, pp. 169–179.
165. Son, S., Lee, J., Byeon, S., *et al.* (2008). Surface chemical analysis of corroded alloys in subcritical and supercritical water oxidation of 2-chlorophenol in continuous anticorrosive reactor system, *Ind. Eng. Chem. Res.*, **47**, pp. 2265–2272.
166. Kritzer, P., Boukis, N. and Dinjus, E. (1999). Corrosion of nickel-base alloy 625 in sub- and supercritical aqueous solutions of $HNO_3$ in the presence of oxygen, *J. Mat. Sci. Lett.*, **18**, pp. 771–773.
167. Hayward, T., Svishchev, I. and Makhija, R. (2003). Stainless steel flow reactor for supercritical water oxidation: corrosion tests, *J. Supercrit. Fluid*, **27**, pp. 275–281.
168. Lee, H., Son, S., Hwang, K., *et al.* (2006). Surface chemical analysis on the corrosion of alloys in the supercritical water oxidation of halogenated hydrocarbon, *Ind. Eng. Chem. Res.*, **45**, pp. 3412–3419.
169. Sridharan, K., Harrington, S., Johnson, A., *et al.* (2007). Oxidation of plasma surface modified zirconium alloy in pressurized high temperature water, *Materials & Design*, **28**, pp. 1177–1185.
170. Nie, S., Chen, Y., Ren, X., *et al.* (2010). Corrosion of alumina-forming austenitic steel Fe-20Ni-14Cr-3Al-0.6Nb-0.1Ti in supercritical water, *J. Nucl. Mater.*, **399**, pp. 231–235.
171. Son, S., Lee, J. and Lee, C. (2007). Corrosion phenomena of alloys by subcritical and supercritical water oxidation of 2-chlorophenol, *J. Supercrit. Fluid*, **44**, pp. 370–378.
172. Botella, P., Frayret, J., Jaszay, T., *et al.* (2003). Experimental study, via current-potential curves, of the anodic behavior of Alloy C-276 and T60 titanium in chlorinated and oxygenated aqueous media under sub- to supercritical conditions, *J. Supercrit. Fluid*, **25**, pp. 269–278.
173. Le Clerq, M. (1996). Ceramic reactor for use with corrosive supercritical fluids, *AIChE J.*, **42**, pp. 1798–1799.

174. Boukis, N., Claussen, N., Ebert, K., *et al.* (1997). Corrosion screening tests of high-performance ceramics in supercritical water containing oxygen and hydrochloric acid, *J. Eur. Ceramic Soc.*, **17**, pp. 71–76.
175. Drews, M., Barr, M., Williams, M., *et al.* (2000). The Corrosion of sol-gel coated type 316SS in chlorinated SC water, Proc. 5th Int. Symp. Supercritical Fluids, Atlanta.
176. Korablov, S. and Yoshimura, M. (2003). Hydrothermal corrosion of TiN PVD films on SUS-304, *Corros. Sci.*, **45**, pp. 531–543.
177. Garcia, K., and Mizia, R. (1995). Corrosion investigation of multilayered ceramics and experimental nickel alloys in SCWO process environments, Proc. 1st Int. Workshop on Supercritical Water Oxidation, WCM forums, Amelia Island, Florida.
178. Hong, G., Killilea, W. and Thomason, T. (1989). Method for solids separation in a wet oxidation type process, *US Patent 4822497.*
179. Dell´Orco, P., Li, L. and Gloyna, E. (1993). The separation of particulates from supercritical water oxidation processes, *Sep. Sci. Technol.*, **28**, pp. 625–642.
180. Marrone, P., Hodes, M., Smith, K., *et al.* (2004). Salt precipitation and scale control in supercritical water oxidation. Part B: commercial/full-scale applications, *J. Supercrit. Fluid*, **29**, pp. 289–312.
181. Calzavara, C., Joussot-Dubien, C., Turc, H., *et al.* (2004). A new reactor concept for hydrothermal oxidation, *J. Supercrit. Fluid*, **31**, pp. 195–206.
182. Fauvel, E., Joussot-Dubien, C., Tanneur, V., *et al.* (2005). A porous reactor for supercritical water oxidation: Experimental results on salty compounds and corrosive solvents oxidation, *Ind. Eng. Chem. Res.*, **44**, pp. 8968–8971.
183. Wellig, B., Weber, M., Lieball, K., *et al.* (2009). Hydrothermal methanol diffusion flame as internal heat source in a SCWO reactor, *J. Supercrit. Fluid*, **49**, pp. 59–70.
184. Bond, L., Mills, C., Whiting, P., *et al.* (1997). Apparatus to remove inorganic scale from a supercritical water oxidation reactor, *J. Cleaner Prod.*, **5**, p. 158.
185. Bermejo, M., Rincon, D., Martin, A., *et al.* (2009). Experimental performance and modeling of a new cooled-wall reactor for the supercritical water oxidation, *Ind. Eng. Chem. Res.*, **48**, pp. 6262–6272.
186. Kawasaki, S., Oe, T., Itoh, S., *et al.* (2007). Flow characteristics of aqueous salt solutions for applications in supercritical water oxidation, *J. Supercrit. Fluid*, **42**, pp. 241–254.
187. Tomita, K. and Oshima, Y. (2004). Stability of manganese oxide in catalytic supercritical water oxidation of phenol, *Ind. Eng. Chem. Res.*, **43**, pp. 7740–7743.
188. Aki, S., Ding, Z. and Abraham, M. (1996). Catalytic supercritical water oxidation: stability of $Cr_2O_3$ catalyst, *AIChE J.*, **42**, pp. 1995–2004.
189. Yu, J. and Savage, P. (2000). Phenol oxidation over $CuO/Al_2O_3$ in supercritical water, *Appl. Catal. B: Environ.*, **28**, pp. 275–288.
190. Ding, Z., Li, L., Wade, D., *et al.* (1998). Supercritical water oxidation of ammonia over a $MnO_2/CeO_2$ catalyst, *Ind. Eng. Chem. Res.*, **37**, pp. 1707–1716.
191. Casal, V. and Schmidt, H. (1998). SUWOX — a facility for the destruction of chlorinated hydrocarbons, *J. Supercrit. Fluid*, **13**, pp. 269–276.
192. Martin, A., Armbruster, U., Schneider, M., *et al.* (2002). Structural transformation of an alumina-supported $MnO_2$-CuO oxidation catalyst by hydrothermal impact of sub- and supercritical water, *J. Mater. Chem.*, **12**, pp. 639–645.
193. Elliott, D., Sealock, L. and Baker, E. (1993). Chemical processing in high-pressure aqueous environments. II. development of catalysts for gasification, *Ind. Eng. Chem. Res.*, **32**, pp. 1542–1548.
194. Kaul, C., Exner, H., Vogel, H., *et al.* (1999). Verhalten von anorganischen katalysatormaterialien gegenüber überkritischen wäßrigen lösungen, *Mat.-Wiss. u. Werkstofftech.*, **30**, pp. 326–331.

195. Webley, P., Tester, J. and Holgate, H. (1991). Oxidation kinetics of ammonia and ammonia-methanol-mixtures in supercritical water in the temperature range 530°C–700°C at 246 bar, *Ind. Eng. Chem. Res.*, **30**, pp. 1745–1754.
196. Yu, J. and Savage, P. (1999). Catalytic oxidation of phenol over $MnO_2$ in supercritical water, *Ind. Eng. Chem. Res.*, **38**, pp. 3793–3801.
197. Oshima, Y., Tomita, K. and Koda, S. (1999). Kinetics of the catalytic oxidation of phenol over manganese oxide in supercritical water, *Ind. Eng. Chem. Res.*, **38**, pp. 4183–4188.
198. Yu, J. and Savage, P. (2000). Kinetics of catalytic supercritical water oxidation of phenol over $TiO_2$, *Environ. Sci. Technol.*, **34**, pp. 3191–3198.
199. Yu, J. and Savage, P. (2001). Catalyst activity, stability, and transformations during oxidation in supercritical water, *Appl. Catal. B: Environ.*, **31**, pp. 123–132.
200. Ding, Z., Aki, S. and Abraham, M. (1995). Catalytic supercritical water oxidation: Phenol conversion and product selectivity, *Environ. Sci. Technol.*, **29**, pp. 2748–2753.
201. Krajnc, M. and Levec, J. (1997). Oxidation of phenol over a transition-metal oxide catalyst in supercritical water, *Ind. Eng. Chem. Res.*, **36**, pp. 3439–3445.
202. Krajnc, M. and Levec, J. (1994). Catalytic Oxidation of Toxic Organics in Supercritical Water, *Appl. Catal. B: Environ.*, **3**, p. L101.
203. Nunoura, T., Lee, G., Matsumura, Y., *et al.* (2002). Modeling of supercritical water oxidation of phenol catalyzed by activated carbon, *Chem. Eng. Sci.*, **57**, pp. 3061–3071.
204. Matsumura, Y., Urase, T., Yamamoto, K., *et al.* (2002). Carbon catalyzed supercritical water oxidation of phenol, *J. Supercrit. Fluid*, **22**, pp. 149–156.
205. Nunoura, T., Lee, G., Matsumura, Y., *et al.* (2003). Reaction engineering model for supercritical water oxidation of phenol catalyzed by activated carbon, *Ind. Eng. Chem. Res.*, **42**, pp. 3522–3531.
206. Nunoura, T., Lee, G., Matsumura, Y., *et al.* (2003). Effect of carbonaceous materials on the oxidation of phenol in supercritical water: A preliminary study, *Ind. Eng. Chem. Res.*, **42**, pp. 3718–3720.
207. Katritzky, A., Barcock, R., Siskin, M., *et al.* (1994). Aqueous high-temperature chemistry of carbo- and heterocycles. 23. reactions of pyridine analogs and benzopyrroles in supercritical water at 460°C, *Energy Fuels*, **8**, pp. 990–1001.
208. Crain, N., Tebbal, S., Li, X., *et al.* (1993). Kinetics and reaction pathways of pyridine oxidation in supercritical water, *Ind. Eng. Chem. Res.*, **32**, pp. 2259–2268.
209. Aki, S. and Abraham, M. (1999). Catalytic supercritical water oxidation of pyridine: comparison of catalysts, *Ind. Eng. Chem. Res.*, **38**, pp. 358–367.
210. Angeles-Hernandez, M., Leeke, G. and Santos, R. (2009). Catalytic supercritical water oxidation for the destruction of quinoline over $MnO_2$/CuO mixed catalyst, *Ind. Eng. Chem. Res.*, **48**, pp. 1208–1214.
211. Jin, L., Ding, Z. and Abraham, M. (1992). Catalytic supercritical water oxidation of 1,4-dichlorobenzene, *Chem. Eng. Sci.*, **47**, pp. 2659–2664.
212. Lin, K. and Wang, H. (2000). Supercritical water oxidation of 2-chlorophenol catalyzed by Cu2+ cations and copper oxide clusters, *Environ. Sci. Technol.*, **34**, pp. 4849–4854.
213. Lin, K. and Wang, H. (1999). Shape selectivity of trace by-products for supercritical water oxidation of 2-chlorophenol effected by CuO/ZSM-48, *Appl. Catal. B: Environ.*, **22**, pp. 261–267.
214. Lin, K., Wang, H. and Yang, Y. (1999). Supercritical water oxidation of 2-chlorophenol effected by $Li^+$ and CuO/Zeolites, *Chemosphere*, **39**, pp. 1385–1396.
215. Lin, K. and Wang, H. (2000). Byproduct shape selectivity in supercritical water oxidation of 2-chlorophenol effected by CuO/ZSM-5, *Langmuir*, **16**, pp. 2627–2631.
216. Krajnc, M. and Levec, J. (1997). The role of catalyst in supercritical water oxidation of acetic acid, *Appl. Catal. B: Environ.*, **13**, pp. 93–103.

217. Yu, J. and Savage, P. (2000). Kinetics of $MnO_2$-catalyzed acetic acid oxidation in supercritical water, *Ind. Eng. Chem. Res.*, **39**, pp. 4014–4019.
218. Park, T., Lim, J., Lee, Y., *et al.* (2003). Catalytic supercritical water oxidation of wastewater from terephthalic acid manufacturing process, *J. Supercrit. Fluid*, **26**, pp. 201–213.
219. Calvo, L. and Vallejo, D. (2002). Formation of organic acids during the hydrolysis and oxidation of several wastes in sub- and supercritical water, *Ind. Eng. Chem. Res.*, **41**, pp. 6503–6509.
220. Yuan, P., Cheng, Z., Jiang, W., *et al.* (2005). Catalytic desulfurization of residual oil through partial oxidation in supercritical water, *J. Supercrit. Fluid*, **35**, pp. 70–75.
221. Guo, Y., Wang, S., Gong, Y., *et al.* (2010). Partial oxidation of municipal sludge with activited carbon catalyst in supercritical water, *J. Haz. Mater.*, **180**, pp. 137–144.
222. Dixon, C. and Abraham, M. (1992). Conversion of methane to methanol by catalytic supercritical water oxidation, *J. Supercrit. Fluid*, **5**, pp. 269–273.
223. Aki, S. and Abraham, M. (1994). Catalytic partial oxidation of methane in supercritical water, *J. Supercrit. Fluid*, **7**, pp. 259–263.
224. Armbruster, U., Martin, A. and Krepel, A. (2001). Partial oxidation of propane in sub- and supercritical water, *J. Supercrit. Fluid*, **21**, pp. 233–243.
225. Richter, T. and Vogel, H. (2001). Die partialoxidation von cyclohexan in überkritischem wasser, *Chem. Ing. Tech.*, **73**, pp. 1165–1168.
226. Richter, T. and Vogel, H. (2002). The partial oxidation of cyclohexane in supercritical water, *Chem. Eng. Technol.*, **25**, pp. 265–268.
227. Fraga-Dubreuil, J., Garcia-Serna, J., Garcia-Verdugo, E., *et al.* (2006). The catalytic oxidation of benzoic acid to phenol in high temperature water, *J. Supercrit. Fluids*, **39**, pp. 220–227.
228. Suresh, A. (1998). Isobutane oxidation in the liquid and supercritical phases: comparison of features, *J. Supercrit. Fluid*, **12**, pp. 165–176.
229. Shah, U., Mahajani, S., Sharma, M., *et al.* (2000). Effect of supercritical conditions on the oxidation of isobutane, *Chem. Eng. Sci.*, **55**, pp. 25–35.
230. Fan, L., Nakayama, Y. and Fujimoto, K. (1997). Air oxidation of supercritical phase isobutane to tert-butyl alcohol, *Chem. Commun.*, **13**, pp. 1179–1180.
231. Mueller-Markgraf, W. (1995). Verfahren zur direktoxidation von propylen zu propylenoxid, *German Patent 19529679A1*.
232. Jessop, P. and Poh, S. (2000). Reactions of supercritical nitrous oxide, Proc. 5th Int. Symp. Supercritical Fluids, Atlanta.
233. Kodra, D. and Vemuri, B. (1994). Autothermal oxidation of dilute aqueous wastes under supercritical conditions, *Ind. Eng. Chem. Res.*, **33**, pp. 575–580.
234. Perrut, M. (2000). Supercritical fluid applications: Industrial developments and economic issues, *Ind. Eng. Chem. Res.*, **39**, pp. 4531–4535.
235. Dunn, J. and Savage, P. (2003). Economic and environmental assessment of high-temperature water as a medium for terephthalic acid synthesis, *Green Chem.*, **5**, pp. 649–655.
236. Licence, P., Ke, J., Sokolova, M., *et al.* (2003). Chemical reactions in supercritical carbon dioxide: from laboratory to commercial plant, *Green Chem.*, **5**, pp. 99–104.
237. Aki, S. and Abraham, M. (1998). An economic evaluation of catalytical supercritical water oxidation: comparison with alternative waste treatment technologies, *Environ. Prog.*, **17**, pp. 246–255.

# Chapter 26

## Unconventional Oxidants for Gas-Phase Oxidations

Patricio RUIZ*, Alejandro KARELOVIC* and Vicente CORTÉS CORBERÁN†

Unconventional oxidants, such as nitric oxide ($N_2O$) and carbon dioxide ($CO_2$), are used to circumvent the main selectivity issue in selective oxidation with molecular oxygen (overoxidation of the desired product) because of their lower oxidizing power and the different nature of the oxygen species they generate. After presenting the most relevant properties on these two oxidants, in this chapter we describe recent advances in the use of $N_2O$ as oxidant for different hydrocarbons (propane, propene, isobutene, methanol, methane, aromatics) and on the use of $CO_2$ in the oxidative dehydrogenation (ODH) of light alkanes and ethylbenzene. Besides, $CO_2$ and $N_2O$ are being used as gas promoters in ODH with oxygen. Their role to master *in situ* the dynamic phenomena at the surface of oxides at work, and their oxidation state, is described. Finally, the role of $CO_2$ in ODH (oxidant or shift of dehydrogenation equilibrium) is discussed.

### 26.1. Nitrous oxide ($N_2O$)

Nitrous oxide is a powerful greenhouse gas that can persist for up to 150 years while it is slowly broken down in the stratosphere. Although $N_2O$ only accounts for around 0.03% of total greenhouse gas emissions, it has a 300-fold greater potential for global warming effects, based on its radiative capacity compared with that of carbon dioxide ($CO_2$). More than two-thirds of the emissions of $N_2O$ come from bacterial and fungal denitrification and nitrification processes in soils. This contribution has been exacerbated through the intensification of agriculture and through the application of synthetic nitrogen-based fertilizers. Since 1997, many non-biological emissions, for example, those associated with the transport industry, have been lowered but emissions from agriculture remain essentially unchanged.[1]

The atmospheric concentration of $N_2O$ has been relatively constant for many centuries (~270 ppbv). The present-day $N_2O$ atmospheric concentration is about 310 ppbv, which means a 9% increase from pre-industrial levels (285 ppbv), at an

*Institute of Condensed Matter and Nanosciences (IMCN), Division Molecules, Solids and Reactivity (MOST), Université catholique de Louvain, Croix du Sud 2/17, B-1348 Louvain-la-Neuve, Belgium.
†Institute of Catalysis and Petroleumchemistry (ICP), CSIC, Calle Marie Curie 2, E-28049 Madrid, Spain.

annual growth rate of 0.2–0.3%. $N_2O$ is the major source of NO$x$ in the stratosphere and an important natural regulator of stratospheric ozone. $N_2O$ contributes to the greenhouse effect, a phenomenon caused by the strong absorbance of infrared radiation in the atmosphere. Although $N_2O$ is not the major contributor to global warming ($\sim$6%), it is much more potent than either of the two most common anthropogenic greenhouse gases, $CO_2$ and methane ($CH_4$).

Chemical production and the burning of organic material and fossil fuels are important sources of $N_2O$ emissions, and those which can be reduced in the short term are associated with chemical production and the energy industry. Emissions from the chemical industry mainly apply to adipic acid, nitric acid, caprolactam, glyoxal, and acrylonitrile production plants, and processes using nitric acid as the oxidizing agent or involving ammonia oxidation. Stationary combustion processes of coal (and fossil fuels in general), biomass, and municipal and industrial waste also involve significant $N_2O$ emissions, and quantification is not accurate. Emissions from vehicles are also not accurately measured.[2,3]

Success in the catalytic oxidation of benzene to phenol has proably been a source of encouragement for the application of $N_2O$ as an oxidant for other reactions. It has been suggested that the dissociation of $N_2O$ allows the formation of a particular oxygen species (called $\alpha$-oxygen) which is able to be inserted into the C–H bond, forming a hydroxyl group. This observation has served as input to study other reactions in the presence of $N_2O$ as oxidant.

In the following sections, recent advances concerning the use of $N_2O$ as an oxidant of different hydrocarbons are described. Very early or extremely recent publications have not been considered. The work presented in these publications continues to be valid and can complement interpretations; this is a useful contribution to improving knowledge so new processes can be developed. Only results presented in the literature in the last few years and describing new advances in this field have been considered.

### 26.1.1. *Oxidation of benzene to phenol* (*OBP*)

Phenol is an important raw material for the synthesis of petrochemicals, agrochemicals, and plastics. Examples of the uses of phenol as an intermediate include the production of bisphenol A, phenolic resins, caprolactam, alkyl phenols, aniline, and other useful chemicals. Today, almost 95% of worldwide phenol production is based on the so-called "cumene process" which is a three-step process (the conversion of benzene and propylene to cumene using supported phosphoric acid catalysts, the conversion of cumene to cumene hydroperoxide with air, and the decomposition of hydroperoxide to phenol and acetone with sulfuric acid). The great interest in the oxidation reaction of benzene to phenol is linked to some disadvantages of the cumene process (environmental impact, production of an explosive hydroperoxide, a multistep process with a high capital investment and high acetone production as a by-product).[4]

The formation of phenol from benzene using $N_2O$ as the oxidant on various metal oxides was demonstrated in the early 1980s. The phenol obtained from benzene oxidation, which is incorporated in the adipic acid production process, can be hydrogenated to cyclohexanone. The nitric acid oxidation of cyclohexanol and cyclohexanone forms $N_2O$ which can be recycled, thus closing the $N_2O$ cycle.[5]

Fe-containing ZSM-5 zeolites are among the most active and most studied catalysts for the OBP process. The process is based on two reactions:

$$N_2O = N_2 + (1/2)O_2 \tag{26.1}$$

and

$$C_6H_6 + N_2O = C_6H_5OH + N_2 \tag{26.2}$$

It has been suggested that extra-lattice complexes of bivalent iron stabilized in the micropore space of the Fe-ZSM-5 matrix are the active sites ($\alpha$-site) of this reaction. These sites are inert towards $O_2$, but they react with $N_2O$ to generate an anion radical oxygen species (called $\alpha$-oxygen $= O_\alpha$). This $\alpha$-oxygen would be the active species that performs the oxidation of benzene to phenol. It has a low binding energy to the surface and a very high reactivity, at room and even lower temperatures. This high reactivity strongly differentiates $O_\alpha$ from the remaining surface oxygen. The most frequently proposed mechanisms in direct $N_2O$ decomposition are:

| | | |
|---|---|---|
| (a) $N_2O + (_*) = N_2 + {}_*O$ | (b) $2N_2O + 2(_*) = 2N_2 + 2{}_*O$ | (c) $N_2O + (_*) = N_2 + {}_*O$ |
| $N_2O + {}_*O = N_2 + O_*O$ | $2{}_*O = O_2 + 2(_*)$ | $N_2O + {}_*O = N_2 + {}_*O_2$ |
| $O_*O = O_2 + (_*)$ | | $N_2O + {}_*O_2 = N_2 + {}_*O_3$ |
| | | ${}_*O_3 = {}_*O + O_2$ |

In Mechanism (b) (Langmuir–Hinshelwood), the migration of oxygen from one active site ($_*$) followed by the recombination with another oxidized site is the rate-determining step. This reaction mechanism requires the active participation of at least two iron centers that are not necessarily located in adjacent positions. In Mechanisms (a) and (c) (Eley–Rideal mechanisms), $N_2O$ decomposition and oxygen evolution occur at the same isolated sites ($*$) after successive collisions between $N_2O$ with ($*$) and ${}_*O$ for (a), and with ($*$), ${}_*O$, and ${}_*O_2$ for (c). The above-described mechanisms are all in agreement with transient response experiments in the 773–848 K interval. The global decomposition reaction is limited by the reaction steps leading to the $O_2$ gas phase. From temporal analysis of products (TAP) it was suggested that Mechanism (a) is the most likely to happen. From the point of view of the oxygen species formed on iron active sites ($*$), the three mechanisms are associated with increasingly complex oxygen species: ${}_*O$, $O_*O$, and $.O_2$ for (a); ${}_*O$ for (b); and ${}_*O$, ${}_*O_2$, and ${}_*O_3$ for (c).[6]

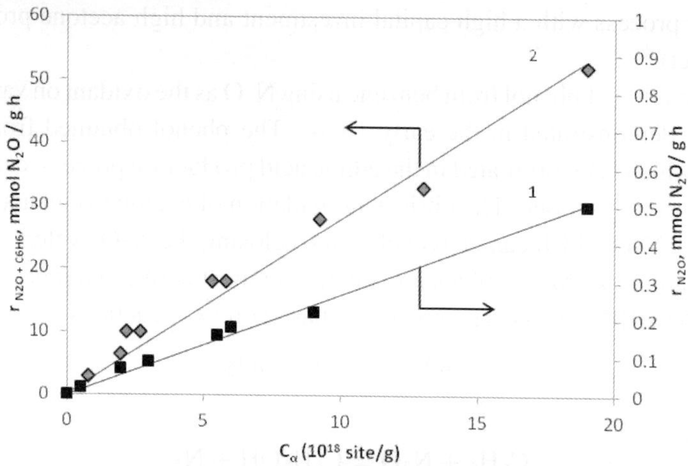


**Figure 26.1.** Rates of $N_2O$ decomposition and benzene oxidation to phenol at 648 K vs $\alpha$-site concentration. Adapted from Ref. 7.

The concentration of $\alpha$-sites can be measured in a static vacuum unit and on the isotopic $^{16}O\alpha/^{18}O_2$ exchange. The rate of OBP is much higher than that of $N_2O$ decomposition, suggesting that benzene accelerates the $N_2O$ decomposition (Fig. 26.1). The oxidation of the $\alpha$-site by $N_2O$ is fast and does not limit the reaction rate. This step proceeds at 473–523 K and has a low activation energy (42–70 kJ/mol). At 698 K, $N_2O$ dissociation with $\alpha$-oxygen should be about 200 times faster than the observed $N_2O$ decomposition. The rate-limiting step is the removal of oxygen. On Fe-ZSM-5 zeolites, the $N_2O$ decomposition proceeds with an activation energy within a range of 200–235 kJ/mol. It seems that both reactions are catalyzed by the same active centers represented by $\alpha$-sites. The activation energy for OBP is about 126 kJ/mol, which is about 90 kJ/mol lower than in the case of $N_2O$ decomposition. The step of $\alpha$-oxygen removal by benzene is much more efficient than that of $N_2O$.

Figure 26.2 presents a general mechanistic scheme for both reactions, which proceed via reversible redox transition $Fe^{II}_{\alpha} = Fe^{III}_{\alpha}$. Bivalent iron is oxidized by $N_2O$. The oxidized site may alternatively react either with another $N_2O$ molecule (decomposition cycle) or with a reducing molecule (reduction cycle). Both routes restore the initial state of the $\alpha$-site.[7]

Using different catalysts in $N_2O$ decomposition (M-zeolites – M-Cu, Co, Fe, etc.) of perovskite-like mixed oxides and supported precious metal (Pd, Rh, etc.), it was concluded that:

(i) Fe-ZSM-5 is always more active than Fe-silicalite that contains the same amount of Fe;

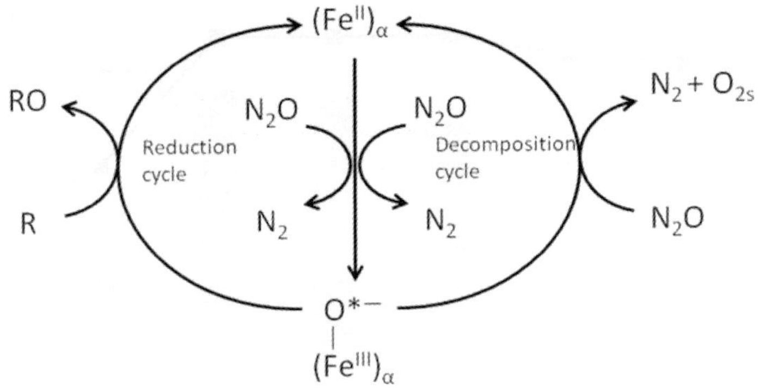


**Figure 26.2.** Catalytic cycles of $N_2O$ decomposition and $N_2O$ reduction on $\alpha$-sites. Adapted from Ref. 7.

(ii) the active sites are formed during activation in an inert atmosphere or *in vacuo*. After this treatment a large fraction of iron is in an extra-framework position and in the divalent state;
(iii) the fraction of active sites present depends upon many experimental factors (impregnation methods, Fe concentration, and $H_2O$ residual pressure);
(iv) the fraction of clusters FexOy and the $Fe^{2+}/Fe^{3+}$ ratio after activation depend on the sample history;
(v) the activity augments with increasing activation temperature in an inert gas;
(vi) at low Fe content, the number of Fe sites where adsorbed oxygen species are formed is roughly identical in Fe-ZSM-5 and Fe-silicalite;
(vii) the activity (calculated per Fe center) increases with the dilution of Fe. The active sites contain a very small number of Fe atoms or, more likely, a single atom. Clustered species, becoming relevant at the highest Fe contents, are characterized by negligible catalytic activity in a $N_2O$ decomposition;
(viii) the number of "$\alpha$-sites" grows with the Fe concentration; and
(ix) all other factors being equal, the number of active sites is at its maximum on catalysts formed by high temperature activation.[6]

The effect of acidity on OBP and the deactivation were studied on ZSM-5 zeolites of different Si/Al ratios (from 14 to 100) and modified with iron cations. Catalysts calcined at 700–900°C showed high oxidative activity for OBP (conversion about 40%) and selectivity towards phenol (about 98%). Samples calcined at 500 and 600°C showed a much lower activity (Fig. 26.3).

The influence of high-thermal treatment could be explained by the enhancement of iron cation mobility, which facilitated the introduction of iron cations into channels resulting in the formation of $\alpha$-sites. The concentration and strength of protonic acidic sites significantly influences the rate of deactivation of Fe-ZSM-5 catalysts.

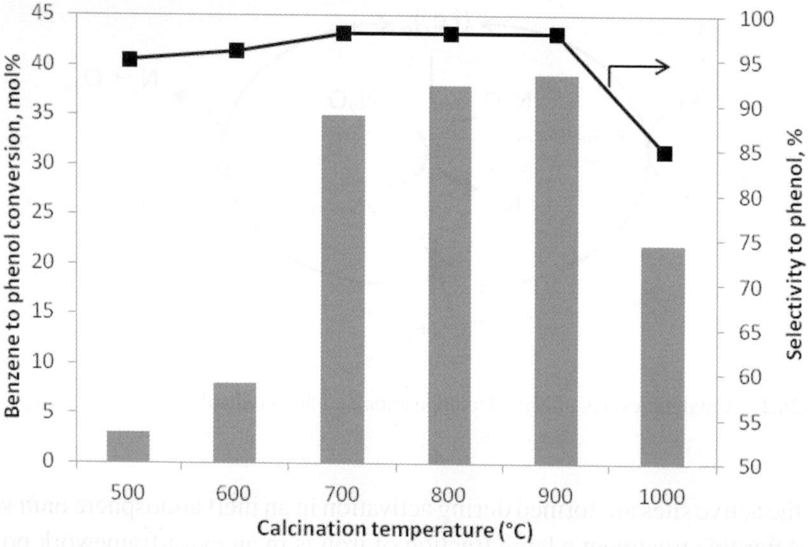


**Figure 26.3.** Benzene conversion and selectivity to phenol over ZSM-5 zeolite modified with iron catalysts and calcined in the temperature range from 500 to 1,000°C for 2 h. Reaction conditions: 450°C; 0.1 g of catalyst; flow rate 42 ml/min; $N_2O$/benzene = 1:1 (activity recorded after 1 h on stream). Adapted from Ref. 8.

The sample calcined at 900°C showed a low number of protonic sites, while in the sample calcined at 550°C, the number of these sites was very high. The strong protonic sites play the predominant role in coke formation. Smaller catalyst crystals led to a higher activity for OBP resulting from less diffusion limitation. Fast diffusion avoids secondary reactions. Coke formation resulting in deactivation of the catalyst was observed to be more severe for the larger crystals. Phenol, thanks to negatively-charged oxygen from the OH group, interacts with acidic protons which results in a strong adsorption, polymerizing in coke deposits.[8]

H-ZSM-5 zeolites were synthesized, with Fe concentrations ranging from a trace level of 30–2,000 ppm and a Si/Al framework composition from 12.5 to 300. The samples were investigated in dehydrated, dehydroxylated, and steamed forms. Together with the changes in concentration of the protonic as well as Al-Lewis sites, changes in the structure and concentration of extra-framework Fe species sites also occurred. The high activity of the H-zeolite can be ascribed to some extra-framework Fe species rather than to the protonic sites. The differences in catalytic activity of the variously treated zeolites should mostly account for the differences in the structure and concentration of Fe species and not for those of the protonic or Al-Lewis sites (Fig. 26.4).[9]

Temperature-programmed desorption (TPD) and Fourier transform infrared spectroscopy (FTIR) have been used to investigate the structure and environment

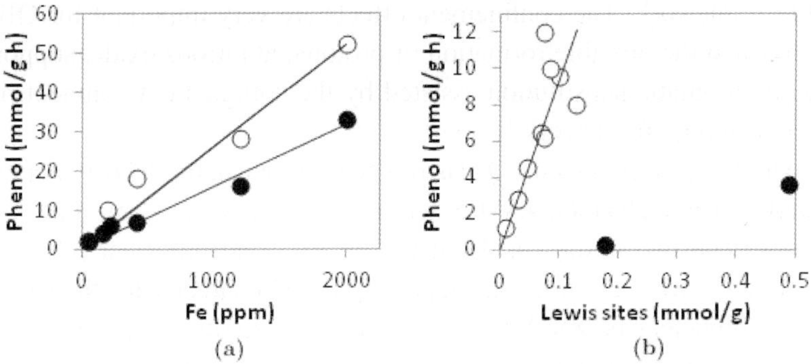


**Figure 26.4.** (a) Rate of phenol formation in benzene oxidation with $N_2O$ on H-ZSM-5 depending on Fe content, (O) dehydrated zeolite, (•) steamed zeolite at 600°C. (b) Rate of phenol formation in benzene oxidation with $N_2O$ on H-ZSM-5 depending on the concentration of Al-Lewis sites on (O) steamed zeolites and on (•) "in situ" dehydroxylated zeolites at 720°C. Adapted from Ref. 9.

of the iron active species of the Fe–MFI catalysts before and after atomic oxygen deposition. $N_2O$ interacts with the Brønsted sites of Fe-ZSM-5 via hydrogen bonding. This type of interaction is nearly absent in Fe-silicalite. Two families of extra-framework iron mononuclear species, FeA and FeB, are present. The difference between these two iron sites is likely due to the number of SiOSi and SiOAl ligands present in the coordination sphere of Fe(II). The FeA site, less coordinated with the MFI zeolitic framework, appears more active and can be associated with the classical $\alpha$-sites. Brønsted sites present in Fe-ZSM-5 interact with $N_2O$ via hydrogen bonding with the formation of stabilized complexes. These complexes are not observed on Fe-silicalite. This difference could partially explain the different activity of the two samples in the $N_2O$ decomposition reaction. Site cooperation between Brønsted and iron-active sites is evidenced in Fe-ZSM-5. The cooperation of Brønsted sites could be associated with their ability to give hydrogen-bonding interactions, a fact that contributes to the increase in concentration of $N_2O$ in the channels of Fe-ZSM-5 with respect to Fe-silicalite. Mononuclear sites characterized by the lowest coordination are the most active in $N_2O$ decomposition. Low or negligible activity is shown by FexOy clusters and $Fe_2O_3$ particles.[6]

The influence of the Brønsted and Lewis acidity on OBP was investigated using a ZSM-5 zeolite. A maximum in catalytic activity and selectivity was reached for steamed samples under mild conditions (30% conversion, 94% selectivity). The Brønsted acid sites play a role of primary importance in OBP as they work in combination with the Lewis acidity. These sites, which are present in the vicinity of the Al framework, are formed during the dealumination of the zeolite. A Langmuir–Hinshelwood mechanism seems to operate. An acid-catalyzed mechanism is proposed, passing through a Wheland-type intermediate stabilized within the zeolite framework. The confinement effects are very important for OBP. DFT study confirmed the possible formation of protonated nitrous oxide, supporting an electrophilic aromatic substitution assisted by the confined environment provided by the active zeolite framework.[10]

The oxidative power toward CO of $\alpha$-oxygen, formed upon $N_2O$ dissociation over an isolated and binuclear Fe-ZSM-5 zeolite, was investigated by means of DFT calculations. The two $\alpha$-sites, isolated $[O–Fe–O]^+$ and binuclear $[O{=}Fe–(\mu O)–(\mu OH)–Fe{=}O]^+$ were considered, suggesting a $Fe^{II}$ character for the isolated and $Fe^{IV}$ for the binuclear Fe-ZSM-5 sites. During the oxidation reaction, the valence state for the isolated iron is II and remains relatively constant while a clear change from IV to II is calculated for the binuclear iron. The reactivity is in line with the high reactivity of $\alpha$-oxygen and the rapid $CO_2$ formation at low temperatures. The reaction would be faster on an isolated $\alpha$-site.[11]

DFT calculations were carried out on a model $(FeO)^{1+}$-ZSM-5 cluster, the $[(SiH_3)_4AlO_4(FeO)]$ cluster, which models the reactivity of $Fe^{3+}$ oxidic clusters. Results were compared to an earlier study on model $Fe^{2+-}$ZSM-5 clusters. The $Fe^{2+}$ site on ZSM-5 is preferred over both $(FeO)^{1+}$ and $Fe^{1+}$ sites for OBP. When $Fe^{1+}$ is the catalytically reactive center, the activation energy of phenol formation is too high, and when $(FeO)^{1+}$ is representative of the reactive center, strongly adsorbed phenolate is the reaction product. The major difference between the two systems appears to be the relative stabilities of the intermediate phenolates. On the $Fe^{3+-}$containing cationic cluster, phenolate appears to be uniquely stable. This result suggests that the experimentally observed preference of $Fe^{2+}$ sites over $(FeO)^{1+}$ on ZSM-5 for OBP is due to the reduced formation of adsorbed phenolate, which is possibly an intermediate for deactivation.[12]

A peculiar behavior of the Fe-ZSM-5 catalyst is that the presence of NO can significantly enhance the catalytic activity of $N_2O$ decomposition, which is opposite to the inhibiting effect of NO observed in the case of noble metal-based catalysts. In adipic and nitric acid plants $N_2O$ is present in a mixture with NO. The NO-assisted $N_2O$ decomposition took place on oligo-nuclear Fe sites by the formation of $NO_2$, which enhances the rate of $O_2$ desorption. The presence of NO could increase the activity and decrease the apparent activation energy of $N_2O$ decomposition. Trace amounts of water vapor result in the hydroxylation of the active Fe sites during $N_2O$ decomposition. The presence of dehydroxylated binuclear Fe sites is directly related to the deactivation of the active sites. The NO treatment of the deactivated Fe- ZSM-5 catalyst could remove the hydroxyl groups by releasing $O_2$ at low temperatures. The promotional effect of NO on $N_2O$ decomposition is explained because NO can catalyze the transformation of the hydroxylated binuclear $Fe^{3+}$ sites into the dehydroxylated binuclear $Fe^{2+}$ sites, which are the active sites for the $N_2O$ decomposition and can promote desorption of $O_2$ during $N_2O$ decomposition.[13]

### 26.1.2. *Propene epoxidation (PEPO)*

The most widely used industrial routes to propylene oxide (PO) are based on the chlorohydrin process or hydroperoxide methods. Much attention has also been directed to processes performed in the presence of hydrogen peroxide in the liquid phase with a TS-1 molecular sieve as the catalyst, iron complexes accommodated in amorphous SBA-15 and MCM-41 modified with alkaline metal salts, and SBA-3 mesoporous molecular sieves doped with transition metal ions (Fe, V, Nb, and Ta).

SBA-3 mesoporous materials modified with vanadium ions show significant activity in PEPO. Among the products, besides PO, propanal and acetone were also observed, whereas the addition of iron ions enhances PEPO performance. The formation of carbonyl compounds resulting from PO isomerization was observed. A PO selectivity of about 20% could be achieved at a propylene conversion of 17% over a mixed Fe/V/SBA-3 catalytic system.[14]

The modification of the FeOx/SBA-15 with alkali metal salts such as KCl shifted the reaction route from allylic oxidation to epoxidation. The increase in the K/Fe ratio increased the $C_3H_6$ conversion (about 7%). The alkali metal takes on following roles: i) increasing the dispersion of FeOx clusters, probably via a surface reaction between FeOx and KCl; ii) decreasing the reactivity of lattice oxygen associated with the FeOx and thus suppressing the allylic oxidation; and iii) eliminating the surface acidity and thus avoiding PO from isomerization.[15]

### 26.1.3. *Propane oxidative dehydrogenation to propylene (ODHP)*

The oxidative dehydrogenation of alkanes to the corresponding alkenes is exothermic and is an alternative to steam cracking, catalytic cracking, and catalytic dehydrogenation, which are all endothermic, require high temperatures, and lead to coke formation.

An attempt to combine the high activity of iron sites in a zeolite matrix with the functionality of the phosphate groups was performed using iron-incorporated aluminophosphate molecular sieves Fe-AlPO4-5 (FAPO). While less effective than steam-activated Fe-ZSM-5, FAPO is found active, selective, and relatively stable in ODHP. Extracting $Fe^{3+}$ species from the framework of FAPO reduces the selectivity to propene. The decrease in the activity is related to the presence of a considerable number of iron clusters and oxides. Carbonates and carboxylates are formed on the surfaces of the FAPO rather than coke. The following elemental steps are proposed to be involved in the reaction:

| | | | |
|---|---|---|---|
| $N_2O + * \rightarrow N_2 + \alpha\text{-}O^*$ | (26.3) | $C_3H_6^* \rightarrow C_3H_6 + {}^*$ | (26.6) |
| $2\,\alpha\text{-}O^* \rightarrow O_2 + 2^*$ | (26.4) | $C_3H_6^* + \alpha\text{-}O^*$ or $N_2O \rightarrow$ carboxylates | (26.7) |
| $C_3H_8 + \alpha\text{-}O* \rightarrow C_3H_6 * + H_2O$ | (26.5) | $C_3H_6^* \rightarrow$ coke | (26.8) |

Step (26.5) is much faster than the recombination and desorption of $\alpha$-oxygen, step (26.4). Two neighboring sites are needed for $O_2$ desorption, while reaction step (26.5) principally requires only one isolated site.[16]

Cr$x$SiBEA zeolites were synthesized by the incorporation of Cr ions into the framework of the zeolite. The activity and selectivity of Cr$x$SiBEA in ODHP depend on the coordination of chromium. The presence of isolated tetrahedral Cr sites with an oxidation state lower than (VI) is essential for maintaining a high selectivity. The propane conversion is about 3%. The selectivity to propene lies between 60 and 80% in the 623–673 K range, and is probably related to the degree of Cr reduction under the reaction conditions and to the nature of surface oxygen formed in the presence of $N_2O$.[17]

The preparation procedure of iron zeolites strongly influences the nature and distribution of iron species in the catalyst and the catalytic performance. Catalysts were prepared by hydrothermal synthesis, liquid-ion exchange, and chemical vapor deposition containing molar Fe/Al ratios in the range of 0.26–1. Activation in steam of the isomorphously substituted iron zeolite leads to superior propylene yields (22–25%) as compared to iron zeolites prepared by liquid-ion exchange and chemical vapor deposition (9–16%), with propylene selectivities around 40%. Iron impurities (170 ppm Fe) in steamed commercial H-ZSM-5 induce relatively low conversions of propene and $N_2O$, but lead to a propylene selectivity of 90%. Mononuclear Fe sites are crucial for ODHP as iron species in large clusters enhance deep oxidation.[18]

Fe-silicalite and Fe-SBA-15 containing almost exclusively isolated $Fe^{3+}$ sites of similar concentration and structure, which are stabilized in markedly different pore geometries, were synthesized. Fe-silicalite was revealed to be much more active than Fe-SBA-15, supporting the fact that the confinement of the iron species in pores of suitable geometry (intersecting channels of *ca.* 0.55 nm diameter) is essential to the creation of high catalytic activity. The large pore zeolites in ordered mesoporous materials apparently do not generate the required intimate contact between potentially active Fe sites and reactant molecules.[19]

### 26.1.4. *Ethane oxidative dehydrogenation to propylene (ODHE)*

The ethane conversion, on a molybdenum-based catalyst (Mo/Si-Ti), varies from 2.4% at 0.46 mg min/cm$^3$ to 4.1% at 1.0 mg min/cm$^3$. Selectivity decreases with increasing contact time. Ethylene selectivities are 75% and above, CO selectivities are 15% and below. $CO_2$ selectivity varies from 2.5% to 3.8% at 2.4% and 4.2% ethane conversion, respectively. The rate of ethylene formation increases with increasing $N_2O$ concentration. The activation energy when using $N_2O$ is 98 kJ/mol, whereas that using $O_2$ is 41 kJ/mol. ODHE performances are attributed to the less oxidized state of the catalyst when using $N_2O$. The catalyst is fully oxidized to

Mo(VI) using $O_2$, whereas the reduction state decreases to an average of +5.8 when using $N_2O$. A more oxidized state would have a greater number of reactive lattice oxygen atoms making ethylene relatively likely to be further oxidized. In the presence of $N_2O$, the ethylene molecule would be near fewer lattice oxygen atoms making ethylene less likely to be oxidized to CO or $CO_2$.[20]

ZSM-5 (of different Si/Al ratios), faujasite Y, and mordenite were used as a matrix for iron(III) ions accommodation using ionic exchange. Amorphous silica, alumina-silica, and silicalite were modified with iron species by the impregnation with an aqueous solution of iron(III) nitrate. Iron-modified zeolites (ZSM-5, mordenite, faujasite) are active for ODHE, while iron compounds impregnated on amorphous silica-alumina or silica did not show any activity. Iron complexes accommodated in channels of ZSM-5 zeolite show the highest activity. Iron complexes generated in the crystalline neutral silicalite of the MFI (mordenite framework inverted) structure and activated under the same conditions as Fe-ZSM-5 did not show any activity. It is suggested that the crystalline structure and the net charge of the zeolite matrix play a significant role in the formation of active iron complexes. Iron-modified ZSM-5 showed high conversion (in the range of 14–28%) and selectivity towards ethene (in the range of 55–87%).[21]

### 26.1.5. *Isobutane oxidative dehydrogenation to isobutene* (*ODHI*)

A highly dispersed surface VOx species, supported over MCM-41, was used in the oxidative dehydrogenation of hydrocarbons, in the presence of $N_2O$ and $O_2$, combining a steady-state catalytic test, steady-state isotopic transient kinetic analysis (SSITKA), and TAP. Corresponding olefins, CO, and $CO_2$ were the main carbon-containing products using $O_2$ and $N_2O$. Minor products were acetaldehyde, acrolein, ethane, and propane. For both oxidants, the initial rate of alkane conversion follows the order: $C_2H_6 < C_3H_8 <$ n-$C_4H_{10} <$ iso-$C_4H_{10}$. The breaking of the weakest C–H bond limits the ODHI reaction independently of the oxidant applied. The oxidant strongly influences the activity and the selectivity. With $N_2O$ the rates of alkane conversion and alkene formation decrease by factor of 3–8 and 1.6–6, while the rates of CO and $CO_2$ formation become 6–30 and 15–70 times lower, respectively. $N_2O$ favors selective olefin production; the alkane activation is the rate-limiting step. $N_2O$ does not decompose over oxidized VOx species in the absence of alkane, while a reversible dissociative $O_2$ adsorption is observed. Reduced VOx species are more reactive in the decomposition of $N_2O$ and the oxygen isotopic exchange than the oxidized ones. The reduced VOx species are re-oxidized by $N_2O$ resulting in gas-phase $N_2$ and restoring the lattice oxygen species. The high activity of reduced VOx species for $N_2O$ decomposition agrees well with DFT calculations on dimeric VOx species over $SiO_2$. Results are consistent with a Mars–van Krevelen mechanism.

This is valid for $O_2$- and $N_2O$-containing feeds. COx formation in the presence of $O_2$ involves oxygen species assumed to be of molecular nature and are not formed from $N_2O$. This may be a reason for the superior performance of $N_2O$ over $O_2$.[22]

### 26.1.6. *Partial oxidation of methanol to produce hydrogen* (*POME*)

Cu/ZnO and Cu/ZnO/$Al_2O_3$ catalysts showed high activity. For catalysts with low copper loading, Activation energy (Ea) equals 482 kJ/mol and Tournover Frequency TOF *ca.* 200 $\text{min}^{-1}$ at 497–499 K, whereas for higher copper contents the Ea and TOF decreased to Ea = 71 kJ/mol and TOF = 160 $\text{min}^{-1}$. Higher selectivities for $H_2$ and $CO_2$ were obtained. The Al is very important for the catalyst stability and longevity. A much higher methanol conversion is obtained, especially at low temperatures, when using $N_2O$ over $Cu_4OZn_{60}$ catalysts. $N_2O$ is a more active oxidizing agent than $O_2$. In contrast with $O_2$, the use of $N_2O$ leads to the formation of large amounts of water and CO. The value of apparent Ea obtained for methanol conversion with $N_2O$ is *ca.* 155 kJ/mol, which is lower than the Ea calculated when $O_2$ is used as the oxidant. The copper metal seems to be active for POME to $H_2$ and $CO_2$.[23]

### 26.1.7. *Methane oxidation to methanol* (*MOMET*)

The following reactions, in the presence of extra-framework oxygen, are involved in MOMET:

$$N_2O + (\,)\alpha = (O)\alpha + N_2 \quad \text{and} \quad (O)\alpha + CH_4 = CH_3OH + (\,)\alpha$$

DFT calculations modeling ZSM-5 as a ($(SiH_3)_4AlO_4M$) (M = Fe, Co) cluster show that in the presence of water, the rate of methanol formation increases. For both clusters, with and without water, the step which has the highest activation barrier is the methanol formation reaction from the hydroxy complex formed on the clusters. The Co-ZSM-5 cluster has a lower activation barrier when compared to that of the Fe-ZSM-5 cluster (49 kcal/mol vs 53 kcal/mol). Water decreases the activation barrier for methanol formation. Activation barrier values decrease to 48 and 39 kcal/mol, for Fe- and Co-ZSM-5 clusters respectively, in the presence of a water molecule adsorbed after the formation of a hydroxyl group.[24]

### 26.1.8. *Oxidative dehydrogenation of ethylbenzene to styrene* (*EBS*)

The EBS process is exothermic so that it can be carried out at a lower temperature compared to that used in the classical ethyl benzene dehydrogenation, and additionally it is not equilibrium-limited. $\gamma$-$Al_2O_3$-supported transition-metal (Fe,Cr and Cu) oxide catalysts led to high conversions of ethyl benzene and $N_2O$. A VOx species was grafted on the SBA-15 surface. Only isolated forms of $V^{5+}$ species

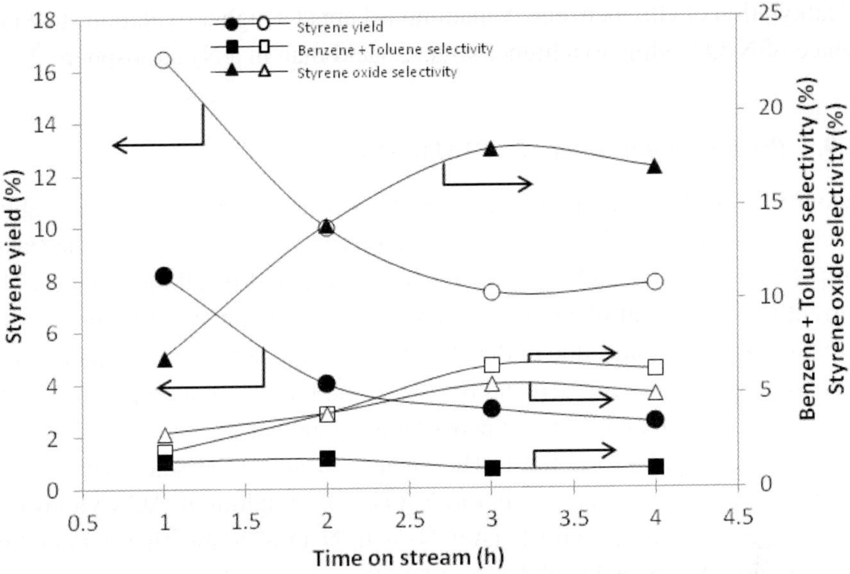


**Figure 26.5.** Yield of styrene and selectivities of benzene, toluene, and styrene oxide in nitrogen and nitrous oxide atmospheres as a function of time on stream. Reaction temperature, 623 K; Weight Hourly Space Velocity (WHSV) of EB, 1; $N_2/N_2O$ flow rate, 1200 $h^{-1}$; 20 VO*x*/$Al_2O_3$. Solid symbols = under $N_2O$ flow; open symbols = under $N_2$ flow. Adapted from Ref. 26.

were observed. The VOx/SBA-15 catalysts were very active in the EBS. Styrene and $CO_2$ were found to be the main carbon containing products formed. Only traces of benzene and toluene as well as CO were observed. The catalyst with a high V loading appeared to be more selective in the total oxidation of aromatics. This effect was attributed to easier reducibility of polymeric $V^{5+}$ species. The highest styrene yield of 38% was achieved at 550°C.[25]

A VO*x*/$\gamma$-$Al_2O_3$ catalyst was prepared by a wet impregnation method. The styrene yield in the presence of $N_2O$ is higher than that in a nitrogen atmosphere (Fig. 26.5).

The yield of styrene decreased with reaction time, however, the yield remained at a much higher level in $N_2O$. A higher selectivity for styrene oxide was observed with $N_2O$. The selectivity for styrene oxide increased with time on stream. The selectivities for benzene and toluene were lower in $N_2O$ than in $N_2$. Dealkylation in the presence of $N_2O$ is higher. The monomeric vanadium (V) species is predominant. With increasing vanadia loading, the formation of polyvanadate species occurs, and at a vanadia loading of 20 wt%, $V_2O_5$ domains are observed along with $AlVO_4$ crystallites. The most active form of the catalyst results when the surface of alumina is covered by two-dimensional polyvanadate species. The monomeric $V^{5+}$ species favors dehydrogenation, whereas the bulk-like $V_2O_5$ preferentially participates in the dealkylation of ethylbenzene. Vanadium is kept at a higher oxidation state in the presence of $N_2O$ leading to a higher styrene yield than in a $N_2$ atmosphere.[26]

### 26.1.9. *Propane ammoxidation* (*PAMMOX*)

Acrylonitrile (ACN) is extensively used for the production of acrylic fibers, rubbers, plastics, and adiponitrile. Today, ACN is produced by a process involving the reaction of propylene, ammonia, and oxygen over complex mixed-metal molybdates. There is currently a great deal of interest in developing catalytic systems for the direct ammoxidation of propane, due to the abundance and lower cost of propane relative to propylene, the increasing demand for ACN, and the risk of a propylene shortage due to increasing demand from the petrochemical industry.

As shown in Table 26.1, the $C_3H_8$ conversion and ACN selectivity over Fe-silicalite at 823 K were rather similar to $N_2O$ or $O_2$, resulting in ACN yields of *ca.* 5%. The selectivity to acetonitrile (AcCN) with $N_2O$ is double that obtained with $O_2$, resulting in AcCN yields of 4.8% and 1.6%, respectively.

A lower COx production is obtained using $N_2O$. The catalyst deactivation is detrimental to the reaction with $N_2O$. Fe-silicalite displayed stable ammoxidation performance with $O_2$ or $N_2O + O_2$ mixtures. Co-feeding of $N_2O$ and $O_2$ induces the best catalytic performance, doubling the conversion of propane as compared to the experiments using the individual oxidants. The selectivity to ACN is very similar in all three cases, while the selectivity to AcCN significantly increased from 10% and 20.5% in $O_2$ and $N_2O$, respectively, to 32% in $N_2O + O_2$. The most striking result for the iron zeolite is the synergetic effect observed in $N_2O + O_2$ mixtures, leading to both increased propane conversions and product yields. To explain this effect, it can be postulated that in a mixture of oxidants, the oxidation state of the active iron sites can be varied to achieve higher ACN and AcCN yields. A similar concept has been proposed to explain the positive effect of $N_2O$ in the ODHP over nickel-molybdate catalysts.[27]

**Table 26.1.** Steady-state performance of steam-activated FeMFI zeolites in propane ammoxidation and comparison to differents catalysts using $O_2$ as the oxidant. Productivity determined as P = Y/[W/Fo($C_3H_8$)]. W/F:($C_3H_8$)/gcat h $mol^{-1}$. Adapted from Ref. 27.

| Catalysts | Oxidant | T/K | W/F | $X(C_3H_8)$ (%) | $S_{ACN}$ (%) | $S_{AcCN}$ (%) | $Y_{ACN}$ (%) | $Y_{AcCN}$ (%) | $P_{ACN}$ | $P_{AcCN}$ |
|---|---|---|---|---|---|---|---|---|---|---|
| Fe-silicalite | $N_2O$ | 823 | 6 | 19 | 27 | 20.5 | 5.1 | 3.9 | 17 | 13 |
| | $O_2$ | 823 | 6 | 16 | 30 | 10 | 4.8 | 1.6 | 16 | 5.3 |
| | $N_2O+O_2$ | 823 | 6 | 42 | 25 | 32 | 10.5 | 13.4 | 35 | 45 |
| | $N_2O+O_2$ | 723 | 6 | 18 | 15 | 30 | 2.7 | 5.4 | 9 | 18 |

### 26.1.10. *Oxidative dehydrogenation of ethanol to acetaldehyde (ODE)*

Acetaldehyde is an important intermediary in organic syntheses and can be obtained as a product of ODE. Carbon-supported copper catalysts, promoted with Pd and Co-Cu/C, Pd-Cu/C, Pd-Co-Cu/C, and Co-Cu/C prepared by impregnation, exhibited significant activity with a selectivity to acetaldehyde near 100% in the range of 150–450°C. The Co-Cu/C and Cu/C catalysts performed the highest conversions of around 80% at 450°C. Acetaldehyde is realized via the formation of surface ethoxide species. $\alpha$-Oxygen plays an important role. The presence of cobalt increased the activity of Co-Cu/C and Pd-Co-Cu/C catalysts. This effect is probably related to the improvement of the redox cycle. A simplified mechanism was proposed for the EOD reaction coupled with the $N_2O$ decomposition reaction. $N_2O$ has an important role in the regeneration of the surface sites and also in the generation of the active $\alpha$-oxygen.[28]

### 26.1.11. *Oxidation of aromatics (OA)*

The catalytic oxidation of toluene with Fe-ZSM-5 catalysts was studied in order to investigate the oxidative competition between aliphatic and aromatic C–H bonds. Fe-ZSM-5 catalysts were prepared by liquid-phase ion exchange, solid-state ion exchange, chemical vapor deposition (CVD), and hydrothermal synthesis. A number of polyaromatic structures were observed, including naphthalene, methylnaphthalene, anthracene, diphenylethane, and large polycyclic structures such as benzanthracene. Xylene isomers were also produced, along with small quantities of phenol and carbon oxides. Benzene was also formed, probably from the cracking of toluene or through a rapid oxidation of the benzylic $CH_3$ structure followed by decarboxylation. The active oxygen species exclusively targets the stronger aromatic hydrogens. The methyl substituent, when attached to a benzene ring, renders it more susceptible to an electrophilic attack at the ortho- and para-positions. The catalytic oxidation of toluene over Fe-ZSM-5 (CVD), yields predominantly para-cresol, with the ortho- and meta- isomers being formed in comparable but smaller quantities, suggesting that the methyl substituent imposes steric restrictions inside the zeolite micropore space, directing hydroxylation towards the more accessible para-position. A rapid deactivation due to coke formation was observed. Aggregate oxide clusters on the surface are not active in the decomposition of $N_2O$. In samples prepared by CVD, the iron species is highly dispersed inside the zeolite channels. The decomposition of $N_2O$ over iron species in extra-framework positions of a MFI zeolite produces a highly electrophilic oxygen species capable of introducing a hydroxyl function into a variety of aromatics substrates. The oxidation of aromatics possessing bulky alkyl groups is strongly influenced by steric restrictions within the micropore space, limiting the yield of hydroxylated product and favoring secondary reactions affecting the side groups. For the oxidation of smaller aromatic substrates, the active oxygen species is influenced by the activating/deactivating nature of the substituent groups.[29]

### 26.1.12. *Use of $N_2O$ as a gas dope in reactions in the presence of oxygen*

Beside the fact that catalytic surfaces often undergo reorganization at the atomic scale when molecules adsorb onto them, the dynamic character of catalysts in oxidation processes mainly comes from their working mechanism, namely the continuous exchange of oxygen atoms between the gas phase and the catalytic surfaces. At the macroscopic level, the dynamic behavior of the oxides materializes as structuration and reconstruction of the catalytic sites which lead to modifications with time-on-stream of the catalysts, of the kinetics, etc. The stabilization of the superficial atoms of the catalytic surface in their most efficient oxidation state can be achieved by tuning the conditions of reaction (partial pressures of oxygen and hydrocarbon) and by adapting the catalyst formulations, e.g. through the addition of elements in tiny quantities. A very promising approach to master the dynamic phenomena *in situ*, at the surface where the oxides are at work, consists of the addition of gaseous "dopes" ("promoters") in the reaction gas feed. The efficiency of the modulation of the oxidation state of the active sites could thus be largely improved thanks to this approach.

Nitrous oxide has been used as a gas dope in the oxidative dehydrogenation of propane (ODP), in the presence of $O_2$, on $NiMoO_4$ (Table 26.2). Comparing it to the test performed in the absence of $N_2O$ (test TR), the addition of 300 ppm of $N_2O$ in the feed (T0.3N2O) induces: i) a decrease in the conversion of propane, ii) an increase in the yield in propylene, iii) an increase in the selectivity of propylene, iv) a decrease in the conversion of oxygen, and v) a decrease in the yield and a weak decrease in the selectivity in $CO_2$.

In the presence of a high amount of $N_2O$, the carbon content increases after the reaction. $Mo^{5+}$ is observed only in the presence of $N_2O$. Under 300 ppm of

**Table 26.2.** Catalytic activity results in the absence or presence of 300 ppm of $N_2O$, on $NiMO_4$. X = conversion of propane; Y = yield in propylene; S = selectivity in propylene. Temperature 723 K. (In parentheses, changes, in %, when compared to test in absence of $N_2O$ (test TR). Adapted from Ref. 30.

| Tests | $N_2O$ | $XC_3$ (%) | $YC_3$ (%) | $SC_3$ (%) | $X\ O_2$ (%) | $Y\ CO_2$ (%) | $S\ CO_2$ (%) |
|---|---|---|---|---|---|---|---|
| TR | 0% | 14.6 | 2.7 | 18.3 | 38.7 | 3.4 | 23.2 |
| T0.3N2O | 300 ppm | 12.4 | 3.0 | 24.2 | 31.5 | 2.7 | 21.4 |
| | | (−15) | (+11) | (+32) | (−19) | (−21) | (−8) |

$N_2O$ in the feed, about 2% of the molybdenum on the surface is present as $Mo^{5+}$ after the catalytic reaction. On the contrary, in the absence of $N_2O$, no reduced molybdenum was observed. When $N_2O$ is added, hydroxyls decrease and the bonds corresponding to M=O increase further. This effect is more marked when the amount of $N_2O$ introduced is higher. In the presence of $O_2$ only, the number of M=O bonds formed increases further. Molybdenum is in a higher oxidation state in the presence of $O_2$. On the contrary, when $N_2O$ is added, Mo is less oxidized. $N_2O$ inhibits the oxidation of molybdenum, promoting its reduction. $N_2O$ could be adsorbed in the same vacancies where $O_2$ adsorption usually occurs. The inhibition in the adsorption of $O_2$ can have as a consequence, to limit (or inhibit) the formation of more electrophilic oxygen species (as $O^-$, $O_2^-$), formed from the dissociation of molecular oxygen. These species could promote the non-selective attack of the hydrocarbons. A higher amount of $N_2O$ deeply reduces the catalysts and promotes the formation of (reduced) sites where carbonaceous products could be formed.[30]

Other results, which complement the information given above, are presented in other references.[31–44]

### 26.1.13. *Final remarks*

The active sites where $O_\alpha$ is formed from $N_2O$ seems to be well identified. This species is highly reactive compared to the rest of the surface oxygen. $O_\alpha$ would be the active species able to be inserted into the C–H bond, forming a hydroxyl group. This rate in OBP is much higher than that of the $N_2O$ decomposition. Both reactions are catalyzed by the same active centers. It could be argued that the exact nature of $O_\alpha$ and the possibility of its migration onto the surface from the sites where it is formed, where it can be incorporated in the benzene molecules, has to be determined more precisely. The role of Brønsted and Lewis acid sites is not definitively established in this reaction and some contradictions remain; the relationship between the nature and number of acid sites, and the structure of the catalytic material would be better understood. In addition, further theoretical approaches seem necessary to advance in this direction.

The general tendency seems to be to apply similar materials and the mechanistic knowledge acquired in OBP with $N_2O$ to activate other hydrocarbons. Advances acquired in the application of $N_2O$ as the oxidant in other reactions are more limited. In some cases, the activity and the yield of the oxygenated products are extremely low. The role of $O_\alpha$ in these reactions, if existing, is not well established or proved. The knowledge in the ODHP process seems to be more advanced than in other processes. Meanwhile, fundamental progress in the understanding of the nature of the active sites as well as in the mechanisms of reaction between $N_2O$ and other hydrocarbons is necessary.

The addition of small amounts of $N_2O$ or NO in the feed seems to be an interesting method to improve selectivity in oxidation. More studies targeted at understanding the mechanism of inhibition of non-selective sites by $N_2O$ and the application to other reactions would be useful for larger applications.

## 26.2. Carbon dioxide ($CO_2$)

Carbon dioxide is a key link in the overall carbon cycle in nature, being the starting material for the photosynthesis of carbon containing compounds and, hence, for most living organisms and fossil fuels. A colorless gas at standard temperature and pressure, its concentration in the Earth's atmosphere is at trace levels.

However, due to the use of fossil fuels, this level has been increasing steadily over the last 150 years, up to around 390 vppm currently. Anthropogenic carbon dioxide annual emissions grew between 1970 and 2004 by about 80%, from 21 to 38 gigatons (Gt), and represented 77% of the total anthropogenic greenhouse gases emissions in 2004.[45] Due to this increase, carbon dioxide has become the most important greenhouse gas, and a global policy has been adapted to limit and reduce its emissions into the atmosphere (the Kyoto Protocol).[46]

All these issues have attracted great interest in the conversion and utilization of $CO_2$. Its possible chemical transformation into other products, reviewed extensively by Arakawa *et al.*,[47] covers a very broad range.

Around 110 megatons (Mt) of $CO_2$ are annually used in commercial synthesis processes, to produce urea, salicylic acid, cyclic carbonates, and polycarbonates. The largest use is for urea production, which reached around 90 Mt/yr in 1997. In addition to these applications, there are a number of promising reactions currently under study in various laboratories, reactions that differ in the extent to which $CO_2$ is reduced during the chemical transformation. They include the synthesis of commodities and intermediates (acetic acid, methanol, carbonates, cyclic carbonates, and lactones), polymers (polyurethanes, polypyrones) and a variety of functionalized carboxylic acids (propenic acid, 3-hexen-1,6-dioic acid). A more detailed description can be found in the cited review.[47]

The production of chemicals from $CO_2$ could have a positive, though very small impact on the global carbon balance, for example, the amount of $CO_2$ generated by a single 500 MW power plant would suffice for current world production of acetic acid. Nevertheless, there are several motivations for such a utilization of $CO_2$:

(i) $CO_2$ is an inexpensive, abundant, and non-toxic feedstock which could replace toxic chemicals;
(ii) it is a renewable feedstock compared to oil or coal;
(iii) its use can lead to new materials and polymers, and new, more efficient routes to existing chemical intermediates;[47]

(iv) there is no loss of latent heat for $CO_2$ because it stays in the gaseous form throughout the reaction; and
(v) $CO_2$ has the highest heat capacity among various typical gases, facilitating heat transfer.

Most of the reactions that use $CO_2$ as a reactant are aimed at the incorporation of a carbonyl group into the product molecule, i.e. they use it as a source of carbon. However, carbon dioxide can also act as an oxygen donor, thus being just an oxidant, it is much milder than molecular oxygen due to its greater chemical stability. This is its role in the reactions of dry reforming, oxidative dehydrogenation (ODH), and partial oxidation. Besides the motivations mentioned above, the use of carbon dioxide as an oxidant also has some advantages. Its use instead of molecular oxygen or air eliminates the risks of flammability of oxygen-fuel reacting mixtures. The reaction is endothermic, thus avoiding the risk of hot spots in the reactor. And the resulting subproduct, carbon monoxide (CO), is a valuable feedstock easily usable with currently available technologies. This allows us to imagine ideal, integrated processes in which 100% of the reactants' carbon content could be converted into useful products.[48] One additional advantage is its ability to oxidize carbon, via the reverse Boudouard reaction ($CO_2 + C \rightleftharpoons 2\,CO$). This can be of relevance in those reactions where the formation of carbonaceous deposits harms the stability of the catalytic performance.

One might argue the drawback of its very low concentration in atmospheric air, but industrial and power plants generate gaseous streams with much higher concentrations[49] (Table 26.3).

Thermodynamics disfavors the dissociation of $CO_2$ to $O_2$ and CO in the gas phase. Under standard conditions, the enthalpy of dissociation is $\Delta H° = +293.0$ kJ/mol. At 427°C, the dissociation constant $K_p$ is only about $10^{-17}$. The literature shows that $CO_2$ could play an oxidant role, but only at high temperatures (usually > 650°C). In fact, the first applications reported on the use of $CO_2$ were those working at very high temperature reactions such as methane oxidative coupling[50] and ethane oxidative dehydrogenation.[51] However, Dury *et al.* reported recently that, in the presence of oxide and noble metal catalysts, $CO_2$ can dissociate and act as an oxidant at much lower temperatures (below 450°C).[52]

The interest in using $CO_2$ for methane oxidative coupling in the 1990s[53] faded away with the rise in the need for hydrogen production for fuel cell applications, and research was diverted to methane dry reforming. As no recent advance on the former has been reported, it will not be further considered.

In the following, the present status of the use of $CO_2$ as an oxidant for different hydrocarbons (alkanes and aromatics), and its use as a gas dope in oxidation reactions in the presence of oxygen, is revised and discussed. Several reviews of these reactions were published prior to the last decade (see below); the more relevant contents will be

**Table 26.3.** Sources and purity of $CO_2$ streams. Adapted from Ref. 49.

| | $CO_2$ emissions, Mt/y |
|---|---|
| Power generation | 7660 |
| Iron and steel production | 1440 |
| Cement production | 1130 |
| Oil refining | 690 |
| Petrochemicals | 520 |
| | $CO_2$ concentration, vol.% |
| Power station flue gas: | |
| Coal fired boiler | 14 |
| Natural gas fired boiler | 8 |
| Natural gas combined cycle | 4 |
| Coal-oxygen combustion | >80 |
| Power station, pre-combustion capture of $CO_2$ | |
| Coal gasification fuel gas | 40 |
| Natural gas partial oxidation fuel gas | 24 |
| Blast furnace gas: | |
| Before combustion | 20 |
| After combustion | 27 |
| Cement kiln off-gas | 14–33 |
| Oil refinery and petrochemical plant fired heaters | 8 |

briefly referred to, and the description of more recent advances will be emphasized. Finally, as $CO_2$ can play different chemical functions in such reactions, its role will be discussed considering the redox properties of the catalyst and the observed kinetics.

### 26.2.1. *Oxidative dehydrogenation of light alkanes*

Anaerobic dehydrogenation of alkanes is an equilibrium-limited process, which endothermically increases with the decrease of the length of the carbon backbone, being maximal for ethane dehydrogenation. To attain economically reasonable conversions, it is operated at high temperatures. This need, together with the further dehydrogenation and oligomerization of the initially formed olefins, causes the loss of reactant by cracking and fast catalyst deactivation by coke. In the last two decades the exothermic ODH with oxygen has been studied as an alternative.[54,55] However, in spite of the huge research effort devoted to the research of ODH of alkanes since then, the problem remains largely unsolved, especially for ethane. This is due to the need to remove heat and to avoid the over-oxidation of olefins to carbon oxides to obtain high olefin selectivity. The use of $CO_2$ as an oxidant helps to tackle both problems. Taking ethane as an example, the thermodynamics of the reactions involved in the overall process are as follows:

$$\text{ODH:} \quad C_2H_6 + CO_2 \rightarrow C_2H_4 + CO + H_2O \quad (\Delta H^{\circ} = +134\,\text{kJ/mol})$$

$$\text{Dehydrogenation (DH):} \quad C_2H_6 \rightarrow C_2H_4 + H_2 \quad (\Delta H^{\circ} = +137\,\text{kJ/mol})$$

$$\text{Dry reforming:} \quad C_2H_6 + 2\,CO_2 \rightarrow 4\,CO + H_2 \quad (\Delta H^{\circ} = +431\,\text{kJ/mol})$$

$$\text{Reverse water-gas shift (RWGS):} \quad H_2 + CO_2 \rightarrow CO + H_2O \quad (\Delta H^{\circ} = +7\,\text{kJ/mol})$$

Similar reactions can describe the process with propane or butanes. Though all these reactions are endothermic, the ODH with $CO_2$ (hereafter $CO_2$-ODH) is the least unfavorable among the reactions of the alkane, ethane.

Krylov's group was the first to report the selective dehydrogenation of ethane by $CO_2$, using mixed Fe-Mn oxide catalysts, at a conference in 1989.[51] The authors later extended the study to the ODH of all the $C_1$-$C_7$ alkanes on several simple transition metal (Fe, Cr, Mn) oxides. It may be underlined that iron oxide and chromium oxide are the topical oxide catalysts for industrial dehydrogenation. The supported manganese oxide catalysts were active, selective, and stable in the conversion of the dry reforming of methane and in the ODH of $C_2$-$C_7$ hydrocarbons and the lower alcohols. Unlike metal catalysts, manganese oxide-based catalysts do not form a carbon layer during the reaction.[56,57] At almost the same time, Wang *et al.* reported the effect of carbon dioxide on the partial oxidation of methane and ethane over $Li^+$/MgO catalysts.[58] Shortly after, Valenzuela *et al.* reported for the first time the ODH of ethane with $CO_2$ on pure ceria[59] and calcium-doped ceria,[60] showing evidence that the reaction with $CO_2$ over ceria-based catalysts is a heterogeneous catalytic reaction. They proposed for the first time that the catalytic reaction is carried out via a redox cycle, where $Ce^{4+}$ is reduced to $Ce^{3+}$ by ethane, producing ethylene, and then $Ce^{3+}$ is oxidized to $Ce^{4+}$ by $CO_2$, producing CO.[60]

To explain the experimental facts (no carbon deposition, CO formation rate always higher than the formation rate of ethylene), the authors suggested that two catalyzed reactions are taking place:

$$CO_2\text{– ODH:} \quad C_2H_6 + CO_2 \rightarrow C_2H_4 + CO + H_2O$$

$$\text{Unselective oxidation:} \quad C_2H_6 + 5CO_2 \rightarrow 7CO + 3H_2O$$

**Table 26.4.** Catalytic performance and reaction conditions of the typical catalysts for the ODH of ethane with $CO_2$. Adapted from ref.[63] and references therein.

| Catalyst | T (°C) | $CO_2/C_2H_6$ ratio | Ethane conversion (mole %) | Ethylene selectivity (mole %) | Ethylene yield (mole %) |
|---|---|---|---|---|---|
| $Ga_2O_3$ | 650 | 5 | 19.6 | 94.5 | 18.6 |
| K-Cr-$MnO_2/SiO_2$ | 850 | 1.5 | 82.6 | 72.8 | 60.1 |
| 6 wt% $Cr_2O_3/SiO_2$-$SO_4^{2-}$ | 650 | 5 | 67.2 | 81.8 | 55.0 |
| 9Fe-9Mn/ Si-2 | 800 | 1 | 68.6 | 92.3 | 63.3 |
| $Na_2WO_4$-Mn/ $SiO_2$ | 800 | 1 | 69.5 | 90.0 | 62.5 |
| CaO-$CeO_2$ | 750 | 2 | 24.2 | 91.0 | 22.0 |
| Cr/H-ZMS-5 (Si/Al = 950) | 600 | 9 | 68.2 | 90.0 | 61.4 |
| Cr-MCM-41 | 700 | 2 | 51.2 | 94.5 | 48.4 |
| Cr/TS-1 (150) | 650 | 4 | 62.2 | 81.0 | 50.4 |

Almost simultaneously, Nakagawa *et al.* reported that gallium oxide ($Ga_2O_3$) is an effective catalyst for this reaction, giving a $C_2H_4$ yield of 18.6% with a selectivity of 94.5% at 650°C.[61]

Most of the further research has been devoted to these catalytic systems, exploring the influence of the synthesis methods, the nature of the support, and its interaction with the active oxidic phase, etc. Publications up to 2003 have been the subject of several reviews,[53,62] from which the reader can get further details.

More recently, Zhang *et al.* summarized the catalytic performance and reaction conditions of topical catalysts for the ODH of ethane with $CO_2$ (Table 26.4).[63] It may be seen that they belong to modifications of the three previously described catalytic systems: transition metal (Mn, Fe, Cr) oxides, ceria-based oxides, and gallium oxide catalysts. The latter have received very little attention, probably due to their high cost. Those based in chromium oxide are the more active at the low temperatures (below 700°C), and hence the most widely studied.

As most of these catalysts have been tested with various $C_2$-$C_4$ alkanes, recent developments will hereafter be grouped by catalytic systems. In general, the main challenges ahead are the need to reduce the temperature of the reaction (increasing the activity) and to increase the stability of the catalytic activity.

#### 26.2.1.1. *Chromium oxide-based catalysts*

Chromium oxide constitutes the active component of the most well-known dehydrogenation catalysts, based on chromia-alumina. Its main drawback for this reaction is the fast deactivation due to coke deposition.

In the case of the ODH with $CO_2$ ($CO_2$-ODH) on chromia catalysts, the goal is to reach higher conversions to be able to reduce the operation temperature. To reach this goal several groups investigated the use of different supports, ranging from conventional oxide supports to mesoporous materials and zeolites.

Using a conventional approach, Ji *et al.* reported the $CO_2$-ODH of ethane over chromia supported on ceria, zirconia, and mixed Zr-Ce oxides.[64] The modification of zirconia with Ce transforms the zirconia phase from a monoclinic to a tetragonal one, which gives a higher specific surface area, a higher number of surface strong basic sites, and improves the on-stream stability of $CrO_x/Ce\text{-}ZrO_2$.

Botavina *et al.*[65] investigated the effect of chromium contents (Cr loadings: from 0.5 to 7.5 wt%) in the $CO_2$-ODH of propane and isobutane over $CrO_x$/silica catalysts prepared by wet impregnation. The catalysts were tested under reaction conditions similar to those required by industrial processes, i.e. looking for high conversions. The introduction of oxygen in the reaction mixture of up to 5.0 vol% resulted in a significant increase of the catalysts' activity and stability, at the expense of some decrease in propene and isobutene selectivity. However, due to the increase of the formation of lighter olefins, the total olefin selectivity increased. The highest activity, selectivity, and stability were observed with a 5.0 wt% Cr loading. Diffuse reflectance (DR) UV-vis studies indicated that Cr(VI) ions in the form of mono-, di-, and polychromates were present in all catalysts, while Cr(III) oxide (alpha-$Cr_2O_3$) was only found for a Cr loading of >3.0 wt%. The catalyst activity seemed to correlate with the more dispersed chromates. Following this observation, a new perspective method of metal vapor synthesis has been investigated quite recently to obtain a high number of dispersed chromium species on the catalyst surface, even at high chromium loadings.[66] In this way, only chromate species were found on the catalyst surface with chromium loadings up to 6.0 wt%; the further increase of the chromium loading results in the appearance of the $\alpha$-$Cr_2O_3$. This catalyst with 6.0 wt% Cr also showed the highest catalytic activity in the $CO_2$-ODH of propane with 69% yield of propene.

Lapidus *et al.*[67] investigated the possibility of increasing the efficiency of silica-supported chromium oxide catalysts in $CO_2$-ODH by the introduction of Li, Na, K, Ca into the catalysts, and the addition of $O_2$ in the reaction mixture. As in the case with the ODH with oxygen on chromia catalysts, potassium has a positive effect,[68] increasing the selectivity to propene and catalyst stability over long duration tests, at a relatively high ratio Cr:K = 20. Co-feeding a small amount of $O_2$ (2%) into the reacting mixture propane-carbon dioxide resulted in the increase of the propene yield and catalyst stability.

The use of ordered mesoporous supports has been investigated in parallel as an alternative strategy to obtain the high dispersion of chromium oxidic species for the ODH with $CO_2$. Bi *et al.* investigated the catalytic behavior for the $CO_2$-ODH of ethane using transition metal-doped M-MCM-41 (M = Ni, Co, Cr) mesoporous materials, prepared by the direct hydrothermal method.[69] Cr-containing catalysts showed the best performance, being active at 450°C and reaching 51.2% conversion with 94.5% ethene selectivity at 700°C. Interestingly, the specific rate of formation of the olefin was higher than that of CO, indicating that besides the ODH reaction some direct dehydrogenation was taking place. At the same conference, Takehira *et al.* reported the $CO_2$-ODH of propane on similarly prepared Cr-MCM-41.[70] The selectivity to olefin was always $> 90\%$, and the propane conversion increased linearly with the Cr content. At 550°C, activity decreased with time-on-stream, but could be recovered not only by oxygen treatment, but also by $CO_2$. The authors concluded that propane is dehydrogenated by $Cr^{6+}O_4$ tetrahedra to form $Cr^{3+}O_8$ octahedra, which in turn was re-oxidized by $CO_2$ to $Cr^{6+}O_4$ tetrahedra.

Similar results have been reported for mesoporous materials MSU-1 and Cr-incorporated in Cr-MSU-1.[71,72] The MSU-x family is a silica-based mesoporous molecular sieve. Its structure does not show long-range ordering, but its high surface area, adjustable uniform pore diameter, and three-dimensional wormlike channels make MSU a promising catalyst support. Supported Cr/MSU-1 prepared by Cr impregnation showed higher activity than Cr-MSU-1 prepared by direct synthesis.[71] The study by x-ray absorption near edge structure (XANES), diffusion reflectance UV-vis, and temperature-programmed reduction with hydrogen ($H_2$-TPR) methods suggested that most Cr species are tetrahedral Cr(VI) in the fresh catalyst. These species were reduced after the reaction. Both a structural deterioration and Cr species reduction lead to the initial decrease of catalyst activity. It is speculated that Cr(VI) species were the active centers at the early reaction stages. With the reduction of the Cr(VI) species, the Cr(III) species transform to generate more stable active centers for the reaction.[72]

Chromium incorporated into a microporous support is also active for the $CO_2$-ODH of ethane.[73] TPR, temperature-programmed oxidation (TPO), FTIR spectroscopy, and X-ray absorption fine-structure (XAFS) were used to analyze the active catalysts, Cr/H-ZSM-5 ($SiO_2/Al_2O_3 > 190$). $Cr^{6+}=O$, or possibly $Cr^{5+}=O$, was the catalytic species on the zeolite support for these catalysts. In contrast, little $Cr^{6+}$ (or $Cr^{5+}$) was detected in the less active catalysts. Recently, Zhao and Wang[74] have compared the performance for the $CO_2$-ODH of ethane using a series of chromium-silicalite-2 molecular sieves (Cr-Si-2), prepared by direct hydrothermal synthesis, with that of chromium oxide supported on the pure silicalite 2 (Cr/Si-2), prepared by conventional impregnation. Results of characterization indicated that monochromate was the dominant chromium species for the group of Cr-Si-2 molecular sieves, and most of the chromium species entered the framework of Cr-Si-2. Samples of equal Cr loading (1.3 wt%) were compared. The direct synthesis sample showed excellent catalytic performance for the $CO_2$-ODH of ethane, giving 45.5% ethylene yield with an 87.9% selectivity at 650°C; the impregnated sample performed less well. Characterization data indicate that the chromium species with a high oxidation state on both samples were partly reduced to aggregated Cr(III) during the ODH, in good agreement with the results of Mimura *et al.*[73] Shi *et al.*[75] investigated the supported Cr/SBA-15 catalysts with different Cr contents prepared by wet impregnation, and the incorporation of cerium by co-impregnation. The activity reached a maximum for a 5 wt% Cr loading, and the performance was improved by the addition of cerium, which increased conversion, decreased the deactivation rate, and favored activity recovery by reoxidation.

Deng *et al.* adopted a different strategy and investigated the effect of the nanosize in chromium oxide.[76,77] Nanosized $Cr_2O_3$ was prepared by the sol-gel method coupled with azeotropic distillation. The obtained nanopowders were characterized by Brunaer-Emmett-Teller specific surface areas (BET), x-ray diffractometry (XRD), transmission electron microscopy (TEM), x-ray photoelectron spectroscopy (XPS), FTIR and $H_2$-TPR techniques. The size distribution of the nanosized oxides, statistically determined by TEM, was narrow, ranging from 20 to 40 nm. The nano-$Cr_2O_3$ catalyst exhibits a much higher ethane conversion and ethylene yield than the normal bulk $Cr_2O_3$ catalyst. At 700°C, the nano-$Cr_2O_3$ catalyst shows 77.1% ethane conversion and 59.0% ethylene yield for the $CO_2$-ODH of ethane.[76] Later, they prepared nanosized composite catalysts $Cr_2O_3/Al_2O_3$, $Cr_2O_3/ZrO_2$, and $Cr_2O_3/MgO$ by co-precipitation coupled with the azeotropic distillation method.[77] The average diameters of these nanosized composite catalysts, calculated using the Scherrer equation, were 8, 12, and 6 nm, respectively, i.e. smaller than that of the bulk nano-$Cr_2O_3$. The catalytic activity of the nanosized composite catalysts varied with the nature of the composites. The $Cr_2O_3/ZrO_2$ catalyst showed the highest ethane conversion and the lowest ethylene selectivity among the nanosized composite catalysts. The ethylene yield over the $Cr_2O_3/Al_2O_3$ catalyst was very low but it exhibited the highest ethylene selectivity. The 10% $Cr_2O_3/MgO$ catalyst exhibited excellent catalytic performance, producing 61.5% ethane conversion and 94.8% ethylene selectivity at 700°C. The authors concluded that the reducibility of chromium and the $Cr^{6+}/Cr^{3+}$ ratio in the nanosized catalysts determine their catalytic performance in the $CO_2$-ODH of ethane. One step further, the $Cr_2O_3/ZrO_2$ (apparently with a 10 wt% Cr) nanocomposite catalysts were modified with Ni, Fe, Co, and Mn oxides, respectively,[78] with 5 wt% of the metal.[79] Each modifier exhibited different effects on catalytic behavior (Table 26.5). The nickel-chromium nanocomposite catalyst mainly favored side reactions (reforming and cracking reactions). But incorporation of Fe, Co, and Mn oxides markedly increased ethylene selectivity. The best performance was observed with the Fe5-Cr10/ZrO2 nanocomposite catalyst, which produced a 50% ethylene yield with 93.17% ethylene selectivity at 650°C.

In a further study,[79] the authors reported the effect of the preparation method (co-precipitation and co-precipitation-impregnation, C-I) on the catalytic performance of Fe-Cr/$ZrO_2$ catalysts. Those catalysts prepared by C-I have higher

**Table 26.5.** Catalytic performance of $Cr_2O_3/ZrO_2$ nanocomposite catalysts. Adapted from Ref. 78.

| Catalyst[a] | T (°C) | Conversion (%) Ethane | Selectivity (%) Ethane | Methane |
|---|---|---|---|---|
| $Cr_2O_3/ZrO_2$ | 600 | 63.4 | 55.3 | 12.1 |
| | 650 | 77.5 | 46.3 | 15.7 |
| $Fe\text{-}Cr_2O_3/ZrO_2$ | 600 | 31.2 | 95.4 | 4.5 |
| | 650 | 53.7 | 93.2 | 6.7 |
| $Ni\text{-}Cr_2O_3/ZrO_2$ | 600 | 95.1 | 3.8 | 15.8 |
| | 650 | 96.3 | 0.0 | 10.7 |
| $Mn\text{-}Cr_2O_3/ZrO_2$ | 600 | 26.4 | 94.1 | 4.7 |
| | 650 | 47.0 | 91.6 | 8.1 |
| $Co\text{-}Cr_2O_3/ZrO_2$ | 600 | 36.3 | 90.3 | 10.7 |
| | 650 | 61.4 | 79.5 | 16.2 |

[a]Reaction conditions: $CO_2$: $C_2H_6$: Ar = 3:1:1; flow rate = 15 ml/min; 0.2 g cat.; P = 0.1MPa; GHSV = 4.5 $Lh^{-1}$ $g^{-1}$.

catalytic stability and higher $CO_2$ conversion, but lower ethylene selectivity than those prepared by co-precipitation. The characterization results indicate that the $Cr^{3+}$ species activate ethane dehydrogenation, and $Fe_3O_4$ is formed during the reaction, which can promote the reverse water-gas-shift (WGS) reaction. The authors propose a complex mechanism, in which chromium activates hydrogen without being reduced, and iron undergoes a redox cycle, as follows. Ethane is activated by the $Cr^{3+}$ species to generate ethylene and H atoms which recombine to form $H_2$ or combine with lattice oxygen to produce $H_2O$, simultaneously reducing $Fe^{3+}$ to $Fe^{2+}$. $CO_2$ dissociates on the active site (denoted by the authors as, but not identified) to produce CO and the active oxygen species ($O^*$), which are absorbed by oxygen vacancies to form adsorbed oxygen species ($O_{ad}$). Then, these adsorbed oxygen species ($O_{ad}$) diffuse into the crystal to create lattice oxygen, which supplements the reduced lattice oxygen used to produce $H_2O$. At the same time, the active oxygen species ($O^*$) reoxidize $Fe^{2+}$ to $Fe^{3+}$, thus completing the redox cycle.

Besides its incorporation into chromia-based catalysts, few recent reports are concerned with iron or manganese oxide catalysts for $CO_2$-ODH. Jin *et al.* reported recently the first example of using cryptomelane-type manganese oxide octahedral molecular sieves (OMS-2) as catalysts for ODH with $CO_2$.[80] The OMS-2 structure consists of one-dimensional tunnels built by 2x2 edge- and corner-sharing $MnO_6$ octahedral chains, forming infinite 3D frameworks with molecule-sized (0.46 nm × 0.46 nm) tunnels. Operating with a short contact time (0.6 s), OMS-2 gave outstanding ethane conversions (70%) and high $C_2H_4$ selectivities (88%) at high temperatures (800°C), whereas the $Cr_2O_3$ (5 wt%) loaded ZSM-5 catalyst only gave 44% selectivity to ethylene. The catalytic performance was stable for 24 h on-line at

T= 800°C. However, the structure of the catalyst was changed to manganosite-type manganese oxide (MnO) as shown by XRD patterns of used samples, which may be due to the lack of oxygen during the reaction. Regeneration of the used catalyst (after 24 h reaction) in oxygen flow at 800°C for 2 h restored OMS-2 structure and the catalytic activity of the regenerated catalyst remained, indicating that the active catalyst is an oxygen-deficient OMS-2 material. In an effort to improve the yield of ethylene, Fe and Mo ions were loaded on the OMS-2 by an incipient wetness impregnation method. The Mo and Fe ion-loaded materials enhanced the conversion of ethane (> 86%) and $C_2H_4$ yield (66%) under the same conditions; the authors claimed that this was the highest ethylene yield reported in the literature for the $CO_2$-ODH of ethane. However, one must be cautious on this point as no data on the contribution of homogeneous gas-phase reactions, really important at such high temperatures,[59,60] is provided and it may be the biggest contribution to the ethylene yield.

#### 26.2.1.2. *Iron oxide and manganese oxide catalysts*

Recently, a new strategy to convert ethane and $CO_2$ to synthesis gas and ethylene in desired relative molar ratios ($C_2H_4/CO/H_2$ = 1/1/1), which can be used directly as a feedstock for hydroformylation to propanal, has been reported.[81] The complex $Na_2WO_4/Mn/SiO_2$ catalytic system has been studied because of its efficiency in giving ethylene by the oxidative coupling of methane. Cobalt-based catalysts are reactive for the partial oxidation of low hydrocarbons to synthesis gas, especially for CO-rich synthesis gas production. Thus, the performance of $Na_2WO_4/Mn/SiO_2$ was modified by the incorporation of cobalt. The goal is a catalyst able to give ethylene by the $CO_2$-ODH of ethane, to produce synthesis gas and ethylene simultaneously. The Co species were reduced into metal Co, which might act as the active phase for CO production. Over Co-promoted $Na_2WO_4/Mn/SiO_2$ catalysts and under the selected conditions ($C_2H_6/CO_2$ = 1/5, F = 60 ml/min, 750°C, 0.3 g catalyst), products with a molar ratio of $C_2H_6/CO/H_2$ = 1/1/1 could be obtained and used directly in the conversion of hydroformylation to propanal.

#### 26.2.1.3. *Ceria-based catalysts*

Ethane conversion, ethylene yield, and selectivity in the $CO_2$-ODH of ethane over Ca-$CeO_2$ catalysts were improved by Valenzuela *et al.*[82] thanks to the synthesis of high specific surface area catalysts by the freeze-drying method. This allowed them to reach a 22% yield with 91% olefin selectivity on nanosized Ca-doped $CeO_2$. However, this catalyst deactivated quickly, and the rate of formation of CO decreased faster than the formation of ethylene. Guío *et al.*[83] studied the kinetics of the catalyzed reactions and its deactivation, by developing an experimental methodology that allows the uncoupling of the catalyst deactivation and catalytic reaction, and simultaneously obtains the kinetic parameters of both processes (i.e. steady-state and deactivation rates, and their apparent activation energies). The deactivation rate of CO formation is one order of magnitude faster than that of ethene formation but both processes show the same apparent activation energy, *ca.* 47 kcal/mol. The apparent activation energy values for the catalytic reactions are $32 \pm 4$ and $26 \pm 2$ kcal/mol for the rates of formation of ethene and CO, respectively.

The efficiency of Ca-doped ceria was attributed to the effect of the solid solution of calcium in the framework of the ceria, which caused an increase in oxygen mobility in the system, as compared to that in the pure stoichiometric ceria.[60] Similar effects can be obtained by ceria modification with zirconium atoms. Based on this rationale, Navas *et al.* investigated nanostructured ceria-zirconia catalysts of variable composition, $Ce_xZr_{1-x}O_2$ ($0 \leq x \leq 1$), and their catalytic performance in the ODH of ethane with $CO_2$.[84] As was observed with Ca doping, the modification of ceria with increasing amounts of zirconium reduces the global activity but increases the selectivity to ethylene. Regardless of the Ce content, all the oxides calcined at 400°C show a nanometric size of 20–40 nm. However, when calcined at 1,000°C, the grain size of pure $CeO_2$ increased to 150–300 nm, while those of Zr-containing samples increased much less, not exceeding 100 nm. In all tests, the formation rate of ethylene was higher than that of CO, but the difference was much reduced over high Zr content samples. The best results were obtained at 750°C with a catalyst composition of $Ce_{0.2}Zr_{0.8}O_2$, reaching 24.4% yield of ethylene, with 90% selectivity. A later work explored if both positive effects (optimal content of cerium and of Ca doping) could be combined to improve the $CO_2$-ODH performance.[85] The incorporation of Ca in the $ZrO_2$-$CeO_2$ network does not modify the structure of $ZrO_2$-$CeO_2$ but improves the performance, showing a maximum at 10 mole % Ca. The addition of 5–10% Ca increased the formation rate by 30% and selectivity to $C_2H_4$ from 70 to 80%.

Shi *et al.* investigated active Ce-based monolithic catalysts, prepared by depositing Ce/SBA-15 samples with different Ce content onto FeCrAl alloy metallic monoliths covered with alumina.[86] They report ethane conversions at 750°C between 54–64% for Ce loadings between 5–12.5 wt% (relative to SBA-15). However, one must be cautious with these results, as the conversions in the presence of an inert (argon) gas instead of $CO_2$ range between 40–44%. This implies that most of the olefin produced comes from the homogeneous gas-phase dehydrogenation reaction, which may be favored by the high void volume ratio of the monolithic support. If one substracts this contribution, as proposed by Valenzuela *et al.*,[60] the catalytic conversion is 14.4 to 20%, not higher than that reported for freeze-drying synthesized CaO-$CeO_2$ catalysts at the same temperature: 21.5% ethane conversion with 91%

selectivity to ethylene.[82] The authors propose a redox mechanism for the reaction, basically identical to that proposed by Valenzuela *et al.*[60]

### 26.2.1.4. *Other catalytic systems*

Co-$BaCO_3$ catalysts exhibit high catalytic performance for the $CO_2$-ODH of ethane.[63] The maximal formation rate of ethylene was 0.264 mmol.min$^{-1}$.g cat.$^{-1}$ (48.0% $C_2H_6$ conversion, 92.2% selectivity, 44.3% $C_2H_4$ yield) on a 7 wt% Co-$BaCO_3$ catalyst at 650°C and 6 L (g cat)$^{-1}$h$^{-1}$.

Ogonowski and Skrzyńska have explored the use of $CO_2$ as an oxidant for the dehydrogenation of isobutane to isobutene over the best-known catalysts for alkanes ODH with oxygen, the V-Mg-O system.[55] The V-Mg-O catalysts were prepared with 20 wt% vanadia, by four different preparation methods: the citrate method, co-precipitation, impregnation, and solid-state reaction. Among them, only those prepared by the citrate method (Cit-VMgO) and co-precipitation were active and selective in the presence of $CO_2$. Magnesium orthovanadate appears to be the most active phase under a $CO_2$ atmosphere.[87] The effects of reaction time and temperature on the conversion, and the role of $CO_2$ in the process, were studied with the Cit-VMgO sample.[88] A decrease of the isobutane conversion was accompanied by the deposition of coke. Active phases of vanadium in the dehydrogenation reaction were believed to be $V^{5+}$ species. Pure $CO_2$ shows only a small ability to oxidize the vanadium species of a lower oxidation state, in agreement with the results of thermodynamic calculations.[89] The authors conducted a thorough thermodynamic analysis of the dehydrogenation of $C_2$–$C_4$ hydrocarbons in the presence of $CO_2$ and compared them with experimental results in an inert gas atmosphere.[90] They concluded that the presence of $CO_2$ enhances the equilibrium conversion of all considered hydrocarbons, and that the physico-chemical properties of the catalyst surface greatly influence its specific activity. Increasing the surface acidity decreases ethane conversion in the presence of $CO_2$, as it requires basic sites for activation. On the other hand, higher hydrocarbons (i.e. propane, *n*-butane and isobutane) are activated on the acid sites, but too strong an interaction with the catalyst surface can promote some undesirable reactions, such as the formation of coke and isomerization.

Vanadium has also been explored as an additive to chromium oxide-based catalysts. V-Cr/SBA-15 catalysts with different V and Cr contents, prepared by the incipient wetness impregnation method, have been explored for the $CO_2$-ODH of propane.[91] The V and Cr bicomponent catalysts exhibited a better performance than those of the monocomponent catalysts. These results are explained as related to the strong interaction between the vanadium oxide and the chromium oxide in the V-Cr/SBA-15 catalysts, shown by TPR results, which remarkably changed the redox properties of the catalysts.

### 26.2.2. $CO_2$ as gas promoter in ODH with oxygen

The objective of using gas promoters in the feed has been discussed above. The influence of the addition of $CO_2$ into the feed has been investigated in the ODH of isobutane over a LaBaSm oxide catalyst in the temperature range 450–600°C, using the feeds $i{-}C_4H_{10}$: $O_2$: He : $CO_2$ = 10 : 5 : $(15-x)$ : $x$ ($x$ = 0, 5, 10, or 15).[92] The catalyst was prepared by the ceramic method. In the absence of $CO_2$ in the feed, the apparent activation energy ($E_A$) for ODH was $20 \pm 1$ kcal/mol (much lower than for anaerobic dehydrogenation, $32 \pm 3$ kcal/mol). When minor amounts of $CO_2$ are co-fed ($x$ = 5, 10), the apparent activation energy decreased to around $13 \pm 3$ kcal/mol, isobutene conversion increased in parallel to the increase of $CO_2$ concentration, selectivity to isobutene increased slightly, and selectivity to propene increased markedly, at the expense of the decreased selectivity to carbon oxides. As a result, an overall selectivity to $C_3$–$C_4$ olefins of between 55 and 80%, with combined yields up to 22% were obtained, values which might be of practical interest as an alternative to mild pyrolysis to obtain light olefins from isobutane.[92] On the contrary, for the highest $CO_2$ content ($x$ = 15), i.e. 50% mole $CO_2$ in the feed, isobutane conversion was lower and isobutene selectivity at (low) isoconversion was much higher (up to 80% vs 30%) than in the absence of $CO_2$, while $E_A$ increased to $26 \pm 1$ kcal/mol, a value much higher than that of ODH and close to that of pure dehydrogenation.

Thus, the influence of $CO_2$ can be attributed to a double effect: on one hand, the faster formation of the more active oxycarbonate phase ($La_2O(CO_3)_2$), predominant in the lower $CO_2$ content; and on the other, the competitive adsorption of molecular $O_2$ and $CO_2$ on the sites which are responsible for total oxidation, predominant in the highest $CO_2$ content.[92]

The influence of the addition of $CO_2$ into the feed has been investigated using $NiMoO_4$ catalysts in the ODH of propane.[52] With respect to the results obtained in the absence of $CO_2$, the main consequences of the addition of $CO_2$ in the gas feed were: i) an increase in the conversion of propane, ii) a decrease in the yield and selectivity of propylene, and iii) an increase in the $CO_2$ yield (Fig. 26.6). These results indicate unambiguously that $CO_2$ can modify the nature of the active and selective sites "during" the selective oxidation reactions. When only a small amount of oxygen (0.5 vol%) is used and no $CO_2$ is added into the feed (Test 0.5O2), $NiMoO_4$ exhibits an extremely high conversion of propane (about 80%) during the first hour of reaction. However, propene is not formed at all and only a small amount of $CO_2$ is produced. After 1 h of reaction, the catalyst deactivates suddenly and its conversion of propane drops completely, to zero. When the catalytic reaction is carried out with a small amount of oxygen, but contains added $CO_2$ (10 vol%) in the feed (Test 0.5O2 + 10CO2), the conversion of propane is low (about 2%),

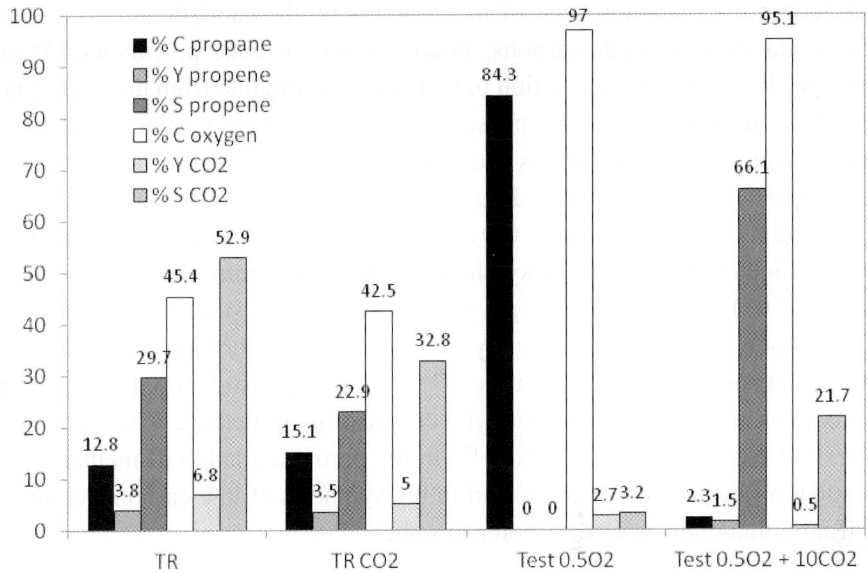


**Figure 26.6.** Conversion of oxygen (%C oxygen), yield of $CO_2$ (%Y $CO_2$) and selectivity (%S $CO_2$) in $CO_2$ obtained with 250 mg of $NiMoO_4$ at 450°C. TR = test in the absence of $CO_2$; $TRCO_2$ = test in the presence of $CO_2$; Test 0.5O2 = only with $O_2$ (0.5 vol.%) but in the absence of $CO_2$; Test 0.5O2+10CO2= with $O_2$ (0.5 vol%) but adding 10 vol% of $CO_2$). From Ref. 52.

but propene is produced with a high selectivity (66%) and with a yield of 1.5%. Moreover, contrary to the behavior observed in the Test 0.5O2, $NiMoO_4$ does not undergo any deactivation with time-on-stream. The catalyst keeps its performance constant for several hours on-stream.

On the other hand, pure $CO_2$ succeeds in oxidizing a reduced oxide (MoOx) much more rapidly and efficiently than pure $O_2$. This allowed the conclusion to be drawn that $CO_2$ can act as a (very powerful) oxidant at moderated temperatures. The results are rationalized by considering that $CO_2$ could dissociate on the surface of the catalyst ($CO_2$(g) = CO(ads) + O(ads)) and that the formed oxygen species, O(ads), is able to induce the change in the oxidation state of molybdenum during the reaction. During the catalytic reaction, $CO_2$ succeeds in maintaining the structure of the catalysts in a rather higher oxidized state ($Mo^{6+}$).[52,93]

The dissociation of $CO_2$ seems difficult at low temperature. Thermodynamics disfavors the gas-phase conversion of $CO_2$ to oxygen and carbon monoxide: $CO_2(g) \rightarrow CO\ (g) + {}^1\!/_2\, O_2(g)$. It is estimated that not more than 2% $CO_2$ transforms into CO(g) and $O_2$(g) at 2,000°C. Then, at temperatures low enough (<450°C), $CO_2$ is generally considered inert. However, in spite of its low reactivity at low temperature, $CO_2$ can dissociate on the surface of a metal oxide or supported metal catalysts into adsorbed CO, and very reactive monoatomic adsorbed oxygen O(ads), which tailors the oxidation state of metal atoms on the catalytic surfaces during reaction and oxidizes hydrocarbons, intermediates, or coke precursors. Oxygen species produced by the dissociation of $CO_2$ can also migrate from the oxide phase where they are formed, to other phases in contact with the former. On the other hand, thermodynamic calculations show that the dissociative adsorption of $CO_2$ to CO(ads) and O(ads) should proceed very spontaneously under low temperature reaction conditions.[94] Dissociation of $CO_2$ to CO and O(ads) has already been shown to play an important role in the synthesis of methanol from synthesis gas in the presence of $CO_2$ and in the water-gas shift reaction under similar temperatures on $Cu/ZnO/Al_2O_3$ catalysts.[94,95] Recently the dissociative adsorption of $CO_2$ has been evidenced at room or low temperatures.[96–99] $CO_2$ may also be used as a valuable gas promoter in the feed for the catalytic combustion of methane (CCM) on Pd/$\gamma$-$Al_2O_3$, $Pd/Ce_{0.21}Zr_{0.79}O_2$ catalysts[100,101] in the partial oxidation of methane,[102] in CO oxidation by gold catalysts,[103] and in the synthesis of low-carbon alcohols on cobalt-modified multi-walled carbon nanotubes.[104]

### 26.2.3. *Oxidative dehydrogenation of ethylbenzene to styrene (ODH of EB)*

Styrene, often quoted as styrene monomer (SM), is one of the largest production bulk monomers nowadays, produced industrially by the steam-aided dehydrogenation of ethylbenzene (EB) over a promoted potassium-iron catalyst. The process suffers from high energy requirements, low per-pass conversions due to its endothermic nature ($\Delta H^{\circ}_{298} = +117.6$ kJ/mol), and equilibrium limitations. A large part of the energy is lost in the steam condensation and cannot be recovered due to practical reasons.[54,105] The ODH of ethylbenzene with pure oxygen has been explored, but it suffers relatively poor selectivity, due to over-oxidation reactions, and safety issues such as the risk of explosion and of hot spots in the catalyst bed due to the highly exothermic nature of the reaction. Due to these issues, and the higher operating costs, this process has not been commercialized. Sato *et al.*[106] first reported the combination of the $CO_2$ RWGS reaction with ethylbenzene dehydrogenation as a way to improve the latter on various catalysts. They later reported that the addition of small amounts of carbon dioxide (ranging from 0.1 to 0.5 mole % of $CO_2$ in the feed) to the reactant feed gas slightly decreased the rate of styrene formation and, to a certain extent, depressed the decay of the catalytic activity of the commercial potassium-promoted iron oxide catalyst.[107]

Mimura and Saito[108,109] discussed the advantages of using $CO_2$ instead of steam for the DH of ethylbenzene to styrene on the basis of thermodynamic considerations. They considered both a one-step pathway, where $CO_2$ plays a direct role in the oxide hydrogenation producing styrene, carbon monoxide, and water, and a two-step pathway, consisting of direct dehydrogenation coupled to the RWGS which removes the hydrogen from the catalytic surface, where styrene, hydrogen, CO, and water are formed:

One-step pathway: $C_6H_5\text{-}CH_2\text{-}CH_3 + CO_2 \rightarrow C_6H_5\text{-}CH{=}CH_2 + CO + H_2O$ (ODH)

Two-step pathway: $C_6H_5\text{-}CH_2\text{-}CH_3 \rightarrow C_6H_5\text{-}CH{=}CH_2 + H_2$ (DH)

$CO_2 + H_2 \rightarrow CO + H_2O$ (RWGS)

They found that the equilibrium yield of styrene for the dehydrogenation in the presence of $CO_2$ by either pathway (one-step or two-step) is much higher than that for the dehydrogenation in the presence of steam: the required temperature to reach the same yield could be around 80°C lower in the presence of $CO_2$. The two-step pathway (i.e. alkane dehydrogenation followed by RWGS) provides a higher equilibrium yield of styrene at a given temperature than the one-step pathway (i.e. direct ODH). They estimated the amount of energy required for the new process using $CO_2$ is much lower than that for a typical commercial process: in the coupled process it is $(1.5\text{–}1.9) \times 10^8$ cal per ton of styrene produced, compared with $1.5 \times 10^9$ cal in the current commercial process using steam.[109]

Early reports on this reaction were focused on iron oxide-based catalysts, supported on different supports: alumina,[108,109] zeolite,[110] and active carbon,[111] among others. Later, hydrotalcite-like compounds,[112] and the topical V-Mg-O catalysts[113] were explored. Park and his group, one of the most active in this field, first reported the use of zirconia as a catalyst,[114] and later on that of ceria-zirconia-based catalysts and vanadium-antimony oxide-based catalysts.[115] Publications up to 2007 have been subject of several reviews.[116–119] Therefore, the most relevant features will be revised hereinafter, as well as more recent publications, and the reader is invited to get further details in the aforementioned reviews.

#### 26.2.3.1. *Vanadium oxide-based catalysts*

Vanadium oxide, combined with or supported on other oxides, is the key component of a great portion of active catalysts for ODH reactions.[55] For example, V-Mg, V-Cr, V-P, and V-Sb oxides are active and selective in oxidation reactions of hydrocarbons such as partial oxidation, ammoxidation, and ODH, including ODH of EB with oxygen. Park's group first studied modified vanadia supported on alumina (V-M/Al, M = Mg, P, Cr, Mo, or Sb), with a 20 wt% content of vanadia. In the $CO_2$-ODH of ethylbenzene, styrene, water, carbon monoxide, and hydrogen were the main products. The addition of MgO, $P_2O_5$, or $MoO_3$ hardly influenced the initial styrene yield (75% at 650°C), but deteriorated the on-stream catalytic activity of V/Al. V/Al and V-Sb/Al catalysts exhibited higher selectivities to styrene (>95%) than those

(81–91%) reported for vanadium oxide loaded on active carbon[111] and MgO[113] supports. The progressive substitution of vanadium by antinomy from 13 to 75 mole % increases the specific surface area, styrene selectivity (up to 96.3%) and, more interestingly, it reduces the decay of activity on-stream from 23% down to 3.3–6.7% conversion loss after one hour on-stream (at 595°C and EB/$CO_2$ molar ratio = 1).[115,120] To neutralize an excessive surface acidity of the alumina support, it was modified with magnesium oxide, using the optimal active phase composition V:Sb = 0.43:0.57.[121] High MgO contents decrease activities of Mg-modified catalysts due to the decrease in their specific surface areas. However, the values of specific yields (styrene yields normalized with respect to BET surface area value, $Y_{ST}/S_{BET}$) are close: 0.8–1.1 %/m$^2$. All V-Sb/$Mg_n$Al catalysts exhibited stable catalytic behavior as compared to the V-Sb/Al, due to the suppression of coking. The modification of alumina with zirconia also has a beneficial effect. The $ZrO_2$ contents at $Al_2O_3$ > 5 wt% increased EB conversion from 55 to 65% and selectivity to styrene from 87 to 92% (at 550°C, with molar ratio $CO_2$/EB = 5, WHSV = 1 h$^{-1}$, time-on-stream = 2 h).

The role of antimony in the active phase seems to be twofold. Its presence avoids the formation of $V_2O_5$ crystallites, observed by XRD in fresh V/Al catalysts, transformed into $V_2O_3$ crystallites after the reaction. Instead of these single vanadium oxides, the XRD patterns of V-Sb/Al catalysts reveal an intermediate phase of mixed V-Sb oxide, $V_{1.1}Sb_{0.9}O_4$. On the other hand, no $V^{3+}$ and a much lower content of $V^{4+}$ species (16.1% vs 33.6%) were found in used V-Sb/Al catalysts. This indicates that antimony plays a key role in stabilizing vanadium in the higher oxidation state, which is the active species.[117]

Recently, Liu *et al.* investigated the effect of vanadium loading in a series of LaVOx/SBA-15 catalysts for the $CO_2$-ODH of EB.[122] The V/La atomic ratio has a prominent influence on the catalytic activity. The 10% $La_2O_3$–15% $V_2O_5$/SBA-15 (wt%) catalyst exhibited the best activity and stability, giving a styrene yield of 74%. The addition of $La^{3+}$ has the effect of hindering carbon deposition and enhancing the stability of the catalysts. The EB conversion and styrene selectivity were 39 and 97%, respectively, at 500°C, and 58 and 80%, respectively, at 600°C. The latter is higher than that observed recently by Burri *et al.* over $CeO_2$-$ZrO_2$/SBA-15 (*ca.* 59% at 600°C).[123] This good performance of the catalyst could be related to the mono-dispersion of the $VO_4^{3-}$ and the synergistic action of lanthanum and vanadium. The same authors studied a similar catalyst with an optimized V/La ratio but supported on MCM-41.[124] The active sites consisted of $V_2O_5$ and $La_2O_3$, and, due to the strong interaction of $V_2O_5$ with MCM-41, there was good $VO_4^{3-}$ dispersion inside the MCM-41 channels. After 4 h on-stream, EB conversion and styrene selectivity were 86.5 and 91.0%, respectively. However, styrene yield decreased monotonously after the initial 2 h on-stream, due to the progressive accumulation of carbonaceous deposits, as evidenced by TEM and high resolution transmission electron microscopy (HRTEM).

Vanadium oxide and cerium oxide have also been used to dope titania–zirconia mixed oxides.[125] Their results will be discussed in the next section.

### 26.2.3.2. *Zirconia-based catalysts*

Various zirconia-based composite oxide catalysts, namely $TiO_2$-$ZrO_2$, $MnO_2$-$ZrO_2$, $CeO_2$-$ZrO_2$, $K_2O/TiO_2$-$ZrO_2$, $B_2O_3/TiO_2$-$ZrO_2$, and $CeO_2$-$ZrO_2$/SBA-15, have been synthesized, characterized by various techniques, and evaluated for the $CO_2$-ODH of ethylbenzene.[118] Among these, the combination of titania-zirconia has attracted considerable attention in recent years as an active catalyst. Both single oxides $ZrO_2$ and $TiO_2$ exhibit excellent catalytic properties for several reactions, and are extensively used as supports. Its combination takes advantage of both, $TiO_2$ (active catalyst and support) and $ZrO_2$ (acid-base properties), and extends their application through the generation of new catalytic sites due to a strong interaction between them. $TiO_2$-$ZrO_2$ composite oxides exhibit a high specific surface area, profound surface acid-base properties, a high thermal stability, and strong mechanical strength.[126] Catalysts for the $CO_2$-ODH of EB require basic properties, to adsorb and activate the $CO_2$ molecule, and specific acidic sites capable of activating EB. This makes the $TiO_2$-$ZrO_2$ mixed oxide combination a most suitable system for this reaction. Park's group investigated the whole range of compositions $(TiO_2)_{1-x}$-$(ZrO_2)_x$, $(0 \leq x \leq 1)$, prepared by a co-precipitation method at pH = 7–8.[127] The XRD pattern of pure $ZrO_2$ revealed a monoclinic phase and the pure $TiO_2$ in the anatase phase, whereas the mixed oxides (40–60 mole % $ZrO_2$) are in an X-ray amorphous state. Ethylbenzene conversion and styrene yield over single $TiO_2$ or $ZrO_2$ were lower, and decreased faster with time-on-stream, than those observed for the 50% $TiO_2$-$ZrO_2$ catalyst. In the absence of $CO_2$, EB conversion on this mixed catalyst at 600°C showed an induction period, reaching a maximum of 53% after 3 h on-stream (using W/F = 16.73 g cat. h/mole; mole ratio $N_2$/EB = 5.1, P = 1 atm). It then decayed with time-on-stream (40% loss in 7 h on-stream), while styrene selectivity remained stable at around 93%. The use of $CO_2$ instead of nitrogen inhibited the conversion decay and kept a stable higher selectivity, close to 100%.[127]

Selectivity can be enhanced by complete suppression of the by-products by means of neutralizing the strong acidic sites of $TiO_2$-$ZrO_2$ by basic oxides such as $K_2O$. Various amounts of $K_2O$ (1–4 wt%) were impregnated over a $TiO_2$-$ZrO_2$ (60:40 mole ratio) composition, which exhibits a better performance than the 50% mixture. Due to the addition of $K_2O$, the $TiO_2$-anatase phase seggregation was clearly observed. The $K_2O$ promoter increases EB conversion and styrene selectivity. $B_2O_3$ was also investigated as a promoter. However, its effect was not positive due to the formation of benzene and toluene as major side products.

Manganese oxide ($MnO_2$) also exhibits redox properties and contributes to the enhancement of the acid-base properties. The $MnO_2$-$ZrO_2$ mixed oxides with 5–50 mole % $MnO_2$ were investigated. The mixed oxides were superior to their individual components: the incorporation of 10% $MnO_2$ to pure $ZrO_2$ in the 10% $MnO_2$-$ZrO_2$ catalyst practically doubled the EB conversion at 600°C, 58–60% vs 25–20% for the single $ZrO_2$ catalyst, with similar selectivity values (95.3% vs 98.9%). The best results were obtained over a 10% $MnO_2$-$ZrO_2$ catalyst at 650°C with an EB conversion of 73% and a product selectivity of 98% to styrene.[118]

Mixed ceria-zirconia oxides have also been explored for the $CO_2$-ODH of ethylbenzene. A mixed oxide with 25 wt% ceria was investigated in two states: bulk oxide and supported on mesoporous SBA-15. The catalytic activity of $CeO_2$-$ZrO_2$ and $CeO_2$-$ZrO_2$/SBA-15 samples were compared with the pure $CeO_2$ and $ZrO_2$ oxides in terms of turnover frequency (TOF) for better comparison. The TOF was defined as the number of moles of EB converted into products per second per mole of the active phase, based on the catalyst composition and weight, and was calculated assuming that all active oxides are in a 4+ oxidation state. Single oxides exhibit similar performance, while the mixed bulk oxide shows better activity,[124] and its spreading onto SBA-15 enhanced its activity by more than tenfold in comparison to its bulk oxides.[118] EB conversion increased from 42 to 56.8% with increasing $CeO_2$-$ZrO_2$ loading from 15 to 25 wt% on the SBA-15 support, but a further loading was not investigated as it could facilitate formation of agglomerated $CeO_2$-$ZrO_2$ particles. However, when the $CeO_2$-$ZrO_2$ loading is increased from 15 to 40 wt%, the selectivity marginally increases from 93.2% to 95.9%.

Titania-zirconia mixed oxides doped with vanadium and cerium oxides were also explored.[125] The $TiO_2$-$ZrO_2$ mixed oxide support, synthesized by a co-precipitation method, showed a high specific surface area (207 $m^2g^{-1}$). A monolayer equivalent (15 wt%) of $V_2O_5$, $CeO_2$, or a combination of both were deposited over the calcined support (XRD amorphous when calcined at 550°C) by wet impregnation or co-impregnation methods to make the $V_2O_5/TiO_2$-$ZrO_2$, $CeO_2/TiO_2$-$ZrO_2$, and $V_2O_5$-$CeO_2/TiO_2$-$ZrO_2$ combined catalysts, respectively. The Raman spectrum of the amorphous support shows the bands usually ascribed to amorphous $ZrTiO_4$. The deposited V and/or Ce oxides are highly dispersed over the support, and the combined acid-base and redox properties of the catalysts play a major role in this reaction. The scanning electron microscopy (SEM) images of all samples reveal spherical-type agglomerates with varying sizes within the nanometer range; the addition of vanadium oxide slightly increases their size, whereas that of ceria has the reverse effect. Initial activity (EB conversion) increases in the order: support < monocomponent-supported oxide < dicomponent-supported oxide; while for selectivity only the dicomponent V-Ce supported oxide showed an increase as compared to that of the bare support. All samples showed stable activity, except the monocomponent vanadia catalyst, as the activity decayed very fast after 2 h on-stream. Thus, the addition of $CeO_2$ to $V_2O_5/TiO_2$-$ZrO_2$ prevented catalyst deactivation.

With the increase of the support calcination temperature to 750°C, crystalline $ZrTiO_4$ was formed. When $V_2O_5$ or $CeO_2$-$V_2O_5$ were supported on this support, few other characteristic lines due to $ZrV_2O_7$ and $CeVO_4$ were observed in addition to $ZrTiO_4$ peaks, and the XPS binding energy of V $2p_{3/2}$ increased from 516.3 to 517.4 eV, indicating the increase of vanadium higher oxidation states, probably due to the newly formed compounds. Among all the catalysts evaluated, the $V_2O_5$-$CeO_2$/$TiO_2$-$ZrO_2$ sample exhibited the highest EB conversion (56%) and styrene selectivity (98%) after 3–4 h on-stream.[128]

### 26.2.4. *The role of $CO_2$ in ODH of alkanes and alkylaromatics*

The presence of $CO_2$ causes various kinetic effects: it accelerates the reaction rate, enhances the selectivity, alleviates the chemical equilibrium, suppresses the unwanted total oxidation products, prevents the hot spots on the catalyst surface, poisons the non-selective sites of the catalysts, and the equilibrium yield of styrene dehydrogenation is much higher in the presence of $CO_2$ than in that of steam.

Since the early reports of the use of $CO_2$ in dehydrogenation reactions, there has been a certain controversy over its role. In other words, is "oxidative" the (oxi)dehydrogenation in the presence of $CO_2$? Some authors expressed this as the alternative between the "one-step" (true ODH, where $CO_2$ acts as soft oxidant), and the "two-steps" (combination of non-oxidative dehydrogenation, DH, and RWGS, where pathway $CO_2$ acts as hydrogen scavenger, shifting the dehydrogenation equilibrium).[108,109]

A key point for the relevance of each of these alternatives is the redox properties of the active phase in relation to $CO_2$. Pure $CO_2$ shows only a small ability for oxidizing vanadium species of a lower oxidation state, in agreement with thermodynamic calculations,[89] but this may be modified by the strong interaction between the vanadium oxide and the chromium[91] or antimony.[121] Pure $CO_2$ succeeds in oxidizing a reduced oxide ($MoO_{3-x}$) much more rapidly and efficiently than pure $O_2$ does, and during the catalytic reaction $CO_2$ keeps the $NiMoO_4$ catalysts in a rather higher oxidized state ($Mo^{6+}$).[52,93] The ability of the oxygen storage capacity (OSC), is usually discussed in terms of the $Ce^{3+}/Ce^{4+}$: cerium may be reduced by CO and reoxidized by $CO_2$.[129] All these data support the idea that, in most of the reducible oxides used for $CO_2$-ODH, carbon dioxide is acting as a real soft oxidant. In those cases where the re-oxidation of the initially reduced active center can not be re-oxidized by $CO_2$, the combined two-step pathway is operating as, for example, in the $CO_2$-ODH of ethane over Fe-Cr/$ZrO_2$ catalysts,[79] where chromium oxide would catalyze DH and iron oxide via the RGWS reaction. Nevertheless, it has been reported in some cases that both pathways could be present simultaneously.

One may conclude that the (oxi)dehydrogenation reaction in the presence of $CO_2$ is a rather complex process in which $CO_2$ is simultaneously playing several important roles that favor the efficiency of the catalysts as compared with dehydrogenation in the same systems. The prevalence of the oxidative or the combined route depends mostly on the nature of the active oxide component and its interaction with $CO_2$.

The use of $CO_2$ in these reactions looks very promising. Nevertheless, further studies are needed to overcome the issues of the stability of catalytic activity in the long term, the improvement of activity to reduce the reaction temperature range to be used, and probably, the integration of energy transfer with some exothermic reaction, to balance the endothermicity of these reactions.

## References


1. Richardson, D., Felgate, H., Watmough, N., *et al.* (2009). Mitigating release of the potent greenhouse gas $N_2O$ from the nitrogen cycle: Could enzymic regulation hold the key, *Trends Biotechnol.*, **27**, pp. 388–397.
2. Pérez-Ramírez, J., Kapteijn, F., Schöffel, K., *et al.* (2003). Formation and control of $N_2O$ in nitric acid production. Where do we stand today? *Appl.Catal. B*: *Environ.*, **44**, pp. 117–151.
3. Pérez-Ramírez, J. (2007). Prospects of $N_2O$ emission regulations in the european fertilizer industry, *Appl. Catal. B*: *Environ.*, **70**, pp. 31–35.
4. Molinari, R. and Poerio, T. (2010). Remarks on Studies for direct production of phenol in conventional and membrane reactors, *Asia Pac. J. Chem. Eng.*, **5**, pp. 191–206.
5. Iwamoto, M., Hirata, J., Matsukami K., *et al.* (1983). Catalytic oxidation by oxide radical ions. 1. One-step hydroxylation of benzene to phenol over group 5 and 6 oxides supported on silica gel, *J. Phys. Chem.*, **87**, pp. 903–905.
6. Rivallan, M., Ricchiardi, G., Bordiga, S., *et al.* (2009). Adsorption and reactivity of nitrogen oxides ($NO_2$, NO, $N_2O$) on Fe–Zeolites, *J. Catal.*, **264**, pp. 104–116.
7. Pirutko, L., Chernyavsky, V., Starokon, E. *et al.* (2009). The role of $\alpha$-sites in $N_2O$ decomposition over FeZSM-5. Comparison with the oxidation of benzene to phenol, *Appl. Catal. B*: *Environ.*, **91**, pp. 174–179.
8. Wac law, A., Nowińska, K. and Schwieger, W. (2004). Benzene to phenol oxidation over iron exchanged zeolite ZSM-5, *Appl. Catal. A*: *Gen.*, **270**, pp. 151–156.
9. Wichterlová, B., Sobalík, Z. and Dědeček, J. (2003). Redox catalysis over Metallo-Zeolites. Contribution to environmental catalysis, *Appl. Catal. B*: *Environ.*, **41**, pp. 97–114.
10. Esteves, P. and Louis, B. (2006). Experimental and DFT study of the partial oxidation of benzene by $N_2O$ over H-ZSM-5: Acid catalyzed mechanism, *J. Phys. Chem. B*, **110**, pp. 16793–16800.
11. Guesmi, H., Berthomieu, D. and Kiwi-Minsker, L. (2010). Reactivity of oxygen species formed upon $N_2O$ dissociation over Fe–ZSM-5 Zeolite: CO Oxidation as a Model, *Catal. Commun.*, **11**, pp. 1026–1031.
12. Ferdi Fellah, M., Onal, I. and van Santen, R. (2010). A density functional theory study of direct oxidation of benzene to phenol by $N_2O$ on a $[FeO]^{1+}$-ZSM-5 Cluster, *J. Phys. Chem. C*, **114**, pp. 12580–12589.
13. Xia, H., Sun, K., Liu, Z., *et al.* (2010). The promotional effect of no on $N_2O$ decomposition over the Bi-nuclear Fe sites in Fe/ZSM-5, *J. Catal.*, **270**, pp. 103–109.
14. Held., A. and Florczak, B. (2009). Vanadium, niobium and tantalum modified mesoporous molecular sieves as catalysts for propene epoxidation, *Catal. Today*, **142**, pp. 329–334.
15. Wang, Y., Yang, W., Yang, L., *et al.* (2006). Iron containing heterogeneous catalysts for partial oxidation of methane and epoxidation of propylene, *Catal. Today*, **117**, pp. 156–162.

16. Wei, W., Moulijn, J. and Mul, G. (2009). FAPO and Fe-TUD-1: Promising catalysts for $N_2O$ mediated selective oxidation of propane? *J. Catal.*, **262**, pp. 1–8.
17. Janas, J., Gurgul, J., Socha, R., *et al.* (2009). Influence of the content and environment of chromium in crsibea zeolites on the oxidative dehydrogenation of propane, *J. Phys. Chem. C*, **113**, pp. 13273–13281.
18. Pérez-Ramírez, J. and Gallardo-Llamas, A. (2005). Impact of the preparation method and iron impurities in Fe-ZSM-5 zeolites for propylene production via oxidative dehydrogenation of propane with $N_2O$, *Appl. Catal. A*: *Gen.*, **279**, pp. 117–123.
19. Santhosh-Kumar, M., Pérez-Ramírez, J., Debbagh, M., *et al.* (2006). evidence of the vital role of the pore network on various catalytic conversions of $N_2O$ over Fe-Silicalite and Fe-SBA-15 with the Same iron constitution, *Appl. Catal. B*: *Environ.*, **62**, pp. 244–254.
20. Woods, M., Mirkelamoglu, B. and Ozkan, U. (2009). Oxygen and nitrous oxide as oxidants: Implications for ethane oxidative dehydrogenation over silica-titania-supported molybdenum, *J. Phys. Chem. C*, **113**, pp. 10112–10119.
21. Held, A., Kowalska, J. and Nowińska, K. (2006). Nitrous oxide as an oxidant for ethane oxydehydrogenation, *Appl. Catal. B*: *Environ.*, **64**, pp. 201–208.
22. Ovsitser, O. and Kondratenko E. (2009). Similarity and differences in the oxidative dehydrogenation of $C_2$–$C_4$ alkanes over nano-sized VOx species using $N_2O$ and $O_2$, *Catal. Today*, **142**, pp. 138–142.
23. Alejo, L., Lago, R., Pefia M., *et al.* (1997). Partial oxidation of methanol to produce hydrogen over Cu-Zn-Based catalysts, *Appl. Catal. A*: *Gen.*, **162**, pp. 281–297.
24. Ferdi-Fellah, M. and Onal, I. (2010). Direct Methane Oxidation to Methanol by $N_2O$ on Fe- and Co-ZSM-5 Clusters with and without Water: A Density Functional Theory Study, *J. Phys. Chem. C*, **114**, pp. 3042–3051.
25. Kustrowski, P., Segura, Y., Chmielarz, L., *et al.* (2006). VOx supported SBA-15 catalysts for the oxidative dehydrogenation of ethylbenzene to styrene in the presence of $N_2O$, *Catal. Today*, **114**, pp. 307–313.
26. Shiju, N., Anilkumar, M., Mirajkar, S., *et al.* (2005). Oxidative dehydrogenation of ethylbenzene over vanadia-alumina catalysts in the presence of nitrous oxide: structure-activity relationship, *J. Catal.*, **230**, pp. 484–492.
27. Pérez-Ramírez, J., Blangenois, N. and Ruiz, P. (2005). Highly efficient Fe-silicalite zeolite in direct propane ammoxidation with $N_2O$ and $O_2$, *Catal. Lett.*, **104**, pp. 163–167.
28. Andrade Sales, E., Oliveira de Souza, T., Costa Santos, R., *et al.* (2005). $N_2O$ decomposition coupled with ethanol oxidative dehydrogenation reaction on carbon-supported copper catalysts promoted by palladium and cobalt, *Catal. Today*, **107–108**, pp. 114–119.
29. Costine, A., O'Sullivan, T. and Hodnett, B. (2005). Oxidative competition between aliphatic and aromatic C–H bonds in the $N_2O$-Fe-ZSM-5 system, *Catal. Today*, **99**, pp. 199–208.
30. Dury, F., Centeno, M., Gaigneaux, E., *et al.* (2003). Interaction of $N_2O$ (as Gas Dope) with nickel molybdate catalysts during the oxidative dehydrogenation of propane to propylene, *Appl. Catal. A*: *Gen.*, **247**, pp. 231–246.
31. Heyden, A., Peters, P., Bell, A., *et al.* (2005). Comprehensive DFT study of nitrous oxide decomposition Over Fe-ZSM-5, *J. Phys. Chem. B*, **109**, pp. 1857–1873.
32. Guesmi, H., Berthomieu, D. and Kiwi-Minsker, L. (2008), Nitrous oxide decomposition on the binuclear $[Fe^{II}(\mu\text{-}O)(\mu\text{-}OH)Fe^{II}]$ Center in Fe-ZSM-5 Zeolite, *J. Phys. Chem. C*, **112**, pp. 20319–20328.
33. Guesmi, H., Berthomieu, D., Bromley, B., *et al.* (2010). Theoretical evidence of the observed kinetic order dependence on temperature during the $N_2O$ Decomposition over Fe-ZSM-5, *Phys. Chem.*, **12**, pp. 2873–2878.
34. Smeets, P., Woertink, J., Sels, B., *et al.* (2010), Transition-metal ions in Zeolites: Coordination and activation of oxygen, *Inorg. Chem.*, **49**, pp. 3575–3583.

35. Prikhod'ko, R., Astrelin, I., Sychev, M., *et al.* (2006). Influence of preparation procedure on the surface chemistry and catalytic characteristics of FeZSM-5 zeolite in selective oxidation of benzene to phenol, *Russ. J. Appl. Chem.*, **79**, pp. 1115–1121.
36. Santiago, M. and Perez-Ramirez, J. (2007). Decomposition of $N_2O$ over hexaaluminate catalysts, *Environ. Sci. Technol.*, **41**, pp. 1704–1709.
37. Bahidsky, M. and Hronec, M. (2005). Direct hydroxylation of aromatics over Copper–Calcium–Phosphates in the gas phase, *Catal. Today*, **99**, pp. 187–192.
38. Pérez-Ramıírez, J. and Gallardo-Llamas, A. (2005). Framework composition effects on the performance of steam-activated FeMFI zeolites in the $N_2O$-mediated propane oxidative dehydrogenation to propylene, *J. Phys. Chem. B*, **109**, pp. 20529–20538.
39. Gallardo-Llamas, A., Mirodatos, C. and Pérez-Ramírez, J. (2005). Cyclic process for propylene production via oxidative dehydrogenation of propane with $N_2O$ over FeZSM-5, *Ind. Eng. Chem. Res.*, **44**, pp. 455–462.
40. Pérez-Ramírez, J., Kondratenko, E. and Debbagh, M. (2005). Transient studies on the mechanism of $N_2O$ activation and reaction with CO and $C_3H_8$ over Fe-silicalite, *J. Catal.*, **233**, pp. 442–452.
41. Díaz-Velásquez, J., Carballo–Suárez, L. and Figueiredo, J. (2006). oxidative dehydrogenation of isobutane over activated carbon catalysts, *Appl. Catal. A*: *Gen.*, **311**, pp. 51–57.
42. Zhu, J., Van Ommen, J. and Lefferts, L. (2006), Partial oxidation of methane by $O_2$ and $N_2O$ to syngas over yttrium-stabilized $ZrO_2$, *Catal. Today*, **112**, pp. 82–85.
43. Kustrowski, P., Chmielarz, L., Surman, J., *et al.* (2005). Catalytic activity of MCM-48-, SBA-15-, MCF-, and MSU-Type mesoporous silicas modified with $Fe^{3+}$ species in the oxidative dehydrogenation of ethylbenzene in the presence of $N_2O$, *J. Phys. Chem. A*, **109**, pp. 9808–9815.
44. Demoulin, O., Seunier, I., Dury, F., *et al.* (2005). Modulation of selective sites by introduction of $N_2O$, $CO_2$ and $H_2$ as gaseous promoters into the feed during oxidation reactions, *Catal. Today*, **99**, pp. 217–226.
45. Intergovernmental Panel on Climate Change. (2007). IPCC Fourth Assessment Report: Climate Change 2007, Synthesis report. http://www.ipcc.ch/publications_and_data/ar4/syr/en/mains2-1.html (accessed March 2011).
46. Kyoto Protocol of the United Nations Framework Convention on Climate Change. (1997). http://unfccc.int/resource/docs/convkp/kpeng.html (accessed March 2011).
47. Arakawa, H., Aresta, M., Armor, J., *et al.* (2001). Catalysis research of relevance to carbon management: progress, challenges, and opportunities. *Chem. Rev.*, **101**, pp. 953–996.
48. Cortés Corberán, V. (2005). Novel approaches for the improvement of selectivity in the oxidative activation of light alkanes, *Catal. Today*, **99**, pp. 33–41.
49. Thambimuthu, K., Davison, J. and Gupta, M. (2003). *Proc. IPCC Workshop on Carbon Dioxide Capture and Storage*, Regina, Canada, pp. 31–52.
50. Nishiyama, T. and Aika. K. (1990). Mechanism of the oxidative coupling of methane using $CO_2$ as an oxidant over PbO-MgO, *J. Catal.*, **122**, pp. 346–351.
51. Mamedov, A., Shiryaev, P., Shashkin, D., *et al.* (1990). Selective dehydrogenation of ethane by carbon dioxide over Fe-Mn oxide catalyst: An in situ study of catalyst phase-composition and structure, in G. Centi and F. Trifiro (eds), *New Developments in Selective Oxidation*(Studies in Surface Science and Catalysis, 55), Elsevier, Amsterdam, pp. 477–482.
52. Dury, F., Gaigneaux, E. and Ruiz. P. (2003). The active role of $CO_2$ at low temperature in oxidation processes: The case of the oxidative dehydrogenation of propane on $NiMoO_4$ catalysts, *Appl. Catal. A*: *Gen.*, **242**, pp. 187–203.
53. Wang, S. and Zhu, Z. (2004). Catalytic conversion of alkanes to olefins by carbon dioxide oxidative dehydrogenations: A review, *Energ. Fuel*, **18**, pp. 1126–1139.
54. Cavani, F. and Trifirò, F. (1995). The oxidative dehydrogenation of ethane and propane as an alternative way for the production of light olefins, *Catal. Today*, **24**, pp. 307–313.

55. Mamedov, E. and Cortés Corberán, V. (1995). Oxidative dehydrogenation of lower paraffins on vanadium oxide-based catalysts: The present state of art and outlooks: A review, *Appl. Catal. A: Gen.*, **127**, pp. 1–40.
56. Krylov, O., Mamedov, A. and Mirzabekova, S. (1994). Catalytic reduction of carbon dioxide by hydrocarbons and other organic compounds, in S. Vic Bellón and V. Cortés Corberán (eds), *New Developments in Selective Oxidation II* (Studies in Surface Science and Catalysis, 82), Elsevier, Amsterdam, pp. 159–166.
57. Krylov, O., Mamedov, A. and Mirzabekova, S. (1995). Catalytic-oxidation of hydrocarbons and alcohols by carbon dioxide on oxide catalysts, *Ind. Eng. Chem. Res.*, **34**, pp. 474–482.
58. Wang, D., Xu, M., Shi, C., *et al.* (1993). Effect of carbon dioxide on the selectivities obtained during the partial oxidation of methane and ethane over Li+/MgO Catalysts, *Catal. Lett.*, **18**, pp. 323–328.
59. Valenzuela, R., Bueno, G., Cortés Corberán, V., *et al.* (1998). Oxidative dehydrogenation of ethane with $CO_2$: An analysis of the heterogeneous reaction contribution on $CeO_2$ Catalysts, *Abstracts 215th ACS Nat. Meeting, Dallas 1998*, COLL-085.
60. Valenzuela, R., Bueno, G., Cortés Corberán, V., *et al.* (2000). Selective oxidehydrogenation of ethane with $CO_2$ over $CeO_2$-based catalysts, *Catal. Today*, **61**, pp. 43–48.
61. Nakagawa, K., Kajita, C., Ide, Y., *et al.* (1998). Dehydrogenation of ethane over gallium oxide in the presence of carbon dioxide, *Chem. Commun.*, **9**, pp. 1025–1026.
62. Krylov, O., Mamedov, A. and Mirzabekova, S. (1995). The regularities in the interaction of alkanes with $CO_2$ on oxide catalysts, *Catal. Today*, **24**, pp. 371–375.
63. Zhang, X., Ye, Q., Xu, B., *et al.* (2007). Oxidative dehydrogenation of ethane over Co-$BaCO_3$ catalysts using $CO_2$ as oxidant: Effects of Co promoter, *Catal. Lett.*, **117**, pp. 140–145.
64. Ji, M., Hong, D., Chang, J., *et al.* (2004). Oxidative dehydrogenation of ethane with carbon dioxide over supported chromium oxide catalysts, in S. Park, J. Chang and K. Lee (eds), *Carbon Dioxide Utilisation for Global Sustainability* (Studies in Surface Science and Catalysis, 153), Elsevier, Amsterdam, pp. 339–342.
65. Botavina, M., Martra, G., Agafonov, Y., *et al.* (2008). Oxidative dehydrogenation of $C_3$-$C_4$ paraffins in the presence of $CO_2$ over $CrO_x/SiO_2$ catalysts, *Appl. Catal. A: Gen.*, **347**, pp. 126–132.
66. Botavina, M., Evangelisti, C., Agafonov, Y., *et al.* (2011). $CrOx/SiO_2$ catalysts prepared by metal vapour synthesis: Physical-chemical characterisation and functional testing in oxidative dehydrogenation of propane, *Chem. Eng. J.*, **166**, pp. 1132–1138.
67. Lapidus, A., Agafonov, Y., Shaporeva, N., *et al.* (2010). Influence of promoters and oxidants on propane dehydrogenation over chromium-oxide catalysts, in *DGMK Tagungsbericht 2010-3*, October 4–6, Berlin, pp. 141–148.
68. Grzybowska-Swierkosz, B. (2002). Effect of additives on the physicochemical and catalytic properties of oxide catalysts in selective oxidation reactions, *Top. Catal.*, **21**, pp. 35–46.
69. Bi, Y., Corberan, V., Zhuang, H., *et al.* (2004). Oxidehydrogenation of ethane with $CO_2$ over transition metal doped MCM-41 mesoporous catalysts, in S. Park, J. Chang and K. Lee (eds), *Carbon Dioxide Utilisation for Global Sustainability* (Studies in Surface Science and Catalysis, 153), Elsevier, Amsterdam, pp. 343–346.
70. Takehira, H., Oishi, Y., Shishido, T., *et al.* (2004). $CO_2$ dehydrogenation of propane over Cr-MCM-41 Catalyst, in S. Park, J. Chang, and K. Lee (eds), *Carbon Dioxide Utilisation for Global Sustainability* (Studies in Surface Science and Catalysis, 153), Elsevier, Amsterdam, pp. 323–328.
71. Liu, L., Li, H., and Zhang Y. (2005). Effect of synthesis parameters on the chromium content and catalytic activities of mesoporous Cr-MSU-x prepared under acidic conditions, *J. Phys. Chem. B*, **110**, pp. 15478–15485.
72. Liu, L., Li, H., and Zhang, Y. (2009). Variations of structure and active species in mesoporous Cr-MSU-x catalyst during the dehydrogenation of ethane with $CO_2$, *Kinet. Catal.*, **50**, pp. 684–690.

73. Mimura, N., Okamoto, M., Yamashita, H., *et al.* (2006). Oxidative dehydrogenation of ethane over Cr/ZSM-5 catalysts using $CO_2$ as an oxidant, *J. Phys. Chem. B*, **110**, pp. 21764–21770.
74. Zhao, X. and Wang, X. (2010). Characterizations and Catalytic properties of chromium silicalite-2 prepared by direct hydrothermal synthesis and impregnation, *Catal. Lett.*, **135**, pp. 233–240.
75. Shi, X., Ji, S. and Wang, K. (2008). Oxidative dehydrogenation of ethane to ethylene with carbon dioxide over Cr–Ce/SBA-15 catalysts, *Catal. Lett.*, **125**, pp. 331–339.
76. Deng, S., Li, H. and Zhang, Y. (2003). Preparation, Characterization and catalytic activity of nanosized chromium oxide, *Chinese J. Inorg. Chem.*, **19**, pp. 825–830.
77. Deng, S. Li, H. and Zhang, Y. (2003). Oxidative dehydrogenation of ethane with carbon dioxide to ethylene over nanosized $Cr_2O_3$ Catalysts, *Chinese J. Inorg. Chem.*, **24**, pp. 744–755.
78. Deng, S., Li, H., Li, S., *et al.* (2007). Activity and characterization of modified $Cr_2O_3/ZrO_2$ Nano-composite catalysts for oxidative dehydrogenation of ethane to ethylene with $CO_2$, *J. Mol. Catal. A*: *Chem.*, **268**, 169–175.
79. Deng, S., Li, H., Li, S., *et al.* (2009). Oxidative dehydrogenation of ethane to ethylene with $CO_2$ over Fe-Cr/$ZrO_2$ catalysts, *Ind. Eng. Chem. Res.*, **48**, pp. 7561–7566.
80. Jin, L., Reutenauer, J., Opembe, N., *et al.* (2009). Studies on dehydrogenation of ethane in the presence of $CO_2$ over Octahedral molecular sieve (oms-2) catalysts, *ChemCatChem*, **1**, pp. 441–444.
81. Zhu, J., Qin, S, Ren, S., *et al.* (2009). $Na_2WO_4/Mn/SiO_2$ catalyst for oxidative dehydrogenation of ethane using $CO_2$ as oxidant, *Catal. Today*, **148**, pp. 310–315.
82. Valenzuela, R., Bueno, G., Solbes, A., *et al.* (2001). Nanostructured ceria-based catalysts for oxydehydrogenation of ethane with $CO_2$, *Top. Catal.*, **15**, pp. 181–188.
83. Guío, M., Prieto, J. and Cortés, C. (2006) Determination of kinetic parameters of the oxide-hydrogenation of ethane with $CO_2$ on nanosized calcium-doped ceria under fast deactivation processes, *Catal. Today*, **112**, pp. 148–152.
84. Navas, J., Sapiña, F., Martínez, E., *et al.* (2007). Nanometric $CeO_2$–$ZrO_2$ catalysts for oxidehydrogenation of ethane with $CO_2$, in *Proceedings of the 8th Natural Gas Convesion Symposium (8 NGCS)*, Brazil 2007, p. 34.
85. Navarro, P., de Melo Monteiro, A., Gómez, A., *et al.* (2010). Nanometric $CeO_2$-$ZrO_2$ catalysts doped with calcium for oxidehydrogenation of ethane with $CO_2$, in *Proceedings of the 9th Novel Gas Conversion Symposium (9 NGCS)*, Lyon, France 2010, pp. 200.
86. Shi X., Ji, S., Wang, K., *et al.* (2008). Oxidative dehydrogenation of ethane over ce-based monolithic catalysts using $co_2$ as oxidant, *Catal. Lett.*, **126**, pp. 426–435.
87. Ogonowski, J. and Skrzyńska, E. (2005). Dehydrogenation of Isobutane in the presence of carbon dioxide, *React. Kinet. Catal. L.*, **86**, pp. 195–201.
88. Ogonowski, J. and Skrzyńska, E. (2007). Dehydrogenation of isobutane with carbon dioxide over a vanadium-magnesium catalyst, *React. Kinet. Catal. L.*, **92**, pp. 267–274.
89. Sakurai, Y., Suzaki,T., Nakagawa, K., *et al.* (2002). Dehydrogenation of ethylbenzene over vanadium oxide-loaded MgO Catalyst: Promoting effect of carbon dioxide, *J. Catal.*, **209**, pp. 16–24.
90. Ogonowski, J. and Skrzyńska, E. (2008). Conversion of lower hydrocarbons in the presence of carbon dioxide: The theoretic analysis and catalytic tests over active carbon supported vanadium oxide, *Catal. Lett.*, **124**, pp. 52–58.
91. Jiang, H. Ji, S., Wu, P., *et al.* (2006). Propane oxidative dehydrogenation with $CO_2$ to propene over V-Cr/SBA-15 catalyst, Ranlia *o Huaxue Xuebao*, **34**, pp. 600–606.
92. Bi, Y., Zhen, K., Valenzuela, R., *et al.* (2000). Oxidative dehydrogenation of isobutene over labasm oxide catalyst: Influence of the addition of $CO_2$ in the feed, *Catal. Today*, **61**, pp. 369–375.

93. Dury, F., Centeno, M., Gaigneaux, E., *et al.* (2003). An attempt to explain the role of $CO_2$ and $N_2O$ as gas dopes in the feed in the oxidative dehydrogenation of propane, *Catal. Today*, **81**, pp. 95–105.
94. Chinchen, G., Spencer, M., Waugh, K., *et al.* (1987). Promotion of methanol synthesis and the water-gas shift reactions by adsorbed oxygen on supported copper catalysts, *J. Chem. Soc. Faraday T.*, **83**, pp. 2193–2212.
95. Dubois, J. and Somorjai, G. (1983). Comments on "Why Carbon Dioxide does not Dissociate on Rhodium at Low Temperature" by W. H. Weinberg, *Surf. Sci.*, **128**, pp. L231–L235.
96. Jacquemin, M., Beuls, A. and Ruiz, P. (2010). Catalytic production of methane from $CO_2$ and $H_2$ at Low Temperature: Insight on the reaction mechanism, *Catal. Today*, **157**, pp. 462–466.
97. Schumacher, N., Andersson, K., Grabow, L., *et al.* (2008). Interaction of carbon dioxide with cu overlayers on Pt(111), *Surf. Sci.*, **602**, pp. 702–711.
98. Staudt, T., Lykhach, Y., Tsud, N., *et al.* (2010). Ceria reoxidation by $CO_2$: A Model Study, *J. Catal.*, **275**, pp. 181–185.
99. Rao, K., Reddy, B. and Park, S. (2010). Novel $CeO_2$ promoted $TiO_2$–$ZrO_2$ nano-oxide catalysts for oxidative dehydrogenation of *p*-diethylbenzene utilizing $CO_2$ as soft oxidant, *Appl. Catal. B: Environ.*, **100**, pp. 472–480.
100. Demoulin, O., Navez, M., Mugabo, J., *et al.* (2007). The Oxidizing role of $CO_2$ at mild temperature on ceria-based catalysts, *Appl. Catal. B: Environ.*, **70**, pp. 284–293.
101. Demoulin, O., Navez, M., Gracia, F., *et al.* (2004). High throughput experimentation applied to the combustion of methane and a comparison with conventional micro-reactor measurements, *Catal. Today*, **91–92**, pp. 85–89.
102. Cellier, C., Le Clef, D., Mateos-Pedrero, C., *et al.* (2005). Influence of the co-feeding of CO, $H_2$, $CO_2$ or $H_2O$ in the Partial Oxidation of Methane over Ni and Rh Supported Catalysts, *Catal. Today*, **106**, pp. 47–51.
103. Mihaylov, M., Ivanova, E., Hao, Y., *et al.* (2008). Gold supported on $La_2O_3$: Structure and reactivity with $CO_2$ and implications for co oxidation catalysis, *J. Phys. Chem. C*, **112**, pp. 18973–18983.
104. Xiao-man, W., Yan-yan, G., Hui. L., *et al.* (2007). Co-Modified multiwall carbon nanotubes as promoter of Co-Mo-K catalyst for low-carbon alcohol synthesis from synthesis gas, *Ziran Kexueban*, **46**, pp. 445–450.
105. James, D. and Castor W. (2011), Styrene, in *Ullmann's Encyclopedia of Industrial Chemistry*, Wiley-VCH Verlag GmbH & Co. KGaA, Weinheim. DOI: 10.1002/14356007.a25_329.pub2
106. Sato, S., Ohhara, M., Sodesawa, T., *et al.* (1988) Combination of ethylbenzene dehydrogenation and carbon dioxide shift-reaction over a sodium oxide/alumina catalyst, *Appl. Catal. A: Gen.*, **37**, pp. 207–215.
107. Matsui, J., Sodesawa, T. and Nozaki F. (1990). Influence of carbon dioxide addition upon decay of activity of a potassium-promoted iron oxide catalyst for dehydrogenation of ethylbenzene, *Appl. Catal. A: Gen.*, **67**, pp. 179–188.
108. Mimura, N. and Saito M. (1999). Dehydrogenation of ethylbenzene to styrene over $Fe_2O_3/Al_2O_3$ catalysts in the presence of carbon dioxide. *Catal. Lett.*, **58**, pp. 59–62.
109. Mimura, N. and Saito M. (2000). Dehydrogenation of ethylbenzene to styrene over $Fe_2O_3/Al_2O_3$ catalysts in the presence of carbon dioxide. *Catal. Today*, **55**, pp. 173–178.
110. Chang, J., Noh, J, Park, S., *et al.* (1998). Effect of carbon dioxide in dehydrogenation of ethylbenzene to styrene over zeolite-supported iron oxide catalyst, *B. Kor. Chem. Soc.*, **19**, pp. 1342–1346.
111. Badstube, T., Papp, H., Kustrowski, P., *et al.* (1998). Oxidative dehydrogenation of ethylbenzene with carbon dioxide on alkali-promoted fe active carbon catalysts, *Catal. Today*, **55**, pp. 169–172.
112. Mimura, N., Takahara, I., Saito, M., *et al.* (2002). Dehydrogenation of ethylbenzene to styrene in the presence of $co_2$ over calcined hydrotalcite-like compounds as catalysts, *Catal. Lett.*, **78**, pp. 125–128.

113. Sakurai, Y., Suzaki, T., Nakagawa, K., *et al.* (2002). Dehydrogenation of ethylbenzene over vanadium oxide-loaded mgo catalyst: promoting effect of carbon dioxide, *J. Catal.*, **209**, pp. 16–24.
114. Park, J., Noh, J., Chang, J., *et al.* (2000). Ethylbenzene to Styrene in the presence of carbon dioxide over zirconia, *Catal. Lett.*, **65**, pp. 75–78.
115. Park, M., Vislovskiy, V., Chang, J., *et al.* (2003). Dehydrogenation of ethylbenzene with carbon dioxide: promotional effect of antimony in supported vanadium-antimony oxide catalyst, *Catal. Today*, **87**, pp. 205–212.
116. Park, S., Chang, J. and Yoo, J. (2002). Carbon Dioxide as a Soft Oxidant: Dehydrogenation of Ethylbenzene into Styrene, in M. Maroto-Valer, C. Song, Y. Soong (eds), *Environmental Challenges and Greenhouse Gas Control for Fossil Fuel Utilization in the 21st Century*, (Symposia "Environmental Challenges for Fossil Fuel Combustion" and "Greenhouse Gas Control and Utilization" held at the ACS National Meeting, San Diego, CA, 1–5 April, 2001) Kluwer Academic/Plenum Publishers, New York, pp. 359–369.
117. Chang, J., Hong, D., Vislovskiy, V., *et al.* (2007). An overview on the dehydrogenation of alkylbenzenes with carbon dioxide over supported vanadium-antimony oxide catalysts, *Catal. Surv. Asia*, **11**, pp. 59–69.
118. Reddy, B., Han, D., Jiang, N., *et al.* (2008). Dehydrogenation of ethylbenzene to styrene with carbon dioxide over $ZrO_2$-based composite oxide catalysts, *Catal. Surv. Asia*, **12**, pp. 56–69.
119. Suzuki, T. (2008). Elucidation of reaction mechanisms over catalyst surface using unsteady-state techniques (in Japanese), *Shokubai*, **50**, pp. 648–653.
120. Chang, J., Vislovskiy, V., Park, M., *et al.* (2003). Utilization of carbon dioxide as soft oxidant in the dehydrogenation of ethylbenzene over supported vanadium-antimony oxide catalystst, *Green Chem.*, **5**, pp. 587–590.
121. Hong, D., Chang, J., Vislovskiy, V., *et al.* (2006). Dehydrogenation of Ethylbenzene with carbon dioxide as oxidant over Mg-Modified alumina-supported V-Sb oxide catalysts, *Chem. Lett.*, **35**, pp. 28–29.
122. Liu, B., Rui, G., Chang, R., *et al.* (2008). Dehydrogenation of ethylbenzene to styrene over LaVOx/SBA-15 catalysts in the Presence of carbon dioxide, *Appl. Catal. A*: *Gen.*, **335**, pp. 88–94.
123. Burri, D., Choi, K., Lee, J., *et al.* (2007). Influence of SBA-15 support on CeO2-ZrO2 catalyst for the dehydrogenation of ethylbenzene to styrene with $CO_2$, *Catal. Commun.*, **8**, pp. 43–48.
124. Liu, B., Rui, G., Jiang, L., *et al.* (2008). Preparation and high performance of $La_2O_3$-$V_2O_5$/MCM-41 Catalysts for Ethylbenzene dehydrogenation in the presence of $CO_2$, *J. Phys. Chem. C*, **112**, pp. 15490–15501.
125. Reddy, B., Lee, S., Han, D., *et al.* (2009). Utilization of carbon dioxide as soft oxidant for oxydehydrogenation of ethylbenzene to styrene over $V_2O_5$-$CeO_2$/$TiO_2$-$ZrO_2$ catalyst, *Appl. Catal. B*: *Environ.*, **87**, pp. 230–238.
126. Reddy, B. and Ataullah, K. (2005). Recent advances on $TiO_2$-$ZrO_2$ Mixed oxides as catalysts and catalyst supports, *Catal. Rev.*, **47**, pp. 257–296.
127. Burri, D., Choi, K., Han, S., *et al.* (2007). Dehydrogenation of ethylbenzene to styrene with $CO_2$ over $TiO_2$-$ZrO_2$ bifunctional catalyst, *B. Kor. Chem. Soc.*, **28**, pp. 53–58.
128. Rao, K., Reddy, B., Abhishek, B., *et al.* (2009). Effect of ceria on the structure and catalytic activity of $V_2O_5$/$TiO_2$-$ZrO_2$ for oxidehydrogenation of ethylbenzene to styrene utilizing $CO_2$ as soft oxidant, *Appl. Catal. B*: *Environ.*, **91**, pp. 649–656.
129. Kaspar, J., Di Monte, R., Fornasiero, P., *et al.* (2001). Dependency of the oxygen storage apacity in zirconia–ceria solid solutions upon textural properties, *Top. Catal.*, **16–17**, pp. 1–4.

# Chapter 27

# Membrane Reactors as Tools for Improved Catalytic Oxidation Processes

Miguel MENÉNDEZ*

This chapter reviews the possibilities that the application of a membrane in a catalytic reactor can improve the selectivity of a catalytic oxidation process to achieve a more compact system or to otherwise increase competitiveness. Classification differentiates between those reactors using dense membranes and those using porous membranes. Dense membranes provide high selectivity towards oxygen or hydrogen and the selective separation of one of these compounds under the reaction conditions is the key element in membrane reactors using such membranes. Porous membranes may have many different operation strategies and the contribution to the reaction can be based on a variety of approaches: reactant distribution, controlled contact of reactants or improved flow. Difficulties for the application of membrane reactors in industrial operation are also discussed.

## 27.1. Introduction

The integration of a membrane and a catalyst in a single piece of equipment to carry out a catalytic process is quite a new idea, and in recent decades researchers have identified many ways in which the synergetic effect achieved by such integration may contribute to developing more selective, safer or more efficient processes. This chapter will show in a systematic way the different combinations of membrane and catalyst in a membrane reactor (MR) that have been described in the literature to carry out catalytic oxidation. Although many reviews on inorganic membrane reactors are available,[1–11] a recent revision of the applications of membrane reactors in catalytic oxidations is not available. Since "catalytic oxidation" is a very wide-ranging concept, this work will exclude two kinds of reactions: biochemical oxidations in MR.[12] (e.g. enzymatic oxidations or the oxidation of organic compounds to $CO_2$ by means of microbes) and electrochemical oxidations in MR.[13] (e.g. catalytic oxidation in the anode of a fuel cell). This restriction leaves the still

*Aragon Institute of Engineering Research (I3A), University of Zaragoza, Mariano Esquillor s/n, 50018 Zaragoza, Spain.

**Table 27.1.** Classification of inorganic membranes most often employed in membrane reactors.

| Porosity | Type of material | Composition | Characteristics |
|---|---|---|---|
| Dense | Ceramic | Mixed oxides | Selectivity to $O_2$ or $H_2$ |
| | Metallic | Pd and Pd alloys | Selectivity to $H_2$ |
| Porous | Microporous | Zeolites, silica, Vycor | May act as molecular sieve, selective in several mixtures |
| | Mesoporous | Alumina, silica, ... | Low permselectivity; can be loaded with a catalyst |

wide field of MRs for catalytic oxidation in chemical and petrochemical processes as the scope of this chapter.

The first step is to classify the large variety of approaches that can distinguish between membrane reactors using dense membranes and those using porous membranes. Although polymeric membranes are more widely used than inorganic ones in most common applications, catalytic oxidations usually require temperatures higher than polymers can withstand and thus inorganic membranes are the usual choice. A classification of the inorganic membranes that have been employed in membrane reactors for catalytic oxidation is shown in Table 27.1.

The shape of the membranes employed in laboratory studies is usually tubular (i.e. tubes with an external diameter of around 1 cm), and this is clearly the preferred option when a catalyst bed is located inside a membrane. However, flat membranes are also used when the material is made in the laboratory and in some pilot plant applications. In some of the most recent works, hollow fiber membranes have been developed. This shape has the advantage of providing a large surface-to-volume ratio and requires a smaller amount of material per unit of surface area, both factors contributing to a lower cost.

## 27.2. Dense Membranes

Dense membranes are those in which the pores of the membrane material do not cross from one side of the membrane to the other. As shown in Table 27.1, two categories of dense membranes are employed for catalytic oxidations: ceramic membranes and metallic membranes.

### 27.2.1. *Ceramic membranes*

Ceramic membranes are composed of metal oxides that have suffered high-temperature treatment. They allow the permeation of oxygen ions by diffusion in the solid phase. Their application in membrane reactors requires the simultaneous

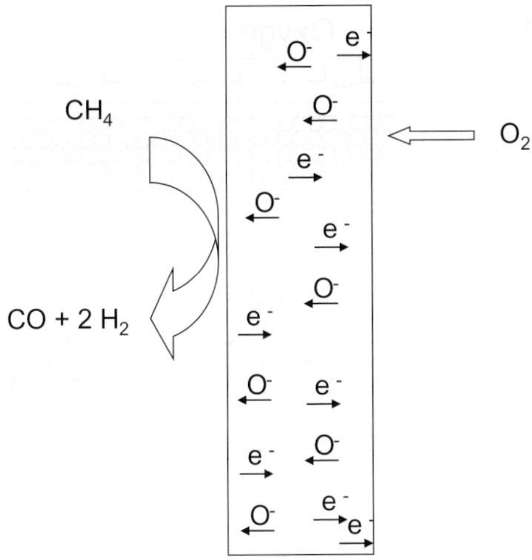


**Figure 27.1.** Flow of oxygen ions and electrons in a ceramic membrane (mixed ion electron conductor).

existence of a flow of electrons as illustrated in Fig. 27.1, and they are thus often referred to as mixed ion electron conducting (MIEC) membranes. If the flow of electrons is carried through an external circuit, the reactor would be a fuel cell. If an external electricity source is added, then it would be an electrochemical reactor. Both of these are beyond the scope of this chapter. Several reviews have discussed the applications of dense ceramic membranes.[9,10,14]

A detailed discussion of the mathematical models of oxygen flow in ceramic membranes is given elsewhere.[15] Typical materials employed in dense ceramic membranes have a brownmillerite or perovskite structure. The most commonly studied application for this kind of membrane is the catalytic partial oxidation of methane (POM) to obtain synthesis gas,

$$CH_4 + \frac{1}{2}O_2 \rightarrow CO + 2H_2$$

although the oxidative coupling of methane,

$$2CH_4 + \frac{1}{2}O_2 \rightarrow C_2H_6 + H_2O$$

and the oxidative dehydrogenation of alkanes,

$$C_nH_{2n+2} + \frac{1}{2}O_2 \rightarrow C_nH_{2n} + H_2O$$

have also been extensively studied in this kind of reactor.

There is a strong industrial interest in a process such as POM using oxygen selective membranes, since this would avoid the use of expensive cryogenic units for the separation of oxygen from air. In addition, the operation temperatures required

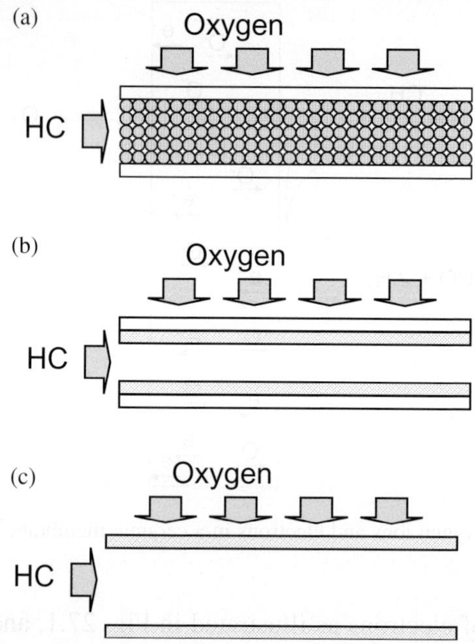


**Figure 27.2.** Three configurations for a catalytic oxidation process with a ceramic membrane: (a) a catalyst bed enclosed inside the membrane; (b) a layer of catalyst attached to the membrane surface; (c) the material of the membrane itself acting as the catalyst.

to achieve a suitable oxygen flow in these membranes (700–900°C) is similar to the reaction temperature. Several configurations have been proposed (Fig. 27.2):

i) a catalyst bed enclosed inside the membrane;
ii) a layer of catalyst attached to the membrane surface;
iii) the material of the membrane itself acting as the catalyst.

The last option has the advantage of simplicity, while the first offers an additional degree of freedom, since the amount of catalyst can be adapted to the permeation flow provided by the membrane. The first option also offers an additional degree of freedom since the amount of catalyst can be adapted to the needs of the membrane (the oxygen flow rate and its consumption rate by the reaction should be similar).

The construction of a pilot plant able to produce 0.1 ton/day of oxygen has been the most significant achievement, obtained by a consortium headed by Air Liquide,[16] although no reports of full-scale application of this kind of membrane have been published. A review of the possibilities in hydrogen production with $CO_2$ capture was carried out by Mundschau *et al.*[17] The application of dense ceramic membranes in membrane reactors for catalytic oxidation has been the target of a large number of research groups worldwide, as may be seen from the list of publications given in

Tables 27.2 (for hydrogen production), 27.3 (for the oxidative coupling of methane) and 27.4 (for the oxidative dehydrogenation of alkanes).

### 27.2.2. *Metallic membranes*

Unlike ceramic membranes, metallic membranes are mainly intended for hydrogen separation. Pd and Pd alloys are the preferred option, although many other metals can be selective to $H_2$.[18] The use of Pd as a hydrogen selective membrane can be traced back to the beginning of the 20th century and the use of such membrane reactors was widely studied by the Gryaznov group in the former USSR.[19] The development of thin Pd membranes supported on a porous material (ceramic or metallic) was a significant step, because it provides a cheaper product (with lower cost per unit of surface and higher hydrogen flux). Although the main application of Pd membrane reactors is to increase the conversion by the removal of hydrogen in reactions with equilibrium-limited conversion, they have also been employed for the partial oxidation of methane (POM),[20–23] providing high purity hydrogen, autothermal reforming of methane,[24] or oxidative reforming of ethanol.[25,26]

A quite recent application of Pd membranes are selective oxidations where the addition of oxygen to one side of the membrane, together with e.g. benzene, and hydrogen to the other side, results in high selectivity to phenol. The Pd membrane is adjacent to a TS-1 membrane. In this case the reaction is similar to the oxidation of benzene with $H_2O_2$, but it is generated *in situ* in the membrane, directly from $H_2$ and $O_2$. In fact it is possible to produce $H_2O_2$ by the oxidation of hydrogen in a Pd membrane reactor.[27] The inconvenience of this process is the low selectivity, since benzene and hydrogen oxidation are favoured.[28]

## 27.3. Porous Membranes

While the role of the membrane is straightforward in membrane reactors using dense membranes (simply to selectively permeate a reactant or product), the use of porous membranes has created a large variety of configurations, depending on the characteristics of the reaction. Therefore, for this kind of membrane, a classification based on the reactor structure rather than on the membrane material is more appropriate, since the same membrane can be used in different ways to improve the performance of different reactions. The following four operating modes, illustrated in Fig. 27.3, will be considered:

(i) membranes as a reactant distributor;
(ii) catalytic membranes as contactors;
(iii) flow-through membrane reactors;
(iv) three-phase reactors.

**Table 27.2.** Membrane reactors using dense ceramic membranes for hydrogen production.

| Material | Remarks | Reference |
|---|---|---|
| Perovskites | $Ni/Al_2O_3$ catalyst increases permeation | 67 |
| $Ba_{0.5}Sr_{0.5}Co_{0.8}Fe_{0.2}O_{3-\delta}$ | Also used for methane coupling | 68 |
| YSZ-$SrCo_{0.4}Fe_{0.6}O_{3-\delta}$ | Found metal redistribution after 7 h operation | 69,70 |
| Brownmillerite | — | 71 |
| $LaGaO_3$ | Electrical power + syngas production | 72 |
| Perovskite | BaSrCoFe as packed catalyst | 73 |
| BaSrCoFeO | Claim stable operation (500 h) LiLaNiO/$Al_2O_3$ packed bed | 74,75,76 |
| Perovskites | — | 77,78 |
| YSZ-$SRCo_{0.4}Fe_{0.6}O_{3-\delta}$ | — | 79 |
| LaSrCoGa or LaNbCo | Mainly total combustion in absence of catalyst | 80 |
| $SrFeCo_{0.5}O_y$ | With a packed bed of Ni/Al2O3 catalyst | 81 |
| $SrFe_{0.7}Al_{0.3}O_{3-\delta}$ | Suitable for POM, other structures cause total oxidation | 82–86 |
| $Ba_{0.5}Sr_{0.5}Co_{0.8}Fe_{0.2}O_3$ | Partial oxidation of heptane | 87,88 |
| BaSrCoFe | — | 89 |
| | Effect of the amount of catalyst | 90 |
| $Sm_{0.4}Ba_{0.6}Fe_{0.8}C_{0.2}O_{3-\delta}$ | — | 91 |
| $Ca_{0.8}Sr_{0.2}Ti_{1-x}Fe_XO_{3-\delta}$ | Packed bed of Ni/$Ca_{0.8}Sr_{0.2}Ti_{1-x}Fe_XO_{3-\delta}$ | 92 |
| YSZ-$SrCo_{0.4}Fe_{0.6}O_{3-\delta}$ | | 93 |
| YSZ-$SrCo_{0.4}Fe_{0.6}O_{3-\delta}$ | Hollow fibre | 94,95,96,97 |
| CeSmO-LaSrCr | Composite membrane; reports 1,000 h operation with high $\Delta P(O_2)$ | 98 |
| $BaCo_xFe_yZr_zO_{3-\delta}$ | Hollow fibre; Ni catalyst | 99,100 |
| Perovskite | Hollow fibre, includes mathematical model | 101,102,103 |
| Complex three layer | Full methane conversion | 104,105 |
| Several perovskites | Fuel oxi-reforming | 106,107,108 |
| $La_{0.6}Sr_{0.4}Co_{0.2}Fe_{0.8}O_{3-\delta}$ | Hollow fibre, total combustion | 109 |
| | Hollow fibre with water splitting | 110 |
| Perovskite | — | 111 |
| $BaCo_{0.7}Fe_{0.2}Nb_{0.1}O_{3-\delta}$ | Simulated coke gas as feed, several catalysts tested | 112,113 |
| $La_{0.8}Sr_{0.2}Fe_{0.7}Ga_{0.3}O_{3-\delta}$ | Work by Air Liquide; membrane supported on $La_{0.8}Ba_{0.2}Fe_{0.7}O_{3-\delta}$ | 114 |
| $BaCo_{0.7}Fe_{0.2}Nb_{0.1}O_{3-\delta}$ | Coke oven gas as feed; LiNiReO/$Al_2O_3$ or Ni/MgO catalyst. Coke formed on catalyst is regenerated by permeated oxygen | 115, 116, 117 |
| $BaCe_{0.1}Co_{0.4}Fe_{0.5}O_{3-\delta}$ | — | 118 |
| $Ba_{0.5}Sr_{0.5}Co_{0.8}Fe_{0.2}O_{3+\delta}$ | Laser ablation of membrane | 119 |
| $Ce_{0.8}Sm_{0.2}O_{2-\delta}$-$La_{0.8}Sr_{0.2}CrO_{3-\delta}$ | $SrTiO_3$ catalyst; phase composite membrane | 120 |

(*Continued*)

**Table 27.2.** (*Continued*)

| Material | Remarks | Reference |
|---|---|---|
| YSZ | Hollow fibre impregnated; Ni catalyst Coke is regenerated by permeated oxygen | 121 |
| | Ta containing perovskite | 122 |
| | Coke oven gas oxidation to syngas | 123,124 |
| $Ba_{0.5}Sr_{0.5}Co_{0.8}Fe_{0.2}O_x$ | Asymmetric membrane | 125 |
| $Ba_{0.5}Sr_{0.5}Co_{0.8}Fe_{0.2}O_x$ | Reforming of $CH_4$ with $CO_2$ | 126 |
| $La_{0.8}Sr_{0.2}F_{0.7}Ga_{0.3}O_{3-\delta}$ $La_{0.8}Sr_{0.2}F_{0.7}Nb_{0.3}O_{3-\delta}$ | 142 h of operation | 127 |

POM: partial oxidation of methane.

**Table 27.3.** Membrane reactors using dense membranes for the oxidative coupling of methane.

| Material | Remarks | References |
|---|---|---|
| Pb/MgO | Pb oxide permeates oxygen | 128 |
| PbO/YSZ | 90% selectivity | 129,130 |
| LaOCl | Textural instability | 131 |
| $Bi_{1.5}Y_{0.3}Sm_{0.2}O_{3-\delta}$ | Better yield than co-feeding | 132 |
| | Li/MgO catalyst; better yield in MR than co-feeding | 133 |
| $Y_2O_3/Bi_2O_3$ | — | 134,135,136 |
| $Bi_{1.5}Y_{0.3}Sm_{0.2}O_{3-\delta}$ | Near surface etching with pure $CH_4$ | 137 |
| $La_{0.8}Sr_{0.2}Co_{0.6}Fe_{0.4}O_{3-\delta}$ | — | 138 |
| Perovskite | Yield 16% | 139,140 |
| — | Some open porosity | 141 |
| BSCF | Membrane has catalytic activity | 74,76 |
| Y-BiO,Sm-Y-BiO | — | 142 |
| $Bi_{1.5}Y_{0.3}Sm_{0.2}O_{3-\delta}$ | Ionic conductivity, 35% yield | 143 |
| BSCF | Improvement with catalyst (LaSr/CaO) | 144 |
| Several mixed oxides | Pt/MgO catalyst | 145 |
| $La_{0.6}Sr_{0.4}Co_{0.2}Fe_{0.8}O_3$ | Hollow fibre | 101 |
| Perovskite | — | 146 |
| $Bi_{1.5}Y_{0.3}Sm_{0.2}O_{3-\delta}$ | Thin catalyst layer, 35% yield | 147 |
| Perovskite | 18% yield | 148 |

In addition, there is a large variety of configurations that cannot be included in the above classifications, often employed in only one or a few exploratory works, that will be reviewed in a fifth (miscellaneous) group.

### 27.3.1. *Membranes as reactant distributors*

In this configuration, often named a packed-bed enclosed membrane reactor (PBMR), the role of the membrane is to provide a controlled flow of oxygen to

**Table 27.4.** Membrane reactors using dense ceramic membranes for the oxidative dehydrogenation (ODH) of alkanes and other selective oxidations.

| Reaction | Membrane | Remarks | Reference |
|---|---|---|---|
| ODH propane | $Ba_{0.5}Sr_{0.5}Co_{0.8}Fe_{0.2}O_{3-\delta}$ | — | 149 |
| ODH ethane | $Ba_{0.5}Sr_{0.5}Co_{0.8}Fe_{0.2}O_{3-\delta}$ | Pd or V/MgO catalyst | 150,151 |
| ODH ethane | | Fluidized bed | 152 |
| ODH ethane | Au\|YSZ\|Pt | $V/Al_2O_3$, electrochemical reactor | 153 |
| ODH ethane | $VLaNi_xV_{1-x}O_{4+d}$ | Testing of materials by pulses | 154 |
| Propane dimerization | $M\text{-}Bi_2O_3$ (M = La, Ce,Eu, Er, V,Nb) | — | 155 |
| Propane to acrolein | $SrFeCo_{0.5}O_y$ | — | 156 |
| $NH_3 \rightarrow NOx$ | Perovskite | 98% selectivity, without noble metal | 157 |
| $NH_3 \rightarrow NOx$ | $Ba_{0.5}Sr_{0.5}Co_{0.8}Fe_{0.2}O_{3-\delta}$ | — | 158 |

a catalyst bed, distributing this feed along it. The dosed distribution of oxygen along the catalyst bed provides a low partial pressure of oxygen in the entire reactor, as opposed to a conventional reactor where most of the reaction occurs near the entry and with a high oxygen partial pressure. For many catalytic oxidations, the reaction order of oxygen for the desired product is smaller than for the formation of carbon oxides, and therefore a distributed dosing of oxygen improves the selectivity to the desired product. For example, in the oxidative coupling of methane, the reaction order of oxygen with many catalysts is smaller for the formation of dimerization products (ethane and ethylene) than for the total oxidation (to CO and $CO_2$).

Although the distribution of oxygen along the reactor is a concept already discussed in chemical reaction engineering textbooks,[29] the application of a porous membrane as a simple way to achieve such a distribution is a more recent development. The first application was used for the oxidative coupling of methane,[30] where a significant improvement in selectivity for a given conversion was found. Obviously the low oxygen concentration has a disadvantage because the reaction is slower and therefore the spatial time required to achieve a given conversion increases. In fact, for very large spatial times, the non-selective contribution of the membrane to the conversion can mask the improvement provided by the oxygen dosage, i.e. below a certain spatial time the increase in selectivity is surpassed by the non-selective conversion caused by the membrane.[31]

The PBMR has been employed successfully in a variety of catalytic selective oxidations:

(i) oxidative dehydrogenation of alkanes (ethane, propane, butane);
(ii) butane oxidation to maleic anhydride;

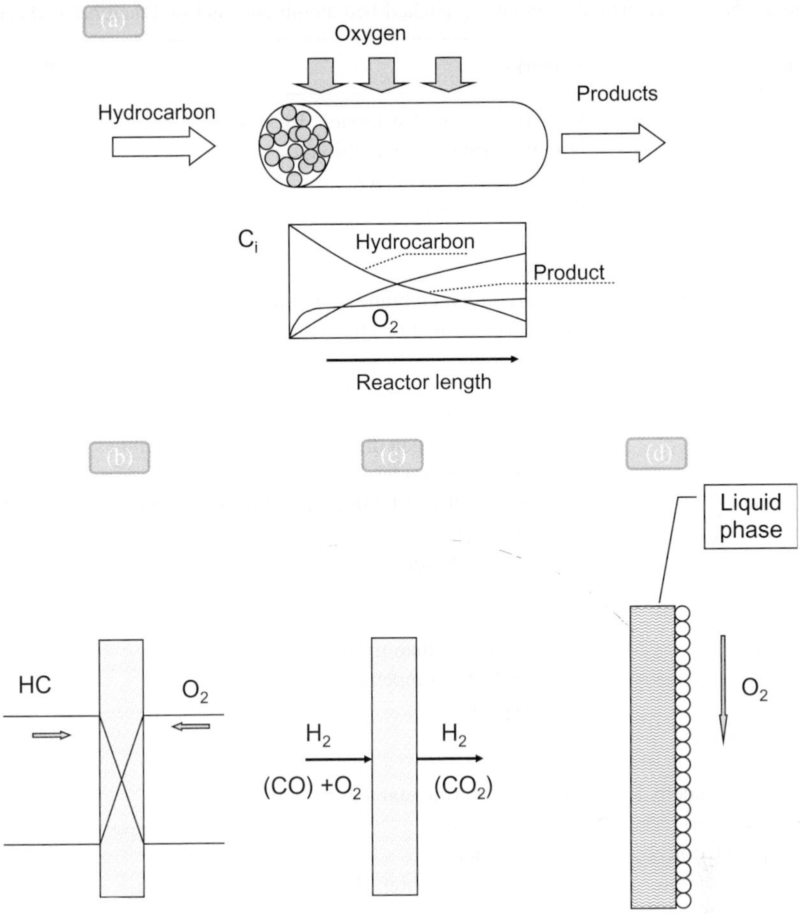


**Figure 27.3.** Four different configurations with a porous membrane in a membrane reactor for selective oxidation: (a) packed-bed enclosed membrane reactor; (b) membrane with catalytic activity as contactor; (c) flow-through configuration; (d) three-phase contactor.

(iii) propane oxidation to acrylic acid;
(iv) ethylene and propylene epoxidation;
(v) styrene oxidation;
(vi) butene oxidation to butadiene;
(vii) methanol to formaldehyde.

A list of experimental works using this configuration is given in Table 27.5.

In general, it may be expected that if the reaction kinetics favours the selectivity to the desired product at a low oxygen concentration, the use of a PBMR will provide an opportunity to improve the performance of the catalyst. Other advantages of this reactor configuration are the possibility of operation with reactant ratios that

**Table 27.5.** Experimental works using packed-bed membrane enclosed membrane reactors.

| Reaction | Remarks | Reference |
|---|---|---|
| MOC | Uniform oxygen distribution, Li/MgO | 30 |
| | Non-uniform oxygen distribution | 159 |
| | Comparison of catalysts | 160 |
| | Vycor® membrane | 36 |
| | $Sm_2O_3$/MgO | 35 |
| ODH C2 | V/MgO | 161 |
| ODH C3 | Using zeolite membrane as distributor | 37,162 |
| ODH C2 | Flames appear in a reactor with co-feeding, but not in MR | 163 |
| ODH C4 | Better selectivity and stability than fixed bed | 164 |
| OHD C3 | Yield optimization | 165 |
| ODH C3 | Non-uniform oxygen distribution; agreement with mathematical model | 166 |
| ODH C3 and C4 | Comparison PBR, PBMR and PBMR with some $O_2$ premixed | 167,168,169 |
| ODH C3 | Membrane in fluidized bed | 170,171 |
| ODH C3 | $Ga_2O_3$ / $MoO_3$ | 171 |
| Butane to maleic anhydride | Effect $O_2$ distribution, MR allows higher butane concentration than fixed bed with co-feeding | 172,173,174,175 |
| | Effect of $CO_2$, compensating reducing atmosphere | 33 |
| Butene oxidation to butadiene | — | 176 |
| Methane to formaldehyde | Includes experimental and modelling | 177 |
| Propane to acrolein | Improved yield | 178 |
| Ethylene epoxidation | Improved yield | 179 |
| Propylene epoxidation | MR safely allows high $H_2$ and $O_2$ concentration | 180 |
| Methanol oxidation to formaldehyde | Improvements in yield, testing of improved safety, reactor modelling and optimization | 181,182,183,184 |

MOC: oxidative coupling of methane; ODH: oxidative dehydrogenation.

would be in the explosion regime if the two reactants were premixed to feed a conventional reactor, and the more homogeneous heat generation that helps avoid hot spots.[32] An example of an operation with reactant concentration beyond the limits of a conventional reactor is the oxidation of butane to maleic anhydride. For this reaction, the butane feed in conventional reactors (fixed or fluidized beds) is kept low (typically below 2%), to keep the mixture outside the explosion limits. With a membrane reactor, higher butane concentrations have been successfully employed by keeping the oxygen concentration low and replenishing the oxygen in order to operate outside the explosion limits (see Fig. 27.4). In this reaction an oxygen concentration which is too low can drastically decrease the selectivity because an

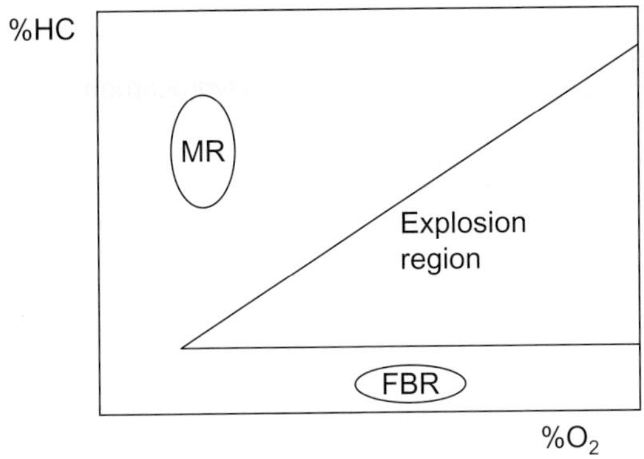


**Figure 27.4.** Operation outside explosion limits in a conventional fixed bed reactor (FBR) and in a membrane reactor (MR).

over-reduced catalyst burns the maleic anhydride, although it has been found that an atmosphere rich in $CO_2$ mitigates this negative effect.[33]

It is worth mentioning here the large variety of materials that have been employed for membranes in PBMR. The requirements are to provide a separation between the oxygen and the hydrocarbon feeds (i.e. to avoid the backpermeation of hydrocarbon to the oxygen side), to provide an even distribution of oxygen along the catalyst bed (although sometimes a non-homogeneous oxygen distribution can improve the yield to the desired product) and to be stable and as inert as possible. The achievement of an even oxygen distribution often requires a significant pressure drop through the membrane because otherwise a preferential permeation towards the end of the bed would occur. A combination of materials is often employed, e.g. an alumina membrane whose pores were filled with silica,[34] an alumina tube in which part of the surface was covered by enamel strips to adapt the permeability to the reaction rate.[35] or Vycor®.[36] Other authors[37] have employed zeolite membranes, which have oxygen permeability and are well suited to the reaction requirements.

### 27.3.2. *Membranes with catalytic activity as contactors*

It is not too difficult to fill the pores of a membrane with a catalytic material, and sometimes the material of the membrane itself (e.g. alumina, silica or zeolite) has suitable catalytic properties for the reaction of interest. In this case the membrane acts as a contactor between two feeds that are fed at opposite sides of the membrane. This was initially proposed[38] as a way to avoid the presence of $NH_3$ in a stream containing $NO_x$ during the reaction of ammonia oxidation. Another option is to preferentially locate the catalyst in one side of the membrane, with the aim of having a low oxygen concentration in contact with the catalyst, as is shown in Fig. 27.5. The possibility of improving the selectivity with this configuration was predicted by a mathematical

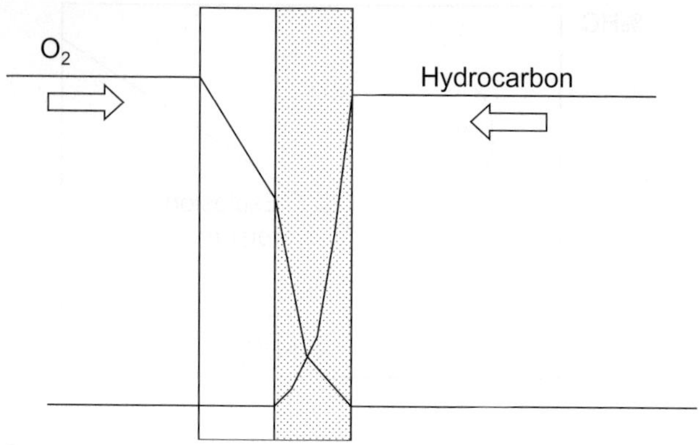


**Figure 27.5.** Concentration profiles in a catalytic membrane contactor with asymmetric catalyst loading.

model[39] and was later observed experimentally for the oxidative dehydrogenation of propane[40,41] and butane.[42]

### 27.3.3. *Flow-through configuration*

In this configuration all the reactants are premixed and permeate through the membrane. This operation mode was proposed for volatile organic compounds (VOC) combustion,[43–47] and it was found that the reaction temperature needed for a given degree of pollutant conversion was lower than in a situation with parallel flow. Since the lower reaction temperature reduces the energy required for heating the feed, this implies an obvious advantage in this process, where the volume of gases to be heated is very large in comparison with the volume of pollutants to be removed. The disadvantage is the pressure drop in the membrane, which may require an additional blower. A recent work proposes the use of this kind of reactor for the destruction of chemical warfare compounds.[48]

A very interesting result was achieved in the selective oxidation of CO in the presence of hydrogen. This reaction removes small amounts of CO remaining in the $H_2$ stream before feeding it to a fuel cell, because CO acts as a poison for the Pt catalyst in proton exchange membrane fuel cells. The main difficulty of this reaction is removing CO below a few ppm without oxidizing hydrogen. Usually an $O_2/CO$ ratio higher than 0.5 is employed as the CO conversion is not large enough if the $O_2/CO$ molar ratio is close to the stoichiometric value. The disadvantage is that the excess of oxygen consumes valuable hydrogen. Hasegawa[49–51] and Sotowa[52] found that a MR in flow-through configuration, using a zeolite Y membrane doped with Pt, can achieve a reduction of CO concentration below 10 ppm with an $O_2/CO$ ratio in the feed of only 1.2, which is an outstanding selectivity. Similar results were achieved by Bernardo,[53,54] thus confirming the interest of this kind of reactor.

Improvements in performance were also obtained in propene epoxidation[55] and the partial oxidation of methane to syngas with Ru.[56]

### 27.3.4. *Three-phase reactions*

In all the reactions considered up to now the reactants and products are gaseous, but porous membranes have also been applied in some three-phase reactions. Reactions tested in this case have included oxidation in the aqueous phase with air or the oxidation of organic substances with $H_2O_2$. The oxidation of organic compounds in the aqueous phase has recently attracted the interest of several researchers[57–60] as a way to remove pollutants. In this kind of reactor the membrane provides an interphase separating the liquid and the gas phase, and in some cases allows the concentration of a gaseous reactant to be handled.

### 27.3.5. *Other miscellaneous configurations*

A quite original approach to selective oxidations using $H_2O_2$ is based on polymeric membranes. In one case the polymer (polydimethyl siloxane) is filled with a zeolite which encapsulates an iron complex, i.e. the homogenous catalyst is immobilized in the membrane. The organic phase and the aqueous phase are fed at opposite sides of the membrane. This system has been used for the oxidation of cyclohexane to cyclohexanol and cyclohexanone, and the products were kept in the organic phase.[61] In addition, Buonomenna used polymeric membranes for the oxidation of benzyl alcohol to benzaldehyde[62] and for the oxidation of cyclohexene,[63] both with $H_2O_2$.

The oxidation with TS-1 zeolite membranes[64] offers the advantage that the separation of the catalyst is easier. A TS-1 zeolite membrane offers good selectivity in the oxidation of alcohols to ketone with $H_2O_2$[65] and styrene oxidation.[66]

## 27.4. Conclusions

It seems quite evident from the numerous examples of MR developed in the literature that they open many opportunities in catalytic oxidation. In some cases the advantage comes from a reduced cost to obtain a main reactant (oxygen), in others the advantage is the improvement in selectivity or a safer operation. The way to the commercial application still seems long and full of hurdles. Some difficulties come from the cost of membranes, other from the resistance to make changes in well-established processes. The risk of membrane breakage is also in the mind of any potential investor. However, the amount of work and the variety of approaches is so impressive, that it is foreseeable that sooner or later one of these developments will generate a successful industrial application. It is also foreseeable that the first application will pave the way for many more.

## References


1. Armor, J. (1998). Applications of Catalytic Inorganic Membrane Reactors to Refinery Products, *J. Membrane Sci.*, **147**, pp. 217–233.
2. Liu, S., Tan, X., Li, K. and Hughes, R. (2001). Methane Coupling Using Catalytic Membrane Reactors, *Catal. Rev.*, **43**, pp. 147–198.
3. Dalmon, J., Cruz-López, A., Farrusseng, D., *et al.* (2007). Oxidation in Catalytic Membrane Reactors, *Appl. Catal. A: Gen.*, **325**, pp. 198–204.
4. Saracco, G., Neomagus, H., Versteeg, G., *et al.*, (1999). High-Temperature Membrane Reactors: Potential and Problems, *Chem. Eng. Sci.*, **54**, pp. 1997–2017.
5. Sanchez-Marcano, J.G. and Tsotsis, T.T. (2002). *Catalytic Membranes and Membrane reactors*; Wiley VCH, Weinheim.
6. Lintz, H. and Reitzmann, A. (2007). Alternative Reaction Engineering Concepts in Partial Oxidations on Oxidic Catalysts, *Catal. Rev.*, **49**, pp. 1–32.
7. Dittmeyer, R., Svajda, K. and Reif, M. (2004). A Review of Catalytic Membrane Layers for Gas/Liquid Reactions, *Top. Catal.*, **29**, pp. 3–27.
8. Sundmacher, K., Rihko-Struckmann, L. and Galvita, V. (2005). Solid Electrolyte Membrane Reactors: Status and Trends, *Catal. Today*, **104**, pp. 185–199.
9. Garagounis, I., Kyriakou, V., Anagnostou, C., *et al.* (2011). Solid Electrolytes: Applications in Heterogeneous Catalysis and Chemical Cogeneration, *Ind. Eng. Chem. Res.*, **50**, pp. 431–472.
10. Yang, W., Wang, H., Zhu, X., *et al.* (2005). Development and Application of Oxygen Permeable Membrane in Selective Oxidation of Light Alkanes, *Top. Catal.*, **35**, pp.155–167.
11. Caro, J. (2006). Membrane Reactors for Catalytic Oxidation, *Chem. Ing. Tech.*, **78**, pp. 899–912.
12. Mazzei, R., Drioli, E. and Giorno, L. (2010). Comprehensive Membrane Science and Engineering, in E. Drioli and L. Giorno (eds), *Biocatalytic Membranes and Membrane Bioreactors*, Elsevier, Oxford, pp. 195–212.
13. Marnellos, G. and Stoukides, M. (2004). Catalytic Studies in Electrochemical Membrane Reactors, *Solid State Ionics*, **175**, pp. 597–603.
14. Garagounis, I., Kyriakou, V., Anagnostou, C., *et al.* (2011). Solid Electrolytes: Applications in Heterogeneous Catalysis and Chemical Cogeneration, *Ind. Eng. Chem. Res.*, **50**, pp. 431–472.
15. Fontaine, M., Norby, T., Larring, Y., *et al.* (2008). Oxygen and Hydrogen Separation Membranes Based on Dense Ceramic Conductors, *Membrane Science and Technology*, **13**, pp. 401–458.
16. Bose, A., Richards, R., Sammells, A., *et al.* (2002). Beyond State-of-the-Art Gas Separation Processes Using Ion-Transport Membranes, *Desalination*, **144**, pp. 91–92.
17. Mundschau, M., Xie, X., Evenson IV, C., *et al.* (2006). Dense Inorganic Membranes for Production of Hydrogen from Methane and Coal with Carbon Dioxide Sequestration, *Catal. Today*, **118**, pp. 12–23.
18. Basile, A., Gallucci, F., Tosti, S. (2008). Synthesis, Characterization, and Applications of Palladium Membranes, in R. Mallada and M. Menendez (eds), *Inorganic Membranes: Synthesis, Characterization and Applications* (Membrane Science and Technology series), Elsevier, Amsterdam, pp. 255–323.
19. Gryaznov, V. (1999). Membrane Catalysis, *Catal. Today*, **51**, pp. 391–395.
20. Basile, A., Paturzo, L. and Laganà, F. (2001). The Partial Oxidation of Methane to Syngas in a Palladium Membrane Reactor: Simulation and Experimental Studies, *Catal. Today*, **67**, pp. 65–75.
21. Basile, A. and Paturzo, L. (2001). An Experimental Study of Multilayered Composite Palladium Membrane Reactors for Partial Oxidation of Methane to Syngas, *Catal. Today*, **67**, pp. 55–64.
22. Paturzo, L. and Basile, A. (2002). Methane Conversion to Syngas in a Composite Palladium Membrane Reactor with Increasing Number of Pd Layers, *Ind. Eng. Chem. Res.*, **41**, pp. 1703–1710.

23. Kikuchi, E. and Chen, Y. (1998). Syngas Formation by Partial Oxidation of Methane in Palladium Membrane Reactor. *Stud. Surf. Sci. Catal.*, **119**, pp. 441–446.
24. Chang, H., Pai, W., Chen, Y., *et al.* (2010). Autothermal Reforming of Methane for Producing High-Purity Hydrogen in a Pd/Ag Membrane Reactor, *Int. J. Hydrogen Energ.*, **35**, pp. 12986–12992.
25. Iulianelli, A., Longo, T., Liguori, S., *et al.* (2009). Oxidative Steam Reforming of Ethanol over Ru-$Al_2O_3$ Catalyst in a Dense Pd-Ag Membrane Reactor to Produce Hydrogen for PEM Fuel Cells, *Int. J. Hydrogen Energ.*, **34**, pp. 8558–8565.
26. Lin, W., Liu, Y. and Chang, H. (2010). Autothermal Reforming of Ethanol in a Pd-Ag/Ni Composite Membrane Reactor, *Int. J. Hydrogen Energ.*, **35**, pp. 12961–12969.
27. Choudhary, V., Gaikwad, A. and Sansare, S. (2001). Non-Hazardous Direct Oxidation of Hydrogen to Hydrogen Peroxide Using a Novel Membrane Catalyst, *Angew. Chem. Int. Edit.*, **40**, pp. 1776–1779.
28. Vulpescu, G., Ruitenbeek, M., Van Lieshout, L., *et al.* (2004). One-Step Selective Oxidation Over a Pd-Based Catalytic Membrane; Evaluation of the Oxidation of Benzene to Phenol as a Model Reaction, *Catal. Commun.*, **5**, pp. 347–351.
29. Levenspiel, O. (1996). *The Chemical Reactor Omnibook.* OSU Bookstores: Corvallis, OR.
30. Coronas, J., Menéndez, M. and Santamaria, J. (1994). Methane Oxidative Coupling Using Porous Ceramic Membrane Reactors. II. Reaction Studies, *Chem. Eng. Sci.*, **49**, pp. 2015–2025.
31. Coronas, J., Gonzalo, A., Lafarga, D., *et al.* (1997). Effect of the Membrane Activity on the Performance of a Catalytic Membrane Reactor, *AIChE J.*, **43**, pp. 3095–3104.
32. Coronas, J., Menéndez, M. and Santamaría, J. (1995). The Porous-Wall Ceramic Membrane Reactor: An Inherently Safer Contacting Device for Gas-Phase Oxidation of Hydrocarbons, *J. Loss Prevent. Proc.*, **8**, pp. 97–101.
33. Mallada, R., Menéndez, M. and Santamaría, J. (2002). On the Favourable Effect of $CO_2$ Addition in the Oxidation of Butane to Maleic Anhydride Using Membrane Reactors, *Appl. Catal. A: Gen.*, **231**, pp. 109–116.
34. Lafarga, D., Santamaria, J. and Menéndez, M. (1994). Methane Oxidative Coupling Using Porous Ceramic Membrane Reactors. I. Reactor Development, *Chem. Eng. Sci.*, **49**, pp. 2005–2013.
35. Tonkovich, A., Jimenez, D., Zilka, J., *et al.* (1996). Inorganic Membrane Reactors for the Oxidative Coupling of Methane, *Chem. Eng. Sci.*, **51**, pp. 3051–3056.
36. Ramachandra, A., Lu, Y., Ma, Y., *et al.* (1996). Oxidative Coupling of Methane in Porous Vycor Membrane Reactors, *J. Membrane Sci.*, **116**, pp. 253–264.
37. Pantazidis, A. Dalmon, J. and Mirodatos, C. (1995). Oxidative Dehydrogenation of Propane on Catalytic Membrane Reactors, *Catal. Today*, **25**, pp. 403–408.
38. Sloot, H., Versteeg, G., Smolders, C., *et al.* (1991). A Non-Permselective Membrane Reactor for the Selective Catalytic Reduction of $NO_x$ with Ammonia, in *Proc. 2nd Int. Conf. on Inorganic Membranes*, Montpellier (France), Trans. Tech. Publ., Zuerich (Switzerland). pp. 261–266.
39. Harold, M., Zaspalis, V., Keizer, K., *et al.* (1993). Intermediate Product Yield Enhancement with a Catalytic Inorganic Membrane. 1. Analytical Model for the Case of Isothermal and Differential Operation, *Chem. Eng. Sci.*, **48**, pp. 2705–2725.
40. Alfonso, M., Julbe, A., Farrusseng, D., *et al.* (1999). Oxidative Dehydrogenation of Propane on V/$Al_2O_3$ Catalytic Membranes. Effect of the Type of Membrane and Reactant Feed Configuration, *Chem. Eng. Sci.*, **54**, pp. 1265–1272.
41. Alfonso, M., Menéndez, M. and Santamaría, J. (2000). Vanadium-Based Catalytic Membrane Reactors for the Oxidative Dehydrogenation of Propane, *Catal. Today*, **56**, pp. 247–252.
42. Alfonso, M., Menéndez, M. and Santamaría, J. (2002). Oxidative Dehydrogenation of Butane on V/MgO Catalytic Membranes, *Chem. Eng. J.*, **90**, pp. 131–138.

43. Pina, M., Menéndez, M. and Santamaría, J. (1996). The Knudsen-Diffusion Catalytic Membrane Reactor: An Efficient Contactor for the Combustion of Volatile Organic Compounds, *Appl. Catal. B: Environ.*, **11**, pp. L19–L27.
44. Pina, M., Irusta, S., Menendez, M., *et al.* (1997). Combustion of Volatile Organic Compounds over Platinum-Based Catalytic Membranes, *Ind. Eng. Chem. Res.*, **36**, pp. 4557–4566.
45. Irusta, S., Pina, M., Menendez, M., *et al.* (1998). Development and Application of Perovskite-Based Catalytic Membrane Reactors, *Catal. Lett.*, **54**, pp. 69–78.
46. Irusta, S., Pina, M., Menendez, M., *et al.* (1998). Catalytic Combustion of Volatile Organic Compounds over La-Based Perovskites, *J. Catal.*, **179**, pp. 400–412.
47. Zalamea, S., Pina, M., Villellas, A., *et al.* (1999). Combustion of Volatile Organic Compounds over Mixed-Regime Catalytic Membranes, *React. Kinet. Catal. L.*, **67**, pp. 13–19.
48. Motamedhashemi, M., Egolfopoulos, F. and Tsotsis, T. (2011). Application of a Flow-Through Catalytic Membrane Reactor (FTCMR) for the Destruction of a Chemical Warfare Simulant, *J. Membrane Sci.*, **376**, pp. 119–131.
49. Hasegawa, Y., Kusakabe, K. and Morooka, S. (2001). Selective Oxidation of Carbon Monoxide in Hydrogen-Rich Mixtures by Permeation through a Platinum-Loaded Y-Type Zeolite Membrane, *J. Membrane Sci.*, **190**, pp. 1–8.
50. Hasegawa, Y., Sotowa, K., Kusakabe, K., *et al.* (2002). The Influence of Feed Composition on CO Oxidation Using Zeolite Membranes Loaded with Metal Catalysts, *Micropor. Mesopor. Mat.*, **53**, pp. 37–43.
51. Hasegawa, Y., Sotowa, K. and Kusakabe, K. (2003). Permeation Behavior during the Catalytic Oxidation of CO in a Pt-Loaded Y-type Zeolite Membrane, *Chem. Eng. Sci.*, **58**, pp. 2797–2803.
52. Sotowa, K., Hasegawa, Y., Kusakabe, K., *et al.* (2002). Enhancement of CO Oxidation by Use of $H_2$-Selective Membranes Impregnated with Noble-Metal Catalysts, *Int. J. Hydrogen Energ.*, **27**, pp. 339–346.
53. Bernardo, P., Algieri, C., Barbieri, G., *et al.* (2006). Catalytic (Pt-Y) Membranes for the Purification of $H_2$-Rich Streams, *Catal. Today*, **118**, pp. 90–97.
54. Bernardo, P., Algieri, C., Barbieri, G., *et al.* (2008). Hydrogen Purification from Carbon Monoxide by Means of Selective Oxidation Using Zeolite Catalytic Membranes, *Sep. Purif. Technol.*, **62**, pp. 629–635.
55. Kobayashi, M., Togawa, J., Kanno, T., *et al.* (2003). Dramatic Innovation of Propene Epoxidation Efficiency Derived from a Forced Flow Membrane Reactor, *J. Chem. Technol. Biotechnol.*, **78**, pp. 303–307.
56. Paturzo, L., Gallucci, F., Basile, A., *et al.* (2003). Partial Oxidation of Methane in a Catalytic Ruthenium Membrane Reactor, *Ind. Eng. Chem. Res.*, **42**, pp. 2968–2974.
57. Gutiérrez, M., Pina, P., Torres, M., *et al.* (2010). Catalytic Wet Oxidation of Phenol Using Membrane Reactors: A Comparative Study with Slurry-Type Reactors, *Catal. Today*, **149**, pp. 326–333.
58. Iojoiu, E., Miachon, S., Landrivon, E., *et al.* (2007). Wet Air Oxidation in a Catalytic Membrane Reactor: Model and Industrial Wastewaters in Single Tubes and Multichannel Contactors, *Appl. Catal. B: Environ.*, **69**, pp. 196–206.
59. Iojoiu, E., Landrivon, E., Raeder, H., *et al.* (2006). The "Watercatox" Process: Wet Air Oxidation of Industrial Effluents in a Catalytic Membrane Reactor. First Report on Contactor CMR Up-Scaling to Pilot Unit, *Catal. Today*, **118**, pp. 246–252.
60. Raeder, H., Bredesen, R., Crehan, G., *et al.* (2003). A Wet Air Oxidation Process Using a Catalytic Membrane Contactor, *Sep. Purif. Technol.*, **32**, pp. 349–355.
61. Langhendries, G., Baron, G., Vankelecom, I., *et al.* (2000). Selective Hydrocarbon Oxidation Using a Liquid-Phase Catalytic Membrane Reactor, *Catal. Today*, **56**, pp. 131–135.
62. Buonomenna, M. and Drioli, E. (2008). Benzyl Alcohol Oxidation to Benzaldehyde in Multiphase Membrane Reactor, *Org. Process Res. Dev.*, **12**, pp. 982–988.

63. Buonomenna, M., Golemme, G., De Santo, M., *et al.* (2010). Direct Oxidation of Cyclohexene with Inert Polymeric Membrane Reactor, *Org. Process Res. Dev.*, **14**, pp. 252–258.
64. Wang, X., Zhang, B., Liu, X., *et al.* (2006). Synthesis of b-Oriented TS-1 Films on Chitosan-Modified $\alpha$-$Al_2O_3$ Substrates, *Adv. Mater.*, **18**, pp. 3261–3265.
65. Chen, P., Chen, X., Tanaka, K., *et al.* (2007). A Novel and Less–Expensive Preparation of Titanium Silicalite–1 Membrane, *Chem. Lett.*, **36**, pp. 1078–1079.
66. Qiu, F., Wang, X., Zhang, X., *et al.* (2009). Preparation and Properties of TS-1 Zeolite and Film Using Sil-1 Nanoparticles as Seeds, *Chem. Eng. J.*, **147**, pp. 316–322.
67. Tsai, C., Dixon, A., Moser, W., *et al.* (1997). Dense Perovskite Membrane Reactors for Partial Oxidation of Methane to Syngas, *AICHE J.*, **43**, pp. 2741–2750.
68. Cong, Y., Shao, Z., Yang, W., *et al.* (2000). Synthesis of Novel Oxygen-Permeable Ceramic Membrane and its Application in Oxidation Conversion of Methane, *Chinese J. Catal.*, **21**, pp. 363–366.
69. Jin, W., Gu, X., Li, S., *et al.* (2000). Experimental and Simulation Study on a Catalyst Packed Tubular Dense Membrane Reactor for Partial Oxidation of Methane to Syngas, *Chem. Eng. Sci.*, **55**, pp. 2617–2625.
70. Jin, W., Li, S., Huang, P., *et al.* (2000). Tubular Lanthanum Cobaltite Perovskite-Type Membrane Reactors for Partial Oxidation of Methane to Syngas, *J. Membrane Sci.*, **166**, pp. 13–22.
71. Sammells, A., Schwartz, M., Mackay, R., *et al.* (2000). Catalytic Membrane Reactors for Spontaneous Synthesis Gas Production, *Catal. Today*, **56**, pp. 325–328.
72. Ishihara, T. and Takita, Y. (2000). Partial Oxidation of Methane into Syngas with Oxygen Permeating Ceramic Membrane Reactors, *Catal. Surv. Jpn*, **4**, pp. 125–133.
73. Dong, H., Shao, Z., Xiong, G., *et al.* (2001). Investigation on POM Reaction in a New Perovskite Membrane Reactor, *Catal. Today*, **67**, pp. 3–13.
74. Shao, Z., Xiong, G., Dong, H., *et al.* (2001). Synthesis, Oxygen Permeation Study and Membrane Performance of a $Ba_{0.5}$ $Sr_{0.5}$ $Co_{0.8}$ $Fe_{0.2}$ $O_{3-\delta}$ Oxygen-Permeable Dense Ceramic Reactor for Partial Oxidation of Methane to Syngas, *Sep. Purif. Technol.*, **25**, pp. 97–116.
75. Shao, H., Zhong, S. and Guo, J. (2004). Pd-Cu, *Chinese J. Catal.*, **25**, pp. 143–148.
76. Shao, Z., Dong, H., Xiong, G., *et al.* (2001). Performance of a Mixed-Conducting Ceramic Membrane Reactor with High Oxygen Permeability for Methane Conversion, *J. Membrane Sci.*, **183**, pp. 181–192.
77. Bouwmeester, H. (2003). Dense Ceramic Membranes for Methane Conversion, *Catal. Today*, **82**, pp. 141–150.
78. Chen, C., Feng, S., Ran, S., *et al.* (2003). Conversion of Methane to Syngas by a Membrane-Based Oxidation-Reforming Process, *Angew. Chem. Int. Edit.*, **42**, pp. 5196–5198.
79. Gu, X., Jin, W., Chen, C., *et al.* (2002). YSZ-$SRCo_{0.4}Fe_{0.6}O_{3-\delta}$ Membranes for the Partial Oxidation of Methane to Syngas, *AIChE J.*, **48**, pp. 2051–2060.
80. Yaremchenko, A., Valente, A., Kharton, V., *et al.* (2003). Oxidation of Dry Methane on the Surface of Oxygen Ion-Conducting Membranes, *Catal. Lett.*, **91**, pp. 169–174.
81. Feng, S., Ran, S., Zhu, D., *et al.* (2004). Synthesis Gas Production from Methane with $SrFeCo_{0.5}O_y$ Membrane Reactor, *Energ. Fuel*, **18**, pp. 385–389.
82. Kharton, V., Yaremchenko, A., Tsipis, E., *et al.* (2004). Characterization of Mixed-Conducting $La_2Ni_{0.9}Co_{0.1}O_{4+\delta}$ Membranes for Dry Methane Oxidation, *Appl. Catal. A: Gen.*, **261**, pp. 25–35.
83. Kharton, V., Sobyanin, V., Belyaev, V. *et al.* (2004). Methane Oxidation on the Surface of Mixed-Conducting $La_{0.3}Sr_{0.7}Co_{0.8}Ga_{0.2}O_{3-\delta}$, *Catal. Commun.*, **5**, pp. 311–316.
84. Kharton, V., Yaremchenko, A., Valente, A., *et al.* (2005). Methane Oxidation over Fe-, Co-, Ni- and V-containing Mixed Conductors, *Solid State Ionics*, **176**, pp.781–791.
85. Kharton, V., Patrakeev, M., Waerenborgh, J., *et al.* (2005). Methane Oxidation over Perovskite-Related Ferrites: Effects of Oxygen Nonstoichiometry, *Solid State Sci.*, **7**, pp. 1344–1352.

86. Kharton, V., Waerenborgh, J., Rojas, D., *et al.* (2005). Mössbauer Spectra and Catalytic Behavior of Perovskite-Like $SrFe_{0.7}Al_{0.3}O_{3-\delta}$, *Catal. Lett.*, **99**, pp. 249–255.
87. Zhu, W., Xiong, G., Han, W., *et al.* (2004). Catalytic Partial Oxidation of Gasoline to Syngas in a Dense Membrane Reactor, *Catal. Today*, **93–95**, pp. 257–261.
88. Zhu, W., Han, W., Xiong, G., *et al.* (2005). Mixed Reforming of Heptane to Syngas in the $Ba_{0.5}Sr_{0.5}Co_{0.8}Fe_{0.2}O_3$ Membrane Reactor, *Catal. Today*, **104**, pp. 149–153.
89. Lu, H., Tong, J., Cong, Y., *et al.* (2005). Partial Oxidation of Methane in $Ba_{0.5}Sr_{0.5}Co_{0.8}Fe_{0.2}O_{3-\delta}$ Membrane Reactor at High Pressures, *Catal. Today*, **104**, pp. 154–159.
90. Zhang, P., Chang, X., Wu, Z., *et al.* (2005). Effect of the Packing Amount of Catalysts on the Partial Oxidation of Methane Reaction in a Dense Oxygen-Permeable Membrane Reactor, *Ind. Eng. Chem. Res.*, **44**, pp. 1954–1959.
91. Ikeguchi, M., Mimura, T., Sekine, Y., *et al.* (2005). Reaction and Oxygen Permeation Studies in $Sm_{0.4}Ba_{0.6}Fe_{0.8}Co_{0.2}O_{3-\delta}$ Membrane Reactor for Partial Oxidation of Methane to Syngas, *Appl. Catal. A: Gen.*, **290**, pp. 212–220.
92. Hamakawa, S., Sato, K., Inoue, T., *et al.* (2006). Design of One-Component Ceramic Membrane-Reactor for Natural Gas Conversion, *Catal. Today*, **117**, pp. 297–303.
93. Hu, J., Xing, T., Jia, Q., *et al.* (2006). Methane Partial Oxidation to Syngas in $YBa_2Cu_3O_{7-x}$ Membrane Reactor, *Appl. Catal. A: Gen.*, **306**, pp. 29–33.
94. Caro, J., Schiestel, T., Werth, S., *et al.* (2006). Perovskite Hollow Fibre Membranes in the Partial Oxidation of Methane to Synthesis Gas in a Membrane Reactor, *Desalination*, **199**, pp. 415–417.
95. Caro, J., Wang, H., Tablet, C., *et al.* (2006). Evaluation of Perovskites in Hollow Fibre and Disk Geometry in Catalytic Membrane Reactors and in Oxygen Separators, *Catal. Today*, **118**, pp. 128–135.
96. Caro, J., Caspary, K., Hamel, C., *et al.* (2007). Catalytic Membrane Reactors for Partial Oxidation Using Perovskite Hollow Fiber Membranes and for Partial Hydrogenation Using a Catalytic Membrane Contactor, *Ind. Eng. Chem. Res.*, **46**, pp. 2286–2294.
97. Kleinert, A., Feldhoff, A., Schiestel, T., *et al.* (2006). Novel Hollow Fibre Membrane Reactor for the Partial Oxidation of Methane, *Catal. Today*, **118**, pp. 44–51.
98. Wang, B., Yi, J., Winnubst, L., *et al.* (2006). Stability and Oxygen Permeation Behavior of $Ce_{0.8}Sm_{0.2}O_{2-\delta}$-$La_{0.8}Sr_{0.2}CrO_{3-\delta}$ Composite Membrane under Large Oxygen Partial Pressure Gradients, *J. Membrane Sci.*, **286**, pp. 22–25.
99. Wang, H., Tablet, C., Schiestel, T., *et al.* (2006). Partial Oxidation of Methane to Syngas in a Perovskite Hollow Fiber Membrane Reactor, *Catal. Commun.*, **7**, pp. 907–912.
100. Wang, H., Feldhoff, A., Caro, J., *et al.* (2009). Oxygen Selective Ceramic Hollow Fiber Membranes for Partial Oxidation of Methane, *AIChE J.*, **55**, pp. 2657–2664.
101. Tan, X., Pang, Z., Gu, Z., *et al.* (2007). Catalytic Perovskite Hollow Fibre Membrane Reactors for Methane Oxidative Coupling, *J. Membrane Sci.*, **302**, pp. 109–114.
102. Tan, X., Thursfield, A., Metcalfe, I., *et al.* (2009). Analysis of a Perovskite Ceramic Hollow Fibre Membrane Reactor for the Partial Oxidation of Methane to Syngas, *Asia Pac. J. Chem. Eng.*, **4**, pp. 251–258.
103. Tan, X. and Li, K. (2009). Design of Mixed Conducting Ceramic Membranes/Reactors for the Partial Oxidation of Methane to Syngas, *AIChE J.*, **55**, pp. 2675–2685.
104. Yin, X., Hong, L. and Liu, Z. (2007). Asymmetric Tubular Oxygen-Permeable Ceramic Membrane Reactor for Partial Oxidation of Methane, *J. Phys. Chem. C*, **111**, pp. 9194–9202.
105. Yin, X., Hong, L. and Liu, Z. (2008). Integrating Air Separation with Partial Oxidation of Methane: A Novel Configuration of Asymmetric Tubular Ceramic Membrane Reactor, *J. Membrane Sci.*, **311**, pp. 89–97.
106. Mundschau, M., Burk, C. and Gribble Jr, D. (2008). Diesel Fuel Reforming Using Catalytic Membrane Reactors, *Catal. Today*, **136**, pp. 190–205.

107. Mundschau, M., Gribble. Jr, D., Henton, L., *et al.* (2010). Reforming Diesel-Fuel Distillates with Membrane Reactors, *Asia-Pacific J. Chem. Eng.*, **5**, pp. 160–168.
108. Mundschau, M., Gribble Jr, D., Plassmeyer, P., *et al.* (2010). Dry Catalytic Partial Oxidation of Diesel-Fuel Distillates into Syngas, *Fuel*, **89**, pp. 1202–1211.
109. Tan, X., Li, K., Thursfield, A., *et al.* (2008). Oxyfuel Combustion Using a Catalytic Ceramic Membrane Reactor, *Catal. Today*, **131**, pp. 292–304.
110. Jiang, H., Wang, H., Werth, S., *et al.* (2008). Simultaneous Production of Hydrogen and Synthesis Gas by Combining Water Splitting with Partial Oxidation of Methane in a Hollow-Fiber Membrane Reactor, *Angew. Chem. Int. Edit.*, **47**, pp. 9341–9344.
111. Zhang, C., Chang, X., Dong, X., *et al.* (2008). The Oxidative Stream Reforming of Methane to Syngas in a Thin Tubular Mixed-Conducting Membrane Reactor, *J. Membrane Sci.*, **320**, pp. 401–406.
112. Cheng, H., Lu, X., Zhang, Y., *et al.* (2009). Hydrogen Production by Reforming of Simulated Hot Coke Oven Gas over Nickel Catalysts Promoted with Lanthanum and Cerium in a Membrane Reactor, *Energ. Fuel*, **23**, pp. 3119–3125.
113. Cheng, H., Lu, X., Liu, X., *et al.* (2009). Partial Oxidation of Simulated Hot Coke Oven Gas to Syngas over Ru-Ni/Mg(Al)O Catalyst in a Ceramic Membrane Reactor, *J. Nat. Gas Chem.*, **18**, pp. 467–473.
114. Julian, A., Juste, E., Geffroy, P., *et al.* (2009). Elaboration of $La_{0.8}Sr_{0.2}Fe_{0.7}Ga_{0.3}O_{3-\delta}$/$La_{0.8}M_{0.2}FeO_{3-\delta}$ (M = Ca, Sr and Ba) Asymmetric Membranes by Tape-Casting and Co-Firing, *J. Membrane Sci.*, **333**, pp. 132–140.
115. Yang, Z., Zhang, Y., Ding, W., *et al.* (2009). Hydrogen Production from Coke Oven Gas over LiNi/$\gamma$-$Al_2O_3$ Catalyst Modified by Rare Earth Metal Oxide in a Membrane Reactor, *J. Nat. Gas Chem.*, **18**, pp. 407–414.
116. Yang, Z., Ding, W., Zhang, Y., *et al.* (2010). Catalytic Partial Oxidation of Coke Oven Gas to Syngas in an Oxygen Permeation Membrane Reactor Combined with NiO/MgO Catalyst, *Int. J. Hydrogen Energ.*, **35**, pp. 6239–6247.
117. Yang, Z., Ding, W., Zhang, Y., *et al.* (2010). Catalytic Partial Oxidation of Coke Oven Gas to Syngas in an Oxygen Permeation Membrane Reactor Combined with NiO/MgO Catalyst, *Renew. Energ.*, **35**, pp. 6239–6247.
118. Li, Q., Zhu, X., He, Y., *et al.* (2010). Partial Oxidation of Methane in $BaCe_{0.1}Co_{0.4}Fe_{0.5}O_{3-\delta}$ Membrane Reactor, *Catal. Today*, **149**, pp. 185–190.
119. Shen, Z., Lu, P., Hu, J., *et al.* (2010). Performance of $Ba_{0.5}Sr_{0.5}Co_{0.8}Fe_{0.2}O_{3+\delta}$ Membrane after Laser Ablation for Methane Conversion, *Catal. Commun.*, **11**, pp. 892–895.
120. Tian, T., Wang, W., Zhan, M., *et al.* (2010). Catalytic Partial Oxidation of Methane over $SrTiO_3$ with Oxygen-Permeable Membrane Reactor, *Catal. Commun.*, **11**, pp. 624–628.
121. Wu, Z., Wang, B. and Li, K. (2010). A novel Dual-Layer Ceramic Hollow Fibre Membrane Reactor for Methane Conversion, *J. Membrane Sci.*, **352**, pp. 63–70.
122. Luo, H., Wei, Y., Jiang, H., *et al.* (2010). Performance of a Ceramic Membrane Reactor with High Oxygen Flux Ta-Containing Perovskite for the Partial Oxidation of Methane to Syngas, *J. Membrane Sci.*, **350**, pp. 154–160.
123. Zhang, Y., Cheng, H., Liu, J., *et al.* (2010). Performance of a Tubular Oxygen-Permeable Membrane Reactor for Partial Oxidation of $CH_4$ in Coke Oven Gas to Syngas, *J. Nat. Gas Chem.*, **19**, pp. 280–283.
124. Zhang, Y., Liu, J., Ding, W., *et al.* (2011). Performance of an Oxygen-Permeable Membrane Reactor for Partial Oxidation of Methane in Coke Oven Gas to Syngas, *Fuel*, **90**, pp. 324–330.
125. Jiang, Q., Nordheden, K. and Stagg-Williams, S. (2009). Reaction Performance of $Ba_{0.5}Sr_{0.5}Co_{0.8}Fe_{0.2}O_x$ Asymmetric Oxygen-Permeable Ceramic Membrane Reactor, AIChE annual meeting, November 2009, Nashville, TN, paper 539a.

126. Jiang, Q., Faraji, S., Nordheden, K., *et al.* (2011). $CO_2$ Reforming Reaction Assisted with Oxygen Permeable Ba0.5Sr0.5Co0.8Fe0.2Ox Ceramic Membranes, *J. Membrane Sci.*, **368**, pp. 69–77.
127. Delbos, C., Lebain, G., Richet, N., *et al.* (2010). Performances of Tubular $La_{0.8}Sr_{0.2}Fe_{0.7}Ga_{0.3}O_{3-\delta}$ Mixed Conducting Membrane Reactor for under Pressure Methane Conversion to Syngas, *Catal. Today*, **156**, pp. 146–152.
128. Fujimoto, K., Asami, K., Omata, K., *et al.* (1991). Selective Oxidative Coupling of Methane with a Membrane Reactor. *Stud. Surf. Sci. Catal.,* **61**, pp. 525–531.
129. Nozaki, T., Yamasaki, O., Omata, K., *et al.* (1992). Selective Oxidative Coupling of Methane with Membrane Reactor, *Chem. Eng. Sci.*, **47**, pp. 2945–2950.
130. Nozaki, T., Hashimoto, S., Omata, K., *et al.* (1993). Oxidative Coupling of Methane with Membrane Reactors Containing Lead Oxide, *Ind. Eng. Chem. Res.*, **32**, pp. 1174–1179.
131. Borges, H., Giroir-Fendler, A., Mirodatos, C., *et al.* (1995). Catalytic Membrane Reactor for Oxidative Coupling of Methane. Part II. Catalytic Properties of LaOCl Membranes, *Catal. Today*, **25**, pp. 377–383.
132. Hibino, T., Sato, T., Ushiki, K., *et al.* (1995). Membrane Reactor for Oxidative Coupling of $CH_4$ with an Oxide Ion-Electron Hole Mixed Conductor, *J. Chem. Soc. Faraday T.*, **91**, pp. 4419–4422.
133. Kao, Y., Lei, L. and Lin, Y. (1997). A Comparative Simulation Study on Oxidative Coupling of Methane in Fixed-Bed and Membrane Reactors, *Ind. Eng. Chem. Res.*, **36**, pp. 3583–3593.
134. Zeng, Y. and Lin, Y. (1997). Catalytic Properties of Yttria Doped Bismuth Oxide Ceramics for Oxidative Coupling of Methane, *Appl. Catal. A: Gen.*, **159**, pp. 101–117.
135. Zeng, Y. and Lin, Y. (1997). Oxidative Coupling of Methane on Oxygen-Semipermeable Yttria-Doped Bismuth Oxide Ceramics in a Reducing Atmosphere, *Ind. Eng. Chem. Res.*, **36**, pp. 277–283.
136. Zeng, Y., Lin, Y. and Swartz, S. (1998). Perovskite-Type Ceramic Membrane: Synthesis, Oxygen Permeation and Membrane Reactor Performance for Oxidative Coupling of Methane, *J. Membrane Sci.*, **150**, pp. 87–98.
137. Xu, S. and Thomson, W. (1998). Stability of $La_{0.6}Sr_{0.4}Co_{0.2}Fe_{0.8}O_{3-\delta}$ Perovskite Membranes in Reducing and Nonreducing Environments, *Ind. Eng. Chem. Res.*, **37**, pp. 1290–1299.
138. Zeng, Y. and Lin, Y. (1999). Stability and Surface Catalytic Properties of Fluorite-Structured Yttria-Doped Bismuth Oxide under Reducing Environment, *J. Catal.*, **182**, pp. 30–36.
139. Lu, Y., Dixon, A., Moser, W., *et al.* (2000). Oxidative Coupling of Methane Using Oxygen-Permeable Dense Membrane Reactors, *Catal. Today*, **56**, pp. 297–305.
140. Lu, Y., Dixon, A., Moser, W., *et al.* (2000). Oxygen-Permeable Dense Membrane Reactor for the Oxidative Coupling of Methane, *J. Membrane Sci.*, **170**, pp. 27–34.
141. Kanno, T., Horiuchi, J. and Kobayashi, M. (2001). Chemical and Physical Modification of a Ceramic Membrane Reactor and the Performance Change in the Oxidative Coupling of Methane, *React. Kinet. Catal. L.*, **72**, pp. 195–200.
142. Zeng, Y. and Lin, Y. (2001). Oxidative Coupling of Methane on Improved Bismuth Oxide Membrane Reactors, *AIChE J.*, **47**, pp. 436–444.
143. Akin, F. and Lin, Y. (2002). Controlled Oxidative Coupling of Methane by Ionic Conducting Ceramic Membrane, *Catal. Lett.*, **78**, pp. 239–242.
144. Wang, H., Cong, Y. and Yang, W. (2005). Oxidative Coupling of Methane in $Ba_{0.5}Sr_{0.5}Co_{0.8}Fe_{0.2}O_{3-\delta}$ Tubular Membrane Reactors, *Catal. Today*, **104**, pp. 160–167.
145. Haag, S., van Veen, A. and Mirodatos, C. (2007). Influence of Oxygen Supply Rates on Performances of Catalytic Membrane Reactors. Application to the Oxidative Coupling of Methane, *Catal. Today*, **127**, pp. 157–164.
146. Taheri, Z., Nazari, K., Safekordi, A., *et al.* (2008). Oxygen Permeation and Oxidative Coupling of Methane in Membrane Reactor: A New Facile Synthesis Method for Selective Perovskite Catalyst, *J. Mol. Catal. A: Chem.*, **286**, pp. 79–86.

147. Bhatia, S., Thien, C. and Mohamed, A. (2009). Oxidative Coupling of Methane (OCM) in a Catalytic Membrane Reactor and Comparison of its Performance with Other Catalytic Reactors, *Chem. Eng. J.*, **148**, pp. 525–532.
148. Olivier, L., Haag, S., Mirodatos, C., *et al.* (2009). Oxidative Coupling of Methane Using Catalyst Modified Dense Perovskite Membrane Reactors, *Catal. Today*, **142**, pp. 34–41.
149. Wang, H., Cong, Y., Zhu, X., *et al.* (2003). Oxidative Dehydrogenation of Propane in a Dense Tubular Membrane Reactor, *React. Kinet. Catal. L.*, **79**, pp. 351–356.
150. Rebeilleau, M., Van Veen, A., Farrusseng, D., *et al.* (2004). Oxidative Activation of Light Alkanes on Dense Ionic Oxygen Conducting Membranes. *Stud. Surf. Sci. Catal.*, **147**, pp. 655–660.
151. Rebeilleau-Dassonneville, M., Rosini, S., Van Veen, A., *et al.* (2005). Oxidative Activation of Ethane on Catalytic Modified Dense Ionic Oxygen Conducting Membranes, *Catal. Today*, **104**, pp. 131–137.
152. Ahchieva, D., Peglow, M., Heinrich, S., *et al.* (2005). Oxidative Dehydrogenation of Ethane in a Fluidized Bed Membrane Reactor, *Appl. Catal. A: Gen.*, **296**, pp. 176–185.
153. Chalakov, L., Rihko-Struckmann, L., Munder, B., *et al.* (2007). Feasibility Study of the Oxidative Dehydrogenation of Ethane in an Electrochemical Packed-Bed Membrane Reactor, *Ind. Eng. Chem. Res.*, **46**, pp. 8665–8673.
154. Crapanzano, S., Babich, I. and Lefferts, L. (2010). Effect of V in $La_2Ni_xV_{1-x}O_{4+\delta}$ on Selective Oxidative Dehydrogenation of Propane, *Appl. Catal. A: Gen.*, **378**, pp. 144–150.
155. Courson, C., Taouk, B. and Bordes, E. (2000). Ion Oxide Conductor as a Catalytic Membrane for Selective Oxidation of Hydrocarbons, *Catal. Lett.*, **66**, pp. 129–138.
156. Zhu, B., Li, H. and Yang, W. (2003). AgBiVMo Oxide Catalytic Membrane for Selective Oxidation of Propane to Acrolein, *Catal. Today*, **82**, pp. 91–98.
157. Pérez-Ramírez, J. and Vigeland, B. (2005). Perovskite Membranes in Ammonia Oxidation: Towards Process Intensification in Nitric Acid Manufacture, *Angew. Chem. Int. Edit.*, **44**, pp. 1112–1115.
158. Sun, S., Rebeilleau-Dassonneville, M., Zhu, X., *et al.* (2010). Ammonia Oxidation in $Ba_{0.5}Sr_{0.5}Co_{0.8}Fe_{0.2}O_{3-\delta}$ Membrane Reactor, *Catal. Today*, **149**, pp. 167–171.
159. Coronas, J., Menéndez, M. and Santamaría, J. (1994). Development of Ceramic Membrane Reactors with a Non-Uniform Permeation Pattern. Application to Methane Oxidative Coupling, *Chem. Eng. Sci.*, **49**, pp. 4749–4757.
160. Miguel, M., Coronas, J., Menéndez, M., *et al.* (1996). Methane Oxidative Coupling over Different Alkali Doped Catalysts: A Comparison of Ceramic Membrane Reactors and Conventional Fixed Bed Reactors, *React. Kinet. Catal. Lett.*, **59**, pp. 277–284.
161. Coronas, J., Menéndez, M. and Santamaría, J. (1995). Use of a Ceramic Membrane Reactor for the Oxidative Dehydrogenation of Ethane to Ethylene and Higher Hydrocarbons, *Ind. Eng. Chem. Res.*, **34**, pp. 4229–4234.
162. Schuurman, Y., Decamp, T., Pantazidis, A., *et al.* (1997). Transient Kinetics of Methane Dehydrogenation and Aromatisation: Experiments and Modelling, *Stud. Surf. Sci. Catal.*, **109**, pp. 351–360.
163. Tonkovich, A., Zilka, J., Jimenez, D., *et al.* (1996). Experimental Investigations of Inorganic Membrane Reactors: A Distributed Feed Approach for Partial Oxidation Reactions, *Chem. Eng. Sci.*, **51**, pp. 789–806.
164. Téllez, C., Menéndez, M. and Santamaría, J. (1997). Oxidative Dehydrogenation of Butane Using Membrane Reactors, *AIChE J.*, **43**, pp. 777–784.
165. Ramos, R.. Menéndez, M. and Santamaría, J. (2000). Oxidative Dehydrogenation of Propane in an Inert Membrane Reactor, *Catal. Today*, **56**, pp. 239–245.
166. Wang, L., Liu, C., Ge, S., *et al.* (2000). Oxidative Dehydrogenation of Butene to Butadiene in Optimum Membrane Reactor, *Chinese J. Catal.*, **21**, pp. 497–499.

167. Ge, S., Liu, C., Fan, Y., *et al.* (2000). Oxidative Dehydrogenation of Butane to Butene and Butadiene Using Inert Membrane Reactor, *Chinese J. Catal.*, **21**, pp. 484–488.
168. Ge, S., Liu, C., Wang, L.J. (2001). Oxidative Dehydrogenation of Butane Using Inert Membrane Reactor with a Non-Uniform Permeation Pattern, *Chem. Eng. J.*, **84**, pp. 497–502.
169. Ge, S., Liu, C., Wang, L., *et al.* (2002). Oxidative Dehydrogenation of Butane to Butadiene and Butene over V-Mg-O catalyst. I. Effect of Reactors, *Acta Petrolei Sinica (Petroleum Processing Section)*, **18**, pp. 59–65.
170. Ahchieva, D., Brzic, D., Peglow, M., *et al.* (2004). Theoretical and Experimental Studies on the Partial Oxidation of Ethane in the Fluidized Bed Membrane Reactor, *Chem. Ing. Tech.*, **76**, pp. 1295–1296.
171. Kotanjac, Ž., van Sint Annaland, M. and Kuipers, J. (2010). A Packed Bed Membrane Reactor for the Oxidative Dehydrogenation of Propane on a $Ga_2 O_3$ / $MoO_3$ Based Catalyst, *Chem. Eng. Sci.*, **65**, pp. 441–445.
172. Mallada, R., Menéndez, M. and Santamaría, J. (2000). Use of Membrane Reactors for the Oxidation of Butane to Maleic Anhydride under High Butane Concentrations, *Catal. Today*, **56**, pp. 191–197.
173. Mallada, R., Pedernera, M., Menéndez, M., *et al.* (2000). Synthesis of Maleic Anhydride in an Inert Membrane Reactor. Effect of Reactor Configuration, *Ind. Eng. Chem. Res.*, **39**, pp. 620–625.
174. Xue, E. and Ross, J. (2000). Use of Membrane Reactors for Catalytic N-Butane Oxidation to Maleic Anhydride with a Butane-Rich Feed, *Catal. Today*, **61**, pp. 3–8.
175. Mota, S. Miachon, S. Volta, J., *et al.* (2001). Membrane Reactor for Selective Oxidation of Butane to Maleic Anhydride, *Catal. Today*, **67**, pp. 169–176.
176. Wang, L., Ge, S., Liu, C., *et al.* (2001). A Novel Porous Membrane Reactor for a Controllable Butene Oxidative Reaction, *J. Porous Media*, **4**, pp. 253–258.
177. Yang, C., Xu, N. and Shi, J. (1998). Experimental and Modeling Study on a Packed-Bed Membrane Reactor for Partial Oxidation of Methane to Formaldehyde, *Ind. Eng. Chem. Res.*, **37**, pp. 2601–2610.
178. Yang, W., Yang, P., Fang, W., *et al.* (2000) Effects of Operation Modes on the Oxidation of Propane to Acrolein in a Membrane Reactor, *Stud. Surf. Sci. Catal.*, **130**, pp. 2705–2710.
179. Al-Juaied, M., Lafarga, D. and Varma, A. (2001). Ethylene Epoxidation in a Catalytic Packed-Bed Membrane Reactor: Experiments and Model, *Chem. Eng. Sci.*, **56**, pp. 395–402.
180. Oyama, S., Bravo-Suárez, J., Lu, J., *et al.* (2007). Novel Directions in Propylene Epoxidation Research: New Catalysts and Feedstocks, and the Application of Membranes, *ACS Division of Petroleum Chemistry, Inc. Preprints*, **52**, pp.263–266.
181. Diakov, V., Lafarga, D. and Varma, A. (2001). Methanol Oxidative Dehydrogenation in a Catalytic Packed-Bed Membrane Reactor, *Catal. Today*, **67**, pp. 159–167.
182. Diakov, V. and Varma, A. (2002). Reactant Distribution by Inert Membrane Enhances Packed-Bed Reactor Stability, *Chem. Eng. Sci.*, **57**, pp. 1099–1105.
183. Diakov, V. and Varma, A. (2003). Methanol Oxidative Dehydrogenation in a Packed-Bed Membrane Reactor: Yield Optimization Experiments and Model, *Chem. Eng. Sci.*, **58**, pp. 801–807.
184. Diakov, V. and Varma, A. (2004). Optimal Feed Distribution in a Packed-Bed Membrane Reactor: The Case of Methanol Oxidative Dehydrogenation, *Ind. Eng. Chem. Res.*, **43**, pp. 309–314.

# Chapter 28

## Structured Catalytic Reactors for Selective Oxidations

Gianpiero GROPPI,* Alessandra BERETTA* and Enrico TRONCONI*

This chapter reviews the application of honeycomb monoliths, open-cell foams (also known as sponges) and other structured catalysts to gas/phase selective oxidation processes.

Monolith catalysts are characterized by well-defined, reproducible geometrical and flow properties, which result in unique performances for heat, mass and momentum transport. As a result, monoliths have become the standard catalyst shape in most applications of environmental catalysis, but they offer potential for optimal design and easy scale-up of catalytic reactors for chemical synthesis processes, too. Recent progress in the fundamental understanding of the above aspects, originating reliable engineering correlations and data as well as improved manufacturing technologies, is herein first summarized. We then discuss a number of exploratory studies at the laboratory and pilot reactor scale, which demonstrate the promising advantages of monolith catalysts over conventional packed-beds of pellets for chemicals production. These are associated primarily with reduced pressure drops, as required, e.g. in short contact time oxidative dehydrogenation processes, and with enhanced conductive heat exchange, that is crucial, e.g. in externally cooled multitublar fixed-bed reactors for strongly exothermic selective oxidations of hydrocarbons.

In comparison to monoliths, applications of open-celled foam structures to the chemical process industry are still at an earlier developmental stage. We report fundamental and applied investigations demonstrating opportunities for the implementation of foam catalysts in the same two main areas of millisecond contact time processes, and of fixed-bed reactors with enhanced heat management.

### 28.1. General Considerations on Structured Catalysts

#### 28.1.1. *Introduction*

For many years, honeycomb monoliths have been the standard catalyst shape in most applications of environmental catalysis. On the other hand, exploration of

*Laboratory of Catalysis and Catalytic Processes, Dipartimento di Energia, Politecnico di Milano, Piazza Leonardo da Vinci, 32 - 20133 Milano, Italy.

the adoption of structured catalysts in other areas of heterogeneous catalysis only started at the beginning of the 1980s, after their successful commercial application to the control of automotive exhausts and the reduction of nitrogen oxides. Particularly attractive were the expectedly lower pressure drop and the potentially smaller size of the reactor as compared to conventional pelletized catalysts in gas-phase processes, but early studies in this field, using methanation and hydrogenation as model reactions, pointed out additional prospective benefits. In a pioneering piece of work, for example, Tucci and Thomson carried out a comparative study of methanation over ruthenium catalysts in both pellet and honeycomb form.[1] In addition to pressure drops lower by two orders of magnitude, they also found significantly higher selectivities (97% vs 83%) over the monolith catalyst, likely resulting from lower internal diffusional resistances. Parmaliana and co-workers[2–5] investigated the hydrogenation of benzene and dehydrogenation of cyclohexane in ceramic monoliths washcoated with alumina impregnated with either Ni or Pt. Again, the low diffusion resistance of monolith catalysts allowed the authors to determine intrinsic kinetic expressions based on an Eley–Rideal mechanism.

In spite of the initially promising indications, however, nearly three decades later the use of monoliths as catalysts or catalyst supports in the processes of the chemical industry is still very limited. Two statements have long discouraged the extensive use of monolithic catalysts outside the well-known environmental applications[6]:

(i) conventional, parallel channel monoliths are virtually adiabatic: this is compatible with the processes for the abatement of pollutants in diluted streams, but would severely limit the control of temperature in many endo- and exothermic chemical processes; and
(ii) the overall load of the catalytically active phase in a monolith catalyst is less than the amount of catalyst in a bed of pellets of comparable volume: again, this is not important for the fast, diffusion limited reactions of environmental catalysis, but would be a clear disadvantage for the reactions under kinetic control usually met in chemical syntheses.

In reality, both such concerns can be overcome by dedicated monolith designs addressing the specific requirements of chemical applications: as presented in Section 28.4, conductive heat exchange in monolith structures can be even more effective than convective heat transfer in packed beds, whereas washcoat catalyst loadings in excess of 25% by volume are well within the range of what is practised with monoliths nowadays. Furthermore, new structured supports, like open-celled foams, are now being considered, which show promising properties in relation to radial heat transfer.

There remain, however, several more practical reasons which hinder the application of structured catalysts and supports to chemical syntheses[7]:

(iii) the many different pelletized catalysts operating in the many processes of the chemical industry are often the result of long and costly development work, their properties are well tailored to the specific process needs and their performances are typically quite satisfactory: accordingly, replacement of the conventional catalyst technology with monolith catalysts requires very significant and proven benefits;
(iv) the production volumes of industrial catalysts are lower by orders of magnitude as compared to the volumes of catalysts for the environment: thus, it is difficult to justify dedicated research efforts as well as capital investment to develop monolithic systems with intrinsic catalytic properties similar to those of conventional systems;
(v) the methods for loading, packaging, sealing and unloading structured catalysts in the synthesis reactors are different from those well established for pellet catalysts, and cannot be directly derived from the experience gained in stationary environmental installations; additional developments in this area are required, too; and
(vi) structured catalysts are intrinsically more expensive than pellet catalysts.

In essence, it appears that substantial improvements are required in order to motivate such a significant change of the catalyst technology.

Notwithstanding such difficulties, however, there are a steadily increasing number of research activities concerning the use of monolithic and other structured catalysts/reactors in the production of chemicals. In fact, after the early phase when only sparse attempts were reported, multiple application areas have now been identified and rationalized in which monolithic catalysts may have intrinsically superior performance characteristics.

One area receiving great attention nowadays in view of its large industrial potential, is the development of novel catalytic oxidation processes using structured reactors with extremely short contact times, whose large flow rates would generate unacceptable pressure drops in packed-bed reactors. Manufacture of olefins via the catalytic oxidative dehydrogenation of light paraffins is an important example of a process in this area for which applications of monolithic catalysts have been envisaged: it is discussed in Section 28.2 of this chapter, with a few more examples.

Again in view of their reduced pressure drop, it has been recognized that monolith structures also hold a good potential for applications as pre- and post-reactors of selective oxidation processes: the related concepts and the existing commercial examples are reviewed in Section 28.3.

An innovative area of development is represented by the use of structured catalysts in chemical processes under non-adiabatic conditions. In fact, as mentioned above, the heat transfer properties of honeycomb monoliths have traditionally been regarded as very poor, but recently, novel monolithic structures and configurations have appeared with interesting characteristics for heat exchange: for example, a new and promising area is the use of honeycomb catalysts with high-thermal conductivity in exothermal selective oxidation processes where multi-tubular reactors are employed. Along similar lines, there is growing interest in the potential of open-celled foam (or sponges) as novel structured catalyst supports with enhanced heat exchange properties. All of these aspects are addressed in Section 28.4.

### 28.1.2. *Geometrical, flow and local transport properties of structured catalysts*

The interest in structured catalysts is mainly related to their peculiar geometrical characteristics, which may result in enhanced flow and transport properties with respect to conventional randomly packed beds of particles. Accordingly, before discussing examples of applications in selective oxidation processes, such properties are briefly outlined in this section for the two main families of structured catalysts, namely honeycombs and foams.

#### 28.1.2.1. *Honeycomb monolith catalysts*

Honeycombs are geometrically regular structures consisting of several identical parallel channels through which the gas flows with segregated parallel streamlines (Fig. 28.1). Honeycombs are usually manufactured starting from both ceramic and metallic materials, either by extrusion or by piling up and rolling corrugated sheets. Accordingly, structures consisting of cells of different shapes (square, hexagonal,

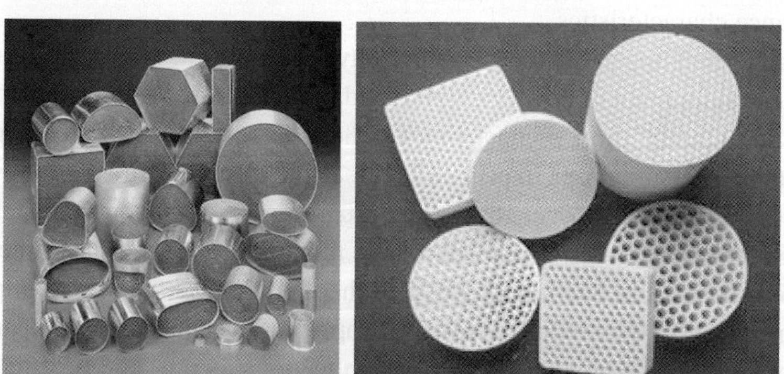

**Figure 28.1.** Examples of metallic and ceramic honeycombs.

**Table 28.1.** Geometrical properties of ceramic honeycombs. Adapted from Ref. 8.

| Monolith type (cell density/ wall thickness) | Open frontal area (OFA) | Hydraulic diameter (mm) | Geometric specific area ($m^2/m^3$) | Bulk density ($kg/m^3$) |
|---|---|---|---|---|
| 400/6.5 | 0.757 | 1.10 | 2,740 | 395 |
| 470/5 | 0.795 | 1.04 | 3,040 | 390 |
| 600/4 | 0.814 | 0.94 | 3,480 | 303 |
| 900/2.5 | 0.856 | 0.78 | 4,370 | 235 |
| 1,200/2.5 | 0.834 | 0.67 | 4,980 | 269 |

triangular, trapezoidal, etc.) and sizes can be obtained. Honeycombs with 900 cpsi (cells per square inch), wall thickness lower than 50 μm and void fraction in excess of 80% are currently commercialized.[8,9] In addition to these outstanding geometrical properties, good mechanical strength, high resistance to thermal shocks and good tolerance to dust make the use of honeycombs ubiquitous in environmental applications which pose severe operating constraints.

From an engineering point of view the extremely regular and periodical structure of honeycombs provides specific advantages with respect to reliable reactor design.

(i) The geometry is completely defined when providing two of the following parameters: pitch (or cell density), open channel size, wall thickness, open frontal area (i.e. void fraction), specific geometric area. For instance, the honeycomb geometry is commercially designated by a cell density (in cells per square inch) and a wall thickness (in mm).[8] Examples of geometrical properties of ceramic honeycombs are reported in Table 28.1.

(ii) With few exceptions,[10] laminar flow conditions usually prevail in the honeycomb channels. Accordingly, well-established and accurate theoretical correlations for laminar flow in ducts can be derived for evaluation of gas-solid heat and mass transfer coefficients. In particular, the effects of the channel shape and of the extent of reaction rate at the catalyst surface can be taken into account even in the presence of developing profiles of velocity, concentration and temperature. As an example, asymptotic values of Nusselt and Sherwood numbers (i.e. dimensionless heat and mass transfer coefficients) for different channel shapes taken from the work of Balakotaiah and co-workers[11] are reported in Table 28.2. Similarly, correlations for the calculation of friction factors to be used in reliable calculations of pressure drops can be derived from the classical heat transfer literature.[12]

(iii) Another important point is that, thanks to the identical geometry of every cell, the description of the behaviour of the single channel is often highly representative of the performances of the whole honeycomb catalyst structure. This allows

**Table 28.2.** Asymptotic values of Nusselt/Sherwood numbers and friction factors for different monolith channel shapes. Adapted from Ref. 11.

| Channel shape | Hydraulic diameter | $Nu(Sh)^{*}_{H2\infty}$ | $Nu(Sh)^{**}_{T\infty}$ | $(f\,Re)_{\infty}$ |
|---|---|---|---|---|
| 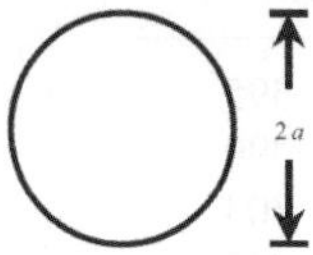 | 2a | 4.364 | 3.656 | 16 |
| 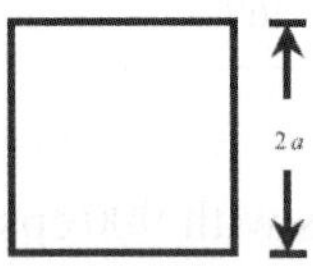 | 2a | 3.089 | 2.977 | 14.23 |
| 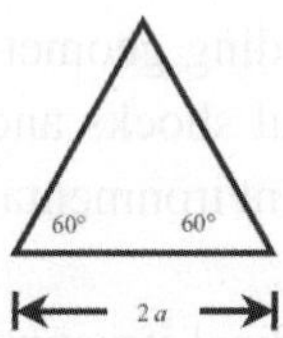 | $\frac{2}{\sqrt{3}}a$ | 1.893 | 2.497 | 13.33 |
| 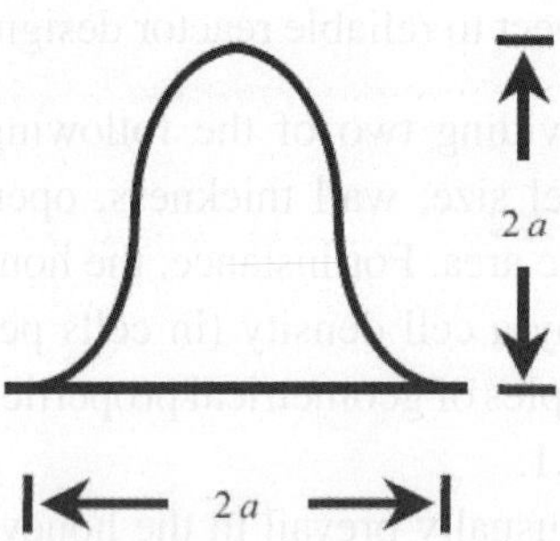 | 1.54a | 2.731 | 2.966 | 13.06 |
| 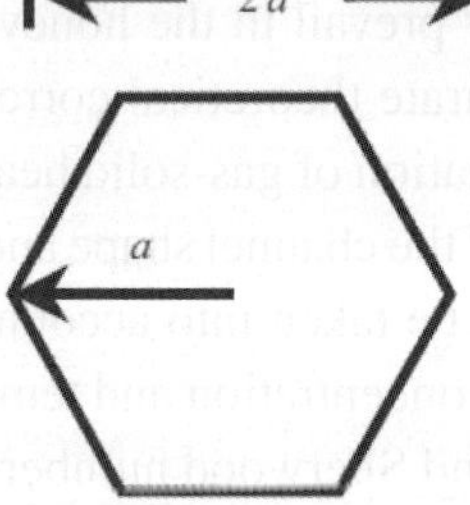 | $\sqrt{3}a$ | 3.861 | 3.340 | 15.05 |

*Asymptotic values for an infinitely slow reaction rate at the catalyst wall.
** Asymptotic values for an infinitely fast reaction rate at the catalyst wall.

for the use of the detailed multidimensional description of flow, temperature and composition fields in the reactor at limited computational cost.

(iv) Though not generally recognized, honeycomb monolith structures can also provide a mechanism for radial and axial heat transfer, namely conduction in the thermally-connected solid matrix, which is not available for random packings of catalyst pellets. Related implications will be thoroughly discussed in the dedicated paragraph in Section 28.4.1 below.

#### 28.1.2.2. *Structured foam catalysts*

The word foam refers to a dispersion of gas bubbles in a liquid, but can also be used to describe a uniform dispersion of a gaseous phase in a solid matrix. Open-celled foams are, in fact, sponge-like structures made of interconnected solid struts which enclose cavities (or pores) communicating through windows. They are commercially available in a variety of metallic and ceramic materials: well-established industrial applications include thermal insulation, energy adsorption, fabrication of noise-reduction devices, filtration of molten metals, purification of hot gases and others.[13]

The use of foams in catalytic processes was first proposed, several years ago, by Twigg and Richardson.[14,15] In fact, foams have interesting structural properties which make them attractive, in principle, as catalyst carriers: they have high porosities (from 70 to 95%), which means low resistance to fluid flow and hence reduced pressure drops, and may have very high surface areas per unit volume, which implies high effectiveness factors and enhanced fluid/solid heat and mass transfer rates with the potential for more compact, efficient and lightweight reactors. Furthermore, in contrast to honeycombs, they allow the radial dispersion of flow, thus favouring an even distribution of the reactants across the catalyst bed. Nevertheless, the study of foam catalysts has been gaining momentum only recently in the research literature, and the number of publications on this subject is still limited.

Indeed, a major obstacle to the development of foam catalysts is the lack of reliable engineering correlations for the relevant morphological and transport properties, which also prevents a conclusive appraisal of their potential in comparison to other more or less conventional structures. Four interlinked areas are of interest in this respect, associated with the description of the foam geometry, as well as of pressure drop, gas/solid mass and heat transfer, and overall (axial and radial) heat transfer in foam catalytic structures, respectively. An exhaustive review of the technical literature on foam catalysts is beyond the scope of the present chapter, so only the most significant published contributions to these topics are summarized in the next paragraphs.

(i) The accurate geometrical description of foams is an inherently difficult task due to the complexity and strong irregularities of the structures resulting from different manufacturing materials and methods (Fig. 28.2). A variety of morphological models, ranging from a simple cubic cell model[16,17] to more sophisticated ones,[18–20] representing the strut network by the packing of a regular polyhedra, have appeared in the engineering literature in an attempt to provide, for

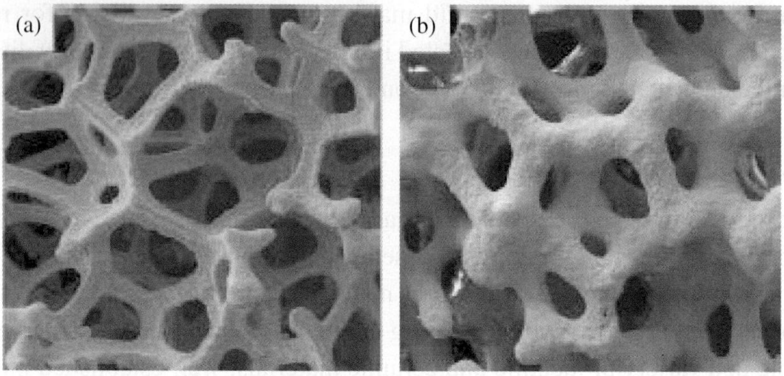

**Figure 28.2.** (a) Metal foam with triangular strut cross section and (b) ceramic foam with circular strut cross section. Reprinted with permission from Ref. 19. Copyright 2009 American Chemical Society.

example, estimates of the specific surface area as a function of experimentally accessible parameters such as porosity and cell diameter, also relying on image analysis techniques[18] or on idealized reconstructed structures.[21] Detailed reviews of the approaches to representation of open-cell foam structures are given in references.[22,23]

(ii) The prediction of pressure drop along foam beds is of course of primary importance for catalytic applications. The published pressure drop data usually follow the Forchheimer equation, indicating that the pressure drop is the sum of viscous and inertial contributions, so that some modified form of the Ergun model can be used to represent these results. This was the approach first adopted by Richardson and co-workers,[24] who correlated $\Delta P$ data over foams with very high porosities (92 and 99.5%) and with different pore densities (10, 30, 45 and 65 pores per inch (PPI), nominal). They also showed that growing washcoat loads incremented the pressure drop, affecting the inertial term of the Ergun equation more, which suggests the importance of the role of the washcoat roughness. Lacroix *et al.*[25] have recently reported the successful application of the Ergun model to the prediction of their pressure drop data in SiC foams, relying on a direct analogy between foams and beds made of spherical particles via the cubic cell geometrical model presented by Giani *et al.*[17] Even though no physical reason can be invoked to explain the extension of the Ergun equation to foams, which have much higher porosities than packed beds of particles, pressure drop data in a variety of foams with different PPI and porosities could be correctly estimated following the analogy: foam → cubic cell model → spherical particles, with the same porosity and surface area as foams, without adjusting the Ergun parameters.

(iii) Concerning fluid/solid mass and heat transfer, even though a number of correlations have been proposed in recent years,[26–30] it is not yet clear whether their predictions of the transport coefficients can be reliably generalized to all the possible foam structures, irrespective of the manufacturing method, the material and the geometry of the foam samples. Notably, according to Ref. 30 one additional structural parameter, related to the foam anisotropy, should be considered in addition to the usual two geometrical properties (e.g. porosity and pore size or strut diameter) utilized to correlate the mass and heat transport characteristics of foams. In order to compare different catalyst carriers, namely pellets, honeycomb monoliths and foams,[17] a defined, dimensionless merit index representing the ratio between the reactant conversion under external diffusion control and the dimensionless pressure drop for typical configurations, was employed and showed the honeycombs performed slightly better than high-porosity foams, both structures being much better than the pellets due to the greatly reduced pressure drop. This is in line with a recent comparative experimental study of CO oxidation over ceramic foams, honeycombs and beads coated with active $Pt/SnO_2$.[31] The results confirmed that foams are superior over particle beds from the viewpoint of combined high mass transfer rates and low pressure drop. This advantage is particularly interesting for the applications of environmental catalysis.

(iv) The radial and axial heat transfer characteristics of foam structures, which are, of course, particularly relevant for their applications as catalyst carriers in selective oxidation processes, are specifically discussed in Section 28.4.3.

## 28.2. Applications of Structured Catalysts in Short Contact Time Processes

The so-called short contact time processes are those processes wherein extremely high throughputs are realized in small reactor volumes, with contact times ranging from milliseconds down to microseconds.[32] These include the partial and selective oxidations of hydrocarbons to produce synthesis gas, olefins and oxygenates.

Metal gauzes, foam monoliths, sponges and traditional honeycomb monoliths have been successfully applied as catalysts or catalyst supports. The main advantages offered by the structured catalysts with high void fractions are represented by the reduced pressure drop and, in the case of foams, by the even distribution of the reactant flow across the fixed bed.

Since the pioneering works in the early 1990s wherein the concept was first proposed and demonstrated, Schmidt and co-workers have widely explored the application of monolithic catalysts for the one-step conversion of natural gas and higher hydrocarbons to valuable products such as $CO/H_2$, ethylene and propylene, at reasonably high conversions rates and selectivity. Since then, several research groups all over the world have worked in this field and have contributed to deepen the comprehension of the governing phenomena. In fact, because the reactors operate under rather severe conditions (high flow rates, high temperatures, fuel-rich feed streams, high pressures) several factors influence the overall performance of short contact time reactors, including strong heat and mass transfer limitations, homogeneous reactions and thermodynamic equilibria. Such factors, along with highly complex flow patterns, make the full understanding of the short contact time processes quite challenging. Debates are still open on the interpretation of the observed results; for instance, the relative roles of heterogeneous and homogeneous reactions in the short contact time oxidative dehydrogenation of alkanes has long been discussed and differently interpreted in the literature.

It is beyond the scope of the present contribution to provide a detailed report on the complexity of the mechanism and kinetics of short contact time processes. Instead, an effort is made to summarize the advancement of the research in this field, trying to focus on the specific characteristics of monolithic structures that are requested and exploited in the short contact time production of chemicals. The focus of this review is on the selective oxidation (or oxidative dehydrogenation) of small alkanes to olefins. Mention is also made of other short contact time oxidation processes, such as the ammoxidation of methane to HCN.

### 28.2.1. *Oxidative dehydrogenation*

A very large number of papers have been based on the study of the oxidative dehydrogenation of paraffins to the corresponding olefins, a sign of the vivid scientific and industrial interest for alternatives to the endothermic technologies of steam cracking and catalytic dehydrogenation. The challenge of developing "new ways to make old chemicals"[33] is made more and more difficult because the existing commercial technologies are well established and continuously subject to incremental improvements; still they suffer from thermodynamic limitations of paraffin conversion (which result in the need to operate at high temperatures, with consequences on coke formation, periodic regenerations and the use of costly materials), and the requirement of large energy inputs (which in turn cause an important environmental impact). On the other hand, the great potential offered by oxidative dehydrogenation, with stoichiometry:

$$C_nH_{2n+2} + 1/2O_2 \rightarrow C_nH_{2n} + H_2O + \text{heat} \tag{28.1}$$

relies on the fact that the reaction is exothermic and not limited by thermodynamic constraints. Great efforts have been made to develop active and selective catalysts for the oxidative dehydrogenation of ethane, propane and *n*-butane.

The issue of selectivity arises from the fact that the desired reaction (28.1) occurs in combination with a large set of possible parallel or consecutive processes, for example, in the case of ethane as the feed fuel:

$$C_2H_6 + 7/2O_2 \rightarrow 2CO_2 + 3H_2O \quad \Delta H = -1429\,\text{kJ/mol} \tag{28.2}$$

$$C_2H_6 + O_2 \rightarrow 2CO + 3H_2 \quad \Delta H = -137\,\text{kJ/mol} \tag{28.3}$$

$$C_2H_6 \rightarrow C_2H_4 + H_2 \quad \Delta H = +136\,\text{kJ/mol} \tag{28.4}$$

$$C_2H_6 + 2H_2O \rightarrow 2CO + 5H_2 \quad \Delta H = +210\,\text{kJ/mol} \tag{28.5}$$

$$C_S + H_2O \rightarrow CO + H_2 \quad \Delta H = +131\,\text{kJ/mol} \tag{28.6}$$

All catalyst formulations, in fact, lead to the unselective formation of deep oxidation products and of syngas ($H_2$ and CO). But cracking processes (here represented by simple dehydrogenation) may also occur at high temperature in the empty volumes of the reactor, with the formation of ethylene, methane, $C_{4+}$ species, but also C (which is an issue when using a Pd-based catalyst, for instance). C gasification, as well as water-gas shift, CO and $H_2$ post combustions, are usually also involved in the process surface kinetics.

A variety of selective oxidation catalysts, based on metal oxides, have been proposed. To date, the best reported performances are by far below the performances of commercial technologies. However, noble metal based catalysts, operating at high temperatures, have also been proposed and the best reported yields to olefins compare well with those of the existing technologies.[34] In the case of both metal oxide based formulations (moderate temperature catalysts) and noble metal catalysts (high temperature catalysts), the use of monolithic reactors has been studied; in the former case it offered the means for improving process selectivity, in the latter case it represented the key for realizing a novel autothermal process concept.

We emphasize that this is not a review of the oxidative dehydrogenation (ODH) of alkanes in general, but attention was paid to those papers which addressed, through experimental or modelling tools, the study of the process kinetics over structured catalysts under high-throughput conditions.

Review papers on the general subject of ODH include those by Fosse Håkonsen and Holmen[35] and Cavani *et al.*[36] The more specific subject of gas-phase syntheses of chemicals on structured catalysts was treated by Groppi *et al.*[37] Comprehensive surveys of the literature on short contact time ODH are also provided by the introductory sections of the papers by Zwerkle *et al.*[38] and by Lange *et al.*[39]

#### 28.2.1.1. *Monolithic catalysts operating at temperatures below 500°C*

Several catalysts based on transition metal oxides have been proposed for the oxidative dehydrogenation of small paraffins, with vanadium oxide being the main component. These systems are active at low to medium temperatures (400–500°C); selectivity is the critical issue and the usually reported performances amount to 20–30% olefin yield. Expert groups in the literature have clarified the intrinsic limit of the process, that is, the decreasing trend of olefin selectivity on increasing paraffin conversion.[40] This is the result of a consecutive kinetic scheme, wherein olefins are highly reactive intermediates which undergo consecutive oxidations to carbon oxides.

Along with efforts in catalyst development for suppressing consecutive reactions, few examples are reported in the literature of attempts to improve the process selectivity by means of reactor design. Membrane reactors and monolithic catalysts operating at short contact time seem to offer room for enhancing the process selectivity. Concerning the use of monoliths, Capannelli *et al.*[41] have compared the performances of a $V_2O_5/\gamma\text{-}Al_2O_3$ catalyst in the oxidative dehydrogenation of propane to propylene in different reactor configurations. They found that by using a single-channel washcoated monolith reactor (obtained by the impregnation of vanadium oxide over a 2 μm thick $\gamma\text{-}Al_2O_3$ layer, deposited on the core side of a tubular ceramic support with an inner diameter (i.d.) of 6.7 mm), much higher propylene selectivities were obtained than those observed in a packed-bed reactor where the catalyst was present in the form of particles with 0.3–0.5 mm diameter (obtained by pelletization of $\gamma\text{-}Al_2O_3$ powders exchanged with vanadium oxide). Reactor performances were compared at equal gas temperature and propylene conversions. Contact times (evaluated with respect to the catalytic phase) were in the order of seconds for the packed bed, and in the order of milliseconds in the case of the monolith. As expected, in the packed-bed reactor, at increasing residence time and reactant conversion, the selectivity of propylene showed a decreasing trend while $CO_X$ were progressively formed; instead, in the monolith-like reactor, propylene selectivity was kept almost constant and a net increase in the olefin yield was obtained with decreasing flow rate.

The authors provided only a qualitative interpretation of their results. They proposed that the better performance of the monolith relied on the beneficial effects of concentration and temperature boundary layers. The expected effect of interphase mass transfer limitations was, in fact, a decrease of $O_2$ concentration at the gas–catalyst interface, with better control of the vanadium valence state and consequently a partial suppression of consecutive oxidation reactions (due to a more important kinetic dependence of deep oxidations on $O_2$ concentration than the desired selective oxidation). The presence of heat transfer limitations was then believed to establish a catalyst temperature higher than the gas-phase temperature, thus promoting heterogeneously-initiated homogeneous reactions at the gas–catalyst interface, with production of peroxo radical species and eventually propylene.[42]

$V_2O_5$-based catalysts have also been tested in the form of honeycombs by the group of Anders Holmen in Trondheim. Despite the fact that this is a traditional "medium temperature catalyst", ODH tests were performed at temperatures higher than 700°C; these are thus reviewed in the following section.

### 28.2.1.2. *Monolithic catalysts operating at temperatures above 500°C*

Very high olefin yields were reported by Schmidt and co-workers,[43–50] who first proposed the oxidative dehydrogenation of light alkanes over insulated noble metal coated monoliths at contact times of a few milliseconds. This new concept of catalytic reactor had been previously applied by the same group to methane partial oxidation and was extended to test the reactivity of $C_2$-$C_6$ alkane/air fuel-rich feeds. Ceramic foam monoliths (with 45 and 80 ppi) were mostly studied as supports of noble metals and bimetallic catalysts.

In typical experiments, the catalyst-impregnated foam monolith was placed and sealed inside a quartz tube; inert alumina-extruded monoliths were placed upstream and downstream from the foam monolith as heat shields. These, along with external insulation, allowed a realization of almost adiabatic conditions; flow rates ranged from 2 to 12 standard liter per minute, corresponding to 13–79 cm/s superficial velocity and contact times from 7 to 40 ms. A moderate pre-heat of flow gases was sufficient to realize rapid light off of the reactor. Within a few seconds, the reactor established at approximately the adiabatic temperature (ranging between 800 and 1,200°C, depending on the alkane nature and the feed composition).

The best olefin yields were observed over Pt-coated monoliths. In the case of ethane/$O_2$ mixtures, selectivities to ethylene up to 65% at 70% ethane conversion and complete $O_2$ conversion were reported.[43] The oxidative dehydrogenation of propane and *n*-butane produced total olefin selectivies of about 60% (mixtures of ethylene and propylene) with high paraffin conversions.[44] Mixtures of ethylene, propylene and 1-butene were observed by the partial oxidation of *n*-pentane and *n*-hexane; ethylene, cyclohexene, butadiene and propylene were the most abundant products of the partial oxidation of cyclohexane.[45]

Further improvements of the selective production of ethylene were obtained by Bodke *et al.*[46,47] by co-feeding $H_2$ to ethane–oxygen mixtures over a Pt-Sn coated monolith; ethylene was produced at 80–85% selectivity with over 70% ethane conversion. Apparently the Pt-Sn alloy could favour the selective oxidation of hydrogen, which thermally drove the selective dehydrogenation of ethane to ethylene.

In spite of the extremely high reaction temperatures, a purely heterogeneous mechanism was originally proposed to explain the formation of olefins on the Pt-surface.[43,44,48] Only in the case of $C_{5+}$ alkanes, was a non-negligible contribution of homogeneous pyrolysis reactions supposed.[45]

However, concerning the effect of catalyst geometry, Huff and Schmidt,[43] and Goetsch and Schmidt[49] found that a single Pt/10%Rh gauze gave results similar to those achieved over a Pt-coated monolith in ethane and propane oxidative dehydrogenations; notably, over the single gauze, the process occurred within 10 to 100 microseconds. The authors interpreted the results by proposing that in the case of monolithic supports, the process also occurred in the same ultra-short timescale, which is at the very entrance of the reactor. Also, in the case of the $H_2$-enriched ethane partial oxidation over a Pt-Sn catalyst, similar performances were observed by running the process over a variety of catalyst supports; it was found[47] that the catalyst geometry did not affect the process, provided that autothermal operation (with high temperature and short contact time) could be guaranteed.

Regarding the effect of catalyst morphology, Bodke *et al.*[50] compared the partial oxidation of several hydrocarbons over Pt in the cases of conventional $\alpha$-$Al_2O_3$ foam monoliths and washcoated monoliths (wherein a 30–50 mm thick layer of $\gamma$-$Al_2O_3$ had been deposited prior to Pt impregnation). They observed that lower ethane conversions and much lower ethylene selectivities were realized by the washcoated monolith; results were interpreted as evidence of the detrimental effect of intraporous mass transfer limitations.

As mentioned above, in recent years several other authors have studied and made contributions to the better comprehension of the oxidative dehydrogenation of light paraffins in short contact time reactors.

Holmen and co-workers[51–58] have studied the partial oxidation of ethane and propane over Pt/10%Rh gauzes, Pt/$\gamma$-$Al_2O_3$ washcoated honeycomb monoliths and, more recently, VMgO/$\gamma$-$Al_2O_3$ coated monoliths and Pt-Sn coated monoliths. Regarding the geometry of supports, cordierite-extruded monoliths from Corning had 400 cells/in$^2$, while the gauze catalyst (from Rasmussen, Hamar, Norway) was woven from 60 $\mu$m diameter wires into 1,024 meshes/cm$^2$. It was confirmed that high yields to olefins could be obtained at high temperature and short contact time over the different catalysts and reactor configurations.

Figure 28.3 reports, for instance, the results of propane oxidative dehydrogenation tests over the VMgO-washcoated monolith. Product distributions are plotted as a function of propane conversion, which was progressively increased by increasing the temperature of an external heating furnace.

At sufficiently high temperatures (>700°C) and high propane conversions (>40%), the different systems behaved very similarly. Also, by extending the comparison to Pt-coated monoliths, only-washcoated monoliths, uncoated monoliths and an empty reactor, high yields to olefins at comparable paraffin conversions were still found. Results from different reactor configurations are reported in Table 28.3. These results indicate unambiguously the great importance of gas-phase reactions.

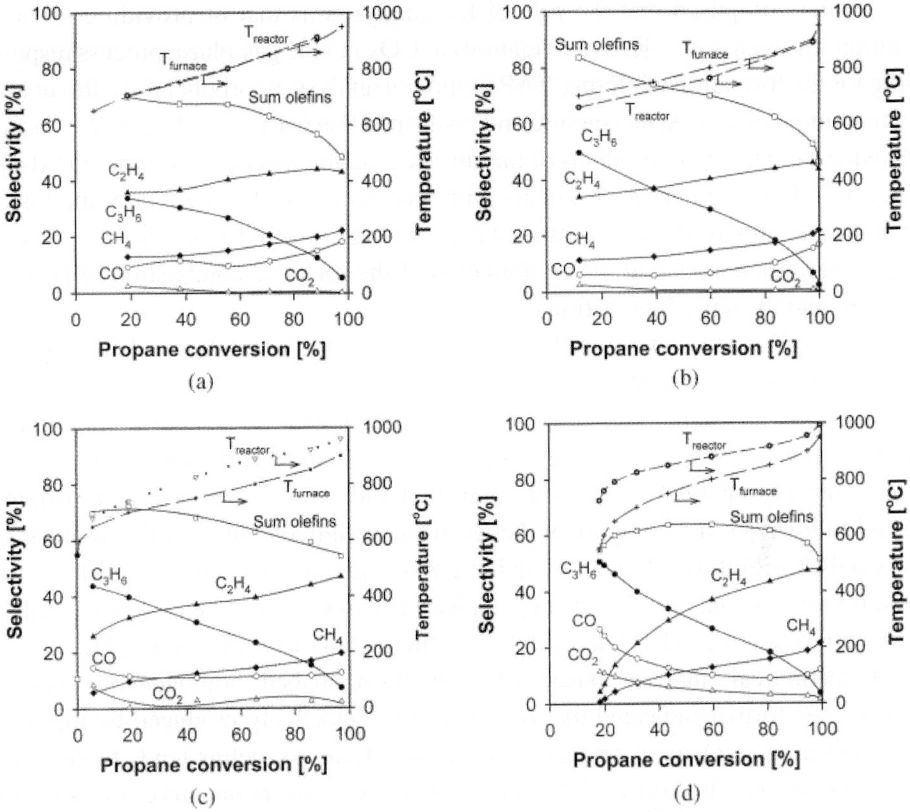


**Figure 28.3.** Oxidative dehydrogenation of propane. Product selectivities and reactor/furnace temperatures as a function of the propane conversion. Feed (Nml/min): propane (308); air (769); Ar (923). Total flow rate: 2,000 Nml/min. (a) The empty reactor; (b) the uncoated honeycomb monolith; (c) the washcoated honeycomb monolith; (d) the washcoated honeycomb monolith impregnated with VMgO. Reprinted from Ref. 53. Copyright 2001 with permission from Elsevier.

**Table 28.3.** Oxidative dehydrogenation of propane over different catalysts (Pt/10%Rh gauze, VMgO and Pt/monolith catalysts) at few milliseconds contact time. Adapted from Ref. 56.

| | | Highest yield % | | |
|---|---|---|---|---|
| Reactor configuration | Propane/$O_2$ ratio | Ethene 950°C | Propene 800–850°C | Sum olefins 800–950°C |
| Empty reactor | 1.9 | 42.9 | 14.9 | 49.9 |
| VMgO/monolith | 1.9 | 47.4 | 15.9 | 53.9 |
| Pt/10%Rh gauze | 1.7 | 47.9 | 16.1 | 53.4 |
| Pt/monolith | 2.0 | 42.7 | 12.8 | 46.9 |

The authors proposed that the role of the catalyst was that of providing thermal ignition (through non-selective oxidations to $CO_X$) to the gas-phase process responsible for the formation of olefins. TAP (temporal analysis of products) studies at low temperatures supported this picture and confirmed that carbon oxides, hydrogen and methane were the main products of the surface reaction mechanism over a $Pt/Al_2O_3$ catalyst.[55] Ethane ODH experiments with a co-feed over $H_2$ were perfomed over Pt-Sn coated monoliths and confirmed the very high selectivity of this process concept toward olefins, due to the combination of the selective combustion of $H_2$ (the co-fuel) and the dehydrogenation of ethane.[57,58]

Beretta, Forzatti and co-workers[59–63] also studied the oxidative dehydrogenation of ethane and propane in short contact time reactors. Experiments were first performed in an isothermal, single channel reactor with an annular configuration (wherein the fluodynamics were simple, mass transfer coefficients were known and catalyst temperature was well controlled and easily measured, as illustrated in the following); the comparison between the results obtained in the absence of the catalyst with those obtained with increasing amounts of catalyst confirmed that olefins were uniquely produced via gas-phase reactions, while the $Pt/Al_2O_3$ catalyst was a non-selective oxidation catalyst (producing only $CO_X$, $H_2$ and $H_2O$).

A theoretical analysis based on a well-established homogeneous kinetic scheme[60,61] further indicated that olefins can be selectively produced by the thermal activation of $O_2$/paraffin mixtures, the selectivity to olefins tends to decrease at increasing paraffin conversion, and the olefins vs conversion curve is practically independent of operating conditions. This implies that, at equal conversion, olefins can be produced at characteristic times of seconds as well as milliseconds, provided that the reaction temperature is low or high enough.

It was thus proposed that in autothermal short contact time reactors, once the fuel-rich feed stream is fed, heterogeneous deep oxidation reactions are initially activated; they, in turn, heat up the gas-phase volume which surrounds the catalytic surface and induce the thermal activation of a homogeneous process.

The principle was demonstrated by realizing autothermal experiments of ethane partial oxidation wherein an oxide-based oxidation catalyst ($BaMnAl_{11}O_{19}$) was used instead of the $Pt/Al_2O_3$ to provide the initial ignition. Indeed, the Pt-containing and the Pt-free reactor behaved similarly, and equally high ethylene yields were produced at contact times of a few milliseconds.[63] The results obtained over the $BaMnAl_{11}O_{19}$ catalyst are reported in Fig. 28.4. It is shown that the measured product distribution was very close to the calculated product distribution of a purely homogeneous adiabatic reactor operating at the same inlet temperature and degree of ethane conversion.

A similar interpretation of the mechanism of olefin production was given by Henning and Schmidt in a later work,[64] wherein the outlet stream was sampled

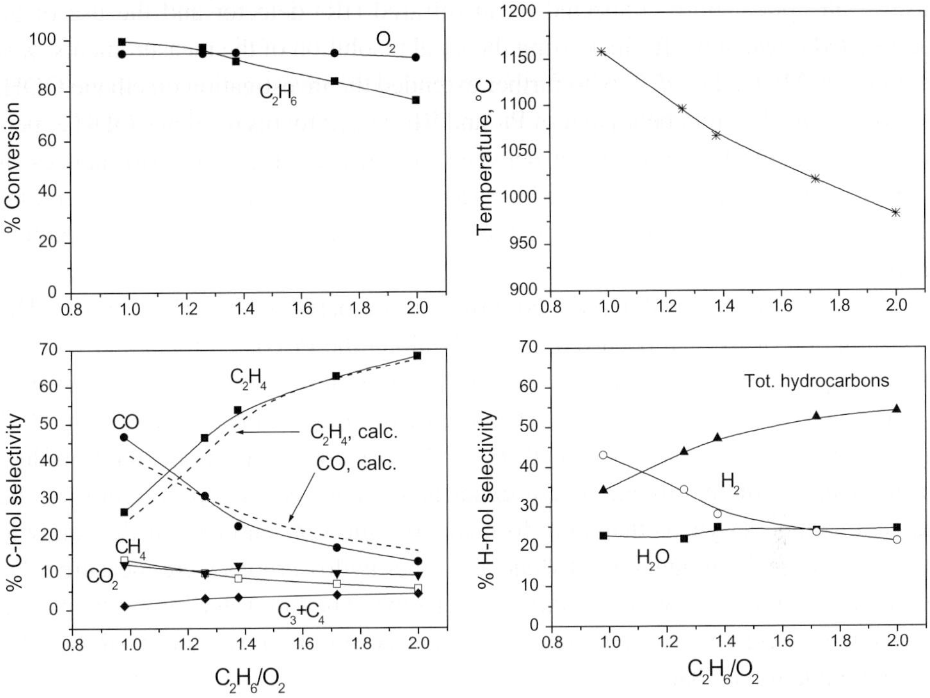


**Figure 28.4.** ODH of ethane in an autothermal reactor in the presence of a $BaMnAl_{11}O_{19}$ catalyst. Effect of ethane/$O_2$ feed ratio. Flow rate: 1NL/min; feed: ethane/air; pre-heat temperature: 500°C. Dashed curves = calculated selectivity of ethylene and CO for a purely homogeneous adiabatic reactor, operating at the same inlet temperature and degree of reactant conversion. Adapted from Ref. 62. Copyright 2001 with permission from Elsevier.

at increasing distance from the outlet monolith section; the analyses allowed the demonstration that a large part of ethylene was produced downstream from the catalytic zone, in the empty and well-heated volumes of the reactor. This work was refined in 2006, when Horn *et al.*[65] were the first to report the application of spatially-resolved sampling techniques for measuring longitudinal profiles of concentration (by capillary sampling and mass spectrometer analysis) and gas-phase temperature (by a sliding thermocouple) in short contact time reactors. Most of the investigations in this and several following papers dealt with the characterization of $CH_4$-CPO reactors with Pt and Rh catalysts. However, the paper by Horn *et al.*[65] also presented the results of an ethane ODH experiment in a Pt-coated foam. The measured concentration profiles showed that $O_2$ and ethane conversions were distributed along the whole catalyst length, with the progressive formation of $CO_X$, $H_2$, $H_2O$ and ethylene. It was confirmed that some conversion of ethane also proceeded downstream of the catalytic zone. Even more sophisticated technical equipment, which also included the optical measurement of the solid temperature through the use of an optical fibre connected to an infrared (IR) detector and the use of an automated linear stage for increasing the axial resolution of the measurements, was applied by Michael *et al.*,[66] who further extended the investigation on ethane ODH, analysing the different behaviour of Pt- and Rh-coated foams in ethane ODH experiments. Notably, the comparison between the different integral performances of the two noble metals, originally addressed by Huff and Schmidt in 1993,[43] represented a sort of building block for the hypothesis of a purely heterogeneous path to ethylene.

The experiments on Pt largely confirmed those performed by Horn *et al.*[65] The consumption of oxygen occurred gradually along the Pt-coated foam and was not complete at the outlet section. Ethylene formation started at some length downstream of the catalyst entrance section. In the case of Rh, the original findings from Huff and Schmidt[43] were also confirmed: $H_2$ and CO were the most abundant product species in the outlet stream. These experiments allowed the appreciation of the large amounts of syngas that were formed from the very entrance of the catalytic foam. Very small amounts of ethylene and other hydrocarbons were measured. $O_2$ was rapidly consumed within a narrow inlet zone, wherein water formation passed through a maximum. These trends resemble very closely those typical of the $CH_4$ partial oxidation on Rh.

The important, novel contribution provided by this recent study was the thermal characterization of the Pt- and Rh-coated monoliths.

Figure 28.5 reports the temperature profiles measured by the pyrometer (assumed representative of the solid surface temperature) and the temperature profiles measured by the thermocouple (representative of the gas-phase temperature) over Pt at varying C/O feed ratios. Progressively increasing temperatures were measured along the axial coordinate in the gas phase as well as on the Pt catalyst surface. A very interesting common feature was observed in the various experiments: ethylene formation started at some distance from the monolith entrance, in correspondence with the reach of about 760°C in the gas phase. This happened progressively downstream from the entrance at an increasing C/O feed ratio. In this case of the analogous experiments with Rh, both the solid and the gas temperatures passed through a maxima (at the very catalyst entrance) followed by a decreasing trend along the monolith length (as in methane partial oxidation).

The overall analysis of results suggested that the behaviour of the Pt-coated foam was dominated by the gas-phase chemistry (characterized by a moderately exothermic oxy-pyrolysis process), while the behaviour of the Rh-coated foam was dominated by the heterogeneous chemistry (characterized by a sequence of highly exothermic oxidation reactions and highly endothermic steam reforming of the hydrocarbons species). The superiority of the Rh catalyst in steam reforming was very evident and confirmed previous results by Beretta and Forzatti.[67] In both

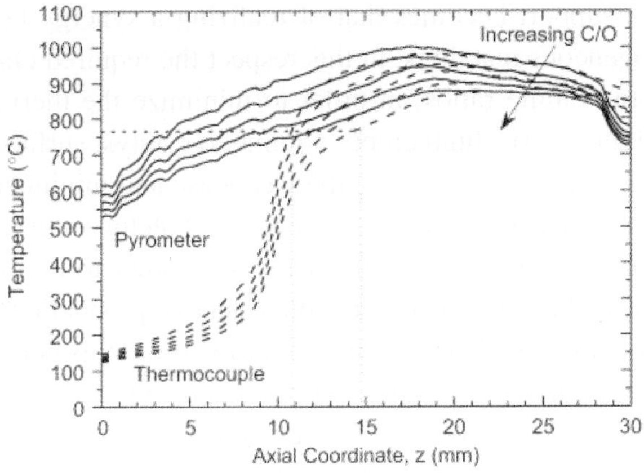


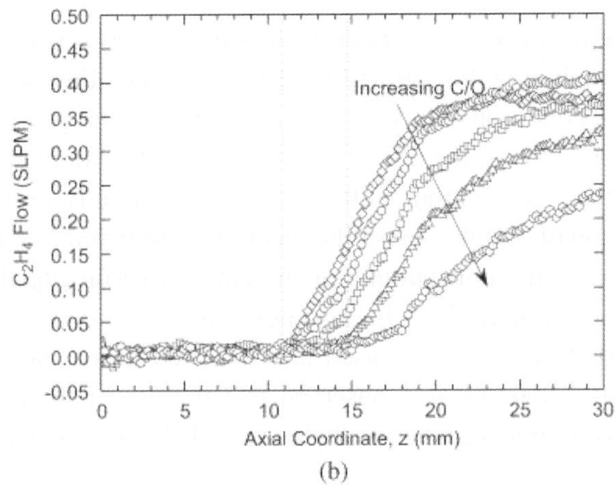


**Figure 28.5.** (a) Comparison of temperature profiles measured by pyrometer and thermocouple with $C_2H_4$ formation at varying C/O feed ratios (from 1.2 to 2.0) over the Pt-coated foam. The dotted lines indicate isotherm/position relationship for $C_2H_4$ formation (b). Reprinted from Ref. 66. Copyright 2010 with permission from Elsevier.

studies, the superiority of Rh in catalysing the steam reforming of hydrocarbon species over Pt was believed to be the key for interpreting the observed differences. In fact, hydrocarbon intermediates formed in the gas phase could, in principle, further react on the surface with the formation of syngas on Rh, but less likely on Pt.

In conclusion, in the case of Pt and the other high temperature catalysts examined so far, several pieces of evidence have been presented in the literature in favour of the prevailing role of homogeneous reactions in the production of olefins at high temperature and short contact time. According to this picture, the role of the catalytic phase is essentially that of providing thermal ignition. In turn, the role of the monolithic support becomes that of realizing a synergy between homogeneous and heterogeneous reactions; in this respect the required characteristics are: (i) high surface to volume ratios, in order to minimize the thermal capacity and allow for rapid light-off; (ii) high empty volume to catalyst surface ratios, in order to guarantee sufficient contact times in the gas phase and minimize undesired heterogeneous steps such as the consecutive oxidation of olefins or the chemical quench of the radical pool; and (iii) high void fractions, in order to realize the necessary short contact times which are needed for the selective production of olefins at the high reaction temperatures of autothermal reactors. From this perspective, the negative effect of washcoating on ethane conversion and ethylene selectivity observed by Bodke *et al.*[47] can be interpreted as the result of the increase of catalyst surface area, up to an unfavourable value of the gas-phase volume to catalyst surface ratio.

It must be noted, however, that efforts have been spent by several research groups with the aim of combining the favourable features of monolithic catalysts operated under autothermal conditions (thus favouring the onset in the gas-phase volumes of selective oxidation paths of the alkane fuel), with the development of intrinsically active and selective formulations able to contribute directly or indirectly to the selective production of olefins. A specific mention is deserved by rare earth oxides-based catalysts wherein contributions of the catalyst surface to the formation of ethyl species were reported.[68–70] Interestingly, through a detailed comparison of the observed performances of Pt-based and $LaMnO_3$-based monoliths in ethane ODH experiments, Donsì *et al.*[71] observed that the use of the perovskite-coated monolith yielded an improvement in ethylene yield. An overview of their results is reported in Fig. 28.6.

The observed trend was attributed to the specific activity of the $LaMnO_3$ catalyst in activating CO oxidation to $CO_2$, thus keeping the high heat release from the unselective surface reactions. In other words, since part of the fuel is inevitably lost to $CO_X$, the authors suggested that a formulation leading to controlled amounts of $CO_2$ was preferable to those, like Pt, which mainly favour the formation of CO. The control of the distribution among the unselective products has a beneficial impact on the thermal behaviour of the reactor and, in particular, on an even distribution of heat. This ultimately would enhance the ethylene yield. This specific feature was also exploited to demonstrate that ethylene yield could even be increased, by co-feeding CO to the standard ODH feed mixture.

Efforts to replace Pt-containing monoliths with Pt-free formulations by exploiting a suitable tailoring of the catalytic formulation was also addressed by Sadikov *et al.*, and Pavlova *et al.*,[72,73] who studied the performances of zirconium phosphates containing Co or Mn in the oxidative dehydrogenation of propane to propylene.

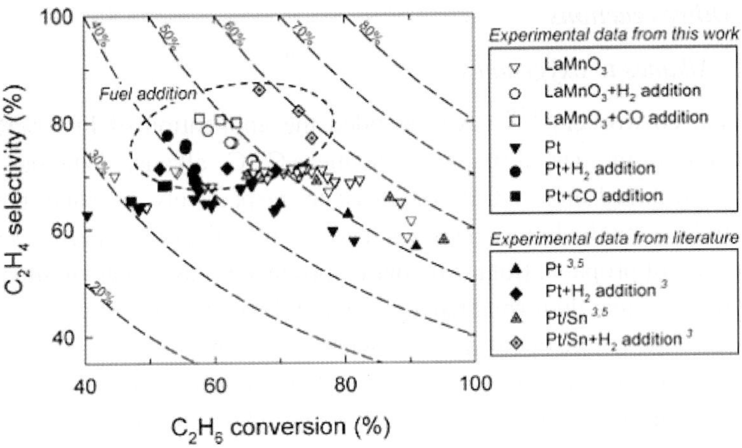


**Figure 28.6.** Ethylene selectivity vs ethane conversion for Pt- and $LaMnO_3$-based monoliths. Experiments with co-feed of $H_2$ and CO. Data concerning the effect of $H_2$ co-feed over Pt and Pt/Sn catalysts (taken from Schmidt and co-workers)[46,47] are also reported in this figure. Reprinted with permission from Ref. 71. Copyright 2005 American Chemical Society.

Though confirming the important role of gas-phase reactions in governing the production of olefins, the authors succeeded in avoiding the use of the noble metal, and largely increased the catalyst resistance to coking, which may become a major issue at high alkane concentrations.

The optimal design of the monolithic catalyst, either in the general case of an unselective catalytic phase like Pt (that is a catalyst leading to the formation of COx only) or in the case of specifically-tuned catalyst formulations, has been sparsely addressed in the literature. It is true, however, that the complexity of the reacting system is extremely high and results in huge computational costs. Few papers[74–78] have addressed the challenge of "merging" the heterogeneous and the homogeneous chemistry of light hydrocarbons into detailed reactor models. The available modelling studies (all dealing with the case of Pt-coated structured catalysts) have differently treated key aspects such as the surface kinetics, the homogeneous chemistry and the reactor model. The comprehensive rationalization of the very complex process is still open, which makes the engineering and optimal design of the structured reactor a relatively unexplored but promising field of investigation. The fluodynamic, physico-chemical and (last but not least) computational complexity has been differently addressed by different authors in the literature, at the expense of some simplifying assumptions, mostly on the reactor model. This explains the partly diverging views of these studies.

### 28.2.2. *Other reactions*

#### 28.2.2.1. *Alkanes to oxygenates*

Schmidt and co-workers[49,79] have extended the application of Pt/10%Rh single gauze reactors to the partial oxidation of linear $C_1$-$C_5$ alkanes. Whereas methane and ethane produced mostly CO and ethylene, respectively, propane, butane and pentane produced both olefins and oxygenates. The number of oxygenates was very low in the case of propane. Butane oxidation produced a significant number of oxygenated products, mainly formaldehyde and acetaldehydes, and oxygenate selectivity improved with a more open gauze. Pentane oxidation gave the highest selectivity to oxygenates, the main products being acetaldehyde and propionaldehyde. Theoretical evaluations showed that the process consisted of a heterogeneously initiated homogeneous reaction, wherein total combustion was primarily catalysed by the Pt surface, and oxygenates and olefins were formed subsequently by gas-phase reactions.

#### 28.2.2.2. *Hydrogen cyanide production*

The catalytic oxidative dehydrogenation of $CH_4$+$NH_3$ (ammoxidation) to produce HCN:

$$CH_4 + NH_3 + 3/2O_2 \rightarrow HCN + 3H_2O + \text{heat} \tag{28.7}$$

is a commercial process (the Andrussow process) carried out at short contact time over 20–50 layers of Pt-10%Rh gauzes, which form a structure a few mm thick and several metres in diameter. In the search of alternative catalyst configurations (a driver being the extremely high cost of the gauzes, largely due to the presence of Rh which is needed to confer ductility rather than for its catalytic properties), Hickman *et al.*[80] investigated the use of foam monoliths, extruded monoliths and metal monoliths coated with Pt. The qualitative behaviour of the monoliths was similar to the industrially used gauzes, with comparable performances. However, even the best Pt monolith catalyst (13.8 wt% Pt-supported on a 6 mm long, 30 ppi $\alpha$-$Al_2O_3$ foam) gave HCN selectivities, based on $CH_4$, consistently significantly lower ($\approx$ 35%) than those on the gauze ($\approx$ 61%), at comparable HCN selectivities based on $NH_3$ ($\approx$ 80%). The authors suggested that $CH_4$ oxidation reactions could be more important on supported catalysts than on gauze, due to the catalytic effect of the support or to differences in the Pt microstructure.

Foam monoliths behaved better than extruded or metal monoliths, which was explained on the basis of the higher mass transfer for the foam compared to straight channelled monoliths. The effect of mass transfer was confirmed by observing that HCN selectivity improved over foam monoliths with smaller pore size and extruded monoliths with higher cell density.

Results were largely confirmed in a later work by Baradwaj and Schmidth,[81] who found that the reaction was highly sensitive to the catalyst microstructure, as activation and differences in performance were observed on catalysts with different support materials.

## 28.3. Applications of Monolithic Catalysts Based on Low Pressure Drop Characteristics

Monolith catalysts are well known to provide outstanding pressure drop performances. As shown in Fig. 28.7,[7] the straightforward passage of gas flow in parallel channels of honeycomb structures under laminar conditions at given gas velocity and specific geometric surface area results in pressure losses lower by one to two orders of magnitude with respect to those associated with the tortuous and often turbulent gas flow passage through packed beds. Lower but still significant advantages in terms of pressure drop can be obtained with foams thanks to their much higher void fraction compared to a packed bed of particles.

This is one of the main reasons for the widespread success of monoliths in environmental applications in which severe limitations on pressure losses are typically posed by strict constraints on energy efficiency. Pressure drops are obviously also an issue in catalytic reactors for chemical process applications, being responsible for compression duties, which can be especially important in the presence of reactant recycling. However, in typical applications, such an aspect can be satisfactorily handled by an appropriate design of the catalytic packed-bed reactor. Some cases still

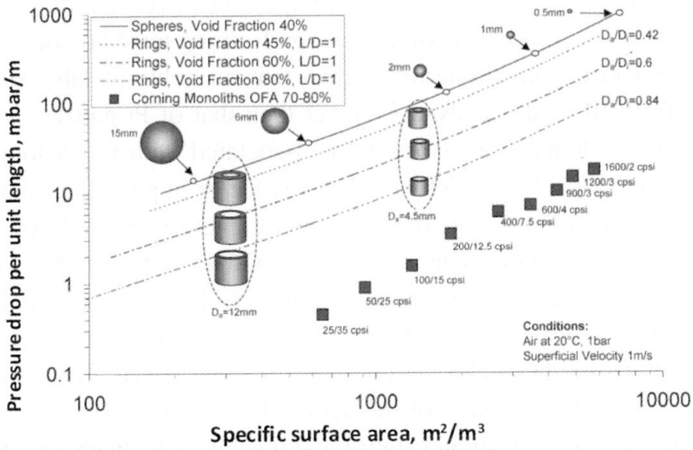


**Figure 28.7.** Comparison of pressure drops in honeycomb monoliths and packed bed of spherical and ring pellets. Reprinted with permission from Ref. 7. Copyright 2004 American Chemical Society.

exist in which almost negligible pressure drops in monolithic catalysts can provide key advantages with respect to existing reactor technologies.

### 28.3.1. *Catalytic post-reactors*

An area of growing interest to the chemical industry is the application of monolithic catalysts as post- or finishing adiabatic reactors downstream of a main conversion reactor. These are typically retrofit installations which must be introduced during plant revamping with minimal modifications to existing piping and compression capacity. In this respect, very low pressure drops, along with the possibility of operating the monolithic reactor in down-, up- and horizontal flow configurations, provide key advantages in the design and operation of the post-reactor. Accordingly, the use of honeycombs as finishing catalysts has been explored in the literature and some commercial applications have been found.

#### 28.3.1.1. *$NH_3$ oxidation*

Selective oxidation of $NH_3$ to NO is the key step in the production of nitric acid. Since the beginning of the last century this process has been based on the use of Pt-based gauzes through which reactants are selectively converted at very short contact times (a few ms). Despite several improvements, including the adoption of Pt-Rh and Pt-Pd-Rh alloys with reduced metal volatility and the use of better gauzes downstream that are made of Pd alloy for metal recovery, Pt losses still remain a major problem, especially in medium and high pressure converters operating at about 900°C.[82]

The cost associated with net Pt losses and with the replacement of exhaust gauzes makes the use of alternative oxide-based catalysts attractive. Despite several years of research, however, the activity of the best-performing mixed oxide catalysts still remain two orders of magnitude lower than that on noble metals. Also, the NO selectivity of mixed oxide catalysts is lower than that of Pt gauzes, possibly due to a major role of homogeneous reactions associated with much longer contact times ($10^{-2}$–$10^{-1}$ s).[83] Since NO yield is a key factor in process economics, such characteristics make unprofitable the adoption of reactors in which low cost oxide catalysts are used instead of noble metal gauzes. On the other hand, the use of dual bed systems consisting of a few platinum gauzes followed by a mixed oxide can represent a viable alternative to conventional technologies. In such a configuration the activity of the oxide catalyst is not a major issue, since upstream Pt gauzes promote ignition and conversion of the main $NH_3$ fraction (about 85%), whereas the second bed acts as a finishing catalyst and must mainly guarantee stability and good selectivity at high temperature (900–950°C in medium- and high-pressure converters), especially with respect to undesired reactions between $NH_3$ and NO.

Several efforts by Russian scientists, which are thoroughly reviewed in Ref. 83, have been devoted to developing catalyst formulations, and design and production methods suitable for such an application. Among the most selective materials are doped iron-aluminum oxides. Retrofitting of existing reactors by the use of a granulated oxide catalyst bed consisting of tablets (5 mm × 5 mm) or extrudates (15 mm length × 5 mm diameter) loaded in special baskets resulted in the following operating problems: (i) pressure drops which were too large (up to 0.1 bar); (ii) uneven flow distribution in thin catalyst layers due to non-uniform bed thickness, which also perturbates the flow pattern in the upstream noble metal gauzes; and (iii) formation of dust associated with thermal cycling of the oxide catalyst particles.

It has been claimed that all these problems can be solved by the replacement of packed beds of catalyst particles with honeycomb monoliths. The technology for extrusion of bulk active monoliths based on promoted iron oxides was developed at the laboratories of the Boreskov Institute in Novosibirsk.[83] Typical honeycomb modules with 5 × 5 mm channels and 1.8–2.2 mm wall thickness were produced with a cross section of 75 × 75 mm and length of 50 mm, and are reported to exhibit 30 $m^2/g$ surface area, 0.3 $g/cm^3$ wall porosity and 8–10 MPa crushing strength.[83,84] Design of monolith channel size and void fraction was obtained by mathematical models including the kinetics of heterogeneous and homogeneous reactions, as well as gas-solid mass transfer effects and the calculation of pressure drop.[85] In addition to the minimal pressure losses and the absence of dust formation due to high mechanical strength and thermal shock resistance, the use of honeycombs was also reported to equalize the flow through the Pt gauzes thanks to the uniform gas permeability through the monolith channels.

Performances during 3,000 h (the normal life of Pt gauzes in high-pressure converters) of operation at 7.3 atm in an industrial unit are reported in Table 28.4. No ammonia slip was observed with a platinum metal loading reduced by 20–25% when replacing three of the twelve gauzes with one honeycomb layer. NO yield decreased from 94.10 to 92.95% during operation, while Pt losses were cut by 20% with respect to those registered before retrofitting (from 0.157 $g/t_{HNO_3}$ to 0.124 $g/t_{HNO_3}$).

**Table 28.4.** Performances of an industrial two bed $NH_3$ converter operated at 7.3 atm. Adapted from Ref. 83.

| Month | Time-on-stream (h) | NO yield (%) | Inlet $NH_3$ (%) | T(°C) | $NH_3$ load ($Nm^3/h$) | $HNO_3$ capacity (t/h) |
|---|---|---|---|---|---|---|
| 1 | 568 | 94.10 | 9.36 | 866 | 6,000 | 14.8 |
| 2 | 601 | 94.07 | 9.15 | 868 | 5,800 | 13.5 |
| 3 | 715 | 93.64 | 9.15 | 868 | 5,700 | 13.0 |
| 4 | 684 | 93.77 | 9.08 | 865 | 5,650 | 14.4 |
| 5 | 708 | 92.95 | 9.13 | 862 | 5,550 | 12.9 |

Also demonstrated was a 1.5 year durability of the monolithic oxide catalyst. Ten commercial reactors adopting dual bed technology with honeycomb catalysts were reported to be under operation in Russia in the year 2000.[83,84]

In an effort to further improve this technology, cordierite-based monoliths were investigated in view of the following properties: (i) high stability towards thermal cycles due to their low thermal expansion coefficient; and (ii) the possibility to extrude honeycombs with higher cell density and improved mass transfer performances. Catalysts prepared via impregnation of the cordierite support with a solution of active perovskite precursors followed by annealing at $T > 500°C$ and calcination at 900°C, showed low activity and selectivity due to interactions between cordierite and the active phase. Some improvement was obtained by the deposition of an inert layer ($ZrO_2$+ mixture of lanthanide oxides) between cordierite and the active perovskites, but stability problems were still observed during ageing tests under real process conditions.[86] A different strategy was attempted based on extrusion of bulk catalysts consisting of cordierite modified with Co-FeV-Bi oxides. Honeycombs with 2.5 mm channel size and 0.4 mm wall thickness were obtained, which are reported to show promising properties for the substitution of a higher number of noble metal gauzes.[84]

#### 28.3.1.2. *Phthalic anhydride production*

Phthalic anhydride, an important feedstock material for the production of plasticizers and plastics, is obtained by the catalytic selective oxidation of *o*-xylene. To reduce investment and utility-specific costs, many efforts have been spent in recent years to enhance the *o*-xylene inlet concentration from 60 g/Nm$^3$ to 100 g/Nm$^3$ and above. Due to the high exothermicity of the process, carried out in multi-tubular, externally cooled reactors, such a concentration raises results with an increase of catalyst thermal loading. As a consequence, the hot spot temperature increases, which results in a progressive decrease of catalyst activity. Deactivation can be partially compensated by an increase of coolant temperature while keeping the catalyst peak temperature constant. However, the deactivation process further proceeds, and eventually the *o*-xylene conversion and phthalic anhydride selectivity performances of the reactor drop below the process constraints, particularly with reference to the quality of the purified product, so that the catalyst must be replaced, thus incurring the high costs associated with the new catalyst loading, the time required by catalyst loading and reactor shutdown and startup procedures.

A solution to overcome this problem is the adoption of catalytic postreactors with the following duties: (i) conversion of unreacted *o*-xylene to phthalic anhydride; (ii) conversion of under-oxidation intermediates (*o*-tolualdheyde, phtalide) to phthalic anhydride; and (iii) extensive destruction by deep oxidation of other side products which affect the quality of phthalic anhydride. As a result an increase of product yield and quality is achieved, particularly when operating with a progressively deactivated catalyst in the main reactor, along with the extension of the operating life of the main catalyst.[87]

Thanks to the reduced thermal load downstream from the main converter, the post-reactor is operated adiabatically, and is controlled only by the regulation of the inlet gas temperature. The use of honeycomb monoliths allows the location of the finishing catalyst bed during retrofit installation, using existing cooling facilities to adjust the inlet gas temperature and to operate the post-reactor with minimum pressure drops, which can be easily handled by existing compressors.

Two post-reactor systems, jointly developed by Lurgi, GEA and Wacker, have been installed in India during plant revamping.[87] The honeycomb catalysts were developed by Lurgi and consist of cordierite monoliths washcoated with a $V_2O_5$-$TiO_2$ active phase.[88] Preferred cell densities are 100–200 cpsi with an active phase loading of 100–150 $kg/m^3$. In one of the two plants, where the post-reactor was installed downstream of a main reactor loaded with an end of life catalyst, an extensive conversion of unreacted *o*-xylene and of intermediate under-oxidation products to phthalic anhydride (*o*-toluic aldehyde, phthalide) was achieved along with a substantial decrease in the production of other undesired side products (benzoquinone) by correct adjustment of the inlet temperature by means of the pre-cooling bundle.

#### 28.3.1.3. *Methanol to formaldehyde*

Similar considerations to those reported in the previous paragraph also apply to the production of formaldehyde by the selective oxidation of methanol. It has been reported that a 25% increase of product yield can be achieved by adopting adiabatic post-reactor technology.[89] For this purpose Haldor Topsoe patented the use of monolith catalysts prepared by impregnation of corrugated silica fibre sheets with a slurry of active catalyst powders consisting of mixed iron-molybdenum oxides with binder additives.[90] The monolith body was then obtained either by rolling a single sheet in a cylindrical shape with straight channels or, preferably, as a cross-corrugated structure by piling up a number of corrugated sheets to form parallel layers with a different orientation of the corrugation among the layers. Fibrous sheets with a thickness of 250 μm and a corrugation height of 2.5 mm, impregnated with an active phase loading of 50–80% w/w, were described in the patent. Comparison of the performances of an adiabatic post-reactor loaded with a monolithic catalyst with those of the same reactor loaded with a crushed (1.0–1.7 mm) conventional formaldehyde catalyst showed that the monolithic system provides a similar increase in conversion and yield as the value provided by the main converter (from 95.6 to >99% and from 91.8 to >94%, respectively) with halved pressure drops.

### 28.3.2. *Reactions for gas generation*

The flow behaviour of monolith catalysts, characterized by a low pressure drop, is also involved in their use as gas generators for rocket thrusters or igniters and for turbine engine restart. These demanding aerospace applications require that a propellant be very rapidly converted to produce large volumes of high-temperature (up to 1,300 K) gas under pressure (up to 2,800 kPa): both the catalytic contact time and the "light-off" time are typically desired to be in the order of milliseconds, and high conversions with large temperature excursions are obtained, much like in short contact time reactors for the production of chemicals. However, more complexity is added by the unsteady-state operation which is usually associated with such applications.

Voecks,[91] and refs therein reported on two proof-of-concept investigations in this area. In the former study, a very high flow rate of liquid propellant consisting of a $H_2$-rich mixture of hydrogen and oxygen was fed to a conventional packed bed of alumina particles, to an alumina-washcoated metal honeycomb monolith and to a washcoated metal sponge monolith. Back pressures up to 1 Mpa were measured across the particle bed, whereas the pressure drop was negligible in the case of the honeycomb and sponge monoliths. All three catalysts exhibited activity at ambient pressure, but only the particles and the sponge monolith were found to be active at the very low pressures typical of space operations, about 14 Pa: this was probably due to the different pressure dependence of the gas-solid mass transfer rates in the honeycomb channels, in the tortuous paths prevailing in the packed bed of particles and in the metal sponge.

## 28.4. Applications of Structured Catalysts Based on Enhanced Heat Exchange

### 28.4.1. *Heat transfer properties of honeycomb monolith structures*

Until recently the use of monolithic catalysts in non-adiabatic reactors was regarded as unfeasible due to poor radial heat transfer properties. Indeed, ceramic honeycomb monoliths are made of essentially insulating materials, a theoretical analysis by Cybulski and Moulijn[92] provided evidence that commercial monolith structures consisting of corrugated metal sheets exhibit modest heat transfer performances too.

Nevertheless, the thermally connected nature of the monolith supports provides, in principle, an alternative mechanism of radial and axial heat transport, namely heat conduction, which is essentially not available for the random packing of catalyst pellets. The conduction within the solid phase of the pellets is, in fact, almost negligible, since only point contacts exist between the pellets, and convection in the gas phase dominates as the primary mechanism for heat exchange. Accordingly, the only practical way of enhancing heat transfer is to increase the flow velocity, but this is limited by the pressure drop, which grows more than linearly with flow rate. By using monolith honeycomb structures with parallel channels as catalyst elements, no radial transfer of gas may exist, but the contribution of thermal conduction through the solid phase (i.e. the monolith matrix) can become quite significant if suitable materials and geometries are adopted.

The effective axial heat conductivity of monolith substrates $k_{e,a}$ is readily estimated as

$$k_{e,a} = k_s(1 - \varepsilon) \tag{28.8}$$

with $k_s$ = intrinsic thermal conductivity of the support material and $\varepsilon$ = monolith void fraction (open frontal area). Early attempts to model radial heat conduction in monoliths, also including comparison with experimental data, were published in Refs. 93–95. Based on a simple analysis of heat conduction in the unit cell of a honeycomb monolith with square channels according to an electrical network analogy, Groppi and Tronconi[96,97] derived the following approximate predictive equation for the effective radial thermal conductivity in washcoated monoliths, $k_{e,r}$:

$$k_{e,r} = \frac{k_s}{(1-\sqrt{\varepsilon+\xi}) + \frac{\sqrt{\varepsilon+\xi}-\sqrt{\varepsilon}}{(1-\sqrt{\varepsilon+\xi})+\frac{k_w}{k_s}\sqrt{\varepsilon+\xi}} + \frac{\sqrt{\varepsilon}}{(1-\sqrt{\varepsilon+\xi})+\frac{k_w}{k_s}(\sqrt{\varepsilon+\xi}-\sqrt{\varepsilon})+\frac{k_g}{k_s}\sqrt{\varepsilon}}} \tag{28.9}$$

where $\varepsilon$ and $\xi$ are the monolith volume fractions of voids and washcoat, whereas $k_s$, $k_g$, $k_w$ are the intrinsic thermal conductivities of solid support, gas phase and washcoat, respectively. A similar equation was also derived for monoliths with equilateral triangular channels.[96] Recently, Hayes and co-workers[98] validated Eq. (28.9) against numerical solutions of the temperature field in honeycomb structures, finding maximum deviations of less than 20% for a typical monolith void fraction of 75%. They also derived an alternative equation, based on a different (parallel) arrangement of the resistance network, which improved somewhat the prediction accuracy.

As Eq. (28.9) shows, the effective conductivity $k_{e,r}$ is directly proportional to the intrinsic thermal conductivity of the support material, $k_s$. Thus, the adoption of highly conductive materials is expectedly highly beneficial for the enhancement of radial heat transfer in monoliths. In Fig. 28.8, estimates of $k_{e,r}$ according to Eq. (28.9) are plotted versus the monolith open frontal area $\varepsilon$ for honeycomb structures made of various metallic and non-metallic materials with different intrinsic thermal conductivity; for the sake of simplicity, the volume fraction of active washcoat, $\xi$, as well as the minor contribution of heat conduction in the gas phase have been neglected in this case. It should be emphasized that when highly conductive materials are used,

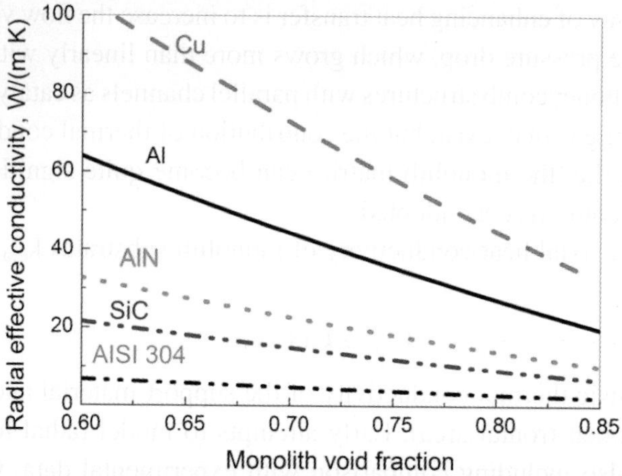


**Figure 28.8.** Effect of material properties and monolith void fraction on estimated radial effective thermal conductivity of honeycomb monoliths with square channels. Adapted from Ref. 97.

the estimates of $k_{e,r}$ in Fig. 28.8 become one order of magnitude greater than the effective radial thermal conductivities in packed beds of catalyst pellets, which are typically in the range 2–5 W/m/K.[99,100] The plot also shows that the radial effective conductivity is adversely affected by large monolith void fractions.

These evaluations point out that heat exchange in monolithic structures can be made efficient (even more efficient than in pellets), but monolith supports with specific designs must be adopted, based on a discerning selection of both the monolith geometry and the material aimed at minimizing resistances to conductive heat transfer.

Notably, the existing commercial monoliths, originally developed for the adiabatic applications of environmental catalysis, were not originally designed for that purpose: in fact, neither the construction material nor the geometry of such supports is optimized for heat conduction. In fact, the intrinsic conductivity of ceramic honeycombs is very low, whereas the available metallic monolith structures are made of poorly conductive alloys (e.g. FeCrAlloy) and are assembled by piling up and rolling corrugated sheets which are in poor thermal contact with each other, thus increasing the overall resistance to heat transfer. Finally, in commercial monoliths, the open frontal area is kept as high as possible, typically 0.7–0.8 for ceramic monoliths and 0.85–0.95 for metallic ones, so as to match the severe pressure drop constraints of environmental processes.

Based on the above considerations, heat conduction in the walls of monolithic structures can be effectively exploited in principle, as an effective mechanism to remove the heat of exothermic selective oxidation reactions. Published studies concerning such applications are discussed in Section 28.4.2.

The potential for enhanced heat transfer, and decreasing radial and axial temperature gradients, is also associated with the adoption of open-celled foams as structured catalyst supports, as outlined in Section 28.4.3. This is a research area of rapidly growing interest nowadays, even though relatively few contributions have appeared to date in the open literature: the relevant publications concerning selective oxidation processes are summarized in Section 28.4.4. Finally, Section 28.4.5 addresses the applications of structured catalysts and reactors of other shapes.

### 28.4.2. *Selective oxidations in conductive honeycomb monoliths*

Groppi and Tronconi have systematically investigated the potential of novel monolithic catalyst supports with high-thermal conductivity in view of replacing conventional packed beds of catalyst pellets in multi-tubular reactors for gas/solid selective oxidations.[96,97,101–104] Starting from the evaluation of effective radial thermal conductivities in monolith structures outlined in Section 28.4.1 and summarized by Fig. 28.8, they predicted that, in principle, the radial heat transfer in fixed-bed gas/solid reactors could be substantially enhanced when changing the dominating heat transfer mechanism from convection to conduction. This would be a very important result, since both the design and the operation of technical packed-bed reactors are limited at present by the removal of the reaction heat, which occurs by convective transport from the randomly packed catalyst pellets to the reactor tube walls: therefore limits on the reactor tube diameter of 1 to 1.5 inches as well as very high gas flow rates are typically required to prevent unacceptable hot spots. Significantly improved radial heat transfer, on the other hand, would bring about reduced risks of thermal runaway, better thermal stability of the catalyst, and possibly improved selectivity, as well as carrying the potential for novel designs of industrial reactors with incremented throughputs and/or enlarged tube diameters, corresponding to reduced investment costs.

In order to assess such prospective advantages, the thermal behaviour of "high conductivity" monolith catalysts in exothermic reactions was investigated both theoretically and experimentally.

#### 28.4.2.1. *Simulation studies*

A preliminary modelling analysis[101] involved the parametric study of a multi-tubular externally-cooled fixed-bed reactor for a generic selective oxidation process, where the catalyst load consisted of cylindrical honeycomb monoliths with washcoated square channels, made of highly conductive supports. In this early work, the attention was focused on the effect of catalyst design. Simulation results were generated by a steady-state, pseudo-continuous 2D monolithic reactor model, where the catalyst is regarded as a continuum consisting of a static, thermally connected solid phase and of a segregated gas phase in laminar flow inside the channels.[101] It was shown that metallic honeycombs are indeed promising for limiting temperature gradients as compared to pellets, to the extent that near-isothermal operation of the catalytic bed can be approached even for strong exothermal duties. In order to take full advantage of heat conduction in monoliths, however, specific honeycomb designs must be developed which include relatively large volume fractions of support made of materials with a high intrinsic conductivity (e.g. copper or aluminum), as well as large loads of active catalytic components. It is worth emphasizing that such designs are structurally different from those of the existing commercial monolithic supports used in environmental catalysis, which carry a relatively small load of active washcoat, since the related reactions are very fast, and exhibit very large open frontal areas; the main goal in these applications is the cutback of the pressure drop.

In a subsequent simulation study, two important industrial selective oxidation processes were addressed in detail, namely the partial oxidation of methanol to formaldehyde and the epoxidation of ethylene to ethylene oxide.[102] In both cases secondary undesired reactions play a significant role, i.e. the combustion of the primary product in the formaldehyde process and the combustion of the ethylene reactant in the ethylene oxide process, so that the study also provided information on how the adoption of "high conductivity" monolith catalysts would affect the selectivity of industrial partial oxidation processes for both a consecutive and a parallel reaction scheme. For both processes intrinsic kinetics applicable to industrial catalysts as well as design and operational parameters for commercial reactors were derived from simulation studies and experimental investigations collected in the literature.

With reference to the formaldehyde reactor, assuming the parameters reported in Table 28.5, the simulations showed that the HCHO molar yield could be incremented from the 93.6%, reported for an optimized packed-bed reactor process,[105] up to over 97% if aluminum-washcoated honeycombs with suitable design were loaded in the original reactor tubes. The optimal performance of the monolith catalysts originates

**Table 28.5.** Simulation parameters for the industrial formaldehyde reactor loaded with "high conductivity" monolith catalysts. Adapted from Ref. 102.

| Reactor | Tube length | L | 0.7 m |
|---|---|---|---|
| | Tube inner diameter | D | 0.0266 m |
| | Catalyst density | $\rho_{cat}$ | 2,000 kg/m$^3$ |
| Operating conditions | Mass velocity | $W_t$ | 2.5 kg/(m$^2$s) |
| | Inlet temperature | $T^\circ$ | 250°C |
| | Inlet pressure | $P^\circ$ | 1.55 ata |
| | Feed $CH_3OH$ mole fraction | $Y^\circ_{CH_3OH}$ | 0.05 |

from: i) a thin catalyst layer, which prevents diffusional limitations from adversely affecting the selectivity; ii) a thickness of the highly conductive monolith walls adequate enough to grant near-isothermal operation, preventing hot spot formation; and iii) a high level of the coolant temperature, which increments the average reactor temperature and hence the overall $CH_3OH$ conversion. In its optimized configuration the monolithic reactor would be virtually isothermal. It is worth noticing that the increment of the coolant temperature can be exploited to compensate for a smaller volume fraction of active catalyst than in packed beds because the process is operated essentially under kinetic control, due to the thin washcoat layers deposited onto the monolith catalysts. A higher temperature level, in fact, does not adversely affect the selectivity in this case since, according to the intrinsic kinetic scheme, the activation energy of the consecutive undesired reaction (combustion of formaldehyde) is lower than that of the primary reaction. Although the optimization of the monolithic reactor was carried out without any constraint on the catalyst temperature, alternative sub-optimal configurations with lower levels of the coolant temperature were investigated too, in view of possible problems with the thermal stability of the catalyst. Even with largely reduced temperature levels, it was still possible to achieve HCHO yields significantly superior to the performance of the industrial packed-bed reactor.

Further simulations showed that a high (>95%) HCHO yield can be achieved even in the case of reactor tubes with a diameter incremented from one, the current industrial standard, to three inches, which would afford important savings in reactor investment costs. However, the volume fraction of the conductive monolith support needs to be incremented by a factor of four to compensate for the greater heat-transfer resistances. For all the simulated conditions the estimated pressure drop was less than 1% of the inlet pressure, versus over 10% in the industrial packed-bed reactor.

In the case of the ethylene oxide reactor, on the other hand, in order to optimize the selectivity it is crucial to prevent hot spots, as the activation energy of the parallel parasitic ethylene combustion is greater than the activation energy of the primary epoxidation reaction. The simulation results[102] confirmed that isothermal operation is also feasible for the ethylene oxide reactor, due to the excellent effective thermal conductivity of the metallic monoliths. For a monolith pitch m = 2 mm, and for monolith volume fractions of support and catalyst both equal to 0.2, the reactor behaviour was found to be identical to that of an ideal isothermal reactor under a variety of conditions, provided that $k_s$, the intrinsic thermal conductivity of the monolith support, was equal to or greater than 50 W/(m K). Again, a reactor design based on larger tubes (D = 76.2 mm instead of 39.25 mm) was also found to be feasible at the expense of a greater volume fraction of catalyst support, but only with $k_s$ greater than 100 W/(m K), which can be provided e.g. by aluminium or copper. Notably, thick catalytic washcoats do not adversely affect the selectivity of this process. Indeed, thick catalyst layers would be desirable to increment the process yield, but the adhesion of thick wash coats onto metallic surfaces may become critical.[106]

On the other hand, increments of the coolant temperature, as adopted for the formaldehyde reactor, are not compatible with the present kinetic scheme, since they would adversely affect the selectivity in this case. For a fixed wash coat thickness (e.g. the greatest one compatible with adhesion requirements), however, the overall catalyst load can still be incremented by incrementing the monolith cell density. Calculations for a 120 μm thick catalytic wash coat shown in Fig. 28.9 suggest that high conductivity honeycombs with small pitch would indeed bring about significant improvements of $C_2H_4$ conversion with only a slight loss of selectivity. While the results presented so far were generated assuming no heat transfer resistance between the monolith catalyst and the coolant, actually a thermal contact resistance can be expected at the interface between the monolith and the inner reactor tube wall, as also detected in the experimental investigations reported below.[104,107] Calculations predict that such a resistance may become critical for the onset of hot spots in the ethylene oxide reactor whenever the corresponding "wall" heat transfer coefficient is less than about 500 W/(m$^2$K).[97] Accordingly, solutions aimed at achieving effective thermal contact between the honeycombs and the reactor tubes ("packaging" methods) represent an important development goal, which must be necessarily pursued in connection with both the manufacturing technologies of monolithic catalysts and with the specific features of the individual catalytic processes.

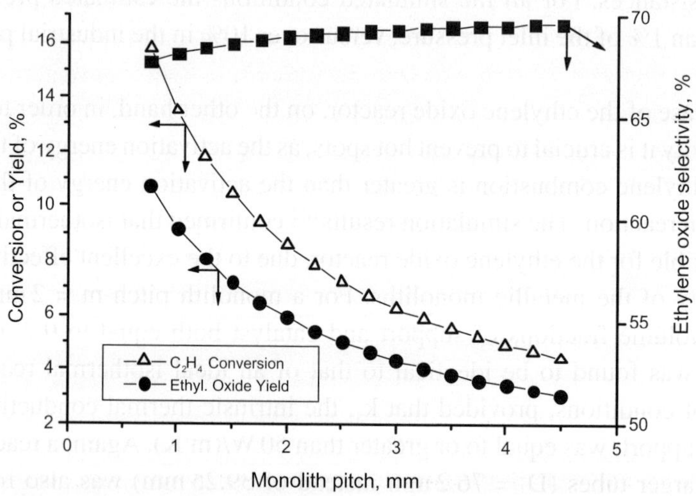


**Figure 28.9.** Epoxidation of ethylene: calculated effect of the monolith pitch on $C_2H_4$ conversion, $C_2H_4O$ selectivity and molar yield. Washcoat thickness: 120 μm; support volume fraction: 0.2; ks: 200 W/(m K); Tcool: 250°C. Tube diameter: 39.2 mm. Reprinted from Ref. 102. Copyright 2001 with permission from Elsevier.

Along similar lines, the application of highly conductive monolithic catalysts in the production of phthalic anhydride by the selective oxidation of *o*-xylene was also simulated.[108] In this study, the characteristics of new prototype honeycomb copper substrates produced at Corning Inc.[109] were assumed, and account was taken of the monolith–reactor contact resistance, included according to the experimental results described in Ref. 107. The simulation results confirm that significant benefits can be expected from the excellent conductive heat transfer properties of these novel substrates. In general, due to a better control of the hot spot, the operating window is enlarged with respect to the conventional packed-bed reactor: for example, it is possible to operate with lower hot spot temperatures, allowing extended catalyst lifetimes. Guidelines are also given for the rational design of the monolithic catalysts with respect to requirements on catalyst inventory and intraporous diffusional limitations. Notably, using a 200/10 cpsi honeycomb, only 28 μm thickness of coating are required to obtain the same loading of active phase, 70 kg/m$^3$, of a conventional packed-bed reactor loaded with $8 \times 5 \times 5$ mm rings coated with an active layer about 100 μm thick. Besides this, using a 400/7 cpsi honeycomb with an active layer thickness of 100 μm, an active phase loading of almost 300 kg/m$^3$ can be obtained, which corresponds to more than 25% of the reactor volume. Such a value, obtained with commercially feasible monolith geometry and washcoat thickness, is indeed able to match the catalyst loading requirements of most of the existing selective oxidation processes. The authors complemented their study with an economic analysis of three operative solutions, which emphasize the different possible strategies in exploiting the advantages of high conductivity monolith supports: (i) operation at equal feed conditions but at a higher coolant temperature to maintain the same hot spot temperature, resulting in a higher yield; (ii) operation at equal air flow but with a higher feed concentration of *o*-xylene, resulting in a higher throughput; and (iii) operation at equal *o*-xylene throughput but with a reduced air flow, which requires less energy. Economic evaluations designated solution (ii) as the most rewarding one by far, however, the related increase in capacity would usually require some investment to de-bottleneck the equipment downstream, while solutions (i) and (iii) appear already very attractive, leading to annual cost reductions in the order of US$1 million for a 45 kt/y unit. Such evaluations, however, did not consider the cost of the novel conductive honeycomb substrates.

#### 28.4.2.2. *Experimental studies*

In parallel with the modelling analyses discussed above, the heat exchange characteristics of "high conductivity" monolith catalysts have also been addressed experimentally,[97,103,104] the goal being to investigate the thermal behaviour of structured metallic catalysts in the presence of a strongly exothermic reaction, with focus specifically on the influence of this catalyst design on parameters such as material,

**Table 28.6.** Characteristics of plate-type catalysts tested in CO oxidation. Adapted from Ref. 119.

| | Sample A | Sample B | Sample C | Sample D | Sample E | Sample F | Sample G |
|---|---|---|---|---|---|---|---|
| Support material | Al | Steel | Steel | Al | Al | Al | Al |
| Support configuration | I | I | I | I | I | II | III |
| No. of slabs | 4 | 4 | 4 | 4 | 4 | 12 | 12 |
| No. of coated slab faces | 6 | 6 | 6 | 6 | 6 | 24 | 24 |
| Slab thickness, mm | 0.5–1 | 0.5–1 | 0.2–0.4 | 0.5–1 | 0.5–1 | 0.5 | 0.5 |
| Gap between slabs, mm | 3 | 3 | 3 | 3 | 3 | 1.5 | 1.5 |
| Slab width, mm | 46 | 46 | 46 | 46 | 46 | 46 | 46 |
| Length, mm | 200 | 200 | 200 | 200 | 200 | 50 | 50 |
| Coating method | I | I | I | II | III | III | III |
| Washcoat load, g | 12.8 | 14.9 | 13.2 | 1.04 | 3.84 | 4.15 | 3.98 |
| Load: 3%Pd $+\gamma$-$Al_2O_3$, g | 2.44 | 2.83 | 2.50 | 0.42 | 3.84 | 4.15 | 3.98 |

configuration and geometry of the structured support, and formulation and load of the catalytic washcoat.

Since commercially available monolithic supports for environmental catalysts are not suitable for this class of applications, as explained above, in the early studies homemade "high conductivity" structured catalysts were prepared by assembling washcoated slabs of aluminium or stainless steel to form plate-type catalytic cartridges, which were also equipped with thermowells for sliding thermocouples in order to monitor the temperature distributions. The washcoat consisted of Pd (3% w/w) on $\gamma$-$Al_2O_3$,[106] and the catalysts were tested in the oxidation of CO, selected as a model exothermic reaction.

Seven samples with different characteristics were prepared and tested, as summarized in Table 28.6. Thus, samples A and B shared the same geometry, but their support was made of aluminium and of stainless steel, respectively, so that a comparison of their thermal behaviour would provide direct information concerning the influence of the intrinsic conductivity of the support material, which is smaller by approximately a factor of ten in the case of steel. Likewise, the other samples were designed and prepared to collect experimental evidence on the role of the thickness of the slabs (sample C), washcoating method (sample D), washcoat load (sample E), volume fraction of metallic support (sample F) and contact thermal resistance at the reactor wall (sample G).

The main results can be summarized as follows:

(i) Over all the samples with Al support, temperature gradients were negligible in the direction transverse to flow, and were moderate along the axial coordinate even under the most severe reaction conditions. Representative measured temperature distributions are displayed in Fig. 28.10: due to the high intrinsic heat conductivity of the Al support, practically all of the overall heat transfer resistance was confined at the interface between the catalyst slabs and the inner reactor wall.

(ii) When tested under the same reaction conditions as sample A, sample B — with a stainless steel support — exhibited much more marked temperature gradients, also along the transverse coordinate (Fig. 28.10). The different behaviours of samples A and B is evidently related to the difference in the intrinsic thermal conductivity of aluminium and steel, which are approximately 200 W/m/K and 20 W/m/K, respectively.

(iii) Sample C, having a support made of steel but with thinner slabs ($s = 0.2$ mm versus 0.5 mm in samples A and B), brought about even stronger gradients as a result of the decreased effective thermal conductivity of the support.

(iv) Different washcoat compositions and washcoat loads (samples D and E) altered the catalytic activity but not the thermal behaviour of the structured systems; particularly, in spite of the incremented load of a more active washcoat, only moderate axial T-gradients and negligible transverse T-gradients were still detected on sample E due to its Al support. Notably, the $CO_2$ productivities measured over sample E were in line with those obtained over a powdered catalyst (with the same composition as its washcoat) loaded in a conventional lab-scale flow microreactor, indicating that the washcoat deposition procedure adopted for the preparation of sample E did not alter the intrinsic catalytic activity. However, the comparison was possible only for limited thermal duties: for example, it was impossible to operate the microflow reactor under the conditions of Fig. 28.10.

(v) Sample F was based on a different design involving a support of aluminium with more densely packed and shorter slabs. However, the geometric exposed area and the overall washcoat load were kept identical to those of sample E. CO oxidation runs at the same operating conditions resulted in very similar temperature profiles over samples E and F, confirming that the thermal behaviour of "high conductivity" structured catalysts is primarily governed by the washcoat-to-support volume ratio, which was, in fact, identical for the two samples.

(vi) Finally, the addition of small fins along the contour of the slabs (sample G) in order to improve the thermal contact at the catalyst–reactor interface resulted in a significant (>30%) decrease in the overall heat transfer resistance with respect to the otherwise identical sample F.

The experimental results were adequately represented by a one-dimensional heterogeneous non-isothermal model of the plate-type structured catalysts, accounting for heat generation by CO oxidation over the catalyst slabs, heat conduction along

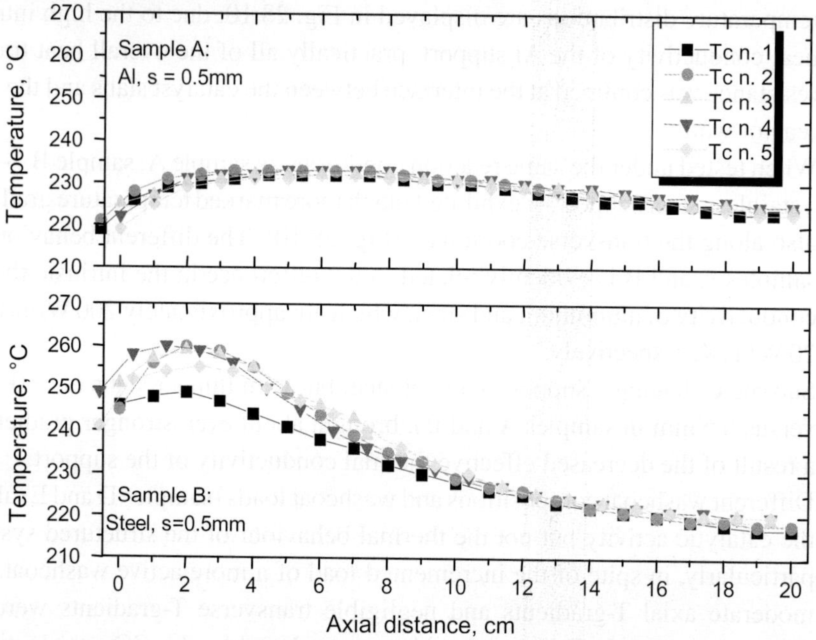


**Figure 28.10.** Comparison of temperature distributions measured over catalyst samples A (Al support) and B (steel support) under the same reaction conditions: CO feed concentration: 5% v/v; feed flow rate: 1,000 $cm^3$/min (STP); oven temperature: 216°C. CO conversion: 100%. Reprinted from Ref. 97. Copyright 2002 with permission from Wiley.

the slabs, and heat exchange to the surrounding heat sink across the catalyst-reactor interface.[97,104] Notice that the assumption of negligible T-gradients in the direction transverse to gas flow (1D approximation) is experimentally verified in the case of highly conductive supports (see e.g. Fig. 28.10). In the model equations, $h_w$ (the wall heat transfer coefficient) was regarded as an adaptive parameter, and was estimated by fitting the calculated axial temperature distributions to a set of experimental catalyst temperature profiles, whereas the rate parameters were estimated from independent regression analysis of CO conversion data in isothermal runs.[104] Satisfactory fits were obtained in the case of samples with Al support (A, D, E, F and G). In all cases the estimated wall heat transfer coefficients $h_w$ were in the order of 100 W/$m^2$/K, the highest value (119 W/$m^2$/K) corresponding to sample G owing to its specific design with fins.[104]

In a later stage of the investigation, Tronconi *et al.*[107] reported an experimental study of the heat transfer properties of monolithic catalysts based on prototype honeycomb copper substrates produced at Corning Inc. from the extrusion of Cu powder, followed by drying and firing.[109] Differently from the conventional manufacturing processes of metallic monoliths based on the piling up and rolling of corrugated sheets, extrusion provides the required continuous and thermally connected matrix which is optimal for conductive heat transfer. The supports were Ni-plated to prevent the oxidation of copper, then washcoated with Pd/$\gamma$-$Al_2O_3$, loaded in a 1" i.d. tube inserted in an oven and tested in both pure heat transfer and reactive experiments, again using CO oxidation as a strongly exothermic model reaction. The axial T-profiles measured by three sliding thermocouples directly inserted into the monolith channels at different radial positions, showed that even under the most severe conditions investigated in reactive runs, corresponding to a radial heat flux exceeding 23 kW/m$^2$, radial temperature gradients were still negligible with respect to the T-differences prevailing between the monolith and the reactor wall, where the controlling heat transfer resistance is confined.

In addition to the characterization of extruded monolithic supports with high intrinsic conductivity suitable for industrial applications, the purpose of this work was also to test the packaging systems aimed at reducing the heat transfer resistance at the monolith–wall interface ("gap" issue). Figure 28.11 compares the temperature differences measured between the monolith centreline and the reactor tube wall in reactive runs at comparable conditions over copper monolith catalysts packaged differently: sample C represents the data for an "advanced" packaging method, whereas sample E was loaded without special measures to ensure good heat transfer across the gap. It is apparent that the introduction of the advanced packaging results in a significant reduction of the temperature gradient across the monolith–tube interface, and hence of the associated thermal resistance, along the entire monolith

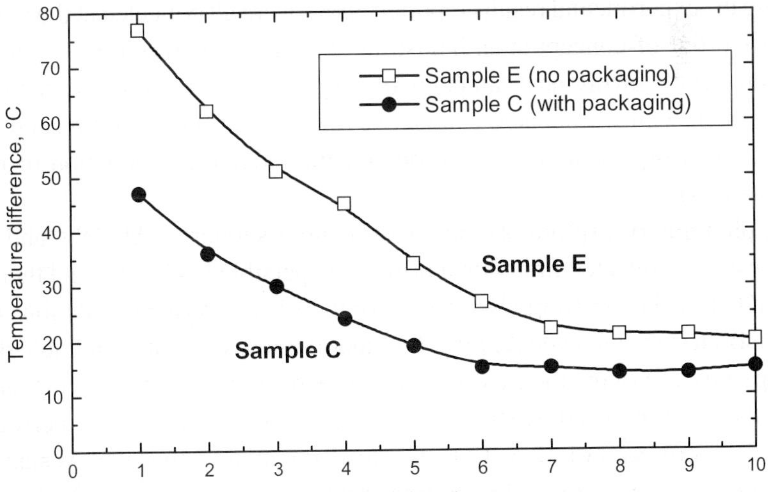


**Figure 28.11.** CO oxidation runs over copper monoliths: effect of packaging on the T-difference between monolith axis and tube wall. Flow rate: 7,000 cm$^3$/min (STP); CO feed: 5% v/v. Toven: 215°C (sample C), 200°C (sample E). Reprinted from Ref. 107. Copyright 2004 with permission from Elsevier.

length. These data confirm that it is critical but also feasible to enhance the wall heat transfer coefficient $h_w$ by a suitable packaging method.

Estimates of the "gap" heat transfer coefficient ($h_w$) were obtained by the regression of temperature profiles in both heat transfer and CO oxidation experiments, based on a heterogeneous one-dimensional monolith reactor model similar to that reported in Ref. 104. The estimated heat transfer coefficient from the monolith to the tube wall was 220 W/($m^2$K) when no special packaging concept was used. With one of several advanced packaging concepts tested, essentially based on controlling the monolith–tube clearance, heat transfer coefficients in the target range of 400–500 W/($m^2$K) were achieved. It was further reported for comparison, that the overall heat transfer coefficients of standard catalyst packings, such as rings, would be well below 100 W/($m^2$K) under the flow rates used in the experiments, and would also be, at best, in the range of 200–250 W/($m^2$K) under typical industrial conditions.

The "gap" resistance was further rationalized in subsequent work.[110,111] It was shown that the associated heat transfer mechanism relies primarily on heat conduction across the stagnant gas film, trapped in the gap between the monolith and the inner reactor tube wall. In fact, the gap resistance was inversely proportional to the gap size evaluated under the reaction conditions (so differential thermal expansion of the monolith and tube materials should be considered),[111] and directly proportional to the gas-phase conductivity, as evidenced by heat transfer measurements with $N_2$-He mixtures of different compositions.[110] Estimates for $h_w$ in excess of 700 W/($m^2$K) were obtained when using pure He.

While the experimental results reported above were all collected at the laboratory scale, one proof-of-concept at an industrial scale has been recently reported in the open literature,[112] involving a campaign of *o*-xylene oxidation runs in a tubular pilot reactor loaded with washcoated conductive (aluminium) honeycomb catalysts and operated under representative conditions for the industrial production of phthalic anhydride (PA).

In a preliminary exploratory phase, structured supports (Al slabs and honeycombs) were washcoated first with a primer (dispersible boehmite) and then with a $V_2O_5/TiO_2$-based precursor powder used for industrial *o*-xylene oxidation catalysts (Polynt). Both steps involved dipping the support into a slurry with suitable rheological properties, removing the excess slurry either by gravity or by blowing, and flash heating the resulting catalyst structure. Isothermal kinetic runs were performed in a lab-scale microreactor (i.d. = 12.6 mm) loaded with a washcoated Al slab shaped in the form of a spiral (3 × 15 cm, total active catalyst mass = 400 mg), covering a range of representative temperatures (320–400°C) and *o*-xylene feed contents (1–3% v/v). The data were analysed by multi-response non-linear regression according to a parallel/sequential scheme of six global reactions associated with power law rate expressions. The kinetic scheme was then incorporated into a 1D heterogeneous

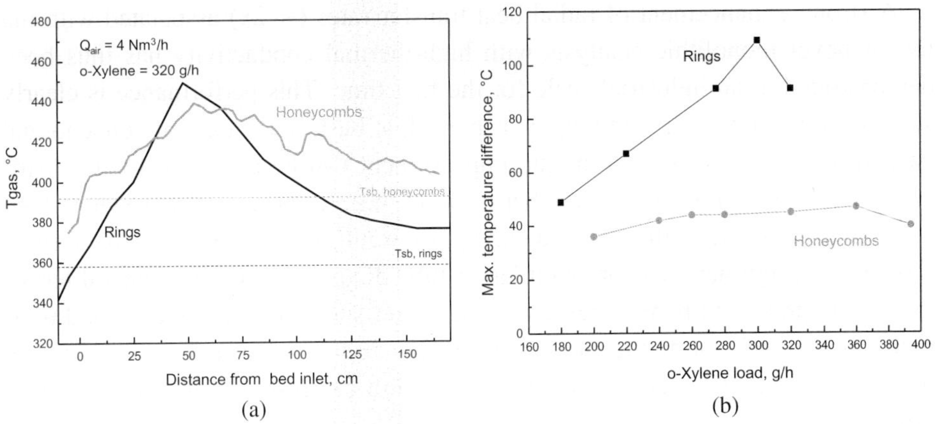


**Figure 28.12.** T-profiles (a) and max. $\Delta T = T_{hs} - T_{sb}$ (b) in pilot runs with rings or Al honeycombs. Adapted from Ref. 112.

simulation model of fixed-bed tubular reactors accounting for the specific heat transfer characteristics of conductive honeycomb catalyst supports.[102,107]

For the pilot reactor runs, 16 Al honeycombs supplied by Corning (26 cpsi, o.d. = 24.4 mm, length = 10 cm)[113] and washcoated with a total catalyst mass of 46 g were loaded in the upper part of an industrial pilot reactor (Polynt), consisting of a single-jacketed tube (length = 3 m, i.d. = 24.6 mm) cooled with molten salts. The tube loading was completed with inert rings. Axial T-profiles were recorded by means of a thermocouple sliding in a 2 mm o.d. thermowell inserted in the central channel of the honeycombs. The pilot reactor was operated continuously for over 1,600 hours. After startup, the air flow rate was kept at 4 $Nm^3/h$ and the *o*-xylene feed load was progressively increased from 120 to 400 g/h while adjusting the salt bath temperature to keep the measured hot spot temperature at around 440°C. Figure 28.12a compares the axial T-profile from one such run at reference industrial conditions for PA production (*o*-xylene load = 320 g/h) with a T-profile measured in the same pilot reactor loaded with conventional catalyst pellets (rings) and operated under the same conditions with a similar hot spot temperature.

The Al honeycomb supports afforded substantially reduced axial T-gradients, and enabled operation of the reactor with a much higher salt bath temperature (392°C vs 358°C in the case of rings): the maximum T-difference with the salt bath was halved (Fig. 28.12b) and the average bed temperature was therefore about 20°C higher. Model-based analysis yielded an estimate of the overall heat transfer coefficient ≈ 415 $W/m^2/K$ (vs 210 for the ring packing), in line with previous lab-scale work. T-gradients were still moderate at *o*-xylene loads in excess of 100 $g/Nm^3$, an upper limit for the current industrial PA packed-bed reactors. The Al honeycombs were successfully unloaded at the end of the pilot reactor runs.

A strong enhancement of radial heat transfer rates ($\approx$ 2x) associated with the use of novel monolithic catalysts with high-thermal conductivity has thus been demonstrated at an industrial scale for the first time. This performance is clearly superior to what could potentially be achieved, at best, in the case of conventional packed-bed reactors. Such a unique improvement can be exploited for intensification of the PA process in a number of ways, e.g. to increase the *o*-xylene feed load $>100$ g/Nm$^3$ (and the PA productivity accordingly) in existing technical reactors, or to design new reactors with larger tube diameters. In more general terms, the results presented herein, being obtained with substrates and under conditions representative of real applications, appear very encouraging in view of practical implementations of "high conductivity" monolith catalysts in selective oxidation processes.

### 28.4.2.3. *Patents*

The adoption of extruded honeycomb catalysts for strongly exothermic gas-phase reactions, and specifically for selective oxidations, is also claimed in industrial patents. US Patent 5,099,085,[114] assigned to Wacker and Degussa, describes the use of honeycomb monolithic catalyst supports for selective chlorination and oxychlorination reactions in multi-tubular fixed-bed reactors. The active phase, $CuCl_2$/KCl on alumina as in commercial pellet catalysts, was deposited onto monolith supports made of ceramic materials, the preferred ones being mullite and cordierite. It was claimed that with such supports the pressure drop across the reactor decreased drastically and the heat dissipation was improved with respect to conventional pellet catalysts having identical composition, and eventually resulting in an incremented selectivity of the reaction. However, based on the investigation of the heat transfer properties of honeycomb supports reported in the previous sections, it appears difficult to rationalize improvements of the heat dissipation when adopting ceramic monolith supports.

European Patent 1110605B and US Patent 7,678,342 to EVC (now Ineos)[115,116] describe the use of catalysts with various metallic honeycomb supports in gas-phase reactions, specifically chlorination/oxychlorination reactions, claiming greater yields and selectivities, avoidance of hot spots, prolonged catalyst lifetime and flexibility in use as compared to conventional catalysts in pellet form.

US Patent Appl. US2009/0176895 to Corning Inc.[117] claims designs and methods to load and operate monolithic catalysts in multi-tubular reactors for use in the chemical processing and/or energy conversion industries. The inventors refer specifically to the use of thermally conductive, metal honeycomb catalyst supports in order to improve thermal uniformity in the reactor tubes for applications to strongly exothermic reactions, such as the partial oxidations of hydrocarbons to products such as ethylene oxide, formaldehyde, phthalic and maleic anhydride, methanol, ethylene dichloride, etc.

Manufacturing methods specific for conductive honeycomb substrates have been patented by Corning Inc. According to US Patent 6,881,703[109] metallic monoliths offering improved heat conductivity are provided from metal powder batches of copper, tin, zinc, aluminium, iron, silver, nickel and their mixtures and alloys by extrusion through a honeycomb die, followed by drying and firing. As a less demanding alternative, US Patent 7,608,344[113] discloses methods to form conductive honeycombs, offering improved heat transfer, by direct extrusion of bulk metal feedstocks in the shape of billets. An example refers to the extrusion of aluminium metal honeycombs of zero wall porosity, similar to those tested in a PA pilot tubular reactor and described in Ref. 112.

### 28.4.3. *Heat transfer properties of foam structures*

The knowledge of the interior thermal properties of chemical reactors is definitely a key characteristic for planning and designing selective oxidation processes, as also discussed in the previous sections. In view of the application of foams as catalyst carriers in fixed-bed tubular reactors, heat transfer with forced convection and radiation in highly porous cellular foams has been studied in recent years by several authors. Some examples are given in Refs. 118–122. A remarkable aspect is that, due to the high contact resistance associated with the point contact between solid particles, foams generally exhibit a higher static thermal conductivity in comparison with packed beds of pellets.[123–127] This is shown, for example, by the experimental measurements of the two-phase thermal conductivity of various sponges (foams) made of alumina, mullite and oxidic-bonded silicon carbide (OBSiC) (in the absence of flow, free convection and radiation) presented in Ref. 127. In all cases, the estimates of thermal conductivities increased with decreasing foam porosity and were nearly constant upon varying the foam pore density (PPI). The estimated conductivities, up to five times greater than in packed beds, in combination with the very low pressure drops, were regarded as particularly promising for industrial applications.

It is worth emphasizing, however, that more work is needed in this very critical area. For instance, Ref. 127 shows the existence of a significant dispersion of the estimates in the literature for foam thermal conductivities; the role of the additional contributions due to convection and radiative heat transfer still need to be assessed; and the role of the thermal resistance at the interface between the foam and the inner reactor tube wall has not been addressed so far.

Recently, the axial two-phase thermal conductivity of ceramic foams has also been measured. The results point out its strong correlation with the superficial flow velocity and the foam porosity.[128]

### 28.4.4. *Selective oxidations over structured foam catalysts*

In spite of their promising properties, particularly in relation to radial heat transfer, only a very limited number of applications of foam catalysts to selective oxidation processes have been reported so far in the open literature.

The state of the art of structured ceramic foam catalysts was reviewed a few years ago by Twigg.[129] According to his analysis, replacing packed beds of traditional catalyst pellets with cartridges of ceramic foams could be beneficial for some industrial processes due to the following primary advantages: (i) the ability to match the shape and size of the reactor, giving easier loading of many long and narrow tubes; (ii) greatly reduced pressure drop, saving process energy costs and enabling larger space velocities; (iii) larger external surfaces; and (iv) enhanced heat transfer, thus avoiding hot spots, e.g. in strongly exothermic oxidation processes, which is the most relevant aspect for the present review. On the other hand, adoption of foams would result in reduced solid loadings compared to packed beds (typically 10–20% versus 45–55% v/v for particles). If the decreased activity per unit reactor volume cannot be recovered by a higher catalyst effectiveness factor (i.e. the reaction is not severely diffusion limited), then the loading of the catalytically active component onto the foams should be incremented. This was the case, for example, in the ethylene epoxidation example discussed by Twigg in Ref. 129. Two-dimensional reactor simulations were performed assuming an incremented silver catalyst loading of the foam in order to achieve the same final conversion and productivity as over the commercial pellets. On this basis, simulated T-profiles revealed a maximum radial temperature difference of 2°C only, versus 20°C in a corresponding packed bed. The enhanced radial heat transfer afforded increased productivity and less recycling. It should be emphasized, however, that such conclusions were based on estimates of the effective radial thermal conductivity and of the wall heat transfer coefficient determined according to correlations[118] derived from very limited data collected on a single 30 ppi foam.

Pestryakov and co-workers[130–132] have pioneered the use of foams for the deep and partial oxidation of organic substances, and specifically alcohols (methanol, ethanol) and ethylene glycol. Thus, bulk foam silver and supported foam Ag/ceramics catalysts were shown to have physico-chemical and catalytic properties exceeding the characteristics of the traditional crystalline silver and supported Ag/pumice catalysts under the same operating conditions. The catalytic properties of copper catalysts supported on ceramic foams were also studied. Due to the three-dimensional open-porous cellular structure, the foam catalysts demonstrated high gas permeability, mechanical strength and low density. Catalytic activity and selectivity of the foam catalysts in the process of the oxidation of methanol to formaldehyde exceeded the characteristics of the conventional crystalline and granulated catalysts of the same composition.

While these early studies were mostly focused on catalytic aspects, recent contributions from the Karlsruhe group of Kraushaar-Czarnetzki appear more focused on assessing the benefits for selective oxidation processes resulting from enhanced heat transfer rates in foams. Reference 133 presents a numerical simulation study comparing the performances of industrial multi-tubular fixed-bed reactors for the oxidation of *o*-xylene with PA, loaded with either conventional spherical catalyst pellets or with ceramic foam monoliths, both coated with the same catalyst. The results point out that foam packings enable a more efficient operation of the PA reactor. Their greater void fractions and surface areas, and above all their enhanced heat transfer properties, allow, in fact, an intensification of the operating conditions while reducing the risks of catalyst deactivation and thermal runaway due to the more controlled T-profiles. According to the authors, replacement of catalyst pellets with foam packings could eventually result in doubling the space-time yield of PA with respect to the current industrial levels. Nevertheless, some caution should be exercised when considering such conclusions, mostly because the heat transfer parameters of the foam catalyst were estimated according to an approximated analogy with a bed of spheres with the same specific surface area per bed volume. Thus, this interesting paper further emphasizes the current need for reliable engineering approaches to the estimation of the heat transfer characteristics of foam catalyst structures, as well as the need for experimental validations of their potential advantages at a representative industrial scale.

Finally, still in relation to the *o*-xylene oxidation process,[134] analyses of the practically relevant problem of how the foam packings should be best fitted into the reactor tubes are required.

### 28.4.5. *Selective oxidations over other structured catalysts*

A pioneering attempt at exploiting conductive heat transfer in structured catalysts in order to improve the temperature control of exothermic chemical processes can be traced to two patents issued in the 1970s to SAES GETTERS.[135,136] They describe catalytic cartridges fabricated by folding aluminium foils on which a catalytic material had been previously deposited. The resulting structure, for which several different shapes were proposed, was inserted into a metal tube, the external wall of which was surrounded by a cooling medium: special care was devoted to securing an effective thermal contact between the conductive sheet and the inner tube wall, leading to optimized configurations with alternating radial circular elements that are associated with large contact areas.[136] Furthermore, a coaxial cylindrical channel was obtained along the centreline of the catalytic module, which could also be exploited for the circulation of a cooling fluid in order to achieve flatter radial and axial temperature profiles. The patents claim a substantially improved efficiency in heat removal due to conduction in the Al foils as compared to pellet catalysts when running strongly exothermic reactions such as selective oxidations. The deposition of the catalytically active phase (e.g. $Ag/Al_2O_3$ for ethylene oxide production) was also based on a proprietary recipe.[137]

Ragaini *et al.* simulated the application of "star" catalytic cartridges to an industrial reactor for ethylene epoxidation.[138] The module consisted of 90 radial elements made of a 0.3 mm thick aluminium foil, with inner and outer diameters of 1 and 3 cm, respectively. Assuming a surface density of 18 mg/cm$^2$ for the active component $Ag/Al_2O_3$ (Ag = 70% w/w) and a reactor 8 m long, the overall load of Ag was 3.63 kg, which is equivalent to the Ag inventory in a single tube of conventional multi-tubular industrial reactors containing pellets with 12.5% Ag w/w.

From the data in the paper it can be estimated that the support volume fraction in the structured reactor was around 60%, whereas the volume fraction of catalytic components was limited to 12%, with a coating thickness slightly over 30 μm. However, more realistic lower contents of Ag in the $Ag/Al_2O_3$ catalytic powder would have resulted in thicker layers. The authors simulated the operation of a single tube reactor externally cooled by either Dowtherm A or boiling water, claiming that a substantial reduction both of the axial temperature gradients and of the pressure drop could be achieved with respect to packed beds. Unfortunately, the assumed operating conditions of the reactor, involving an extremely diluted feed, were quite distant from typical industrial ones, placing uncertainty on these conclusions; nevertheless, this study provides an early example in which the potential advantages of innovative conductive structured catalyst supports for the improvement of strongly exothermic processes are addressed. The major issues in this approach, namely the selection of support materials and the design enhancing both the radial conductive heat transfer and the thermal contact at the support–reactor interface, have already been given careful consideration herein.

A published investigation[139] deals with somewhat similar metal-structured catalyst carriers, made from a thin leaf (0.05–0.3 mm) of a chromium-aluminum steel. Composite structures were stacked in a 26 mm i.d. heated tube, and overall heat transfer coefficients were determined along with pressure drops by feeding air and measuring suitable temperature differences. The results were correlated in terms of Fanning friction factors and Nusselt numbers plotted versus the Reynolds number. Some of the tested structures simultaneously exhibited better heat transfer properties and reduced pressure drop in comparison to ceramic rings and half-rings conventionally used as catalyst supports in random packings, which were also tested for comparison. The improvement was not large, however. At Re $\approx$ 3,000 the best structured carrier gave approximately 18 and 16% higher heat transfer coefficients than ceramic rings and half-rings, respectively, with pressure drops lower by 10 and 40%. Indeed, it is possible that these proposed metal structures did not take full advantage of the conductive heat transfer mechanism, as they were made of thin foils of poorly conductive steel. Reactive experiments were eventually carried out on the best structured carrier identified in Ref. 139. After the deposition of a vanadia-phosphoria catalyst, the strongly exothermic $n$-butane oxidation to maleic anhydride was run with satisfactory results.[140]

For the sake of comprehensiveness, it is worth mentioning here that the intensification of radial heat transfer rates in multi-tubular fixed-bed gas/solid reactors with external cooling for selective oxidations has also been investigated in relation to other classes of structured catalysts, different from honeycomb monoliths with parallel passages. This is the case, for example, with static mixers, which have been proposed as structured catalyst supports in view of their ability to induce divisions and rearrangements of the fluid flow, thus enhancing its turbulence,[141] which is expected in turn to boost heat transfer from the catalyst to the reactor wall. Notably, in this case the concept relies on exploiting convective rather than conductive heat transfer, however, if in the case of structured supports heat transfer is also primarily governed by convection, only modest improvements can be expected over random packings of catalyst pellets. In addition, the analogy between heat and momentum transfer predicts that it may be very hard to decouple increments in heat transfer performances from increments in pressure drop.

Finally, applications of small monolithic metal structures orientated or randomly packed in larger reactors as catalysts for endothermic and exothermic reactions (e.g. selective oxidations) have been reviewed in Ref. 9.

## 28.5. Summary and Conclusions

Monolith catalysts are characterized by regular, well-defined, reproducible, geometric physical and flow properties, which result in unique performances for heat, mass and momentum transport. This also offers unparalleled potential for optimal design and easy scale-up of catalytic reactors for chemical synthesis processes. In the last decade, progress in the fundamental understanding of the above aspects, originating reliable engineering correlations and data, as well as improved manufacturing technologies, leading to new supports with enhanced geometries and made of a wide range of structural and functional materials, have further boosted the known advantages of monoliths with respect to conventional pellet catalysts. As illustrated in this chapter, a good number of exploratory studies have been performed on gas-phase selective oxidation processes, focusing on the specific advantages of monolith catalysts associated with reduced pressure drops, lower diffusional resistances and conductive heat exchange. At this stage, demonstrative realizations are needed addressing practical aspects associated with, for example, loading, sealing, unloading of the monoliths in the reactors, and particularly the economic value of their operational advantages in respect to higher catalyst manufacturing and development costs.

In comparison to monoliths, applications of open-celled foam structures to the chemical process industry are still at an earlier stage of development. Actually, foams do not benefit from the extensive experience of the last three decades in the use of monolith catalysts for environmental processes, and further significant research may thus be required in order to evaluate their potential.

Special opportunities for the industrial implementation of structured catalysts are offered by the growing interest in millisecond contact time processes, in view of the associated requirements on pressure drop and flow distribution to be matched with strict size constraints. In this case, a better control of the complex interplay between heat and mass transfer and heterogeneous/homogeneous reactions granted by structured catalysts would provide guidelines for the design of reactors and processes with optimized performances in terms of selectivity, yield, fast transient response and operational flexibility.

We can also clearly identify another area of growing interest associated with engineering the heat management in structured reactors. The resulting potential for enhanced heat transfer properties in comparison to packed beds of pellets is, of course, very attractive for selective oxidation processes in view of improved selectivity, prolonged catalyst lifetime and safer reactor operation. However, more fundamental and demonstrative work is needed to investigate the thermal transport properties of both honeycombs and foams before such promising performances can be thoroughly assessed in order to overcome the conservative attitude of the chemical process industry.

## References

1. Tucci, E. and Thomson, W. (1979). Monolith Catalyst Favored for Methanation, *Hydrocarb. Process.*, **58**, pp. 123–126.
2. Parmaliana, A., Crisafulli, C., Maggiore, R., *et al.* (1981). Catalytic Activity of Honeycomb Catalysts. I. The Benzene-Cyclohexane (De)Hydrogenation Reaction, *React. Kinet. Catal. Lett.*, **18**, pp. 295–299.
3. Parmaliana, A., Mezzapica, A., Crisafulli, C., *et al.* (1982). Benzene Hydrogenation on Nickel/Honeycomb Catalysts, *React. Kinet. Catal. Lett.*, **19**, pp. 155–160.
4. Parmaliana, A., El Sawi, M., Mento, G., *et al.* (1983). A Kinetic Study of the Hydrogenation of Benzene over Monolithic-Supported Platinum Catalyst, *Appl. Catal.*, **7**, pp. 221–232.
5. Parmaliana, A., El Sawi, M., Fedele, U., *et al.* (1984). A Kinetic Study of Low Temperature Hydrogenation of Benzene over Monolithic-Supported Platinum Catalyst, *Appl. Catal.*, **12**, pp. 49–57.
6. Heck, R., Gulati, S. and Farrauto, R. (2001). The Application of Monoliths for Gas Phase Catalytic Reactions, *Chem. Eng. J.*, **82**, pp. 149–156.
7. Boger, T., Heibel, A. and Sorensen, C. (2004). Monolithic Catalysts for the Chemical Industry, *Ind. Eng. Chem. Res*, **43**, pp. 4602–4611.

8. Gulati, S. (2006). Ceramic Catalyst Supports for Gasoline Fuels, in A. Cybulski and J. Moulijn (eds), *Structured Catalyst and Reactors,* 2nd Edition. CRC Taylor & Francis: Boca Raton, FL, pp. 21–70.
9. Twigg, M. and Webster, D. (2006). Metal and Coated Metal Catalysts, in A. Cybulski and J. Moulijn (eds), *Structured Catalyst and Reactors,* 2nd Edition. CRC Taylor & Francis: Boca Raton, FL, pp. 71–108.
10. Forzatti, P., Groppi, G. and Cristiani, C. (2008). Catalytic Combustion, in G. Ertl, H. Knozinger, F. Schuth, *et al.* (eds), *Handbook of Heterogeneous Catalysis*, Wiley-VCH Verlag GmbH&Co, Weinheim, Germany, pp. 2411–2426.
11. Gundlapally, S. and Balakotaiah, V. (2001). Heat and Mass Transfer Correlations and Bifurcation Analysis of Catalytic Monoliths with Developing Flows, *Chem. Eng. Sci.*, **66**, pp. 1879–1892.
12. Shah, R. and London, A. (1978). *Laminar Forced Convection in Ducts,* Academic Press, New York.
13. Gibson, L. and Ashby, M. (2001). *Cellular Solids: Structure and Properties*, 2nd Edition. Cambridge University Press, Cambridge.
14. Twigg, M. and Richardson, J. (1994). Preparation and Properties of Ceramic Foam Catalyst Supports, in G. Poncelet, J. Martens, B. Delmon, *et al.* (eds), *Preparation of Catalysts VI* (Studies in Surface Science and Catalysis, 91), Elsevier, Amsterdam, pp. 345–359.
15. Twigg, M. and Richardson, J. (2002). Theory and Applications of Ceramic Foam Catalysts, *Chem. Eng. Res. Des.*, **80**, pp. 183–189.
16. Liu, S., Afcan, A. and Masliyah, J. (1994). Steady Incompressible Laminar-Flow in Porous Media, *Chem. Eng. Sci.*, **49** pp. 3565–3586.
17. Giani, L., Groppi, G. and Tronconi, E. (2005). Mass Transfer Characterization of Metallic Foams as Supports for Structured Catalysts, *Ind. Eng. Chem. Res.*, **44**, pp. 4993–5002.
18. Große, J., Dietrich, B., Martin, H., *et al.* (2008). Volume Image Analysis of Ceramic Sponges, *Chem. Eng. Technol.*, **31**, pp. 307–314.
19. Grosse, J., Dietrich, B., Garrido, G., *et al.* (2009). Morphological Characterization of Ceramic Sponges for Applications in Chemical Engineering, *Ind. Eng. Chem. Res.*, **48**, pp. 10395–10401.
20. Smorygo, O., Mikutski, V., Marukovich, A., *et al.* (2011). An Inverted Spherical Model of an Open-Cell Foam Structure, *Acta Mater.*, **59**, pp. 2669–2678.
21. Inayat, A., Schwerdtfeger, J., Freund, H., *et al.* (2011). Periodic Open-Cell Foams: Pressure Drop Measurements and Modeling of an Ideal Tetrakaidecahedra Packing, *Chem. Eng. Sci.*, **66**, pp. 2758–2763.
22. Despois, J. and Mortensen, A. (2005). Permeability of Open-Pore Microcellular Materials, *Acta Mater.*, **53**, pp. 1381–1388.
23. Medraj, M., Baril, E., Loya , V., *et al.* (2007). The Effect of Microstructure on the Permeability of Metallic Foams, *J. Mater. Sci.*, **42**, pp. 4372–4383.
24. Richardson, J., Peng, Y. and Remue, D. (2000). Properties of Ceramic Foam Catalyst Supports: Pressure Drop, *Appl. Catal. A: Gen.*, **204**, pp. 19–32.
25. Lacroix, M., Nguyen, P., Schweich, D., *et al.* (2007). Pressure Drop Measurements and Modeling on SiC Foams, *Chem. Eng. Sci.*, **62**, pp. 3259–3267.
26. Peng, Y. and Richardson, J. (2003). Properties of Ceramic Foam Catalyst Supports: Mass and Heat Transfer, *Appl. Catal. A: Gen.*, **250**, pp. 319–329.
27. Giani, L., Groppi, G. and Tronconi, E. (2005). Heat Transfer Characterization of Metallic Foams, *Ind. Eng. Chem. Res.*, **44**, pp. 9078–9085.
28. Groppi, G., Giani, L. and Tronconi, E. (2007). A Generalized Correlation for Gas/Solid Mass Transfer Coefficients in Metallic and Ceramic Foams, *Ind. Eng. Chem. Res.*, **46**, pp. 3955–3958.
29. Incera Garrido, G., Patcas, F., Lang, S., *et al.* (2008). Mass Transfer and Pressure Drop in Ceramic Foams: A Description for Different Pore Sizes and Porosities, *Chem. Eng. Sci.*, **63**, pp. 5201–5217.

30. Incera Garrido, G. and Kraushaar-Czarnetzki, B. (2010). A General Correlation for Mass Transfer in Isotropic and Anisotropic Solid Foams *Chem. Eng. Sci.*, **65**, pp. 2255–2257.
31. Patcas, F., Garrido, G. and Kraushaar-Czarnetzki, B. (2007). CO Oxidation over Structured Carriers: A Comparison of Ceramic Foams, Honeycombs and Beads, *Chem. Eng. Sci.*, **67**, pp. 3984–3990.
32. Aghalayam, P., Park, Y. and Vlachos, D. (2000). Partial Oxidation of Light Alkanes in Short Contact Time Microreactors, *Catal.*, **15**, pp. 98–137.
33. Schmidt, L., Siddal, J. and Bearden, M. (2000). New Ways to Make Old Chemicals, *AIChE J.*, **46**, pp. 1492–1495.
34. Pereira, C. (1999). New Avenues in Ethylene Synthesis, *Science*, **285**, pp. 670–671.
35. Fosse Håkonsen, S. and Holmen, A. (2008). Oxidative Dehydrogenation of Alkanes, *Handbook of Heterogeneous Catalysis*, J. Wiley, New York, pp. 3384–3400.
36. Cavani, F., Ballarini, N. and Cericola, A. (2007) Oxidative Dehydrogenation of Ethane and Propane: How Far from Commercial Implementation? *Catal. Today*, **127**, pp. 113–131.
37. Groppi, G., Beretta, A. and Tronconi, E. (2006). Monolithic Catalysts For Gas-Phase Synthesis of Chemicals, in A. Cybulski and J. Moulijn (eds), *Structured Catalysts and Reactors*, 2nd Edition. CRC Press Taylor and Francis Group, Boca Raton, FL, pp. 243–310.
38. Zwerkle, D., Allendorf, M., Wolf, M., *et al.* (2000). Understanding Homogeneous and Heterogeneous Contributions to the Platinum-Catalyzed Partial Oxidation of Ethane in Short-Contact-Time Reactors, *J. Catal.*, **196**, pp. 18–32.
39. Lange, J., Schoonebeck, R., Mercera, P., *et al.* (2005). Oxycracking of Hydrocarbons: Chemistry, Technology and Economic Potential, *Applied Catal. A: Gen.*, **283**, pp. 243–253.
40. Cavani, F. and Trifirò, F. (1998). Paraffins as Raw Materials for the Petrochemical Industry, in A. Parmaliana, D. Sanfilippo, F. Frusteri, *et al.* (eds), *Natural Gas Conversion V* (Studies in Surface Science and Catalysis 119), Elsevier, Amsterdam, pp. 561–568.
41. Capannelli, G., Carosini, E., Cavani, F., *et al.* (1996). Comparison of the Catalytic Performance of $V_2O_5/\gamma$-$Al_2O_3$ in the Oxidehydrogenation of Propane to Propylene in Different Reactor Configurations: (i) Packed-Bed Reactor, (ii) Monolith-Like Reactor and (iii) Catalytic Membrane Reactor, *Chem. Eng. Sci.*, **51**, pp. 1817–1826.
42. Cavani, F. and Trifirò, F. (1995). The Oxidative Dehydrogenation of Ethane and Propane as an Alternative Way for the Production of Light Olefins, *Catal. Today*, **24**, pp. 307–313.
43. Huff, M. and Schmidt, L. (1993). Ethylene Formation by Oxidative Dehydrogenation of Ethane over Monoliths at Very Short Contact Times, *J. Phys. Chem.*, **97**, pp. 11815–11822.
44. Huff, M. and Schmidt, L. (1994). Production of Olefins by Oxidative Dehydrogenation of Propane and Butane over Monoliths at Short Contact Times, *J. Catal.*, **149**, pp. 127–141.
45. Faravelli, T., Goldaniga, A., Ranzi, E., *et al.* (1998). Partial Oxidation of Hydrocarbons: Experimental and Kinetic Modeling Study, in A. Parmaliana, D. Sanfilippo, F. Frusteri, *et al.* (eds). *Natural Gas Conversion V* (Studies in Surface Science and Catalysis 119), Elsevier, Amsterdam, pp. 575–580.
46. Bodke, A., Olschki, D., Schmidt, L., *et al.* (1999). High Selectivities to Ethylene by Partial Oxidation of Ethane, *Science*, **285**, pp. 712–715.
47. Bodke, A., Henning, D., Schmidt, L., *et al.* (2000). Oxidative Dehydrogenation of Ethane at Milliseconds Contact Times: Effect of $H_2$ Addition, *J. Catal.*, **191**, pp. 62–74.
48. Huff, M. and Schmidt, L. (1996). Elementary Step Model of Ethane Oxidative Dehydrogenation on Pt-Coated Monoliths, *J. Catal.*, **42**, pp. 3484–3497.
49. Goetsch, D. and Schmidt, L. (1996). Microsecond Catalytic Partial Oxidation of Alkanes, *Science*, **271**, pp. 1560–1562.
50. Bodke, A., Bharadwaj, S. and Schmidt, L. (1998). The Effect of Ceramic Supports on Partial Oxidation of Hydrocarbons over Noble Metal Coated Monoliths, *J. Catal.*, **179**, pp. 138–149.

51. Lodeng, R., Lindvag, O., Kvisle, S., *et al.* (1998). Oxidative Dehydrogenation of Ethane over Pt And Pt/Rh Gauze Catalysts at Very Short Contact Times, in A. Parmaliana, D. Sanfilippo, F. Frusteri, *et al.* (eds). *Natural Gas Conversion V* (Studies in Surface Science and Catalysis 119), Elsevier, Amsterdam, pp. 641–646.
52. Lodeng, R., Lindvag, O., Kvisle, S., *et al.* (1999). Short Contact Time Oxidative Dehydrogenation of $C_2$ and $C_3$ Alkanes over Noble Metal Gauze Catalysts, *Appl. Cat. A: Gen.*, **187**, pp. 25–31.
53. Fathi, M., Lodeng, R., Nilsen, E., *et al.* (2001). Short Contact Time Oxidative Dehydrogenation of Propane, *Catal. Today*, **64**, pp. 113–120.
54. Siberova, B., Fathi, M. and Holmen, A. (2004). Oxidative Dehydrogenation of Ethane and Propane at Short Contact Times, *Appl. Catal. A: Gen.*, **276**, pp. 17–28.
55. Siberova, B., Burch, R., Goguet, A., *et al.* (2003). Low Temperature Oxidation Reactions of Ethane over a $Pt/Al_2O_3$ Catalyst, *J. Catal.*, **219**, pp. 206–213.
56. Siberova, B. (2003). Oxidative Dehydrogenation of Ethane and Propane at Short Contact Times, thesis submitted for the degree of Doktor Ingegnior, NTNU Trondheim, Norway.
57. Hakonsen, S., Silberova, B. and Holmen, A. (2007). Oxidative Dehydrogenation of Ethane on Pt-Sn Impregnated Monoliths, *Top. Catal.* **45**, pp. 61–67.
58. Hakonsen, S., Walmsley, J. and Holmen, A. (2010). Ethene Production by Oxidative Dehydrogenation of Ethane at Short Contact Times over Pt-Sn Coated Monoliths, *Appl. Catal. A: Gen.*, **378**, pp. 1–10.
59. Beretta, A., Piovesan, L. and Forzatti, P. (1999). An Investigation on the Role of $Pt/Al_2O_3$ Catalyst in the Oxidative Dehydrogenation of Propane at Very Short Contact Times, *J. Catal.*, **184**, pp. 455–468.
60. Beretta, A., Forzatti, P. and Ranzi, E. (1999). Production of Olefins via Oxidative Dehydrogenation of Propane in Autothermal Conditions, *J. Catal.*, **184**, pp. 469–478.
61. Beretta A., Ranzi, E. and Forzatti, P. (2001). Oxidative Dehydrogenation of Light Paraffins in Novel Short Contact Reactors. Experimental and Theoretical Investigation, *Chem. Eng. Sci.*, **56**, pp. 779–787.
62. Beretta, A. and Forzatti, P. (2001). High Temperature and Short Contact Time Oxidative Dehydrogenation of Ethane in the Presence of $Pt/Al_2O_3$ and $BaMnAl_{11}O_{19}$ Catalysts, *J. Catal.*, **200**, pp. 45–58.
63. Beretta, A. and Forzatti, P. (2001), Catalyst-Assisted Oxidative Dehydrogenation of Light Paraffins in Short Contact Time Reactors, in J. Spivey, E. Iglesia and T. Fleisch (eds), *Natural Gas Conversion VI* (Studies in Surface Science and Catalysis 136), Elsevier, Amsterdam, pp. 191–196.
64. Henning, D. and Schmidt, L. (2002). Oxidative Dehydrogenation of Ethane at Short Contact Times: Species and Temperature Profiles within and after the Catalyst, *Chem. Eng. Sci.*, **57**, pp. 2615–2625.
65. Horn, R., Degenstein, N., Williams, K., *et al.* (2006). Spatial and Temporal Profiles in Millisecond Partial Oxidation Processes, *Catal. Lett.*, **110**, pp. 169–178.
66. Michael, B., Nare, D. and Schmidt, L. (2010). Catalytic Partial Oxidation of Ethane to Ethylene and Syngas over Rh and Pt Coated Monoliths: Spatial Profiles of Temperature and Composition, *Chem. Eng. Sci.,* **65**, pp. 3893–3902.
67. Beretta, A. and Forzatti, P. (2004). Partial Oxidation of Light Paraffins to Synthesis Gas in Short Contact-Time Reactors, *Chem. Eng. J.*, **99**, pp. 219–226.
68. Mulla, A., Buyeskaya, O. and Baerns, M. (2001). Autothermal Oxidative Dehydrogenation of Ethane to Ethylene Using $Sr_XLa_{1.0}Nd_{1.0}O_Y$ Catalysts as Ignitors, *J. Catal.*, **197**, pp. 43–48.
69. Donsì, F., Pirone, R. and Russo, G. (2002). Oxidative Dehydrogenation of Ethane over a Perovskite-Based Monolithic Reactor, *J. Catal.*, **209**, pp. 51–61.
70. Donsì, F., Pirone, R. and Russo, G. (2004). Catalyst Investigation for Applications of Oxidative Dehydrogenation of Ethane in Short Contact Time Reactors, *Catal. Today*, **91–92**, pp. 285–288.

71. Donsì, F., Cimino, S., Pirone, R., *et al.* (2005). Autothermal Oxidative Dehydrogenation of Ethane on $LaMnO_{3-}$ and Pt-based Monoliths: $H_2$ and CO Addition, *Ind. Eng. Chem. Res.*, **44**, pp. 285–295.
72. Sadykov, V., Pavlova, S., Saputina, N., *et al.* (2000). Oxidative Dehydrogenation of Propane over Monoliths at Short Contact Times, *Catal. Today*, **61**, pp. 93–99.
73. Pavlova, S., Sadykov, V., Frolova, Y., *et al.* (2003). The Effect of the Catalytic Layer Design on Oxidative Dehydrogenation of Propane over Monoliths at Short Contact Times. *Chem. Eng. J.*, **91**, pp. 227–234.
74. Huff, M., Androulakis, I., Sinfelt, J., *et al.* (2000). The Contribution of Gas-Phase Reactions in the Pt-Catalyzed Conversion of Ethane-Oxygen Mixtures, *J. Catal.*, **191**, pp. 46–54.
75. Zerkle, D., Allendorf, M., Wolf, M., *et al.* (2000). Understanding Homogeneous and Heterogeneous Contributions to the Platinum-Catalyzed Partial Oxidation of Ethane in a Short Contact Time Reactor, *J. Catal.*, **196**, pp. 18–39.
76. Donsì, F., Caputo, T., Russo, G., *et al.* (2004). Modeling Ethane Oxy-Dehydrogenation over Monolithic Combustion Catalysts, *AIChE J.*, **50**, pp. 2234–2245.
77. Donsì, F., Williams, K. and Schmidt, L. (2005). A Multistep Surface Mechanism for Ethane Oxidative Dehydrogenation on Pt- and Pt/Sn-Coated Monoliths, *Ind. Eng. Chem. Res.*, **44**, pp. 3453–3470.
78. Vincent, R., Lindstedt, R., Malik, N., *et al.* (2008). The Chemistry of Ethane Dehydrogenation over a Supported Platinum Catalyst, *J. Catal.* **260**, pp. 37–64.
79. Iordanoglou, D. and Schmidt, L. (1998).Oxygenates Formation from N-Butane Oxidation at Short Contact Times: Different Gauze Sizes and Multiple Steady State, *J. Catal.*, **176**, pp. 503–512; Iordanoglou, D., Bodke, A. and Schmidt, L. (1999). Oxygenates and Olefins from Alkanes in a Single-Gauze Reactor at Short Contact Times, *J. Catal.*, **187**, pp. 400–409.
80. Hickman, D., Huff, M. and Schmidt, L. (1993). Alternative Catalyst Supports for HCN Synthesis and $NH_3$ Oxidation, *Ind. Eng. Chem. Res.*, **32**, pp. 809–817.
81. Bharadway, S. and Schmidt, L. (1996). HCN Synthesis by Ammoxidation of Methane and Ethane on Platinum Monoliths, *Ind. Eng. Chem. Res.*, **35**, pp. 1524–1533.
82. Lee, H. and Farrauto, R. (1989). Catalyst Deactivation due to Transient Behaviour in Nitric Acid Production, *Ind. Eng. Chem. Res.*, **28**, pp. 1–5.
83. Sadykov, V., Isupova, L., Zolotarskii, I., *et al.* (2000). Oxide Catalysts for Ammonia Oxidation in Nitric Acid Production: Properties and Perspectives, *Appl. Catal. A: Gen.*, **204**, pp. 59–87.
84. Isupova, L., Sutormina, E., Zakharov, V., *et al.* (2009). Cordierite-Like Mixed Oxide Monolith for Ammonia Oxidation Process, *Catal. Today*, **147S**, pp. 319–323.
85. Zolotarskii, A., Kuzmin, V., Borisova, E., *et al.* (1998). Modelling Two Stage Ammonia Oxidation Performance with the Non Platinum Honeycomb Catalyst, *Abstracts of XIV International Conference on Chemical Reactors CHEMRACTOR-14*, 23–26 June, Tomsk, Russia, pp. 70–71.
86. Isupova, L., Sutormina, N., Kulikovskaia, E., *et al.* (2005). Honeycomb Supported Perovskite Catalysts for Ammonia Oxidation Processes, *Catal. Today*, **105,** pp. 429–435.
87. Eberle, H., Breimair, J., Domes, H., *et al.* (2000). Post Reactor Technology in Phthalic Anhydride Plants, *Petroleum Technology Quarterly*, June.
88. Eberle, H., Helmer, O., Stocksiefen, K., *et al.* (2001). Method of Producing Monolithic Oxidation Catalysts and Their Use in Gas Phase Oxidation of Carbohydrates, *European Patent 1181097A1*.
89. Haldor-Topsoe A/S and Nippon Kasei Chemical Co. (2001). Petrochemical Processes 2001. Formaldehyde, *Hydrocarb. Process.*, **80**, p. 106.
90. Sarup, B., Nielsen, P., Hansen, V., *et al.* (1993). Catalyst for Preparing Aldehyde, *US Patent 5,217,936*.
91. Voecks, G. (1998). Unconventional Utilization of Monolithic Catalysts for Gas-Phase Reactions, in A. Cybulski and J. Moulijn (eds), *Structured Catalyst and Reactors*, Marcel Dekker: New York, pp. 179–208.

92. Cybulski, A. and Moulijn J. (1994). Modeling of Heat Transfer in Metallic Monoliths Consisting of Sinusoidal Cells, *Chem. Eng. Sci.*, **49**, pp. 19–27.
93. Flytzani-Stephanopoulos, M., Voecks, G. and Charng, T. (1986). Modelling of Heat Transfer in Non-Adiabatic Monolith Reactors and Experimental Comparison of Metal Monoliths with Packed Beds, *Chem. Eng. Sci.*, **41**, pp. 1203–1212.
94. Kolaczkowski, S., Crumpton, P. and Spence, A. (1988). Modelling of Heat Transfer in Non-Adiabatic Monolithic Reactors, *Chem. Eng. Sci.*, **43**, pp. 227–231.
95. Kolaczkowski, S., Crumpton, P. and Spence, A. (1989). Channel Interaction in a Non-Adiabatic Monolithic Reactor, *Chem. Eng. J.*, **42**, pp. 167–173.
96. Groppi, G. and Tronconi, E. (1996). Continuous versus Discrete Models of Nonadiabatic Monolith Catalysts, *AIChE J.*, **42**, pp. 2382–2387.
97. Tronconi, E. and Groppi, G. (2002). "High Conductivity" Monolith Catalysts for Gas/Solid Exothermic Reactions, *Chem. Eng. Technol.*, **25**, pp. 743–750.
98. Hayes, R., Rojas, A. and Mmbaga, J. (2009). The Effective Thermal Conductivity of Honeycomb Monolith Structures, *Cat. Today*, **147S**, pp. S113–S119.
99. Doraiswamy, L. and Sharma, M. (1984). *Heterogeneous Reactions: Analysis, Examples and Reactor Design*, Vol. 1, John Wiley, New York.
100. Eigenberger, G. (1997). Reaction Engineering, in G. Ertl, H. Knozinger and J. Weitkamp (eds), *Handbook of Heterogeneous Catalysis*, Vol. 3, J. Wiley-VCH: Weinheim, pp. 1399–1425.
101. Groppi, G. and Tronconi, E. (2000). Design of Novel Monolith Honeycomb Catalyst Supports for Gas/Solid Reactions with Heat Exchange, *Chem. Eng. Sci.*, **55**, pp. 2161–2171.
102. Groppi, G. and Tronconi, E. (2001). Simulation of Structured Catalytic Reactors with Enhanced Thermal Conductivity for Selective Oxidation Reactions, *Catal. Today*, **69**, pp. 63–73.
103. Groppi, G., Airoldi, G., Cristiani, C., *et al.* (2000). Characteristics of Metallic Structured Catalysts with High Thermal Conductivity, *Catal. Today*, **60**, pp. 57–62.
104. Tronconi, E. and Groppi, G. (2000). A Study on the Thermal Behavior of Structured Plate-Type Catalysts With Metallic Supports for Gas/Solid Exothermic Reactions, *Chem. Eng. Sci.*, **55**, pp. 6021–6036.
105. Ray, W., Windes, L. and Schwedock, M. (1989). Steady-State and Dynamic Modelling of a Packed-Bed Reactor for the Partial Oxidation of Methanol to Formaldehyde, *Chem. Eng. Comm.*, **78**, pp. 1–43.
106. Valentini, M., Groppi, G., Cristiani, C., *et al.* (2001). The Deposition of $\gamma$-$Al_2O_3$ Layers on Ceramic and Metallic Supports for the Preparation of Structured Catalysts, *Catal. Today*, **69**, pp. 307–314.
107. Tronconi, E., Groppi, G., Boger, T., *et al.* (2004). Monolithic Catalysts with "High Conductivity" Honeycomb Supports for Gas/Solid Exothermic Reactions: Characterization of the Heat-Transfer Properties, *Chem. Eng. Sci.*, **59**, pp. 4941–4949.
108. Boger, T. and Menegola, M. (2005). Monolithic Catalysts with High Thermal Conductivity for Improved Operation and Economics in the Production of Phthalic Anhydride, *Ind. Eng. Chem. Res.*, **44**, pp. 30–40.
109. Cutler, W., He, L., Olszewski, A., *et al.* (2005). Thermally Conductive Honeycombs for Chemical Reactors, *US Patent 6,881,703*.
110. Groppi, G. and Tronconi, E. (2005). Honeycomb Supports with High Thermal Conductivity for Gas/Solid Chemical Processes, *Catal. Today*, **105**, pp. 297–304.
111. Boger, T. and Heibel, A. (2005). Heat Transfer in Conductive Monolith Structures, *Chem. Eng. Sci.*, **60**, pp. 1823–1835.
112. Groppi, G., Tronconi, E., Cruzzolin, F., *et al.* (2012). Conductive Honeycomb Catalysts: Development and Industrial Pilot Tests for the Oxidation of *o*-Xylene to Phthalic Anhydride, *Ind. Eng. Chem. Res*, **51**, pp. 7590–7596.

113. Abbott, J., Boger, T., He, L., *et al.* (2009). Metal Honeycomb Substrates for Chemical and Thermal Applications, *US Patent 7,608,344*.
114. Strasser, R., Schmidhammer, L., Deller, K., *et al.* (1992). Chlorination Reactions and Oxychlorination Reactions in the Presence of Honeycomb Monolithic Catalyst Supports. *US Patent 5,099,085*.
115. Carmello, D., Marsella, A., Forzatti, P., *et al.* (2008). Metallic Monolith Catalyst Support for Selective Gas Phase Reactions in Tubular Fixed Bed Reactors, *European Patent 1110605B*.
116. Carmello, D., Marsella, A., Forzatti, P., *et al.* (2010). Metallic Monolith Catalyst Support for Selective Gas Phase Reactions in Tubular Fixed Bed Reactors, *US Patent 7,678,342*.
117. Amsden, J., Boulch, G., Heibel, A., *et al.* (2009). Multi-Tubular Reactors with Monolithic Catalysts, U.S. Patent Appl. US2009/0176895.
118. Peng, Y. and Richardson, J. (2004). Properties of Ceramic Foam Catalyst Supports: One-Dimensional and Two-Dimensional Heat Transfer Correlations, *Appl. Catal. A: Gen.*, **266**, pp. 235–244.
119. Lu, T. and Chen, C. (1999). Thermal Transport and Fire Retardance Properties of Cellular Aluminium Alloys, *Acta Mater.*, **47**, pp. 1469–1485.
120. Zhao, C., Kim, T., Lu, T., *et al.* (2004). Thermal Transport in High Porosity Cellular Metal Foams, *J. Thermophys. Heat Tr.*, **18**, pp. 309–317.
121. Younis, L. and Viskanta, R. (1993). Experimental Determination of the Volumetric Heat Transfer Coefficient between Stream of Air and Ceramic Foam, *Int. J. Heat Mass Tran.*, **36**, pp. 1425–1434.
122. Calmidi, V. and Mahajan, R. (2000). Forced Convection in High Porosity Metal Foams, *J. Heat Transf.*, **122**, pp. 557–565.
123. Abramenko, A., Kalinichenko, A., Burster, Y., *et al.* (1999). Determination of the Thermal Conductivity of Foam Aluminum, *J. Eng. Thermophys.*, **72**, pp. 369–373.
124. Singh, R. and Kasana, H. (2004). Computational Aspects of Effective Thermal Conductivity of Highly Porous Metal Foams, *Appl. Therm. Eng.*, **24**, pp. 1841–1849.
125. Paek, J., Kang, B., Kim, S., *et al.* (2000). Effective Thermal Conductivity and Permeability of Aluminium Foam Materials, *Int. J. Thermophys.*, **21**, pp. 453–464.
126. Boomsma, K. and Poulikakos, D. (2001). On the Effective Thermal Conductivity of a Threedimensionally Structured Fluid-Saturated Metal Foam, *Int. J. Heat Mass Tran.*, **44**, pp. 827–836.
127. Dietrich, B., Schell, G., Bucharsky, E.,*et al.* (2010). Determination of the Thermal Properties of Ceramic Sponges, *Int. J. Heat Mass Tran.*, **53**, pp.198–205.
128. Dietrich, B., Kind, M. and Martin, H. (2011). Axial Two-Phase Thermal Conductivity of Ceramic Sponges: Experimental Results and Correlation, *Int. J. Heat Mass Tran.*, **54**, pp. 2276–2282.
129. Twigg, M. and Richardson, J. (2007). Fundamentals and Applications of Structured Ceramic Foam Catalysts, *Ind. Eng. Chem. Res.*, **46**, pp. 4166–4177.
130. Pestryakov, A., Lunin, V., Devochkin, A., *et al.* (2002). Selective Oxidation of Alcohols over Foam-Metal Catalysts, *Appl. Catal. A: Gen.*, **227**, pp. 125–130.
131. Pestryakov, A., Bogdanchikova, N. and Knop-Gericke, A. (2004). Alcohol Selective Oxidation over Modified Foam-Silver Catalysts, *Catal. Today*, **91–92**, pp. 49–52.
132. Pestryakov, A., Petranovskii, V., Pfander, N., *et al.* (2004). Supported Foam–Copper Catalysts for Methanol Selective Oxidation, *Catal. Commun.*, **5**, pp. 777–781.
133. Reitzmann, A., Bareiss, A. and Kraushaar-Czarnetzki, B. (2006). Simulation of a Reactor for the Partial Oxidation of O-Xylene to Phthalic Anhydride Packed with Ceramic Foam Monoliths, *Oil Gas-Eur. Mag.*, **32**, pp. 94–98.
134. Mulheims, P., Marz, S., Muller, S., *et al.* (2011). Fitting Solid Sponges into a Reactor Tube Concerning the Partial Oxidation of *o*-Xylene as an Example, *Chem. Ing. Tech.*, **83**, pp. 286–294.
135. Della Porta, P., Giorgi, T., Cantaluppi, A., *et al.* (1974). Catalyst Cartridge, *US Patent 3,857,680*.

136. Della Porta, P., Ferrario, B., Cantaluppi, A., *et al.* (1975). Catalytic Cartridge, *US Patent 3,890,104*.
137. Della Porta, P., Giorgi, T., Kindl, A., *et al.* (1972). Method of Producing Substrate having a Particulate Metallic Coating, *US Patent 3,652,317*.
138. Ragaini, V., De Luca, G., Ferrario, B., *et al.* (1980). A Mathematical Model for a Tubular Reactor Performing Ethylene Oxidation to Ethylene Oxide by a Catalyst Deposited on Metallic Strips, *Chem. Eng. Sci.*, **35**, pp. 2311–2319.
139. Kolodziej, A., Krajewski, W. and Dubis, A. (2001). Alternative Solution for Strongly Exothermal Catalytic Reactions: A New Metal-Structured Catalyst Carrier, *Catal. Today*, **69**, pp. 115–120.
140. Kołodziej, A., Krajewski, W. and Łojewska, J. (2004). Structured Catalyst Carrier for Selective Oxidation of Hydrocarbons. Modelling and Testing, *Catal. Today*, **91**, pp. 59–65.
141. de Campos, V. and Quinta-Ferreira, R. (2001). Structured Catalysts for Partial Oxidations, *Catal. Today*, **69**, pp. 121–129.

# Index